$$\begin{array}{r} 4327 \\ 11 \\ \hline 4597 \end{array}$$

FOURTH EDITION

College Algebra and Trigonometry with Applications

CHERYL CLEAVES

MARGIE HOBBS

PAUL DUDENHEFER

State Technical Institute at Memphis
Memphis, Tennessee

Prentice Hall
Upper Saddle River, New Jersey ■ *Columbus, Ohio*

0–13–789249–7

Cover Photo: Background, Westlight. Foreground, Tony Stone Images
Editor: Stephen Helba
Developmental Editor: Carol Hinklin Robison
Production Editor: Rex Davidson
Design Coordinator: Julia Zonneveld Van Hook
Text Designer: Rebecca Bobb
Cover Designer: Rod Harris
Production Manager: Patricia A. Tonneman
Illustrations: The Clarinda Company
Marketing Manager: Debbie Yarnell

This book was set in Times Roman and Futura by The Clarinda Company and was printed and bound by Von Hoffmann Press, Inc. The cover was printed by Von Hoffmann Press, Inc.

 © 1998, 1995 by Prentice-Hall, Inc.
Simon & Schuster/A Viacom Company
Upper Saddle River, New Jersey 07458

Portions of this book previously published as *Basic Mathematics for Trades and Technologies* by Cleaves, Hobbs, and Dudenhefer (© 1990), *Introduction to Technical Mathematics* by Cleaves, Hobbs, and Dudenhefer (© 1988), and *Vocational-Technical Mathematics Simplified* by Cleaves, Hobbs, and Dudenhefer (© 1987).

Photo credits: Matthew Brown and Susan Duke/Prentice Hall

Printed in the United States of America

10 9 8 7 6 5 4 3 2 1

ISBN: 0-13-789249-7

Prentice-Hall International (UK) Limited, *London*
Prentice-Hall of Australia Pty. Limited, *Sydney*
Prentice-Hall of Canada, Inc., *Toronto*
Prentice-Hall Hispanoamericana, S. A., *Mexico*
Prentice-Hall of India Private Limited, *New Delhi*
Prentice-Hall of Japan, Inc., *Tokyo*
Simon & Schuster Asia Pte. Ltd., *Singapore*
Editora Prentice-Hall do Brasil, Ltda., *Rio de Janeiro*

Preface

In *College Algebra and Trigonometry with Applications,* Fourth Edition, we have preserved all the features that made the first three editions the most appropriate text on the market for a comprehensive study of mathematics in postsecondary and college technical programs. No other comprehensive text presents such a mathematically sound coverage of arithmetic, algebra, geometry, and trigonometry topics as the Fourth Edition. It remains one of the few such texts to include geometry, data analysis, and statistics. Our emphasis remains on using real-life situations as a context for applied problems.

Readability Level

As in earlier editions, we keep the language simple and have applied two different readability formulas on selected 100-word passages and averaged the results. The readability level generally compares to that of daily newspapers and popular magazines, making our text suitable for a wide range of instructional uses. Of course, when mathematical and technical vocabulary is taken into account, some passages may prove more challenging to read until the definitions are understood. We have carefully examined our definitions and have taken all possible steps to ensure their mathematical and technical soundness as well as their readability.

Content Changes of the Fourth Edition

The Fourth Edition incorporates many valuable suggestions made by users of the earlier editions of the text. Although we continue to use a step-by-step approach, we introduce many topics earlier so that a spiraling development of these topics is possible in later chapters. We have also given greater emphasis to the problem sets and problem solving and have included new material. Specifically, we combined Whole Numbers and Decimals into Chapter 1; introduced graphical representation and function notation in Chapter 2, Integers. We also combined the basic concept of percent, fraction, and decimal equivalents with Chapter 3, Fractions and Percents. Chapter 4, Percent Applications, now focuses on problem-solving strategies for percent applications.

We have written a new Chapter 8 on Symbolic Representation. It includes a more thorough review of signed numbers, an introduction to variable expressions, and translation from words to symbols. A six-step problem-solving strategy is introduced in Chapter 9, Linear Equations, and continues throughout the rest of the text. This strategy employs a major emphasis on critical thinking, estimating, and describing the characteristics of appropriate solutions to applied problems.

Several features have been added to this edition and features that have appeared in earlier editions have been expanded. Each chapter opens with a teamwork project that is appropriate for the content of the chapter. This feature, Good Decisions Through Teamwork, helps students develop teamwork skills while they experience the usefulness of mathematics. Many "Tips and Traps" have been added and each tip includes a title for easy referencing. In addition to the electronic applications, applications covering nursing, horticulture, law enforcement, teaching, and many more careers have been added. To further show the connections between the study of mathematics and the "real-world," articles featuring "Mathematics in the Workplace" have been included.

Commitment to Reform

The authors have been and continue to be active in the reform movements of the American Mathematical Association of Two-Year Colleges (AMATYC), the National Council of Teachers of Mathematics (NCTM), and the Mathematical Association of America (MAA). This edition continues to incorporate reform strategies. We enthusiastically promote the standards and guidelines encouraged by these organizations and the SCANS document. The Instructor's Resource Manual gives some specific references for implementing the Standards in your courses.

We endorse the incremental approach to accomplishing curriculum and pedagogy reform rather than a radical departure from all traditional strategies. We have found that anxiety is increased among both instructors and students when radical changes are incorporated too quickly. The Fourth Edition has accomplished this goal without destroying the "comfort zone" that we have worked so hard to develop.

Calculator Usage

After introducing basic, scientific, and graphics calculators in Chapter 1, we continue with calculator tips throughout the text. These tips are independent of a specific model of calculator and they offer suggestions so that students can logically determine how their calculator operates without referring to the calculator user's manual.

The calculator continues to be emphasized as a tool that facilitates learning and understanding. Assessment strategies that test understanding independently of a calculator are incorporated throughout the text.

Study Strategies and Reference Features

In our experiences as instructors we are aware of the need for students to develop good study habits and good independent learning skills. Students also need a good reference text so they can review mathematical concepts as the need arises. We take great pride in our students' praise of the usefulness of this text. For a detailed description of the features of the text and our suggestions to the students, refer to the "To the Student" portion of the preface.

Additional Resources

Several additional resources are available with the adoption of this text. These resources include the Instructor's Resource Manual, a test-item file and a computerized test item file (PH Custom Test), a Student Solutions Manual, and a "How to Study Technical Mathematics" booklet. Contact your Prentice-Hall representative for more information.

Acknowledgments

A project such as this does not come together without help from lots of people. Our first avenue for input is through our students and fellow instructors at the State Technical Institute at Memphis. We value their comments and suggestions and acknowledge these contributions to the quality of the text. In addition, we are appreciative of the assistance we received in assuring the accuracy of the text and the supplements. Jim and Renee Smith and Kim Collier have spent many hours working every problem and reading every word. However, we take full responsibility for any misprints or errors that may remain in the first printing of the text.

We think the photographs that open each chapter add a human element to the text. Some of the photographs were taken by the authors, but the best ones were taken by Matthew Brown and Susan Duke. We appreciate the care they took in matching the photos to the projects. Brown: Preface, chapters 1, 10, 12, 13, 17, 18, 20, 22b, 24a, 24b, 24c. Duke: chapters 1, 4, 5, 6, 7, 9, 11, 14, 15, 16, 21, 22a, 23, 24d.

We wish to express thanks to all the people who had a part in making this revision a reality. In particular, we thank Steve Helba, Carol Robison, and Rex Davidson at Prentice Hall.

The teaching of mathematics over time produces a wealth of knowledge about instructional strategies and specific content. We are grateful for the many valuable suggestions that were received in these areas. We wish to thank the following individuals who served as reviewers for this Fourth Edition:

Frank Caldwell	York Technical College
LeRoy Mink	ITT Technical Institute—San Antonio
Chuck Coggins	CAD Institute
Carol Maize	Pittsburgh Technical Institute
Susan Poss	Spartanburg Technical College
Jesse Mase	Southern Maine Technical College
Lenora Corbett	Alamance Community College
Larry Blevins	Tyler Jr. College
Bonita M. Markstrom	Front Range Community College
Cindy Frisbie	Louisiana Technical College—Sowela Campus
Jane Fore	Petit Jean Technical College
Catherine Johnson	Alamance Community College
Anita Armfield	York Technical College
Tania McDuffie	Spartanburg Technical College

A very special "thank you" is extended to Cynthia Miller, who supplied many of the special new features in the Fourth Edition. Her input has been invaluable in assuring the most up-to-date and interesting content in many of the chapter openers, applications, and articles. For their contributions, we also extend appreciation to the following individuals:

Good Decisions Through Teamwork

Ch. 15: Patrolman Tim McCarroll, Collierville Police Department, Collierville, TN.   Ch. 16: Janet Callicott, RN, BSN, CPN, LeBonheur Children's Hospital, Memphis, TN.   Ch. 17: Isaac Kullman, Memphis Mensa Chapter, Memphis, TN.   Ch. 18: Dan Snow, CPA, Thompson Financial Group, Memphis, TN.

Career Applications

Ch. 1: Dr. Paul Chodas, Dr. Eleanor Helin, and David Seidel, NASA Jet Propulsion Laboratory, Pasadena, CA.   Ch. 5: Janet Callicott, RN, BSN, CPN, LeBonheur Children's Hospital, Memphis, TN.   Ch. 6: Bettye Yales, owner of Interior Creations, Collierville, TN.   Ch. 12: Don W. Clark, Thomas County Agricultural Extension Office, Thomasville, GA.   Ch. 14: Dan Long, Senior Network Integrator, Chicago, IL.   Ch. 16: Jacquelyn Phillips, Brisbane, Australia.   Ch. 17: George Gibson Jr. and Wayne Raley, Piper Farm Products, Collierville, TN; and Richard Daumgart, Valvoline Instant Oil Change.   Ch. 22: Patrick Gibson, Tornado Expert, Collierville, TN. Ch. 24: Patrolman Tim McCarroll, Collierville Police Department, Collierville, TN; and Tennessee Highway Patrol Office, Memphis, TN.

Ch. 5: Janet Callicott, RN, BSN, CPN, LeBonheur Children's Hospital, Memphis, TN; and Dr. James Erickson, MD, MPH, MS, Civil Air Patrol, Memphis, TN. Ch. 10: George Gibson Jr., Piper Farm Products, Collierville, TN.   Ch. 20: George Gibson Jr., Piper Farm Products, Collierville, TN.

Finally, our extended families continue to be our driving force: First, our spouses, Charles Cleaves, Allen Hobbs, and Gaynell Dudenhefer. Then, our parents, living and deceased, our brothers and sisters, children, grandchildren, nieces and nephews. We have included them and many other friends and acquaintances as the subjects in the application problems.

Cheryl Cleaves
Margie Hobbs
Paul Dudenhefer

To the Student

In almost any career you pursue, the mathematics you learn in this book will serve you well and help you advance your career goals. Anyone can learn to deal with a certain amount of math, even those who have avoided the formal study of mathematics. We have given much thought to the best way to teach mathematics. If you follow the course of the book as we have laid it out, making use of the special features we have put in the text, you will get the most out of this book. The following features are meant to help you learn the mathematics included in this text.

Learning Objectives.   Each section begins with a statement of learning objectives, which lays out for you what you should look for and learn in that section. If you read and think about these before you begin the section, you will know what to look for as you go through the section.

Six-Step Approach to Problem Solving.   This format enables you to take a systematic approach to solving problems. This feature is first found in Chapter 9. You are asked to analyze and compare, and estimate to solve problems.

Rules, Properties, and Formulas.   Boxes for Rules, Properties, and Formulas appear throughout the text to help introduce new procedures and to provide a quick reference for review. To make these procedures as clear as possible, we break them down for you into step-by-step instructions. An example generally follows each rule, property, or formula.

Good Decisions through Teamwork.   Each chapter opens with a suggested class project designed to promote the various facets of teamwork. The collection of projects incorporates a wide variety of team-building strategies. Some general tips for developing effective teams are included in the front matter of the text. Each project involves you, the student, in a unique way. The various projects emphasize computational skills, interpersonal skills, oral or written communication skills, organizational skills, research skills, critical thinking and/or decision-making skills—all skills highly sought by employers. Students submit project reports to a variety of audiences including instructors, peers, employers, and immediate supervisors.

Your instructor may decide to use any or all of the projects, or he or she may organize teams within the class and have each team select a project from a different chapter. Even if a particular project is not used in your class, you will benefit from reading the projects. You will broaden your perception of the usefulness of the mathematics you are studying.

Mathematics in the Workplace.   Excerpts and summaries from many newspaper, magazine, and journal articles are given to help you develop a richer appreciation of the usefulness of mathematics in the workplace and in everyday life. These articles can set the stage for class discussion or they can pique your own interest so that you will go to the library or the Internet for additional information.

Career Applications. At the end of each chapter, you will find one or more career applications. These applications were developed as a result of interviews and research in the workplace to ensure their relevance and accuracy. They demonstrate how widespread math applications are in the workplace and the world around you. In addition to providing a "real world" problem-solving opportunity, they are intended to increase your awareness of the relevance of mathematics to your everyday life and to demonstrate the ways in which you will regularly use the math concepts you are learning.

Calculator Boxes. The use of a calculator is essential in all types of math, and especially college math for technology. In most chapters we have calculator boxes to introduce calculator strategies. They allow multiple calculations to be performed in a continuous series of keystrokes. The boxes discuss how to analyze the procedure and set up a problem for a calculator solution; this is followed by a sample series of keystrokes involved in the solution. In addition, tips are given to help you determine how your specific model of calculator operates without you having to continually refer to your calculator's user manual.

Tips and Traps Boxes. These boxes point out helpful hints or pitfalls involved in doing mathematics procedures. Often, looking at the right and wrong ways of doing something can save you from a costly or time-consuming error down the line. We do not always have to learn from our own mistakes—sometimes we can learn from other people's experience and not make the same mistakes ourselves. These boxes also draw special attention to important observations and connections that you might have missed in an example.

Chapter Summary. Each chapter ends with a summary in the form of a two-column chart. The first column repeats each objective of the section. The second column gives a brief summary and an example or examples to illustrate the objective. A *Words to Know* list is located at the end of each Chapter Summary. This list gives each new vocabulary word in the chapter and the page on which the definition of the word appears. Some definitions are in the body of the text, highlighted by special type. Other definitions are set apart as numbered definitions.

Self-Study Exercises. These short practice sets that are keyed to the learning objectives appear at the end of each section throughout the chapter; they are a signal to you to check yourself to be sure that you understand what you have just read or worked out before you go on to the next section. The answers to every problem are at the end of the text, so you can get immediate feedback on whether you have understood the material.

Assignment Exercises. An extensive set of exercises appears at the end of each chapter to review all the procedures and objectives covered in the chapter. These exercises, organized by section, may be assigned by your instructor as homework, or you may want to work them on your own for extra practice. The answers to the odd-numbered exercises are at the end of the book. Solutions to odd-numbered exercises appear in a separate Student Solutions Manual. Your instructor has the solutions to the even-numbered exercises in the Instructor's Resource Manual.

Concepts Analysis. Too often we focus on the "how to" and overlook the "why" and "where" associated with various mathematical concepts. The Concepts Analysis questions allow you to formalize your understanding of a concept and to connect concepts to other concepts and uses. Error analysis is another way that the understanding of concepts is encouraged.

Trial Test. The trial test at the end of each chapter is designed to enable you to check your understanding of the concepts in the chapter. You should be able to work each problem without referring to any examples in the text or your notes. Take this test yourself before you take the test given by your instructor. The test is located in the last pages of each chapter and the answers to the odd-numbered exercises appear at the end of the book. Solutions to odd-numbered exercises appear in a separate Student Solutions Manual. Your instructor has the solutions to the even-numbered exercises in the Instructor's Resource Manual.

Index. The index is an important part of a math book. You will use the index to cross reference topics and locate topics quickly that relate to a topic you are studying. You will find the index for this book to be very extensive and useful as you study the book.

Table of Contents. The table of contents is like your roadmap for the course. Study the table of contents to determine how the topics are arranged. This will aid you in relating topics to each other.

Student Solutions Manual. This manual can be purchased at your bookstore. It will give you extra "learning insurance" to help you master this material. The manual contains worked-out solutions to the odd-numbered exercises in the Assignment Exercises and the practice test from each chapter of the text. Answers to these exercises appear in the back of your text, but here in the manual, you can study the fully worked-out solutions.

Use of Color in the Text. As you read the text and work through the examples, notice the items that are printed in blue or shaded blue or gray. These shadings will help you follow the logical process of working through the example. Color is also used to highlight important items and features such as the Tips and Traps boxes.

How to Study Technical Mathematics. Your instructor can get free copies of this booklet, which goes over the various learning techniques you can use in class and in preparation for class to make learning mathematics much more efficient and effective.

We wish you much success in your study of mathematics. Many of the improvements for this book were suggested by students such as yourself. If you have suggestions for improving the presentation, please give them to your instructor.

Play Ball, or Work as a Team

"Play ball!" signals the start of a baseball game. But it also signals the start of a team effort to achieve a goal—putting out three players from the other team. And a team effort it is. No one player can achieve this goal alone. All the players together cannot achieve this goal if each works alone. But when all the players work as a team, each carrying out his or her specific function in a coordinated

effort to achieve the goal, then the goal can be attained. The pitcher, the catcher, the base players, the outfielders—all the players—perform their individual tasks cooperatively to strike out a batter or put out a runner sliding into home plate. This is why a ball club is called a team.

This is why businesses, industries, and even colleges more and more are employing teams to achieve goals: A team solution is generally better than a solution reached by one person, even a very knowledgeable person. Why? The world today is so complex, with so many variables, so many aspects to consider, that one person cannot perform very well every task needed to arrive at the best solution possible. But several people, functioning as a team, can divide the tasks among themselves and, as a group, share and discuss their findings, and reach a far better solution than any single individual can.

You may be asked in class to achieve certain goals or carry out a project using teamwork. The team may involve three or more members, maybe even six to eight, and may or may not have a designated leader. If there is no leader designated by team election or by instructor appointment, then different team members who have strengths in the areas of need will likely surface to offer leadership as situations arise. Some typical needs are securing a place to meet, calling the meeting, taking minutes, setting goals and priorities, identifying the tasks to be performed, delegating the tasks among team members, motivating members who get behind, supporting members who lack confidence, helping members work out conflicts, keeping everyone on target, and so on.

Teams work best when tasks and responsibilities are matched to the talents and desires of the individual members. For example, library research is best delegated to team members who are independent workers and interviews are best delegated to team members who are strong in interpersonal relations. In this way each team member can contribute a valuable resource to the group's effort to achieve its goals and feel a sense of satisfaction, importance, and achievement. The effectiveness of the team depends on the effective utilization of the talents of all its members and the recognition by each member of the worth of his or her contribution to the team effort.

For a team to be successful, each member must be committed to the success of the team and to its goals. This commitment, in turn, means that each member helps the team move forward toward its goals, first of all, by completing his or her own share of the work, and then by motivating other members to do the same. This may mean giving a pep talk occasionally or finding something constructive to say when tempted to lash out at a member's backsliding.

At an initial meeting, the team must clarify the nature of the problem to be solved or the task to be performed so that everyone understands the situation the same way. This is the time also for the team to set its priorities so that everyone will know what kinds of solutions will be acceptable ones. If some priorities are more important than others, the team should distinguish between the "must have" priorities and the "nice-to-have" priorities.

Once the problem is clear and the priorities have been set, the team members individually, or possibly in twos, gather the information needed to arrive at a solution. Then the team members meet to discuss the information from individual members and propose solutions to the problem. As a next step they evaluate each solution to see if it meets the priorities agreed to previously. Then the members select the best solution from among the acceptable ones. The best one will be the one that meets all "must have" priorities and as many "nice-to-have" priorities as possible.

The final task is for the team to present its findings in an organized form, such as an oral report by an individual chosen by the team, a group presentation such as a panel discussion, or a written report perhaps with different sections prepared by different members. In any case, the presentation of the findings should include a statement of the problem, the method used to solve the problem, and the solution itself.

Contents

CHAPTER

1

Whole Numbers and Decimals

GOOD DECISIONS THROUGH TEAMWORK

Choose a specific home business in your team's target career field and investigate the cost of starting the business. Begin by deciding whether the home business will market products, such as sports memorabilia, or services, such as desktop publishing. Once your team has agreed on the product or service, investigate the various kinds of equipment, materials, and other required items and their cost. For example, you might search business and trade magazines and interview a professional in the chosen business. Some items to consider are telephones, supplies, furniture, a computer system, fax capability, subscription to an on-line service, and, if necessary, the product inventory.

Now find the total cost to start your home business. Will a loan be needed, or is some other source of funds available? If the budget is tight, can you eliminate or trim any nonessential expenses? Setting priorities early in the planning process is an important part of developing a realistic start-up budget.

When we study a subject for the first time or review in some detail a subject we studied some time ago, we need to begin with the basics. Often, as we examine the basics of a subject, we discover—or rediscover—many bits and pieces of useful information. In this sense, mathematics is no different from any other subject. We begin with a study of whole numbers and decimals, and the basic operations that we perform with them.

1–1 WHOLE NUMBERS, DECIMALS, AND THE PLACE-VALUE SYSTEM

Learning Objectives

1. Identify place values in whole numbers.
2. Read and write whole numbers in words and standard notation.
3. Identify place values in decimal numbers.
4. Read decimal numbers.
5. Write fractions with power-of-10 denominators as decimal numbers.
6. Compare decimal numbers.
7. Round a whole number or a decimal to a place value.
8. Round a whole number or a decimal to a number with one nonzero digit.

Our system of numbers, which is called the *decimal number system*, uses 10 individual figures called *digits*: 0, 1, 2, 3, 4, 5, 6, 7, 8, 9. A *whole number* is made up of one or more digits. When a number contains two or more digits, each digit must be in the correct place for the number to have the value we want it to have. If we mean "ninety-eight," we must place the 9 first and the 8 second to represent 98. If we change the places of these two digits by putting the 8 first and the 9 second, we get a new value (eighty-nine) and a new number (89).

1 Identify Place Values in Whole Numbers.

Each place a digit can occupy in a number has a value called a *place value*. If we know the place value of each digit in a number, we can read the number and understand how much it means. Look at the chart of place values in Fig. 1–1. Notice that each place value *increases* as we move from *right to left* and that each increase is *10 times* the value of the place to the right. For example, the tens place is 10 times the ones place, the hundreds place is 10 times the tens place, and so on.

Billions			Millions			Thousands			Units		
Hundred billions (100,000,000,000's)	Ten billions (10,000,000,000's)	Billions (1,000,000,000's)	Hundred millions (100,000,000's)	Ten millions (10,000,000's)	Millions (1,000,000s)	Hundred thousands (100,000's)	Ten thousands (10,000's)	Thousands (1,000's)	Hundreds (100's)	Tens (10's)	Ones (1's)

Figure 1–1 Whole number place values.

The place values are arranged in *periods*, or groups of three, to make numbers easier to read. The first group of three is called *units*, the second group of three is called *thousands*, the third group is *millions*, and the fourth group is *billions*. Commas are used to mark off these groups in a number. In four-digit numbers, the comma separating the units group from the thousands group is optional. Thus, both 4,575 and 4575 are correct.

> **Rule 1–1** *To identify the place value of digits:*
>
> 1. Mentally position the number on the place-value chart so that the last digit on the right aligns under the ones place.
> 2. Identify the place value of each digit according to its position on the chart.

EXAMPLE 1   In the number 2,472,694,500, identify the place value of the digit 7.

To solve this problem, we align the number on the place-value chart as shown in Fig. 1–2.

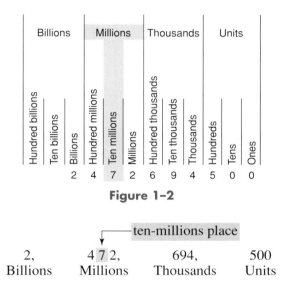

Figure 1–2

2,	4 7 2,	694,	500
Billions	Millions	Thousands	Units

ten-millions place

7 is in the ten-millions place.

2 Read and Write Whole Numbers in Words and Standard Notation.

Now that we have reviewed place values, we can read numbers with little difficulty. All we need to do is use the following rule.

> **Rule 1–2** *To read numbers:*
>
> 1. Mentally position the number on the place-value chart so that the last digit on the right lines up under the ones place.
> 2. Examine the number from right to left, separating each group with commas.
> 3. Identify the leftmost group.
> 4. From the left, read the numbers in each group and the group name. (The group name *units* is not usually read. A group containing all zeros is not usually read.)

EXAMPLE 2   Show how 7543026129 and 2000125 would be read by writing them in words.

Mentally align the digits on the place-value chart (Fig. 1–3). Starting at the right, separate each group of three digits with commas. Identify the leftmost group. The first number is seven *billion,* five hundred forty-three *million,* twenty-six *thousand,* one hundred twenty-nine.

The second number is two *million,* one hundred twenty-five. *Note:* Since the thousands group contains all zeros, it is *not* read.

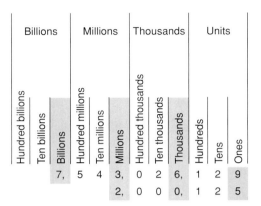

Figure 1–3

Tips and Traps **Special Conventions with Reading and Writing Numbers:**

1. A group name is inserted at each comma.
2. The word *and* should not be used when reading whole numbers.
3. The numbers from 21 to 99 (except 30, 40, 50, and so on) use a hyphen when they are written (forty-three, twenty-six, and so on).

Of course, there are times when we have to write spoken numbers on paper. Numbers written with digits in the appropriate place-value positions are numbers in *standard notation.*

Rule 1–3 *To write numbers in standard notation:*

1. Write the group names in order from left to right, starting with the first group in the number.
2. Fill in the digits in each group. Leave blanks for zeros if necessary.
3. Supply zeros as needed for each group. Each group except the leftmost group must have three digits.
4. Starting at the right, separate the groups with commas as needed.

EXAMPLE 3   The exact cost of a group of airplanes is eight million, two hundred four thousand, twelve dollars. Write this amount as a number in standard notation.

1. Millions	Thousands	Units
2. Millions	Thousands	Units
8	204	_12
3. Millions	Thousands	Units
8	204	0 12
4. Millions	Thousands	Units
8,	204,	012

The cost is $8,204,012.

③ Identify Place Values in Decimal Numbers.

Numbers that are parts of a unit or quantity are called *fractions*. One fraction notation writes one number over another number. The bottom number, called the denominator, represents the number of parts that a whole unit contains. The top number, called the numerator, represents the number of parts being considered. We have already used this notation in expressing division. We will begin our examination of fractions with a special type of fraction called a *decimal fraction*. Other types of fractions will be examined in Chapter 3.

A decimal fraction is a fraction whose denominator is always 10 or some power of 10, such as 100, 1000, and so on. For convenience in this book we use the terms decimal fraction, decimal number, and decimal interchangeably. However, we must keep in mind that decimals are another way of writing fractions. In fraction notation, 3 out of 10 parts would be written as $\frac{3}{10}$. In decimal notation, the denominator 10 is not written but implied by position on the place value chart. A decimal point (.) is used to separate whole amounts on the left and fractional parts on the right. The fraction $\frac{3}{10}$ can be written in decimal notation as 0.3.

Examine the place-value chart shown in Fig. 1–4. We will use the place-value system to understand *decimal numbers*. In Fig. 1–4, to move from *left* to *right*, we divide by 10 to get the value of the next place. For example, to move from the hundreds place to the tens place, we have $100 \div 10 = 10$. To move from the tens place to the ones place, we have $10 \div 10 = 1$.

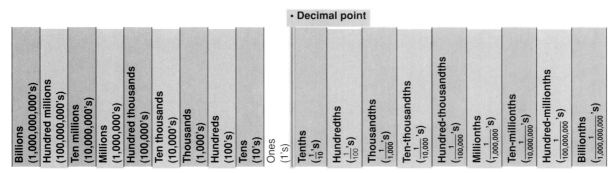

Figure 1–4 Place-value chart.

To extend the place-value chart on the right by moving from the ones place to the next place on the right, we have $1 \div 10 = \frac{1}{10}$. Thus, the place on the right of the ones place is called the *tenths place*. A period (.), called the *decimal point*, is placed between the ones place and tenths place so that we can identify the place value of each digit in a number.

Tips and Traps	*Use of Commas and Periods in Numbers:* The use of a period to separate the whole-number places from the decimal places is not a universally accepted notation. Some cultures use a comma instead. For example, our notation for writing 32,495.8 may be written as 32.495,8 or 32 495,8.

The digits to the right of the ones place are used to represent the numerator of the fractions. The place value of the rightmost digit indicates the denominator.

Rule 1–4 *To identify the place value of digits in decimal fractions:*

1. Mentally position the decimal number on the decimal place-value chart so that the decimal point of the number aligns with the decimal point on the chart.
2. Identify the place value of each digit according to its position on the chart.

EXAMPLE 4   Identify the place value of each digit in 32.4675 by applying Rule 1–4.

Comparing the number 32.4675 to the place-value chart in Fig. 1–5, we see that 3 is in the tens place, 2 is in the ones place, 4 is in the tenths place, 6 is in the hundredths place, 7 is in the thousandths place, and 5 is in the ten-thousandths place.

The place-value chart can be extended on the right side for smaller fractions in the same manner that it is extended on the left for larger numbers.

4 Read Decimal Numbers.

Examine the place-value chart in Fig. 1–5. Notice the *th* on the end of each place value on the right of the decimal point. When reading decimals, this *th* will indicate a decimal number. The word *and* is used to indicate the decimal point.

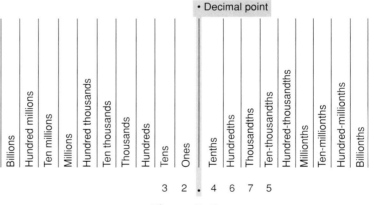

Figure 1–5

To read a decimal number, we begin by reading the whole-number part, then we say *and* to indicate the decimal, then we read the decimal part. In reading the decimal part we use the same procedure as for the whole-number part, and end by reading the place value of the last digit in the decimal part. If the place value of the last digit in the decimal part is two words, these two words are hyphenated. For example, the ten-thousandths, hundred-thousandths, ten-millionths, and hundred-millionths places are all written with hyphens.

Rule 1–5 *To read decimal numbers:*

1. Mentally align the number on the decimal place-value chart so that the decimal point of the number is directly under the decimal point on the chart.
2. Read the whole-number part.

3. Use *and* for the decimal point only if there is a whole-number part.
4. Read the decimal part like the whole-number part.
5. End by reading the *place value* of the last digit in the decimal part.

EXAMPLE 5   Read 52.386 by writing it in words.

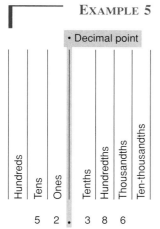

Figure 1–6

1. Align the number on the chart (see Fig. 1–6).
2. Read the whole-number part.

3. Use *and* for the decimal point because there is a whole-number part.

4. Read the decimal part like a whole-number part.

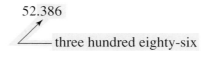

5. End by reading the *place value* of the last digit in the decimal part.

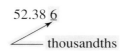

52.386 is "fifty-two and three hundred eighty-six thousandths."

Tips and Traps   *Informal Use of the Word* **Point:**
Informally, the decimal point is sometimes read as "point." Thus, 3.6 is read "three point six."

EXAMPLE 6   Read 0.0162 by writing it in words.

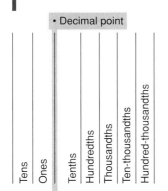

Figure 1–7

1. Align the number on the chart (see Fig. 1–7).
2. There is no whole-number part to read.

0.0162

3. Do *not* read the decimal point.

0.0162

4. Read the decimal part like a whole number.

0.0 162
one hundred sixty-two

5. End by reading the *place value* of the last digit in the decimal part.

0.016 2
ten-thousandths   Be sure to use the hyphen.

0.0162 is "one hundred sixty-two ten-thousandths."

When we read or write whole numbers in words, we do not use the word *and*; that is, 107 is "one hundred seven." When we write whole numbers using numerals, we usually omit the decimal point, even though the decimal point is understood to be at the end of the whole number. Therefore, any whole number, such as 32, can be written without a decimal (32) or with a decimal (32.).

5 Write Fractions with Power-of-10 Denominators as Decimal Numbers.

On the job we may encounter fractions with power-of-10 denominators like 10, 100, 1000, and so on. Examples of these are $\frac{1}{10}$, $\frac{75}{100}$, and so on. These fractions are so quickly changed into decimals that many workers prefer to write them as decimals and perhaps use a calculator to arrive at the answer or solution to the problem. If we encounter fractions like $\frac{1}{10}$ or $\frac{75}{100}$, we may want to use decimals instead. Here is a quick way to convert them.

Any fraction whose denominator is 10, 100, 1000, 10,000, and so on, can be written as a decimal number by writing the numerator and placing the decimal in the appropriate place to indicate the proper place value of the denominator. The decimal number will have the same number of digits in its decimal part as the fraction has zeros in its denominator.

EXAMPLE 7   Write $\frac{3}{10}$, $\frac{25}{100}$, $\frac{425}{100}$, and $\frac{3}{100}$ as decimal numbers.

$\frac{3}{10}$ is written 0.3.

$\frac{25}{100}$ is written 0.25.

$\frac{425}{100}$ is written 4.25.

$\frac{3}{100}$ is written 0.03.

Since $\frac{3}{10} = \frac{30}{100}$, we can see that 0.3 = 0.30. Similarly, 0.7 = 0.70 = 0.700.

Tips and Traps   ***Do Zeros Change the Value of a Decimal Number?***
If we affix zeros on the *right* end of a decimal number, we do not change the value of the number.

$$0.5 = 0.50 = 0.500$$

---

6 Compare Decimal Numbers.

Our employment may sometimes require us to work with close tolerances. For instance, we may be working with wire and need to compare the diameters of several sizes of wire to arrange them in order from largest to smallest or smallest to largest. To do this, we need to be able to compare decimals because wire size is often measured in decimals.

Rule 1–6   *To compare decimal numbers:*
1. Compare whole-number parts.
2. If whole-number parts are equal, compare digits place by place, starting at the tenths place and moving to the right.
3. Stop when two digits in the same place are different.
4. The digit that is larger determines the larger decimal number.

EXAMPLE 8 Compare the two numbers to see which is larger.

32.47   32.48

We begin by looking at the whole-number parts of the two numbers. In this case they are the same. Next, we look at the tenths place for each number. Both numbers have a 4 in the tenths place. Looking at the hundredths place, we see that 32.48 is the larger number because 8 is larger than 7.

EXAMPLE 9 Compare 0.4 and 0.07 to see which is larger.

Since the whole-number parts are the same (0), we compare the digits in the tenths place. 0.4 is larger because 4 is larger than 0.

If we write both numbers in the example above so that they have the same number of digits in the decimal part, they may be easier to compare.

$$0.4 = 0.40$$
$$0.07 = 0.07$$

Since 40 is larger than 7, 0.4 is larger than 0.07. This is equivalent to the process used in comparing common fractions. A common denominator is found and each fraction (or decimal fraction) is changed to an equivalent fraction having the common denominator.

Tips and Traps *Common Denominators in Decimals:*
Decimal fractions have a common denominator if they have the same number of digits to the right of the decimal point.

---

7 Round a Whole Number or a Decimal to a Place Value.

To check calculations made with a calculator or to make mental calculations, it is less cumbersome to use approximate numbers instead of exact numbers. To *round* a number is to express it as an approximation. Whenever a decimal, or any number, is rounded, it becomes less accurate than the original decimal or number. Thus, if we round $3.99 to $4.00, the $4.00 is less accurate than the original $3.99. However, the difference between $3.99 and $4.00 is slight, only 1 cent. On the other hand, if we round $4.25 to $4.00, the difference here is greater and the rounded decimal is much more inaccurate, with a difference of 25 cents. Some job applications will require more accuracy in rounding than others. The rules for rounding numbers with decimals are similar to the rules for rounding whole numbers.

When rounding a number to a certain place value, we must make sure that we are as accurate as our employer wants us to be. Generally, the size of the number and its use will dictate to which decimal place it will be rounded.

Rule 1–7 *To round a whole or decimal number to a given place value:*

1. Locate the digit that occupies the rounding place. Then examine the digit to the immediate right.
2. If the digit to the right of the rounding place is 0, 1, 2, 3, or 4, do not change the digit in the rounding place. If the digit to the right of the rounding place is 5, 6, 7, 8, or 9, add 1 to the digit in the rounding place.

---

3. All digits on the *right* of the digit in the rounding place are replaced with zeros if they are to the left of the decimal point. Digits that are on the right of the digit in the rounding place *and* to the right of the decimal point are dropped.

Tips and Traps *Nine Plus One Still Equals Ten:*
When the digit in the rounding place is 9 and must be rounded up, it becomes 10. The 0 goes in place of the 9 and 1 is carried to the next place to the left.

EXAMPLE 10   Is 24.63 closer to 24.6 or 24.7?

We round to the tenths place. Applying Rule 1–7, we circle the 6, which is in the tenths place, to indicate the rounding place.

24.⑥3

Because the digit to the right of the 6 is 3, we do not change the 6.

24.⑥3
6

Next, we write all digits to the left of the 6 as they are.

24.⑥3
24. 6

Because the 3, which is to the right of 6, is to the right of the decimal, it is dropped. **24.63 is closer to 24.6.**

EXAMPLE 11   Round 46.879 to the hundredths place.

46.8⑦9   7 is in the hundredths place.

46.8⑦9   The next digit to the right is 9, so we add 1 to 7.
8

46.8⑦9   All digits to the left of 7 are written as they are.
46.8 8

46.88   The 9 is dropped because it is to the right of the rounding place and to the right of the decimal.

EXAMPLE 12   Round 32.6 to the tens place.

③2.6   3 is in the tens place.

③2.6   3 is unchanged because 2, the next digit to the right, is less than 5.
3

30   2 is replaced with a zero because it is before the decimal. 6 is dropped because it is to the right of the decimal.

EXAMPLE 13   Round 15.8 to the nearest whole number.

Rounding to the nearest whole number means rounding to the *ones* place.

1⑤.8   5 is in the ones place. 1 is added to 5 and 8 is dropped because it is to the right of
1 6   the decimal.

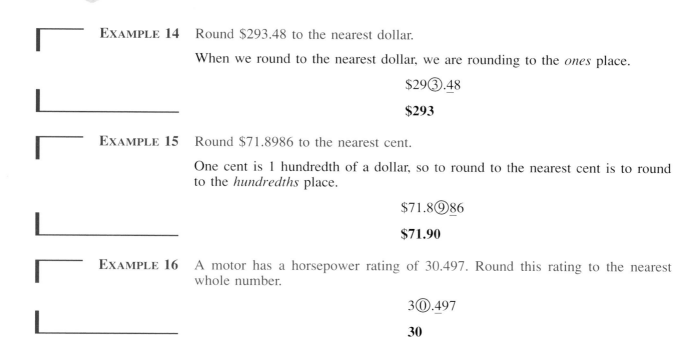

EXAMPLE 14 Round $293.48 to the nearest dollar.

When we round to the nearest dollar, we are rounding to the *ones* place.

$$\$29\textcircled{3}.\underline{4}8$$

$293

EXAMPLE 15 Round $71.8986 to the nearest cent.

One cent is 1 hundredth of a dollar, so to round to the nearest cent is to round to the *hundredths* place.

$$\$71.8\textcircled{9}\underline{8}6$$

$71.90

EXAMPLE 16 A motor has a horsepower rating of 30.497. Round this rating to the nearest whole number.

$$3\textcircled{0}.\underline{4}97$$

30

8 Round a Whole Number or a Decimal to a Number with One Nonzero Digit.

If we want to get a very quick estimate, we sometimes round to one nonzero digit. For example, if we have items priced at $2.39, $3.56, $5.92, and $2.13, we can get a quick estimate of the total by rounding to $2.00, $4.00, $6.00, and $2.00, respectively. Notice that all four rounded amounts have only one digit that is not zero. When we round numbers so that only one digit is not zero, we say that we are rounding to "one nonzero digit."

Look at the numbers below. Notice that the *first* digit on the left is a *nonzero digit* (not a zero); the other digits are all zeros.

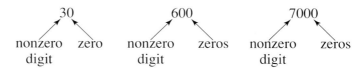

Rule 1–8 *To round to one nonzero digit:*

1. Starting at the *left,* find the *first* digit that is not zero. This digit will be a 1, 2, 3, 4, 5, 6, 7, 8, or 9.
2. Round the number to the place value of the first nonzero digit.
3. All digits to the right of the rounding place are replaced with zeros up to the decimal point. Any digits that are to the right of the digit in the rounding place *and* that follow the decimal point are dropped.

EXAMPLE 17 Round 4.23 to one nonzero digit.

1. Find the first digit from the left that is not zero.

4.23   The 4 is the first nonzero digit.
↑

2. The first nonzero digit, 4, is circled to mark the rounding place.

$\textcircled{4}.23$

3. Round the number to the place value of the first nonzero digit.

④.23   Nothing is added to 4 because the next digit to the right is less than 5.

4. Drop the digits to the right of the decimal point.

4

EXAMPLE 18   Round 78.4 to one nonzero digit.

1. Find the first nonzero digit from the left.

78.4
↑

2. The first nonzero digit is 7.

⑦8.4

3. Round to the place value of the first nonzero digit.

⑦8.4   1 is added to 7 because the next digit to the right is 5 or more.
  8

4. Replace each digit between the rounded digit and the decimal point with a zero. Drop all digits after the decimal point.

80   0 replaces the digit between 8 and the decimal point. Drop the digits after the decimal point.

EXAMPLE 19   Round 0.83 to one nonzero digit.

0.⑧3   Remember to locate the first nonzero digit.
0.8

EXAMPLE 20   If the price of a typewriter is $497.95, estimate this cost to one nonzero digit.

④97.95
$500

Tips and Traps	***Exact Amount Versus Approximate Amount:*** When any amount has been rounded, it is no longer an exact amount. The rounded amount is now an *approximate amount.*

SELF-STUDY EXERCISES 1–1

1   In the number 2,304,976,186, identify the place value of the following digits.

1. 3   **2.** 7   **3.** 1   **4.** 0   **5.** 2

In the number 8,972,069,143, identify the place value of the following digits.

6. 0   **7.** 4   **8.** 7   **9.** 8   **10.** 6

2   Show how these numbers would be read by writing them out as words.

11. 6704   **12.** 89021   **13.** 662900714
14. 3000101   **15.** 15407294376   **16.** 150

Write these words as numbers in standard notation. Use commas when necessary.

17. Seven billion, four hundred   **18.** One million, six hundred twenty-seven thousand, one hundred six

19. Fifty-eight thousand, two hundred one

20. In a telephone conversation a contractor submitted the following bid for a job: "one thousand six dollars." Write this bid as a number.

3

21. What is the place value of the 7 in 32.407?

22. What is the place value of the 8 in 28.396?

23. What is the place value of the 4 in 3.00254?

24. What is the place value of the 3 in 457.2096532?

25. What is the place value of the 7 in 0.0387?

In the following problems, state what digit is in the place indicated.

26. Tens: 46.3079

27. Tenths: 2.0358

28. Thousandths: 520.0765

29. Ten-millionths: 3.002178356

30. Hundredths: 402.3786

4 Write the words for the following decimal numbers.

31. 21.387   **32.** 420.059   **33.** 0.89

34. 0.0568   **35.** 30.02379   **36.** 21.205085

Write the digits for the following numbers.

37. Three and forty-two hundredths

38. Seventy-eight and one hundred ninety-five thousandths

39. Five hundred and five ten-thousandths

40. Seventy-five thousand, thirty-four hundred-thousandths

5 Write the following fractions as decimal numbers.

41. $\dfrac{5}{10}$ **42.** $\dfrac{23}{100}$ **43.** $\dfrac{7}{100}$ **44.** $\dfrac{683}{100}$ **45.** $\dfrac{79}{1000}$

46. $\dfrac{468}{1000}$ **47.** $\dfrac{587}{100}$ **48.** $\dfrac{108}{1000}$ **49.** $\dfrac{603}{100}$ **50.** $\dfrac{400}{100}$

6 Compare the following number pairs and identify the larger number.

51. 3.72, 3.68   **52.** 7.08, 7.06   **53.** 0.23, 0.3

54. 0.56, 0.5   **55.** 2.75, 2.65   **56.** 0.157, 0.2

Compare the following number pairs and identify the smaller number.

57. 8.9, 8.88   **58.** 0.25, 0.3   **59.** 0.913, 0.92

60. 0.761, 0.76   **61.** 5.983, 6.98   **62.** 1.972, 1.9735

Arrange the numbers in order from smaller to larger.

63. 0.23, 0.179, 0.314   **64.** 1.9, 1.87, 1.92   **65.** 72.1, 72.07, 73

66. Two micrometer readings are recorded as 0.837 in. and 0.81 in. Which is larger?

67. A micrometer reading for a part is 3.85 in. The specifications call for a dimension of 3.8 in. Which is larger, the micrometer reading or the specification callout?

68. A washer has an inside diameter of 0.33 in. Will it fit a bolt that has a diameter of 0.325 in.?

69. Aluminum sheeting can be purchased in 0.04-in. thickness or 0.035-in. thickness. Which thickness gives the thicker sheeting?

70. If No. 14 copper wire has a diameter of 0.064 in. and No. 10 wire has a diameter of 0.09 in., which has the larger diameter?

7 Round to the place value indicated.

71. Nearest hundred: 468

72. Nearest hundred: 6248

73. Nearest thousand: 8263

74. Nearest ten thousand: 429,207

75. Nearest thousand: 39,748

76. Nearest ten thousand: 39,748

77. Nearest million: 285,487,412
79. Nearest billion: 82,629,426,021

78. Nearest ten: 468
80. Nearest ten million: 297,384,726

Round the decimals to the nearest whole number.

81. 42.7   **82.** 367.43   **83.** 7.983   **84.** 103.06   **85.** 2.9

Round the decimals to the nearest tenth.

86. 8.05   **87.** 12.936   **88.** 42.574   **89.** 83.23   **90.** 5.997

Round the decimals to the nearest hundredth.

91. 7.036   **92.** 42.065   **93.** 0.792   **94.** 3.198   **95.** 7.773

Round to the nearest thousandth.

96. 0.2173   **97.** 0.0196   **98.** 1.5085   **99.** 4.2378   **100.** 7.0039

Round to the nearest dollar.

101. $219.46   **102.** $82.93   **103.** $507.06   **104.** $2.83   **105.** $5.96

Round to the nearest cent.

106. $8.237   **107.** $0.291   **108.** $0.528   **109.** $5.796   **110.** $238.9238

111. A micrometer measure is listed as 0.7835 in. Round this measure to the nearest thousandth.

112. The diameter of an object is measured as 3.817 in. If specifications call for decimals to be expressed in hundredths, write this measure according to specifications.

113. To the nearest tenth, what is the current of a 2.836-A motor?

8   Round the numbers to numbers with one nonzero digit.

114. 483   **115.** 7.89   **116.** 62.5
119. 0.095   **120.** 3.07   **121.** 52

117. 0.537   **118.** 0.0086
122. 83.09   **123.** 52.8

124. If round steak costs $2.78 per pound, what is the cost per pound to the nearest dollar?

125. An estimate calls for converting 23.077 to an approximate number. What is the approximate number rounded to one nonzero digit?

126. The following amounts of monthly rainfall (in inches) were identified for these months: January, 0.355; March, 1.785; May, 0.45; July, 1.409; September, 0.07; and December, 2.018. Round each amount to one nonzero digit to identify the months with the most similar rainfall.

127. A researcher calculated the average response times (in seconds) as automobile drivers applied the brake when first seeing a road hazard. The following average times were calculated: driver A, 0.0275; driver B, 0.0264; driver C, 0.0234; driver D, 0.0284; and driver E, 0.0379. Round each response time to one nonzero digit to help identify the subjects who were most similar in response time.

1–2   ADDING WHOLE NUMBERS AND DECIMALS

Learning Objectives

1   Add whole numbers.
2   Add decimal numbers.
3   Estimate and check addition.
4   Add using a calculator.

Addition of whole numbers is used on all jobs and occupations at one time or another. Our purpose in this section is to review some ways to gain speed and make fewer mistakes when adding whole numbers.

1 **Add Whole Numbers.**

Commutative and Associative Properties of Addition   It is sometimes easier to discuss the procedures required to work a problem if we name the parts of the problem. In an addition problem, the numbers being added are called *addends* and the answer is called the *sum* or *total*.

Two useful things to know about addition are that it is *commutative* and *associative*. By *commutative* we mean that it does not matter in what *order* we add numbers. We can add 7 and 6 in any order and still get the same answer:

$$7 + 6 = 13 \qquad 6 + 7 = 13$$

■ **DEFINITION 1–1:   Commutative Property.**   The *commutative property* means that values being added (or multiplied) may be added (or multiplied) in any order.

By *associative*, we mean that we can *group* numbers together any way we want when we add and still get the same answer. To add $7 + 4 + 6$, we can group the $4 + 6$ to get 10; then we add the 10 to the 7:

$$7 + (4 + 6) = 7 + 10 = 17$$

Or we can group $7 + 4$ to get 11; then we add the 6 to the 11:

$$(7 + 4) + 6 = 11 + 6 = 17$$

Addition is a *binary operation*; that is, the rules of addition apply to adding *two* numbers at a time. The associative property of addition shows how addition is extended to more than two numbers.

■ **DEFINITION 1–2:   Associative Property.**   The *associative property* means that values being added (or multiplied) may be grouped in any manner.

The commutative and associative properties of addition can help us add whole numbers quickly and accurately. Look for these two aids in the examples that follow.

Tips and Traps   *Symbolic Representation of Rules and Definitions:*
Many rules and definitions can be written symbolically. Symbolic representation allows a quick recall of the rule or definition.

Commutative Property of Addition:

$$a + b = b + a$$

where a and b are numbers.

Associative Property of Addition:

$$a + (b + c) = (a + b) + c$$

where a, b, and c are numbers.
The associative property of addition also allows other possible groupings and extends to more than three numbers.

$$7 + 4 + 6 = 13 + 4 = 17$$
$$3 + 5 + 7 + 9 = 8 + 16 = 24$$

EXAMPLE 1   Add $7 + 3 + 6 + 8 + 2 + 4$ using the commutative and associative properties.

Group $7 + 3$, $8 + 2$, and $6 + 4$. Then add the three groups ($10 + 10 + 10 = 30$).

Another important property of addition is the *zero property of addition*.

Rule 1-9 *Zero property of addition:*

Zero added to any number results in the same number.

$$n + 0 = n \qquad \text{or} \qquad 0 + n = n$$
$$5 + 0 = 5 \qquad \text{or} \qquad 0 + 5 = 5$$

When adding numbers of two or more digits, such as $238 + 456$, we must make sure that the numbers are placed in columns properly. This means that the same place values must be aligned under one another so that all the ones are in the far right column, all the tens in the next column, and so on.

Rule 1-10 *To add numbers of two or more digits:*

1. Arrange the numbers in columns so that the ones place values are in the same column.
2. Add the ones column, then the tens column, then the hundreds column, and so on, until all the columns have been added. *Carry* whenever the sum of any column is more than one digit.

EXAMPLE 2   Shipping fees are sometimes charged by the total weight of the shipment. Find the total weight of the following order: nails, 250 pounds (lb); tacks, 75 lb; brackets, 12 lb; and screws, 8 lb. Arrange in columns and add.

$$
\begin{array}{r}
250 \\
75 \\
12 \\
+\quad 8 \\
\hline
345
\end{array}
$$

The total weight is 345 lb.

② Add Decimal Numbers.

We add whole numbers by adding in columns all digits in the ones place, then adding all digits in the tens place, and so on. To add decimal numbers, we will follow the same procedure. When decimal numbers are aligned in this manner, all decimal points fall in the same vertical line. Aligning decimal points has the same effect as using *like* denominators when adding (or subtracting) fractions.

> **Rule 1–11** *To add decimals:*
>
> Arrange the numbers so that the decimal points are in one vertical line. Then add each column. Align the decimal for the sum in the same vertical line.

EXAMPLE 3   Add 42.3 + 17 + 0.36.

42.3
17   <small>Note that the decimal in 17 is understood to be at the right end.</small>
 0.36

We should be very careful in aligning the digits. Careless writing can cause unnecessary errors. To avoid difficulty in adding the columns, we may write each number so that all have the same number of decimal places by affixing zeros on the right.

$$
\begin{array}{r}
42.30 \\
17.00 \\
0.36 \\
\hline
\mathbf{59.66}
\end{array}
$$

③ Estimate and Check Addition.

Estimating a sum before performing the actual addition gives us an approximate answer. Estimating is a process that is usually done mentally. Estimating sums is important when performing addition mentally or with a calculator.

When the difference between the exact sum and the estimated sum is large, the exact sum may be incorrect and we should add the numbers again.

> **Rule 1–12** *To estimate the answer to an addition problem:*
>
> **1.** Round each addend to the indicated place value or to a number with one nonzero digit.
> **2.** Add the rounded addends.

> **Rule 1–13** *To check an addition problem:*
>
> **1.** Add the numbers a second time and compare to the first sum.
> **2.** Use a different order or grouping if convenient.

EXAMPLE 4   Acme Contractor spent the following amounts on a job: $16,466.15, $23,963.10, and $5,855.20. Estimate the total amount by rounding to thousands. Then find the exact amount and check.

Thousands Place	Estimate	Exact	Check
$16,466.15	$16,000	$16,466.15	$16,466.15
23,963.10	24,000	23,963.10	23,963.10
5,855.20	6,000	5,855.20	5,855.20
	$46,000	**$46,284.45**	**$46,284.45**

The estimate and exact answer are close. The exact answer is reasonable.

The level of accuracy of estimates will vary depending on what place value the addends are rounded to.

4 Add Using a Calculator.

It is important to know the properties of addition so that we can estimate the sums mentally or perform the addition of small numbers mentally. However, in many cases a calculator is used when adding large numbers or when adding several numbers. Several different types of calculators may be used, but we will focus on the most common types of hand-held calculators.

The three most common types of hand-held calculators are the basic, scientific, and graphics calculator. Each type may perform even the simplest calculations differently.

 ■ **General Tips for Using the Calculator**

Addition on the basic and scientific calculators.

1. Enter the first addend. If there is a decimal part, press the decimal key $\boxed{\cdot}$ before the appropriate digit. For numbers less than 1, such as 0.5 or 0.6, it is not necessary to enter the zero before a decimal point.
2. Press the addition operation key $\boxed{+}$.
3. Enter the next addend (as in step 1 above) and press the addition operation key if more addends follow.
4. Continue until all addends have been entered.
5. After entering the final addend, press the addition operation key or the equals key $\boxed{=}$, which will place the sum in the display window.

Both basic and scientific calculators will accumulate the sum of entered addends each time the addition operation key or the equal key is pressed.

 ■ **General Tips for Using the Calculator**

Addition on the graphics calculator.

1. Enter the first addend. If there is a decimal part, press the decimal key $\boxed{\cdot}$ before the appropriate digit. For numbers less than 1, such as 0.5 or 0.6, it is not necessary to enter the zero before a decimal point.
2. Press the addition operation key $\boxed{+}$.
3. Enter the next addend (as in step 1 above) and press the addition operation key if more addends follow.
4. Continue until all addends have been entered.
5. Press the appropriate key to instruct the calculator to perform the calculations in the display window. This key is labeled $\boxed{\text{EXE}}$ for execute on some graphics calculators. Another common label is $\boxed{\text{ENTER}}$.

The graphics calculator does not accumulate the sum of entered addends as operation keys are pressed. All indicated calculations are performed after the execute key is pressed.

EXAMPLE 5 Find the sum of the following numbers using a hand-held calculator.

$$2345 + 3894.745 + 758.05 =$$

Basic or scientific:

$2345 \boxed{+} 3894 \boxed{\cdot} 745 \boxed{+} 758 \boxed{\cdot} 05 \boxed{+} =$ *could be pressed instead of* +

Basic or scientific display: 6997.795
Graphics:

$2345 \boxed{+} 3894 \boxed{\cdot} 745 \boxed{+} 758 \boxed{\cdot} 05 \boxed{EXE}$

Graphics display: 2345 + 3894.745 + 758.05
6997.795

SELF-STUDY EXERCISES 1–2

1 Add.

1.
```
   4
   1
   5
   3
+  2
```

2.
```
   6
   9
   7
   4
+  1
```

3.
```
   3
   5
   2
   4
   7
+  1
```

4.
```
   6
   4
   5
   6
   9
+  6
```

5. In taking inventory, the following $\frac{1}{4}$-inch (in.) hex bolts of various lengths were counted: nine 1 in. long, four $1\frac{1}{4}$ in., seven $1\frac{1}{2}$ in., six 2 in., two $2\frac{1}{4}$ in., and nine $2\frac{1}{2}$ in. How many $\frac{1}{4}$-in. hex bolts were there in all?

6.
```
   1072
   6710
+  1410
```

7.
```
   5273
   4001
+  7682
```

8.
```
   59,718
+  46,567
```

Write in columns and add.

9. $36 + 482 + 961 + 27 + 804$

10. $4582 + 86,724 + 482 + 5826$

11. A mechanic was paid the following for 5 days of work: $86, $124, $67, $85, and $94. How much was the mechanic paid for the 5 days?

12. A truck driver had the following weight slips on three loads of gravel: 8114 lb, 8027 lb, and 8208 lb. What was the total weight of the gravel hauled by the driver?

13. During inventory, the following numbers of slotted-head screws were counted in four different boxes: 84, 63, 72, and 79. How many slotted-head screws were there all together?

14. Four bricklayers were working on the same job. In one day the bricklayers laid the following number of bricks: 1217, 1103, 1039, and 1194. How many bricks did all four lay that day?

15. Canty O'Neal, a purchasing clerk, placed an order for 15 gallons (gal) of white paint, 27 gal of black paint, 5 gal of crimson red paint, and 3 gal of canary yellow paint. How many gallons of paint were ordered?

2 Add.

16. $4.2 + 3.6 + 7.9$

17. $12.8 + 13.52 + 7.86$

18. $3.9 + 4.02 + 0.21$

19. $8.9 + 6.72 + 3.58 + 68.2$

20. $83.37 + 42 + 1.6 + 3$

21. $7 + 4.2 + 14.6 + 0.23$

22. $23.9 + 54.3$

23. $24.5 + 21.2$

24. $205.03 + 56.305$

25. $309.01 + 47.602$

26. $0.784 + 5 + 1.2$

27. $0.225 + 7 + 2.3$

28. $900.75 + 225.85$

29. $300.25 + 113.35$

30. $3.7 + 0.6 + 4.8 + 9$

31. 7.1 + 0.2 + 5.5 + 6

32. 0.70868
        + 0.10937

33. 0.83967
        + 0.33675

34. 51.006
        + 4.507

35. 82.005
        + 3.406

36. 3.487 + 47.5 + 19

37. A pipe that is 0.103 in. thick has an inside diameter of 2.871 in. Find the outside diameter of the pipe.

38. The total current in a parallel circuit is found by adding the individual currents. If a circuit has individual currents of 3.98 A, 2.805 A, and 8.718 A, find the total current.

39. A part-time hourly worker earned $25.97 on Monday, $7.48 on Tuesday, $5.88 on Wednesday, $65.45 on Thursday, and $76.47 on Friday. Find the total week's wages.

40. A residential lot that measures 100.8 ft, 87.3 ft, 104.7 ft, and 98.6 ft on each of its four sides is to be fenced. How many feet of fencing are required?

41. A technician purchased an ac voltage sensor for $11.95, a grounded outlet analyzer for $5.99, and a neon circuit tester for $1.99. How much did she pay for all three items?

42. Julio purchased a $\frac{3}{32}$-in. submini stereo plug adapter for $2.99 and a $\frac{3}{32}$-in. submini mono plug adapter for $1.76. If the sales tax was $0.29, what was his total bill?

3 Estimate the sum by rounding to hundreds. Then find the exact sum and check.

43. 4,256.65
      3,892.10
        576.46
      8,293.00

44. 52,843
      17,497
      13,052
         821

45. 24,003
          5,874
        319,467
         52,855

46. Palmer Associates provided the following prices for items needed to build a sidewalk: concrete, $2,583.45; wire, $43.25; frame material, $18.90; labor, $798. Estimate the cost by rounding each amount to the nearest ten. Find the exact amount.

47. Antonio Juarez expects to spend the following amounts for one semester of college: food, $395; lodging, $1,285; books, $288; supplies, $130; transportation, $162. Estimate the expenditures for one semester by rounding each amount to the nearest hundred. Calculate the exact amount.

Estimate the sums by rounding addends to numbers with one nonzero digit. Then find the exact sum. Finally, check your answer.

48.   940
       + 8299

49.   478.125
       + 146.055

50.   1901
       + 6548

51.   149.25
       + 652.14

52.   16,259
       + 36,542

53.   32,501
       + 16,740

54. A hardware store filled the following order of nails: 25 lb, $2\frac{1}{2}$-in. common; 16 lb, 4-in. common; 12 lb, 2-in. siding; 24 lb, $2\frac{1}{2}$-in. floor brads; 48 lb, 2-in. roofing; and 34 lb, $2\frac{1}{2}$-in. finish. What is the total weight of the order?

55. If four containers have a capacity of 12 gal, 27 gal, 55 gal, and 21 gal, can 100 gal of fuel be stored in these containers? (Find the total capacity of the containers first.)

56. A printer has three printing jobs that require the following numbers of sheets of paper: 185, 83, and 211. Will one ream of paper (500 sheets) be enough to finish the three jobs?

57. How many feet of fencing are needed to enclose the area illustrated in Fig. 1–8?

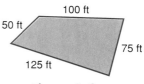

Figure 1–8

4 Estimate the sums by rounding addends to numbers with one nonzero digit. Then use a calculator to find the exact sum.

58. 47,287 + 33,409 + 81,496 + 28,594

59. 387,483 + 879,583 + 592,801

60. 31,592 + 8,584.6 + 13,215.05 + 968

61. 1,328,591 + 35,803,502 + 10,387,921

62. Mario's Restaurant had the following sales daily: Sunday, $3,842.95; Monday, $1,285.68; Tuesday, $1,195.57; Wednesday, $1,843.76; Thursday, $1,526.47; Friday, $2,984.89; and Saturday, $4,359.72. Find the total sales for the week.

1-3 SUBTRACTING WHOLE NUMBERS AND DECIMALS

Learning Objectives

1. Subtract whole numbers.
2. Subtract decimal numbers.
3. Estimate and check subtraction.
4. Subtract using a calculator.

Subtraction of whole numbers is another basic skill required in most occupations. Subtraction is the *inverse operation* of addition. In addition, we add numbers to get their total (such as $5 + 4 = 9$), but to solve the subtraction problem $9 - 5 = ?$, we ask ourselves, What number must be added to 5 to give us 9? The answer is 4 because 4 must be added to 5 to give a total of 9. When we subtract two numbers, the answer is called the *difference* or *remainder*. The initial quantity is the *minuend*. The amount being subtracted from the initial quantity is the *subtrahend*.

1 Subtract Whole Numbers.

In addition, it does not matter in which order numbers are added. But *in subtraction order is important*: $8 - 3 = 5$, but $3 - 8$ does not equal 5; that is, $3 - 8 \neq 5$. Subtraction is *not* commutative, whereas addition is. In addition, grouping does not matter when a problem has three or more numbers. But *in subtracting, grouping is important*.

EXAMPLE 1 Show that $9 - (5 - 1)$ does not equal $(9 - 5) - 1$.

$$9 - (5 - 1) = 9 - 4 = 5, \quad \text{but} \quad (9 - 5) - 1 = 4 - 1 = 3$$

If a subtraction problem contains no parentheses to show the grouping, subtract the first two numbers at the left. Then subtract from that difference the next number in the problem.

EXAMPLE 2 Subtract $8 - 3 - 1$.

$$8 - 3 - 1 = 5 - 1 = \mathbf{4}$$

Rule 1-14 *Zero subtracted from a number results in the same number:*

$$n - 0 = n, \qquad 7 - 0 = 7$$

Tips and Traps *Subtraction and Zeros:*
Subtracting a number from zero is not the same as subtracting zero from a number; that is, $7 - 0 = 7$, but $0 - 7$ does not equal seven.

> **Rule 1–15** *To subtract numbers of two or more digits:*
>
> **1.** Arrange the numbers in columns, with the minuend at the top and the subtrahend at the bottom.
> **2.** Make sure the ones digits are in a vertical line on the right.
> **3.** Subtract the ones column first, then the tens column, the hundreds column, and so on.
> **4.** When subtracting a larger digit from a smaller digit in any given column, borrow 1 from the column to the left. To borrow 1 from the next column to the left is the same as borrowing *one* group of 10; thus, place the borrowed 1 in front of the digit in the minuend.

EXAMPLE 3   Subtract 5327 − 3514.

Arrange in columns.

$$
\begin{array}{r}
{}^{4\,13} \\
5327 \\
-\ 3514 \\
\hline
\mathbf{1813}
\end{array}
$$

Tips and Traps   *Words and Phrases That Imply Subtraction:*
Typical phrases that indicate subtraction in applied problems include: how many are left, how many more, how much less, how much larger, and how much smaller. Also, some applied problems require more than one operation. Many problems requiring more than one operation involve parts and a total. If we know a total and all the parts but one, we must add all the known parts and subtract the result from the total to find the missing part.

EXAMPLE 4   The Randles left Memphis and drove 356 miles on the first day of their vacation. They drove 426 miles on the second day. If they are traveling to Albuquerque, which is 1,050 miles from Memphis, how many more miles do they have to drive?

The phrase, *how many more,* indicates subtraction.

$$1050 \text{ miles} = \text{total miles}$$
$$356 + 426 + \text{miles left to drive} = 1050 \text{ miles}$$

To find the miles left to drive, add 356 + 426 and subtract the result from 1050.

$$356 + 426 = \boxed{782} \qquad 1050 - \boxed{782} = \mathbf{268}$$

2   Subtract Decimal Numbers.

When subtracting decimals, we align the digits just as we do when adding decimals. Again, this is equivalent to finding common denominators for decimal fractions. Aligning the decimal points is sufficient.

> **Rule 1–16** *To subtract decimals:*
>
> Arrange the numbers so that the decimal points are in the same vertical line. Then subtract each column and place the decimal in the difference.

EXAMPLE 5   Subtract 8.29 from 13.76.

$$
\begin{array}{r}
13.76 \\
-\ \ 8.29 \\
\hline
\mathbf{5.47}
\end{array}
$$
Notice how the decimals are aligned.

EXAMPLE 6   Subtract 7.18 from 15.

In this problem we must take care to align the decimals properly. Because 15 is a whole number, its decimal point will be placed after the 5.

$$
\begin{array}{r}
15. \\
-\ \ 7.18 \\
\hline
\end{array}
$$

To subtract, we put zeros in the tenths and hundredths places of 15 and then borrow.

$$
\begin{array}{r}
15.00 \\
-\ \ 7.18 \\
\hline
\mathbf{7.82}
\end{array}
$$

When a worker machines an object using a blueprint as a guide, a certain amount of variation from the blueprint specification is allowed for the machining process. This amount of variation is called *tolerance*. Thus, if a blueprint calls for a part to be 9.47 in. with a tolerance of ±0.05, this means that the actual part can be 0.05 in. *more* than the specification or 0.05 in. *less* than the specification. To find the *largest* possible size of the object, we add $9.47 + 0.05 = 9.52$ in. To find the *smallest* possible size of the object, we subtract $9.47 - 0.05 = 9.42$ in. **The dimensions, 9.52 in. and 9.42 in., are called the *limit dimensions* of the object.**

EXAMPLE 7   Find the limit dimensions of an object with a blueprint specification of 8.097 in. and a tolerance of ±0.005. (This is often written 8.097 ± 0.005.)

$$
\begin{array}{r}
8.097 \\
-\ \ 0.005 \\
\hline
8.092 \text{ in.}
\end{array}
\qquad
\begin{array}{r}
8.097 \\
+\ \ 0.005 \\
\hline
8.102 \text{ in.}
\end{array}
$$

The limit dimensions are 8.092 in. and 8.102 in.

③ Estimate and Check Subtraction.

Estimating a subtraction problem is similar to estimating an addition problem. The numbers in the problem are rounded before the subtraction is performed. The accuracy of the estimate depends on the method used for rounding.

> **Rule 1–17**   *To estimate the difference:*
>
> **1.** Round each number to the indicated place value or to a number with one nonzero digit.
> **2.** Subtract the rounded numbers.

> **Rule 1–18**   *To check a subtraction problem:*
>
> **1.** Add the subtrahend and difference.
> **2.** Compare the result of Step 1 with the minuend. If the two numbers are equal, the subtraction is correct.

EXAMPLE 8   Estimate by rounding to hundreds, then find the exact difference, and check.

$$427.45 - 125$$

	Estimate	Exact	Check
427.45	400	427.45	125.00
− 125.00	− 100	− 125.00	+ 302.45
	300	**302.45**	**427.45**

4 Subtract Using a Calculator.

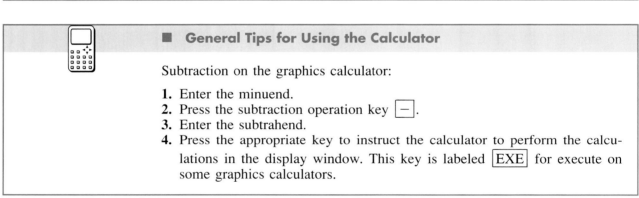

■ **General Tips for Using the Calculator**

Subtraction on the basic and scientific calculators:

1. Enter the minuend.
2. Press the subtraction operation key $\boxed{-}$.
3. Enter the subtrahend.
4. After entering the subtrahend, press the subtraction operation key or the equals key $\boxed{=}$, which will place the difference in the display window.

■ **General Tips for Using the Calculator**

Subtraction on the graphics calculator:

1. Enter the minuend.
2. Press the subtraction operation key $\boxed{-}$.
3. Enter the subtrahend.
4. Press the appropriate key to instruct the calculator to perform the calculations in the display window. This key is labeled $\boxed{\text{EXE}}$ for execute on some graphics calculators.

EXAMPLE 9   Find the difference between $53,943.76 and $34,256.45.

Basic or scientific:

$$53943\boxed{\cdot}76\boxed{-}34256\boxed{\cdot}45\boxed{=}$$

Basic or scientific display: 19687.31
Graphics:

$$53943\boxed{\cdot}76\boxed{-}34256\boxed{\cdot}45\boxed{\text{EXE}}$$

Graphics display: 53943.76 − 34256.45

19687.31

Commas are not entered into the calculator, and most calculator displays do not place commas in the answer.

EXAMPLE 10   Two cuts were made from a 72-in. pipe (see Fig. 1–9). The two lengths cut from the pipe were 28 in. and 15 in. How much of the pipe was left after these cuts were made?

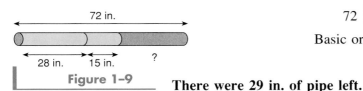

72 ⬜ − ⬜ 28 ⬜ − ⬜ 15 ⬜ = ⬜

Basic or scientific display: 29.

Figure 1–9

There were 29 in. of pipe left.

SELF-STUDY EXERCISES 1–3

1   Do the following subtraction problems.

1. 7 − 2
2. (8 − 3) − 4
3. 8 − (5 − 4)
4. 9 − 5 − 2
5. (7 − 1) − 3
6. 8 − (2 − 1)
7. (8 − 2) − 1
8. 9 − 3 − 4
9. 8 − (5 − 1)
10. 6 − 3 − 2
11. 47 − 23
12. 427 − 26
13. 3672 − 2652

14.   946
       − 831

15.   53,867
       −   831

16. If a mason ordered 75 bags of cement for a job and used only 53, how many bags were left?

17. An inventory sheet shows that 468 outlet boxes were in stock on March 1. Sales during March were 127. How many outlet boxes were left at the end of the month?

18. A reel of cable contained 575 ft. The following amounts were used on three separate jobs: 112 ft, 101 ft, and 41 ft. How much cable was left on the reel?

2   Subtract.

19.   14.86
       − 7.93

20.   20.07
       − 4.236

21. 813.673
       −   9.98

22.   84
       − 27.86

23.   5.079
       − 0.985

24. Subtract 8.8 from 12.7.
25. Subtract 24.38 from 316.2.
26. Subtract 13.5 from 21.
27. Subtract 67.2 from 378.
28. Find the difference between 42 and 37.6.
29. One box of rivets weighs 52.6 lb and another box weighs 37.5 lb. The first box of rivets weighs how much more than the second box?

30. The length of an object is 12.09 in. according to a blueprint. If the tolerance is ±0.01 in., what are the limit dimensions of the object?
31. Two lengths of copper tubing measure 63.6 cm and 3.77 cm. What is the difference in their lengths?

32. Find the limit dimensions of an object whose blueprint dimension is 4.195 in. ± 0.006 in.
33. Steel rods of 0.38 decimeter (dm) and 1.9 dm are cut from a steel rod 2.5 dm long. If the cutting waste is ignored, how long is the piece that is left?

34. If a current of 3.95 A is removed from a 15.5-A parallel circuit, find the remaining current in the circuit.
35. If a hardbound novel costs $27.50 and the softcover edition costs $18.75, how much can be saved by buying the softcover edition?

36. A student buys electronic supplies for an engineering technology class for $27.75 and pays with $30. How much is the change?
37. If the sheet music for a popular song costs $5.25, how much change would be returned if the purchaser pays with a $10 bill? Disregard sales tax.

3   Subtract the following problems. Check your answers.

38.   82
       − 37

39.   961
       − 353

40.   4070
       − 2497

41.   30,021
       − 7,816

Estimate differences by rounding numbers to one nonzero digit. Then find the exact answer.

42.   589,760
        − 498,726

43.   4903.009
        − 2814.125

44.   3070
        − 2896

45. $ 1401.30
        −   802.25

46.   80,096
        − 9,714

47. $ 1801.57
        −   704.42

[4] Estimate differences by rounding numbers to one nonzero digit. Then find the exact answer using your calculator.

48.   8001
        − 3604

49.   3804.621
        − 2617.370

50.   895,740
        − 387,465

51. A fuel tank on one model automobile has a capacity of 75 liters (L). The tank on a different model has a capacity of 92 L. How much larger is the second tank?

52. A bricklayer laid 1283 bricks on one day. A second bricklayer laid 1097 bricks. How many more bricks did the first bricklayer lay?

53. In a stockroom there are 285.8 in. of bar stock and 173.5 in. of round stock. How many inches of bar stock remain after the object in Fig. 1–10 is made from this stock?

54. Referring to Exercise 53, how many inches of round stock remain after the object in Fig. 1–10 is made?

55. Find the missing dimension in Fig. 1–11.

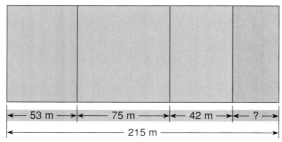

Figure 1–11

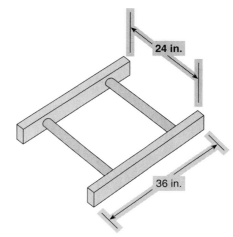

Figure 1–10

1–4   MULTIPLYING WHOLE NUMBERS AND DECIMALS

Learning Objectives

1. Multiply whole-number factors.
2. Multiply decimal factors.
3. Apply the distributive property.
4. Estimate and check multiplication.
5. Multiply using a calculator.

[1] Multiply Whole-Number Factors.

Multiplication is repeated addition. If we have 3 ten-dollar bills, we have $10 + $10 + $10, or $30. Using multiplication, this is the same as 3 times $10, or $30.

When we multiply two numbers, the first number is called the *multiplicand,* and the number we multiply by is called the *multiplier.* Either of the numbers may be referred to as *factors.* The answer or result of multiplication is called the *product.*

$$2 \quad \times \quad 3 \quad = \quad 6$$

multiplicand	multiplier	product
or factor	or factor	

Tips and Traps   *Various Notations for Multiplication:*
Besides the familiar $\times$ or "times" sign, the raised dot ($\cdot$), the asterisk ($*$), and parentheses ( ) may be used to show multiplication.

$$2 \cdot 3 = 6, \qquad 2 * 3 = 6, \qquad 2(3) = 6, \qquad (2)(3) = 6$$

Multiplication is commutative and associative, just like addition. The *commutative property of multiplication* permits two numbers to be multiplied in any order.

$$4 \times 5 = 20, \qquad 5 \times 4 = 20$$

When multiplying more than two numbers, the numbers must be grouped, and the *associative property of multiplication* permits the numbers to be grouped in any way.

$$2 \times (3 \times 5) \qquad \text{or} \qquad (2 \times 3) \times 5$$
$$2 \times \quad 15 \qquad\qquad\qquad 6 \quad \times 5$$
$$30 \qquad\qquad\qquad\qquad 30$$

EXAMPLE 1   Multiply $3 \times 2 \times 9$.

$3 \times 2 \quad \times 9$

$(3 \times 2) \quad \times 9$ Group any two factors.

$\quad 6 \quad \times 9$ Multiply grouped factors.

$\qquad$ **54**   Multiply the factors: 6 and 9.

Another property is the *zero property of multiplication.*

Rule 1–19   *The product of a number and zero is zero:*
$$n \times 0 = 0, \qquad 0 \times n = 0, \qquad 4 \times 0 = 0, \qquad 0 \times 4 = 0$$

EXAMPLE 2   Multiply $2 \times 5 \times 0 \times 7$.

$2 \times 5 \times \ 0 \times 7$

$(2 \times 5) \times (0 \times 7)$ Group factors.

$\quad 10 \quad \times \quad 0$ Multiply each group of factors.

$\qquad\quad$ **0**   Multiply products.

> **Rule 1-20** *To multiply factors of two or more digits:*
>
> **1.** Arrange the factors one under the other.
> **2.** Multiply each digit in the multiplicand by each digit in the multiplier. The product of the multiplicand and each digit in the multiplier gives a *partial product*.
> **a.** Start with the ones digit in the multiplier and multiply the multiplicand from right to left.
> **b.** Align each partial product with its first digit directly under its multiplier digit.
> **3.** Add the partial products.

EXAMPLE 3 Multiply 204 × 103.

```
        1
      204
 ×    103
      612      Multiply: 3 × 204 = 612. Align 612 under 3 in the multiplier.
    0 00       Multiply: 0 × 204 = 000. Align 000 under 0 in the multiplier.
   20 4        Multiply: 1 × 204 = 204. Align 204 under 1 in the multiplier.
   21,012      Add the partial products as they are aligned.
```

Partial products 000 and 204 could be combined on a single line to be 2040, with the rightmost 0 aligned under the 0 in the multiplier and the 4 aligned under the 1 in the multiplier.

Some applied problems require more than one operation to answer the question. The following example requires both multiplication and addition.

EXAMPLE 4 John Paszel is taking inventory and finds 27 unopened boxes of candy. Each of the boxes contains 48 candy bars. In another part of the warehouse, John counts 33 unopened cases of the same candy. These cases each contain 288 bars of candy. How many candy bars should John show on his inventory report?

John must first calculate the number of candy bars in the 27 unopened boxes.

$$27 \times 48 = 1296$$

Next, the number of candy bars in the cases must be calculated.

$$33 \times 288 = 9504$$

Finally, the total number of candy bars is 1296 + 9504, or 10,800.

2 Multiply Decimal Factors.

One way we can multiply decimal numbers is by writing them in their fractional form and then multiplying. To multiply 0.8×0.32, for example, we can write $\frac{8}{10} \times \frac{32}{100}$. Multiplying, we have $\frac{256}{1000}$, which is written as 0.256 in decimal form.

Before continuing, let's use these three numbers to make some observations about decimal fractions and the place-value chart in Fig. 1–12. Recall that the number of zeros in the denominator is the same as the number of places to the right of the decimal point.

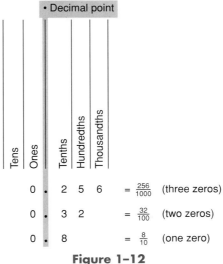

Tens	Ones	Tenths	Hundredths	Thousandths		
0	.	2	5	6	$= \frac{256}{1000}$	(three zeros)
0	.	3	2		$= \frac{32}{100}$	(two zeros)
0	.	8			$= \frac{8}{10}$	(one zero)

Figure 1–12

In 0.8×0.32, we see that 0.8 has one digit to the right of the decimal, while 0.32 has two digits to the right of the decimal. These two factors together have a total of three digits to the right of the decimal. If we compare this total with the number of digits on the right of the decimal in the product, 0.256, we see that they are the same. This fact allows us to multiply decimal numbers without converting them to fractions.

$$
\begin{array}{r}
0.32 \\
\times \quad 0.8 \\
\hline
0.256 \\
\end{array}
\quad
\begin{array}{l}
\text{2 places after decimal} \\
\underline{\text{1 place after decimal}} \\
\text{3 places after decimal}
\end{array}
$$

$$0.256 = \frac{256}{1000}$$

Therefore, we may use the following rule to multiply decimal fractions instead of converting the decimals to common fractions. The results are the same.

Rule 1–21 *To multiply decimal numbers:*

1. Align the numbers as if they were whole numbers and multiply.
2. Then count the total number of digits to the right of the decimal in each factor.
3. Place the decimal in the product so that the number of decimal places is the sum of the numbers of decimal places in the factors.

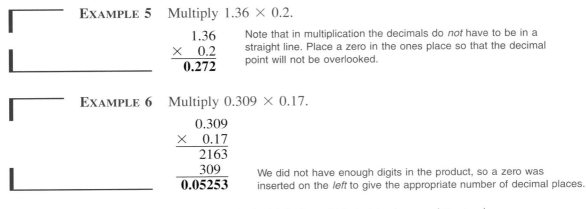

EXAMPLE 5   Multiply 1.36×0.2.

$$
\begin{array}{r}
1.36 \\
\times \quad 0.2 \\
\hline
\mathbf{0.272}
\end{array}
$$

Note that in multiplication the decimals do *not* have to be in a straight line. Place a zero in the ones place so that the decimal point will not be overlooked.

EXAMPLE 6   Multiply 0.309×0.17.

$$
\begin{array}{r}
0.309 \\
\times \quad 0.17 \\
\hline
2163 \\
309 \quad \\
\hline
\mathbf{0.05253}
\end{array}
$$

We did not have enough digits in the product, so a zero was inserted on the *left* to give the appropriate number of decimal places.

EXAMPLE 7   The outside diameter of a pipe is 7.82 cm (see Fig. 1–13). If the pipe wall is 1.56 cm thick, find the inside diameter of the pipe.

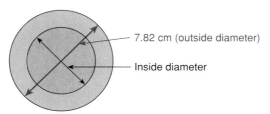

Figure 1–13

We double the thickness (1.56 × 2 = 3.12) to find the total thickness. This total is subtracted from the outside diameter to give the inside diameter.

$$\begin{array}{r} 7.82 \\ -\ 3.12 \\ \hline 4.70 \end{array}$$

The inside diameter is 4.70 cm.

3 Apply the Distributive Property.

Another property of multiplication is called the distributive property. The *distributive property of multiplication* means that multiplying a sum or difference by a factor is equivalent to multiplying each term of the sum or difference by the factor.

Rule 1–22 *Distributive Property of Multiplication:*

1. Add or subtract the numbers within the grouping.
2. Multiply the result of step 1 by the factor outside the grouping.

or

1. Multiply each number inside the grouping by the factor outside the grouping.
2. Add or subtract the products from step 1.

Symbolically,

$$a \times (b + c) = a \times b + a \times c \qquad \text{or} \qquad a(b + c) = ab + ac$$
$$a \times (b - c) = a \times b - a \times c \qquad \text{or} \qquad a(b - c) = ab - ac$$

Tips and Traps   *Other Notations for Multiplication:*
Parentheses show multiplication when the distributive property is used: $a(b + c)$ means $a \times (b + c)$. Letters can be used to represent numbers. Letters that are written together with no operation sign between them show multiplication: ab means $a \times b$; ac means $a \times c$.

EXAMPLE 8   Multiply 3(2 + 4).

Multiplying first gives: Adding first gives:
$$3(2 + 4) =$$ $$3(2 + 4) =$$
$$3(2) + 3(4) =$$ $$3(6) = \mathbf{18}$$
$$6 + 12 = \mathbf{18}$$

EXAMPLE 9   Multiply 2(6 − 5).

Multiplying first gives: Subtracting first gives:

$$2(6 - 5) =$$ $$2(6 - 5) =$$
$$2(6) - 2(5) =$$ $$2(1) = \mathbf{2}$$
$$12 - 10 = \mathbf{2}$$

The distributive property is found in many applied formulas. One example is the formula for the perimeter of a rectangle. A *rectangle* is a four-sided geometric shape whose opposite sides are equal in length, and each corner makes a square corner (see Fig. 1–14). The *perimeter* of a rectangle is the distance around the figure. We would find the perimeter if we were fencing a rectangular yard, installing baseboard in a rectangular room, framing a picture, or outlining a flower bed with landscaping timbers.

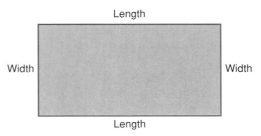

Figure 1–14

The formula for finding the perimeter of a rectangle is

$$P = 2(l + w) \qquad \text{or} \qquad P = 2l + 2w$$

EXAMPLE 10   Find the number of feet of fencing needed to enclose a rectangular pasture that is 1784.6 feet long and 847.3 feet wide.

$$P = 2(l + w) \qquad\qquad \text{or} \qquad P = 2l + 2w$$
$$P = 2(1784.6 + 847.3) \qquad\qquad P = 2(1784.6) + 2(847.3)$$
$$P = 2(2631.9) \qquad\qquad\qquad P = 3569.2 + 1694.6$$
$$P = 5263.8 \qquad\qquad\qquad\qquad P = 5263.8$$

The amount of fencing needed is 5263.8 feet.

4 Estimate and Check Multiplication.

Rule 1–23 *To estimate the answer for a multiplication problem:*

1. Round both factors to a certain place value or to one nonzero digit.
2. Then multiply the rounded numbers.

Rule 1–24 *To check a multiplication problem:*

1. Multiply the numbers a second time and check the product.
2. Interchange the factors if convenient.

EXAMPLE 11   Find the approximate and exact cost of 22 books if each book costs $29.15. Estimate the answer by rounding each factor to a number with one nonzero digit. Then find the exact answer and check your work.

22 rounds to 20 and $29.15 rounds to $30.

$$20 \times \$30 = \$600$$

The approximate cost of the books is $600.

```
      29.15      Check        22
    ×    22                 29.15
      58 30                  1 10
     583 0                   2 2
     ------                   198
     641.30                   44
                            ------
                            641.30
```

The exact cost of the books is $641.30.

Estimating is used in many do-it-yourself projects before exact calculations are made. One such example involves finding the area of a city house lot in the shape of a rectangle. The *area* of a geometric shape is the number of square units needed to cover the shape. Area is used when we are finding the amount of carpet or floor covering needed for a rectangular room, the amount of paint needed to paint a wall, the amount of asphalt needed to pave a parking lot, the amount of fertilizer needed to treat a yard, or the amount of material needed to produce a rectangular sign. Area is measured in *square measures*. One square foot (1 ft^2) indicates a square that measures 1 foot on each side.

5 feet long by 3 feet wide = 15 square feet

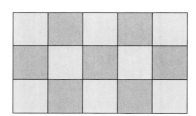

Figure 1–15

To find the area of a rectangle, we multiply the length by the width (see Fig. 1–15). This rule is abbreviated by using a *formula*.

$A = lw$ When two letters are written side-by-side with no operation sign, multiplication is implied.

EXAMPLE 12   Maintenance Consultants needs to apply fertilizer to a customer's lawn. The lawn measures 150.7 feet long and 125.4 feet wide. How many square feet are in the yard?

To estimate the number of square feet in the yard, round the length and width each to one nonzero digit: 150.7 rounds to 200 and 125.4 rounds to 100.

$$A = lw$$
$$A = 200(100)$$
$$A = 20,000$$

The lawn has approximately 20,000 ft^2 of ground to be covered.

To find the exact number of square feet in the yard, use the same formula with the exact measurements.

$$A = lw$$
$$A = (150.7)(125.4)$$
$$A = \mathbf{18,897.78 \ ft^2}$$

5 Multiply Using a Calculator.

■ **General Tips for Using the Calculator**

Multiplication on the basic and scientific calculators:

1. Enter the first factor and press the multiplication operation key $\boxed{\times}$.
2. Enter the next factor and press the multiplication operation key if more factors follow.
3. Continue until all factors have been entered.
4. After entering the final factor, press the multiplication operation key or the equal key $\boxed{=}$, which will place the product in the display window.

Both basic and scientific calculators will accumulate the product of entered factors each time the multiplication operation key or the equal key is pressed.

■ **General Tips for Using the Calculator**

Multiplication on the graphics calculator:

1. Enter the first factor and press the multiplication operation key $\boxed{\times}$.
2. Enter the next factor and press the multiplication operation key if more factors follow.
3. Continue until all factors have been entered.
4. Press the appropriate key to instruct the calculator to perform the calculations in the display window. This key is labeled EXE for execute on some graphics calculators.

The graphics calculator does not accumulate the product of entered factors as operation keys are pressed. All indicated calculations are performed after the execute key is pressed.

EXAMPLE 13   Find the product of the following numbers using a hand-held calculator.

$$3283 \times 34.6 \times 34 =$$

Basic or scientific:

$$3283 \ \boxed{\times} \ 34 \ \boxed{\cdot} \ 6 \ \boxed{\times} \ 34 \ \boxed{=}$$

Basic or scientific display: **3862121.2**

Graphics:

$$3283 \ \boxed{\times} \ 34 \ \boxed{\cdot} \ 6 \ \boxed{\times} \ 34 \ \boxed{\text{EXE}}$$

Graphics display: $3283 \times 34.6 \times 34$

3862121.2

When either or both of the factors of a multiplication problem end in zeros, a short-cut process such as the one in Example 14 can be used.

EXAMPLE 14   Solve the problem 2600×70.

1.
$$\begin{array}{r} 26|00 \\ \times \quad 7|0 \end{array}$$
   Separate the zeros at the end of the factors from the other digits.

2.
$$\begin{array}{r} 26|00 \\ \times \quad 7|0 \\ \hline 182| \end{array}$$
   Multiply the other digits as if the zeros were not there ($26 \times 7 = 182$).

3.
$$\begin{array}{r} 26|00 \\ \times \quad 7|0 \\ \hline 182|000 \end{array}$$
   Affix the zeros to the basic product. Note that the number of zeros affixed to the basic product is the same as the number of zeros at the end of the factors.

This process is sometimes necessary whenever a multiplication problem is too long to fit into a calculator. Many hand-held calculators have only an eight-digit display window for calculations. The problem $26,000,000 \times 3000$ would not fit many basic calculators. This shortcut allows us to work the problem quickly, with or without a calculator.

EXAMPLE 15   Multiply $26,000,000 \times 3000$.

$$\begin{array}{r} 26,000,000 \\ \times \quad 3\,000 \\ \hline 78,000,000,000 \end{array}$$
   Separate ending zeros and multiply 26×3.

SELF-STUDY EXERCISES 1–4

1   Multiply or answer the following.

1. Find the following products.
   (a) 5×3 **(b)** $7 * 8$
   (c) $(9)(7)$ **(d)** $4 \cdot 6$

3. What property of multiplication justifies the statement "$5(3) = 3(5)$"?

2. Jaime Oxnard has 9 wood boxes that he intends to sell for \$7 each. If Jaime sells 6 of the boxes, how much money will he make?

4. Explain the associative property of multiplication and give an example to illustrate the property.

Multiply.

5. $5 \cdot 3 \cdot 0 \cdot 6$ **6.** $3 * 7 * 9 * 2$ **7.** $7 \times 3 \times 4$ **8.** $(3)(2)(0)(8)$

9.
$$\begin{array}{r} 32 \\ \times \quad 7 \end{array}$$
10.
$$\begin{array}{r} 83 \\ \times \quad 7 \end{array}$$
11.
$$\begin{array}{r} 90 \\ \times \quad 7 \end{array}$$
12. 503×204

13. Margaret Johnston purchased 3 cases of candy bars. Each case contained 12 bags and each bag contained 24 pieces of individually wrapped candy. How many individually wrapped candies did Margaret purchase?

14. Chuckee Wright counted 8 unopened boxes of washers. Each box contained 512 washers. What is the total number of washers to be shown on the inventory sheet?

15. Tracie Burke reported reading 6 books a week over a period of 14 weeks. How many books did Tracie read during this time period?

16. Each officer in the Public Safety office wrote, on the average, 25 tickets a week. If there are 7 officers, how many tickets were written over a 4-week period?

17. Jossie Moore is planning to sell candy bars in her Smart Shop. She receives 12 boxes and each box contains 24 candy bars. If Jossie sells the bars for \$1 each, how much money will she get for all the bars?

18. Terry Kelly teaches 5 classes and each class has 46 students. She teaches a sixth class that has 19 students enrolled. How many students is Terry teaching?

2 Multiply.

19. 53.4
        × 0.29

20. 37.7
        × 1.5

21. 9.27
        × 0.35

22. 0.215
        × 0.27

23. 0.271
        × 0.32

24. 3.7 × 1.93

25. 2.78 × 0.03

26. 73.806 × 2.305

27. 1.9067 × 0.2013

28. 8.2037 × 0.602

29. 42 × 0.73

30. 81 × 7.37

31. 81.7 × 0.621

32. 72.6 × 0.532

33. 326 × 1.04

34. 261 × 2.07

35. 6.35 × 40

36. 9.21 × 20

37. 93.07 × 0.01

38. 76.02 × 0.01

39. 586.35 × 0.001

40. 732.65 × 0.001

41. 2.0037 × 4.6

42. 3.0062 × 3.8

43. 7.04 × 0.0025

44. 6.01 × 0.0045

45. 0.457 × 2.003

46. 0.387 × 1.009

47. 0.05 × 0.003

48. 0.07 × 0.008

49. A pipe has an outside diameter of 4.327 in. Find the inside diameter if the pipe wall is 0.185 in. thick.

50. A plastic pipe has an inside diameter of 4.75 in. Find the outside diameter if the pipe wall is 0.25 in. thick.

51. Electrical switches cost $5.70 wholesale. If the retail price is $7.99 each, how much would an electrician save over retail by buying 12 switches at the wholesale price?

52. How much No. 24 electrical wire is needed to cut eight pieces each 18.9 in. long?

53. A retailer purchases 15 cases of potato chips at $8.67 per case. If the chips were sold at $12.95 per case, how much profit did the retailer make?

54. A contractor purchases 500 yd³ of concrete at $21.00 per cubic yard, 24 yd³ of sand at $5.00 per cubic yard, and 36 yd³ of fill dirt at $3.00 per cubic yard. What was the total cost of the materials?

55. A micrometer is designed so that the screw thread makes the spindle advance 0.025 in. in one complete turn. How far will the spindle advance in 15 complete turns?

56. A 4 ft × 8 ft sheet of 1-in.-thick steel weighs 1307 lb and costs $0.30 a pound. Find the cost of five of these sheets.

57. If six 2.5-A devices are connected in parallel, what is the total amperage (A) of the circuit?

58. Sequoia ordered 24 BNC T-adapters at $3.25 each. How much was the total order?

3 Find the product in each of the following by multiplying first, and check your answers by adding first.

59. 5(7 + 4)

60. 3(9 + 11)

61. 14(3 + 11)

62. 4(5 − 2)

63. 1.2(16 − 3)

64. 7(1.8 − 1)

Find the perimeter of the rectangles in Figs. 1–16 and 1–17.

65.

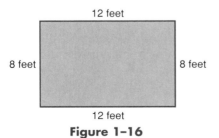

Figure 1–16

66.

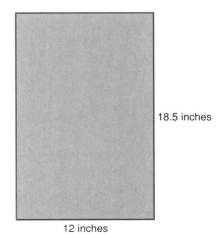

Figure 1–17

67. How much baseboard is required to surround a kitchen if the floor measures 15 feet by 13 feet and no allowance is made for door openings?

68. A dog pen is staked out in the form of a rectangle so that the width is 38 feet and the length is 47 feet. How much fencing must be purchased to build the pen?

4 Complete the following multiplication problems as directed.

69. Find the approximate cost of 24 shirts if each shirt costs $32. What is the exact cost of the shirts?

Find the number of square feet in each of the rectangles illustrated in Figs. 1–18 and 1–19. Check your work.

70.

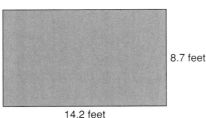

8.7 feet

14.2 feet

Figure 1–18

71.

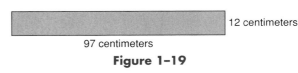

12 centimeters

97 centimeters

Figure 1–19

72. Estimate how many square feet of carpet should be purchased to cover a rectangular room that measures 42.3 feet by 32.5 feet. Find the exact area.

73. A rectangular table top is 12 feet long and 8 feet wide. How many square feet of glass are needed to cover it? Check your work.

5 Use the calculator to find the following products.

74. $5896 \times 47.3 \times 12.83$

75. $52{,}834 * 15 * 29.5$

76. $832 \cdot 703 \cdot 960$

77. $583(294)(627)(0)(8)$

78. Paul Thomas earns $3015 per month. What is Paul's annual salary?

79. Ertha Brower earns $423.20 per week. Calculate her annual salary.

Perform the following multiplications.

80. $3500 \times 42{,}000{,}000$

81. $30{,}800 * 544{,}000$

82. $70{,}000(6000)$

1–5 DIVIDING WHOLE NUMBERS AND DECIMALS

Learning Objectives

1 Use various symbols to indicate division.

2 Divide whole numbers.

3 Divide decimal numbers.

4 Estimate and check division.

5 Find the numerical average.

6 Divide using a calculator.

1 Use Various Symbols to Indicate Division.

Division is the inverse operation for multiplication. If we are trying to determine what number we should multiply 7 by to obtain a product of 28, that would be equivalent to dividing 28 by 7. The result is 4. In division, the number we divide by is the *divisor*. The number being divided is the *dividend*. The answer is the *quotient*.

$$28 \div 7 = 4 \longleftarrow \text{quotient}$$
$$\text{dividend} \longrightarrow \qquad \text{divisor}$$

Besides the "divided by" sign ($\div$), there is also in our culture the symbol $\overline{}$ in problems such as $7\overline{)28}$. The problem $7\overline{)28}$ means the same as $28 \div 7$.

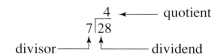

$$\begin{array}{r} 4 \longleftarrow \text{quotient} \\ 7\overline{)28} \end{array}$$
$$\text{divisor} \longrightarrow \qquad \text{dividend}$$

Another division symbol is a bar, such as $\dfrac{28}{7}$, which also means the same as $28 \div 7$.

$$\text{dividend} \longrightarrow \dfrac{28}{7} = 4 \longleftarrow \text{quotient}$$
$$\text{divisor} \longrightarrow$$

Words indicating division include:

1. "divide by" $\longrightarrow$ dividend divided by divisor
2. "goes into" $\longrightarrow$ divisor goes into dividend
3. "divide into" $\longrightarrow$ divisor divided into dividend

EXAMPLE 1   Express 2 goes into 8 as a division using the symbol $\div$, the symbol $\overline{}$, and the bar symbol —. Then solve the problem.

$$8 \div 2 = ? \qquad \text{Because } 4 \times 2 = 8, \quad 8 \div 2 = 4, \quad 2\overline{)8}^{\,4}, \quad \dfrac{8}{2} = 4.$$

2 Divide Whole Numbers.

Some of the problems that arise in division involve *long division*. Study the following example. Long division, like long multiplication, involves using the simple one-digit multiplication facts repeatedly to find the quotient.

EXAMPLE 2   Divide $975 \div 12$.

$12\overline{)975}$ Use long-division symbol.

$\begin{array}{r} 8 \\ 12\overline{)975} \end{array}$ 12 does not divide into 9, but it does go into 97 8 times. Put the 8 over the 7, the last digit of 97.

$\begin{array}{r} 8 \\ 12\overline{)975} \\ \underline{96} \end{array}$ Multiply $8 \times 12 = 96$. Put the 96 under the 97 of the dividend.

$\begin{array}{r} 8 \\ 12\overline{)975} \\ \underline{96} \\ 15 \end{array}$ Subtract $97 - 96 = 1$. This difference must be less than 12, the divisor. Bring down the next digit in the dividend, 5.

$\begin{array}{r} 81 \\ 12\overline{)975} \\ \underline{96} \\ 15 \end{array}$ 12 divides into 15 one time. Put the 1 over the 5 of the dividend.

$\begin{array}{r} 81 \\ 12\overline{)975} \\ \underline{96} \\ 15 \\ \underline{12} \\ 3 \end{array}$ Multiply $1 \times 12 = 12$. Put the 12 under the 15.
Subtract $15 - 12 = 3$. This difference must be less than 12.

$$\begin{array}{r} \underline{\mathbf{81}}\ \ \mathbf{R3} \\ 12\overline{)975} \\ \underline{96} \\ 15 \\ \underline{12} \\ 3 \end{array}$$

The 3 is the remainder.

At times, after subtracting and bringing down the next digit, the divisor will not divide into the resulting number at least one time. In this case, we place a zero in the quotient.

Tips and Traps *Importance of Placing the First Digit Carefully:*
The correct placement of the first digit in the quotient is critical. If the first digit is out of place, all digits that follow will be out of place, giving the quotient too few or too many digits.

EXAMPLE 3 Divide $5\overline{)2535}$.

$$\begin{array}{r} 5 \\ 5\overline{)2535} \\ \underline{25} \\ 03 \end{array}$$

5 divides into 25 five times. Put 5 over the last digit of the 25. Subtract and bring down the 3.

$$\begin{array}{r} \mathbf{5\ 0\ 7} \\ 5\overline{)2\,5\ 3\ 5} \\ \underline{25} \\ 0\,3\,5 \\ \underline{3\,5} \\ 0 \end{array}$$

5 divides into 3 zero times, so put 0 over the 3 of the dividend. Bring down the next digit, which is 5. Divide 5 into the 35: $35 \div 5 = 7$. Put the 7 over the 5 of the dividend. Continue as usual.

When dividing by a one-digit divisor, we can perform the division mentally. This process is usually referred to as *short division*.

EXAMPLE 4 Divide $7\overline{)180}$.

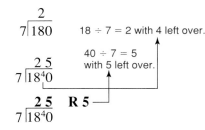

Special properties are associated with division involving zeros, division of a number by itself, and division by 1.

Rule 1–25 *The quotient of zero divided by a nonzero number is zero:*

$$0 \div n = 0, \qquad n\overline{)0}^{\,0}, \qquad \frac{0}{n} = 0$$

$$0 \div 5 = 0, \qquad 5\overline{)0}^{\,0}, \qquad \frac{0}{5} = 0$$

If a number divided by zero is to have an answer, then that answer times zero should equal the number. However, we learned in multiplication that zero times any number is zero. Therefore, a number divided by zero has *no* answer.

$$8 \div 0 = ?$$ What number $\times\ 0 = 8$? Because any number times $0 = 0$, $8 \div 0$ has no answer.

The quotient of a division problem can be only *one* number. Consider the problem $0 \div 0$.

$$0 \div 0 = ?$$ What number $\times\ 0 = 0$? Because *any number* times $0 = 0$, $0 \div 0$ does not have just *one* number as its quotient; that is, the quotient is not *unique*. For this reason, we say that division of zero by zero has no answer, or that division of zero by zero is impossible.

Rule 1–26 *The quotient of a number divided by zero is impossible (undefined):*

$$n \div 0 \text{ is impossible}, \qquad 0\overline{)\,\overset{\text{impossible}}{n}\,}, \qquad \frac{n}{0} \text{ is impossible}$$

$12 \div 0$ is undefined, or impossible

$0 \div 0$ is undefined, or impossible

Rule 1–27 *To divide any nonzero number by itself is 1:*

$$n \div n = 1, \qquad \text{if } n \text{ is not equal to zero;} \qquad 12 \div 12 = 1$$

Rule 1–28 *To divide any number by 1 yields the same number:*

$$n \div 1 = n, \qquad 5 \div 1 = 5$$

In division, as in subtraction, order is important. For example, $12 \div 4 = 3$; however, $4 \div 12$ is not 3. Therefore, division is *not* commutative. When dividing more than two numbers, again as in subtraction, the way the numbers are grouped is important. For example, $(16 \div 4) \div 2 = 4 \div 2 = 2$. If we change the grouping to $16 \div (4 \div 2)$, we have $16 \div 2 = 8$. Notice the different answers: 2 and 8. Therefore, division is *not* associative. If no grouping symbols are included, we divide from left to right.

3 Divide Decimal Numbers.

Dividing decimals is similar to dividing whole numbers. However, as in multiplication, we must be very careful about the placement of the decimal point in the quotient.

To understand the placement of the decimal, consider the example $3.6 \div 1.2$. This division can be written as the fraction $\frac{3.6}{1.2}$. If we multiply the numerator and denominator of the fraction by 10, we make each number a whole number. Because $\frac{10}{10} = 1$, we do not change the value of the fraction.

$$\frac{3.6}{1.2} \times \frac{10}{10} = \frac{3.6 \times 10}{1.2 \times 10} = \frac{36}{12} = 3$$

Now consider the same division problem in long-division form.

$$1.2\overline{)3.6}$$

The decimal point is shifted to the end of the divisor, 1.2, which in this case is one place. (Remember that multiplying by 10 results in moving the decimal one place to the right.) The dividend is also multiplied by 10, which shifts the decimal one place. The decimal in the quotient is written directly above the new decimal in the dividend. The decimal point should be placed in position *before* the division is started.

$$1.2{,}\overline{\smash{)}3.6{,}}\ 3.$$

The reason for moving the decimal point in the divisor is to make the divisor a *whole* number to simplify the division process. This changes the value of the divisor, so the decimal point in the dividend must be moved the same number of places to keep the value of the division problem unchanged.

> **Rule 1-29**   *To divide decimal numbers written in long-division form:*
> 1. Shift the decimal in the divisor so that it is on the right side of all digits. (By shifting the decimal, we are multiplying by 10, 100, 1000, and so on.)
> 2. Shift the decimal in the dividend to the right as many places as the decimal was shifted in the divisor. (Affix zeros if necessary.)
> 3. Write the decimal point in the answer directly above the shifted position of the decimal in the dividend. (Do this *before* dividing.)
> 4. Divide as you would in whole numbers.

EXAMPLE 5   Divide 4.8 ÷ 6.

$$6\overline{\smash{)}4.8}^{\ \cdot}$$ Insert a decimal point above the decimal point in the dividend.

$$6\overline{\smash{)}4.8}^{\ \mathbf{0.8}}$$ Divide.

When the divisor is a whole number, the decimal is understood to be to the right of 6 and is not shifted. When the divisor is a whole number, the decimal point in the dividend is not shifted. The decimal is placed in the quotient directly above the decimal in the dividend.

EXAMPLE 6   Divide 3.12 ÷ 1.2.

$$1.2\overline{\smash{)}3.1{,}2}$$ Shift the decimal in the divisor and dividend.

$$12\overline{\smash{)}31.2}^{\ \cdot}$$ Place the decimal in the quotient, then divide.

$$
\begin{array}{r}
2.6 \\
12\overline{\smash{)}31.2} \\
\underline{24} \\
7\,2 \\
\underline{7\,2}
\end{array}
$$

Not all divisions have a remainder of zero. When divisions have a nonzero remainder, we have two alternatives. We can write the remainder as a fraction by placing it over the divisor and adding the fraction to the quotient, or we can round the quotient to a specified decimal place. We must choose one of these alternatives before dividing.

Rule 1–30 *To express a remainder as a fraction:*

1. Divide to the specified place value.
2. Affix zeros to the dividend after the decimal if needed to carry out the division.
3. Write the remainder over the divisor as a fraction after the last place in the quotient. Reduce the fraction to lowest terms if appropriate (Section 3–5).

EXAMPLE 7   Divide to the hundredths place and write the remainder as a fraction.

$$6.9 \overline{)23}$$

$$
\begin{array}{r}
3.33\frac{23}{69} \\
6.9\,\overline{)23.0\,00} \\
\underline{20\,7} \\
2\,3\,0 \\
\underline{2\,0\,7} \\
2\,30 \\
\underline{2\,07} \\
23
\end{array}
$$

or **$3.33\frac{1}{3}$**

Zeros are affixed so that the division can be carried to the decimal place specified, often hundredths.

Rule 1–31 *To round a quotient to a place value:*

1. Divide to one place past the desired rounding place.
2. Affix zeros to the dividend after the decimal if necessary to carry out the division.
3. Round the quotient to the place specified.

EXAMPLE 8   Divide and round the quotient to the nearest tenth.

$$3.2 \overline{)15.27}$$

$$
\begin{array}{r}
4.77 \\
3.2\,\overline{)15.2\,70} \\
\underline{12\,8} \\
2\,4\,7 \\
\underline{2\,2\,4} \\
2\,30 \\
\underline{2\,24} \\
6
\end{array}
$$

Since we are rounding to the nearest tenth, we divide to the hundredths place.

4.77 rounds to 4.8.

4 **Estimate and Check Division.**

Rule 1–32 *To estimate division:*

1. Round the divisor and dividend to one nonzero digit.
2. Find the first digit of the quotient.
3. Attach a zero in the quotient for each remaining digit in the dividend.

Because division and multiplication are inverse operations, division may be checked by multiplication.

> **Rule 1–33** *To check division:*
>
> **1.** Multiply the divisor by the quotient.
> **2.** Add any remainder to the product in Step 1.
> **3.** The result of Step 2 should equal the dividend.

EXAMPLE 9   Estimate, find the exact answer, and check $913 \div 22$.

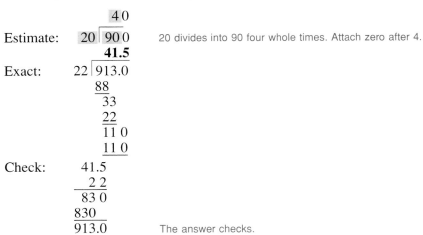

Estimate: 20 divides into 90 four whole times. Attach zero after 4.

The answer checks.

5 Find the Numerical Average.

Sometimes a worker needs to use approximate numbers for one reason or another. Two commonly used approximations are averages and estimates. Averages are often used for comparisons, such as comparing the average mileage different cars get per gallon of gasoline. There are several different kinds of averages. Here we refer to the average known as the arithmetic average or mean. Estimates are used for obtaining a quick but rough idea of what the answer to a problem is *before* working the problem.

In most courses students take, their numerical grade is determined by a process called *numerical averaging*. If the grades for a course are 92, 87, 76, 88, 95, and 96, we can find the average grade by adding the grades and dividing by the number of grades. Because we have six grades, the sum is divided by 6.

$$\frac{92 + 87 + 76 + 88 + 95 + 96}{6} = \frac{534}{6} = 89$$

> **Rule 1–34** *To find the average of a group of numbers or measures:*
>
> Add the numbers or measures and divide the sum by the number of addends.

EXAMPLE 10   A car involved in an energy efficiency study had the following miles per gallon (mpg) listings for five tanks of gasoline: 21.7, 22.4, 26.9, 23.7, and 22.6 mpg. Find the average miles per gallon for the five tanks of gasoline.

$$\frac{21.7 + 22.4 + 26.9 + 23.7 + 22.6}{5} = \frac{117.3}{5} = 23.46$$
$$= 23.5 \text{ mpg} \quad \text{rounded}$$

The average is 23.5 mpg.

Rounding Answers When Averaging:
If the numbers being averaged are expressed to the same place value, the answer is usually rounded to that place value.

6 Divide Using a Calculator.

■ General Tips for Using the Calculator

Division on the basic and scientific calculators:

1. Enter the dividend.
2. Press the division operation key $\boxed{\div}$.
3. Enter the divisor.
4. After entering the divisor, press the equals key $\boxed{=}$, which will place the quotient in the display window.

■ General Tips for Using the Calculator

Division on the graphics calculator:

1. Enter the dividend.
2. Press the division operation key $\boxed{\div}$.
3. Enter the divisor.
4. Press the appropriate key to instruct the calculator to perform the calculation in the display window. This key is labeled EXE for execute on some graphics calculators.

Tips and Traps *Division with a Calculator:*
Division of decimal numbers with a calculator does not require moving any of the decimal points.

EXAMPLE 11 Find the quotient of 9264 ÷ 2.4 using a calculator.

Basic or scientific:

$$9264 \boxed{\div} 2 \boxed{\cdot} 4 \boxed{=}$$

Basic or scientific display: **3860.**

Graphics:

$$9264 \boxed{\div} 2 \boxed{\cdot} 4 \boxed{\text{EXE}}$$

Graphics display: 9264 ÷ 2.4

3860.

Tips and Traps *Division Notation on a Calculator:*
The display on some graphics calculators may use a slash (/) instead of the division symbol (÷).

EXAMPLE 12 Paul Vo shops regularly at Ajax Discount Superstore for items used in his business. In preparing his tax return for the year, Paul has receipts for these amounts: $67.48, $123.76, $47.53, $54.97, $94.50, $46.23, $112.09, and $47.58. What is the amount of his average purchase?

$$\$67.48 + \$123.76 + \$47.53 + \$54.97 + \$94.50 + \$46.23 + \$112.09 + \$47.58 =$$
$$\$594.14 \div 8 = \$74.2675$$

The average amount of the receipts is $74.27.

■ General Tips for Using the Calculator

An average can be found in a continuous sequence on a calculator. One important step is to use the $=$ or $\boxed{\text{EXE}}$ key to find the sum *before* dividing by the number of the measures. Let's apply the calculator to Example 1–12.

Basic or scientific:

67 $\boxed{\cdot}$ 48 $\boxed{+}$ 123 $\boxed{\cdot}$ 76 $\boxed{+}$ 47 $\boxed{\cdot}$ 53 $\boxed{+}$ 54 $\boxed{\cdot}$ 97 $\boxed{+}$ 94 $\boxed{\cdot}$ 50 $\boxed{+}$ 46 $\boxed{\cdot}$

23 $\boxed{+}$ 112 $\boxed{\cdot}$ 09 $\boxed{+}$ 47 $\boxed{\cdot}$ 58 $\boxed{=}$ $\boxed{\div}$ 8 $\Rightarrow$ 74.2675 $74.27, rounded

Graphics:

67 $\boxed{\cdot}$ 48 $\boxed{+}$ 123 $\boxed{\cdot}$ 76 $\boxed{+}$ 47 $\boxed{\cdot}$ 53 $\boxed{+}$ 54 $\boxed{\cdot}$ 97 $\boxed{+}$ 94 $\boxed{\cdot}$ 50 $\boxed{+}$ 46 $\boxed{\cdot}$

23 $\boxed{+}$ 112 $\boxed{\cdot}$ 09 $\boxed{+}$ 47 $\boxed{\cdot}$ 58 $\boxed{\text{EXE}}$ $\boxed{\div}$ 8 $\boxed{\text{EXE}}$ $\Rightarrow$ 74.2675 $74.27, rounded

---

SELF-STUDY EXERCISES 1–5

1 Write the following as divisions using the symbols $\div$, $\boxed{}$, and $-$ for each problem.

1. 4 into 8

2. 3 into 9

3. 6 divided into 24

4. 7 divided into 30

5. 6 divided by 2

6. 8 divided by 4

2 Perform the following long-division problems.

7. 8)600

8. 9)207

9. 6)120

10. 7)21

11. 3)48

12. 8)96

13. 5)215

14. 37)1739

15. 26)312

16. 4)93

17. 48)5982

18. 133)7528

19. In a building where 46 outlets are installed, 1472 ft of cable are used. What is the average number of feet of cable used per outlet?

20. Twelve water tanks are constructed in a welding shop at a total contract price of $14,940. What is the price per tank?

21. Five equally spaced holes are drilled in a piece of $\frac{1}{4}$-in. flat metal stock. The centers of the first and last holes are 2 in. from the end (see Fig. 1–20). What is the distance between the centers of any two adjacent holes? (*Caution:* How many equal center-to-center distances are there?)

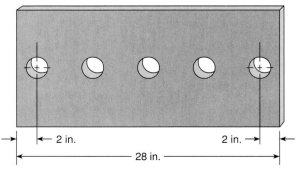

Figure 1–20

[3] Divide.

22. $3\overline{)3.78}$ **23.** $26\overline{)80.34}$ **24.** $21\overline{)1.323}$

25. $1.2\overline{)342}$ **26.** $4.8\overline{)28.32}$ **27.** $7.6\overline{)342}$

28. $6.3\overline{)68.67}$ **29.** $0.23\overline{)0.0437}$ **30.** $0.09\overline{)0.0954}$

31. $0.41\overline{)0.1353}$ **32.** $0.32\overline{)8}$ **33.** $1449 \div 0.07$

34. $8118 \div 0.09$ **35.** $4066 \div 0.38$ **36.** $8.28 \div 4.6$

37. A pipe 7.3 ft long weighs 43.8 lb. What is the weight of 1 ft of pipe?

38. A room requires 770.5 ft^2 of wallpaper, including waste. How many whole single rolls are needed for the job if a roll covers 33.5 ft^2 of surface?

39. The feed per revolution of a drill is 0.012 in. A hole 7.2 in. deep will require how many revolutions of the drill?

40. A piece of channel iron 5.6 ft long is cut into eight pieces. Assuming that there is no waste, what is the length of each piece?

41. If voltage is wattage divided by amperage, find the voltage in a 300-watt circuit drawing 3.2 amps.

42. Exercises 37 to 41 involve measures. In some instances, the answer is a measure and in others, the answer is an amount telling how many. Explain when the answer will be a measure.

Divide to the hundredths place and write the remainder as a fraction.

43. $1.3\overline{)25.8}$ **44.** $2.4\overline{)4.37}$ **45.** $0.06\overline{)156.07}$ **46.** $41.7 \div 21$ **47.** $0.23 \div 2.9$

Divide and round the quotient to the place indicated.

48. Nearest tenth: $0.43\overline{)72.8}$

49. Nearest hundredth: $25\overline{)3.897}$

50. Nearest whole number: $4.1\overline{)34.86}$

51. Nearest cent: $5\overline{)\$4.823}$

52. Nearest dollar: $17\overline{)\$24.98}$

53. If 12 lathes cost $6895, find the cost of each lathe to the nearest dollar.

54. If 12 electrolytic capacitors cost $23.75, find the cost of one capacitor to the nearest cent.

55. To find the depth of an American Standard screw thread, the number of threads per inch is divided into 0.6495. Find the depth of thread of a screw that has 8 threads per inch. Round to the nearest ten-thousandth.

56. A stack of 40 sheets of Fiberglas is 6.9 in. high. What is the thickness of one sheet to the nearest tenth of an inch?

[4] Estimate by rounding to one nonzero digit: then find the exact answer. Check your answers.

57. $7.2\overline{)904.32}$ **58.** $122\overline{)384,512}$

59. $221\overline{)824,604}$ **60.** $7\overline{)59.01}$

61. A landfill job needs 2294 cubic yards of dirt to be delivered. If one truck can carry 18 cubic yards, how many loads will have to be made?

62. If a reel of wire contains 6362.5 ft of wire, how many 50-ft extension cords can be made?

[5] Find the average of the following. Round to the same place value as that used in the problems.

63. Temperature readings: 82.5°, 76.3°, 79.8°, 84.7°, 80.8°, 78.8°, 80.0°

64. Test scores: 86, 73, 95, 85

65. Weight of cotton bales: 515 lb, 468 lb, 435 lb, 396 lb

66. Monthly income: $873.46, $598.21, $293.85, $546.83, $695.83, $429.86, $955.34, $846.95, $1025.73, $1152.89, $957.64, $807.25

67. Average rainfall: 1.25 in., 0.54 in., 0.78 in., 2.35 in., 4.15 in., 1.09 in.

68. Amperes of current: 3.0 A, 2.5 A, 3.5 A, 4.0 A, 4.5 A

[6] Work the following division problems using your calculator. Round the answer to the nearest whole number.

69. $1.58\overline{)85.297}$ **70.** $4\overline{)1203}$ **71.** $5.6\overline{)224.178}$

72. $45\overline{)4,027,500}$ **73.** $32\overline{)1,286,400}$ **74.** $71\overline{)440,271}$

75. Three bricklayers laid 3210 bricks on a job in one day. What was the average number of bricks laid by each bricklayer that day?

77. A shipment of 150 machine parts costs $15,737.50. If this includes a $25 shipping charge, how much does each part cost if shipping charges are not included?

79. A machine operator earns an annual salary of $29,580. What is the operator's monthly salary?

81. An invoice showing the total cost of refrigerator replacement parts indicated $2116.80, which included a $30 freight charge. The catalog price of the parts is $86.95 each. How many parts were ordered?

83. According to a real estate report in the Sunday newspaper, four homes in the Scenic Hills subdivision sold for $86,500, $68,750, $92,780, and $65,990. Find the average selling price of the homes in the subdivision.

85. The weather report listed the daily rainfall for the month of July as 0.2 in., 1.4 in., .05 in., 1.8 in., 0.6 in., 0.8 in., and 0.2 in. Find the average daily rainfall for the month to the nearest tenth of an inch.

76. A developer divided a tract of land into 14 equally valued parcels. If the tract is valued at $147,000, what is the value of each parcel?

78. If a shop manager earns $23,400 annually, what is the monthly salary?

80. A shipment of 288 headlights is billed out at $1329.60, which includes a $48 freight charge. What is the cost per headlight, excluding freight costs?

82. A job requires 95 feet of pipe, which will be cut into 9.3-ft lengths. How many 9.3-ft lengths of pipe will there be?

84. Suprena Anderson paid the following amounts for her textbooks for the 1997 fall semester at Bolton Technical and Community College: English II, $67.50; general science, $92.75; psychology, $42.50; electronic technology I, $94.25; health, $18.50; and introduction to microcomputers, $42.75. Find the average cost of her textbooks for the semester.

1–6   EXPONENTS, ROOTS, AND POWERS OF 10

Learning Objectives

1   Simplify expressions containing exponents.

2   Square numbers and find the square roots of numbers.

3   Use powers of 10 to multiply and divide.

4   Use a calculator to find powers and roots.

1 Simplify Expressions Containing Exponents.

The product of repeated factors can be written in shorter form with the use of whole-number exponents. For example, $4 \times 4 \times 4 = 4^3$. The number 4 is called the *base* and is the repeated factor. The number 3 is called the *exponent* and indicates how many times the repeated factor is used. The expression 4^3 is read "four *cubed*" or "four to the third *power*" or "four raised to the third *power*." In the example $2 \times 2 \times 2 \times 2 \times 2 = 2^5$, the number 2 is the base and the number 5 is the exponent. The expression 2^5 is read "two to the fifth power." In expressions that have 2 as an exponent, such as 4^2, the expression is usually read as "four *squared*"; however, it may also be read as "four to the second power."

To perform the multiplication of 2^5 (or $2 \times 2 \times 2 \times 2 \times 2$) = 32 is to *simplify* the expression. The expression 2^5 is written in *exponential notation*. The number 32 is written in *standard notation*.

$$\text{base} \rightarrow 2^5 \leftarrow \text{exponent}$$

EXAMPLE 1 Identify the base and exponent of the following expressions. Then simplify each expression.

(a) 5^3 (b) 3^2 (c) 2^4 (d) 1.5^2

(a) 5^3 5 is the base; 3 is the exponent.
$$5^3 = 5 \times 5 \times 5 = 125$$

(b) 3^2 3 is the base; 2 is the exponent.
$$3^2 = 3 \times 3 = 9$$

(c) 2^4 2 is the base; 4 is the exponent.
$$2^4 = 2 \times 2 \times 2 \times 2 = 16$$

(d) 1.5^2 1.5 is the base; 2 is the exponent.
$$1.5^2 = 1.5 \times 1.5 = 2.25$$

When a number has 1 for an exponent, the 1 indicates that the number is used as a factor 1 time. For example, $5^1 = 5$. When the number does not have an exponent, the exponent is understood to be 1. For example, the number 8 equals 8^1.

An exponent can be any number, including 1 and 0. Exponents that are not whole numbers will be discussed later. Examine the following expressions in exponential notation.

$$4^3 = 4 \times 4 \times 4 \qquad \text{4 used as a factor 3 times}$$
$$4^2 = 4 \times 4 \qquad \text{4 used as a factor 2 times}$$
$$4^1 = 4 \qquad \text{4 used as a factor 1 time}$$

> **Rule 1–35**   *Any number with an exponent of 1 is that number itself:*
> $$a^1 = a, \qquad \text{for any base } a$$

EXAMPLE 2 Write the following exponential expressions as standard numbers.

(a) 3^1 (b) 8^1 (c) 5^1 (d) 2.3^1

(a) $3^1 = 3$ (b) $8^1 = 8$ (c) $5^1 = 5$ (d) $2.3^1 = 2.3$

EXAMPLE 3 Express the following standard numbers in exponential notation.

(a) 2   (b) 7   (c) 12   (d) 4.5

(a) $2 = 2^1$ (b) $7 = 7^1$ (c) $12 = 12^1$ (d) $4.5 = 4.5^1$

> **Rule 1–36**   *Any number (except zero) with an exponent of 0 equals 1. The expression 0^0 is undefined.*
> $$a^0 = 1, \qquad \text{for any nonzero base } a$$

EXAMPLE 4 Simplify the following exponential expressions.

(a) 9^0 (b) 6^0 (c) 27^0 (d) 2.6^0

(a) $9^0 = 1$ (b) $6^0 = 1$ (c) $27^0 = 1$ (d) $2.6^0 = 1$

2   Square Numbers and Find the Square Roots of Numbers.

The standard number that results from squaring any number is called a *perfect square* or *square*. To square a number is to find the result of using that number as a factor twice. If asked to square 3, we would say $3^2 = 3 \times 3 = 9$. We say 9 is the square of 3.

EXAMPLE 5   Find the square of each of the following.

(a) 2   (b) 7   (c) 3.2

(a) 2; $2^2 = 2 \times 2 = \mathbf{4}$
(b) 7; $7^2 = 7 \times 7 = \mathbf{49}$
(c) 3.2; $3.2^2 = 3.2 \times 3.2 = \mathbf{10.24}$

The process of squaring has an inverse operation, just as addition and multiplication have the inverse operations of subtraction and division, respectively. The inverse operation of squaring is taking the square root of a number. To take the square root of a number (a perfect square) is to determine the number (called the *square root*) which was used as a factor twice to equal that perfect square. The square root of 9 is 3 because 3^2 or $3 \times 3 = 9$.

The *radical sign* $\sqrt{}$ is an operational symbol indicating that the square root is to be taken of the number under the bar. This bar serves as a grouping symbol just like parentheses. The number under the radical sign is called the *radicand*. The entire expression is called a *radical expression*.

radical sign ⟶
$$\sqrt{25} = 5 \leftarrow \text{square root}$$
radicand ⟶

> **Rule 1–37**   *To estimate the square root of a number:*
>
> 1. Select the trial estimate of the square root.
> 2. Square the estimated answer.
> 3. If the square of the estimated answer is less than the original number, adjust the estimated answer to a larger number. If the square of the estimated answer is more than the original number, adjust the estimated answer to a smaller number.
> 4. Square the adjusted answer from Step 3.
> 5. Continue the adjusting process until the square of the trial estimate is the original number.

Square roots of whole numbers that are perfect squares will be whole numbers. The calculator display for the square roots of other numbers will be shown as decimal numbers. When estimating the square root of other numbers, the estimate can show the two whole numbers that the square root is between.

EXAMPLE 6   Find $\sqrt{256}$.

Select 15 as the estimated square root: 15^2 or $15 \times 15 = 225$. The number 225 is less than 256, so the square root of 256 must be larger than 15. We adjust the estimate to 17: 17^2 or $17 \times 17 = 289$. The number 289 is more than 256, so the square root of 256 must be smaller than 17. Adjust the estimate to 16. **Because $16^2 = 256$, 16 is the square root of 256.**

3 Use Powers of 10 to Multiply and Divide.

Powers of 10 are numbers whose only nonzero digit is 1. Thus, 10, 100, 1000, and so on, are powers of 10.

In the following compare the number of zeros in standard notation with the exponent of 10 in exponential notation.

$$1,000,000 = 10^6 \qquad 6 \text{ zeros}$$
$$100,000 = 10^5 \qquad 5 \text{ zeros}$$
$$10,000 = 10^4 \qquad 4 \text{ zeros}$$
$$1000 = 10^3 \qquad 3 \text{ zeros}$$
$$100 = 10^2 \qquad 2 \text{ zeros}$$
$$10 = 10^1 \qquad 1 \text{ zero}$$
$$1 = 10^0 \qquad 0 \text{ zeros}$$

The exponents in powers of 10 indicate the number of zeros used in standard notation.

EXAMPLE 7   Express the following as powers of 10.

(a) 10,000,000   (b) 100,000,000   (c) 100,000,000,000

(a) $1\,0,000,000 = \mathbf{10^{\,7}}$
(b) $1\,00,000,000 = \mathbf{10^{\,8}}$
(c) $1\,00,000,000,000 = \mathbf{10^{\,11}}$

EXAMPLE 8   Express the following in standard notation.

(a) 10^5 (b) 10^{13}

(a) $10^{\,5} = \mathbf{1\,00,000}$
(b) $10^{\,13} = \mathbf{1\,0,000,000,000,000}$

In some applications, powers of 10 are used in multiplication and division problems. Compare each example in the left column with the corresponding example in the right column to find the pattern for multiplying a number by a power of 10.

$5 \times 100 = 500$ $5 \times 10^2 = 500$ Decimal point moved 2 places to the right and 2 zeros attached.

$27 \times 1000 = 27,000$ $27 \times 10^3 = 27,000$ Decimal point moved 3 places to the right and 3 zeros attached.

If 100 sheets of plywood cost $10.57 per sheet, we find the total cost by multiplying: $10.57 \times 100 = 1057. When we solve problems involving multiplying by 10, 100, 1000, and so on, we should be able to perform the calculations mentally by just moving the decimal point. Calculations such as the one above can be done more quickly mentally than if a calculator were used.

Rule 1–38   *To multiply a number by a power of 10:*

1. Move the decimal point in the number to the *right* as many places as the 10, 100, 1000, or so on, has zeros.
2. Attach zeros to the right if necessary.

EXAMPLE 9   Multiply 237×100.

Applying the shortcut rule, we move the understood decimal two places to the *right* and attach two zeros.

$$237.\underset{\curvearrowright}{00} \times 100 = \mathbf{23,700}$$

EXAMPLE 10 Multiply 36.2×1000.

We have three zeros in 1000. We move the decimal point three places to the *right*. Two zeros need to be attached.

$$36.\underset{\underrightarrow{\qquad}}{200} \times 1000 = \mathbf{36,200}$$

Dividing by numbers such as 10, 100, 1000, and so on, can be done very quickly by applying a shortcut rule. Observe the following divisions.

$32 \div 10 = 3.2$ Decimal point moved one place to the *left*.

$78.9 \div 100 = 0.789$ Decimal point moved two places to the *left*.

$52,900 \div 1000 = 52.9$ Decimal point moved three places to the *left*.

If we compare the decimal point in the dividend with the decimal in the quotient, we notice that the decimal point in the quotient has been moved to the left as many places as the divisor (10, 100, 1000, and so on) has zeros.

> **Rule 1–39** *To divide a number by a power of 10:*
>
> **1.** Move the decimal to the *left* as many places as the divisor has zeros.
> **2.** Attach zeros to the left if necessary.
> **3.** Zeros to the right of the decimal point may be dropped following the last nonzero digit of the quotient.

Compare this rule to Rule 1–38 for multiplying decimal numbers by 10, 100, 1000, and so on. For multiplication, the decimal moves to the *right,* whereas for division, the decimal moves to the *left*.

EXAMPLE 11 If 100 lb of floor cleaner costs \$63, what is the cost of 1 lb?

Use the shortcut rule to divide by 100. A whole number has a decimal point understood at the right end.

$$63 \div 100 = 0.\underset{\underleftarrow{\qquad}}{63} = \$0.63$$

The cost for 1 lb is \$0.63.

EXAMPLE 12 If a stack of 100 concrete blocks weighs 1080 lb, find the weight of one concrete block.

$$1080 \div 100 = 10.80 \qquad \text{or} \qquad 10.8 \text{ lb}$$

One block weighs 10.8 lb.

4 Use a Calculator to Find Powers and Roots.

To raise a number to a power using a basic calculator requires repeated multiplication. However, scientific and graphics calculators have keys designed to raise numbers to any power so that the repeated multiplication is done internally. Since scientific or graphics calculators can be used to raise numbers to powers more efficiently, as well as to perform other mathematical operations needed in algebra and other mathematics, additional discussion of calculators will focus on these two types.

Scientific and graphics calculators have special power keys, such as keys for squares and cubes. They also have a "general power" key that can be used for all powers. Even though the "general power" key can be used to square and cube numbers, fewer keystrokes are needed when using the "square" and "cube" keys for these operations.

■ **General Tips for Using the Calculator**

Squaring or cubing on some scientific calculators:

1. Enter the base.
2. Press the square key $\boxed{x^2}$ or cube key $\boxed{x^3}$.

Raising a number to a power on the scientific calculator:

1. Enter the base.
2. Press the general power key $\boxed{x^y}$.
3. Enter the exponent.
4. Press the equals key $\boxed{=}$.

Squaring or cubing on the graphics calculator:

1. Enter the base.
2. Press the square key $\boxed{x^2}$ or cube key $\boxed{x^3}$.
3. Press the $\boxed{\text{EXE}}$ key.

Raising a number to a power on the graphics calculator.

1. Enter the base.
2. Press the general power key $\boxed{x^y}$.
3. Enter the exponent.
4. Press the $\boxed{\text{EXE}}$ key.

Tips and Traps *Labels on Calculator Keys Are Not Universal:*
To make calculator steps easier to follow, we will identify a common label for a function in a box. The exact label will vary with the specific calculator and model. Graphics calculators also provide many functions on menus instead of labels on specific keys. Even though we may use a key notation in this text to show a calculator function, this function may actually appear on a menu. Check your calculator manual for exact location and labeling of functions.

EXAMPLE 13   Use the calculator to find 35^2.

Scientific: 35 $\boxed{x^2}$

Display: **1225.**

Graphics: 35 $\boxed{x^2}$ $\boxed{\text{EXE}}$

Display: 35^2

1225.

EXAMPLE 14   Use the calculator to find 2.3^5.

Scientific: 2 $\boxed{\cdot}$ 3 $\boxed{x^y}$ 5 $\boxed{=}$

Display: 64.36343

Graphics: 2 $\boxed{\cdot}$ 3 $\boxed{x^y}$ 5 $\boxed{\text{EXE}}$

Display: 2.3 x^y 5

64.36343

■ General Tips for Using the Calculator

Finding the square root of a number on most scientific calculators and the graphics calculator:

1. Press the square root key $\boxed{\sqrt{}}$.
2. Enter the radicand.
3. Press the $\boxed{\text{EXE}}$ key or equals key.

Finding the square root of a number on some scientific calculators:

1. Enter the radicand.
2. Press the square root key $\boxed{\sqrt{}}$.

EXAMPLE 15   (a) Evaluate $\sqrt{529}$.

Scientific: $\boxed{\sqrt{}}$ 529 $\boxed{=}$

or

529 $\boxed{\sqrt{}}$

Display: 23.

Graphics: $\boxed{\sqrt{}}$ 529 $\boxed{\text{EXE}}$

Display: $\sqrt{}$ 529

23.

(b) Evaluate $\sqrt{10.89}$.

Scientific: $\boxed{\sqrt{}}$ 10 $\boxed{\cdot}$ 89 $\boxed{=}$

or

10 $\boxed{\cdot}$ 89 $\boxed{\sqrt{}}$

Display: 3.3

Graphics: $\boxed{\sqrt{}}$ 10 $\boxed{\cdot}$ 89 $\boxed{\text{EXE}}$

Display: $\sqrt{}$ 10.89

3.3

SELF-STUDY EXERCISES 1–6

[1]   Give the base and exponent of the following expressions.

1. 4^3 **2.** 9^4 **3.** 2.7^9

Simplify the following exponential expressions.

4. 1.5^3 **5.** 7^2 **6.** 10^3 **7.** 2^4
8. 3.4^2 **9.** 15^1 **10.** 8^1 **11.** 9^0

Express the following with an exponent of 1.

12. 8   **13.** 14.5   **14.** 12   **15.** 23
16. Explain what an exponent of 1 does to the base.
17. Explain what an exponent of 0 does to the base.

[2]   Write in standard notation.

18. 8^2 **19.** 2^2 **20.** 3.5^2
21. 1.4^2 **22.** 13^2 **23.** 1^2
24. 100^2

Square the following numbers.

25. 8   **26.** 9   **27.** 17
28. 18   **29.** 101   **30.** 22

Perform the following operations.

31. $\sqrt{25}$ **32.** $\sqrt{49}$ **33.** $\sqrt{81}$
34. $\sqrt{196}$
35. What does it mean to square a number?
36. What does it mean to take the square root of a number?

3 Multiply the following whole numbers by using powers of 10.

37. 10×10^2
38. 32×1000
39. $3 * 10^4$
40. $2 * 10^4$
41. $102 * 100$
42. 22×10^3

Divide the following whole numbers by using powers of 10.

43. $250 \div 10$
44. $210 \div 10$
45. $\dfrac{300}{10^2}$

46. $\dfrac{900}{10^2}$
47. $2500 \div 10$
48. $120 \div 10^2$

49. What does the exponent of a power of 10 mean when multiplying?

50. What does the exponent of a power of 10 mean when dividing?

4 Use a calculator to evaluate the following.

51. 15^2 ____
52. 7^3 ____
53. 5^7 ____
54. $\sqrt{324}$
55. $\sqrt{784}$
56. $\sqrt{1089}$

1–7 ORDER OF OPERATIONS

Learning Objectives

1 Apply the order of operations to problems containing several operations.

2 Use a calculator to perform a series of operations.

1 **Apply the Order of Operations to Problems Containing Several Operations.**

Whenever several mathematical operations are to be performed, they must follow the proper *order of operations.*

Rule 1–40 *The order of operations follows:*

1. Perform operations within parentheses (or other grouping symbols), beginning with the innermost set of parentheses or apply the distributive property.
2. Evaluate exponential operations and find square roots in order from left to right.
3. Multiply and divide in order from left to right.
4. Add and subtract in order from left to right.

To summarize, use the following key words:

<u>P</u>arentheses (grouping), <u>E</u>xponents (and roots),

<u>M</u>ultiplication and <u>D</u>ivision, <u>A</u>ddition and <u>S</u>ubtraction

---

Tips and Traps *A Memory Aid for the Order of Operations:*
A memory tip for remembering the order of operations is the sentence, "<u>P</u>lease <u>E</u>xcuse <u>M</u>y <u>D</u>ear <u>A</u>unt <u>S</u>ally."

EXAMPLE 1 Evaluate $3(2 + 3)$.

Use parentheses to show both a grouping and multiplication. In this situation the distributive property can also be applied.

$$3(2 + 3) = 3(2) + 3(3)$$
$$= 6 + 9$$
$$= 15$$

or

$$3(2 + 3) = 3(5)$$
$$= 15$$

EXAMPLE 2 Simplify $4^2 - 5(2) \div (4 + 6)$.

$4^2 - 5(2) \div (4 + 6)$	
$4^2 - 5(2) \div 10$	Do operations within parentheses first: $4 + 6 = 10$.
$16 - 5(2) \div 10$	Evaluate exponentiation: $4^2 = 16$.
$16 - 10 \div 10$	Then multiply: $5(2) = 10$.
$16 - 1$	And divide: $10 \div 10 = 1$.
$16 - 1 = \mathbf{15}$	Subtract last.

Tips and Traps   ***Parentheses to Indicate Multiplication, the Distributive Property, or a Grouping:***
As in Example 2, parentheses can be used to indicate multiplication or to indicate an operation that should be done first. If the parentheses contain an operation, the parentheses indicate a grouping. Otherwise, the parentheses indicate multiplication. The expression $5(2)$ indicates multiplication, while $(4 + 6)$ indicates a grouping.

EXAMPLE 3 Evaluate $5 \times \sqrt{16} - 5 + [15 - (3 \times 2)]$.

$5 \times \sqrt{16} - 5 + [15 - (3 \times 2)]$	Work innermost grouping: 3×2.
$5 \times \sqrt{16} - 5 + [15 - 6]$	Work remaining grouping: $15 - 6$.
$5 \times \sqrt{16} - 5 + 9$	Find square root: $\sqrt{16} = 4$.
$5 \times 4 - 5 + 9$	Multiply: 5×4.
$20 - 5 + 9$	Add and subtract from left to right.
$15 + 9 = \mathbf{24}$	

EXAMPLE 4 Evaluate $3.2^2 + \sqrt{21 - 5} \times 2$.

$3.2^2 + \sqrt{21 - 5} \times 2$	Do operations within grouping first; the bar part of the radical symbol is a grouping symbol: $21 - 5 = 16$.
$3.2^2 + \sqrt{16} \times 2$	Then do exponentiation and square root from left to right: $3.2^2 = 10.24$; $\sqrt{16} = 4$.
$10.24 + 4 \times 2$	Then multiply: $4 \times 2 = 8$.
$10.24 + 8$	Add last.
$10.24 + 8 = \mathbf{18.24}$	

2 Use a Calculator to Perform a Series of Operations.

Both the scientific and graphics calculators have been programmed to perform the order of operations in the proper order.

EXAMPLE 5   Evaluate $2.4^2 - 5(2) \div (4 + 6)$.

Scientific: 2 $\boxed{\cdot}$ 4 $\boxed{x^2}$ $\boxed{-}$ 5 $\boxed{\times}$ 2 $\boxed{\div}$ $\boxed{(}$ 4 $\boxed{+}$ 6 $\boxed{)}$ $\boxed{=}$

Display: **4.76**

Graphics: 2 $\boxed{\cdot}$ 4 $\boxed{x^2}$ $\boxed{-}$ 5 $\boxed{(}$ 2 $\boxed{)}$ $\boxed{\div}$ $\boxed{(}$ 4 $\boxed{+}$ 6 $\boxed{)}$ $\boxed{\text{EXE}}$

Display: $2.4^2 - 5(2) \div (4 + 6)$

                                  4.76

■ General Tips for Using the Calculator

The scientific calculator requires that the multiplication operation key be used to indicate the product of 5 and 2. When using the graphics calculator, the expression is entered exactly as it appears in the example or the multiplication operation key can be used. Parentheses keys, $\boxed{(}$ and $\boxed{)}$, are used to show any type of grouping.

EXAMPLE 6   Evaluate $3^4 + \sqrt{77 - 13} \times 2[26 - (45 - 38)]$.

Most scientific and graphics:

3 $\boxed{x^y}$ 4 $\boxed{+}$ $\boxed{\sqrt{}}$ $\boxed{(}$ 77 $\boxed{-}$ 13 $\boxed{)}$ $\boxed{\times}$ 2 $\boxed{(}$ 26 $\boxed{-}$ $\boxed{(}$ 45 $\boxed{-}$ 38 $\boxed{)}$ $\boxed{)}$ $\boxed{\text{EXE}}$

Display: $3x^y4 + \sqrt{} (77 - 13) \times 2$
                 $(26 - (45 - 38))$

                                   385.

Some scientific:

3 $\boxed{x^y}$ 4 $\boxed{+}$ $\boxed{(}$ 77 $\boxed{-}$ 13 $\boxed{)}$ $\boxed{\sqrt{}}$ $\boxed{\times}$ 2 $\boxed{\times}$ $\boxed{(}$ 26 $\boxed{-}$ $\boxed{(}$ 45 $\boxed{-}$ 38 $\boxed{)}$ $\boxed{)}$ $\boxed{=}$

Display: **385.**

SELF-STUDY EXERCISES 1–7

1   Evaluate each problem following the order of operations.

1. $5^2 + 4 - 3$
2. $4^2 + 6 - 4$
3. $4 \times 3 - 9 \div 3$
4. $5 \times 2.9 - 4 \div 2$
5. $25 \div 5 \times 4.8$
6. $64 \div 4 \times 2$
7. $6 \times \sqrt{36} - 2 \times 3$
8. $3 \times \sqrt{81} - 3 \times 4$
9. $4^2 \times 3^2 + (4 + 2) \times 2$
10. $2^2 \times 5^2 + (2 + 1) \times 3$
11. $54 - 3^3 - \dfrac{8}{2}$
12. $156 - 2^3 - \dfrac{9}{3}$
13. $32 - 2.05 \times 4^2 \div 2$
14. $5.2^2 - 3 \times 2^2 \div 6$
15. $4 + 15 \div 3 - \dfrac{0}{7}$
16. $3 + 8 \div 4 - \dfrac{0}{5}$
17. $3 - 2 + 3 \times 3 - \sqrt{9}$
18. $2 - 1 + 4 \times 4 - \sqrt{4}$
19. $2^4 \times (7 - 2) \times 2$
20. $3^4 \times (9 - 3) \times 3$

2   Use the calculator to perform the following operations.

21. $6 \times 9^2 - \dfrac{12}{4}$
22. $7 \times 8^2 - \dfrac{10}{2}$
23. $4^3 + 14 - 8$
24. $2^3 + 12 - 7$
25. $2 \times \sqrt{16} + (8 - \sqrt{25})$
26. $4 \times \sqrt{49} + (9 - \sqrt{64})$
27. $3(2^2 + 1) - 30 \div 3$
28. $4(3^2 + 2) - 60 \div 12$
29. $3 + 10 \div 5 + 2$
30. $4 + 12 \div 4 + 7$
31. $25 - 5^2 \div (7 - 2)$
32. $36 - 6^2 \div (8 - 2)$

33. $2(3.1^2 + 2) - \sqrt{7.29}$ **34.** $3^4 - 2 \times 4.6 \div \dfrac{0}{2}$ **35.** $24(3^2 + 1) - 48.25$

36. State the first operation in the order of operations.

37. State the last operation in the order of operagtions.

ASSIGNMENT EXERCISES

Section 1–1
Write the following fractions as decimal numbers.

1. (a) $\dfrac{3}{10}$ (b) $\dfrac{15}{100}$ (c) $\dfrac{4}{100}$

2. (a) $\dfrac{75}{1000}$ (b) $\dfrac{21}{10}$ (c) $\dfrac{652}{100,000}$

3. What is the place value of 6 in 21.836?

4. What is the place value of 5 in 13.0586?

5. In 13.7213, what digit is in the tenths place?

6. In 15.02167, what digit is in the ten-thousandths place?

7. Identify the place value of the digit 3 in each of the following:
(a) 430   (b) 34,789   (c) 3,456,321

8. Identify the place value of the digit 2 in the following:
(a) 2,785,901   (b) 45,923

9. Write 56,109,110 in words.

10. Write 61,201 in words.

11. Write one million, two hundred sixty-five thousand, four hundred one in standard notation.

12. Write thirty-two thousand, three hundred twenty-one in standard notation.

13. Write the words for 6.803.

14. Write the words for 0.0712.

15. The decimal equivalent of $\frac{5}{8}$ is six hundred twenty-five thousandths. Write this decimal using digits.

16. The thickness of a sheet of aluminum is forty-thousandths of an inch. Write this thickness as a decimal number.

17. Round the following to the indicated places.
(a) 36 to the nearest tens
(b) 74 to the nearest tens
(c) Round 24.237 to the nearest whole number.
(d) Round $42.98 to the nearest dollar.
(e) Round 83.052 to tens.
(f) Round $8.9378 to the nearest cent.
(g) Round $0.9986 to the nearest cent.
(h) Round 0.097032 to hundred-thousandths.

18. Round the following to the indicated places.
(a) Nearest thousand: 65,763
(b) Nearest thousand: 28,714
(c) Nearest ten million: 497,283,016
(d) Nearest hundred: 8236
(e) Nearest ten thousand: 248,217
(f) Nearest hundredth: 7.0893
(g) Nearest thousandth: 1.078834
(h) Nearest tenth: 0.09783

19. Round the following to the indicated places.
(a) Nearest ten: 324
(b) Nearest thousand: 6882
(c) Nearest hundred: 468
(d) Nearest ten thousand: 49,238
(e) Nearest billion: 26,500,000,129
(f) Nearest tenth: 41.378
(g) Nearest hundredth: 6.8957
(h) Nearest ten-thousandth: 23.46097

20. Round the following to numbers with one nonzero digit.
(a) 98   (b) 94   (c) 25,786
(d) 34,786   (e) 12.83   (f) 0.0736
(g) 7.93   (h) 1.876

21. Which of these decimal numbers is larger: 4.783 or 4.79?

22. Which of these decimal numbers is smaller: 0.83 or 0.825?

23. Write these decimal numbers in order of size from smaller to larger: 0.021, 0.0216, 0.02.

24. Two measurements of an object are recorded. If the measures are 4.831 in. and 4.820 in., which measure is larger?

25. The decimal equivalent of $\frac{7}{8}$ is 0.875. The decimal equivalent of $\frac{6}{7}$ is approximately 0.857. Which fraction is larger?

26. Two parts are machined from the same stock. They measure 1.023 in. and 1.03 in. after machining. Which part has been machined more; that is, which part is now smaller?

Section 1–2

Add the following. Use a calculator as directed by your instructor.

27. (a) $6 + 9 + 3 + 5$
(b) $5 + 1 + 6 + 3 + 3$

(c)
```
    8
    5
    3
    6
    2
 +  4
```
(d)
```
    7
    4
    3
    2
    5
 +  4
```

29. An air conditioner uses 10.4 kW, a stove uses 15.3 kW, a washer uses 2.9 kW, and a dryer uses 6.3 kW. What is the total number of killowatts used by the electrical machines?

31. Estimate by rounding to thousands, then find the exact answer. Check the answer.

(a)
```
   $16,742.83
 + $12,349.26
```
(b)
```
   17,402
 + 18,646
```

28. (a) $3.47 + 42.32 + 3.82 + 4.09$
(b) $6.2 + 32.7 + 46.82 + 0.29 + 4.237$
(c) $86.3 + 9.2 + 70.02 + 3 + 2.7$
(d) $42 + 3.6 + 2.1 + 7.83$

30. Add the following.

(a)
```
   3456.08
 + 2147.76
```
(b)
```
   12,467
 + 24,378
```

(c)
```
   23,609
    2200
      76
 +    124
```
(d)
```
   $43,045.36
    5047.47
      87.10
 +   213.08
```

32. A do-it-yourself project requires $57.32 for concrete, $74.26 for fence posts, and $174.85 for fence boards. Estimate the cost by rounding to numbers with one nonzero digit, then find the exact cost. Check the answer.

Section 1–3

33. Subtract the following.
(a) $21.34 - 16.73$
(b) $15.934 - 12.807$
(c) $9 - 7$
(d) $5 - 0$
(e) $8 - 3 - 2 - 3$
(f) $284.73 - 79.831$
(g) $345 - 201$
(h) $13,342 - 1202$

35. A blueprint calls for the length of a part to be 8.296 in. with a tolerance of ± 0.005 in. What are the limit dimensions of the part?

37. Planning a vacation, a family selected a scenic route covering 653 miles and a direct route covering 463 miles. Estimate the difference by converting to round numbers. Then find the exact answer. Check the answer.

Use Fig. 1–21 for Exercises 39 to 42.

39. Find the length of A if $D = 4.237$ in., $B = 1.861$ in., and $C = 1.946$ in.
40. What is the dimension of A if $E = 4.86$ in. and $B = 1.972$ in.?
41. Give the limit dimensions of D if $D = 8.935$ in. with a ± 0.005-in. tolerance.
42. What dimension should be listed for C if D measures 3.7 in. and E measures 1.6 in.?

34. Estimate by rounding to hundreds, then find the exact answer. Check the answer.
(a) $12,346.87 - 4468.63$
(b) $3495 - 3090$
(c) $6767 - 478$
(d) $293.86 - 148$

36. For a moving sale, a family sold a sofa for $75 and a table for $25. If a newspaper advertisement announcing the sale cost $12.75, how much did the family clear on the two items sold?

38. A new foreign car costs $12,677, and a comparable American car costs $11,859. Estimate the difference by rounding to thousands. Then find the exact answer. Check the answer.

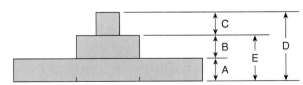

Figure 1–21

Section 1–4

Multiply the following:

43. $2 \times 6 \times 7$
46. 76×5
49. $12,407(270)$

44. $6 \times 3 \times 2 \times 4$
47. 305×45
50. $527 * 342$

45. 127×9
48. $236 \cdot 244$
51. $56,002 \times 7040$

52. A college bookstore sold 327 American history textbooks for $39 each. How much did the bookstore receive for the books?

54. An automotive tire dealer runs a special on heavy-duty, deluxe whitewall truck tires. If the dealer sold 105 tires for $112 each, how much did the dealer take in on the sale?

Perform the indicated operations.

56. 2.4(3 + 1)

57. 5(6 − 2)

59. 6(3 + 7)

60. 3(5 − 3)

62. What number is obtained if the sum of 5 and 2 is multiplied by 6? Use the distributive property.

64. A luxury automotive dealer pays a sound system installer $33.25 per hour. Estimate by rounding to a number with one nonzero digit how much the sound system installer was paid for 37 hours of work. Then find the exact answer. Check the answer.

66. Find the area of a field that is 234.6 feet long by 123.2 feet wide. Estimate the area by rounding to one nonzero digit, then find the exact answer. Check the answer. Express the area in square feet ($A = l \times w$).

68. A piecework employee averages 178.6 pieces per day. If the employee earns $0.28 per item, how much is earned in 5 days?

53. Enhanced 101-key computer keyboards sell through a mail-order supplier for $67 each. How much would a business pay for 21 keyboards?

55. If a wholesaler ordered 144 bare-bones computers for $305 each, how much did the dealer pay for the order?

58. 2(9.2 − 4)

61. 8(7 + 3)

63. A monitor for a computer sells for $687, and a color printer for a computer sells for $523. A multimedia hardware monitor for a computer sells for two times the difference in cost between the $687 monitor and the color printer. How much does a multimedia hardware monitor cost?

65. A worker was offered a job paying $365 per week. If the worker takes the job for 36 weeks, how much will the worker earn? Estimate by rounding to tens, then find the exact answer. Check the answer.

67. A parcel of land measures 1940.7 feet by 620.4 feet. Estimate the area by rounding to hundreds, then find the exact area. Check the answer. Express the area in square feet ($A = l \times w$).

69. If a steel tape expands 0.00014 in. for each inch when heated, how much will a tape 864 in. long expand?

Section 1–5

70. Express 3 divides into 4 using the following symbols.

　(a) ÷　　**(b)** ⌐　　**(c)** ——

Divide the following:

72. 8 ÷ 8

73. 1 ÷ 1

75. 7 ÷ 0

76. 4 ÷ 1

78. 29.25 ÷ 3.6

79. 325 ÷ 25

81. 30,126 ÷ 15

82. 10,160 ÷ 20

83. A group of 27 volunteers will seek contributions to send a first-grade class to the circus. The group leader has 632 envelopes for the collection. If the envelopes are divided equally among the 27 volunteers, how many will each receive? How many will be left over?

85. Divide by short division: 46.7 ÷ 8.

Estimate, find the answer, and check.

87. 475 ÷ 86

71. Express 5 divided by 3 using the following symbols.

　(a) ÷　　**(b)** ⌐　　**(c)** ——

74. 0 ÷ 3

77. 8.43 ÷ 1.6

80. 364.8 ÷ 6

84. A school marching band is in a formation of 7 rows, each with the same number of students. If the band has 56 members, how many are in each row?

86. Divide by short division: 7032 ÷ 9.

88. 35.16 ÷ 1.04

89. If 27 volunteers collected $287 to send a first-grade class to the circus, on the average how much was collected per volunteer? Estimate the answer, then find and check the exact answer.

91. Find the average measure if measures of 42.34 ft, 38.97 ft, 51.95 ft, and 61.88 ft were recorded. Round to the nearest hundredth.

90. A class of 16 students made 768 cookies for a school open house. How many cookies did each student make, assuming each made the same number? Estimate the answer, and then find and check the exact answer.

92. Five light fixtures cost $74.98, $23.72, $51.27, $125.36, and $85.93. Find the average cost of the fixtures to the nearest cent.

Section 1–6

93. Give the base and exponent of each expression. Then simplify each.
(a) 7^3 (b) 2.3^4 (c) 8^4

95. Simplify the following expressions:
(a) 0.9^1 (b) 35^1 (c) 1^1

97. Simplify the following:
(a) 1.7^0 (b) 1^0 (c) 8^0 (d) 149^0

99. Evaluate the following:
(a) 1^2 (b) 125^2 (c) 5.6^2 (d) 21^2

101. Express the following as powers of 10.
(a) 10 (b) 1000 (c) 10,000
(d) 100,000

103. Divide the following by using powers of 10.
(a) $700 \div 100$ (b) $40.56 \div 1000$
(c) $60.5 \div 100$ (d) $23,079 \div 10,000$
(e) $44,582 \div 1000$

94. Give the base and exponent of each expression. Then simplify each.
(a) 5^6 (b) 1.2^2 (c) 10^6

96. Express the following with an exponent of 1.
(a) 904 (b) 76 (c) 0.3

98. Find the value of the following:
(a) 2^2 (b) 1.3^2 (c) 7^2 (d) 12^2

100. Find the square root of the following:
(a) $\sqrt{2500}$ (b) $\sqrt{1.44}$ (c) $\sqrt{289}$
(d) $\sqrt{81}$

102. Multiply the following by using powers of 10.
(a) 3×100 (b) $75 \times 10,000$
(c) 2.2×1000 (d) 5×100
(e) 40.6×10

Section 1–7

Evaluate the following:

104. $2 + (3 + 6) \div 3$

105. $4^2 \times (12 - 7) - 8 + 3$

106. $8^2 - (3 - 1.5) \times 5.2$

107. $4 + 5 - 2 \times 3$

108. $4 + \dfrac{8.6}{2} \times 2$

109. $3.1 \times 4 \times \sqrt{16} - 6^2$

110. $\sqrt{12.25} \times (4 - 2) + 8$

111. $5.2^3 - \sqrt{81} \times (2 + 1)$

112. $2^4 \div 2 - \sqrt{9}$

113. $\dfrac{\sqrt{36}}{1.5} + 9 - 2$

CHALLENGE PROBLEMS

114. You are charged with designing a playground that has 2,160 square yards of space. If the playground is to be rectangular, determine how long and how wide it should be if one of your pieces of equipment requires a space at least 15 yards long. Give whole-number solutions.

116. Use a calculator to evaluate the following powers.
(a) 0.03^2 (b) 0.07^2 (c) 0.005^2
(d) 0.009^2 (e) 0.02^3 (f) 0.004^3

115. Of all the possible ways to design the playground in Problem 114, which rectangle would require the least amount of fencing? Give whole number solutions.

117. Use a calculator to find the square root of the following decimals.
(a) $\sqrt{0.64}$ (b) $\sqrt{0.09}$
(c) $\sqrt{0.0009}$ (d) $\sqrt{0.0625}$
(e) $\sqrt{0.000016}$ (f) $\sqrt{0.4}$

1. Addition and subtraction are inverse operations. Write the following addition problem as a subtraction problem and find the value of the number represented by the letter n: $1.2 + n = 1.7$.

2. Multiplication and division are inverse operations. Write the following multiplication problem as a division problem and find the value of the number represented by the letter n: $5 \times n = 4.5$.

3. Squaring and finding square roots are inverse operations. Write the following square root as a squaring problem and find the value of the number represented by the letter n: $\sqrt{n} = 6$.

4. Give an example illustrating that subtraction is not associative.

5. Give an example illustrating that division is not commutative.

6. Give the steps in the order of operations.

Find and explain the mistake in the following. Rework each problem correctly.

7. $2.5 + 4.9 = $
$$\begin{array}{r} 2.5 \\ +\ 4.9 \\ \hline 6.14 \end{array}$$

8. $2 + 5(4) = $
$7(4) = 28$

9. $\sqrt{9} = 81$

10. How are the procedures for adding whole numbers and adding decimals related?

11. Without making any calculations, do you think 0.004 is a perfect square? Why or why not?

12. Without making any calculations, do you think 0.008 is a perfect cube? Why or why not?

13. Can you find at least one exception to the generalization that a perfect square decimal will have an even number of decimal places? Illustrate your answer.

Objectives	What to Remember with Examples

Section 1–1

1 Identify place values in whole numbers.

Each digit has a place value. Place values are grouped in periods of three digits. Place value names can be found in Fig. 1–1.

> Identify the place value of each of the digits in the number 3628: 3 thousands 6 hundreds 2 tens 8 ones.

2 Read and write whole numbers in words and standard notation.

Standard notation is the usual form of a number or the number written in digits using place values.

Read from the left the numbers in each period and the period name. The period name *units* is not read.

> The number 345,230 is written in standard form.
>
> Write 3462 in words: three thousand, four hundred sixty-two.

3 Identify place values in decimal numbers.

Align the decimal point in the number under the decimal point in the place-value chart. The chart shows the value of each place in the number.

> In 3.75, 3 is in the ones place, 7 is in the tenths place, and 5 is in the hundredths place.

4 Read decimal numbers.

Read whole-number part, *and* for decimal point, decimal part, and place value of rightmost digit.

> Write 32.075 in words: thirty-two and seventy-five thousandths.

5 Write fractions with power-of-10 denominators as decimal numbers.	Write digits on right of decimal as numerator. Denominator will be a power of ten with as many zeros as decimal places.
	Express $\dfrac{53}{1000}$ as a decimal: 0.053.
6 Compare decimal numbers.	Compare decimals by comparing the digits in the same place beginning from the left of each number.
	Which decimal is larger, 0.23 or 0.225? Both numbers have the same digit, 2, in the tenths place. 0.23 is larger because it has a 3 in the hundredths place, while 0.225 has a 2 in that place.
7 Round a whole number or a decimal number to a place value.	Whole numbers: 1. Locate rounding place. 2. Examine digit to the right. 3. Round down if digit is less than 5. 4. Round up if digit is 5 or more. Decimals: Round as in whole numbers; however, digits on the right side of the digit in the rounding place and after the decimal point are dropped rather than replaced with zeros.
	Round 3624 to the nearest tens place: 2 is in the tens place, 4 is the digit to the right; 4 is less than 5, so round down. The rounded value is 3620. Round 5.847 to the nearest tenth: 8 is in the tenths place; 4 is less than 5, so leave 8 as is and drop the 4 and 7. The rounded value is 5.8.
8 Round a whole number or a decimal to a number with one nonzero digit.	Whole numbers: 1. Locate first nonzero digit on the left. 2. Examine the digit to the right. 3. Round down if digit is less than 5. 4. Round up if digit is 5 or more. Decimals: To round decimals to one nonzero digit is to round to the first place from the left that is not zero.
	Round 4789 to one nonzero digit: 4 is the first nonzero digit on the left, 7 is the digit to the right; 7 is 5 or more, so round up. The rounded value is 5000. Round 0.0372 to one nonzero digit: 3 is the first nonzero digit from the left; 7 is 5 or greater, so change 3 to 4 and drop the remaining digits on the right. The rounded value is 0.04.
Section 1–2 **1** Add whole numbers.	Addition is a binary operation that is commutative and associative. Zero property of addition: $0 + n = n + 0 = n$. Arrange numbers in columns of like places. Add each column beginning with the ones place. Carry when necessary.
	$5 + 7 = 7 + 5 = 12$; $(3 + 2) + 6 = 3 + (2 + 6) = 11$ $0 + 3 = 3 + 0 = 3$ $\begin{array}{r} 4824 \\ +\ \ 745 \\ \hline 5569 \end{array}$

2 Add decimal numbers.

Add decimal numbers by arranging the addends so that the decimal points are in the same vertical line.

> Add: 43.35 + 3.7 + 0.462
> $$\begin{array}{r} 43.35 \\ 3.7 \\ +0.462 \\ \hline 47.512 \end{array}$$

3 Estimate and check addition.

Estimate by rounding the addends before finding the sum. Check addition by adding a second time.

> Estimate the sum by rounding to the nearest hundred: 483 + 723
> 500 + 700 = 1200. Exact sum: 1206.

4 Add using a calculator.

Enter each addend, followed by the addition operation key $\boxed{+}$. After the last addend, press $\boxed{=}$ or $\boxed{EXE}$ to display the sum. Use $\boxed{\cdot}$ if there is a decimal point.

> Add: 48.3 + 8492 + 3823.
> Basic or scientific calculator steps:
> 48 $\boxed{\cdot}$ 3 $\boxed{+}$ 8492 $\boxed{+}$ 3823 $\boxed{=}$
> Graphing calculator steps:
> 48 $\boxed{\cdot}$ 3 $\boxed{+}$ 8492 $\boxed{+}$ 3823 $\boxed{EXE}$
> Sum = 12,363.3

Section 1–3

1 Subtract whole numbers.

Subtraction is a binary operation that is *not* commutative or associative. Addition and subtraction are inverse operations.

Zero property of subtraction:
$n - 0 = n$.

Arrange numbers in columns of like places. Subtract each column, beginning with the ones place. Borrow when necessary.

> If 5 + 4 = 9, then 9 − 5 = 4 or 9 − 4 = 5.
> 7 − 0 = 7
> $$\begin{array}{r} 4227 \\ -745 \\ \hline 3482 \end{array}$$

2 Subtract decimal numbers.

Subtract decimal numbers by arranging the minuend and subtrahend so that the decimal points are in the same vertical line.

> Subtract: 53.824 − 4.0423
> $$\begin{array}{r} 53.824 \\ -4.0423 \\ \hline 49.7817 \end{array}$$

3 Estimate and check subtraction.

Estimate by rounding the minuend and subtrahend before finding the difference. Check subtraction by adding the difference and the subtrahend. The check should equal the minuend.

> Estimate the difference by rounding to the nearest hundred: 783 − 423
> 800 − 400 = 400. Exact difference: 783 − 423 = 360. Check: 360 + 423 = 783.

4 Subtract using a calculator.	Enter the minuend, followed by the subtraction operation key $\boxed{-}$. Enter the subtrahend and press $\boxed{=}$ or $\boxed{EXE}$. Use $\boxed{\cdot}$ if there is a decimal point.

> Subtract: 2592 − 1474.
> Basic or scientific calculator steps: 2592 $\boxed{-}$ 1474 $\boxed{=}$
> Graphics calculator steps: 2592 $\boxed{-}$ 1474 $\boxed{EXE}$
> Difference = 1118.

Section 1–4

1 Multiply whole-number factors.	Multiplication is a binary operation that is commutative and associative. Zero property of multiplication: $n \times 0 = 0 \times n = 0$. Arrange numbers in columns of like places. Multiply the multiplicand by each digit in the multiplier. Add the partial products.

> $5 \times 7 = 7 \times 5 = 35;$
> $(3 \times 2) \times 6 = 3 \times (2 \times 6) = 36:$
> $5 \times 0 = 0 \times 5 = 0$
>
> $$\begin{array}{r} 259 \\ \times \quad 23 \\ \hline 777 \\ 518 \\ \hline 5957 \end{array}$$

2 Multiply decimal factors.	To multiply decimal numbers, multiply the numbers as in whole numbers and determine the number of decimal digits in both numbers. Place the decimal in the product to the left of the same number of digits, counting from the right.

> Multiply: 3.25×0.53
>
> $$\begin{array}{r} 3.25 \\ \times \quad 0.53 \\ \hline 9\,75 \\ 1\,62\,5 \\ \hline 1.72\,25 \end{array}$$

3 Apply the distributive property.	$a(b + c) = ab + ac$ $a(b - c) = ab - ac$

> $$\begin{array}{rcl} 3(5 + 6) &=& 3(5) + 3(6) \\ 3(11) &=& 15 \; + 18 \\ 33 &=& 33 \end{array}$$

4 Estimate and check multiplication.	Estimate by rounding the factors before finding the exact product. Check multiplication by multiplying a second time.

> Estimate the product by rounding to one nonzero digit: 483×72; $500 \times 70 = 35{,}000$. Exact product: 34,776.

5 Multiply using a calculator.	Enter each factor, followed by the multiplication operation key $\boxed{\times}$. After the last factor, press $\boxed{=}$ or $\boxed{EXE}$ to display the product.

> Multiply: $48 \times 42 \times 23$.
> Basic or scientific calculator steps: 48 $\boxed{\times}$ 42 $\boxed{\times}$ 23 $\boxed{=}$
> Graphics calculator steps: 48 $\boxed{\times}$ 42 $\boxed{\times}$ 23 $\boxed{EXE}$
> Product = 46,368.

Section 1–5

1 Use various symbols to indicate division.

The division a divided by b can be written as

$$a \div b, \quad b\overline{)a}, \quad \frac{a}{b}, \quad \text{or } a/b$$

The divisor, b, cannot be zero.

> Write 12 divided by 4 in four ways.
> $$12 \div 4, \quad 4\overline{)12}, \quad \frac{12}{4}, \quad 12/4$$

2 Divide whole numbers.

Align the numbers properly in long division.

```
        20   R15
   23)475
       46
       15
        0
       15
```

3 Divide decimal numbers.

When dividing by a decimal number, shift the decimal in the divisor to the right end; shift the decimal in the dividend the same number of places to the right. Place the decimal in the quotient.

Divide one place past the specified place value. Affix zeros in the dividend if necessary. Round to the specified place.

Or divide to specified place and express the remainder as a fraction consisting of remainder over divisor.

> Divide:
> ```
>           3.8
>    2.1)7.98
>        63
>        168
>        168
> ```
>
> Divide and round to tenths:
> ```
>         1.70  ≈ 1.7
>    15)25.60
>       15
>       10 6
>       10 5
>          10
> ```
>
> Divide to tenths and express the remainder as a fraction:
> ```
>          1.7 1/15
>    15)25.6
>       15
>       10 6
>       10 5
>          1
> ```

4 Estimate and check division.

To estimate division, round the dividend and divisor before finding the quotient. Find the first digit of the quotient and add a zero for each remaining digit in the dividend.

To check division, multiply the quotient by the divisor and add the remainder.

> Estimate $2934 \div 42$.
> ```
>         70
>    40)3000
> ```
> Exact quotient = 69   R36
> Check:
> $69 \times 42 = 2898$
> $2898 + 36 = 2934$

5 Find the numerical average.	Find the sum of the values and divide the sum by the number of values.

> Find the average of 74, 65, and 85.
> 74 + 65 + 85 = 224
>   224 ÷ 3 = 74.67 or 75 (rounded)

6 Divide using a calculator.	Enter the dividend followed by the division operation key $\boxed{\div}$. Enter the divisor and press $\boxed{=}$ or $\boxed{\text{EXE}}$. Use $\boxed{\cdot}$ if there is a decimal point.

> Divide: 295 ÷ 15.
> Basic or scientific calculator steps: 295 $\boxed{\div}$ 15 $\boxed{=}$
> Graphics calculator steps: 295 $\boxed{\div}$ 15 $\boxed{\text{EXE}}$
> Quotient = 19.66666667.

Section 1–6

1 Simplify expressions containing exponents.	To simplify an exponential expression, use the base as a factor the number of times indicated by the exponent. $a^1 = a$, for any base a $a^0 = 1$, for any nonzero base

> $5^3 = 5 \times 5 \times 5 = 125$
> $7^1 = 7$
> $9^0 = 1$

2 Square numbers and find square roots of numbers.	Squaring and finding square roots are inverse operations.

> $7^2 = 49$
> $\sqrt{49} = 7$

3 Use powers of 10 to multiply and divide.	To multiply a decimal number by a power of 10, shift the decimal to the *right* as many digits as the power of 10 has zeros. Affix zeros if necessary. To divide a decimal by a power of 10, shift the decimal to the *left*. Note that this process is just the opposite of that for multiplication.

> Multiply:
> $27 \times 10^3 = 27,000$
> $18 \times 100 = 1800$
> $23.52 \times 1000 = 23,520$
>
> Divide: $3.52 \div 10 = 0.352$
> $400 \div 10^2 = 4$
> $23,000 \div 10^3 = 23$

4 Use a calculator to find powers and roots.	Squaring or cubing on some scientific calculators:

1. Enter the base.
2. Press the square key $\boxed{x^2}$ or cube key $\boxed{x^3}$.

Raising a number to a power on some scientific calculators:

1. Enter the base.
2. Press the general power key $\boxed{x^y}$.
3. Enter the exponent.
4. Press the equals key $\boxed{=}$.

Squaring or cubing on graphics calculators and some scientific calculators:

1. Enter the base.
2. Press the square key $\boxed{x^2}$ or cube key $\boxed{x^3}$.
3. Press the $\boxed{\text{EXE}}$ key.

Raising a number to a power on graphics calculators and some scientific calculators:

1. Enter the base.
2. Press the general power key $\boxed{x^y}$.
3. Enter the exponent.
4. Press the $\boxed{\text{EXE}}$ key.

Use a calculator to find 18^2, 21^3, 7^5, and 2.5^4.
18^2 18 $\boxed{x^2}$ $\Rightarrow$ 324 or 18 $\boxed{x^2}$ $\boxed{\text{EXE}}$ $\Rightarrow$ 324
21^3 21 $\boxed{x^3}$ $\Rightarrow$ 9261 or 21 $\boxed{x^3}$ $\boxed{\text{EXE}}$ $\Rightarrow$ 9261
7^5 7 $\boxed{x^y}$ 5 $\boxed{=}$ $\Rightarrow$ 16807 or 7 $\boxed{x^y}$ 5 $\boxed{\text{EXE}}$ $\Rightarrow$ 16807
2.5^4 2.5 $\boxed{x^y}$ 4 $\boxed{=}$ $\Rightarrow$ 39.0625 or 2.5 $\boxed{x^y}$ 4 $\boxed{\text{EXE}}$ $\Rightarrow$ 39.0625

Finding the square root of a number on graphics calculators and most scientific calculators:

1. Press the square root key $\boxed{\sqrt{}}$.
2. Enter the radicand.
3. Press the $\boxed{\text{EXE}}$ or $\boxed{=}$ key.

Finding the square root of a number on some scientific calculators:

1. Enter the radicand.
2. Press the square root key $\boxed{\sqrt{}}$.

Use a calculator to find $\sqrt{729}$ and $\sqrt{33.64}$.
$\sqrt{729}$ 729 $\boxed{\sqrt{}}$ $\Rightarrow$ 27 or $\boxed{\sqrt{}}$ 729 $\boxed{\text{EXE}}$ $\Rightarrow$ 27
$\sqrt{33.64}$ 33.64 $\boxed{\sqrt{}}$ $\Rightarrow$ 5.8 or $\boxed{\sqrt{}}$ 33.64 $\boxed{\text{EXE}}$ $\Rightarrow$ 5.8

Section 1–7

1 Apply the order of operations to problems containing several operations.

Order of operations:
1. Parentheses or groupings, innermost first.
2. Exponential operations and roots from left to right.
3. Multiplication and division from left to right.
4. Addition and subtraction from left to right.

$3^2 + 5(6 - 4) \div 2$
$3^2 + 5(2) \div 2$
$9 + 5(2) \div 2$
$9 + 10 \div 2$
$9 + 5$
14

2 Use a calculator to perform a series of operations.

Both the scientific and graphics calculators will perform the proper order of operations. With the scientific calculator, the times sign must be used with multiplication. With the graphics calculator, the times sign is not required if a parenthesis follows. Use $\boxed{\cdot}$ if there is a decimal point.

$3^2 + 5(6 - 4) \div 2$.
Scientific: 3 $\boxed{x^2}$ $\boxed{+}$ 5 $\boxed{\times}$ $\boxed{(}$ 6 $\boxed{-}$ 4 $\boxed{)}$ $\boxed{\div}$ 2 $\boxed{=}$
Display: 14.
Graphics: 3 $\boxed{x^2}$ $\boxed{+}$ 5 $\boxed{(}$ 6 $\boxed{-}$ 4 $\boxed{)}$ $\boxed{\div}$ 2 $\boxed{\text{EXE}}$
Display: $3^2 + 5(6 - 4) \div 2$
          14.

decimal number system (p. 2)
whole number digits (p. 2)
place value (p. 2)
periods (p. 2)
standard notation (p. 4, 46)
decimal fraction (p. 5)
decimal number (p. 5)
decimal (p. 5)
decimal point (p. 5)
round (p. 9)
nonzero digit (p. 11)
approximate amount (p. 12)
addition (p. 15)
addends (p. 15)
sum (p. 15)
total (p. 15)
commutative property of
 addition (p. 15)
associative property of addition
 (p. 15)
binary operation (p. 15)
carry (p. 16)
estimating (p. 17)

subtraction (p. 21)
inverse operations (p. 21, 36)
inverse (p. 21)
difference (p. 21)
minuend (p. 21)
subtrahend (p. 21)
tolerance (p. 23)
limit dimensions (p. 23)
multiplication (p. 26)
multiplicand (p. 26)
multiplier (p. 27)
factors (p. 26)
product (p. 26)
commutative property of
 multiplication (p. 27)
associative property of
 multiplication (p. 27)
partial product (p. 28)
distributive property of
 multiplication (p. 30)
area (p. 32)
square measures (p. 32)
formula (p. 32)

division (p. 36)
divisor (p. 36)
dividend (p. 36)
quotient (p. 36)
numerical average (p. 42)
base (p. 46)
exponent (p. 46)
cube (p. 46)
cubed (p. 46)
power (p. 46)
squared (p. 46)
simplify (p. 46)
exponential notation (p. 46)
standard notation (p. 46)
perfect square (p. 47)
square (p. 47)
square root (p. 47)
radical sign (p. 48)
radicand (p. 48)
radical expression (p. 48)
powers of 10 (p. 48)
order of operations (p. 53)

TRIAL TEST

1. Write 5,030,102 in words.

2. Write three hundred twenty-four thousand, five hundred twenty in standard notation.

3. Write the digits for seven and twenty-seven thousandths.

4. Write the digits for two hundred four millionths.

5. Round 2743 to hundreds.

6. Round 34,988 to a number with one nonzero digit.

7. Which number is smaller, 5.09 or 5.1?

8. Round 4.018 to the nearest hundredth.

9. Round 48.3284 to the nearest tenth.

10. Round $4.834 to the nearest cent.

Perform the indicated operations.

11. $37 + 158 + 764 + 48$

12. $\$61,532 - \$47,245$

13. $\$13,207 \times 702$

14. $\$25,600 \div 12$

15. $3^2 + 5^3$

16. 46×10^3

17. $3 \times 6^2 - 4 \div 2$

18. $5^3 - (3 + 5) \times \sqrt{9}$

19. $86 \div 10^4$

20. If a stereo costs $495 and a stereo cabinet costs $269, use numbers with one nonzero digit to estimate the total cost. Find the exact cost.

21. If a man has a bill for $165 and his paycheck is $475, estimate how much of the paycheck is left after paying the bill by rounding to tens. Find the exact amount left.

22. A corporation wants to buy 45 VCRs for $335 each. Use numbers with one nonzero digit to estimate the total cost. Find the exact cost.

23. A softball coach paid $126 for 9 pizzas for a party after a successful season. Estimate the cost of each pizza by using numbers with one nonzero digit. Find the exact cost per pizza.

24. A mathematics professor promised to give 2 extra points for each time a student worked a set of self-study exercises. If one student worked 17 sets of self-study exercises, how many points should have been given?

25. Explain the difference between the commutative property of addition and the associative property of addition. Give a numerical illustration to support your explanation.

26. Find the product: 4.08×0.05.

27. Find the product: 42.73×1000.

28. $25\overline{)27.75}$

29. Round answer to the nearest tenth: $7.2\overline{)83.41}$.

30. $52.38 \div 10{,}000$

31. Find the average of these test scores: 82, 95, 76, 84, 72, and 91. Round to the nearest whole number.

32. Estimate the sum by rounding to the nearest whole number: $3.85 + 7.46$.

33. Estimate the difference by rounding each number to the nearest tenth: $0.87 - 0.328$.

34. Estimate the product by rounding each decimal to a number having only one nonzero digit: 42.38×27.9.

35. Estimate the quotient by rounding the divisor and dividend to numbers with one nonzero digit: $37.2\overline{)2987.5}$.

36. The blueprint specification for a machined part calls for its thickness to be 1.485 in. with a tolerance of ± 0.010 in. What are the limit dimensions of the part?

37. Heating oil costs \$1.75 per gallon. What is the cost of 10,000 gal?

38. A construction job requires 16 pieces of reinforced steel, each 7.96 ft in length. What length of steel is required for the job?

2

Integers

GOOD DECISIONS THROUGH TEAMWORK

Each member of your five-member team will assume the role of vice president of operations for each location of your international company. The nature of your business requires a weekly teleconference and each of you schedules working hours between 8:00 A.M. and 5:00 P.M., if possible. The company's headquarters is in Memphis, Tennessee, and it has offices in Toronto, Canada; Frankfurt, Germany; Tokyo, Japan; and San Juan, Puerto Rico.

While you are at the annual meeting in Memphis, your team needs to arrange a time for your weekly teleconference that fits within everyone's normal working hours. If it cannot, select a permanent weekly time or a rotation schedule that will be most convenient for the greatest number of vice presidents. To make these arrangements, begin by assigning the number 0 to the headquarters office and assigning an appropriate integer to the remaining offices based on the difference in time between each location and the headquarters office. Worldwide time zone information can be found on the Internet or a world globe, or in standard reference materials such as a world atlas or encyclopedia. Once you have assigned integers to each location, use teamwork techniques to resolve the problem of establishing a meeting time.

Prepare a report for your classmates that describes your telecommunications plans, lists some of the obstacles to be resolved in establishing the plan, and relates what you have learned from this activity.

Historically, as the need for negative numbers became apparent, the number system was expanded to fill this need. Consider the following. When a temperature is colder than zero degrees, how can we express this temperature numerically? When the selling price of an item is less than the cost of an item, how can we express the profit made on the sale numerically? When a withdrawal on a bank account exceeds the amount of money in the account, how can we express the account balance numerically? These situations demonstrate a need to express values that are less than zero. The first type of number we will explore that is less than zero is the *integer*.

2–1 NATURAL NUMBERS, WHOLE NUMBERS, AND INTEGERS

Learning Objectives

1. Relate integers to natural numbers and whole numbers.
2. Compare integers.
3. Find the absolute value of integers.
4. Find the opposite of integers.
5. Locate points on a rectangular coordinate system.

1 Relate Integers to Natural Numbers and Whole Numbers.

The set of numbers needed first was the set of counting numbers, or *natural numbers*. Natural numbers begin with the numbers 1, 2, 3, 4, 5, 6, 7, and continue indefinitely. As the place value system evolved, the number zero, 0, was added when a certain place value was missing. The natural numbers and zero form a set of numbers called *whole numbers*. The whole numbers are illustrated on the number line in Fig. 2–1. On the number line, zero is to the left of the number one.

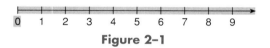

Figure 2–1

Many physical phenomena have values that are less than zero. For instance, when the temperature is 0°, that does not mean that the temperature cannot become any colder than zero. Another type of number is needed to express a temperature colder than zero degrees. When the opposite of each natural number is included, the set of whole numbers can be expanded to form the set of *integers*. Figure 2–2 illustrates the extension of the set of whole numbers to include all integers.

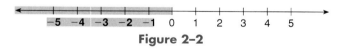

Figure 2–2

The *opposite* of a number is a number that is the same number of units from zero, but in the opposite direction. A number line illustrating integers and their opposites is shown in Fig. 2–3.

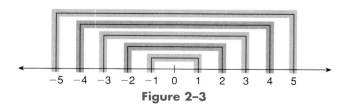

Figure 2–3

As we may notice from the number line, it is not necessary to include the positive sign for positive values. You may sometimes want to include the sign to draw attention to it or to emphasize its direction. Negative signs, however, must always be included. To further illustrate the relationship among the three sets of numbers—natural numbers, whole numbers, and integers—examine Fig. 2–4.

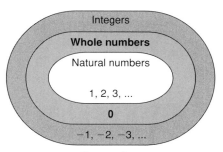

Figure 2–4

EXAMPLE 1   Answer the following questions.

(a) How is the set of natural numbers different from the set of whole numbers?
(b) How is the set of whole numbers different from the set of integers?
(c) Can a number be both an integer and a whole number? If so, give an illustration.
(d) Is any integer also a whole number?

(a) Natural numbers and whole numbers have all the same numbers except for the number zero. Natural numbers do not include zero; whole numbers do include zero.
(b) All whole numbers are included in the set of integers. In addition, integers include the opposite of every natural number.
(c) Yes, every whole number is also an integer. The number 3 is both a whole number and an integer.
(d) An integer is sometimes a whole number. The opposites of natural numbers (the negative integers) are integers but are not whole numbers. The integer −3 is not a whole number.

2   Compare Integers.

If a number is positioned to the left of another number on a number line, it represents a smaller value. Larger values are positioned to the right of a number on the number line. We can show symbolically how two numbers compare in size by using the two symbols, $<$ and $>$. When the first number is smaller than or "is less than" the second number, the symbol that points to the left is used. For example, "5 is less than 7" is written symbolically as $5 < 7$. When the first number is larger than or "is greater than" the second number, the symbol that points to the right is used. For example, "7 is greater than 5" is written symbolically as $7 > 5$.

Tips and Traps   *Memory Tips for Distinguishing Between $<$ and $>$:*
Look at the number line in Fig. 2–5. You will see that numbers get smaller

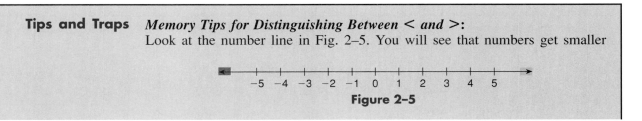

Figure 2–5

as we move to the left. The arrowhead on the left end of the number line has the same direction as the "less than" symbol. In other words, the point of the arrowhead is directed toward the smaller number.

$$-4 < -1 \qquad -3 < 0 \qquad 0 < 4$$

On the other hand, numbers get larger as we move to the right. The arrowhead on the right end of the number line has the same direction as the "greater than" symbol. Again, the point of the arrowhead is directed toward the smaller number.

$$-1 > -4 \qquad 0 > -3 \qquad 4 > 0$$

EXAMPLE 2 Use the symbols $<$ and $>$ to indicate whether the first number is less than or greater than the second number. (a) 7 __ 9 (b) 0 __ -1 (c) -4 __ -2 (d) -2 __ 0

(a) $7 < 9$ 7 is smaller—to the left of 9 on the number line
(b) $0 > -1$ 0 is larger—to the right of -1 on the number line
(c) $-4 < -2$ -4 is smaller—to the left of -2 on the number line
(d) $-2 < 0$ -2 is smaller—to the left of 0 on the number line

3 Find the Absolute Value of Integers.

The distance between each pair of consecutive integers is the same for all integers located on the number line. This concept of distance between numbers on the number line is related to the concept of the *absolute value* of numbers. The absolute value of a number is often described as the number of units of distance the number is from zero. The symbol for absolute value is $| \ |$; $|3|$ is read as "the absolute value of 3."

Because the distance is a physical property that is always expressed as a non-negative amount, the absolute value of a number is always a non-negative value. If a number is a positive number, then the absolute value of the number is the number itself. In symbols, if $a > 0$, then $|a| = a$. If a number is a negative number, then the absolute value of the number is the opposite of the number. In symbols, if $a < 0$, then $|a| = -a$.

The absolute value of zero is zero. In symbols, $|0| = 0$. Study the following examples of absolute values.

$$|-5| = 5 \qquad |27| = 27 \qquad |0| = 0$$

Tips and Traps *Writing "Positive" or "Negative" Symbolically:*
A symbolic way to represent positive numbers is $a > 0$, where a represents any number on the condition that a is greater than zero, or positive.

Symbolically, negative numbers are represented by $a < 0$, which states that any number represented by a is less than zero or negative.

Practice reading in words the symbolic statements that define the absolute value of a number.

■ **DEFINITION 2–1:** **Absolute Value**

$$|a| = a, \text{ for } a > 0$$

The absolute value of a positive number is equal to itself.

$$|a| = -a, \text{ for } a < 0$$

The absolute value of a negative number is equal to its opposite.

$$|0| = 0$$

The absolute value of zero is zero.

EXAMPLE 3   Give the absolute value of the following quantities: (a) $|9|$ (b) $|-4|$ (c) $|0|$

(a) $|9| = 9$ 9 is positive. Its absolute value is the number itself.
(b) $|-4| = 4$ −4 is negative. Its absolute value is the opposite of −4.
(c) $|0| = 0$ 0 is unsigned. Its absolute value is still zero.

Tips and Traps   *Two Components of a Signed Number:*
Integers and other signed numbers tell us two things. In the integer −8, the negative sign tells us the number is to the left of zero and is referred to as the *directional sign* of the number. The 8 tells us how many units the number is from zero and is referred to as the *absolute value* of the number.

---

4 Find the Opposite of Integers.

Every number except zero has an opposite. The opposite of a number is also called the *additive inverse* of the number. A number and its additive inverse (opposite) have the same absolute value but different directional signs.

EXAMPLE 4   Find the opposite of each of the following and illustrate each answer on the number line.   (a) −8   (b) 6   (c) 0

(a) The opposite of −8 is 8, which is illustrated in Fig. 2–6.

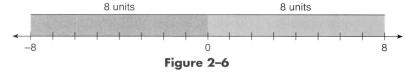

Figure 2–6

(b) The opposite of 6 is −6, which is illustrated in Fig. 2–7.

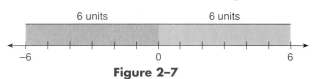

Figure 2–7

(c) Zero has no opposite, which is illustrated in Fig. 2–8.

Figure 2–8

5 Locate Points on a Rectangular Coordinate System.

The number line can be used to represent a value pictorially. This kind of visual representation is called a *one-dimensional graph*. The one dimension represents the distance and direction that a value is from zero. The *rectangular coordinate system* is used to give a graphical representation of two-dimensional values. The rectangular coordinate system is two number lines positioned to form a right angle or square corner (see Fig. 2–9). The number line that runs from left to right is called the *horizontal axis* and is represented by the letter x. The number line that runs from top to bottom is called the *vertical axis* and is represented by the letter y.

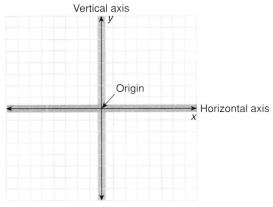

Vertical axis

Origin

Horizontal axis

Figure 2–9

Zero on both number lines is located at the point where the two number lines cross. This point is called the *origin*. Other points are located by horizontal and vertical movements from the origin. Look at the points in Example 5.

EXAMPLE 5   Describe the location of the points in Fig. 2–10 by giving the amount of horizontal and vertical movement from the origin.

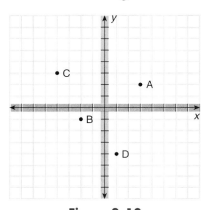

Figure 2–10

Point A: horizontal movement, $+3$; vertical movement, $+2$
Point B: horizontal movement, -2; vertical movement, -1
Point C: horizontal movement, -4; vertical movement, $+3$
Point D: horizontal movement, $+1$; vertical movement, -4

As is common in mathematics, the location of points on a rectangular coordinate system can be written in an abbreviated, symbolic form. Point notation uses two signed numbers to represent the horizontal and vertical movement from the origin to a given point. The value that represents the horizontal movement is called the *x-coordinate*, and the value that represents the vertical movement is called the *y-coordinate*. These signed numbers are separated with a comma and enclosed in parentheses.

Tips and Traps   *Point Notation:*
A symbolic way to represent the location of a point is: (horizontal movement, vertical movement), or (x, y), where x is the horizontal movement from the origin, or the *x*-coordinate, and y is the vertical movement from the *x*-axis, or the *y*-coordinate.

EXAMPLE 6   Write the points in Example 5 using point notation.

Point A = (3, 2)   Point B = (−2, −1)

Point C = (−4, 3)   Point D = (1, −4)

To *plot* a point means to show its location on the rectangular coordinate system.

Rule 2–1   *To plot a point on the rectangular coordinate system:*

1. Start at the origin.
2. Count to the left or right the number of units represented by the first signed number.
3. Starting at the ending point from Step 2, count up or down the number of units represented by the second signed number.

EXAMPLE 7   Plot the following points: Point A = (3, 1), Point B = (−2, 5), Point C = (−3, −2), Point D = (1, −3).

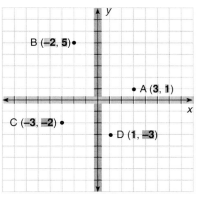

Figure 2–11

The solution is shown in Fig. 2–11.

Suppose a point shows movement in only one direction. The point can still be written in point notation by using the number zero. Points on the horizontal number line have no vertical movement. Points on the vertical number line have no horizontal movement. Examine the points in Fig. 2–12.

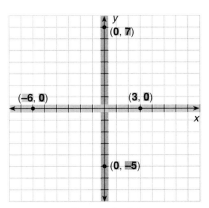

Figure 2–12

EXAMPLE 8 Plot the following points: A = (2, 0), B = (0, 2), C = (−2, 0), D = (0, −2).

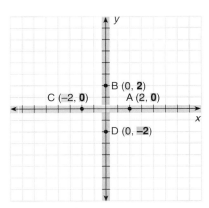

Figure 2–13

The solution is shown in Fig. 2–13.

SELF-STUDY EXERCISES 2–1

1 The following items refer to the number line in Fig. 2–14.

1. On what side of zero are the positive numbers?
2. On what side of zero are the negative numbers?
3. How long does the number line extend in each direction?
4. How many points exist on the number line for each distinct number?
5. What sign (+ or −) is assigned to zero?
6. Arrange these integers in order as they would appear on the number line: −2, 3, 0, −1, −3, 2, 1.

Figure 2–14

7. In what direction on the number line do the numbers increase in value?
8. In what direction on the number line do the numbers decrease in value?

2 Use the "greater than" or "less than" symbol in the following number pairs to make a true statement.

9. 5 __ 8
10. 0 __ −2
11. −4 __ 2
12. −2 __ 0
13. −5 __ −9
14. −5 __ 2
15. 0 __ −5
16. −3 __ 3

Rewrite the following expressions using < or >. When an expression contains one or more letters, the letters stand for numbers.

17. x is less than y
18. a is greater than b
19. 3 + 5 is greater than 6
20. 9 is less than 18 − 6
21. k is greater than t
22. r is less than s

3 Give the value of the following.
23. $|23|$
24. $|0|$
25. $|-10|$
26. -17
27. $|-13|$
28. $|345|$
29. 67
30. $|-61|$

4 Illustrate the following numbers and their opposites on a number line.
31. 7
32. −8
33. −4
34. 12
35. What number has no opposite?
36. Describe the opposite of a positive integer. Describe the opposite of a negative integer.

37. Write the coordinates for the points on the graph in Fig. 2–15.

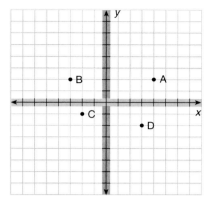

Figure 2–15

A = __   B = __   C = __   D = __

Draw a rectangular coordinate system and locate the points given in Exercises 38 to 43.

38. A = (−7, 4)   **39.** B = (0, −10)   **40.** C = (−2, 0)
41. D = (−7, −7)   **42.** E = (3, 0)   **43.** F = (4, −3)
44. What common property describes the coordinates of the points on the *x*-axis?
45. What common property describes the coordinates of the points on the *y*-axis?
46. Describe the signs of the coordinates of points that lie in the upper left quarter of the graph.
47. Describe the signs of the coordinates of points that lie in the lower right quarter of the graph.

2–2   ADDITION WITH INTEGERS

Learning Objectives

1   Add integers with like signs.

2   Add integers with unlike signs.

Throughout history, numbers have been used for more applications than just counting and recording information. For example, if the low temperature for the day is 3 degrees below zero and it is expected to rise 24 degrees during the day, what will be the high temperature for the day?

1   Add Integers with Like Signs.

Since our earliest experiences with addition, we have added positive numbers. In the addition problem, $3 + 2 = 5$, all three numbers are positive. Numbers are understood to be positive when no sign is given. To emphasize that the numbers are positive, however, we can write the problem as $+3 + +2 = +5$. You can also write the problem using parentheses to separate the two plus signs, $+3 + (+2) = +5$. The plus symbol can serve two purposes in mathematical expressions. It can be used as the directional sign for a positive number and it can be used to show addition. In the expression $+3 + (+2) = +5$, the plus sign in front of the parentheses shows the operation of addition. The plus sign inside the parentheses identifies the number as positive. Addition of integers will build on our knowledge of addition of positive numbers. We will illustrate addition of integers on the number line.

To add 3 and 2, we begin at zero and move three units to the right (positive direction) on the number line (see Fig. 2–16). This move takes us to +3. From the +3 position, we move two units to the right. This second move takes us to the +5 position. In other words, 3 + 2 = 5.

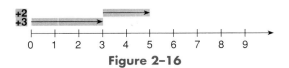

Figure 2-16

On the other hand, to add two negative integers, say, $(-3) + (-2)$, we begin at zero and move three units to the left (negative direction) on the number line (see Fig. 2–17). This takes us to -3 on the number line. Then from -3 we move two units to the left. This second move takes us to -5. Thus, $-3 + (-2) = -5$.

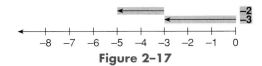

Figure 2-17

When adding the two positive numbers or the two negative numbers, notice that the sum could have been obtained using the following rule.

> **Rule 2-2** *To add numbers with like signs:*
>
> **1.** Add the absolute values of the numbers.
> **2.** Give the sum the common or like sign.

EXAMPLE 1   Add 3 + 6 + 8.

$$3 + 6 + 8 \qquad \text{Add absolute values. Keep common positive sign.}$$
$$9 + 8 = \mathbf{17}$$

Before continuing with additional examples, let's stop and make some observations. First, addition is a *binary operation*. This means that only two numbers are used in the operation. If more than two numbers are involved, two are added and then the next number is added to the sum of the first two. Thus, rules developed for addition apply to two numbers at a time.

EXAMPLE 2   Add $-3 + (-4) + (-1)$.

$$-3 + (-4) + (-1) = \qquad \text{Add absolute values.}$$
$$-7 + (-1) = \mathbf{-8} \qquad \text{Keep common negative sign.}$$

2   Add Integers with Unlike Signs.

Now let's examine the addition of numbers with unlike signs, which can also be illustrated on the number line. When adding numbers with unlike signs, we move in the direction indicated by the sign of each number, positive to the right and negative to the left. To add $-4 + 3$, first move four units to the left of zero (see Fig. 2–18). This takes you to -4. Then from the -4 position, move three units to the right. This takes you to -1. Thus, $-4 + 3 = -1$. This operation can also be expressed as a rule using absolute values.

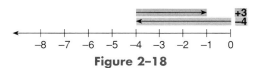

Figure 2-18

> **Rule 2-3** *To add signed numbers with unlike signs:*
> 1. *Subtract* the smaller absolute value from the larger absolute value.
> 2. Give the sum the sign of the number that has the larger absolute value.

EXAMPLE 3   Add $7 + (-12)$.

$7 + (-12) =$ Signs are unlike.

$7 + (-12) = -5$ Subtract absolute values: $12 - 7 = 5$. Keep the sign of the -12 (negative), the larger absolute value.

Tips and Traps   *To Add, You . . . Subtract?*
When addition is extended to integers, phrases we have used in arithmetic, such as "find the sum" or "add," can be confusing to us. Yes, in the operation of addition, sometimes we add absolute values and sometimes we subtract absolute values. We often use the word "combine" to imply the addition of numbers with either like or unlike signs.

EXAMPLE 4   Add $-5 + 7$.

$-5 + 7 =$ Signs are unlike.

$-5 + 7 = 2$ Subtract absolute values: $7 - 5 = 2$. Keep the sign of the 7 (positive), the larger absolute value.

When an addition involves several positive numbers and several negative numbers, it is convenient to group all the positive numbers and all the negative numbers and add each group separately. The sums of each group can then be added. We can do this because of the associative property of addition. The grouping of addends in an addition does not matter. The next example illustrates combining the commutative and associative properties of addition to add several signed numbers.

EXAMPLE 5   Add $8 + (-9) + 13 + (-15)$.

$8 + (-9) + 13 + (-15) =$ Signs are unlike. Group numbers with like signs.

$8 + 13 + [-9 + (-15)] =$ The brackets, [ ], show the grouping and separate the operational plus sign from the directional minus sign of -9. Add groups of numbers with like signs.

$21 + (-24) = -3$ Add resulting sums using the rule for adding numbers with unlike signs.

The properties of addition that involve the number zero apply to integers as well as to whole numbers and decimals. When zero is added to a number, the number is not changed. We say that zero is the *additive identity*. The sum of a number and its opposite is zero. We call the opposite of a number the *additive inverse* of that number. We can also write these definitions using symbols.

■ **DEFINITION 2–2: Additive Identity**   Zero is the *additive identity* because $a + 0 = a$ for all values of a.

■ **DEFINITION 2–3: Additive Inverse**   The opposite of a number is the *additive inverse* of the number because $a + (-a) = 0$ for all values of a.

EXAMPLE 6 Add: (a) $-2 + 0$ (b) $5 + (-5)$ (c) $0 + (-3) + 4$ (d) $5 + (-2) + (-5)$

(a) $-2 + 0 = \mathbf{-2}$ Additive identity
(b) $5 + (-5) = \mathbf{0}$ Additive inverse
(c) $0 + (-3) + 4 =$ Apply additive identity property for $0 + (-3)$.
    $-3 + 4 = \mathbf{1}$ Apply rule for adding numbers with unlike signs.
(d) $5 + (-2) + (-5) =$ Apply commutative and associative properties.
    $5 + (-5) + (-2) =$ Apply additive inverse property.
    $0 + (-2) = \mathbf{-2}$ Apply additive identity property.

SELF-STUDY EXERCISES 2-2

1 Add the following:

1. $7 + 10$
2. $-5 + (-8)$
3. $12 + 87$
4. $-21 + (-38)$
5. $-32 + (-16)$
6. $(-58) + (-103)$

2 Add the following.

7. $-4 + (-2)$
8. $-3 + 7$
9. $4 + (-6)$
10. $-4 + 6$
11. $-18 + 8$
12. $32 + (-72)$
13. $21 + (-14)$
14. $17 + (-4) + 3 + (-1)$
15. $-3 + 2 + (-7)$
16. $47 + (-82) + 2$
17. $14 + (-6) + 1$
18. $-7 + (-3) + (-1)$
19. $4 + 2 + (-3) + 10$
20. $2 + (-1) + 8$
21. $-2 + 1 + (-8) + 12$
22. $15 + (-15)$
23. $-7 + 7$
24. $92 + (-92)$
25. $(-396) + (396)$
26. $503 + (-503)$
27. $0 + 9$
28. $-3 + 0$
29. $0 + (-7)$
30. $18 + 0$
31. $-8 + 0$
32. $0 + (-28)$

33. Write a sum that illustrates the additive inverse.

34. What is the additive identity?

35. Explain why the sum of two positive numbers is positive and the sum of two negative numbers is negative.

36. A stock priced at 42 had the following changes in one week: $-3, +8, -6, -7, +2$. What was the value of the stock at the end of the week?

37. You open a bank account by depositing $242. You then write checks for $21, $32, and $123. What is your new account balance?

2-3   SUBTRACTING INTEGERS

Learning Objectives

1 Subtract integers.

2 Subtract with zero or opposites.

3 Combine addition and subtraction.

Before discussing the subtraction of integers, we need to examine two expressions involving the subtraction of integers, $6 - 9$ and $5 - (-3)$, and to interpret the meaning of the notation. The first expression, $6 - 9$, means positive 9 is subtracted from positive 6. The sign between the 6 and the 9 can be interpreted as subtraction and can be read as "6 subtract 9" or "6 minus 9," just as we would before studying integers. The minus sign is the operational sign telling us to subtract positive 9 from positive 6. The expression could also be written as $6 - (+9)$.

The second expression, $5 - (-3)$, means that negative 3 is subtracted from positive 5. The first sign between 5 and 3 is the operational sign telling us to subtract. The second sign, immediately in front of 3, is the directional sign to show negative 3.

1 Subtract Integers.

Now let's focus on the subtraction of whole numbers. If we let the letter x represent a missing, temporarily unknown number, then the subtraction problem $10 - 7 = x$ can also be interpreted as what number (x) added to 7 gives 10. The number 3 added to 7 gives 10. Another way of stating this relationship is that x is the sum of 10 and -7 since $10 + (-7) = 3$.

We can apply this relationship to subtracting integers on the number line. Let's return to the two expressions, $6 - 9$ and $5 - (-3)$, and subtract them on the number line.

To subtract $6 - 9$, we think of what number added to 9 equals 6, that is, $6 = 9 + x$. Locate 9 on the number line by moving nine units to the right of zero (see Fig. 2–19). From that point, $+9$, we want to get to the $+6$. How far and in what direction should we move? We need to move three units to the left, -3. Thus, $6 = 9 + (-3)$, or $6 - 9 = -3$.

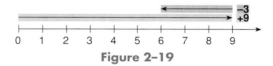

Figure 2–19

Now let's try subtracting $5 - (-3)$. We want to know what number added to -3 equals 5; that is, $5 = -3 + x$. On the number line, locate -3 by moving three units to the left of zero (see Fig. 2–20). To get to $+5$ from -3, how many units and in which direction must we move? We must move eight units to the right, $+8$. Thus, $5 = -3 + 8$, or $5 - (-3) = 8$.

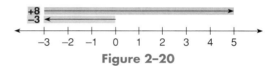

Figure 2–20

Now let's study both solved subtraction problems, $6 - 9 = -3$ and $5 - (-3) = 8$. In the first problem, $6 - 9 = -3$, we see that *to subtract positive 9* is the same as *to add negative 9*. Negative 9 is the opposite, or additive inverse, of positive 9.

In the second case, $5 - (-3) = 8$, we see that *to subtract negative 3* is the same as *to add positive 3*. Positive 3 is the opposite, or additive inverse, of negative 3.

In both cases we added to the minuend (first number) the opposite or additive inverse of the subtrahend (number being subtracted or second number) to obtain the difference (answer).

Rule 2–4   *To subtract signed numbers:*

1. Interpret the problem as adding the opposite of the subtrahend (second number) to the minuend (first number).
2. Apply the appropriate rule for adding signed numbers with like or unlike signs.

EXAMPLE 1   Subtract: (a) $2 - 6$ (b) $-9 - 5$ (c) $12 - (-4)$ (d) $-7 - (-9)$

(a) $2 - 6 =$ Subtract positive 6 from positive 2.
    $2 + (-6) =$ Add the opposite of 6, which is -6, to 2.
    $2 + (-6) = \mathbf{-4}$ Apply the rule for adding numbers with unlike signs.

(b)　$-9 - 5 =$　　　　　　　Subtract positive 5 from negative 9.
　　　$-9 + (-5) =$　　　　　Add the opposite of 5, which is -5, to -9.
　　　$-9 + (-5) = \mathbf{-14}$　　Apply the rule for adding numbers with like signs.

(c)　$12 - (-4) =$　　　　　Subtract negative 4 from positive 12.
　　　$12 + (+4) =$　　　　　Add the opposite of -4, which is $+4$, to 12.
　　　$12 + 4 = \mathbf{16}$　　　　Apply the rule for adding numbers with like signs.

(d)　$-7 - (-9) =$　　　　　Subtract negative 9 from negative 7.
　　　$-7 + (+9) =$　　　　　Add the opposite of -9, which is $+9$, to -7.
　　　$-7 + 9 = \mathbf{2}$　　　　Apply the rule for adding numbers with unlike signs.

2 Subtract with Zero or Opposites.

Subtractions involving zero and opposites will still be interpreted first as equivalent addition problems. To subtract zero from a number is the same as adding zero, which is the original number unchanged: $5 - 0 = 5 + (0) = 5$. To subtract a nonzero number from zero is the same as adding the opposite of the number to zero: $0 - 5 = 0 + (-5) = -5$. Subtracting an opposite from a number is the same as adding a number to itself: $5 - (-5) = 5 + (+5) = 10$.

EXAMPLE 2　Evaluate:　(a) $8 - 0$　(b) $0 - 15$　(c) $32 - (-32)$　(d) $-15 - 15$

(a)　$8 - 0 =$　　　　　　Subtract zero from positive 8.
　　　$8 + (0) =$　　　　　Add zero to positive 8.
　　　$8 + (0) = \mathbf{8}$　　　Zero added to any number is that number.

(b)　$0 - 15 =$　　　　　Subtract positive 15 from zero.
　　　$0 + (-15) =$　　　　Add the opposite of 15, which is -15, to zero.
　　　$0 + (-15) = \mathbf{-15}$　Any number added to zero is that number.

(c)　$32 - (-32) =$　　　Subtract negative 32 from positive 32.
　　　$32 + (+32) =$　　　Add the opposite of -32, which is $+32$, to 32.
　　　$32 + 32 = \mathbf{64}$　　Apply the rule for adding numbers with like signs.

(d)　$-15 - (15) =$　　　Subtract positive 15 from negative 15.
　　　$-15 + (-15) =$　　Add the opposite of 15, which is -15, to -15.
　　　$-15 + (-15) = \mathbf{-30}$　Apply the rule for adding numbers with like signs.

3 Combine Addition and Subtraction.

When we express the sum and difference of integers with all the appropriate operational and directional signs, we have an expression with both an operational sign and a directional sign between every two numbers. We have four different possibilities when two signs are written together. The following Tips and Traps makes some general observations that may reduce any confusion about the four possibilities.

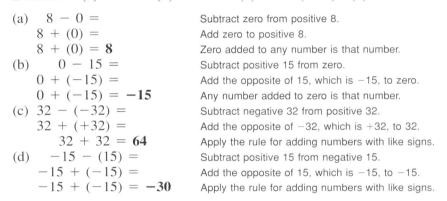

Tips and Traps　*Modifying Notation to Eliminate Two Signs Written Together:*
Writing mathematical expressions that include every operational and directional sign can be cumbersome. In general practice, we omit as many signs as possible. When two signs are written between two numbers, we can write a simplified expression with only one sign.

Plus, Plus: $+3 + (+5)$　　　Add $+3$ and $+5$.　　　$3 + 5$
Minus, Minus: $+3 - (-5)$　　Change to addition.
　　　　　　　　$+3 + (+5)$　　　　　　　$3 + 5$

Plus, Minus: $+3 + (-5)$　　Add $+3$ and -5.　　　$3 - 5$
Minus, Plus: $+3 - (+5)$　　Change to addition.
　　　　　　　　$+3 + (-5)$　　　　　　　$3 - 5$

To generalize, two like signs between integers, whether both plus or both minus, will translate to adding a positive number. Just one plus sign can represent this situation.

Two unlike signs between integers, either a plus then a minus, or a minus then a plus, will translate to adding a negative number. Just one minus sign can represent this situation.

The suggestion in the previous Tips and Traps will facilitate and simplify problems that combine both addition and subtraction of integers.

Rule 2–5 *To add and subtract more than two numbers:*

1. Rewrite the problem so that all integers are separated by only one sign.
2. Interpret the problem as adding a series of signed numbers.

EXAMPLE 3   Evaluate: (a) $3 - (-5) - 6$ (b) $-8 + 10 - (-7)$

(a) $3 - (-5) - 6 =$ Rewrite with only one sign between integers.
      $3 + 5 - 6 =$ Add numbers with like signs.
          $8 - 6 = \mathbf{2}$ Apply the rule for adding numbers with unlike signs.
(b) $-8 + 10 - (-7) =$ Rewrite with only one sign between integers.
       $-8 + 10 + 7 =$ Add numbers with like signs.
          $-8 + 17 = \mathbf{9}$ Apply the rule for adding numbers with unlike signs.

The key to solving applied problems involving integers is to identify which numbers are positive and which are negative. Positive amounts include profits, gains, money in the bank, temperatures above zero, receipts, income, winnings, and so on. Negative amounts include losses, deficits, checks that cleared the bank, temperatures below zero, drops, declines, payments, and so on. Once the positive and negative amounts are known, they may be added or subtracted using the rules for adding or subtracting signed numbers.

EXAMPLE 4   A landscaping business made a profit of $345 one week, a loss of $34 the next week, and a profit of $235 the third week. What was the net profit?

The net profit is the sum of the profits and losses.

$345 \quad - \quad \$34 \quad + \quad \$235$ Interpret profits as positive and losses as negative.
profit   loss   profit

$345 + 235 - 34$ Rearrange and add positives.

$\quad\quad 580 - 34 = 546$ Apply rule for adding numbers with unlike signs.

Thus, the net profit for the   Interpret answer.
three days was $546.

EXAMPLE 5   In a recent year, the highest temperature in Boston was recorded as 98 degrees, and the lowest temperature for Boston in that same year was recorded as -2 degrees. What was the temperature range for the city that year? (The range is the difference between the highest and lowest values.)

$98 - (-2)$ Subtract -2 from 98.

$\quad 98 + 2 = 100$ Change to one sign only and apply the appropriate rule for adding integers.

Thus, the temperature range for Boston   Interpret answer.
in that year was 100 degrees.

1 Subtract.

1. $-3 - 9$	**2.** $8 - 2$	**3.** $9 - 15$	**4.** $-3 - (-7)$
5. $-11 - 14$	**6.** $-6 - (-3)$	**7.** $5 - (-3)$	**8.** $8 - 11$
9. $-8 - 1$	**10.** $11 - (-2)$	**11.** $(-8) - (-7)$	**12.** $(-15) - (-7)$

2 Subtract.

13. $15 - 0$	**14.** $0 - 8$	**15.** $-12 - 0$	**16.** $0 - (-8)$
17. $0 - (-7)$	**18.** $10 - 0$	**19.** $28 - (-28)$	**20.** $-46 - 46$
21. $7 - (-7)$	**22.** $-18 - 18$		

3 Evaluate the following:

23. $-1 + 1 - 4$	**24.** $5 + 3 - 7$	**25.** $7 + 3 - (-4)$
26. $-8 - 2 - (-7)$	**27.** $-3 + 4 - 7 - 3$	**28.** $2 - 4 - 5 - 6 + 8$
29. $8 - 3 + 2 - 1 + 7$	**30.** $-5 - 3 + 8 - 2 + 4$	**31.** $6 - (-3) + 5 - 6 - 9$

32. The temperatures for Bowling Green, Kentucky, ranged from 102 degrees to -5 degrees. What was the temperature range for the city?

33. New Boston, Texas, registered -8 degrees as its lowest temperature one year and 99 degrees as its highest temperature for the same year. What was the temperature range for New Boston?

34. Computing Solutions recorded a profit of $28,296 one quarter (3 months), a loss of $1896 for the second quarter, a profit of $52,597 for the third quarter, and a profit of $36,057 for the fourth quarter. What was the net profit for the year?

35. Explain the difference in subtracting zero from a number and in subtracting a number from zero.

2–4 MULTIPLYING INTEGERS

Learning Objectives

1 Multiply integers.
2 Multiply several integers.
3 Multiply with zero.
4 Evaluate powers of integers.

In multiplying integers, the multiplication of the absolute values of the integers is exactly the same as multiplying whole numbers. However, the rules for multiplying integers must also include the assignment of the proper sign to the product.

1 Multiply Integers.

As for whole numbers, multiplication of integers is commutative and associative; that is, factors may be multiplied in any order and grouped in any manner. Examine these illustrations to see how the signs are assigned to the product of two integers with like signs.

$$4(6) = 24, \qquad (-3)(-7) = 21, \qquad -10(2) = -20, \qquad 8(-9) = -72$$

Rule 2–6 *To multiply two integers:*

1. Multiply the absolute values of the numbers as in whole numbers.
2. If the factors have like signs, the sign of the product is positive.
3. If the factors have unlike signs, the sign of the product is negative.

EXAMPLE 1 Multiply: (a) $-12(-2)$ (b) $10 \cdot 3$ (c) $25(-3)$ (d) $-5 * 7$

(a) $-12(-2) = \mathbf{24}$ Like signs result in a positive product.
(b) $10 \cdot 3 = \mathbf{30}$ Like signs result in a positive product.
(c) $25(-3) = \mathbf{-75}$ Unlike signs result in a negative product.
(d) $-5 * 7 = \mathbf{-35}$ Unlike signs result in a negative product.

When one is multiplied by a number, the result is the number. Thus, one is the *multiplicative identity*.

■ **DEFINITION 2–4: Multiplicative Identity** One is the multiplicative identity because $a \cdot 1 = 1 \cdot a = a$ for all values of a.

2 **Multiply Several Integers.**

Multiplication, like addition, is a binary operation, so when we multiply three or more factors, we multiply two at a time. We apply the appropriate rule for signs each time we multiply. Multiplication is also commutative and associative, so we can change the order and grouping of the factors without affecting the final product.

EXAMPLE 2 Multiply: (a) $4(-2)(6)$ (b) $-3(4)(-5)$ (c) $-2(-8)(-3)$
(d) $-2(-3)(-4)(-1)$

(a) $4(-2)(6)$ Multiply the first two factors and apply the rule for factors with unlike signs.

    $-8(6) = \mathbf{-48}$ Multiply and apply the rule for factors with unlike signs.
(b) $-3(4)(-5)$ Multiply the first two factors and apply the rule for factors with unlike signs.

    $-12(-5) = \mathbf{60}$ Multiply and apply the rule for factors with like signs.
(c) $-2(-8)(-3)$ Multiply the first two factors and apply the rule for factors with like signs.

    $16(-3) = \mathbf{-48}$ Multiply and apply the rule for factors with unlike signs.
(d) $-2(-3)(-4)(-1)$ Multiply the first two and the last two factors.

    $6(4) = \mathbf{24}$ Multiply and apply the rule for factors with like signs.

If we examine the multiplications in Example 2 more closely, we see that the number of negative signs has an effect on the sign of the answer. Part *a* has one negative sign and the product is negative. Part *b* has two negative signs and the product is positive. Part *c* has three negative signs and the product is negative. Part *d* has four negative signs and the product is positive.

> **Rule 2–7** *To determine the sign of the product when multiplying three or more factors:*
>
> **1.** The sign of the product is positive if the number of negative factors is even.
> **2.** The sign of the product is negative if the number of negative factors is odd.

EXAMPLE 3 Multiply: (a) $-2(6)(-1)(-3)$ (b) $(2)(-5)(1)(-3)$

(a) $-2(6)(-1)(-3) = \mathbf{-36}$ Multiply absolute values. The odd number of negative factors makes the product negative.
(b) $(2)(-5)(1)(-3) = \mathbf{30}$ Multiply absolute values. The even number of negative factors makes the product positive.

3 Multiply with Zero.

Multiplications involving zero make use of a special property of zero that extends to integers and, in fact, to all real numbers. The product of zero and any number is zero: $x \cdot 0 = 0$. This is called the *zero property of multiplication*.

Tips and Traps	*Effect of a Zero Factor:*
	If we have two or more factors and one factor is zero, we can immediately write the product as zero without having to work through the steps.

EXAMPLE 4   Multiply: (a) $3(-21)(2)(0)$ (b) $-9(-2)(8)(-1)$

(a) $3(-21)(2)(0) = 0$ Zero is a factor.
(b) $-9(-2)(8)(-1) = -144$ Zero is not a factor.

Application problems may require multiplication of integers.

EXAMPLE 5   In a three-week period, a technology stock declined approximately 2 points each week. How many points did the stock decline in the three weeks?

Because there are equal declines each week, we may multiply the amount of weekly decline times the number of weeks: $(-2)3 = -6$. In interpreting the answer, we can say that the stock declined (negative) a total of 6 points over the three-week period.

4 Evaluate Powers of Integers.

Raising a number to a natural-number power is an extension of multiplication, so determining the sign of the result will be similar to multiplying several integers. Examine the following illustrations to observe the pattern for determining the sign of the result.

$$(+4)^2 = (+4)(+4) = +16 \qquad (-4)^2 = (-4)(-4) = +16$$
$$(+4)^3 = (+4)(+4)(+4) = +64 \qquad (-4)^3 = (-4)(-4)(-4) = -64$$

> **Rule 2–8**   *To raise integers to a natural-number power use the following patterns:*
>
> **1.** A positive number raised to any natural-number power is positive.
> **2.** Zero raised to any natural-number power is zero.
> **3.** A negative number raised to an even natural-number power is positive.
> **4.** A negative number raised to an odd natural-number power is negative.

EXAMPLE 6   Evaluate the following powers: (a) 4^3 (b) 0^8 (c) $(-2)^4$ (d) $(-3)^5$

(a) $4^3 = 4(4)(4) = \mathbf{64}$
(b) $0^8 = \mathbf{0}$
(c) $(-2)^4 = (-2)(-2)(-2)(-2) = \mathbf{16}$
(d) $(-3)^5 = (-3)(-3)(-3)(-3)(-3) = \mathbf{-243}$

EXAMPLE 7   Security systems sometimes have a four-digit code used to activate the system. How many different codes can be made with four digits? Discuss codes that may be impractical.

The number system has 10 digits: 0, 1, 2, 3, 4, 5, 6, 7, 8, and 9, so we can fill each one of the four slots of the code in 10 different ways. To find the total number of different codes, we multiply $10 \cdot 10 \cdot 10 \cdot 10$. This product can be written as 10^4, or 10,000. There are 10,000 ways to make a four-digit security code. Some of these codes, such as 0000, may be impractical.

Tips and Traps *Negative Base Versus an Opposite:*
$(-2)^4$ is not the same expression as -2^4.

$$(-2)^4 = (-2)(-2)(-2)(-2) = 16 \qquad -2^4 = -(2)(2)(2)(2) = -16$$

$(-2)^4$ is a negative base raised to a power. -2^4 is the opposite of 2^4. Even though the values of two expressions may be equal, interpretation may be different. For instance, $(-3)^5$ is not the same expression as -3^5, but both expressions refer to the same value because the exponent is an odd number.

$$(-3)^5 = (-3)(-3)(-3)(-3)(-3) = -243$$
$$-3^5 = -(3)(3)(3)(3)(3) = -243$$

As with whole numbers, raising a nonzero integer to the zero power is equal to one: $x^0 = 1$.

EXAMPLE 8 Evaluate the following powers: (a) $(-2)^0$ (b) $(-7)^0$

(a) $(-2)^0 = \mathbf{1}$ (b) $(-7)^0 = \mathbf{1}$

SELF-STUDY EXERCISES 2–4

1 Multiply.

1. $5 \cdot 8$	**2.** $-4(-3)$	**3.** $7 * 5$	**4.** $(-3)(-7)$
5. $-8(-3)$	**6.** $(-2)(-3)$	**7.** $5(-3)$	**8.** $(-2)(5)$
9. $-4 * 8$	**10.** $-3 \cdot 4$	**11.** $-7 * 8$	**12.** $6(-4)$

13. Holly Hobbs had four checks returned for insufficient funds, and her bank charged her a $28 service charge for each check. Use multiplication of signed numbers to show how these transactions affected her checking account balance.

14. Carolyn Luttrell made 7 withdrawals of $40 each from her checking account. Show with signed numbers how these transactions affected her checking account balance.

15. Using the rules for adding signed numbers, you add or subtract the absolute values depending on whether the signs are alike or unlike. How are the multiplication rules for signed numbers similar or different?

2

16. $5(-2)(3)(2)$	**17.** $6(1)(-3)(-2)$	**18.** $4(0)(-12)(3)$	**19.** $15(-2)(-3)$
20. $5(2)(-3)(0)$	**21.** $-3(2)(-7)(-1)$	**22.** $9(-1)(3)(-2)$	**23.** $(-7)(-5)(-6)$
24. $(-3)(-9)(-12)(-7)$	**25.** $7(-3)(-10)(12)(-8)$		

3

26. $-8(0)$	**27.** $5(0)$	**28.** $0(-12)$	**29.** $(-15)(0)$	**30.** $18(0)$
31. $0 \cdot 3$	**32.** $0(-15)$	**33.** $0(-17)$	**34.** $-28 \cdot 0$	**35.** $46 \cdot 0$

36. Review the definitions for additive inverse and additive identity and write a similar definition for multiplicative inverse.

4 Evaluate.

37. $(-3)^2$ **38.** $(-2)^3$ **39.** $(-5)^2$ **40.** -5^2 **41.** -2^3

42. $(-8)^0$ **43.** $(5)^4$ **44.** 3^4 **45.** 7^0 **46.** $(-42)^0$

47. How many seven-digit telephone numbers can be made with the 10 digits in the number system? Write the expression as a power and evaluate it.

48. Phone companies have recently added a new three-digit, toll-free prefix—888. How many toll-free numbers are now in existence? (*Hint:* See Exercise 47 for the number of seven-digit phone numbers.)

49. Do you think all the phone numbers found in Exercise 47 can be used as legitimate phone numbers? Explain.

50. Many states have automobile license plates with three digits followed by three letters. How many different patterns can be formed with this sequence?

51. How many different license plates can be formed if each license plate has four digits only?

2–5   **DIVIDING INTEGERS**

Learning Objectives

1   Divide integers.

2   Divide with zero.

In dividing integers, the division of the absolute values of the integers is exactly the same as dividing whole numbers. However, the rules for dividing integers must also include the assignment of the proper sign to the quotient.

1   **Divide Integers.**

The rules for determining the sign when dividing integers are similar to the rules for multiplying integers.

> **Rule 2–9**   *To divide two integers:*
>
> **1.** Divide the absolute values of the numbers as in whole numbers.
> **2.** If the values have like signs, the sign of the quotient is positive.
> **3.** If the values have unlike signs, the sign of the quotient is negative.

EXAMPLE 1   Divide: (a) $\dfrac{-8}{-2}$ (b) $\dfrac{6}{3}$ (c) $\dfrac{10}{-2}$ (d) $\dfrac{-9}{1}$

(a) $\dfrac{-8}{-2} = \mathbf{4}$ Like signs give a positive quotient.

(b) $\dfrac{6}{3} = \mathbf{2}$ Like signs give a positive quotient.

(c) $\dfrac{10}{-2} = \mathbf{-5}$ Unlike signs give a negative quotient.

(d) $\dfrac{-9}{1} = \mathbf{-9}$ Unlike signs give a negative quotient.

2   **Divide with Zero.**

Division involving zero applies to integers as it does to whole numbers.

> **Rule 2–10** *To evaluate division with zero:*
>
> Zero divided by any nonzero number is zero.
>
> $$\frac{0}{-5} = 0$$
>
> Division by zero is undefined or impossible.
>
> $$\frac{-5}{0} = \text{undefined or impossible (or infinity)}$$

EXAMPLE 2 Evaluate: (a) $\dfrac{-3}{0}$ (b) $\dfrac{0}{0}$ (c) $\dfrac{0}{-5}$

(a) $\dfrac{-3}{0}$ = **undefined or impossible, or infinity**

(b) $\dfrac{0}{0}$ = **undefined or impossible, or infinity**

(c) $\dfrac{0}{-5}$ = **0**

SELF-STUDY EXERCISES 2–5

1 Divide.

1. $\dfrac{-12}{-6}$　　**2.** $-15 \div 5$　　**3.** $\dfrac{8}{-2}$　　**4.** $\dfrac{-24}{6}$　　**5.** $-28 \div 7$

6. $\dfrac{-32}{-4}$　　**7.** $\dfrac{-50}{-10}$　　**8.** $\dfrac{36}{-6}$　　**9.** $\dfrac{-48}{-6}$　　**10.** $\dfrac{-25}{-5}$

11. You have decreased your house loan balance by $1800 over the past six months. Show this figure as a signed number and find the average monthly decrease.

2 Divide.

12. $\dfrac{12}{0}$　　**13.** $\dfrac{-15}{0}$　　**14.** $\dfrac{0}{+3}$　　**15.** $\dfrac{0}{-12}$　　**16.** $\dfrac{0}{-8}$

17. $\dfrac{-20}{0}$　　**18.** $\dfrac{-100}{0}$　　**19.** $\dfrac{1}{0}$　　**20.** $\dfrac{0}{8}$　　**21.** $\dfrac{-350}{0}$

22. Which two operations have the same rules for handling signs when working with signed numbers?

2–6　ORDER OF OPERATIONS

Learning Objectives　　**1** Use the order of operations for integers.
　　　　　　　　　　　　2 Use calculators to evaluate operations with integers.

1 **Use the Order of Operations for Integers.**

Integers follow the same order of operations as whole numbers.

> **Rule 2-11** *Perform operations in the following order as they appear from left to right:*
>
> 1. Parentheses used as groupings and other grouping symbols.
> 2. Exponents (powers and roots).
> 3. Multiplications and divisions.
> 4. Additions and subtractions.

EXAMPLE 1 Evaluate $(4 + 3) - (3 + 1)$.

$(4 + 3) - (3 + 1)$ Perform operations inside parentheses.

$\quad 7 \quad - \quad 4$ Add or subtract last.

$\quad 7 - 4 = \mathbf{3}$

EXAMPLE 2 Evaluate $8(4 + 6)$.

Evaluate by working within the grouping symbols first.

$8(4 + 6)$ Add $4 + 6 = 10$.

$8(10)$ Multiply 8 by 10.

80

Or, evaluate using the distributive principle.

$8(4 + 6)$ Multiply each term by the factor 8.

$8(4) + 8(6)$

$32 + 48 = \mathbf{80}$ Add $32 + 48$.

Another common grouping symbol is the bar showing division.

EXAMPLE 3 Evaluate $\dfrac{3 - 5}{2} - \dfrac{9}{2 + 1} + 4(5)$.

$\dfrac{3 - 5}{2} - \dfrac{9}{2 + 1} + 4(5)$ Perform operations grouped by the bar or fraction line.

$\dfrac{-2}{2} - \dfrac{9}{3} + 4(5)$ Work multiplications and divisions.

$-1 - 3 + 20$ Work additions and subtractions last.

$-4 + 20 = \mathbf{16}$

EXAMPLE 4 Evaluate $-12 \div 3 - (2)(-5)$.

$-12 \div 3 - (2)(-5)$

$-12 \div 3 - (2)(-5)$ Work multiplications and divisions first.

$\quad\quad -4 \quad - \quad (-10)$ Change to a single sign between the numbers.

$-4 + 10$ Add integers.

$-4 + 10 = \mathbf{6}$

The problem in Example 4 could have been written without parentheses around the 2. The problem could have been expressed as:

$$-12 \div 3 - 2(-5)$$

Again, the multiplications and divisions are done first. But notice how the multiplication can be performed.

$$-12 \div 3 \ - 2(-5)$$

We can consider the minus sign as the sign of the 2. Then we add the result of the division and the result of the multiplication.

$$-4 \qquad +10$$

$$-4 \ + \ 10 \ = \ \mathbf{6}$$ Add integers.

EXAMPLE 5   Evaluate $10 - 3(-2)$.

$$10 \ -3(-2)$$ Think of the multiplication as -3 times -2.

$$10 \ + 6$$

$$10 + 6 = \mathbf{16}$$ Add integers.

Only after parentheses and all multiplications and/or divisions have been taken care of do we perform the final additions and/or subtractions from left to right.

EXAMPLE 6   Evaluate $4 + 5(2 - 8)$.

$$4 + 5\,(2 - 8)$$ First, perform operations in parentheses.

$$4 + 5\,(-6)$$ Work multiplications and/or divisions.

$$4 - 30$$ Finally, perform remaining additions and/or subtractions.

$$4 - 30 = \mathbf{-26}$$

Tips and Traps   ***Order of Operations Is Important*:**
Note what would happen if we proceeded *out of order* in Example 6!

Incorrectly Worked

$$4 + 5(2 - 8)$$ Incorrectly add first instead of last.

$$9(2 - 8)$$ Perform operation in parentheses second instead of first.

$$9(-6)$$ Multiply last instead of second.

$$9(-6) = -54$$

Incorrect Solution

We get two different answers. *The order of operations must be followed to arrive at a correct solution.*

EXAMPLE 7   Evaluate $5 + (-2)^3 - 3(4 + 1)$.

$$5 + (-2)^3 - 3(4 + 1)$$ Perform operation in parentheses.

$$5 + (-2)^3 - 3(5)$$ Raise to a power.

$$5 + (-8) - 3\,(5)$$ Perform multiplication.

$$5 + (-8) - 15$$ Change to one sign only between numbers.

$$5 - 8 - 15$$ Perform first addition of integers.

$$-3 - 15 =$$ Perform remaining addition of integers.

$$\mathbf{-18}$$

2 Use Calculators to Evaluate Operations with Integers.

Calculators with algebraic logic may be used to perform operations with signed numbers by using the $\boxed{+}$, $\boxed{-}$, $\boxed{\times}$, $\boxed{\div}$, and $\boxed{+/-}$ keys. With graphics calculators and some scientific calculators, numbers and signs are entered in the order they are written.

With other scientific calculators, the absolute value of the number is entered first. The $\boxed{+/-}$ key, called the *"sign change key,"* is used to make the number negative. This key is a *toggle key* because it will change the sign of whatever number is showing in the display. For example, if 3 is showing in the display, press $\boxed{+/-}$ to show -3 in the display. Press $\boxed{+/-}$ again to show 3 in the display. Refer to your calculator instruction manual or test your calculator to determine its logic.

Tips and Traps　**Notation for Calculator Illustrations:**
The value following the symbol $\Rightarrow$ indicates the final answer in the calculator display.

EXAMPLE 8　Evaluate the following.

(a) $2 - 7$　　(b) $-7 + 2$

(c) $(-7)(2)$　(d) $\dfrac{-4}{2}$

Common graphics calculator and scientific calculator steps:

(a) $2 - 7$　　$2 \boxed{-} 7 \boxed{EXE} \Rightarrow -5$　($\boxed{EXE}$ may be labeled $\boxed{Enter}$ or $\boxed{=}$).

(b) $-7 + 2$　　$\boxed{-} 7 \boxed{+} 2 \boxed{EXE} \Rightarrow -5$

(c) $(-7)(2)$　　$\boxed{-} 7 \boxed{\times} 2 \boxed{EXE} \Rightarrow -14$

　　　　　　　　　　or

　　　　　　　$\boxed{-} 7 \boxed{(} 2 \boxed{)} \boxed{EXE} \Rightarrow -14$

(d) $\dfrac{-4}{2}$　　$\boxed{-} 4 \boxed{\div} \boxed{2} \boxed{EXE} \Rightarrow -2$

Other scientific calculator steps:

(a) $2 - 7$　　$2 \boxed{-} 7 \boxed{=} \Rightarrow -5$

(b) $-7 + 2$　　$7 \boxed{+/-} \boxed{+} 2 \boxed{=} \Rightarrow -5$

(c) $(-7)(2)$　　$7 \boxed{+/-} \boxed{\times} 2 \boxed{=} \Rightarrow -14$

(d) $\dfrac{-4}{2}$　　$4 \boxed{+/-} \boxed{\div} 2 \boxed{=} \Rightarrow -2$

Tips and Traps　**Getting to Know Your Calculator:**
Although most scientific and graphics calculators follow the order of operations given in Section 2–6, it is a good idea to check the operation of any given calculator to make sure. The owner's manual will be a useful guide to the operation of a particular calculator.

Another strategy is to select an example with numbers that are easy to calculate mentally. Make the mental calculations and then work the problem

---

on the calculator. This mental calculation will help you determine what mathematical concepts are already programmed into the calculator.

Following the order of operations in this text, a scientific or graphics calculator will allow us to enter many operations as they occur from *left to right*. The calculator will impose the correct order of operations for us.

When a bar is used as both a grouping and division symbol, we must instruct the calculator to work the grouping first by enclosing the grouping in parentheses.

EXAMPLE 9 Evaluate the following.

(a) $\dfrac{5 + 1}{3} + 2$ (b) $2(4 - 1) + 3$

Graphics or scientific calculator steps:

(a) $\dfrac{5 + 1}{3} + 2$ $\boxed{(}\ 5\ \boxed{+}\ 1\ \boxed{)}\ \boxed{\div}\ 3\ \boxed{+}\ 2\ \boxed{\text{EXE}}\ \Rightarrow\ 4$

(b) $2(4 - 1) + 3$ $2\ \boxed{(}\ 4\ \boxed{-}\ 1\ \boxed{)}\ \boxed{+}\ 3\ \boxed{\text{EXE}}\ \Rightarrow\ 9$

Other scientific calculator steps:

(a) $\dfrac{5 + 1}{3} + 2$ $\boxed{(}\ 5\ \boxed{+}\ 1\ \boxed{)}\ \boxed{\div}\ 3\ \boxed{+}\ 2\ \boxed{=}\ \Rightarrow\ 4$

(b) $2(4 - 1) + 3$ $2\ \boxed{\times}\ \boxed{(}\ 4\ \boxed{-}\ 1\ \boxed{)}\ \boxed{+}\ 3\ \boxed{=}\ \Rightarrow\ 9$

SELF-STUDY EXERCISES 2–6

1 Evaluate the following problems, paying careful attention to the order of operations.

1. $-12 + 8$

2. $6(-5)$

3. $\dfrac{-10}{5}$

4. $8 - (-9)$

5. $\dfrac{-18}{9}$

6. $3(7) + 9 \div 3$

7. $-6(-2) - 4$

8. $-6(-4)$

9. $7(1 - 4) \div 3 + 2$

10. $\dfrac{2 - 8}{3} + 5(-2) \div 2$

11. $\dfrac{5 - 9}{4} + 4(-3) \div 6$

12. $4 \times 8 - 7 \times 3 + 18 \div 6$

2 Use a scientific or graphics calculator to evaluate the following.

13. $5(3 - 4) - 7(2 - 5) \div (-3)$

14. $142(3 - 21) + 48(27)$

15. $24 - 3(2 + 7) \div 3 + 12$

16. $\dfrac{12 - 18}{3} + 15 - 81 \div 3$

17. $14 - \dfrac{3 + 7}{5} \div 2 + 5(3)$

18. $2(3 - 12) \div 4 \times 6 - 8$

19. $2(4 - 3) - 7 + 4(2 - 8)$

20. $7 - 21 + 138 - 256$

21. Use your calculator to verify that division by zero is impossible. Describe the calculator display that results from this division.

22. Use your calculator to evaluate several addition problems, then change the order of the addends and evaluate. Write a statement about the order of addition.

23. Evaluate several subtraction problems, then change the order of each problem and evaluate. Compare the results for each problem. Write a statement about the order of subtraction.

24. After testing several multiplication examples, write a statement about the order of multiplication.

25. After testing several division examples, write a statement about the order of division.

Consumer Interest: Stock Market Prices

Many savvy savers have recently discovered that their principal will earn much more when invested in the stock market than when invested in savings accounts, certificates of deposit (CDs), mutual funds, or bonds. This trend has also been fueled by many discount stock brokerage firms who charge a fraction of what other firms charge per transaction. The basic idea is simple: you invest by buying shares of publicly traded companies at a specified price, hold it, then sell your shares at a specified price. Ideally you buy at a low price and sell at a high price. Investors who can hold the stock a long time (say, 40 years until retirement, or 18 years until a child goes to college), usually make the most profit. Not only do stock prices usually increase over time, but there is a chance that the stock could "split." This means that one share becomes two shares at half the current price, but the price often rebounds quickly to its pre-split price.

The stock prices you see scrolling along the bottom of financial television channels, news programs, or computer screens report only the change in the price of the stock from the previous day's closing price. This change can be positive, negative, or zero, depending on many factors within the individual company, on politics, or on the financial network.

Let's assume that your child is born in the year 2000, and you sell your restored 1965 Chevrolet Corvair for $10,000 profit to invest for your child's college education. With advice from your certified financial planner and discount stock broker, you decide to invest in Coca-Cola common stock, which sells for $80 per share. You hold these shares for 18 years until your daughter decides she wants to attend college. Your broker advises you to sell within the next month (October) because fewer soft drinks are consumed in the winter and the stock price could fall. The closing price on Friday, September 28, is $341. You record the daily Coca-Cola price change in dollars for the next two weeks (October 1–12, 2018; Monday through Friday):

$$-2, +4, -1, 0, +1, -2, -1, 0, +3, +1$$

Exercises

Use the above information to answer the following exercises. Round all answers to the nearest tenth.

1. Assuming no splits, how many shares of Coca-Cola common stock do you own in 2018?
2. If you sold your stock at the start of the day on October 1, how much money would you receive, and how much profit would you have made? Find the simple interest earned for 18 years by dividing the profit by the principal invested, and multiplying by 100. Dividing this interest by 18 years gives you the simple annual interest. Find it.
3. What is the closing price on October 1, 2018? What would be the difference in the amount of money received if you sold at the end of the day on October 1 instead of at the beginning?
4. Check the price changes given above and guess which day would have been the best to sell on and the worst to sell on. Now list the closing prices on these days and see if your guesses are correct. For the best and worst selling days, what is the difference in price per share, and the amount received from the sale of 125 shares?
5. Assume that when you call in your request to sell on October 12, your broker advises you to hold because she has heard a rumor that the stock may soon split. It does split on October 15 at the closing price of $346. Now how many shares do you hold at what price?
6. If you sell at the beginning of the day on October 16, how much do you receive? If your broker advises you to wait until the end of the month, and

the closing price on October 31 is $185 per share, how much more would you receive?

7. You learn that your smart daughter has received a full academic National Merit scholarship. You decide to hold your stock another year, then sell it and build the vacation/retirement home of your dreams on lakefront property that you own. If the closing price on October 31, 2019, is $238, will you have enough to pay for an $80,000 A-frame home?

8. If you sell your stock on October 31, 2020, and the closing price is $329, will you have enough for your home?

9. Assuming you had to pay $722.50 in broker fees, find the profit earned on your 20-year investment of $10,000 and how much annual simple interest you earned.

10. Compare your interest rate with current savings account, certificate of deposit, money market mutual fund, and bond interest rates. What is the biggest disadvantage to investing in the stock market?

Answers

1. $10,000 divided by $80 per share = 125 shares.
2. You would receive $42,625, or a profit of $32,625, which amounts to 326% simple interest for 18 years (or 18% per year).
3. The closing price on October 1 is $339, which means you would receive $250 less.
4. Closing prices are October 1: $339, October 2: $343, October 3: $342, October 4: $342, October 5: $343, October 8: $341, October 9: $340, October 10: $340, October 11: $343, October 12: $344. The best day to sell is at the end of October 12, and the worst day to sell is at the end of October 1, for a difference of $5 per share, or $625 for 125 shares.
5. You now have 250 shares at $173 per share.
6. At $173, 250 shares gives $43,250, while at $185, 250 shares gives $46,250, or $3000 more.
7. At $238, 250 shares gives $59,500, so you would not have enough.
8. At $329, 250 shares gives $82,250, so you would have enough.
9. Your profit of $71,527.50 ($72,250–$722.50) earned 40% annual simple interest.
10. This interest rate is much higher than if you had invested in savings accounts, certificates of deposit, money market mutual funds, or bonds. The disadvantage is that you are taking a risk that is uninsured by the Federal Deposit Insurance Corporation (FDIC). If the price of the stock falls, you can lose all or part of the investment.

ASSIGNMENT EXERCISES

Section 2–1

The number line in Fig. 2–21 contains 14 positions lettered A through N. Give the letter that corresponds to each of the following numbers.

1. -1	2. $+1$	3. -2
4. $+2$	5. -3	6. $+3$
7. -7	8. 4	9. 5
10. -5	11. 6	12. -6

Figure 2–21

13. Write three examples of natural numbers.

14. Write three examples of integers that are not natural numbers.

15. Is there any number that is a whole number and an integer but not a natural number? If so, what is it? If not, explain why.

16. What is the smallest whole number?

17. What is the smallest natural number?

18. What is the smallest integer?

Use the symbols $>$ and $<$ to show the relationships of the numbers.

19. Is 72°F more than or less than -80°F?

20. Is 0 more than or less than -3?

21. Is 7 more than or less than -5?

22. Is 5 more than or less than 8?

23. Is -9 more than or less than 5?

24. Is -12 more than or less than -8?

Give the value of each of the following numbers:

25. $|5|$ **26.** $|-8|$ **27.** $|-3|$ **28.** $|0|$

29. $|+7|$ **30.** $|+8|$ **31.** $|-11|$ **32.** $|-52|$

Give the opposite of each of the following numbers:

33. -12 **34.** 8 **35.** 15 **36.** 0 **37.** -2

38. -13 **39.** -42 **40.** 156 **41.** 87 **42.** -19

Draw a rectangular coordinate system and locate the following points.

43. A = $(5, -2)$ **44.** B = $(-8, -3)$ **45.** C = $(0, -4)$

46. D = $(3, 7)$ **47.** E = $(-3, 2)$ **48.** F = $(-3, 0)$

49. What are the coordinates of the origin?

50. Which of the four quarters of the coordinate system is used to plot coordinates that are both negative?

51. Write the coordinates for the points on the graph in Fig. 2–22.

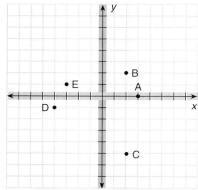

Figure 2–22

Section 2–2

Add.

52. $-3 + (-8)$ **53.** $5 + 12$ **54.** $-7 + 12$ **55.** $(-15) + 8$

56. $7 + (-11)$ **57.** $-5 + (-8)$ **58.** $-6 + 6$ **59.** $8 + (-8)$

60. $7 + 0$ **61.** $0 + (-8)$ **62.** $(-7) + 7$ **63.** $-25 + 0$

64. A publicly traded company had a profit of $256,872 for one year and a loss of $38,956 for the following year. What is the net profit over the two-year period?

65. A football team gained and lost the following yardage during a series of plays beginning with first down: $+4, -5, +9$. If you are the coach, what is your next play? Why?

66. Because of stormy weather, a pilot flying at 35,000 feet descended 8000 feet. What is his new altitude?

67. Agnes opened a checking account by depositing $500. She then wrote checks for $42, $18, and $21. What is her balance after making another deposit of $150?

68. You owe your friend $18 and borrow an additional $42. Use signed numbers to express your debt.

69. Explain the difference between additive identity and additive inverse.

Section 2–3

Evaluate.

70. $8 - 5$

71. $-9 - 4$

72. $-7 - (-2)$

73. $11 - (-3)$

74. $12 + 3 + (-8) - 5$

75. $-6 + 3 - 5 - 7$

76. $8 - 0$

77. $-5 - 0$

78. $0 - 3$

79. $0 - (-2)$

80. $-5 - 5$

81. $18 - (-18)$

82. Temperatures in northern Canada ranged as high as 37 degrees one summer. That same year the temperature was -28 degrees. What is the range of temperatures for the year?

83. What is the difference (or range) in temperatures of 43 degrees above zero and 27 degrees below zero?

84. What is the difference in temperatures 47 degrees below zero and 28 degrees below zero?

Section 2–4

Evaluate.

85. $5(8)$

86. -3×-7

87. 7×-2

88. $(-3)(+2)$

89. $6(-2)$

90. $-7(3)$

91. $-8(0)$

92. $0(5)$

93. $2(3)(-7)(0)$

94. $5(-2)(-1)(-3)$

95. $4(3)(-2)(7)$

96. $(-3)^2$

97. $(7)^3$

98. $(-4)^3$

99. -4^2

100. -2^3

101. On one winter day, the temperature dropped 2 degrees each hour for 5 hours. What was the total drop in temperature?

102. If the temperature in Exercise 101 was 8 degrees originally, what was the temperature at the end of the 5-hour period?

103. You research stock prices and find an article that says, "XYZ stock has dropped 4 points each week for the past 5 weeks." Was the stock price higher or lower 5 weeks ago than it is today? How much higher or lower? Use negative numbers to express drops in prices. Use a signed number to express the amount the stock price decreased or increased.

Section 2–5

Divide.

104. $-8 \div (-4)$

105. $12 \div 3$

106. $\dfrac{18}{9}$

107. $\dfrac{-20}{-5}$

108. $\dfrac{16}{-4}$

109. $\dfrac{-24}{6}$

110. $\dfrac{0}{-8}$

111. $\dfrac{-7}{0}$

112. $\dfrac{-51}{3}$

Section 2–6

Use the order of operations to evaluate the following. Verify the results with a calculator.

113. $7(3 + 5)$

114. $-2(3 - 1)$

115. $\dfrac{15 - 7}{8}$

116. $-20 \div 4 - 3(-2)$

117. $12 - 8(-3)$

118. $7 + 3(4 - 6)$

119. $4 + (-3)^4 - 2(5 + 1)$

120. $(-3)^3 + 1 - 8$

121. $-3 + 2^3 - 7$

122. $4(-6 - 2) - \dfrac{8 + 2}{7 - 5}$

123. $296 - 382(-4)^5$

124. $-71 + 3(-19)$

125. The temperature at 8:00 A.M. is recorded as −3°C. Calculate the temperature at each hour as recorded by the following increases and decreases.

   9:00 A.M.: increase 2°
10:00 A.M.: increase 1°
11:00 A.M.: increase 0°
12:00 P.M.: increase 1°
  1:00 P.M.: no change
  2:00 P.M.: increase 3°
  3:00 P.M.: decrease 4°
  4:00 P.M.: decrease 7°
  5:00 P.M.: decrease 8°
  6:00 P.M.: decrease 12°

126. How many automobile license plates can be formed if the pattern is 3 digits and 3 letters? 6 letters?

1. What two operations for integers use similar rules for handling the signs? Explain the rules for these operations.

3. What operation with 0 is not defined?

5. Illustrate the symbol used for "is greater than" by writing a true statement using the symbol.

7. Explain how to determine the sign of a power if the base is a negative integer. Give an example for an even power and for an odd power.

9. Draw a number line that shows positive and negative integers and zero and place the following integers on the number line: −3, 8, −2, 0, 3, 5.

2. Explain what is meant by "the absolute value of a number." Illustrate your explanation with an example.

4. Describe the process of adding two integers that have different signs.

6. Write the words that describe the correct order for operations with integers.

8. Give an example of multiplying two negative integers and give the product.

10. Find and correct the mistakes in the following problem.

$$(-8)^2 - 3(2)$$
$$16 - 3(2)$$
$$13(2)$$
$$26$$

Objectives	What to Remember with Examples
Section 2–1	
1 Relate integers to natural numbers and whole numbers.	Positive numbers are to the right of zero and negative numbers are to the left of zero on the number line.
	Arrange from smallest to largest: 5, −3, 0, 8, −5 −5, −3, 0, 5, 8
2 Compare integers.	The "greater than" symbol is >. The "less than" symbol is <.
	Use > or < to make a true statement: 5 ? −3; 5 > −3
3 Find the absolute value of integers.	The absolute value of a number is its *distance* from zero without regard to direction.
	Evaluate the following absolute values: $\lvert -3 \rvert, \lvert 5 \rvert$ $\lvert -3 \rvert = 3, \lvert 5 \rvert = 5$

4 Find the opposite of integers.

Opposites are numbers that have the same absolute value but opposite signs.

> Give the opposite of the following:
> 8, −4, −2, +4
> −8, 4, 2, −4

5 Locate points on a rectangular coordinate system.

To plot a point on the rectangular coordinate system: 1. Start at the origin. 2. Count to the left or right the number of units represented by the first signed number. 3. Start at the ending point found in Step 2, count up or down the number of units represented by the second signed number.

> Draw a coordinate system and locate the following points: A (−3, −2), B (4, −1), C (0, −3), D (2, 3).

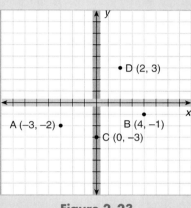

Figure 2–23

Section 2–2

1 Add integers with like signs.

To add integers with like signs, add the absolute values and give the sum the common sign.

> Add: $13 + 7 = 20$
> $-35 + (-13) = -48$

2 Add integers with unlike signs.

To add integers with unlike signs, subtract the smaller absolute value from the larger absolute value. The sign of the sum will be the sign of the number with the larger absolute value.

> Add: $-12 + 7$. Subtract absolute values: $12 - 7 = 5$. Give the 5 the negative sign because -12 has the larger absolute value.
> $-12 + 7 = -5$

When zero is added to a signed number, the result is unchanged: $0 + a = a + 0 = a$.

> Add: $-42 + 0 = -42$
> $0 + 38 = 38$

Section 2–3

1 Subtract integers.

To subtract integers, change the sign of the subtrahend, change subtraction to addition, and use the appropriate rule for adding signed numbers.

> Subtract: $-32 - (-28)$
> $-32 - (-28) = -32 + (+28)$
> $= -4$

| 2 | Subtract with zero or opposites. | To subtract with zero, change the subtraction to addition and use the appropriate rule for addition. |

$$-8 - 0 = -8 + (0) = -8$$
$$0 - (-4) = 0 + (+4) = 4$$

Subtracting an opposite from a number is the same as adding a number to itself.

$$8 - (-8) = 8 + 8 = 16$$
$$-12 - 12 = -24$$

| 3 | Combine addition and subtraction. | Change all subtractions to additions and work the additions from left to right. |

Simplify: $5 - 3 + 2 - (-4)$
$5 + (-3) + 2 + 4 =$
$\qquad 2 + 2 + 4 = 4 + 4 = 8$

Section 2–4

| 1 | Multiply integers. | To multiply integers with like signs, multiply the absolute values and make the sign of the product positive. |

$$-6(-7) = +42$$
$$8 \times 6 = 48$$

To multiply integers with unlike signs, multiply the absolute values and make the sign of the product negative.

$$-7(2) = -14$$
$$8(-3) = -24$$

| 2 | Multiply several integers. | To multiply several integers, perform the multiplication from left to right. |

$$5(-2)(3) = -10(3) = -30$$

| 3 | Multiply with zero. | Any number (including signed numbers) multiplied by zero results in zero: $a \times 0 = 0 \times a = a$ |

$$0 \times 3 = 0$$
$$-5 \times 0 = 0$$
$$0(-7) = 0$$

| 4 | Evaluate powers of integers. | A positive integer raised to a natural-number power results in a positive integer. A negative integer raised to an even-number power results in a positive integer. A negative integer raised to an odd-number power results in a negative integer. |

$$(3)^4 = 81$$
$$(-2)^4 = 16$$
$$(-2)^3 = -8$$
$$-2^4 = -16$$
$$-2^3 = -8$$

Section 2–5

| 1 | Divide integers. | To divide integers, divide the absolute values of the dividend and the divisor. If the dividend and divisor have like signs, the sign of the quotient will be positive. If the dividend and divisor have unlike signs, the sign of the quotient will be negative. |

$$-12 \div (-4) = 3$$
$$15 \div (-3) = -5$$

2 Divide with zero.

Zero divided by any nonzero signed number is zero. A nonzero number *cannot* be divided by zero. $0 \div a = 0$ (if a is not equal to zero); $a \div 0$ is not possible.

$$0 \div 12 = 0 \qquad 0 \div 0 \text{ is not possible or not defined.}$$
$$-7 \div 0 \text{ is not possible or not defined.}$$

Section 2–6

1 Use the order of operations for integers.

Expressions are evaluated using the following order of operations: parentheses, exponents (powers and roots), multiplication and division, addition and subtraction.

Evaluate:
$$5 - 4(7 + 2)^2 - 15 \div 5 =$$
$$5 - 4(9)^2 - 15 \div 5 =$$
$$5 - 4(81) - 15 \div 5 =$$
$$5 - 324 - 15 \div 5 =$$
$$5 - 324 - 3 =$$
$$-319 - 3 =$$
$$-322$$

2 Use calculators to evaluate operations with integers.

Use of the scientific and graphics calculators to evaluate expressions is different.

Use the scientific or graphics calculator to evaluate $5(-3) + 3 - (-2)$.

Most common steps: 5 $\boxed{\times}$ $\boxed{-}$ 3 $\boxed{+}$ 3 $\boxed{-}$ $\boxed{-}$ 2 $\boxed{\text{EXE}}$ $\Rightarrow$ -10

Alternate steps: 5 $\boxed{\times}$ 3 $\boxed{+/-}$ $\boxed{+}$ 3 $\boxed{-}$ 2 $\boxed{+/-}$ $\boxed{=}$ $\Rightarrow$ -10

WORDS TO KNOW

natural numbers (p. 70)
whole numbers (p. 70)
integers (p. 70)
opposites (p. 70)
absolute value (p. 72)
one-dimensional graph
  (p. 73)

rectangular coordinate system
  (p. 73)
horizontal axis (p. 73)
vertical axis (p. 73)
origin (p. 74)
x-coordinate (p. 74)
y-coordinate (p. 74)

additive identity (p. 73, 79)
additive inverse (p. 79)
multiplicative identity (p. 85)
zero property of multiplication
  (p. 86)
sign-change key (p. 92)
toggle key (p. 92)

CHAPTER TRIAL TEST

Use the symbols $>$ and $<$ to write the following as *true* statements.

1. -8 is more than 0

2. 2 is less than 3

3. -5 is more than -1

Answer the following questions.

4. What is the value of $|-12|$?

5. What is the opposite of 8?

Perform the following operations.

6. $-8 - 2$

7. $-3 + 7$

8. $\dfrac{8}{-2}$

9. $2(6)(-4)$

10. $\dfrac{-6}{-2}$

11. $8 + 4 + (-2) + (-7)$

12. $7 - (-3)$

13. $2(-1)(-4)$

14. $-8 + (-3)$

15. $(-8)(3)(0)(-1)$

16. $\dfrac{0}{3}$

17. $0 - 7$

18. $-3 + 5 + 0 + 2 + (-5)$

19. $\dfrac{-7}{0}$

20. $4(-13)$

21. $5 + (-7) - (-3) + 2 - 6$

22. $\dfrac{6}{-1}$

23. $\dfrac{4}{-2}$

24. $8(3) - 2 + 6(7)$

25. $2(3 - 9) \div 2^2 + 7$

26. $\dfrac{10 - 4}{3} - 3$

Perform the following operations. Use a scientific or graphics calculator *as directed by your instructor.*

27. $5(6) - 2 + 10 \div 2$

28. $5(2 - 3) + 16 \div 4$

29. $-2 - 3(-4) + (4)(3)$

30. $\dfrac{-18}{3}$

31. $\dfrac{4 - 2}{2} - 6^2$

32. Write two signed numbers that are opposites. Describe the relationship between the two numbers.

33. How are the multiplicative identity and the additive identity alike? Give examples to illustrate your point.

34. Write the sum of two signed numbers that have unlike signs and whose addition results in a negative number.

35. Temperatures around the world may range from 135 degrees Fahrenheit in Seville, Spain, to -40 degrees Fahrenheit in Fairbanks, Alaska. What is the range (difference) of temperatures?

3

Fractions and Percents

GOOD DECISIONS THROUGH TEAMWORK

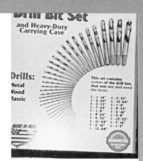

Over the course of one week, take note each time you see a fraction or percent used outside the classroom. Describe the situation and the use made of the fraction or percent. For instance, a computer store advertises a half-off sale on selected peripherals, your bank charges $9\frac{1}{2}\%$ interest on 60-month car loans, or a painter adds $4\frac{1}{2}$ ounces of TSP to a solution to clean a wall before painting it.

With your team, make a master list of situations, eliminating duplications. For each situation on your master list, discuss why a fraction or a percent is used rather than a whole number or a decimal. Discuss too why a fraction is used rather than a percent, or a percent rather than a fraction. On the basis of your discussion, identify major categories for the use of fractions and percents. How many of these categories could apply to a technical setting, such as in science or industry? Give a technology-related example for each category your team judges to be related to a technical setting. Choose a team member to share the results of your discussion with the class.

Learning Objective

1 Identify fraction terminology.

The language of mathematics is important in the study of mathematics. Just as it is important to understand the notation or symbolism used in mathematics, it is important to know the terminology. The first step in our study of fractions will be to learn the words associated with fractions. Thus far, we have reviewed the mathematics of whole numbers and decimals. However, we must often work with numbers called fractions (such as $\frac{1}{2}$, $\frac{1}{4}$, $1\frac{1}{4}$, or $1\frac{1}{2}$) and percents. In this chapter we study how fractions and percents are used to relate to whole numbers and decimals, and how we add, subtract, multiply, and divide fractions.

1 Identify Fraction Terminology.

In Chapter 1, we examined a special type of fraction called a decimal fraction. Now we will look more closely at other types of fractions. A *fraction* is a value that can be expressed as the quotient of two integers.

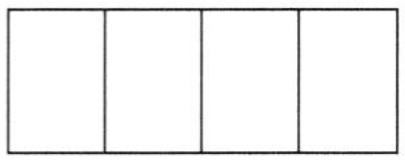

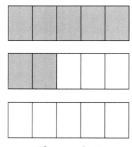

Figure 3–1

If one unit (or amount) is divided into four parts, we can write the fraction $\frac{4}{4}$ to represent this single unit. Figure 3–1 illustrates a unit divided into four equal parts.

The fraction $\frac{4}{4}$ is an example of a *common fraction*. A common fraction consists of two whole numbers. The bottom number, called the *denominator,* indicates *how many equal parts* one whole unit has been divided into. The top number, called the *numerator,* tells *how many of these parts* are being considered. Thus, in Fig. 3–2, one of the four equal parts has been shaded and is represented by the fraction $\frac{1}{4}$. Similarly, the shaded part in Fig. 3–3 is written as the fraction $\frac{3}{7}$.

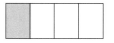

Figure 3–2

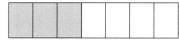

Figure 3–3

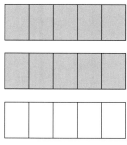

Figure 3–4

The unit is always our *standard* amount when writing fractions. Thus, $\frac{4}{4}$ and $\frac{3}{3}$ represent one unit each, or simply the number 1. Some fractions represent less than one unit (less than 1), for example, $\frac{1}{4}$ or $\frac{3}{4}$. These fractions are called *proper fractions.* Other fractions, however, represent one or more units, for example, $\frac{7}{5}$. These fractions are called *improper fractions.*

Suppose that three units are each divided into five equal parts. If we have seven of these parts, the fraction $\frac{7}{5}$ represents this amount. This fraction is greater than one unit (see Fig. 3–4).

Similarly, if we have 10 of these parts, the fraction $\frac{10}{5}$ represents this amount. If five parts equal one unit, then 10 parts equal two units (see Fig. 3–5). Fifteen parts, $\frac{15}{5}$, equal three units (see Fig. 3–6).

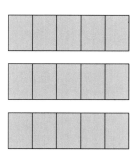

Figure 3–5

Figure 3–6

There is a relationship between fractions and division. We can see this in the fractions $\frac{10}{5}$ (two units) and $\frac{15}{5}$ (three units) because $10 \div 5 = 2$ and $15 \div 5 = 3$. This relationship to division is important to our understanding of fractions. Now we are ready to learn some important definitions concerning fractions.

■ **DEFINITION 3–1: Common Fraction.** A *common fraction* represents the division of one whole number by another.

■ **DEFINITION 3–2: Denominator.** The *denominator* of a fraction indicates the number of parts one unit has been divided into. It is the *bottom* number of a fraction, or the *divisor* of the indicated division.

■ **DEFINITION 3–3: Numerator.** The *numerator* of a fraction indicates the number of the parts being considered. It is the *top* number of a fraction, or the *dividend* of the indicated division.

The numerator and denominator of a fraction are usually separated by a horizontal line, although sometimes a slash is used. This line is called the *fraction line,* and it also serves as a division symbol. The fraction can be read from top to bottom as the numerator *divided by* the denominator. Six divided by two can be written as $2\overline{)6}$, $6 \div 2$, $\frac{6}{2}$, or 6/2.

There are two basic types of common fractions: those whose value is less than 1 and those whose value is equal to or greater than 1.

■ **DEFINITION 3–4: Proper Fraction.** A *proper fraction* is a common fraction whose value is less than one unit; that is, the numerator is less than the denominator. *Examples:*

$$\frac{2}{5}, \quad \frac{3}{7}, \quad \frac{15}{16}, \quad \frac{1}{3}, \quad \frac{7}{10}$$

■ **DEFINITION 3–5: Improper Fraction.** An *improper fraction* is a common fraction whose value is equal to or greater than one unit; that is, the numerator is equal to or greater than the denominator. *Examples:*

$$\frac{4}{4}, \quad \frac{7}{3}, \quad \frac{8}{4}, \quad \frac{100}{10}, \quad \frac{17}{5}$$

When the denominator divides evenly into the numerator, the improper fraction can be written as a whole number, such as $\frac{8}{4} = 2$. But when the value is more than one unit and the denominator cannot divide evenly into the numerator, the improper fraction can be written as a combination of a whole number and a fractional part, such as $\frac{9}{4} = 2\frac{1}{4}$. Such a combination is called a *mixed number*.

■ **DEFINITION 3–6: Mixed Number.** A *mixed number* consists of both a whole number and a fraction. The whole number and fraction are added together. *Example:*

$$3\frac{2}{5} \text{ means three whole units and } \frac{2}{5} \text{ of another unit} \qquad \text{or} \qquad 3\frac{2}{5} = 3 + \frac{2}{5}$$

Sometimes mathematical computations in technical applications become more complex than the fractions and mixed numbers we have looked at so far. We have seen that a fraction like $\frac{2}{7}$ means that a unit has been divided into seven equal parts and we are considering only two of those seven parts. But suppose that our job required us to consider $2\frac{1}{3}$ of those seven parts instead of just two.

In this instance our fraction would be $\frac{2\frac{1}{3}}{7}$. A fraction like $\frac{2\frac{1}{3}}{7}$ is a *complex fraction.*

■ **DEFINITION 3–7: Complex Fraction.** A *complex fraction* has a fraction or mixed number in its numerator or denominator, or both.

Examples of complex fractions:

$$\frac{\frac{1}{2}}{2}, \quad \frac{7}{\frac{3}{4}}, \quad \frac{3\frac{1}{3}}{7}, \quad \frac{5}{4\frac{1}{2}}, \quad \frac{\frac{1}{8}}{\frac{5}{16}}, \quad \frac{6\frac{3}{8}}{4\frac{1}{2}}$$

It is very important to understand that fractions indicate division. The fraction is read from *top to bottom,* with the fraction line read as "divided by." $\frac{\frac{1}{2}}{2}$ is read as $\frac{1}{2}$ divided by 2. $\frac{7}{\frac{3}{4}}$ is read as 7 divided by $\frac{3}{4}$.

As you learned in Chapter 1, many job-related applications make use of another kind of fraction, whose denominator is 10 or a power of 10, such as 100, 1000, and so on. Instead of writing $\frac{3}{10}$, we may write 0.3 (with the zero indicating no whole number, the period or *decimal point* indicating the beginning of the fraction, and the 3 indicating how many $\frac{1}{10}$'s, in this case 3). The mixed number $2\frac{17}{100}$ can be written as 2.17 (with the 2 indicating the whole number, the decimal point indicating the beginning of the fraction, and the 17 indicating how many $\frac{1}{100}$'s, in this case 17).

Tips and Traps **Relating the Decimal Places and the Zeros in the Denominator of the Place Value:**
The number of places after the decimal point indicates the number of zeros in the denominator of the power of 10. Note also that the number after the decimal point indicates the numerator of the fraction.

$$0.015 = \frac{15}{1000} \qquad 2.43 = 2\frac{43}{100}$$

■ **DEFINITION 3–8: Decimal Fraction.** A *decimal fraction* is a fractional notation that uses the decimal point and the place values to its right to represent a fraction whose denominator is 10 or some power of 10, such as 100, 1000, and so on. A decimal fraction is also referred to as a decimal, a decimal number, or a number using decimal notation.

■ **DEFINITION 3–9: Mixed-Decimal Fraction.** A *mixed-decimal fraction* is a notation used to represent a decimal fraction that contains a whole number part as well as a fractional part.

EXAMPLE 1 Rewrite the decimal fractions and mixed-decimal fractions below as proper fractions or mixed numbers:

$$0.4, \quad 0.18, \quad 0.014, \quad 3.7, \quad 4.103$$

The number of places to the right of the decimal indicates the number of zeros in the denominator of the fraction.

$$0.4 = \frac{4}{10} \qquad\qquad 3.7 = 3\frac{7}{10}$$

$$0.18 = \frac{18}{100} \qquad 4.103 = 4\frac{103}{1000}$$

$$0.014 = \frac{14}{1000}$$

To summarize, let's place the various fractions in a diagram like Fig. 3–7 to show the whole picture at a glance.

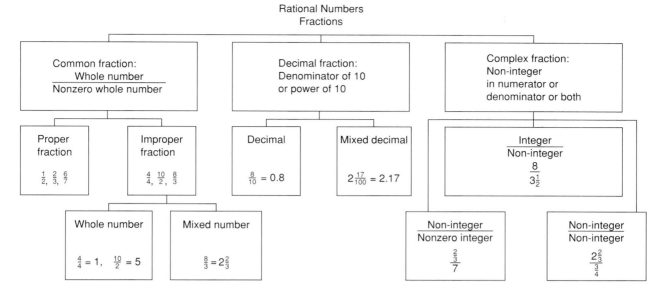

Figure 3–7

Fractions extend the numbers we have studied so far to another level called *rational* numbers. Rational numbers include whole numbers, natural numbers, integers, fractions, and decimal numbers because all these types of numbers can be expressed in fraction form.

■ **DEFINITION 3–10:** **Rational Number.** Any number that can be written in the form of a fraction, with an integer for a numerator and a nonzero integer for a denominator, is a *rational number.*

SELF-STUDY EXERCISES 3–1

1 Write a fraction to represent the shaded portion of Exercises 1 through 7 (Figs. 3–8 to 3–14).

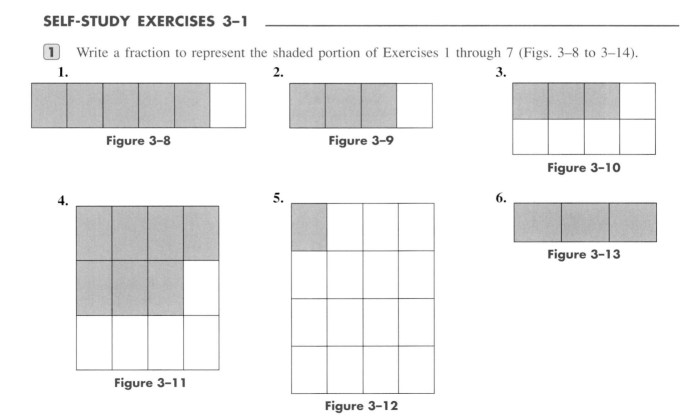

1.

Figure 3–8

2.

Figure 3–9

3.

Figure 3–10

4.

Figure 3–11

5.

Figure 3–12

6.

Figure 3–13

7.

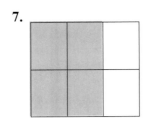

Figure 3–14

Examine each common fraction in Exercises 8 through 11 and answer the following questions.

8. $\dfrac{5}{6}$

9. $\dfrac{8}{8}$

10. $\dfrac{11}{5}$

11. $\dfrac{12}{3}$

(a) One unit has been divided into how many parts?
(b) How many of these parts are used?
(c) The numerator of the fraction is ____.
(d) The denominator of the fraction is ____.
(e) The fraction can be read as ____ divided by ____.
(f) ____ is the divisor of the division.
(g) ____ is the dividend of the division.
(h) Is this common fraction proper or improper?
(i) This fraction represents ____ out of ____ parts.
(j) Does this fraction have a value less than, more than, or equal to 1?

Match the best description with each number given.

____ **12.** $2\dfrac{3}{7}$

____ **14.** $\dfrac{8}{3}$

____ **16.** 6.27

____ **18.** 0.3

____ **20.** $\dfrac{\frac{3}{1}}{\frac{2}{3}}$

____ **22.** $\dfrac{\frac{3}{4}}{\frac{2}{7}}$

____ **13.** 0.689

____ **15.** $14\dfrac{3}{16}$

____ **17.** 3.7

____ **19.** $\dfrac{3}{7}$

____ **21.** $\dfrac{1\frac{1}{2}}{3}$

(a) Decimal fraction
(b) Mixed number
(c) Complex fraction
(d) Mixed-decimal fraction
(e) Proper fraction
(f) Improper fraction

Rewrite the fractions below as proper fractions or mixed numbers.

23. 6.27   **24.** 0.3   **25.** 3.7   **26.** 1.53

3–2   MULTIPLES, DIVISIBILITY, AND FACTOR PAIRS

Learning Objectives

1 Find multiples of a natural number.
2 Determine the divisibility of a number.
3 Find all factor pairs of a natural number.

As we mentioned before, fractions indicate division, and multiplication and division are inverse operations. Before we begin computations with fractions, we will look at some relationships involving multiplication and division.

1 Find Multiples of a Natural Number.

If we count by threes, such as 3, 6, 9, 12, 15, 18, we obtain natural numbers that are *multiples* of 3. Each number above is a multiple of 3 because each is the product of 3 and a natural number; that is,

$$3 = 3 \times 1, \qquad 6 = 3 \times 2, \qquad 9 = 3 \times 3$$
$$12 = 3 \times 4, \qquad 15 = 3 \times 5, \qquad 18 = 3 \times 6$$

■ **DEFINITION 3–11: Multiple.** A *multiple* of a natural number is the product of that number and a natural number.

EXAMPLE 1 Show that 2, 4, 6, 8, and 10 are multiples of 2.

$$2 = 2 \times 1, \qquad 4 = 2 \times 2, \qquad 6 = 2 \times 3$$
$$8 = 2 \times 4, \qquad 10 = 2 \times 5$$

■ **DEFINITION 3–12: Even Number.** Natural numbers that are multiples of 2 are called *even numbers*.

■ **DEFINITION 3–13: Odd Number.** Numbers that are not multiples of 2 are called *odd numbers*.

2 Determine the Divisibility of a Number.

A number is *divisible* by another number if the quotient has no remainder or if the dividend is a multiple of the divisor.

EXAMPLE 2 Is 35 divisible by 7?

35 is divisible by 7 if $35 \div 7$ has no remainder or if 35 is a multiple of 7.

$$35 \div 7 = 5 \qquad \text{or} \qquad 35 = 5 \times 7$$

Yes, 35 is divisible by 7.

EXAMPLE 3 A holiday party is planned for 47 first-grade students. The teachers have a gross (144) of assorted party favors. Can each child receive an equal number of favors with none left over so that no child will receive more than another?

This problem is really asking whether 144 is divisible by 47.

$$144 \div 47 = 3R3 \qquad \text{or} \qquad 144 \div 47 = 3.063829787 \text{ (using a calculator)}$$

The division of $144 \div 47$ has a remainder and the calculator answer is not a natural number; therefore, 144 is not divisible by 47. **No, there will be 3 party favors left over.**

A number of rules or tests can help us decide by inspection if certain numbers are divisible by certain other numbers.

Rule 3–1 *Tests for Divisibility:*

1. A number is divisible by 2 if the last digit is 0, 2, 4, 6, or 8, that is, if the last digit is an even digit.
2. A number is divisible by 3 if the sum of its digits is divisible by 3.
3. A number is divisible by 4 if the last two digits form a number that is divisible by 4.
4. A number is divisible by 5 if the last digit is 0 or 5.
5. A number is divisible by 6 if the number is divisible by *both* 2 *and* 3.
6. A number is divisible by 7 if the division has no remainder.
7. A number is divisible by 8 if the last three digits form a number divisible by 8.
8. A number is divisible by 9 if the sum of its digits is divisible by 9.
9. A number is divisible by 10 if the last digit is 0.

EXAMPLE 4 Use the tests for divisibility to identify which number in each pair of numbers is divisible by the given number.

Numbers	Divisor	Answer
874 or 873	2	**874**; the last digit is an even digit (4).
275 or 270	2	**270**; the last digit is an even digit (0).
427 or 423	3	**423**; the sum of the digits is divisible by 3: $4 + 2 + 3 = 9$.
5912 or 5913	4	**5912**; the last two digits form a number divisible by 4: $12 \div 4 = 3$.
80 or 82	5	**80**; the last digit is 0.
56 or 65	5	**65**; the last digit is 5.
804 or 802	6	**804**; the last digit is even and the sum of the digits is divisible by 3.
58 or 56	7	**56**; it divides by 7 with no remainder.
3160 or 3162	8	**3160**; the last three digits form a number divisible by 8: $160 \div 8 = 20$.
477 or 475	9	**477**; the sum of the digits is divisible by 9; $4 + 7 + 7 = 18$.
182 or 180	10	**180**; the last digit is 0.

③ Find All Factor Pairs of a Natural Number.

The *natural numbers* are the counting numbers, such as 1, 2, 3, 4, 5, 6, 7, 8, 9, 10, 11, 12, and so forth. The natural numbers can also be described as the whole numbers excluding zero.

Any natural number can be expressed as the product of two natural numbers. These two natural numbers are called a *factor pair* of the number.

■ **DEFINITION 3–14: Factor Pair.** A *factor pair* of a natural number consists of two natural numbers whose product equals the given natural number.

Every natural number greater than one has at least one factor pair, one and the number itself.

$$1 \text{ and } 3 \text{ form a factor pair for } 3: 1 \times 3 = 3.$$

$$1 \text{ and } 5 \text{ form a factor pair for } 5: 1 \times 5 = 5.$$

Many natural numbers have more than one factor pair. We will develop a strategy for finding all factor pairs of a natural number.

Rule 3–2 *To find all factor pairs of a natural number:*

1. Write the factor pair of one and the number.
2. Check to see if the number is divisible by 2. If so, write the factor pair of two and the quotient of the beginning natural number and two.
3. Check each natural number in turn for divisibility until you reach a number that has already been found as a quotient in a previous factor pair.

EXAMPLE 5 List all the pairs of factors of 18; then write each distinct factor in order from smallest to largest.

1×18

2×9 18 is divisible by 2: $18 \div 2 = 9$.

3×6 18 is divisible by 3: $18 \div 3 = 6$.
18 is not divisible by 4 or 5.
18 is divisible by 6; 6 was found previously in the factor pair 3×6.

Factors: 1, 2, 3, 6, 9, 18

EXAMPLE 6   List all the pairs of factors of 48; then write each distinct factor in order from smallest to largest.

1×48
2×24 48 is divisible by 2: $48 \div 2 = 24$.
3×16 48 is divisible by 3: $48 \div 3 = 16$.
4×12 48 is divisible by 4: $48 \div 4 = 12$.
48 is not divisible by 5.
6×8 48 is divisible by 6: $48 \div 6 = 8$.
48 is not divisible by 7.
48 is divisible by 8; 8 was found previously in the factor pair 6×8.

Factors: 1, 2, 3, 4, 6, 8, 12, 16, 24, 48

SELF-STUDY EXERCISES 3–2

1 Show that the following numbers are multiples of the first number by listing their factorizations with the first number.

1. 5, 10, 15, 20, 25, 30
2. 6, 12, 18, 24, 30, 36
3. 8, 16, 24, 32, 40, 48
4. 9, 18, 27, 36, 45, 54
5. 10, 20, 30, 40, 50, 60
6. 30, 60, 90, 120, 150, 180

Find 5 multiples of the given numbers by listing their factorizations, starting with the given number times 1.

7. 5
8. 12
9. 7
10. 3
11. 50
12. 4

2 Are the following numbers divisible by the given number? Explain your answer.

13. 2434 by 6
14. 230 by 5
15. 2434 by 4
16. 1221 by 3
17. 756 by 7
18. 920 by 8
19. 621 by 3
20. 426 by 6
21. 1232 by 2

3 Find all the factor pairs for each of the following numbers.

22. 24
23. 36
24. 45
25. 32
26. 16
27. 27
28. 20
29. 30
30. 12
31. 8
32. 4
33. 15
34. 81
35. 64
36. 38
37. 46
38. 51
39. 18
40. 72

3–3   PRIME AND COMPOSITE NUMBERS

Learning Objectives

1 Factor prime and composite numbers.
2 Determine the prime factorization of composite numbers.

1   Factor Prime and Composite Numbers.

When listing all the factor pairs of natural numbers, some numbers may have only one pair of factors, the number itself and 1. These numbers form a special group of numbers called prime numbers. A *prime number* is a whole number greater than 1 that has only one factor pair, the number itself and 1. One is not a prime number because it has only one factor.

EXAMPLE 1 Identify the prime numbers by examining the factor pairs.

(a) 8   (b) 1   (c) 3   (d) 9   (e) 7

(a) **8 is *not* a prime number** because its factor pairs are 1×8 and 2×4.
(b) **1 is *not* a prime number** because a prime must be greater than 1 or because it does not have a factor pair.
(c) **3 is a prime number** because it has one factor pair, 1×3.
(d) **9 is *not* a prime number** because its factor pairs are 1×9 and 3×3.
(e) **7 is a prime number** because it has one factor pair, 1×7.

A *composite number* is a whole number greater than 1 that is not a prime number. Therefore, 8 and 9 in Example 1 are composite numbers. A composite number has a factor pair other than itself and 1.

EXAMPLE 2 Identify the composite numbers by examining the factor pairs of the numbers.

(a) 4   (b) 10   (c) 13   (d) 12   (e) 5

(a) **4 is a composite number** because its factor pairs are 1×4 and 2×2.
(b) **10 is a composite number** because its factor pairs are 1×10 and 2×5.
(c) **13 is *not* a composite number** because its only factor pair is 1×13. It is a prime number.
(d) **12 is a composite number** because its factor pairs are 1×12, 2×6, and 3×4.
(e) **5 is *not* a composite number** because its only factor pair is 1×5. It is a prime number.

Tips and Traps   ***Prime Numbers Less Than 50:***
We can find all the prime numbers that are 50 or less by using a technique developed by an ancient mathematician named Erastosthenes.

Step 1. List the numbers from 1 through 50.

Step 2. Eliminate numbers that are not prime by using the systematic process that follows.

(a) 1 is not prime. Eliminate 1.
(b) 2 is prime. Eliminate all multiples of 2.
(c) 3 is prime. Eliminate all multiples of 3.
(d) 4 has already been eliminated.
(e) 5 is prime. Eliminate all multiples of 5.
(f) 6 has already been eliminated.
(g) 7 is prime. Eliminate all multiples of 7.

Step 3. Circle remaining numbers as prime numbers.

1	②	③	4	⑤	6	⑦	8	9	10
⑪	12	⑬	14	15	16	⑰	18	⑲	20
21	22	㉓	24	25	26	27	28	㉙	30
㉛	32	33	34	35	36	㊲	38	39	40
㊶	42	㊸	44	45	46	㊼	48	49	50

All numbers not already eliminated are prime. Why? Multiples of 11 that are less than 50 have already been eliminated: $11 \times 2 = 22$, $11 \times 3 = 33$, $11 \times 4 = 44$, $11 \times 5 = 55$, which is greater than 50. Similarly, all other composite numbers have already been eliminated.

2 Determine the Prime Factorization of Composite Numbers.

A composite number can be expressed as a product of prime numbers. *Prime factorization* refers to writing a composite number as the product of *only* prime numbers. In this case, the factors are called *prime factors*.

> **Rule 3-3** *To find the prime factors of a composite number:*
>
> 1. Test each prime number in turn to see if the composite number is divisible by the prime.
> 2. Make a factor pair using the first prime number that passes the test in Step 1.
> 3. Carry forward the prime factors and test the remaining factors by repeating Steps 1 and 2.

EXAMPLE 3   Find the prime factorization of 30.

$$30 = 2 \times 15$$
first prime

30 is divisible by 2. Factor 30 into a factor pair using its smallest prime factor, 2.

$$30 = 2 \times 3 \times 5$$
last primes

Carry the prime factor, 2, forward. Factor the composite 15 using its smallest prime factor, 3. Because 5 is also prime, the factoring is complete.

The prime factorization of 30 is 2 × 3 × 5.

EXAMPLE 4   Find the prime factorization of 16.

$$16 = 2 \times 8$$
first prime

Factor 16 into two factors using its smallest prime factor.

$$16 = 2 \times 2 \times 4$$
second primes

Factor 8 into two factors using its smallest prime factor.

$$16 = 2 \times 2 \times 2 \times 2$$
last two primes

Factor 4 into two factors using its smallest prime factor.

The prime factorization of 16 is 2 × 2 × 2 × 2. We can write this expression in exponential notation as 2^4.

SELF-STUDY EXERCISE 3-3

1 List all factor pairs of each number. Then write all the factors in order from smallest to largest.

1. 14   **2.** 22   **3.** 11   **4.** 17   **5.** 18   **6.** 24

2 Find the prime factorization of the following numbers.

7. 50   **8.** 52   **9.** 225   **10.** 125   **11.** 100   **12.** 200
13. 65   **14.** 75   **15.** 121   **16.** 144

Write the prime factorization of the following numbers using exponential notation.

17. 568   **18.** 112   **19.** 124   **20.** 164   **21.** 72   **22.** 900

Learning Objectives

1. Find the least common multiple of two or more numbers.
2. Find the greatest common factor of two or more numbers.

1 Find the Least Common Multiple of Two or More Numbers.

Multiples, especially the *least common multiple,* will be useful when adding and subtracting fractions with different denominators. The least common multiple will be used as the lowest common denominator.

If we count by threes, we obtain natural numbers that are multiples of 3 ($3 \times 1 = 3$, $3 \times 2 = 6$, and so on). If we count by fives, we obtain natural numbers that are multiples of 5 ($5 \times 1 = 5$, $5 \times 2 = 10$, and so on).

Multiples of 3: 3, 6, 9, 12, 15, 18, 21, 24, 27, 30, 33, 36, 39, . . .
Multiples of 5: 5, 10, 15, 20, 25, 30, 35, 40, . . .

The common factors in these lists that are less than 40 are 15 and 30. The smallest multiple common to both sets of numbers above is 15. So 15 is the *least common multiple* of 3 and 5. It is the smallest number divisible by both 3 and 5.

■ **DEFINITION 3–15: Least Common Multiple.** The *least common multiple (LCM)* of two or more natural numbers is the smallest number that is a multiple of each number. It is divisible by each number.

Prime factorization can also be used to find the least common multiple of two or more numbers.

> **Rule 3–4** *The least common multiple of numbers may be found by using the prime factorization of the numbers as follows:*
>
> **1.** List the prime factorization of each number using exponential notation.
> **2.** List the prime factorization of the least common multiple by including each prime factor appearing in *each* number. If a prime factor appears in more than one number, use the factor with the *largest* exponent.
> **3.** Write the resulting expression in standard notation.

EXAMPLE 1   Find the least common multiple of 12 and 40 by prime factorization.

$12 = 2 \times 2 \times 3 \qquad = 2^2 \times 3$ Prime factorization of 12
$40 = 2 \times 2 \times 2 \times 5 = 2^3 \times 5$ Prime factorization of 40
$\text{LCM} = 2^3 \times 3 \times 5$ Prime factorization of LCM
$\textbf{LCM} = \textbf{120}$ LCM in standard notation

2 Find the Greatest Common Factor of Two or More Numbers.

A *common factor* is a factor common to two or more numbers or products. Common factors, especially the *greatest common factor,* will be useful later in simplifying or reducing fractions.

■ **DEFINITION 3–16: Greatest Common Factor.** The *greatest common factor (GCF)* of two or more numbers is the largest factor common to each number. Each number is divisible by the GCF.

Let's take the numbers 30 and 42. The prime factors are

$$30 = 2 \times 3 \times 5, \qquad 42 = 2 \times 3 \times 7$$

The *common* prime factors of both 30 and 42 are 2 and 3, which represent the composite factor 6. The *greatest* common factor is the product of the common prime factors, $2 \times 3 = 6$. So 6 is the GCF of 30 and 42. Each number is divisible by 6.

Rule 3–5 *To find the greatest common factor (GCF) of two or more natural numbers:*

1. List the prime factorization of each number using exponential notation.
2. List the prime factorization of the greatest common factor by including each prime factor appearing in *every* number. If a prime factor appears in more than one number, use the factor with the *smallest* exponent. If there are no common prime factors, the GCF is 1.
3. Write the resulting expression in standard notation.

EXAMPLE 2 Find the greatest common factor of 15, 30, and 45.

$15 = 3 \times 5$	$= 3 \times 5$	Prime factorization of 15
$30 = 2 \times 3 \times 5$	$= 2 \times 3 \times 5$	Prime factorization of 30
$45 = 3 \times 3 \times 5$	$= 3^2 \times 5$	Prime factorization of 45
GCF $= 3 \times 5$		Common prime factors
GCF $= 15$		GCF in standard notation

EXAMPLE 3 Find the greatest common factor of 10, 12, and 13.

$10 = 2 \times 5$	$= 2 \times 5$	Prime factorization of 10
$12 = 2 \times 2 \times 3$	$= 2^2 \times 3$	Prime factorization of 12
$13 = 13$	$= 13$	Prime factorization of 13
GCF $= 1$		No common prime factors

Tips and Traps *LCM Versus GCF:*
Least common multiple (LCM) and greatest common factor (GCF) are concepts that are easily confused. Because multiples of a number are the products of the number and any natural number, multiples will be as large as or larger than the original number. The LCM is the *smallest of the "large or larger" numbers* that are multiples of the original number.

Because factors of a number are the same as or smaller than the given number, the GCF is *the largest of the "small or smaller" numbers* that are factors of the original number.

SELF-STUDY EXERCISES 3–4

[1] Find the least common multiple of the following numbers.

1. 2 and 3	**2.** 5 and 6	**3.** 7 and 8
4. 3 and 4	**5.** 18 and 60	**6.** 10 and 12
7. 12 and 24	**8.** 9 and 18	**9.** 4, 8, and 12

10. 20, 25, and 35   **11.** 3, 9, and 27   **12.** 2, 8, and 16
13. 6, 15, and 18   **14.** 20, 24, and 30   **15.** 12, 18, and 20
16. 6, 10, and 12   **17.** 8, 12, and 32   **18.** 8, 12, and 18
19. 10, 15, and 20   **20.** 30, 50, and 60   **21.** 6, 11, and 33
22. 8, 13, and 39

2 Find the greatest common factor of the following numbers.

23. 18 and 24   **24.** 15 and 25   **25.** 10, 11, and 14
26. 12, 13, and 16   **27.** 6, 8, and 14   **28.** 4, 10, and 18
29. 30 and 45   **30.** 40 and 55   **31.** 36, 60, and 216
32. 18, 30, and 108

MATHEMATICS IN THE WORKPLACE

Astronomy: Planet Alignment

Every body in the universe exerts a gravitational pull on other objects. The more mass a body has, the greater the tug. For example, the Earth has more mass than the moon, so its gravitational pull on the moon is stronger than the moon's gravitational pull on the Earth. As a result, the moon stays in orbit around the Earth, not vice versa. Planets have huge gravitational forces, and when planets are in alignment their combined forces of gravity are tremendous. The strength of their combined gravities on another object (say, an asteroid) depends on four factors:

1. the mass of the planets
2. the planets' distance apart
3. the closest planet's distance from the asteroid
4. the planet's and asteroid's orbital angle from the ecliptic plane

To be in alignment means that the planets and the sun lie on a straight line in a bird's eye view of the ecliptic plane. Planets are usually at their closest point to each other when in alignment. Because some planets' orbits are tilted a few degrees out of the ecliptic plane, the minimal distance apart may not occur when those planets are in alignment.

The least common multiple (LCM) of the planets' orbital periods around the sun is used to calculate how often this alignment occurs. Venus and Earth are in alignment about every 45.0 yr. The mathematical computation of this alignment is found by first finding the prime factorization of their orbital periods. For Earth, 365 days = $5 \cdot 73$. For Venus, 225 days = $3^2 \cdot 5^2$. The

LCM of 365 and 225 is $3^2 \cdot 5^2 \cdot 73 = 16{,}425$ days = 45.0 yr. So about every 45 years Earth, Venus, and the sun are collinear in the ecliptic plane.

Planet alignment is an important factor that NASA and other space agencies consider when developing a spacecraft's launch date window and its trajectory. The combined gravitational forces of aligned planets could greatly affect the spacecraft's flight path, and could even cause a spacecraft to hit a planet.

Space agencies also use planet alignment LCM to conserve spacecraft fuel, flight distance, and time by planning to launch when Earth is closest to the target planet. For example, NASA launched its Mars Global Surveyor and Mars Pathfinder spacecraft on November 7 and December 4, 1996, respectively, to take advantage of the Earth–Mars alignment that occurs about every 2 years. Planet alignment also played a part in choosing the October 6, 1997 launch date and path of the Cassini spacecraft bound for Saturn.

When larger planets align, the combined gravitational effect is *enormous*. Some astronomers believe that Pluto and its moon Charon may have been tugged into orbit around the sun by the alignment of any two or three of the large outer planets: Neptune, Uranus, Saturn, and Jupiter. Because these planets far away from the sun have long orbital periods, alignment of two planets is rare, and alignment of three is even more rare. Voyager II took a rare photo as it left the solar system: seven of our nine planets in the same camera frame. Although not in perfect alignment, even a close alignment of seven planets happens once in a lifetime.

3-5   EQUIVALENT FRACTIONS AND DECIMALS

Learning Objectives

1 Write equivalent fractions with higher denominators.
2 Write equivalent fractions with lowest denominators.
3 Change decimals to fractions.
4 Change fractions to decimals.

116   Chapter 3   Fractions and Percents

Now we are ready to look again at numbers written using fractional form. There are many different ways to express the same value in fractional form. For example, the whole number 1 can be written as $\frac{1}{1}, \frac{4}{4}, \frac{7}{7}, \frac{15}{15}$, and so on.

Fractions are equivalent if they represent the same value. Let's compare the illustrations in Fig. 3–15. In Fig. 3–15, line a is one whole unit divided into only one part ($\frac{1}{1}$). Line b is one unit divided into two parts ($\frac{2}{2}$). Line c is divided into 4 parts ($\frac{4}{4}$), line d is divided into eight parts ($\frac{8}{8}$), and line e is divided into 16 parts ($\frac{16}{16}$).

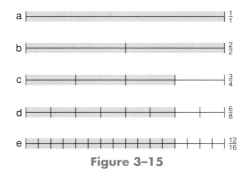

Figure 3–15

Look at the shaded dimensions on lines c, d, and e. What fraction represents this dimension on line c? Three out of four parts, or $\frac{3}{4}$. What fraction represents this dimension on line d? Six out of eight parts, or $\frac{6}{8}$. What fraction represents this dimension on line e? Twelve out of sixteen parts, or $\frac{12}{16}$. Thus, $\frac{3}{4}, \frac{6}{8},$ and $\frac{12}{16}$ are equivalent fractions; that is,

$$\frac{3}{4} = \frac{6}{8} = \frac{12}{16}$$

1 Write Equivalent Fractions With Higher Denominators.

Equivalent fractions are in the same "family of fractions." The first member of the "family" is the fraction in lowest terms; that is, no whole number divides evenly into *both* the numerator and denominator except the number 1. Other "family members" can be found by multiplying both the numerator and denominator by the same number.

EXAMPLE 1   Find five fractions that are equivalent to $\frac{1}{2}$.

$$\frac{1 \times 2}{2 \times 2} \quad \text{or} \quad \frac{1}{2} \times \frac{2}{2} = \frac{2}{4} \qquad\qquad \frac{1 \times 5}{2 \times 5} \quad \text{or} \quad \frac{1}{2} \times \frac{5}{5} = \frac{5}{10}$$

$$\frac{1 \times 3}{2 \times 3} \quad \text{or} \quad \frac{1}{2} \times \frac{3}{3} = \frac{3}{6} \qquad\qquad \frac{1 \times 6}{2 \times 6} \quad \text{or} \quad \frac{1}{2} \times \frac{6}{6} = \frac{6}{12}$$

$$\frac{1 \times 4}{2 \times 4} \quad \text{or} \quad \frac{1}{2} \times \frac{4}{4} = \frac{4}{8}$$

Tips and Traps *Multiplication, Division, and One:*
In Example 1, $\frac{1}{2}$ was multiplied by a fraction whose value is 1, and 1 times any number does not change the value of that number. Written symbolically,

$$\frac{n}{n} = 1 \quad \text{and} \quad 1 \times n = n, \, n \neq 0$$

Many times we will need to find a fraction with a specific denominator that is equivalent to a given fraction. In this case, we first decide what number should be multiplied by the original denominator to give the new denominator. Then we multiply the numerator and denominator by that number. This is allowed because of the **fundamental principle of fractions.**

■ **DEFINITION 3–17: Fundamental Principle of Fractions.** If the numerator and denominator of a fraction are multiplied by the same nonzero number, the value of the fraction remains unchanged.

EXAMPLE 2   Find a fraction equivalent to $\frac{2}{3}$ that has a denominator of 12.

3 times what number is 12? Or 3 divides into 12 how many times? The answer is 4. Then

$$\frac{2 \times 4}{3 \times 4} \qquad \text{or} \qquad \frac{2}{3} \times \frac{4}{4} = \frac{\mathbf{8}}{\mathbf{12}}$$

Rule 3–6   *To change a fraction to an equivalent fraction with a larger specified denominator:*

1. Divide the original denominator into the desired denominator.
2. Multiply the original numerator and denominator by the quotient found in Step 1.

EXAMPLE 3   Change $\frac{5}{8}$ to an equivalent fraction whose denominator is 32.

$$\frac{5}{8} = \frac{?}{32}, \qquad 32 \div 8 = 4 \text{ and } 4 \times 5 = 20$$

$$\frac{5}{8} = \frac{\mathbf{20}}{\mathbf{32}} \qquad \text{or} \qquad \frac{5}{8} \times \frac{4}{4} = \frac{\mathbf{20}}{\mathbf{32}}$$

② Write Equivalent Fractions with Lowest Denominators.

Because any fraction will have an unlimited number of equivalent fractions, we usually want to work with fractions in *lowest terms*. The smaller numbers are easier to work with.

By *lowest terms,* we mean that there is no *whole* number other than 1 that will divide evenly into both the numerator and denominator. Another way of saying this is that the numerator and denominator have no common factors other than 1. When we find an equivalent fraction that has smaller numbers and there are no common factors in the numerator and denominator, we have *reduced to lowest terms.*

Rule 3–7   *To change a fraction to an equivalent fraction with a smaller denominator or to reduce a fraction to lowest terms:*

1. Find a common factor greater than one for the numerator and denominator.
2. Divide both the numerator and denominator by this common factor.
3. Continue until the fraction is in lowest terms or has the desired smaller denominator. The fraction can also be reduced to lowest terms by finding the greatest common factor (GCF) in Step 1.

EXAMPLE 4 Reduce $\frac{8}{10}$ to lowest terms.

Factors of 8: 1, 2, 4, 8
Factors of 10: 1, 2, 5, 10
The greatest common factor (GCF) is two.

$$\frac{8 \div 2}{10 \div 2} \quad \text{or} \quad \frac{8}{10} \div \frac{2}{2} = \frac{4}{5}$$

Tips and Traps **Reducing and Properties of One:**
When reducing the fraction $\frac{8}{10}$, we are dividing by the whole number 1 in the form of $\frac{2}{2}$. A nonzero number divided by itself is 1, and to divide a number by 1 does not change the value of the number. Symbolically,

$$\frac{n}{n} = 1 \quad \text{and} \quad n \div 1 = n; \, n \neq 0$$

EXAMPLE 5 Reduce $\frac{18}{24}$ to lowest terms.

Factors of 18: 1, 2, 3, 6, 9, 18
Factors of 24: 1, 2, 3, 4, 6, 8, 12, 24
The GCF is 6.

$$\frac{18 \div 6}{24 \div 6} \quad \text{or} \quad \frac{18}{24} \div \frac{6}{6} = \frac{3}{4}$$

Tips and Traps **Using the GCF to Reduce to Lowest Terms:**
A fraction can be reduced to lowest terms in the fewest steps by using the *greatest common factor*; however, it can still be reduced correctly in more steps by using any common factor.

$$\frac{18 \div 2}{24 \div 2} = \frac{9}{12}$$

$$\frac{9 \div 3}{12 \div 3} = \frac{3}{4}$$

---

3 Change Decimals to Fractions.

When a decimal is written as a fraction, the number of digits to the right of the decimal point determines the denominator of the fraction. In the decimal 0.23, there are two digits to the right of the decimal, which indicates that the denominator of the fraction will be 100. (One hundred has two zeros.) Decimal numbers with one digit to the right of the decimal have one zero in the denominator. Decimal numbers with three digits to the right of the decimal have three zeros in the denominator, and so on.

> **Rule 3–8** *To convert a decimal number to a fraction or mixed number in lowest terms:*
>
> 1. Write the digits without the decimal point as the numerator.
> 2. Write the denominator as a power of 10 with as many zeros as places to the right of the decimal point.
> 3. Reduce and, if the fraction is improper, convert to a mixed number.

EXAMPLE 6 Write as a fraction 0.4, and 0.075.

$$0.4 = \frac{4}{10} = \frac{2}{5}$$ Tenths indicates a denominator of 10.

$$0.075 = \frac{75}{1000} = \frac{3}{40}$$ Thousandths indicates a denominator of 1000.

4 Change Fractions to Decimals.

With increased calculator use, we may find it more convenient to work with decimals. Therefore, we will often change fractions to decimals when we make certain computations.

The bar separating the numerator and denominator of a fraction indicates division: $\frac{2}{5}$ also means $2 \div 5$ or $5\overline{)2}$.

If we write a decimal after the 2 and affix a zero, we can divide.

$$5\overline{)2.0} \quad \overset{0.4}{}$$

Therefore, $\frac{2}{5} = 0.4$.

Rule 3–9 *To convert a fraction to a decimal number:*

1. Place a decimal point after the numerator.
2. Divide the numerator by the denominator using long division.
3. Affix zeros to the numerator as needed for division.

EXAMPLE 7 Change $\frac{7}{8}$ to a decimal.

$$7 \div 8 \quad \text{or} \quad 8\overline{)7.000} \quad \overset{0.875}{}$$

	6 4
	60
	56
	40
	40

Affix zeros until the division terminates; that is, it has no remainder.

Rule 3–10 *To convert a mixed number to a decimal number:*

1. The whole-number part remains the same.
2. Convert only the fraction part by using Rule 3–9.

EXAMPLE 8 Change $3\frac{2}{5}$ to a mixed decimal.

$$\frac{2}{5} = 5\overline{)2.0} \quad \overset{0.4}{} \quad \text{or} \quad 0.4$$

Then, $3\frac{2}{5} = 3.4$.

When we change some fractions to decimals, the quotient does not terminate.

EXAMPLE 9   Change $\frac{1}{3}$ and $\frac{4}{11}$ to decimals.

$$1 \div 3 \quad \text{or} \quad 3\overline{)1.000} \quad \text{or} \quad \mathbf{0.33\frac{1}{3}}$$

$$
\begin{array}{r}
0.333 \\
3\overline{)1.000} \\
\underline{9} \\
10 \\
\underline{9} \\
10 \\
\underline{9} \\
1
\end{array}
$$

$$4 \div 11 \quad \text{or} \quad 11\overline{)4.000000} \quad \text{or} \quad \mathbf{0.36\frac{4}{11}}$$

$$
\begin{array}{r}
0.363636 \\
11\overline{)4.000000} \\
\underline{3\,3} \\
70 \\
\underline{66} \\
40 \\
\underline{33} \\
70 \\
\underline{66} \\
40 \\
\underline{33} \\
70 \\
\underline{66} \\
4
\end{array}
$$

When fractions are changed into decimals that do not terminate, the decimals are called *repeating decimals*. Repeating decimals can be written with a line over the digit(s) that repeat or three dots at the end to indicate that they repeat. The decimal equivalent can be rounded to any desirable place.

$$\frac{1}{3} = 0.\overline{3} \quad \text{or} \quad 0.333333 \ldots \qquad \frac{4}{11} = 0.\overline{36} \quad \text{or} \quad 0.3636 \ldots$$

 ■ **General Tips for Using the Calculator**

Fractions may be converted to decimal numbers by dividing the numerator by the denominator on a calculator. Remember to enter the dividend (numerator) *first*. The fractions in Example 9 can be converted to decimals on a calculator.

Scientific calculator:

AC 1 ÷ 3 = ⟹ 0.333333333
AC 4 ÷ 11 = ⟹ 0.363636363

Graphics calculator:

AC 1 ÷ 3 EXE ⟹ 0.333333333
AC 4 ÷ 11 EXE ⟹ 0.363636363

SELF-STUDY EXERCISES 3–5

1. Find five fractions that are equivalent to $\frac{4}{5}$.   **2.** Find five fractions that are equivalent to $\frac{7}{10}$.

3. Find a fraction that is equivalent to $\frac{3}{4}$ and has a denominator of 24.

Find the equivalent fractions having the indicated denominators.

4. $\dfrac{3}{8} = \dfrac{?}{16}$ **5.** $\dfrac{4}{7} = \dfrac{?}{21}$ **6.** $\dfrac{9}{11} = \dfrac{?}{44}$

7. $\dfrac{1}{3} = \dfrac{?}{15}$ **8.** $\dfrac{5}{6} = \dfrac{?}{24}$ **9.** $\dfrac{7}{8} = \dfrac{?}{24}$

10. $\dfrac{2}{5} = \dfrac{?}{30}$

2 Reduce the following fractions to lowest terms.

11. $\dfrac{4}{8}$ **12.** $\dfrac{6}{10}$ **13.** $\dfrac{12}{16}$ **14.** $\dfrac{10}{32}$ **15.** $\dfrac{16}{32}$

16. $\dfrac{28}{32}$ **17.** $\dfrac{20}{64}$ **18.** $\dfrac{2}{8}$ **19.** $\dfrac{8}{32}$ **20.** $\dfrac{12}{50}$

21. $\dfrac{10}{16}$ **22.** $\dfrac{4}{16}$ **23.** $\dfrac{24}{32}$ **24.** $\dfrac{12}{64}$ **25.** $\dfrac{14}{64}$

3 Change the following decimals to their fraction or mixed-number equivalents, and reduce answers to lowest terms.

26. 0.5 **27.** 0.1 **28.** 0.2
29. 0.7 **30.** 0.25 **31.** 0.025
32. 3.9 **33.** 4.8 **34.** 0.378
35. 0.875 **36.** 0.375 **37.** 0.625

38. A measure of 0.75 in. represents what fractional part of an inch?

39. What fraction represents a measure of 0.1875 ft?

40. The length of a screw is 2.375 in. Represent this length as a mixed number.

41. An instrument weighs $0.83\frac{1}{3}$ lb. Write this as a fraction of a pound.

42. Some sheet metal is 0.3125 in. thick. What is the thickness expressed as a fraction?

43. A predrilled PC board is 3.125 in. long. Write the length as a mixed number.

4 Change these fractions and mixed numbers to decimals.

44. $\dfrac{2}{5}$ **45.** $\dfrac{3}{10}$ **46.** $\dfrac{7}{8}$ **47.** $\dfrac{5}{8}$ **48.** $\dfrac{9}{20}$

49. $\dfrac{49}{50}$ **50.** $\dfrac{21}{100}$ **51.** $3\dfrac{7}{8}$ **52.** $1\dfrac{7}{16}$ **53.** $4\dfrac{9}{16}$

Write these fractions as repeating decimals.

54. $\dfrac{2}{3}$ **55.** $\dfrac{3}{11}$ **56.** $\dfrac{7}{9}$ **57.** $\dfrac{5}{13}$

Write these fractions as decimals to the hundredths place. Express the remainder as a fraction.

58. $\dfrac{5}{6}$ **59.** $\dfrac{7}{12}$

60. An aerial map shows a building measuring $2\frac{3}{64}$ in. on one side. If decimal measures were used, what would the side of the building measure on the map?

61. The property tax rate in one state is $4\frac{1}{2}\%$ per assessed value. What is the tax rate per assessed value expressed as a mixed-decimal percent?

62. A plan specifies allowing a gap of $\frac{1}{8}$ in. between vinyl flooring and the wall for expansion. What should the gap be in decimal notation?

3-6 IMPROPER FRACTIONS AND MIXED NUMBERS

Learning Objectives

1 Convert improper fractions to whole or mixed numbers.

2 Convert mixed numbers and whole numbers to improper fractions.

1 Convert Improper Fractions to Whole or Mixed Numbers.

As we recall from Section 3–1, an improper fraction is a fraction whose value is equal to or greater than one unit, such as $\frac{6}{3}$, $\frac{15}{7}$, or $\frac{5}{5}$. Problems involving improper fractions are sometimes easier to solve if we change the improper fractions into whole numbers or mixed numbers.

> **Rule 3–11** *To convert an improper fraction to a whole or mixed number:*
>
> Perform the division indicated. Any remainder of the division is expressed as a fraction or decimal equivalent.

EXAMPLE 1   Convert $\frac{15}{3}$ to a whole or mixed number.

$$\frac{15}{3} \quad \text{means} \quad 15 \div 3 \quad \text{or} \quad 3\overline{\smash{\big)}15} \quad \begin{array}{r} 5 \\ \underline{15} \\ 0 \end{array}$$

$$\frac{15}{3} = 5$$

If there is no remainder from the division, the improper fraction converts to a whole number.

Tips and Traps — *Converting to a Whole or Mixed Number Is Different from Reducing:* Converting an improper fraction to a whole or mixed number should not be confused with reducing to lowest terms. An improper fraction is in lowest terms if its numerator and denominator have no common factor. Therefore, the improper fraction $\frac{10}{7}$ is in lowest terms. The improper fraction $\frac{10}{4}$ is not in lowest terms. It will reduce to $\frac{5}{2}$, which is in lowest terms.

When converting an improper fraction to a whole or mixed number, the fractional part should be in lowest terms. We can reduce to lowest terms either before dividing or after dividing.

EXAMPLE 2   Convert $\frac{28}{8}$ to a mixed number.

$$\frac{28}{8} = \frac{7}{2} \qquad 2\overline{\smash{\big)}7} \quad \begin{array}{r} 3 \\ \underline{6} \\ 1 \end{array} = 3\frac{1}{2} \qquad \text{Fraction is reduced before dividing.}$$

or

$$\frac{28}{8} \qquad 8\overline{\smash{\big)}28} \quad \begin{array}{r} 3 \\ \underline{24} \\ 4 \end{array} = 3\frac{4}{8} = 3\frac{1}{2} \qquad \text{Fraction is reduced after dividing.}$$

2 Convert Mixed Numbers and Whole Numbers to Improper Fractions.

It is sometimes useful to convert a mixed number to an improper fraction when solving certain problems. In the improper fraction, the numerator indicates the total number of parts considered. The denominator indicates how many parts one unit has been divided into.

> **Rule 3–12** *To convert a mixed number to an improper fraction:*
> 1. Multiply the denominator of the fractional part by the whole number.
> 2. Add the numerator of the fractional part to the product; this sum becomes the numerator of the improper fraction.
> 3. The denominator of the improper fraction is the same as the denominator of the fractional part of the mixed number.

EXAMPLE 3 Change $6\frac{2}{3}$ to an improper fraction.

$$6\frac{2}{3} = \frac{(3 \times 6) + 2}{3} = \frac{20}{3}$$

Whole numbers can also be written as improper fractions. Any whole number can be written as a fraction with a denominator of 1. Then the whole number divided by 1 can be changed to an improper fraction having any denominator by using the procedures described in Section 3–5.

EXAMPLE 4 Change 8 to fifths.

$$8 = \frac{8}{1} = \frac{8 \times 5}{1 \times 5} = \frac{40}{5}$$

SELF-STUDY EXERCISES 3–6

[1] Convert the following improper fractions to whole or mixed numbers.

1. $\frac{12}{5}$ 2. $\frac{10}{7}$ 3. $\frac{12}{12}$ 4. $\frac{32}{7}$ 5. $\frac{24}{6}$

6. $\frac{15}{7}$ 7. $\frac{23}{9}$ 8. $\frac{47}{5}$ 9. $\frac{86}{9}$ 10. $\frac{38}{21}$

11. $\frac{57}{15}$ 12. $\frac{64}{4}$ 13. $\frac{72}{10}$ 14. $\frac{19}{2}$ 15. $\frac{36}{4}$

[2] Change the following mixed numbers to improper fractions.

16. $2\frac{1}{3}$ 17. $3\frac{1}{8}$ 18. $1\frac{7}{8}$ 19. $6\frac{5}{12}$ 20. $9\frac{5}{8}$

21. $3\frac{7}{8}$ 22. $7\frac{5}{12}$ 23. $6\frac{7}{16}$ 24. $8\frac{1}{32}$ 25. $1\frac{5}{64}$

26. $7\frac{3}{10}$ 27. $8\frac{2}{3}$ 28. $33\frac{1}{3}$ 29. $66\frac{2}{3}$ 30. $12\frac{1}{2}$

Change each of the following whole numbers to an equivalent fraction having the indicated denominator.

31. $5 = \frac{?}{3}$ 32. $9 = \frac{?}{2}$ 33. $7 = \frac{?}{8}$ 34. $8 = \frac{?}{4}$ 35. $3 = \frac{?}{16}$

3-7 **FINDING COMMON DENOMINATORS AND COMPARING FRACTIONS AND DECIMALS**

Learning Objectives
[1] Find common denominators.
[2] Compare fractions and decimals.

1 Find Common Denominators.

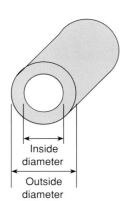

Inside
diameter

Outside
diameter

Figure 3–16

Is it possible to have a pipe with an outside diameter of $\frac{5}{8}$ in. and an inside diameter of $\frac{21}{32}$ in. (see Fig. 3–16)? To answer this question, it is necessary to compare the size of two fractions. To do this, we first select a *common denominator;* that is, we find a denominator that each denominator will divide into evenly. A common denominator can always be found by multiplying the two denominators together. However, many times a smaller number can be used as a common denominator. We should use the least common multiple (LCM) as our *least common denominator* (LCD). Because 8 divides evenly into 32, we can use 32 for our common denominator. Next, we change $\frac{5}{8}$ to an equivalent fraction with a denominator of 32.

$$\frac{5}{8} = \frac{5 \times 4}{8 \times 4} = \frac{20}{32}$$

Now to answer our original question, is $\frac{20}{32}$, which is equivalent to $\frac{5}{8}$, larger than $\frac{21}{32}$? No. Then $\frac{5}{8}$ in. cannot be the outside diameter of a pipe with an inside diameter of $\frac{21}{32}$ in.

With many fractions, the least common denominator can be found by inspection. For fractions with larger denominators, you may need to use the procedure for finding the LCM discussed in Section 3–4.

Rule 3–13 *To find the least common denominator (LCD) of two or more fractions:*

Find the least common multiple (LCM) of the denominators of the fractions.

EXAMPLE 1 Find the least common denominator for the fractions: $\frac{5}{12}, \frac{4}{15}, \frac{3}{8}$. Write the prime factorization of each denominator.

$$\begin{array}{ccc}
\underline{12} & \underline{15} & \underline{8} \\
2 \times 6 & \mathbf{3 \times 5} & 2 \times 4 \\
2 \times 2 \times 3 & & 2 \times 2 \times 2 \\
\mathbf{2^2 \times 3} & & \mathbf{2^3}
\end{array}$$

$$\text{LCM or LCD} = 2^3 \times 3 \times 5$$
$$= 8 \times 3 \times 5$$
$$= \mathbf{120}$$

Tips and Traps *An Alternate Procedure for Finding the LCM or LCD:*
Another strategy for finding the LCM or LCD of several fractions is to divide out duplicated factors.

1. Arrange the denominators horizontally.
2. Divide out any prime factor that will divide evenly into at least two denominators.
3. The LCM or LCD will be the product of the primes and remaining factors.

Look at the denominators from Example 1.

	8	12	15	
2				Divide out a factor of 2.
2	4	6	15	Divide out another factor of 2.
3	2	3	15	Divide out a factor of 3.
	2	1	5	

Primes: 2, 2, 3
Factors: 2, 1, 5
LCM = $2 \cdot 2 \cdot 3 \cdot 2 \cdot 1 \cdot 5$
     = 120

2 Compare Fractions and Decimals.

To compare quantities, you must compare like amounts. Thus, when comparing fractions, we must have fractions with the same denominator. To compare $\frac{3}{7}$ and $\frac{5}{7}$, which have the same denominators, we compare the numerators. Therefore, $\frac{3}{7}$ is smaller than $\frac{5}{7}$.

To compare fractions with different denominators, we must first change the fractions to equivalent fractions, which have a common denominator.

Rule 3–14 *To compare fractions:*

1. Find the least common denominator (the *smallest* number each denominator will divide into evenly), which is also the least common multiple.
2. Change each fraction to an equivalent fraction having the least common denominator as its denominator.
3. Compare the numerators.

Tips and Traps *Comparing Fractions:*
Fractions can be compared by changing them to equivalent fractions having *any* common denominator, but using the least common denominator gives us smaller numbers to work with.

EXAMPLE 2   Two drill bits have diameters of $\frac{3}{8}$ in. and $\frac{5}{16}$ in., respectively. Which drill bit will make the larger hole?

The least common denominator is 16.

$$\frac{3}{8} = \frac{6}{16}, \qquad \frac{5}{16} = \frac{5}{16}$$

Now that each fraction has been changed to an equivalent fraction with the same denominator, we compare the numerators: $\frac{6}{16}$ is larger than $\frac{5}{16}$ because 6 is larger than 5, so $\frac{3}{8}$, which is equivalent to $\frac{6}{16}$, is larger than $\frac{5}{16}$. **The drill bit with a $\frac{3}{8}$-in. diameter will drill the larger hole.**

Sometimes we are asked to compare a fraction and a decimal. This comparison requires changing one of the numbers so that both numbers are in fraction format or both are in decimal format. Unless otherwise specified, you may use either format.

EXAMPLE 3   An engineering rule is used to measure the length of a computer motherboard as 15.8 inches. The computer case that will hold the board is measured at $15\frac{7}{8}$ inches. Will the board fit inside the case?

Change $15\frac{7}{8}$ to a decimal.

$$7 \div 8 = 0.875$$

$$15\frac{7}{8} = 15.875$$

The case measurement, 15.875, is greater than 15.80, so the board will fit into the case.

Our employment may sometimes require us to work with close tolerances. For instance, we may be working with wire and need to compare the diameters of several sizes of wire to arrange them in order from larger to smaller or smaller to larger. To do this, we need to be able to compare decimals because wire size is often measured in decimals.

Rule 3–15 *To compare decimal numbers:*

1. Compare whole-number parts.
2. If whole-number parts are equal, compare digits place by place, starting at the tenths place and moving to the right.
3. Stop when two digits in the same place are different.
4. The larger digit determines the larger decimal number.

EXAMPLE 4   Compare the two numbers to see which is larger.

$$32.47 \qquad 32.48$$

We begin by looking at the whole-number parts of the two numbers. In this case they are the same. Next, we look at the tenths place for each number. Both numbers have a 4 in the tenths place. **After looking at the hundredths place, we see that 32.48 is the larger number because 8 is larger than 7.**

EXAMPLE 5   Compare 0.4 and 0.07 to see which is larger.

Since the whole-number parts are the same (0), we compare the digits in the tenths place. **0.4 is larger because 4 is larger than 0.**

If we write both numbers in the example above so that they have the same number of digits in the decimal part, they may be easier to compare.

$$0.4 = 0.40$$

$$0.07 = 0.07$$

Because 40 is larger than 7, 0.4 is larger than 0.07. This is equivalent to the process used in comparing common fractions. A common denominator is found and each fraction (or decimal fraction) is changed to an equivalent fraction having the common denominator.

Tips and Traps   *Decimal Fractions with Common Denominators:*
Decimal fractions have a common denominator if they have the same number of digits to the right of the decimal point.

EXAMPLE 6   A measuring instrument called a micrometer is used to measure two objects. The measurements taken are 0.386 in. and 0.388 in. Which object is larger?

To compare the measurements above, we note that the digits in the tenths and hundredths places are the same. **Comparing the digits in the thousandths places, we see that the object measuring 0.388 in. is larger.**

SELF-STUDY EXERCISES 3–7

1   Find the least common denominator for the following fractions.

1. $\dfrac{5}{8}, \dfrac{4}{9}$ 2. $\dfrac{3}{10}, \dfrac{4}{15}$ 3. $\dfrac{9}{10}, \dfrac{4}{25}$ 4. $\dfrac{7}{12}, \dfrac{9}{16}, \dfrac{5}{8}$ 5. $\dfrac{2}{3}, \dfrac{5}{12}, \dfrac{7}{8}$

2 Show which fraction is larger.

6. $\frac{2}{3}, \frac{3}{5}$ 7. $\frac{5}{12}, \frac{7}{16}$ 8. $\frac{8}{9}, \frac{7}{8}$ 9. $\frac{5}{8}, \frac{11}{16}$ 10. $\frac{15}{32}, \frac{29}{64}$

11. $\frac{7}{12}, \frac{9}{16}$ 12. $\frac{3}{8}, \frac{4}{5}$ 13. $\frac{7}{11}, \frac{9}{10}$ 14. $\frac{4}{15}, \frac{3}{16}$ 15. $\frac{1}{2}, \frac{7}{16}$

16. Is the thickness of a $\frac{3}{16}$-in. sheet of metal greater than the length of a $\frac{15}{64}$-in. sheet metal screw?

17. Will a pipe with a $\frac{5}{16}$-in. outside diameter fit inside a pipe with a $\frac{3}{8}$-in. inside diameter?

18. A hollow-wall fastener has a grip range up to $\frac{7}{16}$ in. Is it long enough to fasten a thin sheet metal strip to a plywood wall if the combined thickness is $\frac{3}{8}$ in.?

19. A range top is $29\frac{1}{8}$ in. long by $19\frac{1}{2}$ in. wide. Is it smaller than an existing opening $29\frac{3}{16}$ in. long by $19\frac{9}{16}$ in. wide?

20. A wrench is marked $\frac{5}{8}$ at one end and $\frac{19}{32}$ at the other. Which end is larger?

21. Charles Bryant has a wrench marked $\frac{25}{32}$, but it is too large. Would a $\frac{7}{8}$-in. wrench be smaller?

22. For a do-it-yourself project, Brenda Jinkins needs to cut a piece of sheet metal slightly longer than the required $10\frac{21}{32}$ in. of the plans and trim it down to size. Brenda cut the piece $10\frac{3}{4}$ in. Was it cut too short?

23. A plastic anchor for a No. 6 $\times$ 1-in. screw requires that at least a $\frac{3}{16}$-in. diameter hole be drilled. Will a $\frac{1}{4}$-in. drill bit be large enough?

24. The plastic anchor in Exercise 23 requires a minimum hole depth of $\frac{7}{8}$ in. Is a $\frac{3}{4}$-in. hole deep enough?

25. Is a $\frac{3}{8}$-in. wrench too large or too small for a $\frac{1}{2}$-in. bolt?

Show which decimal is larger.

26. 0.023 or 0.03

27. 0.38 or 0.392

28. 5.38 or 5.375

29. 4.207 or 4.71

30. A nurse measures and records temperatures of 96.5°, 97.3°, 98.6°. Which temperature is highest?

31. A hip replacement medical implant is measured by two quality control technicians, and measures of 0.0256 cm and 0.0249 cm are recorded. Which measure is larger?

32. Two cities base the property tax rate on mills. One city has a rate of 0.973 mills and the other has a rate of 0.975 mills. Which city has the lower tax rate?

33. Describe an instance when you needed to compare two or more decimals or fractions. How did you do this?

3-8   ADDING FRACTIONS AND MIXED NUMBERS

Learning Objectives

1 Add fractions.

2 Add mixed numbers.

So far we have been studying what fractions and mixed numbers represent and how to convert them to their equivalents. Now it is time to put that knowledge to use to solve the kinds of problems commonly found on the job in the various technologies. Let's look first at problems using addition.

1 Add Fractions.

The most important rule to remember when adding fractions is that all fractions to be added must have the same denominator. Trying to add unlike fractions is like trying to add unlike objects or measures. Before unlike fractions can be added, we change the fractions to equivalent fractions having a common denominator and then add the fractions. The following rule is used when adding fractions.

Rule 3–16 *To add fractions:*

1. If the denominators are not the same, find the least common denominator.
2. Change each fraction not already expressed in terms of the common denominator to an equivalent fraction having the common denominator.
3. Add the numerators only.
4. The common denominator will be the denominator of the sum.
5. Reduce the answer to lowest terms and change improper fractions to whole or mixed numbers.

EXAMPLE 1   Find the sum of $\dfrac{3}{8} + \dfrac{1}{8}$.

Because the denominators are the same, start with Step 3 of Rule 3–16.

$$\frac{3}{8} + \frac{1}{8} = \frac{4}{8} = \mathbf{\frac{1}{2}}$$

Tips and Traps   *Shortening Written Steps:*
We are beginning to work some of the basic steps mentally. For example, $\frac{4}{8} = \frac{4 \div 4}{8 \div 4} = \frac{1}{2}$ was written as $\frac{4}{8} = \frac{1}{2}$. The understanding of basic principles is important; however, we do not want to burden ourselves with unnecessary written procedures.

EXAMPLE 2   Add $\dfrac{5}{32} + \dfrac{3}{16} + \dfrac{7}{8}$.

The least common denominator may be found by inspection. Both 8 and 16 divide evenly into 32, so 32 may be used as the common denominator. Next, change each fraction to an equivalent fraction whose denominator is 32.

$$\frac{5}{32} = \frac{5}{32}, \qquad \frac{3}{16} = \frac{6}{32}, \qquad \frac{7}{8} = \frac{28}{32}$$

Then add the numerators.

$$\frac{5}{32} + \frac{6}{32} + \frac{28}{32} = \frac{39}{32}$$

Finally, change to a mixed number.

$$\frac{39}{32} = \mathbf{1\frac{7}{32}}$$

Notice that in a problem of this type nearly all our fraction skills are used.

EXAMPLE 3   A plumber uses a $\frac{9}{16}$-in.-diameter copper tube wrapped with $\frac{5}{8}$-in. insulation. What size hole must be bored in the stud for the insulated pipe to be installed?

From Fig. 3–17, we can see that the $\frac{5}{8}$-in. insulation increases the diameter on each side of the pipe. To get the total diameter of the pipe and insulation, we add $\frac{9}{16} + \frac{5}{8} + \frac{5}{8}$. The thickness of the insulation is added twice because it counts in the total diameter of the pipe and insulation two times.

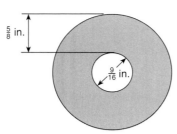

$$\frac{9}{16} = \frac{9}{16}$$ The LCD is 16.

$$\frac{5}{8} = \frac{10}{16}$$

$$+ \frac{5}{8} = \frac{10}{16}$$

$$\frac{29}{16} = 1\frac{13}{16}$$

Figure 3–17

The total diameter is $1\frac{13}{16}$ in. and the diameter of the hole must be at least this measure.

2 Add Mixed Numbers.

There are two popular methods for adding mixed numbers. One method is to add the whole numbers and fractions separately and then combine the sums. The other method is to change each mixed number to an improper fraction and then follow the rules for adding fractions. The first method is more appropriate for most situations.

EXAMPLE 4   Add $5\frac{2}{3} + 7\frac{3}{8} + 4\frac{1}{2}$.

It is usually more convenient to arrange the problem vertically when adding. Using the first method, we have

$$5\frac{2}{3} = 5\frac{16}{24}$$ The LCD is 24.

$$7\frac{3}{8} = 7\frac{9}{24}$$

$$+ 4\frac{1}{2} = 4\frac{12}{24}$$

$$16\frac{37}{24} \qquad \frac{37}{24} = 1\frac{13}{24}$$

$$16 + 1\frac{13}{24} = 17\frac{13}{24}$$

Using the second method, we have

$$5\frac{2}{3} = \frac{17}{3} = \frac{136}{24}$$

$$7\frac{3}{8} = \frac{59}{8} = \frac{177}{24}$$

$$+ 4\frac{1}{2} = \frac{9}{2} = \frac{108}{24}$$

$$\frac{421}{24} = 17\frac{13}{24}$$

EXAMPLE 5   A carpenter cut the following pieces from a metal rod: $3\frac{3}{8}$ in., $9\frac{5}{16}$ in., and $7\frac{3}{4}$ in. What was the total length cut from the rod?

To find the total length cut from the rod, we add the mixed numbers.

$$3\frac{3}{8} = 3\frac{6}{16}$$ The LCD is 16.

130   CHAPTER 3   Fractions and Percents

$$9\frac{5}{16} = 9\frac{5}{16}$$

$$+ \ 7\frac{3}{4} = 7\frac{12}{16}$$

$$19\frac{23}{16} = 20\frac{7}{16} \qquad \frac{23}{16} = 1\frac{7}{16}, \ 19 + 1\frac{7}{16} = 20\frac{7}{16}$$

$20\frac{7}{16}$ in. were cut from the rod.

Tips and Traps **Writing Whole Numbers in Mixed Number Form:**
When adding mixed numbers and whole numbers, the whole number can be thought of as a mixed number with zero as the numerator of the fraction.

$$5 + 3\frac{1}{3} = \quad 5\frac{0}{3}$$
$$+ \ 3\frac{1}{3}$$

Recall the addition property of zero. Zero added to any number does not change the value of the number. Thus, $5 + 3\frac{1}{3} = 8\frac{1}{3}$.

EXAMPLE 6   Find the largest permissible measurement of a part if the blueprint calls for the part to be 2 in. long and the tolerance is $\pm\frac{1}{8}$ in.

Tolerance is the amount a part can vary from the blueprint specification. To find the largest permissible measure, we add.

$$2 + \frac{1}{8} = 2\frac{1}{8}$$

The largest permissible measure is $2\frac{1}{8}$ in.

Tips and Traps **What are Tolerance and Limit Dimensions?**
Tolerance is an often-used concept in technical applications. The "plus or minus" symbol ($\pm$) indicates that the actual measure of an object can vary by being no more than a specified amount larger (plus) or no less than a specified amount smaller (minus). If you are asked to find the largest possible acceptable measure, *add* the specified amount. If you are asked to find the smallest possible acceptable measure, *subtract* the specified amount. If you are asked to find the *limits* or *limit dimensions*, find *both* the largest and smallest acceptable measures.

SELF-STUDY EXERCISES 3–8

1   Add; reduce answers to lowest terms and convert any improper fractions to whole or mixed numbers.

1. $\dfrac{5}{16} + \dfrac{1}{16}$

2. $\dfrac{1}{2} + \dfrac{1}{8} + \dfrac{3}{4}$

3. $\dfrac{1}{8} + \dfrac{1}{2}$

4. $\dfrac{3}{8} + \dfrac{5}{32} + \dfrac{1}{4}$

5. $\dfrac{5}{16} + \dfrac{1}{4}$

6. $\dfrac{15}{16} + \dfrac{1}{2}$

7. $\dfrac{3}{32} + \dfrac{5}{64}$

8. $\dfrac{7}{8} + \dfrac{3}{5}$

9. $\dfrac{3}{4} + \dfrac{8}{9}$

10. $\dfrac{7}{8} + \dfrac{5}{24}$

11. What is the thickness of a countertop made of $\frac{7}{8}$-in. plywood and $\frac{1}{16}$-in. Formica?

12. Three pieces of steel are joined together. What is the total thickness if the pieces are $\frac{1}{2}$ in., $\frac{7}{16}$ in., and $\frac{29}{32}$ in.?

13. Three books are placed side by side. They are $\frac{5}{16}$ in., $\frac{7}{8}$ in., and $\frac{3}{4}$ in. wide. What is the total width of the books if they are polywrapped in one package?

14. Find the outside diameter of a pipe (see Fig. 3–18) whose wall is $\frac{1}{2}$ in. thick if its inside diameter is $\frac{7}{8}$ in.

15. What length bolt is needed to fasten two pieces of metal each $\frac{7}{16}$ in. thick if a $\frac{1}{8}$-in. lockwasher is used and a $\frac{1}{4}$-in. nut is used?

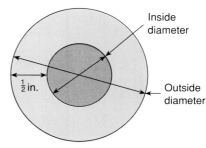

Figure 3–18

---

[2] Add; reduce answers to lowest terms and convert any improper fractions to whole or mixed numbers.

16. $2\dfrac{3}{5} + 4\dfrac{1}{5}$

17. $1\dfrac{5}{8} + 2\dfrac{1}{2}$

18. $3\dfrac{3}{4} + 7\dfrac{3}{16} + 5\dfrac{7}{8}$

19. $\dfrac{1}{6} + \dfrac{7}{9} + \dfrac{2}{3}$

20. $2\dfrac{1}{4} + 2\dfrac{9}{16}$

21. $1\dfrac{5}{16} + 4\dfrac{7}{32}$

22. $3\dfrac{1}{4} + 1\dfrac{7}{16}$

23. $4\dfrac{1}{2} + 9$

24. $\dfrac{2}{3} + 3\dfrac{4}{5}$

25. $5\dfrac{1}{8} + 3 + 4\dfrac{9}{16}$

26. The studs of an outside wall are $5\frac{3}{4}$ in. thick. The inside wall board is $\frac{7}{8}$ in. thick and the outside covering is $2\frac{3}{16}$ in. thick. What is the total thickness of the wall?

27. A blueprint calls for a piece of bar stock $3\frac{7}{8}$ in. long. If a tolerance of $\pm\frac{1}{16}$ in. is allowed, what is the longest permissible measurement for the bar stock?

28. If $4\frac{3}{8}$ gal of water are used to dilute $7\frac{1}{4}$ gal of acid, how many gallons are in the mixture?

29. How much bar stock is needed to make bars of the following lengths: $10\frac{1}{4}$ in., $8\frac{7}{16}$ in., $5\frac{15}{32}$ in.? Disregard waste.

30. Three pieces of I-beam each measuring $7\frac{5}{8}$ in. are needed to complete a job. How much I-beam is needed?

3–9 SUBTRACTING FRACTIONS AND MIXED NUMBERS

Learning Objectives

[1] Subtract fractions.

[2] Subtract mixed numbers.

[1] Subtract Fractions.

The steps for subtracting fractions and mixed numbers are very similar to the steps for adding fractions and mixed numbers.

Rule 3–17 *To subtract fractions:*

1. If the denominators are not the same, find the least common denominator.
2. Change each fraction not expressed in terms of the common denominator to an equivalent fraction having the common denominator.
3. Subtract the numerators.
4. The common denominator will be the denominator of the difference.
5. Reduce the answer to lowest terms.

Fractions must have the same denominator before they can be subtracted. Chapter 1 emphasized a very important contrast between addition and subtraction. In addition, the order of the terms (addends) does not matter; that is, $3 + 4$ is the same as $4 + 3$. In subtraction, however, the order of the terms is very important because subtraction is not commutative. We must be careful when arranging a subtraction problem.

EXAMPLE 1   Subtract $\dfrac{3}{8} - \dfrac{7}{32}$.

$$\dfrac{3}{8} = \dfrac{12}{32} \qquad \text{Change } \tfrac{3}{8} \text{ to an equivalent fraction with a denominator of 32.}$$

$$-\dfrac{7}{32} = \dfrac{7}{32} \qquad \text{Subtract numerators and keep the common denominator.}$$

$$\mathbf{\dfrac{5}{32}}$$

2 Subtract Mixed Numbers.

As in addition, there is more than one method for subtracting mixed numbers. Because the method that requires changing each mixed number to an improper fraction is generally not the most practical method, the examples below use the method that considers the whole numbers and fractional parts separately.

EXAMPLE 2   Subtract $15\dfrac{7}{8} - 4\dfrac{1}{2}$.

$$15\dfrac{7}{8} = 15\dfrac{7}{8}$$

$$-\ 4\dfrac{1}{2} = 4\dfrac{4}{8}$$

$$\mathbf{11\dfrac{3}{8}}$$

An additional procedure is used when the fractional part of the subtrahend (number being subtracted) is larger than the fractional part of the minuend (first number). This causes us to have to borrow from the whole-number part of the minuend.

> **Rule 3-18** *To borrow when subtracting mixed numbers:*
>
> 1. If the fractional parts of the mixed numbers do not have the same denominator, change them to equivalent fractions having a common denominator.
> 2. When the fraction in the subtrahend is larger than the fraction in the minuend, borrow one whole number from the whole-number part of the minuend. That makes the whole number 1 less.
> 3. Change the one whole number borrowed to an improper fraction having the common denominator. For example, $1 = \frac{3}{3}$, $1 = \frac{8}{8}$, $1 = \frac{n}{n}$, where n is the common denominator.
> 4. Add the borrowed fraction ($\frac{n}{n}$) to the fraction already in the minuend.
> 5. Subtract the fractional parts and the whole-number parts.
> 6. Reduce the answer to lowest terms.

EXAMPLE 3 Subtract $15\frac{3}{4}$ from $18\frac{1}{2}$.

$$18\frac{1}{2} = 18\frac{2}{4} = 17\frac{4}{4} + \frac{2}{4} = 17\frac{6}{4} \qquad 18 - 1 = 17, \ 1 = \frac{4}{4}, \ \frac{4}{4} + \frac{2}{4} = \frac{6}{4}$$

$$-\ 15\frac{3}{4} = 15\frac{3}{4} = 15\frac{3}{4} \qquad\qquad = 15\frac{3}{4}$$

$$2\frac{3}{4}$$

EXAMPLE 4 A metal block weighing $127\frac{1}{2}$ lb is removed from a flatbed truck that carried a payload of $433\frac{3}{8}$ lb. How many pounds remain on the truck?

$$433\frac{3}{8} = 433\frac{3}{8} = 432\frac{11}{8} \qquad 433 - 1 = 432, \ 1 = \frac{8}{8}, \ \frac{8}{8} + \frac{3}{8} = \frac{11}{8}$$

$$-\ 127\frac{1}{2} = 127\frac{4}{8} = 127\frac{4}{8}$$

$$305\frac{7}{8}$$

$305\frac{7}{8}$ lb remain on the truck.

Tips and Traps *Think of Whole Numbers in Mixed Number Form Before Subtracting:*
As in addition, when subtracting whole numbers and mixed numbers, consider the whole number to have zero fractional parts. Then follow the same procedures as before. Borrow when necessary.

EXAMPLE 5 Subtract 27 from $45\frac{1}{3}$.

$$45\frac{1}{3} = 45\frac{1}{3}$$

$$-\ 27 = 27\frac{0}{3}$$

$$18\frac{1}{3}$$

EXAMPLE 6 How many feet of wire are left on a 100-ft roll if $27\frac{1}{4}$ ft are used from the roll?

$$100 \quad = 99\frac{4}{4}$$
$$- \quad 27\frac{1}{4} = 27\frac{1}{4}$$
$$\overline{\qquad\qquad\quad 72\frac{3}{4}}$$

There are $72\frac{3}{4}$ ft left on the roll.

EXAMPLE 7 Three lengths measuring $5\frac{1}{4}$ in., $7\frac{3}{8}$ in., and $6\frac{1}{2}$ in. are cut from a 64-in. bar of angle iron. If $\frac{3}{16}$ in. is wasted on each cut, how many inches of angle iron remain?

Visualize the problem by making a sketch (see Fig. 3–19). Then find the total amount of angle iron used. This includes the three lengths and the waste for three cuts.

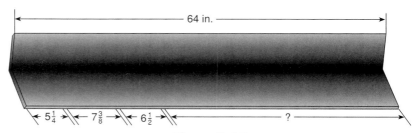

Figure 3–19

$$5\frac{1}{4} + 7\frac{3}{8} + 6\frac{1}{2} + \frac{3}{16} + \frac{3}{16} + \frac{3}{16} =$$

$$5\frac{4}{16} + 7\frac{6}{16} + 6\frac{8}{16} + \frac{3}{16} + \frac{3}{16} + \frac{3}{16} = 18\frac{27}{16}$$

$$18\frac{27}{16} = 18 + \frac{27}{16} = 18 + 1\frac{11}{16} = 19\frac{11}{16} \qquad \text{The total amount removed is } 19\frac{11}{16} \text{ inches.}$$

Next we will subtract to find the amount of angle iron remaining.

$$64 - 19\frac{11}{16} =$$

$$63\frac{16}{16} - 19\frac{11}{16} = 44\frac{5}{16}$$

There are $44\frac{5}{16}$ in. of angle iron remaining.

SELF-STUDY EXERCISES 3–9

1 Subtract; reduce when necessary.

1. $\dfrac{7}{8} - \dfrac{5}{8}$

2. $\dfrac{9}{16} - \dfrac{3}{8}$

3. $\dfrac{7}{16} - \dfrac{3}{8}$

4. $\dfrac{5}{8} - \dfrac{1}{2}$

5. $\dfrac{5}{32} - \dfrac{1}{64}$

6. $\dfrac{7}{8} - \dfrac{3}{4}$

2

7. $9\dfrac{11}{16} - 5$

8. $23\dfrac{3}{16} - 5\dfrac{7}{16}$

9. $9\dfrac{1}{4} - 4\dfrac{5}{16}$

10. $9\frac{1}{32} - 3\frac{3}{8}$

11. A length of bar stock $16\frac{3}{8}$ in. long is cut so that a piece only $7\frac{9}{16}$ in. long remains. What is the length of the cutoff piece? Disregard waste.

12. A concrete foundation includes $7\frac{7}{8}$ in. of base fill. If the foundation is to be 18 in. thick, how thick must the concrete be?

13. A casting is machined so that $22\frac{1}{5}$ lb of metal remain. If the casting weighed $25\frac{3}{10}$ lb, how many pounds were removed by machine?

14. A bolt $2\frac{5}{8}$ in. long fastens two pieces of wood 1 in. and $1\frac{7}{32}$ in. thick. If a $\frac{3}{32}$-in.-thick lockwasher and a $\frac{1}{8}$-in.-thick washer are used, what thickness is the nut if it is flush with the bolt? When giving the measure of a bolt, the length does not include the bolt head.

15. Find the missing length in Fig. 3–20.

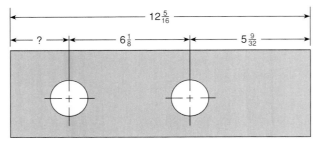

Figure 3–20

3-10   MULTIPLYING FRACTIONS AND MIXED NUMBERS

Learning Objectives

1   Multiply fractions.

2   Multiply mixed numbers.

We saw in Chapter 1 that the multiplication of whole numbers is a shortcut for addition and therefore a time saver on the job. For example, we saw that multiplying 2×4 to get 8 was quicker and more efficient than adding $2 + 2 + 2 + 2$ to get the same number 8. When multiplying a whole number by a whole number, we are increasing a given number of units by a certain number of times. Thus, in $2 \times 7 = 14$, we are increasing 2 by seven times. However, when multiplying a fraction by a fraction, we are doing something different.

When multiplying a fraction by a fraction, we are trying to find *a part of a part*. For instance, $\frac{1}{2} \times \frac{1}{2}$ is $\frac{1}{2}$ of $\frac{1}{2}$. The word "of" is the clue that we must multiply to find the part we are looking for. The examples in Fig. 3–21 illustrate what is meant by finding a part of a part.

$$\tfrac{1}{2} \times \tfrac{1}{2}$$

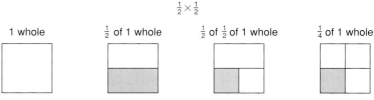

Figure 3–21

When adding or subtracting fractions and mixed numbers, it is necessary to have a common denominator. Otherwise, the fractional parts cannot be added or subtracted. However, this is not the case when multiplying. In multiplying fractions, we do *not* change fractions to equivalent fractions having a common denominator. Look at two more examples of taking a part of a part (Figs. 3–22 and 3–23).

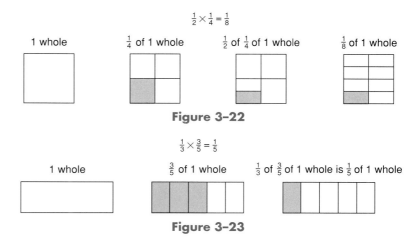

$$\tfrac{1}{2} \times \tfrac{1}{4} = \tfrac{1}{8}$$

1 whole $\tfrac{1}{4}$ of 1 whole $\tfrac{1}{2}$ of $\tfrac{1}{4}$ of 1 whole $\tfrac{1}{8}$ of 1 whole

Figure 3–22

$$\tfrac{1}{3} \times \tfrac{3}{5} = \tfrac{1}{5}$$

1 whole $\tfrac{3}{5}$ of 1 whole $\tfrac{1}{3}$ of $\tfrac{3}{5}$ of 1 whole is $\tfrac{1}{5}$ of 1 whole

Figure 3–23

1   Multiply Fractions.

Figures 3–21, 3–22, and 3–23 show graphically how we can multiply a fraction by a fraction, that is, take a part of a part. On the job, however, we may not want to stop and draw pictures or figures. Instead, we need rules to follow so that we can make our computations quickly.

Rule 3–19   *To multiply fractions:*

1. Multiply the numerators of the fractions for the numerator of the product.
2. Multiply the denominators for the denominator of the product.
3. Reduce the product to lowest terms.

EXAMPLE 1   Find $\dfrac{1}{2}$ of $\dfrac{1}{4}$.

$$\frac{1}{2} \times \frac{1}{4} = \mathbf{\frac{1}{8}}$$

EXAMPLE 2   Find $\dfrac{1}{3}$ of $\dfrac{3}{5}$.

$$\frac{1}{3} \times \frac{3}{5} = \frac{3}{15} = \mathbf{\frac{1}{5}}$$

Tips and Traps   *Reduce or Cancel Before Multiplying:*

In Example 2, reducing was necessary. When multiplying fractions, it is desirable to reduce common factors before they are multiplied.

$$\frac{1}{3} \times \frac{3}{5} = \frac{1 \times \overset{1}{\cancel{3}}}{\underset{1}{\cancel{3}} \times 5} = \frac{1}{5}$$

Because both a numerator and a denominator have a common factor of 3, the common factor can be reduced before multiplying. Remember from Section 3–5 that reducing applies the principles $\frac{n}{n} = 1$ and $1 \times n = n$. This process is also referred to as *canceling*.

EXAMPLE 3 Find $\dfrac{3}{9}$ of $\dfrac{2}{7}$.

$$\dfrac{\overset{1}{\cancel{3}}}{\underset{3}{\cancel{9}}} \times \dfrac{2}{7} = \dfrac{\mathbf{2}}{\mathbf{21}}$$ 3 is a common factor of both a numerator and a denominator. Reduce before multiplying.

EXAMPLE 4 Multiply $\dfrac{2}{3} \times \dfrac{5}{9} \times \dfrac{1}{6}$.

$$\dfrac{\overset{1}{\cancel{2}}}{3} \times \dfrac{5}{9} \times \dfrac{1}{\underset{3}{\cancel{6}}} = \dfrac{\mathbf{5}}{\mathbf{81}}$$ 2 is a common factor of both a numerator and a denominator. Reduce before multiplying.

Tips and Traps *Cancel from Any Numerator to Any Denominator:*
Common factors that are reduced can be diagonal to each other, one above the other, or separated by another fraction, but one factor *must* be in the numerator and the other in the denominator.

EXAMPLE 5 Multiply $\dfrac{2}{5} \times \dfrac{10}{21} \times \dfrac{6}{12}$.

$$\dfrac{\overset{1}{\cancel{2}}}{\underset{1}{\cancel{5}}} \times \dfrac{\overset{2}{\cancel{10}}}{21} \times \dfrac{\overset{1}{\cancel{6}}}{\underset{\underset{1}{2}}{\cancel{12}}} = \dfrac{\mathbf{2}}{\mathbf{21}}$$ 5 and 10 are diagonal to each other. 6 and 12 are above each other. 2 and 2 are separated by another fraction.

Other patterns of reducing could have been done.

2 Multiply Mixed Numbers.

Multiplying mixed numbers or a combination of whole numbers, fractions, and mixed numbers is based on the same rules as for multiplying fractions. When multiplying mixed numbers or any combination of whole numbers, fractions, and mixed numbers, we first change each whole number or mixed number to an improper fraction. Then we can proceed as in multiplying fractions.

Rules 3–20 *To multiply mixed numbers, fractions, and whole numbers:*

1. Change each mixed number or whole number to an improper fraction.
2. Reduce as much as possible.
3. Multiply numerators.
4. Multiply denominators.
5. Convert the answer to a whole or mixed number if possible.

EXAMPLE 6 Multiply $2\dfrac{1}{2} \times 5\dfrac{1}{3}$.

$$2\dfrac{1}{2} \times 5\dfrac{1}{3} = \dfrac{5}{2} \times \dfrac{16}{3} = \dfrac{5}{\underset{1}{\cancel{2}}} \times \dfrac{\overset{8}{\cancel{16}}}{3} = \dfrac{40}{3} = \mathbf{13\dfrac{1}{3}}$$

EXAMPLE 7 An engineering technician needed 10 pieces of wire each $3\frac{5}{8}$ in. long. What total length will be needed for the wire pieces?

$$3\frac{5}{8} \times 10 = \frac{29}{8} \times \frac{\overset{5}{\cancel{10}}}{1} = \frac{145}{4} = 36\frac{1}{4}$$

$36\frac{1}{4}$ in. of wire will be needed.

EXAMPLE 8 Bricks that are $2\frac{1}{4}$ in. thick are used to lay a brick wall with $\frac{3}{8}$-in. mortar joints. What will be the height of the wall after nine courses?

A *course* of bricks means a horizontal row (see Fig. 3–24). If nine horizontal rows are laid, how many mortar joints are included? There will be eight mortar joints between nine rows of bricks and one mortar joint between the first row of bricks and the foundation. The total height of the wall after nine courses of bricks will include nine rows of bricks at $2\frac{1}{4}$ in. per row and nine mortar joints at $\frac{3}{8}$ in. per mortar joint.

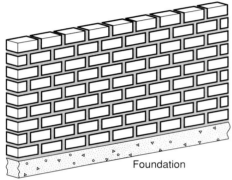

Foundation

Figure 3–24

$$\left(9 \times 2\frac{1}{4}\right) + \left(9 \times \frac{3}{8}\right)$$

$$\left(\frac{9}{1} \times \frac{9}{4}\right) + \left(\frac{9}{1} \times \frac{3}{8}\right)$$

$$\frac{81}{4} + \frac{27}{8}$$

$$20\frac{1}{4} + 3\frac{3}{8}$$

$$20\frac{2}{8} + 3\frac{3}{8} = 23\frac{5}{8}$$

The wall will be $23\frac{5}{8}$ in. high.

SELF-STUDY EXERCISES 3–10

1 Multiply and reduce answers to lowest terms.

1. $\frac{3}{4} \times \frac{1}{8}$ **2.** $\frac{1}{2} \times \frac{7}{16}$ **3.** $\frac{5}{8} \times \frac{7}{10}$

4. $\frac{2}{3} \times \frac{7}{8}$ **5.** $\frac{1}{2} \times \frac{3}{4} \times \frac{8}{9}$ **6.** $\frac{3}{8} \times \frac{5}{6} \times \frac{1}{2}$

7. $\frac{7}{8} \times \frac{2}{5} \times \frac{4}{21}$ **8.** $\frac{11}{12} \times \frac{9}{10} \times \frac{8}{15}$ **9.** $\frac{3}{10} \times \frac{4}{15}$

10. $\frac{15}{16} \times \frac{7}{10}$

2 Multiply and reduce answers to lowest terms. Convert improper fractions to whole or mixed numbers.

11. $7 \times 3\frac{1}{8}$ **12.** $\frac{3}{5} \times 125$ **13.** $2\frac{3}{4} \times 1\frac{1}{2}$

14. $9\frac{1}{2} \times 3\frac{4}{5}$ **15.** $\frac{1}{5} \times 7\frac{5}{8}$ **16.** $\frac{2}{3} \times 3\frac{1}{4}$

17. A fuel tank that holds 75 liters (L) of fuel is $\frac{1}{4}$ full. How many liters of fuel are in the tank?

18. If steps are 12 risers high and each riser is $7\frac{1}{2}$ in. high, what is the total rise of the steps?

19. A piece of sheet metal is $\frac{1}{16}$ in. thick. How thick would a stack of 250 pieces be?

20. An alloy, which is a substance composed of two or more metals, is $\frac{11}{16}$ copper, $\frac{7}{32}$ tin, and $\frac{3}{32}$ zinc. How many kilograms (kg) of each metal are needed to make 384 kg of alloy?

3-11 DIVIDING FRACTIONS AND MIXED NUMBERS

Learning Objectives

1. Find reciprocals.
2. Divide fractions.
3. Divide mixed numbers.
4. Simplify complex fractions.

Once we can multiply fractions, we can easily handle the division of fractions and mixed numbers. The reason for this is that multiplication and division are inverse operations.

Let's make a comparison. If six units are divided into two equal parts, how many units will be in each part? $6 \div 2 = ?$ The answer is 3. But at the same time, how many units are $\frac{1}{2}$ of 6 units? $\frac{1}{2} \times 6 = ?$ Again, the answer is 3.

In the two examples above, $6 \div 2$ and $\frac{1}{2} \times 6$ represent three units. To divide by two, then, is the same as taking half of a quantity.

1 Find Reciprocals.

If we compare $12 \div 3$ and $\frac{1}{3} \times 12$, we would find that both answers are the same, 4. This means that $12 \div 3 = \frac{1}{3} \times 12$ or $12 \times \frac{1}{3}$. So, not only is there a relationship between multiplication and division, there is also a relationship between 3 and $\frac{1}{3}$. Pairs of numbers like $\frac{1}{2}$ and 2 or $\frac{1}{3}$ and 3 are called *reciprocals*.

■ **DEFINITION 3–18: Reciprocals.** Two numbers are *reciprocals* if their product is 1. Thus, $\frac{1}{2}$ and 2 are reciprocals because $\frac{1}{2} \times 2 = 1$, and $\frac{2}{3}$ and $\frac{3}{2}$ are reciprocals because $\frac{2}{3} \times \frac{3}{2} = 1$.

> **Rule 3–21** *To find the reciprocal of a number:*
>
> 1. Write the number in fractional form.
> 2. Interchange the numerator and denominator so that the numerator becomes the denominator and the denominator becomes the numerator.

Tips and Traps *Reciprocals and Inverting:*
Interchanging the numerator and denominator of a fraction is commonly referred to as *inverting* the fraction.

EXAMPLE 1 Find the reciprocals of $\frac{2}{3}$, $\frac{4}{7}$, $\frac{1}{5}$, 3, 1, $2\frac{1}{2}$, 0.8, 3.5, and 0.

The reciprocal of $\frac{2}{3}$ is $\frac{3}{2}$.
The reciprocal of $\frac{4}{7}$ is $\frac{7}{4}$.
The reciprocal of $\frac{1}{5}$ is $\frac{5}{1}$ or 5.
The reciprocal of 3 is $\frac{1}{3}$. ($3 = \frac{3}{1}$)
The reciprocal of 1 is 1. ($1 = \frac{1}{1}$)
The reciprocal of $2\frac{1}{2}$ is $\frac{2}{5}$. ($2\frac{1}{2} = \frac{5}{2}$)
Write 0.8 as a fraction: $\frac{8}{10} = \frac{4}{5}$. The reciprocal is $\frac{5}{4}$ or 1.25.
Write 3.5 as an improper fraction: $3\frac{5}{10} = 3\frac{1}{2} = \frac{7}{2}$. The reciprocal is $\frac{2}{7}$ or 0.285714285.
0 has no reciprocal because division by 0 is impossible. $0 = \frac{0}{1}$, $\frac{1}{0}$ is undefined.

2 Divide Fractions.

Let's review the parts of a division problem.

$$15 \div 3 = 5$$

dividend   divisor   quotient

To help identify the divisor, remember that the symbol $\div$ is always read "divided by."

Rule 3–22 *To divide fractions:*

1. Change the division to an equivalent multiplication by replacing the divisor with its reciprocal and replacing the division sign ($\div$) with a multiplication sign ($\times$).
2. Perform the resulting multiplication.

EXAMPLE 2   Find $\dfrac{5}{8} \div \dfrac{2}{3}$.

$$\frac{5}{8} \div \frac{2}{3} = \frac{5}{8} \times \frac{3}{2} = \frac{15}{16}$$

Tips and Traps   *Invert and Multiply:*
A more common way of stating the division of fractions rule is "invert the divisor and multiply."

EXAMPLE 3   Divide $\dfrac{7}{8}$ by $\dfrac{14}{15}$.

$$\frac{7}{8} \div \frac{14}{15} = \frac{7}{8} \times \frac{15}{\overset{1}{\underset{2}{14}}} = \frac{15}{16}$$

EXAMPLE 4   An auger bit will advance $\frac{1}{16}$ in. for each turn (see Fig. 3–25). How many turns are needed to drill a hole $\frac{5}{8}$ in. deep? ($\frac{5}{8}$ in. can be divided into how many $\frac{1}{16}$-in. parts?)

$$\frac{5}{8} \div \frac{1}{16} = \frac{5}{\underset{1}{8}} \times \frac{\overset{2}{16}}{1} = 10$$

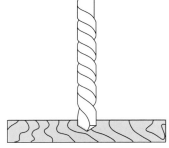

Figure 3–25

Ten turns are needed.

3 Divide Mixed Numbers.

Mixed numbers and whole numbers can be divided by first writing the mixed numbers or whole numbers as improper fractions and then following the rules for dividing fractions.

Rule 3–23 *To divide mixed numbers, fractions, and whole numbers:*

1. Change each mixed number or whole number to an improper fraction.
2. Convert to an equivalent multiplication problem by using the reciprocal of the divisor.
3. Multiply according to the rule for multiplying fractions.

EXAMPLE 5   Find $2\frac{1}{2} \div 3\frac{1}{3}$.

$$2\frac{1}{2} \div 3\frac{1}{3} = \frac{5}{2} \div \frac{10}{3} = \frac{\overset{1}{5}}{2} \times \frac{3}{\underset{2}{10}} = \frac{3}{4}$$

EXAMPLE 6   Find $5\frac{3}{8} \div 3$.

$$5\frac{3}{8} \div 3 = \frac{43}{8} \div \frac{3}{1} = \frac{43}{8} \times \frac{1}{3} = \frac{43}{24} = 1\frac{19}{24}$$

EXAMPLE 7   A developer subdivided $5\frac{1}{4}$ acres into lots, each containing $\frac{7}{10}$ of an acre. How many lots were made?

$$5\frac{1}{4} \div \frac{7}{10} = \frac{21}{4} \div \frac{7}{10} = \frac{\overset{3}{21}}{\underset{2}{4}} \times \frac{\overset{5}{10}}{\underset{1}{7}} = \frac{15}{2} = 7\frac{1}{2}$$

Seven lots can be made so that each is $\frac{7}{10}$ of an acre. The $\frac{1}{2}$ lot is left over.

EXAMPLE 8   How many pieces of cable, each $1\frac{3}{4}$ ft long, can be cut from a reel containing 100 ft of cable? See Fig. 3–26.

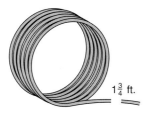

$1\frac{3}{4}$ ft.

Figure 3–26

$$100 \div 1\frac{3}{4} = \frac{100}{1} \div \frac{7}{4} = \frac{100}{1} \times \frac{4}{7} = \frac{400}{7} = 57\frac{1}{7}$$

Fifty-seven pieces of cable can be cut to the desired length. The extra $\frac{1}{7}$ of the desired length is considered waste.

EXAMPLE 9   A pipe that is $21\frac{1}{2}$ in. long is cut into four equal pieces. If $\frac{3}{16}$ in. is wasted on each cut, how long will each piece be?

Figure 3–27

How many cuts will need to be made? See Fig. 3–27. Three cuts are to be made and each cut wastes $\frac{3}{16}$ in., so first find the amount wasted.

$$\frac{3}{16} \times 3 = \frac{9}{16}$$

Next, find how much pipe will be left to divide equally into four parts.

$$21\frac{1}{2} - \frac{9}{16}$$

$$21\frac{1}{2} = 21\frac{8}{16} = 20\frac{24}{16}$$ Align vertically.

$$-\quad\frac{9}{16} = \quad\frac{9}{16} = \quad\frac{9}{16}$$

$$20\frac{15}{16} \text{ in.}$$

Now find the length of each part.

$$20\frac{15}{16} \div 4 = \frac{335}{16} \div \frac{4}{1} = \frac{335}{16} \times \frac{1}{4} = \frac{335}{64} = 5\frac{15}{64}$$

Each part will be $5\frac{15}{64}$ in. long.

4 Simplify Complex Fractions.

Recall from Section 3–1, a complex fraction is one in which either the numerator or the denominator or both contain a number other than a whole number. Examples of complex fractions are

$$\frac{\frac{3}{4}}{2}, \qquad \frac{6}{\frac{1}{2}}, \qquad \frac{\frac{2}{3}}{\frac{1}{5}}, \qquad \frac{2\frac{1}{2}}{7}, \qquad \frac{1\frac{2}{3}}{3\frac{5}{8}}, \qquad \frac{4}{1\frac{1}{2}}$$

Recall also that fractions indicate division as we read from top to bottom. The large fraction line is read as "divided by." In the complex fraction $\frac{2\frac{1}{2}}{7}$, for example, we would read "$2\frac{1}{2}$ divided by 7." The complex fraction $\frac{4}{1\frac{1}{2}}$ would be read "4 divided by $1\frac{1}{2}$." If we think of fractions as divisions, then complex fractions are merely another way of writing problems like the ones we worked in the previous three objectives.

> **Rule 3–24** *To simplify a complex fraction:*
>
> Perform the indicated division.

EXAMPLE 10 Simplify $\dfrac{6\frac{3}{8}}{4\frac{1}{2}}$.

$$\frac{6\frac{3}{8}}{4\frac{1}{2}} = 6\frac{3}{8} \div 4\frac{1}{2} = \frac{51}{8} \div \frac{9}{2} = \frac{\overset{17}{\cancel{51}}}{\underset{4}{\cancel{8}}} \times \frac{\overset{1}{\cancel{2}}}{\underset{3}{\cancel{9}}} = \frac{17}{12} = 1\frac{5}{12}$$

EXAMPLE 11   Simplify $\dfrac{\frac{4}{5}}{6}$.

$$\frac{\frac{4}{5}}{6} = \frac{4}{5} \div 6 = \frac{4}{5} \div \frac{6}{1} = \frac{\overset{2}{\cancel{4}}}{5} \times \frac{1}{\underset{3}{\cancel{6}}} = \frac{2}{15}$$

Complex fractions are sometimes needed to change mixed decimals to fractions. For example, $0.33\frac{1}{3}$ contains the fraction $\frac{1}{3}$. To convert these decimal numbers to fractions, we write them as complex fractions.

EXAMPLE 12   Write $0.33\frac{1}{3}$ as a fraction.

$$0.33\frac{1}{3} = \frac{33\frac{1}{3}}{100} = 33\frac{1}{3} \div 100$$

Count places for digits only; that is, do not count the fraction $\frac{1}{3}$ as a place.

Following our rule for dividing mixed numbers, write $33\frac{1}{3}$ as an improper fraction. Then invert the divisor 100 and multiply.

$$\frac{\overset{1}{\cancel{100}}}{3} \times \frac{1}{\underset{1}{\cancel{100}}} = \frac{1}{3}$$

SELF-STUDY EXERCISES 3–11

1 Give the reciprocal of the following:

1. $\dfrac{3}{7}$ 2. 8 3. $2\dfrac{1}{5}$ 4. $\dfrac{1}{5}$ 5. 7

2 Divide and reduce answers to lowest terms. Convert improper fractions to whole or mixed numbers.

6. $\dfrac{4}{5} \div \dfrac{8}{9}$ 7. $\dfrac{11}{32} \div \dfrac{3}{8}$ 8. $\dfrac{3}{4} \div \dfrac{3}{8}$

9. $\dfrac{7}{10} \div \dfrac{2}{3}$ 10. $\dfrac{5}{6} \div \dfrac{2}{5}$ 11. $\dfrac{7}{8} \div \dfrac{3}{16}$

12. $\dfrac{1}{5} \div \dfrac{1}{2}$

13. One-half inch is divided into $\frac{1}{16}$-in. segments. How many $\frac{1}{16}$-in. segments are in $\frac{1}{2}$ in.?

3 Divide and reduce answers to lowest terms. Convert improper fractions to whole or mixed numbers.

14. $10 \div \dfrac{3}{4}$ 15. $8 \div \dfrac{1}{4}$ 16. $2\dfrac{1}{2} \div 4$ 17. $7\dfrac{3}{8} \div \dfrac{1}{2}$ 18. $30\dfrac{1}{3} \div 4\dfrac{1}{3}$

19. Lumber is given in rough dimensions. Rough lumber that is 2 in. thick will dress out to $1\frac{5}{8}$ in. How many dressed 2 × 4's are in a stack that is $29\frac{1}{4}$ in. high?

20. How many $4\frac{5}{8}$-ft lengths can be cut from a 50-ft length of conduit?

21. A truck will hold 21 yd³ (cubic yards) of gravel. If an earth mover has a shovel capacity of $1\frac{3}{4}$ yd³, how many shovelfuls will be needed to fill the truck?

22. Three shelves of equal length are to be cut from a 72-in. board. If $\frac{1}{8}$ in. is wasted on each cut, what is the maximum length each shelf can be? (Two cuts will be made to divide the entire board into three equal lengths.)

23. If $\frac{1}{8}$ in. represents 1 ft on a drawing, find the dimensions of a room that measures $2\frac{1}{2}$ in. by $1\frac{7}{8}$ in. on the drawing. (How many $\frac{1}{8}$'s are there in $2\frac{1}{2}$; how many $\frac{1}{8}$'s are there in $1\frac{7}{8}$?)

24. A segment of I-beam is $10\frac{1}{2}$ ft long. Into how many whole $2\frac{1}{4}$-ft pieces can it be divided? Disregard waste.

25. How many $17\frac{5}{8}$-in. strips of quarter-round molding can be cut from a piece $132\frac{3}{4}$ in. long? Disregard waste.

26. How many $9\frac{1}{4}$-in. drinking straws can be cut from a $216\frac{1}{2}$-in. length of stock? How much stock will be left over?

27. It takes $12\frac{1}{2}$ minutes to sand a piece of oak stock. If Florence Randle worked 100 minutes, how many pieces did she sand?

4 Divide and reduce answers to lowest terms. Convert improper fractions to whole or mixed numbers.

28. $\dfrac{\frac{1}{2}}{\frac{2}{5}}$

29. $\dfrac{\frac{2}{2}}{\frac{2}{3}}$

30. $\dfrac{\frac{7}{1}}{1\frac{1}{4}}$

31. $\dfrac{2\frac{2}{5}}{5}$

32. $\dfrac{3\frac{1}{4}}{9\frac{3}{4}}$

33. $\dfrac{12\frac{1}{2}}{2\frac{1}{2}}$

34. $\dfrac{\frac{7}{1}}{\frac{1}{4}}$

35. $\dfrac{1\frac{1}{3}}{\frac{3}{6}}$

36. $\dfrac{\frac{10}{1}}{3\frac{1}{3}}$

37. $\dfrac{33\frac{1}{3}}{100}$

Write as a fraction.

38. $0.16\frac{2}{3}$

39. $0.83\frac{1}{3}$

40. $0.66\frac{2}{3}$

3–12 SIGNED FRACTIONS AND DECIMALS

Learning Objectives

1 Change a signed fraction to an equivalent signed fraction.

2 Perform basic operations with signed fractions.

3 Perform basic operations with signed decimals.

1 **Change a Signed Fraction to an Equivalent Signed Fraction.**

A fraction has three basic signs, the sign of the fraction, the sign of the numerator, and the sign of the denominator. The fraction $\frac{2}{3}$ expressed as a *signed fraction* is $+\frac{+2}{+3}$. When a signed fraction has negative signs, it is sometimes convenient to change the signed fraction to an equivalent signed fraction.

The rules for operating with integers can be extended to apply to signed fractions. Applying the rules of subtraction and division allows us to manipulate the signs of a fraction.

> **Rule 3–25** *To find an equivalent signed fraction:*
>
> Change any two of the three signs to the opposite sign.

EXAMPLE 1 Change $-\frac{-2}{-3}$ to three equivalent signed fractions.

$$-\frac{-2}{-3} = +\frac{+2}{-3}$$ Change the sign of the fraction and the sign of the numerator.

$$-\frac{-2}{-3} = +\frac{-2}{+3}$$ Change the sign of the fraction and the sign of the denominator.

$$-\frac{-2}{-3} = -\frac{+2}{+3}$$ Change the sign of the numerator and the sign of the denominator.

Changing the signs of a fraction is a manipulation tool that will allow us to simplify our work when performing basic operations with signed fractions.

2 Perform Basic Operations With Signed Fractions.

The same rules for performing the basic operations with integers can be used with signed fractions. In applying these rules, we will also change the signs of a fraction whenever it will give us a simpler problem to work.

EXAMPLE 2   Add $\dfrac{-3}{4} + \dfrac{5}{-8}$.

$$\frac{-3}{4} + \frac{-5}{8}$$ Change the signs of the numerator and denominator in the second fraction so that both denominators are positive.

$$\frac{-6}{8} + \frac{-5}{8}$$ Change to equivalent fractions with a common denominator.

$$\frac{-11}{8}$$ Add numerators, applying the rule for adding numbers with like signs.

$$-1\frac{3}{8}$$ Change to mixed number. The sign of the mixed number is determined by the rule for dividing numbers with unlike signs.

EXAMPLE 3   Subtract $\dfrac{-3}{7} - \dfrac{-5}{7}$.

$$\frac{-3}{7} + \frac{5}{7}$$ Change the signs of the second fraction by changing the sign of the fraction and the numerator.

$$\frac{2}{7}$$ Apply the rule for adding numbers with unlike signs.

Tips and Traps   *Manipulating Signs:*
Manipulating the signs of a fraction is used to simplify the steps of a problem. Generally, we use this tool to create an equivalent problem with fewer negative signs.

EXAMPLE 4   Multiply $\left(\dfrac{-4}{5}\right)\left(\dfrac{3}{-7}\right)$.

$$\frac{-4}{5} \cdot \frac{3}{-7}$$ Multiply numerators and denominators.

$$\frac{-12}{-35}$$ Apply rule for dividing numbers with like signs.

$$\frac{12}{35}$$

EXAMPLE 5   Simplify $\left(\dfrac{-2}{3}\right)^3$.

$\left(\dfrac{-2}{3}\right)^3$ Cube numerator and cube denominator.

$\dfrac{(-2)^3}{3^3}$ Apply rule for raising a negative number to an odd power.

$\dfrac{-8}{27}$ or $-\dfrac{8}{27}$

3 Perform Basic Operations with Signed Decimals.

The same rules for performing the basic operations with integers can be used with signed decimals.

EXAMPLE 6   Add -5.32 and -3.24.

$\begin{array}{r} -5.32 \\ -3.24 \\ \hline \mathbf{-8.56} \end{array}$ Align the decimals and use the rule for adding numbers with like signs.

EXAMPLE 7   Subtract -3.7 from 8.5.

$$8.5 - (-3.7) = 8.5 + 3.7 = \mathbf{12.2}$$

EXAMPLE 8   Multiply 3.91 and -7.1.

$\begin{array}{r} 3.91 \\ -\ 7.1 \\ \hline 391 \\ 27\ 37 \\ \hline \mathbf{-27.761} \end{array}$ Use the rule for multiplying numbers with unlike signs.

Or use the calculator.

Scientific Calculator:

$3 \boxed{\cdot} 9\ 1 \boxed{\times} 7 \boxed{\cdot} 1 \boxed{+/-} \boxed{=} \Rightarrow -27.761$

Graphics Calculator:

$3 \boxed{\cdot} 9\ 1 \boxed{\times} \boxed{-} 7 \boxed{\cdot} 1 \boxed{\text{EXE}} \Rightarrow -27.761$

EXAMPLE 9   Divide: $(-1.2) \div (-0.4)$.

Determine the sign of the quotient according to the rule for dividing signed numbers. Division of two negative numbers results in a positive quotient. Perform the division of decimals.

$0.4\overline{)1.2}$ with quotient $3.$ Shift decimal points.

SELF-STUDY EXERCISES 3–12

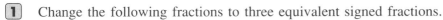

1 Change the following fractions to three equivalent signed fractions.

1. $+\dfrac{+5}{+8}$

2. $-\dfrac{3}{4}$

3. $\dfrac{-2}{-5}$

4. $-\dfrac{-7}{-8}$

5. $\dfrac{7}{8}$

2 Perform the indicated operations.

6. $\dfrac{-7}{8} + \dfrac{5}{8}$ 7. $\dfrac{-4}{5} + \dfrac{-3}{10}$ 8. $\dfrac{1}{2} - \dfrac{-3}{5}$ 9. $\dfrac{-3}{5} \times \dfrac{10}{-11}$ 10. $-\dfrac{5}{8} \div \dfrac{4}{5}$

3 Perform the indicated operations with signed decimals.

11. $5.823 - 32.12$
14. $34.6 \times (-3.2)$
17. $83.2 \div (-3)$
20. $4.23 - 4.2/(-1.2)$

12. $-8.32 + 7.21$
15. -7.2×8.2
18. $-0.826 \div -2$

13. $-84.23 - 7.21$
16. $-83.1 \times (-4.1)$
19. $-3.2 + 7.8(-3.2 + 0.2)$

3-13   CALCULATORS WITH FRACTION KEY

Learning Objective   **1** Use the fraction key on a calculator to perform operations with fractions.

1 Use the Fraction Key on a Calculator to Perform Operations with Fractions.

Some scientific and graphics calculators have a special key for making calculations with fractions. On these calculators, numbers can be entered as fractions and results can be displayed either in fraction or decimal form (or mixed number form for improper fractions). The *fraction key* is generally labeled ⌐a b/c⌐. The numerator and denominator are separated with a special symbol. The fraction $\frac{2}{3}$ appears as 2 ⌐ 3. The mixed number $3\frac{1}{5}$ appears as 3 ⌐ 1 ⌐ 5.

> **Rule 3–26**   *To reduce the fraction using a calculator:*
>
> To reduce the fraction, enter the numerator, press the fraction key, and then enter the denominator. To display the fraction in lowest terms, press the equals key (on the scientific calculator) or the EXE key (on the graphics calculator).

EXAMPLE 1   Reduce $\dfrac{30}{36}$.

Scientific: 30 a b/c 36 = ⇒ 5 ⌐ 6
Graphics: 30 a b/c 36 EXE ⇒ 5 ⌐ 6

Although we have already examined the process for finding the decimal equivalent of a fraction by dividing, we will now find decimal equivalents with the fraction key.

> **Rule 3–27**   *To find the decimal equivalent of a fraction with a calculator:*
>
> When a fraction or mixed number is displayed on the calculator and after the equal or EXE key has been pressed, press the fraction key.

EXAMPLE 2   Use the fraction key to show the decimal equivalent of $\frac{5}{6}$.

5 a b/c 6 = a b/c ⇒ 0.833333333   or EXE for =

Improper fractions can be found by accessing the function $\boxed{\text{d/c}}$ above the fraction key, which appears on the calculator as $\boxed{\text{a b/c}}^{\text{d/c}}$.

Tips and Traps   *Calculator Shift Keys:*
A special key is used to access calculator functions that are written above the keys. This key is labeled differently on various calculators. Frequently used labels are $\boxed{\text{Shift}}$ $\boxed{\text{INV}}$ and $\boxed{\text{2nd}}$. Many times, these keys and the labels above the keys are color coded to make the functions and the access key easier to locate.

EXAMPLE 3   Change $2\frac{7}{8}$ to an improper fraction using the improper fraction key.

$\boxed{2}$ $\boxed{\text{a b/c}}$ $\boxed{7}$ $\boxed{\text{a b/c}}$ $\boxed{8}$ $\boxed{\text{Shift}}$ $\boxed{\text{d/c}}$ Shift key may be labeled differently.

$\dfrac{23}{8}$

Tips and Traps   *Calculator Toggle Keys:*
Some keys on a calculator act as a toggle in certain situations. With $\frac{23}{8}$ in the display, press the fraction key several times. The display alternates among the improper fraction form, the decimal equivalent, and the mixed number. The $\boxed{\text{a b/c}}$ key is a toggle key in this instance.

The fraction key can also be used to add, subtract, multiply, or divide fractions.

Rule 3–28   *To perform calculations of fractions on the calculator:*
1. Enter the fraction or mixed number using the fraction key.
2. Use the operation keys as usual.

EXAMPLE 4   Simplify $\dfrac{3}{4} - \dfrac{2}{3} + 3\dfrac{1}{2}$.

$\boxed{3}$ $\boxed{\text{a b/c}}$ $\boxed{4}$ $\boxed{-}$ $\boxed{2}$ $\boxed{\text{a b/c}}$ $\boxed{3}$ $\boxed{+}$ $\boxed{3}$ $\boxed{\text{a b/c}}$ $\boxed{1}$ $\boxed{\text{a b/c}}$ $\boxed{2}$ $\boxed{=}$ or EXE

$3\dfrac{7}{12}$

SELF-STUDY EXERCISES 3–13

1   Use a calculator with a fraction key to perform the following operations.

1. $\dfrac{5}{8} + \dfrac{7}{12}$

2. $3\dfrac{1}{7} - 5\dfrac{3}{5}$

3. $2\dfrac{7}{8} \times 5\dfrac{1}{12}$

4. $\dfrac{3}{5} \div \dfrac{12}{35}$

5. $3\dfrac{1}{5} \div 4$

6. $\dfrac{-1}{4} + \dfrac{3}{4}$

7. $\dfrac{-5}{8} + \dfrac{7}{12}$ **8.** $15\dfrac{3}{5} \times 18\dfrac{1}{12}$ **9.** $3\dfrac{4}{5} + 7\dfrac{3}{8}$

10. $-\dfrac{7}{8} \div -\dfrac{2}{3}$

3-14 FINDING NUMBER AND PERCENT EQUIVALENTS

Learning Objectives

1. Change any number to its percent equivalent.
2. Change any percent to its numerical equivalent.

In our study of fractions and decimals, we have learned some very useful ways to express parts of quantities and to compare quantities. In this chapter we continue by standardizing our comparison of parts of quantities. We do this by expressing quantities in relation to a standard unit of 100. We can use this relationship, called a *percent,* to solve many different types of technical problems.

The word *percent* means "per hundred" or "for every hundred." 35 percent means 35 per hundred, or 35 out of every hundred, or $\frac{35}{100}$. 100 percent represents one hundred out of 100 parts, or $\frac{100}{100}$, or 1 whole quantity. The symbol % is used to represent "percent."

■ **DEFINITION 3–19:   Percent.**   A *percent* is a fractional part of one hundred expressed with a percent sign (%).

1 Change Any Number to Its Percent Equivalent.

It often happens on the job that we need to use percents, but the numbers we have are not expressed as percents. For example, suppose that an electronic parts supply house wants to keep its inventory of general-purpose soldering irons always at 1200. After a large order was filled, the inventory dropped to 800 soldering irons, and the parts manager ordered one-half of that amount to bring the inventory back up to 1200. We can express this $\frac{1}{2}$ as a percent by multiplying by 1 in the form of 100%.

$$\frac{1}{2} \times 100\% = \frac{1}{2} \times \frac{\overset{50}{\cancel{100}}\%}{\underset{1}{1}} = 50\%$$

> **Rule 3–29**   *To change any number to its percent equivalent:*
> Multiply by 1 in the form of 100%.

This rule can be applied to change any type of number—such as fractions, decimals, whole numbers, or mixed numbers—to a percent equivalent.

EXAMPLE 1   Change the following fractions to percent equivalents: $\frac{1}{4}$, $\frac{1}{3}$, $\frac{3}{8}$, $\frac{4}{7}$, $\frac{1}{200}$, and $\frac{3}{1000}$.

$$\frac{1}{4} \times 100\% = \frac{1}{4} \times \frac{\overset{25}{\cancel{100}}\%}{\underset{1}{1}} = \mathbf{25\%}$$

$$\frac{1}{3} \times 100\% = \frac{1}{3} \times \frac{100\%}{1} = \frac{100\%}{3} = \mathbf{33\dfrac{1}{3}\%}$$

$$\frac{3}{8} \times 100\% = \frac{3}{8} \times \frac{\overset{25}{\cancel{100\%}}}{1} = \frac{75\%}{2} = \mathbf{37\frac{1}{2}\%}$$

$$\frac{4}{7} \times 100\% = \frac{4}{7} \times \frac{100\%}{1} = \frac{400\%}{7} = \mathbf{57\frac{1}{7}\%}$$

$$\frac{1}{200} \times 100\% = \frac{1}{\underset{2}{\cancel{200}}} \times \frac{\overset{1}{\cancel{100\%}}}{1} = \mathbf{\frac{1}{2}\%}$$ $\frac{1}{2}\%$ means $\frac{1}{2}$ of every hundredth, or $\frac{1}{2}$ of 1%.

$$\frac{3}{1000} \times 100\% = \frac{3}{\underset{10}{\cancel{1000}}} \times \frac{\overset{1}{\cancel{100\%}}}{1} = \mathbf{\frac{3}{10}\%}$$ $\frac{3}{10}\%$ means $\frac{3}{10}$ of every hundredth, or $\frac{3}{10}$ of 1%.

EXAMPLE 2 Change the following decimals to percent equivalents: 0.3, 0.25, 0.07, 0.006, and 0.0025.

We will use the shortcut procedure to multiply by 100: move the decimal point two places to the right.

$$0.3 \times 100\% = 030.\% = \mathbf{30\%}$$
$$0.25 \times 100\% = 025.\% = \mathbf{25\%}$$
$$0.07 \times 100\% = 007.\% = \mathbf{7\%}$$
$$0.006 \times 100\% = 000.6\% = \mathbf{0.6\%}$$ 0.6% means 0.6 of every hundredth, or 0.6 of 1%.

$$0.0025 \times 100\% = 000.25\% = \mathbf{0.25\%}$$ 0.25% means 0.25 of every hundredth, or 0.25 of 1%.

EXAMPLE 3 Change the following whole numbers to their percent equivalents: 1, 3, 7.

$$1 \times 100\% = \mathbf{100\%}$$ 100 out of 100 or all of something

$$3 \times 100\% = \mathbf{300\%}$$

When we talk of more than 100%, we are talking about more than one whole quantity. Thus, 300% is three whole quantities, or three times a quantity.

$$7 \times 100\% = \mathbf{700\%}$$ 7 whole quantities, or 7 times a quantity

EXAMPLE 4 Change the following mixed numbers and decimals to their percent equivalents: $1\frac{1}{4}$, $3\frac{2}{3}$, 5.3, and 5.12.

$$1\frac{1}{4} \times 100\% = \frac{5}{4} \times \frac{\overset{25}{\cancel{100\%}}}{1} = \mathbf{125\%}$$

$$3\frac{2}{3} \times 100\% = \frac{11}{3} \times \frac{100\%}{1} = \frac{1100\%}{3} = \mathbf{366\frac{2}{3}\%}$$

$$5.3 \times 100\% = 530.\% = \mathbf{530\%}$$
$$5.12 \times 100\% = 512.\% = \mathbf{512\%}$$

2 Change Any Percent to Its Numerical Equivalent.

Percents are used as a convenient way of expressing the relationship of any quantity to 100. They are excellent time-savers in making comparisons or stating problems in the technologies. However, we cannot use percents as such when solving a problem. Instead, we need to first convert the percents to fractional, decimal, whole, or mixed-number equivalents.

Rule 3–30 *To change a percent to a numerical equivalent:*

Divide by 100%.

The numerical equivalent of a percent can be expressed in fractional or decimal form. In the following example, several percents are changed to both their fractional and decimal equivalents.

EXAMPLE 5 Change the following percents to their fractional and decimal equivalents: 75%, 38%, and 5%.

	Fractional equivalent	*Decimal equivalent*
75%	75% ÷ 100%	75% ÷ 100%

$$\frac{\overset{3}{\cancel{75}\%}}{1} \times \frac{1}{\underset{4}{\cancel{100}\%}} = \frac{3}{4} \qquad .75 = \mathbf{0.75}$$

Use the shortcut procedure for dividing by 100%: move the decimal point two places to the left.

38%	38% ÷ 100%	38% ÷ 100%

$$\frac{\overset{19}{\cancel{38}\%}}{1} \times \frac{1}{\underset{50}{\cancel{100}\%}} = \frac{19}{50} \qquad .38 = \mathbf{0.38}$$

5%	5% ÷ 100%	5% ÷ 100%

$$\frac{\overset{1}{\cancel{5}\%}}{1} \times \frac{1}{\underset{20}{\cancel{100}\%}} = \frac{1}{20} \qquad .05 = \mathbf{0.05}$$

Some percents will change more conveniently to one numerical equivalent than to another. In solving problems, we normally change the percent to the most convenient numerical equivalent for solving the problem. In the following examples, both the fractional and decimal equivalents of the percent will be given, and you can judge for yourself when fractional equivalents are more convenient than decimal equivalents, and vice versa.

EXAMPLE 6 Change the following percents to their fractional and decimal equivalents: $33\frac{1}{3}\%$, $37\frac{1}{2}\%$.

	Fractional equivalent	*Decimal equivalent*
	$33\frac{1}{3}\% \div 100\%$	$33\frac{1}{3}\% \div 100\%$

$$\frac{\overset{1}{\cancel{100}\%}}{3} \times \frac{1}{\underset{1}{\cancel{100}\%}} = \frac{1}{3} \qquad .33\frac{1}{3} = \mathbf{0.33\frac{1}{3}}$$

Remember, a decimal point separates whole quantities from fractional parts. Therefore, there is an *unwritten* decimal between 33 and $\frac{1}{3}$. Because $\frac{1}{3}$ will not change to a terminating decimal equivalent, using the decimal equivalent of $33\frac{1}{3}\%$ will create some difficult calculations.

$$\begin{array}{cc} \textit{Fractional} & \textit{Decimal} \\ \textit{equivalent} & \textit{equivalent} \end{array}$$

$$37\frac{1}{2}\% \div 100\% \qquad 37\frac{1}{2}\% \div 100\% = \mathbf{0.37\frac{1}{2}}$$

$$\frac{\overset{3}{\cancel{75}}}{2}\% \times \frac{1}{\cancel{100}\%}_{4} = \frac{3}{8}$$

Decimal equivalents are desirable when computations are made on a calculator. However, the decimal $0.37\frac{1}{2}$ would still not be adaptable to most calculators because of the fractional part that remains. For that reason, when a mixed-number percent is changed to decimal form, first change the fractional part of the mixed number to its decimal equivalent.

$$\frac{1}{2} = 0.5 \qquad 2\overline{)1.0}\;{}^{0.5}$$

Thus, $37\frac{1}{2}\% = 37.5\%$. Then, divide by 100%.

$$37.5\% \div 100\% = \mathbf{0.375}$$

In this form, the decimal equivalent may be used on a calculator.

EXAMPLE 7 Change 5.25% to its fractional and decimal equivalents.

$$\begin{array}{cc} \textit{Fractional} & \textit{Decimal} \\ \textit{equivalent} & \textit{equivalent} \end{array}$$

$$5.25\% = 5\frac{25}{100}\% = 5\frac{1}{4}\% \qquad 5.25\% = 5.25\% \div 100\% = \mathbf{0.0525}$$

$$5\frac{1}{4}\% = 5\frac{1}{4}\% \div 100\%$$

$$= \frac{21}{4}\% \times \frac{1}{100\%}$$

$$= \frac{\mathbf{21}}{\mathbf{400}}$$

EXAMPLE 8 Change the following percents to their fractional and decimal equivalents: $\frac{1}{2}\%$ and 0.25%.

$$\begin{array}{cc} \textit{Fractional} & \textit{Decimal} \\ \textit{equivalent} & \textit{equivalent} \end{array}$$

$$\frac{1}{2}\% = \frac{1}{2}\% \div 100\% = \frac{1}{2}\% \times \frac{1}{100\%} = \frac{\mathbf{1}}{\mathbf{200}} \qquad \frac{1}{2}\% = 0.5\% = 0.5\% \div 100\% = \mathbf{0.005}$$

$$0.25\% = \frac{25}{100}\% = \frac{1}{4}\% \qquad\qquad\qquad 0.25\% = 0.25\% \div 100\% = \mathbf{0.0025}$$

$$\frac{1}{4}\% = \frac{1}{4}\% \div 100\%$$

$$= \frac{1}{4}\% \times \frac{1}{100\%}$$

$$= \frac{\mathbf{1}}{\mathbf{400}}$$

When a quantity is 100% or more, the fractional and decimal equivalents will be equal to or more than the whole number 1. The numerical equivalents will be whole numbers, mixed numbers, or mixed decimals. Again, when solving problems, use the more convenient numerical equivalent.

EXAMPLE 9 Change the following percents to their whole-number equivalents or to both their mixed-number and mixed-decimal equivalents: 700%, 375%, $233\frac{1}{3}\%$, and $462\frac{1}{2}\%$.

$$700\% \div 100\% = \frac{\overset{7}{\cancel{700\%}}}{1} \times \frac{1}{\cancel{100\%}} = \mathbf{7}$$

or

$$700\% \div 100\% = \mathbf{7}$$

Mixed-number
equivalent

$$375\% \div 100\%$$

$$\frac{\overset{15}{\cancel{375\%}}}{1} \times \frac{1}{\cancel{100\%}} = \frac{15}{4} = \mathbf{3\frac{3}{4}}$$

Mixed-decimal
equivalent

$$375\% \div 100\% = \mathbf{3.75}$$

Mixed-number
equivalent

$$233\frac{1}{3}\% \div 100\% = \frac{\overset{7}{\cancel{700\%}}}{3} \times \frac{1}{\cancel{100\%}} = \frac{7}{3} = \mathbf{2\frac{1}{3}}$$

To allow more mental calculation, consider $233\frac{1}{3}\%$ to be $200\% + 33\frac{1}{3}\%$.

$$200\% = 2 \quad \text{and} \quad 33\frac{1}{3}\% = \frac{1}{3}$$

Thus, $200\% + 33\frac{1}{3}\% = 2 + \frac{1}{3} = \mathbf{2\frac{1}{3}}$.

Mixed-decimal
equivalent

$$233\frac{1}{3}\% \div 100\% = \mathbf{2.33\frac{1}{3}}$$

Mixed-number
equivalent

$$462\frac{1}{2}\% \div 100\% = \frac{\overset{37}{\cancel{925}}}{2} \times \frac{1}{\cancel{100}} = \frac{37}{8} = \mathbf{4\frac{5}{8}}$$

Again, it is easier to consider $462\frac{1}{2}\%$ to be $400\% + 62\frac{1}{2}\%$.

$$400\% = 4 \quad \text{and} \quad 62\frac{1}{2}\% \div 100\% = \frac{\overset{5}{\cancel{125}}}{2} \times \frac{1}{\cancel{100}} = \frac{5}{8}$$

Then, $462\frac{1}{2}\% = 4 + \frac{5}{8} = \mathbf{4\frac{5}{8}}$.

*Mixed-decimal
equivalent*

$$462\frac{1}{2}\% = 462.5\%$$

$$462.5\% \div 100\% = \mathbf{4.625}$$

Because many percents, fractions, and decimals are frequently used, it would be helpful to know these equivalents from memory. Knowing these will save much time on the job. Table 3–1 is a list of the most commonly used percents and equivalents. The equivalents were determined by the procedures used in Rules 3–29 and 3–30.

TABLE 3–1 Common Percent, Fraction, and Decimal Equivalents

Percent	Fraction	Decimal	Percent	Fraction	Decimal
10%	$\frac{1}{10}$	0.1	60%	$\frac{3}{5}$	0.6
20%	$\frac{1}{5}$	0.2	$66\frac{2}{3}\%$	$\frac{2}{3}$	$0.66\frac{2}{3}$ or 0.667^{a}
25%	$\frac{1}{4}$	0.25	70%	$\frac{7}{10}$	0.7
30%	$\frac{3}{10}$	0.3	75%	$\frac{3}{4}$	0.75
$33\frac{1}{3}\%$	$\frac{1}{3}$	$0.33\frac{1}{3}$ or 0.333^{a}	80%	$\frac{4}{5}$	0.8
40%	$\frac{2}{5}$	0.4	90%	$\frac{9}{10}$	0.9
50%	$\frac{1}{2}$	0.5	100%	$\frac{1}{1}$	1.0

[a]These decimals can be expressed with fractions or rounded decimals. Their fraction equivalents are exact amounts, and rounded decimals are approximate amounts.

To help us remember these equivalents, we will group them differently and suggest some mental calculations as memory aids.

$$50\% = \frac{1}{2} = 0.5$$

This is one of the easiest to learn because of its easy comparison to money. One dollar is 100 cents. One-half dollar is 50 cents. One-half dollar is $0.50 in dollar-and-cent notation.

$$25\% = \frac{1}{4} = 0.25$$

$$75\% = \frac{3}{4} = 0.75$$

Again, relating these to money, one-fourth of a dollar is 25 cents or $0.25, and three-fourths of a dollar is 75 cents or $0.75.

Ten percent and multiples of 10 percent are usually easy to remember.

$$10\% = \frac{1}{10} = 0.1 \qquad 60\% = \frac{6}{10} = \frac{3}{5} = 0.6$$

$$20\% = \frac{2}{10} = \frac{1}{5} = 0.2 \qquad 70\% = \frac{7}{10} = 0.7$$

$$30\% = \frac{3}{10} = 0.3 \qquad\qquad 80\% = \frac{8}{10} = \frac{4}{5} = 0.8$$

$$40\% = \frac{4}{10} = \frac{2}{5} = 0.4 \qquad\qquad 90\% = \frac{9}{10} = 0.9$$

$$50\% = \frac{5}{10} = \frac{1}{2} = 0.5 \qquad\qquad 100\% = \frac{1}{1} = 1$$

When 3 is divided into 100, the result is $33\frac{1}{3}$.

$$\begin{array}{r} 33\frac{1}{3} \\ 3\overline{)100} \end{array}$$

$$33\frac{1}{3}\% = \frac{1}{3} = 0.33\frac{1}{3}$$

Two times $33\frac{1}{3}$ is $66\frac{2}{3}$.

$$66\frac{2}{3}\% = \frac{2}{3} = 0.66\frac{2}{3}$$

SELF-STUDY EXERCISES 3–14

1

1. If $\frac{2}{5}$ of the electricians in a city are self-employed, what percent are self-employed?

2. If $\frac{7}{10}$ of the bricklayers in a city are male, what percent are male?

Change the following numbers to their percent equivalents.

3. $\frac{5}{8}$ **4.** $\frac{7}{9}$ **5.** $\frac{7}{1000}$ **6.** $\frac{1}{350}$ **7.** 0.2

8. 0.14   **9.** 0.007   **10.** 0.0125   **11.** 5   **12.** 8

13. $1\frac{1}{3}$ **14.** $3\frac{1}{2}$ **15.** $4\frac{3}{10}$ **16.** $2\frac{1}{5}$ **17.** 3.05

18. 7.2   **19.** 15.1   **20.** 36.25

2 Change the following to both fractional and decimal equivalents.

21. 36%   **22.** 45%   **23.** 20%

24. 75%   **25.** $6\frac{1}{4}\%$ **26.** 62.5%

27. $66\frac{2}{3}\%$ **28.** 0.6%   **29.** $\frac{1}{5}\%$

30. 0.05%   **31.** $8\frac{1}{3}\%$ **32.** 18.75%

Change to equivalent whole numbers.

33. 800%   **34.** 400%

Change to both mixed-number and mixed-decimal equivalents.

35. 250%   **36.** 425%   **37.** 176%

38. 380%   **39.** $137\frac{1}{2}\%$ **40.** 387.5%

Change to mixed-number equivalents.

41. $166\frac{2}{3}\%$ **42.** $316\frac{2}{3}\%$

Change to mixed-decimal equivalents.

43. 115.3% **44.** $212\frac{1}{2}\%$ **45.** $106\frac{1}{4}\%$

Fill in the missing percents, fractions, or decimals *from memory*. Work this exercise only after memorizing the equivalents in Table 3–1.

	Percent	*Fraction*	*Decimal*		*Percent*	*Fraction*	*Decimal*
46.	10%			**47.**		$\frac{1}{4}$	
48.			0.2	**49.**		$\frac{1}{3}$	
50.	50%			**51.**		$\frac{4}{5}$	
52.			0.75	**53.**	$66\frac{2}{3}\%$		
54.			1	**55.**		$\frac{3}{10}$	
56.	40%			**57.**			0.7
58.		$\frac{9}{10}$		**59.**	60%		
60.		$\frac{1}{1}$		**61.**			0.25
62.		$\frac{2}{3}$		**63.**	70%		
64.			0.5	**65.**		$\frac{3}{5}$	
66.			0.1	**67.**	20%		
68.			$0.33\frac{1}{3}$	**69.**			$0.66\frac{2}{3}$
70.		$\frac{3}{4}$		**71.**	30%		
72.			0.9	**73.**		$\frac{1}{5}$	
74.			0.6	**75.**	100%		

CAREER APPLICATION

Aerospace Technology: Asteroid and Comet Impacts on Earth

In January 1997 NASA tracked 115+ asteroids in resonance with Earth, 7,000+ numbered asteroids (well-defined orbits), and 15,000+ unnumbered asteroids. Most asteroids revolve around the sun in a more elliptical (oval) orbit than Earth's nearly circular orbit, which means some asteroids from the Asteroid Belt cross Earth's path. These Earth-crossing asteroids are called Apollo Asteroids. An asteroid is said to be "in resonance" with the Earth if it is periodically "close" to Earth. A collision could eventually result because Earth's gravitational pull would tug the asteroid even closer.

For obvious reasons, predicting when an asteroid or comet might hit the Earth is of great interest to NASA, the U.S. National Security Council, and the general public. Recent technological developments in computer simulation software, electronic CCD cameras (mounted onto telescopes), and electronic transmission of data have made it possible to detect asteroids more easily, define their orbits more quickly, and predict with reasonable confidence any close approaches to Earth. A close approach is generally regarded as 0.15 astronomical units (AU). One AU is the average distance of the earth from the sun, approximately 93,000,000 miles. Thus, a close approach would be approximately 0.15 × 93,000,000 or 13,950,000 miles.

You can mathematically predict how often an Apollo Asteroid will come close to Earth (and possibly collide with Earth) by finding the least common multiple (LCM) of their orbiting periods around the sun. For example, an asteroid named Toro in a highly elliptical orbit around the sun comes close periodically to Venus's almost circular orbit. Toro orbits the sun in about 584 days (approximately 1.6 years or ~1.6 years), and Venus orbits in about 224 days (~0.6152 years). The LCM of these orbiting periods is 2920 days (~8 years). In 2920 days, Toro orbits the sun 5 times while Venus orbits the sun 13 times. We say Toro is in "5 to 13 resonance" with Venus, and collision is possible. When the collision may occur depends on other factors such as their minimal distance apart in their current orbits, Toro's orbital inclination to the ecliptic plane, and other planets in resonance with Toro that might intersect it first.

Exercises

Use the information above to answer the following. Assume that 1 year = 365 days.

1. Toro is also in resonance with the Earth. Find the LCM of Toro's and Earth's orbiting periods (584 days and 365 days, respectively).
2. How many orbits will Toro and Earth each complete during this time? State the resonance of Toro to Earth.
3. Earth's collision with Toro is not expected for at least 1 million years because Toro's orbit is tipped slightly out of the ecliptic plane. Draw a diagram showing the orbits of Earth and Toro around the sun looking down on the ecliptic plane. Now draw another diagram showing the same information but from a side view of the ecliptic plane.
4. NASA's definition of a *close approach* is when an asteroid or comet comes within 13,950,000 miles (0.15 AU) of Earth. Comet Toutaki is predicted to be ~3,255,000 miles (0.035 AU) from Earth on December 30, 2000. It will have its next close approach to Earth in 2004, which means Comet Toutaki is in a 1 to 4 resonance with Earth. What is the LCM of Comet Toutaki's and Earth's orbiting periods in years? In days?
5. Do you have enough information to find Comet Toutaki's orbiting period around the sun? If so, what is its period in years? In days? Is this the only possible answer?
6. NASA's Near Earth Asteroid Rendezvous (NEAR) Mission will reach Eros, a 22-mi long asteroid, in January 1999 and orbit it for at least 6 months. Scientists hope to learn its dimensions, mass, density, spin, chemical composition, and geology. Eros orbits the sun 4 times in the time Earth orbits the sun 7 times, so Eros is in a 4 to 7 resonance with Earth. Find Eros' orbiting period in days and in years.
7. If NASA is unable to reach Eros in 1999, when is the next time Eros and Earth will be positioned about the same as in 1999?
8. Scientists have found gaps in the Asteroid Belt corresponding to missing asteroids whose mathematically calculated orbiting periods would have predicted them to be in resonance with Jupiter. Give one possible explanation of what happened to the missing asteroids.
9. If a missing asteroid's orbiting period was mathematically predicted to be ~3650 days, and Jupiter's orbiting period is ~12 years, find the LCM

that predicts how often the asteroid and Jupiter will be close to each other.

10. Use Exercise 9 to find the resonance of the asteroid to Jupiter.

Answers

1. LCM = 2920 days, or 8 yr.
2. Toro will complete 5 orbits, and Earth will complete 8 orbits. Toro is in a 5 to 8 resonance with Earth.
3. Top view: Side view:

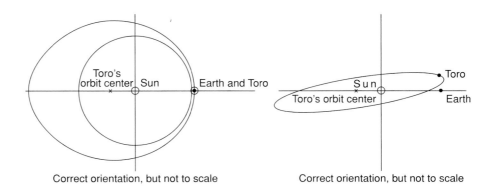

Correct orientation, but not to scale   Correct orientation, but not to scale

4. LCM = 4 yr = ~1460 days.
5. Yes. Comet Toutaki's orbiting period is 4 yr, or 1460 days. This is the only possible answer.
6. Eros' orbiting period is ~639 days, or ~1.75 yr.
7. The next time will be in 2008.
8. The missing asteroids were mathematically predicted to be in resonance with Jupiter, so they were frequently close to Jupiter. Jupiter's gravitational pull would have eventually sucked in the missing asteroids, leaving "negative rings," or gaps, in the Asteroid Belt.
9. 3650 days = 10 yr. The LCM of 10 years and 12 years is 60 years. So the asteroid and Jupiter would have been close every 60 years.
10. In 60 years the asteroid orbited the sun 6 times and Jupiter orbited the sun 5 times. So the asteroid was in a 5 to 6 resonance with Jupiter.

ASSIGNMENT EXERCISES

Section 3-1
Write a fraction to represent the shaded portion of Figs. 3–28 to 3–32.

1.

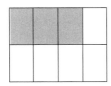

Figure 3–28

2.

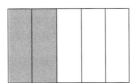

Figure 3–29

3.

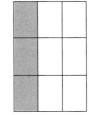

Figure 3–30

4.

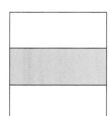

Figure 3–31

5.

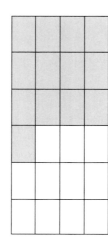

Figure 3–32

Answer the following questions for the numbers in Exercises 6 through 8.

6. $\frac{3}{7}$

7. $\frac{9}{4}$

8. $\frac{2}{2}$

(a) What is the numerator of the fraction?
(b) What is the denominator of the fraction?
(c) One unit has been divided into how many parts?
(d) How many of these parts are used?
(e) The fraction can be read as _____ divided by _____.
(f) What is the divisor of the division?
(g) What is the dividend of the division?
(h) Is this fraction proper or improper?
(i) Does this fraction have a value less than, more than, or equal to 1?
(j) This fraction represents _____ out of _____ parts.

Match the best description with each fraction given.

_____ **9.** $\frac{5}{2}$

_____ **10.** 0.172

_____ **11.** $6\frac{1}{7}$

_____ **12.** 4.59

_____ **13.** $\frac{7}{11}$

_____ **14.** $\frac{\frac{2}{3}}{\frac{1}{4}}$

_____ **15.** $33\frac{1}{3}$

(a) Mixed-decimal fraction
(b) Decimal fraction
(c) Complex fraction
(d) Mixed number
(e) Proper fraction
(f) Improper fraction

Rewrite the decimals below as proper fractions or mixed numbers.

16. 4.273   **17.** 0.87   **18.** 1.002
19. 2.03   **20.** 1.1

Section 3–2
Find five multiples of each of the following numbers.

21. 2   **22.** 9   **23.** 21   **24.** 14
25. 7   **26.** 11   **27.** 8   **28.** 15

Are the following numbers divisible by the given number? Explain.

29. 153 by 3
30. 8234 by 4
31. 8726 by 6
32. 5986 by 5

List all factor pairs for each number. Then write the factors in order from smallest to largest.

33. 48
34. 50
35. 51
36. 63

Section 3–3

Find the prime factorization of each number. Write in factored form and then in exponential notation.

37. 44
38. 128
39. 216
40. 98

Section 3–4

Find the least common multiple of the following numbers:

41. 18 and 40
42. 12 and 18
43. 12, 18, and 30
44. 6, 10, and 12

Find the greatest common factor of the following numbers:

45. 10 and 12
46. 12 and 18
47. 12, 18, and 30
48. 4, 9, and 16

Section 3–5

Find the equivalent fractions having the denominators indicated.

49. $\dfrac{5}{8} = \dfrac{?}{24}$
50. $\dfrac{3}{7} = \dfrac{?}{35}$
51. $\dfrac{5}{12} = \dfrac{?}{60}$

52. $\dfrac{4}{5} = \dfrac{?}{40}$
53. $\dfrac{2}{3} = \dfrac{?}{15}$
54. $\dfrac{4}{9} = \dfrac{?}{18}$

55. $\dfrac{3}{4} = \dfrac{?}{32}$
56. $\dfrac{1}{6} = \dfrac{?}{30}$
57. $\dfrac{1}{5} = \dfrac{?}{55}$

58. $\dfrac{7}{8} = \dfrac{?}{64}$

Reduce the following fractions to lowest terms.

59. $\dfrac{6}{12}$
60. $\dfrac{8}{10}$
61. $\dfrac{4}{32}$

62. $\dfrac{26}{64}$
63. $\dfrac{2}{8}$
64. $\dfrac{8}{32}$

65. $\dfrac{34}{64}$
66. $\dfrac{16}{64}$
67. $\dfrac{12}{32}$

68. $\dfrac{45}{90}$
69. $\dfrac{6}{8}$
70. $\dfrac{75}{100}$

Change the decimals to fractions in lowest terms.

71. 0.7
72. 0.83
73. 0.95
74. 0.25
75. 0.872
76. 0.081
77. 0.02
78. 0.005

Change the fractions to decimals. If necessary, round to the nearest thousandth.

79. $\dfrac{1}{5}$
80. $\dfrac{1}{10}$
81. $\dfrac{5}{8}$
82. $\dfrac{3}{7}$
83. $\dfrac{9}{11}$

Section 3–6

Write the following improper fractions as whole or mixed numbers.

84. $\dfrac{35}{7}$
85. $\dfrac{18}{5}$
86. $\dfrac{27}{6}$

87. $\dfrac{39}{8}$
88. $\dfrac{21}{15}$
89. $\dfrac{43}{8}$

90. $\dfrac{22}{7}$

91. $\dfrac{175}{2}$

92. $\dfrac{135}{3}$

Write the following whole or mixed numbers as improper fractions.

93. 8

94. $10\dfrac{1}{2}$

95. $7\dfrac{1}{8}$

96. $5\dfrac{7}{12}$

97. $9\dfrac{3}{16}$

98. $7\dfrac{8}{17}$

99. $4\dfrac{3}{5}$

100. $9\dfrac{1}{9}$

101. 12

102. $16\dfrac{2}{3}$

Change each of the following whole numbers to an equivalent fraction having the denominator indicated.

103. $2 = \dfrac{?}{10}$

104. $6 = \dfrac{?}{4}$

105. $11 = \dfrac{?}{3}$

106. $7 = \dfrac{?}{5}$

Section 3–7

Find the least common denominator for the fractions.

107. $\dfrac{1}{4}, \dfrac{1}{3}, \dfrac{1}{5}$

108. $\dfrac{7}{8}, \dfrac{2}{3}$

109. $\dfrac{3}{4}, \dfrac{1}{16}$

110. $\dfrac{1}{12}, \dfrac{3}{4}$

111. $\dfrac{5}{12}, \dfrac{3}{10}, \dfrac{13}{15}$

112. $\dfrac{1}{12}, \dfrac{3}{8}, \dfrac{15}{16}$

113. Is a $\frac{5}{8}$-in. wrench larger or smaller than a $\frac{9}{16}$-in. wrench?

114. An alloy contains $\frac{2}{3}$ metal A and the same quantity of another alloy contains $\frac{3}{5}$ metal A. Which alloy contains more metal A?

115. Is a $\frac{3}{8}$-in.-thick piece of plaster board thicker than a $\frac{1}{2}$-in.-thick piece?

116. A $\frac{9}{16}$-in. tube must pass through a wall. Is a $\frac{3}{4}$-in.-diameter hole large enough?

117. Is a $\frac{19}{32}$-in. wrench larger or smaller than a $\frac{7}{8}$-in. bolt head?

Determine which fraction is smaller.

118. $\dfrac{3}{7}, \dfrac{1}{9}$

119. $\dfrac{3}{8}, \dfrac{4}{8}$

120. $\dfrac{5}{9}, \dfrac{4}{9}$

121. $\dfrac{1}{4}, \dfrac{3}{16}$

122. $\dfrac{5}{8}, \dfrac{11}{16}$

123. $\dfrac{7}{8}, \dfrac{27}{32}$

124. $\dfrac{7}{64}, \dfrac{1}{4}$

125. $\dfrac{1}{2}, \dfrac{9}{19}$

Section 3–8

Add; reduce sums to lowest terms and convert any improper fractions to mixed numbers or whole numbers.

126. $\dfrac{1}{8} + \dfrac{5}{16}$

127. $\dfrac{3}{16} + \dfrac{9}{64}$

128. $\dfrac{3}{14} + \dfrac{5}{7}$

129. $\dfrac{3}{5} + \dfrac{5}{6}$

130. $3\dfrac{7}{8} + 7 + 5\dfrac{1}{2}$

131. $2\dfrac{7}{16} + 6\dfrac{5}{32}$

132. $9\dfrac{7}{8} + 5\dfrac{3}{4}$

133. $3\dfrac{7}{8} + 5\dfrac{3}{16} + 1\dfrac{7}{32}$

134. $2\dfrac{1}{4} + 3\dfrac{7}{8}$

135. A pipe is cut into two pieces measuring $7\frac{5}{8}$ in. and $10\frac{7}{16}$ in. How long was the pipe before it was cut? Disregard waste.

136. Find the total thickness of a wall if the outside covering is $3\frac{7}{8}$ in. thick, the studs are $3\frac{7}{8}$ in., and the inside covering is $\frac{5}{16}$-in. paneling.

137. If $7\frac{5}{16}$ in. of a piece of square bar stock is turned so that it is cylindrical and $5\frac{9}{32}$ in. remains square, what is the total length of the bar stock?

138. Two castings weigh $27\frac{1}{2}$ lb and $20\frac{3}{4}$ lb. What is the total weight of the two castings?

139. Three metal rods $3\frac{1}{8}$ in., $5\frac{3}{32}$ in., and $7\frac{9}{16}$ in. were welded together end to end. How long is the welded rod?

140. In Fig. 3–33, what is the length of side A?

141. In Fig. 3–33, what is the length of side B?

142. A hollow-wall fastener has a grip range up to $\frac{3}{4}$ in. Is it long enough to fasten three sheets of metal $\frac{5}{16}$ in. thick, $\frac{3}{8}$ in. thick, and $\frac{1}{16}$ in. thick?

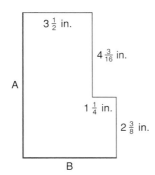

Figure 3–33

143. Figure 3–34 shows $\frac{1}{2}$-in. copper tubing wrapped in insulation. What is the distance across the tubing and insulation?

144. In Exercise 143, what would be the overall distance across the tubing and insulation if $\frac{3}{8}$-in.-ID tubing were used?

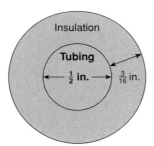

Figure 3–34

Section 3–9

Subtract; reduce to lowest terms when necessary.

145. $\dfrac{5}{9} - \dfrac{2}{9}$

146. $\dfrac{11}{32} - \dfrac{5}{64}$

147. $3\dfrac{5}{8} - 2$

148. $7 - 4\dfrac{3}{8}$

149. $8\dfrac{7}{8} - 2\dfrac{29}{32}$

150. $7 - 2\dfrac{9}{16}$

151. $12\dfrac{11}{16} - 5$

152. $48\dfrac{5}{12} - 12\dfrac{11}{15}$

153. A bolt 2 in. long fastens a piece of $\frac{7}{8}$-in.-thick wood to a piece of metal. If a $\frac{3}{16}$-in.-thick lock washer, a $\frac{1}{16}$-in. washer, and a $\frac{7}{16}$-in.-thick nut are used, what is the thickness of the metal if the nut is flush with the bolt after tightening?

154. Pins of $2\frac{3}{8}$ in. and $3\frac{7}{16}$ in. were cut from a drill rod 12 in. long. If $\frac{1}{16}$ in. of waste is allowed for each cut, how many inches of drill rod are left?

155. A piece of tapered stock has a diameter of $2\frac{5}{16}$ in. at one end and a diameter of $\frac{55}{64}$ in. at the other end. What is the difference in the diameters?

156. Four lengths measuring $6\frac{1}{4}$ in., $9\frac{3}{16}$ in., $7\frac{1}{8}$ in., and $5\frac{9}{32}$ in. are cut from 48 in. of copper tubing. How much copper tubing remains? Disregard waste.

Section 3–10

Multiply and reduce answers to lowest terms. Convert improper fractions to whole or mixed numbers.

157. $\dfrac{1}{3} \times \dfrac{7}{8}$ **158.** $\dfrac{2}{5} \times \dfrac{7}{10}$ **159.** $\dfrac{7}{9} \times \dfrac{3}{8}$

160. $\dfrac{2}{3} \times \dfrac{5}{8} \times \dfrac{3}{16}$ **161.** $\dfrac{15}{16} \times \dfrac{4}{5} \times \dfrac{2}{3}$ **162.** $5 \times \dfrac{3}{4}$

163. $\dfrac{7}{16} \times 18$ **164.** $\dfrac{3}{16} \times 184$ **165.** $1\dfrac{1}{2} \times \dfrac{4}{5}$

166. In a concrete mixture, $\frac{4}{7}$ of the total volume is sand. How much sand is needed for 135 cubic yards (yd^3) of concrete?

167. Concrete blocks are 8 in. high. If a $\frac{3}{8}$-in. mortar joint is used, how high will a wall of 12 courses of concrete blocks be? (*Hint:* There are 12 rows of mortar joints.)

168. An adjusting screw will move $\frac{3}{64}$ in. for each full turn. How far will it move in four turns?

169. A chef is making a dessert that is $\frac{3}{4}$ the original recipe. How much flour should be used if the original recipe calls for $3\frac{2}{3}$ cups of flour?

170. If an alloy is $\frac{3}{5}$ copper and $\frac{2}{5}$ zinc, how many pounds of each metal are in a casting weighing $112\frac{1}{2}$ lb?

Section 3–11

Divide and reduce answers to lowest terms. Convert improper fractions to whole or mixed numbers.

171. $\dfrac{7}{8} \div \dfrac{3}{4}$ **172.** $\dfrac{4}{9} \div \dfrac{5}{16}$ **173.** $\dfrac{7}{8} \div \dfrac{3}{32}$

174. $8 \div \dfrac{2}{3}$ **175.** $18 \div \dfrac{3}{4}$ **176.** $35 \div \dfrac{5}{16}$

177. $5\dfrac{1}{10} \div 2\dfrac{11}{20}$ **178.** $27\dfrac{2}{3} \div \dfrac{2}{3}$ **179.** $7\dfrac{1}{5} \div 12$

180. On a house plan, $\frac{1}{4}$ in. represents 1 ft. Find the dimensions of a porch that measures $4\frac{1}{8}$ in. by $6\frac{1}{2}$ in. on the plan. (How many $\frac{1}{4}$'s are there in $4\frac{1}{8}$; how many $\frac{1}{4}$'s are there in $6\frac{1}{2}$?)

181. A pipe that is 12 in. long is to be cut into four equal parts. If $\frac{3}{16}$ in. is wasted per cut, what is the maximum length that each pipe can be? (It will take three cuts to divide the entire length into four equal parts.)

182. A stack of $\frac{5}{8}$-in. plywood is $21\frac{7}{8}$ in. high. How many sheets of plywood are in the stack?

183. A rod $1\frac{1}{8}$ yd long is cut into 6 equal pieces. What is the length of each piece? Disregard waste.

184. If $7\frac{1}{2}$ gallons of liquid are distributed equally among five containers, what is the average number of gallons per container?

Divide and reduce answers to lowest terms. Convert improper fractions to whole or mixed numbers.

185. $\dfrac{\frac{1}{3}}{6}$ **186.** $\dfrac{\frac{4}{4}}{5}$ **187.** $\dfrac{8}{1\frac{1}{2}}$

188. $\dfrac{3\frac{1}{4}}{5}$ **189.** $\dfrac{2\frac{1}{5}}{8\frac{4}{5}}$ **190.** $\dfrac{16\frac{2}{3}}{3\frac{1}{3}}$

191. $\dfrac{12\frac{1}{2}}{100}$ **192.** $\dfrac{37\frac{1}{2}}{100}$

Section 3–12

Change each fraction to three equivalent signed fractions.

193. $-\dfrac{3}{8}$

194. $\dfrac{-5}{9}$

195. $\dfrac{-7}{-8}$

Perform the indicated operations.

196. $\dfrac{-7}{8} + \dfrac{-3}{8}$

197. $\dfrac{5}{9} - \dfrac{-3}{7}$

198. $\dfrac{-5}{8} * \dfrac{-2}{3}$

199. $\dfrac{-4}{5} \div -\dfrac{7}{15}$

200. $3.23 + (-4.61)$

201. 0.27×0.13

202. $-4.36 + (-7.23)$

203. $-12.4 \div 0.2$

Section 3–13

Use a calculator to perform the following operations.

204. $\dfrac{-11}{12} - \dfrac{-7}{8}$

205. $-\dfrac{7}{8} + \left(-\dfrac{5}{12}\right)$

206. $1\dfrac{3}{5} \div \left(-7\dfrac{5}{8}\right)$

207. $-2\dfrac{5}{8} \times 4\dfrac{1}{2}$

Section 3–14

Change to percent equivalents.

208. $\dfrac{87}{100}$

209. 0.7

210. $4\dfrac{1}{3}$

211. 125

212. $\dfrac{5}{6}$

213. 17.3

214. 18

Change to both decimal and fraction equivalents.

215. 72%

216. 40%

217. $12\dfrac{1}{2}\%$

218. $16\dfrac{2}{3}\%$

219. $\dfrac{2}{3}\%$

220. $\dfrac{3}{5}\%$

221. 275%

222. 124%

223. $112\dfrac{1}{2}\%$

224. $183\dfrac{1}{3}\%$

Change to decimal equivalents.

225. 227.2%

226. 73.8%

227. 9.275%

228. 275%

229. 340%

▰▰▰ CHALLENGE PROBLEM ▰▰▰

230. Len Smith has 180 ft of fencing and needs to build two square or rectangular holding yards for your two sheltie dogs. How should he design the two yards to get the largest area fenced in for each dog by using the 180 ft of wire?

▰▰▰ CONCEPTS ANALYSIS ▰▰▰

1. What two operations require a common denominator?

2. Explain how to find the reciprocal of a fraction.

3. What steps must be followed to find the reciprocal of a mixed number?

4. What number can be written as any fraction that has the same numerator and denominator?

5. What operation requires the use of the reciprocal of a fraction?

6. Name the operation that has each of the following for an answer: sum? difference? product? quotient?

7. What operation must be used to solve an applied problem if the total and one of the parts are given?

8. What does the denominator of a fraction indicate?

9. What does the numerator of a fraction indicate?

10. What kind of fraction has a value less than one?

Find, explain, and correct the mistakes in the following problems.

11. $\dfrac{5}{8} + \dfrac{1}{8} = \dfrac{6}{16} = \dfrac{3}{8}$

12.
$$\begin{array}{r} 12 \\ - \ 5\dfrac{3}{4} \\ \hline 7\dfrac{3}{4} \end{array}$$

13. $\dfrac{3}{5} \times 2\dfrac{1}{5} = 2\dfrac{3}{25}$

14. $\dfrac{5}{8} \div 4 = \dfrac{5}{8} \times \dfrac{4}{1} = \dfrac{5}{2} = 2\dfrac{1}{2}$

15.
$$\begin{array}{r} 12\dfrac{3}{4} = 12\dfrac{6}{8} = 11\dfrac{16}{8} \\ - \ 4\dfrac{7}{8} = 4\dfrac{7}{8} = 4\dfrac{7}{8} \\ \hline 7\dfrac{9}{8} = 7 + 1\dfrac{1}{8} = 8\dfrac{1}{8} \end{array}$$

16. $0.3\% = 0.3\% \times 100\%$
$ = 30$

CHAPTER SUMMARY

| Objectives | What to Remember with Examples |

Objectives

What to Remember with Examples

Section 3–1

1 Identify fraction terminology.

The numerator (top number) of a fraction represents the number of parts of the whole amount we are considering. The denominator (bottom number) of a fraction represents the number of parts a whole amount has been divided into. Proper fractions are less than 1. Improper fractions are equal to or larger than 1. A mixed number consists of a whole number and a common fraction written together and indicates addition of the whole number and fraction. A complex fraction has a fraction or mixed number in the numerator, or denominator, or both. The fraction line for complex fractions also indicates division. A decimal fraction is a number written with a decimal point to represent a fractional part whose denominator is 10 or a power of 10.

> Identify the following as proper fractions, improper fractions, or mixed numbers: $\frac{5}{37}, \frac{7}{7}, \frac{12}{5}, 8\frac{5}{9}.$ $\frac{5}{37}$, proper fraction; $\frac{7}{7}$, improper fraction; $\frac{12}{5}$, improper fraction; $8\frac{5}{9}$, mixed number.
>
> Write the following in decimal notation: $\frac{3}{10}, \frac{5}{100}, 4\frac{23}{100}.$ $\frac{3}{10} = 0.3, \frac{5}{100} = 0.05,$ $4\frac{23}{100} = 4.23$

Section 3–2

1 Find multiples of a natural number.

Multiply a number by a natural number to find a multiple.

> Find the first five multiples of 7: $7 \times 1 = 7, 7 \times 2 = 14, 7 \times 3 = 21,$ $7 \times 4 = 28, 7 \times 5 = 35$. The first five multiples of 7 are 7, 14, 21, 28, and 35.

2 Determine the divisibility of a number.

Apply the appropriate divisibility test or perform the division to see if the division has no remainder.

> Is 2195 divisible by 6? No, the number should be even and the sum of the digits should be divisible by 3.

3 Find all factor pairs of a natural number.

Start with the pair of factors using the number and 1. Examine other natural numbers to find other pairs of factors, if they exist. Continue until the factors begin to repeat but are in the opposite order.

> Find all the factors of 24: $1 \times 24, 2 \times 12, 3 \times 8, 4 \times 6$; 5 is not a factor, 6×4 is a repeat. Factors are 1, 2, 3, 4, 6, 8, 12, 24.

Section 3–3

1 Factor prime and composite numbers.

Identify the factors of the number. Numbers whose only factors are the number and 1 are prime. Other numbers are composite.

> Is 18 prime or composite? Factors of 18 are 1, 2, 3, 6, 9, 18; therefore, 18 is composite.

2 Determine the prime factorization of composite numbers.

Find the smallest prime factor of the composite number. Retain prime factors and continue factoring composite factors until all factors are prime.

> Find the prime factorization of 28:
> $$28 = 2 \times 14$$
> $$= 2 \times 2 \times 7 \text{ or } 2^2 \times 7$$

Section 3–4

1 Find the least common multiple of two or more numbers.

1. Find the prime factorization of each number in exponential notation.
2. The LCM includes each prime factor appearing in any of the numbers the largest number of times that it appears in any factor.
3. Write the result in standard notation.

> Find the LCM for 12, 15, and 30:
> $12 = 2^2 \times 3, 15 = 3 \times 5, 30 = 2 \times 3 \times 5$
> $\text{LCM} = 2^2 \times 3 \times 5 \text{ or } 60$

2 Find the greatest common factor of two or more numbers.

1. List the prime factorization of each number in exponential notation.
2. The GCF includes factors common to each number.
3. Write the result in standard notation.

> Find the GCF of 12, 15, 30:
> $12 = 2^2 \times 3, 15 = 3 \times 5, 30 = 2 \times 3 \times 5$
> $\text{GCF} = 3$

Section 3–5

1 Write equivalent fractions with higher denominators.

Equivalent fractions can be made by multiplying both the numerator and denominator by the same number.

> Write three fractions equivalent to $\frac{7}{8}$.
> $$\frac{7}{8} \times \frac{2}{2} = \frac{14}{16} \qquad \frac{7}{8} \times \frac{3}{3} = \frac{21}{24} \qquad \frac{7}{8} \times \frac{4}{4} = \frac{28}{32}$$

2 Write equivalent fractions with lowest denominators.

Fractions in which the numerator and denominator have a common factor can be reduced by dividing both the numerator and denominator by the greatest common factor.

> Reduce $\frac{12}{16}$.
> $$\frac{12}{16} = \frac{12}{16} \div \frac{4}{4} = \frac{3}{4}$$

| **3** Change decimals to fractions. | To change a decimal to a fraction (or mixed number) in lowest terms, write the number without the decimal as the numerator and make the denominator have the same number of zeros as the decimal has decimal places. |

Write 0.23 as a fraction:

$$\frac{23}{100}$$

| **4** Change fractions to decimals. | To change any fraction to a decimal, divide the denominator into the numerator by placing a decimal after the last digit of the numerator and affixing zeros as needed. |

Change $\frac{5}{8}$ to a decimal:

```
      .625
  8 ⟌ 5.000
      4 8
        20
        16
        40
        40
```

Section 3–6

| **1** Convert improper fractions to whole or mixed numbers. | To convert an improper fraction to a whole or mixed number, divide the numerator by the denominator. Write any remainder as a fraction having the original denominator as its denominator. |

Convert the following to whole or mixed numbers: $\frac{18}{6}, \frac{15}{4}$.

$$\frac{18}{6} = 3; \quad \frac{15}{4} = 3\frac{3}{4}$$

| **2** Convert mixed numbers and whole numbers to improper fractions. | To convert a mixed number to an improper fraction:
1. Multiply the whole number by the denominator.
2. Add the numerator to the result of Step 1.
3. Place the sum from Step 2 over the original denominator. |

Change $4\frac{7}{8}$ to an improper fraction.

$$4\frac{7}{8} = \frac{(8 \times 4) + 7}{8} = \frac{39}{8}$$

Section 3–7

| **1** Find common denominators. | To find the lowest common denominator, use the process shown in Section 3–3 for finding the least common multiple. |

Find the lowest common denominator for $\frac{7}{18}$ and $\frac{5}{24}$.

$18 = 2 \cdot 3 \cdot 3 = 2 \cdot 3^2$
$24 = 2 \cdot 2 \cdot 2 \cdot 3 = 2^3 \cdot 3$
$\text{LCD} = 2^3 \cdot 3^2 = 2 \cdot 2 \cdot 2 \cdot 3 \cdot 3 = 72$

The smallest number that can be divided evenly by both 18 and 24 is 72.

| **2** Compare fractions and decimals. | To compare fractions:
1. Write the fractions as equivalent fractions with common denominators.
2. Compare the numerators. The larger numerator indicates the larger fraction.

To compare mixed numbers: |

1. Compare the whole number parts, if different.
2. If the whole number parts are equal, write the fractions with common denominators.
3. Compare the numerators.

To compare decimals:
1. Compare the digits in the same place beginning from the left of each number.

Which fraction is smaller, $\frac{2}{5}$ or $\frac{5}{12}$?

$$\frac{2}{5} = \frac{24}{60}$$

$$\frac{5}{12} = \frac{25}{60}$$

Since $\frac{24}{60}$ is smaller than $\frac{25}{60}$, $\frac{2}{5}$ is smaller than $\frac{5}{12}$.

Which decimal is larger, 0.23 or 0.225? Both numbers have the same digit, 2, in the tenths place. 0.23 is larger because it has a 3 in the hundredths place, while 0.225 has a 2 in that place.

Section 3–8

1 Add fractions.

To add fractions, find the common denominator and convert each fraction to an equivalent fraction with the common denominator. Add the numerators and place the sum over the common denominator. Reduce the sum if possible.

Add: $\frac{1}{7} + \frac{3}{7} + \frac{2}{7} = \frac{6}{7}$

Add: $\frac{5}{8} + \frac{3}{4} = \frac{5}{8} + \frac{6}{8} = \frac{11}{8} = 1\frac{3}{8}$

2 Add mixed numbers.

To add mixed numbers:
1. Convert each fractional part to an equivalent fraction with the LCD.
2. Place the sum of the numerators over the LCD.
3. Add the whole number parts.
4. Write the improper fraction from Step 2 as a whole or mixed number; add the result to the whole number from Step 3.
5. Simplify if necessary.

Add $4\frac{3}{4} + 5\frac{2}{8} + 1\frac{1}{2}$.

$$4\frac{3}{4} = 4\frac{6}{8}$$

$$5\frac{2}{8} = 5\frac{2}{8}$$

$$1\frac{1}{2} = 1\frac{4}{8}$$

$$\overline{\qquad 10\frac{12}{8}}$$

$$\frac{12}{8} = 1\frac{4}{8} \text{ and } 10 + 1\frac{4}{8} = 11\frac{1}{2}$$

Section 3–9

1 Subtract fractions.

To subtract fractions:
1. Convert each fraction to an equivalent fraction that has the LCD as its denominator.
2. Subtract the numerators.
3. Place the difference over the LCD.
4. Reduce if possible.

Subtract $\frac{5}{8} - \frac{7}{16}$.

$$\frac{5}{8} = \frac{10}{16}$$

$$\frac{7}{16} = \frac{7}{16}$$

$$\frac{3}{16}$$

2 Subtract mixed numbers.

To subtract mixed numbers:
1. Write fractions as equivalent fractions with common denominators.
2. Borrow from the whole number and add to the fraction if the fraction in the minuend is smaller than the fraction in the subtrahend.
3. Subtract the fractions.
4. Subtract the whole numbers.
5. Simplify if necessary.

Subtract $5\frac{3}{8} - 3\frac{9}{16}$.

$$5\frac{3}{8} = 5\frac{6}{16} = 4\frac{22}{16}$$

$$3\frac{9}{16} = 3\frac{9}{16} = 3\frac{9}{16}$$

$$1\frac{13}{16}$$

Section 3–10

1 Multiply fractions.

To multiply fractions:
1. Reduce the numerator and denominator that have a common factor.
2. Multiply the numerators for the numerator of the product.
3. Multiply the denominators for the denominator of the product.
4. Be sure the product is reduced.

Multiply $\frac{4}{5} \times \frac{7}{10} \times \frac{15}{35}$.

$$\overset{2}{\underset{1}{\cancel{\frac{4}{5}}}} \times \overset{1}{\underset{5}{\cancel{\frac{7}{10}}}} \times \overset{3}{\underset{5}{\cancel{\frac{15}{35}}}} = \frac{6}{25}$$

2 Multiply mixed numbers.

To multiply fractions, whole numbers, and mixed numbers:
1. Write whole numbers as fractions with denominators of 1.
2. Write mixed numbers as improper fractions.
3. Reduce as much as possible.
4. Multiply the numerators for the numerator of the product.
5. Multiply the denominators for the denominator of the product.
6. Write the product as a whole number, mixed number, or fraction in lowest terms.

Multiply $4 \times 3\frac{1}{5} \times \frac{2}{7}$.

$$\frac{4}{1} \times \frac{16}{5} \times \frac{2}{7} = \frac{128}{35} = 3\frac{23}{35}$$

Section 3–11

1 Find reciprocals.

To write the reciprocal of a number, first express the number as a fraction. Then interchange the numerator and denominator.

Find the reciprocal of $\frac{3}{5}$, 6, $2\frac{3}{4}$.

The reciprocal of $\frac{3}{5}$ is $\frac{5}{3}$; the reciprocal of 6 is $\frac{1}{6}$; the reciprocal of $2\frac{3}{4}$ or $\frac{11}{4}$ is $\frac{4}{11}$.

2 Divide fractions.

To divide fractions, replace the divisor with its reciprocal and change the division to multiplication. Then multiply.

Divide $\frac{4}{5} \div \frac{8}{9}$.

$$\frac{4}{5} \div \frac{8}{9} = \frac{\overset{1}{4}}{5} \times \frac{9}{\underset{2}{8}} = \frac{9}{10}$$

3 Divide mixed numbers.

To divide mixed numbers:
1. Write each mixed number as an improper fraction.
2. Invert the divisor and change the division to multiplication.
3. Simplify if possible.
4. Multiply.

Divide $4\frac{2}{3} \div 1\frac{1}{6}$.

$$4\frac{2}{3} \div 1\frac{1}{6} = \frac{14}{3} \div \frac{7}{6} = \frac{\overset{2}{14}}{\underset{1}{3}} \times \frac{\overset{2}{6}}{\underset{1}{7}} = \frac{4}{1} \text{ or } 4.$$

4 Simplify complex fractions.

To simplify a complex fraction:
1. Write mixed numbers and whole numbers as improper fractions.
2. Write the division (as indicated by the fraction bar) as multiplication and invert the divisor.
3. Simplify if possible.
4. Multiply.
5. Convert the improper fraction to a whole or mixed number if necessary.

Simplify the following complex fraction.

$$\frac{4\frac{1}{2}}{3\frac{3}{5}} = \frac{\frac{9}{2}}{\frac{18}{5}} = \frac{9}{2} \div \frac{18}{5} = \frac{\overset{1}{9}}{2} \times \frac{5}{\underset{2}{18}} = \frac{5}{4} = 1\frac{1}{4}$$

Section 3–12

1 Change a signed fraction to an equivalent signed fraction.

If any two of the three signs of a signed fraction are changed, the value of the fraction is not changed.

Write three equivalent signed fractions for $-\frac{7}{8}$.

$$-\frac{7}{8} = -\frac{+7}{+8} = \frac{+7}{-8} \text{ or } -\frac{-7}{-8} \text{ or } \frac{-7}{+8}$$

2 Perform basic operations with signed fractions.

To add or subtract signed fractions, write equivalent fractions that have positive integers as denominators and have common denominators. Add or subtract the numerators using the rules for adding signed numbers. To multiply or divide signed fractions, multiply or divide the fractions, paying attention to the rules for multiplying or dividing signed numbers.

> Add $\frac{-5}{8} + \frac{7}{8}$.
>
> $$\frac{-5}{8} + \frac{7}{8} = \frac{2}{8} = \frac{1}{4}$$

3 Perform basic operations with signed decimals.

Use the same rules to add or subtract decimals that were used to perform basic operations with integers.

> Add $4.37 + (-2.91)$
>
> $4.37 - 2.91 = 1.46$

Section 3–13

1 Use the fraction key on a calculator to perform operations with fractions.

Calculators that have a fraction key usually identify it as $\boxed{\text{a b/c}}$. To enter fractions or mixed numbers, enter each part, followed by the fraction key. Use the four operation keys in the usual way to perform operations with fractions on the calculator.

> Multiply $\frac{3}{4} \times \frac{7}{8}$.
>
> 3 $\boxed{\text{a b/c}}$ 4 $\boxed{\times}$
>
> 7 $\boxed{\text{a b/c}}$ 8 $\boxed{=}$ $\Rightarrow$ 21 ⌐ 32

Section 3–14

1 Change any number to its percent equivalent.

Multiply a number by 100% to change a number to a percent.
For a shortcut, move the decimal two places to the right.

> Change $\frac{1}{2}$, 1.2, and 7 to percents.
>
> $\frac{1}{2} \times 100\% = \frac{100\%}{2} = 50\%$
>
> $1.2 \times 100\% = 120\%$
>
> $7 = 700\%$

2 Change any percent to its numerical equivalent.

Divide by 100% to change a percent to a number. Reduce if possible.
For a shortcut, move the decimal two places to the left.

> Change 7% to a fraction.
>
> $7\% \div 100\% = \dfrac{7}{100}$
>
> Change $1\frac{1}{4}\%$ to a fraction.
>
> $1\frac{1}{4}\% \div 100\% = \dfrac{5\%}{4} \div 100\% = \dfrac{5\%}{4} \times \dfrac{1}{100\%} = \dfrac{5}{400} = \dfrac{1}{80}$
>
> Convert 3.5% to a decimal.
>
> $3.5\% \div 100\% = 0.035 = 0.035$
>
> Convert 245% to a mixed number.
>
> $245\% \div 100\% = \dfrac{245\%}{100\%} = 2\dfrac{45}{100} = 2\dfrac{9}{20}$
>
> Convert 124.5% to a decimal.
>
> $124.5\% \div 100\% = 1.245 = 1.245$

Convert $100\frac{1}{4}\%$ to a decimal.

$$100\frac{1}{4}\% = 100.25\%$$

$$100.25\% \div 100\% = 1.0025$$

WORDS TO KNOW

fraction (p. 104)
common fraction (p. 105)
denominator (p. 105)
numerator (p. 105)
proper fraction (p. 105)
improper fraction (p. 105)
mixed number (p. 105)
complex fraction (p. 105)
decimal fraction (p. 106)
mixed-decimal fraction (p. 106)
rational number (p. 107)
multiple (p. 109)
even number (p. 109)

odd number (p. 109)
divisible (p. 109)
factor pair (p. 110)
prime number (p. 111)
composite number (p. 112)
prime factorization (p. 113)
prime factors (p. 113)
least common multiple (p. 114)
greatest common factor
    (p. 114)
equivalent fractions (p. 125)
fundamental principle of
    fractions (p. 118)

reduce to lowest terms (p. 118)
common denominator (p. 125)
least common denominator
    (LCD) (p. 125)
tolerance (p. 131)
canceling (p. 137)
reciprocals (p. 140)
inverting (p. 141)
signed fractions (p. 145)
fraction key (p. 148)
percent (p. 150)

CHAPTER TEST

Write a fraction to represent the following.

1. 3 out of 4 people in a survey

2. $7 \div 9$

Convert the following to mixed or whole numbers.

3. $\dfrac{9}{3}$

4. $\dfrac{14}{9}$

Convert the following to improper fractions.

5. $4\dfrac{6}{7}$

6. $3\dfrac{1}{10}$

Write the prime factors of the following.

7. 96

8. 132

Perform the following operations. When possible, simplify first. Make sure your answers are in lowest terms.

9. $\dfrac{5}{6} \times \dfrac{3}{10}$

10. $\dfrac{3}{7} \times \dfrac{2}{9}$

11. $2\dfrac{2}{9} \times 1\dfrac{3}{4}$

12. $7 \times \dfrac{1}{3}$

13. $7\dfrac{1}{2} \div \dfrac{5}{9}$

14. $\dfrac{4\frac{2}{3}}{2\frac{1}{2}}$

15. $\dfrac{7}{12} + \dfrac{5}{6}$

16. $\dfrac{5}{12} \div \dfrac{5}{6}$

17. $2\dfrac{3}{7} + 5 + \dfrac{1}{2}$

18. $\dfrac{3}{32} + 4 + 1\dfrac{3}{4}$

19. $\dfrac{7}{9} - \dfrac{2}{3}$

20. $6\dfrac{1}{4} - 2\dfrac{3}{4}$

21. $\dfrac{5\frac{2}{3}}{1\frac{1}{9}}$

Determine which is larger. Show your work.

22. $\dfrac{7}{8}, \dfrac{11}{12}$

23. $\dfrac{7}{32}, \dfrac{5}{16}$

Arrange the fractions in order, beginning with the smallest. Show your work.

24. $\dfrac{5}{7}, \dfrac{10}{21}, \dfrac{3}{4}$

Write the percent equivalent.

25. $\dfrac{3}{5}$

26. $\dfrac{5}{8}$

Solve the following problems.

27. Two of the seven security employees at the local community college received safety awards from the governor. Write a fraction to represent what part of the total number of employees received an award.

28. A candy-store owner mixed $1\frac{1}{2}$ pounds of caramels, $\frac{3}{4}$ pound of chocolates, and $\frac{1}{2}$ pound of candy corn. What was the total weight of the mixed candy?

29. A homemaker had $5\frac{1}{2}$ cups of sugar on hand to make a batch of cookies requiring $1\frac{2}{3}$ cup of sugar. How much sugar was left?

30. If $6\frac{1}{4}$ ft of wire is needed to make one electrical extension cord, how many extension cords can be made from $68\frac{3}{4}$ feet of wire?

31. A costume maker figures one costume requires $2\frac{2}{3}$ yards of red satin material. How many yards of red satin material would be needed to make three costumes?

32. Will a $\frac{5}{8}$-in.-wide drill bit make a hole wide enough to allow a $\frac{1}{2}$-in. (outside diameter) copper tube to pass through?

4

Problem Solving
with Percents

Your team of four to six members is employed in the same department of a local industry, and your supervisor has assigned you to distribute annual raises that must average 4% per department among the team members. No team member can get exactly 4%. And the raise must be at least 2% and no more than 6%. You may establish your own criteria for distributing the raises, but you are given the years of experience and the rating on annual

performance reviews for each member. A performance rating of 1 is the lowest rating possible and a rating of 5 is the highest. Each team member should assume the role of one of the following employees for purposes of this project: Employee 1 has 4 years experience, a performance rating of 4, and a salary of $28,500; Employee 2 has 3 years experience, a performance rating of 3, and a salary of $22,800; Employee 3 has 10 years experience, a performance rating of 4, and a salary of $32,700; Employee 4 has 7 years experience, a performance rating of 3, and a salary of $31,400; Employee 5 has 15 years experience, a performance rating of 3, and a salary of $34,600; Employee 6 has 12 years experience, a performance rating of 5, and a salary of $32,400.

As a team, decide the amount of increase each of you will recommend to your supervisor. Prepare a report for your supervisor that includes a justification for your recommendations; a table showing the original salary, the amount of increase, the new salary, and the percent of increase for each employee; and calculations to verify that the total amount of increases is exactly 4% of the total original salaries except for rounding discrepancies. Will the percents of change also average 4%? Why or why not?

Problems involving percents are one of the most common applications of mathematics in the workplace and in everyday life. Everyone needs to understand percents to be a well-informed employee or citizen.

4-1 PERCENTAGE PROPORTION

Learning Objectives

1. Solve a proportion for any missing element.
2. Identify the rate, base, and percentage in percent problems.
3. Solve the percentage proportion for any missing element.

One of the most practical methods for solving percent problems is to solve the percentage proportion. This will be the method that we use most often at this point in the text because it applies to numerous situations and it requires very few formula manipulation skills. As we increase our experience with manipulating formulas, we will examine some of the other methods.

1 Solve a Proportion for Any Missing Element.

The equivalent fractions $\frac{5}{10}$ and $\frac{1}{2}$ can be written as the proportion $\frac{5}{10} = \frac{1}{2}$. Each fraction is also called a *ratio,* and two ratios that are equal or equivalent form a *proportion.*

■ **DEFINITION 4-1: Ratio.** A *ratio* is a fraction comparing a quantity or measure in the numerator to a quantity or measure in the denominator.

■ **DEFINITION 4-2: Proportion.** A *proportion* is a mathematical statement that shows two fractions or ratios are equal.

A property of proportions that helps us in solving proportions with a missing element is that the *cross products* in a proportion are equal.

■ **DEFINITION 4-3: Cross Products.** In a proportion, the *cross products* are the product of the numerator of the first fraction times the denominator of the second, and the product of the denominator of the first fraction times the numerator of the second. In the proportion $\frac{a}{b} = \frac{c}{d}$, the cross products are $a \times d$ and $b \times c$.

> **Rule 4-1** *Property of Proportions:*
>
> The cross products in a proportion are equal. Symbolically,
>
> If $\frac{a}{b} = \frac{c}{d}$, then $a \times d = b \times c$, provided that b and d are not equal to zero.
>
> Also, if $a \times d = b \times c$, then $\frac{a}{b} = \frac{c}{d}$.

We can apply this property to show that $\frac{5}{10} = \frac{1}{2}$ is a true statement. The first cross product is $5 \times 2 = 10$. The second cross product is $10 \times 1 = 10$. The cross products are equal; thus, the fractions are equal.

This property can also be used to find a missing element when three of the four elements are known.

EXAMPLE 1 Find the value for a in $\dfrac{a}{12} = \dfrac{3}{9}$.

$$\frac{a}{12} = \frac{3}{9}$$

$$a \times 9 = 12 \times 3$$ Find the cross products.

$$a \times 9 = 36$$ When the product and one factor are known, divide the product by the known factor to find the missing factor.

$$a = \frac{36}{9}$$

$$a = 4$$

Thus, $\dfrac{4}{12} = \dfrac{3}{9}$.

2   Identify the Rate, Base, and Percentage in Percent Problems.

All problems involving percents will have three basic elements: the rate, the base, and the percentage. Knowing what each element is and how all three are related helps us solve problems with percents.

The *rate* (R) is the percent; the *base* (B) is the number that represents the original or total amount; the *percentage* (P) represents part of the base. In the statement, 50% of 80 is 40, the rate is 50%, the base is 80, and the percentage is 40.

EXAMPLE 2   Identify the given and missing elements for

(a) 20% of 75 is what number?
(b) What percent of 50 is 30?
(c) Eight is 10% of what number?

Use the identifying key words for rate (*percent* or %), base (*total, original,* associated with the word *of*), and percentage (*part,* associated with the word *is*).

      R B P
(a) 20% of 75 is what number?
    percent total   part
         R B P
(b) What percent of 50 is 30?
      percent   total part
     P R B
(c) Eight is 10% of what number?
    part   percent   total

3   Solve the Percentage Proportion for Any Missing Element.

It is possible to consolidate all three elements into one formula called the *percentage proportion.* The relationship among the rate, percentage, base, and the standard unit of 100 can be represented by two fractions that are equal to each other.

Formula 4–1   *Percentage Proportion:*

$$\frac{R}{100} = \frac{P}{B}$$

where R = rate or percent

P = percentage or part

B = base or total

In a percent problem, if we know any two of the elements, we can find the third element by using the property of proportions and cross multiplication. This is called *solving* the proportion.

Rule 4–2 *To solve the percentage proportion:*

1. Cross multiply to find the cross products.
2. Divide the product of the two known factors by the factor with the letter.

EXAMPLE 3   What is 20% of 75?

The rate is 20%, the base is 75, and the percentage is missing.

$$\frac{R}{100} = \frac{P}{B}$$ Set up proportion.

$$\frac{20}{100} = \frac{P}{75}$$ Substitute the known elements.

$$20 \times 75 = 100 \times P$$ Cross multiply.

$$1500 = 100 \times P$$

$$\frac{1500}{100} = P$$ Divide to find P.

$$15 = P$$

Therefore, the percentage is 15.

Tips and Traps   *Multiply, Then Divide:*
The solution of the percentage proportion always involves one multiplication calculation and one division calculation. One cross product will be two numbers. They will be multiplied. The other cross product will be one number and one letter. Divide the cross product with two numbers by the one number in the other cross product.

$$\frac{20}{100} = \frac{P}{75}$$

$$20 \times 75 = 100 \times P$$

Multiply $20 \times 75 = 1500$. Divide 1500 by 100. $1500 \div 100 = 15$. So $P = 15$.

   ■ **General Tips for Using the Calculator**

Proportions can be solved using a calculator with a continuous series of steps.

$$\frac{20}{100} = \frac{P}{75}$$

$$20 \ \boxed{\times} \ 75 \ \boxed{\div} \ 100 \ \boxed{=} \ \Rightarrow 15$$

EXAMPLE 4 What percent of 50 is 30?

The rate is missing, the base = 50, the percentage = 30.

$$\frac{R}{100} = \frac{P}{B}$$ Set up proportion.

$$\frac{R}{100} = \frac{30}{50}$$ Substitute known elements.

$R \times 50 = 100 \times 30$ Cross multiply.

$R \times 50 = 3000$

$$R = \frac{3000}{50}$$ Divide to find R.

$R = 60$

Since R is the rate, 60 represents 60%.

EXAMPLE 5 Eight is 10% of what number?

The rate = 10%, the base is missing, the percentage = 8.

$$\frac{R}{100} = \frac{P}{B}$$ Set up proportion.

$$\frac{10}{100} = \frac{8}{B}$$ Substitute known elements.

$10 \times B = 8 \times 100$ Cross multiply.

$10 \times B = 800$

$$B = \frac{800}{10}$$ Divide to find B.

$B = 80$

The base is 80.

The percentage proportion has several advantages. The obvious advantage is that we can use only one formula, the percentage proportion, to solve problems for percentage, rate, or base. Another advantage is that we can standardize our approach to percentage problems—have one basic way to solve them that works in all cases.

This general approach to percentage problems has other advantages, also. One is that the standard 100 takes care of having to convert decimal numbers to percents or to convert percents to decimal equivalents, as is done in traditional approaches to these problems. This approach also simplifies the steps. For example, all problems involve one multiplication (cross multiplication) and one division step (division by whichever number the letter is multiplied by). This approach also includes either multiplication by 100 or division by 100, which can be done mentally.

SELF-STUDY EXERCISES 4–1

1 Find the value of the letter in each proportion.

1. $\dfrac{x}{10} = \dfrac{1}{2}$ **2.** $\dfrac{b}{16} = \dfrac{2}{8}$ **3.** $\dfrac{3}{a} = \dfrac{4}{12}$ **4.** $\dfrac{4}{c} = \dfrac{2}{5}$

5. $\dfrac{2}{3} = \dfrac{x}{6}$ **6.** $\dfrac{2}{5} = \dfrac{a}{20}$ **7.** $\dfrac{1}{2} = \dfrac{3}{x}$ **8.** $\dfrac{2}{4} = \dfrac{9}{b}$

9. $\dfrac{d}{3} = \dfrac{1}{2}$ **10.** $\dfrac{2}{x} = \dfrac{3}{5}$ **11.** $\dfrac{4}{9} = \dfrac{c}{2}$ **12.** $\dfrac{8}{3} = \dfrac{2}{b}$

2 Identify the given and missing elements for each of the following as R (rate), B (base), and P (percentage).

13. What percent of 10 is 2?

14. Two is 20% of what number?

15. 20% of 10 is what number?

16. Three books is what percent of four books?

17. 15% of how many dollars is $9?

18. What percent of 25 students is 5 female students?

19. 6 of 15 motorists is what percent?

20. How many nurses is 20% of the 15 nurses on duty?

21. 35% of how many pieces of sod is 70 pieces?

22. What percent of a total bill of $45 is $3.15?

3

23. Find the percentage if the base is 75 and the rate is 5%.

24. Find the percentage if the base is 25 and the rate is 2.5%.

25. What is the rate if the base is 10.5 and the percentage is 7?

26. Find the rate when the base is 80 and the percentage is 30.

27. If the percentage is 4.75 and the rate is $33\frac{1}{3}$%, find the base.

28. Find the base when the rate is 15% and the percentage is 52.5.

29. If the rate is $12\frac{1}{2}$% and the base is 75, find the percentage.

30. If the percentage is 11 and the rate is 5%, find the base.

31. Find the rate when the percentage is 15 and the base is 75.

32. What is the base if the percentage is 35 and the rate is 17.5%?

4–2 SOLVING PERCENTAGE PROBLEMS

Learning Objectives

1 Use the percentage proportion to find the percentage.

2 Use the percentage proportion to find the rate.

3 Use the percentage proportion to find the base.

4 Use the percentage proportion to solve applied problems.

Let's examine some additional examples in which problems are solved using the percentage proportion. In these examples, pay particular attention to the clues that suggest whether the missing element is the rate, percentage, or base. Also notice the suggestions for using decimal versus fractional equivalents, as well as making calculations by hand versus by calculator.

1 Use the Percentage Proportion to Find the Percentage.

EXAMPLE 1 25% of 180 is what number?

25% is the rate, and the only other decision that has to be made is whether 180 is the base or the percentage. The missing number is 25% of, or part of, 180. Therefore, 180 is the base. The word *of* is a key word to look for in identifying the base. We are finding part *of* a quantity (the base).

Now, set up the proportion.

$$\frac{25}{100} = \frac{P}{180}$$

$$25 \times 180 = 100 \times P \qquad \text{Cross multiply.}$$

$$4500 = 100 \times P$$

$$\frac{4500}{100} = P \qquad \text{Divide by 100.}$$

$$45 = P$$

45 is 25% of 180.

Knowing fractional equivalents can save time when making calculations by hand. Let's substitute $\frac{1}{4}$ for 25% or 25 hundredths ($\frac{25}{100}$) and rework this example.

$$\frac{1}{4} = \frac{P}{180} \qquad \text{The left ratio is the fractional equivalent of } \tfrac{25}{100} \text{ reduced to lowest terms, } \tfrac{1}{4}.$$

$$1 \times 180 = 4 \times P$$

$$180 = 4 \times P$$

$$\frac{180}{4} = P$$

$$45 = P$$

If a calculator is available, the decimal equivalent for the percent can be used to save calculation time: $25\% = 0.25$. To employ our same technique of using proportions in problems involving percents, we can write 0.25 as $\frac{0.25}{1}$. (To divide by 1 does not change the value of a number.)

$$\frac{0.25}{1} = \frac{P}{180} \qquad \text{The left ratio employs the decimal equivalent of } \tfrac{25}{100}, \text{ that is, 0.25.}$$

$$0.25 \times 180 = P \qquad 1 \times P = P. \text{ One times a number does not change the value of the number.}$$

$$45 = P$$

Again, the percentage is 45.

The next example shows how the fractional equivalent of a mixed number percent can be used.

EXAMPLE 2 $33\frac{1}{3}\%$ of 282 is what number?

The rate is $33\frac{1}{3}\%$. The key word *of* tells us that 282 is the base. The percentage is missing.

$$\frac{33\frac{1}{3}}{100} = \frac{P}{282}$$

If the fractional equivalent for $33\frac{1}{3}\%$ or $\frac{33\frac{1}{3}}{100}$ is known from memory, we can simplify the calculations by substituting the fractional equivalent for $\frac{33\frac{1}{3}}{100}$ in the proportion. $\left(\frac{33\frac{1}{3}}{100} = \frac{1}{3}; \text{ see Example 6 on page 152.}\right)$

$$\frac{1}{3} = \frac{P}{282}$$

$$1 \times 282 = 3 \times P \qquad \text{Cross multiply.}$$

$$\frac{282}{3} = P \qquad \text{Divide.}$$

$$94 = P$$

Thus, the percentage is 94.

Even with a calculator, you may prefer to use the fractional equivalent because the decimal equivalent of $33\frac{1}{3}\%$ is a repeating decimal, and requires rounding. Examine the effect of rounding to various places.

$$0.3 \times 282 = 84.6$$

$$0.33 \times 282 = 93.06$$

$$0.333 \times 282 = 93.906$$

$$0.3333 \times 282 = 93.9906$$

$$0.333333333 \times 282 = 93.99999991$$

Because the desired degree of accuracy may vary depending on the technical application, it is advisable to find the exact answer on the calculator by using the fractional equivalent. A close approximate answer can be found by using the full calculator value for nonterminating decimal equivalents.

Percents that are less than 1% or more than 100% follow the same procedures and cautions as any other percents. However, there are some time-savers we can use to make our work easier and faster when dealing with these percents. Again note the suggestions about using calculators versus solving problems by hand and using fractional versus decimal equivalents.

It is always advisable to anticipate the approximate size of an answer before making any calculations. This approximation will help you discover many careless mistakes, especially when you are using a calculator. 1% of any number can be found by making mental calculations.

$$1\% \text{ of } 175 \text{ is}$$

$$\frac{1}{100} = \frac{P}{175}$$

$$175 = 100 \times P$$

$$\frac{175}{100} = P$$

$$1.75 = P$$

To find 1% of a number, divide the number by 100 or move the decimal two places to the left.

Because 1% of a number can be found mentally, when working with a percent that is less than 1%, we will first find 1% of the number. The answer we are seeking must be less than 1% of the number. This estimating procedure will be very useful in checking the decimal placement. For instance, let's find $\frac{1}{4}\%$ of 875. $\frac{1}{4}\%$ means $\frac{1}{4}$ of 1%. 1% of 875 is 8.75 (move the decimal two places to the left). $\frac{1}{4}\%$ of 875 would be $\frac{1}{4}$ of 8.75, or a little more than the whole number 2. We can use this estimate to check the work in the following example.

EXAMPLE 3 $\frac{1}{4}\%$ of 875 is what number?

$\frac{1}{4}\%$ and 0.25% are equivalent. Either can be used in solving this problem. Because we are given the rate and the base, we need to find the percentage.

$$\frac{\frac{1}{4}}{100} = \frac{P}{875} \qquad\qquad \frac{0.25}{100} = \frac{P}{875}$$

$$\frac{1}{4} \times 875 = 100 \times P \qquad 0.25 \times 875 = 100 \times P \qquad \text{Cross multiply.}$$

$$\frac{875}{4} = 100 \times P \qquad\qquad 218.75 = 100 \times P$$

$$\frac{\frac{875}{4}}{100} = P \qquad\qquad \frac{218.75}{100} = P \qquad \text{Divide by 100.}$$

$$\frac{875}{4} \div 100 = P \qquad\qquad 2.1875 = P$$

$$\frac{\overset{35}{\cancel{875}}}{4} \times \frac{1}{\underset{4}{\cancel{100}}} = P$$

$$\frac{35}{16} = P$$

$$2\frac{3}{16} = P$$

Therefore, the percentage is $2\frac{3}{16}$, or 2.1875. Use your calculator to verify that $\frac{3}{16} = 0.1875$ (3 ÷ 16 = 0.1875).

100% of a number is 1 times the number, or the number itself. When working with percents that are larger than 100%, we can also quickly make a rough estimate of the answer. For instance, 325% of 86 is what number?

325% of 86 is more than three times 86. 100% of 86 = 86. Thus, 325% of 86 must be more than 3 times 86, or more than 258. Thus, our estimate for Example 4 is more than 258.

EXAMPLE 4   325% of 86 is what number?

Again, we are finding the percentage. The rate is 325% and the base is 86.

Option 1   Option 2

$$\frac{325}{100} = \frac{P}{86} \qquad\qquad \frac{13}{4} = \frac{P}{86} \left(325\% = 3\frac{1}{4} = \frac{13}{4} \right)$$

$$325 \times 86 = 100 \times P \qquad 13 \times 86 = 4 \times P \qquad \text{Cross multiply.}$$

$$\frac{27950}{100} = P \qquad\qquad \frac{1118}{4} = P \qquad \text{Divide.}$$

$$279.5 = P \qquad\qquad 279.5 = P$$

Thus, the percentage is 279.5 or $279\frac{1}{2}$.

2 Use the Percentage Proportion to Find the Rate.

The rate is the easiest of the three parts of a percentage problem to identify. When neither of the given numbers in a problem has a percent symbol or is followed by the word *percent,* the rate is missing.

EXAMPLE 5 What percent of 48 is 24?

The missing element in this problem is the rate. Is 48 the percentage (part) or the base (total)? 48 follows the key word *of* and is the base. 24 follows the key word *is* and is the percentage.

Set up the proportion.

$$\frac{R}{100} = \frac{24}{48}$$

$R \times 48 = 100 \times 24$ Cross multiply.

$R \times 48 = 2400$

$$R = \frac{2400}{48}$$ Divide.

$R = 50$

Since R is the rate, 50 represents 50%.

EXAMPLE 6   What percent is 2 out of 600?

The rate is missing. The 2 represents the part or percentage and 600 is the base.

$$\frac{R}{100} = \frac{2}{600}$$

$R \times 600 = 2 \times 100$ Cross multiply.

$R \times 600 = 200$

$$R = \frac{200}{600}$$ Divide or reduce.

$$R = \frac{1}{3}$$

Thus, the rate is $\frac{1}{3}$%.

EXAMPLE 7   261 is what percent of 87?

The rate is missing in this problem. Does 261 represent the percentage or the base? 261 is associated with the key word *is* and represents the percentage or part. 87 is associated with the key word *of* and represents the base (one total amount). Because the percentage is larger than the base, the rate will be more than 100%; that is, the base is used more than once.

$$\frac{R}{100} = \frac{261}{87}$$

$26,100 = 87 \times R$ Cross multiply.

$$\frac{26,100}{87} = R$$ Divide.

$300 = R$

Thus, the rate is 300%.

Tips and Traps   ***Alternate Percentage Formulas:***
The two most common types of percent problems involve finding the percentage or finding the rate.

An alternative percentage formula is often used to find the percentage. To find the percent of a number (base), such as 20% of 15, change the percent or rate to a decimal and multiply it and the number.

$$\text{Percentage} = \text{Rate} \times \text{Base}$$

$$P = 20\% \times 15 \qquad \text{Change rate to decimal equivalent.}$$

$$P = 0.20 \times 15 \qquad \text{Multiply.}$$

$$P = 3$$

Another alternative percentage formula is often used to find the rate. To find what percent one number (part) is of another (base), such as 3 is what percent of 15, divide the part by the base. Then move the decimal point two places to the right.

$$\text{Rate} = \frac{\text{Percentage}}{\text{Base}}$$

$$R = \frac{3 \text{ (part)}}{15 \text{ (base)}} \qquad \text{Divide.}$$

$$R = 0.2$$

$$R = 20\% \qquad \text{Change quotient to percent.}$$

■ General Tips for Using the Calculator

Some scientific calculators and graphics calculators have a percent key, but it is very common to work a percent problem without using the percent key. Percent problems are performed efficiently on calculators by entering the percent in decimal notation when finding the percentage or base, or changing the calculator display to percent notation when finding the rate. Even when a calculator has a percent key, the use of the $\boxed{\%}$ key may vary. The most common practice is to mentally convert percent notation to decimal notation or vice versa. The percent key is most often used in place of the equal key. In addition to performing the same purpose as the equal key, the % key also uses the decimal equivalent of the rate when the rate is used in the problem.

28% of 37 is what number?

Calculator using $\boxed{\%}$ *key:* The $\boxed{\%}$ key automatically converts the rate to a decimal.

$$\boxed{\text{AC}} \ 28 \ \boxed{\times} \ 37 \ \boxed{\%} \Rightarrow 10.36$$

Calculator using $\boxed{=}$ *or* $\boxed{\text{EXE}}$:

$$\boxed{\text{AC}} \ \boxed{\cdot} \ 28 \ \boxed{\times} \ 37 \ \boxed{=} \Rightarrow 10.36$$

15 is what percent of 52?

Calculator using $\boxed{\%}$ *key:* The $\boxed{\%}$ key automatically converts the decimal answer to a percent.

$$\boxed{\text{AC}} \ 15 \ \boxed{\div} \ 52 \ \boxed{\%} \Rightarrow 28.84615385 \qquad \text{a percent}$$

15 is approximately 28.8% of 52.

Calculator using $\boxed{=}$ *or* $\boxed{\text{EXE}}$: The $\boxed{=}$ or $\boxed{\text{EXE}}$ key gives a decimal answer. You must change the decimal to a percent by mentally multiplying by 100%.

$$\boxed{\text{AC}} \ 15 \ \boxed{\div} \ 52 \ \boxed{\text{EXE}} \Rightarrow 0.2884615385 \qquad \text{a decimal}$$

0.2884615385 is approximately 28.8%.

3 Use the Percentage Proportion to Find the Base.

EXAMPLE 8 20% of what number is 45?

This time we know the rate, 20%, and the percentage, 45. We are looking for the base, as indicated by the key word *of*.

Option 1	Option 2	Option 3
$\dfrac{20}{100} = \dfrac{45}{B}$	$\dfrac{1}{5} = \dfrac{45}{B}$	$\dfrac{0.2}{1} = \dfrac{45}{B}$
$20 \times B = 100 \times 45$	$B = 5 \times 45$	$45 = 0.2 \times B$
$20 \times B = 4500$	$\boxed{B = 225}$	$\dfrac{45}{0.2} = B$
$B = \dfrac{4500}{20}$		$\boxed{225 = B}$
$\boxed{B = 225}$		

Thus, the base is 225.

EXAMPLE 9 $\frac{3}{4}$% of what number is 11.25?

The rate is $\frac{3}{4}$% and the percentage is 11.25, as signaled by the key word *is*. The base is missing.

Option 1	Option 2	
$\dfrac{\frac{3}{4}}{100} = \dfrac{11.25}{B}$	$\dfrac{0.75}{100} = \dfrac{11.25}{B}$	
$\dfrac{3}{4} \times B = 1125$	$0.75 \times B = 1125$	Cross multiply.
$B = \dfrac{1125}{\frac{3}{4}}$	$B = \dfrac{1125}{0.75}$	Divide.
	$\boxed{B = 1500}$	
$B = \dfrac{\overset{375}{\cancel{1125}}}{1} \times \dfrac{4}{\underset{1}{\cancel{3}}}$		
$\boxed{B = 1500}$		

Thus, the base is 1500.

To mentally check that the answer is reasonable, find 1% of 1500. 1% of 1500 = 15. $\frac{3}{4}$% of 1500 should be less than 15. 11.25 is less than 15, so the answer is reasonable.

EXAMPLE 10 398.18 is 215% of what number?

Here, we are looking for the base, as the key word *of* lets us know. We are given the percentage and the rate.

$$\frac{215}{100} = \frac{398.18}{B}$$

$$215 \times B = 39{,}818 \qquad \text{Cross multiply.}$$

$$B = \frac{39,818}{215} \qquad \text{Divide.}$$

$$B = 185.2 \quad \text{or} \quad 185\frac{1}{5}$$

Therefore, the base is 185.2 or 185$\frac{1}{5}$.

Because the rate is more than 100%, we expected the base to be smaller than the percentage. Thus, the answer is reasonable.

When solving applied problems, the most difficult task is identifying the two parts that are given and determining which part is missing. Then the proportion can be set up and solved. Let's examine several examples of applied problems involving percents.

4 Use the Percentage Proportion to Solve Applied Problems.

EXAMPLE 11 If a type of solder contains 55% tin, how many pounds of tin are needed to make 10 lb of solder?

First, let's be sure we understand the word *solder*. Solder is a mixture of metals. In this problem, the *total amount* or the *base* is the 10 lb of solder. We know that 55% of this 10 lb of solder is tin; that is, the *rate* of tin in the solder is 55%. We want to find how much or what part of the 10 lb of solder is tin. This means that we are trying to find the *percentage* or *part*.

We write the proportion.

$$\frac{55}{100} = \frac{P}{10}$$

$$55 \times 10 = 100 \times P$$

$$550 = 100 \times P$$

$$\frac{550}{100} = P$$

$$5\frac{1}{2} = P$$

Thus, there are 5$\frac{1}{2}$ lb of tin in 10 lb of solder.

To check the reasonableness of the answer, 55% is a little more than $\frac{1}{2}$. 5$\frac{1}{2}$ lb is a little more than $\frac{1}{2}$ of 10.

EXAMPLE 12 If a 150-horsepower (hp) engine delivers only 105 hp to the driving wheels of a car, what is the efficiency of the engine?

Efficiency means the *percent* the output (105 hp) is of the total amount (150 hp) the engine is capable of delivering. So the base amount is 150 hp, the part or percentage delivered is 105 hp, and the percent of 150 represented by 105 is the rate or efficiency.

$$\frac{R}{100} = \frac{105}{150}$$

$$R \times 150 = 100 \times 105$$

$$R \times 150 = 10,500$$

$$R = \frac{10,500}{150}$$

$$R = 70$$

The engine is 70% efficient.
    Because the engine was not operating at full capacity (150 hp), we expected the efficiency to be less than 100%.

EXAMPLE 13   The effective value of current or voltage in an ac circuit is 71.3% of the maximum voltage. If a voltmeter shows a voltage of 110 volts (V) in a circuit, what is the maximum voltage?

71.3% of the maximum voltage is 110 V. The maximum voltage is the *base,* and the amount of voltage shown in the voltmeter is 110 V or the *percentage.*

$$\frac{71.3}{100} = \frac{110}{B}$$

$$71.3 \times B = 100 \times 110$$

$$71.3 \times B = 11{,}000$$

$$B = \frac{11{,}000}{71.3}$$

$$B = 154.2776999$$

or

$$B = 154 \text{ V}$$ to the nearest volt

Thus, the maximum voltage is 154 V.
    Because the rate was less than 100%, we expected the base, the maximum voltage, to be larger than the percentage.

SELF-STUDY EXERCISES 4–2

[1]
1. 20% of 375 is what number?
3. $66\frac{2}{3}$% of 309 is what number?
5. $\frac{3}{4}$% of 90 is what number?
7. What number is 134% of 115?
9. 400% of 231 is what number?

2. 75% of 84 is what number?
4. 34.5% of 336 is what number?
6. 0.2% of 470 is what number?
8. Find 275% of 84.
10. $37\frac{1}{2}$% of 920 is what number?

[2]
11. What percent of 348 is 87?
13. 72 is what percent of 216?

15. 37.8 is what percent of 240?
17. 32 is what percent of 4000?
19. What percent of 125 is 625?

12. What percent of 350 is 105?
14. 28 is what percent of 85 (to the nearest tenth percent)?
16. What percent of 175 is 28?
18. What percent is 2 out of 300?
20. 173.55 is what percent of 156?

[3]
21. 50% of what number is 36?
23. $12\frac{1}{2}$% of what number is 43?
25. $\frac{2}{3}$% of what number is $2\frac{2}{5}$?
27. 150% of what number is $112\frac{1}{2}$?
29. 92 is 500% of what number?

22. 60% of what number is 30?
24. 15.87 is 34.5% of what number?
26. 0.3% of what number is 0.825?
28. 43% of what number is 107.5?
30. $133\frac{1}{3}$% of what number is 348?

[4]   Solve the following problems.
31. Cast iron contains 4.25% carbon. How much carbon is contained in a 25-lb bar of cast iron?

32. 3645 rolls of landscape fabric are manufactured during one day. After being inspected, 121 of these rolls were rejected as imperfect. What percent of the rolls was rejected? (Round to the nearest whole percent.)

33. An engine operating at 82% efficiency transmits 164 hp. What is the engine's maximum capacity in horsepower?

35. The voltage of a generator is 120 V. If 6 V are lost in a supply line, what is the rate of voltage loss?

37. A certain ore yields an average of 67% iron. How much ore is needed to obtain 804 lb of iron?

39. 385 defective alcohol swabs were produced during a day. If 4% of the alcohol swabs produced were defective, how many alcohol swabs were produced in all?

34. If wrought iron contains 0.07% carbon, how much carbon is in a 30-lb bar of wrought iron?

36. A contractor figures it costs $\frac{1}{2}$% of the total cost of a job to make a bid. What would be the cost of making a bid on a $115,000 job?

38. A contractor makes a profit of $12,350 on a $115,750 job. What is the percent of profit? (Round to the nearest whole percent.)

40. In a welding shop, 104,000 welds are made. If 97% of them are acceptable, how many are acceptable?

4-3 INCREASES AND DECREASES

Learning Objectives

1. Find the amount of increase or decrease in percent problems.
2. Find the new amount directly in percent problems.
3. Find the rate or the base in increase or decrease problems.

Percents are often used in problems dealing with increases or decreases. For instance, if a TV repair shop is advised that its recent order for 250 power cords will be reduced by 14% because of a shortage of copper wire, the shop needs to find out how many power cords this decrease will amount to. The shop may have to order additional cords from another supplier.

1 Find the Amount of Increase or Decrease in Percent Problems.

When working with increases or decreases, the *original amount* (250 power cords) will be the *base*. The *percentage* (power cords *not* received) will be the amount of *change (increase* or *decrease)*. The *new amount* will be the original amount plus or minus the amount of change.

EXAMPLE 1 Pipefitters are to receive a 9% increase in wages per hour. If they were making $9.25 an hour, what will be the *amount of increase per hour* (to the nearest cent)? Also, what will be the *new wage per hour*? The original wage per hour is the base, and we want to find the amount of increase (percentage).

$$\frac{9}{100} = \frac{P}{9.25}$$

P represents amount of increase and 9% is the rate of increase.

$$9 \times 9.25 = 100 \times P$$

$$83.25 = 100 \times P$$

$$\frac{83.25}{100} = P$$

$$\$0.8325 = P \quad \text{or} \quad \$0.83 \text{ to the nearest cent is the amount of increase.}$$

The pipefitters will receive an $0.83 per hour increase in wages.

$$\$9.25 + \$0.83 = \$10.08 \qquad \text{New amount} = \text{original amount} + \text{amount of increase.}$$

Their new hourly wage will be $10.08.

EXAMPLE 2 Molten iron shrinks 1.2% while cooling. What is the cooled length of a piece of iron if it is cast in a 24-cm pattern?

First, we will find the amount of shrinkage (amount of decrease, percentage). The original amount, 24 cm, is the base.

$$\frac{1.2}{100} = \frac{P}{24}$$

P represents amount of decrease and 1.2% is the rate of decrease.

$$1.2 \times 24 = 100 \times P$$

$$28.8 = 100 \times P$$

$$\frac{28.8}{100} = P$$

$$0.288 = P$$

The amount of shrinkage is 0.288 cm. Then the length of the cooled piece (new amount) is

24 − 0.288 = 23.712 cm   New amount = original amount − amount of decrease.

EXAMPLE 3 Uncut earth is hard, packed soil. As it is dug, the volume increases or swells. A contractor figures that there will be a 20% earth swell when a mixture of uncut loam and clay soil is excavated. If 150 cubic yards (yd^3) of uncut earth are to be removed, taking into account the earth swell, how many cubic yards will have to be hauled away?

To find the amount of earth swell (amount of increase, percentage), find 20% of 150 yd^3. 150 yd^3 is the original amount or base.

Option 1

$$\frac{20}{100} = \frac{P}{150}$$

$$20 \times 150 = 100 \times P$$

$$3000 = 100 \times P$$

$$\frac{3000}{100} = P$$

$$30 = P$$

Option 2

$$\frac{1}{5} = \frac{P}{150}$$

$$150 = 5 \times P$$

$$\frac{150}{5} = P$$

$$30 = P$$

P represents the amount of increase and 20%, or $\frac{1}{5}$, or 0.2, is the rate of increase.

Option 3

$$\frac{0.2}{1} = \frac{P}{150}$$

$$0.2 \times 150 = P$$

$$30 = P$$

Thus, the earth will swell 30 yd^3 when cut. To find the amount of earth to be hauled away, add 150 yd^3 and 30 yd^3.

150 + 30
  = 180 yd³ to be hauled away   New amount = original amount + amount of increase.

2 Find the New Amount Directly in Percent Problems.

When knowing the amount of increase is not necessary, the new amount can be figured directly. Remember, all of a quantity is 100%. If a quantity is to be increased by 20%, then the new amount will be 100% + 20% or 120% of the original amount.

If 150 yd^3 of uncut earth (base) increases by 20% when cut, the new amount (percentage) will be 120% of 150 yd^3.

Option 1   Option 2

$$\frac{120}{100} = \frac{P}{150}$$ $$\frac{1.2}{1} = \frac{P}{150}$$ P represents the new amount because 120% is the new rate; that is, 120% of 150 is the new amount.

$$\frac{6}{5} = \frac{P}{150}$$ $$1.2 \times 150 = P$$

$$6 \times 150 = 5 \times P$$ $$180 = P$$

$$900 = 5 \times P$$

$$\frac{900}{5} = P$$

$$180 = P$$

Thus, we can find the 180 yd^3 to be hauled away directly.

EXAMPLE 4   A drying process causes a 2% weight loss in a casting. If the wet casting weighs 130 kg, how much will the dried casting weigh?

The amount of weight loss (amount of decrease) is 2% of 130 kg.

Option 1

$$\frac{2}{100} = \frac{P}{130}$$ P represents the amount of decrease because 2% is the rate or percent of decrease.

$$2 \times 130 = 100 \times P$$

$$260 = 100 \times P$$

$$\frac{260}{100} = P$$

$$2.6 = P$$

Thus, the casting will have a weight loss of 2.6 kg. The dried casting will weigh

$$130 - 2.6 = 127.4 \text{ kg}$$ New amount = original amount − amount of decrease.

Because knowing the amount of weight loss is not necessary, the dried casting weight could have been calculated directly. The original weight equals 100%. There will be 2% weight loss, so the dried weight will be 100% − 2%, or 98% of the original weight.

Option 2

$$\frac{98}{100} = \frac{P}{130}$$ P represents the new amount because 98% is the percent that the new amount is of the original weight; that is, 98% of 130 is the new amount.

$$98 \times 130 = 100 \times P$$

$$\frac{12,740}{100} = P$$

$$127.4 = P$$

Thus, the weight of the dried casting is 127.4 kg.

EXAMPLE 5   A 3% error is acceptable for a machine part to be usable. If the part is intended to be 57 cm long, what is the range of measures that is acceptable for this part?

The machine part can be ±3% from the ideal length of 57 cm. The range of acceptable measures is found by finding the smallest acceptable measure and the largest acceptable measure. The symbol $\pm$ is read "plus or minus." It means that the item can be more $(+)$ or less $(-)$ than the designated amount. In this case, the part can be 3% longer than 57 cm or 3% shorter than 57 cm and still be usable. The smallest acceptable value is 97% of the ideal length $(100\% - 3\%)$.

$$\frac{97}{100} = \frac{P}{57}$$

P represents the smallest acceptable amount because 97% is the *smallest acceptable percent.*

$$97 \times 57 = 100 \times P$$

$$\frac{5529}{100} = P$$

$$55.29 = P$$

Thus, the smallest acceptable measure is 55.29 cm.

The largest acceptable value is 103% of the ideal length $(100\% + 3\%)$.

$$\frac{103}{100} = \frac{P}{57}$$

P represents the largest acceptable amount because 103% is the *largest acceptable percent.*

$$5871 = 100 \times P$$

$$\frac{5871}{100} = P$$

$$58.71 = P$$

Thus, the largest acceptable value is 58.71 cm. The range of acceptable measures is from 55.29 cm to 58.71 cm.

We can also work this problem by finding the amount of acceptable error. Then subtract this amount from the intended length of the machine part to get the smallest acceptable measure. Add to get the largest acceptable measure.

3% of 57 is the amount of acceptable error.

$$\frac{3}{100} = \frac{P}{57}$$

P represents the acceptable error because 3% is the *acceptable error percent.*

$$171 = 100 \times P$$

$$\frac{171}{100} = P$$

$$1.71 = P$$

57 ± 1.71 represents the range of acceptable measures.

$$57 - 1.71 = 55.29$$

Smallest acceptable amount = original amount − amount of acceptable error.

The smallest acceptable value is 55.29 cm.

$$57 + 1.71 = 58.71$$

Largest acceptable amount = original amount + amount of acceptable error.

The largest acceptable value is 58.71 cm.

3 Find the Rate or the Base in Increase or Decrease Problems.

There are many kinds of increase or decrease problems that involve finding either the rate or the base.

EXAMPLE 6 A worn brake lining is measured to be $\frac{3}{32}$ in. thick. If the original thickness was $\frac{1}{4}$ in., what is the percent of wear?

First, the amount of wear is $\frac{1}{4} - \frac{3}{32}$.

$$\frac{8}{32} - \frac{3}{32} = \frac{5}{32}$$ Find the common denominator. Subtract.

The amount of wear (decrease) is the percentage. The base is the original amount.

$$\frac{R}{100} = \frac{\frac{5}{32}}{\frac{1}{4}}$$ R represents the percent of wear. $\frac{5}{32}$ is the amount of wear.

$$\frac{1}{4} \times R = \frac{\overset{25}{\cancel{100}}}{1} \times \frac{5}{\underset{8}{\cancel{32}}}$$

$$\frac{1}{4} \times R = \frac{125}{8}$$

$$R = \frac{125}{8} \div \frac{1}{4}$$

$$R = \frac{125}{\underset{2}{\cancel{8}}} \times \frac{\overset{1}{\cancel{4}}}{1}$$

$$R = 62\frac{1}{2}$$

The percent of wear is $62\frac{1}{2}\%$.

EXAMPLE 7 During the month of May, an electrician made a profit of $1525. In June, the same electrician made a profit of $1708. What is the percent of increase?

The amount of increase is $1708 − $1525 = $183. Then $183 is what percent of the *original* amount?

$$\frac{R}{100} = \frac{183}{1525}$$ R represents the percent of increase. 183 is the amount of increase.

$$R \times 1525 = 18,300$$

$$R = \frac{18,300}{1525}$$

$$R = 12$$

Thus, the percent of increase is 12%.

In the applications used in this section on increase and decrease, we can see how important it is to understand what 100% of something really means. Many applications of percents use the fact that 100% of a quantity is the entire quantity. This knowledge lets us compute many quantities directly, without having to first figure the increase or decrease separately.

SELF-STUDY EXERCISES 4–3

1. Find the amount of increase if 432 is increased by 25%.

2. If 78 is increased by 40%, what is the new amount?

3. Find the amount of decrease if 68 is decreased by 15%.

2

5. Steel rods shrink 10% when cooled from furnace temperature to room temperature. If a tie rod is 30 in. long at furnace temperature, how long is the cooled tie rod?

7. If 17% extra flooring is needed to allow for waste when the boards are laid diagonally, how much flooring should be ordered to cover 2045 board feet of floor? Answer to the nearest whole board foot.

9. When making an estimate on a job, the contractor wants to make a 10% profit. If all the estimated costs are $15,275, what is the total bid of cost and profit for the job?

3

11. The cost of a pound of nails increased from $2.36 to $2.53. What was the percent of increase to the nearest whole-number percent?

13. A chicken farmer bought 2575 baby chicks. Of this number 2060 lived to maturity. What percent of loss was experienced by the chicken farmer?

15. An engine that has a 4% loss of power has an output of 336 hp. What is the input (base) horsepower of the engine?

17. A floor in a doctor's office normally requiring 2580 board feet is to be laid diagonally. If a 17% waste allowance is necessary for flooring laid diagonally, how much flooring must be ordered? (Round to the nearest whole number.)

19. Steel bars shrink 10% when cooled from furnace temperature to room temperature. If a cooled steel bar is 36 in. long, how long was it when it was formed?

21. An output of 141 hp is required for an engine. If there is a 6% loss of power, what amount of input horsepower (or base) is needed?

4. If 135 is decreased by 75%, what is the new amount?

6. 1650 board feet of 1-in. $\times$ 8-in. common boards are needed to subfloor a house. If 10% extra flooring is needed to allow for waste when the boards are laid square, how much flooring should be ordered?

8. 25,400 bricks are required for a construction job. If 2% more bricks are needed to allow for breakage, how many bricks must be ordered?

10. Rock must be removed from a highway right of way. If 976 cubic yards (yd^3) of unblasted rock are to be removed, how many cubic yards is this after blasting? Blasting causes a 40% swell in volume.

12. A landscape contractor estimated materials, shrubs, saplings, and labor for a job to cost $5385. An estimate one year later for the same job was $7808, due to inflation. Find the percent of increase due to inflation to the nearest whole number.

14. An electrician recorded costs of $1297 for a job. If she received $1232 for the job, what was the percent of money lost on the job? Round to the nearest whole number.

16. A contractor figures that 10 yd^3 of sand are needed for a job. If a 5% allowance for waste must be included, how much sand must be ordered?

18. A shop manager records a 14% loss on rivets for waste. If the shop needs 25 lb of rivets, how many pounds must be ordered to compensate for loss due to waste?

20. The cost of No. 1 pine studs increased from $3.85 each to $4.62 each. Find the percent of increase.

4–4 **BUSINESS APPLICATIONS**

Learning Objectives

1 Calculate sales tax and payroll deductions.

2 Calculate discount and commission.

3 Calculate interest on loans and investments.

Employment in any profession will, at some time or other, require using mathematics for business purposes. Every company, plant, organization, corporation, partnership, or proprietorship is itself a business of some sort and is concerned with matters such as ordering and paying for supplies and equipment, figuring trade discounts on materials purchased, and processing the payroll accurately and on time.

1 Calculate Sales Tax and Payroll Deductions.

One of the most common applications of percents in business settings is that of *sales tax*. Most states, counties, and cities add to the purchase price of many items a certain percent for sales tax. Then the total amount that a purchaser pays is the price of the item plus the sales tax.

EXAMPLE 1 A 5% sales tax is levied on an order of building supplies costing $127.32. What is the amount of sales tax to be paid? What is the total bill?

Option 1
To find the amount of sales tax to be paid, we need to find 5% of $127.32. The amount of sales tax is the percentage, and the cost of the supplies is the base.

$$\frac{5}{100} = \frac{P}{127.32}$$

P represents the amount of sales tax. 5% is the percent or sales tax rate.

$$5 \times 127.32 = 100 \times P$$

$$636.60 = 100 \times P$$

$$\frac{636.60}{100} = P$$

$$\$6.366 = P$$

When dealing with money, the amounts are usually rounded to the nearest cent. **Thus, the sales tax is $6.37.**
   To find the total amount to be paid, we add the cost of the supplies and the sales tax.

$$\$127.32 + \$6.37 = \mathbf{\$133.69}$$

Total bill = cost of supplies + sales tax.

Option 2
We could have found the total bill directly by considering the cost of the supplies as 100%. Because a 5% sales tax will be added, the total bill will be 105% of the cost of the supplies. Again, $127.32 is the base.

$$\frac{105}{100} = \frac{P}{127.32}$$

P represents the total bill because 105% is the total rate.

$$105 \times 127.32 = 100 \times P$$

$$13368.6 = 100 \times P$$

Zero in the hundredths place can be dropped.

$$\frac{13368.6}{100} = P$$

$$\$133.686 = P$$

$$\mathbf{\$133.69} = P$$

Total bill to the nearest cent

   Other forms of taxes that involve most workers are withholding tax (income tax) and social security tax (FICA). These two taxes are normally deducted from a worker's paycheck.

EXAMPLE 2 If the rate of social security tax (FICA) is 6.2% of the first $62,700 of earnings in a given year, how much social security tax is withheld on a weekly paycheck of $425?

The amount of pay before any deductions are made is called the *gross pay*, and it is the base. The amount of social security tax (FICA) will be the percentage. The rate is 6.2%.

$$\frac{6.2}{100} = \frac{P}{425}$$

P represents the amount of FICA deduction.

$$6.2 \times 425 = 100 \times P$$

$$2635 = 100 \times P$$

$$\frac{2635}{100} = P$$

$$\$26.35 = P$$

Therefore, $26.35 will be withheld for FICA.

Many employers take various *payroll deductions* from employees' paychecks. Some are required, such as withholding tax, FICA, and retirement contributions. Others are made as a convenience to the employee, such as insurance payments, union dues, and credit union or bank deposits.

The employee's total earnings before any deductions are made are called *gross pay*. The "take-home" pay or the amount after deductions is called *net pay*.

EXAMPLE 3 An x-ray technician's total weekly earnings are $475 and the take-home pay is $365.75. What percent of the gross pay are the total deductions?

There are two ways to approach this problem. One way is to first subtract the net pay from the gross pay to find the amount of the deductions.
Option 1

$$\$475 - 365.75 = \boxed{109.25}$$ total deductions

Then we find the percent that the deductions are of the gross pay. The amount of deductions is the percentage, and the gross pay is the base.

$$\frac{R}{100} = \frac{109.25}{475}$$

R represents the percent of deductions. $109.25 is the amount of the deductions.

$$R \times 475 = 100 \times 109.25$$

$$R \times 475 = 10,925$$

$$R = \frac{10,925}{475}$$

$$R = 23$$

Thus, the deductions are 23% of the gross pay.
Option 2
Another way of approaching this problem is to find the percent the net pay (percentage) is of the gross pay (base).

$$\frac{R}{100} = \frac{365.75}{475}$$

R represents the percent of net pay. $365.75 is the percentage or the amount of net pay.

$$R \times 475 = 100 \times 365.75$$

$$R \times 475 = 36,575$$

$$R = \frac{36{,}575}{475}$$

$$R = 77$$

The net pay is 77% of the gross pay. Because the gross pay is 100% and the net pay is 77%, the deductions are 100% − 77%, or **23% of the gross pay.**

Tips and Traps   *Matching Descriptions for Rate and Percentage:*
In the two approaches given in Example 3, notice that the descriptions of the rate and percentage match in both cases. In the first approach, you had the rate or percent *of deductions* and the amount *of deductions*. In the second approach, you had the rate or percent *of net pay* and the amount *of net pay*.

2 Calculate Discount and Commission.

Many businesses give customers discounts. These discounts may be given as incentives for the customers to pay cash or to pay within a certain time period. *Discounts* are also given to increase sales, to reduce inventories, or to move seasonal stock. In most cases the discounted price is determined by deducting a certain percent of the original price from the original price. The original price is the base.

EXAMPLE 4   A landscape contractor is given a 3% discount for paying cash for the horticultural supplies that are bought. If the total bill before the discount is $143.38, what is the amount that the landscape contractor will pay in cash?

Again, this problem can be approached two different ways. We can find the cash discount (percentage) by finding 3% of 143.38.

Option 1

$$\frac{3}{100} = \frac{P}{143.38}$$ P represents the amount of discount and 3% is the rate of discount.

$$3 \times 143.38 = 100 \times P$$

$$430.14 = 100 \times P$$

$$\frac{430.14}{100} = P$$

$$4.3014 = P$$

$$\$4.30 = P$$ to the nearest cent

Thus, the landscape contractor pays $143.38 − $4.30 or $139.08.

Option 2

Another approach in solving this problem is to again utilize our knowledge of 100%. The total bill is 100% and a 3% cash discount is given. Therefore, the discounted price is 97% of the original price (base).

$$\frac{97}{100} = \frac{P}{143.38}$$ P represents the discounted price and 97% is the rate of the discounted price.

$$97 \times 143.38 = 100 \times P$$

$$13{,}907.86 = 100 \times P$$

$$\frac{13{,}907.86}{100} = P$$

$$139.0786 = P$$

$$\mathbf{\$139.08 = P} \qquad \text{Amount to be paid to the nearest cent}$$

Persons in the sales profession are sometimes paid a salary based on the amount of sales made. This salary is usually a certain percent of the total sales and is called a *commission.*

EXAMPLE 5   A salesperson receives a 6% commission on all sales made. If this salesperson sells $15,575 in merchandise during a given pay period, what is the commission?

The amount of total sales during the pay period ($15,575) is the base. The commission or part of the total sales that the salesperson receives in wages is the percentage.

$$\frac{6}{100} = \frac{P}{15{,}575} \qquad \text{P represents the amount of commission and 6\% is the rate of commission.}$$

$$6 \times 15{,}575 = 100 \times P$$

$$93{,}450 = 100 \times P$$

$$\frac{93{,}450}{100} = P$$

$$\$934.50 = P$$

The salesperson's commission for the pay period is $934.50.

EXAMPLE 6   An automotive parts salesperson earns a salary of $125 per week and 8% commission on all sales over $2500 per week. The sales during one week were $4875. What is the salesperson's salary for that week?

First, we need to determine the amount of sales on which the salesperson will earn a commission.

$$\$4875 - 2500 = \$2375$$

The salesperson receives an 8% commission on $2375 (base).

$$\frac{8}{100} = \frac{P}{2375} \qquad \text{P represents the amount of commission earned. \$2375 is the base on which the commission is earned.}$$

$$8 \times 2375 = 100 \times P$$

$$19{,}000 = 100 \times P$$

$$\frac{19{,}000}{100} = P$$

$$\$190.00 = P \qquad \text{commission}$$

The salesperson will receive a $125 base salary and $190 in commission.

$$\$125 + 190 = \$315$$

The salary for that week will be $315.

3   Calculate Interest on Loans and Investments.

Interest is the amount charged for borrowing or loaning money, or the amount of money earned when money is saved or invested. The *amount of interest* is the *percentage,* the total amount invested or borrowed is the *principal* or *base,* and the *percent of interest* is the *rate.*

The rate of interest is always expressed as a percent per time period. For example, the rate of interest on a loan may be 12% per year or per annum. The rate of interest or finance charge on a charge account may be $1\frac{1}{2}$% per month.

There are many different ways of figuring interest. Most banks or loan institutions use compound interest; some institutions figure interest using the exact time of a loan; some use an approximate time of a loan such as 30 days per month or 360 days per year. However, the Truth-in-Lending Law requires that all businesses equate their interest rate to an annual simple interest rate known as the *annual percentage rate* (APR). This allows consumers to compare rates of various institutions and to understand exactly what rate they are earning or are being charged.

Simple interest for one time period can be found by using our basic percentage proportion.

EXAMPLE 7   Find the interest for 1 year on a loan of $5000 if the interest rate is 15% per year.

$5000 is the principal of the loan or the base. We are looking for the interest (percentage).

$$\frac{15}{100} = \frac{P}{5000}$$

P represents the amount of interest. 15% is the interest rate.

$$15 \times 5000 = 100 \times P$$

$$75{,}000 = 100 \times P$$

$$\frac{75{,}000}{100} = P$$

$$\$750 = P$$

The interest on the loan is $750.

EXAMPLE 8   A credit-card service charges a finance charge (interest) of $1\frac{1}{2}$% per month on the average daily balance of the account. If the average daily balance on an account is $157.48, what is the finance charge for the month?

$$\frac{1\frac{1}{2}}{100} = \frac{P}{157.48}$$

P represents the finance charge, or interest, for one month. Remember, 1.5 can be substituted for $1\frac{1}{2}$ if desired.

$$1.5 \times 157.48 = 100 \times P$$

$$236.22 = 100 \times P$$

$$\frac{236.22}{100} = P$$

$$2.3622 = P$$

$$\$2.36 = P \qquad \text{to the nearest cent}$$

The finance charge is $2.36.

EXAMPLE 9 $1250 is invested for 6 months at an interest rate of $8\frac{1}{2}$% per year. Find the simple interest earned.

First, we will find the simple interest for one time period, which is one year.

$$\frac{8\frac{1}{2}}{100} = \frac{P}{1250}$$

P represents the interest for one year because $8\frac{1}{2}$% is the rate for one year. Remember, 8.5 may be substituted for $8\frac{1}{2}$.

$$8\frac{1}{2} \times 1250 = 100 \times P$$

$$10,625 = 100 \times P$$

$$\frac{10,625}{100} = P$$

$$\$106.25 = P$$

The interest for a full year is $106.25. Now we will find what portion of a year is represented by 6 months. There are 12 months in a year, so 6 months is $\frac{6}{12}$ or $\frac{1}{2}$ year.

To find the interest for 6 months, we find $\frac{1}{2}$ of the interest for a full year, or $\frac{1}{2}$ of $106.25.

$$\frac{1}{2} \times \frac{106.25}{1} = \frac{106.25}{2} = \$53.125 \text{ or } \$53.13$$ to the nearest cent
Do not forget this step when the amount of interest is figured for more or less time than the time period of the interest rate.

The interest earned is $53.13.

EXAMPLE 10 For one month a business received $3.07 in interest on an account balance of $245.75. What is the monthly rate of interest?

$3.07 is the interest or percentage, and $245.75 is the principal or base.

$$\frac{R}{100} = \frac{3.07}{245.75}$$ R represents the *monthly rate* of interest because $3.07 is the *monthly* interest.

$$R \times 245.75 = 100 \times 3.07$$

$$R \times 245.75 = 307$$

$$R = \frac{307}{245.75}$$

$$R = 1.24923703$$

$$R = 1.25\%$$ to the nearest hundredth of a percent

The monthly rate is 1.25%.

SELF-STUDY EXERCISES 4–4

1. Find the sales tax and the total bill on an order of office supplies costing $75.83 if the tax rate is 6%. Round to the nearest cent.
2. Materials to landscape a property total $785.84. What is the total bill if the sales tax rate is $5\frac{1}{2}\%$? Round to the nearest cent.
3. If the rate of social security tax is 6.2%, find the tax on gross earnings of $375.80. Round to the nearest cent.
4. An employee's gross earnings for a pay period are $895.65. The net pay for this salary is $675.23. What percent of the gross pay are the total deductions? Round to the nearest whole percent.

5. An employee has a net salary of $576.89 and a gross salary of $745.60. What percent of the gross salary is the total of the deductions? Round to the nearest whole percent.

6. A manufacturer will give a 2% discount to customers paying cash. If a parts store paid cash for an order totaling $875.84, what amount was saved? Calculate to the nearest cent.

8. What is the commission earned by a salesperson who sells $18,890 in merchandise if a 5% commission is paid on all sales?

10. A parts house manager is paid a salary of $2153 monthly plus a bonus of 1% of the net earnings of the business. Find the total salary for a month when the net earnings of the business were $105,275.

7. Find the cash price for an order of hospital supplies totaling $3985.57 if a 3% discount is offered for cash orders. Calculate to the nearest cent.

9. A manufacturer's representative is paid a salary of $140 per week and 7% commission on all sales over $3200 per week. The sales for a given week were $7412. What is the representative's salary for that week?

3 Solve the following problems.

11. An electrician purchases $650 worth of electrical materials. A finance charge of $1\frac{1}{2}\%$ per month is added to the bill. What is the finance charge for 1 month?

13. A nurse is charged $10.24 in interest on a credit-card account with a $584.87 average daily balance. What is the rate of interest? Round to the nearest hundredth of a percent.

15. Find the interest on a loan of $5840 at 12% per year for 2 years, 6 months.

12. $10,000 is invested for 3 months at 12% per year. How much interest is earned?

14. Find the interest on a loan of $2450 at 15% per year for 1 year.

CAREER APPLICATION

Electronics: Percentage Error

Technicians are often asked to find the percentage error in components and in parts of circuits.

When you calculate a value (called the *theoretical value* or the *predicted value*) and then measure that value (called the *measured* or *actual value*), the results are seldom exactly the same. It is not enough to say that measured values were close or not close to the predicted values. You need to give a percentage answer. This is called percentage error or percentage difference.

$$\% \text{ error} = \frac{\text{actual value} - \text{theoretical value}}{\text{theoretical value}} \times 100$$

or

$$\% \text{ error} = \frac{\text{measured value} - \text{coded value}}{\text{coded value}} \times 100$$

Suppose that you have a circuit in which you have calculated and then measured three voltage drops, or potential differences in voltage. These voltage drops are called V_1 and V_2 and V_3, and they are read as "V sub 1" and "V sub 2" and "V sub 3." They mean "voltage drop number one" and "voltage drop number two" and "voltage drop number three." The subscripts are numbers written "sub" or "under" the line and are there simply for identification purposes. Subscripts are an easy way of making complicated sets of numbers easier to read. By using sequential numbers, we can arrange the numbers in an order that is easy to use.

As long as you understand what is meant, you can also write V1 and V2 and V3. The convention is that a letter coming first means the number following the letter is meant to be a type of subscript or identifier. This practice is particularly important when programming computers.

Suppose your calculations show that you should get

$$V_1 = 20 \text{ V}, \qquad V_2 = 30 \text{ V}, \qquad V_3 = 40 \text{ V}$$

When measurements are made using a voltmeter, the results are

$$V_1 = 21 \text{ V}, \qquad V_2 = 29 \text{ V}, \qquad V_3 = 43 \text{ V}$$

First consider V_1. Since $V_{1\_meas} = 21$ V and $V_{1\_coded} = 20$ V,

$$\% \text{ error} = \frac{21 \text{ V} - 20 \text{ V}}{20 \text{ V}} \times 100 = \frac{1}{20} \times 100 = \frac{100}{20} = +5\%$$

Next consider V_2. Since $V_{2\_meas} = 29$ V and $V_{2\_coded} = 30$ V,

$$\% \text{ error} = \frac{29 \text{ V} - 30 \text{ V}}{30 \text{ V}} \times 100 = \frac{-1}{30} \times 100 = \frac{-100}{30} = -3.3\%$$

Next consider V_3. Since $V_{3\_meas} = 43$ V and $V_{3\_coded} = 40$ V,

$$\% \text{ error} = \frac{43 \text{ V} - 40 \text{ V}}{40 \text{ V}} \times 100 = \frac{300}{40} = 7.5\%$$

Your writeup could say that all measurements were within 10% of what was predicted, which means they were very close indeed.

Now calculate the % error for three more voltage drops, each with an actual difference of 3 volts. What is the % error if $V_{4\_meas} = 403$ V and $V_{4\_coded} = 400$ V?

$$\% \text{ error} = \frac{403 \text{ V} - 400 \text{ V}}{400 \text{ V}} \times 100 = \frac{300}{400} = 0.75\%, \text{ or less than } 1\%$$

Very close indeed.

What is the % error if $V_{5\_meas} = 3997$ V and $V_{5\_coded} = 4000$ V? The actual measurement difference is only 3 volts. But

$$\% \text{ error} = \frac{3997 \text{ V} - 4000 \text{ V}}{4000 \text{ V}} \times 100 = \frac{-300}{4000}$$

$$= -0.075\%, \text{ or less than } 1\%$$

which is extremely close.

What is the % error if $V_{6\_meas} = 1$ V and $V_{6\_coded} = 4$ V? The actual measurement difference is only 3 volts. But

$$\% \text{ error} = \frac{1 \text{ V} - 4 \text{ V}}{4 \text{ V}} \times 100 = \frac{-300}{4} = -75\%, \text{ or extremely high error}$$

Notice that when the measured value was more than the theoretical value, as with V_1, V_3, and V_4, the % error ended as positive. And when the measured value was less than the theoretical value, as with V_2, V_5, and V_6, the % error ended as negative.

Exercises

Fill in the % errors in the following data table and then decide if each error is high, low, or in the middle. Low is closer to zero, middle is closer to -10 or $+10$, and high is less than -10 or greater than $+10$.

Sample Data Table for Recording Voltages

Voltage Drop	Theoretical Value (V_c)	Measured Value (V_m)	% Error	High, Low, or Middle?
1. V_1	50	55		
2. V_2	50	46		
3. V_3	500	505		
4. V_4	500	480		
5. V_5	250	200		
6. V_6	250	260		
7. V_7	25	26		
8. V_8	25	23		
9. V_9	2	1.5		
10. V_{10}	2	2.2		

Answers for Exercises

Voltage Drop	Theoretical Value (V_c)	Measured Value (V_m)	% Error	High, Low, or Middle?
1. V_1	50	55	10	Middle
2. V_2	50	46	−8	Middle
3. V_3	500	505	1	Low
4. V_4	500	480	−4	Low
5. V_5	250	200	−20	High
6. V_6	250	260	4	Low
7. V_7	25	26	4	Low
8. V_8	25	23	−8	Middle
9. V_9	2	1.5	−25	High
10. V_{10}	2	2.2	10	Middle

ASSIGNMENT EXERCISES

Section 4–1

Find the value of the letter in each proportion.

1. $\dfrac{1}{2} = \dfrac{a}{9}$ 2. $\dfrac{x}{7} = \dfrac{3}{4}$ 3. $\dfrac{7}{16} = \dfrac{21}{y}$ 4. $\dfrac{3}{a} = \dfrac{2}{5}$

5. $\dfrac{2}{9} = \dfrac{x}{6}$ 6. $\dfrac{3}{16} = \dfrac{a}{4}$ 7. $\dfrac{2}{3} = \dfrac{4}{c}$ 8. $\dfrac{1}{4} = \dfrac{2}{a}$

9. $\dfrac{d}{5} = \dfrac{1}{2}$ 10. $\dfrac{3}{x} = \dfrac{5}{4}$ 11. $\dfrac{2}{9} = \dfrac{c}{3}$ 12. $\dfrac{7}{3} = \dfrac{10}{x}$

Identify the given and missing elements for each of the following as *R* (rate), *B* (base), and *P* (percent).

13. What percent of 25 is 5?
14. Six is 40% of what number?
15. 5% of 180 is what number?
16. $15 is what percent of $120?
17. 45% of how many dollars is $36?
18. What percent of 10 syringes is two syringes?
19. Six is what percent of 25 sacks of grass seed?
20. How many landscape contractors is 15% of 40 landscape contractors?
21. 18% of 150 pieces of sod is how many?
22. What percent of a total of 28 students is 8 students?

Solve the following problems involving percents.

23. 5% of 480 is what number?
24. $62\frac{1}{2}$% of 120 is what number?
25. $\frac{1}{4}$% of 175 is what number?
26. $233\frac{1}{3}$% of 576 is what number?
27. 39 is what percent of 65?
28. What percent of 118 is 42.48?
29. What percent of 65 is 162.5?
30. 80% of what number is 116?

31. 24% of what number is 19.92?

33. 260% of what number is 395.2?

35. 38.25 is what percent of 250?

37. What percent of 26 is 130?

39. $10\frac{1}{3}\%$ of what number is 8.68?

32. 7.56 is $6\frac{3}{4}\%$ of what number?

34. 3 is 0.375% of what number?

36. 83% of 163 is what number?

38. 4.75% of 348.2 is what number?

Section 4–2

40. Specifications for bronze call for 80% copper. How much copper is needed to make 300 lb of bronze?

42. When laying a subfloor using common 1-in. × 8-in. boards laid diagonally, 17% is allowed for waste. How many board feet will be wasted out of 1250 board feet?

44. The voltage loss in a line is 2.5 V. If this is 2% of the generator voltage, what is the generator voltage?

46. If a freshman class of 1125 college students is made up of 8% international students, how many international students are in the class?

48. A survey studied 600 people for their views on nuclear power plants near their towns. Of these, 75 people said that they approved of nuclear power plants near their towns. What percent approved?

50. It is estimated that only 19% of the licensed big game hunters on a state wildlife management area are successful. If 95 big game hunters were successful, what was the total number of big game hunters?

41. 84 lb of 224 lb of an alloy is zinc. What percent of the alloy is zinc?

43. 27 out of 2374 pieces produced by a particular machine were defective. What percent (to the nearest hundredth of a percent) were defective?

45. If a family spends 28% of its income on food, how much of a $950 paycheck goes for food?

47. A city prosecuted 1475 individuals with traffic citations. If 36,875 individuals received traffic citations, what percent were prosecuted?

49. In a certain college 67 students made the dean's list. If this was 33.5% of the student body, what was the total number of students in the college?

Section 4–3

51. A rough casting weighs 32.7 kg. After finishing on a lathe, it weighs 29.3 kg. Find the percent of weight loss to the nearest whole number percent.

53. A contractor ordered 800 board feet of lumber for a job that required 750 board feet. What percent, to the nearest whole number, of the required lumber was ordered for waste?

55. A lathe costing $600 was sold for $516. What was the percent of decrease in the price of the lathe?

57. A wet casting weighing 145 kg has a 2% weight loss in the drying process. How much will the dried casting weigh?

59. A blueprint specification for a part lists its overall length as 62.5 cm. If a tolerance of ±0.8% is allowed, find the limit dimensions of the length of the part.

61. A computer disk that once sold for $2.25 now sells for 25% less. How much does the computer disk now sell for?

52. A steel beam expands 0.01% of its length when exposed to the sun. If a beam measures 49.995 ft after being exposed to the sun, what is its cooled length?

54. A brickmason received a 12% increase in wages, amounting to $35.40. Find the amount of wages received before the increase. Find the amount of wages received after the increase.

56. According to specifications, a machined part may vary from its specified measure by ±0.4% and still be usable. If the specified measure of the part is 75 in. long, what is the range of measures acceptable for the part?

58. A mixture of uncut loam and clay soil will have a 20% earth swell when it is excavated. If 300 cubic yards of uncut earth are to be removed, how many cubic yards will have to be hauled away if the earth swell is taken into account?

60. A book that used to sell for $18.50 now sells for 20% more. How much does the book now sell for?

62. A paperback dictionary originally sold for $4. It now sells for $1 more. What is the percent of the increase?

63. A laptop computer that was originally $2400 is now $300 more. What is the percent of the increase?

64. When earth is dug up, it usually increases in bulk or expands by 20%. How much earth will a contractor have to haul away if 150 cubic feet are dug up?

65. A shipping carton is rated to hold 50 pounds. A stronger carton that can hold 15% more weight will be used for added safety. How much weight will the new carton hold?

66. A 20-inch bar of iron measured 20.025 inches when it was heated. What was the percent of increase?

67. A pearl necklace measuring 18 inches was exchanged for one measuring 24 inches. What was the percent of increase?

68. A casting weighed 130 ounces when first made. After it dried, it weighed 127.4 ounces. What was the percent of weight loss caused by drying?

69. A dieter went from 168 pounds to 160 pounds in one week. To the nearest tenth of a percent, what was the percent of weight loss?

70. Workers took a 10% pay reduction to help their company stay open during economic hard times. What was the reduced annual salary of a worker who originally earned $35,000?

71. A motorist traded an older car with 350 horsepower for a new car with 17.4% less horsepower. What is the horsepower of the new car to the nearest whole number?

Section 4–4

Solve the following problems.

72. Discontinued paneling is sale priced at 25% off the regular price. If the regular price is $12.50 per sheet, what is the sale price per sheet rounded to the nearest cent?

73. A parts distributor is paid a weekly salary of $250 plus an 8% commission on all sales over $2500 per week. What is the distributor's salary for a given week if the sales for that week totaled $4873?

74. An employee's gross earnings for a month are $1750. If the net pay is $1237, what percent of the gross pay is the total of the deductions? Round to the nearest whole percent.

75. An order of lumber totaled $348.25. If a 5% sales tax is added to the bill, what is the total bill? Round to the nearest cent.

76. A builder purchased a concrete mixer for $785. The builder did not know the sales tax rate, but the total bill was $828.18. Find the sales tax rate. Round to the nearest tenth of a percent.

77. An employer must match employees' social security contributions. If the weekly payroll is $27,542 and if the FICA rate is 6.2%, what are the employer's contributions? (All employees' year-to-date salaries were under maximum salary subject to the FICA tax.)

78. A contractor can receive a 2% discount on a monthly sand and gravel bill of $1655.75 if the bill is paid within 10 days of the statement date. How much is saved by paying the bill within the 10-day period? Figure to the nearest cent.

79. A businessperson is charged a $4.96 monthly finance charge on a bill of $283.15. What is the monthly interest rate on the account? Round to the nearest hundredth of a percent.

80. Find the interest on a loan of $3200 if the annual interest rate is 16% and the loan is for 9 months.

81. A property owner sold a house for $127,500 and paid off all outstanding mortgages. $52,475 cash was left from this transaction. If the money was invested at 14% interest for 18 months, how much interest was earned on this investment?

82. The sales tax in one city is 8.75% of the purchase price. How much is the sales tax on a purchase of $78.56?

83. A small town charges 3.25% of the purchase price for sales tax. What is the sales tax on a purchase of $27.45?

84. Interest for one year on a loan of $2400 was $396. What was the interest rate?

85. A $2000 certificate of deposit earned $170 interest in one year. What was the interest rate?

86. If a loan for $500 at 18.75% interest per annum (year) is paid in 3 months, how much is the interest?

87. What is the interest on a business loan for $6500 at 9.75% interest per annum (year) paid in 7 months?

88. A salesperson earns a commission of 15% of the total monthly sales. If the salesperson earned $2145, how much were the total sales?

89. A salesclerk in a store is paid a salary plus 3% commission. If the salesclerk earned $10.65 commission for a weekend, how much were the sales?

■■■■■ CHALLENGE PROBLEMS ■■■■■

90. A homeowner with an annual family income of $35,500 spends in a year $6900 for a home mortgage, $950 for property taxes, $380 for homeowner's insurance, $2400 for utilities, and $200 for maintenance and repair. To the nearest tenth, what percent of the homeowner's annual income is spent for housing?

91. A motorist with an annual income of $18,250 spends each year $3420 on automobile financing, $652 on gasoline, $625 on insurance, and $150 on maintenance and repair. To the nearest tenth, what percent of the motorist's annual income is used for automotive transportation?

92. A 78-pound alloy of tin and silver contains 69.3 pounds of tin. Find the percent of silver in the alloy to the nearest tenth of a percent.

93. There are 25 women in a class of 35 students. Find the percent of men in the class to the nearest tenth of a percent.

94. A librarian checked out 25 books in a bookmobile. The books included 5 mysteries, 2 science fiction novels, 8 biographies, and 10 classics. Mysteries and biographies accounted for what percent of the books checked out?

■■■■■ CONCEPTS ANALYSIS ■■■■■

1. Under what conditions are two fractions proportional?

2. Solving a proportion with one missing term requires two computations. In the proportion $\dfrac{R}{100} = \dfrac{65}{26}$, what are the two computations that are to be performed to find the value of R?

3. Give some clues for determining if a value in a percent problem represents a rate.

4. Give some clues for determining if a value in a percent problem represents a percentage.

5. Give some clues for determining if a value in a percent problem represents a base.

6. Is the amount of sales tax required on a purchase in your state determined by a percent? What is the sales tax rate in your state?

7. If the total bill, including sales tax, on a purchase is 107% of the original amount, what is the sales tax rate?

8. If a dress is marked 25% off the original price, what percent of the original price does the buyer pay?

9. If a quantity increases 50%, is the new amount twice the original amount? Explain your answer.

10. If the cost of an item decreases 50%, is the new amount half the original amount? Explain your answer.

Find and explain any mistakes in the following. Rework the incorrect problems correctly.

11.
$$\frac{R}{100} = \frac{4}{5}$$
$$\frac{R}{25} = \frac{1}{5}$$
$$5 \times R = 25 \times 1$$
$$5 \times R = 25$$
$$R = 25 \div 5$$
$$R = 5\%$$

12. $3\% = 0.3$

13. What percent of 25 is 75?
$$\frac{R}{100} = \frac{25}{75}$$
$$75 \times R = 100 \times 25$$
$$75 \times R = 2500$$
$$R = 2500 \div 75$$
$$R = 33\frac{1}{3}\%$$

14. 26 is 0.5% of what number?

$$\frac{0.5}{100} = \frac{26}{B}$$

$$0.5 \times 26 = 100 \times B$$

$$13 = 100 \times B$$

$$\frac{13}{100} = B$$

$$0.13 = B$$

15. If the cost of a $15 shirt increases 10%, what is the new cost of the shirt?

$$\frac{10}{100} = \frac{P}{15}$$

$$100 \times P = 10 \times 15$$

$$100 \times P = 150$$

$$P = \frac{150}{100}$$

$$P = 1.5$$

The shirt increased $1.50 in price.

MATHEMATICS IN THE WORKPLACE

Health Sciences: Ovarian Cancer Survivability

According to a recent study published in the *New England Journal of Medicine,* women who inherit a genetic susceptibility to ovarian cancer have greater survival rates from this disease than women without the genetic flaw. Women who are born with a flaw in a gene called BRCA1 face a 65% lifetime risk of getting ovarian cancer and an 85% risk of getting breast cancer. There is a 95% chance they will get either ovarian or breast cancer.

In this study conducted at the University of Pennsylvania Medical Center, doctors investigated whether the ovarian cancer in women with the genetic flaw is lethal as quickly as it is in women who have normal BRCA1 genes. They found that ovarian cancer patients with the bad gene survived an average of six years, compared with about two years for other patients. This is one small bit of good news for women with the BRCA1 flaw, most common in Jewish women of Eastern European descent.

Experts believe that about 10% of all ovarian cancer results from an inherited predisposition, and 80% to 90% of this is caused by BRCA1. The American Cancer Society predicts that about 26,700 cases of ovarian cancer will be reported in 1997 in the United States.

This study did not investigate the survivability of women with breast cancer resulting from the flawed BRCA1, but preliminary evidence from other research suggests that this disease is also less aggressive than in normal BRCA1 women. Women who get sick with these inherited cancers can be a little bit encouraged that their survival rate is better than that of the typical ovarian or breast cancer patient. Genetic screening tests for BRCA1 flaws are now available.

CHAPTER SUMMARY

Objectives	What to Remember with Examples

Section 4–1

1 Solve a proportion for any missing element.

Multiply diagonally across the equal sign to find the cross products. Then divide the product of the two known factors by the unknown factor of the other product.

$$\frac{3}{y} = \frac{2}{5}$$

$$3 \times 5 = 2 \times y$$

$$15 = 2 \times y$$

$$\frac{15}{2} = y$$

$$7\frac{1}{2} = y \text{ or } y = 7\frac{1}{2}$$

2 Identify the rate, base, and percentage in percent problems.

Use the identifying key words for rate (*percent* or *%*), base (*total, original;* associated with the word *of*), and percentage (*part;* associated with the word *is*).

$$
\begin{array}{ccc}
& R & B \quad P \\
\text{(a)} & \text{What percent of } 10 \text{ is } 5?
\end{array}
$$

$$
\begin{array}{ccc}
& R & B \qquad P \\
\text{(b)} & 25\% \text{ of what number is } 3?
\end{array}
$$

$$
\begin{array}{ccc}
& P \quad R & B \\
\text{(c)} & \text{Three is } 20\% \text{ of what number?}
\end{array}
$$

3 Solve the percentage proportion for any missing element.

Multiply diagonally across the equal sign to find the cross products. Divide the product of the two numbers by the factor with the *R*, *P*, or *B*.

$$
\frac{R}{100} = \frac{5}{25}
$$
$$
25 \times R = 5 \times 100
$$
$$
25 \times R = 500
$$
$$
R = \frac{500}{25}
$$
$$
R = 20\%
$$

Section 4–2

1 Use the percentage proportion to find the percentage.

Use key words like *of,* signaling base, and *is,* signaling percentage, to help identify the missing element. Then set up the percentage proportion with *P* for percentage. The percentage is a part of the whole amount.

What amount is 5% of $200? *Of* identifies $200 as the base. The percent or rate is given. The missing element must be the percentage.

$$
\frac{5}{100} = \frac{P}{200}
$$
$$
5 \times 200 = P \times 100
$$
$$
1000 = P \times 100
$$
$$
\frac{1000}{100} = P
$$
$$
\$10 = P
$$

Use a shortcut. Multiply base by decimal equivalent of the percent.

$$
5\% \times 200 = P
$$
$$
0.05 \times 200 = 10
$$

2 Use the percentage proportion to find the rate.

Use key words like *of,* signaling base, and *is,* signaling percentage, to help identify the missing element. Then set up the percentage proportion with *R* for rate. The rate is a percent.

What percent of 6 is 2? *Is* suggests 2 is the percentage. *Of* identifies 6 as the base.

$$
\frac{R}{100} = \frac{2}{6}
$$
$$
6 \times R = 2 \times 100
$$
$$
6 \times R = 200
$$

$$R = \frac{200}{6}$$

$$R = 33\frac{1}{3}\%$$

Use a shortcut. Divide percentage by the base.

$$\frac{2}{6} = 0.33\frac{1}{3} = 33\frac{1}{3}\%$$

3 Use the percentage proportion to find the base.

Use key words like *of,* signaling base, and *is,* signaling percentage, to help identify the missing element. Then set up the percentage proportion with B for base. The base is the total amount.

12 is 24% of what number? The percent is given. 12 is the part or percentage and is suggested by *is. Of* identifies "what number" as the missing base.

$$\frac{24}{100} = \frac{12}{B}$$
$$24 \times B = 12 \times 100$$
$$24 \times B = 1200$$
$$B = \frac{1200}{24}$$
$$B = 50$$

4 Use the percentage proportion to solve applied problems.

Use key words like *of,* signaling base, and *is,* signaling percentage, to help identify the missing element. Then set up the percentage proportion with R, P, or B as the missing element. The rate is a percent. The percentage is a part of the whole amount. The base is the total amount.

LaQuita sold 12 boxes of candy for a school project. If she started with 25 boxes, what percent of the candy did she sell? The percent is missing. 25 boxes is the total amount, or base. 12 is the part or percentage.

$$\frac{R}{100} = \frac{12}{25}$$
$$25 \times R = 12 \times 100$$
$$25 \times R = 1200$$
$$R = \frac{1200}{25}$$
$$R = 48\%$$

Section 4–3

1 Find the amount of increase or decrease in percent problems.

The original amount is the base. The new amount is the original amount plus the increase or the original amount less the decrease. Subtract the new amount and the original amount to find the amount of increase or decrease.

Julio made $15.25 an hour but took a pay cut to keep his job. If he now makes $12.20 an hour, what was the percent of the pay cut? $15.25 − $12.20 = $3.05 (amount of decrease). $3.05 is the part (percentage) being considered. $15.25 is the original amount, or base.

$$\frac{R}{100} = \frac{3.05}{15.25}$$
$$15.25 \times R = 3.05 \times 100$$

$$15.25 \times R = 305$$
$$R = \frac{305}{15.25}$$
$$R = 20\%$$

2 Find the new amount directly in percent problems.

Add the percent of increase to 100% or subtract the percent of decrease from 100%. Use this new percent in the percentage proportion.

A project requires 5 lb of galvanized nails. If 15% of the nails will be wasted, how many pounds must be purchased? 5 lb is the base, the original amount. The percent is 100% + 15%, or 115%. The amount of nails needed is the percentage, or part.

$$\frac{115}{100} = \frac{P}{5}$$
$$100 \times P = 5 \times 115$$
$$100 \times P = 575$$
$$P = \frac{575}{100}$$
$$P = 5.75 \text{ lb}$$

3 Find the rate or the base in increase or decrease problems.

Subtract the original amount and the new amount to find the amount of increase or decrease. Then set up the percentage proportion, using R or B for the missing element.

A 5-in. power edger blade now measures $4\frac{3}{4}$ in. What is the percent of wear? $4\frac{4}{4} - 4\frac{3}{4} = \frac{1}{4}$. The part or percentage is $\frac{1}{4}$. The base or original amount is 5 in.

$$\frac{R}{100} = \frac{\frac{1}{4}}{5}$$
$$5 \times R = \frac{1}{4} \times 100$$
$$5 \times R = 25$$
$$R = \frac{25}{5}$$
$$R = 5\%$$

A PC lost 20% of its memory when a set of RAM chips was removed. If the PC now has 5.12G of memory, what was the original memory? Current memory is 100% − 20% = 80% of original memory, or base. 5.12G is the part, or percentage.

$$\frac{80}{100} = \frac{5.12}{B}$$
$$80 \times B = 5.12 \times 100$$
$$80 \times B = 512$$
$$B = \frac{512}{80}$$
$$B = 6.4\text{G}$$

Section 4–4

1 Calculate sales tax and payroll deductions.

The rate is the percent of tax or deduction. The base is the amount of the purchase or the gross pay. The percentage is the amount of tax or deduction. Solve with a percentage proportion.

If the sales tax rate is 7%, find the tax and total bill on a purchase of $30.

$$\frac{7}{100} = \frac{P}{30}$$
$$7 \times 30 = 100 \times P$$
$$210 = 100 \times P$$
$$\frac{210}{100} = P$$
$$\$2.10 = P$$

Total cost is $30 + \$2.10 = \$32.10$.

If 8% of gross pay is deducted for retirement, how much is deducted for retirement from a monthly salary of $1650? What is the net pay?

$$\frac{8}{100} = \frac{P}{1650}$$
$$8 \times 1650 = 100 \times P$$
$$13{,}200 = 100 \times P$$
$$\frac{13{,}200}{100} = P$$
$$\$132 = P \qquad \text{deducted for retirement}$$

Net pay is $1650 - \$132 = \$1518$.

2 Calculate discount and commission.

The rate is the percent of discount or commission. The base is the price or amount of sales. The percentage is the amount of the discount or commission. Solve with a percentage proportion using P as the missing element.

A $45 dress is on sale at 20% off. How much is the discount? What is the sale price?

$$\frac{20}{100} = \frac{P}{45}$$
$$100 \times P = 20 \times 45$$
$$100 \times P = 900$$
$$P = \frac{900}{100}$$
$$P = \$9 \qquad \text{discount}$$

The sale price is $45 - \$9 = \$36$.

Chang sold $5000 of furniture and earned a 3% commission. How much is his commission?

$$\frac{3}{100} = \frac{P}{5000}$$
$$100 \times P = 3 \times 5000$$
$$100 \times P = 15{,}000$$
$$P = \frac{15{,}000}{100}$$
$$P = \$150$$

3 Calculate interest on loans and investments.

The rate is the percent of interest. The base is the amount of the loan or investment. The percentage is the amount of interest. Solve with a percentage proportion using P as the missing element.

Boris borrowed $2500 at 8.5% interest for one year. How much did he have to repay?

$$\frac{8.5}{100} = \frac{P}{2500}$$

$$100 \times P = 8.5 \times 2500$$

$$100 \times P = 21{,}250$$

$$P = \frac{21{,}250}{100}$$

$$P = \$212.50 \qquad \text{interest}$$

He repaid $2500 + $212.50 = $2712.50.

Interest on an investment is found the same way.

WORDS TO KNOW

ratio (p. 176)
proportion (p. 176)
cross products (p. 176)
rate (p. 177)
base (p. 177)
percentage (p. 177)

percentage proportion (p. 177)
new amount (p. 189)
percent of increase or decrease (p. 189)
sales tax (p. 195)
FICA (p. 195)

gross pay (p. 196)
payroll deductions (p. 196)
discount (p. 197)
commission (p. 197)
interest (p. 198)
annual percentage rate (p. 199)

CHAPTER TRIAL TEST

Find the value of the letter in each proportion.

1. $\dfrac{a}{10} = \dfrac{4}{5}$

2. $\dfrac{2}{8} = \dfrac{3}{x}$

3. $\dfrac{2}{c} = \dfrac{1}{6}$

Identify the given and missing elements for each of the following as R (rate), B (base), and P (percent).

4. $10 is what percent of $35?

5. 40% of 10 x-ray technicians is how many?

6. What percent of 24 syringes is 8?

7. 9 is what percent of 27 dogwood trees?

8. How many books is 30% of 40 books?

9. 12% of 50 grass plugs is how many?

Solve the following problems involving percents.

10. 10% of 150 is what number?

11. What number is $6\frac{1}{4}\%$ of 144?

12. What percent of 275 is 33?

13. 45.75 is 15% of what number?

14. 55 is what percent of 11?

15. 250% of what number is 287.5?

16. What percent of 360 is 1.2?

17. 245% of what number is 164.4? Round to the nearest hundredth.

18. 5.4% of 57 is what number?

19. $15,000 is invested at 14% per year for 3 months. How much interest is earned on the investment?

20. A casting measuring 48 cm when poured shrinks to 47.4 cm when cooled. What is the percent of decrease?

21. An electronic parts salesperson earned $175 in commission. If the commission is 7% of sales, how much did the salesperson sell?

22. Electronic parts increased 15% in cost during a certain period, amounting to an increase of $65.15 on one order. How much would the order have cost before the increase? Round to the nearest cent.

23. In 1996, an area vocational school had an enrollment of 325 men and 123 women. In 1997 there were 149 women. What was the percent of increase of women students? Round to the nearest hundredth.

24. The payroll for an electrical shop for one week was $1500. If federal income and FICA taxes averaged 28%, how much was withheld from the $1500?

25. During one period, a bakery rejected 372 items as unfit for sale. In the following period, the bakery rejected only 323 items, a decrease in unfit bakery items. What was the percent of decrease? Round to the nearest hundredth.

26. Materials to landscape a new home cost $643.75. What is the amount of tax if the rate is 6%? Round to the nearest cent.

27. A casting weighed 36.6 kg. After milling, it weighs 34.7 kg. Find the percent of weight loss to the nearest whole percent.

28. After soil was excavated for a project, it swelled 15%. If 275 yd^3 were excavated, how many cubic yards of soil were there after excavation?

29. The total bill for machinist supplies was $873.92 before a discount of 12%. How much was the discount? Round to the nearest cent.

30. A business paid a $5.58 finance charge on a monthly balance of $318.76. What was the monthly rate of interest? Round to the nearest hundredth.

5

Direct Measurement

GOOD DECISIONS THROUGH TEAMWORK

Think about what would happen if the metric system completely replaced the US customary system in the United States. Would people still refer to an *inch*worm or the 50 *yard* line? Would they still use sayings like, "An *ounce* of prevention is worth a *pound* of cure"? What about Robert Frost's poem "Stopping by Woods on a Snowy Evening," which ends with "And *miles* to go before I sleep"?

Spend a week observing examples of both systems of measurement in everyday use. Look for things like product packaging, tools, hardware items, sporting events, clothing sizes, language use (like Frost's poem), and so on. Record the examples of the measures and the situations in which they were used. Then investigate the advantages and disadvantages of using each system of measurement in different situations. You might investigate library sources or interview instructors and other professionals in contrasting fields like physical science and social science, or English and engineering.

With your team members, discuss the advantages and disadvantages of each type of measurement you have observed or investigated. Also discuss the status of the gradual conversion to the metric system in the United States and what impact this conversion is having on the way people speak, work, play, and so on. Choose a team member to present the team's findings to your class.

Measurements are used in business, industry, health care, and our everyday lives. Increased international trade has created a need for a worldwide standard of measurements. The International System of Units (SI), more commonly called the *metric system*, has been adopted by most nations of the world, including the United States, and a conversion to this system of measurement is gradually taking place.

However, the United States has used a nonmetric system of measurement for many years, and this system is still the customary system of measurement for many businesses and industries. This system is called the *English* or the *U.S. customary system of measurement*, and we begin our study of measurements with it.

5–1 THE U.S. CUSTOMARY SYSTEM OF MEASUREMENT

Learning Objectives

1. Identify uses of U.S. customary system measures of length, weight, and capacity.
2. Convert from one U.S. customary unit of measure to another.
3. Write mixed U.S. customary measures in standard notation.

The *U.S. customary* or *English system* of measurement has always been used in the United States. Many of the units of measure within this system have become obsolete through the years. In our discussion of the U.S. customary system of measurement, we will include only those measures currently in use.

1 Identify Uses of U.S. Customary System Measures of Length, Weight, and Capacity.

There are three basic units in the U.S. customary system commonly used to measure length. They are the inch, the foot, and the mile. Table 5–1 gives the relationships among these measurements of length.

TABLE 5–1 U.S. Customary Units of Length or Distance

12 inches (in.)[a] = 1 foot (ft)[b]	36 inches (in.) = 1 yard (yd)
3 feet (ft) = 1 yard (yd)	5280 feet (ft) = 1 mile (mi)

[a]The symbol ″ means inches (8″ = 8 in.) or seconds (60″ = 60 seconds).

[b]The symbol ′ means feet (3′ = 3 ft) or minutes (60′ = 60 minutes).

Figure 5–1

Inch: An inch is slightly less than the width of a quarter coin (Fig. 5–1). Historically, an inch was considered to be the width of a man's thumb. Inches are used to measure lengths that vary by increments the size of the width of a man's thumb or the width of a U.S. quarter. Some examples of lengths measured in inches are the sizes of men's trousers, belts, shirts, and jackets. Also, the diagonal distance in inches from corner to opposite corner of a television screen determines its size, such as 13-in. or 19-in. television screens. Photographs and picture frames, such as 5 × 7 or 8 × 10, are measured in inches. Lumber sizes, like 2 × 4's or 2 × 12's, are indicated in inches.

Foot: A foot is 12 inches, or about the width of a placemat for a dining table (Fig. 5–2). The measure was originally based on the length of a human foot. Feet are used to measure larger lengths varying in size by increments of 12 inches. Lumber, for instance, is sold by the foot. A 2 × 4 is usually sold in lengths of 8 ft, 12 ft, 16 ft, or 20 ft. Human heights are also often measured in feet (and inches), such as 5 ft, 6 in.; 5 ft, 11 in.; or 6 ft, 2 in. Elevation, such as the height of mountains and the altitude of airplanes, is designated in feet.

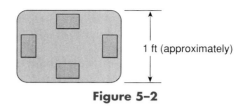

Figure 5-2

Mile: A mile is 5280 feet, or the length of approximately eight city blocks (Fig. 5–3). The mile was originally used by the Romans, who considered a mile to be 1000 paces of 5 feet each. Long distances are measured in miles, such as the distances between cities and the lengths traveled on a road, street, or highway. The mile is also used to designate the speed of a vehicle and speed limits, such as 55 miles per hour (mph) or 30 mph.

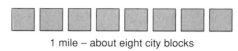

1 mile – about eight city blocks

Figure 5-3

Another length measure is the *yard*. The yard is three feet or 36 inches. Yards are used in similar circumstances as feet, as well as measures for fabric and football fields.

Many goods are exchanged or sold according to their amount of weight or mass. There are three commonly used measuring units for weight or mass in the U.S. customary system. They are the ounce, pound, and ton. Table 5–2 gives the relationships among these measures of weight or mass.

TABLE 5-2 U.S. Customary Units of Weight or Mass

16 ounces (oz) = 1 pound (lb)
2000 pounds (lb) = 1 ton (T)

1 ounce — approximately three packets of artifical sweetener

Figure 5-4

1 lb – about one can of cut green beans

Figure 5-5

Ounce: The ounce is $\frac{1}{16}$ of a pound and is used to measure objects or products that vary in weight by increments of this weight. It is the approximate weight of three individual packages of artificial sweetener (Fig. 5–4). (The troy ounce, which will not be used in this text, measures precious metals and is $\frac{1}{12}$ of a pound. *Ounce* is derived from a Latin word for $\frac{1}{12}$.) Ounces are used for measures of lightweight items like first-class letters, tubes of toothpaste or medicinal ointments, canned goods, and dry-packaged foods like pasta, candy, and gravy mixes.

Pound: A pound is equal to 16 ounces. It is used to measure heavier items that vary by increments of this weight. The term is derived from an old English word for weight. A pound is the weight of a 4.25-in. tall can of cut green beans or a paper container of coffee (Fig. 5–5). Many small, medium, and large items of merchandise are measured in pounds, like grocery and market items, people's weights, and air pressure in automotive tires. The pound is also used to calculate shipping charges for merchandise and parcel-post packages.

Ton: A ton is 2000 pounds. The word originally meant a weight or measure. The typical American-made station wagon may help you to visualize this measure somewhat (Fig. 5–6) because an American-made station wagon may weigh 3 tons (although such vehicle weights are usually expressed in pounds). The ton is used for extremely heavy items, such as very large volumes of grain, the weights of huge animals like elephants (4 to 7 tons), and the weights by which coal and iron are sold.

3 tons – approximately

Figure 5–6

The U.S. customary system of measurement has several units commonly used to measure capacity or volume. The system includes measuring units for both liquid and dry capacity measures; however, the dry capacity measures are seldom used. It is more common to express dry measures in terms of weight than in terms of capacity.

The U.S. customary units of measure for capacity or volume are the ounce, cup, pint, quart, and gallon. Table 5–3 gives the relationships among the liquid measures for capacity or volume. In the U.S. customary system, the term *ounce* is used to represent both a weight measure and a liquid capacity measure. The measures are different and have no common relationship. The context of the problem will suggest whether the unit for weight or capacity is meant.

TABLE 5–3   U.S. Customary Units of Liquid Capacity or Volume

8 ounces (oz) = 1 cup (c)	4 cups (c) = 1 quart (qt)
2 cups (c) = 1 pint (pt)	4 quarts (qt) = 1 gallon (gal)
2 pints (pt) = 1 quart (qt)	

Ounce: A liquid ounce is a volume, not a weight. It is about the quantity of two small bottles of fingernail polish or perfume (Fig. 5–7). Liquid ounces are used for bottled medicine, canned or bottled carbonated beverages, baby bottles and formula, and similar-size quantities.

Cup: A cup is equal to 8 ounces. The term is derived from an old English word for tub, a kind of container. This measure is about the volume of a coffee cup (Fig. 5–8). Cups are most often used for 8-oz quantities in cooking. Measuring cups used for cooking usually divide their contents into cups and portions of cups.

1 ounce – approximately two small bottles of fingernail polish

Figure 5–7

1 cup – approximately one large coffee cup

Figure 5–8

Pint: A pint is two cups. Its name comes from an Old English word meaning the spot that marks a certain level in a measuring device. It is the quantity of a *medium*-size paper container of milk (Fig. 5–9). Liquids like milk, automobile motor additives, produce like strawberries, ice cream, and similar quantities may be packaged in pint containers. In some localities, shucked oysters are sold by the pint.

Quart: A quart is two pints, or four cups, or 32 ounces. The term is derived from an Old English word for fourth. It is a fourth of a gallon. Milk and various citrus juices are often sold in quart containers (Fig. 5–10). Motor oil and ice cream are also packaged in quart containers. Quarts are also used in measuring liquid quantities in cooking and in measuring the capacities of cookware like mixing bowls, pots, and casserole dishes.

Gallon: A gallon is four quarts. It is based on the "wine gallon" of British origin. Paint, varnish, and stain are often sold in gallon cans (Fig. 5–11). Motor fuel and heating oil are usually sold by the gallon. Large quantities of liquid propane and various chemicals are also sold by the gallon. Often, statistics on liquid consumption, such as water or alcoholic beverages, are reported in gallons consumed per individual during a specified time period.

1 pint – one large single-serving container of milk

Figure 5–9

1 quart – one container of milk

Figure 5–10

PAINT
1 GALLON

1 gallon – one large can of paint

Figure 5–11

EXAMPLE 1   Indicate the most reasonable measuring units for the following:

1. Package of rice
2. Height of Doctor Washington
3. A hippopotamus
4. Sugar for a cake
5. Expensive French perfume
6. Distance between Memphis and New Orleans
7. Bottle of suntan lotion
8. Container of eggnog
9. Sherbet in grocery store freezer
10. Tank of pesticide
11. Man's dress shirt size
12. Material for draperies

1. Pounds or ounces
2. Feet, or feet and inches
3. Tons or pounds
4. Cups
5. Ounces
6. Miles

7. Ounces
8. Gallon, quart, or pint
9. Gallon or pint
10. Gallons
11. Inches
12. Yards

2 Convert from One U.S. Customary Unit of Measure to Another.

Technical applications of units of measure often call for changing from one unit to another. Converting from one unit of measure to another in the U.S. customary system can be done by deciding whether the beginning unit of measure is larger or smaller than the desired unit of measure. When changing to a larger unit of measure, there will be fewer units and a division step is required. When changing to a smaller unit of measure, there will be more units and a multiplication step is required. This process is often confusing, so an alternative process that employs a *unity ratio* can be used.

■ **DEFINITION 5–1:   Unity Ratio.**   A *unity ratio* is a ratio of measures whose value is 1.

A ratio is a fraction. A unity ratio, then, is a fraction with one unit of measure in the numerator and a different, but equivalent, unit of measure in the denominator. Some examples of unity ratios are

$$\frac{12 \text{ in.}}{1 \text{ ft}}, \quad \frac{1 \text{ ft}}{12 \text{ in.}}, \quad \frac{3 \text{ ft}}{1 \text{ yd}}, \quad \frac{1 \text{ mi}}{5280 \text{ ft}}$$

In each unity ratio, the value of the numerator equals the value of the denominator. When we have a ratio with the numerator and denominator equal, the value of the ratio is 1. We call these ratios *unity ratios* because their value is 1. The word *unity* means 1. When we convert from one unit of measure to another, we use a unity ratio that contains the original unit and the new unit.

> **Rule 5–1** *To change from one U.S. customary unit of measure to another:*
>
> 1. Set up the original amount as a fraction with the original unit of measure in the numerator.
> 2. Multiply this by a unity ratio with the original unit in the denominator and the new unit in the numerator.
> 3. Reduce like units of measure and all numbers wherever possible.

EXAMPLE 2   Find the number of inches in 5 ft.

To make this conversion, we multiply 5 ft by a unity ratio that contains both inches and feet so that these are the units involved.

Because 5 ft is a whole number, it is written with 1 as the denominator. The original unit is placed with the 5 in the *numerator* of the first fraction.

$$\frac{5 \text{ ft}}{1}\left(\frac{\quad}{\quad}\right)$$

The unit of measure we are changing *from* (feet) is placed in the *denominator* of the unity ratio. This placement will allow the original unit to reduce later.

$$\frac{5 \text{ ft}}{1}\left(\frac{\quad}{\text{ft}}\right)$$

The unit we are changing *to* (inches) is placed in the *numerator* of the unity ratio.

$$\frac{5 \text{ ft}}{1}\left(\frac{\text{in.}}{\text{ft}}\right)$$

Now we place in the unity ratio the numerical values that make these two units of measure equivalent (1 ft = 12 in.). Then we complete the calculation, reducing wherever possible.

$$\frac{5 \cancel{\text{ft}}}{1}\left(\frac{12 \text{ in.}}{1 \cancel{\text{ft}}}\right) = 60 \text{ in.}$$

Thus, 5 ft = 60 in.

EXAMPLE 3   How many pints are in 4.5 quarts?

$$\frac{4.5 \cancel{\text{qt}}}{1}\left(\frac{2 \text{ pt}}{1 \cancel{\text{qt}}}\right) = 4.5 \, (2 \text{ pt}) = 9 \text{ pt} \qquad \text{from quart (denominator) to pint (numerator)}$$

Thus, 4.5 qt = 9 pt.

Sometimes it is necessary to convert a U.S. customary unit to some U.S. unit other than the next larger or smaller unit of measure. For example, suppose that a dressmaker needs to know how many yards are in a certain number of inches of fabric. One way to solve this problem is to convert inches to feet with one unity ratio and then convert feet to yards with another unity ratio.

Tips and Traps *Changing to Any Larger or Smaller Unit:*
To change from one U.S. customary unit to one other than the next larger or smaller unit, proceed as before, but multiply the original amount by as many unity ratios as needed to attain the new U.S. customary unit.

EXAMPLE 4 How many inches are in $2\frac{1}{3}$ yd?

Write $2\frac{1}{3}$ as an improper fraction.

$$2\frac{1}{3}\ \text{yd} = \frac{7}{3}\ \text{yd}$$

Multiply the improper fraction by two unity ratios. One unity ratio is needed to convert yards to feet, and the other is needed to convert feet to inches. Note the original unit in the *numerator* of the improper fraction, $\frac{7}{3}$.

$$\frac{7\ \text{yd}}{3}\left(\frac{\text{ft}}{\text{yd}}\right)\left(\frac{\text{in.}}{\text{ft}}\right)$$

Insert the numerical values that make the units equivalent in each unity ratio (3 ft = 1 yd; 12 in. = 1 ft). Then complete the calculation, reducing wherever possible.

$$\frac{7\ \cancel{\text{yd}}}{\cancel{3}}\left(\frac{\overset{1}{\cancel{3}}\ \cancel{\text{ft}}}{1\ \cancel{\text{yd}}}\right)\left(\frac{12\ \text{in.}}{1\ \cancel{\text{ft}}}\right) = 84\ \text{in.}$$

Thus, $2\frac{1}{3}$ yd = 84 in.

Alternative method
If we use the conversion factor 36 in. = 1 yd, we could eliminate one unity ratio from the calculation; that is, we could convert $2\frac{1}{3}$ yd ($\frac{7}{3}$ yd) to inches as follows:

$$\frac{7\ \text{yd}}{3}\left(\frac{36\ \text{in.}}{1\ \text{yd}}\right)$$ Set up the unity ratio using the fact that 36 in. = 1 yd.

$$\frac{7\ \cancel{\text{yd}}}{\cancel{3}}\left(\frac{\overset{12}{\cancel{36}}\ \text{in.}}{1\ \cancel{\text{yd}}}\right) = 84\ \text{in.}$$ Reduce units and numbers; then multiply.

Thus, $2\frac{1}{3}$ yd = 84 in.

EXAMPLE 5 Find the number of ounces in 2 gallons.

Four unity ratios can be employed to convert gallons to quarts, quarts to pints, pints to cups, and cups to ounces. The values are obtained from Table 5–3.

$$\frac{2\ \text{gal}}{1}\left(\frac{\text{qt}}{\text{gal}}\right)\left(\frac{\text{pt}}{\text{qt}}\right)\left(\frac{\text{c}}{\text{pt}}\right)\left(\frac{\text{oz}}{\text{c}}\right) =$$

$$\frac{2\ \cancel{\text{gal}}}{1}\left(\frac{4\ \cancel{\text{qt}}}{1\ \cancel{\text{gal}}}\right)\left(\frac{2\ \cancel{\text{pt}}}{1\ \cancel{\text{qt}}}\right)\left(\frac{2\ \cancel{\text{c}}}{1\ \cancel{\text{pt}}}\right)\left(\frac{8\ \text{oz}}{1\ \cancel{\text{c}}}\right) =$$ Reduce measures.

$$2(4)(2)(2)(8)\ \text{oz} = 256\ \text{oz}$$ Multiply.

Thus, 2 gal = 256 oz.

Alternative method
We could work this same problem with fewer unity ratios if we made some preliminary calculations with the conversion factors from Table 5–3. For example, 4 c = 1 qt and 8 oz = 1 c, and we can multiply 4 (cups) by 8 (ounces per cup) to find the number of ounces in a quart; that is, $4 \times 8 = 32$ oz. Thus, 32 oz = 1 qt. We can now find the number of ounces in 2 gallons with fewer unity ratios.

$$\frac{2 \text{ gal}}{1}\left(\frac{4 \text{ qt}}{1 \text{ gal}}\right)\left(\frac{32 \text{ oz}}{1 \text{ qt}}\right)$$

Set up unity ratios using 4 qt = 1 gal and 32 oz = 1 qt.

$$\frac{2 \cancel{\text{gal}}}{1}\left(\frac{4 \cancel{\text{qt}}}{1 \cancel{\text{gal}}}\right)\left(\frac{32 \text{ oz}}{1 \cancel{\text{qt}}}\right) = 256 \text{ oz}$$

Reduce and multiply.

Thus, 2 gal = 256 oz.

Tips and Traps   *Two Shortcuts for Changing Units of Measure:*

1. To change a U.S. customary unit to a desired *smaller* unit: *Multiply* the number of larger units by the number of desired smaller units that equals 1 larger unit.

   2 yd = _____ ft   The desired smaller unit is feet: 3 ft = 1 yd.

   2×3 ft = 6 ft

   Multiply number of yards by 3 ft.
   Dimension analysis: $\frac{2 \text{ yd}}{1} \times \frac{3 \text{ ft}}{1 \text{ yd}} = 6$ ft

   Thus, 2 yd = 6 ft.

   Let's reduce this shortcut to key word clues.

   To smaller unit → obtain more units → multiply

2. To change a U.S. customary unit to a desired *larger* unit: *Divide* the original measure by the number of smaller units that equals 1 desired larger unit.

   12 ft = _____ yd   The desired larger unit is yards: 1 yd = 3 ft.

   $$\frac{12 \text{ ft}}{3 \text{ ft}} = \frac{\overset{4}{\cancel{12} \text{ ft}}}{\underset{1}{\cancel{3} \text{ ft}}} = 4 \text{ yd}$$

   Divide original measure by measure having number of smaller units equal to 1 larger unit.
   Dimension analysis: $\frac{12 \text{ ft}}{1} \times \frac{1 \text{ yd}}{3 \text{ ft}} = 4$ yd

   Thus, 12 ft = 4 yd.

   Using key word clues:

   To larger unit → obtain fewer units → divide

---

3 **Write Mixed U.S. Customary Measures in Standard Notation.**

Some measures are expressed by using two or more units of measure. For example, the weight of an object may be recorded as 5 lb 3 oz.

■ **DEFINITION 5–2: Mixed Measures.** Measures expressed by using two or more units of measure are called *mixed measures.*

■ **DEFINITION 5–3: Standard Notation.** *Standard notation* means that, in a mixed measure, each unit of measure is converted to the next larger unit of the mixed measure whenever possible.

If an answer is 3 ft 15 in., the 15 in. is converted to 1 ft 3 in. and the 1 ft is added to the 3 ft.

3 ft 15 in. = 3 ft + 12 in. + 3 in. = 3 ft + 1 ft + 3 in. = 4 ft 3 in.

Thus, in standard notation, 3 ft 15 in. is 4 ft 3 in.

EXAMPLE 6 Express (a) 8 lb 20 oz and (b) 1 gal 5 qt in standard notation.

(a) 8 lb 20 oz   20 oz = 1 lb 4 oz

Thus, 8 lb 20 oz = 9 lb 4 oz in standard notation.

(b) 1 gal 5 qt   5 qt = 1 gal 1 qt

Thus, 1 gal 5 qt = 2 gal 1 qt.

If the mixed measure contains three or more different measures, then conversion to standard notation may require two steps.

EXAMPLE 7 Express 2 yd 4 ft 16 in. in standard notation.

2 yd 4 ft 16 in. = 2 yd 4 ft 12 in. + 4 in. = 2 yd 4 ft + 1 ft + 4 in. =

2 yd 5 ft 4 in. = 2 yd 3 ft + 2 ft 4 in. = 2 yd + 1 yd 2 ft 4 in. =

3 yd 2 ft 4 in.

Thus, 2 yd 4 ft 16 in. = 3 yd 2 ft 4 in.

SELF-STUDY EXERCISES 5–1

1 Identify the appropriate U.S. customary measure for each of the following.

1. Shipping weight of a sofa
2. Liquid medicine in a bottle
3. Package of dried beans
4. Large home aquarium
5. Container of motor oil
6. Shipment of coal
7. Sack of potatoes
8. Size of a casserole dish
9. Height of a hospital patient
10. Milk for a cake recipe
11. Man's belt size
12. Cloth for a Mardi Gras costume
13. Hourly speed of a train
14. Weight of a first-class letter
15. Shaving lotion
16. Distance from home to a doctor's office across town
17. Package of macaroni
18. Pork roast
19. Parcel-post package
20. Tube of ointment

2 Using unity ratios, convert the given measures to the new units of measure.

21. 4 ft = _____ in.
22. 7 yd = _____ ft
23. $2\frac{1}{2}$ mi = _____ yd

24. Find the number of yards in 28 ft.
25. If a car with 0 miles on the odometer was driven 10,560 ft, how many miles would register on the odometer?

Using any method, convert the given measures to the new units of measure.

26. How many ounces are in 3 lb?
27. Find the number of pounds in $36\frac{4}{5}$ oz.
28. How many ounces are in 45.8 lb?
29. A can of vegetables weighs 1.2 lb. How many ounces is this?
30. The net weight of a can of corn is 17 oz. If a case contains 24 cans, what is the net weight of a case in ounces? in pounds?
31. How many quarts are in 5 gal?
32. How many pints are in $6\frac{1}{2}$ qt?
33. Find the number of gallons in 6 qt.
34. How many pints are in $7\frac{1}{2}$ gal?
35. How many gallons are in 72 pt?
36. How many ounces are in 2 gal?
37. How many yards are in 54 in.?
38. How many inches are in 3 yd?
39. How many feet are in $1\frac{1}{3}$ mi?
40. How many cups are in $1\frac{1}{2}$ gal?

3 Express the following measures in standard notation.

41. 2 ft 20 in.	**42.** 1 mi 6375 ft	**43.** 2 lb $19\frac{1}{2}$ oz	**44.** 1 gal 5 qt
45. 1 gal 3 qt 2 pt	**46.** 1 T 2500 lb	**47.** 2 yd 1 ft 23 in.	**48.** 1 qt 5 c 10 oz
49. 2 ft 10 in.	**50.** 5 lb 25 oz	**51.** 3 gal 5 qt 48 oz	**52.** 6 qt 20 oz
53. 1 mi 265 yd 4500 ft 25 in.	**54.** 2 ft 40 in.	**55.** 2 T 2600 lb 15 oz	**56.** 3 lb 21 oz
57. 2 lb 5 oz	**58.** 1 yd 4 ft 16 in.	**59.** 2 yd 2 ft 11 in.	**60.** 1 mi 875 ft 6 in.

5–2 ADDING AND SUBTRACTING U.S. CUSTOMARY MEASURES

Learning Objectives

1 Add U.S. customary measures.

2 Subtract U.S. customary measures.

Once we can convert from one U.S. customary unit to another, we can perform addition, subtraction, multiplication, and division in the U.S. customary system.

1 Add U.S. Customary Measures.

We can add U.S. customary measures *only* when their units are the same; that is, measures with the same units are *like measures*. *Unlike measures* are measures with different units. Before we can add 3 ft and 2 in., we need to change one of the measures so that they are both expressed as inches or as feet.

Rule 5–2 *To add unlike U.S. customary measures:*

1. Convert to a common U.S. customary unit.
2. Add.

EXAMPLE 1 Add 3 ft + 2 in.

Because 2 in. is a fraction of a foot, we can avoid working with fractions by converting the measures with the larger unit of measure (feet) to the smaller unit of measure (inches).

$$3 \text{ ft} = \frac{3 \text{ ft}}{1}\left(\frac{12 \text{ in.}}{1 \text{ ft}}\right) = 36 \text{ in.}$$

Because 3 ft = 36 in., then 3 ft + 2 in. =

$$36 \text{ in.} + 2 \text{ in.} = 38 \text{ in.}$$

Rule 5–3 *To add mixed U.S. customary measures:*

1. Align the measures vertically so that the common units are written in the same vertical column.
2. Add.
3. Express the sum in standard notation.

EXAMPLE 2 Add and write the answer in standard form: 6 lb 7 oz and 3 lb 13 oz.

$$
\begin{array}{r}
6 \text{ lb} \ \ 7 \text{ oz} \\
+ \ 3 \text{ lb } 13 \text{ oz} \\
\hline
9 \text{ lb } 20 \text{ oz}
\end{array}
$$
(20 oz = 1 lb 4 oz)

Thus, 9 lb 20 oz = 10 lb 4 oz.

2 Subtract U.S. Customary Measures.

To add measures, we must express different units of measure in a common unit. To subtract measures, we must also express different units in a common unit of measure.

Rule 5–4 *To subtract unlike U.S. customary units:*

1. Convert to a common U.S. customary unit.
2. Subtract.

EXAMPLE 3 Subtract 15 in. from 2 ft.

Because changing 15 in. to feet will give us a mixed number, it is easier if we convert 2 ft to inches.

$$2 \text{ ft} = 24 \text{ in.}$$
$$24 \text{ in.} - 15 \text{ in.} = \mathbf{9 \text{ in.}}$$

Rule 5–5 *To subtract mixed U.S. customary measures:*

1. Align the measures vertically so that the common units are written in the same vertical column.
2. Subtract.

EXAMPLE 4 Subtract 5 ft 3 in. from 7 ft 4 in.

Align the measures that are alike in a vertical line and then subtract.

$$\begin{array}{r} 7 \text{ ft } 4 \text{ in.} \\ - \ 5 \text{ ft } 3 \text{ in.} \\ \hline \mathbf{2 \text{ ft } 1 \text{ in.}} \end{array}$$

Sometimes, when subtracting mixed measures, we must subtract a larger unit from a smaller unit. In this case we must apply to the mixed measures our knowledge of borrowing.

Rule 5–6 *To subtract a larger U.S. customary unit from a smaller unit in mixed measures:*

1. Align the common measures in vertical columns.
2. Borrow one unit from the next larger unit of measure in the minuend, convert to the equivalent smaller unit of measure, and add it to the smaller unit of measure.
3. Subtract the measures.

EXAMPLE 5 Subtract 3 lb 12 oz from 7 lb 8 oz.

$$\begin{array}{r} 7 \text{ lb } \ \ 8 \text{ oz} \\ - \ 3 \text{ lb } 12 \text{ oz} \end{array}$$

We always begin subtracting with the smallest unit of measure, which should be on the *right*. In this example, we notice that 12 oz cannot be subtracted from

8 oz. To perform this subtraction, we rewrite 7 lb as 6 lb 16 oz. Then we combine the 16 oz with 8 oz to increase the 8 oz to 24 oz.

$$
\begin{array}{ll}
7 \text{ lb } 8 \text{ oz} = 6 \text{ lb } 16 \text{ oz} + 8 \text{ oz} = & 6 \text{ lb } 24 \text{ oz} \\
3 \text{ lb } 12 \text{ oz} = & -\ 3 \text{ lb } 12 \text{ oz} \\
& \overline{\mathbf{3 \text{ lb } 12 \text{ oz}}}
\end{array}
$$

SELF-STUDY EXERCISES 5–2

[1] Add and write the answer in standard notation.

1. 5 oz + 2 lb

2. 4 ft + 7 in.

3.   8 lb 2 oz
   + 7 lb 9 oz

4.   5 ft 45 in.
   + 7 ft 30 in.

5.   5 qt 1 pt
   + 2 qt $1\frac{1}{2}$ pt

6.   8 gal 3 qt
   + 5 gal 2 qt

7.   7 yd 2 ft
   + 1 yd 2 ft

8.   3 c 6 oz
   + 1 c 3 oz

9.   1 ft 13 in.
   + 2 ft 15 in.

10.   4 yd 2 ft 7 in.
   + 2 yd 1 ft 10 in.

Solve the following problems. When necessary, express the answers in standard notation.

11. A plumber has a 2-ft length of copper tubing and a 7-in. length of copper tubing. What is the total?

12. How long are two extension cords together if one is 6 ft and the other is 60 in.?

13. A mixture for hamburgers contains 2 lb 8 oz of ground round steak and 3 lb 7 oz of regular ground beef. How much does the hamburger mixture weigh?

14. According to hospital maternity records, one infant twin weighed 6 lb 1 oz and the other weighed 5 lb 15 oz at birth. What was their total weight?

[2] Subtract and write the answer in standard form.

15. 2 ft − 18 in.

16. 2 qt − 3 pt

17. 3 yd − 7 ft

18. 6 lb − 18 oz

19.   3 lb 12 oz
   − 2 lb 6 oz

20.   12 lb 7 oz
   − 5 lb 12 oz

21.   3 ft 8 in.
   − 2 ft 9 in.

22.   19 lb 12 oz
   − 8 lb 15 oz

23.   2 ft 30 in.
   − 1 ft 40 in.

24.   5 gal 3 qt 1 pt
   − 1 gal 3 qt $1\frac{1}{2}$ pt

25. A seamstress had a length of cloth 5 ft long. What was its length after cutting off 10 in.?

26. A cabinet maker cut 9 in. from a 3-ft shelf in a medical lab. How long was the shelf after it was cut?

27. If a computer with a pentium® processor can sort a list of names in 1 min 30 sec and a computer with a math coprocessor can do the same job in 45 sec, how much time is saved by using the computer with the math coprocessor?

28. A car with a 2.0-L engine can accelerate a certain distance in 1 min 27 sec. A car with a 5.0-L engine can accelerate the same distance in 48 sec. How much faster is the car with the 5.0-L engine?

29. A package containing a laser printer weighed 74 lb 3 oz. The container and packing material weighed 4 lb 12 oz. How much did the laser printer weigh?

30. To weigh a pet poodle, Stacey held the dog while standing on a scale. If the scale showed 132 lb 6 oz and Stacey weighs 115 lb 8 oz, how much does the poodle weigh?

5-3 MULTIPLYING AND DIVIDING U.S. CUSTOMARY MEASURES

Learning Objectives

[1] Multiply a U.S. customary measure by a number.

[2] Multiply a U.S. customary measure by a measure.

[3] Divide a U.S. customary measure by a number.

4 Divide a U.S. customary measure by a measure.

5 Change from one U.S. customary rate measure to another.

1 Multiply a U.S. Customary Measure by a Number.

In addition to adding and subtracting U.S. customary measures, we often need to multiply or divide those measures by some number. Problems often arise in various fields that require us to multiply a measure by a given number. For example, a carpenter may buy cypress siding in lengths of 18 ft 6 in. To find the total length of six pieces of the siding, we multiply 18 ft 6 in. by 6. We do this by multiplying 6 by the numbers associated with each unit of measure.

Rule 5–7 *To multiply a U.S. customary measure by a given number:*

1. Multiply the numbers associated with each unit of measure by the given number.
2. Write the resulting measure in standard notation.

EXAMPLE 1 A container has a capacity of 21 gal 3 qt of weed killer. What is the total capacity of eight such containers?

$$
\begin{array}{r}
21 \text{ gal } 3 \text{ qt} \\
\times \qquad 8 \\
\hline
168 \text{ gal } 24 \text{ qt}
\end{array}
\qquad (24 \text{ qt} = 6 \text{ gal})
$$

168 gal 24 qt = 168 gal + 6 gal = 174 gal in standard notation

The eight containers have a combined capacity of 174 gal.

2 Multiply a U.S. Customary Measure by a Measure.

Length measures can be multiplied by like length measures to produce square measures. Square measures indicate areas (Fig. 5–12).

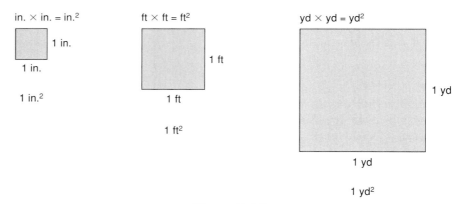

Figure 5–12

Rule 5–8 *To multiply a length measure by a like length measure:*

1. Multiply the numbers associated with each like unit of measure.
2. The product will be a square unit of measure.

EXAMPLE 2   A desktop is 2 ft × 3 ft (Fig. 5–13). What is the number of square feet in the surface?

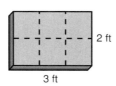

2 ft

3 ft

Figure 5–13   2 ft × 3 ft = **6 ft²**

3   Divide a U.S. Customary Measure by a Number.

We are frequently required to divide measures by a number. For example, if a recipe calls for 2 gal 2 qt of a liquid ingredient and the recipe is halved (divided by 2), what amount of this ingredient should be used? To solve this problem, we divide the measure by 2.

> **Rule 5–9**   *To divide a U.S. customary measure by a given number that divides evenly into each measure:*
>
> 1. Divide the numbers associated with each unit of measure by the given number.
> 2. Write the resulting measure in standard notation.

EXAMPLE 3   How much milk is needed for a half-recipe if the original recipe calls for 2 gal 2 qt?

$2\overline{)2 \text{ gal } 2 \text{ qt}}$ We divide each measure by 2.

$\dfrac{1 \text{ gal } 1 \text{ qt}}{2\overline{)2 \text{ gal } 2 \text{ qt}}}$

The half-recipe requires 1 gal 1 qt of the ingredient.

If a given number does not divide evenly into each measure, there will be a remainder. For instance, if we divided 5 gal by 2, the 2 would divide into the 5 gal 2 times with 1 gal left over as a remainder.

> **Rule 5–10**   *To divide U.S. customary measures by a given number that does not divide evenly into each measure:*
>
> 1. Set up the problem and proceed as in long division.
> 2. When a remainder occurs after subtraction, convert the remainder to the same unit used in the next smaller measure and add it to the next smaller measure.
> 3. Then continue by dividing the given number into this next smaller measure.
> 4. If a remainder occurs when the smallest unit is divided, express the remainder as a fractional part of the smallest unit.

EXAMPLE 4 Divide 5 gal 3 qt 1 pt by 3.

$$
\begin{array}{r}
1\ \text{gal} \qquad 3\ \text{qt} \qquad 1\tfrac{2}{3}\ \text{pt} \\
\hline
3)\overline{5\ \text{gal} \qquad 3\ \text{qt} \qquad 1\ \text{pt}} \\
\underline{3\ \text{gal}} \\
2\ \text{gal} = \underline{\ 8\ \text{qt}\ } \\
11\ \text{qt} \\
\underline{\ 9\ \text{qt}\ } \\
2\ \text{qt} = \underline{4\ \text{pt}} \\
5\ \text{pt} \\
\underline{3\ \text{pt}} \\
2\ \text{pt}
\end{array}
$$

3 qt + 8 qt = 11 qt

11 qt ÷ 3   =   3 qt, remainder 2 qt

1 pt + 4 pt = 5 pt

5 pt ÷ 3   =   1 pt, remainder 2

Write the remainder, 2, as a fraction, $\tfrac{2}{3}$, and add to 1 pint to get $1\tfrac{2}{3}$ pints.

Thus, 1 gal 3 qt $1\tfrac{2}{3}$ pt is the solution.

4 Divide a U.S. Customary Measure by a Measure.

In addition to dividing measures by a number, we are often called on to divide measures by a measure. For instance, if tubing is manufactured in lengths of 8 ft 4 in. and a part is 10 in. long, how many parts can be cut from the length of tubing if we do not account for waste? To solve such problems, we must express both measures in the same unit, just as we did when we added and subtracted measures. We generally convert to the *smallest* unit used in the problem.

> **Rule 5–11** *To divide a U.S. customary measure by a U.S. customary measure:*
>
> **1.** Convert both measures to the same unit if they are different.
> **2.** Write the division as a fraction, including the common unit in the numerator and the denominator.
> **3.** Reduce the units and divide the numbers.

EXAMPLE 5 Divide 8 ft 4 in. by 10 in.

Converting 8 ft 4 in. to inches, we have

$$8\ \text{ft} = \frac{8\ \cancel{\text{ft}}}{1}\left(\frac{12\ \text{in.}}{1\ \cancel{\text{ft}}}\right) = 96\ \text{in.}$$

$$8\ \text{ft}\ 4\ \text{in.} = 96\ \text{in.} + 4\ \text{in.} = 100\ \text{in.}$$

Dividing, we have 100 in. ÷ 10 in.

If we write this division in fraction form, we can see more easily that the common units will reduce. In other words, the answer will be a number (not a measure) telling *how many times* one measure divides into another.

$$\frac{100\ \cancel{\text{in.}}}{10\ \cancel{\text{in.}}} = 10$$

Therefore, we see that 10 parts of equal length can be cut from the tubing.

5 Change from One U.S. Customary Rate Measure to Another.

A *rate measure* is a ratio of two different kinds of measures. It is often referred to simply as a *rate*. Some examples of rates are 55 miles per hour, 20 cents per mile, and 3 gallons per minute. In each of these rates, the word *per* means *divided by*.

The rate 55 miles per hour means 55 miles ÷ 1 hour or $\frac{55 \text{ mi}}{1 \text{ hr}}$. The rate 20 cents per mile means 20 cents ÷ 1 mile or $\frac{20 \text{ cents}}{1 \text{ mi}}$. The rate 3 gallons per minute means 3 gallons ÷ 1 minute or $\frac{3 \text{ gal}}{1 \text{ min}}$.

Notice that every rate measure requires two units of measure, one in the numerator and one in the denominator. Sometimes it is necessary to express the numerator and/or denominator of a rate measure in other units of measure. We can do this by using unity ratios. Measures of time are universally accepted. All major countries use the same units of measure for time. The basic units of time are the year, month, week, day, hour, minute, and second. Table 5–4 gives the relationships among the units of measure for time. You will need these measures when working with rates.

TABLE 5–4 Time Measures

1 year (yr) = 12 months (mo)	1 day (da) = 24 hours (hr)
1 year (yr) = 365 days (da)	1 hour (hr) = 60 minutes (min)[a]
1 week (wk) = 7 days (da)	1 minute (min) = 60 seconds (sec)[b]

[a]The symbol ′ means feet (3′ = 3 ft) or minutes (60′ = 60 minutes).

[b]The symbol ″ means inches (8″ = 8 in.) or seconds (60″ = 60 seconds).

Rule 5–12 *To convert one U.S. customary rate measure to another:*

1. Set up the original rate measure equal to the new rate measure without its numerical values.
2. Compare the units of both numerators and both denominators to determine which original units will change.
3. Multiply each original measure that changes by a unity ratio containing the new unit so that the unit to be changed will reduce.

The conversion procedure is the same as for converting one measure to another measure. A separate unity ratio is used for each part of the rate that will change. For example, when both the numerator and denominator of a rate measure change, we need to multiply by at least two unity ratios to make the conversion. We first multiply by a unity ratio that will reduce the numerator of the original rate measure. Then we multiply by a unity ratio that will reduce the denominator of the original rate measure.

EXAMPLE 6 Change $\dfrac{3 \text{ pt}}{\text{sec}}$ to $\dfrac{\text{qt}}{\text{min}}$.

Because we are converting the numerator from pints to quarts, we use a unity ratio with quarts in the numerator and pints in the denominator to reduce pints in the original problem.

$$\frac{3 \text{ pt}}{\text{sec}} \left(\frac{1 \text{ qt}}{2 \text{ pt}} \right)$$

To complete the conversion, we multiply by the unity ratio that will change *seconds* in the denominator to *minutes*. Seconds must be reduced, so our unity ratio must have seconds in the *numerator*.

$$\frac{3 \text{ pt}}{\text{sec}} \left(\frac{1 \text{ qt}}{2 \text{ pt}} \right) \left(\frac{60 \text{ sec}}{1 \text{ min}} \right)$$

$$\frac{3 \cancel{\text{ pt}}}{\cancel{\text{sec}}} \left(\frac{1 \text{ qt}}{\cancel{2} \cancel{\text{ pt}}_{1}} \right) \left(\frac{\overset{30}{\cancel{60}} \cancel{\text{ sec}}}{1 \text{ min}} \right) = 90 \frac{\text{qt}}{\text{min}}$$

Thus, 3 $\frac{\text{pt}}{\text{sec}}$ equals 90 $\frac{\text{qt}}{\text{min}}$.

Occasionally, a rate measure requires more than two unity ratios to convert it to another rate measure.

SELF-STUDY EXERCISES 5–3

1 Multiply and write answers for mixed measures in standard notation.

1. 12 mi
    5

2. 18 gal
    6

3. 21 lb
    12

4. 3 qt 1 pt
    4

5. 7 lb 3 oz
    8

6. 5 gal 2 qt
    7

7. 8 gal 3 qt
    5

8. 7 ft 3 in.
    8

9. Tuna is packed in 1 lb 8 oz cans. If a case contains 24 cans, how much does a case weigh?

10. A car used 1 qt 1 pt of oil on each of 5 days. Find the total oil used.

2 Multiply.

11. 5 in. × 7 in.

12. 12 ft × 9 ft

13. 15 yd × 12 yd

14. 4 mi × 27 mi

15. A room is to be covered with square linoleum tiles that are 1 ft by 1 ft. If the room is 18 ft by 21 ft, how many tiles (square feet) are needed?

16. A horticulturist stores a stock solution of fertilizer in two tanks, each with a capacity of 23 gal 9 oz. How much liquid fertilizer is needed to fill both tanks?

17. Latonya has three containers, each containing 1 qt 3 pt of photographic solution. How much total photographic solution is in all three containers?

18. A package of nails weighs 1 lb 4 oz. How much would five packages weigh?

19. If a history textbook weighs 2 lb 8 oz, how much would three of the same books weigh?

3 Divide.

20. 12 gal ÷ 2

21. 3 days 6 hr ÷ 2

22. 20 yd 2 ft 6 in. ÷ 2

23. 4 yd 1 ft 9 in. ÷ 3

24. 4 gal 3 qt 1 pt ÷ 6

25. 60 gal ÷ 9 (express in gallons only)

26 5 hr ÷ 3 (express in hours and minutes)

27. If eight trucks require 42 qt of oil for each to get a complete oil change, how many quarts are required for each truck?

28. Sixty feet of wire are required to complete eight jobs. If each job requires an equal amount of wire, find the amount of wire required for one job. (Express in feet and inches.)

29. A vat holding 10 gal 2 qt of defoliant is emptied equally into three tanks. How many gallons and quarts are in each tank?

30. How many pieces of $\frac{1}{2}$-in. OD (outside diameter) plastic pipe 8 in. long can be cut from a piece 72 in. long?

31. A roll of No. 14 electrical cable 150 ft long is divided into 30 equal sections. How long is each section?

32. A greenhouse attendant has a container with 6 gal 2 qt 10 oz of potassium nitrate solution that will be stored in two smaller containers. How much solution will be stored in each smaller container?

33. For a family picnic, Mr. Sonnier prepared 96 lb 12 oz of boiled crawfish. He brought the crawfish to the picnic site in four containers containing equal amounts. How much did the crawfish in each container weigh?

4 Divide.

34. 36 ft ÷ 12 ft　　　　　　**35.** 51 in. ÷ 3 in.　　　　　　　　**36.** 2 ft 8 in. ÷ 4 in.

37. 6 lb 12 oz ÷ 6 oz　　　　**38.** 2 ft 6 in. ÷ 10 in.

39. How many 6-in. pieces can be cut from 48 in. of pipe?

40. How many 2-lb boxes can be filled from 18 lb of nails?

41. How many 15-oz cans are in a case if the case weighs 22 lb 8 oz?

42. How many 8-dollar tickets can be purchased for 72 dollars?

43. How many pieces of wood 5 in. long can be cut from a piece 45 in. long?

5 Work the following problems.

44. $\dfrac{60 \text{ qt}}{\text{sec}} = \underline{\quad} \dfrac{\text{gal}}{\text{sec}}$

45. $\dfrac{45 \text{ lb}}{\text{hr}} = \underline{\quad} \dfrac{\text{lb}}{\text{min}}$

46. $\dfrac{3 \text{ mi}}{\text{hr}} = \underline{\quad} \dfrac{\text{ft}}{\text{hr}}$

47. $\dfrac{144 \text{ lb}}{\text{min}} = \underline{\quad} \dfrac{\text{oz}}{\text{min}}$

48. $\dfrac{30 \text{ gal}}{\text{min}} = \underline{\quad} \dfrac{\text{qt}}{\text{sec}}$

49. $\dfrac{30 \text{ lb}}{\text{day}} = \underline{\quad} \dfrac{\text{oz}}{\text{hr}}$

50. $\dfrac{8 \text{ in}}{\text{sec}} = \underline{\quad} \dfrac{\text{ft}}{\text{min}}$

51. $\dfrac{5 \text{ mi}}{\text{min}} = \underline{\quad} \dfrac{\text{ft}}{\text{sec}}$

52. A car traveling at the rate of 30 mph (miles per hour) is traveling how many feet per second?

53. A pump that can pump $45\frac{\text{gal}}{\text{hr}}$ can pump how many quarts per minute?

54. A pump can dispose of sludge at the rate of $3200\frac{\text{lb}}{\text{hr}}$. How many pounds can be disposed of per minute?

55. If water flows through a pipe at the rate of $50\frac{\text{gal}}{\text{min}}$, how many gallons will flow per second?

5–4　INTRODUCTION TO THE METRIC SYSTEM

Learning Objectives

1 Identify uses of metric measures of length, weight, and capacity.

2 Convert from one metric unit of measure to another.

3 Make calculations with metric measures.

In the *metric system,* or the International System of Units (SI), there is a standard unit for each type of measurement. The *meter* is used for length or distance, the *gram* is used for weight or mass, and the *liter* is used for capacity or volume. A series of prefixes are added to the standard units to indicate measures greater than the standard units or less than the standard units.

Many consider the metric system to be easier to use than the U.S. customary system. One reason for this is that all measures greater than or less than the standard unit are in powers of 10. Changing from larger to smaller or from smaller to larger measures only requires multiplying or dividing by 10 or a power of 10. Another reason the metric system is easier to use than the U.S. customary system is that the prefixes used with the standard unit represent the power of 10 by which the standard unit is divided or multiplied.

■ **DEFINITION 5–4: Metric System.** The *metric system* is an international system of measurement that uses standard units and prefixes to indicate their powers of 10.

1 Identify Uses of Metric Measures of Length, Weight, and Capacity.

Before we start our study of metric measurements, look at some of the prefixes used in the metric system. Keep in mind that the prefix will have the same meaning no matter which unit (meter, gram, or liter) the prefix is attached to.

The prefixes used in this chapter for *smaller* units than the standard unit are as follows:

$$\textbf{deci-}\ \frac{1}{10}\ \text{of} \qquad \textbf{centi-}\ \frac{1}{100}\ \text{of} \qquad \textbf{milli-}\ \frac{1}{1000}\ \text{of}$$

The prefixes used in this chapter for *larger* units than the standard unit are as follows:

$$\textbf{deka-}\ 10\ \text{times} \qquad \textbf{hecto-}\ 100\ \text{times} \qquad \textbf{kilo-}\ 1000\ \text{times}$$

Other prefixes are used with very large and very small measures.

The prefixes can be related to our decimal system of numeration. Let's compare our decimal place-value chart with these prefixes (Fig. 5–14). The standard unit (whether meter, gram, or liter) corresponds to the *ones* place. All the places to the left are powers of the standard unit. That is, the value of *deka-* (some textbooks use *deca-*) is 10 times the standard unit; the value of *hecto-* is 100 times this unit; the value of *kilo-* is 1000 times this unit; and so on. All the places to the right of the standard unit are subdivisions of the standard unit. That is, the value of *deci-* is $\frac{1}{10}$ of the standard unit; the value of *centi-* is $\frac{1}{100}$ of this unit; the value of *milli-* is $\frac{1}{1000}$ of this unit; and so on.

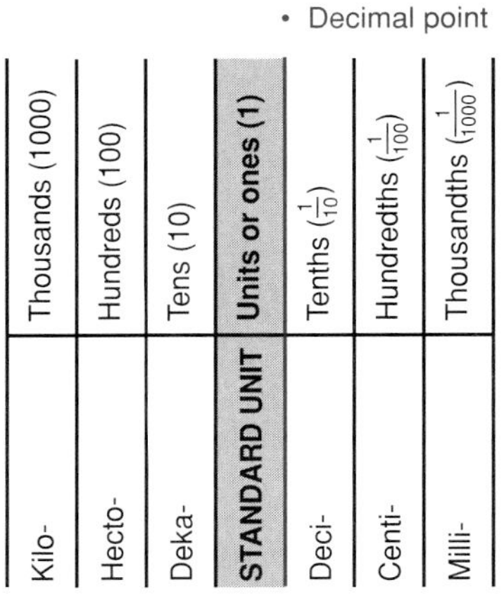

• Decimal point

Thousands (1000)	Hundreds (100)	Tens (10)	Units or ones (1)	Tenths ($\frac{1}{10}$)	Hundredths ($\frac{1}{100}$)	Thousandths ($\frac{1}{1000}$)
Kilo-	Hecto-	Deka-	STANDARD UNIT	Deci-	Centi-	Milli-

Figure 5–14

EXAMPLE 1 Give the value of the following metric units in terms of the prefix and the standard unit (gram, liter, or meter).

(a) Kilogram (kg) = 1000 times 1 gram or 1000 g
(b) Deciliter (dL) = $\frac{1}{10}$ of 1 liter or 0.1 L
(c) Hectometer (hm) = 100 times 1 meter or 100 m
(d) Dekaliter (dkL) = 10 times 1 liter or 10 L
(e) Milliliter (mL) = $\frac{1}{1000}$ of 1 liter or 0.001 L
(f) Centigram (cg) = $\frac{1}{100}$ of 1 gram or 0.01 g

Remember that all metric measurements of length, weight, and volume are expressed either as one of the standard units alone or with a suitable prefix.

Meter: The *meter* is the standard unit for measuring length. Both *meter* and *metre* are acceptable spellings for this unit of measure; however, we will use the spelling *meter* throughout this book. A meter is about 39.37 in., or 3.37 in. (approximately $3\frac{1}{3}$ in.) longer than a yard (Fig. 5–15). The meter is the appropriate unit to measure lengths and distances like room dimensions, land dimensions, lengths of poles, heights of mountains, and heights of buildings. The abbreviation for meter is *m*.

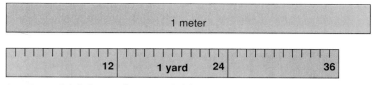

1 meter – slightly longer than a yardstick
(36 inches = 1 yard) (39.37 inches = 1 meter)

Figure 5–15

Kilometer: To measure long distances, a larger measuring unit is needed. A *kilometer* is 1000 meters and is used for these longer distances (Fig. 5–16). The abbreviation for kilometer is *km.* The prefix *kilo* attached to the word *meter* means 1000. The distance from one city to another, one country to another, or one landmark to another would be measured in kilometers. Driving at a speed of 55 miles (or 90 km) per hour, we would travel 1 km in about 40 seconds (sec). An average walking speed is 1 km in about 10 minutes (min).

1 kilometer – about five city blocks
1 mile – about eight city blocks

Figure 5–16

Centimeter: When measuring objects less than 1 m long, the most common unit used is the *centimeter (cm).* The prefix *centi* means "$\frac{1}{100}$ of," and a centimeter is one hundredth of a meter. A centimeter is about the width of a thumbtack head, somewhat less than $\frac{1}{2}$ in. (Fig. 5–17). Centimeters are used to measure medium-sized objects such as tires, clothing, textbooks, and television pictures.

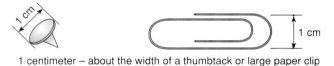

1 centimeter – about the width of a thumbtack or large paper clip

Figure 5–17

Millimeter: Many objects are too small to be measured in centimeters, so an even smaller unit of measure is needed. A *millimeter (mm)* is "$\frac{1}{1000}$ of" a meter. It is about the thickness of a plastic credit card or a dime (Fig. 5–18). Certain film sizes, bolt and nut sizes, the length of insects, and similar items are measured in millimeters.

Other units and their abbreviations are decimeter, dm; dekameter, dkm; and hectometer, hm.

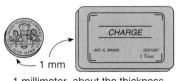

1 millimeter–about the thickness
of a dime or a charge card

Figure 5–18

Gram: The standard unit for measuring weights in the metric system is the gram (Fig. 5–19). A gram is described as the weight of 1 cubic centimeter (cm^3) of water. A cubic centimeter is a cube that has each edge equal to 1 centimeter in length. It is a cube a little smaller than a sugar cube. The abbreviation *g* is used for gram. The gram is used for measuring small or light objects such as a paper clip, a cube of sugar, a coin, and a bar of soap.

1 gram – about the weight of two paper clips

Figure 5–19

1 kilogram – about the weight of an average book with a one-inch spine

Figure 5–20

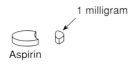

Aspirin

Figure 5–21

Kilogram: A *kilogram (kg)* is 1000 grams (Fig. 5–20). Since a cube 10 cm on each edge can be divided into 1000 cm³, the weight of the amount of water required to fill this cube would be 1000 grams or 1 kilogram. A kilogram is approximately 2.2 lb. The kilogram is probably the most often used unit for measuring weight. It is used for the weight of people, meat, sacks of flour, automobiles, and so on.

Milligram: Because the standard unit for measuring weight, the gram, is used only to measure small objects, the *milligram* (1/1000 of a gram) is used to measure *very* small objects. Milligrams are much too small for ordinary uses; however, pharmacists use milligrams *(mg)* when measuring small amounts of drugs, vitamins, and medications (Fig. 5–21).

Other units and their abbreviations are decigram, dg; centigram, cg; dekagram, dkg; and hectogram, hg.

Liter: A *liter (L)* is the volume of a cube 10 cm on each edge. It is the standard metric unit of capacity (Fig. 5–22). Like the meter, it may be spelled *liter* or *litre,* but we will use the spelling *liter.* A cube 10 cm on each edge filled with water weighs approximately 1 kg, so 1 L of water weighs about 1 kg. One liter is just a little larger than a liquid quart (a U.S. customary unit for measuring capacity). Soft drinks are often sold in 2-liter bottles, gasoline is sold by the liter at some service stations, and numerous other products are sold in liter containers.

Milliliter: A liter is 1000 cm³, so $\frac{1}{1000}$ of a liter, or a *milliliter,* has the same capacity or volume as a cubic centimeter (Fig. 5–23). Most liquid medicine is labeled and sold in milliliters *(mL)* or cubic centimeters (cc or cm³). Medicines, perfumes, and other very small quantities are measured in milliliters.

1 liter – about the volume of a quart of milk or a soft drink in a plastic bottle

Figure 5–22

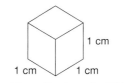

1 cubic centimeter = 1 milliliter

Figure 5–23

Other units and their abbreviations are deciliter, dL; centiliter, cL; dekaliter, dkL; hectoliter, hL; and kiloliter, kL.

There are other multiples and standard units in the metric system, but those discussed here are the ones frequently used.

EXAMPLE 2 Choose the most reasonable metric measure for each of the following.

1. Distance from Jackson, Mississippi, to New Orleans, Louisiana
  (a) 322 m   **(b)** 322 km   **(c)** 322 cm   **(d)** 322 mm

2. Weight of an adult woman
  (a) 56 g   **(b)** 56 mg   **(c)** 56 kg   **(d)** 56 dkg
3. Width of a color slide film
  (a) 35 cm   **(b)** 35 m   **(c)** 35 mm   **(d)** 35 km
4. Bottle of eye drops
  (a) 30 dL   **(b)** 30 dkL   **(c)** 30 L   **(d)** 30 mL
5. Weight of a pill
  (a) 352 mg   **(b)** 352 dg   **(c)** 352 g   **(d)** 352 kg

1. **(b)** 322 km (about 200 mi)
2. **(c)** 56 kg (about 120 lb)
3. **(c)** 35 mm (roll of film for automatic cameras)
4. **(d)** 30 mL (about 1 oz)
5. **(a)** 352 mg (one regular-strength aspirin)

2 Convert From One Metric Unit of Measure to Another.

To understand the processes involved in changing one metric unit into another, let's arrange the units into a place-value chart like the one we used for decimals (Fig. 5–24). The units are arranged from left to right and from largest to smallest.

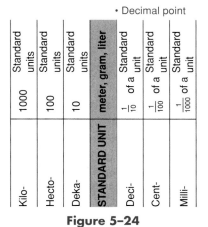

Figure 5–24

As we move from any place in the chart one place to the *right,* the metric unit changes to the next smaller unit. In effect, the larger unit is broken down into 10 smaller units. In other words, we are *multiplying* the larger unit by 10 when we move one place to the right.

For example, suppose that we wanted to change 2 m to decimeters. Because a decimeter is $\frac{1}{10}$ of a meter, there must be 10 dm in 1 meter. If it takes 10 dm to make 1 m, how many decimeters does it take to make 2 m? It takes twice as many as it does to make 1 m.

$$2 \times 10 = 20$$

Therefore, 2 m = 20 dm.

Instead of multiplying by 10, we can use the metric-value chart as directed in the following rule because, when changing to a smaller unit, we will always be moving to the right on the chart. Suppose that we needed to change 4 L to centiliters. The first step to the right would change 4 L to 40 dL. The next step to the right would change the 40 dL to 400 cL.

> **Rule 5–13** *To change from one metric unit to* any smaller *metric unit:*
>
> 1. Mentally position the measure on the metric-value chart so that the decimal is immediately *after* the original measuring unit.
> 2. Move the decimal to the right so that it *follows* the new measuring unit. (Affix zeros if necessary.)

EXAMPLE 3   43 dkm = _____ cm?

Place 43 dkm on the chart so that the last digit is in the dekameters place; that is, the understood decimal that follows the 3 will be *after* the dekameters place (see Fig. 5–25). To change to centimeters, move the decimal point so that it falls *after* the centimeters place. Fill in the empty places with zeros (see Fig. 5–26). Recall that a shortcut to multiplication by powers of 10 is to move the decimal point one place to the right for each time 10 is used as a factor.

Thus, 43 dkm = 43,000 cm.

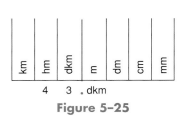

Figure 5–25

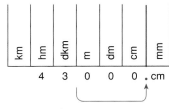

Figure 5–26

EXAMPLE 4   2.5 dg _____ mg? (Note the decimal location.)

Place the number 2.5 on the chart so that the decimal point falls *after* the decigrams place (see Fig. 5–27). (Note the decimal after the decigrams place.) To change to milligrams, shift the decimal two places to the right so that it falls *after* the milligrams place (see Fig. 5–28). (Note the decimal after the milligrams place.)

Thus, 2.5 dg = 250 mg.

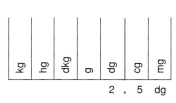

Figure 5–27

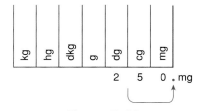

Figure 5–28

As we move one place to the *left* on the metric-value chart, the metric unit changes to the next larger unit. In effect, the smaller units are consolidated into one larger unit 10 times larger than each smaller unit. In other words, we are *dividing* the smaller unit by 10 when we move one place to the left.

For instance, suppose that we wanted to change 20 mm to centimeters. Because 1 millimeter is $\frac{1}{10}$ of a centimeter, there are 10 mm in 1 cm. If 1 cm equals 10 mm, how many centimeters are there in 20 mm? There are as many centimeters as there are 10's in 20 mm (20 ÷ 10 = 2). Therefore, 20 mm = 2 cm.

Because we are dividing by 10, we can use the metric-value chart as directed in the following rule. When changing to a larger unit, we will always move to the left on the chart.

> **Rule 5–14** *To change from one metric unit to* **any** larger *metric unit:*
>
> 1. Mentally position the measure on the metric-value chart so that the decimal is immediately *after* the original measuring unit.
> 2. Move the decimal to the left so that it *follows* the new measuring unit. (Affix zeros if necessary.)

EXAMPLE 5   3495 L = _____ kL?

The number is placed on the chart so that the digit 5 is in the liters place; that is, the decimal point falls after the liters place (see Fig. 5–29). (Note the understood decimal point after the liters place.) To change to kiloliters, move the decimal *three* places to the left so that the decimal falls *after* the kiloliters place (see Fig. 5–30). Recall that a shortcut to division by powers of 10 is to move the decimal point one place to the left for each time 10 is used as a divisor.
   Thus, 3495 L = 3.495 kL.

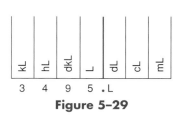

Figure 5–29

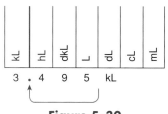

Figure 5–30

EXAMPLE 6   2.78 cm = _____ dkm?

Place the number on the chart so that the decimal is *after* the centimeters place (see Fig. 5–31). (Note the decimal position after the centimeters place.) Move the decimal three places to the left so that it falls *after* the dekameters place (see Fig. 5–32). (Note the decimal position after the dekameters place.)
   Therefore, 2.78 cm = 0.00278 dkm.

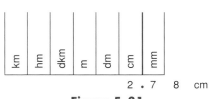

Figure 5–31

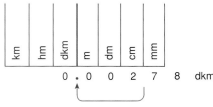

Figure 5–32

To summarize, we may change from one metric unit to another metric unit of measure of the same type by positioning the measure on the metric-value chart so that the decimal is immediately *after* the original measuring unit. Then we move the decimal to the right or left as necessary so that it *follows* the new measuring unit. Zeros may be affixed if necessary.

We may also change from one metric unit to another without thinking of the number positioned on the metric-value chart. Instead we mentally locate the original measuring *unit* on the metric-value chart and move to the desired new measuring *unit*. As we move, we carefully note the number of units we move past and the direction in which we move. Then we locate the decimal point in the original metric measure and shift the decimal in the same direction and the same number of units.

Tips and Traps   *How Far and Which Way?*

To determine the movement of the decimal point when changing from one metric unit to another, answer the questions:

1. *How far* is it from the original unit to the new unit (how many places)?
2. *Which way* is the movement on the chart?

Change 28.392 cm to m.

How far is it from cm to m (Fig. 5–33)?   Two places
Which way?   Left

Figure 5–33

Move the decimal in the original measure *two places to the left.*

$$28.392 \text{ cm} = 0.28392 \text{ m}$$

③ Make Calculations With Metric Measures.

Up to this point, we have been comparing metric units and changing from one unit to a larger or smaller unit. Now we will use metric units in problems involving addition, subtraction, multiplication, and division. Our employment will often require us to perform these basic mathematical operations.

When adding or subtracting measures, it is important that we remember to add or subtract only *like* or *common* measures. By this we mean that the same measuring unit is used. For instance, 5 cm and 3 cm are *like* measures. In contrast, 17 kg and 4 hg are *unlike* measures.

Rule 5–15   *To add or subtract* **like** *or* **common** *metric measures:*

Add or subtract the numerical values. The answer will have the common unit of measure. *Examples:*

$$7 \text{ cm} + 4 \text{ cm} = 11 \text{ cm}$$
$$15 \text{ dkg} - 3 \text{ dkg} = 12 \text{ dkg}$$

Rule 5–16   *To add or subtract* **unlike** *metric measures:*

First change the measures to a common unit of measure. Then add or subtract the numerical values. The answer will have the common unit of measure.

EXAMPLE 7   Add 9 mL + 2 cL.

Change cL to mL; that is 2 cL = 20 mL. Thus, 9 mL + 20 mL = **29 mL.**
   Look at this example again:
$$9 \text{ mL} + 2 \text{ cL}$$
We could have changed milliliters to centiliters:
$$9 \text{ mL} = 0.9 \text{ cL}$$

Then,

0.9 cL + 2 cL =   Remember, in addition and subtraction, decimals must be in a straight
  0.9 cL   line. There is an understood decimal after the addend 2.
  2   cL
  ─────
  2.9 cL

Does 2.9 cL = 29 mL? Yes. Therefore, our answers are equivalent.

EXAMPLE 8   Subtract: 14 km − 34 hm.

Change hm to km; that is, 34 hm = 3.4 km.

  14.0 km
− 3.4 km   Caution: Watch alignment of decimals.
  ─────────
  10.6 km

The difference is 10.6 km, or 106 hm.

Tips and Traps   *Incompatible Measures:*
Look at this problem: 2 cm + 5 g. Can 2 cm be added to 5 g? Can a common unit of measure be found for centimeters and grams? Centimeters measure length and grams measure weight (Fig. 5–34); therefore, no common unit of measure can be found. Thus, we cannot add 2 cm + 5 g.

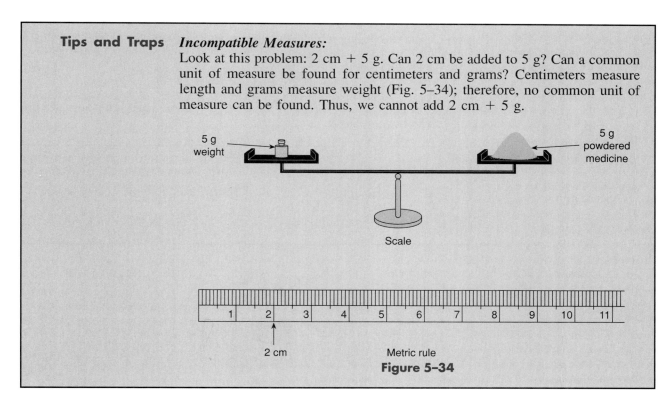

Figure 5–34

Multiplication is a shortcut for repeated addition. If we wanted to know the total length of three pieces of landscape timber that are each 2.7 m long, we can multiply 2.7 m by 3; that is, 2.7 m × 3 = 8.1 m.

Rule 5–17   *To multiply a metric measure by a number:*

Multiply the numerical values. The answer will have the same unit of measure as the original measure.

Suppose we had a tube that was 30 cm long and we needed it cut into five equal parts (Fig. 5–35). How long would each part be? To solve this problem, we need to divide 30 cm by 5.

$$\frac{30 \text{ cm}}{5} = 6 \text{ cm}$$

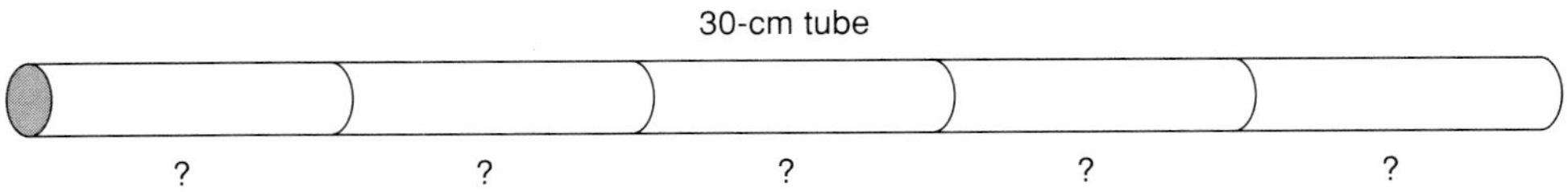

30-cm tube

?   ?   ?   ?   ?

Figure 5–35

Rule 5–18   *To divide a metric measure by a number:*

Divide the numerical values. The answer will have the same unit of measure as the original measure.

All types of measures can be divided by a number because we are simply dividing a quantity of parts.

There are situations where it becomes necessary to divide a measure by a measure. To give an example, suppose that we had to administer a 250-mg dose of ascorbic acid in 100-mg tablets. How many tablets would be needed per dose?

$$250 \text{ mg} \div 100 \text{ mg} \quad \text{or} \quad \frac{250 \text{ mg}}{100 \text{ mg}} = 2.5 \qquad \text{Units reduce.}$$

Notice that the answer, 2.5, does not have a units label. That is because we are looking for *how many* tablets are needed.

Rule 5–19   *To divide a metric measure by a like measure:*

Divide the numerical values. The answer will be a number that tells how many parts there are. If unlike measures are used, we must first change to like measures.

EXAMPLE 9   Solve the following problems.

  (a) Four micrometers each weighing 752 g will fit into a shipping carton. If the shipping carton and filler weigh 217 g, what is the total weight of the shipment?

  (b) A 5-m-long board is to be cut into four equal parts. How long is each part?

  (c) How many 25-g packages of seed can be made from 2 kg of seed?

  (a) Total weight = carton and filler + four micrometers

$$= 217 \text{ g} + (4 \times 752 \text{ g})$$

$$= 217 \text{ g} + 3008 \text{ g}$$

$$= 3225 \text{ g, or } 3.225 \text{ kg}$$

Thus, the total weight is 3225 g, or 3.225 kg.

  (b) $\dfrac{5 \text{ m}}{4} = 1.25 \text{ m}$ Divide length by number of parts.

Thus, each part is 1.25 m long.

  (c) $\dfrac{2 \text{ kg}}{25 \text{ g}} = \dfrac{2000 \text{ g}}{25 \text{ g}}$ Convert kg to g; then divide. Reduce units.

$$= 80$$

Thus, 80 packages of seed can be made.

[1]

1. Give the value of the following metric units.
   (a) kilometer (km)
   (b) dekaliter (dkL)
   (c) decigram (dg)
   (d) millimeter (mm)
   (e) hectogram (hg)
   (f) centiliter (cL)

Choose the most reasonable metric measure for each of the following.

2. Height of a 6-year-old pediatric patient
   (a) 1.5 km   (b) 1.5 m   (c) 1.5 cm   (d) 1.5 mm
3. Diameter of a dime
   (a) 1.5 m   (b) 1.5 cm   (c) 1.5 mm   (d) 1.5 km
4. Distance from Los Angeles to San Francisco
   (a) 800 m   (b) 800 mm   (c) 800 km   (d) 800 cm
5. Length of a pencil
   (a) 20 km   (b) 20 cm   (c) 20 m   (d) 20 mm
6. Overnight accumulation of snowfall
   (a) 8 cm   (b) 8 m   (c) 8 km
7. Width of home videotape
   (a) 8 m   (b) 8 km   (c) 8 cm   (d) 8 mm
8. Weight of the average male adult
   (a) 70 mg   (b) 70 kg   (c) 70 g
9. Weight of a can of tuna
   (a) 184 mg   (b) 184 kg   (c) 184 g
10. Weight of a teaspoonful of sugar
    (a) 2 g   (b) 2 kg   (c) 2 mg
11. Weight of a vitamin C tablet
    (a) 250 g   (b) 250 mg   (c) 250 kg
12. Weight of a dinner plate
    (a) 350 kg   (b) 350 mg   (c) 350 g
13. Weight of a sack of dog food
    (a) 25 g   (b) 25 kg   (c) 25 mg
14. Volume of a tank of pesticide
    (a) 60 L   (b) 60 mL
15. Dose of cough syrup
    (a) 100 L   (b) 100 mL
16. Glass of juice
    (a) 0.12 mL   (b) 0.12 L

[2] Change to the measure indicated. When using the metric-value chart, the decimal is placed immediately *after* the measuring unit.

17. 4 m = ____ dm
18. 7 g = ____ dg
19. 58 km = ____ hm
20. 8 hL = ____ dkL
21. 0.25 km = ____ hm
22. 21 dkL = ____ L
23. 8.5 cm = ____ mm
24. 14.2 dg = ____ cg
25. How many milliliters are there in 15.3 cL?
26. How many meters are there in 46 dkm?
27. How many dekagrams are there in 7.5 hg?
28. How many millimeters are there in 16 cm?

Change each measure to the unit of measure indicated. When using the metric-value chart, the decimal is placed immediately *after* the measuring unit.

29. 4L = ____ cL
30. 8 m = ____ mm
31. 58 km = ____ m
32. 8 hg = ____ g
33. 0.25 km = ____ m
34. 21 dkg = ____ dg
35. 10.25 hm = ____ cm
36. 8.33 L = ____ mL
37. 2 km = ____ mm
38. 0.7 g = ____ cg
39. Change 2.36 hL to liters.
40. Change 0.467 dkm to centimeters.
41. Change 3.8 kg to decigrams.
42. Change 13 dkm to centimeters.

Change each measure to the unit indicated. When using the metric-value chart, the decimal is placed immediately *after* the measuring unit.

43. 28 m = _____ dkm

44. 238 hL = _____ kL

45. 101 mg = _____ cg

46. 60 hm = _____ km

47. 29 dkL = _____ hL

48. 192.5 g = _____ dkg

49. 17 cm = _____ dm

50. How many decimeters are in 4389 cm?

51. How many grams are in 47 dg?

52. How many deciliters are in 2.25 cL?

Change each measure to the measure indicated. When using the metric-value chart, the decimal is placed immediately *after* the measuring unit.

53. 2743 mm = _____ m

54. 385 g = _____ kg

55. 15 dkm = _____ km

56. 8 cL = _____ L

57. 296,484 m = _____ hm

58. 29.83 dg = _____ dkg

59. 0.3 cm = _____ dkm

60. 40 dL = _____ kL

61. 2857 mg = _____ kg

62. 15,285 m = _____ km

63. Change 297 cm to hectometers.

64. Change 0.03 mL to liters.

3 Add or subtract these measures.

65. 3 m + 8 m

66. 7 hL + 5 hL

67. 15 cg − 9 cg

68. 2 dm + 4 cm

69. 5 cL + 9 mL

70. 4 m + 2 L

71. 14 kL − 39 hL

72. 1 g − 45 cg

73. 3 cg − 5 mL

74. 7 km + 2 m

75. A patient absorbs 175 ml of fluid through an IV. If the IV bag has 825 ml left, how much fluid was in the bag to begin with?

76. 653 dkL of orange juice concentrate are removed from a vat containing 8 kL of the concentrate. How much concentrate remains in the vat?

Multiply.

77. 43 m × 12

78. 3.4 m × 12

79. 50.32 dm × 3

80. A plot of ground is divided into seven plots, each with road frontage of 138.5 m. What is the total road frontage of the plot of ground?

Divide.

81. 39 m ÷ 3

82. $\dfrac{54 \text{ cL}}{6}$

83. A block of silver weighing 978 g is cut into six equal pieces. How much does each piece weigh?

84. Two pieces of steel each 12 m long are cut into a total of 60 equal pieces. How long is each piece?

85. 2.5 cg ÷ 0.5 cg

86. 3 m ÷ 10 cm

87. How many 250-mL prescriptions can be made from a container of 4 L of decongestant?

88. How many 500-g containers are needed to contain 40 kg of grass seed?

Solve the following.

89. Add 4.6 cL + 5.28 dL of photographic developer.

90. Add 3 m + 2 dkm of fabric for draperies.

91. Subtract 19.8 km − 32.3 hm of paved highway.

92. Subtract 13 kL − 39 hL of stored liquid.

93. Multiply 0.25 cL of cologne by 5.

94. Multiply a 35-mm film size by 2.

95. A length of satin fabric 30 dm long is cut into 4 equal pieces. How long is each piece?

96. An IV bag holding 250 ml of an antibiotic is calibrated (marked off) into 5 equal sections. How many milliliters are represented by each section?

97. How many containers of jelly can be made from 8500 L of jelly if each container holds 4 dL of jelly?

98. How many 2-kg vials of hydrochloric acid (HCl) can be obtained from 38 kg of HCl?

METRIC–U.S. CUSTOMARY COMPARISONS

Learning Objective   **1**   Convert between U.S. customary measures and metric measures.

Learning to convert from the U.S. customary system to the metric system, or vice versa, is important because both systems are used in the United States. These conversions from U.S. customary to metric or metric to U.S. customary are more challenging than working entirely within one system. A given industry often uses one system exclusively, making conversions from one system to another uncommon. However, an industry that commonly uses the U.S. customary system in the United States may need to make conversions to the metric system to market its products in other countries.

1 Convert Between U.S. Customary Measures and Metric Measures.

To convert measures from the U.S. customary system to the metric system, and vice versa, we need only one *conversion factor* for each type of measurement (length, weight, and capacity). The following factors have been rounded to the nearest hundredth of a unit.

For length:
    1 m = 1.09 yd
For weight:
    1 kg = 2.20 lb
For capacity:
    (dry measure)
        1 L = 0.91 dry qt
    (liquid measure)
        1 L = 1.06 liquid qt

Dry quarts are less common than liquid quarts; if the type of quart (dry or liquid) is not specified, use liquid quarts.

Other conversion factors, such as centimeters to inches and kilometers to miles, can be derived from these factors using unity ratios. Naturally, the more conversion factors we have before us, the better; but one per type of measurement is all that is necessary.

Other conversion factors are listed in Table 5–5. The numbers in parentheses are rounded to ten-thousandths and can be used if greater accuracy is desired. Calculators sometimes have conversion factors with 10 or more decimal places programmed into the calculator for even greater accuracy.

TABLE 5–5   Metric–U.S. Customary Conversion Factors

Measure of Length	
Metric to U.S. customary units	*U.S. customary to metric units*
1 meter = 39.37 inches	1 inch = 25.4 millimeters
1 meter = 3.28 feet (3.2808)	1 inch = 2.54 centimeters
1 meter = 1.09 yards (1.0936)	1 inch = 0.0254 meter
1 centimeter = 0.39 or 0.4 inch (0.3937)	1 foot = 0.3 meter (0.3048)
1 millimeter = 0.04 inch (0.03937)	1 yard = 0.91 meter (0.9144)
1 kilometer = 0.62 mile (0.6214)	1 mile = 1.61 kilometers (1.6093)

Capacity: Liquid Measure	
Metric to U.S. customary units	*U.S. customary to metric units*
1 liter = 1.06 liquid quarts (1.0567)	1 liquid quart = 0.95 liter (0.9463)

TABLE 5–5 Metric–U.S. Customary Conversion Factors *continued*

Capacity: Dry Measure	
Metric to U.S. customary units	*U.S. customary to metric units*
1 liter = 0.91 dry quart (0.9081)	1 dry quart = 1.1 liters (1.1012)

Measure of Weight	
Metric to U.S. customary units	*U.S. customary to metric units*
1 gram = 0.04 ounce (0.0353)	1 ounce = 28.35 grams (28.3495)
1 kilogram = 2.2 pounds (2.2046)	1 pound = 0.45 kilogram (0.4536)

Tips and Traps   *To Memorize or Not:*

We do *not* need to memorize these conversion factors. They are available in numerous resource materials and are generally accessible in the workplace. The important thing is to remember to use the appropriate unity ratio.

Rule 5–20   *To convert metric measures to U.S. customary measures or U.S. customary measures to metric measures:*

1. Select the appropriate conversion factor. (For convenience, select the conversion factor that allows the new measure to be in the *numerator* of the unity ratio and places a 1 in the denominator.)
2. Set up a unity ratio so that the measure we are changing from reduces out and the new measure remains.

EXAMPLE 1   A hospital patient weighs 90 lb. How many kilograms does the patient weigh?

$$\frac{90 \text{ lb}}{1}\left(\frac{0.45 \text{ kg}}{1 \text{ lb}}\right)$$ Using 1 kg = 2.2 lb involves division. Answers will vary slightly.

$$\frac{90 \text{ lb}}{1}\left(\frac{0.45 \text{ kg}}{1 \text{ lb}}\right) = 90(0.45 \text{ kg}) = 40.5 \text{ kg}$$

Thus, the patient weighs about 40.5 kg.

EXAMPLE 2   A vinyl landscape pond that has a capacity of 76 liters is about to be filled with water. How many gallons is its capacity?

$$\frac{76 \text{ L}}{1}\left(\frac{1.06 \text{ qt}}{1 \text{ L}}\right)\left(\frac{1 \text{ gal}}{4 \text{ qt}}\right)$$ Table 5–5 does not give a conversion factor for gallons, so use two unity ratios.

$$\frac{\overset{19}{\cancel{76}} \text{ \cancel{L}}}{1}\left(\frac{1.06 \text{ qt}}{1 \text{ \cancel{L}}}\right)\left(\frac{1 \text{ gal}}{\underset{1}{4} \text{ qt}}\right) = 20.14 \text{ gal or } 20 \text{ gal}$$

Thus, the pond is a 20-gal pond.

SELF-STUDY EXERCISES 5–5

1   Change to the units indicated.

1. 9 m to inches
2. 120 m to yards
3. 42 km to miles
4. 6 L to liquid quarts
5. 10 dry qt to liters
6. 27 kg to pounds
7. 50 lb to kilograms
8. 7 in. to centimeters
9. 18 ft to meters

10. 39 mL to ounces
(*Hint:* 1 liquid qt = 32 oz.)

11. $5\frac{3}{4}$ gal to liters
(*Hint:* 1 gal = 4 liquid qt.)

12. A spool of wire contains 100 ft of wire. How many meters of wire are on the spool?

13. A sheet of metal weighing 60 lb weighs how many kilograms?

14. Two cities 150 mi apart are how many kilometers apart?

15. A road 30 m wide is how many yards wide?

16. A tourist in Europe traveled 200 km, 60 km, and 120 km by car. How many total miles was this?

17. A patient in therapy jogged 5 km, 4 km, and 3 km. How many miles did the patient jog?

18. A container holds 12 dry quarts. How many liters will the container hold?

19. A spool of electrical wire contains 100 m of wire. How many feet of wire are on the spool?

MATHEMATICS IN THE WORKPLACE

Health Sciences: Body Mass Index

Overweight Americans now outnumber those who are not overweight, according to the National Center for Health Statistics. Body mass index (BMI) is the standard measure used by health practitioners to measure a person's degree of obesity or emaciation. BMI is body weight in kilograms (kg) divided by height in meters squared, or BMI = w/h^2.

According to federal guidelines, a BMI greater than 25 means that you are overweight. Surveys show that 59% of men and 49% of women have BMIs greater than 25. People in their fifties tend to be the fattest—73% of men and 64% of women in this group have BMIs of 25+. Extreme obesity is defined as a BMI greater than 40. To calculate your BMI:

1. Multiply your weight by 0.45 to convert to kilograms.
2. Convert your height to inches.
3. Multiply the inches by 0.025 to get meters.
4. Square this number.
5. Divide your weight in kilograms by this number, and round to the nearest whole number. The result is your BMI.

Persons with high or low BMIs can expect to pay higher life and health insurance premiums. They may even be denied life and health insurance. BMI is used in many medical applications. For example, overweight persons needing certain surgeries (especially orthopedic) may be forced to wait until they have lost weight before they can receive their operations due to the health risk.

BMI is also used to find a person's body surface area (BSA) in square meters. Persons with heart problems are tested for heart efficiency. A person's cardiac index (CI) measures heart efficiency through volume of blood rate flow and their body surface area according to this formula:

$$CI = \frac{\text{cardiac output}}{\text{BSA}}$$

$$= \frac{\text{liters of blood per minute}}{\text{BSA}} \; (\text{L/min/m}^2)$$

For a patient with a low cardiac index, doctors usually operate to correct the low blood flow problem, and patients are typically instructed to lose weight. A low CI can result from low cardiac output, high body surface area, or both. A very low CI is life threatening and demands immediate treatment.

For these weight-related problems and many others, health officials strongly recommend maintaining a reasonable weight throughout your entire life.

5–6 READING INSTRUMENTS USED TO MEASURE LENGTH

Learning Objectives

1. Read the English rule.
2. Read the metric rule.

1 Read the English Rule.

Most of us are familiar with the *English rule;* in elementary and high school we called it simply a *ruler*. Although we used it for drawing straight lines, it was intended primarily for measuring lengths.

The most common English rule uses an inch as the standard unit. Each inch is subdivided into fractional parts, usually 8, 16, 32, or 64. Let's examine carefully the portion of the English rule illustrated in Fig. 5–36.

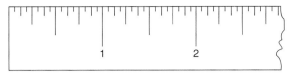

Figure 5–36

Each inch is divided into 16 equal parts; thus, each part is $\frac{1}{16}$ in.; that is, the first mark from the left edge represents $\frac{1}{16}$ in. The left end of the rule represents zero (0). (Some rules leave a small space between zero and the end of the rule.)

The second mark from the left edge of the rule represents $\frac{2}{16}$ or $\frac{1}{8}$ in. This mark is slightly longer than the first mark. Look at Fig. 5–37, which has each of the division marks labeled.

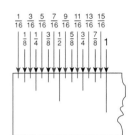

Figure 5–37

The fourth mark from the left is labeled $\frac{1}{4}$; that is, $\frac{4}{16} = \frac{1}{4}$. In each case, fractions are always reduced to lowest terms. Notice that the $\frac{1}{4}$ mark is slightly longer than the $\frac{1}{8}$ mark.

On the rule the division marks are made different lengths to make the rule easier to read. The shortest marks represent fractions that, in lowest terms, are sixteenths ($\frac{1}{16}, \frac{3}{16}, \frac{5}{16}, \frac{7}{16}, \frac{9}{16}, \frac{11}{16}, \frac{13}{16}, \frac{15}{16}$). The marks representing fractions that reduce to eighths are slightly longer than the sixteenths marks ($\frac{1}{8}, \frac{3}{8}, \frac{5}{8}, \frac{7}{8}$). Next, the marks representing fractions that reduce to fourths are slightly longer than the eighths marks ($\frac{1}{4}, \frac{3}{4}$). The marks representing one-half ($\frac{1}{2}$) are longer than the fourths marks, and the inch marks are longer than the one-half marks. Look at Fig. 5–37 again and pay particular attention to the lengths of the division marks.

In many of the examples and exercises that follow, we will measure *line segments* whose beginning and end are identified by capital letters, such as line segment *AB* in Fig. 5–38.

EXAMPLE 1 Measure line segment *AB* (Fig. 5–38).

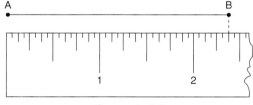

Figure 5–38

Align point *A* with zero. Line segment *AB* goes past the 2-in. mark, but not up to the 3-in. mark. Therefore, the measure of *AB* will be a mixed number between 2 and 3. Point *B* is $\frac{3}{8}$ in. past 2. **Thus, *AB* is $2\frac{3}{8}$ in.**

Tips and Traps	***Judge to the Closest Mark:*** A line segment may not always line up exactly with a division mark. If this is the case, we must use eye judgment and decide which mark is closer to the end of the line segment.

EXAMPLE 2   Measure line segment *CD* (Fig. 5–39).

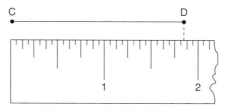

Figure 5–39

Point *D* lines up between $1\frac{13}{16}$ and $1\frac{7}{8}$. Remember, measurements are always approximations; using our best eye judgment, point *D* seems closer to $1\frac{13}{16}$ than $1\frac{7}{8}$. **We will say *CD* is $1\frac{13}{16}''$ to the nearest sixteenth of an inch.**

In practice, measurements are considered acceptable if they are within a desired *tolerance*. In the example above, the smallest division is $\frac{1}{16}$, so the desired tolerance would normally be plus or minus one-half of one-sixteenth, or $\pm\frac{1}{32}$ ($\frac{1}{2}$ of $\frac{1}{16} = \frac{1}{32}$). That is, the acceptable measure can be $\frac{1}{32}$ more than or $\frac{1}{32}$ less than the ideal measure.

If the ideal measure is $1\frac{13}{16}$ in. and the tolerance is $\frac{1}{32}$ in., the *range* of acceptable values is from $1\frac{13}{16} - \frac{1}{32}$ to $1\frac{13}{16} + \frac{1}{32}$.

$$1\frac{13}{16} = 1\frac{26}{32}$$

$$1\frac{26}{32} - \frac{1}{32} = 1\frac{25}{32} \qquad 1\frac{26}{32} + \frac{1}{32} = 1\frac{27}{32}$$

The acceptable range is from $1\frac{25}{32}$ to $1\frac{27}{32}$.

2   Read the Metric Rule.

The *metric rule* illustrated in Fig. 5–40 shows centimeters as the major divisions, represented by the longest lines. Each centimeter is divided into 10 millimeters, represented by the shortest lines. A line slightly longer than the millimeter line divides each centimeter into two equal parts of 5 mm each.

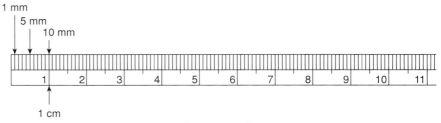

Figure 5–40

The metric rule is read like the English rule with the exception that only two measures are indicated, centimeters and millimeters. Other measures are calculated in relation to millimeters or centimeters. Thus, every 10 cm or 100 mm make 1 decimeter (dm).

EXAMPLE 3   Find the length of line segment *AB* (Fig. 5–41) to the nearest millimeter.

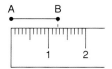

Figure 5–41

The line segment *AB* extends to between 12 mm and 13 mm. It appears to be closer to the 12-mm mark by eye judgment. **Thus, line segment *AB* is 12 mm to the nearest millimeter. The measure could also be written as 1.2 cm.**

EXAMPLE 4   Find the length of the line segment *CD* (Fig. 5–42) if the acceptable tolerance is ± 0.5 mm.

Figure 5–42

The end of line segment *CD* falls approximately halfway between the 35-mm mark and the 36-mm mark, by eye judgment, measuring about 35.5 mm. Because the tolerance is ± 0.5 mm, both 35 and 36 mm would be acceptable measures of the line segment *CD*. **Thus, 35 mm, 35.5 mm, and 36 mm would be acceptable approximations for line segment *CD*.**

SELF-STUDY EXERCISES 5–6

1   Measure the line segments 1 through 10 in Fig. 5–43 to the nearest sixteenth of an inch (tolerance = $\pm \frac{1}{32}$ in.).

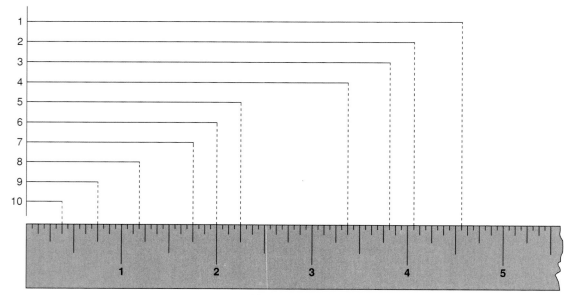

Figure 5–43

2 Measure the line segments 11 through 20 in Fig. 5–44 to the nearest millimeter. Measures can be expressed in millimeters or centimeters.

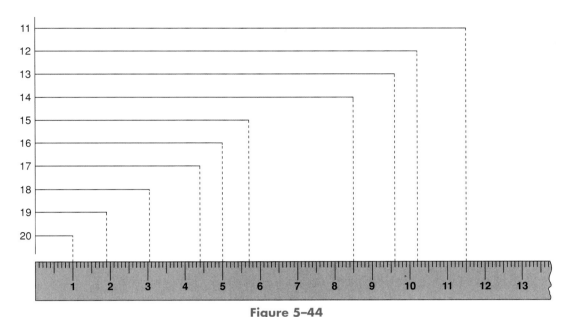

Figure 5–44

Nursing: Drug Dosage and IV Concentration

When administering patient drugs, most of a nurse's time is spent calculating the correct dosage and concentration, both of which depend on the patient's body weight. Because even minor errors in dosage or concentration can have drastic (or even life-threatening) consequences, it is imperative that the mathematics be calculated correctly. Drug manufacturers recommend dosages based on milligrams of medicine per kilogram of body weight (mg/kg).

For example, the recommended dosage of *Lasix,* a common drug used in congestive heart failure, is 1 mg of *Lasix* per kg of body weight (1 mg/kg). A patient's weight is typically measured in pounds, so nurses must convert lb to kg by dividing the weight in lb by 2.2 (1 kg = 2.2 lb). Conversely, to change a patient's weight from kg to lb, nurses multiply the weight in kg by 2.2.

Although in the hospital nurses administer metric doses, many times they must convert to the U.S. customary system for patient at-home-care instructions. For example, *Children's Tylenol,* a common drug used to treat fever and pain, contains 80 mg of medicine per 5 cubic centimeters (cc) of volume. Knowing that 5 cc equals 1 teaspoon (tsp), nurses calculate how many tsp should be given at home to a patient whose body weight requires a stronger or weaker dose of *Tylenol.*

In the hospital, some medicines are given through an intravenous (IV) drip. The rate of the IV drip determines the concentration of medicine given. For example, an infusion of *Dobutamine,* a cardiac drug used to improve heart muscle contractility for a patient, at 1.0 cc/hr equals a concentration of 10 micrograms of medicine per kilogram of body weight per minute (10 mcg/kg/min). Nurses could then calculate how fast the drip would need to be if the doctor increased the dosage to 14 mcg/kg/min. Doctors can also order the IV concentration changed, which means the amount of medicine dissolved in the IV liquid is increased (for a higher concentration) or decreased (for a lower concentration).

Exercises

Use the above information to answer the following exercises. Round all answers to the nearest tenth.

1. A premature baby boy weighing 4.5 lb needs a dose of *Lasix*. Convert his weight to kg and calculate how many mg of the drug he should receive.
2. Find the dosage of *Lasix* needed for his 3.4 lb twin sister.
3. Calculate the dosage of *Lasix* needed for a 67-year-old man weighing 192 lb.
4. The recommended dosage of *Children's Tylenol* is 10 mg/kg. How many mg of *Tylenol* should be given to an 18-month-old child who weighs 24 lb and has an ear infection?
5. Use your answer from Exercise 4 to tell the parent this child's dosage in tsp.
6. If a *Dobutamine* IV drip is going at 1.0 cc/hr, which equals 10 mcg/kg/min, how fast would it need to go if the doctor ordered it increased to 20 mcg/kg/min?
7. For a *Dobutamine* IV drip going at 1.0 cc/hr (10 mcg/kg/min), how fast would it need to go if the doctor ordered it decreased to 8 mcg/kg/min?
8. If a doctor ordered the concentration of a *Dobutamine* IV drip doubled, how would a nurse change the amount of medicine added to a bag of IV liquid?

Answers

1. For a weight of about 2 kg, 2 mg of *Lasix* should be given.
2. For a weight of about 1.5 kg, 1.5 mg of *Lasix* should be given.
3. For about 87.3 kg of weight, 87.3 mg of *Lasix* should be given.
4. For about 10.9 kg of weight, 109.0 mg of *Tylenol* should be given.
5. Because *Children's Tylenol* comes 80 mg/1 tsp, about 1.4 or $1\frac{1}{2}$ tsp should be given.
6. The concentration has been doubled, so the IV should drip at twice the rate, or 2.0 cc/hr.
7. The IV should drip at 0.8 cc/hr.
8. The amount of medicine added to the bag of IV liquid should be doubled.

ASSIGNMENT EXERCISES

Section 5–1

Identify the appropriate U.S. customary measure for each of the following.

1. Package of spaghetti
2. Tank of cotton defoliant
3. Container of motor oil
4. Distance from work to the hospital
5. Package of taco shells
6. Porterhouse steak
7. Shipment of iron
8. Sack of ammonium nitrate
9. Size of an aluminum pot
10. Height of a tree
11. Sugar for a pie recipe
12. Man's shirt size
13. Cloth for a pair of kitchen curtains
14. Hourly speed of an aircraft
15. Shaving cream

Using unity ratios, convert the given measures to the new units of measure.

16. 8 ft = _____ in.
17. 12 ft = _____ yd
18. 11 yd = _____ ft
19. $1\frac{1}{5}$ mi = _____ ft
20. How many feet of wire are needed to put a fence along a property line $2\frac{1}{4}$ mi long?
21. How many ounces are in 5 lb?
22. An object weighing $57\frac{3}{5}$ lb weighs how many ounces?
23. Find the number of pounds in 680 oz.

24. A can of fruit weighs 22.4 oz. How many pounds is this?

25. The net weight of a can of peas is 19 oz. If a case contains 16 cans, what is the net weight of a case in ounces? in pounds?

26. How many quarts are in 8 pt?

27. How many pints are in $7\frac{1}{2}$ qt?

28. Find the number of gallons in 15 qt.

29. Find the number of pints in 3 gal.

30. How many gallons are in 36 pt?

31. A cereal company has 12 T of prepared grain and wants to market it as packages of several ounces each. How many ounces of prepared grain is this?

32. A sack of concrete mix weighs 70 lb. How many ounces does a sack of concrete mix weigh?

33. How many feet of wire are needed to fence a property line $1\frac{1}{4}$ mi long?

34. A cook has 1 qt of vegetable oil. The recipe used requires 2 c of oil. How many recipes can be made from the quart of oil?

Express the following measures in standard notation.

35. 6 ft 17 in.

36. 1 mi 5375 ft

37. 12 lb $17\frac{1}{2}$ oz

38. 2 gal 7 qt

39. 1 gal 2 qt 5 pt

40. 2 T 3100 lb

41. 3 yd 2 ft 16 in.

42. 1 qt 3 c 12 oz

43. $3\frac{1}{4}$ ft 10 in.

44. 2 lb 21 oz

45. 1 gal 3 qt 48 oz

Section 5–2

Add or subtract as indicated. Write answers in standard form.

46. 12 oz + 2 lb

47. 8 ft − 49 in.

48. 4 gal + 3 qt

49. 2 ft 9 in.
  + 8 ft 2 in.

50. 7 lb 8 oz
  + 5 lb 9 oz

51. 5 gal 3 qt
  + 2 gal 3 qt

52. 7 ft 9 in.
  − 4 ft 6 in.

53. 4 lb 9 oz
  − 3 lb 11 oz

54. 4 yd 1 ft 8 in.
  − 2 yd 2 ft 11 in.

55. A rug 12 ft 6 in. long must fit in a room whose length is 10 ft 9 in. How much should be trimmed from the rug to make it fit the room?

56. Two packages to be sent air express each weigh 5 lb 4 oz. What is the shipping weight of the two packages?

57. A water hose purchased for an RV was 2 ft long. What was its length after cutting off 7 in.?

58. A vinyl flooring installer cut 19 in. from a piece of vinyl 13 ft long. How long was the vinyl piece after it was cut?

Section 5–3

Multiply and write answers for mixed measures in standard form.

59. 42 ft
  12 ft

60. 8 lb 3 oz
  9

61. 9 in.
  7 in.

62. 10 gal 3 qt
  7

Divide.

63. 20 yd 2 ft 6 in. ÷ 2

64. 5 gal 3 qt 2 pt ÷ 6

65. 65 ft ÷ 12 (Write answer in feet.)

66. 21 ft ÷ 4 (Write answer in feet and inches.)

67. If 18 lb of candy are divided equally into four boxes, express the weight of the contents of each box in pounds and ounces.

68. If 32 equal lengths of pipe are needed for a job and each length is to be 2 ft 8 in., how many feet of pipe are needed for the job?

Work the following problems.

69. 14 ft ÷ 4 ft

70. 2 mi 120 ft ÷ 15 ft

71. 400 lb ÷ 90 lb

72. 6 yd 2 ft ÷ 5 ft

73. $5\dfrac{\text{mi}}{\text{min}} = \underline{\hspace{1cm}} \dfrac{\text{mi}}{\text{hr}}$

74. $2520\dfrac{\text{gal}}{\text{hr}} = \underline{\hspace{1cm}} \dfrac{\text{qt}}{\text{hr}}$

75. $88\dfrac{\text{ft}}{\text{sec}} = \underline{\hspace{1cm}} \dfrac{\text{mi}}{\text{hr}}$

76. $18\dfrac{\text{mi}}{\text{gal}} = \underline{\hspace{1cm}} \dfrac{\text{ft}}{\text{gal}}$

77. A pump that can move water at the rate of $75\dfrac{\text{gal}}{\text{hr}}$ can move how many gallons per minute?

78. A plane that travels at the rate of 240 mph is traveling how many feet per second?

79. How many quarts of milk are needed for a recipe that calls for 3 pt of milk?

80. How many $\frac{1}{2}$-oz servings of jelly can be made from a $1\frac{1}{2}$-lb container of jelly?

Section 5–4

Give the prefix that relates each of the following numbers to the standard unit in the metric system.

81. 1000 times

82. $\dfrac{1}{10}$ of

83. $\dfrac{1}{1000}$ of

84. 10 times

85. $\dfrac{1}{100}$ of

86. 100 times

Give the value of the prefixes of the following units of measure based on a standard measuring unit.

87. dekameter (dkm)

88. hectogram (hg)

89. milligram (mg)

90. centigram (cg)

91. kiloliter (kL)

92. deciliter (dL)

Choose the most reasonable answer.

93. Height of the Washington Monument
    (a) 200 m   **(b)** 200 cm   **(c)** 200 mm   **(d)** 200 km

94. Height of Mt. Rushmore
    (a) 1.6 km   **(b)** 1.6 m   **(c)** 1.6 cm   **(d)** 1.6 mm

95. Weight of an egg
    (a) 50 g   **(b)** 50 kg   **(c)** 50 mg

96. Weight of a saccharin tablet
    (a) 50 kg   **(b)** 50 mg   **(c)** 50 g

97. Weight of a man's shoe
    (a) 0.25 g   **(b)** 0.25 mg   **(c)** 0.25 kg

98. Carton of milk
    (a) 4 L   **(b)** 4 mL

99. Bottle of medicine
    (a) 50 L   **(b)** 50 mL

Change to the unit indicated.

100. 0.4 dkm = ____ hm

101. 67.1 m = ____ dkm

102. 4 m = ____ dm

103. 2.3 m = ____ mm

104. 5 cm = ____ mm

105. 0.123 hm = ____ mm

106. How many millimeters are in 0.432 km?

107. 23 dkm = ____ mm

108. 42.7 cm = ____ dkm

109. 41,327 dkm = ____ km

110. A board is 1.82 m long. How many centimeters are in the board?

111. 394.5 g = ____ hg

112. 2.7 hg = ____ dg

113. 3,000,974 cg = ____ kg

Perform the operations indicated.

114. 25 mm − 14 mm

115. 12 g + 5 m

116. 17 mg − 8 mL

117. 8 g − 52 cg

118. 4.3 dkg × 7

119. 6.83 cg × 9

120. $\dfrac{18 \text{ cm}}{9}$

121. 7.5 kg ÷ 0.5 kg

122. $\dfrac{8 \text{ hL}}{20 \text{ L}}$

123. 34 hL ÷ 4

124. 2.4 m ÷ 5 cm

125. Fabric must be purchased to make seven garments, each requiring 2.7 m of fabric. How much fabric must be purchased?

126. Candy weighing 526 g is mixed with candy weighing 342 g. What is the weight of the mixture?

127. A recipe calls for 5 mL of vanilla flavoring and 24 cL of milk. How much liquid is this?

128. Twenty boxes, each weighing 42 kg, are to be moved. How much weight must be moved?

129. A metal rod 42 m long is cut into seven equal pieces. How long is each piece?

130. Thirty-two kilograms of a chemical are distributed equally among 16 students. How many kilograms of chemical does each student receive?

131. A serving of punch is 25 cL. How many servings can be obtained from 25 L of punch?

132. How many containers of jelly can be made from 8548 L of jelly if each container holds 4 dL of jelly?

133. A bolt of fabric contains 6.8 dkm. If a shirt requires 1.7 m, how many shirts can be made from the bolt?

Section 5-5
Make the following conversions.

134. 7 m = _____ inches

135. 215 m = _____ yards

136. 69 km = _____ miles

137. 15 L = _____ liquid quarts

138. 12 dry qt = _____ liters

139. 32 kg = _____ pounds

140. 10 lb = _____ kilograms

141. 9 in. = _____ centimeters

142. 21 ft = _____ meters

143. 14.8 dkL = _____ quarts

144. $3\frac{1}{2}$ gal = _____ liters

145. How many meters are in 200 ft of pipe?

146. Concrete weighing 90 lb weighs how many kilograms?

147. Two cities 175 mi apart are how many kilometers apart?

148. A room 10 m wide is how many feet wide?

Section 5-6
Measure line segments 149 through 158 in Fig. 5–45 (tolerance = $\pm\frac{1}{32}$ in.).

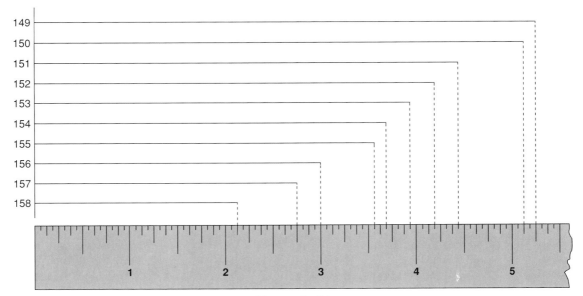

Figure 5–45

Measure line segments 159 through 168 in Fig. 5–46 to the nearest millimeter.

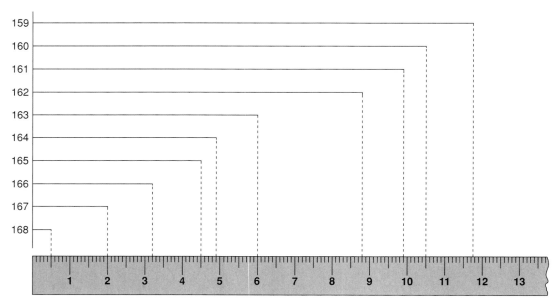

Figure 5–46

CHALLENGE PROBLEMS

169. Compare and contrast the U.S. customary system of measurement and the metric system of measurement.

170. Why do you think the metric system has not fully replaced the U.S. customary system in the United States?

CONCEPTS ANALYSIS

1. Give four items that would be measured with pounds.

2. Give four items that would be measured with feet.

3. If you were building a house, what units of linear measure would you be likely to use?

4. Explain how metric units of measure of length, weight, and capacity are similar.

5. Which measure is longer, a yard or a meter?

6. Which metric measure would you use if you were dispensing a liquid medicine?

7. If your medicine bottle reads 5 mg, is the measure more likely to be a liquid or a capsule?

Identify the mistake, explain why it is wrong, and correct the mistake in each of the following.

8. 3.252 dkL = 325.2 kL

9. 418 yd = _____ ft?

$$418 \text{ yd} \times \frac{3 \text{ ft}}{1 \text{ yd}} = 1234 \text{ yd}$$

10. 24 pt = _____ gal?

$$24 \text{ pt} \times \frac{1 \text{ gal}}{4 \text{ pt}} = 6 \text{ gal}$$

11. A stack of plywood is 4 ft high. If each sheet of plywood is $\frac{3}{4}$ in. thick, how many pieces of plywood are in the stack?

4 ft = 48 in. The stack of plywood is 48 in. tall. $48 \times \frac{3}{4} = 36$. There are 36 sheets of plywood in the stack.

Objectives	What to Remember with Examples

Section 5–1

1 Identify uses of U.S. customary system measures of length, weight, and capacity.

Associate small, medium, and large objects with units designating small, medium, and large amounts.

> Nose drops, oz; sofa, lb; coal, T; travel, mi; belt, in.; window, ft or in.; cloth, yd; oil, qt; fuel, gal; cooking, c

2 Convert from one U.S. customary unit of measure to another.

1. Write the original measure in the numerator of a fraction with 1 in the denominator.

2. Multiply by a unity ratio with the original unit of measure in the denominator and the new unit in the numerator.

Multiplication can be used to change to smaller units.
Division can be used to change to larger units.
A unity ratio (or ratios) can be used to change to any unit.

> Change 5 ft to inches.
>
> $$\frac{5 \text{ ft}}{1} \times \frac{12 \text{ in.}}{1 \text{ ft}} = 60 \text{ in.}$$
>
> Change 285 ft to yards.
>
> $$\frac{285 \text{ ft}}{1} \times \frac{1 \text{ yd}}{3 \text{ ft}} = 95 \text{ yd}$$
>
> Change $3\frac{1}{2}$ quarts to cups.
>
> $$3\frac{1}{2} \text{ quarts} = \frac{7}{2}$$
>
> $$\frac{7 \text{ qt}}{2} \times \frac{2 \text{ pt}}{1 \text{ qt}} \times \frac{2 \text{ c}}{1 \text{ pt}} = 14 \text{ c}$$

3 Write mixed U.S. customary measures in standard notation.

Each unit is converted to the next larger unit when possible.

> 2 ft 16 in. = 3 ft 4 in.
> (16 in. = 1 ft 4 in.)

Section 5–2

1 Add U.S. customary measures.

To add or subtract U.S. customary units:
1. Convert the measures to the same unit of measure.
2. Add or subtract.

> Add 5 lb and 7 oz.
>
> $$\frac{5 \text{ lb}}{1} \times \frac{16 \text{ oz}}{1 \text{ lb}} = 80 \text{ oz}$$
>
> 80 oz + 7 oz = 87 oz

2 Subtract U.S. customary measures.

To add or subtract mixed U.S. customary measures:
1. Align like measures in columns.
2. Add or subtract, carrying or borrowing as necessary.

> Subtract 2 ft 8 in. from 7 ft.
>
> $\quad$ 6 ft 12 in. $\qquad$ Rewrite 7 ft as 6 ft 12 in.
> $\underline{- \text{ 2 ft } \text{ 8 in.}}$
> $\quad$ 4 ft   4 in.

Section 5-3

1 Multiply a U.S. customary measure by a number.

To multiply a measure by a number:
1. Multiply each measure with a different unit by the number.
2. Express the answer in standard notation.

> Multiply 2 gal 3 qt by 5.
>
> $$\begin{array}{r} 2 \text{ gal } 3 \text{ qt} \\ \times \qquad 5 \\ \hline 10 \text{ gal } 15 \text{ qt or} \\ 13 \text{ gal } \;\; 3 \text{ qt} \end{array}$$

2 Multiply a U.S. customary measure by a measure.

To multiply a length measure by a like length measure:
1. Multiply the numbers associated with each like unit of measure.
2. The product will be a square unit of measure.

> Multiply 5 yd by 12 yd.
>
> $5 \text{ yd} \times 12 \text{ yd} = 60 \text{ yd}^2$

3 Divide a U.S. customary measure by a number.

To divide a measure by a number:
1. Divide the largest unit by the number.
2. Convert any remainder to the next smaller unit.
3. Repeat Steps 1 and 2 until no other measures are left.
4. Express any remainder as a fraction of the smallest measure.
5. The quotient will be a measure.

> Divide 7 lb 5 oz by 5.
>
> $$\begin{array}{r} \quad 1 \text{ lb} \qquad 7\frac{2}{5} \text{ oz} \\ \hline 5\,\overline{)\,7 \text{ lb} \qquad 5 \text{ oz}} \\ \underline{5 \text{ lb}} \qquad \qquad \\ 2 \text{ lb} = \underline{32 \text{ oz}} \\ 37 \text{ oz} \\ \underline{35 \text{ oz}} \\ 2 \end{array}$$

4 Divide a U.S. customary measure by a measure.

To divide a measure by a measure:
1. Convert the measures to the same unit.
2. Divide. The quotient is a number that indicates how many.

> How many 5-oz glasses of juice can be poured from 1 gallon of juice?
>
> $$\frac{1 \text{ gal}}{1} \times \frac{4 \text{ qt}}{1 \text{ gal}} \times \frac{2 \text{ pt}}{1 \text{ qt}} \times \frac{2 \text{ c}}{1 \text{ pt}} \times \frac{8 \text{ oz}}{1 \text{ c}} = 128 \text{ oz}$$
>
> $$\frac{128 \text{ oz}}{5 \text{ oz}} = 25\frac{3}{5} \text{ glasses}$$

5 Change from one U.S. customary rate measure to another.

1. Set up the original rate equal to the new rate without its numerical values.
2. Decide which units will change.
3. Multiply the original rate by the unity ratio or ratios with the new unit to reduce the old unit.

256   CHAPTER 5   Direct Measurement

Change 10 mi/hr to ft/hr.

$$\frac{10 \text{ mi}}{1 \text{ hr}} \text{ to } \frac{\text{ft}}{\text{hr}}$$

$$\frac{10 \text{ mi}}{1 \text{ hr}} \left(\frac{5280 \text{ ft}}{1 \text{ mi}} \right)$$

52,800 ft/hr

Section 5–4

1 Identify uses of metric measures of length, weight, and capacity.

Associate small, medium, and large objects with units designating small, medium, and large amounts. Powers of 10 prefixes are used, such as *kilo* for 1000.

Perfume, mL; soda, L; travel, km; racetrack, m; pill, mg; potatoes, kg; eye drops, mL

2 Convert from one metric unit of measure to another.

Move the decimal point in the original measure to the left or right as many places as necessary to move from the original measuring unit to the new unit on the metric chart of prefixes.

Change 5.04 cl to liters. Move the decimal two places to the left.

5.04 cL = 0.0504 L

3 Make calculations with metric measures.

To add or subtract metric measures:
1. Change measures to measures with like units if necessary.
2. Add or subtract.

To multiply or divide a metric measure by a number, multiply or divide the numbers and keep the same measuring unit.

To divide a metric measure by a measure:
1. Change measures to measures with like units if necessary.
2. Divide the numbers, canceling the units of measure. The answer will be a number.

7 km + 34 m =
7000 m + 34 m = 7034 m
or
7 km + 0.034 m = 7.034 km

7 mg × 3 = 21 mg
5 m ÷ 25 cm

$$\frac{500 \text{ cm}}{25 \text{ cm}} = 20$$

Section 5–5

1 Convert between U.S. customary measures and metric measures.

1. Set up the original amount as a fraction with the original unit of measure in the numerator and 1 in the denominator.
2. Multiply by a unity ratio with the original unit of measure in the denominator and the new unit in the numerator.

How many feet are in 14 meters?

$$\frac{14 \text{ m}}{1} \times \frac{3.28 \text{ ft}}{1 \text{ m}} = 45.92 \text{ ft}$$

Section 5–6

1 Read the English rule.

See Fig. 5–47. Align the rule along the object. Count the number of inches and fractional parts ($\frac{1}{32}$'s, $\frac{1}{16}$'s, $\frac{1}{8}$'s, $\frac{1}{4}$'s, and so on) to determine the approximate length. Use eye judgment to estimate closeness of the object to a mark on the rule.

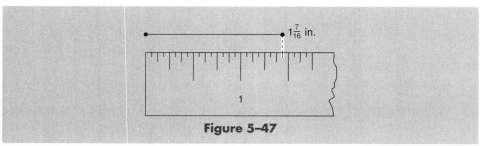

Figure 5–47

2 Read the metric rule.

See Fig. 5–48. Align the rule along the object. Count the number of millimeters and/or centimeters (10 mm = 1 cm) to determine the approximate length. Use eye judgment to estimate closeness of the object to a mark on the rule.

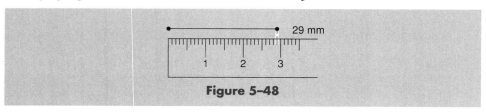

Figure 5–48

WORDS TO KNOW

English system (p. 215)
U.S. customary system (p. 215)
unity ratio (p. 218)
mixed measures (p. 221)
standard notation (p. 221)
unlike measures (p. 223)
like measures (p. 223)
rate measure (p. 228)
metric system (p. 231)

meter (p. 232)
deci- (p. 232)
centi- (p. 232)
milli- (p. 232)
deka- (p. 232)
hecto- (p. 232)
kilo- (p. 232)
kilometer (p. 233)
centimeter (p. 233)

millimeter (p. 233)
gram (p. 233)
kilogram (p. 234)
milligram (p. 234)
liter (p. 234)
milliliter (p. 234)
conversion factors (p. 243)
English rule (p. 245)
metric rule (p. 247)

CHAPTER TRIAL TEST

Change to the measure indicated (see Tables 5–1 to 5–4 for conversion factors).

1. 3 ft = _____ in.

2. 36 oz = _____ lb

3. 32 qt = _____ gal

4. Add: 2 yd 6 ft 10 in. + 3 ft 7 in. Write your answer in standard notation.

5. $60\dfrac{\text{gal}}{\text{min}} = \underline{\quad}\dfrac{\text{qt}}{\text{sec}}$

6. 21 in. ÷ 3 in.

7. A 495-ft section of highway is to be resurfaced. How many yards is this?

8. How many quarts are contained in a 55-gal drum?

9. If an automobile travels at 55 mph, how many feet is it traveling per second?

10. A spark plug wire kit contains $18\frac{1}{2}$ ft of wire in a coil. The directions call for cutting off individual lengths of 28 in. each. How many 28-in. wires can be cut from the coil?

11. Find the measure of the line segment AB in Fig. 5–49 (tolerance = $\pm\frac{1}{32}$ in.).

12. Select the appropriate U.S. customary measure for the contents of a swimming pool.

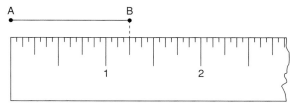

Figure 5–49

13. Write 1 gal 3 qt 6 c 20 oz in standard notation.

14. Which is the appropriate metric measure for eye drops: 28 mL, 28 dL, 28 L, or 28 kL?

Give the metric prefix for each of the following.

15. $\frac{1}{10}$ of standard unit

16. 10 × standard unit

Change to the metric unit indicated.

17. 298 m = _____ km

18. 8 dm = _____ mm

19. 5.2 dL of liquid are poured from a container holding 10 L. How many liters of liquid remain in the container?

20. A bar of soap weighs 175 g. How much do 15 bars of soap weigh (in kilograms)?

21. 75 mi = _____ km

22. 25 kg = _____ lb

23. 4 L = _____ pt

24. How many liters of weed killer are contained in a 55-gal drum?

25. A 0.243-caliber bullet travels 2450 $\frac{ft}{sec}$ for the first 200 yd. Using conversion factors, change 2450 $\frac{ft}{sec}$ to $\frac{meters}{sec}$.

CHAPTER

6

Area and Perimeter

GOOD DECISIONS THROUGH TEAMWORK

You and your team are designing options for your customer to place a single-family home on the lot purchased by the customer. The subdivision covenant where the lot is located requires that no per-manent structure be located closer than 25 feet from the front curb and 10 feet from the side and back property lines. The minimum amount of living space must be 2800 square feet, with at least 1800 square feet on the ground level. The lot measures $103\frac{1}{2}$ ft wide $\times$ $121\frac{3}{4}$ ft deep.

Make three different scale drawings of the lot and house, as viewed from above, that meet the subdivision covenants. Prepare a written document for your customer that presents the strengths and weaknesses of each option.

Pair with another team and have each team alternately playing the role of the customer. In an open discussion, the customer team will decide which, if any, of the options is suitable for its needs. If no option presented meets the needs of the customer, construct a new scale drawing that will meet the needs of the customer, or show the customer why his or her needs cannot be accommodated within the requirements of the subdivision covenants.

There are occasions in everyday life and in almost every occupation when we need to figure how much surface space a straight-sided, flat figure, called a polygon, occupies or what is the combined length of its sides. Most of us will have to work with such geometric figures and their measurements at one time or another. It is to our advantage to be familiar with them and the formulas that can save both time and money on the job and at home.

6-1 SQUARES, RECTANGLES, AND PARALLELOGRAMS

Learning Objectives

1. Find the perimeter and area of a square.
2. Find the perimeter and area of a rectangle.
3. Find the perimeter and area of a parallelogram.

A *polygon* is a plane or flat closed figure described by straight line segments and angles. Among the most familiar and easy-to-work-with polygons are the *parallelogram*, the *rectangle*, and the *square*. We will begin with a definition of terms.

■ **DEFINITION 6-1: Parallelogram.** A *parallelogram* is a four-sided polygon whose opposite sides are parallel (Fig. 6-1).

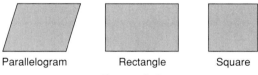

| Parallelogram | Rectangle | Square |

Figure 6-1

■ **DEFINITION 6-2: Rectangle.** A *rectangle* is a parallelogram whose angles are all right angles (Fig. 6-1).

■ **DEFINITION 6-3: Square.** A *square* is a parallelogram whose sides are all of equal length and whose angles are all right angles (Fig. 6-1).

1 Find the Perimeter and Area of a Square.

Two of the most useful concepts associated with polygons are *area* and *perimeter*. It is important that we can distinguish between these concepts.

■ **DEFINITION 6-4: Perimeter.** The *perimeter* is the total length of the sides of a plane figure (Fig. 6-2).

■ **DEFINITION 6-5: Area.** The *area* is the amount of surface of a plane figure (Fig. 6-3).

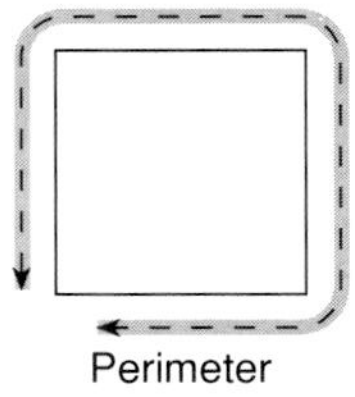

Perimeter

Figure 6-2

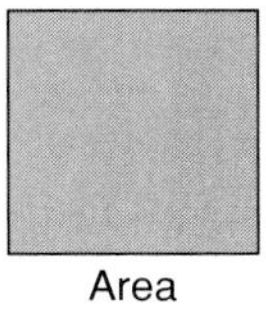

Area

Figure 6-3

When do you need to find the perimeter of a polygon and when do you need to find the area? If a carpenter had to install baseboard molding in a newly constructed den or family room, he or she would need to know how many feet of baseboard molding were needed. The carpenter needs to know the distance

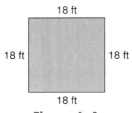

18 ft

18 ft   18 ft

18 ft

Figure 6–4

around the room, that is, the perimeter of the room. According to the plans, the room is square and measures 18 ft on each side (Fig. 6–4).

The perimeter, or distance around the room, would be figured by adding together the length of each wall: 18 ft + 18 ft + 18 ft + 18 ft = 72 ft. Because all sides are equal and there are four sides, we can multiply 4 × 18.

Formula 6–1   *Perimeter of a square:*

$$P_{\text{square}} = 4s$$

where *s* is the length of one side

An expression like 4*s* means 4 times the value represented by *s*.

Tips and Traps   *Subscripts:*
Subscripted words, letters, or numbers are a handy way of providing additional information. Because we will be examining the formulas for the perimeter and area of several different shapes, we may sometimes use subscripts to distinguish among them. For instance, the perimeter of a square, rectangle, or parallelogram could be indicated as P_{square}, $P_{\text{rectangle}}$, or $P_{\text{parallelogram}}$, respectively.

EXAMPLE 1   How much aluminum edge molding is needed to surround a stainless steel kitchen sink that measures 40 cm on each side (Fig. 6–5)?

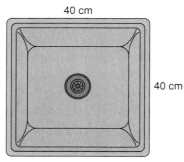

40 cm

40 cm

Figure 6–5

$P_{\text{square}} = 4s$

$P_{\text{square}} = 4(40 \text{ cm})$ Substitute measurement of one side. Multiply.

$P_{\text{square}} = 160 \text{ cm}$

The sink requires 160 cm of molding.

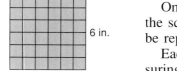

6 in.

6 in.

Figure 6–6

On the other hand, the area of a square is the entire amount of surface that the square occupies. If we had a square that measured 6 in. on a side, it would be represented as in Fig. 6–6.

Each small square in the larger figure is a *square inch,* that is, a square measuring 1 in. on each side. If we count the number of square inches in the larger figure, we find 36. This is the area of the square in question. The area can be expressed as 36 sq in. or 36 in.^{2}. Instead of counting each square inch in the larger square, we could have arrived at 36 sq in. or 36 in.^{2} by multiplying one side by the other (6 in. × 6 in. = 36 in.^{2}). From this we get the formula for the area of a square.

Representing Square Measuring Units:

The exponent 2 following a unit of measure indicates *square measure* or area. Thus, 5 in.^{2} = 5 square inches, 23 ft^2 = 23 square feet, 120 cm^2 = 120 square centimeters, and so on. This is a shortcut way of expressing a measure that has *already been "squared."*

Formula 6–2 *Area of a square:*

$$A_{\text{square}} = s^2$$

where s is the length of one side

The expression s^2 means $s \times s$, just as 5^2 means 5×5.

EXAMPLE 2 Lynn Fly, who installs vinyl flooring, discovered that one of the squares in the flooring pattern was damaged. He decided to cut out the damaged square and replace it. The damaged square measured 20 cm on each side. What is the area of the square to be replaced?

$A_{\text{square}} = s^2$

$A_{\text{square}} = (20 \text{ cm})^2$ Substitute measurement of one side. Square the measurement.

$A_{\text{square}} = 400 \text{ cm}^2$

The vinyl square measures 400 cm^2.

Tips and Traps   *Dimension Analysis:*

When evaluating a formula, the measuring units are often omitted from the written steps. However, it is very important to analyze the measuring units of the problem and to determine the measuring unit of the solution.

Example 1

$P_{\text{square}} = 4s$

$P_{\text{square}} = 4(40)$ The measuring unit is centimeters and a measure is multiplied by a number.

$P_{\text{square}} = \textbf{160 cm}$ The measuring unit of the solution is centimeters.

Perimeter is always a linear measure, which means the measuring unit should be to the first power.

Example 2

$A_{\text{square}} = s^2$

$A_{\text{square}} = (20)^2$ The measuring unit is centimeters and a measure is squared or a measure is multiplied by a measure.

$A_{\text{square}} = \textbf{400 cm}^2$ The measuring unit of the solution is square centimeters.

Area is always a square measure, which means that measures of area will have an exponent of two.

---

2 **Find the Perimeter and Area of a Rectangle.**

Finding the perimeter of a rectangle is similar to finding the perimeter of a square. We add the measures of the four sides, the longer sides being called the *length* (*l*) and the shorter sides being called the *width* (*w*), as shown in Fig. 6–7.

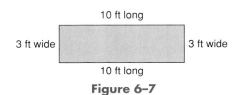

Figure 6–7

$$P = 10 + 3 + 10 + 3 = 26 \text{ ft}$$

We notice that the perimeter is composed of two lengths and two widths:

$$P = 2l + 2w$$

The distributive property shows that $2(l + w)$ and $2l + 2w$ are equivalent.

Formula 6–3 *Perimeter of a rectangle:*

$$P_{\text{rectangle}} = 2(l + w)$$

where l is the length and w is the width

Formula 6–3 shows that *both* the length *and* the width are multiplied by 2.

EXAMPLE 3

A shop that makes custom picture frames has an order for a frame whose outside measurements will be 42 in. by 30 in. (Fig. 6–8). How many inches of picture frame molding will be needed for the job?

$P_{\text{rectangle}} = 2(l + w)$

$P_{\text{rectangle}} = 2(42 + 30)$ Substitute. Perform operations.

$P_{\text{rectangle}} = 2(72)$

$P_{\text{rectangle}} = 144$

42 in.

30 in.

Figure 6–8

When analyzing the dimensions, inches are added to inches, giving the results in inches. Then the inches are multiplied by a number and the final result is inches. **Thus, 144 in. are needed for the project.**

In some applications, we may need to decrease the total perimeter to account for doorways and other openings in the perimeter.

EXAMPLE 4

A chain-link fence is to be installed around a yard measuring 25 m by 30 m (Fig. 6–9). A gate 3 m wide will be installed to allow entrance of an automobile. The gate comes preassembled from the manufacturer. How much fencing does the installer need to put up the fence?

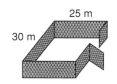

25 m

30 m

Figure 6–9

$P_{\text{rectangle}} = 2(l + w) - 3$ Subtract the length of the gate.

$P_{\text{rectangle}} = 2(30 + 25) - 3$ Substitute. Perform operations.

$P_{\text{rectangle}} = 2(55) - 3$

$P_{\text{rectangle}} = 110 - 3$

$P_{\text{rectangle}} = 107 \text{ m}$ Perimeter is a linear measure.

The job requires 107 m of fencing.

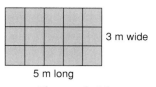

3 m wide

5 m long

Figure 6–10

The area of a rectangle can be found by dividing it into smaller squares and counting the number of squares (Fig. 6–10). By count, there are 15 square meters, or 15 m^2, in the rectangle. Multiplying length by width also gives us 15 m^2.

Formula 6–4   *Area of a rectangle:*

$$A_{\text{rectangle}} = lw$$

where *l* is the length and *w* is the width

The expression *lw* represents length times width.

EXAMPLE 5   A carpet installer needs to carpet a room measuring 16 ft by 20 ft. Projecting out from one wall is a fireplace whose hearth measures 3 ft by 6 ft (Fig. 6–11). How many square yards of carpet does the installer require for the job? How much is wasted?

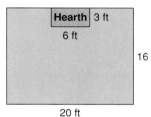

Hearth   3 ft

6 ft

16 ft

20 ft

Figure 6–11

This problem has several steps. First, compute the area of the room without considering the area of the hearth. This is the amount of carpet needed for the job. Second, compute the area of the fireplace hearth. This is the amount wasted. Finally, analyze the dimensions to see if the result is expressed using the desired measuring unit. If not, we will need to convert to the desired measuring unit.

Let A_{room} = the area of the room and A_{hearth} = the area of the hearth.

$A_{\text{room}} = lw$	
$A_{\text{room}} = 20 \times 16$	Substitute and multiply.
$A_{\text{room}} = 320 \text{ ft}^2$	Area is a square measure.
$A_{\text{hearth}} = lw$	
$A_{\text{hearth}} = 3 \times 6$	Substitute and multiply.
$A_{\text{hearth}} = 18 \text{ ft}^2$	Area is a square measure.

Because carpet is sold in square yards, we must convert the square footage to square yards. Since 9 ft^2 = 1 yd^2, we can use a *unity ratio* to convert square feet to square yards.

A unity ratio, as we recall from Chapter 5, is a fraction with one unit of measure in the numerator and a different, but equivalent, unit of measure in the denominator. A unity ratio contains the original unit and the new unit. Now we can complete the problem using the unity ratio $\frac{1 \text{ yd}^2}{9 \text{ ft}^2}$.

$$\frac{320 \text{ ft}^2}{1} \times \frac{1 \text{ yd}^2}{9 \text{ ft}^2} = \frac{320 \text{ yd}^2}{9} = 35\frac{5}{9} \text{ yd}^2 \text{ carpet needed,}$$

$$\frac{\overset{2}{\cancel{18}} \text{ ft}^2}{1} \times \frac{1 \text{ yd}^2}{\underset{1}{\cancel{9}} \text{ ft}^2} = 2 \text{ yd}^2 \text{ carpet wasted}$$

This job requires $35\frac{5}{9}$ yd^2 of carpet. Of this, 2 yd^2 is waste that can be used for carpeting smaller areas, such as closets.

Example 5 brings up some interesting real-life questions. Can a fraction of a yard of carpet be purchased? If carpet can be purchased only in 12-ft or 15-ft widths, will there be additional waste? Is it more practical to purchase extra carpet that will be wasted or to spend extra labor costs in installing the carpet with several seams? You may want to investigate common practices in the industry.

In other applications, we may find that additional calculations are necessary for finding one or more areas. For instance, in estimating the amount of paint, wallcovering, tongue-and-groove flooring, roofing, shingles, siding, floor tiles, bricks, and similar building materials, consideration must be given to the decrease or increase in area coverage of these materials on installation. Generally, such decreases and increases are standardized by the industry and made known to the user. For example, a paint manufacturer will indicate on the product label the number of square feet of coverage on bare wood and a different number of square feet of coverage on previously painted surfaces. A roofing shingle manufacturer will indicate how many bundles of shingles are needed per 100 ft² of roof depending on the amount of shingle overlap desired.

EXAMPLE 6   If a $\frac{1}{2}$-in. mortar joint is used to build a wall of 2-in. × 4-in. × 8-in. bricks, six bricks will cover one square foot. If a $\frac{1}{4}$-in. mortar joint is used, seven bricks will be needed to cover 1 square foot. How many bricks would be needed to build a 50-ft × 14-ft wall with $\frac{1}{2}$-in. joints? With $\frac{1}{4}$-in. joints?

$$A_{\text{wall}} = 50 \times 14$$

$$A_{\text{wall}} = 700 \text{ ft}^2$$

$$\frac{700 \text{ ft}^2}{1} \times \frac{6 \text{ bricks}}{1 \text{ ft}^2} = 4200 \text{ bricks} \qquad \text{Find number of bricks at 6 per ft}^2\text{, using a unity ratio.}$$

$$\frac{700 \text{ ft}^2}{1} \times \frac{7 \text{ bricks}}{1 \text{ ft}^2} = 4900 \text{ bricks} \qquad \text{Find number of bricks at 7 per ft}^2\text{, using a unity ratio.}$$

Using $\frac{1}{2}$-in. mortar joints, 4200 bricks are needed. Using $\frac{1}{4}$-in. mortar joints, 4900 bricks are needed.

EXAMPLE 7   Asphalt roofing shingles are sold in bundles. The number of bundles needed to cover a square (100 ft²) depends on the amount of overlap when the shingles are installed. If the overlap allows 4 in. of each shingle to be exposed, then four bundles are needed per square. With a 5-in. exposure, 3.2 bundles are needed per square. Figure the number of bundles for a 4-in. exposure and for a 5-in. exposure for a roof measuring 30 ft × 20 ft.

$$A_{\text{roof}} = 30 \times 20$$

$$A_{\text{roof}} = 600 \text{ ft}^2$$

$$\frac{\overset{6}{600} \text{ ft}^2}{1} \times \frac{1 \text{ square}}{\underset{1}{100} \text{ ft}^2} = 6 \text{ squares} \qquad \text{Find the number of squares (100 ft}^2 \text{ areas) using a unity ratio.}$$

$$6 \times 4 = 24 \text{ bundles} \qquad \text{Find the number of bundles for a 4-in. exposure.}$$

$$6 \times 3.2 = 19.2 \text{ } or \text{ } 20 \text{ bundles} \qquad \text{Find the number of bundles for a 5-in. exposure.}$$

Therefore, 24 bundles of shingles are required for a 4-in. exposure, and 20 bundles (from 19.2 bundles) are required for a 5-in. exposure.

3   Find the Perimeter and Area of a Parallelogram.

The perimeter of a parallelogram, like the perimeter of a rectangle, is the sum of its four sides. Instead of naming the sides of the parallelogram *length* and *width,* we will use the terms *base* and *adjacent side.* Notice the locations of the base and adjacent side in Fig. 6–12.

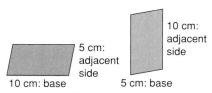

Figure 6-12

■ **DEFINITION 6–6: Base.** The *base* of any polygon is a horizontal side or a side that would be horizontal if the polygon's orientation were modified.

■ **DEFINITION 6–7: Adjacent Side.** The *adjacent side* of any polygon is a side that has an end point in common with the base.

The formula for the perimeter of a parallelogram is twice the sum of the base and adjacent side.

> **Formula 6-5** *Perimeter of a parallelogram:*
>
> $$P_{\text{parallelogram}} = 2(b + s)$$
>
> where b is the base and s is the adjacent side

Formula 6–5 is similar to the formula for the perimeter of a rectangle. However, for the area of a parallelogram, we must use still other terminology and a somewhat different formula than we used for the area of a rectangle. Study the parallelogram in Fig. 6–13.

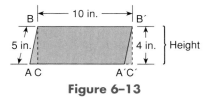

Figure 6-13

The *height* of a parallelogram or of any other geometric figure is an important dimension. Sometimes height is called *altitude*.

■ **DEFINITION 6–8: Height.** The *height* of a polygon is the perpendicular distance from the base to the highest point of the polygon above the base.

The height of the parallelogram in Fig. 6–13 is 4 in. Notice that the height and the adjacent side are different lengths. The blue-shaded portions of Fig. 6–13 are triangles.

If triangle ABC ($\triangle ABC$) in Fig. 6–13 were transposed to the right side of the parallelogram, we would have a rectangle. The area, therefore, would be the product of the *base* times the *height;* that is, $10 \times 4 = 40$ in.^{2}.

> **Formula 6-6** *Area of a parallelogram:*
>
> $$A_{\text{parallelogram}} = bh$$
>
> where b is the base and h is the height

EXAMPLE 8 Find the perimeter and the area of a parallelogram with a base of 16 in., an adjacent side of 8 in., and a height of 7 in. (Fig. 6–14).

Visualize the parallelogram.

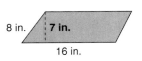

8 in. 7 in.
16 in.

Figure 6–14

$P_{\text{parallelogram}} = 2(b + s)$ $A_{\text{parallelogram}} = bh$

$P_{\text{parallelogram}} = 2(16 + 8)$ $A_{\text{parallelogram}} = 16 \times 7$ Substitute.

$P_{\text{parallelogram}} = 2(24)$ $A_{\text{parallelogram}} = \mathbf{112 \ in.^2}$

$P_{\text{parallelogram}} = \mathbf{48 \ in.}$

 ■ **General Tips for Using the Calculator**

Calculations for area and perimeter require one or more operations. We prefer to make all calculations in a continuous sequence of steps using a calculator.

Area of a Square

Assume a square with 5-in. sides.

Some *scientific calculators:*

$$\boxed{AC} \ 5 \ \boxed{x^2} \Rightarrow 25$$

Other scientific and *graphics calculators:*

$$\boxed{AC} \ 5 \ \boxed{x^2} \ \boxed{EXE} \Rightarrow 25$$

Also, some calculators may require the use of another key to activate the $\boxed{x^2}$ key, particularly if the symbol x^2 is written above the key, $\overset{x^2}{\boxed{\ }}$. The key that activates the operations above a key may be labeled shift, 2nd, or INV.

Perimeter of a Rectangle

Assume a rectangle is 10 cm long and 5 cm wide.

Enter the sum first:

$$\boxed{AC} \ 5 \ \boxed{+} \ 10 \ \boxed{=} \ \boxed{\times} \ 2 \ \boxed{=} \Rightarrow 30$$

Use parentheses:

$$\boxed{AC} \ 2 \ \boxed{\times} \ \boxed{(} \ 5 \ \boxed{+} \ 10 \ \boxed{)} \ \boxed{=}$$

or

$$\boxed{AC} \ 2 \ \boxed{(} \ 5 \ \boxed{+} \ 10 \ \boxed{)} \ \boxed{EXE} \Rightarrow 30$$

SELF-STUDY EXERCISES 6–1

1

1. Find the perimeter and the area of Fig. 6–15.

2. Name the shape in Fig. 6–15.

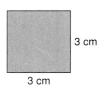

3 cm

3 cm

Figure 6–15

Solve the following problems involving perimeter and area.

3. Madison Duke will wallpaper a laundry room 8 ft by 8 ft by 8-ft high. How many square feet of paper will be needed if there are 63 ft² of openings in the room?

4. The square parking lot of a doctor's office is to have curbs built on all four sides. If the lot is 150 ft on each side, how many feet of curb are needed? Allow 10 ft for a driveway into the parking lot.

5. Making no allowances for bases, the pitcher's mound, or the home plate area, how many square yards of artificial turf are needed to resurface an infield at an indoor baseball stadium? The infield is 90 ft on each side. (9 ft² = 1 yd².)

6. Ted Davis is a farmer who wants to apply fertilizer to a 40-acre field with dimensions $\frac{1}{4}$ mi × $\frac{1}{4}$ mi. Find the area in square miles.

7. If $\frac{1}{4}$ mi is 1320 ft, how many feet of fencing would be needed to enclose the field in Exercise 6, assuming that a 12-ft steel gate will be installed?

8. A 36-in. × 36-in. ceramic tile shower stall will be installed. How many 4-in. × 4-in. tiles will be needed to cover the floor? Disregard the drain opening and grout spaces.

9. A border of 4-in. × 4-in. wall tiles surrounds the floor of the shower in Exercise 8. How many tiles are needed for this border? Disregard spaces for grout.

10. A 20-in. × 20-in. central heating and air conditioning return air vent will be installed in a wall. Find the perimeter and area of the wall opening.

11. Find the perimeter and the area of the shape in Fig. 6–16.

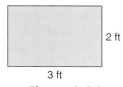

2 ft

3 ft

Figure 6–16

12. Name the shape in Fig. 6–16.

Solve the following problems involving perimeter and area.

13. A rectangular parking lot is 340 ft by 125 ft. Find the number of square feet in the parking lot.

14. A room is 15 ft by 12 ft. How many square feet of flooring are needed for the room?

15. Wallpaper will be used to paper a small kitchen 9 ft by 10 ft by 8-ft high. How many square feet of paper will be needed if there are 63 ft² of openings in the kitchen?

16. How many feet of quarter-round molding are needed to finish around the baseboard after sheet vinyl flooring is installed if the room is 16 ft by 18 ft and there are three 3-ft-wide doorways?

17. The dimensions of a sun porch are 9 ft 6 in. by 15 ft. How many board feet of 1-in.-thick tongue-and-groove flooring are needed for the porch? (Board feet for 1-in.-thick lumber is the same number as ft².) Disregard waste.

18. The swimming pool in Fig. 6–17 measures 32 ft by 18 ft. How much fencing is needed, including material for a gate, if the fence is to be built 7 ft from each side of the pool?

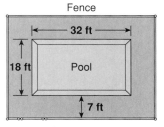

Figure 6–17

19. A Formica tabletop measures 40 in. by 62 in. How many feet of edge trim would be needed (12 in. = 1 ft)?

20. A countertop requires rolled edging to be installed on all four sides. How much rolled edging material is needed if the countertop measures 25 in. × 40 in.?

3

21. Find the perimeter and area of the shape in Fig. 6–18.

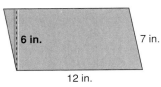
6 in.   7 in.
12 in.

Figure 6–18

22. Name the shape in Fig. 6–18.

Solve the following problems involving perimeter and area.

23. An illuminated sign indicating the main entrance of a hospital is shaped like a parallelogram with a base of 48 in. and an adjacent side of 30 in. How many feet of aluminum molding are needed to frame the sign?

24. Find the area of a parking lot shaped like a parallelogram 200 ft long if the perpendicular distance between the sides is 45 ft (Fig. 6–19).

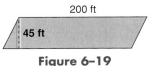
200 ft
45 ft

Figure 6–19

25. A customized van will have a window cut in each side in the shape of a parallelogram with a base of 20 in. and an adjacent side of 11 in. How many inches of trim are needed to surround the two windows?

26. A design in a marble floor will feature parallelograms with a base of 48 in. and a height of 24 in. If the figure will be made of 8-in. × 8-in. marble tiles, how many tiles will be needed for each figure, assuming no waste?

27. A machine cuts sheet metal into parallelograms measuring 6 cm along the base and 3 cm in height. How many can be cut from a piece of sheet metal 100 cm × 200 cm, assuming no waste?

28. A contemporary building will have a window in the shape of a parallelogram with a base of 50 in. and an adjacent side of 30 in. How many inches of trim are needed to surround the window?

29. A table for a reading lab will have a top in the shape of a parallelogram with a base of 36 in. and an adjacent side of 18 in. How many inches of edge trim are needed to surround the tabletop?

30. Signs in the shape of parallelograms with a base of 24 in. and a height of 10 in. will be cut from sheet metal. How many whole signs can be cut from a piece of sheet metal 48 in. × 48 in.? Illustrate your answer with a drawing.

MATHEMATICS IN THE WORKPLACE

Engineering: Computer Analysis of Tornado

Engineering Analysis, Inc., specializes in severe weather threat assessment for business and government. According to a new computer software program written by engineers at Engineering Analysis, Inc., Mississippi ranked first as the nation's most tornado-prone state. Mississippi beat out states traditionally perceived as tornado-prone, such as Kansas, Nebraska, Oklahoma, and Indiana, which rounded out the top five. Arkansas was sixth, followed by Iowa, Alabama, Illinois, and Wisconsin. Texas ranked twentieth. The least likely tornado targets were Idaho and Hawaii.

The new computer software program ranks states according to the amount of *land area* damaged by twisters in each state. Instead of simply counting the number of tornadoes per state each year, this new software divides the yearly total land area damaged in each state by that state's total land area. The resulting number represents the average amount of land damaged by tornadoes in that state over the previous year.

Engineering Analysis, Inc., obtained National Weather Service data that shows the starting point, ending point, and width of every U.S. tornado from 1950 to 1995. Using *geometry* and *trigonometry,* the computer program then calculated the distance each tornado traveled on land, and multiplied the distance by its width ($A = l \times w$) to determine how much land area the twister passed over. Developed by a National Weather Service forecaster, this index is a reliable indicator of tornado threat because it considers both the number of tornadoes and the severity of each storm.

AREA AND CIRCUMFERENCE OF A CIRCLE ▬▬▬▬▬▬▬

Learning Objectives   **1** Find the circumference of a circle.
                           2 Find the area of a circle.
                           3 Solve applied problems involving circles.

Tanks, rings, MRI machines, pipes, pillars, arches, gears, pulleys, wheels, and similar items are all based on the *circle* in one way or another. In many cases the circle or some part of the circle is found in composite figures in combination with one or more polygons. In this section we will study circles, their circumferences, and their areas.

Most of us are familiar with circles and some of their properties. However, for technical applications, it is necessary to clarify what a circle is and what some of its properties are. Refer to Fig. 6–20.

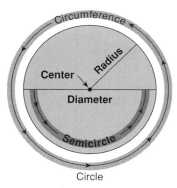

Figure 6–20

■ **DEFINITION 6–9: Circle.** A *circle* is a closed curved line whose points lie in a plane and are the same distance from the *center* of the figure.

■ **DEFINITION 6–10: Center.** The *center* of a circle is the point that is the same distance from every point on the circumference of the circle.

■ **DEFINITION 6–11: Radius.** A *radius* (plural: *radii*, pronounced "ray · dē · ī") is a straight line segment from the center of a circle to a point on the circle. It is half the diameter.

■ **DEFINITION 6–12: Diameter.** The *diameter* of a circle is the straight line segment from a point on the circle through the center to another point on the circle.

■ **DEFINITION 6–13: Circumference.** The *circumference* of a circle is the perimeter or length of the closed curved line that forms the circle.

■ **DEFINITION 6–14: Semicircle.** A *semicircle* is one-half a circle and is created by drawing a diameter.

1 **Find the Circumference of a Circle.**

The circle is a geometric form that has a special relationship between its circumference and its diameter. If we divide the circumference of any circle by the diameter, the quotient is always the same number. This number is a nonrepeating, nonterminating decimal approximately equal to 3.1415927 to seven decimal places. The Greek letter π (pronounced "pie") is used to represent this value. Convenient approximations often used in calculations involving π are $3\frac{1}{7}$ and 3.14. Because the quotient of the circumference and the diameter is π and because division and multiplication are inverse operations, the circumference equals π times the diameter. Because the diameter is twice the radius, the circumference is also equal to π times twice the radius.

Formula 6–7   *Circumference of a circle:*

$$C = \pi d \quad \text{or} \quad C = 2\pi r$$

where *d* is the diameter and *r* is the radius

Tips and Traps   *Calculator Values of π:*
All calculations involving π will be approximations. Many calculators include a π function where π to seven or more decimal places can be used by pressing a single key. Other calculators may require use of more than one key to activate the $\boxed{\pi}$ key. However, in making calculations by hand or with a calculator not having the $\boxed{\pi}$ function, 3.14 is sometimes adequate. **This text uses the calculator value 3.141592654 for π in all examples and exercises.**

EXAMPLE 1   What is the circumference of a circle whose diameter is 1.3 m (Fig. 6–21)? Round to tenths.

Figure 6–21

$$C = \pi d = \pi(1.3) = 4.08407045$$

The $\boxed{\pi}$ key on scientific or graphics calculators is used for the value of π.

The circumference is 4.1 m (rounded).   Circumference is a linear measure.

Tips and Traps   *Fractions Versus Decimals:*
For ease in using a calculator, convert mixed U.S. customary linear measurements to their decimal equivalents. For instance, if a diameter is 7 ft 6 in., convert it to 7.5 ft (from $7\frac{6}{12}$ ft, in which $\frac{6}{12} = 0.5$).

EXAMPLE 2   Find to the nearest hundredth the circumference of a circle whose radius is 1 ft 9 in. (Fig. 6–22).

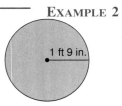

Figure 6–22

$$C = 2\pi r$$

$$C = 2\pi(1.75)$$ Substitute for π and r. $1\frac{9}{12}$ ft = 1.75 ft

$$C = 10.99557429$$

$$C = 11.00 \text{ ft} \quad \text{(rounded)}$$ Circumference is a linear measure.

2 Find the Area of a Circle.

The area of a circle, like the circumference, is obtained from the relationships within the circle itself. If we divide a circle into two semicircles and then subdivide each semicircle into a number of pie-shaped pieces, we have something like *A* in Fig. 6–23. If we then spread out the upper pie-shaped pieces and lower pie-shaped pieces, we have a figure like *B* in Fig. 6–23. Now if we pushed the upper pieces and lower pieces together, the result would approximate a rectangle like C in Fig. 6–23, whose length is $\frac{1}{2}$ the circumference and whose width is the radius. Thus, the area of the circle would be approximately the area of the rectangle, that is, length times width. Since the length of the rectangle is one-half the circumference and the width is the radius, the area of a circle equals one-half the circumference times the radius.

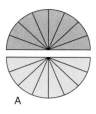 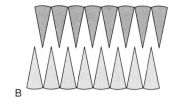 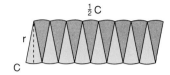

A   B   C

Figure 6–23

$A = \dfrac{1}{2}C \times r$ The formula for circumference is $C = 2\pi r$, so we can substitute $2\pi r$ for C.

$A = \dfrac{1}{2}(2\pi r)(r)$

$A = \dfrac{1}{2}(2\pi r)(r)$ Multiply. Reduce where possible.

$A = \pi r^2$

Formula 6-8 *Area of a circle:*

$$A_{\text{circle}} = \pi r^2$$

where r is the radius

Recall that r^2 means $r \times r$.

EXAMPLE 3   Find the area of a circle whose radius is 8.5 m (Fig. 6–24). Round to tenths.

8.5 m

Figure 6–24

$A = \pi r^2 = \pi(8.5)^2$

$= \pi(72.25) = 226.9800692 \text{ m}^2$ Area is a square measure.

The area of the circle is 227.0 m².

Recall in the order of operations that powers precede multiplication.

Tips and Traps   *Use a Continuous Calculator Sequence:*
When mixed U.S. customary measures do not convert to convenient, terminating decimal equivalents, a continuous calculator sequence may be preferred. For example,

$$3 \text{ ft } 4 \text{ in.} = 3 + \frac{4}{12} \text{ ft}$$

We can reduce and calculate with the mixed number or the improper fraction equivalent.

$$3 + \frac{4}{12} = 3 + \frac{1}{3} = 3\frac{1}{3} = \frac{10}{3}$$

Or we can make a continuous sequence using $3 + \frac{4}{12}$.

$$3 \boxed{+} 4 \boxed{\div} 12 \boxed{=} \Rightarrow 3.333333333$$

Remember, a scientific or graphics calculator applies the order of operations.

EXAMPLE 4   Find the area to the nearest hundredth of the top of a circular tank with a diameter of 12 ft 8 in.

If we are using a calculator, we can calculate the area using continuous steps as shown.

$A = \pi r^2$

$A = (\pi)\left(\dfrac{12 + \dfrac{8}{12}}{2}\right)^2$ Follow the order of operations. 8 in. $= \frac{8}{12}$ ft. The diameter, $12 + \frac{8}{12}$, is divided by 2 to equal the radius.

$A = 126.012772$ ft^2 calculator result

$A = 126.01$ ft^2 (rounded)   Area is a square measure.

The area of the top of the tank is 126.01 ft^2 or 125.95 ft^2 (basic calculator using $\pi = 3.14$).

   ■ **General Tips for Using the Calculator**

The calculator sequence for Example 4 follows:

Some *scientific calculators:*

$\boxed{AC}$ 12 $\boxed{+}$ 8 $\boxed{\div}$ 12 $\boxed{=}$ $\boxed{\div}$ 2 $\boxed{=}$ $\boxed{x^2}$ $\boxed{\times}$ $\boxed{\pi}$ $\boxed{=}$ $\Rightarrow$ 126.012772

Other scientific and *graphics calculators:*

$\boxed{AC}$ 12 $\boxed{+}$ 8 $\boxed{\div}$ 12 $\boxed{EXE}$ $\boxed{\div}$ 2 $\boxed{EXE}$ $\boxed{x^2}$ $\boxed{\times}$ $\boxed{\pi}$ $\boxed{EXE}$ $\Rightarrow$ 126.0127721

An alternative sequence of keystrokes will result if the parentheses keys are used. Experiment with your calculator to find other calculator sequences.

$\boxed{\pi}$ $\boxed{\times}$ $\boxed{(}$ $\boxed{(}$ 12 + 8 ÷ 12 $\boxed{)}$ $\boxed{\div}$ 2 $\boxed{)}$ $\boxed{x^2}$ $\boxed{=}$ $\Rightarrow$ 126.0127721

③ Solve Applied Problems Involving Circles.

EXAMPLE 5   A 15-in.-diameter wheel has a 3-in. hole in the center. Find the area of a side of the wheel to the nearest tenth (Fig. 6–25).

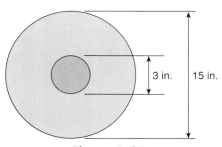

Figure 6–25

We are asked to find the area of the shaded portion of the wheel in Fig. 6–25. To do so, we find A_{outside}, the area of the larger circle (diameter 15 in.), and *subtract* from it the area of A_{inside}, the smaller circle (diameter 3 in.). The shaded portion is called a *ring*.

$$A_{\text{outside}} = \pi r^2 \quad \left(r = \frac{15}{2} = 7.5\right) \qquad A_{\text{inside}} = \pi r^2 \quad \left(r = \frac{3}{2} = 1.5\right)$$

$$A_{\text{outside}} = \pi(7.5)^2 \qquad\qquad\qquad A_{\text{inside}} = \pi(1.5)^2$$

$$A_{\text{outside}} = \pi(56.25) \qquad\qquad\quad A_{\text{inside}} = \pi(2.25)$$

$$A_{\text{outside}} = 176.7145868 \text{ in.}^2 \qquad A_{\text{inside}} = 7.068583471 \text{ in.}^2$$

Area of wheel (ring) $= A_{\text{outside}} - A_{\text{inside}}$

$$A_{\text{wheel}} = 176.7145868 - 7.068583471 = 169.6460033 \quad \text{or} \quad 169.6 \text{ in.}^2$$

The area of the wheel (ring) is 169.6 in.^{2}.

EXAMPLE 6 A bandsaw has two 25-cm wheels spaced 90 cm between centers (Fig. 6–26). Find the length of the continuous saw blade.

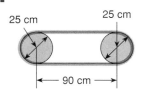

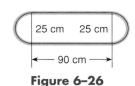

Figure 6–26

This layout is a composite figure. The figure consists of a semicircle at each end and a rectangle in the middle. It is called a *semicircular-sided* figure. The two semicircles equal one whole circle, so we need to find the circumference of one circle (wheel) and add it to the lengths of the two sides of the rectangle.

$C = \pi d$ Total length of blade $= C + 2l$

$C = \pi(25)$ Total length of blade $= 78.53981634 + 2(90)$

$C = 78.53981634$ cm   Total length of blade $= 258.5$ cm   (rounded)

The bandsaw blade is 258.5 cm in length.

Many formulas that have a specific application for a particular industry may include geometric formulas for area, perimeter, or circumference.

EXAMPLE 7 Find the cutting speed of a lathe if a piece of work 7 in. in diameter turns on a lathe at 75 rpm.

The *cutting speed* is the speed of a tool that passes over the work, such as the speed of a sander as it sands (passes over) a piece of wood. If the cutting speed is too fast or too slow, safety and quality are impaired. The formula for cutting speed is CS $= C$ (in feet) $\times$ rpm.

CS $=$ cutting speed

$C =$ circumference or one revolution (in feet)

rpm $=$ revolutions per minute

Cutting speed is measured in *feet per minute* (ft/min).

$C = \pi d$

$C = \pi(7)$

$C = 21.99114858$ in.   Convert 21.99114858 in. into feet using a unity ratio.

$C = 1.832595715$ ft $\dfrac{21.99114858 \text{ in.}}{1} \times \dfrac{1 \text{ ft}}{12 \text{ in.}} = 1.832595715 \text{ ft}$

CS $= C \times$ rpm $= 1.832595715 \times 75 = 137$ ft/min.   (rounded)

The cutting speed of the lathe is approximately 137 ft/min.

[1] Find the circumference of circles with the following dimensions. Round to tenths.

1. Diameter = 8 cm
2. Radius = 3 in.
3. Radius = 1.5 ft
4. Diameter = 5.5 m

[2] Find the blue shaded area of Figs. 6–27 to 6–30 to the nearest tenth.

5.

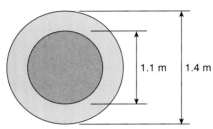

1.1 m   1.4 m

Figure 6–27

6.

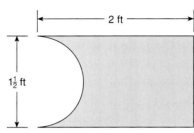

2 ft

$1\frac{1}{2}$ ft

Figure 6–28

7.

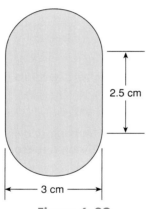

2.5 cm

3 cm

Figure 6–29

8.

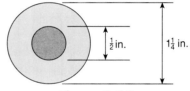

$\frac{1}{2}$ in. $1\frac{1}{4}$ in.

Figure 6–30

[3] Solve the following applied problems.

9. A swimming pool is in the form of a semicircular-sided figure. Its width is 20 ft and the parallel portions of the sides are each 20 ft (Fig. 6–31). What is the area of a 5-ft-wide walk surrounding the pool?

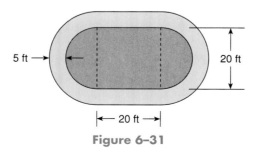

5 ft   20 ft

20 ft

Figure 6–31

10. A belt connecting two 9-in.-diameter drums on a conveyor system needs replacing. How many inches must the new belt be if the centers of the drums are 10 ft apart (Fig. 6–32)?

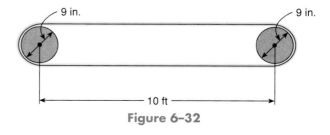

9 in.   9 in.

10 ft

Figure 6–32

11. Find the area of the ring formed by the cross-cut section of Fig. 6–33.

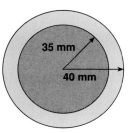

Figure 6–33

12. The wall of a galvanized water pipe is 3.5 mm thick. If the outside circumference is 68 mm, what is the area (to nearest tenth) of the shaded inside cross-cut section (Fig. 6–34)?

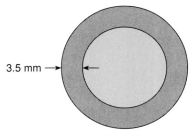

Figure 6–34

13. A 2-in.-inside-diameter pipe and a 4-in.-inside-diameter pipe empty into a third pipe whose inside diameter is 5 in. (Fig. 6–35). Is the third pipe large enough for the combined flow? (Justify your answer.)

14. A large pipe whose interior cross-sectional area is 20 in.² empties into two smaller pipes that each have an interior diameter of 4 in. (Fig. 6–36). Are the smaller pipes together large enough to carry off the flow from the larger pipe? (Justify your answer.)

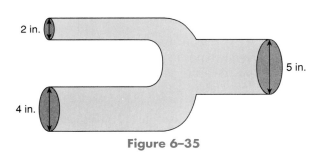

Figure 6–35

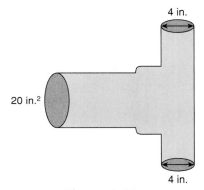

Figure 6–36

15. Cutting speed, when applied to a grinding wheel, is called *surface speed*. What is the surface speed in ft/min of a 9-in.-diameter grinding wheel revolving at 1200 rpm? (Surface speed = circumference in feet × rpm.)

17. A lamp lights up effectively an area 10′ in diameter (Fig. 6–37). How many square feet does the lamp light effectively?

16. A polishing wheel 12 in. in diameter revolves at 500 rpm. What is the surface speed? (See Exercise 15 for formula.)

18. A steel sleeve with an outside diameter of 3 in. measures $2\frac{1}{2}$ in. across the inside. The sleeve will be babbitted to fit a 1.604-in.-diameter motor shaft. What must the thickness of the babbitt be (in hundredths)? The babbitt is

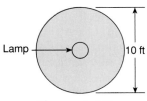

Figure 6–37

shaded blue in Fig. 6–38. A babbitt is a lining of babbitt metal, a soft antifriction alloy, that allows two metal parts to fit snugly together.

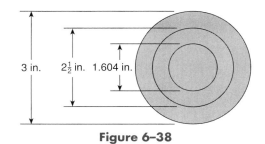

Figure 6–38

19. Two 12-cm-diameter drums are connected by a belt to form a conveyor system. The centers of the drums are 2 m apart (Fig. 6–39). How long must a replacement belt be (to the nearest tenth of a meter)?

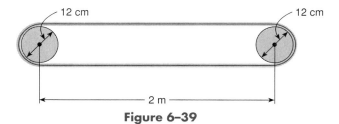

Figure 6–39

Interior Design: Paint, Wallpaper, and Border

While earning a technical college degree, you work for an interior design company. After the decorator has assisted the customer in choosing paint, wallpaper, and border, your job is to go to the site, measure the room(s), and order the correct amounts. Your company sometimes orders an extra gallon of paint, one extra single roll of wallpaper, and 20 extra feet of wallpaper border.

Your first job is a child's rectangular room measuring 12 ft by 16 ft with 8-ft ceilings. Three of its walls will be covered with wallpaper and the fourth short wall will be painted with chalkboard paint. A ceiling border 6 in. wide will circle the room. The non-chalkboard short wall and one of the long walls each have a centered window 6 ft wide by 5 ft tall. The other long wall has a 30 in. by 80 in. entry door and a 60 in. by 80 in. closet door. Neither the windows nor the doors will be papered or painted.

Exercises

Use the above measurements to answer the following. Show all computations. Round up all answers.

1. Using a ruler, draw a two-dimensional diagram of this room looking down from the ceiling. Label the length of each side, and place the doors and windows correctly.
2. Using a ruler, draw a two-dimensional diagram of each of the two short walls of this room looking at the walls while standing inside the room. Label the length of each side, and place the window correctly.
3. Using a ruler, draw a two-dimensional diagram of each of the two long walls of this room looking at the walls while standing inside the room. Label the length of each side, and place the doors and the window correctly.
4. Using the diagram from Exercise 2, calculate the amount of chalkboard paint needed to paint two coats on the short wall without the window. Each gal of chalkboard paint covers 300 ft^2.

5. Using the diagram from Exercise 1, calculate the length of ceiling wallpaper border needed to circle the entire room. Keep in mind that the doors and windows do not reach the ceiling. If the wallpaper border comes only in 20-ft rolls, how many rolls will you need? If the customer wants border seams only at the corners of the room, how many rolls will it take?

6. Use the diagrams from Exercises 2 and 3 to calculate the area to be covered with wallpaper on the two long walls and the remaining short wall. Assume that the wallpaper is put up first, then the border covers the wallpaper at the ceiling. Don't forget to subtract the area of the doors and windows. (*Note:* $1 \text{ ft}^2 = 144 \text{ in.}^2$.)

7. Assume that the customer chose a uniform wallpaper pattern that does not require matching. How many rolls of wallpaper will it take if each single roll of wallpaper covers 28 ft^2? If this pattern is sold only in double rolls, how many double rolls will you need?

8. Most wallpaper today comes in 27-in.-wide rolls. Paperers cut panels equal to the height of the room and cut out doors and windows from each panel. To find the number of single rolls needed, professional wallpaper hangers typically divide the total area to be covered by 22 ft^2. This technique ensures that enough wallpaper will be available to have the minimum number of seams after allowing for door and window waste and pattern matching. How many single and double rolls will be needed using this method of calculation?

9. Place the order for the chalkboard paint, rolls of wallpaper border, and double rolls of wallpaper.

Answers

1. Ceiling-view diagram:

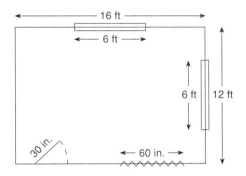

2. Short-wall diagrams:

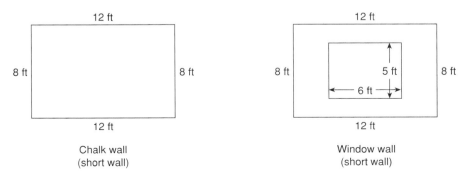

Chalk wall (short wall)   Window wall (short wall)

3. Long-wall diagrams:

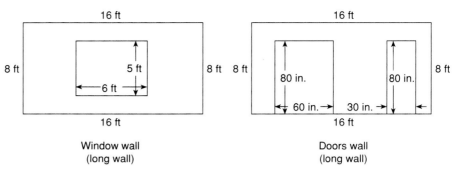

Window wall
(long wall)

Doors wall
(long wall)

4. The chalkboard wall's area is 96 ft². Double that for two coats equals 192 ft², so 1 gal is plenty.
5. a. 56 ft of wallpaper border are needed.
    b. 3 rolls are needed for seams anywhere.
    c. 4 rolls are needed for corner seams only.
6. Short wall with window: 66 ft².
    Long wall with window: 98 ft².
    Long wall with doors: 78 ft².
    Total wallpaper area: 242 ft².
7. 9 single rolls; 5 double rolls
8. 11 single rolls; 6 double rolls
9. Order 2 gal of chalkboard paint, 5 rolls of wallpaper border, and 6 double rolls of wallpaper.

Thanks to Bettye Yates, owner of Interior Creations, Collierville, TN.

ASSIGNMENT EXERCISES

Section 6–1
Find the perimeter and the area of Figs. 6–40 to 6–42.

1.

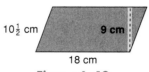

$10\frac{1}{2}$ cm   **9 cm**
18 cm

Figure 6–40

2.

$2\frac{1}{2}$ ft

$2\frac{1}{2}$ ft

Figure 6–41

3.

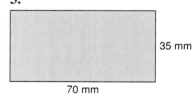

35 mm

70 mm

Figure 6–42

4. If a parking lot for a new hospital measures 275 ft by 150 ft, how many square feet need to be blacktopped?
6. A hall wall with no windows or doors measures 25 ft long by 8 ft high. Find the number of square feet to be covered if paneling is installed on the two walls.
8. Vincent Ores, a contractor, is to brick the storefront of a landscape service that has a doorway measuring 7 ft by 6 ft. How many bricks are needed if the storefront is 20 ft by 12 ft and the bricks cover at the rate of 6 per square foot using $\frac{1}{2}$-in. mortar joints?

5. In Exercise 4, find the amount of curbing needed to surround the parking lot if 14 feet are allowed for a driveway.
7. A den 18 ft by $16\frac{1}{2}$ ft is to be carpeted. How many square yards of carpeting are needed?
9. A roof measuring 16 ft by 20 ft is to be covered with asphalt roofing cement. How much would the project cost if the asphalt roofing cement spreads at the rate of 150 square feet per gallon and costs $4.75 per gallon? The cement is purchased by the gallon only.

10. Estimate the number of feet of roll fencing needed to fence a square storage area measuring $15\frac{1}{2}$ ft on a side. A preassembled gate 4 ft wide will be installed.

11. A dining room wall 12 ft long by 8 ft high will be wallpapered. How many single rolls (24 in. by 20 ft) of wallpaper will be needed? Assume there is no waste.

12. A square area of land is 10,000 m². If a baseball field must be at least 99.1 m along each foul line to the park fence, is the square adequate for regulation baseball?

13. Debbie Murphy is building a contemporary home with four front windows, each in the form of a parallelogram. If each window has a 5-ft base and a height of 2 ft, how many square feet of the 25-ft × 11-ft wall will require stain?

14. If the stain used on the wall in Exercise 13 is applied at a cost of $2.75 per square yard, find the cost to the nearest dollar of staining the front wall.

Section 6–2
Find the perimeter (or circumference) and the area of Figs. 6–43 to 6–46. Round to hundredths.

15.

Figure 6–43

16.

Figure 6–44

17.

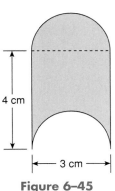

Figure 6–45

18.

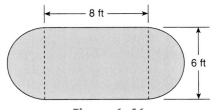

Figure 6–46

19. A circle has a diameter of 2.5 m. If its diameter were increased 1.2 m, what would be the difference in the areas (to the nearest tenth) of the two circles?

20. If a circular flower bed has a radius of 24 in., how many $5\frac{3}{4}$-in.-long flat bricks would be needed to enclose it at the rate of two bricks per 12 in.?

21. Three water hoses, whose interior cross-sectional areas are 1 in.², $1\frac{1}{2}$ in.², and 2 in.², empty into a larger hose. What inside diameter must the larger hose be to carry away the combined flow of the three smaller hoses? Round to tenths.

22. Two 35-mm-diameter drums connect a conveyor belt. If the centers of the drums are 70 cm apart, how many cm long is the conveyor belt? Round to tenths.

23. A $\frac{1}{4}$-in. electric drill with variable speed control turns as slowly as 25 rpm. If an abrasive disk with a 5-in. diameter is attached to the drill drive shaft, what is the disk's slowest cutting speed in ft/min? Round to the nearest whole number. (Cutting speed = circumference in feet × rpm.)

24. What is the cross-sectional area of the opening in a round flue tile whose inside diameter is 8 in.? Round any part of an inch to the next tenth of an inch.

25. Triple-wall galvanized chimney pipe for pre-fabricated sheet-metal fireplaces has a circumference of 47 in. What is the inside diameter of the fire-stop spacer through which the chimney pipe passes as it goes through the ceiling? Round to the nearest tenth.

27. Find the area of the blue shaded portion of the sidewalk in Fig. 6–47. The blue shaded portion is one-fourth of a circle. Express the answer in square yards rounded to tenths. ($144 \text{ in.}^2 = 1 \text{ ft}^2$, $9 \text{ ft}^2 = 1 \text{ yd}^2$.)

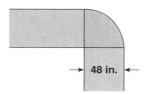

48 in.

Figure 6–47

29. If a cogwheel makes one complete revolution and its radius is 9.4 in., how long is the path traveled by any one point on the cogwheel? Round to tenths.

26. Find the area of the cross-section of a wire whose diameter is $\frac{1}{16}$ in. (Express the answer to the nearest thousandth.)

28. Find the area of the composite layout shown in Fig. 6–48. Round the answer to the nearest tenth.

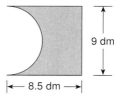

9 dm

8.5 dm

Figure 6–48

CHALLENGE PROBLEMS

30. Lou Ferrante plans to build a dog pen. He has 360 feet of fencing and would like to enclose as much area as possible for his dog. What length and width should he make the dog pen?

32. Given that the formula for finding the area of a triangle (Fig. 6–49) is $A = \frac{1}{2}bh$, find the area of the shaded triangle in Fig. 6–51 for Problem 6 of the Concepts Analysis.

31. Tosha Riddle, an interior designer, needs to order a tablecloth to cover a circular table that is 24 inches from the floor and 18 inches across the top. The cloth must drape the table so that it just touches the floor. Determine the shape needed for the cloth. Find the size of the fabric needed for the cloth.

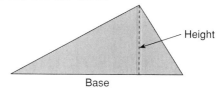

Height

Base

Figure 6–49

CONCEPTS ANALYSIS

1. Describe five activities or jobs that would require you to find the perimeter of a shape.

3. Draw five parallelograms of different sizes and cut out each parallelogram. Cut each parallelogram into 2 pieces by cutting through two corners or vertices. Describe similarities and/or differences between the two pieces of each parallelogram.

2. Describe five activities or jobs that would require you to find the area of a shape.

4. Each piece of the parallelogram in Problem 3 forms a triangle. Use the comparisons from Problem 3 to write a formula to find the area of a triangle as it relates to a parallelogram.

5. Draw a shape that has the same relationship to a parallelogram as a square has to a rectangle (Fig. 6–50).

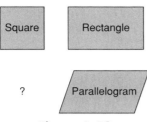

Figure 6–50

6. Explain how you could find the area of the following composite figure (Fig. 6–51).

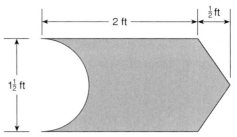

Figure 6–51

7. Discuss the similarities and differences between a square and a rectangle.

9. Drawing freely on the formulas for the circumference of a circle, devise your own formulas to find the radius and the diameter of a circle.

8. Discuss the similarities and differences between a rectangle and a parallelogram.

10. The new shape in Problem 5 is called a rhombus. List the properties of a rhombus. Write a formula for finding the area and perimeter of a rhombus.

CHAPTER SUMMARY

| Objectives | What to Remember with Examples |

Section 6–1

1 Find the perimeter and area of a square.

Formula for perimeter: $P_{\text{square}} = 4s$: s is the length of a side.
Formula for area: $A_{\text{square}} = s^2$.

> Find the perimeter of a square 6.5 cm on a side.
>
> $P_{\text{square}} = 4s$
> $P_{\text{square}} = 4(6.5)$
> $P_{\text{square}} = 26$ cm
>
> Find the area of a square 1.5 in. on a side.
>
> $A_{\text{square}} = s^2$
> $A_{\text{square}} = 1.5^2$
> $A_{\text{square}} = 2.25$ in.^{2}

2 Find the perimeter and area of a rectangle.

Formula for perimeter: $P_{\text{rectangle}} = 2(l + w)$: l is length of long side; w is width, or length of short side.
Formula for area: $A_{\text{rectangle}} = lw$.

> A rectangular flower bed is 8 ft × 3 ft. How many feet of edging are needed to surround the bed?
>
> $P_{\text{rectangle}} = 2(l + w)$
> $P_{\text{rectangle}} = 2(8 + 3)$
> $P_{\text{rectangle}} = 2(11)$
> $P_{\text{rectangle}} = 22$ ft
>
> How much landscaping fabric is needed to cover the bed?
>
> $A_{\text{rectangle}} = lw$
> $A_{\text{rectangle}} = 8 \times 3$
> $A_{\text{rectangle}} = 24$ ft^2

3 Find the perimeter and area of a parallelogram.

Formula for perimeter: $P_{\text{parallelogram}} = 2(b + s)$: b is the base, or length of a long side; s is the length of a short side.

Formula for area: $A_{\text{parallelogram}} = bh$: b is the base; h is the height, or length of a perpendicular distance between the bases.

Find the perimeter of a parallelogram with a base of 1 m and a side of 0.5 m.

$P_{\text{parallelogram}} = 2(b + s)$
$P_{\text{parallelogram}} = 2(1 + 0.5)$
$P_{\text{parallelogram}} = 2(1.5)$
$P_{\text{parallelogram}} = 3$ m

Find the area of a sign shaped like a parallelogram with a base of 36 in. and a height of 12 in.

$A_{\text{parallelogram}} = bh$
$A_{\text{parallelogram}} = 36(12)$
$A_{\text{parallelogram}} = 432$ in.^{2}

Section 6-2

1 Find the circumference of a circle.

Formula for circumference: $C = \pi d$, or $C = 2\pi r$: d is the diameter or distance across the center of a circle; r is the radius, or half the diameter; π is approximated on a calculator as 3.141592654.

Find the circumference of a circle with a 3-cm diameter.

$C = \pi d$
$C = \pi(3)$
$C = 9.4$ cm (rounded)

What is the distance around a circle with an 18-in. radius?

$C = 2\pi r$
$C = 2\pi(18)$
$C = 113.1$ in. (rounded)

2 Find the area of a circle.

Formula for area: $A_{\text{circle}} = \pi r^2$.

Find the area of a circle whose diameter is 3 m. First, find the radius: $3 \div 2 = 1.5$.

$A_{\text{circle}} = \pi r^2$
$A_{\text{circle}} = \pi(1.5)^2$
$A_{\text{circle}} = 7.07$ m^2 (rounded)

3 Solve applied problems involving circles.

Some problems may involve composite figures and require several areas or circumferences or perimeters to be used.

A belt moving around two pulleys 5 in. in diameter must be replaced. How long must the belt be if the centers of the pulleys are 18 in. apart? This setup forms a semicircular-sided figure (Fig. 6–52). Both semicircles form one full circle. Find its circumference and add the upper and lower distances ($2 \times 18 = 36$).

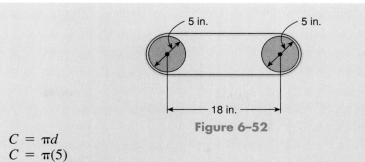

Figure 6–52

$C = \pi d$

$C = \pi(5)$

$C = 15.70796327$

$15.70796327 + 36 = 51.7$ in. (rounded). The belt must be 51.7 in. long.

WORDS TO KNOW

parallelogram (p. 261)
rectangle (p. 261)
square (p. 261)
perimeter (p. 261)
area (p. 261)
length (p. 263)

width (p. 263)
base (p. 267)
adjacent side (p. 267)
height (p. 267)
circle (p. 271)
center (p. 271)

radius (p. 271)
diameter (p. 271)
circumference (p. 271)
semicircle (p. 271)

CHAPTER TRIAL TEST

Find the perimeters and areas of Figs. 6–53 to 6–56.

1.

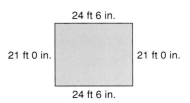

Figure 6–53

2.

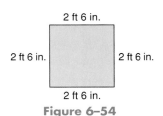

Figure 6–54

3.

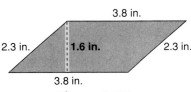

Figure 6–55

4.

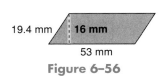

Figure 6–56

5. A company charges $3.00 labor per square yard to install carpet and padding. If a dining room is 12 ft × 13 ft, what would the labor cost to install the padding and carpet (9 ft² = 1 yd²)?

7. How many square feet of floor space are there in the plan shown in Fig. 6–57?

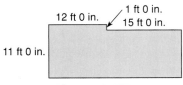

Figure 6–57

6. How many squares (100 ft²) of siding would be needed for four sides of a garage 18 ft × 15 ft × 9 ft high if there are 125 ft² of openings? Allow one-fourth extra siding for overlap and waste.

8. A machine stamps 1-in. squares from sheet metal. How many can it cut from a 3-ft × 4-ft sheet of metal?

Use Fig. 6–58 to identify the numbered parts of the circle.

9. ____
10. ____
11. ____
12. ____

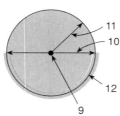

Figure 6–58

Solve these problems and round, when necessary, to hundredths.

13. Find the circumference of a circle whose radius is 6 in.

14. Find the area of the circle in Exercise 13.

15. What is the perimeter of a roller-skating rink with the dimensions shown in Fig. 6–59?

16. Find the area to the nearest square foot of the skating rink in Exercise 15.

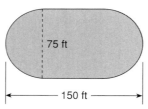

75 ft

150 ft

Figure 6–59

17. A water pipe with an inside diameter of $\frac{5}{8}$ in. and an outside diameter of $\frac{7}{8}$ in. is cut off at one end. What is the cross-sectional area of the ring formed by the wall of the pipe?

18. Find the area of the blue shaded portion of the tiled walk that surrounds a rectangular swimming pool (Fig. 6–60).

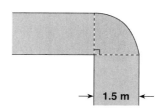

1.5 m

Figure 6–60

19. Find the area of a washer with an inside diameter of $\frac{1}{4}$ in. and an outside diameter of $\frac{1}{2}$ in.

20. If it costs $1.25 to sod each square foot of a circular lawn (Fig. 6–61), how much would it cost, to the nearest cent, if the lawn is $6\frac{1}{2}$ ft wide and surrounds a round goldfish pond having a diameter of 12 ft?

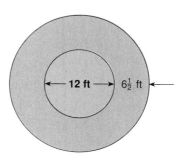

12 ft $6\frac{1}{2}$ ft

Figure 6–61

7

Interpreting and Analyzing Data

GOOD DECISIONS THROUGH TEAMWORK

Your team has been hired to conduct market research for a major consumer magazine. Your assignment is to choose an area of interest, conduct a survey, and prepare a report of your findings. Begin by determining a suitable multiple-choice survey question, for example, "Which long-distance telephone company do you prefer?" "What single piece of exercise equipment would you most want to purchase?" Then identify five to ten possible responses, including "None of the above."

Next, determine your team's survey methods. How many responses do you feel would be adequate to make the results reliable? When and where will you obtain the responses? On campus? At a mall? Through the mail? Discuss how the time of day and location can affect survey results.

Conduct the survey and record your responses. Then tabulate, calculating the total number of respondents and the number choosing each possible response. Illustrate your findings with a circle, a bar, and a line graph.

Write a report documenting your methods, results, and conclusions. Include the tabulation of responses and the summary graphs. Keep in mind that a high-quality report could mean another high-paying market research project for your team.

A *graph* is a means of showing information visually. Some graphs show how our tax dollars are divided among various government services, trace the fluctuations in a patient's temperature, or illustrate regional planting seasons. Other graphs are used to show equations, inequalities, and their solutions. Tables, on the other hand, usually list data, such as income tax tables showing taxes due on different incomes.

7-1 READING CIRCLE, BAR, AND LINE GRAPHS

Learning Objectives

1. Read circle graphs.
2. Read bar graphs.
3. Read line graphs.

Many graphs give us useful information at a glance. However, we must be able to read, or interpret, the graphs properly to use them to our benefit. Three common graphs used to represent data are the circle graph, the bar graph, and the line graph.

1 Read Circle Graphs.

■ **DEFINITION 7–1: Circle Graph.** A *circle graph* uses a divided circle to show pictorially how a total amount is divided into parts.

The complete circle represents the total amount of one whole quantity. Then the circle is divided into parts so that the sum of all the parts equals the whole quantity. These parts can be expressed as fractions, decimals, or percents. Figure 7–1 illustrates a circle graph.

Figure 7–1 Distribution of wholesale price of $425 color television receiver.

To "read" a graph is to examine the information on the graph.

Rule 7–1 *To read a circle, bar, or line graph:*

1. Examine the title of the graph to find out what information is shown.
2. Examine the parts to see how they relate to one another and to the whole amount.
3. Examine the labels for each part of the graph and any explanatory remarks that may be given.
4. Use the given parts to calculate additional numerical amounts.

EXAMPLE 1 Answer the following questions using Fig. 7–1.

(a) What percent of the wholesale price is the cost of labor?

(b) What percent of the wholesale price is the cost of materials?

(c) What would the wholesale price be if no tariff (tax) was paid on imported parts?

(a) $\dfrac{R}{100} = \dfrac{130}{425}$ *R* represents the percent of wholesale price that is labor cost.

$R \times 425 = 13,000$

$R = \dfrac{13,000}{425}$

$R = 30.58823529$

$R = 30.6\%$ (labor)

(b) $\dfrac{R}{100} = \dfrac{125}{425}$ *R* represents the percent of wholesale price that is materials cost.

$R \times 425 = 12,500$

$R = \dfrac{12,500}{425}$

$R = 29.41176471$

$R = 29.4\%$ (materials)

(c) **Price − tariff = 425 − 40 = \$385 (cost without tariff)**

2 Read Bar Graphs.

Each type of graph developed historically because of a need for the kind of information it made accessible to the user. The bar graph is no exception.

■ **DEFINITION 7–2: Bar Graph.** A *bar graph* uses two or more bars to show pictorially how two or more amounts compare to each other rather than to a total.

Numerical amounts are represented by the lengths of the bars. The bars can be drawn either horizontally or vertically.

The axis (*reference line*) along the length of the bars is scaled with numerical amounts. Then the other axis (reference line) along the base of the bars will be used to label the bars. Figure 7–2 illustrates a bar graph with horizontal bars.

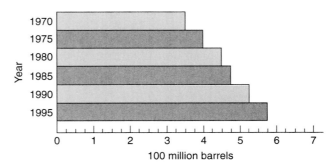

Figure 7–2 Company oil production.

EXAMPLE 2   Answer the following questions using Fig 7–2.

(a) How many more 100 million barrels of oil are indicated for the company in 1995 than in 1975?

(b) Judging from the graph, should company oil production in 2000 be more or less than in 1995?

(c) How many 100 million barrels of oil did the company produce in 1975?

(a) 1995 production − 1975 production = 5.75 − 4.00 = 1.75 hundred million barrels.

(b) More, because the trend has been toward greater production.

(c) Four hundred million barrels in 1975.

3 Read Line Graphs.

A third kind of graph we encounter in our reading of industrial reports, handbooks, and the like, is the line graph.

■ **DEFINITION 7–3: Line Graph.** A *line graph* uses one or more lines to show changes in data.

The horizontal axis or reference line in a line graph usually represents periods of time or specific times. The vertical axis or reference line is usually scaled to represent numerical amounts. Line graphs are used to pictorially represent trends in data and to quickly show high values and low values. Figure 7–3 illustrates a line graph.

EXAMPLE 3 Answer the following questions regarding a patient's temperature using Fig. 7–3.

(a) On what date and time of day did the patient's temperature first drop to within 0.2 degrees of normal (98.6 degrees)?

(b) On which post-op (post-operative) days was the patient's temperature within 0.2 degrees of normal?

(c) What was the highest temperature recorded for the patient?

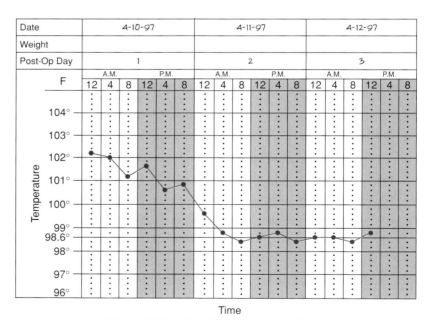

Figure 7–3 Graphic temperature chart.

(a) **4-11-97 at 4 A.M.** (Each "dot" is 0.2 degrees, so the temperature was 98.8 degrees.)

(b) **Post-op days 2 and 3** (beginning at 4 A.M. on day 2)

(c) **102.2 degrees** (recorded at 12 A.M. on 4-10-97)

1. Use Fig. 7–4 to answer Exercises 1 to 3.

1. What percent of the gross salary goes into savings? Round to tenths.
2. What percent of the take-home pay is federal income tax? Round to tenths.
3. What percent of the gross pay is the take-home pay? Round to tenths.

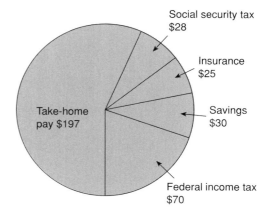

Figure 7–4 Distribution of weekly salary of $350.

2. Use Fig. 7–5 to answer Exercises 4 to 6.

4. What expenditure is expected to be the same next year as this year?
5. What two expenditures are expected to increase next year?
6. What two expenditures are expected to decrease next year?

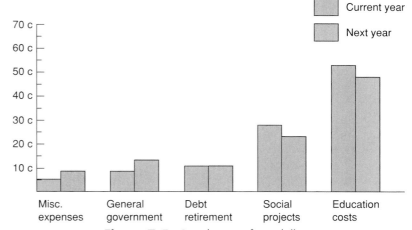

Figure 7–5 Distribution of tax dollars.

3. Use Fig. 7–6 to answer Exercises 7 to 10.

7. How many amperes of current are produced by 50 V when the resistance is 10 ohms?
8. How many volts are needed to produce 2 A of current when the resistance is 25 ohms?
9. Approximately how many volts are required to produce a current of 3.5 A when the resistance is 10 ohms?
10. Find the resistance when 100 V are needed to produce a current of 4 A.

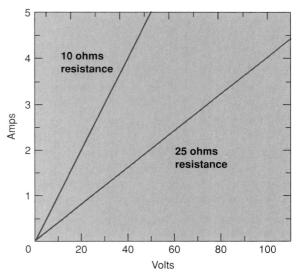

Figure 7–6 Amperage produced by voltage across two resistances.

FREQUENCY DISTRIBUTIONS, HISTOGRAMS, AND FREQUENCY POLYGONS

Learning Objectives

1. Interpret and make frequency distributions.
2. Interpret and make histograms.
3. Interpret and make frequency polygons.

1 Interpret and Make Frequency Distributions.

Suppose for a class of 25 students the instructor recorded the following grades:

76 91 71 83 97 87 77 88 93 77 93 81 63

79 74 77 76 97 87 89 68 90 84 88 91

It would be difficult for the instructor to make sense of all these numbers as they appear above. But the instructor can arrange the scores into several smaller groups, called *class intervals*. The word *class* means a special group, just as a group of students in a course is a class.

These scores can be grouped into class intervals such as 60–64, 65–69, 70–74, 75–79, 80–84, 85–89, 90–94, and 95–99. Each of these class intervals has an odd number of scores. The *middle score* of each interval is a *class midpoint*.

The instructor can now *tally* the number of scores that fall into each class interval to get a *class frequency,* the number of scores in each class interval.

A compilation of class intervals, midpoints, tallies, and class frequencies is called a *frequency distribution.*

EXAMPLE 1 Answer the following questions about the frequency distribution in Table 7–1.

(a) How many students scored 70 or above?

$$2 + 6 + 3 + 5 + 5 + 2 = 23$$

23 students scored 70 or above.

(b) How many students made A's (90 or higher)?

$$5 + 2 = 7$$

7 students made A's (90 or higher).

(c) What percent of the total grades were A's (90's)?

$$\frac{7 \text{ A's}}{25 \text{ total}} = \frac{7}{25} = 0.28 = 28\% \text{ A's}$$

(d) Were the students prepared for the test or was the test too difficult?

The relatively high number of 90's (7) compared to the relatively low number of 60's (2) suggests that the **test was not too difficult for the class as a whole and some students were not prepared.**

TABLE 7–1 Frequency Distribution of 25 Scores

Class Interval	Midpoint	Tally	Class Frequency
60–64	62	/	1
65–69	67	/	1
70–74	72	/ /	2
75–79	77	++++ /	6
80–84	82	/ / /	3
85–89	87	++++	5
90–94	92	++++	5
95–99	97	/ /	2

(e) What is the ratio of F's (60's) to A's (90's)?

$$\frac{2 \text{ F's}}{7 \text{ A's}} = \frac{2}{7}$$

The ratio is $\frac{2}{7}$.

EXAMPLE 2   Students in a history class reported their credit-hour loads as shown. Make a frequency distribution of their credit hours. Credit hours carried: 3, 12, 15, 3, 6, 6, 12, 9, 12, 9, 6, 3, 12, 18, 6, 9.

To have a class interval with an easy-to-find midpoint, use an odd-numbered interval. Here, an interval of 5 is used; that is, 0–4 contains five possibilities: 0, 1, 2, 3, and 4. The middle number is the midpoint, 2. Make a tally mark for each time the credit hours of a student falls in the interval. Then count the tally marks to get the class frequency (Table 7–2).

TABLE 7–2

Class Interval	Midpoint	Tally	Class Frequency
0–4	2	/ / /	3
5–9	7	++++ / /	7
10–14	12	/ / / /	4
15–19	17	/ /	2

2️⃣ **Interpret and Make Histograms.**

A *histogram* is a bar graph in which the two scales are class intervals and class frequencies. The frequency distribution from Example 2 can be made into a histogram. The frequencies in this histogram form the vertical scale. The intervals form the horizontal scale.

EXAMPLE 3   Answer the following questions about the histogram in Figure 7–7.

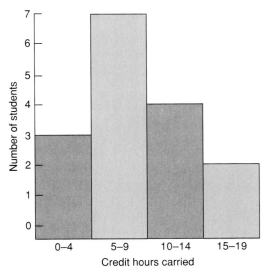

Figure 7–7 Credit hours carried by 16 students in history class.

(a) How many students carried 5 to 9 hours?
   7 carried 5 to 9 hours.
(b) How many students carried less than 15 hours?
   $3 + 7 + 4 =$ **14 carried less than 15 hours.**

(c) What percent of the total carried 10 to 14 hours?

$$\frac{4}{16} = \frac{1}{4} = 0.25 = \mathbf{25\%}$$

(d) What is the ratio of students carrying 0 to 4 hours to those carrying 5 to 9 hours?

$$\frac{3 \text{ with } 0 \text{ to } 4}{7 \text{ with } 5 \text{ to } 9} = \frac{3}{7}$$

The ratio is $\frac{3}{7}$.

EXAMPLE 4   A hospital gives vacation leave based on employee years of service. The employees fall into four categories: 0–2 years, 8 employees; 3–5 years, 6 employees; 6–8 years, 4 employees; 9–11 years, 2 employees. Make a histogram showing this information.

The years of service are already arranged in class intervals and so may be used as given. The class frequency or number of employees in each interval is also provided and so may be used as given. Therefore, it is *not* necessary to make a frequency distribution. The class intervals form the horizontal scale and the frequencies form the vertical scale (see Fig. 7–8).

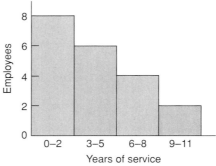

Figure 7–8   Years of service of 20 employees.

3   Interpret and Make Frequency Polygons.

A *frequency polygon* is a line graph made by identifying each class midpoint, marking it with a dot, and connecting the dots with a line. The histogram from Example 4 may be used to make a frequency polygon. The class midpoints may be considered an average for each class interval.

EXAMPLE 5   Make a frequency polygon for the histogram in Example 4.

First, identify the class midpoints with a dot, as in Fig. 7–9. Connect the dots with a line, as in Fig. 7–10. Write the line graph without the bars, as in Fig. 7–11.

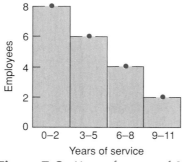

Figure 7–9   Years of service of 20 employees.

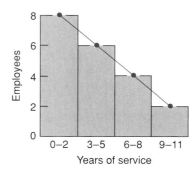

Figure 7–10   Years of service of 20 employees.

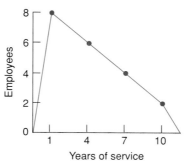

Figure 7-11 Years of service of 20 employees.

EXAMPLE 6 Answer the following questions using the frequency polygon in Example 5.

(a) How many employees have 5 or fewer years of service?
$8 + 6 = $ **14 with 5 or fewer years of service.**

(b) What is the ratio of employees with 2 or fewer years of service to employees with 9 or more years of service?

$$\frac{8 \text{ with } 0 \text{ to } 2 \text{ years}}{2 \text{ with } 9 \text{ or more}} = \frac{8}{2} = \frac{4}{1}$$

The ratio is $\frac{4}{1}$.

(c) What percent of the number of employees with 2 or fewer years of service is the number of employees with 9 or more years of service?

$$\frac{2}{8} = \frac{1}{4} = 0.25 = \textbf{25\%}$$

(d) What is the average number of years of service for employees in the 6 to 8 years of service bracket?
7 is the midpoint or average for the interval.

SELF-STUDY EXERCISES 7-2

1 Use the frequency distribution in Table 7–3 to answer the following questions. The distribution shows the ages of 25 college students in a landscaping class.

1. How many students are 22 or younger?
2. How many students are older than 34?
3. What is the ratio of the number of students 38 to 40 to the number of students 17 to 19?
4. What is the ratio of the smallest class frequency to the largest class frequency?
5. What percent of the total class are students age 17 to 19?
6. What percent of the total class are students age 20 to 22?
7. What two age groups make up the smallest number of students in the class?
8. What two age groups make up the largest number of students in the class?

TABLE 7-3 Frequency Distribution of 25 Ages

Class Interval	Midpoint	Tally	Class Frequency
38–40	39	/	1
35–37	36	/	1
32–34	33	/ /	2
29–31	30	/ / /	3
26–28	27	/ /	2
23–25	24	++++ /	6
20–22	21	++++ / /	7
17–19	18	/ / /	3

9. How many students are over age 28?
10. How many students are under age 26?

Use the following hourly pay rates (rounded to the nearest whole dollar) for 33 support employees in a private college to complete a frequency distribution using the format shown in Table 7–4.

TABLE 7–4 Pay Rates of 33 Support Employees

	Class Interval	Midpoint	Tally	Class Frequency
11.	$14–16	_____	_____	_____
12.	$11–13	_____	_____	_____
13.	$8–10	_____	_____	_____
14.	$5–7	_____	_____	_____

$6 $6 $10 $7 $6 $6 $6

$6 $7 $7 $8 $8 $6 $6

$11 $10 $7 $11 $8 $16 $6

$6 $9 $6 $7 $9 $6 $6

$12 $13 $7 $15 $5

Use the following 40 test scores of two physics classes to complete a frequency distribution using the format in Table 7–5.

TABLE 7–5 Test Scores of 40 Physics Students

	Class Interval	Midpoint	Tally	Class Frequency
15.	91–95	_____	_____	_____
16.	86–90	_____	_____	_____
17.	81–85	_____	_____	_____
18.	76–80	_____	_____	_____
19.	71–75	_____	_____	_____
20.	66–70	_____	_____	_____
21.	61–65	_____	_____	_____
22.	56–60	_____	_____	_____

57 91 76 89 82 59 72 88

76 84 67 59 77 66 56 76

77 84 85 79 69 88 75 58

85 65 67 66 93 83 69 81

80 64 78 76 72 90 79 90

2 Complete the following.

23. Use the information from Exercises 11 through 14 to make a histogram. Use the reference lines in Fig. 7–12 as guides.

24. Use the information from Exercises 15 through 22 to make a histogram.

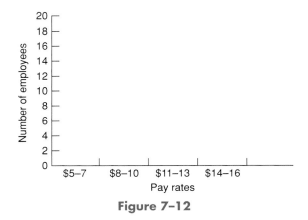

Figure 7–12

3

25. Use the information from Exercises 11 through 14 to make a frequency polygon. Use the reference lines in Fig. 7–12 as guides.

26. Use the information from Exercises 15 through 22 to make a frequency polygon.

7–3   FINDING STATISTICAL MEASURES

Learning Objectives

1 Find the arithmetic mean or arithmetic average.

2 Find the median, the mode, and the range.

In this age of information explosion and with the increased availability of computers, we have massive amounts of data available to us. To make this data useful in the decision-making process, we need to examine the data and summarize key trends and characteristics. This summary is generally in the form of statistical measurements. A *statistical measurement* or *statistic* is a standardized, meaning-

296   CHAPTER 7   Interpreting and Analyzing Data

ful measure of a set of data that reveals a certain feature or characteristic of the data.

☐1 Find the Arithmetic Mean or Arithmetic Average.

An *average* is an approximate number that represents a quantity considered typical or representative of several related quantities. There are several different types of averages. The most common is the *arithmetic mean* or *arithmetic average*. This average is the sum of the quantities in the data set divided by the number of quantities.

> **Rule 7–2** *To find the arithmetic mean or average:*
>
> **1.** Add the quantities.
> **2.** Divide their sum by the number of quantities.

EXAMPLE 1   Find the average of each group of quantities.

(a) Pulse rates: 68, 84, 76, 72, 80
Because there are 5 pulse rates, find their sum and divide by 5:

$$\frac{68 + 84 + 76 + 72 + 80}{5} = \frac{380}{5} = 76$$

The average pulse rate is 76.

(b) Pounds: 21, 33, 12.5, 35.2 (to nearest hundredth)
Because there are 4 weights, find their sum and divide by 4.

$$\frac{21 + 33 + 12.5 + 35.2}{4} = \frac{101.7}{4} = 25.425$$

The average weight is 25.43 lb (to nearest hundredth).

EXAMPLE 2   An automobile used 41 gallons of regular gasoline on a trip of 876 miles. What was the average miles per gallon to the nearest tenth?

The number of miles for each of the 41 gallons of gasoline has already been added (876), so we divide by 41.

$$\frac{876}{41} = 21.4 \qquad \text{to nearest tenth}$$

The car averaged 21.4 miles per gallon on the trip.

EXAMPLE 3   A horticulture student made the following final grades for the past term: B, A, A, C. Find the student's QPA (quality point average) to the nearest hundredth. (See Table 7–6.)

TABLE 7–6 Quality Points for Final Grades

Subject	Grade	Hours		Points per Hour		Total Points
Algebra 101	B	3	×	3	=	9
Spray chemicals 102	A	3	×	4	=	12
Landscape 301	A	3	×	4	=	12
English 101	C	4	×	2	=	8
		13 hours				41 points

To find the QPA for a term, the quality points for the letter grade of each course are multiplied by the credit hours of each course to obtain the total quality points for each course. This total is divided by the total credit hours earned. The quality points awarded are A = 4, B = 3, and C = 2.

$$\frac{41}{13} = 3.153846154 \quad \text{or} \quad 3.15 \quad \text{to nearest hundredth}$$

Thus, the student's QPA for the term is 3.15.

EXAMPLE 4   A community college student has the following grades in Physics 101: 73, 84, 80, 62, and 70. What grade is needed on the last test for a C or 75 average?

One way to find out the needed grade is to assume that each grade is 75 for the 6 tests:

$$\frac{75 + 75 + 75 + 75 + 75 + 75}{6} = \frac{450}{6} = 75$$

Then find the sum of the first five tests actually taken:

$$73 + 84 + 80 + 62 + 70 = 369$$

Subtract to find the difference between this sum and 450. The difference is the needed score.

$$450 - 369 = 81$$

The student must earn a score of 81. To check, assume 81 on the last test and find the average.

$$\frac{73 + 84 + 80 + 62 + 70 + 81}{6} = \frac{450}{6} = 75$$

Thus, 81 is the score needed on the last test.

2   Find the Median, the Mode, and the Range.

Besides the mean or arithmetic average, we sometimes use the *median* and the *mode* to describe data.

The *median* is the middle quantity when the given quantities are arranged in order of size.

> **Rule 7-3**   *To find the median:*
> 1. Arrange the quantities in order of size.
> 2. If the number of quantities is odd, the median is the middle quantity.
> 3. If the number of quantities is even, the median is the average of the two middle quantities.

EXAMPLE 5   A TPR chart shows a patient's temperature, pulse rate, and respiration rate. The following pulse rates were recorded on a TPR chart: 68, 88, 76, 64, 72. What is the median pulse rate?

In order of size: 88
                     76
                     72 ← median or middle value in odd number of values
                     68
                     64

The median pulse rate is 72.

EXAMPLE 6 The following temperatures were recorded: 56°, 48°, 66°, and 62°. What is the median temperature?

Because the number of temperatures is even, find the average of the two middle values.

In order of size:
$$48$$
$$\left.\begin{array}{c}56 \\ 62\end{array}\right\} \frac{56 + 62}{2} = \frac{118}{2} = 59$$
$$66$$

Thus, the median temperature is 59°.

The *mode* is the most frequently occurring quantity among the quantities considered.

Rule 7–4 *To find the mode:*

1. Identify the quantity that occurs most frequently as the mode.
2. If no quantity occurs more than another quantity, there is no mode for this data set.
3. If more than one quantity occurs with the same frequency that is the greatest frequency, the set of quantities or scores will have more than one mode.

EXAMPLE 7 The following hourly pay rates are used at fast-food restaurants: cooks, $5.50; servers, $5.15; bussers, $5.15; dishwashers, $5.25; managers, $7.50. Find the mode.

The hourly pay rate of $5.15 occurs more than any other rate. It is the mode.

EXAMPLE 8 The following daily shifts are worked by employees at a mall clothing store: 4 hours, 6 hours, 8 hours. Find the mode.

No shift occurs more than another, so there is no mode.

The mean, the median, and the mode are also referred to as *measures of central tendency.* Another group of statistical measures are *measures of variation or dispersion.* One of these measures is the *range.* The range is the difference between the highest quantity and the lowest quantity in a set of data.

EXAMPLE 9 Find the range for the data described in Example 7.

The high value is $7.50. The low value is $5.15.

$$\text{range} = \$7.50 - \$5.15 = \mathbf{\$2.35}$$

Another measure of variation or dispersion is *standard deviation.* Standard deviation is introduced in a Career Application at the end of the chapter.

Tips and Traps | *Use More Than One Statistical Measure:*
A common mistake when making conclusions or inferences from statistical measures is to examine only one statistic, such as the mean. To obtain a complete picture of the data requires looking at more than one statistic.

---

1 Find the mean of the given quantities. Round to hundredths if necessary.

1. 12, 14, 16, 18, 20
4. 85, 68, 77, 65
7. 32°F, 41°F, 54°F
10. $32, $43, $22, $63, $36
13. Respiration rates: 16, 24, 20

2. 13, 15, 17, 19, 21
5. 37.6, 29.8
8. 10°C, 13°C, 15°C
11. 11 in., 17 in., 16 in.
14. Pulse rates: 68, 84, 76

3. 68, 54, 73, 69
6. 65.3, 67.9
9. $27, $32, $65, $29, $21
12. 9 in., 7 in., 8 in.

Solve the following problems.

15. A baseball player made 276 home runs over a 16-year period. What was the average number of home runs per year to the nearest tenth?

16. An automobile salesperson sold 163 new cars over a 6-month period. What was the average number of cars sold per month to the nearest tenth?

17. An automobile used 32 gallons of regular gasoline on a trip of 786 miles. What was the average miles per gallon to the nearest tenth?

18. A pickup truck used 25 gallons of regular gasoline on a trip of 256 miles. What was the average miles per gallon to the nearest tenth?

2 Find the median for each group of numbers.

19. 32, 56, 21, 44, 87
22. 21, 33, 18, 32, 19, 44

20. 78, 23, 56, 43, 38
23. $22, $35, $45, $30, $29

21. 12, 21, 14, 18, 15, 16
24. $66, $54, $76, $55, $69

25. The following hourly pay rates are used at fast-food restaurants: cooks, $5.15; servers, $5.25; bussers, $5.15; dishwashers, $5.25; managers, $8.25. Find the median pay rate.

26. The following hourly pay rates are used at a locally owned store: clerks, $5.45; bookkeepers, $6.25; operators, $5.15; assistant managers, $7.95. Find the median pay rate.

Find the mode for each group of numbers.

27. 2, 4, 6, 2, 8, 2
29. 21, 32, 67, 34, 23, 22
31. $56, $67, $32, $78, $67, $20, $67, $56

28. 5, 12, 5, 5, 20
30. 32, 45, 41, 23, 56, 77
32. $32, $87, $67, $32, $32, $87, $77, $22

33. The following weekend work shifts are in effect at a mall clothing store: 4 hours in A.M., 6 hours in P.M., 4 hours in P.M. Find the mode for the number of hours.

34. The following prices are in effect at a fast-food restaurant: $1.75, hamburgers; $1.97, hot ham sandwiches; $2.38, chicken fillet sandwiches; $1.97, roast beef sandwiches. Find the mode.

Find the range for each group of numbers.

35. 22, 36, 41, 41, 17
37. 10, 23, 12, 17, 13, 16
39. $25, $15, $25, $40, $19
41. 23°F, 37°F, 29°F, 54°F, 46°F, 71°F, 67°F

36. 28, 33, 36, 13, 28
38. 23, 23, 18, 32, 29, 14
40. $36, $44, $26, $52, $19

7–4 **COUNTING TECHNIQUES AND SIMPLE PROBABILITIES**

Learning Objectives

1 Count the number of ways objects in a set can be arranged.

2 Determine the chance of an event occurring if an activity is repeated over and over.

1 **Count the Number of Ways Objects in a Set Can Be Arranged.**

A *set* is a well-defined group of objects or *elements*. The numbers 2, 4, 6, 8, and 10 can be a set of even numbers between 1 and 12. Women, men, and children can be a set of people. A, B, and C can be a set of three capital letters.

Counting, in this section, refers to determining all the possible ways the elements in a set can be arranged. One way to find out is to *list* all possible arrangements and then count the number of arrangements.

EXAMPLE 1 List and count the ways the elements in the set *A*, *B*, and *C* can be arranged.

When *A* is first, there are only two possibilities, *BC* and *CB:*

ABC
ACB

When *B* is first, there are only two possibilities, *AC* and *CA:*

BAC
BCA

When *C* is first, there are only two possibilities, *AB* and *BA:*

CAB
CBA

Therefore, *A*, *B*, and *C* can be arranged in six ways. Each of these ways can also be called a set.

If more than three elements are in the set, the procedure becomes more challenging. It may be helpful to use a *tree diagram,* which allows each new set of possibilities to branch out from a previous possibility.

EXAMPLE 2 List and count the ways the elements in the set *W*, *X*, *Y*, and *Z* can be arranged.

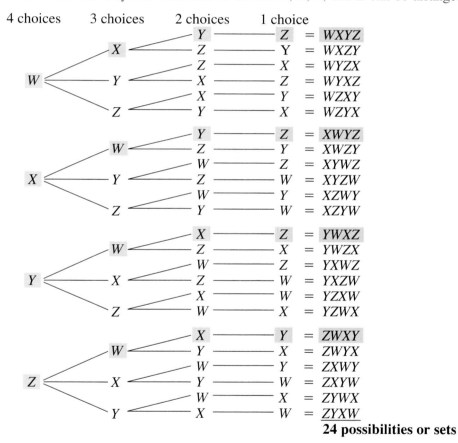

24 possibilities or sets

As evident in Example 2, the greater the number of elements in a set, the greater the complexity and time required to list all possible arrangements.

A faster way to obtain a count of the possible ways the elements in any set can be arranged is by using logic and common sense.

In Example 2, we have four possibilities for the first letter, *W*, *X*, *Y*, or *Z*. For each of the four possible first letters, we have three choices left. For each of these three letters, we have two choices left. Finally, for each of these two letters, we have one choice left.

By multiplying the number of possible letter choices for each position, we can determine the total number of possibilities without listing them: $4 \cdot 3 \cdot 2 \cdot 1 = 24$.

EXAMPLE 3   A coin is tossed three times. With each toss, the coin falls heads up or tails up. How many combinations of heads and tails are there with three tosses of the coin?

There are three tosses. Each toss has only two possibilities, heads or tails; that is,

1st toss	2nd toss	3rd toss
2 possibilities	2 possibilities	2 possibilities

By multiplying the number of possible outcomes for each toss, we get $2 \cdot 2 \cdot 2 = 8$. So 8 combinations are possible.

Tosses: 1st   2nd   3rd

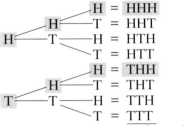

This tree diagram illustrates the possible outcomes of three tosses.

8 possibilities or sets

EXAMPLE 4   Henry has three ties: a red tie, a blue tie, and a green tie. He also has three shirts: a white shirt, a pink shirt, and a yellow shirt. How many combinations of shirts and ties are possible?

If we start with the shirts, there are three possibilities (white, pink, and yellow). For each shirt, there are three possible ties (red, blue, and green). **So we have $3 \cdot 3 = 9$ possible combinations.**

EXAMPLE 5   Given the digits 1, 2, 3, 4, 5, and 6, how many three-digit numbers can be formed without repeating a digit?

The numbers to be formed contain three digits, so there are three positions to fill. We have six digits to work with. The first digit can be one of six. For each of these six digits, there are five possible second digits. For each of the five second digits, there are four possible third digits.

Positions:	1st	2nd	3rd
Possibilities:	6	5	4

By multiplying the possibilities for each position, we get $6 \cdot 5 \cdot 4 = \mathbf{120}$ **possible combinations or sets.**

2 Determine the Chance of an Event Occurring if an Activity is Repeated Over and Over.

Probability means the chance of an event occurring if an activity is repeated over and over. The probability of an event occurring is expressed as a ratio or a percent.

Weather forecasters generally use percents, such as when they forecast a 60% chance of rain or a 20% chance of snow. This text will use ratios like $\frac{3}{5}$ for 3 chances out of 5 or $\frac{2}{3}$ for 2 chances out of 3.

When a coin is tossed, two outcomes are possible, heads or tails. But only one side will be on top. The probability of tossing heads is one out of two, or $\frac{1}{2}$.

> **Rule 7–5** *To express the probability of event A occurring:*
>
> 1. Make a ratio with the number of *A*'s in the set divided by the total number of elements in the set.
> 2. Express the ratio in lowest terms.

EXAMPLE 6   When a die is rolled, what is the probability that a three will appear?

A die has six sides numbered by dots to represent 1 through 6. Each side is an element in the set of six sides. So there is a total of six elements in the set. Only one element is a three. **The probability of rolling a three is $\frac{1}{6}$.**

EXAMPLE 7   A holiday gift shopper wrapped eight men's ties in separate boxes. There were two solid-color ties and six striped ties. If the gift boxes were given at random to eight men, what is the probability of a man receiving a solid-color tie?

The total elements in the set are eight. There are two solid-color elements. **The probability of receiving a solid-color tie is $\frac{2}{8}$, which reduces to $\frac{1}{4}$.**

EXAMPLE 8   A practical nurse has a box of 144 syringes in individual sterile packets. Suppose three of the syringes will have torn packets. What is the probability that the first syringe picked will have a torn packet? If the packet for the first syringe selected is torn, what is the probability of picking a second syringe with a torn packet?

On the first pick, the probability of picking one syringe with a torn packet is $\frac{3}{144}$, which reduces to $\frac{1}{48}$. We assume that the first pick was a syringe with a torn packet, so this leaves a total of 143 syringes and now only two have torn packets. **On the second pick, the probability of picking a syringe with a torn packet is $\frac{2}{143}$.**

SELF-STUDY EXERCISES 7–4

1   Complete the following exercises on counting techniques.

1. List and count all the combinations possible for Keaton, Brienne, and Renee to be seated in three adjacent seats at a basketball game.

2. List and count all the combinations possible for arranging books A, B, C, and D on a shelf.

3. Count all the combinations possible for Jim Riddle to arrange a T-shirt, a sport shirt, a dress shirt, and a sweater on a shelf for display.

4. How many combinations are possible for arranging containers of cotton balls, gauze pads, swabs, tongue depressors, and adhesive tape in a row on a shelf in a doctor's examining room?

5. Part of a landscaping process involves five steps. The steps can be arranged in any order. The landscaping company efficiency officer wants to determine the most efficient order for the five steps. How many combinations of steps are possible?

2 Complete the following exercises on simple probability.

6. A drawing will be held to award door prizes. If the names of 24 people are in the pool for the drawing, what is the probability that David's name will be pulled at random for the first prize? If his name is pulled, what is the probability that Gaynell's name will be pulled next?

8. Mimi tosses 21 pet food coupons, 16 dishwashing detergent coupons, and 11 cereal coupons into a container. What is the probability of reaching in and picking a cereal coupon?

10. A TV quiz program puts all questions in a box. If the box contains five hard questions, five average questions, and five easy questions, what is the probability of being asked an easy question?

7. A box of greeting cards contains 20 friendship cards, 10 get-well cards, and 10 congratulations cards. What is the probability of picking a get-well card at random?

9. A jar holds 10 lock washers and 15 flat washers. What is the probability of drawing a lock washer at random?

CAREER APPLICATION

Teaching: Using Statistics to Examine Class Performance

Teachers often use a variety of statistical measures in evaluating the understanding of a particular concept in a class. When examining a set of data, measures of central tendency such as the mean, median, and mode tell only part of the story. *Measures of variation or dispersion* are also important in examining a set of data. The spread between the highest and lowest amounts in the set of data is the *range*. While this statistical measure gives us some information about dispersion, it does not tell us whether the highest or lowest values are typical of the set of data or they are extreme outliers. We can get a clearer picture of the set of data by examining how much each data point differs or deviates from the mean of the set of data. If we attempt to average the deviations from the mean, we experience a problem. Consider the set of data including the amounts 45, 63, 87, and 91. First, we will find the mean and the deviation from the mean of each amount. We will introduce some new symbols: x_i, $\sum$, and $\bar{x}$. Using these symbols, we can write the formula for finding the mean as $\bar{x} = \frac{\sum x_i}{n}$. This can be read as "the mean (or x bar) is equal to the sum of each value of x (x sub i) divided by the number of values (n)." The Greek capital letter *sigma*, $\sum$, is called a summation symbol and indicates the addition of a group of values. The notation, x_i, identifies each data value by using a subscript: x_1 is the first value, x_2 is the second value, and x_n is the nth value.

$$\bar{x} = \frac{45 + 63 + 87 + 91}{4} = \frac{286}{4} = 71.5$$

Next, we will define the deviation from the mean as $\bar{x} - x_i$. That is, each value of x is subtracted from the mean. We will arrange these values in a table and find the sum of the deviations, $(\bar{x} - x_i)$.

i	x_i	$\bar{x} - x_i$
1	45	$71.5 - 45 = 26.5$
2	63	$71.5 - 63 = 8.5$
3	87	$71.5 - 87 = -15.5$
4	91	$71.5 - 91 = -19.5$
Total	286	0

As we might expect, the sum of the deviations equals zero because the sum of the positive deviations $(26.5 + 8.5 = 35)$ equals the sum of the negative deviations $(-15.5 + -19.5 = -35)$. To avoid this situation, mathematicians have accepted a statistical measure called the *standard deviation* that uses the square of each deviation from the mean. The square of a negative value will always be positive. There are other variations of the formula for finding the standard deviation of a set of values, but we will examine only one formula.

$$s = \sqrt{\frac{\sum(\bar{x} - x_i)^2}{n - 1}}$$

To calculate the standard deviation manually, it is helpful to arrange our calculations in a table. The formula for the standard deviation directs us to average the squared deviations by using one less than the number of values and then find the square root of this average. We are using the formula for standard deviation that deals with a sample of an entire population. In this formula instead of dividing by n, the number of data values in the sample, we divide by $n - 1$. In very large sets of data, there is very little difference in the calculations using n versus $n - 1$. The logical basis for using $n - 1$ assumes that one of the data values is exactly the mean (which may or may not be true) and there are $n - 1$ data values that deviate from the mean.

i	x_i	$\bar{x} - x_i$	$(\bar{x} - x_i)^2$
1	45	$71.5 - 45 = 26.5$	$26.5^2 = 702.25$
2	63	$71.5 - 63 = 8.5$	$8.5^2 = 72.25$
3	87	$71.5 - 87 = -15.5$	$(-15.5)^2 = 240.25$
4	91	$71.5 - 91 = -19.5$	$(-19.5)^2 = 380.25$
Total	286	0	1395

$$s = \sqrt{\frac{1395}{3}} = \sqrt{465} = 21.563859 \ or \ 21.6$$

In interpreting the meaning of standard deviation, a smaller value indicates that the mean is a typical value in the data set. A large standard deviation indicates that the mean is not typical and other statistical measures should be examined to get a better understanding of the characteristics of the data set. In this example, the mean of 71.5 with a standard deviation of 21.6 indicates that the data values do not cluster closely around the mean.

Exercises

Chloe Duke teaches two classes of Technical Math. She has given each class the same quiz on probability. In examining the mean and range for each class, she discovered that they were exactly the same, but from a visual inspection of the scores it appeared that the morning class had a better understanding of the topic than the afternoon class. She proceeded to justify her speculation by calculating the standard deviation for each set of grades.

The quiz scores for the morning class are: 79, 80, 80, 59, 81, 80, 80, 35, 80, 85.

The quiz scores for the afternoon class are: 66, 100, 64, 99, 50, 99, 69, 63, 67, 62.

1. Find the mean score for each class.
2. Find the range for each class.
3. Find the median for each class.
4. Find the mode for each class.

5. Find the standard deviation for each class.

6. Explain why you agree or disagree with Ms. Duke's speculation.

Answers

1. Mean of morning class = 73.9; mean of afternoon class = 73.9.
2. Range of morning class = 50; range of afternoon class = 50.
3. Median of morning class = 80; median of afternoon class = 66.5
4. Mode for morning class = 80; mode for afternoon class = 99.
5. Standard deviation of morning class = 15.4; standard deviation of afternoon class is 18.3.
6. The high range for each class indicates that the scores have a wide dispersion. In examining the median for each set of scores, you see the median for the morning class is significantly higher than that of the afternoon class. The mode for the afternoon class is not at all descriptive of class scores as a whole. The standard deviation for the afternoon class is larger than that of the morning class, which indicates that the scores of the afternoon are not as closely clustered about the mean as those of the morning class. In looking at the total picture, you see the median and the standard deviation give us the most useful information in agreeing with Ms. Duke that the morning class as a whole has a better understanding of probability than the afternoon class.

ASSIGNMENT EXERCISES

Section 7–1

Use Fig. 7–13 to answer Exercises 1 to 4.

1. In what year(s) did women use more sick days than men?
2. In what year(s) did men use about five sick days?
3. In what year(s) did men use more sick days than women?
4. What was the greatest number of sick days for men?

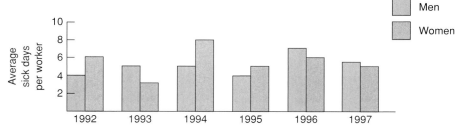

Figure 7–13 Comparison of sick days for men and women.

Use Fig. 7–14 to answer Exercises 5 to 7.

5. What percent of the total cost is the cost of the lot? Round to the nearest tenth.
6. What percent of the total cost is the cost of the house? Round to the nearest tenth.
7. The cost of the lot and landscaping represents what percent of the total cost? Round to the nearest tenth.

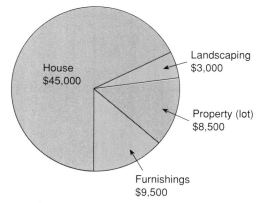

Figure 7–14 Distribution of costs for a $66,000 home.

Use Fig. 7–15 to answer Exercises 8 to 10.

8. What was the patient's highest pulse rate? The highest respiration rate?

9. On what date and time were the patient's pulse rate and respiration rate at their highest level?

10. On what days was the patient's respiration rate below 25?

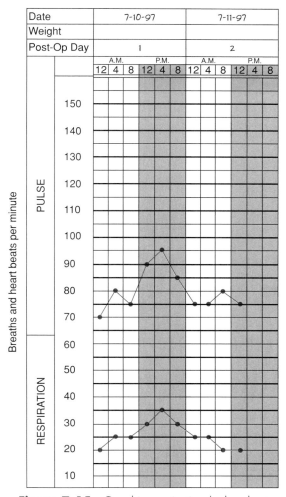

Figure 7–15 Graphic respiration/pulse chart.

Section 7–2

Use the histogram in Fig. 7–16 to answer the questions that follow about a public school system's classroom computers and their memory expressed in megabytes (MB).

11. How many computers have 4 MB to 8 MB of memory?

12. How many computers have 24 MB to 32 MB of memory?

13. How many computers does the system have?

14. How many computers have more than 8 MB of memory?

15. What is the ratio of the number of 24 MB–32 MB computers to the number of 12 MB–16 MB computers?

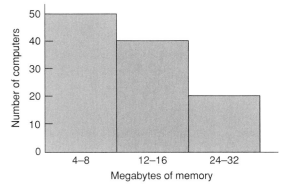

Figure 7–16 Public school system's classroom computers and memory.

A regional horticultural association is composed of 54 members of several clubs. Make a frequency distribution of the members' ages: 17, 18, 20, 21, 21, 24, 24, 29, 29, 29, 31, 31, 33, 33, 34, 35, 35, 38, 38, 38, 39, 41, 42, 43, 43, 43, 43, 45, 45, 47, 47, 48, 48, 48, 49, 50, 51, 51, 52, 56, 56, 58, 58, 60, 60, 62, 64, 64, 65, 66, 68, 70, 71, 71. Use the format shown in Table 7–7.

TABLE 7–7 Ages of Club Members of Regional Horticultural Association

	Class Interval	Midpoint	Tally	Class Frequency
16.	66–75	_____	_____	_____
17.	56–65	_____	_____	_____
18.	46–55	_____	_____	_____
19.	36–45	_____	_____	_____
20.	26–35	_____	_____	_____
21.	16–25	_____	_____	_____

22. Make a histogram of the members' ages from the previous frequency distribution. Use the reference lines in Fig. 7–17 as guides.

23. Make a frequency polygon from the histogram in Fig. 7–17.

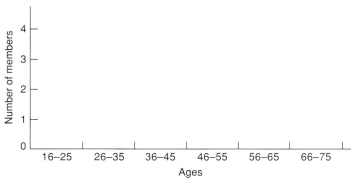

Figure 7–17 Ages of members of horticultural association.

Answer Exercises 24 to 31 based on the information in Exercises 22 and 23.

24. In what age group is the least number of members?

25. In what age group is the greatest number of members?

26. How many members are under age 36?

27. How many members are over age 55?

28. What is the ratio of the number of members age 66 to 75 to members age 16 to 25?

29. What is the ratio of the number of members age 66 to 75 to the number of members age 46 to 55?

30. What percent (to the nearest tenth of a percent) of the members age 46 to 55 are the members age 16 to 25?

31. What percent (to the nearest tenth of a percent) of the total number of members are the members age 46 to 55?

32. Make a frequency distribution of the following scores on an English language test: 68, 70, 74, 77, 78, 82, 82, 84, 86, 86, 86, 88, 89, 90, 90, 93.

33. Make a frequency distribution of the following mileages (miles per gallon, or mpg) reported one week by customers to an automotive rental company for six-cylinder cars: 22, 22, 23, 23, 23, 24, 24, 25, 26, 26, 27, 29, 29, 30, 30, 31, 31.

34. Make a histogram with the information from Exercise 33.

35. Make a frequency polygon based on the histogram you made for Exercise 34.

Section 7-3

Solve the following problems.

36. Jim Smith made 176 baskets over a 16-game period. What was the average number of baskets per game to the nearest tenth?

37. Ben Duke sold 63 new cars over a 5-month period. What was the average number of cars sold per month to the nearest tenth?

38. An automobile used 22 gallons of regular gasoline on a trip of 358 miles. What was the average miles per gallon to the nearest tenth?

39. A delivery truck used 21 gallons of regular gasoline on a trip of 289 miles. What was the average miles per gallon to the nearest tenth?

40. The following hourly pay rates are used at fast-food restaurants: cooks, $6.25; servers, $6.95; bussers, $5.20; dishwashers, $5.20; managers, $8.25. Find the median pay rate.

42. The following weekend work shifts are in effect at a mall clothing store: 3 hours in A.M., 6 hours in P.M., 3 hours in P.M. Find the mode.

44. Kim Collier, a student at a technical college, made the final grades listed in Table 7–8 for the past term. Find Kim's QPA (quality point average) to the nearest hundredth.

TABLE 7–8 Grade Distribution

Subject	Grade	Hours	Points per Hour
Electronics 101	A	4	4
Circuits 201	A	4	4
Algebra 101	B	4	3

46. Janice Van Dyke has the following grades in Algebra 102: 98, 82, 87, 72, and 82. What grade is needed on the last test for a B or 85 average?

48. Find the mean, median, and mode for these test scores: 67, 87, 76, 89, 70, 69, 82.

Section 7–4

49. Susan Duke's new clothes consist of three blazers, four skirts, and two sweaters. How many three-piece outfits can she make from the new clothes?

51. There are three magazines on horses and ten magazines on fashion in Dr. Nelson Campany's waiting room. If Shirley Riddle sends her toddler to get two magazines, what is the probability that the child will randomly pick a fashion magazine on the first draw? If the child is successful, what is the probability of the child picking a fashion magazine on the second draw?

53. If three coins are tossed, what is the number of possible combinations of heads and tails?

55. Suppose Nurse Lee has five 3.5-inch floppy disks. He knows one is filled but cannot remember which one. If he picks one and puts it in the computer, what is the probability of his picking the filled disk?

41. The following hourly pay rates are used at a locally owned store: office assistants, $5.85; bookkeepers, $6.20; cashiers, $5.45; assistant managers, $7.90. Find the median pay rate.

43. The following prices are in effect at a fast-food restaurant: $1.85, hamburgers; $1.98, hot ham sandwiches; $2.28, chicken sandwiches; $1.85, roast beef sandwiches. Find the mode.

45. Tami Murphy made the final grades listed in Table 7–9 for the past term. Find Tami's QPA (quality point average) to the nearest hundredth.

TABLE 7–9 Grade Distribution

Subject	Grade	Hours	Points per Hour
English 201	A	3	4
History 202	B	3	3
Philosophy 101	A	3	4

47. Sarah Smith has the following grades in American History: 79, 73, 71, 78, and 86. What grade is needed on the last test for a C or 75 average?

50. For her first professional job, Martha Deskin, a recent college graduate, purchased four pairs of shoes, three business suits, and five shirts. How many clothing combinations are possible?

52. Two coins are tossed three times. Count the number of combinations of heads and tails with three tosses of the two coins.

54. Cy Pipkin is puzzled over a true-false question on a test and does not know the answer. What is the probability that he will pick the right answer by chance?

56. A box of wrapped candies is said to contain 5 cherry cordials, 10 cremes, and 15 chocolate-covered caramels. What is the probability of picking a cherry cordial?

57. A box of stationery contains 12 sheets imprinted with an animal scene, 24 sheets imprinted with a floral design, and 12 plain sheets. What is the probability of picking a sheet with an animal scene at random?

CHALLENGE PROBLEM

One complete circle is made up of 360 degrees. A measuring instrument called a *protractor* is used to draw angles of various sizes.

58. Use a protractor to make a circle graph of the sources of a construction company's yearly income of $150,000. Show 50% from small business construction, 25% from home remodeling, 15% from local government jobs, and 10% from miscellaneous work.

CONCEPTS ANALYSIS

1. What type of information does a circle graph show?

2. Give a situation where it would be appropriate to organize the data in a circle graph.

3. What type of information does a bar graph show?

4. Give a situation where it would be appropriate to organize the data in a bar graph.

5. What type of information does a line graph show?

6. Give a situation where it would be appropriate to organize the data in a line graph.

7. Explain the similarities and differences between a histogram and a frequency polygon.

8. Explain the differences among the three types of averages: the mean, the median, and the mode.

9. Show by example how using a tree diagram for counting gives the same result as multiplying the number of choices for each position in the set.

10. If there are 5 red marbles and 7 blue marbles in a bag, is the probability of drawing a red marble $\frac{5}{7}$? Explain your answer.

CHAPTER SUMMARY

Objectives	What to Remember with Examples

Section 7–1

1 Read circle graphs.

A circle graph compares parts to a whole.

If the total monthly revenue at a used car dealership is $75,000, what is the revenue from trucks? (See Fig. 7–18.)

35% of 75,000 =
0.35 × 75,000 = $26,250

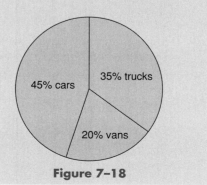

Figure 7–18

2 Read bar graphs. Bar graphs compare items to each other.

What is the ratio of men's salaries to women's salaries in the pants department? (See Fig. 7–19.)

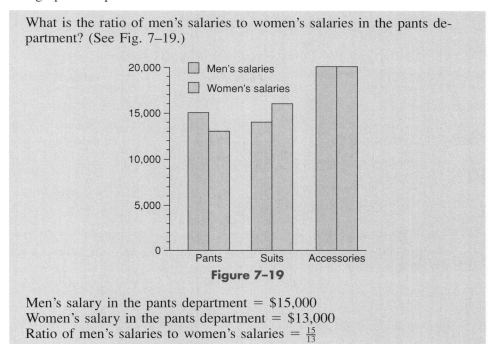

Figure 7–19

Men's salary in the pants department = $15,000
Women's salary in the pants department = $13,000
Ratio of men's salaries to women's salaries = $\frac{15}{13}$

3 Read line graphs. A line graph shows how an item changes with time.

Use the line graph of the average prices for textbooks to find the year when textbooks averaged $47 per book. (See Fig. 7–20.)

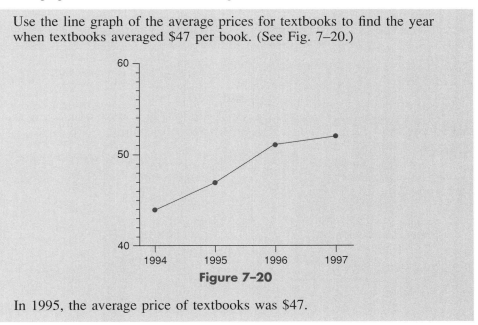

Figure 7–20

In 1995, the average price of textbooks was $47.

Section 7–2
1 Interpret and make frequency distributions. To make a frequency distribution, determine the appropriate interval for classifying the data. Tally the data.

Make a frequency distribution with the following data, indicating leave days for State College employees. (See Table 7–12.)

2 2 4 4 4 5 5 6 6 8 8 8 9 12 12 12 14 15 20 20

TABLE 7–12 Annual Leave Days of 20 State College Employees

Class Interval	Tally	Class Frequency
16–20	//	2
11–15	++++	5
6–10	++++ /	6
1–5	++++ //	7

2 Interpret and make histograms.

A histogram is a bar graph representing data in a frequency distribution.

Make a histogram using the data for State College. (See Fig. 7–21.)

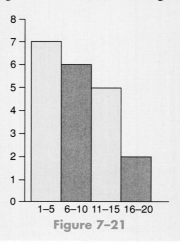

Figure 7–21

3 Interpret and make frequency polygons.

A frequency polygon is a line graph connecting the midpoints of each class interval in a frequency distribution.

Draw a frequency polygon using the data for State College. (See Fig. 7–22.)

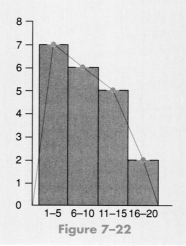

Figure 7–22

Section 7–3

1 Find the arithmetic mean or arithmetic average.

Add the quantities; divide by the number of quantities.

> Find the arithmetic mean of the following test scores: 76, 86, 93, 87, 68, 76, 88
>
> $$76 + 86 + 93 + 87 + 68 + 76 + 88 = 574$$
> $$574 \div 7 = 82$$
>
> The mean is 82.

2 Find the median, the mode, and the range.

The median of an odd number of quantities is the middle quantity when the quantities are arranged in order of size. For an even number of quantities, average the two middle quantities.

The mode is the quantity that occurs most frequently. A set of quantities may have no mode or more than one mode.

The range is the difference between the largest and smallest quantity.

> Find the median, mode, and range of the set of test scores:
>
> 68, 76, 76, $\underline{86}$, 87, 88, 93
>
> The median is 86.
>
> The mode is 76.
>
> The range is $93 - 68 = 25$.

Section 7–4

1 Count the number of ways objects in a set can be arranged.

Multiply the number of positions possible for each object in the set.

> Renee Smith's closet has two new blazers (navy and red) and four new skirts (gray, black, tan, and brown). How many outfits can she make from the new clothes?
>
> 1 blazer + 1 skirt = 1 outfit
>
> $$\begin{pmatrix} \text{blazer} \\ \text{choices} \end{pmatrix} \cdot \begin{pmatrix} \text{skirt} \\ \text{choices} \end{pmatrix} = \begin{pmatrix} \text{possible} \\ \text{outfits} \end{pmatrix}$$
> $$2 \quad \cdot \quad 4 \quad = \quad 8$$

2 Determine the chance of an event occurring if an activity is repeated over and over.

The chance of an event occurring is the ratio of the number of possible successful outcomes to the number of possible outcomes.

> A box in a doctor's office contains 30 $\frac{1}{2}$-in. adhesive strips and 70 $\frac{3}{4}$-in. adhesive strips. What is the probability of picking a $\frac{3}{4}$-in. adhesive strip?
>
> $$\frac{70}{100} = \frac{7}{10}$$
>
> The probability of picking a $\frac{3}{4}$-in. adhesive strip is $\frac{7}{10}$.

WORDS TO KNOW

graph, (p. 288)
circle graph, (p. 288)
bar graph, (p. 289)
line graph, (p. 290)
class intervals, (p. 292)
class midpoint, (p. 292)
tally, (p. 292)
class frequency, (p. 292)
frequency distribution, (p. 292)
histogram, (p. 293)

frequency polygon, (p. 294)
statistical measurements,
  (p. 296–297)
average, (p. 297)
arithmetic mean or arithmetic
  average, (p. 297)
median, (p. 298)
mode, (p. 298)
measures of central tendency,
  (p. 299)

measures of variation or
  dispersion, (p. 299)
range, (p. 299)
standard deviation, (p. 299)
set, (p. 300)
elements, (p. 300)
counting, (p. 301)
tree diagram, (p. 301)
probability, (p. 302)

CHAPTER TRIAL TEST

1. A _____ graph is used to show how different data items are related to each other.
2. A _____ graph is used to show how a whole quantity is related to its parts.
3. A _____ graph is used to show how an item or items change over time.

Answer Problems 4 and 5 from the line graph of Fig. 7–23.

4. How many degrees warmer was it indoors at midnight than outdoors?
5. What was the change in outdoor temperature between 11:00 A.M. and noon?

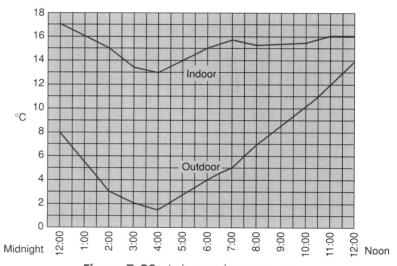

Figure 7–23 Indoor-outdoor temperature.

Answer the following questions about the manufacturing costs for the electronic game as shown in the circle graph in Fig. 7–24.

6. What percent of the total cost is materials?
7. What would the profit be if there were no tariff or tax on imported parts?
8. How much are overhead and materials together?
9. What percent of the total cost is the profit?
10. What is the ratio of labor to total cost?
11. What is the ratio of the tariff to total cost?

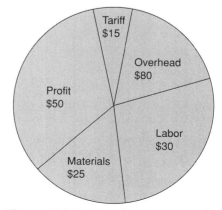

Figure 7–24 Manufacturing costs of a $200 electronic game.

Use the bar graph in Fig. 7–25 to answer the questions about the academic-year starting salaries of women and men college professors in various academic departments of a certain college.

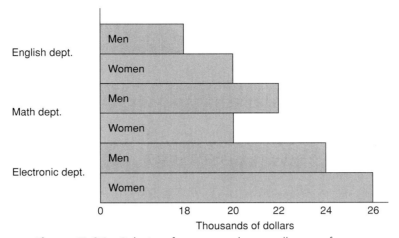

Figure 7–25 Salaries of women and men college professors.

12. In what departments do men make more than women?

13. In what departments do women make more than men?

14. What percent of women's salaries is men's salaries in the English Department (to the nearest tenth of a percent)?

15. What percent of men's salaries is women's salaries in the Electronics Department (to the nearest tenth of a percent)?

16. What is the ratio of men's salaries to women's salaries in the Math Department?

17. What is the ratio of men's salaries to women's salaries in the English Department?

18. Make a line graph to illustrate the following numerical information about the average prices of two-, three-, and four-bedroom homes in a certain subdivision. Use the reference lines in Fig. 7–26 as guides.

    2-bedroom homes, average $80,000
    3-bedroom homes, average $90,000
    4-bedroom homes, average $100,000

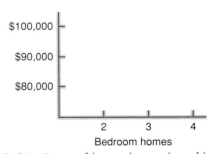

Figure 7–26 Prices of homes by number of bedrooms.

Use the frequency distribution shown in Table 7–13 to answer the questions that follow. The distribution shows the number of correct answers on a 30-question test in a science class.

19. How many students scored less than 16 correct?

20. How many students scored more than 25 correct?

21. What is the ratio of the number of students scoring 6 to 10 correct to those scoring 21 to 25 correct?

22. What percent of the total scored 21 to 30 correct?

TABLE 7–13 Frequency Distribution of Correct Answers

Class Interval	Midpoint	Tally	Class Frequency
26–30	28	/ / /	3
21–25	23	/ / / /	4
16–20	18	┼┼┼ / /	7
11–15	13	/ / /	3
6–10	8	/ /	2
1–5	3	/	1

Use the following ages of 24 children in a day-care center to complete a frequency distribution, using the format shown in Table 7–14.

TABLE 7-14 Ages of 24 Day-Care Children

	Class Interval	Midpoint	Tally	Class Frequency
23.	4–6	_____	_____	_____
24.	1–3	_____	_____	_____

```
1  1  1  1  1  1  1  2  2  2
3  3  3  3  3  4  4  4  4  5
5  5  5  6
```

Use the histogram in Fig. 7–27 to answer questions 25 to 27 about a company's software programs and the number of employees trained to use them.

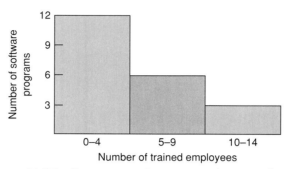

Figure 7-27 Company employees trained to use software.

25. How many employees at the most can use only three programs?

26. How many programs does the company have?

27 How many employees does the company have?

28. Find the mean, median, and mode for these test scores: 77, 87, 77, 89, 70, 69, 82.

29. Find the mean, median, and mode for these test scores: 81, 78, 69, 75, 81, 93, 68.

30. List and count the ways the elements in the set *L, M, N,* and *O* can be arranged.

31. Rayford has two ties: a red tie and a blue tie. He also has three shirts: a white shirt, a green shirt, and a yellow shirt. How many combinations of shirts and ties are possible?

32. Millie has in her purse three green eye shadows, four white eye shadows, and two black eye shadows. What is the probability of her pulling out a black eye shadow?

33. An envelope contains the names of two men and three women to be interviewed for a promotion at Washington's Landscape Service. The interviewer wants to pull names to determine the order of the interviews. What is the probability of pulling a woman's name first?

34. If a small boy has one red marble and three yellow marbles in his pocket, what is the probability of his pulling out the red one on the first try?

8

Symbolic Representation

GOOD DECISIONS THROUGH TEAMWORK

Symbolic representation is used in everyday life to aid in overcoming language barriers and to allow for concepts to be processed more rapidly. There is an instantaneous association of a concept with a picture or symbol. Even preschool children who haven't yet learned to read can distinguish between the symbols indicating male and female restroom facilities. A symbol can be mentally processed much more quickly than the words represented by the symbol. For example, the human brain can process the symbol much faster than it can process the words *no left turn*.

In your teams use a flip chart or board to write the words and draw the nonmathematical symbols that any team member can recall. One team member will serve as recorder and another will serve as reporter. Discuss the components of an effective symbolic representation of a concept. Next, design signs that contain no words to distinguish a walking path from a bicycle path, to direct persons to the information office on your campus, to direct persons to keep off the grass. In addition, identify one additional communication need on your campus and design a "no words" sign to meet this need.

Compare the symbolic representations your team designed with those designed by other teams in your class. Share the "no words" sign you designed to meet the additional communication need on your campus and measure its effectiveness by having classmates interpret the symbolic representation. Extend this activity beyond your classroom by researching international symbols for signs and investigate the history of the development of this means of communication.

Early in the development of our number system, symbols were useful for expressing numbers and mathematical concepts. These symbols allow mathematical expressions to be written in a concise form that overcomes many barriers created by differences in languages and cultures. For symbolism to be effective, there must be a widespread understanding of the numbers and concepts represented by the symbols. We will review many mathematical concepts by examining the most common symbolism used in mathematics. As we continue our study, some of these initial concepts will be expanded and additional mathematical symbolism will be introduced.

8–1 RATIONAL NUMBERS

Learning Objectives

1. Express powers with negative integer exponents in exponential form and in ordinary form.
2. Perform basic operations with rational numbers.

In Chapters 1, 2, and 3, we examined various symbolic representations for rational numbers. To review the relationship of these rational numbers, examine Fig. 8–1. Since the definition of a rational number is any number that can be expressed as the quotient of any integer and a nonzero integer, the illustration also shows how the various types of rational numbers can be written in rational form.

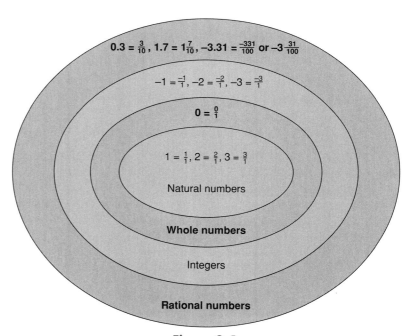

$$0.3 = \frac{3}{10}, \ 1.7 = 1\frac{7}{10}, \ -3.31 = \frac{-331}{100} \text{ or } -3\frac{31}{100}$$

$$-1 = \frac{-1}{1}, \ -2 = \frac{-2}{1}, \ -3 = \frac{-3}{1}$$

$$0 = \frac{0}{1}$$

$$1 = \frac{1}{1}, \ 2 = \frac{2}{1}, \ 3 = \frac{3}{1}$$

Natural numbers

Whole numbers

Integers

Rational numbers

Figure 8–1

1 Express Powers with Negative Integer Exponents in Exponential Form and in Ordinary Form.

Rational numbers are manipulated through basic operations to represent different conditions and relationships among the numbers. We have already become familiar with rational numbers expressed as natural numbers, whole numbers, integers, fractions, mixed numbers, decimals, and mixed decimals. Now let's examine another way to represent certain fractions.

A special property of fractions useful in several manipulations is the property of *reciprocals*. A number times its reciprocal equals one: $n \times \frac{1}{n} = 1$. Another symbolic way to represent a reciprocal is by using the exponent -1. When an exponent is different from a natural number, the symbolism takes on a different meaning than repeated multiplication.

To show a reciprocal using negative exponents, the base remains the same and the exponent is the opposite of the original exponent. Using negative exponents, the reciprocal of 10^1 is 10^{-1}. The reciprocal of 10^2 is 10^{-2}. Because the reciprocal of 10^1 or 10 is also $\frac{1}{10}$, then $\frac{1}{10}$ and 10^{-1} must represent the same number. Similarly, $\frac{1}{100}$ is the same as 10^{-2}.

EXAMPLE 1   Write the following by using positive exponents, then write them as ordinary numbers.

(a) 10^{-4} (b) 5^{-2} (c) 4^{-3}

(a) $10^{-4} = \left(\dfrac{1}{10}\right)^4$ or $\dfrac{1}{10,000}$ (b) $5^{-2} = \left(\dfrac{1}{5}\right)^2$ or $\dfrac{1}{25}$

(c) $4^{-3} = \left(\dfrac{1}{4}\right)^3$ or $\dfrac{1}{64}$

EXAMPLE 2   Write the following as equivalent expressions with negative exponents.

(a) $\dfrac{1}{9}$ (b) $\dfrac{1}{16}$ (c) $\left(\dfrac{1}{8}\right)^3$

(a) $\dfrac{1}{9} = \left(\dfrac{1}{3}\right)^2 = 3^{-2}$ (b) $\dfrac{1}{16} = \left(\dfrac{1}{4}\right)^2 = 4^{-2}$ (c) $\left(\dfrac{1}{8}\right)^3 = 8^{-3}$

or $\dfrac{1}{16} = \left(\dfrac{1}{2}\right)^4 = 2^{-4}$

We can use the calculator to verify the concept of negative exponents.

■ Calculator Tips for Reciprocals and Exponents

Reciprocals

Most scientific and graphics calculators have a reciprocal function. The reciprocal key or function is usually labeled $1/x$ or x^{-1}. To find the reciprocal of a number on the calculator, enter the number and press the reciprocal key. Some scientific calculators display the reciprocal in decimal form instantly. Graphics and scientific calculators may require you to press the *execute*, *equal*, or *enter* key before the reciprocal is displayed.

To find the reciprocal of $\frac{1}{4}$, enter $\frac{1}{4}$ as a fraction, or enter its decimal equivalent of 0.25, then press the reciprocal key. The reciprocal of $\frac{1}{4}$ is 4.

Exponents

The general exponent key on the calculator is usually labeled x^y. This key is also referred to as the general power key. The base is entered first, followed by the general power key, and then the exponent. Because the exponent can be more than one digit, a positive or negative value, or various types of numbers or expressions, you must instruct the calculator that the entry of the exponent is complete by pressing the *equal* or *execute* key.

Find 5^2 using the calculator. 5 $\boxed{x^y}$ 2 $\boxed{=}$ yields the result of 25. To be sure how a new key on the calculator operates, first work some examples for which you know the result.

Find 5^{-2} using the calculator. 5 $\boxed{x^y}$ 2 $\boxed{+/-}$ $\boxed{=}$ yields the result of 0.04. Is this the result we expected? Working the example without the general exponent key, we have $5^{-2} = (\frac{1}{5})^2 = \frac{1}{25}$. Is $\frac{1}{25}$ equal to 0.04? Yes, $1 \div 25 = 0.04$.

If these keystrokes did not yield the appropriate result or if the calculator failed to make any calculation at all, try a different sequence of keystrokes. For example, try 5 $\boxed{x^y}$ $\boxed{-}$ 2 $\boxed{=}$. When learning to use different features of a calculator, it is important to consult the instruction manual and/or investigate several possible options.

EXAMPLE 3 Using the general power key on the calculator, verify the following statement: $6^{-3} = (\frac{1}{6})^3$.

To verify a statement, show that the value of the expression on the left is equal to the value of the expression on the right.

$$6^{-3} \quad 6 \ \boxed{x^y} \ 3 \ \boxed{+/-} \ \boxed{=} \Rightarrow 0.004629629$$

$$\left(\frac{1}{6}\right)^3 \quad 1 \ \boxed{\div} \ 6 \ \boxed{x^y} \ 3 \ \boxed{=} \Rightarrow 0.004629629$$

A noncontinuous sequence for $(\frac{1}{6})^3$ can be used. For example,

$$6^3 = 216 \qquad 1 \div 216 = \Rightarrow 0.004629629$$

② Perform Basic Operations with Rational Numbers.

An important part of symbolic representation is the manipulating and combining of numbers through basic operations. A careful development of the basic operations of integers was presented in Chapter 2. We will review these operations before we expand our number system to include additional types of numbers and symbolic representations.

Rule 8–1 *Adding and Subtracting signed numbers:*

To *add* signed numbers having *like* signs:

1. Add their absolute values.
2. Give the answer the common sign.

To *add* signed numbers having *unlike* signs:

1. Subtract the smaller absolute value from the larger absolute value.
2. Give the answer the sign of the number with the larger absolute value.

To *subtract* signed numbers:

1. Change the subtraction to an equivalent addition by changing the sign of the subtrahend to its opposite.
2. Use the appropriate rule for adding signed numbers.

EXAMPLE 4 Perform the indicated operations.

(a) $-5 - 7$ (b) $-6 + 9$ (c) $4 - (-2)$ (d) $5 - 4.7 + 2.6 - 9$

(a) $-5 - 7 = \mathbf{-12}$ Interpret as -5 plus -7. Apply rule for adding numbers with like signs.

(b) $-6 + 9 = \mathbf{3}$ Apply rule for adding numbers with unlike signs.

(c) $4 - (-2) =$ Change to an equivalent addition.

$$4 + 2 = 6$$ Apply rule for adding numbers with like signs.

(d) $5 - 4.7 + 2.6 - 9 =$ Change to an equivalent addition.

$5 + (-4.7) + 2.6 + (-9) =$ Apply rule for adding numbers with unlike signs.

$0.3 + 2.6 + (-9) =$ Apply rule for adding numbers with like signs.

$2.9 + (-9) = -6.1$ Apply rule for adding numbers with unlike signs.

Rule 8–2 *Multiplying and dividing signed numbers:*

To *multiply* two signed numbers having *like* signs:

1. Multiply their absolute values.
2. Make the sign of the product positive.

To *multiply* two signed numbers having *unlike* signs:

1. Multiply their absolute values.
2. Make the sign of the product negative.

To *divide* signed numbers having *like* signs:

1. Divide their absolute values.
2. Make the sign of the quotient positive.

To *divide* signed numbers having *unlike* signs:

1. Divide their absolute values.
2. Make the sign of the quotient negative.

EXAMPLE 5   Perform the indicated operations.

(a) $-3(7)$ (b) $-\dfrac{3}{8}\left(-\dfrac{5}{6}\right)$ (c) $\dfrac{-18}{-3}$ (d) $\dfrac{84}{-2.5}$

(a) $-3(7) = -21$ Apply the rules for multiplying numbers with unlike signs.

(b) $-\dfrac{\overset{1}{3}}{8}\left(-\dfrac{5}{\underset{2}{6}}\right) = \dfrac{5}{16}$ Apply the rules for multiplying numbers with like signs.

(c) $\dfrac{-18}{-3} = 6$ Apply the rules for dividing numbers with like signs.

(d) $\dfrac{84}{-2.5} = -33.6$ Apply the rules for dividing numbers with unlike signs.

Tips and Traps   *Memory Tips:*

It is common to remember clue words to help recall important rules. However, just as symbols are meaningless if you don't know what they represent, these clue words are meaningless if you do not understand the rules they represent. Addition:

likes . . . add . . . keep common sign

unlikes . . . subtract . . . keep sign of larger absolute value

Subtraction: Change double signs to single sign and use rules for addition of likes or unlikes.

$$+\ +\ \Rightarrow\ + \qquad +-\ \Rightarrow\ -$$
$$-\ -\ \Rightarrow\ + \qquad -\ +\ \Rightarrow\ -$$

Multiplication/division: multiply or divide as indicated.

likes . . . +

unlikes . . . −

SELF-STUDY EXERCISES 8–1

[1] Write the following by using positive exponents, then write them as ordinary numbers.

1. 10^{-8} **2.** 7^{-3} **3.** 2^{-4}

4. 6^{-2} **5.** 4^{-3}

Write the following as equivalent expressions with positive exponents, then as equivalent expressions with negative exponents.

6. $\dfrac{1}{5}$ **7.** $\dfrac{1}{25}$ **8.** $\dfrac{1}{49}$

9. $\dfrac{1}{27}$ **10.** $\dfrac{1}{64}$

Use the general power key on the calculator to verify the following statements.

11. $8^{-2} = \left(\dfrac{1}{8}\right)^{2}$ **12.** $5^{-3} = \left(\dfrac{1}{5}\right)^{3}$ **13.** $12^{-7} = \left(\dfrac{1}{12}\right)^{7}$

14. $15^{-3} = \left(\dfrac{1}{15}\right)^{3}$ **15.** $\left(\dfrac{1}{3}\right)^{-2} = 3^{2}$

[2] Perform the indicated operations. Round to hundredths if necessary.

16. $8 + (-2)$ **17.** $7 + (-3)$ **18.** $15 + (-21)$

19. $28 + (-63)$ **20.** $-8 + 5$ **21.** $-7 + 12$

22 $-12 + (-4)$ **23.** $-11 + (-16)$ **24.** $-16 - 21$

25. $-42 - 18$

26. The temperature in Juneau, Alaska, was -13 degrees, then dropped 8 degrees in 2 hours. What was the temperature after the drop?

27. Latrica already owed \$15 to LaQuinta when she borrowed \$46. Express the amount Latrica now owes as a signed number.

28. $-5 + 3 - 7$ **29.** $-14 - 17 + 21$ **30.** $5(-3)$

31. $-4(-2)$ **32.** $\dfrac{-3}{8}\left(\dfrac{-5}{6}\right)$ **33.** $(-4.2)(-7.6)$

34. $6.7(-5.2)$ **35.** $\dfrac{-42}{6}$ **36.** $\dfrac{48}{-8}$

37. $\dfrac{-72}{(-9)}$ **38.** $\left(\dfrac{-5}{8}\right) \div \left(\dfrac{-3}{4}\right)$ **39.** $-6.7 \div (-5.3)$

40. $-2.1 \div 0.18$

8–2   REAL NUMBERS

Learning Objectives

[1] Examine the relationship between powers and roots using exponential and root notation.

[2] Position irrational numbers on the number line.

3 Estimate the value of irrational numbers.

4 Simplify numerical expressions using the order of operations.

So far our number system has been developed through rational numbers. As we examine the roots of all rational numbers, we will find it necessary to expand our study of numbers to include another type of number.

1 **Examine the Relationship Between Powers and Roots Using Exponential and Root Notation.**

In Section 1–6 we examined the use of exponents to indicate powers of numbers and roots of perfect powers. We will now extend our use of exponents to cover roots of nonperfect powers. Before we examine roots, we will look at *perfect powers*. Rather than give a formal definition at this time, we will illustrate the concept.

The opposite or inverse operation for raising to powers is finding roots. For example, a *square root* is the number used as a factor two times to give a certain number. A *cube root* is the number used as a factor three times to give a certain number. A *fourth root* is the number used as a factor four times to give a certain number. Using the values in Table 8–1, we can say that the square root of 4 is 2, the square root of 9 is 3, and the square root of 16 is 4. Similarly, the cube root of 8 is 2, the cube root of 27 is 3, and the cube root of 64 is 4. The fourth root of 16 is 2, the fourth root of 81 is 3, and the fourth root of 256 is 4. Exponents can be used to represent roots symbolically. Keep in mind that symbolic representations are useful only if you understand what is being represented. The choice of symbols is meaningless if the appropriate interpretation is not made.

TABLE 8–1

Powers of 2	Powers of 3	Powers of 4	Powers of 5
$2^1 = 2$	$3^1 = 3$	$4^1 = 4$	$5^1 = 5$
$2^2 = 4$	$3^2 = 9$	$4^2 = 16$	$5^2 = 25$
$2^3 = 8$	$3^3 = 27$	$4^3 = 64$	$5^3 = 125$
$2^4 = 16$	$3^4 = 81$	$4^4 = 256$	$5^4 = 625$
$2^5 = 32$	$3^5 = 243$	$4^5 = 1024$	$5^5 = 3125$
$2^6 = 64$	$3^6 = 729$	$4^6 = 4096$	$5^6 = 15,625$

Square roots are indicated by using an exponent of $\frac{1}{2}$, cube roots are indicated by using an exponent of $\frac{1}{3}$, and fourth roots are indicated by using an exponent of $\frac{1}{4}$. To generalize, the root of a number is indicated by the denominator of a fractional exponent.

EXAMPLE 1   Write the following roots using fractional exponents.

(a) Square root of 25   (b) Cube root of 125   (c) Fourth root of 625

(a) Square root of 25 $= 25^{1/2}$

(b) Cube root of 125 $= 125^{1/3}$

(c) Fourth root of 625 $= 625^{1/4}$

Another common notation for roots is the *radical sign*, $\sqrt{}$. The number for which a root is to be found is written under the radical sign and is called the *radicand.* For cube roots, the number 3 is written in the "v" portion of the radical sign. Fourth roots are indicated by placing a 4 in the "v" portion, and so on. The root indication is called the *index* of the root.

$$\text{Square root of } 25 = 25^{1/2} = \sqrt{25}$$
$$\text{Cube root of } 125 = 125^{1/3} = \sqrt[3]{125}$$
$$\text{Fourth root of } 625 = 625^{1/4} = \sqrt[4]{625}$$

Because of the availability of scientific and graphics calculators, the radical notation, which once was the most popular notation for roots, is now being replaced by the fractional exponent notation for roots.

■ Calculator Tips for Roots

There are numerous ways to find the root of a number using a scientific or graphics calculator. By examining several different ways, we can strengthen our understanding of the notation for roots and the use of the calculator. Some scientific calculators have special function keys for the most common roots: square roots and cube roots. To use these keys, enter the base or radicand and then press the appropriate special root key; or first press the special root key, enter the base or radicand, and then press the *equal* (=) or *execute* (EXE) key. Some graphics calculators have a menu choice of special roots. From the menu, select the appropriate root, then enter the base or the radicand. To see the value of the root, press the execute or enter key. To find roots in general, other calculator functions are necessary.

Using the general power key, $\boxed{x^y}$, on a calculator, find the fourth root of 16.

Exponent as fraction: 16 $\boxed{x^y}$ 1 $\boxed{\text{a b/c}}$ 4 $\boxed{=}$ ⇒ 2.
$$16^{1/4} = 2$$

Exponent as decimal equivalent: 16 $\boxed{x^y}$ $\boxed{\cdot}$ 25 $\boxed{=}$ ⇒ 2.
$$16^{0.25} = 2$$

Exponent as indicated division: 16 $\boxed{x^y}$ $\boxed{(}$ 1 $\boxed{\div}$ 4 $\boxed{)}$ $\boxed{=}$ ⇒ 2.
$$16^{1/4} = 2$$

Using the general root key, $\boxed{x^{1/y}}$, on a calculator, find the fourth root of 16. Enter the base or radicand, press the general root key, enter the index of the root, and then press equal. Some calculators have a general root key in radical notation, $\boxed{\sqrt[y]{x}}$. Some typical sequences of keys are:

16 $\boxed{x^{1/y}}$ 4 $\boxed{=}$ ⇒ 2   Enter index second.

16 $\boxed{\sqrt[y]{x}}$ 4 $\boxed{=}$ ⇒ 2

4 $\boxed{\sqrt[y]{x}}$ 16 $\boxed{=}$ ⇒ 2   Enter index first.

Graphics Calculators: Most graphics calculators have the same notations as scientific calculators; however, you may need to access these functions through a menu. Check with your calculator's instruction manual for specific details.

Test various sequences of steps on your calculator to determine the different options you have for finding the roots of numbers.

EXAMPLE 2   Using the calculator, find the following roots.

(a) $144^{1/2}$ (b) $343^{1/3}$ (c) $\sqrt[4]{1,296}$

(a) $144^{1/2}$ 144 $\boxed{x^{1/y}}$ 2 $\boxed{=}$ ⟹ 12.   Other options will also yield correct results.

(b) $343^{1/3}$ 343 $\boxed{x^{1/y}}$ 3 $\boxed{=}$ ⟹ 7.

(c) $\sqrt[4]{1296}$ 1296 $\boxed{x^{1/y}}$ 4 $\boxed{=}$ ⟹ 6.

2 Position Irrational Numbers on the Number Line.

To continue our discussion of roots, at this point we will limit our focus to square roots. As you would expect, not all numbers are perfect powers. Perfect squares that are less than or equal to 100 are 1, 4, 9, 16, 25, 36, 49, 64, 81, and 100. All other integers between 1 and 100 are not perfect squares. If you examine the square roots of some of these numbers on the calculator, you will see that the numbers appear to be nonterminating, nonrepeating decimals. These values are actually approximations because they have been rounded: $2^{1/2} = 1.414213562$, and $3^{1/2} = 1.732050808$. The exact amounts are called *irrational numbers*. This new type of number expands our number system beyond rational numbers to *real numbers*. Real numbers include both rational and irrational numbers (see Fig. 8–2). Other types of irrational numbers will be introduced later.

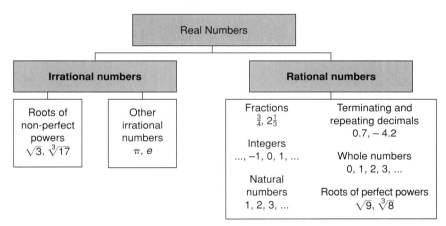

Figure 8–2

Roots are ordered and have a position on the number line similar to the ordering of rational numbers. Let's position some irrational numbers on the number line.

EXAMPLE 3   Use calculator values to find the approximate positions of the following square roots on the number line (Fig. 8–3).

(a) $2^{1/2}$ (b) $3^{1/2}$ (c) $5^{1/2}$ (d) $6^{1/2}$

(a) $2^{1/2} \approx 1.414213562$ (b) $3^{1/2} \approx 1.732050808$

(c) $5^{1/2} \approx 2.236067977$ (d) $6^{1/2} \approx 2.449489743$

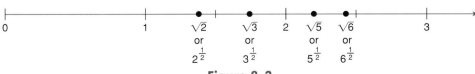

Figure 8–3

TABLE 8–2

Number	Square Root of Number
1	1
2	Between 1 and 2
3	Between 1 and 2
4	2
5	Between 2 and 3
6	Between 2 and 3
7	Between 2 and 3
8	Between 2 and 3
9	3
10	Between 3 and 4

3 Estimate the Value of Irrational Numbers.

To develop a number sense with irrational numbers, practice estimating irrational numbers. Again, we use the property that roots (irrational numbers) are ordered and have a position on the number line similar to the ordering of rational numbers. Looking at the results of the calculations in Example 3, we see that 2 and 3 are between the whole numbers 1 and 4. The square root of 1 is 1 and the square root of 4 is 2. The square roots of 2 and 3 are between the square roots of 1 and 4, or between 1 and 2. Similarly, 5 and 6 are between 4 and 9. The square root of 4 is 2 and the square root of 9 is 3. Thus, the square roots of 5 and 6 should be between 2 and 3 (see Table 8–2).

EXAMPLE 4

Determine the two whole numbers between which the following square roots will fall.

(a) $11^{1/2}$ (b) $15^{1/2}$ (c) $75^{1/2}$ (d) $120^{1/2}$

Make a list of perfect squares that will include the examples.

1, 4, 9, 16, 25, 36, 49, 64, 81, 100, 121

(a) $11^{1/2}$ 11 is between 9 and 16; $11^{1/2}$ is between 3 and 4.

(b) $15^{1/2}$ 15 is between 9 and 16; $15^{1/2}$ is between 3 and 4.

(c) $75^{1/2}$ 75 is between 64 and 81; $75^{1/2}$ is between 8 and 9.

(d) $120^{1/2}$ 120 is between 100 and 121; $120^{1/2}$ is between 10 and 11.

Now we will summarize the different types of numbers that we have used as exponents so far.

Tips and Traps   *Types of Numbers Used as Exponents:*

Exponent number type	Indication
Natural number exponents greater than 1	Repeated multiplication
$4^3 = 4 \times 4 \times 4 = 64$	
Exponent of 1	Expression equals the base
$7^1 = 7$	
Exponent of 0	Expression equals 1
$5^0 = 1$	
Negative integral exponents	Reciprocals
$3^{-2} = \left(\dfrac{1}{3}\right)^2 = \dfrac{1}{9}$	
Fractional exponents	Roots
$9^{1/2} = \sqrt{9} = 3$	

---

4 Simplify Numerical Expressions Using the Order of Operations.

When more than one operation is included in a problem, we must perform the operations in a specified order. We will briefly review the order of operations. For more thorough coverage, refer to Section 2–6.

> **Rule 8-3** *Order of operations for signed numbers:*
>
> Perform operations in the following order as they appear from left to right:
>
> 1. Parentheses used as groupings and other grouping symbols
> 2. Exponents (powers and roots)
> 3. Multiplications and divisions
> 4. Additions and subtractions

EXAMPLE 5 Evaluate the following expressions.

(a) $3 + 2(5 - 8)$ (b) $7 - (4 - 9)$ (c) $4(-5) - \dfrac{8}{-2} + 2(5 + 2)$

(d) $4 + 3(5 - 7)^2 - 6$

(a) $3 + 2(5 - 8) =$ Perform operation inside parentheses.
    $3 + 2(-3) =$ Multiply numbers with unlike signs.
      $3 + -6 = \mathbf{-3}$ Add numbers with unlike signs.

(b) $7 - (4 - 9) =$ Perform operation inside parentheses.
    $7 - (-5) =$ Change to equivalent addition.
    $7 + 5 = \mathbf{12}$ Add numbers with like signs.

(c) $4(-5) - \dfrac{8}{-2} + 2(5 + 2) =$ Perform operation inside parentheses.

   $4(-5) - \dfrac{8}{-2} + 2(7) =$ Perform multiplications and divisions as they appear from left to right.

   $-20 - (-4) + 14 =$ Change to equivalent addition.
     $-20 + 4 + 14 = \mathbf{-2}$ Perform additions and subtractions as they appear from left to right.

(d) $4 + 3(5 - 7)^2 - 6 =$ Perform operations inside parentheses.
   $4 + 3(-2)^2 - 6 =$ Raise to power.
    $4 + 3(4) - 6 =$ Multiply.
     $4 + 12 - 6 =$ Add from left to right.
      $16 - 6 = \mathbf{10}$

SELF-STUDY EXERCISES 8-2

[1] Write the roots using fractional exponents.

1. Square root of 36
2. Square root of 81
3. Cube root of 64
4. Fourth root of 81
5. Seventh root of 48
6. Fifth root of 32

Use a calculator to find the roots.

7. $196^{1/2}$
8. $529^{1/2}$
9. $216^{1/2}$
10. $2197^{1/3}$
11. $16,807^{1/5}$

[2] Use a calculator to find the approximate value of the roots and locate each one on a number line.

12. $7^{1/2}$
13. $8^{1/2}$
14. $5^{1/3}$
15. $12^{1/2}$

[3] Determine the two whole numbers between which each square root lies.

16. $18^{1/3}$
17. $7^{1/2}$
18. $48^{1/2}$
19. $152^{1/2}$

Write each number in a different form based on your understanding of exponents.

20. 18^5
21. 7^4
22. 183^1
23. 14^0
24. 196^0
25. 8^{-2}
26. 7^{-2}
27. 15^{-1}
28. $27^{1/3}$
29. $49^{1/2}$

4 Evaluate the following expressions.

30. $5 + 7(8 - 12)$

31. $14 - (2 - 7)$

32. $15((-2) - \left(\dfrac{16}{-4}\right) + 5(3 + 1)$

33. $5 + 7(2 - 9)^2 - 13$

34. $8 - 7(3 - 8)^{-3} - 14$

8-3 ■ VARIABLE NOTATION

Learning Objectives

1 Identify equations, terms, factors, constants, variables, and coefficients.

2 Write verbal interpretations of symbolic statements.

3 Translate verbal statements into symbolic statements using variables.

4 Simplify variable expressions.

An important part of symbolic representation is the ability to represent an unknown value. Such notation will allow us to develop rules of operations with these symbols and thus find the unknown value.

1 Identify Equations, Terms, Factors, Constants, Variables, and Coefficients.

Before we begin to work with equations to solve problems, we need to understand some of the basic concepts and terminology of equations. In mathematics we often use the symbol "=" to show that quantities are "equal to" each other. We can write the statement "five is equal to two plus three" in symbols as $5 = 2 + 3$. This symbolic statement is called an *equation*. "Symbolic" means we are using symbols like "=" and "+."

■ **DEFINITION 8–1: Equation.** An *equation* is a symbolic statement that two expressions or quantities are equal in value. This statement may be true or false.

EXAMPLE 1 Verify that the following statements are true equations.

(a) $3(8) = 24$ (b) $12 - 3 = 9$

To verify that these statements are true equations, we must *evaluate* or find the value of the expression, or quantity, on each side of the equals sign. To be a true equation, the value of the left side of the equation must equal the value of the right side of the equation.

(a) $3(8) = 24$ 3 times 8 is 24.
       $\mathbf{24 = 24}$ The equation is true.

(b) $12 - 3 = 9$ 12 minus 3 is 9.
       $\mathbf{9 = 9}$ The equation is true.

Some types of equations contain a *missing* number or *unknown* number. We *solve* an equation by finding the missing number that makes the equation *true*. The missing number is represented in the equation by a letter such as x, a, or z. In each equation, the letter has a certain (unknown) value. The same letter may be used in several equations, and its value in each equation depends on the other numbers and relationships in the equation. Because the value of the letter varies with each equation, the letter is often called a *variable*. Because the value of the letter is unknown until the equation is solved, the letter is also called the *unknown* in the equation. The *solution*, also called the *root*, of an equation is the value of the variable that makes the equation true.

■ **DEFINITION 8–2:** **Variable** or **Unknown.** A letter that represents an unknown or missing number is called a *variable* or *unknown*.

■ **DEFINITION 8–3:** **Root** or **Solution.** The *root* or *solution* of an equation is the value of the variable that makes the equation true.

EXAMPLE 2 Solve the following equations.

(a) $x = 3 + 8$ (b) $\dfrac{18}{3} = n$ (c) $y = 3(4) - 5$

Because the value of one side of an equation must equal the value of the other side of the equation, we solve the equation, that is, find the missing number or variable by evaluating the opposite side of the equation.

(a) $x = 3 + 8$ 8 added to 3 is 11.
   $\mathbf{x = 11}$

(b) $\dfrac{18}{3} = n$ 18 divided by 3 is 6.
   $\mathbf{6 = n}$

(c) $y = 3(4) - 5$ Remember, multiplication is worked first in the order of operations: 3 times 4 is 12.
   $y = 12 - 5$ Add 12 and −5.
   $\mathbf{y = 7}$ 12 plus − 5 = 7.

Because both sides of an equation are equivalent, it does not matter which side comes first in the equation. In these examples, for instance, $x = 3 + 8$ means the same as $3 + 8 = x$; $\frac{18}{3} = n$ means the same as $n = \frac{18}{3}$; and $y = 3(4) - 5$ means the same as $3(4) - 5 = y$. In other words, $n = 5$ is equivalent to $5 = n$, and so on. The unknown, or variable, may appear on either the left or right side of an equation.

Rule 8–4 *Symmetric property of equality:*

If the sides of an equation are interchanged, equality is maintained.

If $a = b$, then $b = a$.

Numbers multiplied together are called *factors*. Because division is the inverse operation of multiplication, it can be rewritten as an equivalent product. $\frac{3}{y}$ can be rewritten as $3 \cdot \frac{1}{y}$. Thus, division can be expressed as multiplication of factors. An algebraic expression, or quantity, can be written showing a known factor multiplied times a variable, or unknown factor, such as $5x$. The number 5 is the known factor multiplied times some unknown number that we represent with the letter x.

If an algebraic expression contains only factors, then we refer to this expression as a *term*.

$2xy$ is a term containing the factors 2, x, and y.

$\dfrac{2x}{y}$ is a term containing the factors 2, x, and $\dfrac{1}{y}$.

The expression $2(x + 3)$ is a term containing the factors 2 and $(x + 3)$. Because $(x + 3)$ is grouped, it is one quantity and is considered as a factor. However, if the distributive principle is applied, and each term within parentheses is multiplied by 2, then $2(x + 3)$ becomes $2x + 6$. In this expression there are *two* terms, $2x$ and 6.

In algebra it is important to be able to distinguish between terms and factors.

■ **DEFINITION 8–4: Factors.** Algebraic expressions that are multiplied are called *factors*.

■ **DEFINITION 8–5: Terms.** Algebraic expressions that are single quantities or quantities that are added or subtracted are called *terms*.

A term can be a single letter; a single number; the product of numbers and letters; the product of numbers, letters, and/or groupings; or the quotient of numbers, letters, and/or groupings. The fraction line implies that the numerator and/or the denominator is grouped.

EXAMPLE 3 Identify the terms in each expression by drawing a box around each term.

(a) $a - 2$ (b) $3x + 5$ (c) $2x - 4(x + 7)$

(d) $2ab + 4a - b + 2(a + b)$ (e) $\dfrac{x}{2}$ (f) $\dfrac{2a + 1}{3}$ (g) $\dfrac{2a}{3} + 1$

(a) $\boxed{a} - \boxed{2}$ Terms are separated by "+" and "−" signs that are not within a grouping.

(b) $\boxed{3x} + \boxed{5}$

(c) $\boxed{2x} - \boxed{4(x + 7)}$

(d) $\boxed{2ab} + \boxed{4a} - \boxed{b} + \boxed{2(a + b)}$

(e) $\boxed{\dfrac{x}{2}}$

(f) $\boxed{\dfrac{2a + 1}{3}}$ The numerator or denominator of a fraction is a grouping, such as $2a + 1$.

(g) $\boxed{\dfrac{2a}{3}} + \boxed{1}$

Terms that contain only numbers are called *number terms* or *constants*, and terms that contain only letters or that contain both numbers and letters are called *letter terms* or *variable terms*.

■ **DEFINITION 8–6: Number Terms** or **Constants.** A term that contains only numbers is called a *number term* or *constant*.

■ **DEFINITION 8–7: Letter Term** or **Variable Term.** A term that contains only one letter, several letters used as factors, or a combination of letters and numbers used as factors is called a *letter term* or *variable term*.

If a term contains more than one factor, each factor is the *coefficient* of the other factors. Any factor(s) in a term containing more than one factor may be grouped as the coefficient of the remaining factor(s). In the term $2ab$:

- 2 is the *coefficient* of ab.
- a is the *coefficient* of $2b$.
- b is the *coefficient* of $2a$.
- $2b$ is the *coefficient* of a.
- $2a$ is the *coefficient* of b.
- ab is the *coefficient* of 2.

■ **DEFINITION 8–8: Coefficient.** Each factor of a term containing more than one factor is the *coefficient* of the remaining factor(s).

In algebraic expressions, we are usually interested in the *numerical coefficient* of a term. In the term $2ab$, 2 is the *numerical coefficient* of ab.

Recall that mathematics makes use of signs, both positive and negative. If a term is positive, such as $2a$, there is no need to write the "+" sign before the term. But if the term is negative, such as $-5b$, then the "−" sign *must* be expressed. In the term $2a$, the numerical coefficient is 2, which is positive. In the term $-5b$, the numerical coefficient is negative, (-5).

If the term is of the type $-\frac{a}{3}$ or $\frac{2b}{7}$, the term is the product of a fractional coefficient and a letter; that is, the term $-\frac{a}{3} = -\frac{1}{3}\left(\frac{a}{1}\right)$ or $-\frac{1}{3}a$ and the term $\frac{2b}{7} = \frac{2}{7}\left(\frac{b}{1}\right)$ or $\frac{2}{7}b$. The *numerical coefficient* of a is $-\frac{1}{3}$. The *numerical coefficient* of b is $\frac{2}{7}$. Similarly, in the term $\frac{4n}{5}$ the *numerical coefficient* of n is $\frac{4}{5}$.

■ **DEFINITION 8–9: Numerical Coefficient.** The numerical factor of a term is the *numerical coefficient*. Unless otherwise specified, the word *coefficient* will be used to mean the *numerical coefficient*. The numerical factor should be written *in front* of the variable.

EXAMPLE 4 Identify the numerical coefficient in each term.

(a) $2x$ (b) $-3ab$ (c) $4(x + 3)$ (d) $-\dfrac{n}{3}$ (e) $\dfrac{2b}{5}$

(a) The coefficient of x is **2.**
(b) The coefficient of ab is **−3.**
(c) The coefficient of $(x + 3)$ is **4.**
(d) The term $-\dfrac{n}{3}$ is the same as $-\dfrac{1}{3}n$. Therefore, the coefficient of n is $-\dfrac{1}{3}$.

(e) $\dfrac{2b}{5}$ is the same as $\dfrac{2}{5}\left(\dfrac{b}{1}\right)$ or $\dfrac{2}{5}b$. Thus, the coefficient of b is $\dfrac{2}{5}$.

When a letter term has no written numerical coefficient, the coefficient is understood to be 1; that is, the coefficient of x is 1: $x = 1x$. Similarly, the coefficient of ab is 1: $ab = 1ab$. Likewise, the numerical coefficient of $-x$ is -1: $-x = -1x$. The coefficient of $-bc$ is -1: $-bc = -1bc$.

2 Write Verbal Interpretations of Symbolic Statements.

In early studies of mathematics, unknown values were represented by using boxes, circles, underlining, or other symbols. For example, a problem might be to find the missing value that will make the following statement true: 5 + ____ = 8. The underline identifies the position of the missing value and the other numbers and symbols give the conditions of the problem.

We commonly use letters of the alphabet to show the position of the missing value. For instance, the same problem that was stated before could be written as $5 + x = 8$ or $5 + y = 8$ or $5 + a = 8$. The choice of letters is not significant. The position of the letter and the other conditions of the statement are important. This letter is called a *variable* or *unknown*. The interpretation of the mathematical statement can be phrased in several ways. $5 + x = 8$ can be interpreted as "What number added to 5 will give 8?" It can also be interpreted as "5 is added to a number and the result is 8. What is the missing number?"

Another convention in using variable notation is that multiplication symbols can be omitted when a number and a variable are multiplied. For instance, $6x$ means 6 times the value of x. If there is more than one missing number, different variables or notations are used to represent each different number. To write a symbolic statement to represent the problem that two different numbers are multiplied, we can use two different letters such as xy or the same letter with subscripts, such as x_1x_2. In both cases the two numbers are multiplied. For clarity, the raised dot or parentheses may be used to indicate multiplication, even though they are not always necessary. We avoid the use of the times sign ($\times$) so that we do not confuse it with a missing value labeled x. A raised dot or parentheses can be used with subscripted variables. For example, $x_1 \cdot x_2$ or $(x_1)(x_2)$ means that two variables are multiplied.

EXAMPLE 5   Write a verbal interpretation of the following statements.

(a) $x - 7 = 4$ (b) $\dfrac{x}{5} = 3$ (c) $2x + 3 = 15$ (d) $2(x + 3) = 14$

Various phrases can describe the conditions of each statement.

(a) $x - 7 = 4$.   **When 7 is subtracted from a number, the answer is 4.**

(b) $\dfrac{x}{5} = 3$.   **A number divided by 5 is 3.**

(c) $2x + 3 = 15$.   **Fifteen is the result when 3 is added to 2 times a number.**

(d) $2(x + 3) = 14$.   **If the sum of a number and 3 is doubled, the result is 14.**

3 Translate Verbal Statements into Symbolic Statements Using Variables.

As we stated earlier, symbolic representation of verbal statements or real-life situations will allow us to apply a systematic process for manipulating these symbols to determine the value of missing amounts. We are making a translation similar to translating from one language to another. To make these symbolic translations, we need to watch for key words or phrases that imply or suggest specific operations. The following are some key words or phrases that distinguish between addition and subtraction.

Addition:	the sum of, plus, increased by, more than, added to, exceeds, longer, total, heavier, older, wider, taller, gain, greater than, more, expands
Subtraction:	less than, decreased by, subtracted from, the difference between, diminished by, take away, reduced by, less, minus, shrinks, younger, lower, shorter, narrower, slower, loss

EXAMPLE 6   Translate the following statements into symbols.

(a) The sum of 12, 23, and a third number is 52.

(b) When a number is subtracted from 45, the result is 17.

(a) The sum of 12, 23, and a third number is 52: $12 + 23 + x = 52.$

(b) When a number is subtracted from 45, the result is 17: $45 - x = 17.$

Words or phrases that imply multiplication and division are:

Multiplication:	times, multiply, of, the product of, multiplied by
Division:	divide, divided by, divided into, how big is each part, how many parts can be made from

Some words that imply multiplication or division may indicate a specific number in the multiplication or division. Some examples are twice (2 times), double (2 times), triple (3 times), and half of (1/2 times or divided by 2).

EXAMPLE 7   Translate the following statements into symbols.

(a) If a value is tripled, the result is 63.

(b) How many shelves that are 3 feet long can be made from a board that is 12 feet long?

(a) If a value is tripled, the result is 63: $3x = 63.$

(b) How many shelves that are 3 feet long can be made from a board that is 12 feet long? $\frac{12}{3} = x.$

EXAMPLE 8   Explain the difference between the following two symbolic statements.

Four times the difference between a number and 8 is 24.
The difference between four times a number and 8 is 24.

Four times the difference between a number and 8 is 24. This statement can be written symbolically as $4(x - 8) = 24$. It means that the difference is taken first and then the result is multiplied by 4.

The difference between four times a number and 8 is 24. This statement can be written symbolically as $4x - 8 = 24$. A number is multiplied by 4 and then 8 is subtracted from the result.

4 Simplify Variable Expressions.

We saw in adding and subtracting fractions that we added or subtracted fractions with like or common denominators. Similarly, in adding and subtracting measures we needed like units of measure. With variable or letter terms, we can only add or subtract like terms.

■ **DEFINITION 8–10: Like Terms.** Terms are *like terms* if they are number terms or if they are letter terms with exactly the same letter factors.

The terms $4y$ and $2y$ are like terms because both contain the same letter (y). Similarly, 3 and 1 are like terms because both are numbers. We *cannot* combine 3 and $4y$ because they are unlike terms (number term and letter term). We *cannot* combine $4y$ and $2x$ because they are unlike terms (different letters).

Rule 8–5 *To combine numerical terms:*

1. Change subtraction to addition if appropriate.
2. Add the numerical terms using the appropriate rule for adding signed numbers.
3. The sum will be a number term.

Rule 8–6 *To combine like letter terms:*

1. Change subtraction to addition if appropriate.
2. Add the numerical coefficients of the like letter terms using the appropriate rule for adding signed numbers.
3. The sum will have the same letter factors as the like terms being added.

EXAMPLE 9 Simplify the following expressions by combining like terms.

(a) $5a + 2a - a$ (b) $3x + 5y + 8 - 2x + y - 12$

(a) $5a + 2a - a = 6a$ Add coefficients: $5 + 2 - 1 = 6$.

(b) $3x + 5y + 8 - 2x + y - 12$ Add like terms.
$= x + 6y - 4$
$3x - 2x = x$
$5y + y = 6y$
$8 - 12 = -4$

The *distributive principle* can be extended to include multiplying number terms and letter terms. For now, we will only examine situations in which we are multiplying number and letter terms by number terms.

EXAMPLE 10 Simplify by applying the distributive principle.

(a) $3(5x - 2)$ (b) $-7(x + y - 3z)$ (c) $-(-3x + 4)$

(a) $3(5x - 2) = 3(5x) - 3(2)$ Apply the distributive
$= 15x - 6$ principle.

(b) $-7(x + y - 3z) = -7(x) - 7(y) - 7(-3z)$ Apply the distributive
$= -7x - 7y + 21z$ principle.

(c) $-(-3x + 4) = -1(-3x) - 1(+4)$ Apply the distributive
$= 3x - 4$ principle.

SELF-STUDY EXERCISES 8-3

1 Verify that the following statements are true.

1. $5(-3) = -15$ **2.** $17 - 6 = 11$ **3.** $12 - 5(2) = -7 + 9$

4. $8 - 3(5) = 2(-1 - 3) + 1$

Solve the following equations.

5. $n = 4 + 7$ **6.** $m = 8 - 2$ **7.** $5 - 9 = y$

8. $\dfrac{12}{2} = x$ **9.** $p = 2(5) - 1$ **10.** $b = 6 - 3(5)$

Identify the terms in each expression by drawing a box around each term.

11. $7 + c$ **12.** $4a - 7$ **13.** $3x - 2(x + 3)$

14. $\dfrac{a}{3}$ **15.** $7xy + 3x - 4 + 2(x + y)$ **16.** $14x + 3$

17. $\dfrac{7}{(a + 5)}$ **18.** $\dfrac{4x}{7} + 5$

19. Write an algebraic expression that contains 3 terms.

Identify the numerical coefficient of each term.

20. $5x$ **21.** $-4xy$ **22.** $\dfrac{n}{5}$ **23.** $\dfrac{2a}{7}$

24. $6(x + y)$ **25.** $\dfrac{-4}{5x}$

26. Write an expression that has one term and a numerical coefficient of -15.

2   Write a verbal interpretation of the following symbolic statements.

27. $x + 4 = 7$ **28.** $x - 5 = 2$ **29.** $3x = 15$ **30.** $3x + 1 = 7$

3   Write a symbolic interpretation for the following statements.

31. Five more than a number is 12.   **32.** A number divided by 6 is 9.

33. Four times the difference of a number and 3 is 12.   **34.** Three less than four times a number is 12.

35. The sum of 12, 7, and a third number is 17.   **36.** Seven more than twice a number is 21.

37. If 15 degrees is added to a temperature, it would be 48 degrees. What is the temperature?   **38.** If 15 milliliters of water are added to a medicine, there are 45 milliliters in all. What is the volume of the medicine?

39. How many 5-foot pieces of I-beam can be cut from a piece that is 45 feet long?   **40.** A piece of oak flooring that is 18 feet long has an unacceptable flaw that requires 3 feet to be trimmed from one end. How many 5-foot boards can be cut from the remaining length of flooring?

4   Simplify by combining like terms.

41. $3a - 7a + a$ **42.** $-8x + y - 3y$ **43.** $5x - 3y + 2x + y$

44. $-4a + b + 9 + a - b - 3$ **45.** $2x + 8 - x + 4 - x - 2$ **46.** $3a + 5b + 8c + 1 + b$

Simplify by applying the distributive principle and combining like terms if appropriate.

47. $5(2x - 4)$ **48.** $4 + 3(2x + 3)$ **49.** $3 - (4x - 2)$ **50.** $5 - 2(6a + 1)$

8-4   **FUNCTION NOTATION**

Learning Objectives   **1**   Examine number patterns by using function notation.

                     2   Investigate the effects of changing input values in function notation.

                     3   Graph independent and dependent values of functions.

In the study of mathematics, you will often gain valuable insights into basic principles by examining patterns. One of the most useful notations to facilitate this examination is function notation.

1 Examine Number Patterns by Using Function Notation.

Suppose you wish to rent a car and the salesperson quotes the rate of $25 per day plus $4 for each 100 miles (or any part of 100) that the car is driven. You would like to calculate the total rental charge for several different mileage totals. You can best examine the pattern developed for changing mileage by using function notation. In *function notation* we have two types of variables, an *independent variable* and a *dependent variable*. As the words imply, one variable is dependent on the other. In our car rental example, the total cost of the rental (dependent variable) is dependent on the number of miles the car is driven (independent variable). We can let different letters represent the two variables, such as c for the total cost of the rental and m for the miles driven. Using these variables, a function relating these two variables would be $c = \$25 + \$4m$ or $c = 25 + 4m$. To reveal patterns that develop from different values of the independent variable, it is common to arrange the information in tables and to graph the results (see Table 8–3).

Table 8–3

Car Rental Costs

Exact Miles	Hundred Miles, m	Calculations, $c = 25 + 4m$	Cost, c
278	3	$c = 25 + 4(3)$ $c = 25 + 12$ $c = 37$	$37
346	4	$c = 25 + 4(4)$ $c = 25 + 16$ $c = 41$	$41
412	5	$c = 25 + 4(5)$ $c = 25 + 20$ $c = 45$	$45

Another common function notation expresses the independent and dependent variables in the following way: $f(x) = 25 + 4x$, or $C(m) = 25 + 4m$. In the first representation, $f(x) = 25 + 4x$, a standard notation called the "function of x" is used; x represents the independent variable and $f(x)$ represents the dependent variable. We can think of the values of the independent variable as the *input values*. Calculations are performed. Then the values of the dependent variable are the results of the calculations, or the *output values*.

In the second representation, $C(m) = 25 + 4m$, the letters chosen serve as reminders for the interpretation of the variables. $C(m)$ is the dependent variable and represents the cost of the car rental. The independent variable m represents the number of miles (in hundreds or parts of hundreds) that the car is driven. Table 8–4 shows how these alternative notations are used.

EXAMPLE 1   Examine Table 8–3 or 8–4 and answer the following questions.

(a) Without making the calculations, what would you expect the cost of the car rental to be for 194 miles?

(b) Would the rental for 304 miles and 400 miles be the same? Explain.

(a) The rate would be calculated for 200 miles and would be $4 less than the rate for 300 miles, or $8 more than the basic rate. The rate would be $37 − $4 = $33, or $25 + $8 = $33.

(b) Both rates would be calculated at 400 miles. The 4 miles over 300 miles would count as a portion of the next 100 miles.

TABLE 8-4

| | | Car Rental Costs | |
Exact Mile	Hundred Miles, x	Calculations, $f(x) = 25 + 4x$	Cost, $f(x)$
278	3	$f(3) = 25 + 4(3)$ $f(3) = 25 + 12$ $f(3) = 37$	$37
346	4	$f(4) = 25 + 4(4)$ $f(4) = 25 + 16$ $f(4) = 41$	$41
412	5	$f(5) = 25 + 4(5)$ $f(5) = 25 + 20$ $f(5) = 45$	$45

EXAMPLE 2   Evaluate the function $f(x) = x - 1$ for the following values.

(a) $f(4)$ (b) $f(8)$ (c) f(0)   (d) $f(-1)$

(a) $f(4) = 4 - 1$
$\quad f(4) = \mathbf{3}$

(b) $f(8) = 8 - 1$
$\quad f(8) = \mathbf{7}$

(c) $f(0) = 0 - 1$ 1 less than or to the left of 0 is −1.
$\quad f(0) = \mathbf{-1}$

(d) $f(-1) = -1 - 1$ 1 less than or to the left of −1 is −2.
$\quad f(-1) = \mathbf{-2}$

2 Investigate the Effects of Changing Input Values in Function Notation.

A function shows a relationship between independent and dependent variables. As values change for the input or independent variable, values of the output or dependent variable will change accordingly.

Depending on the context of the problem, there may be some values that cannot be considered. For example, in the car rental problem in Example 1, an input value of 0 would mean that the car was driven no miles. Even if the car was driven 1 mile, the customer would be charged for 100 miles. Is this situation likely to occur? Also, can a car be driven a negative number of miles? The only appropriate values of the independent variable would be positive values. With the way the rental rate is defined, only natural numbers, which represent the number of hundreds of miles, or fraction thereof, that the car was driven, are used. Therefore, we can say that, in this situation, the values of the independent variable are restricted to natural numbers. Is there a maximum number that is appropriate? Can the car be driven 2000 miles in one day? Probably not. Thus, even though we may not know the exact maximum, we know that there is a reasonable amount that can be considered as the maximum. These suitable values for the independent variable make up a set of values called the *domain* of the function. As you can see, the values that make up the domain are considered within the context of the situation.

Once the values of the domain (input) have been determined, then the possible values of the dependent variable (output) can be examined. These output values make up a set of numbers called the *range*. In the car rental example, the range would include numbers like $29, $33, $37, $41, $45, $49, and so on. The upper limit of the range would be determined by a reasonable maximum number of miles the car could be driven in one day.

The concept of functions will be expanded in later chapters, but for now we want to look for patterns that develop in the output values (range) as input values (domain) are changed.

EXAMPLE 3   Anticipate the characteristics of the output values for various input values in the following.

(a) $f(x) = 3x$ for whole number values in the domain.

(b) $f(x) = x - 5$ for any integer.

(a) The output values (range) will be whole numbers. Also, the output values will be multiples of 3.

(b) The output values (range) will be integers. Also, the output value will be a natural number if the input value for x is an integer greater than 5, the output value will be 0 if the input value is exactly 5, and the output value will be a negative integer if the input value is an integer less than 5.

3 Graph Independent and Dependent Values of Functions.

A visual representation can be made of the data found in evaluating functions. Let's take the data from the car rental problem of Example 1 and graph the findings. Consider each pair of input and output values to be a point on a rectangular coordinate system.

EXAMPLE 4   Graph the data from Example 1 on a rectangular coordinate system.

The values of the independent variable (miles driven per day) will be represented on the horizontal or x-axis. The corresponding values of the dependent variable (rental cost per day) will be represented on the vertical or y-axis (Fig. 8–4).

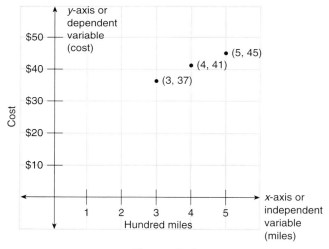

Figure 8–4

When the number of miles driven per day *increases*, the rental cost per day also *increases*. Such a function is called an *increasing function*.

SELF-STUDY EXERCISES 8–4

1   Evaluate the function for the indicated value.

1. $f(x) = x + 5; f(2), f(-3), f(0)$

2. $f(x) = 3x - 7; f(-4), f(5), f\left(\dfrac{1}{6}\right)$

3. $f(x) = 4x - \dfrac{3}{5}; f(-0.2), f\left(\dfrac{3}{4}\right), f\left(-\dfrac{3}{5}\right)$

2 Describe the characteristics of the output values for various input values.

4. $f(x) = 5x$ for whole numbers in the domain

5. $f(x) = 2x - 3$ for any integer in the domain

6. $f(x) = x^{-1/2}$ for any integer that is a perfect square

7. $f(x) = x + 8$ for any negative integer

3 Evaluate the functions and graph the data on a rectangular coordinate system.

8. $f(x) = x$; find $f(-2), f(-1), f(0), f(1), f(2)$

9. $f(x) = 2x$; find $f(-2), f(-1), f(0), f(1), f(2)$

10. Ace Rental Company charges $20 plus $12 per day (or any part of a day) to rent a wheelchair. Identify the dependent variable and the independent variable. Make a table to show the pattern that develops from using various values of the independent variable. Write a function that represents the cost of renting the wheelchair for any number of days. Draw a graph that represents the function.

CAREER APPLICATION

Horticulture: Leaf Chlorophyll Production

Green plants use carbon dioxide and water in the presence of sunlight to make carbohydrate energy in a process called *photosynthesis.* Plants give off oxygen as a by-product of photosynthesis. Humans and other living organisms use oxygen and water in a process called *respiration*, which gives off carbon dioxide as a by-product. Horticulturists and other scientists are interested in maximizing chlorophyll production to give plants a lush green appearance, and to provide humans with plenty of oxygen to breathe. Chlorophyll is the green pigment produced in leaves, which receive the sunlight needed for photosynthesis. The more sunlight received → the more chlorophyll cells produced → the greater the photosynthesis rate → the greener the leaves and the more oxygen produced.

Botanists found that during the months of June through September, the maximum possible number of chlorophyll cells in a small elm leaf is about 500, which is added to 100 times the number of months of direct sunlight received so far that year. The number of hours of sunlight are greatest, and the sun's rays are most direct, in the months of June, July, August, and September. For example, at the end of June, a small elm leaf has received 1 month of direct sunlight. One hundred times 1 is 100, and 500 added to 100 equals 600. So the maximum number of chlorophyll cells in a small elm leaf at the end of June is about 600 cells.

Exercises

Use the information above to answer the following.

1. Horticulturists who know algebra can write the number of months of direct sunlight received so far in a year using symbolic representation. Choose a variable to represent the months of sunlight. (*Remember*: You do not have to choose x or y; sometimes a different letter related to the meaning of the variable is better.) Define your variable in words.

2. Find the value of the input variable you defined in Exercise 1 for these dates:
   (a) July 31 (b) September 30 (c) September 15 (d) June 1

3. Based on what you know so far, guess which dates will have the maximum chlorophyll cell production.

4. Now pick a different variable to represent the output variable or the maximum possible number of chlorophyll cells in a small elm leaf. Choose a letter appropriate to the meaning of the variable. Define your variable in words.

5. Use the values of the input variable found in Exercise 2 and the botanist's prediction to find the value of the output variable you defined in Exercise 4 on these dates:
   (a) July 31   (b) September 30   (c) September 15   (d) June 1

6. Use your answers from Exercise 5 to predict the date of maximum chlorophyll production in a small elm leaf. If this is different from your initial guess in Exercise 3, think of a reason why. What happens after this date to make chlorophyll-cell production decrease?

7. Use the variables you chose in Exercises 1 and 4 to write an equation showing how the number of chlorophyll cells is found when you know the number of months of direct sunlight received.

8. How should the equation in Exercise 7 be altered to predict chlorophyll production in an elm leaf three times the size of the small one?

9. How should the equation in Exercise 7 be altered to predict chlorophyll production in an elm leaf half the size of the small one?

10. Explain in words how the equation in Exercise 7 should be altered to predict chlorophyll production in a variegated privet leaf the size of the small elm leaf. The variegated leaf has green and white stripes of equal size.

11. Why would a NASA scientist working on the development of a human colony on the moon be interested in predicting chlorophyll production?

12. Give at least two reasons why remote tropical rain forests grow such lush, thick vegetation when there are few humans there to breathe out carbon dioxide.

13. Should a person with asthma choose a yellow-leaf coleus plant, a green-leaf philodendron plant, or a variegated spider plant to put on his or her office desk? Why?

14. In the mid-1980s an experimental, self-contained, sealed glass habitat named Biosphere I was built in the Arizona desert. For four years it was to be home to seven adult humans and several types of animals used for food, milk, etc. The project ended early because of several unforeseen problems, one of which was lack of oxygen. Give at least two possible reasons for the lack of oxygen.

Answers

1. Answers vary, but m = months of direct sunlight is a good choice.
2. (a) $m = 2$ (b) $m = 4$ (c) $m = 3.5$ (d) $m = 0$
3. Answers vary, but July or August are the usual answers.
4. Answers vary, but C = the number of chlorophyll cells in a small elm leaf is a good choice.
5. (a) $C = 700$ (b) $C = 900$ (c) $C = 850$ (d) $C = 500$
6. September 30 appears to be the date of maximum chlorophyll production in a small elm leaf, perhaps because the ground is warmer in September than in July or August, while the air is slightly cooler in September. After this date, it gets much cooler, the sun's rays are not as direct, and elm leaves turn yellow and fall off the tree.
7. $C = 500 + 100m$
8. $C = 3(500 + 100m) = 1500 + 300m$ because the elm leaf size is tripled.
9. $C = \frac{1}{2}(500 + 100m) = 250 + 50m$ because the elm leaf size is halved.
10. $C = 250 + 50m$ because the variegated leaf has only half the green chlorophyll pigment of a solid green leaf.
11. Our moon has no natural oxygen, carbon dioxide, living organisms, or green plants. A NASA moon colony could produce all these items by in-

corporating humans and plants in a careful balance of photosynthesis and respiration.

12. Answers may include the following: most tropical rain forests have numerous living organisms other than humans that respire; tropical rain forests are near the equator and receive the most direct sunlight as well as plenty of sunlight year round; rain forests have an abundant supply of water necessary for photosynthesis.

13. A person with asthma should choose the green-leaf philodendron plant because, of the three, it has the greatest amount of chlorophyll cells. (Before modern asthma treatment, people with asthma would plant a balsam tree forest in their yards because this tree gives off tremendous amounts of oxygen.)

14. Answers include the following: Biosphere I could have had too many living organisms competing for the oxygen; there were too few green plants to give off oxygen; temperatures were either too cold or too hot for ideal plant growth; or too few nutrients were available for plant development.

ASSIGNMENT EXERCISES

Section 8–1

Write the reciprocal of the following. Express your answer with a positive exponent and write it as an ordinary number.

1. 2^{-3} **2.** 5^2 **3.** $\left(\dfrac{1}{3}\right)^2$ **4.** $\left(\dfrac{1}{4}\right)^{-3}$

Write equivalent expressions using positive exponents, then write them as ordinary numbers. Verify your answers with a calculator.

5. 4^{-3} **6.** 5^{-4} **7.** 2^{-6} **8.** 10^{-7}

Write equivalent expressions with positive exponents, then write equivalent expressions with 1 as the denominator. Verify your answers with a calculator.

9. $\dfrac{1}{7}$ **10.** $\dfrac{1}{81}$ **11.** $\dfrac{1}{121}$ **12.** $\dfrac{1}{125}$

Perform the indicated operations. Round to hundredths if necessary.

13. $12 - 5 + 9$ **14.** $(-3.2)(7.5)$ **15.** $\left(-\dfrac{5}{8}\right)\left(\dfrac{4}{9}\right)$ **16.** $\dfrac{-2.8}{-1.5}$

17. $\left(-\dfrac{3}{7}\right) \div \left(-\dfrac{9}{28}\right)$

18. Penna's Landscape Company purchased 60 water oaks for \$22 each. Forty-seven of the trees were sold for \$38 each and the remaining trees died before they could be sold. Use signed numbers to express the amount of money the company made (or lost) on the trees.

Section 8–2

Estimate the following roots by identifying the two whole numbers between which the root lies. Verify each estimate with a calculator.

19. $53^{1/2}$ **20.** $72^{1/2}$ **21.** $130^{1/2}$ **22.** $68^{1/2}$

23. $\sqrt{145}$ **24.** $\sqrt{210}$

Use a calculator to locate the following roots on a number line.

25. $11^{1/2}$ **26.** $12^{1/2}$ **27.** $13^{1/2}$ **28.** $14^{1/2}$

Evaluate the following expressions.

29. $8(2 - 7) + 3^2 - 49^{1/2}$

30. $-5 + (-2 + 8)^2 - 2$

Section 8–3

Identify the terms in the following expressions by drawing a box around each term.

31. $15x - \dfrac{3a}{7} + \dfrac{(x - 7)}{5}$

32. $5x - 8 + \dfrac{3}{y}$

Write a verbal interpretation for the following statements.

33. $x + 5 = 2$

34. $x - 7 = 11$

35. $\dfrac{x}{8} = 7$

36. $3x + 7 = -3$

37. $3(x + 7) = -3$

Write a symbolic statement for each of the following verbal statements.

38. A number decreased by 5 is 8. What is the number?

39. Seven more than twice a number is 11. Find the number.

40. A certain stock listed on the New York Stock Exchange closed at $42\frac{3}{8}$, a decrease of $3\frac{1}{8}$ points from the opening price. What is the opening price?

41. Twice the sum of a number and 8 is 40. What is the number?

42. The print shop used 31 cases of copy paper during one month. End-of-the-month inventory indicated 172 cases on hand. How many cases of paper were on hand at the beginning of the month? Write an equation containing addition to solve the problem. Then write an equation containing subtraction that will also solve the problem.

Section 8–4

Evaluate each of the following functions for the indicated values of the independent variable.

43. $f(x) = 2x - 1$. Find $f(-2)$; $f(-1)$; $f(0)$; $f(1)$; $f(2)$.

44. $f(x) = -2x + 5$. Find $f(-3)$; $f(-2)$; $f(-1)$; $f(0)$.

45. Describe the pattern formed by the independent and dependent variables in Exercise 43.

46. Describe the pattern formed by the independent and dependent variables in Exercise 44.

47. Use the number pairs formed in Exercise 43 to make a visual representation on a rectangular coordinate system.

CHALLENGE PROBLEMS

TABLE 8–5

Hours	Pay
5	$30.95
8	$49.52
12	$74.28
8	$49.52

48. Table 8–5 represents hours worked and earnings that Jason made for one week on his part-time job. What is the independent variable and what is the dependent variable? Write a function that represents the information about Jason's earnings. (You will need to determine his hourly pay.) Predict Jason's earnings for other hours worked. Graph the function.

49. Examine Table 8–6 to determine a pattern and write a function using the pattern. Then use the function to find at least two additional sets of data. Graph the data.

TABLE 8-6

Independent Variable	Dependent Variable
−1	−5
3	7
5	13
7	19

50. Write a function and use it to create a set of data. Swap data sets with a classmate. Examine the data from your classmate to determine the function represented by the data. Determine if your classmate wrote the correct function for your data.

CONCEPTS ANALYSIS

1. Describe the five numbers or number types used as exponents. Explain what each means and give an example to illustrate each explanation.

2. What is the basic requirement for two numbers to be reciprocals? Illustrate your response.

3. Describe reciprocals written with exponents.

4. Give an example of a number and its reciprocal written in exponential form. Demonstrate that they are reciprocals.

5. Identify the error, describe why it is an error, and rework the problem correctly.

$$8 - 3(2 + 7)^2 - 49^{1/2} =$$
$$5(2 + 7)^2 - 49^{1/2} =$$
$$5(9)^2 - 49^{1/2} =$$
$$5(81) - 49^{1/2} =$$
$$5(81) - 7 =$$
$$405 - 7 =$$
$$398$$

6. Describe like terms and explain how they are combined. Give an example and combine the terms.

7. Explain the difference between *factors* and *terms*. Give an example of each.

8. If you solved an equation and got $7 = x$, but wanted to write your root as $x = 7$, what mathematical property would allow you to do so?

9. Describe and give examples of the various forms of rational numbers.

10. Write a paragraph explaining how the natural numbers, integers, rational numbers, and real numbers are related. Draw a diagram that illustrates these numbers and give examples of each.

CHAPTER SUMMARY

Objectives	What to Remember with Examples

Section 8-1

1 Express powers with negative integer exponents in exponential form and in ordinary form.

To show a reciprocal of a number written with an exponent, the base remains the same and the exponent is the opposite of the original exponent.

> Write the reciprocal of 10^{-3}, 5^4, 3^{-5}. The reciprocal of 10^{-3} is 10^3, the reciprocal of 5^4 is 5^{-4}, the reciprocal of 3^{-5} is 3^5.
> Write the following with positive exponents and as ordinary numbers.
>
> $$7^{-2} = \left(\frac{1}{7}\right)^2 = \frac{1}{49} \qquad 2^{-5} = \left(\frac{1}{2}\right)^5 = \frac{1}{32}$$

2 Perform basic operations with rational numbers.

To add signed numbers with like signs, add the absolute values and use the common sign.

> $$-5 + (-3) = -8; \qquad 8 + 2 = 10$$

To add signed numbers with unlike signs, subtract the absolute values and use the sign of the larger absolute value.

$$(-5) + 3 = -2; \qquad 12 + (-7) = +5$$

To subtract signed numbers, convert to addition by adding the opposite of the subtrahend.

$$12 - (-8) = 12 + 8 = 20; \quad -7 - (-2) = -7 + 2 = -5;$$
$$-8 - 12 = -8 + (-12) = -20$$

To multiply or divide signed numbers with like signs, multiply or divide the absolute values. The product or quotient will be positive.

$$-3(-8) = +24; \quad \frac{-40}{-10} = +4; \quad 5(-3) = -15; \quad \frac{20}{-5} = -4$$

Section 8–2

1 Examine the relationship between powers and roots using exponential and root notation.

The root of a number is indicated by the denominator of a fractional exponent.

$$16^{1/2} = \sqrt{16} = 4; \qquad 125^{1/3} = \sqrt[3]{125} = 5; \qquad 32^{1/5} = \sqrt[5]{32} = 2$$

2 Position irrational numbers on the number line.

Position irrational numbers on the number line (Fig. 8–5) by approximating the value of the irrational number mentally or with a calculator.

$$6^{1/2} = 2.449489743 \qquad 7^{1/2} = 2.645751311 \qquad 8^{1/2} = 2.828427125$$
$$9^{1/2} = 3.000000000 \qquad 10^{1/2} = 3.16227766$$

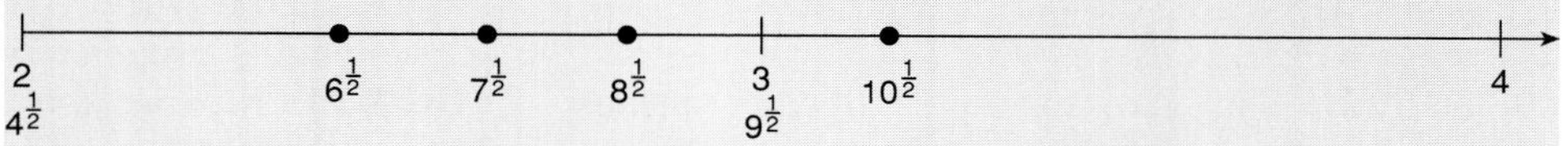

Figure 8–5

3 Estimate the value of irrational numbers.

To estimate the square root of a number, identify the two perfect squares between which the number lies. The square root of the number will be between the square roots of the two perfect squares.

15 is between 9 and 16, so $15^{1/2}$ is between 3 and 4. 38 is between 36 and 49, so $38^{1/2}$ is between 6 and 7.

4 Simplify numerical expressions using the order of operations.

Perform operations in the following order as they appear from left to right:
1. Parentheses used as groupings and other grouping symbols
2. Exponents (powers and roots)
3. Multiplications and divisions
4. Additions and subtractions

$$5 - 3(2 + 5^2) - 36^{1/2} + 2$$
$$5 - 3(2 + 25) - 36^{1/2} + 2$$
$$5 - 3(27) - 36^{1/2} + 2$$
$$5 - 3(27) - 6 + 2$$
$$5 - 81 - 6 + 2$$
$$-76 - 6 + 2$$
$$-82 + 2$$
$$-80$$

1 Identify equations, terms, factors, constants, variables, and coefficients.

An *equation* is a statement that two quantities are equal. A *variable* is a letter that represents an unknown value. A *root* or *solution* of an equation is the value of the variable that makes the equation a true statement.

$x = 5 + 2$ is an *equation*. x is the *variable*. 7 is the *root* or *solution*.

Factors are expressions of multiplication.

$5a$ means 5 *times a*. 5 is a *factor* of $5a$. a is a *factor* of $5a$.

Terms are algebraic expressions that are added or subtracted.

In the expression $5a + 3b + 7$, $5a$ is a *term* and $3b$ is a *term*. 7 is a *term*.

Constant terms are terms that contain only numbers.

In $5a + 3b + 7$, the *constant* term is 7.

Variable terms are terms that have at least one letter.

In $5a + 3b + 7$, $5a$ and $3b$ are *variable* terms.

A *coefficient* is one factor as it relates to the remaining factors of a term.

In $5a + 7$, 5 is the *coefficient* of a and a is the *coefficient* of 5. 7 has no coefficient. The coefficient 5 is also called the *numerical coefficient*.

2 Write verbal interpretations of symbolic statements.

Mathematical symbols can be translated into verbal phrases and statements.

$2x - 3 = 5$ can be translated into "Three less than twice a number is five."

3 Translate verbal statements into symbolic statements using variables.

Verbal phrases can be translated into mathematical symbols.

The statement "A number increased by 13 results in 52" is translated into $x + 13 = 52$.

4 Simplify variable expressions.

Combine like terms by adding or subtracting the coefficients of the terms and by using the same letter or letters for the sum or difference.

$5m + 4m = (5 + 4)m = 9m$

Apply the distributive principle by multiplying the factor outside the parentheses by each term inside the parentheses.

$-4(x + 2y - 5) = -4x - 8y + 20$

Distributing -1 changes only the sign of each term inside the parentheses.

$-(2x - 3) = -1(2x - 3) = -2x + 3$

Section 8–4

1 Examine number patterns by using function notation.

A function can be used to show patterns that develop from using various values of the independent variable.

> Universal Works charges $50 as a flat fee and $18 per hour (or any part of an hour) for leveling and clearing ground. Write a function that expresses the cost of service in relation to the number of hours of work. Describe the pattern. The total cost includes $50 in addition to the hourly cost of $18 times the number of hours. Thus, the function is $f(x) = 18x + 50$.

2 Investigate the effects of changing input values in function notation.

As values change for the input or independent variable, values of the output or dependent variable change accordingly.

> Find the cost for 2 hours of work, 4 hours of work, and 10 hours of work by Universal Works.
>
> | | | |
> |---|---|---|
> | $f(x) = 18x + 50$ | $f(x) = 18x + 50$ | $f(x) = 18x + 50$ |
> | $f(2) = 18(2) + 50$ | $f(4) = 18(4) + 50$ | $f(10) = 18(10) + 50$ |
> | $f(2) = 36 + 50$ | $f(4) = 72 + 50$ | $f(10) = 180 + 50$ |
> | $f(2) = \$86$ | $f(4) = \$122$ | $f(10) = \$230$ |

3 Graph independent and dependent values of functions.

Functions can be represented graphically on a coordinate system. The independent variable is represented on the horizontal or x-axis and the dependent variable is represented on the vertical or y-axis.

> Figure 8–6 represents the function used by Universal Works for leveling and clearing ground.

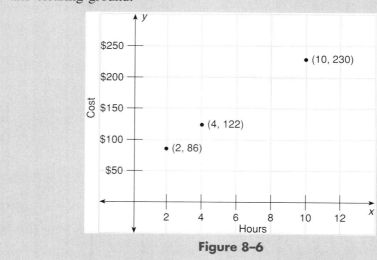

Figure 8–6

WORDS TO KNOW

reciprocals (p. 319)
perfect power (p. 323)
square root (p. 323)
cube root (p. 323)
fourth root (p. 323)
radical sign (p. 323)
radicand (p. 323–324)
index (p. 324)
irrational number (p. 325)
real number (p. 325)
equation (p. 328)

variable or unknown (p. 329)
root or solution (p. 329)
symmetric property of equality (p. 329)
factor (p. 329, 330)
term (p. 329, 330)
number term or constant (p. 330, 331)
letter term or variable term (p. 330, 331)
coefficient (p. 331)

numerical coefficient (p. 331)
like terms (p. 334)
distributive principle (p. 334)
function notation (p. 336)
independent variable (p. 336)
dependent variable (p. 336)
input values (p. 336)
output values (p. 336)
domain (p. 337)
range (p. 337)
increasing function (p. 338)

Write equivalent expressions using positive exponents, then write them as ordinary numbers.

1. 12^{-2}

2. 5^{-4}

Write equivalent expressions with positive exponents, then write the expressions with a denominator of 1.

3. $\dfrac{1}{10}$

4. $\dfrac{1}{144}$

5. Write and illustrate the rule for adding signed numbers with unlike signs.

Estimate the roots by determining between which two whole numbers each root lies.

6. $88^{1/2}$

7. $148^{1/2}$

8. $\sqrt{96}$

Evaluate the following expressions.

9. $5 - 32 \cdot 6 - 5^2 \cdot 16^{1/2}$

10. $7 - (8 - 3) - 3(2) + 18 \div 9$

Identify the terms in each expression by drawing a box around each.

11. $5x^2$

12. $7(3x - 1) + 5x - 3$

Write a symbolic statement for each verbal statement.

13. Five added to a number gives a result of 35.

14. Seven times the difference of 8 and a number is 11.

Write a verbal expression for the following symbolic statements.

15. $x + 8 = 21$

16. $3x - 8 = 14$

Evaluate each of the following functions for the indicated value of the independent variable.

17. $f(x) = 5x + 2$. Find $f(-4)$, $f(-2)$, $f(0)$, $f(2)$, $f(4)$.

18. $f(x) = -3x - 5$. Find $f(-3)$, $f(-1)$, $f(0)$, $f(1)$, $f(3)$.

19. Graph the function described in Exercise 17.

20. Graph the function described in Exercise 18.

21. Write a function that represents the data shown below. The number of hours and cost of rental is for a computer that is rented by the hour (or any part of an hour).

Hours Rented	Cost of Rental
4	$16
6	$22
7	$25
8	$28

9

Linear Equations

GOOD DECISIONS THROUGH TEAMWORK

Form teams in which members have an interest in a similar type of business, industry, or technology. Examples of businesses, industries, or technologies to investigate are accounting, transportation, manufacturing, catering, engineering, and health-related services. A textbook on spreadsheets is a good source of other applications to investigate.

With your team, find samples of spreadsheets used to perform a specific function in the selected business, industry, or technology. Gather examples of the spreadsheets from library research, interviews with professionals or educators in the selected field, or other appropriate sources. Select one of the spreadsheets to analyze further. This analysis should include the purpose of the spreadsheet, how the information is collected, what additional information is derived through calculations, and what equations, or formulas, are used to make those calculations.

Present your team's findings to the class and be prepared to answer questions from class members.

Equations and formulas play an important role in the mathematics used in the workplace. We have found missing values in the percentage proportion, $\frac{R}{100} = \frac{P}{B}$, and in various area and perimeter formulas like $P = 2(l + w)$ or $A = \pi r^2$. We will now investigate how equations can be used to model numerous problem-solving situations.

9–1   SOLVING BASIC EQUATIONS

Learning Objectives

1. Solve basic equations.
2. Check the solutions of equations.

Algebraic equations relate to many real-life situations. For example, suppose we know that five packages of copier paper cost $15. We could represent the cost of the five packages as $5x$, which is 5 times the unknown cost of each package (x). Since the five packages together cost a total of $15, the $5x$ must equal $15. We can write an equation to represent this relationship: $5x = 15$. This kind of equation is called a basic equation and is the subject of Section 9–1.

1   Solve Basic Equations.

■ **DEFINITION 9–1: Basic Equation.** An equation that consists of only a variable and its coefficient on one side of the equal sign and only a number term on the other side of the equal sign is called a *basic equation*.

Some examples of basic equations are:

$$5x = 15 \qquad -2 = 4b$$
$$2x = 11 \qquad -3y = 6$$
$$8 = 3a$$

A basic equation states that a number times a variable equals a product. In the equation $5x = 15$, we have the number 5 times the variable x equal to the product 15. We can determine mentally that the missing number must be 3; however, we want to learn a procedure for solving even the simplest of equations so that we can apply the same procedure to more complicated equations.

An equation is *solved* when the letter is alone on one side of the equation; that is, the coefficient of the letter term is 1. The number on the side opposite the letter is called the *root* or *solution* of the equation.

In the equation $5x = 15$, the numerical coefficient of the letter term is 5. Since we would like the numerical coefficient of x to be 1, *we multiply both sides of the equation by the reciprocal of 5*, which is $\frac{1}{5}$. Recall that multiplication of a number by the reciprocal, or multiplicative inverse, of that number gives us 1.

$$\left(\frac{1}{5}\right)5x = 15\left(\frac{1}{5}\right)$$

$$\left(\frac{1}{\cancel{5}}\right)\cancel{5}x = \cancel{15}^{\,3}\left(\frac{1}{\cancel{5}}\right) \qquad \text{Reduce where possible.}$$

$$1x = 3 \qquad \text{Work multiplication on each side.}$$

$$x = 3$$

The basic principle that underlies this procedure and all other procedures for solving equations is to preserve the equality of the two sides of the equation.

Rule 9–1 *To preserve equality:*

If we perform an operation on one side of the equation, we must perform the same operation on the other side.

For example, if we had a scale with 1 oz on one side and 1 oz on the other side, the scale would be balanced. If we increased or decreased the weight on one side, we would need to do the same on the other side or one side would be heavier than the other and the equality of both sides would be lost (see Fig. 9–1).

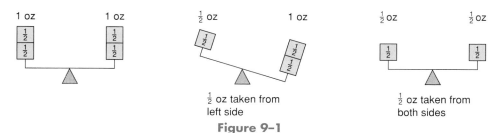

Figure 9–1

Let's look at the equation $5x = 15$ again. Notice that the numerical coefficient 5 is "attached" to the unknown, x, by multiplication. Since division is the inverse operation of multiplication, we can solve the equation by *dividing both sides of the equation by 5*. This procedure also gives us a numerical coefficient of 1 for x.

$$\frac{5x}{5} = \frac{15}{5} \qquad \text{Divide both sides by 5.}$$

$$\frac{\overset{1}{\cancel{5}}x}{\underset{1}{\cancel{5}}} = \frac{\overset{3}{\cancel{15}}}{\underset{1}{\cancel{5}}} \qquad \text{Reduce within each fraction.}$$

$$1x = 3$$

$$x = 3$$

In short, we can solve a basic equation by multiplication or by division, as indicated in Rule 9–2.

Rule 9–2 *To solve a basic equation:*

Divide both sides of the equation by the coefficient of the letter term;
or
Multiply both sides of the equation by the reciprocal of the coefficient of the letter term.

This rule is based on the *multiplication axiom*.

■ **DEFINITION 9–2: Multiplication Axiom.** Both sides of an equation may be multiplied by the same *nonzero* quantity without changing the equality of the two sides.

If $a = b$ and $c \neq 0$, then $ac = bc$.

Since multiplication and division are inverse operations, to divide by a number is the same as to multiply by its multiplicative inverse—its reciprocal.
Let us now solve several basic equations using multiplication or division.

EXAMPLE 1   Solve the equation $4x = 48$.

$$4x = 48$$

$$\left(\frac{1}{4}\right)4x = 48\left(\frac{1}{4}\right)$$ Multiply both sides of the equation by the reciprocal of the coefficient of the letter term.

$$x = 12$$

The same equation can be solved by dividing both sides of the equation by the coefficient of the letter term.

$$4x = 48$$

$$\frac{4x}{4} = \frac{48}{4}$$ Divide both sides by 4.

$$1x = 12$$ This step is usually performed mentally and omitted from the written steps.

$$x = 12$$

EXAMPLE 2   Solve the equation $45 = 0.5x$.

$$\frac{45}{0.5} = \frac{0.5x}{0.5}$$ Divide both sides by 0.5.

$$90 = x$$

Multiplication and division are effective no matter which side of the equation contains the letter term. Once the equation is solved, however, many people prefer to put the letter on the left ($x = 90$ instead of $90 = x$). Either way is correct.

Tips and Traps   ***Improper Fractions Versus Mixed Numbers:***
The solution of many equations will not be a whole number. When this is the case, the solution can be left as a *proper* or *improper fraction* in lowest terms. The solution can also be changed to its decimal or mixed-number equivalent. When applied problems are solved, we will interpret the answer in an appropriate form. In the meantime, we will leave the solutions as fractions in lowest terms.

EXAMPLE 3   Solve the equation $\frac{1}{4}n = 7$.

Using multiplication: $\frac{1}{4}n = 7$

$$\left(\frac{4}{1}\right)\frac{1}{4}n = 7\left(\frac{4}{1}\right)$$ The reciprocal of $\frac{1}{4}$ is $\frac{4}{1}$.

$$n = 28$$

Using division: $\frac{1}{4}n = 7$

$$\frac{\frac{1}{4}n}{\frac{1}{4}} = \frac{7}{\frac{1}{4}}$$ $7 \div \frac{1}{4} = \frac{7}{1} \times \frac{4}{1} = 28$

$$n = 28$$

EXAMPLE 4   Solve the equation $-n = 25$.

The numerical coefficient is -1, so the equation is not solved. The numerical coefficient must be $+1$ for the equation to be solved.

$$\frac{-1n}{-1} = \frac{25}{-1}$$

$$n = -25$$

2   Check the Solutions of Equations.

When an equation has been solved, the *solution*, or *root*, can be checked to avoid careless mistakes.

Rule 9-3   *To check the solution or root of an equation:*

1. Substitute the solution in place of the letter in the original equation.
2. Perform all indicated operations.
3. If the solution is correct, the value of the left side should equal the value of the right side.

EXAMPLE 5   The solution for the equation $4x = 48$ (Example 1) was found to be 12. Check or verify this root.

$4x = 48$ Substitute 12 for *x*.

$4(12) = 48$ Perform the multiplication.

$48 = 48$ The root is verified if the final equality is true.

SELF-STUDY EXERCISES 9-1

1   Solve the following equations. Use multiplication or division. Show your work.

1. $7x = 56$ **2.** $2x = 18$ **3.** $15 = 3x$ **4.** $36 = 1.2n$
5. $-3a = 27$ **6.** $-18 = 9b$ **7.** $-7c = -49$ **8.** $-72 = 8x$
9. $5x = 32$ **10.** $36 = 8y$ **11.** $-6n = 15$ **12.** $-28 = -3b$

13. $-x = 7$ **14.** $-2 = -y$ **15.** $\frac{1}{3}x = 5$ **16.** $4 = \frac{1}{2}y$

17. $\frac{3}{5}x = 2$ **18.** $-\frac{3}{4}y = 5$ **19.** $-\frac{1}{2}x = -6$ **20.** $21 = \frac{3}{8}n$

2   **21.–30.** Check or verify each solution or root for the even-numbered Exercises 2–20.

ISOLATING THE VARIABLE IN SOLVING EQUATIONS ████████

Learning Objectives

1 Solve equations with like terms on the same side of the equation.
2 Solve equations with like terms on opposite sides of the equation.

Equations are not always basic equations. Sometimes they are more involved and must be *simplified* into the form of the basic equation so that they can be solved. This section will show how to simplify or change into basic equations certain types of equations that are not basic equations.

1 Solve Equations with Like Terms on the Same Side of the Equation.

Suppose a family of four went to lunch at a fast-food restaurant. The mother had only $7 and the father had only $8, which they planned to combine for the four lunches. At the last minute the daughter decided to go to a friend's house for lunch and left the restaurant. An algebraic equation can be used to represent this scenario and find out how much each family member can spend on lunch after the daughter left.

Let x equal the amount of money each family member can spend. There are four members, so the total amount to be spent is 4 times x, or $4x$. Because the daughter left, we must subtract the amount (x) she would have spent. This gives us $4x - x$. The total amount of money still equals the mother's $7 plus the father's $8. These relationships can be expressed in equation form: $4x - x = 7 + 8$.

Recall that a basic equation has a single letter term for one side and a single number term for the opposite side. If an equation has more than one letter term on one side or more than one number term on the opposite side, we *combine* the terms on each side to form a basic equation.

We can combine only *like* terms. By "combine," we mean adding them according to the signed number rules.

EXAMPLE 1 Solve the equation $4x - x = 7 + 8$.

$4x - x = 7 + 8$ The coefficient of x is -1.

$\boxed{4x - x} = \boxed{7 + 8}$ Think of $4x$ plus $-1x = 3x$. $7 + 8 = 15$.

$\qquad 3x = \boxed{15}$ We now have a basic equation.

The basic equation can be solved as before with multiplication or division on both sides. Division is more convenient with whole numbers.

$\dfrac{3x}{3} = \dfrac{15}{3}$ We divide both sides by the numerical coefficient of x to "undo" multiplication.

$\qquad x = 5$

2 Solve Equations with Like Terms on Opposite Sides of the Equation.

In many equations, like terms do not appear on the same side of the equal sign. In such cases the like terms must be manipulated so that they appear on the same side of the equal sign *before* they can be combined. The basic purpose is to *isolate* the variable or letter terms on one side of the equation and thereby obtain a basic equation to be solved.

Terms can be moved from one side of an equation to the other by applying the *addition axiom,* which states that the same amount can be added to both sides of an equation without changing the equality of the two sides.

EXAMPLE 2   Solve the equation $2x + 4 = 8$.

The like terms 4 and 8 are on opposite sides of the equal sign. To isolate the $2x$, we need to eliminate 4. Therefore, if we had some way to move the number term 4 to the right side of the equation, all the number terms would be on one side and the letter term would be isolated on the opposite side.

We can move the 4 from the letter term side to the opposite side by adding the *opposite* of 4 (that is, -4) to *both* sides. Remember, whatever we do to one side, we must do to the other side.

$$2x + 4 = 8$$

$$2x \boxed{+ 4 - 4} = \boxed{8 - 4} \qquad \text{Add } -4 \text{, the opposite of 4, to both sides.}$$

$$2x \boxed{+ 0} = \boxed{4} \qquad 4 - 4 = 0, \ 8 - 4 = 4$$

$$2x = 4 \qquad 2x + 0 = 2x$$

Now we have a *basic equation* that we can solve with multiplication or division on both sides. Here, division is used.

$$\frac{2x}{2} = \frac{4}{2} \qquad \text{We divide both sides by the numerical coefficient.}$$

$$\boldsymbol{x = 2}$$

Example 2 applied the *addition axiom* to solve the equation.

■ **DEFINITION 9–3:   Addition Axiom.**   The same quantity can be added to both sides of an equation without changing the equality of the two sides.

$$\text{If } a = b, \text{ then } a + c = b + c.$$

EXAMPLE 3   Solve the equation $9 - 4x = 8x$.

In this equation the like terms $-4x$ and $8x$ are on opposite sides of the equation. We want to manipulate the terms in this equation so that both letter terms are isolated on one side of the equation and the number term is on the other side. The simplest way of accomplishing this is to eliminate $-4x$ from the left side of the equation by adding its opposite to both sides.

$$9 - 4x = 8x$$

$$9 \boxed{- 4x + 4x} = \boxed{8x + 4x} \qquad \text{Add } 4x \text{, the opposite of } -4x \text{, to both sides.}$$

$$9 \boxed{+ 0} = \boxed{12x} \qquad -4x + 4x = 0, \ 8x + 4x = 12x$$

$$9 = 12x \qquad 9 + 0 = 9$$

$$\frac{9}{12} = \frac{12x}{12} \qquad \text{Divide both sides by the coefficient of } x.$$

$$\frac{9}{12} = x \qquad \text{Reduce to lowest terms.}$$

$$\frac{3}{4} = x \qquad \textbf{or} \qquad x = \frac{3}{4}$$

EXAMPLE 4   Solve the equation $7 - 5x = 12$.

The like terms are 7 and 12. Add -7 to both sides of the equation to isolate the variable term.

$$7 - 5x = 12$$

$$\boxed{7 - 7} - 5x = \boxed{12 - 7} \qquad 7 - 7 = 0; \qquad 12 - 7 = 5$$

$$\boxed{0} - 5x = \boxed{5} \qquad 0 - 5x = -5x$$

$$-5x = 5$$

$$\frac{-5x}{-5} = \frac{5}{-5} \qquad \text{Note that both sides are divided by } -5 \text{ because the coefficient is negative.}$$

$$x = -1$$

Notice in the preceding examples that, each time we add to both sides the opposite of a term, it is eliminated on one side of the equation and appears as the *opposite* on the other side of the equation. Knowing this, we can *transpose* as a shortcut for adding opposites to both sides. *Transpose* means "move across as the opposite of."

Tips and Traps   ***Transposing or Sorting Terms:*** A shortcut to applying the addition axiom for moving terms is called *transposing*. When the opposite of a term is to be added to both sides of an equation, the term will be omitted from one side of the equation and written as its opposite on the other side. That is, the term is *transposed* to the other side of the equation as its opposite. Another way to describe this process is to *sort* the terms so that like terms are on the same side of the equation.

In other words, we can *mentally* add the opposite term to both sides but show only part of the procedure. Let's rework Examples 2 and 3 by transposing like terms to isolate the variable.

$$2x \boxed{+\ 4} = 8 \qquad\qquad 9 \boxed{-\ 4x} = 8x$$

$$2x = 8 \boxed{-\ 4} \qquad\qquad 9 = 8x \boxed{+\ 4x}$$

$$2x = 4 \qquad\qquad 9 = 12x$$

$$\frac{2x}{2} = \frac{4}{2} \qquad\qquad \frac{9}{12} = \frac{12x}{12}$$

$$x = 2 \qquad\qquad \frac{3}{4} = x$$

EXAMPLE 5   Solve the equation $9x + 2 = 6x - 10$.

We must first decide on which side of the equation to isolate the letter terms. Once that decision has been made, the number terms must be collected on the opposite side. We will solve this equation two ways, by isolating the letter terms on the left and by isolating the letter terms on the right. The main point is that, no matter which side is chosen for the letter terms, the number terms must be on the opposite side.

    Letter Terms Isolated on Left   **Letter Terms Isolated on Right**

$$9x \boxed{+\ 2} = \boxed{6x} - 10 \qquad\qquad \boxed{9x} + 2 = 6x \boxed{-\ 10}$$

$$9x \boxed{-\ 6x} = -10 \boxed{-\ 2} \qquad\qquad 2 \boxed{+\ 10} = 6x \boxed{-\ 9x}$$

$$3x = -12 \qquad\qquad\qquad 12 = -3x$$

$$\frac{3x}{3} = \frac{-12}{3} \qquad\qquad\qquad \frac{12}{-3} = \frac{-3x}{-3}$$

$$x = -4 \qquad\qquad\qquad\qquad -4 = x$$

Sequence of Strategies for Solving Equations:
1. Combine like terms on each side of the equation.
2. Transpose terms to isolate the variable or letter terms on one side and the number terms on the other.
3. Combine like terms on each side to get a basic equation.
4. Solve the basic equation by multiplication or division.

EXAMPLE 6 Solve the equation $2x - 3 + 5x = 8 - 6x$.

$$2x - 3 + 5x = 8 - 6x$$

$$7x - 3 = 8 - 6x \qquad \text{Combine the like terms } 2x \text{ and } 5x \text{ on the left side.}$$

$$7x + 6x = 8 + 3 \qquad \text{Transpose to collect like terms.}$$

$$13x = 11 \qquad \text{Combine the like terms to get a basic equation.}$$

$$\frac{13x}{13} = \frac{11}{13} \qquad \text{Divide by the coefficient of } x.$$

$$x = \frac{11}{13}$$

Number terms and coefficients in equations can also be fractions and decimals.

EXAMPLE 7 Solve $9 + \frac{1}{4}b = \frac{3}{4}b$.

$$9 + \frac{1}{4}b = \frac{3}{4}b \qquad \text{Note that the letter terms are on opposite sides.}$$

$$9 = \frac{3}{4}b - \frac{1}{4}b \qquad \text{Transpose. The denominators of the letter terms are the same.}$$

$$9 = \frac{2}{4}b \qquad \text{Combine the letter terms.}$$

$$9 = \frac{1}{2}b \qquad \text{Reduce.}$$

$$\left(\frac{2}{1}\right)9 = \left(\frac{2}{1}\right)\frac{1}{2}b \qquad \text{Multiply both sides by the reciprocal of } \frac{1}{2}.$$

$$\mathbf{18 = b}$$

EXAMPLE 8 Solve $1.1 = 3.4 + R$.

$$1.1 = 3.4 + R \qquad \text{Transpose.}$$

$$1.1 - 3.4 = R \qquad \text{Combine terms.}$$

$$\mathbf{-2.3 = R}$$

SELF-STUDY EXERCISES 9–2

1 Solve the following equations.

1. $3x + 7x = 60$

2. $42 = 8m - 2m$

3. $5a - 6a = 3$

4. $3m - 9m = 3$

5. $y + 3y = 32$

6. $0 = 2x - x$

2 Solve the following equations.

7. $b + 6 = 5$	**8.** $1 = x - 7$	**9.** $5t - 18 = 12$
10. $10 - 2x = 4$	**11.** $2y + 7 = 17$	**12.** $3x + 7 = x$
13. $3a - 8 = 7a$	**14.** $10x + 18 = 8x$	**15.** $4x + 7 = 8$
16. $4x = 5x + 8$	**17.** $2t + 6 = t + 13$	**18.** $12 + 5x = 6 - x$

19. $4y - 8 = 2y + 14$ **20.** $8 - 7y = y + 24$ **21.** $\dfrac{2x}{3} = 18$

22. $\dfrac{y + 1}{2} = 7$ **23.** $\dfrac{8 - R}{76} = 1$ **24.** $P = \dfrac{1}{2} + \dfrac{1}{3}$

25. $x + \dfrac{1}{7}x = 16$ **26.** $\dfrac{2}{5} - x = \dfrac{1}{2}x + \dfrac{4}{5}$ **27.** $\dfrac{3}{7}m - \dfrac{1}{2} = \dfrac{2}{3}$

28. $\dfrac{1}{4}s = \dfrac{1}{4} + \dfrac{1}{10} + \dfrac{1}{20}$ **29.** $m = 2 + \dfrac{1}{4}m$ **30.** $\dfrac{7}{9} + 3 = \dfrac{1}{2}T$

31. $2.3x = 4.6$ **32.** $0.8R = 0.6$ (round to nearest tenth)

33. $0.33x + 0.25x = 3.5$ (round to nearest hundredth) **34.** $0.04x = 0.08 - x$ (round to nearest hundredth)

35. $0.47 = R + 0.4R$ (round to nearest hundredth)

9–3	**APPLYING THE DISTRIBUTIVE PROPERTY IN SOLVING EQUATIONS**

Learning Objective   **1**   Solve equations that contain parentheses.

When an equation contains an addition or subtraction in parentheses and that quantity in parentheses is multiplied by another factor, we have an example of the *distributive property*. (See Chapter 1.)

1 **Solve Equations That Contain Parentheses.**

Imagine that a restaurant hired you to fill a vacated position. The manager told you that you would be paid $2 more per hour than the previous worker. You forgot to ask about the hourly rate of the previous worker, but in three hours you earned a gross pay of $18. You can use algebra to figure the pay rate of the previous worker. Just let x equal the previous worker's hourly pay rate and add $2 to it, which is $x + 2$. Then multiply that amount by three and make it equal to your gross pay of $18. The equation would look like this: $3(x + 2) = 18$.

If an equation contains a distributive multiplication, such as $3(x + 2)$, the first step in solving the equation is to apply the distributive property.

EXAMPLE 1   Solve the equation $3(x + 2) = 18$ for x.

$3(x + 2) = 18$	Each term in parentheses must be multiplied by 3.
$3x + 6 = 18$	
$3x = 18 - 6$	Transpose.
$3x = 12$	Combine.
$\dfrac{3x}{3} = \dfrac{12}{3}$	Divide.
$x = 4$	

EXAMPLE 2   Solve the equation $5x + 7 = 2(3 - x) - 13$ for x.

$5x + 7 = 2(3 - x) - 13$	Each term in parentheses must be multiplied by 2.
$5x + 7 = 6 - 2x - 13$	
$5x + 7 = -7 - 2x$	Combine.

$$5x + 2x = -7 - 7 \qquad \text{Transpose.}$$

$$7x = -14 \qquad \text{Combine.}$$

$$\frac{7x}{7} = \frac{-14}{7} \qquad \text{Divide.}$$

$$x = -2$$

Recall from Section 9–1 that the root of an equation can be checked by substituting the root for the letter in each place it appears in the equation and then evaluating both sides of the equation. If the root (solution) is correct, both sides of the equation will equal the same number.

Let's check the root $x = -2$ for Example 2.

$$5x + 7 = 2(3 - x) - 13$$

$$5(-2) + 7 = 2(3 - (-2)) - 13 \qquad \text{Substitute } -2 \text{ for } x. \text{ Because the grouping } 2(3 - (-2)) \text{ has two negatives, be careful with the signs.}$$

$$5(-2) + 7 = 2(3 + (+2)) - 13 \qquad \text{Evaluate the grouping first.}$$

$$5(-2) + 7 = 2(5) - 13 \qquad \text{Perform all multiplications next.}$$

$$-10 + 7 = 10 - 13 \qquad \text{Add last.}$$

$$-3 = -3$$

Since $-3 = -3$, the solution checks.

Tips and Traps	***The Sign of a Term:*** In solving any equation, it is very important to be careful with the sign of each term. The algebraic sign of any term is the sign that comes *before* the term.

EXAMPLE 3 Solve the equation $6 - (x + 3) = 2x$ for x.

Since $(x + 3)$ is in parentheses, it is a grouping that is handled as one term. The sign of the term is negative and the numerical coefficient is understood to be -1.

The first step in solving this equation is to apply the distributive property by multiplying $(x + 3)$ by its understood coefficient, -1.

$$6 - (x + 3) = 2x$$

$$6 - 1(x + 3) = 2x \qquad \text{-1 is understood before } (x + 3). \text{ Distribute.}$$

$$6 - x - 3 = 2x \qquad \text{Transpose.}$$

$$6 - 3 = 2x + x \qquad \text{Combine.}$$

$$3 = 3x \qquad \text{Divide.}$$

$$\frac{3}{3} = \frac{3x}{3} \qquad \text{Divide.}$$

$$1 = x$$

> **Rule 9–4** *To multiply a sum or difference in parentheses by a coefficient using a shortcut:*
>
> **1.** Consider the sign between terms as the sign of the second term.
> **2.** Multiply each term in the parentheses by the coefficient using the rules for multiplying signed numbers.

EXAMPLE 4 Solve and check the equation $28 = 7x - 3(x - 4)$ for x.

$$28 = 7x - 3(x - 4)$$ Each term in parentheses is multiplied by -3.

$$28 = 7x - 3x + 12$$ Distribute.

$$28 - 12 = 4x$$ Combine like terms on the right and then transpose.

$$16 = 4x$$ Combine.

$$\frac{16}{4} = \frac{4x}{4}$$ Divide.

$$\mathbf{4 = x}$$

As equations get more involved, the importance of checking becomes more apparent. To check the root 4, we will substitute 4 for x in the equation.

$$28 = 7x - 3(x - 4)$$

$$28 = 7(4) - 3(4 - 4)$$ Substitute 4 for x. Then simplify the grouping $4 - 4 = 0$.

$$28 = 7(4) - 3(0)$$ Multiply first; then subtract.

$$28 = 28 - 0$$

$$\mathbf{28 = 28}$$

The following rule summarizes some strategies for solving equations.

Rule 9–5 *To solve equations use the following strategies as needed:*

1. Apply the distributive property.
2. Combine like terms on each side of the equation.
3. Transpose terms to collect the variable or letter terms on one side and number terms on the other.
4. Combine like terms on each side of the equation.
5. Solve the basic equation by using multiplication by the reciprocal of the coefficient of the variable term or division by the coefficient of the variable term.

Tips and Traps *Key Words for Strategies:*
The following key words can be used as memory tools:

1. Distribute.
2. Combine.
3. Transpose or sort.
4. Combine.
5. Multiply or divide.

SELF-STUDY EXERCISES 9–3

1 Solve the following equations.

1. $3(7 + x) = 30$ **2.** $15 = 5(2 - y)$ **3.** $6(3x - 1) = 12$

4. $4t = 2(7 + 3t)$ **5.** $2x = 7 - (x + 6)$ **6.** $6x - 2(x - 3) = 30$

7. $4a + 5 = 3(2 + a) - 4$ **8.** $8x - (3x - 2) = 12$ **9.** $15 - 3(2x + 2) = 6$

10. $(2b + 1) = 5(b - 4) + 1$

STRATEGIES FOR PROBLEM SOLVING ▬▬▬▬▬

Learning Objective   **1** Solve applied problems using equations.

The three previous sections began with a problem that could be represented as an algebraic equation and solved. Each section showed how to solve specific kinds of linear equations exemplified by each problem. In this section, however, we present the following six-step process to help you interpret a word problem or other real-life problem and set up an equation to solve that problem.

1. **Unknown facts** means "What fact or facts am I trying to find?" Usually you will choose a letter to represent the unknown fact or one of the unknown facts.
2. **Known facts** means "What relevant facts are known or given in the problem?"
3. **Relationships** means "How are the known and unknown facts related?" and "What two quantities are equal?" The relationships should be expressed first in words and then symbolically in an equation or formula.
4. **Estimation** means "Approximately what should be the result of the solution or the 'answer' to the problem?" Your estimation may sometimes be expressed as "more than . . ." or "less than . . ." or "between . . . and . . . ."
5. **Solution** means the solution of the equation or formula for the unknown facts.
6. **Interpretation** means expressing the answer in terms of the context of the problem. For example, if the problem asks for the number of truckloads of gravel needed for a job, the answer is expressed in truckloads of gravel.

The six steps are applied in the following examples.

EXAMPLE 1 Cher Hattier was taken to lunch by a friend. The order included a 12-inch turkey sub, a 6-inch turkey sub, and two iced teas. Cher didn't notice the price of the subs but did recall that the 12-inch sub cost $1.50 more than the smaller sub and that two teas cost $2.50. The total cost of lunch was $10. How much did each sub cost?

Unknown facts The cost of each sub is requested.

Known facts The 12-inch sub was $1.50 more than the 6-inch sub. The two teas cost $2.50. The total cost of the meal was $10.

Relationships Let x = the cost of the 6-inch sub. Then $x + \$1.50$ is the cost of the 12-inch sub. The cost of the two subs plus $2.50 for the iced teas equals $10; that is, $x + x + 1.50 + 2.50 = 10$.

Estimation The subs together cost $7.50 because the teas cost $2.50 ($10 − $2.50 = $7.50). So the subs cost roughly half of $7.50, or around $3.00 or $4.00 each.

Solution
$$x + x + 1.50 + 2.50 = 10$$

$$2x + 4 = 10 \qquad \text{\small $x + x = 2x$; 1.50 + 2.50 = 4.00, or 4}$$

$$2x = 10 - 4 \qquad \text{\small Transpose.}$$

$$2x = 6$$

$$\frac{2x}{2} = \frac{6}{2} \qquad \text{\small Solve for } x.$$

$$x = 3 \qquad \text{\small Cost of 6-inch sub.}$$

Interpretation **Thus, the 6-inch sub cost $3.00, and the 12-inch sub cost $4.50 ($3 + $1.50).**

Check:
$$x + x + 1.50 + 2.50 = 10$$
$$3 + 3 + 1.50 + 2.50 = 10$$
$$10 = 10$$

EXAMPLE 2 A student ordered a student stethoscope and two pairs of support hose. The total bill was $41.50, including a $2.50 shipping charge. If the student stethoscope cost twice as much as one pair of support hose, how much did she pay for the stethoscope and each pair of hose?

Unknown facts
We are asked for the cost of the stethoscope and each pair of hose.

Known facts
The stethoscope cost twice what a pair of hose cost. The shipping was $2.50. The total bill was $41.50.

Relationships
Let x = the cost of a pair of hose. Then $2x$ = the cost of the stethoscope. Thus, the cost of each pair of hose ($x + x$) plus the cost of the stethoscope ($2x$) plus the shipping ($2.50) equals $41.50. The equation would be: $x + x + 2x + 2.50 = 41.50$.

Estimation
Since the cost of the stethoscope is twice that of a pair of hose, and there are two pairs of hose, the stethoscope must cost around $20 and each pair of hose must cost about $10.

Solution

$x + x + 2x + 2.50 = 41.50$	Combine.
$4x + 2.50 = 41.50$	Transpose.
$4x = 41.50 - 2.50$	Combine like terms.
$\dfrac{x}{4} = \dfrac{39}{4}$	Divide.
$x = 9.75$	

Interpretation
One pair of hose cost $9.75. Since the stethoscope cost twice that amount ($2x$), the stethoscope cost 2(9.75), or $19.50.

Check:
$$x + x + 2x + 2.50 = 41.50$$
$$9.75 + 9.75 + 2(9.75) + 2.50 = 41.50 \qquad \text{Substitute.}$$
$$41.50 = 41.50$$

EXAMPLE 3 A nursing student purchased a new pathology text, a used history text, and a notebook. When he checked his receipt for $132 plus tax, he discovered that the new pathology text cost twice as much as the used history text and notebook combined. If the notebook cost $3.00, how much did he pay for the used history text? For the new pathology text?

Unknown facts
How much does each text cost?

Known facts
The notebook cost $3.00. The total cost is $132.00. The pathology text cost twice the cost of the history text and notebook combined.

Relationships
Let x = the cost of the history text. The notebook cost $3. The cost of the pathology text is twice the cost of the history text and the notebook, that is, $2(x + 3)$. The cost of the history text plus the pathology text plus the notebook equals $132. Therefore, the equation is $x + 2(x + 3) + 3 = 132$.

Estimation
Ignoring the notebook, the pathology text cost twice the cost of the history text, so the ratio is 2 to 1. If these two books cost about $132, a ratio of 2 to 1 would be $88 to $44 (132/3 = 44). So the pathology text is around $88 and the history text is around $44.

Solution

$x + 2(x + 3) + 3 = 132$	
$x + 2x + 6 + 3 = 132$	Distribute.
$3x + 9 = 132$	Combine like terms.
$3x = 132 - 9$	Transpose.
$3x = 123$	

$$\frac{3x}{3} = \frac{123}{3}$$ Solve for x.

$$x = 41$$

Interpretation **The history text cost \$41. The pathology text cost $2(x + 3)$. Thus, $2(41 + 3) = 82 + 6 = \$88$.**

Check:
$$x + 2(x + 3) + 3 = 132$$
$$41 + 2(41 + 3) + 3 = 132$$
$$41 + 82 + 6 + 3 = 132$$
$$132 = 132$$

EXAMPLE 4 A horticulturist has marked off a 32-m wide rectangular nursery plot with 158 m of fencing. Because the plants must be properly spaced from one another, she needs to know the length of the plot. Find the length in meters.

Unknown facts She needs to know the length of the nursery plot in meters.

Known facts The nursery plot is rectangular. The width is 32 m. The fencing gives us the perimeter of 158 m.

Relationships The formula for the perimeter of a rectangle includes the relationships needed: the perimeter is twice the sum of the length and width, or $p = 2(l + w)$.

Estimation The length must be more than the width of 32 m and less than half the perimeter, that is, less than 79 m.

Solution
$$p = 2(l + w)$$
$$158 = 2(l + 32) \quad \text{Substitute in formula.}$$
$$158 = 2(l) + 2(32) \quad \text{Distribute.}$$
$$158 = 2l + 64$$
$$158 - 64 = 2l \quad \text{Transpose.}$$
$$94 = 2l \quad \text{Combine like terms.}$$
$$\frac{94}{2} = \frac{2l}{2} \quad \text{Divide.}$$
$$47 = l$$

Interpretation **Thus, the length of the nursery plot is 47 m.**

Check:
$$p = 2(l + w)$$
$$158 = 2(47 + 32)$$
$$158 = 94 + 64$$
$$158 = 158$$

EXAMPLE 5 A machine part weighs 2.7 kg and is to be shipped in a carton weighing x kg. If the total weight of three packaged machine parts is 9.3 kg, how much does each carton weigh? (See Fig. 9–2.)

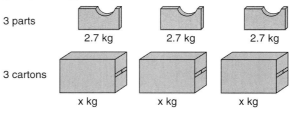

3 parts   2.7 kg   2.7 kg   2.7 kg

3 cartons   x kg   x kg   x kg

Figure 9–2

Unknown facts	How much does each empty carton weigh?
Known facts	There are three parts and three cartons. Each part weighs 2.7 kg. The three packaged parts (in cartons) weigh a total of 9.3 kg.
Relationships	Let x equal the weight of one empty carton. Each packaged part weighs the sum of the part (2.7 kg) plus the weight of a carton (x kg). The sum of the three packaged parts equals 9.3 kg; that is, $3(2.7 + x) = 9.3$ or $(2.7 + x) + (2.7 + x) + (2.7 + x) = 9.3$.
Estimation	Three packaged parts weigh 9.3 kg, and $\frac{9.3}{3} = 3.1$ kg. Thus, one carton weighs more than 0.1 kg and well under 1 kg.

Solution

$$3(2.7 + x) = 9.3$$

$$3(2.7) + 3(x) = 9.3 \qquad \text{Distribute.}$$

$$8.1 + 3x = 9.3 \qquad \text{Make calculations.}$$

$$3x = 9.3 - 8.1 \qquad \text{Transpose.}$$

$$3x = 1.2$$

$$\frac{3x}{3} = \frac{1.2}{3} \qquad \text{Divide.}$$

$$x = 0.4$$

Interpretation

Each carton weighs 0.4 kg.

Check:
$$3(2.7 + \boxed{x}) = 9.3$$
$$3(2.7 + \boxed{0.4}) = 9.3$$
$$9.3 = 9.3$$

Tips and Traps   *Problem-Solving Tips:*

1. Read the problem carefully. Read it several times and phrase by phrase.
2. Understand all the words in the problem.
3. Analyze the problem:
   What are you asked to find?
   What facts are given?
   What facts are implied?
4. Visualize the problem.
5. Describe the conditions or relationships of the problem "symbolically."
6. Examine the options.
7. Develop a *plan* for solving the problem.
8. Write your *plan* "symbolically"; that is, write an equation.
9. Anticipate the characteristics of a reasonable solution.
10. Solve the equation.
11. Verify your answer with the conditions of the problem.

Information is often displayed in a table of rows and columns called a *spreadsheet*. These rows and columns may display results of calculations, such as totals or percents, in addition to the original data. Many computer software programs such as Lotus 123, Excel, and Quatro help you build an *electronic spreadsheet* and draw appropriate graphs through the use of formulas and equations. The calculations are made automatically once the formulas are defined in the spreadsheet. Key information is placed in a specific location on the spreadsheet called an *address,* and formulas are developed using these addresses as the variables. These spreadsheet programs are useful because key information can be changed while retaining the basic formulas of the spreadsheet. This process allows a person to see quickly how various changes in the key data will affect the result.

The 7th Inning Sports Memorabilia Shop is developing an annual operating budget. The budget categories and the projected amount of expense for each category are shown in the spreadsheet in Fig. 9–3. Additional information needed is the total operating budget for the year and the percent of total annual budget for each category in the projected annual budget. Formulas must be developed to calculate this information.

	A	B	C
1	The 7th Inning Budgeted Operating Expenses		
2			
3	Expense	Budget Amount	Percent of Total Budget
4			
5	Salaries	$42,000.00	
6	Rent	$36,000.00	
7	Depreciation	$9,000.00	
8	Utilities and Phone	$14,500.00	
9	Taxes and Insurance	$14,000.00	
10	Advertising	$3,500.00	
11	Purchases	$120,000.00	
12	Other	$500.00	
13			
14	Total		

Figure 9–3

The spreadsheet program labels the columns with the letters A, B, and C, and the rows with numbers 1 through 14. The person developing the spreadsheet uses row 1 for the title of the spreadsheet, row 3 to label the columns of data, and column A to label the rows of data. Each position on the spreadsheet is called a *cell,* and the program identifies each cell by its column letter and row number. For example, the amount budgeted for taxes and insurance is in cell B9 and is $14,000.00.

Formulas are now developed to make the needed calculations. Each spreadsheet program gives various shortcuts for writing formulas and formats for giving instructions unique to that program; however, we will write the basic concepts used, and program-specific conventions can be added as appropriate.

EXAMPLE 6   Write the formulas and make the calculations to complete the spreadsheet in Figure 9–3. To find the total to be placed in cell B14, we need to add the amounts in cells B5 through B12. We represent this in a formula by giving the addresses of the cells to be added: B14 = B5 + B6 + B7 + B8 + B9 + B10 + B11 + B12.

To calculate the percent of the total budget, the specific amount is divided by the total budget and then multiplied by 100. We will write a formula for each line of data. Most programs use an asterisk (*) to show multiplication and a forward slash (/) to show division.

$$C5 = B5/B14*100 \qquad C6 = B6/B14*100 \qquad C7 = B7/B14*100$$

$$C8 = B8/B14*100 \qquad C9 = B9/B14*100 \qquad C10 = B10/B14*100$$

$$C11 = B11/B14*100 \qquad C12 = B12/B14*100$$

There are two ways to determine the value for cell C14. If we use the percent method, the percentage and the base would be the same amount, so the total percent would be 100%. To build in a check against the spreadsheet formulas, however, it is advisable to find the total percent by adding the calculated percents. It is easy to make a typing error in the formulas or to place the formula in the wrong cell. The total should be 100% or extremely close. There may be a small discrepancy due to the effects of rounding. C14 = C5 + C6 + C7 + C8 + C9 + C10 + C11 + C12. The spreadsheet with the completed calculations is shown in Fig. 9–4.

	A	B	C
1	The 7th Inning Budgeted Operating Expenses		
2			
3	Expense	Budget Amount	Percent of Total Budget
4			
5	Salaries	$42,000.00	17.5
6	Rent	$36,000.00	15.0
7	Depreciation	$9,000.00	3.8
8	Utilities and Phone	$14,500.00	6.1
9	Taxes and Insurance	$14,000.00	5.8
10	Advertising	$3,500.00	1.5
11	Purchases	$120,000.00	50.1
12	Other	$1500.00	0.2
13			
14	Total	$239,500.00	100.0

Figure 9–4

SELF-STUDY EXERCISES 9–4

1 Write the following statements as equations and solve.

1. The sum of x and 4 equals 12. Find x.

2. 4 less than 2 times a number is 6. Find the number.

3. Three parts totaling 27 lb are packaged for shipping. Two parts weigh the same. The third part weighs 3 lb less than each of the two equal parts. Find the weight of each part.

5. A wet casting weighing 4.03 kg weighs 3.97 kg after drying. Write and solve an algebraic equation to find the weight loss due to drying.

7. An engineering student purchased a new circuits text, a used Spanish text, and a graphics calculator. She remembered that the calculator cost twice as much as the Spanish book and that the circuits text cost $70.00. The total before tax was $235.00. What was the cost of the calculator and the Spanish text?

9. Phil Chu, owner of Chu's Landscape Service, knows that one of his fertilizer tanks holds twice as many gallons of liquid fertilizer as a second tank. The two tanks together hold 325 gallons. If both tanks are filled to capacity, how many gallons of fertilizer does each tank hold?

11. Develop formulas to complete the spreadsheet in Fig. 9–5 to show data for the actual expenses for the 7th Inning Sports Memorabilia Shop.

4. How many gallons of water must be added to 24 gal of pure cleaner to make 60 gal of diluted cleaner?

6. A plumber needs three times as much perforated pipe as solid pipe to lay a drain line 400 ft long. How much of each type of pipe is needed?

8. Mary Jefferson purchased a home on a square lot 150 feet on each side. She wants to enclose the entire lot with a cedar fence. One estimate for the job was for $14.00 per linear foot. How much will the fence cost?

10. Carolina Villa hired a stone mason to build a triangular flower bed at one end of her patio. She will need enough mulch to cover 60 square feet, the area of the flower bed. If the flower bed has an altitude of 10 feet, how long is the base?

12. Use the formulas to complete the spreadsheet for the 7th Inning Sports Memorabilia Shop.

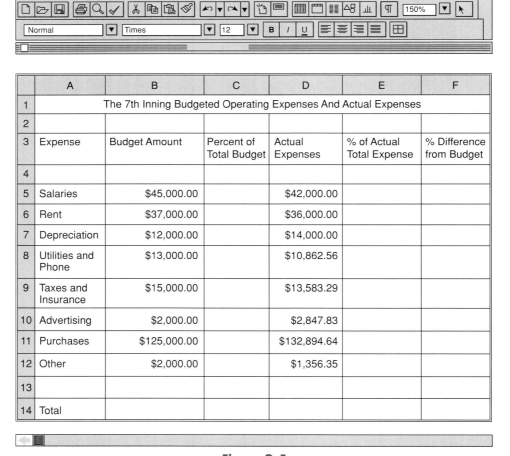

	A	B	C	D	E	F
1	The 7th Inning Budgeted Operating Expenses And Actual Expenses					
2						
3	Expense	Budget Amount	Percent of Total Budget	Actual Expenses	% of Actual Total Expense	% Difference from Budget
4						
5	Salaries	$45,000.00		$42,000.00		
6	Rent	$37,000.00		$36,000.00		
7	Depreciation	$12,000.00		$14,000.00		
8	Utilities and Phone	$13,000.00		$10,862.56		
9	Taxes and Insurance	$15,000.00		$13,583.29		
10	Advertising	$2,000.00		$2,847.83		
11	Purchases	$125,000.00		$132,894.64		
12	Other	$2,000.00		$1,356.35		
13						
14	Total					

Figure 9–5

Electronics: Kirchhoff's Laws

Kirchhoff's current law (KCL) and Kirchhoff's voltage law (KVL) form the basis of all electronics. The only difficult thing about the two laws is spelling Kirchhoff—notice that there are two h's and two f's. The rest is easy.

Kirchhoff's current law (KCL) says that *the sum of all currents at a node equals zero.* A node is like an intersection near your house. At 3 A.M. there are no cars in the intersection. If each entering car counts $+1$, and each exiting car counts -1, then the sum of all the cars will equal zero. KCL works the same way, only with current, which is measured in amperes (A). Because an ampere is rather large, measurements are often in milliamperes (mA), or thousandths of an ampere. Figure 9–6 illustrates a node.

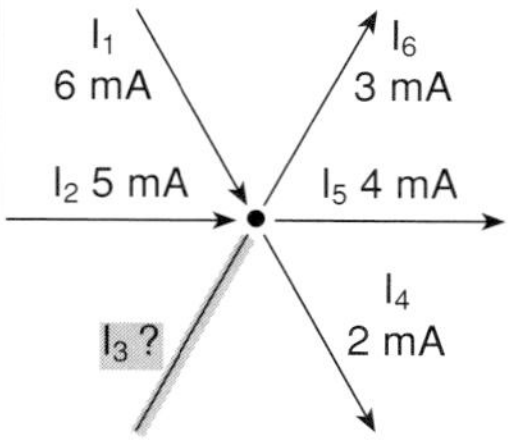

Arrows are often used to indicate current. *Let each entering arrow have a + sign and each exiting arrow have a − sign.* Assume that there is a node with two currents entering the node ($I_1 = 6$ mA, $I_2 = 5$ mA), an unknown current called I_3, and three currents exiting the node ($I_4 = 2$ mA, $I_5 = 4$ mA, and $I_6 = 3$ mA). See Fig. 9–6, circuit 1. As an equation,

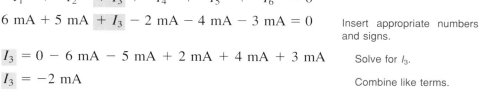

$$I_1 \ + \ I_2 \ + I_3 + \ I_4 \ + \ I_5 \ + \ I_6 \ = 0$$

$$6 \text{ mA} + 5 \text{ mA} + I_3 - 2 \text{ mA} - 4 \text{ mA} - 3 \text{ mA} = 0 \qquad \text{Insert appropriate numbers and signs.}$$

$$I_3 = 0 - 6 \text{ mA} - 5 \text{ mA} + 2 \text{ mA} + 4 \text{ mA} + 3 \text{ mA} \qquad \text{Solve for } I_3.$$

$$I_3 = -2 \text{ mA} \qquad \text{Combine like terms.}$$

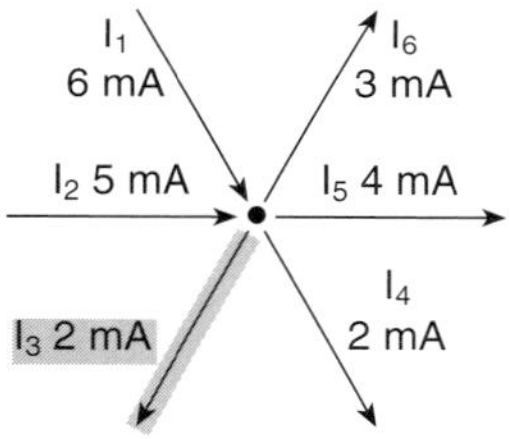

This means that I_3 *is exiting the node and equals 2 mA,* as shown in Fig. 9–7. To prove this, the numbers need to be reinserted into the original equation.

$$I_1 \ + \ I_2 \ + \ I_3 \ + \ I_4 \ + \ I_5 \ + \ I_6 \ = 0$$

$$6 \text{ mA} + 5 \text{ mA} - 2 \text{ mA} - 2 \text{ mA} - 4 \text{ mA} - 3 \text{ mA} = 0 \qquad \text{It works!}$$

Kirchhoff's voltage law (KVL) says that *the sum of all voltages in any loop equals zero.* Picture a bug walking around any loop in a circuit. Voltage is always measured as a drop in potential across something, so there is a plus sign at one end of the something and a minus at the other end. As the bug walks around the circuit, *the first sign encountered is the one used.* The equations for KVL look like those for KCL. The only difference is that a closed loop, rather than a node, is used.

Look at circuit 2 in Fig. 9–8, which has two voltage sources ($V_1 = 8$ V, $V_2 = 7$ V), an unknown voltage called V_3, and three voltage drops ($V_4 = 3$ V, $V_5 = 4$ V, $V_6 = 2$ V). Assume you start at the top of the circuit and go down the left side to the bottom and then up the right side.

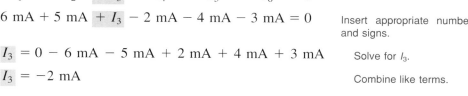

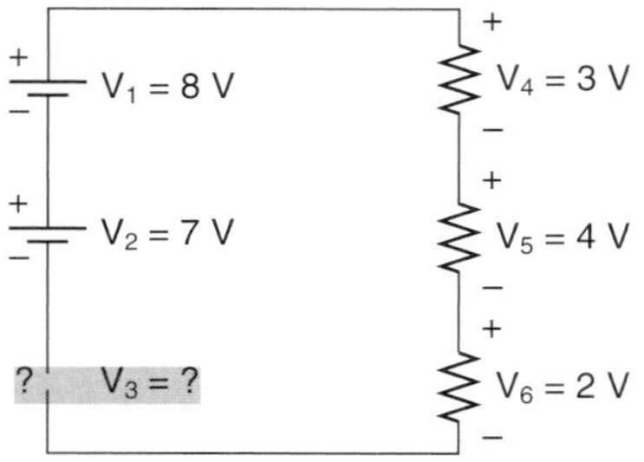

Circuit 1

Figure 9–6

Completed circuit 1

Figure 9–7

Circuit 2

Figure 9–8

As an equation then:

$$V_1 + V_2 + V_3 + V_6 + V_5 + V_4 = 0$$
$$8\text{ V} + 7\text{ V} + V_3 - 2\text{ V} - 4\text{ V} - 3\text{ V} = 0 \qquad \text{Insert appropriate numbers and signs.}$$
$$V_3 = 0 - 8\text{ V} - 7\text{ V} + 2\text{ V} + 4\text{ V} + 3\text{ V} \qquad \text{Solve for } V_3.$$
$$V_3 = -6\text{ V}$$

This means that V_3 *is a voltage drop of 6 V* as shown in Fig. 9–9.

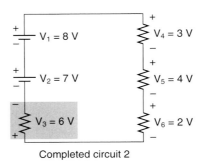

Completed circuit 2

Figure 9–9

To prove this, the numbers need to be reinserted into the original equation.

$$V_1 + V_2 + V_3 + V_4 + V_5 + V_6 = 0$$
$$8\text{ V} + 7\text{ V} - 6\text{ V} - 3\text{ V} - 4\text{ V} - 2\text{ V} = 0 \qquad \text{It works!}$$

Exercises

Redraw each circuit in Figs. 9–10 to 9–15. Find the missing currents and voltages in all the circuits and draw the correct component on the circuit. For each circuit, write a complete equation using all the values to show that the algebraic sum of the currents at a node or of the voltages in a loop equals zero.

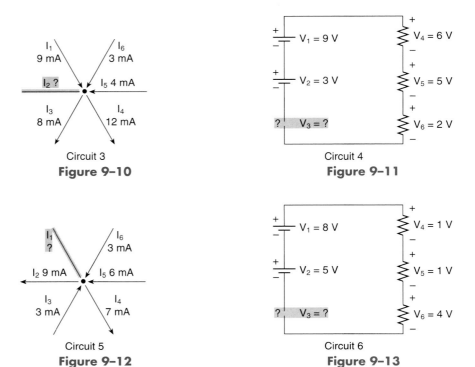

Circuit 3

Figure 9–10

Circuit 4

Figure 9–11

Circuit 5

Figure 9–12

Circuit 6

Figure 9–13

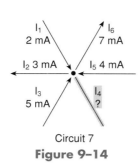

Circuit 7

Figure 9–14

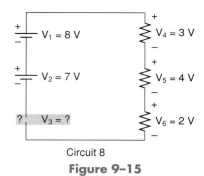

Circuit 8

Figure 9–15

Answers for Exercises

$$I_1 \quad + \quad I_2 \quad + \quad I_3 \quad + \quad I_4 \quad + \quad I_5 \quad + \quad I_6 \quad = 0$$
$$9 \text{ mA} + 4 \text{ mA} - 8 \text{ mA} - 12 \text{ mA} + 4 \text{ mA} + 3 \text{ mA} = 0$$

$$V_1 \quad + \quad V_2 \quad + \quad V_3 \quad + \quad V_4 \quad + \quad V_5 \quad + \quad V_6 \quad = 0$$
$$9 \text{ V} + 3 \text{ V} + 3 \text{ V} - 4 \text{ V} - 5 \text{ V} - 6 \text{ V} = 0$$

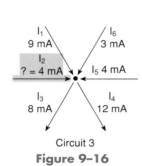

Circuit 3

Figure 9–16

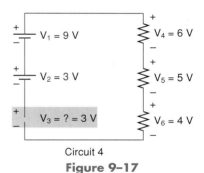

Circuit 4

Figure 9–17

$$I_1 \quad + \quad I_2 \quad + \quad I_3 \quad + \quad I_4 \quad + \quad I_5 \quad + \quad I_6 \quad = 0$$
$$4 \text{ mA} - 9 \text{ mA} + 3 \text{ mA} - 7 \text{ mA} + 6 \text{ mA} + 3 \text{ mA} = 0$$

$$V_1 \quad + \quad V_2 \quad + \quad V_3 \quad + \quad V_4 \quad + \quad V_5 \quad + \quad V_6 \quad = 0$$
$$8 \text{ V} + 5 \text{ V} - 7 \text{ V} - 4 \text{ V} - 1 \text{ V} - 1 \text{ V} = 0$$

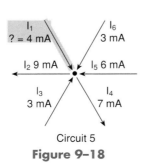

Circuit 5

Figure 9–18

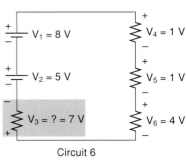

Circuit 6

Figure 9–19

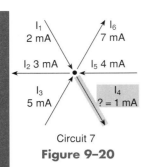

Circuit 7
Figure 9–20

$$I_1 \ + \ I_2 \ + \ I_3 \ + \ \boxed{I_4} \ + \ I_5 \ + \ I_6 \ = 0$$
$$2 \text{ mA} - 3 \text{ mA} + 5 \text{ mA} \ \underline{-1 \text{ mA}} + 4 \text{ mA} - 7 \text{ mA} = 0$$

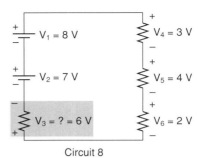

Circuit 8
Figure 9–21

$$V_1 \ + \ V_2 \ + \ \boxed{V_3} \ + \ V_4 \ + \ V_5 \ + \ V_6 \ = 0$$
$$8 \text{ V} + 7 \text{ V} \ \boxed{-6 \text{ V}} \ - 2 \text{ V} - 4 \text{ V} - 3 \text{ V} = 0$$

▮ ASSIGNMENT EXERCISES ▮

Section 9–1

Solve the following equations using multiplication or division. Check the answers.

1. $3x = 21$
2. $4x = -28$
3. $-15 = 2b$
4. $-5y = 30$
5. $-7y = -49$
6. $-5 = -m$
7. $8 = -x$
8. $0.6 = -a$
9. $3 = \frac{1}{5}x$
10. $-\frac{2}{7}x = 8$
11. $-\frac{3}{8}x = -24$
12. $\frac{1}{2}x = -5$
13. $42 = -\frac{6}{7}x$
14. $-\frac{5}{8}x = -10$
15. $\frac{1}{7}x = 12$

Section 9–2

Solve the following equations.

16. $5y - 7y = 14$
17. $4x + x = 25$
18. $36 = 9a - 5a$
19. $2b - 7b = 10$
20. $0 = 4t - t$
21. $21 = x + 2x$
22. $8x - 3x = 6 + 9$
23. $20 - 4 = 2x - 6x$
24. $13 - 27 = 3x - 10x$
25. $x + 7 = 10$
26. $x - 5 = 3$
27. $x - 3 = -4$
28. $3 + y = -5$
29. $1 = a - 4$
30. $t + 7 = 12$
31. $3x + 4 = 19$
32. $4x - 3 = 9$
33. $15 - 3x = -6$
34. $5 = 3x - 7$
35. $-7 = 6x - 31$
36. $4 = 7 - 4x$
37. $-12 = -8 - 2x$
38. $2x + 6 = x$
39. $5x - 12 = 9x$
40. $12x + 27 = 3x$
41. $3x + 9 = 10$
42. $7x = 8x + 4$
43. $4y + 8 = 3y - 4$
44. $10 + 4x = 5 - x$
45. $7 - 4y = y + 22$
46. $7x - 1 = 4x + 17$
47. $4x - 3 = 2x + 6$
48. $y - 5 = 6y + 30$
49. $8 - 2y = 15 - 3y$
50. $6 - 7x = 15 - x$
51. $5x - 12 = 2x + 15$
52. $18x - 21 = 15x + 33$
53. $7x - 5 + 2x = 3 - 4x + 12$
54. $2x - 3 + 15 = 7x - 8 - 6x$
55. $3x - 5x + 2 = 6x - 5 + 12x$
56. $\frac{R}{7} - 6 = -R$
57. $0 = \frac{8}{9}c + \frac{1}{4}$
58. $\frac{2}{15}P - P = 4$

59. $6.7y - y = 8.4$ (round to tenths)

60. $0.9R = 0.3$ (round to tenths)

61. $0.86 = R + 0.4R$ (round to hundredths)

62. $0.04y = 0.02 - y$ (round to hundredths)

Section 9–3
Solve the following equations.

63. $18 = 6(2 - y)$

64. $4(6 + x) = 36$

65. $3x = 3(9 + 2x)$

66. $4(5x - 1) = 16$

67. $7x - 3(x - 8) = 28$

68. $4a = 8 - (a + 7)$

69. $5x = 7 + (x + 5)$

70. $3(x + 2) - 5 = 2x + 7$

71. $4(3 - x) = 2x$

72. $3(2x - 4) = 4x - 6$

73. $-2(4 - 2x) = -16$

74. $3(2x + 1) = -3$

75. $5(3 - 2x) = -5$

76. $-16 = -2(-2x + 4)$

77. $8 = 6 - 2(3x - 1)$

78. $4x - (x + 3) = 3$

79. $3(x - 1) = 18 - 2(x + 3)$

80. $-(x - 1) = 2(x + 7)$

81. $-(2x + 1) = -7$

82. $2 + 3(x - 4) = 2x - 5$

83. $7 = 3 + 4(x + 2)$

84. $7(x + 2) = -6 + 2x$

85. $3(4x + 3) = 3 - 4(x - 1)$

86. $3(2 - x) - 1 = 4(3 - x)$

Section 9–4
Write the following statements as equations and solve.

87. The difference between x and 6 equals 8. Find x.

88. Twice a number increased by 5 is 17. Find the number.

89. Five times the sum of x and 6 is 42 more than x. Find x.

90. How many gallons of water must be added to 46 gal of pure alcohol to make 100 gal of alcohol solution?

91. If one technician works 3 hours less than another and the total of their hours is 51, how many hours has each technician worked?

92. The shorter side of a carpenter's square in the shape of an L is 6 in. shorter than the longer side. If the total length of the L-shaped tool is 24 in., what is the measure of each side of the square?

93. A 6-inch square of sheet metal is formed into a circular tube. Will the tube fit into a round hole 1.5 inches in diameter?

94. LaShaundria worked 14 hours in one week and then received a $0.50 per hour pay raise. She then worked 21 hours the remainder of the week at the new pay rate. She earned $220.50 for the week. How much was her old pay rate? Her new pay rate?

95. Ms. Galendez's back yard is a rectangle whose length is twice the width. If the perimeter of the yard is 720 feet, what are the dimensions of her yard?

96. Develop spreadsheet formulas for the Samaritan Home Health Company payroll sheet shown in Table 9–1. The company pays time and a half for any hours in excess of 40 during any work week.

97. Make the calculations to complete the spreadsheet for Samaritan Home Health Company for the given week.

TABLE 9–1

Employee	Date	Hourly Rate	Hours Worked for Week	Regular Pay	Overtime Pay	Total Gross Pay
Gayden, Bertha	8/19	$ 9.25	40			
Harrover, Roy	8/19	$13.60	42			
Harshman, Carol	8/19	$ 7.80	38			
Kearney, Claude	8/19	$16.50	48			
Morgan, Relbue	8/19	$12.75	30			
Oswalt, Alisa	8/19	$ 6.45	40			
Stapleton, Iven	8/19	$ 9.15	45			
Total Gross Payroll						

Make up a word problem or situation that would use the following equations in the solution.

98. $2x + 7 = 47$

99. $3(x + 5) + 2 = 80$

1. What is a basic equation and how do you solve it?

2. Write the following in the proper sequence for solving equations. An item can be used more than once.
   (a) Apply the addition axiom to arrange the equation so that number terms are on one side and letter terms are on the other.
   (b) Apply the multiplication axiom by dividing both sides by the coefficient of the letter term or by multiplying both sides by the reciprocal of the coefficient of the letter term.
   (c) Apply the distributive property to remove any parentheses.
   (d) Combine like terms that are on the same side of the equal sign.

3. Write an example of an equation and solve it to illustrate the use of the multiplication axiom. If several steps are in the solution, identify the step that applies the multiplication axiom.

4. Write an example of an equation and solve it to illustrate the use of the addition axiom. If several steps are in the solution, identify the step that applies the addition axiom.

5. In a basic equation like $\frac{3}{4}x = \frac{5}{8}$, solve the equation in two different ways. Why are the two ways equivalent? What property of equality was used?

6. Write an example of an equation and solve it to illustrate the distributive property. If several steps are in the solution, identify the step that applies the distributive property.

7. Explain the Estimation step used in solving applied problems. Why is this step important in the problem-solving process?

Identify and explain the first mistake encountered in each problem. Then work the problem correctly.

8.
$$7(x - 3) + 2 = 4 + 2x$$
$$7x - 3 + 2 = 4 + 2x$$
$$7x - 1 = 4 + 2x$$
$$5x - 1 = 4$$
$$5x = 5$$
$$x = 1$$

9.
$$5 - 3(x + 2) = 3(x + 1)$$
$$2(x + 2) = 3(x + 1)$$
$$2x + 4 = 3x + 3$$
$$-x = -1$$
$$x = 1$$

10.
$$5x - (3x - 6) = 18$$
$$5x - 3x - 6 = 18$$
$$2x - 6 = 18$$
$$2x = 18 + 6$$
$$2x = 24$$
$$x = 12$$

Objectives	What to Remember with Examples

Section 9–1

1 Solve basic equations.

Multiply both sides of the equation by the reciprocal of the coefficient of the letter term *or* divide both sides of the equation by the coefficient of the letter term. This process applies the multiplication axiom.

Solve:

$$\frac{x}{5} = 8$$
$$\frac{5}{1}\left(\frac{x}{5}\right) = 8\left(\frac{5}{1}\right)$$
$$x = 40$$

$$-8x = 24$$
$$\frac{-8x}{-8} = \frac{24}{-8}$$
$$x = -3$$

$$\frac{1}{3}x = \frac{4}{9}$$
$$\left(\frac{3}{1}\right)\left(\frac{1}{3}x\right) = \left(\frac{4}{9}\right)\left(\frac{3}{1}\right)$$
$$x = \frac{4}{3}$$

$$7.5 = 2.5x$$
$$\frac{7.5}{2.5} = \frac{2.5x}{2.5}$$
$$3 = x$$

$$\frac{x}{0.6} = 2.9$$
$$0.6\left(\frac{x}{0.6}\right) = 0.6(2.9)$$
$$x = 1.74$$

2 Check the solutions of equations.

Substitute the value of the variable for the variable in each place that it appears in the equation. Perform operations on both sides of the equation. The two sides of the equation should be equal.

Verify that $x = 3$ is the solution for the equation $2x - 1 = 5$.

$$2(3) - 1 = 5$$
$$6 - 1 = 5$$
$$5 = 5$$

Section 9–2

1 Solve equations with like terms on the same side of the equation.

Combine the like terms on the same side of the equation. Solve the remaining basic equation.

Solve $3x - 5x = 12$ for x.

$$3x - 5x = 12$$
$$-2x = 12$$
$$\frac{-2x}{-2} = \frac{12}{-2}$$
$$x = -6$$

2 Solve equations with like terms on opposite sides of the equation.

Use the addition axiom to move letter terms to one side of the equation and number terms to the other. Combine like terms. Solve the remaining basic equation.

Solve $x - 5 = 7$ for x.

$$x - 5 = 7$$
$$x - 5 + 5 = 7 + 5$$
$$x = 12$$

Solve $7.2 = x - 3.5$ for x.

$$7.2 = x - 3.5$$
$$7.2 + 3.5 = x$$
$$10.7 = x$$

Solve $x - \frac{3}{8} = \frac{5}{8}$ for x.

$$x - \frac{3}{8} = \frac{5}{8}$$

$$x = \frac{5}{8} + \frac{3}{8}$$

$$x = \frac{8}{8}$$

$$x = 1$$

Solve $3x - 5 = 5x + 7$ for x.

$$3x - 5 = 5x + 7$$
$$3x - 5x = 7 + 5$$
$$-2x = 12$$
$$\frac{-2x}{-2} = \frac{12}{-2}$$
$$x = -6$$

Section 9–3

1 Solve equations that contain parentheses.

Remove parentheses using the distributive property. Continue, solving the equation.

Solve $3(2x - 5) = 8x + 7$ for x.

$$3(2x - 5) = 8x + 7$$
$$6x - 15 = 8x + 7$$
$$6x - 8x = 7 + 15$$
$$-2x = 22$$
$$\frac{-2x}{-2} = \frac{22}{-2}$$
$$x = -11$$

Section 9–4

1 Solve applied problems using equations.

Six-Step Problem-Solving Strategy

A shipment of textbooks to a college bookstore was sent in two boxes weighing 37 lb total. One box weighed 9 lb more than the other. What was the weight of each box?

Unknown facts
Weight of each of two boxes.

Known facts
Total weight of two boxes is 37 lb.
One box weighs 9 lb more.

Relationships

$x = $ weight of one box $x + x + 9 = 37$
$x + 9 = $ weight of other box

Estimation
If the boxes were the same weight, each would weigh $17\frac{1}{2}$ lb.
Thus, one box will be less than and one box will be more than $17\frac{1}{2}$ lb.

Solution

$$x + x + 9 = 37$$
$$2x + 9 = 37$$
$$2x = 37 - 9$$
$$2x = 28$$
$$\frac{2x}{2} = \frac{28}{2}$$
$$x = 14 \text{ lb}$$
$$x + 9 = 23 \text{ lb}$$

Interpretation
The two boxes weigh 14 and 23 pounds.

CHAPTER TRIAL TEST

Solve the following equations.

1. $5x = 30$

2. $-8y = 72$

3. $\dfrac{x}{2} = 5$

4. $-\dfrac{3}{5}m = -6$

5. $x = 5 - 8$

6. $5x + 2x = 49$

7. $y - 2y = 15$

8. $2x = 3x - 7$

9. $5 - 2x = 3x - 10$

10. $5x + 3 - 7x = 2x + 4x - 11$

11. $3(x + 4) = 18$

12. $2(x - 7) = -10$

13. $7 - (x - 3) = 4x$

14. $3x + 2 = 4(x - 1) - 1$

15. $4.5x - 3.4 = 2.1x - 0.4$

16. 5 added to x equals 16. Find x.

17. Twice the sum of a number and 3 is 10. Find the number.

18. One tank holds four times as many gallons as a smaller tank. If the tanks hold 2580 gal together, how much does each tank hold?

19. The state wildlife service reported that the deer population in 1997 on the Brinks Wildlife Management Area was estimated to be 125 animals greater than the deer population on the Henderson Wildlife Management Area. The total deer population for both wildlife management areas (WMAs) was estimated to be 3125 animals. How many deer were estimated for each WMA in 1997?

20. Helping Medical Corporation operates two hospitals in one small county. The chief executive officer reported for one month that an increase in patient use of the out-patient center at one of the hospitals was twice that of the other hospital's out-patient center. If the number of out-patients served by both out-patient facilities for the month was 2550, how many out-patients used each facility for the month?

CHAPTER
10
Equations with Fractions and Decimals

GOOD DECISIONS THROUGH TEAMWORK

The proportionality of the sides of similar triangles has given rise to the classic problem of computing the unknown height of an object, such as a build-

ing, if the building casts a shadow on the ground and a nearby object whose height is known also casts a shadow on the ground.

Your team is assigned the task of computing the height of a measurable object, such as a post holding a stop sign, by using the similar triangles formed by a team member and his or her shadow and by the object and its shadow. Measure the length of the shadow of a team member of known height or a measuring device of known height and compare to the length of the shadow cast by the object of unknown height.

Solve the similar triangles. What is the calculated height of the object chosen? Now verify that height by direct measurement. How close is the computed length to the actual length? Is there a difference? Why or why not? If there is a difference, what is the percent of difference?

Repeat the process with two other measurable objects. Find the percent of difference between the calculated amount and the measured amount. Next, find the height of an object that cannot be measured directly. Establish a reasonable range of values that your team expects the actual measure to be.

Present your team's findings to the class. Include a demonstration of the method used and the solutions of the similar right triangles. Make a chart or transparency to aid in the presentation. Defend your estimate of the unmeasurable object.

Thus far, we have used mostly integers in equations. However, real-life situations will require us to deal with various fractional and decimal quantities in solving equations.

10-1 SOLVING FRACTIONAL EQUATIONS BY CLEARING THE DENOMINATORS

Learning Objectives

1. Solve fractional equations by clearing the denominators.
2. Solve applied problems involving rate, time, and work.

1. Solve Fractional Equations by Clearing the Denominators.

A technique for solving equations that contain fractions is to *clear* the equation of all denominators in the first step. Then the resulting equation, which contains no fractions, can be solved using previously learned techniques.

This process, also called clearing the fractions, is another application of the multiplication axiom. First, we will look at an example that contains only one fraction.

EXAMPLE 1 Solve $2x + \dfrac{3}{4} = 1$ by clearing the equation of fractions first.

We can *clear* an equation of a single fraction by multiplying the *entire* equation by the denominator of the fraction. *This means multiplying each term by the denominator.*

$$2x + \frac{3}{4} = 1$$

$$4(2x) + 4\left(\frac{3}{4}\right) = 4(1) \qquad \text{Multiply each term by the denominator 4.}$$

$$4(2x) + \overset{1}{4}\left(\frac{3}{\underset{1}{4}}\right) = 4(1) \qquad \text{Reduce and multiply.}$$

$$8x + 3 = 4 \qquad \text{Use methods from Chapter 9 to finish the solution.}$$

$$8x = 4 - 3$$

$$8x = 1$$

$$\frac{8x}{8} = \frac{1}{8}$$

$$x = \mathbf{\frac{1}{8}}$$

Check: $\qquad 2x + \dfrac{3}{4} = 1$

$$\overset{1}{2}\left(\frac{1}{\underset{4}{8}}\right) + \frac{3}{4} = 1$$

$$\frac{1}{4} + \frac{3}{4} = 1$$

$$1 = 1$$

> **Rule 10–1** *To clear an equation of a single fraction:*
>
> Apply the multiplication axiom by multiplying the entire equation by the denominator of the fraction.

EXAMPLE 2 Solve $\dfrac{4d}{3} = 12$.

$$\frac{4d}{3} = 12$$

$$(\overset{1}{\cancel{3}})\frac{4d}{\underset{1}{\cancel{3}}} = (3)(12)$$ Multiply both sides by the denominator 3. Reduce where possible.

$$4d = 36$$

$$\frac{4d}{4} = \frac{36}{4}$$

$$d = \mathbf{9}$$

Check: $\dfrac{4(\overset{3}{\cancel{9}})}{\underset{1}{\cancel{3}}} = 12$

$$12 = 12$$

An advantage of this rule is that it can be applied to equations when the unknown is in the *denominator* of the fraction. For example, we can use this procedure to solve $\frac{3}{7x} = 4$. Note that the term $\frac{3}{7x}$ is *not* the product of $\frac{3}{7}$ and x, since $\frac{3}{7}(\frac{x}{1}) = \frac{3x}{7}$. Therefore, the numerical coefficient of x is *not* $\frac{3}{7}$. We solve the equation by clearing the fraction.

EXAMPLE 3 Solve $\dfrac{10}{x} = 2$.

$$\frac{10}{x} = 2$$

$$(\overset{1}{\cancel{x}})\frac{10}{\underset{1}{\cancel{x}}} = (x)2$$ Multiply both sides by denominator x. Reduce where possible.

$$10 = 2x$$

$$\frac{10}{2} = \frac{2x}{2}$$

$$\mathbf{5} = x$$

Check: $\dfrac{\overset{2}{\cancel{10}}}{\underset{1}{\cancel{5}}} = 2$

$$2 = 2$$

When we solve equations with a variable in the denominator of a fraction, we may get a "root" that will not make a true statement when substituted in the original equation. Solutions or roots that do not make a true statement in the original equation are called **extraneous roots. When solving an equation with the variable in the denominator, we *must* check the root to see if it makes a true statement in the original equation.**

One situation that will produce an extraneous root is a value that will cause the denominator of any fraction to be zero. Any such value can be called an **excluded value.** Example 4 illustrates this situation.

EXAMPLE 4 Solve $\dfrac{0}{x} = 5$.

$$\frac{0}{x} = 5$$

$$x\left(\frac{0}{x}\right) = 5(x)$$

$$\frac{0}{5} = \frac{5x}{5}$$

$$0 = x$$

Check: $\quad \dfrac{0}{0} = 5 \qquad$ Does not check.

Therefore, the equation $\frac{0}{x} = 5$ has no solution.

Another type of equation containing fractions is one in which the fraction contains more than one term in its numerator or denominator. Rule 10–1 applies in this case also.

EXAMPLE 5 Solve $\dfrac{12}{Q + 6} = 1$. Identify any excluded values.

The denominator will be zero when $Q + 6 = 0$ or when $Q = -6$. Thus, -6 is an excluded value.

$$\frac{12}{Q + 6} = 1$$

$$(Q + 6) \; \frac{12}{Q + 6} = (Q + 6)\,1 \qquad \text{Multiply both sides of the equation by the denominator, the quantity } Q + 6. \text{ Reduce where possible.}$$

$$12 = Q + 6$$

$$12 - 6 = Q \qquad \text{Transpose.}$$

$$6 = Q \qquad \text{Because the numerical coefficient of } Q \text{ is already 1, the equation is solved.}$$

Check: $\quad \dfrac{12}{6 + 6} = 1$

$$\frac{12}{12} = 1$$

$$1 = 1$$

In the next example, we will look at an equation that has more than one fraction.

EXAMPLE 6 Solve $-\frac{1}{4}x = 9 - \frac{2}{3}x$ by clearing all fractions first.

The two fractions have denominators of 4 and 3. To clear *both* fractions or denominators at the same time, we multiply each term of the *entire* equation by both denominators or by the product of 4 and 3. This product is called the **least common multiple** (LCM) of 4 and 3 and is found the same way as the least common denominator.

$$-\frac{1}{4}x = 9 - \frac{2}{3}x$$

$$(4)(3)\left(-\frac{1}{4}x\right) = (4)(3)(9) - (4)(3)\left(\frac{2}{3}x\right)$$ Multiply each term by both denominators.

$$(\overset{1}{\cancel{4}})(3)\left(-\frac{1}{\underset{1}{\cancel{4}}}x\right) = (4)(3)(9) - (4)(\overset{1}{\cancel{3}})\left(\frac{2}{\underset{1}{\cancel{3}}}x\right)$$ Reduce and multiply the remaining factors.

$$-3x = 108 - 8x \qquad \text{This equation contains no fractions.}$$

$$-3x + 8x = 108$$

$$5x = 108$$

$$\frac{5x}{5} = \frac{108}{5}$$

$$x = \boxed{\frac{108}{5}}$$

Check using a calculator:

$$-\frac{1}{4}x = 9 - \frac{2}{3}x$$

$$-\frac{1}{4}\left(\frac{108}{5}\right) = 9 - \frac{2}{3}\left(\frac{108}{5}\right)$$

Left Side:

1 $\boxed{\text{a b/c}}$ 4 $\boxed{+/-}$ $\boxed{\times}$ 108 $\boxed{\text{a b/c}}$

5 $\boxed{=}$ $\Rightarrow$ $-5\frac{2}{5}$ or -5.4

Right Side:

9 $\boxed{-}$ 2 $\boxed{\text{a b/c}}$ 3 $\boxed{\times}$ 108 $\boxed{\text{a b/c}}$

5 $\boxed{=}$ $\Rightarrow$ $-5\frac{2}{5}$ or -5.4

> **Rule 10–2** *To clear an equation of fractions:*
>
> Multiply each term of the *entire* equation by the least common multiple (LCM) of the denominators of the equation.

Clearing fractions creates larger numbers in the new equation, but all numbers will be integers instead of fractions.

EXAMPLE 7 Solve $\frac{1}{R} = \frac{1}{4} + \frac{1}{10} + \frac{1}{20}$ by clearing fractions first.

$R = 0$ is the excluded value.

$$\frac{1}{R} = \frac{1}{4} + \frac{1}{10} + \frac{1}{20}$$

The LCM is 20*R* because *R*, 4, 10, and 20 will all divide evenly into **20*R*.**

$$20R\left(\frac{1}{R}\right) = \overset{5}{\cancel{20R}}\left(\frac{1}{\cancel{4}}\right) + \overset{2}{\cancel{20R}}\left(\frac{1}{\cancel{10}}\right) + \overset{1}{\cancel{20R}}\left(\frac{1}{\cancel{20}}\right)$$

Multiply each term in the *entire* equation by 20R and reduce.

$$20 = 5R + 2R + R$$

Combine like terms.

$$\frac{20}{8} = \frac{8R}{8}$$

Divide by the coefficient 8.

$$\frac{20}{8} = R$$

$$\boxed{\frac{5}{2}} = \boldsymbol{R}$$

in lowest terms

Check: $\dfrac{1}{R} = \dfrac{1}{4} + \dfrac{1}{10} + \dfrac{1}{20}$

Substitute $\dfrac{5}{2}$ for *R*.

$$\frac{1}{\frac{5}{2}} = \frac{1}{4} + \frac{1}{10} + \frac{1}{20}$$

Left Side:

$$\frac{1}{\frac{5}{2}} = 1\left(\frac{2}{5}\right) = \frac{2}{5}$$

Right Side:

$$\frac{1}{4} + \frac{1}{10} + \frac{1}{20} =$$

$$\frac{5}{20} + \frac{2}{20} + \frac{1}{20} = \frac{8}{20} = \frac{2}{5}$$

Tips and Traps *Fractions Versus Decimals:*

Sometimes applied problems using fractional equations have their solutions expressed as decimal numbers rather than as fractions. In these cases we may want to perform the indicated division represented by the fractional solution.

The equation in Example 7 is derived from the formula for finding total resistance in a parallel dc circuit with three branches rated at 4, 10, and 20 ohms. Since ohms are expressed in decimal numbers, our fractional solution could be converted to a decimal equivalent.

$$R = \frac{5}{2} \quad \text{or} \quad 2.5 \text{ ohms}$$

---

2 Solve Applied Problems Involving Rate, Time, and Work.

Our knowledge of fractional equations now enables us to solve various types of applied problems. The first type of applied problem using fractions that we will examine deals with the amount of work done, the rate of work, and the time worked.

Formula 10–1 *Formula for amount of work:*

Amount of work done = rate of work × time worked

As with percents and other rates we have studied, a rate measure involves a unit of time. If car *A* travels 50 miles in 1 hour, then car *A*'s *rate of work* (travel) is 50 miles per 1 hour, or $\frac{50 \text{ mi}}{1 \text{ hr}}$ expressed as a fraction. If car *A* travels for 3 hours, then the rate times the time ($\frac{50 \text{ mi}}{1 \text{ hr}} \times 3$ hr) equals the total amount of work (travel) completed. Car A traveled (worked) 150 miles:

$$\text{Amount of work} = \text{rate} \times \text{time}$$

$$= \frac{50 \text{ mi}}{1 \text{ hr}} \times 3 \text{ hr}$$

The unit of time used in the rate measure must be compatible with the time measure.

$$= 150 \text{ mi}$$

EXAMPLE 8 A carpenter can install 1 door in 3 hours. How many doors can the carpenter install in 30 hours?

Known facts Rate of work = 1 door per 3 hours, $\frac{1}{3}$ door per hour, or $\frac{1 \text{ door}}{3 \text{ hr}}$
Time worked = 30 hours

Unknown facts W = amount of work or number of doors installed

Relationships Amount of work = rate of work × time worked

Estimation More than one door can be installed in 30 hr.

Solution

$$W = \frac{1 \text{ door}}{\overset{}{\underset{1}{\cancel{3} \text{ hr}}}} \times \overset{10}{\cancel{30} \text{ hr}}$$

$$W = 10 \text{ doors}$$

Interpretation **Thus, 10 doors can be installed in 30 hours.**

If two workers or machines do a job together, we find the amount of work done by each worker or machine. Combined, the amounts equal 1 total job. The basic formula for solving this kind of work problem is shown in Formula 10–2.

Formula 10–2 *Formula for completing one job when A and B are working together:*

$$\left(\begin{array}{c} A\text{'s} \\ \text{amount of} \\ \text{work} \end{array} \right) + \left(\begin{array}{c} B\text{'s} \\ \text{amount of} \\ \text{work} \end{array} \right) = 1 \text{ completed job}$$

or

$$\left(\begin{array}{c} A\text{'s} \\ \text{rate of} \times \text{time} \\ \text{work} \quad \text{worked} \end{array} \right) + \left(\begin{array}{c} B\text{'s} \\ \text{rate of} \times \text{time} \\ \text{work} \quad \text{worked} \end{array} \right) = 1 \text{ completed job}$$

EXAMPLE 9 Pipe 1 can fill a tank in 6 min and pipe 2 can fill the same tank in 8 min. How long would it take for both pipes together to fill the tank? See Fig. 10–1.

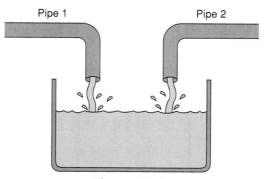

Figure 10–1

Known facts	Pipe 1 fills the tank at a rate of 1 tank per 6 min, $\frac{1}{6}$ tank per minute, or $\frac{1 \text{ tank}}{6 \text{ min}}$. Pipe 2 fills the tank at a rate of 1 tank per 8 min, $\frac{1}{8}$ tank per minute, or $\frac{1 \text{ tank}}{8 \text{ min}}$.
Unknown facts	T = time (in minutes) for both pipes together to fill the tank.
Relationships	Amount of work of Pipe 1 = $\frac{1 \text{ tank}}{6 \text{ min}} (T)$. Amount of work of Pipe 2 = $\frac{1 \text{ tank}}{8 \text{ min}} (T)$. Amount of work together = Pipe 1's work + Pipe 2's work.
Estimation	Both pipes together should fill the tank more quickly than the faster rate, or in less than 6 minutes.

Solution

$$\frac{1 \text{ tank}}{6 \text{ min}} (T \text{ min}) + \frac{1 \text{ tank}}{8 \text{ min}} (T \text{ min}) = 1 \text{ tank} \qquad \text{The LCM is } 24.$$

$$(24)\left(\frac{1}{6}T\right) + (24)\left(\frac{1}{8}T\right) = (24)(1)$$

$$(\overset{4}{24})\left(\frac{1}{6}T\right) + (\overset{3}{24})\left(\frac{1}{8}T\right) = (24)(1) \qquad \text{Reduce and multiply.}$$

$$4T + 3T = 24$$

$$7T = 24$$

$$\frac{7T}{7} = \frac{24}{7}$$

$$T = \frac{24}{7}\left(\text{or } 3\frac{3}{7}\right)\text{min}$$

Interpretation · **Both pipes working together can fill the tank in $3\frac{3}{7}$ min.**

Tips and Traps · *Improper Fractions Versus Mixed Numbers:*
In solving ordinary equations, a root that is an improper fraction, like $\frac{24}{7}$, is left as an improper fraction. However, when solving applied problems such as work problems, it is better to change improper fractions like $\frac{24}{7}$ into mixed numbers like $3\frac{3}{7}$ because $3\frac{3}{7}$ min is easier to understand than $\frac{24}{7}$ min.

Now, continuing with work problems, suppose two pipes, one a faucet and the other a drain, had opposite functions; that is, suppose Pipe 1 is filling the tank. Pipe 2 is a drain and is emptying the tank. In this case, we would subtract the work done by the drain from the work done by the faucet. This combined action, if the faucet fills at a faster rate than the drain empties, will result in a full tank. The formula for solving this type of work problem is shown in Formula 10–3.

Formula 10–3 · *Formula for completing one job when working in opposition:*

$$\begin{pmatrix} A\text{'s} \\ \text{amount of} \\ \text{work} \end{pmatrix} - \begin{pmatrix} B\text{'s} \\ \text{amount of} \\ \text{work} \end{pmatrix} = 1 \text{ completed job}$$

or

$$\begin{pmatrix} A\text{'s} \\ \text{rate of} \times \text{time} \\ \text{work} \quad \text{worked} \end{pmatrix} - \begin{pmatrix} B\text{'s} \\ \text{rate of} \times \text{time} \\ \text{work} \quad \text{worked} \end{pmatrix} = 1 \text{ completed job}$$

Here, we *subtract* the smaller amount of work from the larger.

EXAMPLE 10   Working alone, a faucet can fill a tank in 6 min. A drain working alone can empty the tank in 8 min (Fig. 10–2). If both the faucet and drain are open, in how many minutes will the tank start to overflow if the faucet is not turned off?

Faucet

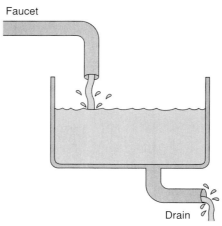

Drain

Figure 10–2

Known facts   Faucet's rate of work = 1 tank filled per 6 min, $\frac{1}{6}$ tank per min, or $\frac{1 \text{ tank}}{6 \text{ min}}$.
Drain's rate of work = 1 tank emptied per 8 min, $\frac{1}{8}$ tank per min, or $\frac{1 \text{ tank}}{8 \text{ min}}$.

Unknown facts T = time (min) until tank overflows with faucet and drain both working.

Relationships

$$\text{Amount of work of faucet} = \frac{1 \text{ tank}}{6 \text{ min}} (T)$$

$$\text{Amount of work of drain} = \frac{1 \text{ tank}}{8 \text{ min}} (T)$$

$$\begin{array}{l}\text{Amount of work when both faucet}\\ \text{and drain are open}\end{array} = \text{faucet's work} - \text{drain's work}$$

Estimation   With both the faucet and drain open, it should take longer to fill the tank than if the faucet were open and the drain closed, or longer than 6 min.

Solution

$$\frac{1 \text{ tank}}{6 \text{ min}} (T \text{ min}) - \frac{1 \text{ tank}}{8 \text{ min}} (T \text{ min}) = 1 \text{ tank}$$

$$\frac{1}{6}T - \frac{1}{8}T = 1$$

$$(24)\left(\frac{1}{6}T\right) - (24)\left(\frac{1}{8}T\right) = (24)(1)$$

$$(\overset{4}{24})\left(\frac{1}{6}T\right) - (\overset{3}{24})\left(\frac{1}{8}T\right) = (24)(1)$$

$$4T - 3T = 24$$

$$T = 24 \text{ min}$$

Interpretation   **With both pipes working, one filling and the other emptying, the tank will be full in 24 min.**

1 Solve the following fractional equations. Identify excluded values as appropriate.

1. $\dfrac{2}{7}x = 8$

2. $-7 = \dfrac{21}{33}p$

3. $\dfrac{1}{3}r = \dfrac{6}{7}$

4. $0 = -\dfrac{2}{5}c$

5. $-\dfrac{5}{3}m = 9$

6. $-9m = \dfrac{5}{3}$

7. $\dfrac{5}{8}t = 1$

8. $-\dfrac{5}{7}p = -\dfrac{11}{21}$

9. $10 = -\dfrac{1}{35}t$

10. $\dfrac{5}{12}z = 20$

11. $\dfrac{2x}{3} = 18$

12. $\dfrac{7}{Q} = 21$

13. $\dfrac{y+1}{2} = 7$

14. $\dfrac{7}{p-4} = -8$

15. $0 = \dfrac{x}{4}$

16. $-\dfrac{8}{P} = -72$

17. $\dfrac{P}{-8} = -72$

18. $-8 = \dfrac{4B}{B-6}$

19. $\dfrac{3P}{7} = 12$

20. $\dfrac{8-R}{76} = 1$

21. $\dfrac{2}{9}c + \dfrac{1}{3}c = \dfrac{3}{7}$

22. $-\dfrac{1}{4}x = 9 - \dfrac{2}{3}x$

23. $\dfrac{2}{7}y + \dfrac{3}{8} = \dfrac{1}{7}y + \dfrac{5}{3}$

24. $\dfrac{1}{3}x + \dfrac{1}{2}x = \dfrac{20}{3}$

25. $\dfrac{7}{R} - \dfrac{2}{R} = -1$

26. $S = \dfrac{1}{15} + \dfrac{1}{5} + \dfrac{1}{30}$

27. $18 - \dfrac{1}{4}x = \dfrac{1}{2}$

28. $\dfrac{1}{7}H - \dfrac{1}{3}H = 0$

29. $\dfrac{7}{16}h + \dfrac{1}{9} = \dfrac{1}{3}$

30. $x + \dfrac{1}{4}x = 8$

31. $3y + 9 = \dfrac{1}{4}y$

32. $18 = \dfrac{4}{3x} - \dfrac{3}{2x}$

33. $\dfrac{2}{7}p + 1 = \dfrac{1}{3}p$

34. $S = \dfrac{1}{10} + \dfrac{1}{25} + \dfrac{1}{50}$

35. $-x + \dfrac{1}{7} = \dfrac{1}{2}x$

36. $0 = 1 + \dfrac{2}{9}c - c$

37–42. Check or verify the roots of Exercises 12, 14, 16, 18, 25, and 32.

2 Set up an equation for each of the following. Solve each equation.

43. A licensed electrician can install 8 light fixtures in 2 hr. How many light fixtures can be installed in 20 hr?

44. A printing press can produce 1 day's newspaper in 4 hr. A higher-speed press can do 1 day's newspaper in 2 hr. How much time would it take both presses to do the 1 job?

45. One machine can pack 1 day's salmon catch in 8 hr. A second machine can pack 1 day's catch in only 5 hr. How much time would it require for 1 day's catch to be packed if both machines were used together?

46. A painter can paint a house in 6 days. Another painter takes 8 days to paint the same house. If they work together, how much time would it take them to paint the house?

47. A tank has two pipes entering it and one leaving it. Pipe 1 fills the tank in 3 min. Pipe 2 takes 7 min to fill the same tank. Pipe 3, however, empties the tank in 21 min. How much time does it take to fill the tank with all three pipes operating at the same time?

10–2 **SOLVING DECIMAL EQUATIONS**

Learning Objectives

1 Solve decimal equations.

2 Solve applied problems with decimal equations.

1 Solve Decimal Equations.

A technique similar to clearing fractions or denominators is to clear the equation of all decimals before starting to solve the equation. Like the equation cleared of all fractions, the equation cleared of all decimals contains only integers. It can then be solved using techniques previously studied.

> **Rule 10–3** *To clear an equation of decimals:*
>
> Multiply each term of the *entire* equation by the least common multiple (LCM) of the fractional amounts following the decimal points.

EXAMPLE 1   Solve $1.1 = 3.4 + R$ by first clearing the equation of decimals.

In $1.1 = 3.4 + R$, the LCM is 10 because both decimals are in tenths (have understood denominators of 10).

$$1.1 = 3.4 + R$$
$$10(1.1) = 10(3.4) + 10R \qquad \text{Multiply entire equation by } 10.$$
$$11 = 34 + 10R$$
$$11 - 34 = 10R$$
$$-\frac{23}{10} = \frac{10R}{10}$$
$$-2.3 = R$$

Check: $1.1 = 3.4 + R$
$$1.1 = 3.4 + (-2.3)$$
$$1.1 = 1.1$$

Tips and Traps   *LCM for Decimals:*
Digits to the right of the decimal point represent fractional amounts whose denominators are determined by place value position. The LCM for all the decimal numbers in an equation will be the denominator of the place value indicated by the fractional amount with the most digits to the right of the decimal.

This procedure of clearing decimals allows us to avoid having to divide an amount by a decimal, which can be a common source of error.

EXAMPLE 2   Solve $0.38 + 1.1y = 0.6$ by first clearing the equation of decimals.

$$0.38 + 1.1y = 0.6 \qquad \text{The LCM is } 100.$$
$$100(0.38) + 100(1.1y) = 100(0.6)$$
$$38 + 110y = 60$$
$$110y = 60 - 38$$
$$\frac{110y}{110} = \frac{22}{110}$$
$$y = 0.2$$

2 Solve Applied Problems with Decimal Equations.

A number of applied problems are solved with decimal equations. One common problem involves interest on a loan or on an investment.

A basic tool for interest problems is the interest formula. It resembles the percentage formula, $P = RB$, but it includes the time over which the money was borrowed or invested. If we know three of the four elements, we can find the fourth.

Formula 10–4 *Formula for Simple Interest:*

$$I = PRT$$

where I = **interest**, P = **principal**, R = **rate** or percent in decimal form, and T = **time.**

For comparison to the percentage formula, interest is the part or percentage and the principal is the base.

EXAMPLE 3   A $1000 investment is made for $2\frac{1}{2}$ years at 8.25%. Find the amount of interest.

In using the interest formula, change the percent to a decimal equivalent. Express the time in the same time measure as the rate. Then substitute the values in the formula.

$I = PRT$

$I = \$1000 \times 0.0825 \times 2.5$ years   8.25% = 0.0825; $2\frac{1}{2}$ years = 2.5 years

$I = \$206.25$

The interest for $2\frac{1}{2}$ years is $206.25.

Tips and Traps   *Clearing Decimals Versus Using a Calculator:*
Example 3 illustrates that, in some cases, clearing the decimals is not the most efficient means to solve an equation with decimals. The procedure may produce extremely large, cumbersome numbers and may not clear all the decimals. A calculator will give the solution more quickly and more efficiently if we proceed as if the equation contained whole numbers. The calculator should be used as a tool for solving mathematical problems.

Let's use the interest formula to solve problems when facts other than the interest are missing.

EXAMPLE 4   Aetna Photo Studio borrowed $3500 for some darkroom equipment and had to pay $1890 in interest over a three-year period. What was the interest rate?

Known facts   Principal = $3500, interest = $1890, and time = 3 years.

Unknown facts R = rate.

Relationships $I = PRT$

Estimation   One year's interest is about $\frac{1}{3}$ of the total interest, or about $600.

10% of $3500 = $350

20% of $3500 = $350 $\times$ 2 = $700

Solution

Therefore R is more than 10% and less than 20%.

$$1890 = 3500 \times R \times 3$$
$$1890 = 10{,}500R$$
$$\frac{1890}{10{,}500} = \frac{10{,}500R}{10{,}500}$$
$$0.18 = R \qquad\qquad \text{0.18 = 18\%}$$

Interpretation

The interest rate was 18%.

Many other real-life situations may be solved with formulas that use decimal numbers. Let's examine a few.

EXAMPLE 5

The distance formula is distance = rate × time. If a car was driven 82.5 mi at 55 miles per hour (mph), how long did the trip take?

Known facts

Distance = 82.5 mi. Rate = $\dfrac{55 \text{ mi}}{\text{hr}}$.

Unknown facts

T = time in hours

Relationships

Distance = rate × time

Estimation

At 55 mi per hr, a car would travel 110 miles in 2 hours. It would take less than 2 hours to travel 82.5 miles.

Solution

$$82.5 = 55 \times T$$
$$82.5 = 55T$$
$$\frac{82.5}{55} = \frac{55T}{55}$$
$$1.5 = T$$

Interpretation

The trip took 1.5 or $1\frac{1}{2}$ hours.

EXAMPLE 6

The formula for electrical power is $V = \frac{W}{A}$: voltage (V) equals wattage (W) divided by amperage (A). Find the voltage to the nearest hundredth needed for a circuit of 1280 W with a current of 12.23 A.

Estimate: $1200 \div 12 = 100$.

$$V = \frac{W}{A}$$
$$V = \frac{1280}{12.23}$$
$$V = 104.6606705 \qquad \text{or} \qquad 104.66 \text{ V}$$

The voltage is 104.66 V.

EXAMPLE 7

Juan earned $30.38 working 6.75 hr at a fast-food restaurant. What was his hourly wage?

Known facts

Pay = $30.38. Hours = 6.75.

Unknown facts

Wage = W

Relationships

Hourly wage × hours worked = amount of pay

Estimation

7 hr at $4 = $28. 7 hr at $5 = $35. Hourly wage is between $4 and $5.

Solution	$W \times 6.75 = 30.38$
	$6.75W = 30.38$
	$\dfrac{6.75W}{6.75} = \dfrac{30.38}{6.75}$
	$W = 4.500740741$ or \$4.50 per hr
Interpretation	**The hourly wage is \$4.50 per hr.**

SELF-STUDY EXERCISES 10–2

1   Solve the following equations.

1. $2.3x = 4.6$

2. $0.8R = 0.6$ (round to nearest tenth)

3. $0.33x + 0.25x = 3.5$ (round to nearest hundredth)

4. $0.3a = 4.8$

5. $1.5p = 7$ (round to nearest tenth)

6. $0.04x = 0.08 - x$ (round to nearest hundredth)

7. $0.4p = 0.014$

8. $0.47 = R + 0.4R$ (round to nearest hundredth)

9. $2.3 = 5.6 + y$

10. $4.3 = 0.3x - 7.34$

11. $2x + 3.7 = 10.3$

12. $0.16 + 2.3x = -0.3$

13. $1.5x + 2.1 = 3$

14. $3.82 - 2.5y = 1$

15. $0.15p = 2.4$

2   Solve the following problems using decimal equations.

16. If the formula for force is force = pressure × area, how many pounds of force are produced by a pressure of 35 psi on a piston whose surface area is 2.5 in.²? Express the pounds of force as a decimal number.

17. The circumference of a circle equals the product of π times the diameter. If a steel rod has a diameter of 1.5 in., what is the circumference of the rod to the nearest hundredth? (Circumference is the distance around a circle.)

18. The distance formula is distance = rate × time. If a trucker drove 682.5 mi at 55 miles per hour (mph), how long did she drive? (Answer to the nearest whole number.)

19. The distance formula is distance = rate × time. If a tractor-trailer rig was driven 422.5 mi at 65 miles per hour (mph) on interstate highways, how long did the trip take?

20. Electrical resistance in ohms (Ω) is voltage (V) divided by amperes (A). Find the resistance to the nearest tenth for a motor with a voltage of 12.4 volts requiring 1.5 amps.

21. Rita earned \$39.75 working 7.5 hr at a college bookstore. What was her hourly wage?

22. The formula for electrical power is $V = \frac{W}{A}$: voltage (V) equals wattage (W) divided by amperage (A). Find the voltage to the nearest hundredth needed for a circuit of 500 W with a current of 3.2 A.

23. Find the interest paid on a loan of \$2400 for 1 year at an interest rate of 11%.

24. Find the interest paid on a loan of \$800 at $8\frac{1}{2}\%$ interest for 2 years.

25. Find the total amount of money (maturity value) that the borrower will pay back on a loan of \$1400 at $12\frac{1}{2}\%$ simple interest for 3 years.

26. Find the rate of interest on an investment of \$2500 made by Nurse Honda for a period of 2 years if she received \$612.50 in interest.

27. Maddy Brown needed start-up money for her landscape service. She borrowed \$12,000 for 30 months and paid \$360 interest on the loan. What interest rate did she pay?

28. Raul Fletes needed money to buy lawn equipment. He borrowed \$500 for 7 months and paid \$53.96 in interest. What was the rate of interest?

29. Linda Davis agreed to lend money to Alex Luciano at a special interest rate of 9%, on the condition that he borrow enough that he would pay her \$500 in interest over a 2-year period. What was the minimum amount Alex could borrow?

30. Rob Thweatt needed money for medical school. He borrowed $6000 at 12% interest. If he paid $360 interest, what was the duration of the loan?

10-3 USING PROPORTIONS TO SOLVE PROBLEMS

Learning Objectives

1. Solve fractional equations that are proportions.
2. Solve problems with direct proportions.
3. Solve problems with inverse proportions.
4. Solve problems that involve similar triangles.

1 Solve Fractional Equations that Are Proportions.

Another common type of equation that contains fractions is a *proportion*. An example of a proportion is

$$\frac{N}{5} = \frac{1}{12}$$

In a proportion, *each side* of the equation is a fraction or ratio. (See Chapter 4 for an introduction to ratios and proportions.)

The term *ratio* is often used when the fraction represents quantities or measures in solving (word) problems. We will examine this use of ratios and proportions in objective 2. For now, our concern is with examining a technique for solving an equation expressed as a proportion.

The relationship of the terms of the ratios or fractions in a proportion such as

$$\frac{6}{9} = \frac{2}{3}$$

can also be stated as

"6 is to 9 as 2 is to 3"

Still another way of representing a proportion is

$6 : 9 = 2 : 3,$ which is read   "6 is to 9 as 2 is to 3"

As noted in Chapter 4, the *cross products* are equal in a proportion. A cross product is the product of the numerator of one ratio and the denominator of the other. We state this relationship symbolically in the following property.

Property 10-1   *Property of Proportions:*

$$\text{If } \frac{a}{b} = \frac{c}{d}, \text{ then } ad = bc, \qquad b, d \neq 0$$

Remember, properties do not apply when denominators are zero.

That is,

$$\text{if } \quad \frac{6}{9} = \frac{2}{3}$$

$$\text{then} \quad 6(3) = 9(2)$$

$$\text{or} \quad 18 = 18.$$

Another way of stating this is that the product of the *extremes* (end factors *a* and *d*) equals the product of the *means* (middle factors *b* and *c*).

We can use this property when solving equations in the form of a proportion.

EXAMPLE 1 Solve the following proportions.

(a) $\dfrac{x}{4} = \dfrac{9}{6}$ (b) $\dfrac{4x}{5} = \dfrac{17}{20}$ (c) $\dfrac{x-2}{x+8} = \dfrac{3}{5}$ (d) $\dfrac{3}{x} = 7$

(a) $\dfrac{x}{4} = \dfrac{9}{6}$

$6x = 36$ Cross multiply. $x(6) = 6x$; $4(9) = 36$

$\dfrac{6x}{6} = \dfrac{36}{6}$ Solve for *x*.

$x = 6$

(b) $\dfrac{4x}{5} = \dfrac{17}{20}$

$80x = 85$ Cross multiply. $4x(20) = 80x$; $5(17) = 85$

$\dfrac{80x}{80} = \dfrac{85}{80}$ Solve for *x*.

$x = \dfrac{85}{80}$ Reduce.

$x = \dfrac{17}{16}$

(c) $\dfrac{x-2}{x+8} = \dfrac{3}{5}$

$5(x-2) = 3(x+8)$ Cross multiply.

$5x - 10 = 3x + 24$ Distribute.

$5x - 3x = 24 + 10$ Transpose.

$2x = 34$ Combine terms.

$\dfrac{2x}{2} = \dfrac{34}{2}$ Solve for *x*.

$x = 17$

(d) $\dfrac{3}{x} = 7$ $7 = \dfrac{7}{1}$

$\dfrac{3}{x} = \dfrac{7}{1}$

$3 = 7x$ Cross multiply.

$\dfrac{3}{7} = \dfrac{7x}{7}$ Solve for *x*.

$\dfrac{3}{7} = x$

2 Solve Problems With Direct Proportions.

Many circumstances in problem solving involve two pairs of data that are directly related.

■ **DEFINITION 10–1: Direct Proportion.** A *direct proportion* is one in which the quantities being compared are directly related so that as one quantity increases (or decreases), the other quantity also increases (or decreases). This relationship is also called a *direct variation*.

Data are often arranged in data tables so that these relationships can be examined. If 4 apples cost $1, let's set up a data table to examine related costs.

Apples	Cost
4	$1
8	$2
12	$3
16	$4

Another way to use this relationship is to find the cost of a specified number of apples or to determine how many apples can be purchased with a specified amount of money.

Rule 10–4 *To set up a direct proportion:*

1. Establish two pairs of related data.
2. Write one pair of data in the numerators of the two ratios.
3. Write the other pair of data in the denominators of the two ratios.
4. Form a proportion using the two ratios.

EXAMPLE 2 If 4 apples cost $1, (a) find the cost of 10 apples and (b) how many apples can be purchased for $10?

(a) Pair 1: 4 apples cost $1.
   Pair 2: 10 apples cost c dollars.

$$\frac{4 \text{ apples}}{10 \text{ apples}} = \frac{\$1}{\$c}$$ Pair 1 is the numerator of each ratio.
   Pair 2 is the denominator of each ratio.

$$4c = 10$$

$$\frac{4c}{4} = \frac{10}{4}$$

$$c = 2.50$$ to nearest cent

Ten apples cost $2.50.

(b) Pair 1: 4 apples cost $1.
   Pair 2: a apples cost $10.

$$\frac{4 \text{ apples}}{a \text{ apples}} = \frac{\$1}{\$10}$$ Pair 1
   Pair 2

$$40 = a$$

Forty apples can be bought for $10.

Pairs of data that are directly related can also be set up by making each pair a ratio. In Example 2(a), a direct proportion could be written as:

$$\frac{4 \text{ apples}}{\$1} = \frac{10 \text{ apples}}{\$c}$$

The solution would be the same.

$$4c = 10$$
$$c = \$2.50$$

A third way to use data that are directly related is to identify a pair of data in which both values are known and then use that relationship to find a conversion factor to find additional information. For example, if 4 apples cost \$1, how much does 1 apple cost?

$$\$1 \div 4 \text{ apples} = \$0.25 \text{ per apple}$$

A conversion factor is used to multiply the number of items. If each apple costs \$0.25, then 10 apples would be $10 \times \$0.25$, or \$2.50. This conversion factor can also be called a *constant of direct variation*. The direct variation formula is $y = kx$, where x is the independent variable and y is the dependent variable. Another way to express this is $k = \frac{y}{x}$.

Rule 10–5 *To find the constant of direct variation:*

1. Identify a pair of related data in which both values are known.
2. Write the related pair as a ratio. The units in the denominator of the ratio should match the units of the independent variable.
3. Leave the ratio as a fraction or change it to a decimal equivalent.

Any of these three methods can be used in working with data that are directly related. However, we will continue with the first method.

EXAMPLE 3 A truck will travel 102 mi on 6 gal of gasoline. How far will it travel on 30 gal of gasoline?

Pair 1:   102 mi uses 6 gal of gasoline.

Pair 2: m mi uses 30 gal of gasoline.

Dimension Analysis

$$\frac{102 \text{ mi}}{m \text{ mi}} = \frac{6 \text{ gal}}{30 \text{ gal}} \qquad \frac{\text{distance}_1}{\text{distance}_2} = \frac{\text{gasoline}_1}{\text{gasoline}_2}$$

$$\frac{102}{m} = \frac{6}{30} \qquad \frac{\text{mi}}{\text{mi}} = \frac{\text{gal}}{\text{gal}}$$

$$102(30) = 6m \qquad \text{Cross multiply. mi(gal)} = \text{gal (mi)}$$

$$\frac{3060}{6} = \frac{6m}{6} \qquad \text{Divide by gal. Reduce. } \frac{\text{mi(gal)}}{\text{gal}} = \frac{\text{gal(mi)}}{\text{gal}}$$

$$510 = m \qquad m \text{ is expressed in miles.}$$

Thus, 510 mi can be traveled on 30 gal of gasoline.

EXAMPLE 4   If a metal rod tapers 1 in. for every 24 in. of length, what is the amount of taper of a 30-in. piece of rod? (see Fig. 10–3.)

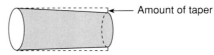

← Amount of taper

Figure 10–3

Pair 1:   1-in. taper for 24-in. length

Pair 2: x-in. taper for 30-in. length

$$\frac{1\text{-in. taper}}{x\text{-in. taper}} = \frac{24\text{-in. length}}{30\text{-in. length}}$$

$$\frac{1}{x} = \frac{24}{30}$$

$$1(30) = 24x$$

$$30 = 24x$$

$$\frac{30}{24} = \frac{24x}{24}$$

$$\frac{5}{4} = x \qquad \text{or} \qquad 1\frac{1}{4} = x$$

The amount of taper for a 30-in. length of rod is $1\frac{1}{4}$ in.

EXAMPLE 5   The ratio of chicory to coffee in a New Orleans coffee mixture is 1 : 8. If the coffee company uses 75 lb of chicory for a batch of the coffee mixture, how many pounds of coffee are needed?

Pair 1:   1 lb chicory for 8 lb coffee

Pair 2:   75 lb chicory for x lb coffee

$$\frac{1\text{ lb chicory}}{75\text{ lb chicory}} = \frac{8\text{ lb coffee}}{x\text{ lb coffee}}$$

$$\frac{1}{75} = \frac{8}{x}$$

$$1x = 75(8) \qquad \text{Cross multiply.}$$

$$x = 600$$

Thus, the coffee mixture requires 600 lb of coffee.

③ Solve Problems With Inverse Proportions.

Now that we have seen how to set up a direct proportion, let's turn our attention to relationships, where two quantities are *inversely proportional*.

■ **DEFINITION 10–2: Inverse Proportion.**   An *inverse proportion* is one in which the quantities being compared are inversely related so that as one quantity increases the other quantity decreases, or as one decreases the other increases. This relationship is also called *inverse variation*.

For example, as we *increase* pressure on materials such as foam rubber, gases, loose dirt, and so on, the materials become compressed and so *decrease* in size

394   CHAPTER 10   Equations with Fractions and Decimals or volume. Or, as we *decrease* pressure on these same materials, they *increase* in size or volume as they expand. This relationship is the opposite, or inverse, of direct proportion.

Another example is that of having 3 workers frame a house in 2 weeks. If the contractor *increases* the number of workers to 6, the time it takes would *decrease* to 1 week, assuming that the workers work at the same rate of speed. In other words, the time required is *inversely proportional* to the number of workers on the job.

Unlike the directly related ratios in a proportion, inversely related ratios do not allow us the flexibility we had in setting up ratios of unlike measures. As we proceed, we will analyze dimensions to show the problems that arise when using ratios with unlike measures.

Rule 10–6   *To set up an inverse proportion:*

1. Establish two pairs of related data.
2. Arrange one pair as the numerator of one ratio and the denominator of the other.
3. Arrange the other pair so that each ratio contains like measures.
4. Form a proportion using the two ratios.

EXAMPLE 6   If the intensity of a light illuminating a wall is 30 candelas when the light source is 10 ft from the wall, to what level does the intensity in candelas decrease if the light source is 15 ft from the wall and if the intensity is inversely proportional to the square of the distance from its source?

The relationship here is one of *inverse variation*. As the distance between the wall and light source *increases*, the intensity *decreases*.

Pair 1:   30 candelas at 10 ft

Pair 2: x candelas at 15 ft

$$\frac{30 \text{ candelas at } 10 \text{ ft}}{x \text{ candelas at } 15 \text{ ft}} = \frac{(15 \text{ ft})^2}{(10 \text{ ft})^2}$$

Dimension Analysis
$$\frac{\text{intensity}_1}{\text{intensity}_2} = \frac{\text{distance}^2}{\text{distance}^1}$$

$$\frac{30}{x} = \frac{15^2}{10^2}$$

$\frac{\text{candelas}}{\text{candelas}} = \frac{\text{ft}^2}{\text{ft}^2}$

$$30(100) = 225x$$

Cross multiply. candelas(ft^2) = ft^2(candelas)

$$3000 = 225x$$

$$\frac{3000}{225} = x$$

Divide by ft^2: $\frac{\text{candelas}(\cancel{ft^2})}{\cancel{ft^2}}$ = candelas

$$x = \textbf{13.33 candelas at 15 ft}$$

Tips and Traps   *Setting up Inverse Proportions:*
Suppose we had arranged the data so that each pair formed a ratio and then we inverted one of the ratios. Why can't we form ratios of unlike measures in our problem? Look at the value to see if the answer conforms to what we expect the answer to be.

$$\frac{30 \text{ candelas at } 10 \text{ ft}}{(10 \text{ ft})^2} = \frac{(15 \text{ ft})^2}{x \text{ candelas}}$$

$$30x = 22{,}500$$

$$x = 750$$

Did we expect the intensity to increase? No. Analyze the dimensions.

$$\frac{\text{candelas}}{\text{ft}^2} = \frac{\text{ft}^2}{\text{candelas}}$$

$$\text{candelas(candelas)} = \text{ft}^2(\text{ft}^2)$$

Is this a true statement? No. These measures are not equal.

EXAMPLE 7   If 5 machines take 12 days to complete a job, how long will it take for 8 machines to do the job?

As the number of machines *increases,* the amount of time required to do the job *decreases.* Thus, the quantities are *inversely proportional.*

Pair 1:   5 machines finish in 12 days .

Pair 2:   8 machines finish in x days .

$$\frac{5 \text{ machines}}{8 \text{ machines}} = \frac{x \text{ days for 8 machines}}{12 \text{ days for 5 machines}}$$

Each ratio uses like measures. The pairs are arranged inversely.

Dimension Analysis

$$\frac{5}{8} = \frac{x}{12}$$

$$\frac{\text{machines}}{\text{machines}} = \frac{\text{days}}{\text{days}}$$

$$5(12) = 8x$$

Cross multiply: machines(days) = machines(days)

$$60 = 8x$$

$$\frac{60}{8} = x$$

Divide by machines.
$$\frac{\text{machines (days)}}{\text{machines}} = \text{days}$$

$$x = \frac{15}{2} \quad \text{or} \quad 7\frac{1}{2} \text{ days}$$

It will take $7\frac{1}{2}$ days for 8 machines to do the job.

Inverse variation has a conversion factor similar to the constant of direct variation. The inverse variation formula is $y = \frac{k}{x}$, where x is the independent variable and y is the dependent variable. This formula can also be written as $k = xy$.

Rule 10–7   *To find the constant of indirect variation:*

1. Identify a pair of related data in which both values are known.
2. Write the related pair as a product.

The constant of indirect variation for Example 7 would be 5 machines times 12 days. To finish the example using the inverse variation formula,

$$y = \frac{k}{x}$$

$$y = \frac{(5 \text{ machines})(12 \text{ days})}{8 \text{ machines}}$$

$$y = \frac{60 \text{ days}}{8}$$

$$y = 7\frac{1}{2} \text{ days}$$

Gears and pulleys involve *inverse* relationships. Suppose that a large gear and a small gear are in mesh or a large pulley is connected by a belt to a smaller pulley. The larger gear or pulley has the *slower* speed, and the smaller gear or pulley has the *faster* speed. A bicycle is an example. The larger gear or pulley is turned by the pedals, and this motion is transmitted to the smaller gear or pulley turning the rear wheel.

If the diameter (greatest distance across a circular object) of the larger gear or pulley is increased, the larger size would mean that the larger gear or pulley would take even longer to make a complete revolution. However, it would turn the smaller gear or pulley even faster. Thus, the larger gear or pulley would revolve *more slowly* while the smaller one would revolve *more quickly*.

EXAMPLE 8 A gear measuring 10 in. across is in mesh with a gear measuring 5 in. across (Fig. 10–4). If the larger gear has a speed of 25 revolutions per minute (rpm), at how many rpm does the smaller gear turn?

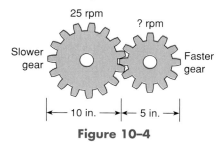

Figure 10–4

Because gears in mesh are *inversely* related, we set up an inverse proportion. Each ratio uses like measures and the ratios are in inverse order.

Pair 1:   25 rpm of larger gear for 10-in. size of larger gear.

Pair 2:   x rpm of smaller gear for 5-in. size of smaller gear.

Estimate: We expect the speed of the smaller gear to be faster than the 25-rpm speed of the larger gear.

$$\frac{25 \text{ rpm of larger gear}}{x \text{ rpm of smaller gear}} = \frac{5 \text{ in. of smaller gear}}{10 \text{ in. of larger gear}}$$

$$\frac{25}{x} = \frac{5}{10}$$

$$25(10) = 5x \qquad\qquad \text{Cross multiply.}$$

$$250 = 5x$$

$$\frac{250}{5} = x$$

$$50 = x$$

Thus, the smaller gear turns at the faster speed of 50 rpm.

4 Solve Problems That Involve Similar Triangles.

Many applications involving proportions are based on the properties of similar triangles. Before looking at applied problems using similar triangles, we will examine the definitions of similar and congruent triangles. *Congruent triangles* have the same size and shape. *Similar triangles* have the same shape but not the same size.

Every triangle has six parts: three angles and three sides. Each angle or side of one similar or congruent triangle has a corresponding angle or side in the other similar or congruent triangle. The symbol for showing congruency is $\cong$ and is read "is congruent to." The symbol for a triangle is $\triangle$.

Examine the triangles in Fig. 10–5.

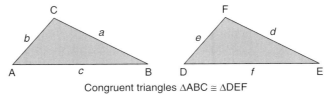

Congruent triangles $\triangle ABC \cong \triangle DEF$

Figure 10–5

Corresponding angles of congruent triangles are equal in measure.	Corresponding sides of congruent triangles are equal in measure.
Angle A = angle D	Side a = side d
Angle B = angle E	Side b = side e
Angle C = angle F	Side c = side f

The symbol for showing similarity is $\sim$ and is read "is similar to." Examine Fig. 10–6, which shows two similar triangles.

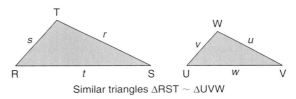

Similar triangles $\triangle RST \sim \triangle UVW$

Figure 10–6

Corresponding angles of similar triangles are equal in size.	Corresponding sides of similar triangles are directly proportional.
Angle R = angle U	Side r corresponds to side u
Angle S = angle V	Side s corresponds to side v
Angle T = angle W	Side t corresponds to side w

■ **DEFINITION 10–3: Congruent Triangles.** Two triangles are *congruent* if they have the same size and shape. Each angle of one triangle is equal to its corresponding angle in the other triangle. Each side of one triangle is equal to its corresponding side in the other triangle.

■ **DEFINITION 10–4: Similar Triangles.** Two triangles are *similar* if they have the same shape, but different size. Each angle of one triangle is equal to its corresponding angle in the other triangle. Each side of one triangle is directly proportional to its corresponding side in the other triangle.

In Fig. 10–7, $\triangle ABC$ is similar to $\triangle MNQ$. The angles that correspond are $\angle A$ and $\angle M$, $\angle B$ and $\angle N$, $\angle C$ and $\angle Q$. (The symbol for angle is $\angle$.) The angles correspond because each pair is equal. Because the triangles are similar, their corresponding sides are proportional. Corresponding sides are opposite equal angles. The corresponding sides in Fig. 10–7 are $\overline{BC}$ and $\overline{NQ}$, $\overline{AC}$ and $\overline{MQ}$, $\overline{AB}$ and $\overline{MN}$. It is common to label the angles of a triangle with capital letters and the sides opposite these angles with the corresponding lowercase letters. Because the corresponding sides are proportional, their ratios are equal. For example, we can write the proportion

$$\frac{BC}{NQ} = \frac{AC}{MQ} = \frac{AB}{MN} \qquad \text{or} \qquad \frac{a}{m} = \frac{b}{n} = \frac{c}{q}$$

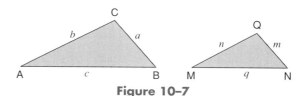

Figure 10–7

Property 10-2   *Similar triangles:*

If two triangles are *similar*, the ratios of their corresponding sides are equal.

We can use this property of similar triangles to solve many problems.

EXAMPLE 9   Find the missing side in Fig. 10–8 if $\triangle ABC \sim \triangle DEF$.

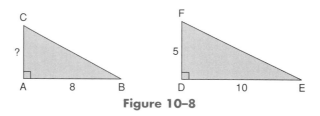

Figure 10–8

Write ratios of corresponding sides in a proportion. For convenience, single lowercase letters may be used to identify sides in similar triangles. Side a is opposite $\angle A$, side b is opposite $\angle B$, and so on.

Pair 1:   8 from $\triangle ABC$ corresponds and is proportional to 10 from $\triangle DEF$.

Pair 2: b from $\triangle ABC$ corresponds and is proportional to 5 from $\triangle DEF$.

$\dfrac{8}{b} = \dfrac{10}{5}$ Substitute values and find missing side *b*.

$10b = 8(5)$ Cross multiply.

$10b = 40$

$b = 4$

Side b is 4 units long.

EXAMPLE 10   A tree surgeon must know the height of a tree to determine which way it should fall so that it will not endanger lives, traffic, or property. A 6-ft pole makes a 4-ft shadow on the ground when the tree makes a 20-ft shadow (Fig. 10–9). What is the height of the tree?

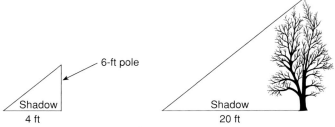

Figure 10–9

Because the triangles formed are similar, we can use a proportion to solve the problem. Let x = the height of the tree.

Pair 1:   6-ft pole casts a 4-ft shadow.

Pair 2: x-ft tree casts a 20-ft shadow.

$$\frac{6 \text{ (height of pole)}}{x \text{ (height of tree)}} = \frac{4 \text{ (shadow of pole)}}{20 \text{ (shadow of tree)}}$$

$$6(20) = 4x \qquad \text{Cross multiply.}$$
$$120 = 4x$$
$$30 = x$$

The tree is 30 ft tall.

EXAMPLE 11   A building lies between points A and B, so the distance between these points cannot be measured directly by a surveyor. Find the distance using the similar triangles shown in Fig. 10–10.

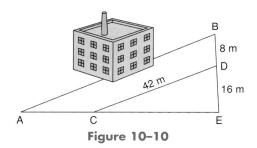

Figure 10–10

$\triangle ABE \sim \triangle CDE$. Note that CD must be made parallel to AB for the triangles to be similar. Parallel means the lines are the same distance apart from end to end.

Visualize the triangles as separate triangles (Fig. 10–11). Using the lowercase letter e to represent the missing measure of AB could lead to confusion. In $\triangle CDE$, side CD also could be thought of as side e.

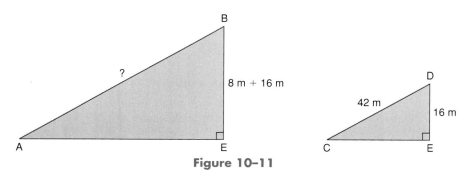

Figure 10–11

CHAPTER 10   Equations with Fractions and Decimals

Pair 1: AB from $\triangle ABE$ corresponds to CD from $\triangle CDE$.

Pair 2: BE from $\triangle ABE$ corresponds to DE from $\triangle CDE$.

$$\frac{AB}{BE} = \frac{CD}{DE} \qquad BE = BD + DE = 8 + 16 = 24$$

$$\frac{AB}{24} = \frac{42}{16}$$

$$16AB = (24)42$$

$$16AB = 1008$$

$$AB = 63 \text{ m}$$

Thus, the distance from A to B is 63 m.

SELF-STUDY EXERCISES 10–3

1 Solve the following proportions.

1. $\dfrac{x}{5} = \dfrac{9}{15}$

2. $\dfrac{3x}{16} = \dfrac{3}{8}$

3. $\dfrac{x-1}{x+6} = \dfrac{4}{5}$

4. $\dfrac{5}{x} = 8$

5. $\dfrac{2}{7} = \dfrac{x-4}{x+3}$

6. $\dfrac{2x+1}{8} = \dfrac{3}{7}$

7. $\dfrac{3x-2}{3} = \dfrac{2x+1}{3}$

8. $\dfrac{5}{2x-2} = \dfrac{1}{8}$

9. $\dfrac{8}{3x+2} = \dfrac{8}{14}$

10. $\dfrac{2x}{8} = \dfrac{3x+1}{7}$

2 Solve the following problems using proportions.

11. If 7 cans of dog food sell for \$4.13, how much will 10 cans sell for?

12. If 6 cans of coffee sell for \$22.24, how much will 20 cans sell for?

13. A mechanic took 7 hours to tune up 9 fuel-injected engines. At this rate, how many fuel-injected engines can be tuned up in 37.5 hours? Round to the nearest whole number.

14. A costume maker took 9 hours to make 4 head-pieces for a Mardi Gras ball. At this rate, how many complete headpieces can be made in 35 hours?

15. How far can a family travel in 5 days if it travels at the rate of 855 miles in 3 days?

16. How far can a tractor-trailer rig travel in 8 days if it travels at the rate of 1680 miles in 4 days?

17. How much crystallized insecticide does 275 acres of farmland need if the insecticide treats 50 acres per 100 pounds?

18. How much fertilizer does 2625 square feet of lawn need if the fertilizer treats 1575 square feet per gallon? Express the answer to the nearest tenth of a gallon.

3 Solve the following problems using proportions.

19. The fan pulley and alternator pulley are connected by a fan belt on an automobile engine. The fan pulley is 225 cm in diameter and the alternator pulley is 125 cm in diameter. If the fan pulley turns at 500 rpm, how many revolutions per minute does the faster alternator pulley turn?

20. The volume of a certain gas is inversely proportional to the pressure on it. If the gas has a volume of 160 in.^{3} under a pressure of 20 pounds per square inch (psi), what will the volume be if the pressure is decreased to 16 psi?

21. A pulley that measures 15 in. across (diameter) turns at 1600 rpm and drives a larger pulley at the rate of 1200 rpm. What is the diameter of the larger pulley in this inverse relationship?

22. Six painters can trim the exterior of all the new brick homes in a subdivision in 9 weeks. The contractor wants to have the homes ready in just 3 weeks and so needs more painters. How many painters are needed for the job? Assume that the painters all work at the same rate.

23. Two groundskeepers take 25 hr to prepare a golf course for a tournament. How long would it take 5 groundskeepers to prepare the golf course?

24. A gear measures 5 in. across. It turns another gear 2.5 in. across. If the larger gear has a speed of 25 revolutions per minute (rpm), what is the rpm of the smaller gear?

25. A small pulley 3 in. in diameter turns 250 revolutions per minute (rpm) and drives a larger pulley at 150 rpm. What is the diameter (distance across) of the larger pulley?

26. Two painters working at the same speed can paint 800 ft² of wall space in 6 hr. If a third painter paints at the same speed, how long will it take the three of them to paint the same wall space?

27. Three machines can complete a printing project in 5 hr. How many machines would be needed to finish the same project in 3 hr?

28. A gear measures 6 in. across. It is in mesh with another gear with a diameter of 3 in. If the larger gear has a speed of 60 revolutions per minute (rpm), what is the rpm of the smaller gear?

29. Nurse Lee can prepare dosages for her patients in 30 min. If she gets help from assistants who work at her rate and together they complete the preparation in 6 min, how many helpers did she get?

30. A 4.5-in. pulley turning at 1000 revolutions per minute (rpm) is belted to a large pulley turning at 500 rpm. What is the size of the larger pulley?

4 Write the corresponding parts not given for the congruent triangles in problems 31–33 (Figs. 10–12 to 10–14).

31. $AB = FE$
$BC = DE$
$AC = DF$

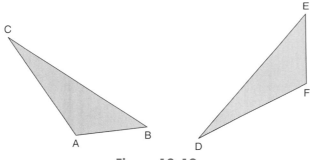

Figure 10–12

32. $\angle R$ and $\angle U = 90°$
$QR = TU$
$RP = SU$

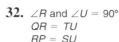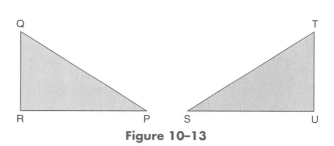

Figure 10–13

33. $JK = NM$
$\angle J = \angle M$
$\angle K = \angle N$

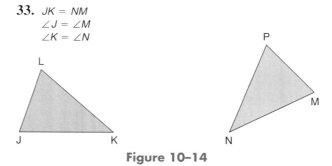

Figure 10–14

34. $\triangle ABC \sim \triangle EDF$. $\angle A = \angle E$, $\angle C = \angle F$. Find a and d. Use Fig. 10–15.

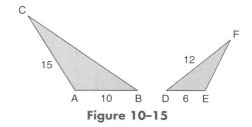

Figure 10–15

35. $\triangle ABC \sim \triangle DEC$. $\angle 1$ and $\angle 2$ have the same measure. Find DC and DE. (*Hint:* Let $DC = x$ and $AC = x + 3$. Use Fig. 10–16.)

36. Find the height of a tree that makes a 30-ft shadow when a 6-ft 6-in. pole makes a shadow of 3 ft.

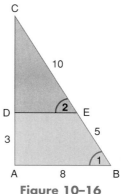

Figure 10–16

CAREER APPLICATION

Electronics: Voltage-Divider Equation

One handy equation often used in electronics is the voltage-divider equation. This equation enables you to calculate directly the voltage at a particular place in the circuit without first finding the current. In a simple circuit with one source, V_t, and two voltage drops, V_1 (across R_1) and V_2 (across R_2), the equation for V_2 is

$$V_2 = \frac{R_2 V_t}{R_1 + R_2}$$

Solve this equation for R_2. R_2 is highlighted so that you can follow it throughout the process.

$(R_1 + R_2)V_2 = \dfrac{(R_1 + R_2)R_2 V_t}{R_1 + R_2}$ Multiply both sides by the denominator.

$(R_1 + R_2)V_2 = R_2 V_t$ Distribute.

$R_1 V_2 + R_2 V_2 = R_2 V_t$ Collect terms containing R_2 on right.

$R_1 V_2 = R_2 V_t - R_2 V_2$

$R_1 V_2 = R_2(V_t - V_2)$ Reverse distributive property.

$\dfrac{R_1 V_2}{(V_t - V_2)} = \dfrac{R_2(V_t - V_2)}{(V_t - V_2)}$ Divide both sides by $(V_t - V_2)$.

$\dfrac{R_1 V_2}{(V_t - V_2)} = R_2$ or $R_2 = \dfrac{R_1 V_2}{V_t - V_2}$

Does this equation work? Let's say that $R_1 = 51\ \Omega$, $V_2 = 7$ V, and $V_t = 10$ V. Then

$$R_2 = \frac{(51\ \Omega)(7\ \text{V})}{10\ \text{V} - 7\ \text{V}}$$

$$R_2 = \frac{(51\ \Omega)7\ \text{V}}{3\ \text{V}}$$

$$R_2 = 119\ \Omega$$

Volts cancel, leaving only ohms. Now check in the original formula.

$$V_2 = \frac{R_2 V_t}{R_1 + R_2};\ 7\ \text{V} = \frac{(119\ \Omega)(10\ \text{V})}{51\ \Omega + 119\ \Omega}$$ Correct

Exercises

For each of the following, first solve using letters to get the general equation. Then prove that the new equation is correct by using the given numbers to find a numerical answer. Prove the numerical answer obtained by putting the values back in the original equation.

1. Solve for R_1 in the equation $V_1 = \dfrac{R_1 V_t}{R_1 + R_2}$. What is R_1 if $V_1 = 7$ V, $V_t = 9$ V, and $R_2 = 8$ kΩ?

2. Solve for R_2 in the equation $V_2 = \dfrac{R_2 V_t}{R_1 + R_2}$. What is R_2 if $V_2 = 3$ V, $V_t = 8$ V, and $R_1 = 120$ Ω?

Note: The next problems use the current-divider formula, which uses G (measured in siemens) and I (measured in amperes).

3. Solve for G_1 in the equation $I_1 = \dfrac{G_1 I_t}{G_1 + G_2}$. What is G_1 if $I_1 = 5$ mA, $I_t = 11$ mA, and $G_2 = 8$ mS?

4. Solve for G_2 in the equation $I_2 = \dfrac{G_2 I_t}{G_1 + G_2}$. What is G_2 if $I_2 = 15$ mA, $I_t = 35$ mA, and $G_1 = 4$ mS?

Answers for Exercises

1. $R_1 = \dfrac{-R_2 V_1}{V_1 - V_t}$ or $R_1 = \dfrac{R_2 V_1}{(V_t - V_1)}$; $28 \text{ k}\Omega = \dfrac{8 \text{ k}\Omega \, (7 \text{ V})}{(9 \text{ V} - 7 \text{ V})}$

Proof: $V_1 = \dfrac{R_1 V_t}{R_1 + R_2}$; $7 \text{ V} = \dfrac{(28 \text{ k}\Omega)(9 \text{ V})}{28 \text{ k}\Omega + 8 \text{ k}\Omega}$;

2. $R_2 = \dfrac{-R_1 V_2}{V_2 - V_t}$ or $R_2 = \dfrac{R_1 V_2}{(V_t - V_2)}$; $72 \text{ }\Omega = \dfrac{120 \text{ }\Omega \, (3 \text{ V})}{(8 \text{ V} - 3 \text{ V})}$

Proof: $V_2 = \dfrac{R_2 V_t}{R_1 + R_2}$; $3 \text{ V} = \dfrac{(72 \text{ Ohms})(8 \text{ V})}{120 \text{ Ohms} + 72 \text{ Ohms}}$;

3. $G_1 = \dfrac{-G_2 I_1}{I_1 - I_t}$ or $G_1 = \dfrac{G_2 I_1}{(I_t - I_1)}$; $6.67 \text{ mS} = \dfrac{8 \text{ mS} \, (5 \text{ mA})}{(11 \text{ mA} - 5 \text{ mA})}$

Proof: $I_1 = \dfrac{G_1 I_t}{G_1 + G_2}$; $5 \text{ mA} = \dfrac{(6.67 \text{ mS})(11 \text{ mA})}{6.67 \text{ mS} + 8 \text{ mS}}$;

4. $G_2 = \dfrac{-G_1 I_2}{I_2 - I_t}$ or $G_2 = \dfrac{G_1 I_2}{(I_t - I_2)}$; $3 \text{ mS} = \dfrac{4 \text{ mS}(15 \text{ mA})}{(35 \text{ mA} - 15 \text{ mA})}$

Proof: $I_2 = \dfrac{G_2 I_t}{G_1 + G_2}$; $15 \text{ mA} = \dfrac{(3 \text{ mS})(35 \text{ mA})}{4\text{mS} + 3 \text{ mS}}$

ASSIGNMENT EXERCISES

Section 10–1

Solve the following fractional equations.

1. $P = \dfrac{1}{2} + \dfrac{1}{3}$ **2.** $x + \dfrac{1}{7}x = 16$ **3.** $\dfrac{2}{5} - x = \dfrac{1}{2}x + \dfrac{4}{5}$

4. $\dfrac{R}{7} - 6 = -R$

5. $\dfrac{3}{7}m - \dfrac{1}{2} = \dfrac{2}{3}$

6. $0 = \dfrac{8}{9}c + \dfrac{1}{4}$

7. $\dfrac{2}{15}P - P = 4$

8. $\dfrac{1}{4}S = \dfrac{1}{4} + \dfrac{1}{10} + \dfrac{1}{20}$

9. $m = 2 + \dfrac{1}{4}m$

10. $\dfrac{7}{9} + 3 = \dfrac{1}{2}T$

11–15. Check or verify Exercises 2, 4, 6, 8, and 10.

Solve the following equations.

16. $6x + \dfrac{1}{4} = 5$

17. $2x + 4 = \dfrac{1}{2}$

18. $\dfrac{2}{5}x + 6 = \dfrac{2}{3}x$

19. $\dfrac{3}{10}x = \dfrac{1}{8}x + \dfrac{2}{5}$

20. $\dfrac{2}{5} + 3x = \dfrac{1}{10} - x$

21. $\dfrac{1}{x} = \dfrac{2}{5} + \dfrac{3}{10}$

22. $\dfrac{1}{R} = \dfrac{1}{10} + \dfrac{1}{3} + \dfrac{1}{6}$

23. $\dfrac{2}{P} = \dfrac{1}{2} + \dfrac{1}{4} - \dfrac{5}{12}$

24. $\dfrac{3}{x} + 4 = \dfrac{1}{5} - 7$

25. $\dfrac{5}{12}x - \dfrac{3}{4} = \dfrac{1}{9} - \dfrac{2}{3}x$

Set up a fractional equation for each of the following. Solve each equation.

26. A plastic tube fills a container in 10 min. A drain in the container, however, empties the container in 30 min. If the drain is open, how long will it take to fill the container?

27. Melissa can complete a landscape project in 3 hr. Henry can complete the same project in 7 hr. How long would it take Melissa and Henry working together to complete the landscape project?

28. One optical scanner can read a stack of sheets in 20 min. A second scanner can read the same stack in 12 min. How much time would it take for both scanners together to process the one stack of sheets?

29. An apprentice electrician can install 5 light fixtures in 2 hr. How many light fixtures can be installed in 10 hr?

30. A brick mason can erect a retaining wall in 6 hr. The brick mason's apprentice can do the same job in 10 hr. How much time would it take both of them working together to erect a retaining wall?

Section 10–2

Solve the following equations.

31. $3.4 = 1.5 + T$

32. $2y + 2.9 = 11.7$

33. $2.3x - 4.1 = 0.5$

34. $0.22 + 1.6x = -0.9$

35. $6.8 = 0.2y - 8.64$

36. $1.4x - 7.2 = 3.5x - 4.3$

37. $0.3x - 2.15 = 0.8x + 3.75$

38. $1.3x + 2 = 8.6x - 3.24$

39. $2.7 - x = 5 + 2x$

40. $4x - 3.2 + x = 3.3 - 2.4x$

41. $6.7y - y = 8.4$ (round to tenths)

42. $0.9R = 0.3$ (round to tenths)

43. $\dfrac{4x}{0.7} = \dfrac{3}{1.2}$ (round to hundredths)

44. $0.86 = R + 0.4R$ (round to hundredths)

45. $0.04y = 0.02 - y$ (round to hundredths)

46. $\dfrac{x}{6} = \dfrac{1.8}{3}$

47. $\dfrac{2.1}{x} = \dfrac{4.3}{7}$ (round to tenths)

48. The distance formula is distance = rate × time. If a portable MRI unit traveled 350.8 mi to and from a rural hospital at 50 miles per hour (mph), how long to the nearest hour did the trip to and from the hospital take?

49. Electrical resistance in ohms (Ω) is voltage (V) divided by amperes (A). Find the resistance of a small motor with a voltage of 8.5 volts requiring 0.5 amps.

50. Lester earned \$29.69 working $4\frac{3}{4}$ hr for a landscape service. What was his hourly wage?

51. The formula for electrical power is $V = \frac{W}{A}$: voltage (V) equals wattage (W) divided by amperage (A) or $V = \frac{W}{A}$. Find the voltage to the nearest hundredth needed for a circuit of 385 W with a current of 3.5 A.

52. If the formula for simple interest is interest = principal × rate × time, find the interest to the nearest cent on a principal of $1000 at a 19.5% rate for 1.5 years.

53. If the formula for force is force = pressure × area, how many pounds of force are produced by a pressure of 30 psi on a piston whose surface area is 12.5 in.^{2}?

Section 10–3

Solve the following proportions.

54. $\dfrac{x}{6} = \dfrac{5}{3}$

55. $\dfrac{3x}{8} = \dfrac{3}{4}$

56. $\dfrac{x-3}{x+6} = \dfrac{2}{5}$

57. $\dfrac{7}{x} = 6$

58. $\dfrac{2}{3} = \dfrac{x+3}{x-7}$

59. $\dfrac{4x+3}{15} = \dfrac{1}{3}$

60. $\dfrac{3x-2}{3} = \dfrac{2x+1}{4}$

61. $\dfrac{5}{4x-3} = \dfrac{3}{8}$

62. $\dfrac{8}{3x-2} = \dfrac{2}{3}$

63. $\dfrac{4x}{7} = \dfrac{2x+3}{3}$

64. $\dfrac{2x}{3x-2} = \dfrac{5}{8}$

65. $\dfrac{5x}{3} = \dfrac{2x+1}{4}$

66. $\dfrac{5}{9} = \dfrac{x}{2x-1}$

67. $\dfrac{7}{x} = \dfrac{5}{4x+3}$

68. $\dfrac{3}{5} = \dfrac{2x-3}{7x+4}$

Solve the following problems using proportions.

69. If 15 machines can complete a job in 6 weeks, how many machines are needed to complete the job in 4 weeks?

70. In preparing a banquet for 30 people, Cedric Henderson uses 9 lb of potatoes. How many pounds of potatoes will be needed for a banquet for 175 people?

71. A car with a speed control device travels 100 mi at 50 mi per hour (mph). The trip takes 2 hr. If the car traveled at 40 mph, how much time would the driver need to reach the same destination?

72. A 6-ft landscape engineer casts a 5-ft shadow on the ground. How tall is a nearby tree that casts a 30-ft shadow?

73. There are 25 women in a class of 35 students. If this is typical of all classes in the college, how many women are enrolled in the college if it has 6300 students all together?

74. It takes five people 7 days to clear an acre of land of debris left by a tornado; inversely, more people can do the job in less time. How long would it take seven people all working at the same rate?

75. A gear whose diameter is 45 cm is in mesh with a gear whose diameter is 30 cm. If the larger gear turns at 1000 rpm, how many revolutions does the smaller, faster gear turn per minute?

76. A certain cloth sells at a rate of 3 yd for $7.00. How many yards can Clemetee Whaley buy for $35.00?

77. Three workers take 5 days to assemble a shipment of microwave ovens; inversely, more workers can do the job in less time. How long will it take five workers to do the same job?

78. An architect's drawing is scaled at $\frac{3}{4}$ in. = 6 ft. What is the actual height of a door that measures $\frac{7}{8}$ in. on the drawing?

79. A CAD program scales a blueprint so that $\frac{5}{8}$ inch = 2 feet. What is the actual measure of a wall that is shown as $1\frac{5}{16}$ inches on the blueprint?

80. A 10-in. pulley makes 900 revolutions every minute. It drives a larger pulley at 500 rpm. What is the diameter of the larger pulley in this inverse relationship?

81. On recent trips to rural health centers, a portable mammography unit used 81.2 gal of unleaded gasoline. If the travel involved a total of 845 mi, how many gallons of gasoline would be used for travel of 1350 mi? Round to tenths.

82. A pulley whose diameter is 3.5 in. is belted to a pulley whose diameter is 8.5 in. In this inverse relationship, if the smaller, faster pulley turns at the rate of 1200 rpm, what is the rpm of the slower pulley? Round to the nearest whole number.

83. A contractor estimates that a painter can paint 300 ft^2 of wall space in 3.5 hr. How many hours should the contractor estimate for the painter to paint 425 ft^2? Express the answer to the nearest tenth.

84. A coffee company mixes 1.6 lb of chicory with every 3.5 lb of coffee. At this ratio, how many pounds of chicory are needed to mix with 2500 lb of coffee? Round to the nearest whole number.

85. A wire 825 ft long has a resistance of 1.983 ohms (Ω). How long is a wire of the same diameter if the resistance is 3.247 ohms? Round to the nearest whole foot.

86. The ceramic tile ordered for a job weighs 510 lb. If the average weight of the ceramic tile is 4.25 lb per square foot, how many square feet are estimated for this job?

87. A gear is 4 in. across. It turns a smaller gear 2 in. across. If the larger gear has a speed of 30 revolutions per minute (rpm), what is the rpm of the smaller gear?

88. A small pulley gear with a 6 in. diameter turns 350 revolutions per minute (rpm) and drives a larger pulley at 150 rpm. What is the diameter (distance across) of the larger pulley?

89. Two machines can complete a printing project in 6 hr. How many machines would be needed to finish the same project in 4 hr?

90. Waylon can install a hard drive in 10 PCs in 6 hr. He gets help from assistants who work at his rate and together they complete the installation in 2 hr. How many helpers did he get?

91. Write proportions for the similar triangles in Fig. 10–17.

92. $\triangle LMN \sim \triangle XYZ$, $\angle L = \angle X$, $\angle M = \angle Y$. Find m and x (see Fig. 10–18).

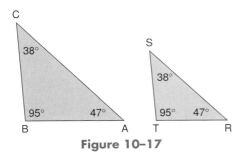

Figure 10–17

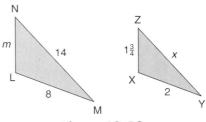

Figure 10–18

93. Find AB if $\triangle DEC \sim \triangle AEB$ (see Fig. 10–19).

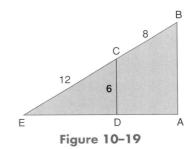

Figure 10–19

94. Make up a word problem that can be solved with a direct proportion. Include a lawn mower, tanks of gasoline, and acres to be mowed.

95. A gear turning at 130 rpm has 50 teeth. It is in mesh with another gear that turns at 65 rpm. How many teeth does the other gear have?

96. Make up a word problem that can be solved with an inverse proportion. Include a belt, pulleys, rpm's, and diameters of the pulleys.

97. The ratio of water to antifreeze in a mixture of radiator solution is 2 to 5. If the radiator is filled with 10 gallons of liquid, how much is water and how much is antifreeze?

1. What is the rule for clearing an equation of fractions or denominators?

2. When the variable is in the denominator of a fraction, why should the root always be checked?

3. What is an extraneous root?

4. How can the LCM for a decimal equation be identified easily?

5. What is the rule for clearing an equation of decimals?

6. Is it always wise to clear an equation of decimals? Explain.

7. State the simple interest formula in words and symbols.

8. Express in symbols the property of a proportion.

9. Explain the difference between a direct proportion and an inverse proportion.

10. Explain how to set up a direct proportion and an inverse proportion.

11. Give some examples of situations that are directly proportional.

12. Give some examples of situations that are inversely proportional.

Find a mistake in the following examples. Explain the mistake and correct it.

13. $\dfrac{2}{3}x = \dfrac{4}{5}$

$\left(\dfrac{3}{2}\right)\dfrac{2}{3}x = \left(\dfrac{4}{5}\right)\left(\dfrac{2}{3}\right)$

$x = \dfrac{8}{15}$

14. $0.5x + x = 3.7$

$0.6x = 3.7$

$\dfrac{0.6x}{0.6} = \dfrac{3.7}{0.6}$

$x = 6.167 \quad \text{(rounded)}$

15. $3(0.2x - 3.1) = 4.7$

$10[3(0.2x - 3.1)] = 10[4.7]$

$30(2x - 31) = 47$

$60x - 930 = 47$

$60x = 930 + 47$

$60x = 977$

$\dfrac{60x}{60} = \dfrac{977}{60}$

$x = 16.283 \quad \text{(rounded)}$

CHAPTER SUMMARY

Objectives	What to Remember with Examples
Section 10–1 **1** Solve fractional equations by clearing the denominators.	To clear the equation of denominators: multiply the *entire* equation by the least common multiple (LCM) of the denominators of the equation.

$\dfrac{3}{a} + \dfrac{1}{3} = \dfrac{2}{3a} \qquad \text{LCM} = 3a$

$(3a)\left(\dfrac{3}{a}\right) + (3a)\left(\dfrac{1}{3}\right) = (3a)\left(\dfrac{2}{3a}\right)$

$(\overset{3}{\cancel{3a}})\left(\dfrac{3}{\cancel{a}}\right) + (\overset{a}{\cancel{3a}})\left(\dfrac{1}{\cancel{3}}\right) = (\overset{1}{\cancel{3a}})\left(\dfrac{2}{\cancel{3a}}\right)$

$9 + a = 2$

$a = 2 - 9$

$a = -7$

If a variable is in a denominator, check the root to see if it makes a true statement and is not an *extraneous root*.

Check:

$$\frac{3}{-7} + \frac{1}{3} = \frac{2}{3(-7)} =$$

$$\frac{3}{-7} + \frac{1}{3} = \frac{2}{-21} \qquad \text{LCM} = 21$$

$$(21)\left(\frac{3}{-7}\right) + (21)\left(\frac{1}{3}\right) = (21)\left(\frac{2}{-21}\right)$$

$$(\overset{-3}{\cancel{21}})\left(\frac{3}{\cancel{-7}}\right) + (\overset{7}{\cancel{21}})\left(\frac{1}{\cancel{3}}\right) = (\overset{-1}{\cancel{21}})\left(\frac{2}{\cancel{-21}}\right)$$

$$-9 + 7 = -2$$
$$\qquad -2 = -2 \qquad \text{The root makes a true statement.}$$

2 Solve applied problems involving rate, time, and work.

Amount of work: rate $\times$ time = amount of work

A *rate* is a ratio such as 1 card per 15 min or $\frac{1 \text{ card}}{15 \text{ min}}$. Cancel units when solving.

Lashonda can install a PC video card in 15 min. How many cards can she install in $1\frac{1}{2}$ hours?

$$\text{rate} \times \text{time} = \text{amount of work } (W)$$

$$\frac{1 \text{ card}}{15 \text{ min}} \times 90 \text{ min} = W \qquad 1\frac{1}{2} \text{ hr} = 90 \text{ min}$$

$$6 \text{ cards} = W$$

Completing one job when working together:
(A's rate $\times$ time) + (B's rate $\times$ time) = 1 job
Let T = time. Let 1 = 1 completed job.

Galenda assembles a product in 10 min and Marcus assembles the product in 15 min. How long will it take them together to assemble one product? Galenda's time $\times$ rate: $\frac{1}{10}T$. Marcus's time $\times$ rate: $\frac{1}{15}T$

$$\frac{1}{10}T + \frac{1}{15}T = 1$$
$$\text{LCM} = 30$$

$$(30)\left(\frac{1}{10}T\right) + (30)\left(\frac{1}{15}T\right) = (30)(1)$$

$$(\overset{3}{\cancel{30}})\left(\frac{1}{\cancel{10}}T\right) + (\overset{2}{\cancel{30}})\left(\frac{1}{\cancel{15}}T\right) = (30)(1)$$

$$3T + 2T = 30$$
$$5T = 30$$
$$\frac{5T}{5} = \frac{30}{5}$$
$$T = 6 \text{ min}$$

Completing one job when working in opposition:
(A's rate $\times$ time) $-$ (B's rate $\times$ time) $= 1$ job

Solve as above but *subtract* the two amounts of work.

> An inlet valve can fill a vat in 2 hr. A drain valve can empty the vat in 5 hr. With both valves open, how long will it take for the vat to fill? Inlet valve's rate $\times$ time: $\frac{1}{2}T$. Drain valve's rate $\times$ time: $\frac{1}{5}T$
>
> $$\frac{1}{2}T - \frac{1}{5}T = 1$$
> $$\text{LCM} = 10$$
>
> $$(10)\left(\frac{1}{2}T\right) - (10)\left(\frac{1}{5}T\right) = (10)1$$
>
> $$(\overset{5}{\cancel{10}})\left(\frac{1}{\cancel{2}}T\right) - (\overset{2}{\cancel{10}})\left(\frac{1}{\cancel{5}}T\right) = (10)1$$
>
> $$5T - 2T = 10$$
> $$3T = 10$$
> $$\frac{3T}{3} = \frac{10}{3}$$
> $$T = 3\frac{1}{3}\text{ hr}$$

Section 10-2

1 Solve decimal equations.

To clear the equation of decimals: multiply the *entire* equation by the least common multiple (LCM) of the fractional amounts following the decimal point. (The fractional amount with the most places after the decimal point will be the LCM.)

> $$3.5x + 2.75 = 10$$
> $$\text{LCM} = 100$$
>
> $$(100)(3.5x) + (100)(2.75) = (100)(10)$$
>
> $$350x + 275 = 1000$$
> $$350x = 1000 - 275$$
> $$350x = 725$$
> $$\frac{350x}{350} = \frac{725}{350}$$
> $$x = 2.071428571$$
> $$x = 2.07 \qquad \text{(rounded)}$$

2 Solve applied problems with decimal equations.

In many cases, decimal equations set up to solve problems may be treated as if the decimals were whole numbers.

> John worked 2.5 hr at \$8.50 per hour. What was his pay? Let $P = $ pay.
>
> $$P = 2.5(8.50)$$
> $$P = \$21.25$$

Interest formula: $I = PRT$, where I is interest, R is rate, P is principal, and T is time. If any three elements are given, the fourth one can be found. A calculator is useful in making the calculations.

Sequoia earned $300 on an investment of $1500 for 2 years. What interest rate was she paid? Given: principal ($1500), interest ($300), and time (2 years). The rate (R) is missing.

$$I = PRT$$
$$300 = 1500 \times R \times 2$$
$$300 = 3000R$$
$$\frac{300}{3000} = \frac{3000R}{3000}$$
$$0.1 = R$$

Change decimal to a percent: $0.1 = 10\%$ (Move decimal 2 places to the right).

Section 10–3

1 Solve fractional equations that are proportions.

Property of proportions: If $\frac{a}{b} = \frac{c}{d}$, then $ad = bc$ ($b, d \neq 0$)

To solve:
1. Find the cross products.
2. Divide both sides by the coefficient of the letter term.

$$\frac{4}{2} = \frac{x}{6}$$
$$(2)(x) = (4)(6)$$
$$2x = 24$$
$$\frac{2x}{2} = \frac{24}{2}$$
$$x = 12$$

2 Solve problems with direct proportions.

Verify that the problem involves direct proportion: as one amount increases (decreases), the other amount increases (decreases).

Set up and solve direct proportion:
1. Write numerator and denominator of each ratio or fraction in *same* order.
2. Make the two ratios or fractions equal to each other.

Kinta makes $72 for 8 hr of work in a hospital business office. How many hours must he work to make $150?
Proportion is direct: as hours increase, pay increases.

$$\frac{\$72}{\$150} = \frac{8 \text{ hr}}{x \text{ hr}}$$
$$\frac{72}{150} = \frac{8}{x}$$
$$(72)(x) = (8)(150)$$
$$72x = 1200$$
$$\frac{72x}{72} = \frac{1200}{72}$$
$$x = 16\frac{2}{3} \text{ hr}$$

3 Solve problems with inverse proportions.

Verify that the problem involves inverse proportion: as one amount increases (decreases), the other amount decreases (increases).

Set up and solve the inverse proportion:
1. Write each ratio or fraction in *like* measures.
2. Write each ratio or fraction in *inverse* order.
3. Make the two ratios or fractions equal to each other.

Jane can paint a room in 4 hr. If she has two helpers who also can paint a room in 4 hr., how long will it take all three to paint the same room? Proportion is inverse: as number of painters increases, time decreases.

$$\frac{1 \text{ painter}}{3 \text{ painters}} = \frac{x \text{ hr}}{4 \text{ hr}} \text{ (inverse proportion)}$$

$$\frac{1}{3} = \frac{x}{4}$$

$$(3)(x) = (1)(4)$$

$$3x = 4$$

$$\frac{3x}{3} = \frac{4}{3}$$

$$x = 1\frac{1}{3}\text{hr}$$

4 Solve problems that involve similar triangles.

Each angle of one triangle is equal to its corresponding angle in the other triangle. Each side of one triangle is proportional to its corresponding side in the other triangle.

Find the height of a building that makes a shadow of 25 m when a meter stick makes a shadow of 1.5 m.

$$\frac{x\text{-m building}}{1\text{-m stick}} = \frac{25\text{-m building shadow}}{1.5\text{-m stick shadow}}$$

$$\frac{x}{1} = \frac{25}{1.5}$$

$$1.5\,x = 25$$

$$x = \frac{25}{1.5}$$

$$x = 16.67 \text{ m}$$

WORDS TO KNOW

excluded value (p. 379)
extraneous root (p. 379)
least common multiple (LCM)
  (p. 380)
rate of work (p. 381)
simple interest (p. 387)

principal (p. 387)
time (p. 387)
proportions (p. 390)
ratio (p. 390)
property of proportions
  (p. 390)
constant of direct variation
  (p. 393)

constant of indirect variation
  (p. 396)
direct proportion (p. 392)
inverse proportion (p. 394)
congruent triangles (p. 398)
similar triangles (p. 398)

CHAPTER TRIAL TEST

Solve the following equations containing fractions.

1. $\frac{3}{8}y = 6$

2. $4 = \frac{1}{3}x + 2$

3. $\frac{3a}{7} = 9$

4. $\frac{R}{7} = \frac{2}{5}$

5. $\frac{3 + Q}{1} = \frac{4}{5}$

6. $\frac{3}{y + 2} = \frac{2}{3}$

7. $\frac{8}{y + 2} = -7$

8. $\frac{4}{5}z + z = 8$

9. $\frac{2}{7}x = \frac{1}{2}x + 4$

10. $-\frac{2}{3}x = \frac{1}{4}x - 11$

11. $5x + \frac{3}{5} = 2$

12. $3x + 2 = \frac{2}{3}$

13. $\frac{3}{5}x + \frac{1}{10}x = \frac{1}{3}$

14. $\frac{1}{x} = \frac{1}{3} + \frac{5}{6}$

Solve the following equations containing decimal numbers. Round to hundredths when necessary.

15. $1.3x = 8.02$

16. $4.5y + 1.1 = 3.6$

17. $\frac{1.2}{x} = 4.05$

18. $\frac{3.8}{6} = \frac{0.05}{R}$

19. $0.18x = 300 - x$

20. $4.3 = 7.6 + x$

21. $3x + 1.4 = 8.9$

22. $7.9 = 0.5x - 8.35$

23. $0.23 + 7.1x = -0.8$

Solve the following problems involving fractions, decimal numbers, and proportions.

24. A pipe can fill 1 tank in 4 hr. If a second pipe can empty 1 tank in 6 hr, how long will it take for the tank to fill with both pipes operating?

25. A 9-in. gear is in mesh with a 4-in. gear. If the larger gear makes 75 rpm, how many revolutions per minute does the smaller gear make in this inverse relationship?

26. One employee can wallpaper a room in 2 hr. Another employee can wallpaper the same room in 3 hr. How long will it take both employees to wallpaper the room when working together?

27. Using the formula, pressure $= \frac{force}{area}$, how much pressure does a force of 32.75 lb exert on a surface of 24.65 in.^{2} in a hydraulic system? Answer in pounds per square inch (psi) rounded to hundredths.

28. A one-year loan for the purchase of an electronically controlled assembly machine in a factory cost the management $2758 in simple interest. If the interest rate was 19% how much did the machine cost (principal)? Round to the nearest dollar. Interest = principal × rate × time.

29. If a compact car used 62.5 L of unleaded gasoline to travel 400 mi, how many liters of gasoline would the driver use to travel 350 mi? Round to tenths.

30. If three workers take 8 days to complete a job, how many workers would be needed to finish the same job in only 6 days if each worked at the same rate? (More workers take fewer days.)

31. Resistance in a parallel dc circuit equals voltage divided by amperes. What is the resistance (in ohms) if the voltage is 40 V and amperage is 3.5 A? Express the answer as a decimal rounded to thousandths.

32. The ratio of men to women in technical and trade occupations is estimated to be 3 to 1, that is, $\frac{3}{1}$. If 56,250 men are employed in such occupations in a certain city, how many employees are women?

33. If an ice maker produces 75 lb of ice in $3\frac{1}{2}$ hr, how many pounds of ice would it produce in 5 hr? Round to the nearest whole number.

34. Find HI if $\triangle ABC \sim \triangle GHI$ (see Fig. 10–20).

35. Find DB if $\triangle ABE \sim \triangle CDE$, $CD = 9$, $AB = 12$, $DE = 15$ (see Fig. 10–21).

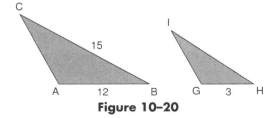

Figure 10–20

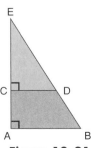

Figure 10–21

36. A model of a triangular part is shown in Fig. 10–22. Find side x if the part is to be similar to the model.

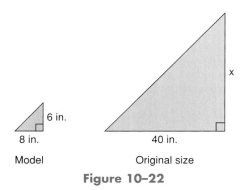

6 in.

8 in.

Model

40 in.

Original size

Figure 10–22

MATHEMATICS IN THE WORKPLACE

Business Management: Financial Ratios

Financial ratios are formed by dividing the amount of one business component by another. Typical business components include capital, revenue, costs (fixed/variable, materials/labor), profit, advertising, assets, liabilities, production levels, sales, commissions, accounts receivable, accounts payable, and net worth.

Financial ratios are used for various reasons, primarily to assess profit level (and to find profit leaks), lack of capital, areas of over- or underinvestment, and trends toward insolvency. Companies may compare their ratios with published ratios of others in their industry to find out how they compare. A company's ranking against industry standards of ratio values is also useful when a bank is deciding whether to lend them money, and when creditors are deciding whether to sell them materials.

The numeric values of financial ratios vary depending on the industry. For example, one could expect to find a high proportion of the total assets in equipment (equipment assets/total assets) for a machine-tool or chemical-processing plant, and a low proportion for a radio-assembly plant that needs little fixed assets. Financial ratios are not absolute measures; what's good for one business may spell disaster for another. A 60% ratio of fixed assets to net worth (fixed assets/net worth) for a corrugated-container plant may be satisfactory, but could result in bankruptcy for a men's suits manufacturer.

Some ratio values are viewed differently by different people. For example, the ratio

$$\text{Net-Worth Turnover} = \frac{\text{net sales}}{\text{net worth}}$$

has several meanings, and even different desired values for the following interests. From an *economic viewpoint*, this ratio measures the activity of the money that stock-holders have put up as risk. From the investors' standpoint, the higher the ratio, the better the risk. From a *financial viewpoint*, the ratio measures the velocity of the turnover of the invested capital. From a *commercial viewpoint*, it indicates overtrading or undertrading. *Overtrading* refers to a high level of sales for the capital invested, which invariably requires higher credit so that the creditors have more capital at stake than the stockholders. As the related ratio of liabilities to net worth increases, the creditors' margin of safety is reduced and the company may be in serious difficulty, with a drop in earnings or a trend toward price reduction. *Undertrading* refers to a low level of sales for the capital invested. This insufficient volume of trade can mean cash flow problems and could lead to insolvency.

Because financial ratios are approximate, even among similar companies, there is always a range of numeric values for each ratio. Within this range exists a powerful leverage for increased profits and growth. Just a few percentage points difference in some ratios holds the clue to perhaps just as many percentage points difference in profits. However, the isolated value of a ratio is not as important as its change from period to period, and its interrelation with other ratios.

In conclusion, financial ratios are used not only for management decision making, but also as barometers of financial conditions within the company and within the industry. When ratios are combined with valid and balanced judgment, they are benchmarks against which decisions can be tested. The maximum profits to be earned on the total capital employed are the result of optimum levels or optimum changes in the areas of pricing, costing, facility use, productivity, and balancing sources of investment.

CHAPTER

11

Powers and Logarithms

Powers and logarithms are used every day to describe situations of nonlinear increases or decreases. For example, investments that earn compound interest grow more than investments that earn simple interest. Formulas with variable exponents are used to determine the future accumulated amount of the investment, while formulas with logarithms can be used to determine how long it takes an investment to reach a specified goal. The formula for continuous compound interest is $A = Pe^{rt}$, where A is the accumulated amount of the investment, P is the initial amount of the investment, r

is the annual rate of interest, t is time in years, and e is a constant that is approximately equal to 2.718. The logarithmic formula, $t = (\ln A - \ln P)/r$, is used to determine how long it takes an investment to reach a specified goal.

Form pairs of teams, with three or four in each team. One team will represent an investment company trying to sell investment plans. The other team will represent consumers shopping for an investment plan for something like a child's college education, the purchase of a first home, starting a business, or planning for retirement.

The selling team will prepare and deliver a sales presentation showing examples of how the growth of a fixed investment will vary with different investment rates and times. The presentation should also show when the investor can expect a fixed investment to double or triple for various investment rates. This team should sell the consumer on the benefits of establishing an investment early in one's life.

The consumer team will determine the purpose and goal of its investment and prepare for the sales presentation by becoming knowledgeable about how the growth of a fixed investment will vary and when an investment can be expected to double or triple for various investment rates. This team should be prepared to ask questions when the selling team makes its presentation, and team members should reach a group decision on acceptance or rejection of the selling team's investment plan.

In Chapter 8, we saw that repeated multiplication formed *powers* whenever values have a positive integral exponent. In this chapter, we will extend our study of powers to include variable powers and logarithms. Powers are found in numerous applications that use scientific notation and occur frequently in formulas used to solve career-related problems.

11-1 LAWS OF EXPONENTS

Learning Objectives

1. Multiply powers with like bases.
2. Divide powers with like bases.
3. Find a power of a power.

We have seen in various equations and formulas that a variable can be all types of numbers: natural numbers, whole numbers, integers, fractions, decimals, and signed numbers. When raising a variable to a power, we must consider all the types of numbers that the variable can be. Also, as we study other types of numbers, we will look at their powers. Examine the chart of values for x^2, x^3, and x^4 in Table 11–1.

TABLE 11-1

x	x^2	x^3	x^4
5	$(5)(5) = 25$	$(5)(5)(5) = 125$	$(5)(5)(5)(5) = 625$
-3	$(-3)(-3) = 9$	$(-3)(-3)(-3) = -27$	$(-3)(-3)(-3)(-3) = 81$
$\dfrac{1}{3}$	$\left(\dfrac{1}{3}\right)\left(\dfrac{1}{3}\right) = \dfrac{1}{9}$	$\left(\dfrac{1}{3}\right)\left(\dfrac{1}{3}\right)\left(\dfrac{1}{3}\right) = \dfrac{1}{27}$	$\left(\dfrac{1}{3}\right)\left(\dfrac{1}{3}\right)\left(\dfrac{1}{3}\right)\left(\dfrac{1}{3}\right) = \dfrac{1}{81}$
0.7	$(0.7)(0.7) = 0.49$	$(0.7)(0.7)(0.7) = 0.343$	$(0.7)(0.7)(0.7)(0.7) = 0.2401$
-2.1	$(-2.1)(-2.1) = 4.41$	$(-2.1)(-2.1)(-2.1) = -9.261$	$(-2.1)(-2.1)(-2.1)(-2.1) = 19.4481$

As we begin our study of powers, we will first examine expressions with natural-number exponents. Then we will include exponents that are whole numbers, integers, fractions, decimals, and other types of numbers.

There are several laws of exponents that apply to terms that have *like* bases. The *laws of exponents* are contained in the *rules* and *definitions* presented in this section.

1 Multiply Powers With Like Bases.

To multiply 2^3 by 2^2 using repeated multiplication, we have $2(2)(2)$ times $2(2)$ or $2(2)(2)(2)(2)$. Another way of writing this is $2^3 \times 2^2 = 2^5$. We can multiply the terms x^4 and x^2. Even though the value of x is not known, $x^4(x^2)$ is $x(x)(x)(x)$ times $x(x)$ or $x(x)(x)(x)(x)(x)$.

$$x^4(x^2) = x^6$$

The following rule is a shortcut for using repeated multiplication.

> **Rule 11-1** *To multiply powers that have like bases:*
>
> The exponent of the product is the *sum* of the exponents of the powers and the base of the product remains unchanged. This rule can be stated symbolically as
>
> $$a^m(a^n) = a^{m+n}$$
>
> where a, m, and n are various types of numbers.

EXAMPLE 1   Write the products of the following expressions.

   (a) $y^4(y^3)$ (b) $a(a^2)$ (c) $b(b)$ (d) $x(x^3)(x^2)$ (e) $m(m)(m^2)$

   (a) $y^4(y^3) = y^{4+3} = \boldsymbol{y^7}$

   (b) $a(a^2) = a^{1+2} = \boldsymbol{a^3}$ Remember, when there is no written exponent, the exponent is 1.

   (c) $b(b) = b^{1+1} = \boldsymbol{b^2}$

   (d) $x(x^3)(x^2) = x^{1+3+2} = \boldsymbol{x^6}$

   (e) $m(m)(m^2) = m^{1+1+2} = \boldsymbol{m^4}$

Tips and Traps   ***Do You Always Add Exponents When Multiplying?***
Rule 11–1 applies *only* to expressions with *like* bases. Thus, $x^2(y^3)$ can only be written as x^2y^3. No other simplification can be made.

---

2   **Divide Powers With Like Bases.**

When reducing fractions, any factors common to both the numerator and denominator will cancel or reduce to a factor of 1. This concept is used when powers that have like bases are divided.

EXAMPLE 2   Reduce or simplify the following fractions; that is, perform the division indicated by each fraction.

   (a) $\dfrac{x^5}{x^2}$ (b) $\dfrac{a^2}{a}$ (c) $\dfrac{b^7}{b^5}$ (d) $\dfrac{m^3}{m^3}$ (e) $\dfrac{y^3}{y^4}$

   (a) $\dfrac{x^5}{x^2} = \dfrac{x(x)(x)\,(\cancel{x})(\cancel{x})}{\cancel{x}(\cancel{x})} = \boldsymbol{x^3}$

   (b) $\dfrac{a^2}{a} = \dfrac{a\,(\cancel{a})}{\cancel{a}} = \boldsymbol{a}$ (or a^1)

   (c) $\dfrac{b^7}{b^5} = \dfrac{b(b)\,(\cancel{b})(\cancel{b})(\cancel{b})(\cancel{b})(\cancel{b})}{(\cancel{b})(\cancel{b})(\cancel{b})(\cancel{b})(\cancel{b})} = \boldsymbol{b^2}$

   (d) $\dfrac{m^3}{m^3} = \dfrac{(\cancel{m})(\cancel{m})(\cancel{m})}{(\cancel{m})(\cancel{m})(\cancel{m})} = \boldsymbol{1}$ All factors of m reduce to 1.

   (e) $\dfrac{y^3}{y^4} = \dfrac{(\cancel{y})(\cancel{y})(\cancel{y})}{(y)\,(\cancel{y})(\cancel{y})(\cancel{y})} = \dfrac{\boldsymbol{1}}{\boldsymbol{y}}$ The denominator has more factors of y than the numerator.

Rule 11–2   *To divide powers that have like bases:*

The exponent of the quotient is the *difference* between the exponents of the dividend and divisor, and the base of the quotient remains unchanged. This rule can be stated symbolically as

$$\frac{a^m}{a^n} = a^{m-n}$$

where a, m, and n are various types of numbers except that $a \neq 0$.

Now let's apply Rule 11–2 to the problems in Example 2.

(a) $\dfrac{x^5}{x^2} = x^{5-2} = x^3$ (d) $\dfrac{m^3}{m^3} = m^{3-3} = m^0 = 1$

(b) $\dfrac{a^2}{a} = a^{2-1} = a^1$ or a (e) $\dfrac{y^3}{y^4} = y^{3-4} = y^{-1} = \dfrac{1}{y}$

(c) $\dfrac{b^7}{b^5} = b^{7-5} = b^2$

Because expressions with positive integral exponents are evaluated by using repeated multiplication, it is preferable to rewrite expressions with negative exponents as equivalent expressions with positive exponents. This manipulation is accomplished by applying the definition of negative exponents and exponents of zero.

EXAMPLE 3 Write the following quotients using positive exponents.

(a) $\dfrac{x^5}{x^8}$ (b) $\dfrac{a}{a^4}$ (c) $\dfrac{b}{b^2}$ (d) $\dfrac{y^{-3}}{y^2}$ (e) $\dfrac{x^3}{x^{-5}}$

(a) $\dfrac{x^5}{x^8} = x^{5-8} = x^{-3} = \dfrac{1}{x^3}$ (d) $\dfrac{y^{-3}}{y^2} = y^{-3-2} = y^{-5} = \dfrac{1}{y^5}$

(b) $\dfrac{a}{a^4} = a^{1-4} = a^{-3} = \dfrac{1}{a^3}$ (e) $\dfrac{x^3}{x^{-5}} = x^{3-(-5)} = x^{3+5} = x^8$

(c) $\dfrac{b}{b^2} = b^{1-2} = b^{-1} = \dfrac{1}{b}$

Tips and Traps *Negative and Positive Exponents:*
Again, the inverse relationship of multiplication and division will allow us flexibility in applying the laws of exponents.
Look again at Example 3, part (a).

$$\dfrac{x^5}{x^8} \quad \text{is} \quad x^5 \div x^8 \quad \text{or} \quad x^5 \cdot \dfrac{1}{x^8} \quad \text{or} \quad x^5 \cdot x^{-8}$$

Then, the multiplication law of exponents can be used.

$$x^5 \cdot x^{-8} = x^{5+(-8)} = x^{-3}$$

A factor can be moved from a numerator to a denominator (or vice versa) by changing the sign of the exponent.

$$x^{-2} = \dfrac{1}{x^2}, \qquad \dfrac{1}{x^{-3}} = x^3$$

This property *does not apply* to a term that is part of a numerator or denominator that has two or more terms.

$$\dfrac{x^{-2} + 1}{3} \quad \textit{does not equal} \quad \dfrac{1}{3x^2}.$$

Another way to express this caution is that *only factors* of the entire numerator or denominator can be moved. If more than one term is in the numerator, each term is divided by the denominator.

$$\dfrac{x^{-2} + 1}{3} = \dfrac{x^{-2}}{3} + \dfrac{1}{3} = \dfrac{1}{3x^2} + \dfrac{1}{3}$$

3 Find a Power of a Power.

In the term $(2^3)^2$, we have a power raised to a power. 2^3 is the base of the expression and 2 is the exponent.

$$(2^3)^2 \quad \text{is} \quad (2^3)(2^3) = 2^{3+3} = 2^6 \quad \text{or} \quad 64$$

Let's look at some examples of raising variable powers to a power.

EXAMPLE 4   Find the following powers of powers by first expressing the term as repeated multiplication and then applying the multiplication law of exponents.

(a) $(3^2)^3$ (b) $(x^3)^4$ (c) $(a^2)^2$ (d) $(y^3)^2$ (e) $(n^3)^5$

(a) $(3^2)^3 = (3^2)(3^2)(3^2) = 3^{2+2+2} = 3^6 = \mathbf{729}$

(b) $(x^3)^4 = (x^3)(x^3)(x^3)(x^3) = x^{3+3+3+3} = \mathbf{x^{12}}$

(c) $(a^2)^2 = (a^2)(a^2) = a^{2+2} = \mathbf{a^4}$

(d) $(y^3)^2 = (y^3)(y^3) = y^{3+3} = \mathbf{y^6}$

(e) $(n^3)^5 = (n^3)(n^3)(n^3)(n^3)(n^3) = n^{3+3+3+3+3} = \mathbf{n^{15}}$

From Example 4 we can see a pattern developing. This pattern leads us to the following rule.

Rule 11–3   *To raise a power to a power:*

Multiply exponents and keep the same base.

$$(a^m)^n = a^{mn}$$

Applying this rule to the problems in Example 4, we have

(a) $(3^2)^3 = 3^{2(3)} = 3^6 = 729$

(b) $(x^3)^4 = x^{3(4)} = x^{12}$

(c) $(a^2)^2 = a^{2(2)} = a^4$

(d) $(y^3)^2 = y^{3(2)} = y^6$

(e) $(n^3)^5 = n^{3(5)} = n^{15}$

There are other laws of exponents that are extensions or applications of the three laws we have already examined. While these additional laws are not necessary, they are useful tools for simplifying expressions containing exponents. These laws are expressed in the rules that follow.

Rule 11–4   *To raise a fraction or quotient to a power:*

Raise both the numerator and denominator to that power.

$$\left(\frac{a}{b}\right)^n = \frac{a^n}{b^n}, \quad b \neq 0$$

EXAMPLE 5   Raise the following fractions to the indicated power.

(a) $\left(\dfrac{2}{3}\right)^2$ (b) $\left(\dfrac{x}{y^3}\right)^3$ (c) $\left(\dfrac{-3}{4}\right)^2$ (d) $\left(\dfrac{-1}{3}\right)^3$

(a) $\left(\dfrac{2}{3}\right)^2 = \dfrac{2^2}{3^2} = \dfrac{4}{9}$

(b) $\left(\dfrac{x}{y^3}\right)^3 = \dfrac{x^3}{(y^3)^3} = \dfrac{x^3}{y^9}$

(c) $\left(\dfrac{-3}{4}\right)^2 = \dfrac{(-3)^2}{4^2} = \dfrac{9}{16}$ $(-3)(-3) = +9$

(d) $\left(\dfrac{-1}{3}\right)^3 = \dfrac{(-1)^3}{3^3} = \dfrac{-1}{27}$ $(-1)(-1)(-1) = -1$

Rule 11–5   *To raise a product to a power:*

Raise each factor to the indicated power.

$$(ab)^n = a^n b^n$$

EXAMPLE 6   Raise the following products to the indicated powers.

(a) $(ab)^2$ (b) $(a^2b)^3$ (c) $(xy^2)^2$ (d) $(3x)^2$ (e) $(2x^2y)^3$ (f) $(-5xy)^2$

(a) $(ab)^2 = a^{1(2)}b^{1(2)} = \mathbf{a^2 b^2}$

(b) $(a^2b)^3 = a^{2(3)}b^{1(3)} = \mathbf{a^6 b^3}$

(c) $(xy^2)^2 = x^{1(2)}y^{2(2)} = \mathbf{x^2 y^4}$

(d) $(3x)^2 = 3^{1(2)}x^{1(2)} = 3^2 x^2 = \mathbf{9x^2}$

(e) $(2x^2y)^3 = 2^{1(3)}x^{2(3)}y^{1(3)} = 2^3 x^6 y^3 = \mathbf{8x^6 y^3}$

(f) $(-5xy)^2 = (-5)^{1(2)}x^{1(2)}y^{1(2)} = (-5)^2 x^2 y^2 = \mathbf{25x^2 y^2}$

In applying the laws of exponents, it is very important to recognize important differences.

Tips and Traps   *Limitations of the Laws of Exponents:*

It is very important to understand what the laws of exponents *do not* include.

- Rule 11–5 applies to factors.

$$(a + b)^3 \quad \text{does not equal} \quad a^3 + b^3$$

- Rule 11–1 applies to like bases.

$$a^2(b^3) \quad \text{does not equal} \quad ab^5 \text{ or } (ab)^5$$

- An exponent affects only the one factor or grouping immediately to the left.

$$3x^2 \quad \text{and} \quad (3x)^2 \quad \text{are not equal}$$

In the term $3x^2$, the numerical coefficient 3 is multiplied times the square of x. In the term $(3x)^2$, $3x$ is squared. Thus, $(3x)^2 = 3^2 x^2 = 9x^2$.

- A negative coefficient of a base is not affected by the exponent.

$$-x^3 \quad \text{means} \quad -(x)(x)(x)$$

If $x = 4$, $-x^3 = -(4)^3$ or $-(64) = -64$. If $x = -4$, $-x^3 = -(-4)^3$ or $-(-64) = 64$.

---

1   Write the products of the following.

1. $x^3(x^4)$ **2.** $m(m^3)$ **3.** $a(a)$

4. $x^2(x^3)(x^5)$ **5.** $y(y^2)(y^3)$ **6.** $a^2(b)$

2   Write the quotients of the following. Write the answers with positive exponents.

7. $\dfrac{y^7}{y^2}$ **8.** $\dfrac{x^5}{x}$ **9.** $\dfrac{a^3}{a^4}$

10. $\dfrac{b^6}{b^5}$ **11.** $\dfrac{m^2}{m^2}$ **12.** $\dfrac{x}{x^3}$

13. $\dfrac{y^5}{y}$ **14.** $\dfrac{n^2}{n^{-5}}$ **15.** $\dfrac{x^{-4}}{x^6}$

16. $\dfrac{n^2}{n^3}$ **17.** $\dfrac{x^7}{x^0}$ **18.** $\dfrac{x^{-3}}{x^2}$

19. $\dfrac{x^{-4}}{x^{-2}}$

3   Raise the following terms to the indicated power.

20. $(x^3)^2$ **21.** $(y^7)^0$ **22.** $\left(-\dfrac{1}{2}\right)^3$

23. $\left(-\dfrac{2}{7}\right)^2$ **24.** $\left(\dfrac{a}{b}\right)^4$ **25.** $(2m^2n)^3$

26. $\left(\dfrac{x^2}{y}\right)^3$ **27.** $\left(\dfrac{3}{5}\right)^2$ **28.** $(-2a)^2$

29. $(x^2y)^3$

11–2	BASIC OPERATIONS WITH ALGEBRAIC EXPRESSIONS CONTAINING POWERS

Learning Objectives   **1**   Add or subtract like terms containing powers.

                             2   Multiply or divide algebraic expressions containing powers.

1   **Add or Subtract Like Terms Containing Powers.**

Now that variables with exponents have been introduced, we must broaden our concept of *like terms*. For letter terms to be like terms, the letters as well as the exponents of the letters must be exactly the same; that is, $2x^2$ and $-4x^2$ are like terms because the x's are both squared. But $2x^2$ and $4x$ are not like terms because the x's do not have the same exponents.

EXAMPLE 1   Tell whether the following pairs of terms are like terms.

     (a) $3a^2b$ and $-\frac{2}{3}a^2b$ (b) $-8xy^2$ and $7x^2y$ (c) $10ab^2$ and $(-2ab)^2$
     (d) $5x^4y^3$ and $-2y^3x^4$ (e) $2x^2$ and $3x^3$

     (a) **$3a^2b$ and $-\frac{2}{3}a^2b$ are like terms.** All letters and their exponents are the same.
     (b) **$-8xy^2$ and $7x^2y$ are *not* like terms.** In $-8xy^2$ the exponent of x is 1, and in $7x^2y$ the exponent of x is 2. Also, the exponents of y are not the same.

(c) **$10ab^2$ and $(-2ab)^2$ are *not* like terms.** In the term $10ab^2$ only the b is squared. In $(-2ab)^2$ the entire term is squared and the result of the squaring is $4a^2b^2$.

(d) **$5x^4y^3$ and $-2y^3x^4$ are like terms.** Both x factors have an exponent of 4 and both y factors have an exponent of 3. Because multiplication is commutative, the order of factors does not mattter.

(e) **$2x^2$ and $3x^3$ and *not* like terms.** The exponents of x are not the same.

When an algebraic expression contains several terms, we simplify the expression as much as possible by combining any like terms.

Rule 11–6 *To combine like terms:*

1. Combine the coefficients.
2. The letter factors and exponents do not change.

EXAMPLE 2 Simplify the following algebraic expressions by combining like terms.

(a) $5x^3 + 2x^3$ (b) $x^5 - 4x^5$ (c) $a^3 + 4a^2 + 3a^3 - 6a^2$
(d) $3m + 5n - m$ (e) $y + y - 5y^2 + y^3$

(a) $5x^3 + 2x^3 = (5+2)x^3 = 7x^3$

$5x^3$ and $2x^3$ are like terms; therefore, add the coefficients 5 and 2. The answer will have the same letter factor and exponent as the like terms.

(b) $x^5 - 4x^5 = (1-4)x^5 = -3x^5$

x^5 and $-4x^5$ are like terms. $1 - 4 = -3$. The answer will have the same letter factor and exponent as the like terms.

(c) $a^3 + 4a^2 + 3a^3 - 6a^2 = 4a^3 - 2a^2$

a^3 and $3a^3$ are like terms. $4a^2$ and $-6a^2$ are like terms.

(d) $3m + 5n - m = 2m + 5n$

$3m$ and $-m$ are like terms.

(e) $y + y - 5y^2 + y^3 = 2y - 5y^2 + y^3$

y and y are like terms.

When an algebraic expression contains a grouping preceded immediately by a minus sign, this means we are subtracting the entire grouping. Recall that the grouping is multiplied by -1, the implied coefficient of the grouping. This causes *each* sign within the grouping to be changed to its opposite and at the same time removes the parentheses. If the grouping is preceded by a plus sign or an unexpressed positive sign, parentheses are removed without changing signs, as if each term in the grouping were multiplied by $+1$, the implied coefficient.

EXAMPLE 3 Simplify the following expressions.

(a) $y^2 + 2y - (3y^2 + 5y)$
(b) $(m^3 - 3m^2 - 5m + 4) - (4m^3 - 2m^2 - 5m + 2)$

(a) $y^2 + 2y - (3y^2 + 5y) =$ Distribute implied coefficient of -1.

$y^2 + 2y - 1(3y^2 + 5y) =$

$y^2 + 2y - 3y^2 - 5y =$ Combine like terms.

$-2y^2 - 3y$

(b) $(m^3 - 3m^2 - 5m + 4) \; \boxed{-} \; (4m^3 - 2m^2 - 5m + 2) =$ Distribute the implied coefficient of -1.

$(m^3 - 3m^2 - 5m + 4) \; \boxed{-1} \; (4m^3 - 2m^2 - 5m + 2) =$

$\qquad m^3 - 3m^2 - 5m + 4 - 4m^3 + 2m^2 + 5m - 2 =$ Combine like terms.

$-3m^3 - m^2 + 2$

2   Multiply or Divide Algebraic Expressions Containing Powers.

> **Rule 11–7**   *To multiply or divide algebraic expressions:*
> **1.** The coefficients are multiplied or divided using the rules for signed numbers.
> **2.** The letter factors are multiplied or divided using the laws of exponents.

Look at the following examples.

EXAMPLE 4   Multiply or divide as indicated. Express answers with positive exponents.

(a) $(4x)(3x^2)$ (b) $(-6y^2)(2y^3)$ (c) $(-a)(-3a)$ (d) $\dfrac{2x^4}{x}$

(e) $\dfrac{-6y^5}{2y^3}$ (f) $\dfrac{-4x}{4x^2}$ (g) $\dfrac{-5x^4}{15x^2}$ (h) $\dfrac{3x}{12x^3}$

(a) $(4x)(3x^2) = \boxed{4(3)} \, \boxed{(x^{1+2})} = \textbf{12 } \boldsymbol{x^3}$

(b) $(-6y^2)(2y^3) = \boxed{-6(2)} \, \boxed{(y^{2+3})} = \boldsymbol{-12 \, y^5}$

(c) $(-a)(-3a) = \boxed{-1(-3)} \, \boxed{(a^{1+1})} = \textbf{3 } \boldsymbol{a^2}$
The coefficient of $-a$ is -1. The exponent of $-a$ is 1.

(d) $\dfrac{2x^4}{x} = \boxed{\dfrac{2}{1}} \, \boxed{(x^{4-1})} = \textbf{2 } \boldsymbol{x^3}$
The coefficient of x in the denominator is 1. The coefficients of the letter factors are divided using the rules for dividing signed numbers. The exponent of x is 1. The letter factors are divided using the laws of exponents.

(e) $\dfrac{-6y^5}{2y^3} = \boxed{\dfrac{-6}{2}} \, \boxed{(y^{5-3})} = \boldsymbol{-3 \, y^2}$

(f) $\dfrac{-4x}{4x^2} = \boxed{\dfrac{-4}{4}} \, \boxed{(x^{1-2})} = \boldsymbol{-1 \, x^{-1}} \qquad$ or $\qquad -\dfrac{1}{x}$
This can also be written as $\dfrac{1}{-x}$ or $\dfrac{-1}{x}$ to have a positive exponent.

(g) $\dfrac{-5x^4}{15x^2} = \boxed{\dfrac{-5}{15}} \, \boxed{(x^{4-2})} = \boldsymbol{-\dfrac{1}{3}x^2} \qquad$ or $\qquad -\dfrac{x^2}{3}$
Remember that $-\dfrac{1}{3}x^2$ is the same as $-\dfrac{1}{3}\left(\dfrac{x^2}{1}\right) \qquad$ or $\qquad -\dfrac{x^2}{3}$.

(h) $\dfrac{3x}{12x^3} = \dfrac{3}{12}(x^{1-3}) = \dfrac{1}{4}x^{-2} = \dfrac{1}{4\,x^2}$

Since $\dfrac{1}{4}x^{-2} = \dfrac{1}{4}\left(\dfrac{1}{x^2}\right)$, this can be written as $\dfrac{1}{4x^2}$.

If factors are to be multiplied times more than one term, the distributive property is applied.

EXAMPLE 5 Perform the multiplications.

(a) $2x(x^2 - 4x)$ (b) $-2y^3(2y^2 + 5y - 6)$ (c) $4a(3a^3 - 2a^2 - a)$

(a) $2x\,(x^2 - 4x) = 2x^3 - 8x^2$

(b) $-2y^3\,(2y^2 + 5y - 6) = -4y^5 - 10y^4 + 12y^3$

(c) $4a\,(3a^3 - 2a^2 - a) = 12a^4 - 8a^3 - 4a^2$

When more than one term is divided by a single term, we must divide *each* term in the dividend (numerator) by the divisor (denominator). Actually, each term in the numerator represents the numerator of a separate fraction with the given denominator.

EXAMPLE 6 Perform the divisions.

(a) $\dfrac{4x^2 - 2x + 8}{2}$ (b) $\dfrac{18a^4 + 15a^3 - 9a^2 - 12a}{3a}$

(c) $\dfrac{3x^3 - x^2}{x^3}$ (d) $\dfrac{6x^3 + 2x^2}{2x^2}$

(a) $\dfrac{4x^2 - 2x + 8}{2} = \dfrac{\overset{2}{4x^2}}{\underset{1}{2}} - \dfrac{\overset{1}{2x}}{\underset{1}{2}} + \dfrac{\overset{4}{8}}{\underset{1}{2}} = 2x^2 - x + 4$

(b) $\dfrac{18a^4 + 15a^3 - 9a^2 - 12a}{3a} = \dfrac{18a^4}{3a} + \dfrac{15a^3}{3a} - \dfrac{9a^2}{3a} - \dfrac{12a}{3a}$

$= 6a^3 + 5a^2 - 3a - 4$

(c) $\dfrac{3x^3 - x^2}{x^3} = \dfrac{3x^3}{x^3} - \dfrac{x^2}{x^3} = 3 - x^{-1}$ or $3 - \dfrac{1}{x}$

(d) $\dfrac{6x^3 + 2x^2}{2x^2} = \dfrac{6x^3}{2x^2} + \dfrac{2x^2}{2x^2} = 3x + 1$

SELF-STUDY EXERCISES 11–2

1 Simplify the following expressions.

1. $3a^2 + 4a^2$

2. $5x^3 - 2x^3$

3. $b^2 + 3a^2 + 2b^2 - 5a^2$

4. $3a - 2b - a$

5. $x - 3x - 2x^2 - 3x^2$

6. $3a^2 - 2a^2 + 4a^2$

7. $x^2 + 3y - (2x^2 + 5y)$

8. $4m^2 - 2n^2 - (2m^2 - 3n^2)$

9. $7a + 3b + 8c + 2a - (b - 2c)$

10. $5x + 3y - (7x - 2z)$

2 Perform the multiplication or division as indicated.

11. $7x(2x^2)$

12. $(-2m)(-m^2)$

13. $(-3m)(7m)$

14. $(-y^3)(2y^3)$

15. $\dfrac{6x^4}{3x^2}$

16. $\dfrac{-5a^2}{10a}$

17. $\dfrac{-7x}{-14x^3}$

18. $\dfrac{-9x^5}{12x^2}$

19. $3x(x - 6)$

20. $4x(3x^2 - 7x + 8)$

21. $-4x(2x - 3)$

22. $2x^2(5 + 2x)$

23. $\dfrac{6x^2 - 4x}{2x}$

24. $\dfrac{12x^5 - 6x^3 - 3x^2}{3x^2}$

25. $\dfrac{7x^4 - x^2}{x^3}$

26. $\dfrac{8x^4 + 6x^3}{2x^2}$

11–3   POWERS OF 10 AND SCIENTIFIC NOTATION

Learning Objectives

1. Multiply and divide by powers of 10.
2. Change a number from scientific notation to ordinary notation.
3. Change a number from ordinary notation to scientific notation.
4. Multiply and divide numbers in scientific notation.

1   Multiply and Divide by Powers of 10.

We have learned that our number system is based on the number 10; that is, each place value in our number system is a *power of 10*. Look at the place-value chart in Fig. 11–1 to see how our number system relates to powers of 10.

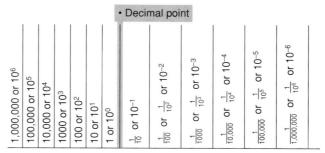

Figure 11–1

The ones place is 10^0. This is consistent with the definition of zero exponents. The tenths place is $\frac{1}{10}$ or 10^{-1}, which is consistent with the definition of negative exponents.

Notice the relationship between the exponent of 10 and the number of zeros in the ordinary numbers listed below.

$$1,000,000 = 10^6 \qquad \frac{1}{10} = 10^{-1}$$

$$100,000 = 10^5 \qquad \frac{1}{100} = 10^{-2}$$

$$10,000 = 10^4 \qquad \frac{1}{1000} = 10^{-3}$$

$$1000 = 10^3 \qquad \frac{1}{10,000} = 10^{-4}$$

$$100 = 10^2 \qquad \frac{1}{100,000} = 10^{-5}$$

$$10 = 10^1 \qquad \frac{1}{1,000,000} = 10^{-6}$$

$$1 = 10^0$$

The absolute value of the exponent equals the number of zeros in the ordinary number.

We learned in arithmetic that we could quickly multiply or divide by 10, 100, 1000, and so on, by shifting the decimal point. When multiplying or dividing by a power of 10, the exponent of 10 tells how many places the decimal is to be shifted and in what direction.

Rule 11–8   *To multiply by a power of 10:*

1. If the exponent is positive, shift the decimal point to the *right* the number of places indicated by the *positive* exponent.
2. If the exponent is negative, shift the decimal point to the *left* the number of places indicated by the *negative* exponent.

Rule 11–9   *To divide by a power of 10:*

1. Change the division to an equivalent multiplication.
2. Use Rule 11–8 for multiplying by a power of 10.

EXAMPLE 1   Perform the multiplications or divisions by using powers of 10.

(a) 275×10 (b) 0.18×100 (c) 2.4×1000 (d) 43×0.1
(e) $3.14 \div 10$ (f) $0.48 \div 100$ (g) $20.1 \div 1000$

(a) $275 \times 10 = 275 \times 10^1 = \mathbf{2750}$
   Because the exponent is $+1$, move the decimal one place to the right.
(b) $0.18 \times 100 = 0.18 \times 10^2 = \mathbf{18}$
   Because the exponent is $+2$, move the decimal two places to the right.
(c) $2.4 \times 1000 = 2.4 \times 10^3 = \mathbf{2400}$
   Because the exponent is $+3$, move the decimal three places to the right.
(d) $43 \times 0.1 = 43 \times 10^{-1} = \mathbf{4.3}$
   Because the exponent is -1, move the decimal one place to the left.

(e) $3.14 \div 10 = 3.14 \times \dfrac{1}{10}$

   Change the division to an equivalent multiplication.
   $= 3.14 \times 10^{-1}$
   Express the fraction $\frac{1}{10}$ as a power of 10. Because the exponent of 10 is -1, move the decimal one place to the *left*.
   $= \mathbf{0.314}$

(f) $0.48 \div 100 = 0.48 \times \dfrac{1}{100} = 0.48 \times 10^{-2} = \mathbf{0.0048}$

(g) $20.1 \div 1000 = 20.1 \times \dfrac{1}{1000} = 20.1 \times 10^{-3} = \mathbf{0.0201}$

A power of 10 can be multiplied or divided by another power of 10 by following the laws of exponents.

EXAMPLE 2   Multiply or divide by using the laws of exponents.

(a) $10^5(10^2)$ (b) $10^{-1}(10^2)$ (c) $10^0(10^3)$ (d) $\dfrac{10}{10^3}$

(e) $\dfrac{10^5}{10^4}$ (f) $\dfrac{10^2}{10^2}$ (g) $\dfrac{10^{-2}}{10^3}$

(a) $10^5(10^2) = 10^{5+2} = \mathbf{10^7}$ (b) $10^{-1}(10^2) = 10^{(-1+2)} = \mathbf{10^1}$
                                                            or **10**

(c) $10^0(10^3) = 10^{0+3} = \mathbf{10^3}$

(d) $\dfrac{10}{10^3} = 10^{1-3} = \mathbf{10^{-2}}$ **or** $\dfrac{\mathbf{1}}{\mathbf{10^2}}$

(e) $\dfrac{10^5}{10^4} = 10^{5-4} = \mathbf{10^1}$ **or** $\mathbf{10}$

(f) $\dfrac{10^2}{10^2} = 10^{2-2} = \mathbf{10^0}$ **or** $\mathbf{1}$

(g) $\dfrac{10^{-2}}{10^3} = 10^{-2-3} = \mathbf{10^{-5}}$ **or** $\dfrac{\mathbf{1}}{\mathbf{10^5}}$

2 Change a Number from Scientific Notation to Ordinary Notation.

Powers of 10 are used in many applications in a special form that saves time when performing operations with very large or very small numbers. This special form for writing such numbers is called *scientific notation*.

To write a number in scientific notation, the decimal is shifted so that the number has a numerical value of at least 1 but less than 10; that is, the whole-number part of the value is a single, nonzero digit. Then that value is multiplied by the appropriate power of 10 so that the entire expression has the same value as the original number.

■ **DEFINITION 11–1: Scientific Notation.** A number is expressed in *scientific notation* if it is the product of two factors and the absolute value of the first factor is a number greater than or equal to 1 but less than 10 and the second factor is a power of 10.

Tips and Traps   *Characteristics of Scientific Notation:*

- Numbers between 0 and 1 require negative exponents when written in scientific notation.
- The first factor in scientific notation will always have only one nonzero digit to the left of the decimal.
- The use of the times sign ($\times$) for multiplication is the most common representation for scientific notation.

EXAMPLE 3   Which of the following terms are expressed in scientific notation?

    (a) 4.7×10^2 (b) 0.2×10^{-1} (c) -3.4×5^2 (d) 2.7×10^0
    (e) 34×10^4 (f) 8×10^{-6} (g) $-2.8 \div 10^3$

(a) 4.7 is more than 1 but less than 10. 10^2 is a power of 10. Multiplication is indicated. **Thus, the term is in scientific notation.**

(b) Even though 10^{-1} is a power of 10, **this term is not in scientific notation** because the first factor (0.2) is less than 1 and therefore is *not* more than or equal to 1.

(c) The absolute value of -3.4 is more than 1 and less than 10, but 5^2 is not a power of 10. **Thus, this term is not in scientific notation.**

(d) 2.7 is more than 1 and less than 10. 10^0 is a power of 10. **Thus, this term is in scientific notation.**

(e) 34 is greater than 10. Even though 10^4 is a power of 10, **this term is not in scientific notation** because the first factor (34) is 10 or more.

(f) 8 is more than 1 and less than 10. 10^{-6} is a power of 10. **Thus, this term is in scientific notation.**

(g) The absolute value of -2.8 is more than 1 and less than 10, but division rather than multiplication is the indicated operation. **This term is not in scientific notation.**

When a number is written strictly according to place value, it is called an *ordinary number*.

> **Rule 11–10** *To change from a number written in scientific notation to an ordinary number:*
>
> 1. Perform the indicated multiplication by moving the decimal point in the first factor the appropriate number of places. Insert zeros as necessary.
> 2. Omit the power-of-10 factor.

Remember, when multiplying by a power of 10, the exponent of 10 tells us how many places and in which direction to move the decimal.

EXAMPLE 4 Change the following to ordinary numbers.

(a) 3.6×10^4 (b) 2.8×10^{-2} (c) 1.1×10^0 (d) 6.9×10^{-5}
(e) 9.7×10^6

(a) $3.6 \times 10^4 = 36000. = \mathbf{36,000}$
The decimal is moved four places to the right.

(b) $2.8 \times 10^{-2} = .028 = \mathbf{0.028}$
The decimal is moved two places to the left.

(c) $1.1 \times 10^0 = \mathbf{1.1}$
The decimal is moved no (zero) places.

(d) $6.9 \times 10^{-5} = .000069 = \mathbf{0.000069}$
The decimal is moved five places to the left.

(e) $9.7 \times 10^6 = 9700000. = \mathbf{9,700,000}$
The decimal is moved six places to the right.

③ Change a Number from Ordinary Notation to Scientific Notation.

If we want to express an ordinary number in scientific notation, we are basically reversing the procedures we used before. Shifting the decimal point changes the value of a number. The power-of-ten factor is used to offset or balance this change. It is important to maintain the original value of the number.

> **Rule 11–11** *To change from a number written in ordinary notation to scientific notation:*
>
> 1. Indicate where the decimal should be positioned in the ordinary number so that the number is valued at 1 or between 1 and 10 by inserting a caret ($\wedge$) in the proper place.
> 2. Determine how many places and in which direction the decimal shifts *from* the new position (caret) *to* the old position (decimal point). This number will be the exponent of the power of 10.

Tips and Traps *A Balancing Act:*
Moving the decimal in the ordinary number will change the value of the number unless you balance the effect of the move in the power-of-10 factor. When a decimal is moved in the first factor, the value is changed. To offset

this change, an opposite change must be made in the power-of-10 factor. From the new to the old position indicates the proper number of places and the direction (positive or negative) for balancing with the power-of-10 factor.

Remember the word "NO." Count from "New" to "Old."

$$3800 = 3.8 \times 10^3 \ (3_{\wedge}800. \times 10^3 \qquad\qquad N \rightarrow O = +3)$$
$$0.0045 = 4.5 \times 10^{-3} \ (0.004_{\wedge}5 \times 10^{-3} \qquad N \rightarrow O = -3)$$

EXAMPLE 5   Express the following numbers in scientific notation.

(a) 285   (b) 0.007   (c) 9.1   (d) 85,000   (e) 0.00074

(a) $285 \rightarrow 2_{\wedge}85 = \mathbf{2.85 \times 10^2}$
   The unwritten decimal is after the 5. Place the caret between 2 and 8 so that the number 2.85 is between 1 and 10. Count *from* the caret *to* the decimal to determine the exponent of 10. A move two places to the right represents the exponent +2.

(b) $0.007 \rightarrow 0.007_{\wedge} = \mathbf{7 \times 10^{-3}}$
   7 is between 1 and 10. Count *from* the caret *to* the decimal. A move three places to the left represents the exponent −3.

(c) $9.1 = \mathbf{9.1 \times 10^0}$
   9.1 is already between 1 and 10, so the decimal does not move; that is, the decimal moves zero places.

(d) $85,000 \rightarrow 8_{\wedge}5000 = \mathbf{8.5 \times 10^4}$
   From the caret *to* the decimal is four places to the right.

(e) $0.00074 \rightarrow 0.0007_{\wedge}4 = \mathbf{7.4 \times 10^{-4}}$
   From the caret *to* the decimal is four places to the left.

Occasionally a number will be in power-of-10 notation, but it will not be in scientific notation because the first factor is not equal to 1 or is not between 1 and 10. When this is the case, shift the decimal to the proper place and adjust the original power-of-10 factor appropriately.

Tips and Traps   *Between 1 and 10:*
The expression "between 1 and 10" means any number that is more than 1 but less than 10. The first factor in scientific notation must be equal to 1 *or* between 1 and 10.

EXAMPLE 6   Express the following in scientific notation.

(a) 37×10^5 (b) 0.03×10^3

(a) $3_{\wedge}7 \times 10^5 = 3.7 \times 10^1 \times 10^5 \qquad N \rightarrow O = +1$
   $\qquad\qquad = \mathbf{3.7 \times 10^6}$

(b) $0.03_{\wedge} \times 10^3 = 3. \times 10^{-2} \times 10^3 \qquad N \rightarrow O = -2$
   $\qquad\qquad = \mathbf{3 \times 10^1}$

[4] Multiply and Divide Numbers in Scientific Notation.

Multiplication and division can be performed with numbers expressed in scientific notation without having to convert the numbers in scientific notation to ordinary numbers.

> **Rule 11-12** *To multiply numbers in scientific notation:*
>
> 1. Multiply the first factors using the rules of signed numbers.
> 2. Multiply the power-of-10 factors by using the laws of exponents.
> 3. Examine the first factor of the product (step 1) to see if its value is equal to 1 or between 1 and 10.
>   (a) If so, write the results of steps 1 and 2.
>   (b) If not, shift the decimal so that the first factor is equal to 1 or is between 1 and 10, and adjust the exponent of the power-of-10 factor accordingly.

EXAMPLE 7 Perform the multiplications.

(a) $(4 \times 10^2)(2 \times 10^3)$ (b) $(3.7 \times 10^3)(2.5 \times 10^{-1})$
(c) $(8.4 \times 10^{-2})(5.2 \times 10^{-3})$

(a) $(4 \times 10^2)(2 \times 10^3) = \mathbf{8 \times 10^5}$
Because 8 is between 1 and 10, we do not make any adjustments.
(b) $(3.7 \times 10^3)(2.5 \times 10^{-1}) = \mathbf{9.25 \times 10^2}$
Because 9.25 is between 1 and 10, we do not make any adjustments.
(c) $(8.4 \times 10^{-2})(5.2 \times 10^{-3}) = 43.68 \times 10^{-5}$
43.68 is not between 1 and 10. Therefore we must make adjustments.

$$43.68 \rightarrow 4_\wedge 3.68 \quad \text{or} \quad 4.368 \times 10^1$$

Now we multiply 4.368×10^1 times 10^{-5}.

$$4.368 \times 10^1 \times 10^{-5} = 4.368 \times 10^{1-5}$$
$$= \mathbf{4.368 \times 10^{-4}}$$

Division involving numbers in scientific notation follows a similar procedure.

> **Rule 11-13** *To divide numbers in scientific notation:*
>
> 1. Divide the factors using the rules of signed numbers.
> 2. Divide the power-of-10 factors by using the laws of exponents.
> 3. Examine the first factor of the quotient (step 1) to see if its value is equal to 1 or between 1 and 10.
>   (a) If so, write the results of steps 1 and 2.
>   (b) If not, shift the decimal so that the first factor is equal to 1 or is between 1 and 10, and adjust the exponent of the power-of-10 factor accordingly.

EXAMPLE 8 Perform the divisions.

(a) $\dfrac{3 \times 10^5}{2 \times 10^2}$ (b) $\dfrac{1.44 \times 10^{-3}}{6 \times 10^{-5}}$ (c) $\dfrac{9.6 \times 10^{29}}{3.2 \times 10^{111}}$ (d) $\dfrac{1.25 \times 10^3}{5}$

(a) $\dfrac{3 \times 10^5}{2 \times 10^2} = \dfrac{3}{2} \times 10^{5-2} = \mathbf{1.5 \times 10^3}$
Because 1.5 is between 1 and 10, no adjustments are necessary.

(b) $\dfrac{1.44 \times 10^{-3}}{6 \times 10^{-5}} = \dfrac{1.44}{6} \times 10^{-3-(-5)} = 0.24 \times 10^{-3+5} = 0.24 \times 10^2$
0.24 is less than 1, so adjustments are necessary.

$$0.24 \rightarrow 0.2_\wedge 4 = 2.4 \times 10^{-1}$$

Then $2.4 \times 10^{-1} \times 10^2 = \mathbf{2.4 \times 10^1}$.

(c) $\dfrac{9.6 \times 10^{29}}{3.2 \times 10^{111}} = \dfrac{9.6}{3.2} \times 10^{29-111} = \mathbf{3} \times \mathbf{10^{-82}}$

(d) $\dfrac{1.25 \times 10^3}{5}$ Remember, 5 is the same as 5×10^0.

$\dfrac{1.25 \times 10^3}{5 \times 10^0} = \dfrac{1.25}{5} \times 10^{3-0} = 0.25 \times 10^3$ 0.25 is less than 1.

$0.25 \to 0.2_\wedge 5 = 2.5 \times 10^{-1}$

Then, $2.5 \times 10^{-1} \times 10^3 = 2.5 \times 10^{-1+3} = \mathbf{2.5 \times 10^2}$

■ General Tips for Using the Calculator

Power-of-10 Key

The power-of-10 key, labeled $\boxed{\text{EXP}}$ or $\boxed{\text{EE}}$ on most calculators, is a shortcut key for entering the following keys:

$$\boxed{\times}\ 10\ \boxed{x^y}$$

This key is used only for power-of-10 factors and only the *exponent* of 10 is entered. If you enter $\boxed{\times}$ 10, then $\boxed{\text{EXP}}$, your answer will have one extra factor of 10.

Parts (a), (b), and (d) of Example 8 can be worked using the power-of-10 key on most calculators.

(a) $3\ \boxed{\text{EXP}}\ 5\ \boxed{\div}\ 2\ \boxed{\text{EXP}}\ 2\ \boxed{=}\ \Rightarrow 1500$

This result will then need to be expressed in scientific notation if desired: $1500 = 1.5 \times 10^3$.

Part (c) cannot be done with all calculators because some calculators will accept no more than two digits for an exponent.

The internal program of a calculator has predetermined how the output of a calculator will be displayed. For example, even if you would like an answer displayed in scientific notation, it may fall within the guidelines for display as an ordinary number. You must make the conversion to scientific notation. The reverse may also be true.

We will now apply our knowledge of scientific notation to solving applied problems.

EXAMPLE 9   A star is 5.7 light-years from Earth. If 1 light-year is 5.87×10^{12} miles, how many miles from Earth is the star?

To solve this problem, we express the given information in a direct proportion of two fractions relating light-years to miles.

Pair 1: 1 light year = 5.87×10^{12} mi
Pair 2: 5.7 light years = x mi

$\dfrac{1 \text{ light-year}}{5.7 \text{ light-years}} = \dfrac{5.87 \times 10^{12} \text{ mi}}{x \text{ mi}}$ Pair 1 becomes the two numerators.
Pair 2 becomes the two denominators.

$\dfrac{1}{5.7} = \dfrac{5.87 \times 10^{12}}{x}$ Cross multiply.

$x = (5.87 \times 10^{12})(5.7)$ $5.7 = 5.7 \times 10^0$

$x = 33.459 \times 10^{12}$ Perform scientific notation adjustment.

$x = 3.3459 \times 10^1 \times 10^{12}$

$$x = 3.3459 \times \boxed{10^{13}}$$

or $\quad 3.3 \times 10^{13}$ First factor is rounded to tenths.

The star is 3.3×10^{13} miles from Earth.

EXAMPLE 10 An angstrom unit is 1×10^{-7} mm. How many millimeters are in 14.82 angstrom units?

We will set up a direct proportion of fractions relating angstrom units to millimeters.

Pair 1: 1 angstrom $= 1 \times 10^{-7}$ mm
Pair 2: 14.82 angstroms $= x$ mm

$$\frac{1 \text{ angstrom}}{14.82 \text{ angstroms}} = \frac{1 \times 10^{-7} \text{ mm}}{x \text{ mm}}$$ Pair 1 becomes the two numerators.
Pair 2 becomes the two denominators.

$$\frac{1}{14.82} = \frac{1 \times 10^{-7}}{x}$$ Cross multiply.

$$x = 14.82 \times 10^{-7}$$

$$x = 1.482 \times \boxed{10^{1} \times 10^{-7}}$$ Adjust.

$$x = 1.482 \times \boxed{10^{-6}} \text{ mm}$$

There are 1.482×10^{-6} millimeters in 14.82 angstrom units.

EXAMPLE 11 One coulomb (C) is approximately 6.28×10^{18} electrons. How many coulombs are in 2.512×10^{21} electrons?

Pair 1: 1 C $= 6.28 \times 10^{18}$ electrons
Pair 2: x C $= 2.512 \times 10^{21}$ electrons

Setting up the direct proportion, we have

$$\frac{1 \text{ C}}{x \text{ C}} = \frac{6.28 \times 10^{18} \text{ electrons}}{2.512 \times 10^{21} \text{ electrons}}$$ Pair 1 becomes the two numerators.
Pair 2 becomes the two denominators.

$$\frac{1}{x} = \frac{6.28 \times 10^{18}}{2.512 \times 10^{21}}$$ Cross multiply.

$$(6.28 \times 10^{18})(x) = 2.512 \times 10^{21}$$ Divide by the coefficient of x.

$$x = \frac{\boxed{2.512 \times 10^{21}}}{\boxed{6.28 \times 10^{18}}}$$

$$x = \boxed{0.4 \times 10^{3}}$$ Adjust.

$$x = 4 \times 10^{-1} \times 10^{3}$$

$$x = 4 \times 10^{2}$$ Write as an ordinary number.

$$x = 400 \text{ C}$$

There are 400 C in 2.512×10^{21} electrons.

SELF-STUDY EXERCISES 11-3

1 Multiply or divide as indicated.

1. 0.37×10^{2} **2.** 1.82×10^{3} **3.** 5.6×10^{-1} **4.** 142×10^{-2}

5. 78×10^{4} **6.** 62×10^{0} **7.** $4.6 \div 10^{4}$ **8.** $6.1 \div 10$

9. $7.2 \div 10^{1}$ **10.** $42 \div 10^{0}$ **11.** $10^{4}(10^{6})$ **12.** $10^{-3}(10^{-4})$

13. $10^{0}(10^{-3})$ **14.** $10^{-3}(10^{4})$ **15.** $10(10^{2})$ **16.** $\dfrac{10^{4}}{10^{2}}$

17. $\dfrac{10}{10^4}$ **18.** $\dfrac{10^4}{10^4}$ **19.** $\dfrac{10^{-2}}{10^3}$ **20.** $\dfrac{10^0}{10^1}$

2 Write the following as ordinary numbers.

21. 4.3×10^2 **22.** 6.5×10^{-3} **23.** 2.2×10^0 **24.** 7.3×10
25. 9.3×10^{-2} **26.** 8.3×10^4 **27.** 5.8×10^{-3} **28.** 8×10^4
29. 6.732×10^0 **30.** 5.89×10^{-3}

3 Express the following in scientific notation.

31. 392 **32.** 0.02 **33.** 7.03 **34.** 42,000
35. 0.081 **36.** 0.0021 **37.** 23.92 **38.** 0.101
39. 1.002 **40.** 721

Write in scientific notation.

41. 42×10^4 **42.** 32.6×10^3 **43.** 0.213×10^2
44. 0.0062×10^{-3} **45.** $56,000 \times 10^{-3}$

4 Perform the indicated operations. Express the answers in scientific notation.

46. $(6.7 \times 10^4)(3.2 \times 10^2)$ **47.** $(1.6 \times 10^{-1})(3.5 \times 10^4)$ **48.** $(5.0 \times 10^{-3})(4.72 \times 10^0)$

49. $(8.6 \times 10^{-3})(5.5 \times 10^{-1})$ **50.** $\dfrac{3.15 \times 10^5}{4.5 \times 10^2}$ **51.** $\dfrac{4.68 \times 10^3}{7.2 \times 10^7}$

52. $\dfrac{4.55 \times 10^{-1}}{6.5 \times 10^{-4}}$ **53.** $\dfrac{7.84 \times 10^{-2}}{9.8 \times 10^0}$

54. A star is 5.5 light years from Earth. If one light year is 5.87×10^{12} miles, how many miles from Earth is the star?

55. An angstrom (Å) unit is 1×10^{-7} mm. How many angstrom units are in 4.2×10^{-5} mm?

11–4 POLYNOMIALS

Learning Objectives

1 Identify polynomials, monomials, binomials, and trinomials.

2 Identify the degree of terms and polynomials.

3 Arrange polynomials in descending order.

1 **Identify Polynomials, Monomials, Binomials, and Trinomials.**

A special type of algebraic expression that contains variables and exponents is a *polynomial*.

■ **DEFINITION 11–2: Polynomial.** A *polynomial* is an algebraic expression in which the exponents of the variables are nonnegative integers.

EXAMPLE 1 Identify which expressions are polynomials. If an expression is not a polynomial, explain why.

(a) $5x^2 + 3x + 2$ (b) $5x - \dfrac{3}{x}$ (c) 9 (d) $-\dfrac{1}{2}x + 3x^{-2}$

(a) $\mathbf{5x^2 + 3x + 2}$ **is a polynomial.**
(b) $\mathbf{5x - \frac{3}{x}}$ **is not a polynomial** because the term $-\frac{3}{x}$ is equivalent to $-3x^{-1}$. A polynomial cannot have a variable with a negative exponent.
(c) 9 is a polynomial because it is equivalent to $9x^0$ and the exponent, zero, is a nonnegative integer.
(d) $-\frac{1}{2}x + 3x^{-2}$ is not a polynomial because $3x^{-2}$ has a negative exponent.

Some polynomials have special names, depending on the number of terms contained in the polynomial.

■ **DEFINITION 11–3: Monomial.** A *monomial* is a polynomial containing one term.

$$3, \quad -2x, \quad 5ab, \quad 7xy^2, \quad \frac{3a^2}{4} \quad \text{are monomials.}$$

■ **DEFINITION 11–4: Binomial.** A *binomial* is a polynomial containing two terms.

$$x + 3, \quad 2x^2 - 5x, \quad x + \frac{y}{4}, \quad 3(x - 1) + 2 \quad \text{are binomials.}$$

■ **DEFINITION 11–5: Trinomial.** A *trinomial* is a polynomial containing three terms.

$$a + b + c, \quad x^2 - 3x + 4, \quad x + \frac{2a}{7} - 5 \quad \text{are trinomials.}$$

EXAMPLE 2 Identify each of the following expressions as a polynomial. Then state whether each one is a monomial, binomial, or trinomial.

(a) $x^2y - 1$ (b) $4(x - 2)$

(c) $\dfrac{2x + 5}{2y}$ (d) $3x^2 - x + 1$

(e) $3x^2 - (x + 1)$

(a) $x^2y - 1$	**Binomial**
(b) $4(x - 2)$	**Monomial.** If the distributive property is applied, the expression will become a binomial: $4(x - 2) = 4x - 8$.
(c) $\dfrac{2x + 5}{2y}$	This one-term expression is **not a monomial because it is not a polynomial** (the exponent of y is -1 if it is in the numerator).
(d) $3x^2 - x + 1$	**Trinomial**
(e) $3x^2 - (x + 1)$	**Binomial**

Tips and Traps *Counting Terms:*
Terms in an algebraic expression are separated by plus or minus signs that are not in a grouping.

$$5(x + 3) \boxed{-} 4 \qquad \text{2 terms}$$
$$(3x - 4)(x + 1) \qquad \text{1 term}$$

---

② Identify the Degree of Terms and Polynomials.

■ **DEFINITION 11–6: Degree of a Term.** The *degree of a term* that has only one variable with a nonnegative exponent is the same as the exponent of the variable.

$$3x^4, \quad \text{fourth degree} \qquad -5x, \quad \text{first degree}$$

Number terms other than zero have a degree of 0. A variable to the zero power is implied.

$$5 = 5x^0, \quad \text{degree zero}$$

If a term has more than one variable, the degree of the term is the sum of the exponents of the variable factors.

$$2xy = 2x^1 y^1, \quad \text{second degree} \qquad -2ab^2 = -2a^1 b^2, \quad \text{third degree}$$

EXAMPLE 3 Identify the degree of each term in the polynomial $5x^3 + 2x^2 - 3x + 3$.

$5x^3$, degree 3 (or third degree)
$2x^2$, degree 2 (or second degree)
$-3x$, degree 1 (or first degree)
3, degree 0

Special names are associated with terms of degree 0, 1, 2, and 3. A *constant term* has degree 0. A *linear term* has degree 1. A *quadratic term* has degree 2. A *cubic term* has degree 3.

■ **DEFINITION 11–7: Degree of a Polynomial.** The *degree of a polynomial* that has only one variable and only positive integral exponents is the degree of the term with the largest exponent.

A *linear polynomial* has degree 1. A *quadratic polynomial* has degree 2. A *cubic polynomial* has degree 3.

$5x^3 - 2$ has a degree of 3 and is a cubic polynomial.

$x + 7$ has a degree of 1 and is a linear polynomial.

$7x^2 - 4x + 5$ has a degree of 2 and is a quadratic polynomial.

EXAMPLE 4 Identify the degree of the following polynomials.

(a) $5x^4 + 2x - 1$ (b) $3x^3 - 4x^2 + x - 5$ (c) 7   (d) $x - \dfrac{1}{2}$

(a) $5x^4 + 2x - 1$ has a **degree of 4.**
(b) $3x^3 - 4x^2 + x - 5$ has a **degree of 3 and is a cubic polynomial.**
(c) 7 has a **degree of 0 and is a constant.**
(d) $x - \frac{1}{2}$ has a **degree of 1 and is a linear polynomial.**

③ Arrange Polynomials in Descending Order.

The terms of a polynomial are customarily arranged in order based on the degree of each term of the polynomial. The terms can be arranged beginning with the term with the highest degree (*descending order*) or beginning with the term with the lowest degree (*ascending order*).

Tips and Traps *Most Common Arrangement of Polynomials:*
Polynomials are most often arranged in descending order so that the degree of the polynomial is the degree of the first term.

The first term of a polynomial arranged in descending order is called the *leading term* of the polynomial. The coefficient of the leading term of a polynomial is called the *leading coefficient*.

EXAMPLE 5 Arrange each polynomial in descending order and identify the degree, the leading term, and the leading coefficient of the polynomial.

(a) $5x + 3x^3 - 7 + 6x^2$ (b) $x^4 - 2x + 3$
(c) $4 + x$ (d) $x^2 + 5$

(a) $5x + 3x^3 - 7 + 6x^2 = \mathbf{3x^3 + 6x^2 + 5x - 7}$.
   Third degree, leading term is $3x^3$, leading coefficient is 3.

(b) $\mathbf{x^4 - 2x + 3}$ is already in descending order.
   Fourth degree, leading term is x^4, leading coefficient is 1.

(c) $4 + x = \mathbf{x + 4}$.
   First degree or linear polynomial, leading term is x, leading coefficient is 1.

(d) $\mathbf{x^2 + 5}$ is already in descending order.
   Second degree or quadratic polynomial, leading term is x^2, leading coefficient is 1.

Tips and Traps   *Coefficients of Zero:*
A polynomial arranged in descending order can be written with every successive degree represented. Missing terms have a coefficient of 0.
In Example 5(b), $x^4 - 2x + 3$ is the same as $x^4 + 0x^3 + 0x^2 - 2x^1 + 3x^0$.

SELF-STUDY EXERCISES 11–4

1   Identify each of the following expressions as a monomial, binomial, or trinomial.

1. $2x^3y - 7x$ **2.** $5xy^2 + 8y$ **3.** $3xy$
4. $7ab$ **5.** $5(3x - y)$ **6.** $(x - 5)(2x + 4)$
7. $5x^2 - 8x + 3$ **8.** $7y^2 + 5y - 1$ **9.** $\dfrac{4x - 1}{5}$
10. $\dfrac{x}{6} - 5$ **11.** $4(x^2 - 2) + x^3$ **12.** $3x^2 - 7(x - 8)$

2   Identify the degree of each term in each polynomial.

13. $6x$ **14.** $8x^2$ **15.** $6x^2 - 8x + 12$
16. $7x^3 - 8x + 12$ **17.** $x - 12$ **18.** $3x^2 - 8$
19. 15 **20.** 21 **21.** $2x - \dfrac{1}{4}$
22. $8x^2 + \dfrac{5}{6}$

Identify the degree of the following polynomials.

23. $5x^2 + 8x - 14$ **24.** $x^3 - 8x^2 + 5$ **25.** $9 - x^3 + x^6$
26. $12 - 15x^2 - 7x^5$ **27.** $2x - \dfrac{4}{5}x^2$ **28.** $\dfrac{7}{8} - x$

3   Arrange each polynomial in descending order and identify the degree, leading term, and leading coefficient of the polynomial.

29. $5x - 3x^2$ **30.** $7 - x^3$ **31.** $4x - 8 + 9x^2$
32. $5x^2 + 8 - 3x$ **33.** $7x^3 - x + 8x^2 - 12$ **34.** $7 - 15x^4 + 12x$
35. $-7x + 8x^6 - 7x^3$ **36.** $15 - 14x^8 + x$

EXPONENTIAL EXPRESSIONS, EQUATIONS, AND FORMULAS

Learning Objectives

1. Evaluate formulas with at least one exponential term.
2. Evaluate formulas that contain a power of the natural exponential, e.
3. Solve exponential equations in the form $b^x = b^y$, where $b > 0$ and $b \neq 1$.

Many scientific, technical, and business phenomena have the property of exponential growth; that is, the growth rate does not remain constant as certain physical properties increase. Instead, the growth rate increases exponentially. For example, notice the difference between $2x$ and 2^x when x increases. If we write the expressions in function notation, we have two different functions of x.

$$f(x) = 2x$$
$$g(x) = 2^x$$

Examine the values in Table 11–2 and the graphical representation in Fig. 11–2.

TABLE 11–2

x	$f(x) = 2x$	$g(x) = 2^x$
1	$2(1) = 2$	$2^1 = 2$
2	$2(2) = 4$	$2^2 = 4$
3	$2(3) = 6$	$2^3 = 8$
4	$2(4) = 8$	$2^4 = 16$
5	$2(5) = 10$	$2^5 = 32$
6	$2(6) = 12$	$2^6 = 64$

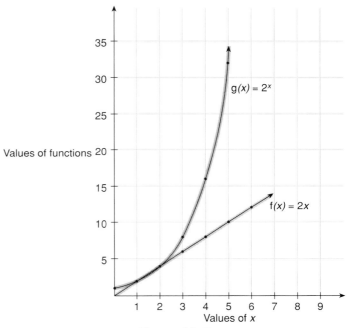

Figure 11–2

2^x is said to increase *exponentially*.

Before the easy availability of scientific and graphics calculators, logarithms were used extensively to make numerical calculations. Exponential and logarithmic expressions are found in many formulas. Even with the use of a scientific or graphics calculator, an understanding of exponential and logarithmic expressions is necessary. In this section, we will evaluate expressions containing exponential expressions with the scientific or graphics calculator.

1 Evaluate Formulas With at Least One Exponential Term.

Many formulas have terms that contain exponents. An *exponential expression* is an expression that contains at least one term that has a *variable exponent*. A variable exponent is an exponent that has at least one letter factor. Exponential expressions can be evaluated on a scientific calculator by using the general power key $\boxed{x^y}$. The base of the term is entered first, then $\boxed{x^y}$, followed by the value of the exponent. An *exponential equation* or formula contains at least one term that has a variable exponent.

A commonly used formula that contains a variable exponent is the formula for calculating the compound amount for compound interest. *Compound interest* is the interest calculated at the end of each period and then added to the principal for the next period. The *compound amount* is the principal and interest accumulated over a period of time. The accumulated amount A that a principal P will be worth at the end of t years, when invested at an interest rate r and compounded n times a year, is given by the following formula.

Formula 11–1 *Compound Amount:*

$$A = P\left(1 + \frac{r}{n}\right)^{nt}$$

where A = accumulated amount
$\quad P$ = original principal
$\quad t$ = time in years
$\quad r$ = rate per year
$\quad n$ = compounding periods per year

EXAMPLE 1 Using the formula $A = P(1 + \frac{r}{n})^{nt}$ and a scientific or graphics calculator, find the accumulated amount on an investment of \$1500, invested at an interest rate of 9% for 3 years, if the interest is compounded quarterly.

$$A = P\left(1 + \frac{r}{n}\right)^{nt}$$

P = \$1500; r = 9% or 0.09; n = quarterly or 4 times a year; t = 3 years

$$A = 1500\left(1 + \frac{0.09}{4}\right)^{(4)(3)}$$

$$A = 1500(1 + 0.0225)^{12}$$

$$A = 1500(1.0225)^{12}$$

1.0225 $\boxed{x^y}$ 12 $\boxed{=}$ 1.30604999 $\Rightarrow$

$$A = 1500(1.30604999)$$

$$A = 1959.07$$

Rounded

The accumulated amount of the \$1500 investment after 3 years is \$1959.07 to the nearest cent.

■ **General Tips for Using the Calculator**

Should we do every step on the calculator? The following sequence of keystrokes is one option for performing all the calculations in Example 1 in a continuous series.

1500 $\boxed{\times}$ $\boxed{(}$ 1 $\boxed{+}$ .09 $\boxed{\div}$ 4 $\boxed{)}$ $\boxed{x^y}$ $\boxed{(}$ 4 $\boxed{\times}$ 3 $\boxed{)}$ $\boxed{=}$

Whenever possible, it is advisable to do some calculations mentally. This can greatly decrease the complexity of the calculator sequences.

1500 $\boxed{\times}$ $\boxed{(}$ 1 $\boxed{+}$ .09 $\boxed{\div}$ 4 $\boxed{)}$ $\boxed{x^y}$ 12 $\boxed{=}$ $\Rightarrow$ 1959.074985

Try each sequence on your calculator.

---

2 Evaluate Formulas that Contain a Power of the Natural Exponential, e.

In many applications involving circles, the irrational number π (approximately equal to 3.14159) is used. Another irrational number, e, arises in the discussion of many physical phenomena. Many formulas will contain a power of the natural exponential e.

Exponential change is an interesting phenomenon. Let's look at the value of the expression $(1 + \frac{1}{n})^n$ as n gets larger and larger. See Table 11–3.

TABLE 11–3

n	$(1 + \frac{1}{n})^n$	Result
1	$(1 + \frac{1}{n})^n$	2
2	$(1 + \frac{1}{n})^n$	2.25
3	$(1 + \frac{1}{n})^n$	2.37037037
10	$(1 + \frac{1}{n})^n$	2.59374246
100	$(1 + \frac{1}{n})^n$	2.704813829
1000	$(1 + \frac{1}{n})^n$	2.716923932
10,000	$(1 + \frac{1}{n})^n$	2.718145927

The value of the expression changes very little as the value of n gets larger. We can say that the value approaches a given number. We will call the number e, the *natural exponential*. The natural exponential, e, like π, is an irrational number and will never terminate or repeat as more decimal places are examined.

■ **DEFINITION 11–8: Natural Exponential, e.** The *natural exponential*, e, is the limit that the value of the expression $(1 + \frac{1}{n})^n$ approaches as n gets larger and larger without bound.

To evaluate formulas containing the natural exponential e, we will use a scientific or graphics calculator.

■ General Tips for Using the Calculator

Natural Exponential Key

The natural exponential key is generally labeled $\boxed{e^x}$. On most calculators the exponent is entered after pressing the natural exponential key. To find $e^{2.3}$, enter the following sequence:

$$\boxed{e^x} \ 2.3 \ \boxed{=} \ \Rightarrow 9.974182455$$

On other calculators, the exponent is entered before pressing the natural exponential key. Try your calculator. A good test for the $\boxed{e^x}$ key is to find e^0.

EXAMPLE 2   The formula for the atmospheric pressure (in millimeters of mercury) is $P = 760e^{-0.00013h}$, where h is the height in meters. Find the atmospheric pressure at 100 meters above sea level ($h = 100$).

$P = 760e^{-0.00013h}$ $h = 100$

$P = 760e^{-0.00013(100)}$

$P = 760e^{-0.013}$ Multiply -0.00013 times 100 mentally.

$P = 760(0.987084135)$ $e^{-0.013} = 0.987084135$

$P = 750.1839427$ One calculator sequence: $760 \ \boxed{\times} \ \boxed{e^x} \ \boxed{-} \ .013 \ \boxed{=}$

Thus, the atmospheric pressure at 100 meters above sea level is 750.18 mm.

3 Solve Exponential Equations in the Form $b^x = b^y$, where $b > 0$ and $b \neq 1$.

To illustrate the properties of exponential equations, we will look at the equation $2^x = 32$. In the equation, the value of x will be the power of 2 that gives a result of 32. We can rewrite 32 as 2^5. Thus, $2^x = 2^5$. To solve the equation for x, we apply Rule 11–14.

Rule 11–14 *To solve an exponential equation in the form of $b^x = b^y$:*

Apply the following property and solve for x. If $b^x = b^y$, and $b > 0$ and $b \neq 1$, then $x = y$.

Tips and Traps *Excluded Bases:*
It is important to notice that this property is appropriate only when the bases are alike, positive, and not equal to one. Why must we exclude $b = 1$? If $1^5 = 1^8$, then does $5 = 8$?

EXAMPLE 3   Solve the equation $2^x = 32$.

$2^x = 32$ Rewrite 32 as a power of 2.

$2^x = 2^5$

If $2^x = 2^5$,   then $x = 5$.   Apply Rule 11–14.

Check to see if the solution is appropriate.   Does $2^5 = 32$? Yes, so 5 is the correct solution.

EXAMPLE 4   Solve the equation, $3^{x+1} = 27$.

$3^{x+1} = 27$ Rewrite 27 as a power of 3.

$3^{x+1} = 3^3$

If $3^{x+1} = 3^3$, then $x + 1 = 3$.

$x + 1 = 3$ Solve for x.

$\quad x = 3 - 1$

$\quad x = 2$

Check to see if the solution, 2, is correct.   Does $3^{2+1} = 27$?

$$3^3 = 27$$

Thus, the solution, $x = 2$, is correct.

It will not always be possible to rewrite an exponential equation as an equation with like bases. In such cases, other methods for solving the exponential equation are used.

SELF-STUDY EXERCISES 11–5

1   Using the compound amount formula, $A = P(1 + \frac{r}{n})^{nt}$, find the accumulated amount for the following.

1. Principal = \$1500, rate = 10%, compounded annually, time = 5 years.

2. Principal = \$1750, rate = 8%, compounded quarterly, time = 2 years.

3. The number of grams of a chemical that will dissolve in a solution is given by the formula $C = 100e^{0.02t}$, where t = temperature in degrees Celsius. Evaluate when:
(a) $t = 10$ (c) $t = 25$
(b) $t = 20$ (d) $t = 30$

4. The compound amount when an investment is compounded continually (every instant) is expressed by the formula $A = Pe^{ni}$, where A = compounded amount, P = principal, n = number of years, and i = interest rate per year. Find the compound amount when:
(a) Principal = $1000, interest = 9%, for 2 years.
(b) Principal = $1500, interest = 10%, for 6 months.

Evaluate using a scientific or graphics calculator.

5. 4^3 **6.** 3^{-5} **7.** 5^{10} **8.** 8^{-3} **9.** $9^{2.5}$ **10.** $10^{-\frac{5}{2}}$

[2] A formula for electric current is $i = 1.50e^{-200t}$, where t is the time in seconds. Calculate the current for the following times. Express the answers in scientific notation.

11. 1 second   **12.** 1.1 seconds   **13.** 0.5 second

Evaluate using a scientific or graphics calculator. Round to hundredths.

14. e^2 **15.** e^{-3} **16.** $e^{0.21}$ **17.** $e^{-3.5}$

[3] Solve for x.

18. $3^x = 3^7$ **19.** $5^x = 5^{-3}$ **20.** $3^{x+4} = 3^6$
21. $2^{x-3} = 2^7$ **22.** $6^x = 6^7$ **23.** $3^x = 27$

24. $2^x = 64$ **25.** $3^x = \dfrac{1}{81}$ **26.** $2^x = \dfrac{1}{64}$

27. $4^{3x} = 128$ **28.** $3^{2x} = 243$ **29.** $4^{3-x} = \dfrac{1}{16}$

30. $2^{4-x} = \dfrac{1}{16}$

11-6 LOGARITHMIC EXPRESSIONS

Learning Objectives

[1] Write exponential expressions as equivalent logarithmic expressions.
[2] Write logarithmic expressions as equivalent exponential expressions.
[3] Evaluate common and natural logarithmic expressions using a calculator.
[4] Evaluate logarithms with a base other than 10 or e.
[5] Evaluate formulas containing at least one logarithmic term.
[6] Simplify logarithmic expressions by using the properties of logarithms.

Logarithms were first introduced as a relatively fast way of carrying out lengthy calculations. The scientific calculator has diminished the importance of logarithms as a computational device; however, the importance of logarithms in advanced mathematics, electronics, and theoretical work is more evident than ever. Many formulas use logarithms to show the relationships of physical properties. Also, logarithms are used to solve many exponential equations.

[1] Write Exponential Expressions as Equivalent Logarithmic Expressions.

In Chapter 1, we looked at powers of whole numbers and roots of perfect powers. To expand on that information, we will extend our study to logarithmic expressions. A *logarithmic expression* is an expression that contains at least one term with a logarithm. The three basic components of a power are the base, exponent, and result of exponentiation or power. Since the word *power* is often used to mean more than one concept, we will use the **result of exponentiation** terminology.

$$\overset{\text{exponent}}{3^{\,4}} = 81 \leftarrow \text{result of exponentiation (power)}$$
$$\underset{\text{base}}{}$$

When finding a power or the result of exponentiation, you are given the base and exponent. When finding a root, you know the base (radicand) and the index of the root (exponent). A third type of calculation will be to find the exponent when you are given the base and the result of exponentiation. The exponent in this process is called the *logarithm*. In the logarithmic form $\log_b x = y$, b is the *base*, y is the *exponent* or *logarithm*, and x is the *result of exponentiation*. This equation is read, "The log of x to the base b is y." In the exponential form $x = b^y$, b is also the base, y is the exponent, and x is the result of the exponentiation.

Rule 11–15 *To convert an exponential expression to a logarithmic equation:*

If $x = b^y$, then $\log_b x = y$, provided that $b > 0$ and $b \neq 1$.

1. The exponent in the exponential form is the variable solved for in the logarithmic form.
2. The base in the exponential form is the base in the logarithmic form.
3. The variable solved for in the exponential form is the result of exponentiation in logarithmic form. This result is sometimes referred to as the *argument*.

EXAMPLE 1 Rewrite the following in logarithmic form.

(a) $2^4 = 16$ (b) $3^2 = 9$ (c) $2^{-2} = \dfrac{1}{4}$

(a) $2^4 = 16$ converts to $\mathbf{\log_2 16 = 4}$ Base = 2, exponent = 4, result of exponentiation = 16.

(b) $3^2 = 9$ converts to $\mathbf{\log_3 9 = 2}$ Base = 3, exponent = 2, result of exponentiation = 9.

(c) $2^{-2} = \dfrac{1}{4}$ converts to $\mathbf{\log_2 \dfrac{1}{4} = -2}$ Base = 2, exponent = −2, result of exponentiation = $\frac{1}{4}$.

---

2 Write Logarithmic Expressions as Equivalent Exponential Expressions.

The inverse of the process in the preceding objective will change a logarithmic expression to an exponential expression.

Rule 11–16 *To convert a logarithmic equation to an exponential equation:*

$\log_b x = y$ converts to $x = b^y$, provided that $b > 0$ and $b \neq 1$.

1. The dependent variable (solved for variable) in the logarithmic form is the exponent in the exponential form.

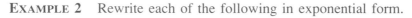

2. The base in the logarithmic form is the base in the exponential form.

3. The result of the exponentiation in logarithmic form is the dependent variable (solved for variable) in exponential form.

EXAMPLE 2   Rewrite each of the following in exponential form.

(a) $\log_2 32 = 5$ (b) $\log_3 81 = 4$ (c) $\log_5 \dfrac{1}{25} = -2$

(d) $\log_{10} 0.001 = -3$

(a) $\log_2 32 = 5$ converts to $2^5 = 32$

Base = 2, exponent = 5, result of exponentiation = 32.

(b) $\log_3 81 = 4$ converts to $3^4 = 81$

Base = 3, exponent = 4, result of exponentiation = 81.

(c) $\log_5 \dfrac{1}{25} = -2$ converts to $5^{-2} = \dfrac{1}{25}$

Base = 5, exponent = -2, result of exponentiation = $\frac{1}{25}$.

(d) $\log_{10} 0.001 = -3$ converts to $10^{-3} = 0.001$

Base = 10, exponent = -3, result of exponentiation = 0.001.

3 **Evaluate Common and Natural Logarithmic Expressions Using a Calculator.**

When the base of a logarithm is 10, the logarithm is referred to as a *common logarithm*. If the base is omitted in a logarithmic expression, the base is assumed to be 10. Thus, $\log_{10} 1000 = 3$ is normally written as $\log 1000 = 3$. On a calculator, expressions containing common logarithms can be evaluated using the $\boxed{\log}$ key.

A logarithm with a base of e is a *natural logarithm* and is abbreviated as *ln*. Scientific or graphics calculators normally have an $\boxed{\text{ln}}$ key. Thus, $\log_e 1 = 0$ is normally written as *ln* $1 = 0$. Expressions containing natural logarithms can be evaluated using the $\boxed{\text{ln}}$ key.

Calculators are used regularly when calculating with logarithms.

■ General Tips for Using the Calculator

Scientific and graphics calculators generally have two logarithm keys, $\boxed{\log}$ for *common logarithms* and $\boxed{\text{ln}}$ for *natural logarithms*.

Evaluate log 2 and ln 2.

Most Calculators:
To find the common log of 2, press the $\boxed{\log}$ key, followed by the result of exponentiation, 2.

$$\boxed{\log}\ 2\ \boxed{=}\ \Rightarrow 0.3010299957$$

To find the natural log of 2, press the $\boxed{\text{ln}}$ key, followed by the result of exponentiation, 2.

$$\boxed{\text{ln}}\ 2\ \boxed{=}\ \Rightarrow 0.6931471806$$

Some calculators require that the result of exponentiation be entered, followed by the $\boxed{\log}$ or $\boxed{\text{ln}}$ key. Experiment with your calculator to determine the necessary key strokes. A good test is to verify that log 1 = 0 and ln 1 = 0.

EXAMPLE 3 Evaluate the following using a calculator.

(a) log 10,000   (b) log 0.000001   (c) $\log_4 256$ (d) $\log_3 \frac{1}{27}$

(a) log 10,000 = **4**   [log] 10000 [=].

(b) log 0.000001 = **−6**   Enter [log] .000001 [=].

(c) Rewrite the logarithmic equation as an exponential equation with like bases and solve.

$\log_4 256 = x$ is $4^x = 256$ $256 = 4^4$

$4^x = 4^4$

$x = 4$

(d) Rewrite the logarithmic equation as an exponential equation with like bases and solve.

$\log_3 \frac{1}{27} = x$ is $3^x = \frac{1}{27}$ $27 = 3^3; \frac{1}{27} = 3^{-3}$

$3^x = 3^{-3}$

$x = -3$

EXAMPLE 4 Evaluate the following using a calculator. Express the answers to the nearest ten-thousandth.

(a) ln 5   (b) ln 4.5   (c) ln 948

(a) ln 5 = **1.6094**   (b) ln 4.5 = **1.5041**   (c) ln 948 = **6.8544**

4 Evaluate Logarithms with a Base Other Than 10 or e.

To find the logarithm for a base other than 10 or e using a calculator, a conversion formula is necessary.

> **Rule 11–17** *To evaluate a logarithm with a base* **b** *other than 10 or* **e***:*
>
> $$\log_b a = \frac{\log a}{\log b}$$
>
> A similar process can be used with natural logarithms.
>
> $$\log_b a = \frac{\ln a}{\ln b}$$

EXAMPLE 5 Find $\log_7 343$.

$\log_7 343 = \dfrac{\log 343}{\log 7} = 3$ [log] 343 [÷] [log] 7 [=] ⇒ 3

5 Evaluate Formulas Containing at Least One Logarithmic Term.

Many formulas in science make use of both common and natural logarithms.

EXAMPLE 6   The loudness of sound is measured by a unit called a *decibel*. A very faint sound, called the *threshold sound*, is assigned an intensity of I_o. Other sounds have an intensity of I, which is a specified number times the threshold sound ($I = nI_o$). Then the decibel rate is given by the formula $bel = 10 \log \frac{I}{I_o}$. Find the decibel rating for sounds having the following intensities (I).

(a) A whisper, $110I_o$
(b) A speaking voice, $230I_o$
(c) A busy street, $9{,}000{,}000I_o$
(d) Loud music, $875{,}000{,}000{,}000I_o$
(e) A jet plane at takeoff, $109{,}000{,}000{,}000{,}000I_o$

(a) A whisper, $I = 110I_o$

$$bel = 10 \log \frac{110I_o}{I_o} \qquad\qquad \frac{I_o}{I_o} = 1$$

$$bel = 10 \log 110 \qquad\qquad \log 110 = 2.041392685$$

$$bel = \mathbf{20} \text{ (rounded)}$$

(b) A speaking voice, $I = 230I_o$

$$bel = 10 \log \frac{230I_o}{I_o} \qquad\qquad \frac{I_o}{I_o} = 1$$

$$bel = 10 \log 230 \qquad\qquad \log 230 = 2.361727836$$

$$bel = \mathbf{24} \quad \text{(rounded)}$$

(c) A busy street, $I = 9{,}000{,}000I_o$

$$bel = 10 \log \frac{9{,}000{,}000I_o}{I_o}$$

$$bel = 10 \log 9{,}000{,}000 \qquad\qquad \log 9{,}000{,}000 = 6.954242509$$

$$bel = \mathbf{70} \quad \text{(rounded)}$$

(d) Loud music, $I = 875{,}000{,}000{,}000I_o$

$$bel = 10 \log \frac{875{,}000{,}000{,}000I_o}{I_o}$$

$$bel = 10 \log 875{,}000{,}000{,}000$$

$$bel = 10 \log (8.75 \times 10^{11}) \qquad 10 \boxed{\log} 8.75 \boxed{\text{EXP}} 11 \boxed{=} \Rightarrow 119.4200805$$

$$bel = \mathbf{119} \quad \text{(rounded)}$$

(e) A jet plane at takeoff, $I = 109{,}000{,}000{,}000{,}000I_o = (1.09 \times 10^{14})I_o$

$$bel = 10 \log \frac{(1.09 \times 10^{14})I_o}{I_o}$$

$$bel = 10 \log (1.09 \times 10^{14})$$

$$bel = 10(14.0374265)$$

$$bel = \mathbf{140} \text{ (rounded)}$$

⑥ Simplify Logarithmic Expressions by Using the Properties of Logarithms.

Many applications of logarithms and exponential expressions require the understanding of the properties of logarithms. Because a logarithm is an exponent, the laws of exponents are appropriate in simplifying logarithmic expressions. A few manual manipulations can be made before using the calculator. Similar laws are appropriate for both common and natural logarithms and logarithms with bases other than 10 and e.

Property 11-1 *Properties of Logarithms:*

$$\log_b mn = \log_b m + \log_b n$$

$$\log_b \frac{m}{n} = \log_b m - \log_b n$$

$$\log_b m^n = n \log_b m$$

$$\log_b b = 1$$

We can illustrate these laws with examples using a calculator.

EXAMPLE 7 Show that the following statements are true by using a calculator:

(a) $\log 6 = \log 2 + \log 3$ (b) $\log 2 = \log 6 - \log 3$
(c) $\log 2^3 = 3 \log 2$ (d) $\log 20 = 1 + \log 2$

(a) $\log 6 = \log 2(3) = \log 2 + \log 3$

$\boxed{\log}\ 6\ \boxed{=}\ \Rightarrow 0.77815125, \qquad \boxed{\log}\ 2\ \boxed{+}\ \boxed{\log}\ 3\ \boxed{=}\ \Rightarrow 0.77815125$

(b) $\log 2 = \log \dfrac{6}{3} = \log 6 - \log 3$

$\boxed{\log}\ 2\ \boxed{=}\ \Rightarrow 0.301029995, \qquad \boxed{\log}\ 6\ \boxed{-}\ \boxed{\log}\ 3\ \boxed{=}\ \Rightarrow 0.301029995$

(c) $\log 2^3 = 3 \log 2$

$\boxed{\log}\ \boxed{(}\ 2\ \boxed{x^y}\ 3\ \boxed{)}\ \boxed{=}\ \Rightarrow 0.903089987, \qquad 3\ \boxed{\times}\ \boxed{\log}\ 2\ \boxed{=}\ \Rightarrow 0.903089987$

(d) $\log 20 = \log 10(2) = \log 10 + \log 2 = 1 + \log 2$

$\boxed{\log}\ 20\ \boxed{=}\ \Rightarrow 1.301029996, \qquad 1 + \boxed{\log}\ 2\ \boxed{=}\ \Rightarrow 1.301029996$

SELF-STUDY EXERCISES 11–6

[1] Rewrite the following as logarithmic equations.

1. $3^2 = 9$

2. $2^5 = 32$

3. $9^{\frac{1}{2}} = 3$

4. $16^{\frac{1}{4}} = 2$

5. $4^{-2} = \dfrac{1}{16}$

6. $3^{-4} = \dfrac{1}{81}$

[2] Rewrite the following as exponential equations.

7. $\log_3 81 = 4$

8. $\log_{12} 144 = 2$

9. $\log_2 \dfrac{1}{8} = -3$

10. $\log_5 \dfrac{1}{25} = -2$

11. $\log_{25} \dfrac{1}{5} = -0.5$

12. $\log_4 \dfrac{1}{2} = -0.5$

Solve for x by using an equivalent exponential expression.

13. $\log_4 64 = x$

14. $\log_3 x = -4$

15. $\log_6 36 = x$

16. $\log_7 \dfrac{1}{49} = x$

17. $\log_5 x = 4$

18. $\log_4 \dfrac{1}{256} = x$

3 Evaluate the following with a calculator. Express the answer to the nearest ten-thousandth.

19. log 3
20. log 6
21. log 2.4
22. log 4.2
23. log 150
24. log 0.0012
25. *ln* 4
26. *ln* 2.5
27. *ln* 0.15
28. *ln* 275
29. *ln* 100

4 Evaluate the following logarithms with a calculator.

30. $\log_5 125$
31. $\log_3 729$
32. $\log_{\frac{1}{2}} 0.03125$
33. $\log_7 49$
34. $\log_8 56$

5 The intensity of an earthquake is measured on the *Richter scale* by the formula

$$\text{Richter scale rating} = \log \frac{I}{I_o}$$

where I_o is the measure of the intensity of a very small (faint) earthquake.

Find the Richter scale ratings of earthquakes having the following intensities:

35. $1000 I_o$
36. (b) $100,000 I_o$
37. $100,000,000 I_o$

6 Solve for x by using an equivalent logarithmic expression and the facts that $\log_3 2 = 0.631$ and $\log_3 5 = 1.465$.

38. $\log_3 10 = x$
39. $\log_3 8 = x$
40. $\log_3 6 = x$

CAREER APPLICATION

Electronics: Prefixes Used in Technical Fields

Technical people must often deal with numbers that are very large or very small. It is difficult to read and to write numbers that have many decimal places in them, both to the left and to the right of the decimal point. To make things easier, these numbers are usually changed to the power-of-10 form, that is, the number times 10 to some power, which shows how many places you are away from the decimal point. Then the power-of-10 factor can be replaced with a letter that stands for that value.

$$9.876 \times 10^3 = 9.876 \times 1000 = 9876$$

The 3 is the exponent and can be read as +3, which means the decimal point moves three places to the right from where it is in the original 9.876; so you end up with 9876. The 10^3 can be replaced with k for kilo, which means "thousand." If the exponent is +6, then the 10^6 can be replaced with M for mega (million), and the decimal point moves 6 places to the right. If the exponent is +9, then the 10^9 can be replaced with G for giga (billion), and the decimal point moves 9 places to the right.

$$9.876 \times 10^3 = 9.876 \text{ k} = 9.876 \times 1000 = 9876$$

$$9.876 \times 10^6 = 9.876 \text{ M} = 9.876 \times 1,000,000 = 9,876,000$$

$$9.876 \times 10^9 = 9.876 \text{ G} = 9.876 \times 1,000,000,000 = 9,876,000,000$$

$$9.876 \times 10^{12} = 9.876 \text{ T} = 9.876 \times 1,000,000,000,000$$
$$= 9,876,000,000,000$$

$$9.876 \times 10^{15} = 9.876 \text{ P} = 9.876 \times 1,000,000,000,000,000$$
$$= 9,876,000,000,000,000$$

$$10^{15} = P = peta = quadrillion$$
$$10^{12} = T = tera = trillion$$
$$10^{9} = G = giga = billion$$
$$10^{6} = M = mega = million$$
$$10^{3} = k = kilo = thousand$$
$$10^{0} = (unit = no\ prefix)$$
$$10^{-3} = m = milli = thousandth$$
$$10^{-6} = \mu = micro = millionth$$
$$10^{-9} = n = nano = billionth$$
$$10^{-12} = p = pico = trillionth$$
$$10^{-15} = f = femto = quadrillionth$$

If the exponent is negative, then the decimal point must move to the left. The prefix that replaces 10^{-3} is m for milli, and the prefix that replaces 10^{-6} is μ for micro. (The Greek letter mu, μ, was used to get an "m" sound for the prefix micro.)

$$9.876 \times 10^{-3} = 9.876\ m = 9.876 \times 0.001 = 0.009876$$
$$9.876 \times 10^{-6} = 9.876\ \mu = 9.876 \times 0.000001 = 0.000009876$$
$$9.876 \times 10^{-9} = 9.876\ n = 9.876 \times 0.000000001 = 0.000000009876$$
$$9.876 \times 10^{-12} = 9.876\ p = 9.876 \times 0.000000000001$$
$$= 0.000000000009876$$
$$9.876 \times 10^{-15} = 9.876\ f = 9.876 \times 0.000000000000001$$
$$= 0.000000000000009876$$

Our technology is moving so fast that it is important to learn the words to keep up with developments. Measurements are being made today in many fields in femto seconds; 1 femto second (1 f sec) means that the 1 is in the fifteenth place. Fourteen zeros are between the decimal point and 1. If it takes 1 femto second to do a calculation, then 1 peta of calculations can be done in 1 sec.

Don't try to memorize the prefixes, but be able to use them.

Exercises on Prefixes

$$P = 10^{15} \qquad T = 10^{12} \qquad G = 10^{9} \qquad M = 10^{6} \qquad k = 10^{3}$$
$$m = 10^{-3} \qquad \mu = 10^{-6} \qquad n = 10^{-9} \qquad p = 10^{-12} \qquad f = 10^{-15}$$

Convert each of the following numbers into a basic unit with no prefix. Follow the examples. The letter that follows the prefix is a unit that you keep. These are common units used in electronics.

1. 9753 kW = _____9,753,000_____ W
2. 15.68 μA = _____0.00001568_____ A
3. 4.567 mA = _____ A
4. 3.471 MW = _____ W
5. 65.42 μS = _____ S
6. 4892 pF = _____ F
7. 15.89 km = _____ m
8. 3.51 GV = _____ V

9. 53.2 kΩ = _____		Ω
10. 1.30 fS = _____		S
11. 5.69 nH = _____		H
12. 75.6 pF = _____		F
13. 24.56 mm = _____		m
14. 876.54 μS = _____		S

Answers for Exercises on Prefixes

3. 4.567 mA =	0.004567	A
4. 3.471 MW =	3,471,000	W
5. 65.42 μS =	0.00006542	S
6. 4892 pF =	0.000000004892	F
7. 15.89 km =	15,890	m
8. 3.51 GV =	3,510,000,000	V
9. 53.2 kΩ =	53,200	Ω
10. 1.30 fS =	0.0000000000000130	S
11. 5.69 nH =	0.00000000569	H
12. 75.6 pF =	0.0000000000756	F
13. 24.56 mm =	0.02456	m
14. 876.54 μS =	0.00087654	S

ASSIGNMENT EXERCISES

Section 11–1
Perform the indicated operations. Write the answers with positive exponents.

1. $x^5 \cdot x^5$ **2.** $x^2(x^4)$ **3.** $x^{-1}(x^7)$

4. $\dfrac{x^7}{x^4}$ **5.** $\dfrac{x^8}{x^5}$ **6.** $\dfrac{x^3}{x^5}$

7. $(x^3)^4$ **8.** $(-x^3)^3$ **9.** $(x^{-3})^{-5}$

Section 11–2
Simplify the following.

10. $4x^3 + 7x - 3x^3 - 5x$ **11.** $8x - 2x^4 - 3x^3 + 5x - x^3$ **12.** $5x - 3x + 7x^2 - 8x$

13. $4x^2 - 3y^2 + 7x^2 - 8y^2$ **14.** $4x^3(-3x^4)$ **15.** $-7x^8(-3x^{-2})$

16. $\dfrac{12x^5}{6x^3}$ **17.** $\dfrac{12x^7}{-18x^4}$ **18.** $\dfrac{11x^4}{22x^7}$

19. $5x(2x^2 + 3x - 4)$ **20.** $8x^3(2x - 6)$ **21.** $\dfrac{6x^3 - 12x^2 + 21x}{3x}$

22. $\dfrac{25y^5 - 85y^3 + 70y^2}{-5y}$

Section 11–3
Perform the indicated operations. Express as ordinary numbers.

23. $10^5 \cdot 10^7$ **24.** $10^{-2} \cdot 10^8$ **25.** $10^7 \cdot 10^{-10}$

26. 4.2×10^5 **27.** $8.73 \div 10^{-3}$ **28.** $5.6 \div 10^{-2}$

Change to scientific notation.

29. 52,000   **30.** 160   **31.** 0.00017

Perform the indicated operations and write the result in scientific notation.

32. $(4.2 \times 10^5)(3.9 \times 10^{-2})$

33. $(7.8 \times 10^{53})(5.6 \times 10^{72})$

34. $\dfrac{5.2 \times 10^8}{6.1 \times 10^5}$

35. $\dfrac{1.25 \times 10^3}{3.7 \times 10^{-8}}$

Section 11–4

36. Describe a polynomial and give an example.

37. Is the expression $5x^3 - 3x^{-2}$ a polynomial? Why or why not?

38. Arrange the following polynomials in descending order and identify the degree of each polynomial, the leading term, and the leading coefficient.
(a) $5x + 3x^3 - 8 + x^2$
(b) $3y^5 - 7y - 8y^4 + 12$

Section 11–5

A formula for electric current is $i = 1.50e^{-200t}$, where t is time in seconds. Calculate the current for the following times. Express the answers in scientific notation.

39. 0.07 second

40. 0.2 second

41. 0.4 second

42. The number of grams of a chemical that will dissolve in a solution is given by the formula $C = 100e^{0.05t}$, where t = temperature in degrees Celsius. Evaluate when:
(a) $t = 10$ (b) $t = 20$
(c) $t = 45$ (d) $t = 50$.

43. The compound amount for an investment is expressed by the formula $A = P(1 + \frac{r}{n})^{nt}$, where A = compounded amount, P = principal, n = number of compounding periods per year, r = interest rate per year, and t = number of years. Find the compound amount when:
(a) Principal = \$2000, interest = 8%, for 3 years and compounded quarterly.
(b) Principal = \$500, interest = 12%, for 9 months and compounded monthly.

Solve for x.

44. $2^x = 2^6$

45. $3^x = 3^{-2}$

46. $2^{x+3} = 2^7$

47. $4^{x-2} = 4^2$

48. $5^{2x-1} = 5^2$

49. $6^{3x+2} = 6^{-3}$

50. $2^x = 16$

51. $3^x = 81$

52. $3^x = \dfrac{1}{9}$

53. $2^x = \dfrac{1}{32}$

54. $4^{2x} = 64$

55. $5^{3x} = 125$

56. $3^{4-x} = \dfrac{1}{27}$

57. $6^{2-x} = \dfrac{1}{36}$

Evaluate.

58. e^3

59. e^{-4}

60. $e^{-0.12}$

61. e^{-10}

Section 11–6

Rewrite the following as logarithmic equations.

62. $5^2 = 25$

63. $3^4 = 81$

64. $81^{\frac{1}{2}} = 9$

65. $27^{\frac{1}{3}} = 3$

66. $5^{-3} = \dfrac{1}{125}$

67. $4^{-3} = \dfrac{1}{64}$

68. $8^{-\frac{1}{3}} = \dfrac{1}{2}$

69. $9^{-\frac{1}{2}} = \dfrac{1}{3}$

70. $121^{\frac{1}{2}} = 11$

71. $12^{-2} = \dfrac{1}{144}$

72. Write a true exponential expression; convert it to logarithmic form.

Rewrite the following as exponential equations. Verify if the equation is true.

73. $\log_{11} 121 = 2$

74. $\log_3 81 = 4$

75. $\log_{15} 1 = 0$

76. $\log_{25} 5 = \dfrac{1}{2}$

77. $\log_7 7 = 1$

78. $\log_3 3 = 1$

79. $\log_4 \dfrac{1}{16} = -2$

80. $\log_2 \dfrac{1}{16} = -4$

81. $\log_9 \dfrac{1}{3} = -0.5$

82. $\log_{16} \dfrac{1}{4} = -0.5$

Evaluate the following with a scientific calculator. Express the answers to the nearest ten-thousandth.

83. $\log 5$

84. $\log 3.8$

85. $\log 180$

86. $\log 0.0015$

87. $\log 0.4$

88. $\ln 12$

89. $\ln 270$

90. $\ln 0.134$

91. $\ln 0.8$

92. $\ln 80$

93. $\log_5 30$

94. $\log_7 120$

Solve for x by using an equivalent exponential expression.

95. $\log_4 16 = x$

96. $\log_7 49 = x$

97. $\log_7 x = 3$

98. $\log_5 x = -2$

99. $\log_6 \dfrac{1}{36} = x$

100. $\log_4 \dfrac{1}{64} = x$

101. The intensity of an earthquake is measured on the Richter scale by the formula

$$\text{Richter scale rating} = \log \frac{I}{I_o}$$

where I_o is the measure of the intensity of a very small (faint) earthquake. Find the Richter scale ratings of the earthquakes having the following intensities;

(a) $100 I_o$ (b) $10,000 I_o$ (c) $150,000,000 I_o$

Solve for x by using an equivalent logarithmic expression and the facts that $\log_2 3 = 1.585$ and $\log_2 7 = 2.807$.

102. $\log_2 21$

103. $\log_2 9$

104. $\log_2 6$

CHALLENGE PROBLEM

105. The Environmental Protection Agency (EPA) monitors atmosphere and soil contamination by dangerous chemicals. When possible, the chemical contamination is decomposed by using microorganisms that change the chemicals so they are no longer harmful. A particular microorganism can reduce the contamination level to about 65% of the existing level every 30 days. A soil test for a contaminated site shows 72,000,000 units per cubic meter of soil.

(a) Write a formula for determining the contamination level after x 30-day periods: the initial contamination times the percent reduction (65% or 0.65) raised to the xth power.

(b) What is the level of contamination after 60 days? after 150 days?

(c) A "safe level" is 60,000 units of contamination per cubic meter of soil. Estimate how long it would take for the soil to reach this safe level. Discuss your method of arriving at the estimate.

CONCEPTS ANALYSIS

Write the following laws of exponents and properties in your own words. Also, give an example illustrating each.

1. $a^m \cdot a^n = a^{m+n}$

2. $\dfrac{a^m}{a^n} = a^{m-n}, \quad a \neq 0$

3. $a^0 = 1, \quad a \neq 0$

4. $a^{-n} = \dfrac{1}{a^n}, \quad a \neq 0$

5. $(a^m)^n = a^{mn}$

6. $\left(\dfrac{a}{b}\right)^n = \dfrac{a^n}{b^n}, \quad b \neq 0$

7. $(ab)^n = a^n b^n$

8. What two conditions must be satisfied before a number is in scientific notation?

9. Explain in words what is meant by the following mathematical statement: If $b^x = b^y$, $b > 0$ and $b \neq 1$, then $x = y$. Write an equation that meets these conditions and solve it.

10. Show symbolically the relationship between an exponential equation and a logarithm. Give an example illustrating this relationship.

11. Explain the difference between a common logarithm and a natural logarithm.

Find the mistake in the following examples. Explain the mistake and correct it.

12. $\dfrac{9x^3 - 12x^2 + 3x}{3x} = 3x^2 - 4x$

13. $x^5(x^3) = x^{15}$

14. $\dfrac{3.4 \times 10^5}{2 \times 10^{-2}} = 1.7 \times 10^3$

15. $\begin{aligned} 2^{x-3} &= 4^2 \\ x - 3 &= 2 \\ x &= 2 + 3 \\ x &= 5 \end{aligned}$

CHAPTER SUMMARY

Objectives	What to Remember with Examples

Section 11-1

1 Multiply powers with like bases.

To multiply powers with like bases, add the exponents and keep the common base as the base of the product.

> Multiply. $x^5(x^{-7}) =$
> $$x^{5+(-7)} = x^{-2} \text{ or } \frac{1}{x^2}$$

2 Divide powers with like bases.

To divide powers with like bases, subtract the exponents and keep the common base as the base of the quotient.

> Divide. $\dfrac{a^7}{a^4} =$
> $$a^{7-4} = a^3$$

3 Find a power of a power.

To find a power of a power, multiply the exponents and keep the same base.

Simplify. $(y^5)^4 =$
$y^{5(4)} = y^{20}$

Section 11–2

1 Add or subtract like terms containing powers.

Like terms are terms that not only have the same letter factors, but also have the same exponent. Like terms are added or subtracted by adding or subtracting the coefficients of the like terms and keeping the variable factors and their exponents exactly the same.

Simplify. $3x^4 + 8x^2 - 7x^2 + 2x^4 = 5x^4 + x^2$

2 Multiply or divide algebraic expressions containing powers.

To multiply or divide algebraic expressions containing powers, multiply or divide the coefficients; then add or subtract the exponents of like bases.

Simplify. $(3x^4y^5)(7x^2yz) = 21x^6y^6z$

Section 11–3

1 Multiply and divide by powers of 10.

To multiply by powers of 10, add the exponents and keep the base of 10. To divide by powers of 10, subtract the exponents and keep the base of 10.

Multiply. $10^6(10^7) = 10^{13}$ Divide. $10^3 \div 10^5 = 10^{-2}$

2 Change a number from scientific notation to ordinary notation.

To change a number from scientific notation to ordinary notation, perform the indicated multiplication by moving the decimal point in the first factor the appropriate number of places. Insert zeros as necessary. Omit the power of 10 factor.

Write 3.27×10^{-4} as an ordinary number.
Shift the decimal 4 places to the *left*. 0.000327

3 Change a number from ordinary notation to scientific notation.

To change a number written in ordinary notation to scientific notation, insert a caret in the proper place to indicate where the decimal should be positioned so that the number is valued at 1 or between 1 and 10. Then determine how many places and in which direction the decimal shifts from the new position (caret) to the old position (decimal point). This number will be the exponent of the power of 10.

Write 54,000 in scientific notation.
$5_\wedge 4000. = 5.4 \times 10^4$

4 Multiply and divide numbers in scientific notation.

To multiply numbers in scientific notation, multiply the first factors; then multiply the powers of 10 by using the laws of exponents. Next, examine the first factor of the product to see if its value is equal to 1 or is between 1 and 10. If the factor is 1 or is between 1 and 10, the process is complete. If the factor is not 1 or not between 1 and 10, shift the decimal so that the factor is equal to 1 or is between 1 and 10 and adjust the exponent of the power of 10 accordingly.

Multiply.
$(4.5 \times 10^{89})(7.5 \times 10^{36}) =$
$33.75 \times 10^{125} =$
$3_\wedge 375 \times 10^1 \times 10^{125}$
3.375×10^{126}

To divide numbers in scientific notation, use steps similar to multiplication, but apply the rule for the division of signed numbers and the laws of exponents for division.

> Divide.
> $$(3 \times 10^{-3}) \div (4 \times 10^2) =$$
> $$0.75 \times 10^{-5} =$$
> $$07_\wedge 5 \times 10^{-1} \times 10^{-5} =$$
> $$7.5 \times 10^{-6}$$

Section 11–4

1 Identify polynomials, monomials, binomials, and trinomials.

Polynomials are algebraic expressions in which the exponents of the variable are nonnegative integers. A monomial is a polynomial with a single term. A binomial is a polynomial containing two terms. A trinomial is a polynomial containing three terms.

> Give an example of a polynomial, a monomial, a binomial, and a trinomial.
>
> polynomial: $4x^3 + 6x^2 - x + 3$ monomial: $8x$
> binomial: $6x + 3$ trinomial: $8x^2 - 4x - 3$

2 Identify the degree of terms and polynomials.

The degree of a term containing one variable is the exponent of the variable. The degree of a polynomial that has only one variable is the degree of the term that has the largest exponent.

> What is the degree of the polynomial listed in the previous example? The highest exponent is 3; thus, the degree of the polynomial is 3.

3 Arrange polynomials in descending order.

To arrange polynomials in descending order, list the term that has the highest degree first, the term that has the next highest degree second, and so on, until all terms have been listed.

> Arrange the following polynomial in descending order.
>
> $$4 - 2x + 7x^5 - 3x^2 + x^3$$
>
> Descending order:
>
> $$7x^5 + x^3 - 3x^2 - 2x + 4$$

Section 11–5

1 Evaluate formulas with at least one exponential term.

A scientific or graphics calculator can be used along with the order of operations to evaluate formulas with at least one exponential term.

> Use the formula for compound interest to find the compound amount for a loan of $5000 for 3 years at an annual interest rate of 6% if the principal is compounded semiannually.
>
> $$A = P\left(1 + \frac{r}{n}\right)^{nt}$$
>
> $$A = 5000\left(1 + \frac{0.06}{2}\right)^{2(3)}$$
>
> $$A = 5000(1.03)^6$$
> $$A = 5000(1.194052297)$$
> $$A = \$5970.26 \text{ (rounded)}$$

2 Evaluate formulas that contain a power of the natural exponential, e.

Use a scientific or graphics calculator to evaluate formulas containing the natural exponential, e.

Use the formula $P = 760e^{-0.00013h}$ for atmospheric pressure to find P at 50 meters above sea level (h).

$$P = 760e^{-0.00013(50)}$$
$$P = 760e^{-0.0065}$$
$$P = 760(0.9935210793)$$
$$P = 755.08 \quad \text{(rounded)}$$

3 Solve exponential equations in the form $b^x = b^y$, where $b > 0$ and $b \neq 1$.

To solve an exponential equation in the form $b^x = b^y$, apply the property: when the bases are equal, the exponents are equal. If the bases are not equal, rewrite the bases as powers so that the bases are equal. The exponents will then be equal. This property does not apply unless b is positive and not equal to one.

Solve the equation $4^{3x+1} = 8$. Rewrite the bases: $4 = 2^2$ and $8 = 2^3$

$$2^{2(3x+1)} = 2^3$$

So,

$$2(3x + 1) = 3$$
$$6x + 2 = 3$$
$$6x = 1$$
$$x = \frac{1}{6}$$

Section 11–6

1 Write exponential expressions as equivalent logarithmic expressions.

To write exponential expressions as equivalent logarithmic expressions, use the following format: $x = b^y$ converts to $\log_b x = y$.

Write $16 = 2^4$ in logarithmic form. $\log_2 16 = 4$

2 Write logarithmic expressions as equivalent exponential expressions.

To write logarithmic expressions as equivalent exponential expressions, use the following format: $\log_b x = y$ converts to $x = b^y$.

Write $\log_5 125 = 3$ in exponential form. $5^3 = 125$

3 Evaluate common and natural logarithmic expressions using a calculator.

Use the $\boxed{\log}$ key to find common logarithms and the $\boxed{\ln}$ key to find natural logarithms. Some scientific calculators require the number to be entered, followed by the $\boxed{\log}$ or $\boxed{\ln}$ key. Most graphics and scientific calculators require the $\boxed{\log}$ or $\boxed{\ln}$ key to be entered, followed by the number.

Use a calculator to find log 25. Scientific calculator steps:

$\boxed{\log}$ 25 $\boxed{=}$ or 25 $\boxed{\log}$ $\Rightarrow$ 1.397940009

4 Evaluate logarithms with a base other than 10 or e.

To evaluate logarithms with bases other than 10 or e, divide the common or natural log of the number by the common or natural log of the base.

Evaluate $\log_4 64$:

$$\log_4 64 = \frac{\log 64}{\log 4} = 3 \qquad \boxed{\log}\ 64 \div \boxed{\log}\ 4\ \boxed{=} \Rightarrow 3$$

5 Evaluate formulas containing at least one logarithmic term.

To evaluate formulas containing at least one logarithmic term, use a calculator and follow the order of operations.

Use the formula $bel = 10 \log \left(\dfrac{I}{I_o} \right)$ to find the decibel rating for a sound that is 350 times the threshold sound ($350I_o$).

$$bel = 10 \log \left(\dfrac{350I_o}{I_o} \right)$$
$$bel = 10 \log 350$$
$$bel = 25.44 \quad \text{(rounded)}$$

10 $\boxed{\times}$ $\boxed{\log}$ 350 $\boxed{=}$

6 Simplify logarithmic expressions by using the properties of logarithms.

The laws of exponents also apply to expressions with logarithms.
$$\log_b mn = \log_b m + \log_b n$$
$$\log_b \frac{m}{n} = \log_b m - \log_b n$$
$$\log_b m^n = n \log_b m$$
$$\log_b b = 1$$

Write log (3)(8) in another way. Use a calculator to verify that the equation is true.

$$\log (3)(8) = \log 3 + \log 8$$
$\boxed{\log}$ $\boxed{(}$ 3 $\boxed{\times}$ 8 $\boxed{)}$ $\boxed{=}$ $\Rightarrow 1.380211242$
$\boxed{\log}$ 3 $\boxed{+}$ $\boxed{\log}$ 8 $\boxed{=}$ $\Rightarrow 1.380211242$

WORDS TO KNOW

CHAPTER TRIAL TEST

Perform the indicated operations. Write the answers with positive exponents.

1. $(x^4)(x)$

2. $x^7(x^5)$

3. $\dfrac{x^6}{x^3}$

4. $\dfrac{x^0}{x^2}$

5. $\left(\dfrac{4}{7} \right)^2$

6. $(6a^2b)^2$

7. $\left(\dfrac{x^2}{y} \right)^2$

8. $3x(4x^3)$

9. $(-2a^4)(3a^2)$

10. $\dfrac{12x^2}{4x^3}$

11. $4a(3a^2 - 2a + 5)$

12. $\dfrac{60x^3 - 45x^2 - 5x}{5x}$

13. $(10^3)^2$

14. $\dfrac{10^{-5}}{10^3}$

Write the following as ordinary numbers.

15. 42×10^3

16. 0.83×10^2

17. 420×10^{-2}

18. 21×10^{-3}

19. $42 \div 10^3$

20. $8.4 \div 10^{-2}$

Write the following in scientific notation.

21. 240

22. 5.2301

23. 0.00086

24. 39×10^5

25. 783×10^{-5}

Perform the indicated operations. Express the answers in scientific notation.

26. $(5.9 \times 10^5)(3.1 \times 10^4)$

27. $(7.2 \times 10^{-3})(4.1 \times 10^2)$

28. $\dfrac{2.87 \times 10^5}{3.5 \times 10^7}$

29. $\dfrac{5.25 \times 10^4}{1.5 \times 10^2}$

30. A star is 5.9 light-years from Earth. If 1 light-year is 9.45×10^{12} km, how many kilometers from Earth is the star?

31. The total resistance (in ohms) of a dc series circuit equals the total voltage divided by the total amperage. If the total voltage is 3×10^3 V and the total amperage is 2×10^{-3} A, find the total resistance (in ohms) expressed as an ordinary number.

Evaluate using a calculator.

32. 5^{-8}

33. $15^{\frac{3}{2}}$

34. $e^{-0.25}$

35. 12^5

Solve for x.

36. $4^x = 4^{-3}$

37. $2^{x-4} = 2^5$

38. $3^x = \dfrac{1}{9}$

39. $2^{2x-1} = 8$

Rewrite the following as logarithmic equations.

40. $2^8 = 256$

41. $4^{-\frac{1}{2}} = \dfrac{1}{2}$

Rewrite the following as exponential equations.

42. $\log_5 625 = 4$

43. $\log_3 \dfrac{1}{27} = -3$

Evaluate the following with a calculator. Express the answer to the nearest ten-thousandth.

44. $\log 4.8$

45. $\ln 32$

Solve for x by using an equivalent exponential expression and a calculator.

46. $\log_4 x = -2$

47. $\log_6 216 = x$

Use a calculator to evaluate the following. Round to ten-thousandths.

48. $\log_3 5$

49. $\log_2 6$

50. $\log_7 2$

51. $\log_8 21$

52. The formula for the population growth of a certain species of insect in a controlled research environment is

$$P = 1{,}000{,}000e^{0.05t}$$

where t = time in weeks. Find the projected population after (a) 2 weeks and (b) 3 weeks.

54. The height in meters of the male members of a certain group is approximated by the formula

$$h = 0.4 + \log t$$

where t represents age in years for the first 20 years of life ($1 \le t \le 20$). Find the projected height of a 10-year-old male.

53. The revenue in thousands of dollars from sales of a product is approximated by the formula

$$S = 125 + 83 \log (5t + 1)$$

where t is the number of years after a product was marketed. Find the projected revenue from sales of the product after 3 years.

55. Find the accumulated amount on an investment of $5000 at a rate of 5.8% per year, compounded annually for 2 years. Use the formula $A = P(1 + \frac{r}{n})^{nt}$.

12

Roots and Radicals

GOOD DECISIONS THROUGH TEAMWORK

One of the most important advantages of working in a team is the opportunity to benefit from the work of fellow team members. In our everyday lives, time is one of our most important resources. For example, if you are interested in gathering information about one or more topics, the tasks can be divided among team members and your team can have all the information in a fraction of the time. This team project involves gathering infor-

mation on roots, irrational numbers, and imaginary numbers.

The concept of taking roots introduces a new type of number called an irrational number. For this project, refer to historical records to answer the following questions. When was this type of number first recorded and what purpose did it serve? When was a radical symbol first used for indicating a root of a number? When did it become standard practice for fractional or rational exponents to be used for indicating a root of a number? Do you think this notation for roots will become more or less common in the future? Why or why not? Find examples of formulas that use radicals or rational exponents. Explain what each formula can be used to find.

When planning team strategies, each team member should work on some portion of the project, and the team as a whole should discuss some resources for finding the desired information. A decision must be made on how much detail each team member should record for his or her findings and how the team will share and compile the information.

Prepare a written report using the information the team has compiled. This project could be extended to include a similar investigation of imaginary and complex numbers.

In Chapter 8, we examined the symbolic representation for irrational numbers that resulted from taking roots of numbers that are not perfect powers. We will extend our study of roots to include roots of variables.

As we saw before, there are two common notations for representing roots: fractional or rational exponents and radical notation. We will look at both types of notation and the conventions used with each.

12-1 ROOTS AND NOTATION CONVENTIONS

Learning Objectives

1. Write powers and roots using rational exponent and radical notations.
2. Use laws of exponents to simplify and evaluate expressions with rational exponents.

In Chapter 8 we illustrated by using a calculator that the relationship between the square root of a number and that number raised to the one-half power was the same.

$$\sqrt{8} = 2.828427125 \quad 8^{\frac{1}{2}} = 2.828427125$$

Similarly, the cube root of a number and a number raised to the one-third power is the same.

$$\sqrt[3]{8} = 2 \qquad 8^{\frac{1}{3}} = 2$$

Thus, we determined that the index or order of a root is indicated by the number in the "$\sqrt{\ }$" portion of the radical symbol or the number in the denominator of a rational exponent. By combining our knowledge of roots, the laws of exponents, and the arithmetic of rational numbers, we can extend our study of powers and roots to include variables.

1. Write Powers and Roots Using Rational Exponent and Radical Notations.

For our discussion of powers and roots of variables, unless otherwise specified, we will consider variables to represent positive values. Suppose that $\sqrt{x}$ is raised to the fourth power. In rational exponent notation, $\sqrt{x}$ is written as $x^{\frac{1}{2}}$.

$$(x^{\frac{1}{2}})^4 = x^{\frac{4}{2}} = x^2 \qquad \text{Multiply exponents and reduce.}$$

We apply the laws of exponents and arithmetic properties of fractions.

What happens if $\sqrt{x}$ is raised to the third power (cubed)?

$$(x^{\frac{1}{2}})^3 = x^{\frac{3}{2}}$$

This exponent can be left in improper fraction form, and it brings up the need for further interpretation of the meaning of a rational or fractional exponent.

> **Rule 12-1** *To convert between rational exponent notation and radical notation:*
>
> The numerator of a rational exponent represents the power and the denominator of the rational exponent represents the index or order of the root.
>
>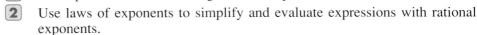

Then, is $(\sqrt{x})^2$ the same as $\sqrt{x^2}$? *Yes,* for non-negative values of x or $x \geq 0$.

$$(x^{\frac{1}{2}})^2 = x^{\frac{2}{2}} = x^1 = x, \qquad (x^2)^{\frac{1}{2}} = x^{\frac{2}{2}} = x^1 = x$$

Let's expand this concept to show the relationships between powers and roots.

Property 12–1   *Powers and roots—inverse operations:*

Raising to powers and extracting roots are inverse operations for non-negative values. The order in which the operations are performed does not matter.

$$(x^{\frac{1}{n}})^n = x \qquad (\sqrt[n]{x})^n = x$$
$$(x^n)^{\frac{1}{n}} = x \qquad \sqrt[n]{x^n} = x$$

for $x \geq 0$.

EXAMPLE 1   Write the following in both rational exponent and radical notations.

(a) The fourth root of the square of x.
(b) The square of the fourth root of x.
(c) The square root of the cube of x.
(d) The cube of the square root of x.

	Rational Exponent Notation	**Radical Notation**
(a)	$(x^2)^{\frac{1}{4}}$	$\sqrt[4]{x^2}$
(b)	$(x^{\frac{1}{4}})^2$	$(\sqrt[4]{x})^2$
(c)	$(x^3)^{\frac{1}{2}}$	$\sqrt{x^3}$
(d)	$(x^{\frac{1}{2}})^3$	$(\sqrt{x})^3$

Even though the calculator has made the pencil-and-paper method of taking roots obsolete, it is still helpful to perform some mental manipulations on expressions containing powers and roots before evaluating these expressions using the calculator. One common manipulation is to simplify rational exponents and radical expressions by applying the laws of exponents.

Rule 12–2   *To simplify radicals using fractional exponents and the laws of exponents:*

1. Convert the radicals to equivalent expressions using fractional exponents.
2. Apply the laws of exponents and the arithmetic of fractions.
3. Convert simplified expressions back to radical notation if desired.

EXAMPLE 2   Convert the following radical expressions to equivalent expressions using fractional exponents and simplify if appropriate.

(a) $\sqrt[3]{x}$ (b) $\sqrt[5]{2y}$ (c) $(\sqrt{ab})^3$ (d) $\sqrt[4]{16b^8}$ (e) $(\sqrt[3]{27xy^5})^4$

(a) $\sqrt[3]{x} = \boldsymbol{x^{\frac{1}{3}}}$
(b) $\sqrt[5]{2y} = \boldsymbol{(2y)^{\frac{1}{5}}}$ **or** $\boldsymbol{2^{\frac{1}{5}}y^{\frac{1}{5}}}$
(c) $(\sqrt{ab})^3 = \boldsymbol{(ab)^{\frac{3}{2}}}$ **or** $\boldsymbol{a^{\frac{3}{2}}b^{\frac{3}{2}}}$
(d) $\sqrt[4]{16b^8} = (2^4 b^8)^{\frac{1}{4}} = (2^4)^{\frac{1}{4}}(b^8)^{\frac{1}{4}} = \boldsymbol{2b^2}$

     The fourth root of 16 is the number used as a factor 4 times to equal 16. By inspection (or by using a calculator), $\sqrt[4]{16} = 2$ and $2^4 = 16$.

(e) $\left(\sqrt[3]{27xy^5}\right)^4 = (27x^1y^5)^{\frac{4}{3}} = (27)^{\frac{4}{3}}x^{\frac{4}{3}}(y^5)^{\frac{4}{3}} = 27^{\frac{4}{3}}x^{\frac{4}{3}}y^{\frac{20}{3}} = \mathbf{81x^{\frac{4}{3}}y^{\frac{20}{3}}}$

27 is a perfect cube; $3^3 = 27$. Therefore, $27^{\frac{4}{3}} = (3^3)^{\frac{4}{3}} = (3)^4 = 81$.

Tips and Traps   *Simplifying Coefficients:*
When coefficients of variable terms have rational exponents, the numerical equivalent can be determined if desired. Usually, if the coefficient is a perfect power of the indicated root, we will evaluate the coefficient. Otherwise, we may leave the coefficient with the rational exponent.

---

2 Use Laws of Exponents to Simplify and Evaluate Expressions with Rational Exponents.

Other laws of exponents are applied to fractional exponents. Remember, the laws of exponents apply to factors having *like bases*. The following example illustrates other laws of exponents applied to fractional exponents.

EXAMPLE 3   Perform the following operations and simplify. Express answers with positive exponents in lowest terms.

(a) $(x^{\frac{3}{2}})(x^{\frac{1}{2}})$ (b) $(3a^{\frac{1}{2}}b^3)^2$ (c) $\dfrac{x^{\frac{1}{2}}}{x^{\frac{1}{3}}}$ (d) $\dfrac{10a^3}{2a^{\frac{1}{2}}}$

(a) $(x^{\frac{3}{2}})(x^{\frac{1}{2}}) = x^{\frac{3}{2}+\frac{1}{2}} = x^{\frac{4}{2}} = \mathbf{x^2}$

(b) $(3a^{\frac{1}{2}}b^3)^2 = 3^2ab^6 = \mathbf{9ab^6}$

(c) $\dfrac{x^{\frac{1}{2}}}{x^{\frac{1}{3}}} = x^{\frac{1}{2}-\frac{1}{3}} = \mathbf{x^{\frac{1}{6}}}$ $\qquad \dfrac{1}{2} - \dfrac{1}{3} = \dfrac{3}{6} - \dfrac{2}{6} = \dfrac{1}{6}$

(d) $\dfrac{10a^3}{2a^{\frac{1}{2}}} = 5a^{3-\frac{1}{2}} = \mathbf{5a^{\frac{5}{2}}}$ $\qquad 3 - \dfrac{1}{2} = \dfrac{3}{1} - \dfrac{1}{2} = \dfrac{6}{2} - \dfrac{1}{2} = \dfrac{5}{2}$

Now, let's evaluate some expressions both before and after we apply some simplification procedures.

EXAMPLE 4   Evaluate the expressions in Example 3 for $x = 2$, $a = 3$, and $b = 4$ both before and after simplifying.

	Before	**After**

(a) $(x^{\frac{3}{2}})(x^{\frac{1}{2}}) = (2^{\frac{3}{2}})(2^{\frac{1}{2}})$

$\boxed{2}\ \boxed{x^y}\ 3\ \boxed{a\,b/c}\ 2\ \boxed{\times}\ 2\ \boxed{x^y}\ 1$
$\boxed{a\,b/c}\ 2\ \boxed{=}\ \Rightarrow \mathbf{4}$

$x^2 = 2^2$

$2\ \boxed{x^2}\ \Rightarrow \mathbf{4}$

(b) $(3a^{\frac{1}{2}}b^3)^2 = (3 \cdot 3^{\frac{1}{2}} \cdot 4^3)^2$

$\boxed{(}\ 3\ \boxed{\times}\ 3\ \boxed{x^y}\ 1\ \boxed{a\,b/c}\ 2\ \boxed{\times}\ 4$
$\boxed{x^y}\ 3\ \boxed{)}\ \boxed{x^y}\ 2\ \boxed{=}\ \Rightarrow \mathbf{110,592}$

$9ab^6 = 9 \cdot 3 \cdot 4^6$

$9\ \boxed{\times}\ 3\ \boxed{\times}\ 4\ \boxed{x^y}\ 6\ \boxed{=}\ \Rightarrow \mathbf{110,592}$

(c) $\dfrac{x^{\frac{1}{2}}}{x^{\frac{1}{3}}} = \dfrac{2^{\frac{1}{2}}}{2^{\frac{1}{3}}}$

$2\ \boxed{x^y}\ 1\ \boxed{a\,b/c}\ 2\ \boxed{\div}\ 2\ \boxed{x^y}\ 1$
$\boxed{a\,b/c}\ 3\ \boxed{=}\ \Rightarrow \mathbf{1.122462048}$

$x^{\frac{1}{6}} = 2^{\frac{1}{6}}$

$2\ \boxed{x^y}\ 1\ \boxed{a\,b/c}\ 6\ \boxed{=}\ \Rightarrow$
$\mathbf{1.122462048}$

(d) $\dfrac{10a^3}{2a^{\frac{1}{2}}} = \dfrac{10 \cdot 3^3}{2 \cdot 3^{\frac{1}{2}}}$

$5a^{\frac{5}{2}} = 5 \cdot 3^{\frac{5}{2}}$

$$\boxed{(}\ 10\ \boxed{\times}\ 3\ \boxed{x^y}\ 3\ \boxed{)}\ \div\ \boxed{(}\ 2 \qquad\qquad 5\ \boxed{\times}\ 3\ \boxed{x^y}\ 5\ \boxed{\text{a b/c}}\ 2\ \boxed{=}\ \Rightarrow$$
$$\boxed{\times}\ 3\ \boxed{x^y}\ 1\ \boxed{\text{a b/c}}\ 2\ \boxed{)} \qquad\qquad\qquad\qquad \mathbf{77.94228634}$$
$$\boxed{=}\ \Rightarrow\ \mathbf{77.94228634}$$

Exponents can also be expressed as decimals: $\sqrt{x} = x^{\frac{1}{2}} = x^{0.5}$.

EXAMPLE 5 Perform the following operations. Express the answers with positive exponents.

(a) $a^{2.3}(a^3)$ (b) $(5a^{3.5}b^{0.5})^2$

(a) $a^{2.3}(a^3) = a^{2.3+3} = \mathbf{a^{5.3}}$

What does $a^{5.3}$ mean? If written as an improper fraction, $a^{5.3} = a^{5\frac{3}{10}} = a^{\frac{53}{10}}$. This means we take the tenth root of a to the 53rd power or $\sqrt[10]{a^{53}}$.

(b) $(5a^{3.5}b^{0.5})^2 = 5^2 a^{3.5(2)} b^{0.5(2)}$
$$= \mathbf{25a^7 b}$$

SELF-STUDY EXERCISES 12-1

1 Write the following in both rational exponent and radical notations.

1. The fifth root of the cube of x.
2. The cube of the fifth root of x.
3. The square root of the fourth power of x.
4. The square of the sixth root of x.
5. The fourth root of the cube of x.
6. The fourth power of the square root of x.

Convert the following radical expressions to equivalent expressions using fractional exponents and simplify.

7. $\sqrt[4]{a}$
8. $\sqrt[3]{m}$
9. $\sqrt[5]{3x}$
10. $\sqrt[4]{7a}$
11. $(\sqrt{xy})^4$
12. $\sqrt[3]{27a^6}$
13. $\sqrt[3]{32x^7}$
14. $(\sqrt{36x^3y^4})^4$
15. $\left(\sqrt[3]{16a^4b^7}\right)^2$

2 Perform the following operations and simplify. Express the answers with positive exponents in lowest terms.

16. $(a^{\frac{5}{2}})(a^{\frac{1}{2}})$
17. $(b^{\frac{3}{5}})(b^{\frac{1}{5}})$
18. $(4x^{\frac{1}{3}}y^4)^3$
19. $(2m^{\frac{1}{5}}n^2)^5$
20. $\dfrac{a^{\frac{3}{5}}}{a^{\frac{1}{5}}}$
21. $\dfrac{b^{\frac{1}{5}}}{b^{\frac{3}{5}}}$
22. $\dfrac{m^{\frac{7}{8}}}{m^{\frac{3}{4}}}$
23. $\dfrac{x^{\frac{1}{3}}}{x^{\frac{5}{6}}}$
24. $\dfrac{a^2}{a^{\frac{1}{3}}}$
25. $\dfrac{14a^4}{7a^{\frac{3}{5}}}$
26. $\dfrac{8a^{\frac{2}{3}}}{24a^5}$
27. $\dfrac{6a^{\frac{7}{8}}}{27a^2}$

Evaluate the following expressions for $a = 2$, $b = 4$, and $c = 5$ both before and after simplifying. Verify that the result is the same.

28. $(a^{\frac{1}{3}})(a^{\frac{4}{3}})$
29. $(4a^{\frac{1}{3}}b^2)^3$
30. $\dfrac{c^{\frac{3}{5}}}{c^{\frac{1}{4}}}$
31. $\dfrac{5b^2}{10b^{\frac{1}{3}}}$

Perform the following operations and express the answers with positive exponents.

32. $(a^{3.2})(a^4)$
33. $\dfrac{a^{5.6}}{a^7}$
34. $(3a^{1.2}b^2)^3$
35. $(4c^{1.3}d^{0.3})^{10}$

12-2 SIMPLIFYING SQUARE-ROOT EXPRESSIONS

Learning Objectives

1 Find the square root of variables.

2 Simplify square-root radical expressions containing perfect-square factors.

We have already simplified variable expressions with rational exponents. Now we will simplify variable expressions with variables in radical notation. While radical notation is used less and less, during this transition phase there is a significant number of situations where you will still encounter radical notation.

1   Find the Square Root of Variables.

The square root of a variable is best understood by using the rational exponent notation and the laws of exponents. At this point we will consider only real numbers and we will *assume* that the variables represent positive values.

$$\sqrt{x^2} = (x^2)^{\frac{1}{2}} = x^1 = x$$
$$\sqrt{x^4} = (x^4)^{\frac{1}{2}} = x^2$$
$$\sqrt{x^6} = (x^6)^{\frac{1}{2}} = x^3$$

Tips and Traps   *Perfect Square Variable Factors:*
A positive variable with an even exponent is a perfect square. To find the square root of a variable factor, take $\frac{1}{2}$ of the exponent.

EXAMPLE 1   Find the square root of the following positive variables.

(a) $\sqrt{x^8}$ (b) $-\sqrt{x^{12}}$ (c) $\pm\sqrt{x^{18}}$ (d) $\sqrt{\dfrac{a^2}{b^4}}$

(a) $\sqrt{x^8} = (x^8)^{\frac{1}{2}} = \boldsymbol{x^4}$ (b) $-\sqrt{x^{12}} = -(x^{12})^{\frac{1}{2}} = \boldsymbol{-x^6}$

(c) $\pm\sqrt{x^{18}} = \pm(x^{18})^{\frac{1}{2}} = \boldsymbol{\pm x^9}$ (d) $\sqrt{\dfrac{a^2}{b^4}} = \dfrac{(a^2)^{\frac{1}{2}}}{(b^4)^{\frac{1}{2}}} = \dfrac{\boldsymbol{a}}{\boldsymbol{b^2}}$

2   Simplify Square-Root Radical Expressions Containing Perfect-Square Factors.

Some radicands will be made up of factors that are perfect squares, while others will not be perfect squares. In the latter case, it may be useful *to factor* the radicand.

■   **DEFINITION 12–1:   To factor.**   To *factor an algebraic expression* is to write it as the indicated product of two or more factors, that is, as a multiplication.

Some radicands that are not perfect squares can be *factored* into a perfect square times another factor. We simplify radicals by taking the square root of as many perfect-square factors as possible. The factors that are not perfect squares are left under the radical sign, and the square root of the perfect-square factors will be outside the radical: $\sqrt{4x} = 2\sqrt{x}$; $\sqrt{3x^2} = x\sqrt{3}$; that is, when the square root of a factor is actually taken, the result is no longer under the radical symbol.

A number such as 8 is not a perfect square; however, it does have a perfect square factor: 4 is a factor of 8 and 4 is a perfect square. Thus, $\sqrt{8}$ can be written as $\sqrt{4 \cdot 2} = 2\sqrt{2}$.

Similarly, x^3 is not a perfect square because the exponent is not even. However, any power having an odd exponent larger than 1 has a perfect square factor.

Using the laws of exponents, $a^3 = a^2 a^1$. Then $\sqrt{a^3}$ can be simplified as $\sqrt{a^2 \cdot a^1} = a\sqrt{a}$. To find a^2 and a^1 as factors of the power a^3, we subtract the exponent 1 from the odd exponent 3.

We will simplify any radical expression by taking the square root of *all* perfect-square factors. This procedure can be summarized as a rule.

Rule 12–3 *To simplify square-root radicals containing perfect-square factors:*

1. If the radicand is a perfect square, express it as a square root without the radical sign.
2. If the radicand is *not* a perfect square, factor the radicand into as many perfect-square factors as possible. The square roots of the perfect-square factors will appear *outside* the radical and the other factors will stay *inside* (under) the radical sign.

If the radicand is *not* a perfect square and *cannot* be factored into one or more perfect-square factors, it is already in simplified form.

EXAMPLE 2   Simplify the following radicals.

(a) $\sqrt{21ab^2}$ (b) $\sqrt{32}$ (c) $\sqrt{y^7}$ (d) $\sqrt{18x^5}$
(e) $\sqrt{75xy^3z^5}$ (f) $\sqrt{7x}$

When factoring coefficients, some perfect squares that can be used are 4, 9, 16, 25, 36, 49, 64, 81.

(a) $\sqrt{21ab^2} = \sqrt{21}\left(\sqrt{a}\right)\left(\sqrt{b^2}\right) = \sqrt{21}\left(\sqrt{a}\right)\left(b\right) = \boldsymbol{b\sqrt{21a}}$

Factors removed from under the radical sign are written before the radical. Because the factors 21 and a are not perfect squares and have no perfect-square factors, they are written under the same radical sign.

(b) What is the *largest* perfect-square factor of 32? 4 is a factor of 32, but 16 is also a factor of 32. Use the *largest* perfect-square factor.

$$\sqrt{32} = \sqrt{16 \cdot 2} = \sqrt{16}\left(\sqrt{2}\right) = \boldsymbol{4\sqrt{2}}$$

(c) The largest perfect square factor of y^7 is y^{7-1}, or y^6.

$$\sqrt{y^7} = \sqrt{y^6 \cdot y^1} = \sqrt{y^6}\left(\sqrt{y}\right)$$
$$= \boldsymbol{y^3\sqrt{y}}$$

(d) $\sqrt{18x^5} = \sqrt{9 \cdot 2 \cdot x^4 \cdot x^1}$
$= \sqrt{9}\left(\sqrt{2}\right)\left(\sqrt{x^4}\right)\left(\sqrt{x}\right)$
$= 3\left(\sqrt{2}\right)\left(x^2\right)\sqrt{x}$
$= \boldsymbol{3x^2\sqrt{2x}}$

In this example we showed each step in our simplifying process. However, we customarily do most of these steps mentally.

$$\sqrt{18x^5} = \sqrt{9 \cdot 2 \cdot x^4 \cdot x} = \boldsymbol{3x^2\sqrt{2x}}$$

Write the square roots of the perfect-square factors outside the radical sign. The other factors are written under the radical sign.

(e) $\sqrt{75xy^3z^5} = \sqrt{25 \cdot 3 \cdot x \cdot y^2 \cdot y \cdot z^4 \cdot z}$
$= \boldsymbol{5yz^2\sqrt{3xyz}}$

(f) $\sqrt{7x} = \boldsymbol{\sqrt{7x}}$ 7 and x contain no perfect-square factors.

Tips and Traps   *Finding Perfect-Square Factors:*

- Whole-number perfect squares: 1, 4, 9, 16, 25, 36, 49, 64, 81, 100, 121, 144, ...
- One is a factor of any number: $8 = 8 \cdot 1$. To factor using the perfect square 1 does not simplify a radicand.
- Any variable with an exponent higher than 1 is a perfect square or has a perfect-square factor.
- Perfect-square variables:

$$x^2, x^4, x^6, x^8, x^{10}, \ldots$$

- Variables with a perfect-square factor:

$$x^3 = x^2 \cdot x^1, \qquad x^5 = x^4 \cdot x^1$$
$$x^7 = x^6 \cdot x^1, \qquad x^9 = x^8 \cdot x^1$$

- A convenient way to keep up with perfect-square factors is to circle them. The square roots of circled factors are written outside the radical sign. The uncircled factors stay in the radicand as is.

$$\sqrt{75ab^4c^3} = \sqrt{\textcircled{25} \cdot 3 \cdot a \cdot \textcircled{b^4} \cdot \textcircled{c^2} \cdot c^1}$$
$$= 5b^2c\sqrt{3ac}$$

EXAMPLE 3   Simplify the following.

    (a) $\sqrt{7x^2}$ (b) $\sqrt{9a}$ (c) $\sqrt{32m^5n^6}$

    (a) $\sqrt{7\textcircled{x^2}} = \boldsymbol{x\sqrt{7}}$

    (b) $\sqrt{\textcircled{9}a} = \boldsymbol{3\sqrt{a}}$

    (c) $\sqrt{32m^5n^6} = \sqrt{\textcircled{16} \cdot 2 \cdot \textcircled{m^4} \cdot m \cdot \textcircled{n^6}} = \boldsymbol{4m^2n^3\sqrt{2m}}$

SELF-STUDY EXERCISES 12–2

1 Find the square root of the following positive variables.

1. $\sqrt{x^{10}}$ **2.** $\sqrt{x^6}$ **3.** $\pm\sqrt{x^{16}}$ **4.** $\sqrt{\dfrac{x^{14}}{y^{24}}}$

5. $\sqrt{a^6b^{10}}$ **6.** $-\sqrt{a^2b^4c^{12}}$ **7.** $\sqrt{\dfrac{x^2y^4}{z^{10}}}$ **8.** $\sqrt{\dfrac{a^4}{b^{10}c^{12}}}$

2 Simplify.

9. $\sqrt{24}$ **10.** $\sqrt{98}$ **11.** $\sqrt{48}$

12. $\sqrt{x^9}$ **13.** $\sqrt{y^{15}}$ **14.** $\sqrt{12x^3}$

15. $\sqrt{56a^5}$ **16.** $\sqrt{72a^3x^4}$ **17.** $\sqrt{44x^5y^2z^7}$

18. Create a square-root radical with the product of a constant factor that is not a perfect square, but has a perfect-square factor, and a variable factor with an odd exponent greater than 1. Then simplify.

12–3 BASIC OPERATIONS WITH SQUARE-ROOT RADICALS

Learning Objectives   **1** Add or subtract square-root radicals.

                   2 Multiply square-root radicals.

3 Divide square-root radicals.

4 Rationalize a denominator.

1   Add or Subtract Square-Root Radicals.

As you recall, when adding or subtracting measures or algebraic terms, we can only add or subtract like quantities. Similarly, only *like* radicals can be added or subtracted. Square-root radicals are *like* radicals whenever the radicands are identical.

■ **DEFINITION 12–2: Like Radicals.** When the radicands are identical and the radicals have the same order or index, the radicals are *like radicals*.

> **Rule 12–4** *To add or subtract like square-root radicals:*
>
> Add or subtract the coefficients of the radicals. The common radicand will be used in the solution.
>
> $$a\sqrt{b} + c\sqrt{b} = (a + c)\sqrt{b}$$

EXAMPLE 1   Add or subtract the following radicals when possible.

(a) $3\sqrt{7} + 2\sqrt{7}$ (b) $4\sqrt{2} + \sqrt{2}$ (c) $5\sqrt{3} + 7\sqrt{5} + 2\sqrt{3} - 4\sqrt{5}$

(d) $3 + \sqrt{3}$ (e) $\dfrac{7}{8}\sqrt{5} - \dfrac{3}{8}\sqrt{5}$ (f) $3\sqrt{11} - 3\sqrt{11}$

(g) $2\sqrt{2} - \sqrt{3}$

(a) $3\sqrt{7} + 2\sqrt{7} = \mathbf{5\sqrt{7}}$

(b) $4\sqrt{2} + \sqrt{2} = \mathbf{5\sqrt{2}}$

When no coefficient is written in front of a radical, the coefficient is 1.

(c) $5\sqrt{3} + 7\sqrt{5} + 2\sqrt{3} - 4\sqrt{5} = \mathbf{7\sqrt{3} + 3\sqrt{5}}$

(d) $\mathbf{3 + \sqrt{3}}$

No addition can be performed. These are not like radicals.

(e) $\dfrac{7}{8}\sqrt{5} - \dfrac{3}{8}\sqrt{5} = \dfrac{4}{8}\sqrt{5} = \dfrac{1}{2}\left(\dfrac{\sqrt{5}}{1}\right) = \dfrac{\mathbf{\sqrt{5}}}{\mathbf{2}}$

(f) $3\sqrt{11} - 3\sqrt{11} = 0\sqrt{11} = \mathbf{0}$

Zero times any number is zero.

(g) $\mathbf{2\sqrt{2} - \sqrt{3}}$

The terms cannot be combined. These are not like radicals.

We can add or subtract square-root radical expressions only if they have *like* radicands. However, when radicals are not in simplest form, you are not able to recognize like radicals. We may be able to simplify the radical expressions and obtain like radicands. If we obtain like radicands, then we can add or subtract.

EXAMPLE 2   Add or subtract the following radical expressions.

(a) $12\sqrt{5} + 3\sqrt{20}$ (b) $\sqrt{3} - \sqrt{27}$ (c) $6\sqrt{3} + 2\sqrt{8}$

(a) $12\sqrt{5} + 3\sqrt{20}$ $\sqrt{20}$ can be simplified.

    $12\sqrt{5} + 3\sqrt{4 \cdot 5}$

    $12\sqrt{5} + 3 \cdot 2\sqrt{5}$

    $12\sqrt{5} + 6\sqrt{5}$

    $\mathbf{18\sqrt{5}}$

(b) $\sqrt{3} - \sqrt{27}$

$\quad \sqrt{3} - \sqrt{9 \cdot 3}$

$\quad \sqrt{3} - 3\sqrt{3}$

$\quad \mathbf{-2\sqrt{3}}$

$\sqrt{27}$ can be simplified.

(c) $6\sqrt{3} + 2\sqrt{8}$

$\quad 6\sqrt{3} + 2\sqrt{4 \cdot 2}$

$\quad 6\sqrt{3} + 2 \cdot 2\sqrt{2}$

$\quad \mathbf{6\sqrt{3} + 4\sqrt{2}}$

$\sqrt{8}$ can be simplified.

Terms cannot be combined. Radicals are still unlike radicals.

2 Multiply Square-Root Radicals.

When multiplying two square-root radicals, the expressions under the radical signs (radicands) are multiplied together. Numbers in front of the radical signs are coefficients of the radicals and are multiplied separately.

Rule 12–5 *To multiply square-root radicals:*

1. The coefficients are multiplied to give the coefficient of the product.
2. The radicands are multiplied to give the radicand of the product.
3. Simplify if possible.

$$a\sqrt{b} \cdot c\sqrt{d} = ac\sqrt{bd}$$

EXAMPLE 3 Multiply the following radicals.

(a) $\sqrt{3} \cdot \sqrt{5}$ (b) $\sqrt{\dfrac{7}{8}} \cdot \sqrt{\dfrac{2}{3}}$ (c) $3\sqrt{2} \cdot 4\sqrt{3}$ (d) $\sqrt{3} \cdot \sqrt{12}$

(a) $\sqrt{3} \cdot \sqrt{5} = \mathbf{\sqrt{15}}$

(b) $\sqrt{\dfrac{7}{8}} \cdot \sqrt{\dfrac{2}{3}} = \sqrt{\dfrac{14}{24}} = \sqrt{\mathbf{\dfrac{7}{12}}}$

Reduce.

(c) $3\sqrt{2} \cdot 4\sqrt{3} = \mathbf{12\sqrt{6}}$

Multiply coefficients 3 and 4. Then multiply radicands 2 and 3.

(d) $\sqrt{3} \cdot \sqrt{12} = \sqrt{36} = \mathbf{6}$

36 is a perfect square.

3 Divide Square-Root Radicals.

When dividing square-root radicals, we follow a similar procedure. Coefficients and radicands are divided (or reduced) separately.

Rule 12–6 *To divide square-root radicals:*

1. The coefficients are divided to give the coefficient of the quotient.
2. The radicands are divided to give the radicand of the quotient.
3. Simplify if possible.

$$\dfrac{a\sqrt{b}}{c\sqrt{d}} = \dfrac{a}{c}\sqrt{\dfrac{b}{d}} \qquad c \text{ and } d \neq 0$$

EXAMPLE 4 Divide the following radicals.

(a) $\dfrac{\sqrt{12}}{\sqrt{4}}$ (b) $\dfrac{\sqrt{\frac{2}{3}}}{\sqrt{\frac{7}{4}}}$ (c) $\dfrac{3\sqrt{6}}{6}$ (d) $\dfrac{5\sqrt{20}}{\sqrt{10}}$

(a) $\dfrac{\sqrt{12}}{\sqrt{4}} = \sqrt{\dfrac{12}{4}} = \sqrt{3}$

(b) $\dfrac{\sqrt{\frac{2}{3}}}{\sqrt{\frac{7}{4}}} = \sqrt{\dfrac{\frac{2}{3}}{\frac{7}{4}}} = \sqrt{\dfrac{2}{3}\left(\dfrac{4}{7}\right)}$ $\left(\dfrac{2}{3} \div \dfrac{7}{4} = \dfrac{2}{3} \cdot \dfrac{4}{7}\right)$

$\sqrt{\dfrac{8}{21}} = \sqrt{\dfrac{4 \cdot 2}{21}} = \dfrac{2\sqrt{2}}{\sqrt{21}}$

(c) $\dfrac{3\sqrt{6}}{6}$

The 6 in the denominator *will not* divide into the 6 in the numerator. The 6 in the numerator is a radicand and the 6 in the denominator is not. However, the 3 and the denominator 6 can be divided (or reduced) because they are both outside the radical. Remember, $\sqrt{6}$ is not 6, but 2.449 (the square root of 6).

$\dfrac{\overset{1}{\cancel{3}}\sqrt{6}}{\underset{2}{\cancel{6}}} = \dfrac{\sqrt{6}}{2}$ A coefficient of 1 does not have to be written in front of the radical.

(d) $\dfrac{5\sqrt{20}}{\sqrt{10}}$

The 5 and 10 are not divided or reduced because the 10 is a radicand and the 5 is not. However, the 20 and 10 can be divided because they are both square-root radicands.

$$\dfrac{5\sqrt{20}}{\sqrt{10}} = 5\sqrt{2}$$

4 **Rationalize a Denominator.**

Radicals are often rewritten in an equivalent form whenever they appear in the *denominator* of a fraction. Before the common use of calculators, this was a popular manipulation. Dividing by a rational number, and most often a whole number, was easier and more accurate than dividing by a rounded decimal approximation of the radical. Even now, finding common denominators and other procedures are easier if all denominators contain only rational numbers. Thus, $\frac{1}{\sqrt{3}}$ and $\sqrt{\frac{2}{5}}$ are generally rewritten so that the denominator is a rational number.

This procedure is called *rationalizing the denominator*.

Rule 12–7 *To rationalize a denominator:*

1. Remove perfect-square factors from all radicands.
2. Multiply the denominator by another radical so that the resulting radicand is a perfect square.

> **3.** To preserve the value of the fraction, multiply the numerator by the same radical. Thus, we multiply by an equivalent of 1.
> **4.** Reduce the resulting fraction, if possible.

EXAMPLE 5 Rationalize all denominators. Simplify the answers if possible.

(a) $\dfrac{1}{\sqrt{3}}$ (b) $\dfrac{5}{\sqrt{7}}$ (c) $\dfrac{2}{\sqrt{x}}$ (d) $\dfrac{4}{\sqrt{8}}$ (e) $\dfrac{5\sqrt{2}}{x^2\sqrt{3x}}$ (f) $\sqrt{\dfrac{2}{3}}$

(a) $\dfrac{1}{\sqrt{3}} = \dfrac{1}{\sqrt{3}} \cdot \dfrac{\sqrt{3}}{\sqrt{3}} = \dfrac{\sqrt{3}}{3}$ $\sqrt{3} \cdot \sqrt{3} = (\sqrt{3})^2 = 3$

   Remember, a radical times itself is the same as squaring a radical.

(b) $\dfrac{5}{\sqrt{7}} = \dfrac{5}{\sqrt{7}} \cdot \dfrac{\sqrt{7}}{\sqrt{7}} = \dfrac{5\sqrt{7}}{7}$ $\sqrt{7} \cdot \sqrt{7} = (\sqrt{7})^2 = 7$

(c) $\dfrac{2}{\sqrt{x}} = \dfrac{2}{\sqrt{x}} \cdot \dfrac{\sqrt{x}}{\sqrt{x}} = \dfrac{2\sqrt{x}}{x}$ $\sqrt{x} \cdot \sqrt{x} = (\sqrt{x})^2 = x$

(d) $\dfrac{4}{\sqrt{8}} = \dfrac{4}{\sqrt{4 \cdot 2}} = \dfrac{4}{2\sqrt{2}} = \dfrac{2}{\sqrt{2}}$ Simplify perfect-square factors, then rationalize the denominator.

$$= \dfrac{2}{\sqrt{2}} \cdot \dfrac{\sqrt{2}}{\sqrt{2}} = \dfrac{2\sqrt{2}}{2} = \sqrt{2}$$

(e) $\dfrac{5\sqrt{2}}{x^2\sqrt{3x}} = \dfrac{5\sqrt{2}}{x^2\sqrt{3x}} \cdot \dfrac{\sqrt{3x}}{\sqrt{3x}} = \dfrac{5\sqrt{6x}}{x^2 \cdot 3x} = \dfrac{5\sqrt{6x}}{3x^3}$

(f) $\sqrt{\dfrac{2}{3}} = \dfrac{\sqrt{2}}{\sqrt{3}} \cdot \dfrac{\sqrt{3}}{\sqrt{3}} = \dfrac{\sqrt{6}}{3}$

When is rationalizing a denominator practical? Before calculators and computers, a radical in the denominator meant dividing by a decimal approximation of an irrational number. Calculations were more difficult and answers were generally less accurate. Are calculator values less accurate if the denominator is not rationalized? Examine the next example.

EXAMPLE 6 Compare the approximate value of the following expressions:

$$\dfrac{1}{\sqrt{2}} \quad \text{and} \quad \dfrac{\sqrt{2}}{2}$$

$1 \div \boxed{\sqrt{}}\ 2\ \boxed{\text{EXE}} \Rightarrow 0.7071067812$

$\boxed{\sqrt{}}\ 2 \div 2\ \boxed{\text{EXE}} \Rightarrow 0.7071067812$

The radical expressions are equivalent.

Tips and Traps **Why Do We Still Rationalize Denominators?**
In reality, the importance of rationalizing radical expressions in technical applications is minimal. Approximate values and calculators and computers are most often used.

The most important use of rationalizing is to find alternate, but equivalent, representations of expressions. These alternate representations may make it easier to identify like radicals and other properties.

Traditionally, radical expressions are in simplest form if:

1. There are no perfect-square factors in any radicand.
2. There are no fractional radicands.
3. The denominator of a radical expression is rational or contains no radicals.

A logical sequence for simplifying radical expressions follows in Rule 12–8.

Rule 12–8 *To simplify a radical expression:*

1. Reduce radicands and coefficients whenever possible.
2. Simplify expressions with perfect-square factors in all radicands.
3. Reduce radicands and coefficients whenever possible.
4. Rationalize denominators that contain radicals.
5. Reduce radicands and coefficients whenever possible.

EXAMPLE 7 Perform the operation and simplify if possible.

(a) $\sqrt{y^3} \cdot \sqrt{8y^2}$ (b) $\sqrt{\dfrac{1}{3}} \cdot \sqrt{\dfrac{8}{3x}}$

(a) $\sqrt{y^3} \cdot \sqrt{8y^2} = \sqrt{8y^5} = \sqrt{4 \cdot 2 \cdot y^4 \cdot y} = \mathbf{2\,y^2\sqrt{2y}}$

(b) $\sqrt{\dfrac{1}{3}} \cdot \sqrt{\dfrac{8}{3x}} = \sqrt{\dfrac{8}{9x}} = \dfrac{\sqrt{8}}{\sqrt{9x}} = \dfrac{\sqrt{4 \cdot 2}}{\sqrt{9 \cdot x}} = \dfrac{2\sqrt{2}}{3\sqrt{x}} \cdot \dfrac{\sqrt{x}}{\sqrt{x}} = \dfrac{\mathbf{2\sqrt{2x}}}{\mathbf{3x}}$

SELF-STUDY EXERCISES 12–3

[1] Add or subtract. Simplify radicals where necessary.
1. $5\sqrt{3} + 7\sqrt{3}$
2. $8\sqrt{5} - 12\sqrt{5}$
3. $4\sqrt{7} + 3\sqrt{7} - 5\sqrt{7}$
4. $2\sqrt{3} - 8\sqrt{5} + 7\sqrt{3}$
5. $9\sqrt{11} - 3\sqrt{6} + 4\sqrt{6} - 12\sqrt{11}$
6. $44\sqrt{2} + \sqrt{3} - \sqrt{2} + 5\sqrt{3}$
7. $2\sqrt{3} + 5\sqrt{12}$
8. $7\sqrt{5} + 2\sqrt{45}$
9. $4\sqrt{63} - \sqrt{7}$
10. $3\sqrt{6} - 2\sqrt{54}$
11. $8\sqrt{2} - 3\sqrt{28}$
12. $2\sqrt{3} + \sqrt{48}$
13. $3\sqrt{5} + 4\sqrt{180}$
14. $7\sqrt{98} - 2\sqrt{2}$
15. $6\sqrt{40} - 2\sqrt{90}$
16. $\sqrt{12} - \sqrt{27}$

[2] Multiply and simplify if possible.
17. $5\sqrt{2} \cdot 3\sqrt{5}$
18. $8\sqrt{3} \cdot 5\sqrt{12}$
19. $5\sqrt{3x} \cdot 4\sqrt{5x^2}$
20. $2x\sqrt{3x^4} \cdot 7x^2\sqrt{8x}$

Perform the indicated operations and simplify if possible.
21. $\sqrt{x^2} \cdot \sqrt{3x}$
22. $\sqrt{\dfrac{2}{3}} \cdot \sqrt{\dfrac{4}{5y}}$
23. $\sqrt{\dfrac{1}{x}} \cdot \sqrt{\dfrac{8}{7x}}$
24. $\sqrt{\dfrac{4}{9y^2}} \cdot \sqrt{\dfrac{1}{2y}}$
25. $\sqrt{\dfrac{8x}{3}} \cdot \sqrt{\dfrac{2x^2}{3}}$
26. $\sqrt{7} \cdot \sqrt{x^2}$
27. $\sqrt{\dfrac{4x^2}{x^3}} \cdot \sqrt{12x}$
28. $\sqrt{\dfrac{1}{2}} \cdot \sqrt{\dfrac{x^2}{3}}$
29. $\sqrt{\dfrac{y^3}{2}} \cdot \sqrt{\dfrac{2}{7y}}$
30. $\sqrt{8x} \cdot \sqrt{x^2}$

[3] Divide and simplify.
31. $\dfrac{4\sqrt{12}}{2\sqrt{6}}$
32. $\dfrac{15\sqrt{24}}{9\sqrt{2}}$
33. $\dfrac{12x^2\sqrt{8x}}{15x\sqrt{6x^3}}$
34. $\dfrac{6x^4\sqrt{25x^3}}{2x\sqrt{16x^2}}$

[4] Rationalize the denominator and simplify.
35. $\dfrac{5}{\sqrt{3}}$
36. $\dfrac{6}{\sqrt{5}}$
37. $\dfrac{1}{\sqrt{8}}$
38. $\dfrac{\sqrt{5}}{\sqrt{11}}$

39. $\dfrac{\sqrt{8}}{\sqrt{12}}$ **40.** $\dfrac{5x}{\sqrt{3x}}$ **41.** $\dfrac{2x^2}{\sqrt{7x^2}}$ **42.** $\dfrac{4x^5}{\sqrt{12x^3}}$

12–4 COMPLEX AND IMAGINARY NUMBERS

Learning Objectives

1 Write imaginary numbers using the letter *i*.

2 Raise imaginary numbers to powers.

3 Write real and imaginary numbers in complex form, $a + bi$.

4 Combine complex numbers.

1 **Write Imaginary Numbers Using the Letter *i*.**

Taking the square root of a negative number introduces a new type of number. This new type of number is called an *imaginary number*.

■ **DEFINITION 12–3: Imaginary Number.** An *imaginary number* has a factor of $\sqrt{-1}$, which is represented by *i* ($i = \sqrt{-1}$).

Imaginary numbers and real numbers combine to form the set of complex numbers (Fig. 12–1).

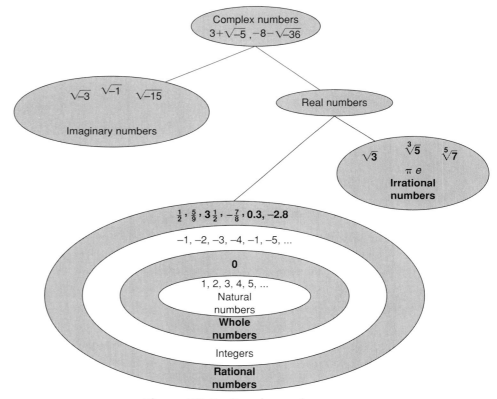

Figure 12–1 Complex number system.

■ **DEFINITION 12–4: Complex Number.** A *complex number* is a number that can be written in the form $a + bi$, where a and b are real numbers and i is $\sqrt{-1}$.

The square root of a negative number, say, $\sqrt{-16}$, can be simplified as $\sqrt{-1 \cdot 16}$ or $4\sqrt{-1}$. Because the square root of negative one is an *imaginary* factor and 4 is a real factor, we will use the letter i to represent $\sqrt{-1}$ and rewrite $4\sqrt{-1}$ as $4i$. Similarly, $\sqrt{-4} = \sqrt{-1 \cdot 4} = 2\sqrt{-1} = 2i$.

> **Tips and Traps** *Is i Different from j?*
> Electronics and other applications of imaginary numbers may use j rather than i to represent $\sqrt{-1}$.

EXAMPLE 1 Rewrite the following imaginary numbers using the letter i for $\sqrt{-1}$.

(a) $\sqrt{-9}$ (b) $\sqrt{-25}$ (c) $\sqrt{-7}$

(a) $\sqrt{-9} = \sqrt{-1 \cdot 9} = 3\sqrt{-1} = 3i$

(b) $\sqrt{-25} = \sqrt{-1 \cdot 25} = 5\sqrt{-1} = 5i$

(c) $\sqrt{-7} = \sqrt{-1 \cdot 7} = \sqrt{7} \cdot \sqrt{-1} = \sqrt{7}i$ or $i\sqrt{7}$.

> **Tips and Traps** *Are Numerical Coefficients Always First?*
> $\sqrt{7}i$ and $\sqrt{7i}$ do not represent the same amount. In the first term, $\sqrt{7}i$, i is not under the radical symbol. In the second term, $\sqrt{7i}$, i is under the radical symbol. Because it is easy to confuse the two terms, when the coefficient of an imaginary number is an irrational number, it is appropriate to write the i factor first: $\sqrt{7}i = i\sqrt{7}$

2 Raise Imaginary Numbers to Powers.

Some powers of imaginary numbers are real numbers. Examine the pattern that develops with powers of i.

$$i = \sqrt{-1}$$
$$i^2 = (\sqrt{-1})^2 = -1$$
$$i^3 = i^2 \cdot i^1 = -1i = -i$$
$$i^4 = i^2 \cdot i^2 = -1(-1) = 1$$
$$i^5 = i^4 \cdot i^1 = 1(i) = i$$
$$i^6 = i^4 \cdot i^2 = 1(-1) = -1$$
$$i^7 = i^4 \cdot i^3 = 1(-i) = -i$$
$$i^8 = i^4 \cdot i^4 = 1(1) = 1$$
$$i^9 = i^4 \cdot i^4 \cdot i^1 = 1(1)(i) = i$$
$$i^{10} = i^4 \cdot i^4 \cdot i^2 = 1(1)(-1) = -1$$
$$i^{11} = i^4 \cdot i^4 \cdot i^3 = 1(1)(-i) = -i$$
$$i^{12} = i^4 \cdot i^4 \cdot i^4 = 1(1)(1) = 1$$

> **Rule 12–9** *To simplify a power of* **i**:
>
> 1. Divide the exponent by 4 and examine the **remainder**.
> 2. The power of i will simplify as follows, based on the remainder in step 1.

$$\text{Remainder of } 0 \Rightarrow 1$$

$$\text{Remainder of } 1 \Rightarrow i$$

$$\text{Remainder of } 2 \Rightarrow -1$$

$$\text{Remainder of } 3 \Rightarrow -i$$

EXAMPLE 2 Simplify the following powers of i.

(a) i^{15} (b) i^{20} (c) i^{33} (d) i^{18}

(a) $i^{15} = -i$ $15 \div 4 = 3$ R3, remainder of $3 \Rightarrow -i$

(b) $i^{20} = 1$ $20 \div 4 = 5$, remainder $0 \Rightarrow 1$

(c) $i^{33} = i$ $33 \div 4 = 8$ R1, remainder $1 \Rightarrow i$

(d) $i^{18} = -1$ $18 \div 4 = 4$ R2, remainder $2 \Rightarrow -1$

Tips and Traps Powers of i:

All powers of i can be simplified to one of four values: i, -1, $-i$, or 1. Even powers of i can be simplified to a real number, either 1 or -1.

3 Write Real and Imaginary Numbers in Complex Form, *a* + *bi*.

A complex number has two parts, a real part and an imaginary part. In $a + bi$, if $a = 0$, then $a + bi$ is the same as $0 + bi$ or bi, which is an imaginary number. If $b = 0$, then $a + bi$ is the same as $a + 0 \cdot i$ or a, which is a real number. Thus, real numbers and imaginary numbers are also complex numbers.

EXAMPLE 3 Rewrite the following in the form $a + bi$.

(a) 5   (b) $\sqrt{-36}$ (c) $-3i^2$ (d) $6 - \sqrt{-5}$

(a) $5 = 5 + 0i$ $a = 5$, $b = 0$

(b) $\sqrt{-36} = 6i = 0 + 6i$ $a = 0$, $b = 6$

(c) $-3i^2 = -3(-1) = 3 = 3 + 0i$ $a = 3$, $b = 0$

(d) $6 - \sqrt{-5} = 6 - \sqrt{(5)(-1)} = 6 - i\sqrt{5}$

4 Combine Complex Numbers.

Complex numbers are combined by adding *like* parts. Real parts are added together and imaginary parts are added together.

Rule 12–10 *To combine complex numbers:*

1. Add real parts for the real part of the answer.
2. Add imaginary parts for the imaginary part of the answer.

$$(a + bi) + (c + di) = (a + c) + (b + d)i$$

EXAMPLE 4   Combine the following:

(a) $(3 + 5i) + (8 - 2i)$ (b) $(-3 + i) - (7 - 4i)$ (c) $-2 + (5 + 6i)$

(a) $(3 + 5i) + (8 - 2i) = \mathbf{11} + \mathbf{3}i$ $3 + 8 = 11$; $5i - 2i = 3i$
(b) $(-3 + i) - (7 - 4i) =$ Distribute -1.
        $(-3 + i) + (-7 + 4i) = \mathbf{-10} + \mathbf{5}i$ $-3 - 7 = -10$; $i + 4i = 5i$
(c) $-2 + (5 + 6i) = \mathbf{3} + \mathbf{6}i$ $-2 + 5 = 3$

SELF-STUDY EXERCISES 12–4

1 Write the following imaginary numbers using the letter i. Simplify if possible.

1. $\sqrt{-25}$ **2.** $\sqrt{-36}$ **3.** $\sqrt{-64x^2}$ **4.** $\sqrt{-32y^5}$

2 Simplify the following powers of i.

5. i^{17} **6.** i^{20} **7.** i^{24} **8.** i^{32}

3 Write the following real and imaginary numbers in complex form.

9. 15   **10.** $33i$ **11.** $5 + \sqrt{-4}$ **12.** $8 + \sqrt{-32}$
13. $7 - \sqrt{-3}$ **14.** $-7i^3$ **15.** $4i^6$

4 Combine the following complex numbers.

16. $(4 + 3i) + (7 + 2i)$ **17.** $\sqrt{12} - 5\sqrt{-3} + (\sqrt{8} - \sqrt{-27})$
18. $12 - 3i - (8 - 4i)$ **19.** $15 + 8i - (3 - 12i)$
20. $(4 + 7i) - (3 - 2i)$

12–5   EQUATIONS WITH SQUARES AND SQUARE ROOTS

Learning Objectives   **1** Solve equations with squared letter terms.
                           2 Solve equations with square-root radical terms.

1 Solve Equations with Squared Letter Terms.

In solving equations with squared letter terms, we will use the same procedures for solving equations that we used in Chapter 9. When squared letter terms are involved, we need to take one more step. After the equation is simplified to a basic equation with a coefficient of $+1$, we will take the square root of both sides of the equation to find the roots. In doing so, the solutions will be square roots expressed with $\pm$ to show both the negative and the positive square-root values. Equations that we will examine in this section are *quadratic equations*. In Chapter 15, we will define quadratic equations and examine several types of quadratic equations.

Rule 12–11   *To solve an equation containing only squared variable terms and number terms:*

1. Perform all normal steps to isolate the squared variable and to obtain a numerical coefficient of 1.
2. Take the square root of both sides.
3. Identify both roots of the equation.

EXAMPLE 1   Solve the following equations:

(a) $x^2 = 4$ (b) $x^2 + 9 = 25$ (c) $5x^2 + 10 = 30$

(d) $\dfrac{3}{25} = \dfrac{x^2}{48}$ (e) $x^2 + 16 = 0$

(a) $x^2 = 4$ Take the square root of both sides.

$x = \pm 2$

(b) $x^2 + 9 = 25$ Isolate the squared variable term.

$x^2 = 25 - 9$ Combine like terms.

$x^2 = 16$ Take the square root of both sides.

$x = \pm 4$

(c) $5x^2 + 10 = 30$ Isolate the squared variable term.

$5x^2 = 30 - 10$ Combine like terms.

$5x^2 = 20$ Divide by the coefficient of 5.

$\dfrac{5x^2}{5} = \dfrac{20}{5}$

$x^2 = 4$

$x = \pm 2$

(d) $\dfrac{3}{25} = \dfrac{x^2}{48}$ To solve a proportion, cross multiply.

$3(48) = 25x^2$

$144 = 25x^2$

$\dfrac{144}{25} = \dfrac{25x^2}{25}$

$\dfrac{144}{25} = x^2$

$\pm\dfrac{12}{5} = x$

(e) $x^2 + 16 = 0$ Isolate the squared variable term.

$x^2 = -16$ Take the square root of both sides.

$x = \pm\sqrt{-16}$ Express the roots as imaginary numbers using i.

$x = \pm 4i$

Tips and Traps   *Imaginary Roots Versus No Real Roots:*
Sometimes within the context of an applied problem, imaginary or complex roots are unacceptable. Often a problem will specify that only real roots are acceptable.

---

2 **Solve Equations with Square-Root Radical Terms.**

In this discussion we consider only equations containing a radical isolated on either or both sides of the equation. Whenever this is the case, we solve the equation by first squaring both sides of the equation. This will rid the equation of radicals. Then we can use previously learned procedures to finish solving the equation.

> **Rule 12–12** *To solve an equation containing square-root radicals that are isolated on either or both sides of the equation:*
>
> 1. Square both sides to eliminate any radicals.
> 2. Perform all normal steps to isolate the variable and solve the equation.

EXAMPLE 2 Solve the following equations.

(a) $\sqrt{x} = \dfrac{3}{4}$ (b) $\sqrt{x-2} = 5$ (c) $\sqrt{5x^2 - 36} = 8$

(d) $\sqrt{1.7x^2} = \sqrt{6.8}$

(a) $\quad \sqrt{x} = \dfrac{3}{4}$ Square both sides of the equation.

$\quad (\sqrt{x})^2 = \left(\dfrac{3}{4}\right)^2$

$\quad\quad x = \dfrac{9}{16}$

(b) $\quad \sqrt{x-2} = 5$ Square both sides of the equation.

$\quad (\sqrt{x-2})^2 = 5^2$ Any square-root radical squared equals the radicand.

$\quad\quad x - 2 = 25$

$\quad\quad\quad x = 25 + 2$

$\quad\quad\quad \mathbf{x = 27}$

(c) $\quad \sqrt{5x^2 - 36} = 8$

$\quad (\sqrt{5x^2 - 36})^2 = 8^2$ Do this step mentally.

$\quad\quad 5x^2 - 36 = 64$

$\quad\quad\quad 5x^2 = 64 + 36$

$\quad\quad\quad 5x^2 = 100$

$\quad\quad\quad \dfrac{5x^2}{5} = \dfrac{100}{5}$ Do this step mentally.

$\quad\quad\quad x^2 = 20$

$\quad\quad\quad x = \pm\sqrt{20}$

$\quad\quad\quad \mathbf{x = \pm 2\sqrt{5}}$ Exact roots

$\quad\quad\quad \mathbf{x \approx \pm 4.472}$ Approximate roots

(d) $\sqrt{1.7x^2} = \sqrt{6.8}$

$\quad\quad 1.7x^2 = 6.8$

$\quad\quad x^2 = \dfrac{6.8}{1.7}$

$\quad\quad x^2 = 4$

$\quad\quad \mathbf{x = \pm 2}$

EXAMPLE 3 Shirley Riddle needs to prepare a flower bed for planting. She needs a square bed with an area of 128 ft². How long should each side of the bed be?

$\quad\quad A = s^2$ Formula for the area of a square

$\quad 128 = s^2$

$\quad \sqrt{128} = s$

$\quad \mathbf{8\sqrt{2} = s}$ Exact root

$\quad \mathbf{11.3 \approx s}$ Approximate root

Each side of the bed should be 11.3 ft.

Tips and Traps *Examining Roots:*

Exact Roots Versus Approximate Roots

A root is an exact amount when no rounding has been done. If an exact root is desired, give the root in simplest form.

   If the root has been rounded, it becomes an approximate root. Approximate roots are often more meaningful than exact roots, especially in applied problems.

One Root Versus Two Roots

Look again at the four problems in Example 2. Parts (a) and (b) have variables to the first power after the radical is removed. Therefore, there will be *at most* one solution. Parts (c) and (d) have variables that are squared when the radical is removed. These equations will have *at most* two roots or solutions.

Extraneous Roots

The techniques of squaring both sides of an equation and taking the square root of both sides of an equation are equivalent to multiplying or dividing by a variable, thus introducing the possibility of an extraneous root. Be sure to check your roots.

SELF-STUDY EXERCISES 12–5

[1] Solve the following equations. Use a calculator to evaluate to the nearest thousandth when necessary.

1. $y^2 = 25$

2. $-9 + 3x^2 = 18$

3. $3 = \dfrac{36}{x^2}$

4. $\dfrac{3}{81} = \dfrac{x^2}{12}$

5. $3 + R^2 = 12$

6. $\dfrac{1}{2}P^2 = \dfrac{3}{7}$

7. $\dfrac{4}{9} = \dfrac{x^2}{3}$

8. $16 - T^2 = 0$

9. $5x^2 - 10 = 20$

10. $x^2 + 25 = 0$

[2] Solve the following equations. Use a calculator if necessary. Evaluate to the nearest thousandth.

11. $\sqrt{x} = 9$

12. $\sqrt{\dfrac{y}{3}} = 8$

13. $8 = \sqrt{y - 2}$

14. $\sqrt{3x^2 - 3} = 9$

15. $\sqrt{\dfrac{3}{x + 2}} = 12$

16. $\sqrt{x^2} = 2$

17. $10 = \sqrt{2x}$

18. $\sqrt{\dfrac{2}{3}x^2} = 4$

19. $\sqrt{1.3y^2} = 2.4$

20. $\sqrt{y^2} = 2.9$

CAREER APPLICATION

Horticulture: Insect Mating

Thomasville, Georgia, is known as the "rose capital of the world" for the abundant native and imported roses that bloom almost yearround due to its mild climate and excellent iron-rich soil. In preparation for its annual Rose Festival each April, local horticulturists control as many variables as they can to maximize the quality and quantity of April roses.

One big problem is aphids, which are tiny sucking insects that secrete a sweet juice ants love to eat. Ants will even carry around and nurse aphids just to get their honeydew. While the aphids eat the new growth on leaves, stems, and buds, the ants loosen soil around roots, causing rose plants to wilt and die. Aphids and ants can destroy a rose garden in a few days.

A horticulture researcher studies aphid mating in the field and in the laboratory. When equal numbers of male and female aphids are in close proximity, the number of their matings per hour depends on the temperature according to the following formula:

$$M = 5t^{\frac{3}{2}} - 3t^{\frac{1}{2}}$$

where M is the number of group matings per hour and t is the temperature in degrees Celsius (°C). This formula approximates the matings for temperatures between 4° and 25°C, inclusive.

Knowing the temperatures at the maximum and minimum number of matings can help horticulturists control aphids and ants before they begin to affect the quality and quantity of rose production. Aphids can be eradicated with natural predators (for example, ladybugs which eat aphids), manual labor (for example, washing them off with soap), or insecticide (for example, diazinon or malathion spray). The type of treatment depends on the degree of infestation, how much time is available for eradication, treatment cost, and one's attitudes toward environmental protection.

Exercises

Use the above information to answer the following. Round answers to the nearest unit.

1. Approximate the number of matings at the following Celsius temperatures: (a) 9° (b) 16° (c) 6° (d) 20.5° (e) 36°
2. Convert the exponential mating formula to radical form. Now use the square root calculator key to perform the same calculations in Exercise 1. If your answers differ, explain why. Which form uses fewer calculator key strokes?
3. Use your answers from Exercise 1 to state the trend of mating frequency as temperature increases. Do you think this trend will continue for 36°C? Explain why or why not.
4. At what temperatures do the maximum and minimum numbers of matings occur? State your answers in degrees Celsius and degrees Fahrenheit. What are the maximum and minimum number of matings for this temperature range?
5. If Thomasville's average February temperature of 51.5°F was predicted to be 60°F this year, use the formula to calculate the expected percentage increase or decrease in matings this year compared to typical years. How should horticulturists plan to alter their aphid-control measures?
6. If Thomasville's average October temperature of 66.0°F was predicted to be 57°F this year, use the formula to calculate the expected percentage increase or decrease in matings this year compared to typical years. How should horticulturists plan to alter their aphid-control measures?

Answers

1. (a) 126 matings (b) 308 matings (c) 66 matings (d) 451 matings (e) Unknown because 36° is outside the temperature range.
2. The answers should be the same. Radical form uses fewer key strokes.
3. As temperature increases, the number of matings increases. The trend is hard to predict. At 36°C (97°F), the matings will probably decrease or stop completely (especially if the humidity is also high). Matings might actually increase, however, in a frantic effort to reproduce before death.

4. The maximum number of matings is about 610, which occurs at 25°C (77°F). The minimum number of matings is about 34, which occurs at 4°C (39°F).
5. At 51.5°F (10.8°C), 168 matings are predicted. At the higher temperature of 60°F (16°C), 308 matings are expected. This 83% increase in aphid mating means that horticulturists should be prepared to increase their aphid-control measures.
6. At 66°F (19°C), 401 matings are predicted. At the lower temperature of 57°F (14°C), 251 matings are expected. This 37% decrease in aphid mating means that horticulturists should be prepared to decrease their aphid-control measures.

ASSIGNMENT EXERCISES

Section 12–1
Write in both rational exponent and radical notations.

1. The square root of the seventh power of x.
2. The cube root of the fourth power of x.
3. The square of the cube root of x.
4. The fifth power of the fifth root of x.

Write the following using fractional exponents.

5. $\sqrt{x}$
6. $\sqrt[3]{x^5}$
7. $\sqrt[5]{x^4}$
8. $\sqrt[4]{9x}$
9. $(\sqrt[3]{xy})^4$
10. $\sqrt[3]{64x^{10}}$
11. $\sqrt{7}$

Write the following in radical format.

12. $x^{\frac{5}{8}}$
13. $y^{\frac{3}{5}}$
14. $a^{\frac{1}{4}}$

Convert the following radicals to equivalent expressions using fractional exponents.

15. $\sqrt[3]{x}$
16. $\sqrt[4]{p}$
17. $\sqrt[5]{4y}$
18. $(\sqrt{ab})^6$
19. $\sqrt[3]{8b^{12}}$
20. $(\sqrt{49x^2y^3})^4$

Perform the following operations. Express the answers with positive exponents in lowest terms.

21. $(a^{\frac{1}{2}})(a^{\frac{3}{2}})$
22. $(x^{\frac{4}{3}})(x^{\frac{2}{3}})$
23. $y^{\frac{3}{4}} \cdot y^{\frac{1}{4}}$
24. $y^{\frac{5}{8}} \cdot y^{\frac{1}{8}}$
25. $(3x^{\frac{1}{4}}y^2)^3$
26. $(2x^{\frac{3}{4}}y)^2$
27. $(4ax^{\frac{1}{2}})^3$
28. $(x^{\frac{1}{2}})^{\frac{1}{3}}$
29. $\dfrac{x^{\frac{3}{4}}}{x^{\frac{1}{4}}}$
30. $\dfrac{x^{\frac{1}{6}}}{x^{\frac{5}{6}}}$
31. $\dfrac{a^{\frac{5}{6}}}{a^{-\frac{1}{3}}}$
32. $\dfrac{a^{\frac{7}{10}}}{a^{\frac{2}{5}}}$
33. $\dfrac{x^{\frac{5}{8}}}{x^{\frac{3}{4}}}$
34. $\dfrac{y^{\frac{1}{3}}}{y^{\frac{5}{6}}}$
35. $\dfrac{a^3}{a^{\frac{1}{3}}}$

Evaluate the expressions in 36 to 42 for $a = 2$, $b = 1$, and $x = 3$ both before and after simplifying.

36. $\dfrac{a^2}{a^{\frac{3}{5}}}$
37. $\dfrac{12a^4}{6a^{\frac{1}{2}}}$
38. $\dfrac{27x^3}{9x^{\frac{2}{3}}}$
39. $\dfrac{15a^{\frac{3}{5}}}{10a^5}$
40. $\dfrac{14a^{\frac{5}{6}}}{24a^2}$
41. $a^{2.3}(a^4)$
42. $(3a^{1.2}b^2)^3$

Section 12–2
Find the square root of the following if all variables represent positive numbers.

43. $\sqrt{y^{12}}$
44. $\sqrt{a^{10}}$
45. $-\sqrt{b^{18}}$
46. $\pm\sqrt{\dfrac{x^4}{y^6}}$

Simplify the following expressions.

47. $(\sqrt{5})^2$ **48.** $(\sqrt{x^5})^2$ **49.** $\sqrt{x^2}$ **50.** $(\sqrt{8})^2$

51. $\sqrt{9P^3}$ **52.** $\sqrt{8^2}$ **53.** $\sqrt{18a^2b}$ **54.** $\sqrt{12x^2y^3}$

55. $\sqrt{32x^5y^2}$ **56.** $\sqrt{63x^4y^7}$ **57.** $\sqrt{75x^{10}y^9}$ **58.** $\sqrt{125xy^3}$

Section 12–3

Add or subtract the following radicals.

59. $5\sqrt{3} - 7\sqrt{3}$ **60.** $4\sqrt{2} + 3\sqrt{5} - 8\sqrt{2} + 6\sqrt{5}$

61. $3\sqrt{7} - 2\sqrt{28}$ **62.** $\sqrt{2} - \sqrt{8}$ **63.** $2\sqrt{6} + 3\sqrt{54}$

64. $3\sqrt{5} - 2\sqrt{45}$ **65.** $4\sqrt{3} - 8\sqrt{48}$ **66.** $\sqrt{40} + \sqrt{90}$

67. $5\sqrt{8} - 3\sqrt{50}$ **68.** $5\sqrt{7} - 4\sqrt{63}$ **69.** $3\sqrt{2} - 5\sqrt{32}$

Multiply the following radicals and simplify.

70. $\sqrt{6} \cdot \sqrt{3}$ **71.** $2\sqrt{8} \cdot 3\sqrt{6}$ **72.** $2\sqrt{a} \cdot \sqrt{b}$

73. $5\sqrt{3} \cdot 8\sqrt{7}$ **74.** $2\sqrt{3} \cdot 5\sqrt{18}$ **75.** $-8\sqrt{5} \cdot 4\sqrt{30}$

Divide the following radicals and simplify.

76. $\dfrac{4\sqrt{8}}{2}$ **77.** $\dfrac{3\sqrt{5}}{2\sqrt{20}}$ **78.** $\dfrac{2\sqrt{90}}{\sqrt{5}}$ **79.** $\dfrac{6\sqrt{18}}{8\sqrt{12}}$

80. $\dfrac{14\sqrt{56}}{7\sqrt{7}}$ **81.** $\dfrac{5\sqrt{48}}{20\sqrt{20}}$ **82.** $\dfrac{\sqrt{9x}}{\sqrt{3x}}$ **83.** $\dfrac{\sqrt{3y^3}}{\sqrt{y^3}}$

84. $\left(\sqrt{\dfrac{25}{36}}\right)^2$ **85.** $\left(\sqrt{\dfrac{9}{16}}\right)^2$ **86.** $\sqrt{\dfrac{9c^4}{25y^6}}$

Rationalize the denominator and simplify.

87. $\dfrac{5}{\sqrt{17}}$ **88.** $\dfrac{1}{\sqrt{8}}$ **89.** $\dfrac{\sqrt{7}}{\sqrt{12}}$ **90.** $\dfrac{\sqrt{3}}{\sqrt{7x}}$

91. $\dfrac{\sqrt{3}}{\sqrt{8}}$ **92.** $\dfrac{\sqrt{7}}{5\sqrt{18}}$ **93.** $\dfrac{5\sqrt{3}}{\sqrt{24}}$ **94.** $\dfrac{\sqrt{15}}{5\sqrt{7}}$

Section 12–4

Write the following numbers using the letter i. Simplify if possible.

95. $\sqrt{-100}$ **96.** $-\sqrt{-16x^2}$ **97.** $\pm\sqrt{-24y^7}$

Simplify the powers of i.

98. i^5 **99.** i^{14} **100.** i^{98} **101.** i^{77}

Write the following as complex numbers in simplified form.

102. 5 **103.** $15i$ **104.** $3 + \sqrt{-9}$

105. $-12i^5$ **106.** $-6i^{11}$

Simplify.

107. $(5 + 3i) + (2 - 7i)$ **108.** $(4 - i) - (3 - 2i)$

109. $(7 - \sqrt{-9}) + (4 + \sqrt{-16})$

Section 12–5

Solve the following equations containing squares and radicals. Use a calculator if needed. Evaluate to the nearest thousandth.

110. $q^2 = 81$ **111.** $x^2 - 36 = 0$ **112.** $3x^2 - 2 = 7$

113. $x^2 - 4 = 0$ **114.** $7 + P^2 = 107$ **115.** $x^2 + 4 = 0$

116. $x^2 + 81 = 0$

117. $18 = 2x^2$

118. $\sqrt{P + 2} = 12$

119. $\sqrt{\dfrac{27}{2}} = x$

120. $\sqrt{3 + x} = 14$

121. $\sqrt{q + 3} = 7$

122. $\sqrt{\dfrac{1}{4x}} = 2$

123. $\sqrt{1.3x^2} = 11.7$

124. $\sqrt{x^2 + 1} = 5$

125. $\sqrt{x^2 + 2} = 9$

126. $\sqrt{3 + y^2} = 10$

127. $\sqrt{Q^2 - 1} = 0$

128. $0 = \sqrt{z^2 - 4}$

129. $\sqrt{2 + y^2} = 8$

CHALLENGE PROBLEMS

130. List all whole numbers less than 100 that are not perfect squares, but have perfect square factors. Then write the square root of each number and simplify the square root.

Perform the indicated operation and express in simplest form.

131. $5i(3 - 4i)$

132. $12(7 + 2i)$

CONCEPTS ANALYSIS

Write the following rules in words. Assume that all radicands represent positive values.

1. $a\sqrt{b} \cdot c\sqrt{d} = ac\sqrt{bd}$

2. $\dfrac{a\sqrt{b}}{c\sqrt{d}} = \dfrac{a}{c}\sqrt{\dfrac{b}{d}}; \quad c, d \neq 0$

3. $a\sqrt{b} + c\sqrt{b} = (a + c)\sqrt{b}$

4. $(\sqrt{x})^2 = x$ or $\sqrt{x^2} = x$, for positive values of x.

5. List the conditions for a radical expression to be in simplest form.

6. What does it mean to *rationalize* a denominator? What calculations or manipulations with fractions are easier if the denominator is a rational number?

7. What property of equality can be used to solve an equation that contains a squared variable term?

8. When should the procedure of taking the square root of both sides of an equation be used?

9. What property of equality can be used to solve an equation that contains a square root radical?

10. When should the procedure of squaring both sides of an equation be used?

11. Write the following property in words.

$$x^{\frac{1}{n}} = \sqrt[n]{x}$$

12. Write the following property in words.

$$x^{\frac{m}{n}} = \sqrt[n]{x^m} \text{ or } (\sqrt[n]{x})^m$$

Illustrate with numerical examples the following properties of radicals by using the laws of exponents and fractional exponents. Assume that all radicands are positive values.

13. $\sqrt[n]{xy} = \sqrt[n]{x} \cdot \sqrt[n]{y}$

14. $\sqrt[m]{\sqrt[n]{x}} = \sqrt[n]{\sqrt[m]{x}} = \sqrt[mn]{x}$

15. $\sqrt[n]{\dfrac{x}{y}} = \dfrac{\sqrt[n]{x}}{\sqrt[n]{y}}$

16. $x^{-\frac{m}{n}} = \dfrac{1}{x^{\frac{m}{n}}}$

Objectives	What to Remember with Examples

Section 12-1

1 Write powers and roots using rational exponent and radical notations.

To write a radical expression as an expression with fractional exponents, the radicand is the base of the expression. The exponent of the radicand is the numerator, and the index of the root is the denominator of the fractional exponent.

> Write in fractional exponent form.
> $$\sqrt[5]{3^2} = 3^{\frac{2}{5}}$$
>
> Write in radical form:
> $$5^{\frac{1}{2}} = \sqrt{5}; \quad 7^{\frac{3}{5}} = \sqrt[5]{7^3} \text{ or } \left(\sqrt[5]{7}\right)^3$$

2 Use laws of exponents to simplify and evaluate expressions with rational exponents.

All the rules governing exponents that were used for integral exponents apply to expressions with rational exponents.

> Simplify.
> $$x^{\frac{1}{3}} \cdot x^{\frac{2}{3}} = x^{\frac{1}{3}+\frac{2}{3}} = x^{\frac{3}{3}} = x; \quad \frac{x^{\frac{7}{8}}}{x^{\frac{3}{4}}} = x^{\frac{7}{8}-\frac{3}{4}} = x^{\frac{7}{8}-\frac{6}{8}} = x^{\frac{1}{8}}$$

Section 12-2

1 Find the square root of variables.

Variable factors are perfect squares if the exponent is divisible by 2. To find the square root of a perfect square variable with an exponent, take half of the exponent and keep the same base.

> Give the square root of the following. x^6, x^{10}, x^{24}
> $$\sqrt{x^6} = x^3, \quad \sqrt{x^{10}} = x^5, \quad \sqrt{x^{24}} = x^{12}$$

2 Simplify square-root radical expressions containing perfect-square factors.

To simplify square-root radical expressions containing perfect-square factors, factor constants and variables using the largest possible perfect-square factor of the constant and of each variable. Note that the largest perfect square of a variable will be written with the largest possible even-numbered exponent.

> Simplify the following radical expressions. $\sqrt{98}, \sqrt{x^{13}}, \sqrt{72y^9}, 5\sqrt{12}$
> $$\sqrt{98} = \sqrt{49(2)} = 7\sqrt{2}; \quad \sqrt{x^{13}} = \sqrt{x^{12}(x)} = x^6\sqrt{x}$$
> $$\sqrt{72y^9} = \sqrt{36 \cdot 2 \cdot y^8 \cdot y} = 6y^4\sqrt{2y}$$
> $$5\sqrt{12} = 5\sqrt{4 \cdot 3} = 5 \cdot 2\sqrt{3} = 10\sqrt{3}$$

Section 12-3

1 Add or subtract square-root radicals.

To add or subtract square-root radicals, first simplify all radical expressions; then add or subtract the coefficients of like radical terms. Like radical terms are terms that have exactly the same factors under the radical and have the same index.

> Add or subtract. $3\sqrt{5x} + 7\sqrt{5x}; 2\sqrt{18} - 5\sqrt{8}$
> $$3\sqrt{5x} + 7\sqrt{5x} = 10\sqrt{5x}$$
> $$2\sqrt{18} - 5\sqrt{8} = 2\sqrt{9 \cdot 2} - 5\sqrt{4 \cdot 2} = 2 \cdot 3\sqrt{2} - 5 \cdot 2\sqrt{2} =$$
> $$6\sqrt{2} - 10\sqrt{2} = -4\sqrt{2}$$

| **2** Multiply square-root radicals. | To multiply radicals that have the same index, multiply the coefficients and write as the coefficient of the product, and multiply the radicands and write as the radicand of the product; then simplify the radical. |

$$\text{Multiply.} \quad 5\sqrt{7} \cdot 8\sqrt{14} = 40\sqrt{98} = 40\sqrt{49 \cdot 2} = 40 \cdot 7\sqrt{2} = 280\sqrt{2}$$

| **3** Divide square-root radicals. | To divide radicals that have the same index, divide the coefficients for the coefficient of the quotient; then divide the radicands for the radicand of the quotient. Simplify any remaining radical expressions that can be simplified; then simplify any resulting coefficients that can be simplified. |

$$\text{Divide.} \quad \frac{12\sqrt{75}}{8\sqrt{6}} = \frac{3\sqrt{3 \cdot 25}}{2\sqrt{3 \cdot 2}} = \frac{3\sqrt{25}}{2\sqrt{2}} = \frac{3 \cdot 5}{2\sqrt{2}} = \frac{15}{2\sqrt{2}}$$

| **4** Rationalize a denominator. | To rationalize a denominator of a radical expression, remove perfect-square factors from all radicands. Multiply the denominator by another radical so that the resulting radicand is a perfect square. Then multiply the numerator by the same radical the denominator was multiplied by; simplify the resulting radicals, and reduce the resulting fraction, if possible. |

Rationalize the denominator.

$$\sqrt{\frac{3}{5}} = \frac{\sqrt{3}}{\sqrt{5}} \cdot \frac{\sqrt{5}}{\sqrt{5}} = \frac{\sqrt{15}}{5}$$

Section 12–4

| **1** Write imaginary numbers using the letter i. | The letter i is used to represent $\sqrt{-1}$. Thus, the square root of negative numbers can be expressed as imaginary numbers and simplified using the letter i. Be careful to distinguish when the negative is *outside* the radical and when it is *under* the radical. |

$$\text{Simplify.} \quad \sqrt{-48} = \sqrt{-1 \cdot 16 \cdot 3} = 4i\sqrt{3}$$

| **2** Raise imaginary numbers to powers. | Imaginary numbers can be raised to powers by examining the remainder when the exponent is divided by 4. Review Rule 12–9. |

$$\text{Simplify:} \quad i^{17} = i^{16} \cdot i = i; \quad i^{42} = i^{40} \cdot i^2 = i^2 = -1$$

| **3** Write real and imaginary numbers in complex form, $a + bi$. | A complex number is a number that can be written in the form $a + bi$, where a and b are real numbers and i is $\sqrt{-1}$. Either a or b can be zero. If a is zero, the number is an imaginary number; if b is zero, the number is a real number. |

Rewrite the following as complex numbers. $\sqrt{-81}$, 38, $\sqrt{9}$, $\sqrt{-19}$

$$\sqrt{-81} = 9i = 0 + 9i; \quad 38 = 38 + 0i; \quad \sqrt{9} = 3 = 3 + 0i$$
$$\sqrt{-19} = i\sqrt{19} = 0 + i\sqrt{19}$$

| **4** Combine complex numbers. | To combine complex numbers, add the real parts for the real part of the result; then add the imaginary parts for the imaginary part of the result. Be careful with signs when subtracting complex numbers. |

$$\text{Simplify:} \quad (5 + 3i) + (7 - 8i) = (5 + 7) + (3 - 8)i = 12 - 5i;$$
$$(3 - 8i) - (4 + 6i) = (3 - 4) + (-8 - 6)i = -1 - 14i$$

Section 12–5

1 Solve equations with squared letter terms.

To solve an equation containing only squared variable terms and number terms, perform all normal steps to isolate the squared variable and to obtain a numerical coefficient of 1 for the squared variable. Then take the square root of both sides. *Note:* when taking the square root of both sides, be sure to use both the positive and the negative square roots of the constant. Thus, two solutions are possible. Simplify radicals for exact solutions; use a calculator for approximate solutions.

$$\text{Solve:} \quad 3x^2 + 5 = 29$$
$$3x^2 = 24$$
$$x^2 = 8$$
$$x = \pm\sqrt{8}$$
$$\text{Exact:} \quad x = \pm 2\sqrt{2}$$
$$\text{Approximate:} \quad x = \pm 2.828$$

2 Solve equations with square-root radical terms.

To solve a basic equation containing square-root radicals, square both sides to eliminate any radical; then perform all normal steps to isolate the variable and solve the equation.

$$\text{Solve:} \quad \sqrt{2x + 1} = 5$$
$$(\sqrt{2x + 1})^2 = 5^2$$
$$2x + 1 = 25$$
$$2x = 24$$
$$x = 12$$

WORDS TO KNOW

rational exponents, (p. 460)
decimal exponents, (p. 463)
factor an algebraic expression, (p. 464)
like radicals, (p. 467)
rationalize a denominator, (p. 469)

imaginary numbers, (p. 472)
complex numbers, (p. 472)
the letter i, (p. 472)
remainder, (p. 473)
$a + bi$, (p. 474)
real part of a complex number, (p. 474)

imaginary part of a complex number, (p. 474)
quadratic equation, (p. 475)

CHAPTER TRIAL TEST

Perform the indicated operations. Simplify if possible. Rationalize the denominators if needed.

1. $2\sqrt{7} \cdot 3\sqrt{2}$

2. $\dfrac{\sqrt{8}}{\sqrt{2}}$

3. $4\sqrt{3} + 2\sqrt{3}$

4. $3\sqrt{8} - 4\sqrt{8}$

5. $\dfrac{4\sqrt{2}}{\sqrt{3}}$

6. $\sqrt{3y^3} \cdot \sqrt{15y^2}$

7. $\dfrac{6\sqrt{8}}{2\sqrt{3}}$

8. $\dfrac{3\sqrt{a}}{\sqrt{b}}$

9. $\dfrac{3\sqrt{5}}{2} \cdot \dfrac{7}{\sqrt{3x}}$

Solve the following equations containing square-root radicals or squared terms. Evaluate to the nearest thousandth.

10. $x^2 = 144$

11. $\sqrt{5 - x^2} = 2$

12. $8 = y^2 - 1$

13. $7 = \sqrt{Q}$ **14.** $\sqrt{\dfrac{x^2}{2}} = 6$ **15.** $\sqrt{\dfrac{6}{y}} = \sqrt{\dfrac{2}{3}}$

16. $\sqrt{4x} = 20$ **17.** $x^2 + 49 = 0$

Convert the following radical expressions to equivalent expressions using fractional exponents and simplify.

18. $\sqrt[6]{x}$ **19.** $\sqrt[3]{27x^{15}}$ **20.** $\left(\sqrt{5x^4}\right)^6$ **21.** $\sqrt[5]{x^{10}y^{15}z^{30}}$

Perform the following operations. Express answers with positive exponents in lowest terms.

22. $a^{\frac{4}{5}} \cdot a^{\frac{1}{5}}$ **23.** $\left(125x^{\frac{1}{2}}y^6\right)^{\frac{1}{3}}$ **24.** $\dfrac{b^{\frac{3}{4}}}{b^{\frac{1}{4}}}$

25. $\dfrac{12x^{\frac{3}{5}}}{6x^{-\frac{2}{5}}}$ **26.** $\dfrac{r^{-\frac{1}{5}}s^{\frac{1}{3}}}{r^{\frac{3}{5}}s^{-\frac{5}{3}}}$

Simplify.

27. i^{23} **28.** i^{88}

Add or subtract.

29. $(5 + 3i) - (8 - 2i)$ **30.** $(7 - i) + (4 - 3i)$

13

Formulas and Applications

GOOD DECISIONS THROUGH TEAMWORK

Formulas are equations that have been used so frequently in certain applications that they have become standard in those applications. Examples are the simple interest formula, $I = PRT$, and Ohm's law in electronics, $E = IR$. Many fields, like ballistics, accounting, manufacturing, photography, real estate, physics, mechanics, construction estimation, statistics, medicine, engineering, etc., use formulas to make calculations quickly and conveniently.

Your team project is to have team members investigate their major field or some other field of interest to determine what formulas are used, if any. Collect three to five formulas for each field, including the meaning of the symbols, what the formulas are used for, variations of the formulas, and how to solve the formulas or variations. You might interview instructors, recognized experts, and other professionals for this purpose.

Discuss with your team members the advantages of using the formulas, any difficulties in using them, and whether computers are used with them. Discuss also why some fields may not use formulas. Then present your team's findings to your class.

In Chapters 4, 6, and 10, we solved a number of applied problems with percent, area, perimeter, and interest *formulas*, although we did not always call them formulas. The procedure for work problems, rate × time = amount of work, was a formula. Another formula was the one used to find force: force = $\frac{\text{pressure}}{\text{area}}$.

Formulas are really procedures that have been used so frequently to solve certain types of problems that they have become the accepted means of solving these problems. Most formulas are expressed with one or more letter terms rather than words, and these procedures are written in an abbreviated, symbolic equation, such as $A = \pi r^2$ and $P = 2(l + w)$, the formulas for the area of a circle and the perimeter of a rectangle, respectively. Electronic spreadsheets and calculator and computer programs are developed through formulas.

13-1 FORMULA EVALUATION

Learning Objective

1 Evaluate formulas for a given variable.

The most common use of formulas is for finding missing values. If we know values for all but one variable of a formula, we can find the unknown value.

1 Evaluate Formulas for a Given Variable.

To *evaluate* a formula is to substitute known numerical values for some variables and perform the indicated operations to find the value of the remaining variable in question. In so doing, we may need to use any and all of the steps and procedures we learned previously for solving equations.

Many formulas involve more than one operation. When evaluating such formulas, we follow the rules for the order of operations.

> **Rule 13-1** *Order of operations:*
>
> Perform operations in the following order as they appear from left to right.
>
> 1. Parentheses used as groupings and other grouping symbols
> 2. Exponents (powers and roots)
> 3. Multiplications and divisions
> 4. Additions and subtractions

This section will illustrate several formula evaluations to help develop a concept of formula evaluation. Subsequent sections will focus on more specific formulas and their evaluations.

The percentage proportion studied previously (Chapter 4) is actually a formula and shows how any formula written as a proportion may be evaluated. This may be a good place to begin formula evaluation.

EXAMPLE 1 Evaluate the formula $\frac{R}{100} = \frac{P}{B}$ if $R = 20$ and $P = 10$.

$$\frac{R}{100} = \frac{P}{B}$$

$$\frac{20}{100} = \frac{10}{B} \qquad \text{Substitute values.}$$

$$20(B) = 10(100) \qquad \text{Cross multiply.}$$

$$20B = 1000 \qquad \text{Divide.}$$

$$\frac{20B}{20} = \frac{1000}{20}$$

$$B = 50 \qquad \text{Interpret solution.}$$

Thus, the base is 50.

The formulas in Examples 2 to 4 are variations of the percentage proportion. In the first formula, the rate is isolated.

EXAMPLE 2 Evaluate the formula $R = \frac{100P}{B}$ if $P = \$75$ and $B = \$125$.

$$R = \frac{100P}{B} \qquad \text{Rate} = 100 \text{ times the percentage divided by the base.}$$

$$R = \frac{100(75)}{125} \qquad \text{Substitute values.}$$

$$R = \frac{7500}{125} \qquad \text{Perform calculations.}$$

$$R = 60 \qquad \text{Interpret solution.}$$

Thus, the rate is 60%.

Sometimes the letter we are asked to solve for is not isolated. In this case, we substitute the given values and isolate the variable as in any equation. The next example shows the variable in the numerator of a fraction.

EXAMPLE 3 Solve the formula $P = \frac{RB}{100}$ for B if $P = 45$ and $R = 30$.

$$P = \frac{RB}{100} \qquad \text{Percentage} = \text{rate times base divided by 100.}$$

$$45 = \frac{30B}{100} \qquad \text{Substitute values.}$$

$$45(100) = \frac{30B}{100}(100) \qquad \text{Multiply to eliminate denominator.}$$

$$4500 = \frac{30B}{\cancel{100}}(\cancel{100}) \qquad \text{Reduce whenever possible.}$$

$$\frac{4500}{30} = \frac{30B}{30} \qquad \text{Divide.}$$

$$150 = B \qquad \text{Interpret solution.}$$

Therefore, the base is 150.

The variable may also be in the denominator of a fraction.

EXAMPLE 4 Evaluate the formula $B = \frac{100P}{R}$ if $P = 60$ m and $B = 200$ m.

$$B = \frac{100P}{R} \qquad \text{Base} = 100 \text{ times percentage divided by the rate.}$$

$$200 = \frac{100(60)}{R} \qquad \text{Substitute values.}$$

$$200 = \frac{6000}{R}$$ Multiply in numerator.

$$200(R) = \frac{6000}{R}(R)$$ Multiply to eliminate denominator.

$$200R = \frac{6000}{\cancel{R}}(\cancel{R})$$ Reduce.

$$200R = 6000$$

$$\frac{200R}{200} = \frac{6000}{200}$$ Divide.

$$R = 30$$ Interpret solution.

Therefore, the rate is 30%.

Rule 13–2   *To evaluate a formula:*

1. Write the formula.
2. Rewrite the formula substituting known values for variables of the formula.
3. Solve the equation from Step 2 for the missing variable.
4. Interpret the solution within the context of the formula.

Let's look at the formula for the perimeter of a rectangle (Chapter 6). Again, we will solve for a letter that is not isolated. In this example, the variable is within parentheses and must be removed from the parentheses.

EXAMPLE 5   Solve the formula $P = 2(l + w)$ for w if $P = 12$ ft and $l = 4$ ft.

$$P = 2(l + \boxed{w})$$ Perimeter of a rectangle = two times the sum of the length and the width.

$$12 = 2(4 + \boxed{w})$$ Substitute values.

$$12 = 8 + 2\boxed{w}$$ Apply the distributive property.

$$12 - 8 = 2\boxed{w}$$ Isolate the term with the variable.

$$4 = 2\boxed{w}$$ Combine like terms.

$$\frac{4}{2} = \frac{2w}{2}$$ Divide.

$$2 = w$$ Interpret solution.

Therefore, the width is 2 ft.

The formula for the area of a circle (Chapter 6) contains a squared variable. In the next example, we will solve for the squared variable.

EXAMPLE 6   Evaluate the formula $A = \pi r^2$ for r if $A = 144$ in.^{2}.

$$A = \pi r^2$$ Area of a circle = π times the radius squared.

$$144 = \pi r^2$$ Substitute values.

$$\frac{144}{\pi} = \frac{\pi r^2}{\pi}$$ Isolate the variable by dividing both sides by π.

$$\frac{144}{\pi} = r^2$$

$$r = \sqrt{\frac{144}{\pi}}$$ Take the square root of both sides. For convenience, rewrite with r on the left side.

$$r = \sqrt{45.83662361}$$ Only the principal square root is appropriate.

$$r = 6.770275003$$ Interpret solution.

Thus, the radius is 6.77 in. (rounded).

A variation of the same formula for the area of a circle allows us to solve for the variable under a square root radical.

EXAMPLE 7   Evaluate the formula $r = \sqrt{\frac{A}{\pi}}$ for A if $r = 6.25$ cm.

$$r = \sqrt{\frac{A}{\pi}}$$ Radius = square root of the area of the circle divided by π.

$$6.25 = \sqrt{\frac{A}{\pi}}$$ Substitute values.

$$(6.25)^2 = \left(\sqrt{\frac{A}{\pi}}\right)^2$$ Square both sides.

$$39.0625 = \frac{A}{\pi}$$ Perform calculations.

$$39.0625(\pi) = \frac{A}{\overset{1}{\underset{1}{\pi}}}(\overset{1}{\pi})$$ Eliminate denominator.

$$39.0625\pi = A$$ Evaluate using the calculator value for π.

$$122.718463 = A$$ Interpret solution.

Thus, the area is 122.72 cm^2 (rounded).

The interest formula (Chapter 4) may be used to illustrate a variable that is one of several factors.

EXAMPLE 8   Evaluate the formula $I = PRT$ for P if $I = \$94.50$, $R = 21\%$, and $T = \frac{1}{2}$ year.

For convenience in using a calculator, convert $\frac{1}{2}$ year to 0.5 year. In this formula, the rate is expressed as a decimal equivalent, $21\% = 0.21$.

$$I = PRT$$

$$94.50 = P(0.21)(0.5)$$ Substitute values.

$$94.50 = 0.105P$$ Multiply 0.21 and 0.5.

$$\frac{94.50}{0.105} = \frac{0.105P}{0.105}$$ Divide.

$$900 = P$$ Interpret solution.

Therefore, the principal is $900.

1 Evaluate the following formulas. Use $A = 5$, $B = 6$, and $C = 8$.

1. $X = B + (3C - A)$

2. $X = A^2 - (2C - B)$

3. $X = C(B^2 + 4AC) - 3$

4. $X = \dfrac{3(C - A)}{3} + B$

5. $X = \dfrac{2(C - A)^2}{A - 1}$

Evaluate the interest formula $I = PRT$ using the following values.

6. Find the interest if $P = \$800$, $R = 15.5\%$ and $T = 2\frac{1}{2}$ years.

7. Find the rate if $I = \$427.50$, $P = \$1500$, and $T = 2$ years.

8. Find the time if $I = \$236.25$, $P = \$750$, and $R = 10.5\%$.

9. Find the principal if $I = \$838.50$, $R = 21\frac{1}{2}\%$, and $T = 1\frac{1}{2}$ years.

Evaluate the percentage formula $P = \frac{RB}{100}$ using the following values.

10. Find the percentage if $R = 15\%$ and $B = 600$ lb.

11. Find the rate if $P = 24$ kg and $B = 300$ kg.

12. Find the base if $P = \$250$ and $R = 7.4\%$. Round to hundredths.

13. Evaluate the rate formula $R = \frac{100P}{B}$ for P if $R = 16\%$ and $B = 85$.

14. Evaluate the base formula $B = \frac{100P}{R}$ for R if $B = \$2200$ and $P = \$374$.

15. Find the length of a rectangular work area if the perimeter is 180 in. and the width is 24 in. Use the formula $P = 2(l + w)$.

16. Find the radius of a circle whose area is 254.5 cm^2 using the formula $r = \sqrt{\frac{A}{\pi}}$. Round to the nearest tenth.

17. Find the cost (C) if the markup (M) on an item is \$5.25 and the selling price (S) is \$15.75. Use the formula $M = S - C$.

18. Using the formula for the side of a square, $s = \sqrt{A}$, find the length of a side of a square field whose area is $\frac{1}{16}$ mi^2.

19. Evaluate the formula for the area of a circle, $A = \pi r^2$, if $r = 7$ in. Round to the nearest tenth.

20. Evaluate the formula for the area of a square, $A = s^2$, if $s = 2.5$ km.

21. Using the markup formula $M = S - C$, find the selling price (S) if the markup (M) is \$12.75 and the cost (C) is \$36.

13–2 FORMULA REARRANGEMENT

Learning Objective

1 Rearrange formulas to solve for a given variable.

We saw in Section 13–1 that we could solve a formula for any missing variable, no matter where the missing information was located in the formula. In developing formulas for spreadsheets or in programming calculators or computers, the missing values must be indicated by a variable that is isolated on the left side of an equation. This often requires that the formula be rearranged.

1 Rearrange Formulas to Solve for a Given Variable.

Formula rearrangement generally refers to isolating a letter term other than the one already isolated (if any) in the formula. Solving formulas in this manner shortens our work when doing repeated formula evaluations. Since we will not evaluate the formulas after we rearrange them, we will not go into what each formula represents. Here we are interested only in the techniques used in the rearrangement of the formulas. After the formulas are solved for the desired variable, the formula is rewritten with the variable on the left side for convenience.

The following formulas require transposition and careful handling of signs.

EXAMPLE 1 Solve the formula $M = S - C$ for S.

$$M = \boxed{S} - C$$ Transpose to isolate S.

$$M + C = \boxed{S}$$ The coefficient of S is positive, so the formula is solved.

$$\boxed{S} = M + C$$ S is rewritten on the left for convenience.

Tips and Traps | ***Where is the Variable or Unknown in a Formula?***
It may help to think of the letter term we are solving for as the unknown or variable, and to think of the other letter terms *as if* they were the number terms in an ordinary equation.

EXAMPLE 2 Solve the formula $M = S - C$ for C.

$$M = S \boxed{- C}$$ Transpose to isolate C.

$$M - S = \boxed{-C}$$

$$\frac{M - S}{-1} = \frac{\boxed{-C}}{-1}$$ The coefficient of C is negative, so we need to divide both sides by -1.

$$-M + S = \boxed{C}$$ Note effect on signs after division by -1.

$$S - M = \boxed{C}$$ Write the positive term first.

$$\boxed{C} = S - M$$ C is rewritten on the left for convenience.

Sometimes a formula contains a term of several factors and we need to solve for one of those factors. In this case, treat the factor being solved for as the variable and the other factors as its coefficient.

Rule 13–3 *To rearrange a formula:*

1. Determine which variable of the formula will be isolated (solved for).
2. Highlight or mentally locate all instances of the variable to isolate.
3. Treat all other variables of the formula as you would numbers in an equation and perform normal techniques for solving an equation.
4. If the isolated variable is on the right side of the equation, interchange the sides so that it appears on the left side.

EXAMPLE 3 Solve for R in the formula $I = PRT$.

$$I = P\boxed{R}T$$ Since we are solving for R, PT is its coefficient.

$$\frac{I}{PT} = \frac{P\cancel{R}T}{\cancel{PT}}$$ Divide both sides by the coefficient of the variable.

$$\frac{I}{PT} = \boxed{R}$$ Rewrite with R on the left.

$$\boxed{R} = \frac{I}{PT}$$

Sometimes formulas contain addition (or subtraction) and multiplication in which we must use the distributive property to solve for a particular letter. In the formula $A = P(1 + ni)$, we must use the distributive property to solve for either n or i because each appears inside the grouping.

EXAMPLE 4 Solve for n in $A = P(1 + ni)$.

$$A = P(1 + ni)$$ Identify the variable to be isolated.

$$A = P + Pni$$ Use the distributive principle to remove n from parentheses.

$$A - P = Pni$$ Transpose P to isolate the term with n.

$$\frac{A - P}{Pi} = \frac{Pni}{Pi}$$ Divide both sides by the coefficient of n.

$$\frac{A - P}{Pi} = n$$ Rewrite.

$$n = \frac{A - P}{Pi}$$

Sometimes we will use the distributive principle *in reverse* to rearrange formulas. If we wish to solve the formula $S = FC + VC$ for C, we must first rewrite the formula by using the distributive principle because the C appears in two terms. This is the same as expressing $FC + VC$ as the product of C and $(F + V)$.

EXAMPLE 5 Solve for C in $S = FC + VC$.

$$S = FC + VC$$ Since C appears in two terms, use the distributive principle to express the terms as the product of C and $(F + V)$.

$$S = C(F + V)$$

$$\frac{S}{F + V} = \frac{C(F + V)}{F + V}$$ Since we are solving for C, divide both sides by the coefficient of C, which is $(F + V)$.

$$\frac{S}{F + V} = C$$

$$C = \frac{S}{F + V}$$

When the formula contains division, the denominator is eliminated before further steps are taken.

EXAMPLE 6 Solve the formula $R = \frac{V}{I}$ for V.

$$R = \frac{V}{I}$$ Identify the variable to be isolated.

$$(I)R = \frac{V}{I}(I)$$ Multiply both sides by the denominator I to eliminate it.

$$IR = V$$

$$V = IR$$

Tips and Traps   *Subscripted Variables:*
Many formulas express relationships between two similar measurements by noting the different measurements with *subscripts*.

The average temperature for a 3-hour period is found by adding the temperatures for each of the periods and dividing by 3.

$$\text{Average temp.} = \frac{\text{temp. 1st hour} + \text{temp. 2nd hour} + \text{temp. 3rd hour}}{3}$$

Using an abbreviated form of the formula, we have

$$t_{av} = \frac{t_1 + t_2 + t_3}{3}$$

Since we are referring to temperatures throughout the formula, we use subscripts to distinguish the various temperatures. To read this formula with subscripts we say "t sub av equals the sum of t sub one plus t sub two plus t sub three, all divided by three."

When we wish to rearrange formulas written in the form of a proportion, we can use the property of proportions that allows cross multiplication. Once we have cross multiplied, we use methods previously discussed to solve for the appropriate letter.

EXAMPLE 7   Solve for T_1 in the formula $\dfrac{T_1}{T_2} = \dfrac{V_1}{V_2}$.

$\dfrac{T_1}{T_2} = \dfrac{V_1}{V_2}$ Identify the variable to be isolated. Watch the subscripts.

$T_1 V_2 = T_2 V_1$ Cross multiply.

$\dfrac{T_1 V_2}{V_2} = \dfrac{T_2 V_1}{V_2}$ Divide both sides by the coefficient of T_1 to solve for T_1.

$T_1 = \dfrac{T_2 V_1}{V_2}$

To solve a formula for a letter that is squared, we first solve for the squared letter and then take the square root of both sides of the formula.

EXAMPLE 8   Solve for a in the formula $c^2 = a^2 + b^2$.

$c^2 = a^2 + b^2$ Identify the variable to be isolated.

$c^2 - b^2 = a^2$ Transpose b^2 to isolate the letter term being solved for.

$\sqrt{c^2 - b^2} = \sqrt{a^2}$ Take the square root of both sides.

$\sqrt{c^2 - b^2} = a$

$a = \sqrt{c^2 - b^2}$

To rearrange a formula with a single term on each side, when one side is a square root, we *begin* by squaring both sides of the formula. Then we solve for the desired letter by using the same techniques used earlier to rearrange formulas.

EXAMPLE 9   Solve for A in the formula $s = \sqrt{A}$.

$s = \sqrt{A}$ Identify the variable to be isolated.

$s^2 = (\sqrt{A})^2$ Square both sides to eliminate the $\sqrt{}$.
Notice that $(\sqrt{A})^2 = \sqrt{A} \cdot \sqrt{A} = A$.

$s^2 = A$

$A = s^2$

1 Rearrange the following formulas.

1. Solve $E = X + I$ for X.

2. Solve $A = M - K$ for K.

3. Solve $S = 2\pi rh$ for r.

4. Solve $m = bx + by$ for b.

5. Solve $A = m(x + 2y)$ for y.

6. Solve $\dfrac{T_1}{T_2} = \dfrac{V_1}{V_2}$ for T_2.

7. Solve $V = \pi r^2 h$ for r.

8. Solve $m = \sqrt{XY}$ for X.

9. Solve $\dfrac{R}{100} = \dfrac{P}{B}$ for R.

10. Solve $P = 2(b + s)$ for b.
11. Solve $C = 2\pi r$ for r.
12. Solve $A = lw$ for l.
13. Solve $R = AC - BC$ for C.
14. Solve $A = s^2$ for s.
15. Solve $D = RT$ for R.
16. Solve $S = P - D$ for D.
17. Solve $A = \pi r^2$ for r.
18. Solve $a^2 + b^2 = c^2$ for c.

19. The formula for finding the amount of a re-payment on a loan is $A = I + P$, where A is the amount of the repayment, I is the interest, and P is the principal. Solve the formula for interest.

20. The formula for finding interest is $I = PRT$, where I represents interest, P represents principal, R represents rate, and T represents time. Rearrange the formula to find the time.

13-3 TEMPERATURE FORMULAS

Learning Objectives

1 Make Celsius/Kelvin temperature conversions.
2 Make Rankine/Fahrenheit temperature conversions.
3 Convert Fahrenheit to Celsius temperatures.
4 Convert Celsius to Fahrenheit temperatures.

Many technical applications require conversions between different temperature scales. We will be concerned with temperature conversions involving the Celsius, Kelvin, Rankine, and Fahrenheit scales.

1 **Make Celsius/Kelvin Temperature Conversions.**

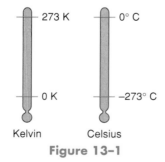

Kelvin   Celsius

Figure 13-1

One scale used to measure temperature in the metric system of measurement is called the *Kelvin* scale. Units on this scale are abbreviated with a capital K (without the symbol ° because these units are called *Kelvins*) and are measured from absolute zero, the temperature at which *all* heat is said to be removed from matter. Another metric temperature scale is the *Celsius* scale (abbreviated °C), which has as its zero the freezing point of water. The Kelvin and Celsius scales are related such that absolute zero on the Kelvin scale is the same as $-273°C$ on the Celsius scale. Each unit of change on the Kelvin scale is equal to 1 degree of change on the Celsius scale; that is, the size of a Kelvin and a Celsius degree is the same on both scales (see Fig. 13-1).

> **Formula 13-1** *To convert Celsius to Kelvin:*
> $$K = °C + 273$$

Although temperature conversions can be made using unity ratios, we generally use formulas rather than unity ratios to make temperature conversions.

EXAMPLE 1 A temperature of 21°C on the Celsius scale is the same as what temperature on the Kelvin scale?

According to the formula $K = °C + 273$, we can find the Kelvin temperature by *adding* 273 to the Celsius temperature.

$$K = 21 + 273$$
$$K = 294$$

Thus, 21°C = 294 K.

Formula 13–2 *To convert Kelvin to Celsius:*

$$°C = K - 273$$

EXAMPLE 2 A temperature of 300 K on the Kelvin scale corresponds to what temperature on the Celsius scale?

To find the Celsius temperature, we *subtract* 273 from the Kelvin reading.

$$°C = \boxed{K} - 273$$
$$°C = \boxed{300} - 273$$
$$°C = 27$$

Thus, 300 K = 27°C.

② Make Rankine/Fahrenheit Temperature Conversions.

The U.S. customary system temperature scale that starts at absolute zero is called the *Rankine* scale. It is related to the more familiar *Fahrenheit* scale, which places the freezing point of water at 32°. The Rankine (°R) and Fahrenheit (°F) scales have the same relationship as the Kelvin and Celsius scales in the metric system; that is, 1 degree of change on the Rankine scale is equal to 1 degree of change on the Fahrenheit scale. Absolute zero (the zero for the Rankine scale) corresponds to 460 degrees *below* zero on the Fahrenheit scale.

Formula 13–3 *To convert Fahrenheit to Rankine:*

$$°R = °F + 460$$

EXAMPLE 3 What temperature on the Rankine scale corresponds to 40°F on the Fahrenheit scale?

To find the Rankine temperature, we *add* 460 to the Fahrenheit temperature reading.

$$°R = \boxed{°F} + 460$$
$$°R = \boxed{40} + 460$$
$$°R = 500$$

Thus, 40°F = 500°R.

Formula 13–4 *To convert Rankine to Fahrenheit:*

$$°F = °R - 460$$

EXAMPLE 4 What temperature on the Fahrenheit scale corresponds to 650°R on the Rankine scale?

To find the Fahrenheit temperature, we *subtract* 460 from the Rankine temperature.

$$°F = \boxed{°R} - 460$$
$$°F = \boxed{650} - 460$$
$$°F = 190$$

Thus, 650°R = 190°F.

3 Convert Fahrenheit to Celsius Temperatures.

The Celsius and Fahrenheit scales are the most common temperature scales used for reporting air and body temperatures. Since we still use both scales, we need to be able to convert Fahrenheit temperatures to Celsius and Celsius temperatures to Fahrenheit. The formulas used for converting temperatures using these two scales are more complicated than the previous ones because 1 degree of change on the Celsius scale does *not* equal 1 degree of change on the Fahrenheit scale.

Formula 13–5   *To convert Fahrenheit to Celsius:*

$$°C = \frac{5}{9}(°F - 32)$$

If we are given a temperature reading in the Fahrenheit scale and need to change it to the Celsius scale, we use the formula $°C = \frac{5}{9}(°F - 32)$.

EXAMPLE 5   Change 212°F (the boiling point of water) to degrees Celsius.

We begin by writing 212 in place of the °F in the formula.

$$°C = \frac{5}{9}(\boxed{°F} - 32)$$

$$°C = \frac{5}{9}(\boxed{212} - 32)$$

The first step in the order of operations is to work groupings, so we need to subtract $212 - 32$, which makes the value of the grouping 180.

Recall that a number, $\frac{5}{9}$, written in front of a grouping (°F $-$ 32) means that we *multiply* the number $\frac{5}{9}$ by the value of the grouping.

$$°C = \frac{5}{9}(180) \qquad \tfrac{5}{9}(180) \text{ means } \tfrac{5}{9} \times 180$$

The next step in the order of operations is to multiply or divide.

$$°C = \frac{5}{9}\left(\frac{180}{1}\right) \qquad \frac{5}{\overset{}{\underset{1}{9}}} \times \frac{\overset{20}{\cancel{180}}}{1} = 100$$

$$°C = 100$$

Thus, 212°F = 100°C.

The order of these operations is reviewed in Chapter 1, Chapter 8, and Section 13–1.

EXAMPLE 6   According to Dr. Shotwell, an antifungal powder containing tolnaftate may be stored at a room temperature of 77°F. Change 77°F to degrees Celsius.

$$°C = \frac{5}{9}(\boxed{°F} - 32)$$

$$°C = \frac{5}{9}(\boxed{77} - 32)$$ Write 77 in place of °F.

$$°C = \frac{5}{9}(45)$$ Work grouping.

$$°C = \frac{5}{9} \times \frac{\overset{5}{\cancel{45}}}{1}$$ Multiply. Reduce if possible.

$$°C = 25$$

Thus, 77°F = 25°C.

4 Convert Celsius to Fahrenheit Temperatures.

When a temperature is expressed in Celsius and needs to be expressed in Fahrenheit, we use the formula $°F = \frac{9}{5}°C + 32$ and follow the rules of the order of operations as indicated in the examples. When no groupings are present, multiplication is done before addition.

> **Formula 13-6** *To convert Celsius to Fahrenheit:*
>
> $$°F = \frac{9}{5}°C + 32$$

EXAMPLE 7   Change 100°C to degrees Fahrenheit.

$$°F = \frac{9}{5}\boxed{°C} + 32$$ A number, $\frac{9}{5}$, written in front of a letter, C, indicates multiplication.

$$°F = \frac{9}{5}(\boxed{100}) + 32$$ Write 100 in place of C.

$$°F = \frac{9}{\cancel{5}_{1}} \times \frac{\overset{20}{\cancel{100}}}{1} + 32$$ Multiply first, reducing if possible.

$$°F = 180 + 32$$ Add.

$$°F = 212$$

Thus, 100°C = 212°F.

EXAMPLE 8   The label on a dropper bottle of ofloxacin opthalmic solution warns that the medicine must not be stored at a temperature above 25°C. Change 25°C to degrees Fahrenheit.

$$°F = \frac{9}{5}\boxed{°C} + 32$$

$$°F = \frac{9}{5}(\boxed{25}) + 32$$ Write 25 in place of C.

$$°F = \frac{9}{\cancel{5}_{1}} \times \frac{\overset{5}{\cancel{25}}}{1} + 32$$ Multiply first, reducing if possible.

$$°F = 45 + 32 \qquad \text{Add.}$$

$$°F = 77$$

Thus, 25°C = 77°F.

Tips and Traps   *Unnecessary Duplication of Formulas?*
In this section, we have given six formulas instead of the necessary three. Each even-numbered formula could have been derived by rearranging the odd-numbered formula that preceded it.

Formula 13–1: $K = °C + 273$ Formula 13–2: $°C = K - 273$

Formula 13–3: $°R = °F + 460$ Formula 13–4: $°F = °R - 460$

Formula 13–5: $°C = \dfrac{5}{9}(°F - 32)$ Formula 13–6: $°F = \dfrac{9}{5}°C + 32$

Argument for Unnecessary Formulas

If formulas are easily accessible and in a format that does not cause you to lose valuable time *searching* for the appropriate formula, then all variations of a formula are handy. You can save calculation time and use calculators and computers most efficiently by selecting the formula with the variable that matches your missing information on the left.

Argument Against Unnecessary Formulas

If formulas are not easily accessible and if you will often need to commit formulas to memory, memorize the fewest number of formulas possible.

SELF-STUDY EXERCISES 13–3

1   Make the following temperature conversions.
1. $82°C = \_\_\_\_ K$ **2.** $438 K = \_\_\_\_ °C$ **3.** $17°C = \_\_\_\_ K$
4. $273 K = \_\_\_\_ °C$ **5.** $98 K = \_\_\_\_ °C$ **6.** $71°C = \_\_\_\_ K$
7. $60°C = \_\_\_\_ K$ **8.** $192 K = \_\_\_\_ °C$

2   Make the following temperature conversions.
9. $460°F = \_\_\_\_ °R$ **10.** $98°F = \_\_\_\_ °R$ **11.** $710°R = \_\_\_\_ °F$
12. $920°R = \_\_\_\_ °F$ **13.** $180°F = \_\_\_\_ °R$ **14.** $212°F = \_\_\_\_ °R$
15. $350°R = \_\_\_\_ °F$ **16.** $600°R = \_\_\_\_ °F$ **17.** $0°R = \_\_\_\_ °F$
18. $32°F = \_\_\_\_ °R$

3   Change the following Fahrenheit temperatures to Celsius.
19. $95°F$ **20.** $32°F$ **21.** $113°F$
22. $41°F$ **23.** $59°F$ **24.** $50°F$
25. $149°F$ **26.** $122°F$ **27.** $176°F$
28. $248°F$

4   Change the following Celsius temperatures to Fahrenheit.
29. $70°C$ **30.** $15°C$ **31.** $45°C$
32. $50°C$ **33.** $20°C$ **34.** $215°C$
35. $310°C$ **36.** $410°C$ **37.** $185°C$
38. $0°C$

Learning Objectives

1. Find the perimeter and area of a trapezoid.
2. Find the perimeter and area of a triangle.
3. Use the Pythagorean theorem to find the missing side of a right triangle.
4. Find the area of prisms and cylinders.
5. Find the volume of prisms and cylinders.
6. Find the area and volume of a sphere.
7. Find the area and volume of a cone.

Geometry involves the study and measurement of shapes. We have already seen in Chapter 6 four geometric figures: the square, the rectangle, the parallelogram, and the circle. We used formulas to find their perimeters and areas. These formulas are listed next.

Formulas from Chapter 6

Square
Perimeter: $P = 4s$ Area: $A = s^2$ s is the measure of one side

Rectangle
Perimeter: $P = 2(l + w)$ Area: $A = lw$ l is length and w is width

Parallelogram
Perimeter: $P = 2(b + s)$ Area: $A = bh$ b is base, h is height, and s is adjacent side

Circle
Circumference: $C = \pi d$ or $2\pi r$ Area: $A = \pi r^2$ π is an irrational constant, r is the radius, and d is the diameter

---

Geometric layouts and figures used in technical occupations are not limited to squares, rectangles, and parallelograms. Other common figures are *trapezoids* and *triangles*.

1. **Find the Perimeter and Area of a Trapezoid.**

Trapezoids are used in the construction of many objects such as picture frames (illustrated in Fig. 13–2), table tops, windows, and roofs.

■ **DEFINITION 13–1: Trapezoid.** A *trapezoid* is a four-sided polygon having only two parallel sides.

Unlike a parallelogram, the trapezoid's four sides may all be of unequal size, as illustrated in Fig. 13–3.

Figure 13–2

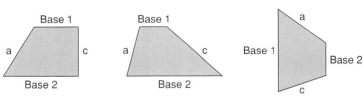

Figure 13–3

Note that the *parallel* sides are called *bases* (b_1 and b_2) with the subscripts 1 and 2 used to distinguish them. The nonparallel sides are designated side a and side c. The perimeter is the sum of the lengths of all four sides, or $b_1 + b_2 + a + c$. Since all four sides may be unequal, neither the bases nor the nonparallel sides may be combined to simplify the formula.

Formula 13-7 *Perimeter of a trapezoid:*

$$P = b_1 + b_2 + a + c$$

where b_1, b_2, a, and c are sides of the trapezoid.

EXAMPLE 1   The rear window of a pickup truck is a trapezoid whose shorter base is 40 in., whose longer base is 48 in., and whose nonparallel sides are each 16 in. How many inches of rubber gasket are needed to surround the window?

$P = b_1 + b_2 + a + c$

$P = 40 + 48 + 16 + 16$ Substitute in formula.

$P = 120$ in.

The rubber gasket to surround the rear window must be 120 in. long.

The area of a trapezoid can be represented by dividing it into squares (Fig. 13–4). However, if we recognize that by placing two congruent trapezoids end to end we form a parallelogram, then the area of one trapezoid is half the area of the larger parallelogram. Since the area of a parallelogram is the height times the base, the area of a trapezoid is *one-half* the height times the "base" of the parallelogram formed by joining two equal trapezoids end to end.

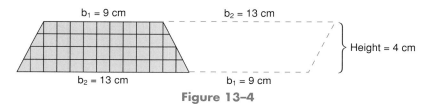

Figure 13–4

Formula 13-8 *Area of a trapezoid:*

$$A = \frac{1}{2}h(b_1 + b_2)$$

b_1 is one of two parallel sides, b_2 is the other parallel side, and h is the perpendicular distance between the bases.

EXAMPLE 2   A trapezoidal table top for special use in a reading classroom measures 60 cm on the shorter base and 90 cm on the longer base. If the height (perpendicular distance between the bases) is 50 cm, how much Formica is needed to re-cover the top?

$A = \frac{1}{2}h(b_1 + b_2)$

$A = \frac{1}{2}(50)(60 + 90)$ Substitute in formula.

$$A = \frac{1}{2}(\overset{25}{\underset{1}{\cancel{50}}})(150)$$ Reduce where possible.

$$A = 3750 \text{ cm}^2$$

Thus, 3750 cm² of Formica are needed.

2   Find the Perimeter and Area of a Triangle.

■ **DEFINITION 13–2: Triangle.**   A *triangle* is a polygon that has three sides.

It also has three vertices. The sides of a triangle form three angles. The *degree* is a unit used to measure angles. A more thorough discussion of the degree used as an angle measure is found in Chapter 21.

Tips and Traps	*Angle Measures of Any Triangle:* In any triangle, no matter what its shape or size, the sum of the measures of its three angles is 180°.

Classifying Triangles by Angles
There are two primary separations of triangles according to angles. These are triangles that have one right (90°) angle and triangles that do not have a right angle. For this discussion, see Fig. 13–5.

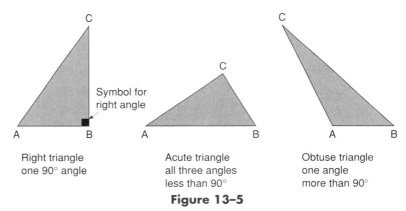

Figure 13–5

■ **DEFINITION 13–3: Right Triangle.**   A *right triangle* is a triangle that has one right (90°) angle. The sum of the other two angles is 90°.

One side of a right triangle is used quite frequently in mathematics and deserves special attention. This side is the *hypotenuse*.

■ **DEFINITION 13–4: Hypotenuse.**   The side of a right triangle that is opposite the right angle is called the *hypotenuse*.

The other sides are called *legs,* or the vertical side may be called the *altitude* and the horizontal side the *base*. The right angle is usually marked ⌐ or ⌐ .

Right triangles are used in many technologies. Surveyors, carpenters, and navigators are but a few who frequently use right triangles to solve problems involving relationships among lines and angles.

■ **DEFINITION 13–5: Oblique Triangle.**   An *oblique triangle* is any triangle that does not contain a right angle.

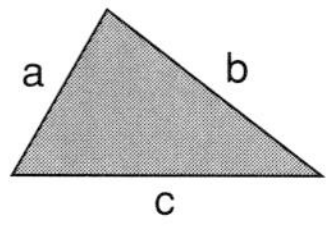

Figure 13-6

There are two kinds of oblique triangles, *acute* and *obtuse*. Acute triangles have each angle less than 90°. Obtuse triangles have one angle more than 90°. For a more thorough discussion of acute and obtuse angles and triangles, see Chapter 21.

The perimeter of a triangle is the sum of the lengths of the three sides (Fig. 13–6).

Formula 13-9 *Perimeter of a triangle:*

$$P = a + b + c$$

a, *b*, and *c* are sides of a triangle.

EXAMPLE 3

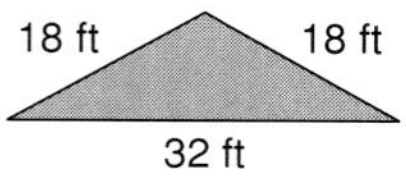

18 ft   18 ft

32 ft
Figure 13-7

The triangular gable ends of an apartment unit in a complex under construction will be outlined in contrasting trim. If each gable is 32 ft wide and 18 ft on each side, how many feet of contrasting wood trim are needed for each gable? Disregard overlap at the corners (Fig. 13–7).

$P = a + b + c$

$P = 18 + 32 + 18$ Substitute in formula.

$P = 68$ ft

Each gable will require 68 linear feet of trim.

The area of a triangle, like the area of a trapezoid, can be figured by dividing it into square units (Fig. 13–8). Notice that one side is designated as the base and that the height is the perpendicular distance from the base to the top. Like the trapezoid, the triangle forms a parallelogram if an identical triangle is placed beside the original triangle. A triangle, then, is half a parallelogram, and its area can be expressed as half the area of a parallelogram, that is, one-half the base times the height. Any side can serve as the base of a triangle. However, the height of a triangle is the perpendicular distance from the selected base to the vertex of the angle opposite the base. This vertex opposite the base is often referred to as the *apex* of the triangle.

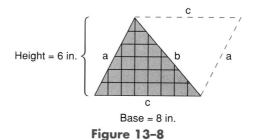

Figure 13-8

Formula 13-10 *Area of a triangle:*

$$A = \frac{1}{2}bh$$

b is one side of the triangle, *h* is the height.

EXAMPLE 4

A gable end of a house has a rise of 7 ft 6 in. and a span of 30 ft 6 in. What is its area? The rise is the height of the gable and the span is the base of the gable (Fig. 13–9).

Figure 13-9

$$A = \frac{1}{2}bh$$

$$A = \frac{1}{2}(30.5)(7.5)$$

$$A = 0.5(30.5)(7.5)$$

$$A = 114.375 \text{ ft}^2$$

Substitute in formula after converting to one common unit of measure. 7 ft 6 in. = 7.5 ft, 30 ft 6 in. = 30.5 ft because 1 ft = 12 in.

The area of a gable is 114.375 ft.²

In a right triangle, the sides adjacent to the hypotenuse or the sides that form the right angle are perpendicular to each other. Thus, one is the height and the other is the base. If a triangle is known to be a right triangle, the height and the base can be identified even if they are not labeled as such.

EXAMPLE 5 Find the area of a piece of sheet metal cut as a right triangle. One side is 12 in., one is 16 in., and the third is 20 in. (Fig. 13–10).

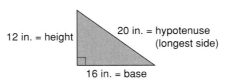

Figure 13-10

Designate one side as the base and another as the height. The longest side is the hypotenuse.

$$A = \frac{1}{2}bh$$

$$A = \frac{1}{2}(\overset{8}{\cancel{16}})(12) \qquad \text{Substitute in formula.}$$

$$A = 96 \text{ in.}^2$$

The area of the sheet metal is 96 in.².

The height of a triangle need not be measured along one side such as in a right triangle or inside the triangle such as in the gable of Example 4. In some triangles, the height is measured along an imaginary line outside the triangle from the base to the highest point of the triangle. The triangles illustrated in Fig. 13–11 are examples of triangles with the height measured outside the triangle itself. The area is calculated using the standard formula. Let's figure the area of $\triangle DEF$. ($\triangle$ means triangle.)

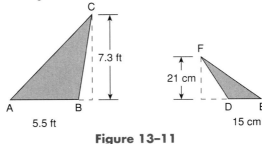

Figure 13-11

EXAMPLE 6 The base of $\triangle DEF$ in Fig. 13–11 is 15 cm and the height is 21 cm. Find the area.

$$A = \frac{1}{2}bh$$

$$A = \frac{1}{2}(15)(21) \qquad \text{Substitute in formula.}$$

$$A = \frac{1}{2}(315)$$

$$A = 157.5 \text{ cm}^2$$

The area of $\triangle DEF$ is 157.5 cm².

3 Use the Pythagorean Theorem to Find the Missing Side of a Right Triangle.

One of the most famous and useful theorems in mathematics is the Pythagorean theorem. It is named for the Greek mathematician, Pythagoras.

■ **DEFINITION 13–6: Pythagorean Theorem.** The *Pythagorean theorem* states that the square of the hypotenuse of a right triangle is equal to the sum of the squares of the two legs of the triangle.

In symbols, the Pythagorean theorem may be expressed as Formula 13–11.

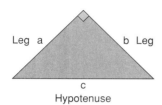

Leg a   b **Leg**

c
Hypotenuse

Figure 13–12

> **Formula 13–11** *Pythagorean Theorem*
>
> $$c^2 = a^2 + b^2$$
>
> c is the hypotenuse of a right triangle; a and b are the legs.

This theorem, illustrated in Fig. 13–12, has numerous applications in many technical fields. When working with the Pythagorean theorem, we will know two sides of a right triangle and can expect to find the remaining side.

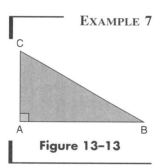

Figure 13–13

EXAMPLE 7 In $\triangle ABC$ (Fig. 13–13), if $AB = 4$ ft and $AC = 3$ ft, find BC.

$(BC)^2 = (AC)^2 + (AB)^2$ State theorem symbolically.

$(BC)^2 = 3^2 + 4^2$ Substitute values.

$(BC)^2 = 9 + 16$

$(BC)^2 = 25$

$BC = \sqrt{25}$ Take the square root of both sides.

$BC = 5$ ft   Use only the positive square root.

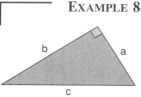

Figure 13–14

EXAMPLE 8 If $a = 8$ mm and $c = 17$ mm, find b (Fig. 13–14).

$c^2 = a^2 + b^2$ State theorem symbolically.

$17^2 = 8^2 + b^2$ Substitute values.

$289 = 64 + b^2$

$289 - 64 = b^2$ Transpose to isolate.

$225 = b^2$

$\sqrt{225} = b$ Take the square root of both sides.

15 mm $= b$

Many times we are confronted with a problem that contains "hidden" triangles. In these cases, we need to visualize the triangle or triangles in the problem. Often, we have to draw one or more of the sides of the "hidden" triangle as an aid in solving the problem.

506

EXAMPLE 9   Find the center-to-center distance between pulleys A and C (Fig. 13–15).

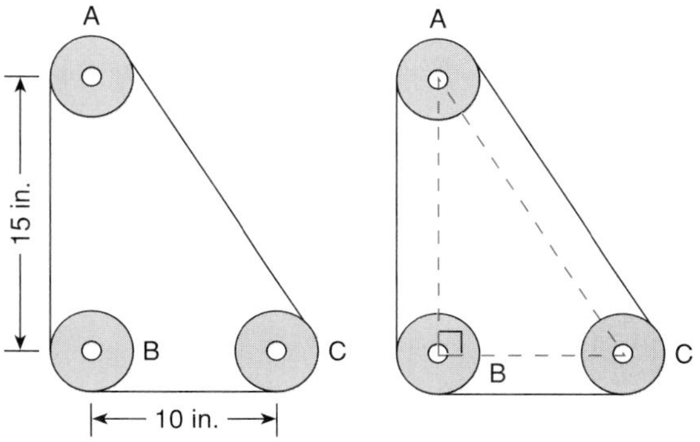

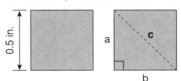

Figure 13–15

Connect the center points of the three pulleys to form a right triangle. The hypotenuse is the distance between pulleys A and C. We use the Pythagorean theorem and substitute the given values for the two known sides.

$(AC)^2 = (AB)^2 + (BC)^2$ State theorem symbolically.

$(AC)^2 = 15^2 + 10^2$ Substitute values.

$(AC)^2 = 225 + 100$

$(AC)^2 = 325$ Take the square root of both sides.

$AC = \sqrt{325}$

$AC = 18.0277564$

Thus, the distance from pulley A to pulley C is 18.0 in. (to the nearest tenth).

EXAMPLE 10   The head of a bolt is a square 0.5 in. on a side (distance across flats). What is the distance from corner to corner (distance across corners)? (See Fig. 13–16.)

Figure 13–16

The sides of the bolt head form the legs of a triangle if a diagonal line is drawn from one corner to the opposite corner. This diagonal forms the hypotenuse of the right triangle. Because the legs of the triangle are known to be 0.5 in. each, we can substitute in the Pythagorean theorem to find the diagonal line, the hypotenuse, which is the distance across corners.

$c^2 = a^2 + b^2$ State theorem symbolically.

$c^2 = 0.5^2 + 0.5^2$ Substitute values.

$c^2 = 0.25 + 0.25$

$c^2 = 0.5$ Take the square root of both sides.

$c = \sqrt{0.5}$

$c = 0.707106781$

Thus, the distance across corners is 0.71 in. (to the nearest hundredth).

Sometimes right triangles are used in technical fields to represent certain relationships, such as forces acting on an object at right angles or electrical and electronic phenomena related in the way the three sides of a right triangle are related. Let's look at an example.

EXAMPLE 11

Forces A and B come together at a right angle to produce force C (Fig. 13–17). If force A is 74.8 lb and the resulting force C is 91.5 lb, what is force B?

Figure 13–17

Since the forces are related in the way the sides of a right triangle are related, we may use the Pythagorean theorem to find the missing force B, a leg of the triangle.

$A^2 + B^2 = C^2$	State theorem symbolically.
$74.8^2 + B^2 = 91.5^2$	Substitute values.
$5595.04 + B^2 = 8372.25$	Transpose to isolate.
$B^2 = 8372.25 - 5595.04$	
$B^2 = 2777.21$	Take the square root of both sides.
$B = \sqrt{2777.21}$	
$B = 52.7$	To the nearest tenth

Thus, force B is 52.7 lb.

4 Find the Area of Prisms and Cylinders.

Common household items like ice cubes, cardboard storage boxes, and toy building blocks are examples of solid geometric figures classified generally as *prisms*. Cans and pipes are examples of *cylinders*.

■ **DEFINITION 13–7: Prism.** A *prism* is a solid whose bases (ends) are parallel, congruent polygons and whose faces (sides) are parallelograms, rectangles, or squares. In a *right prism* the faces are perpendicular to the bases.

■ **DEFINITION 13–8: Cylinder.** A right circular *cylinder* is a solid with a curved surface and two circular bases such that the height is perpendicular to the bases.

■ **DEFINITION 13–9: Height.** The *height* of a solid with two bases is the shortest distance between the two bases.

In right circular cylinders and in right prisms, the height is the same as the length of a side or face. However, in oblique prisms the height is the perpendicular distance between the bases and is different from the length of a side or face (see Fig. 13–18).

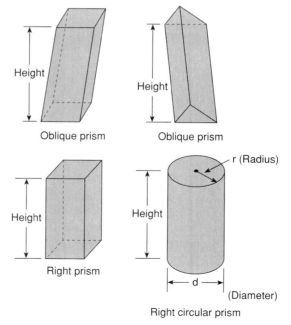

Figure 13–18

A Solid May Not Be Solid at All:
When we refer to a figure as a solid, we mean the figure is three-dimensional; it has length, width, and height. A plane figure, on the other hand, is two dimensional. An empty box or can is still referred to as a "solid" geometrical shape or figure.

Some jobs require us to find the area of a solid figure. The area of a solid figure can refer to just the area of the *sides* of the figure. Or area can refer to the overall area, including the bases along with the sides.

■ **DEFINITION 13–10: Lateral Surface Area.** The *lateral surface area* (LSA) of a solid figure is the area of its sides only.

■ **DEFINITION 13–11: Total Surface Area.** The *total surface area* (TSA) of a solid figure is the area of the sides plus the area of its base or bases.

To find the lateral surface area, all we need to do is find the sum of the areas of each side using the formulas from Chapter 6. However, we can also find the lateral surface area (LSA) of a prism or cylinder by multiplying the perimeter of the base times the height of the solid. In the formulas and examples that follow, we will consider only *right* prisms and *right* circular cylinders.

Formula 13–12 *Lateral surface area of a right prism or cylinder:*

$\quad$ LSA $= ph,$ $\quad p =$ perimeter of base, $h =$ height of the solid figure

To get the total surface area (TSA), we add to the lateral surface area the areas of the two bases.

Formula 13–13 *Total surface area of a right prism or cylinder:*

$\quad\quad$ TSA $= ph + 2B,$ $\quad\quad\quad h =$ height of solid figure,

$\quad\quad p =$ perimeter of base, $B =$ area of base

EXAMPLE 12 Find the lateral surface area of a rectangular shipping carton measuring 24 in. in length, 12 in. in width, and 20 in. in height (Fig. 13–19).

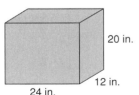
24 in.
20 in.
12 in.
Figure 13–19

LSA $= ph$

LSA $= (2l + 2w)h$ $\quad\quad$ Since the base is a rectangle, its perimeter is $2l + 2w$.

LSA $= [2(24) + 2(12)]20$ $\quad\quad$ Substitute numerical values.

LSA $= [72]20$

LSA $= 1440$ in.^{2}

The lateral surface area of the carton is 1440 in.^{2}.

EXAMPLE 13 How many square centimeters of sheet metal are required to manufacture a can whose radius is 4.5 cm and whose height is 9 cm? Assume no waste or overlap.

TSA $= ph + 2B$ $\quad\quad$ Total surface area is needed.

TSA $= 2\pi rh + 2\pi r^2$ $\quad\quad$ $p = 2\pi r, B = \pi r^2.$

$$TSA = 2(\pi)(4.5)(9) + 2(\pi)(4.5)^2 \qquad \text{Substitute values.}$$

$$TSA = 254.4690049 + 127.2345025$$

$$TSA = 381.70 \text{ cm}^2 \qquad \text{Rounded}$$

The can requires 381.70 cm² of sheet metal.

EXAMPLE 14 Find the total surface area of the triangular prism shown in Fig. 13–20.

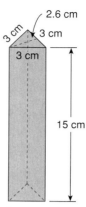

Figure 13–20

$$TSA = ph + 2B$$

$$TSA = 9(15) + (2)\left(\frac{1}{2}\right)(3)(2.6) \qquad \begin{array}{l}\text{Perimeter of triangular base is } 3 + 3 + 3 = 9 \text{ cm.}\\\text{Area of triangular base is } \frac{1}{2}bh, \text{ or } \frac{1}{2}(3)(2.6).\end{array}$$

$$TSA = 135 + 7.8$$

$$TSA = 142.8 \text{ cm}^2$$

The total surface area of the triangular prism is 142.8 cm².

5 Find the Volume of Prisms and Cylinders.

At times we need to know the *volume* of an object, such as a container, to estimate, for example, how many of these containers can be loaded into a given size storage area or shipped in a tractor-trailer rig of certain dimensions.

■ **DEFINITION 13–12: Volume.** The *volume* of a solid geometric figure is the amount of space it occupies, measured in terms of three dimensions (length, width, and height).

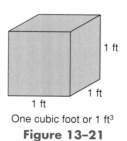

1 ft

1 ft

1 ft

One cubic foot or 1 ft³

Figure 13–21

If we had a rectangular box measuring 1 ft long, 1 ft wide, and 1 ft high (Fig. 13–21), it would be a cube representing 1 cubic foot (ft³). Its volume is calculated by multiplying length × width × height, or $1 \times 1 \times 1 = 1$ ft³. This cubic measure is often indicated by the exponent 3 following the unit of measure, meaning that the measure has *already been "cubed."* It is merely a shortcut for indicating such cubic measures as ft³ = cubic feet, in.³ = cubic inches, cm³ = cubic centimeters, and so on.

From this concept of 1 ft³ or 1 cubic foot comes the formula for the volume of a rectangular box as length × width × height or $V = lwh$. Note that $l \times w$ is actually the formula for the area of the rectangle (or square) that forms the base of the rectangular box. If the base of the prism were a triangle, pentagon, hexagon, or other polygon, we would have to use the appropriate formula for its area. In the case of a cylinder, we would use the formula for the area of a circle because its base is a circle. Thus, the general formula for the volume of *any* right prism or cylinder is as follows.

> **Formula 13–14** *Volume of right prism or cylinder:*
>
> $V = Bh$, B = area of base, h = height of prism or cylinder

EXAMPLE 15

Find the volume of the triangular prism in Example 14, whose height is 15 cm and whose bases are triangles 3 cm on a side and 2.6 cm in height.

$V = Bh_2$ h_1 = 2.6 cm (height of prism base), h_2 = 15 cm (height of prism)

$V = \left(\dfrac{1}{2}bh_1\right)h_2$ Substitute the formula for the area of the triangular base.

$V = \left[\dfrac{1}{2}(3)(2.6)\right]15$ Substitute values.

$V = 58.5 \text{ cm}^3$

The volume of the prism is 58.5 cm³.

EXAMPLE 16

What is the cubic-inch displacement (space occupied) of a cylinder whose diameter is 5 in. and whose height is 4 in.?

$V = Bh$

$V = \pi r^2 h$ Substitute the formula for the area of the circular base.

$V = \pi(2.5)^2(4)$ Substitute values; $r = \frac{1}{2}$ diameter, or 2.5.

$V = 78.5 \text{ in.}^3$ Rounded

The cylinder displacement is 78.5 in.³.

6   Find the Area and Volume of a Sphere.

Soccer balls, golf balls, tennis balls, baseballs, and ball bearings are *spheres*. Spheres are also used as tanks to store gas and water because spheres hold the greatest volume for a specified amount of surface area.

■ **DEFINITION 13–13:** **Sphere.** A *sphere* is a solid formed by a curved surface whose points are all equidistant from a point inside called the *center* (Fig. 13–22).

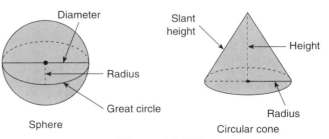

Figure 13–22

■ **DEFINITION 13–14:** **Great Circle.** The *great circle* divides the sphere in half at its greatest diameter and is formed by a plane through the center of the sphere.

A sphere does not have bases like prisms and cylinders. The surface area of a sphere includes *all* the surface and so there is only one formula. Because of the relationship of the sphere to the circle, the formula includes elements of the formula for the area of a circle. The total surface area of the sphere is four times the area of a circle with the same radius.

Formula 13–15 *Total surface area of a sphere:*

$$\text{TSA} = 4\pi r^2$$

where *r* equals the radius.

The formula for the volume of a sphere also contains elements found in formulas for a circle. The volume of a sphere is calculated with Formula 13–16.

Formula 13–16 *Volume of a sphere:*

$$V = \frac{4\pi r^3}{3}$$

where *r* equals the radius.

Note that the radius is *cubed,* or raised to the power of 3, which indicates volume.

EXAMPLE 17   Find the surface area and volume of a sphere whose diameter is 90 cm.

$\text{TSA} = 4\pi r^2$

$\text{TSA} = 4(\pi)(45)^2$ Substitute values. $\frac{1}{2}d = r$, so $r = 45$.

$\textbf{TSA} = \textbf{25,447 cm}^2$ Rounded

$V = \frac{4\pi r^3}{3}$

$V = \frac{4(\pi)(45)^3}{3}$ Substitute values.

$V = \textbf{381,704 cm}^3$ Rounded

7   Find the Area and Volume of a Cone.

One example of an object based on the *cone* is the funnel or the circular rain cap placed on top of stove vent pipes extending through the roofs of many homes.

■ **DEFINITION 13–15: Cone.** A right *cone* is a solid whose base is a circle and whose side surface tapers to a point, called the *vertex* or *apex,* and whose height is a perpendicular line between the base and apex (Fig. 13–22).

In Fig. 13–22, note that in addition to the perpendicular height of the cone, there is also the *slant height* of a cone.

■ **DEFINITION 13–16: Slant Height.** The *slant height* of a cone is the distance along the side from the base to the apex.

The lateral surface area of a cone equals the circumference of the base times $\frac{1}{2}$ the slant height, or LSA $= 2\pi r \frac{s}{2}$, which is simplified as follows.

> **Formula 13-17** *Lateral surface area of a cone:*
> $$\text{LSA} = \pi rs, \qquad r = \text{radius}, \; s = \text{slant height}$$

The total surface area, then, is the lateral surface area plus the area of the base.

> **Formula 13-18** *Total surface area of a cone:*
> $$\text{TSA} = \pi rs + \pi r^2$$
> where r = radius of circular base, s = slant height, πr^2 = area of base

The volume of a cone is equal to one-third the area of the base times the *height of the cone* (*not* the slant height).

> **Formula 13-19** *Volume of a cone:*
> $$V = \frac{\pi r^2 h}{3}$$
> where πr^2 = area of circular base, h = height of cone

EXAMPLE 18

Find the lateral surface area, total surface area, and volume of a cone whose diameter is 8 cm, height is 6 cm, and slant height is 7 cm. Round to hundredths.

$\text{LSA} = \pi rs$

$\text{LSA} = (\pi)(4)(7)$ Substitute values: $r = \frac{1}{2}d$, or 4.

$\textbf{LSA} = \textbf{87.9645943 cm}^2$

$\text{TSA} = \pi rs + \pi r^2$

$\text{TSA} = 87.9645943 + (\pi)(4)^2$ Substitute values.

$\textbf{TSA} = \textbf{138.23 cm}^2$ Rounded

$V = \dfrac{\pi r^2 h}{3}$

$V = \dfrac{(\pi)(4)^2(6)}{3}$ Substitute values.

$\textbf{\textit{V}} = \textbf{100.53 cm}^3$ Rounded

EXAMPLE 19

Find the weight of the cast-iron solid shown in Fig. 13–23 if cast iron weighs 0.26 lb per cubic inch. Round to the nearest whole pound.

The solution requires finding the volume of the cone that forms the top of the solid, the volume of the cylinder that forms the middle portion of the solid, and the volume of the hemisphere (half sphere) that forms the bottom of the solid.

$$V_{\text{cone}} = \frac{\pi r^2 h}{3} \qquad V_{\text{cylinder}} = \pi r^2 h \qquad V_{\text{hemisphere}} = \frac{1}{2}\left(\frac{4\pi r^3}{3}\right)$$

$$V_{\text{cone}} = \frac{(\pi)(4.5)^2(8)}{3} \qquad V_{\text{cylinder}} = (\pi)(4.5)^2(10) \qquad V_{\text{hemisphere}} = \frac{1}{2}\left(\frac{4(\pi)(4.5)^3}{3}\right)$$

$$V_{\text{cone}} = 169.6460033 \text{ in.}^3 \qquad V_{\text{cylinder}} = 636.1725124 \text{ in.}^3 \qquad V_{\text{hemisphere}} = 190.8517537 \text{ in.}^3$$

$$\text{Total volume} = V_{\text{cone}} + V_{\text{cylinder}} + V_{\text{hemisphere}}$$

$$\text{Total volume} = 996.6702694 \text{ in.}^3$$

Convert to pounds:

$$\frac{996.6702694 \text{ in.}^3}{1} \times \frac{0.26 \text{ lb}}{1 \text{ in.}^3} = 259 \text{ lb.} \qquad \text{Rounded}$$

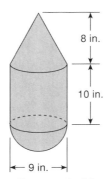

Figure 13–23

To the nearest whole pound, the solid cast-iron figure weighs 259 lb.

SELF-STUDY EXERCISES 13–4

1 Find the perimeter and the area of Figs. 13–24 and 13–25. Round to hundredths.

1.

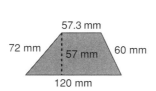

Figure 13–24

2.

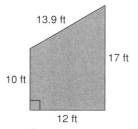

Figure 13–25

Solve the following problems involving perimeter and area.

3. The six glass panes in a kitchen light fixture each measure $4\frac{1}{2}$ in. along the top and 10 in. along the bottom. The top and bottom are parallel. The height of each pane is 8 in. What is the combined area of the six trapezoidal panes?

4. A section of a hip roof is a trapezoid measuring 38 ft at the bottom, 14 ft at the top, and 10 ft high. Find the area of this section of the roof in square feet.

5. A lot in an urban area is 60 ft wide. The sides are 120 ft and 154 ft, and they are perpendicular to the width of the lot. Find the area of the trapezoidal property.

6. A swimming pool is fashioned in a trapezoidal design. The parallel sides are $18\frac{1}{2}$ ft and 31 ft. The other sides are $24\frac{1}{2}$ ft and 24 ft. What is the perimeter of the pool?

514

2 Find the perimeter and area of the triangles in Figs. 13–26 and 13–27. Round to hundredths.

7.

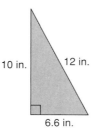

10 in.   12 in.

6.6 in.

Figure 13–26

8.

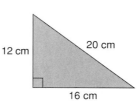

12 cm   20 cm

16 cm

Figure 13–27

9. Find the area of the triangle in Fig. 13–28.

3.5 cm

9 cm

Figure 13–28

Solve the following problems involving perimeter and area.

10. Max Cisneros is planning a patio that will adjoin the sides of his L-shaped home. One side of the home is 24 ft and the other is 18 ft. The shape of the patio is triangular. Draw a representation of the patio and find the perimeter of the triangle if its hypotenuse is 30 ft. Find the number of square feet of surface area that needs to be covered with concrete.

11. A louver, or triangular vent, for a gable roof measures 6 ft 6 in. wide and stands 3 ft high. What is the area in square feet of the vented portion of the gable?

12. A mason lays tile in the form of a triangle. The height of the triangle is 12 ft and the base is 5 ft. What is the area?

13. A metal worker cuts a triangular plate with a base of 11 in. and a height of $4\frac{1}{2}$ in. from a piece of metal. What is the area of the plate?

14. If aluminum siding costs $6.75 a square yard installed, how much would it cost to put the siding on the two triangular gable ends of a roof under construction? Each gable has a span (base) of 30 ft 6 in. and a rise (height) of 7 ft 6 in. Any portion of a square yard is rounded to the next highest square yard.

15. Find the height of a triangle if its base is 15 ft 6 in. and its area is 62 ft^2.

16. Find the perimeter of a triangle whose sides are 2 ft 6 in., 1 yd 8 in., and 4 ft 6 in.

3 Find the missing side of the right triangle. Round the final answers to the nearest thousandth. Use Fig. 13–29 for Exercises 17 through 19.

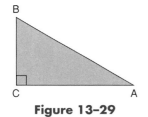

B

C   A

Figure 13–29

17. $AC = ?$
$BC = 7$ cm
$AB = 25$ cm

18. $AC = 24$ mm
$BC = ?$
$AB = 26$ mm

19. $AC = 15$ yd
$BC = 8$ yd
$AB = ?$

Use Fig. 13–30 for Exercises 20 through 22. Find the missing dimensions. Round to thousandths where appropriate.

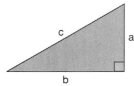
Figure 13–30

20. $a = 5$ cm
$b = 4$ cm
$c = ?$

21. $a = ?$
$b = 9$ m
$c = 11$ m

22. $a = 9$ ft
$b = ?$
$c = 15$ ft

Solve the following problems. Round the final answers to the nearest thousandth if necessary.

23. A light pole will be braced with a wire that is to be tied to a stake in the ground 18 ft from the base of the pole, which extends 26 ft above the ground. If the wire is attached to the pole 2 ft from the top, how much wire must be used to brace the pole? (See Fig. 13–31.)

24. Find the center-to-center distance between holes A and C in a sheet metal plate if the distance between the centers of A and B is 16.5 cm and the distance between the centers of B and C is 36.2 cm (Fig. 13–32).

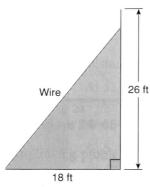
Figure 13–31

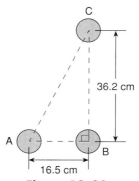
Figure 13–32

25. Find the length of a rafter that has a 10-in. overhang if the rise of the roof is 10 ft and the joists are 48 ft long (Fig. 13–33.)

26. A stair stringer is 8 ft high and extends 10 ft from the wall (Fig. 13–34). How long will the stair stringer be?

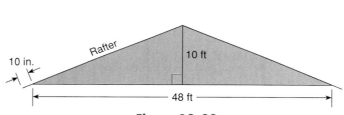
Figure 13–33

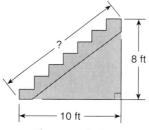
Figure 13–34

27. A machinist wishes to strengthen an L bracket that is 5 cm by 12 cm by welding a brace to each end of the bracket (Fig. 13–35). How much metal rod is needed for the brace?

28. To make a rectangular table more stable, a diagonal brace is attached to the underside of the table surface. If the table is 27 dm by 36 dm, how long is the brace?

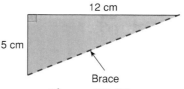
Figure 13–35

29. Find the length of the side of the largest square nut that can be milled from a piece of round stock whose diameter is 15 mm (Fig. 13–36).

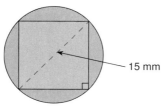

Figure 13–36

30. A rigid length of electrical conduit must be shaped as shown in Fig. 13–37 to clear an obstruction. What is the total length of the conduit needed? (*Hint:* Do not forget to include *AB* and *CD* in the total length.)

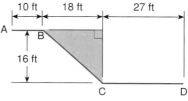

Figure 13–37

31. The vector diagram in Fig. 13–38 is used in electrical applications. Find the voltage of E_a.

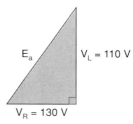

Figure 13–38

[4] Find the lateral surface area and total surface area of the solids in Figs. 13–39 and 13–40. Round to the nearest hundredth if necessary.

32.

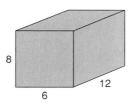

Figure 13–39

33.

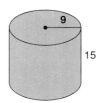

Figure 13–40

Solve the following problems.

34. What is the total surface area of a cylindrical oil storage tank 35 ft in diameter and 12 ft tall? Round to the nearest whole number.

35. What is the total surface area of a triangular prism whose height is 6 in. and whose triangular base measures 2 in. on each side with a height of 1.7 in.?

36. A solid chocolate candy bar in the form of a prism is enclosed in a clear wrapper. The manufacturer wants to wrap the sides of the bar with paper indicating the company name, ingredients, and weight of the bar. What is the lateral area to be wrapped if the bar measures as shown in Fig. 13–41?

37. Find the number of square inches needed to make a paper label for an aluminum can $2\frac{1}{2}$ in. in diameter and $4\frac{3}{4}$ in. tall. Assume no waste or overlap. Round to tenths.

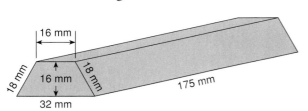

Figure 13–41

38. A wall-mounted three-way speaker is to be covered with wood-grained vinyl. What is the lateral area to be covered if the speaker is 19 in. high, 11 in. wide, and 10 in. deep? (*Hint:* Assume that the 11×19 rectangle is the base.)

39. If the side of a 1-lb coffee can will be imprinted to show the manufacturer and contents, how many square inches of lateral surface may be imprinted if the can is 4 in. across and $5\frac{1}{2}$ in. high? Round to tenths.

5 Find the volume of the solids in Figs. 13–42 and 13–43. Round to the nearest hundredth if necessary.

40.

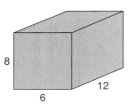

Figure 13–42

41.

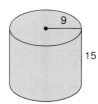

Figure 13–43

Solve the following problems.

42. How many cubic inches are in an aluminum can $2\frac{1}{2}$ in. in diameter and $4\frac{3}{4}$ in. tall? Round to tenths.

44. A right pentagonal prism is 10 cm high. If the area of each pentagonal base is 32 cm^2, what is the volume of the prism?

46. A cylindrical water well is 1200 ft deep and 6 in. across. How much soil and other material were removed? Round to the nearest cubic foot. (*Hint:* Convert measures to a common unit.)

43. What is the volume of a cylindrical oil storage tank 40 ft in diameter and 15 ft tall? Round to the nearest whole number.

45. What is the volume of a triangular prism whose height is 8 in. and whose triangular base measures 4 in. on each side with a height of 3.46 in.? Round to hundredths.

6 Solve the following problems. Round to tenths.

47. Find the surface area of a sphere with a radius of 5 cm.

49. How many square feet of steel are needed to manufacture a spherical water tank with a diameter of 45 ft?

51. A spherical propane tank has a diameter of 4 ft. How many square feet of surface area need to be painted?

48. Find the volume of a sphere with a radius of 6 in.

50. If 1 ft^3 = 7.48 gal, how many gallons can the water tank in Exercise 49 hold?

52. If a propane tank is filled to 90% of its capacity, how many gallons of propane will the tank in Exercise 51 hold? (1 ft^3 = 7.48 gal.)

7 Solve the following problems. Round to tenths.

53. Find the lateral surface area, total surface area, and volume of the cone in Fig. 13–44.

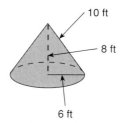

Figure 13–44

54. How many cubic feet are in a conical pile of sand 30 ft in diameter and 20 ft high?

55. How many square centimeters of sheet metal are needed to form a conical rain cap 25 cm in diameter if the slant height will be 15 cm?

57. Find the total surface area of a conical tank whose radius is 15 ft and whose slant height is 20 ft.

59. A cylindrical water tower with a conical top and hemispheric bottom (see Fig. 13–45) needs to be painted. If the cost is $2.19 per square foot, how much would it cost (to the nearest dollar) to paint the tank?

56. A cone-shaped storage container holds a photographic chemical. If the container is 80 cm wide and 30 cm high, how many liters of the chemical does it hold if 1 L = 1000 cm³?

58. Find the height of a conical tank whose volume is 261.67 ft³ and whose radius is 5 ft. (*Hint:* Rearrange the volume formula to find the height.)

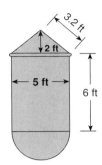

Figure 13–45

13–5 MISCELLANEOUS TECHNICAL FORMULAS

Learning Objective

1 Evaluate miscellaneous technical formulas.

In addition to temperature conversion formulas and geometric formulas, several other formulas are used in technical applications. In this section, we will present a miscellaneous assortment of technical formulas and show how they are evaluated and, if necessary, rearranged.

1 Evaluate Miscellaneous Technical Formulas.

EXAMPLE 1 Evaluate the formula $E = \frac{I - P}{I}$ if $I = 24{,}000$ calories (cal) and $P = 8600$ cal.

$E = \dfrac{I - P}{I}$ Engine efficiency = the difference between heat input and output divided by heat input.

$E = \dfrac{24{,}000 - 8600}{24{,}000}$ Substitute given values.

$E = \dfrac{15{,}400}{24{,}000}$ Perform calculations in numerator grouping.

$E = 0.642$ Nearest thousandth

Thus, the engine efficiency is 0.642 (64.2% efficient).

EXAMPLE 2 Evaluate the formula $R_T = \frac{R_1 R_2}{R_1 + R_2}$ if $R_1 = 10$ ohms and $R_2 = 6$ ohms.

$R_T = \dfrac{R_1 R_2}{R_1 + R_2}$ Total resistance = product of first resistance and second resistance divided by sum of first and second resistances.

$R_T = \dfrac{10(6)}{10 + 6}$ Substitute given values.

$R_T = \dfrac{60}{16}$ Perform calculations in numerator and denominator groupings.

$$R_T = 3.75$$

Thus, the total resistance in the circuit is 3.75 ohms.

In the next formula, do not forget to square the D.

EXAMPLE 3 Evaluate the formula $H = \frac{D^2N}{2.5}$ if $D = 4$ and $N = 8$.

$$H = \frac{D^2N}{2.5}$$ Horsepower = the diameter of the cylinder squared times the number of cylinders divided by 2.5.

$$H = \frac{4^2(8)}{2.5}$$ Perform power operation.

$$H = \frac{16(8)}{2.5}$$ Perform multiplication.

$$H = \frac{128}{2.5}$$

$$H = 51.2$$

The engine is rated at 51.2 hp.

Sometimes we are asked to solve for a letter term that is not isolated. Let's look again at the horsepower formula and solve for something other than horsepower.

EXAMPLE 4 Solve the formula $H = \frac{D^2N}{2.5}$ for D (in inches) if $H = 45$ and $N = 6$.

$$H = \frac{D^2N}{2.5}$$

$$45 = \frac{D^2(6)}{2.5}$$

$$(2.5)45 = \frac{D^2(6)}{2.5}(2.5)$$ Multiply to eliminate denominator.

$$112.5 = 6D^2$$

$$\frac{112.5}{6} = D^2$$

$$18.75 = D^2$$

$$D = 4.330$$ Rounded

Thus, the diameter of the piston is 4.330 in.

Tips and Traps *Interpret the Solution Within the Context of the Formula:*
In practical problems where a negative square root is unrealistic, we use only the positive square root. In the preceding example, for instance, we cannot have a negative diameter of a piston.

Caution: The next formula includes both powers and roots. Because the process for solving these equations involves multiplying and dividing by variables, the solution should be checked for extraneous roots.

EXAMPLE 5 In the formula $Z = \sqrt{R^2 + X^2}$, solve for R (in ohms) if $Z = 12.4$ ohms and $X = 12$ ohms.

$$Z = \sqrt{R^2 + X^2}$$ Impedance = the square root of the sum of the resistance squared plus the reactance squared.

$$12.4 = \sqrt{R^2 + (12)^2}$$

$$(12.4)^2 = (\sqrt{R^2 + 144})^2 \qquad \text{Eliminate radical.}$$

$$153.76 = R^2 + 144 \qquad \text{Isolate } R^2.$$

$$153.76 - 144 = R^2$$

$$9.76 = R^2$$

$$R = 3.124 \qquad \text{Rounded}$$

Thus, the resistance is 3.1 ohms (rounded to tenths).

Note the many steps in the following formula and the importance of observing the order of operations. The formula is for finding the length of a belt connecting two pulleys.

EXAMPLE 6 Evaluate $L = 2C + 1.57(D + d) + \frac{D + d}{4C}$ if $C = 24$ in., $D = 16$ in., and $d = 4$ in. See Fig. 13–46.

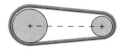

Figure 13–46

L = length of belt joining two pulleys,
C = distance between centers of pulleys,
D = diameter of large pulley,
d = diameter of small pulley.

$$L = 2C + 1.57(D + d) + \frac{D + d}{4C}$$

$$L = 2(24) + 1.57(16 + 4) + \frac{16 + 4}{4(24)}$$

$$L = 2(24) + 1.57(20) + \frac{20}{96} \qquad \text{Work groupings in parentheses, numerator, and denominator.}$$

$$L = 48 + 31.4 + 0.20833333 \qquad \text{Work multiplication and division.}$$

$$L = 79.61 \qquad \text{Rounded}$$

The pulley belt is 79.61 in. long.

SELF-STUDY EXERCISES 13–5

1 Complete the following.

1. Use the formula $R_t = \frac{R_1 R_2}{R_1 + R_2}$ to find the total resistance (R_t) if one resistance (R_1) is 12 ohms and the second resistance (R_2) is 8 ohms.

2. What is the percent efficiency (E) of an engine if the input (I) is 25,000 calories and the output (P) is 9600 calories? Use the formula $E = \frac{I - P}{I}$.

3. Distance is rate times time, or $D = RT$. Find the rate if the distance traveled is 140 mi and the time traveled is 4 hr.

4. The formula for voltage (Ohm's law) is $E = IR$. Find the amperes of current (I) if the voltage (E) is 120 V and the resistance (R) is 80 ohms.

5. According to Boyle's law, if temperature is constant, the volume of a gas is inversely proportional to the pressure on it. Find the final volume (V_2) of a gas using the formula $\frac{V_1}{V_2} = \frac{P_2}{P_1}$ if the original volume (V_1) is 15 ft^3, the original pressure (P_1) is 60 lb per square inch (psi), and the final pressure (P_2) is 150 psi.

6. The formula for power (P) in watts (W) is $P = I^2 R$. Find the current (I) in amperes if a device draws 63 W and the resistance (R) is 7 ohms.

7. Use the formula $H = \frac{D^2 N}{2.5}$ to find the number of cylinders (N) required in an engine of 3.2 hp (H) if the cylinder diameter (D) is 2 in.

8. The formula for the speed (s) of a driven pulley in revolutions per minute (rpm) is $s = \frac{DS}{d}$. Find the speed of a driven pulley with a diameter (d) of 5 in. if the diameter (D) of the driving pulley is 10 in. and its speed (S) is 800 rpm.

9. If the distance (C) between the centers of the pulleys in Exercise 8 is 24 in., find the length (L) of the belt connecting them using the formula

$$L = 2C + 1.57(D + d) + \frac{D + d}{4C}$$

Round to hundredths.

10. Find the reactance (X) in ohms using the formula $Z = \sqrt{R^2 + X^2}$ if the impedence (Z) is 10 ohms and the resistance (R) is 9 ohms. Round to tenths.

CAREER APPLICATION

Electronics: Ohm's Law

Ohm's law is derived from Kirchhoff's laws to form the basis for much of the work done in electronics. In its simplest forms, Ohm's law is represented by three equations:

Ohm's law: $E = IR$, $P = IE$, $G = \frac{1}{R}$

E represents potential difference, which is measured in volts (V).

I represents current, which is measured in amperes (A).

P represents power, which is measured in watts (W).

R represents resistance, which is measured in ohms (Ω).

G represents conductance, which is measured in siemens (S).

First consider $E = IR$. The R can be replaced by $\frac{1}{G}$, so $E = \frac{I}{G}$. This can be rewritten as $I = GE$, which can be rearranged to get $G = \frac{I}{E}$.

The equation $P = IE$ can be rearranged to get $I = \frac{P}{E}$ and $E = \frac{P}{I}$. Also, IR can be substituted for E in the equation $P = IE$ to get $P = IIR = I^2 R$, and $\frac{1}{G}$ can be substituted for R to get $P = \frac{I^2}{G}$.

Assume that you start with $P = \frac{I^2}{G}$ and wish to solve for I or G. Eliminate denominators (multiply through by G) to get $PG = I^2$. This gives both $G = \frac{I^2}{P}$ and also $I = \sqrt{PG}$.

Suppose that $P = 4$ mW and $G = 7$ μS. Then $I = 167$ μA.

Proof: $P = \frac{I^2}{G} = \frac{(167 \text{ μA})^2}{7 \text{ μS}} = 3984$ W or 4 mW   Rounded

Exercises

Solve for each of the following first in letters (for the general equation) and then prove your equation by finding the numerical answer. Prove the numerical answer obtained by substituting back into the original equation. Be sure to use the given prefixes.

1. Solve for I in $P = I^2 R$. What is I if $P = 8$ W and $R = 2$ Ω?
2. Solve for G in $E = \frac{I}{G}$. What is G if $E = 12$ V and $I = 4$ mA?
3. Solve for P in $G = \frac{I^2}{P}$. What is P if $G = 8$ mS and $I = 6$ mA?
4. Solve for R in $P = I^2 R$. What is R if $I = 4$ mA and $P = 8$ mW?

5. Solve for E in $P = \frac{E^2}{R}$. What is E if $P = 9$ W and $R = 63$ Ω?
6. Solve for E in $P = E^2G$. What is E if $P = 8$ W and $G = 2$ mS?
7. Solve for R_3 in $R_t = R_1 + R_2 + R_3$. What is R_3 if $R_t = 1410$ Ω, $R_1 = 420$ Ω, and $R_2 = 440$ Ω?
8. Solve for G_3 in $G_t = G_1 + G_2 + G_3$. What is G_3 if $G_t = 92$ mS, $G_1 = 22$ mS, and $G_2 = 24$ mS?
9. Solve for R_1 in $V_1 = \frac{R_1 V_t}{R_1 + R_2}$. What is R_1 if $V_1 = 8$ V, $V_t = 12$ V, and $R_2 = 41$ Ω?
10. Solve for G_2 in $I_2 = \frac{G_2 I_t}{G_1 + G_2}$. What is G_2 if $I_2 = 5$ mA, $I_t = 8$ mA, and $G_1 = 6$ mS?

Answers for Exercises

1. Solve for I in $P = I^2R$. What is I if $P = 8$ W and $R = 2$ Ω?

$$I = \sqrt{\frac{P}{R}} = \sqrt{\frac{8}{2}} = 2 \text{ A} \qquad \text{Proof:} \quad 2^2 \times 2 = 8$$

2. Solve for G in $E = \frac{I}{G}$. What is G if $E = 12$ V and $I = 4$ mA?

$$G = \frac{I}{E} = \frac{4 \text{ mA}}{12 \text{ V}} = 0.333 \text{ mS} = 333 \text{ }\mu\text{S} \qquad \text{Proof:} \quad \frac{4 \text{ mA}}{0.333 \text{ mS}} = 12 \text{ V}$$

3. Solve for P in $G = \frac{I^2}{P}$. What is P if $G = 8$ mS and $I = 6$ mA?

$$P = \frac{I^2}{G} = \frac{(6 \text{ mA})^2}{8 \text{ mS}} = 4.5 \text{ mW} \qquad \text{Proof:} \quad \frac{(6 \text{ mA})^2}{4.5 \text{ mW}} = 8 \text{ mS}$$

4. Solve for R in $P = I^2R$. What is R if $I = 4$ mA and $P = 8$ mW?

$$R = \frac{P}{I^2} = \frac{8 \text{ mW}}{(4 \text{ mA})^2} = 0.5 \text{ k}\Omega = 500 \text{ }\Omega \qquad \text{Proof:} \quad (4 \text{ mA})^2(500 \text{ }\Omega) = 8 \text{ mW}$$

5. Solve for E in $P = \frac{E^2}{R}$. What is E if $P = 9$ W and $R = 63$ Ω?

$$E = \sqrt{PR} = \sqrt{9 \times 63} = 23.8 \text{ V} \qquad \text{Proof:} \quad \frac{(23.8 \text{ V})^2}{63 \text{ }\Omega} = 9 \text{ W}$$

6. Solve for E in $P = E^2G$. What is E if $P = 8$ W and $G = 2$ mS?

$$E = \sqrt{\frac{P}{G}} = \sqrt{\frac{8 \text{ W}}{2 \text{ mS}}} = \sqrt{\frac{8 \text{ W}}{0.002 \text{ S}}} = 63.2 \text{ V}$$

$$\text{Proof:} \quad (63.2 \text{ V})^2(2 \text{ mS}) = 7988.48 \text{ mW} = 8 \text{ W}$$

7. Solve for R_3 in $R_t = R_1 + R_2 + R_3$. What is R_3 if $R_t = 1410$ Ω, $R_1 = 420$ Ω, and $R_2 = 440$ Ω?

$$R_3 = R_t - R_1 - R_2 = 1410 \text{ }\Omega - 420 \text{ }\Omega - 440 \text{ }\Omega = 550 \text{ }\Omega$$
Proof: $420 \text{ }\Omega + 440 \text{ }\Omega + 550 \text{ }\Omega = 1410 \text{ }\Omega$

8. Solve for G_3 in $G_t = G_1 + G_2 + G_3$. What is G_3 if $G_t = 92$ mS, $G_1 = 22$ mS, and $G_2 = 24$ mS?

$$G_3 = G_t - G_1 - G_2 = 92 \text{ mS} - 22 \text{ mS} - 24 \text{ mS} = 46 \text{ mS}$$
Proof: $22 \text{ mS} + 24 \text{ mS} + 46 \text{ mS} = 92 \text{ mS}$

9. Solve for R_1 in $V_1 = \frac{R_1 V_t}{R_1 + R_2}$. What is R_1 if $V_1 = 8$ V, $V_t = 12$ V, and $R_2 = 41$ Ω?

$$R_1 = \frac{V_1 R_2}{V_t - V_1} = \frac{8 \text{ V } (41 \text{ }\Omega)}{12 \text{ V} - 8 \text{ V}} = 82 \text{ }\Omega$$

$$\text{Proof:} \quad \frac{(82 \text{ }\Omega)(12 \text{ V})}{82 \text{ }\Omega + 41 \text{ }\Omega} = 8 \text{ V}$$

10. Solve for G_2 in $I_2 = \dfrac{G_2 I_t}{G_1 + G_2}$. What is G_2 if $I_2 = 5$ mA, $I_t = 8$ mA, and $G_1 = 6$ mS?

$$G_2 = \frac{I_2 G_1}{I_t - I_2} = \frac{5 \text{ mA} (6 \text{ mS})}{8 \text{ mA} - 5 \text{ mA}} = 10 \text{ mS}$$

Proof: $\dfrac{(10 \text{ mS})(8 \text{ mA})}{6 \text{ mS} + 10 \text{ mS}} = 5$ mA

■ ASSIGNMENT EXERCISES ■

Section 13–1

Evaluate the following formulas. Use $D = 2.5$, $E = 3$, and $F = 4$.

1. $Y = E + (3F - D)$

2. $Y = F(E^2 + 4DF) - 3$

3. $Y = \dfrac{2(F - D)^2}{D - 1}$

4. $Y = \dfrac{3(F - D)}{3} + E$

5. $Y = D^2 - (2F - E)$

Evaluate the percentage formula $P = \dfrac{RB}{100}$ using the following values.

6. Find the percentage if $R = 10\%$ and $B = 300$ lb.

7. Find the rate if $P = 12$ kg and $B = 125$ kg.

8. Find the base if $P = \$28.05$ and $R = 8.5\%$.

9. Evaluate the rate formula $R = \dfrac{100P}{B}$ for P if $R = 12\%$ and $B = 90$.

10. Evaluate the base formula $B = \dfrac{100P}{R}$ for R if $B = \$5000$ and $P = \$450$.

Evaluate the interest formula $I = PRT$ using the following values.

11. Find the interest if $P = \$440$, $R = 16\%$, and $T = 2\frac{3}{4}$ years.

12. Find the rate if $I = \$2484$, $P = \$4600$, and $T = 3$ years.

13. Find the time if $I = \$387.50$, $P = \$1550$, and $R = 12.5\%$.

14. Find the principal if $I = \$1665$, $R = 18\frac{1}{2}\%$, and $T = 1\frac{1}{2}$ years.

15. Find the length of a rectangular work area if the perimeter is 160 in. and the width is 30 in. Use the formula $P = 2(l + w)$.

16. Find the radius of a circle whose area is 132.7 mm² using the formula $r = \sqrt{\frac{A}{\pi}}$. Round to the nearest tenth.

17. Find the cost (C) if the markup (M) on an item is $\$25.75$ and the selling price (S) is $\$115.25$. Use the formula $M = S - C$.

18. Using the formula for the side of a square, $s^2 = A$, find the length of a side of a square field whose area is $\frac{1}{4}$ mi².

19. Evaluate the formula for the area of a circle, $A = \pi r^2$, if $r = 5.5$ in. Round to the nearest tenth.

20. Evaluate the formula for the area of a square, $A = s^2$, if $s = 3.25$ km.

Section 13–2

Solve the following formulas as indicated.

21. $D = F - (m + n)$ for F.

22. $H = C - S$ for S.

23. $V = lwh$ for w.

24. $R = h(p + 3q)$ for q.

25. $c = ah + ab$ for a.

26. $K = \dfrac{m + n}{p}$ for m.

27. $B = cr^2x$ for r.

28. $r = \sqrt{s^2 - t^2}$ for t.

29. $PB = A$ for B.

30. $V = lwh$ for h.

31. $I = Prt$ for r.

32. $s = c + m$ for c.

33. $s = r - d$ for r.

34. $s = r - d$ for d.

35. $v = v_0 - 32t$ for t.

36. $P = 2(l + w)$ for w.

37. $A = P(1 + rt)$ for t.

38. $V = \frac{1}{3} Bh$ for h.

39. The formula for finding the sale price on an item is $S = P - D$, where S is the sale price, P is the original price, and D is the discount. Solve the formula for the original price.

40. The formula for finding tax is $T = RM$, where T represents tax, R represents the tax rate, and M represents the marked price. Rearrange the formula to find the marked price.

Section 13–3

Make the following temperature conversions.

41. 78°C = _____ K

42. 410 K = _____ °C

43. 12°F = _____ °R

44. 720°R = _____ °F

45. 95°C = _____ °F

46. 86°F = _____ °C

47. The label on a container of the acid-reducer famotidine states that storage above 40°C should be avoided. What is 40°C on the Fahrenheit scale?

48. The freezing point of benzene is 5°C. What is this temperature on the Fahrenheit scale?

49. Candy that should reach a temperature of 365°F should reach what temperature on the Celsius scale?

50. Summer road surface temperatures of 122°F would show as what reading on the Celsius scale?

Section 13–4

Find the perimeter and the area of Figs. 13–47 to 13–50.

51.

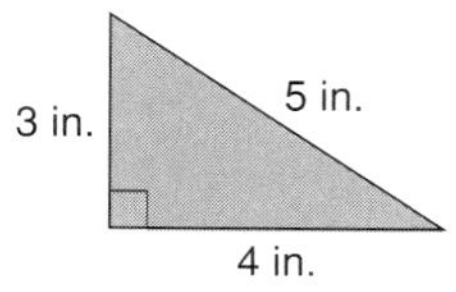

![3 in., 5 in., 4 in. right triangle]

Figure 13–47

52.

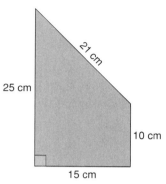

Figure 13–48

53.

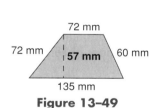

Figure 13–49

54.

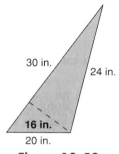

Figure 13–50

55. Find the area of the triangle in Fig. 13–51.

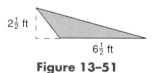

Figure 13–51

Solve the following problems involving perimeter and area.

56. A trapezoidal wall section of a contemporary home has a height of 9 ft, a top length of 7 ft, and a bottom length of 14 ft. Find the area of the wall section.

57. The sides of a triangle are $2\frac{1}{2}$ ft, $3\frac{3}{4}$ ft, and 5 ft. If the triangle is to be made into an advertising sign for a restaurant, how many feet of contrasting trim would be needed to outline the sign?

58. Using your knowledge of formula rearrangement and triangles, find the base of a triangle whose height is 26.25 ft and whose area is 236.25 ft^2.

59. If a triangular louver whose height is 3 ft and whose base is 6 ft 6 in. is installed in each gable end of a house, how many square feet of ventilation would the two louvers provide?

60. One trapezoidal side of a hip roof measures $21\frac{1}{2}$ ft along the ridge line (upper base) and $45\frac{1}{3}$ ft at the bottom (lower base). The height of the side is 16 ft. How many bundles of roofing shingles are necessary to cover the side at a 5-in. exposure if 3.2 bundles cover each square (100 ft^2)?

62. A gable end of a house has a span (base) of 30 ft and a rise (height) of $5\frac{1}{2}$ ft. If the gable is to be covered with 8-in. wide bevel siding with a $6\frac{1}{2}$-in. exposure to the weather, how many square feet of siding are needed if we assume a 23% loss for lap and waste? (Round any portion of a square foot to the next highest square foot.)

64. A trapezoidal flower bed has its longer base along the entire length of a 45-ft outside wall of a building. If the shorter base is 25 ft and the adjacent sides are each 11.5 ft, how many feet of garden edging material are needed to surround the flower bed if no edging is used against the building wall?

61. A stairway to the upstairs portion of a house has a run (base) of 10 ft and a rise (height) of $7\frac{1}{2}$ ft. How many sheets of 4-ft × 8-ft paneling are needed to finish the exposed side? Disregard waste.

63. If the three sides of the gable end of a roof are each 16 ft, how many feet of trim would be needed to surround the gable end?

65. If the flower bed in Exercise 64 projects 5.7 ft from the building wall, how many square feet of surface are available for flowering plants?

Use Fig. 13–52 to solve the following exercises. Round the final answers to the nearest thousandth if necessary.

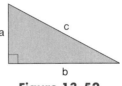

Figure 13–52

66. $a = 3$ m	**67.** $a = 9$ in.	**68.** $a = 8$ cm	**69.** $a = 7$ ft	**70.** $a = 8$ mm
$b = ?$	$b = 12$ in.	$b = 15$ cm	$b = ?$	$b = ?$
$c = 5$ m	$c = ?$	$c = ?$	$c = 10$ ft	$c = 17$ mm
71. $a = ?$	**72.** $a = ?$	**73.** $a = 11$ mi	**74.** $a = 10$ in.	**75.** $a = ?$
$b = 15$ yd	$b = 12$ km	$b = 17$ mi	$b = 24$ in.	$b = 40$ cm
$c = 17$ yd	$c = 15$ km	$c = ?$	$c = ?$	$c = 50$ cm

Solve the following problems. Round the final answers to the nearest thousandth if necessary.

76. If the base of a ladder is placed on the ground 4 ft from a house, how tall must the ladder be (to the nearest foot) to reach the chimney top that extends $18\frac{1}{2}$ ft from the ground?

77. In an automobile, there are three pulleys connected by one belt. The center-to-center distance between the pulleys farthest apart cannot be measured conveniently. The other center-to-center distances are 12 in. and 18 in. (Fig. 13–53). Find the distance between the pulleys farthest apart.

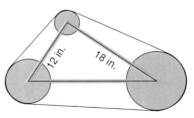

Figure 13–53

78. A central vacuum outlet is installed in one corner of a rectangular room that measures 9′ × 12′. How long must the nonelastic hose be to reach all parts of the room?

79. Find the diameter of a piece of round steel from which a 3-in. square nut can be milled.

80. Find the distance across the corners of a square nut that is 7.9 mm on a side (Fig. 13–54).

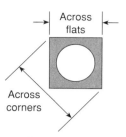

Figure 13–54

81. Find the volume of the triangular prism of Fig. 13–55.

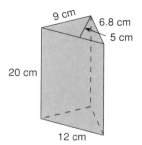

Figure 13–55

82. Find the lateral surface area of the triangular prism of Fig. 13–55.

83. Find the total surface area of the triangular prism of Fig. 13–55.

84. Find the lateral surface area of the cylinder of Fig. 13–56.

85. Find the total surface area of the cylinder of Fig. 13–56.

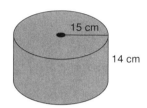

Figure 13–56

86. Find the volume of the cylinder of Fig. 13–56.

87. How many cubic yards of top soil are needed to cover an 85-ft by 65-ft area for landscaping if the topsoil will be 6 in. deep? Round to the nearest whole number. ($27 \text{ ft}^3 = 1 \text{ yd}^3$.)

88. If concrete weighs 160 lb per cubic foot, what is the weight of a concrete circular slab 4 in. thick and 15 ft across?

89. What is the lateral surface area of a hexagonal column 20 ft high that measures 6 in. on a side?

90. An interstate highway was repaired in one section 48 ft across, 25 ft long, and 8 in. deep. If concrete costs $25.50 per cubic yard, what is the cost of the concrete needed to repair the highway rounded to the nearest dollar? ($27 \text{ ft}^3 = 1 \text{ yd}^3$.)

91. A pipeline to carry oil between two towns 5 mi apart has an inside diameter of 18 in. If 1 mi = 5280 ft and $1 \text{ ft}^3 = 7.48$ gal, how many gallons of oil will the pipeline hold (to the nearest gallon)?

92. How many barrels of oil will a tank hold if its height is $65\frac{1}{2}$ ft and its radius is 20 ft? Round to the nearest whole barrel. (31.5 gal = 1 barrel and $1 \text{ ft}^3 = 7.48$ gal.)

Solve the following problems. Round the final answer to the nearest tenth unless otherwise specified.

93. Find the total surface area of a sphere whose radius is 9 m.

94. Find the total surface area of a sphere whose diameter is 20 cm.

95. Find the volume of a sphere whose radius is 12 ft.

96. Find the volume of a sphere whose diameter is 30 cm.

97. Find the lateral surface area of a cone whose radius is 6 cm and whose slant height is 9 cm.

98. Find the total surface area of a cone whose radius is 4 m and whose slant height is 8 m.

99. Find the volume of a cone whose radius is 6 in. and whose height is 10 in.

100. If 1 gal = 231 in.^{3}, find the number of gallons that a conical oil container 18 in. high and 23 in. across can hold.

101. The entire exterior surface of a conical tank whose slant height is 12 ft and whose diameter is 18 ft will be painted. If the paint covers at a rate of 350 ft² per gallon, how many gallons of paint are needed for the job? Round any fraction of a gallon to the next whole gallon.

103. How many square feet of plastic material are needed to devise a wind-tunnel cone whose base is 6 ft across and whose height is 12 ft (Fig. 13–57). (*Hint:* To find the slant height, consider it to be the hypotenuse of a right triangle.)

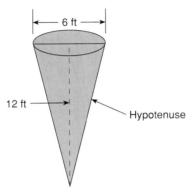

Figure 13–57

102. A hopper deposited sand in a cone-shaped pile with a diameter of 9′ 6″ and a height of 8′ 3″. To the nearest cubic foot, how much sand was deposited?

104. To the nearest square foot, how much nylon material is needed to construct a conical tent-like pavilion 30 ft in diameter and 5 ft from the apex to the base? (*Hint:* See Exercise 103).

105. A No. 5 soccer ball has a diameter of 8.8 in. How much leather is needed to cover the surface?

107. A spherical tank is anchored halfway in the ground. How many cubic feet of earth had to be excavated to install the tank? The tank measures 20 ft across. Round to the nearest whole cubic foot.

106. How much does a 4-in. lead ball weigh if lead weighs 1 lb per 2.4 in.³?

Section 13–5

108. Use the formula $R_t = \dfrac{R_1 R_2}{R_1 + R_2}$ to find the total resistance (R_t) if one resistance (R_1) is 10 ohms and the second resistance (R_2) is 9 ohms. Round to tenths.

110. Distance is rate times time, or $D = RT$. Find the rate if the distance traveled is 260 mi and the time traveled is 4 hr.

112. What is the percent efficiency (E) of an engine if the input (I) is 22,600 calories and the output (P) is 5600 calories? Use the formula $E = \dfrac{I - P}{I}$. Round to the nearest tenth of a percent.

109. The formula for voltage (Ohm's law) is $E = IR$. Find the amperes of current (I) if the voltage (E) is 220 V and the resistance (R) is 80 ohms.

111. According to Boyle's law, if temperature is constant, the volume of a gas is inversely proportional to the pressure on it. Find the final volume (V_2) of a gas using the formula $\dfrac{V_1}{V_2} = \dfrac{P_2}{P_1}$ if the original volume (V_1) is 30 ft³, the original pressure (P_1) is 75 lb per square inch (psi), and the final pressure (P_2) is 225 psi.

113. The formula for power (P) in watts (W) is $P = I^2 R$. Find the current (I) in amperes if a device draws 80 W and the resistance (R) is 8 ohms. Round to tenths.

114. Find the impedance (Z) in ohms using the formula $Z = \sqrt{R^2 + X^2}$ if the resistance (R) is 4 ohms and the reactance (X) is 7 ohms. Round to tenths.

115. The formula for the speed (S) of a driven pulley in revolutions per minute (rpm) is $S = \frac{DS}{d}$. Find the speed of a driven pulley with diameter (d) of 3 in. if the diameter (D) of the driving pulley is 7 in. and its speed (S) is 600 rpm.

116. If the distance (C) between the centers of the pulleys in Exercise 115 is 30 in., find the length (L) of the belt connecting them using the formula

$$L = 2C + 1.57(D + d) + \frac{D + d}{4C}$$

Round to hundredths.

117. Use the formula $H = \frac{D^2 N}{2.5}$ to find the number of cylinders (N) required in an engine of 5 hp (H) if the cylinder diameter (D) is 2.5 in.

CHALLENGE PROBLEMS

118. Devise your own formulas for the following relationships.
   (a) An electrical power company computes the monthly charges by multiplying the kilowatts of power used times the cost per kilowatt and adds to that a fixed monthly fee.
   (b) A store figures ending balance on a charge account by multiplying the interest rate times the previous unpaid balance and then adding the previous balance and purchases and subtracting payments.
   (c) Profit on the sale of a certain type of item is the product of the number of items sold and the difference between the selling price of the item and its cost to the seller.

119. Explain the usefulness of formula rearrangement in solving applied problems. Use at least one formula to illustrate.

CONCEPTS ANALYSIS

1. Briefly describe the procedure for evaluating a formula when numerical values are given for every variable in the formula except one.

2. Give some instances when it would be desirable to rearrange a formula.

3. Describe the similarities and differences in the formulas for the area of a parallelogram and a triangle.

4. Describe the similarities and differences in the formulas for the area of a parallelogram and a trapezoid.

5. State the Pythagorean theorem in words. Illustrate the Pythagorean theorem by drawing a right triangle, labeling its parts, and writing the theorem symbolically.

6. Make up a practical application that can be solved using the Pythagorean theorem.

Find the mistakes in the following problems. Explain the mistakes and rework the problem correctly.

7. In the formula $a = 3b - c$, find b if $a = 5$ and $c = 1$.

$$5 = 3b - 1$$
$$5 + 1 = 3b$$
$$6 = 3b$$
$$3 = b$$

8. Solve the following formula for y: $3x + 2y = 6$

$$3x = -2y + 6$$
$$x = \frac{-2y + 6}{3}$$
$$x = -\frac{2}{3}y + 2$$

9. In the right triangle ABC (Fig. 13–58), find b if $a = 8$ mm and $c = 17$ mm.

$$c^2 = a^2 + b^2$$
$$17^2 = 8^2 + b^2$$
$$289 = 64 + b^2$$
$$289 - 64 = b^2$$
$$\sqrt{225} = b$$
$$15 = b$$

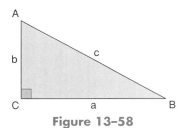

Figure 13–58

10. Solve the formula for force (F) if the diameter of the piston (D) is 3.5 in. and the pressure (P) is 125 lb per in.^{2} (psi).

$$F = \frac{PD^2}{1.27}$$

$$F = \frac{(125)(3.5)^2}{1.27}$$

$$F = \frac{437.5^2}{1.27}$$

$$F = \frac{191,406.25}{1.27}$$

$$F = 150,713.5827 \text{ lb}$$

CHAPTER SUMMARY

Objectives

What to Remember with Examples

Section 13–1

1 Evaluate formulas for a given variable.

Substitute values. Solve using rules for solving equations and/or the order of operations.

> Find the area (A) if the length (l) is 4 ft and the width (w) is 2 ft.
>
> $A = lw$
> $A = (4)(2)$
> $A = 8 \text{ ft}^2$

Sometimes the letter to solve for must be isolated.

> If the perimeter (P) of a square is 12 in., find the length of a side.
>
> $P = 4s$
> $12 = 4s$
> $\dfrac{12}{4} = \dfrac{4s}{4}$
> $s = 3 \text{ in.}$

Section 13–2

1 Rearrange formulas to solve for a given variable.

Isolate the desired variable so that it appears on the left. This often makes evaluation simpler. Apply all rules for solving equations.

> Solve the formula $S = \frac{R + P}{2}$ for R.
>
> $(2)S = \dfrac{R + P}{2}(2)$
> $2S = R + P$
> $2S - P = R$
> $R = 2S - P$

Section 13-3

1 Make Celsius/Kelvin temperature conversions.

Use formulas:
$K = °C + 273$

> Change 20°C to K.
>
> $K = °C + 273$
> $K = 20 + 273$
> $K = 293$

$°C = K - 273$

> Change 281 K to °C.
>
> $°C = K - 273$
> $°C = 281 - 273$
> $°C = 8$

2 Make Rankine/Fahrenheit temperature conversions.

Use formulas: $°R = °F + 460$

> Change 50°F to °R.
>
> $°R = °F + 460$
> $°R = 50 + 460$
> $°R = 510$

$°F = °R - 460$

> Change 750°R to °F.
>
> $°F = °R - 460$
> $°F = 750 - 460$
> $°F = 290$

3 Convert Fahrenheit to Celsius temperatures.

Use formula: $°C = \dfrac{5}{9}(°F - 32)$

> Change 14°F to °C.
>
> $°C = \dfrac{5}{9}(°F - 32)$
>
> $°C = \dfrac{5}{9}(14 - 32)$
>
> $°C = \dfrac{5}{9}(-18)$
>
> $°C = -10$

4 Convert Celsius to Fahrenheit temperatures.

Use formula: $°F = \dfrac{9}{5}°C + 32$

> Change 28°C to °F.
>
> $°F = \dfrac{9}{5}°C + 32$
>
> $°F = \dfrac{9}{5}(28) + 32$
>
> $°F = \dfrac{252}{5} + 32$
>
> $°F = 50.4 + 32$
>
> $°F = 82.4$

Section 13-4

1 Find the perimeter and area of a trapezoid.

Use formulas: $P = b_1 + b_2 + a + c$

> Find the perimeter (P) of a trapezoid whose bases are 10 cm and 8 cm and whose other sides (a, c) are each 7 cm.
>
> $P = b_1 + b_2 + a + c$
> $P = 10 + 8 + 7 + 7$
> $P = 32$ cm

$A = \frac{1}{2}h(b_1 + b_2)$

> Find the area of a trapezoid whose bases are 11 and 18 in. and whose height is 12 in.
>
> $A = \frac{1}{2}h(b_1 + b_2)$
>
> $A = \frac{1}{2}(12)(11 + 18)$
>
> $A = \frac{1}{2}(12)(29)$
>
> $A = 174$ in.^{2}

2 Find the perimeter and area of a triangle.

Use formulas: $P = a + b + c$

> What is the perimeter (P) of a triangular roof vent whose sides a, b, and c are 6 ft, 4 ft, and 4 ft?
>
> $P = a + b + c$
> $P = 6 + 4 + 4$
> $P = 14$ ft

$A = \frac{1}{2}(bh)$

> A triangle has a base of 3 m and a height of 2 m. Find its area.
>
> $A = \frac{1}{2}(bh)$
>
> $A = \frac{1}{2}(3)(2)$
>
> $A = 3$ m^2

3 Use the Pythagorean theorem to find the missing side of a right triangle.

The square of the hypotenuse (of a right triangle) equals the sum of the squares of the other two sides. $c^2 = a^2 + b^2$

> The hypotenuse (c) of a right triangle is 10 in. If side b is 6 in., find side a.
>
> $$c^2 = a^2 + b^2$$
> $$10^2 = a^2 + 6^2$$
> $$100 = a^2 + 36$$
> $$100 - 36 = a^2$$
> $$64 = a^2$$
> $$\sqrt{64} = a$$
> $$a = 8 \text{ in.}$$

4 Find the area of prisms and cylinders.

Use the formula for lateral surface area (area of sides): LSA = ph, where p is the perimeter of the base and h is the height.

> Find the lateral surface area of a triangular prism whose base is 4 in. on each side and whose height is 10 in.
>
> LSA = ph
> LSA = $(4 + 4 + 4)(10)$
> LSA = $(12)(10)$
> LSA = 120 in.^{2}

Use the formula for total surface area (sides plus bases): TSA = $ph + 2B$, where p is the perimeter of the base, h is the height, and B is the area of a base.

> Find the total surface area of the preceding triangle if the height of each base is 3.464 in.
>
> TSA = $ph + 2B$
>
> TSA = $120 + 2\left(\dfrac{1}{2}\right)(4)(3.464)$
>
> TSA = $120 + 13.856$
> TSA = 133.856 in.^{2}

5 Find the volume of prisms and cylinders.

Use formula: $V = Bh$, where B is the area of the base and h is the height.

> Find the volume of a cylinder whose diameter is 20 mm and whose height is 80 mm.
>
> $V = Bh$
> $V = \pi r^2 h$
>
> $V = \pi(10)^2(80)$ $\qquad r = \dfrac{1}{2}d = 10$ mm
>
> $V = \pi(100)(80)$
> $V = 25{,}132.74123$ mm^3

6 Find the area and volume of a sphere.

Use formula for total surface area: TSA = $4\pi r^2$, where r is the radius.

> What is the surface area of a sphere 6 in. in diameter?
>
> TSA = $4\pi r^2$
>
> TSA = $4\pi(3)^2$ $\qquad r = \dfrac{1}{2}d = 3$ in.
>
> TSA = $4\pi(9)$
> TSA = 113.10 in.^{2} rounded

Use formula for volume: $V = \dfrac{4\pi r^3}{3}$, where r is the radius.

> A spherical gas tank is 10 ft in diameter. Find its volume.
>
> $V = \dfrac{4\pi r^3}{3}$
>
> $V = \dfrac{4\pi(5)^3}{3}$ $\qquad r = \dfrac{1}{2}d = 5$ ft
>
> $V = \dfrac{4\pi(125)}{3}$
>
> $V = 523.60$ ft^3 rounded

7 Find the area and volume of a cone.

Use formula for lateral surface area: $LSA = \pi r s$, where r is the radius and s is the slant height.

> A conical pile of gravel has a diameter of 30 ft and a slant height of 40 ft. What is the lateral surface area?
>
> $LSA = \pi r s$
> $LSA = \pi(15)(40)$
> $LSA = 1884.96 \text{ ft}^2$ rounded

Use formula for total surface area: $TSA = \pi r s + \pi r^2$, where r is the radius and s is the slant height. πr^2 is the area of the base.

> Find the total surface area of the preceding cone.
>
> $TSA = \pi r s + \pi r^2$
> $TSA = 1884.955592 + \pi(15)^2$
> $TSA = 1884.955592 + \pi(225)$
> $TSA = 1884.955592 + 706.8583471$
> $TSA = 2591.81 \text{ ft}^3$ rounded

Use formula for volume: $V = \dfrac{\pi r^2 h}{3}$, where r is the radius and h is the height.

> Find the volume of a cone whose radius is 6 in. and whose height is 12 in.
>
> $V = \dfrac{\pi r^2 h}{3}$
> $V = \dfrac{\pi(6)^2(12)}{3}$
> $V = \dfrac{\pi(36)(12)}{3}$
> $V = 452.39 \text{ in.}^3$ rounded

Section 13-5

1 Evaluate miscellaneous technical formulas.

Apply all techniques for solving equations.

> The formula for the distance (d) an object falls is $d = \frac{1}{2}gt^2$. Find the distance if gravity (g) is 32 ft per second squared and seconds (t) is 3.
>
> $d = \dfrac{1}{2}gt^2$
> $d = \dfrac{1}{2}(32)(3)^2$
> $d = \dfrac{1}{2}(32)(9)$
> $d = 144 \text{ ft}$

WORDS TO KNOW

formula (p. 488)
evaluation (p. 488)
formula rearrangement (p. 492)
subscripts (p. 494–495)
Kelvin (p. 496)
Celsius (p. 496)
Rankine (p. 497)
Fahrenheit (p. 497)
geometry (p. 501)
trapezoid (p. 501)

parallel (p. 502)
base (p. 502)
triangle (p. 503)
right triangle (p. 503)
hypotenuse (p. 503)
leg (p. 503)
altitude (p. 503)
oblique triangle (p. 503)
Pythagorean theorem (p. 506)
prism (p. 508)

cylinder (p. 508)
height (p. 508)
lateral surface area (p. 509)
total surface area (p. 509)
volume (p. 510)
sphere (p. 511)
great circle (p. 511)
cone (p. 512)
slant height (p. 512)
horsepower (p. 520)

CHAPTER TRIAL TEST

1. The electrical resistance of a wire is found from the formula $R = \frac{PL}{A}$. Rearrange the formula to find the length L of the wire.

2. The formula for the volume (V) of a solid rectangular figure is $V = lwh$ (length × width × height). If the volume of a mailing container must be 7.5 cm^3 and its length 1.5 cm and width 0.5 cm, what must its height be?

3. Engine displacement d is found from the formula $d = \pi r^2 sn$. Solve to find r (the radius of the bore).

4. Using $d = 351$ in.^{3}, $s = 3.5$ in. stroke, and $n = 8$ cylinders, calculate to the nearest tenth the radius of the bore with the rearranged formula in Problem 3.

Use the following formulas for Problems 5 to 8:

$$K = °C + 273 \quad °C = K - 273 \quad °R = °F + 460 \quad °F = °R - 460 \quad °C = \frac{5}{9}(°F - 32) \quad °F = \frac{9}{5}°C + 32$$

5. 88°C = _____ K

6. 104°F = _____ °C

7. 195°C = _____ °F

8. 65°F = _____ °R

9. A section of a hip roof is a trapezoid measuring 35 ft at the bottom, 15 ft at the top, and 10 ft high. Find the area of this section of the roof in square feet.

10. The gable end of a roof is a triangle whose rise (height) is 7 ft and whose span (base) is 24 ft. The gable end contains a window 24 in. × 36 in. How much of the gable area will need to be painted?

11. A metal rod is welded to a metal support (Fig. 13–59). If the two sides of the metal support are 8 dm and 6 dm, find the length of the metal rod needed to make the brace.

12. Find the perimeter of the polygons shown in the following figures (Fig. 13–60).

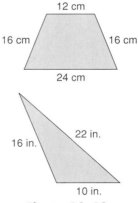

Figure 13–60

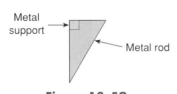

Figure 13–59

Solve the following problems.

13. A right pentagonal prism measures 1 in. on each side of its base and has a height of 10 in. What is its lateral surface area?

14. How much space is required to store 30 microcomputers crated in boxes that measure 62 cm by 70 cm by 50 cm?

15. A spherical tank 12 ft in diameter can hold how many gallons of fluid if 1 ft^3 = 7.48 gal? Answer to the nearest whole gallon.

16. The bases of a brass prism are equilateral triangles whose bases are 3 in. and whose heights are 2.6 in. If the prism's height is 8 in., what is the volume of the brass prism?

17. Hard coal broken into small pieces is dumped into a pile shaped like a cone. The base is 35 ft across and the pile stands 12 ft tall. How many cubic feet of coal are in the pile? Round to tenths.

18. A spherical gas storage tank 2.5 m wide needs to be sandblasted, primed, and refinished. The owner received an estimate of $8.50 per square meter. How much should the job cost the owner, to the nearest dollar?

19. A sheet metal worker wants to make a tin cone that will later be modified for use. If the base of the cone is to be 30 cm across and its height is 25 cm, what is the total surface area required to the nearest tenth? (*Hint:* Use your knowledge of a right triangle to find the slant height.)

20. A cold water pipe whose outside diameter is $\frac{7}{8}$ in. runs 23 ft in an attic. How many whole rolls of insulating wrap are needed if one roll covers $5\frac{1}{2}$ ft^2 and no allowance is made for overlap. (*Hint:* $\frac{7}{8}$ in. is $\frac{7}{8} \times \frac{1}{12} = 0.0729$ ft.)

21. A steel rod has a diameter of 20 in. Find the volume of steel in a 5-ft length of the rod in cubic feet to the nearest tenth.

22. Find the total surface area of a cylindrical storage tank if it is 30 ft tall and has a diameter of 12 ft. Round to hundredths.

23. Using the formula $P = \frac{1.27F}{D^2}$, calculate force F (in pounds) if the pressure P is 180 psi and the piston diameter is 3.25 in. Round to nearest hundredth.

24. Use the formula $R_t = \frac{R_1 R_2}{R_1 + R_2}$ to find the total resistance (R_t) if one resistance (R_1) is 9 ohms and the second resistance (R_2) is 8 ohms. Round to tenths.

25. If the efficiency (E) of an engine is 70% and the input (I) is 40,000 calories, find the output (P) in calories. Use the formula $E = \frac{I - P}{I}$.

CHAPTER

14

Products and Factors

In Chapter 14, we will study methods of factoring, like removing common factors and grouping, and even factoring of special products, like the difference of two perfect squares and perfect-square trinomials. When faced with one or more problems to solve, however, there are always the same questions: Do I have to factor? If I do, where do I start? Is there a way to select the most appropriate method of factoring for a given problem?

Your team should answer these questions. Some of your team members might use the library to examine several mathematics textbooks and, if possible, solutions manuals to discover the approaches of different authors. Other team members might discuss these questions with mathematics instructors at your school, at nearby colleges, and in local high schools. Then as a team, discuss your findings and use them to formulate answers to the three questions. In some cases, there may be more than one right answer.

Make an oral presentation to your class of the questions, how you researched the answers to the questions, and the answers themselves. You may want to experiment with overhead transparencies for this presentation. Be prepared to answer questions on your methods and solutions.

Throughout our study of mathematics, we have examined products and factors. To reduce fractions, we looked for factors common to both the numerator and denominator. In examining perfect squares, we looked for two identical factors to find the square root of a value. We looked for perfect-square factors of a radicand so that the radical could be written in a simpler form. In Chapter 14, we will again find it useful to examine products and factors.

14–1 THE DISTRIBUTIVE PROPERTY AND COMMON FACTORS

Learning Objective

1 Factor an expression containing a common factor.

Applying the distributive property and finding common factors have been introduced earlier in the text and were applied in different contexts. In this section, rather than using the distributive property to multiply and obtain a product, we will start with a product and regenerate the factors that produce the product. In other words, we want to undo the multiplication. Factoring will have some similarities to division, which is the inverse operation of multiplication.

1 Factor an Expression Containing a Common Factor.

The multiplication problem $7a(3a + 2)$ is the indicated product of $7a$ and the grouped quantity $3a + 2$. This form of the expression is called the *factored* form. After the expression is multiplied, we have two terms written as the indicated sum $21a^2 + 14a$. This is called the *expanded* form. To rewrite the expression $21a^2 + 14a$ as the indicated product $7a(3a + 2)$ is to *factor* it.

Let's look at a general example of the distributive property:

$$a(x + y) = ax + ay$$

Notice that a appears as a factor in both terms on the right side of the equal sign. When a factor appears in each of several terms, it is called a *common factor* of the terms. The distributive property in reverse can be used to write the addition as a multiplication. In other words, we can *factor* the expression.

EXAMPLE 1 Write $3a + 3b$ in factored form.

Because 3 is the common factor in both terms, we can use the distributive property to factor the expression.

$$3a + 3b = \mathbf{3}(\textbf{\textit{a}} + \textbf{\textit{b}})$$

The distributive property also applies if we have more than two terms.

EXAMPLE 2 Write $3ab + 9a + 12b$ in factored form.

Because 3 is the only factor that appears in all three terms of the expression, it is the common factor.

$$3ab + 9a + 12b = \mathbf{3}(\textbf{\textit{ab}} + \textbf{3\textit{a}} + \textbf{4\textit{b}})$$

$$3 \cdot 3 \qquad 3 \cdot 4$$

When looking for a common factor, we always look for *all* common factors.

EXAMPLE 3   Factor $10a^2 + 6a$ completely.

Because 2 and a are common factors, the greatest common factor is $2a$.

$$10a^2 + 6a = \mathbf{2a(5a + 3)} \qquad \frac{10a^2}{2a} = 5a, \frac{6a}{2a} = 3$$

We should always check our factoring by multiplying.

$$2a(5a + 3) = 10a^2 + 6a$$

This process of checking can be done mentally.

EXAMPLE 4   Factor $2x^2 + 4x^3$ completely.

Because 2 and x^2 are common factors, the greatest common factor is $2x^2$.

$$2x^2 + 4x^3 = \mathbf{2x^2(1 + 2x)} \qquad \left(\frac{2x^2}{2x^2} = 1, \frac{4x^3}{2x^2} = 2x \right)$$

Note that a 1 remains in the grouping because $2x^2$ is the product of $2x^2$ and 1.

Tips and Traps   ***When Is It Necessary to Write a 1?***

We have found that it is not always necessary to write the number 1. When is this the case? Because of the multiplicative property of 1, $1 \cdot n = n$; when 1 is a *factor*, the 1 does not have to be written. Because of the definition of the exponent of 1, $a^1 = a$; when the exponent of a factor is 1, the 1 does not necessarily have to be written. When 1 is a term instead of a factor, it must be written. Look at Example 4 again.

$$2x^2 + 4x^3 = 2x^2(1 + 2x)$$

Within the factor $(1 + 2x)$, there are two terms, 1 and $2x$. The term of 1 must be written. In Example 1, examine the factor $(a + b)$. This factor resulted from the division of $\frac{3a}{3}$ and $\frac{3b}{3}$. In each case the quotient was $1a$ and $1b$, respectively; but the 1's are factors of a term instead of separate terms. Thus, the factors of 1 do not have to be written.

SELF-STUDY EXERCISES 14–1

[1]   Factor completely. Check.

1. $7a + 7b$
2. $m^2 + 2m$
3. $6x^2 + 3x$
4. $5ab + 10a + 20b$
5. $4ax^2 + 6a^2x$
6. $5a - 7ab$
7. $12a^2 - 15a + 6$
8. $3x^3 - 9x^2 - 6x$
9. $8a^2b + 14ab^3$
10. $3m^2 - 6m^3$

Learning Objectives

1. Use the FOIL method to multiply two binomials.
2. Multiply polynomials to obtain three special products.

In this section, we will multiply an expression by another expression rather than by a group of factors. The basic concept involves using the distributive property more than once. Procedures that encourage doing some of the steps mentally will then be introduced.

1. Use the FOIL Method to Multiply Two Binomials.

As we expand our experiences with multiplication, we will again need to use the appropriate terminology associated with polynomials. Recall:

A *polynomial* is an expression with constants and variables with whole number exponents and contains one or more terms with at least one variable term.
A *monomial* is a polynomial that contains only one term, such as $5x^2$.
A *binomial* is a polynomial that contains only two terms, such as $3a + 4$.
A *trinomial* is a polynomial that contains only three terms, such as $4x^2 + x - 2$.

We can multiply two binomials by using the distributive property. According to the distributive property, each term of the first factor is multiplied by each term of the second factor. In the case where the first factor is a binomial, we use the distributive property more than once.

EXAMPLE 1 Multiply $(x + 4)(x + 2)$.

$$(x + 4)(x + 2) = x(x + 2) + 4(x + 2) \qquad \text{Distributive property}$$
$$= x^2 + 2x + 4x + 8 \qquad \text{Distributive property}$$
$$= x^2 + 6x + 8 \qquad \text{Combine like terms.}$$

FOIL Method

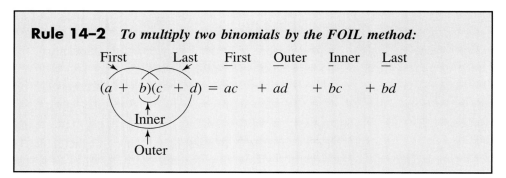

Rule 14–2 *To multiply two binomials by the FOIL method:*

$$\underbrace{\text{First}} \quad \underbrace{\text{Last}} \quad \underbrace{\text{First}} \quad \underbrace{\text{Outer}} \quad \underbrace{\text{Inner}} \quad \underbrace{\text{Last}}$$
$$(a + b)(c + d) = ac + ad + bc + bd$$
$$\text{Inner}$$
$$\text{Outer}$$

To help us remember a systematic way of accomplishing this multiplication, we can use the word **FOIL**, where F refers to the product of the *first* terms in each factor, O refers to the product of the two *outer* terms, I refers to the product of the *inner* terms, and L refers to the product of the *last* terms in each factor. If the inner and outer products are like terms, they should be combined.

EXAMPLE 2 Use the FOIL method to multiply $(2x - 3)(x + 1)$.

$$\qquad\qquad \underline{F} \quad \underline{O} \quad \underline{I} \quad \underline{L}$$
$$(2x - 3)(x + 1) = 2x^2 + 2x - 3x - 3$$
$$= 2x^2 - x - 3$$

2 Multiply Polynomials to Obtain Three Special Products.

When we multiply the sum of two terms by the difference of the same two terms, we can make some special observations about the product.

$$\underset{0}{(x + 3)(x - 3) = x^2 \underbrace{- 3x + 3x}_{} - 9 = x^2 - 9}$$

F   O   I   L

Examine the final product. Notice that the product has only *two* terms and is a *difference*. The sum of the outer and inner products is zero. Also notice that the terms of the product are *perfect squares*. Now compare the product with its factors. The first term of the product is the *square of the first term* in either of its factors. The second term in the product is the *square of the second term* in either of its factors.

We call pairs of factors like $(a + b)(a - b)$ the sum and difference of the same two terms or *conjugate pairs*. The product is called "the difference of two perfect squares."

Rule 14–3   *To mentally multiply the sum and difference of the same two terms:*

1. Square the first term.
2. Insert a minus sign.
3. Square the second term.

EXAMPLE 3   Find the products mentally.

(a) $(x + 2)(x - 2)$ (b) $(a - 4)(a + 4)$
(c) $(2m + 3)(2m - 3)$ (d) $(2a + y)(2a - y)$

Since each of these is the sum and difference of the same two terms, the product will be the difference of two perfect squares.

(a) $(x + 2)(x - 2) = x^2 - 4$

(b) $(a - 4)(a + 4) = a^2 - 16$

(c) $(2m + 3)(2m - 3) = 4m^2 - 9$

(d) $(2a + y)(2a - y) = 4a^2 - y^2$

When we multiply the same two binomials, we can also make some special observations. Because $x \cdot x = x^2$, $(x + 3)(x + 3) = (x + 3)^2$. The quantity $(x + 3)^2$ is called a *binomial square*.

EXAMPLE 4   Find the products of the binomial squares.

(a) $(x + 2)^2$ (b) $(x - 3)^2$ (c) $(2x - 5)^2$ (d) $(3a - 2b)^2$

To find the products, write each binomial square as two factors; then use the FOIL method to multiply.

(a) $(x + 2)^2 = (x + 2)(x + 2) = x^2 + 4x + 4$ $2x + 2x = 4x$

(b) $(x - 3)^2 = (x - 3)(x - 3) = x^2 - 6x + 9$ $-3x - 3x = -6x$

(c) $(2x - 5)^2 = (2x - 5)(2x - 5)$
$ = \mathbf{4x^2 - 20x + 25}$ $-10x - 10x = -20x$

(d) $(3a - 2b)^2 = (3a - 2b)(3a - 2b)$
$ = \mathbf{9a^2 - 12ab + 4b^2}$ $-6ab - 6ab = -12ab$

Notice that each product is a trinomial (three terms). These are special trinomials called *perfect square trinomials*. In each, the first and last terms are positive and perfect squares, and result from squaring the terms of the binomial. To get the middle term of the trinomial, we combine the outer and inner products. Because the outer and inner products are the same, the middle term can be found more easily by doubling the product of the two terms of the binomial. Let's apply these observations to an example.

EXAMPLE 5   Find the product of $(2x + 5)^2$.

$(2x + 5)^2 = 4x^2 + \underline{} + 25$ Square binomial terms.

$(\boxed{2x} + \boxed{5})^2 = \mathbf{4x^2 + 20x + 25}$ Double product of binomial terms.
 $2x(5) = 10x;\ 2(10x) = 20x$.

Rule 14–4   *To mentally square a binomial:*

1. Square the first term.
2. Double the product of the two terms.
3. Square the second term.

EXAMPLE 6   Square the following binomials using the steps given in Rule 14–4.

(a) $(x + 2)^2$ (b) $(2x - 3)^2$ (c) $(5x + 1)^2$ (d) $(7x + 2)^2$

	Square first term	Double product of terms	Square second term
(a) $(x + 2)^2 =$	$(x)^2$ x^2	$2 \cdot (2x)$ $+\ 4x$	$(2)^2$ $+\ 4$
(b) $(2x - 3)^2 =$	$(2x)^2$ $4x^2$	$2 \cdot (-6x)$ $-\ 12x$	$(-3)^2$ $+\ 9$
(c) $(5x + 1)^2 =$	$(5x)^2$ $25x^2$	$2 \cdot (5x)$ $+\ 10x$	$(1)^2$ $+\ 1$
(d) $(7x + 2)^2 =$	$(7x)^2$ $49x^2$	$2 \cdot (14x)$ $+\ 28x$	$(2)^2$ $+\ 4$

Now let's look at the special products of the following two polynomials:

$$(a + b)(a^2 - ab + b^2)$$

The binomial is the sum of two terms and the trinomial's first and last terms are the squares of the two terms of the binomial. The middle term is the product of the two terms of the binomial preceded by a negative sign.

$$(a - b)(a^2 + ab + b^2)$$

The binomial is the difference of two terms and the trinomial's first and last terms are the squares of the two terms of the binomial. The middle term is the product of the two terms of the binomial preceded by a positive sign.

Tips and Traps　*Recognizing Patterns:*

Special products are products that fit a specific pattern. If the product fits the pattern, then the product is predictable and can be done mentally. Let's look carefully at the pattern developed by the special products:

$$(a + b)(a^2 - ab + b^2) \qquad \text{and} \qquad (a - b)(a^2 + ab + b^2)$$

In addition to the first and last terms of the trinomial being the squares of the two terms of the binomial and the middle term being the product of the two terms, a key element in these two special cases is that, when the binomial is a *sum* ($+$), the middle term of the trinomial is *negative* ($-$), and when the binomial is a *difference* ($-$), the middle term of the trinomial is *positive* ($+$).

Observe the products when the binomial and trinomial are multiplied by an extension of the FOIL method or repeated applications of the distributive principle.

EXAMPLE 7　Multiply $(a + b)(a^2 - ab + b^2)$.

Multiply each term in the trinomial by each term in the binomial.

$$(a + b)(a^2 - ab + b^2) = a^3 \quad - a^2b + ab^2$$
$$+ a^2b - ab^2 \quad + b^3$$
$$\overline{a^3 \qquad\qquad\qquad + b^3}$$

First　Middle　Last

The four middle terms add to zero.

Thus, $(a + b)(a^2 - ab + b^2) = \boldsymbol{a^3 + b^3}$, the *sum of two perfect cubes.*

EXAMPLE 8　Multiply $(a - b)(a^2 + ab + b^2)$.

Multiply each term in the trinomial by each term in the binomial.

$$(a - b)(a^2 + ab + b^2) = a^3 \quad + a^2b + ab^2$$
$$- a^2b - ab^2 \quad - b^3$$
$$\overline{a^3 \qquad\qquad\qquad - b^3}$$

First　Middle　Last

The four middle terms add to zero.

Thus, $(a - b)(a^2 + ab + b^2) = \boldsymbol{a^3 - b^3}$, the *difference of two perfect cubes.*

Rule 14–5　*To mentally multiply a binomial and a trinomial of the types $(a + b)(a^2 - ab + b^2)$ and $(a - b)(a^2 + ab + b^2)$:*

1. Cube the first term of the binomial.
2. If the binomial is a sum, insert a plus sign. If it is a difference, insert a minus sign.
3. Cube the second term of the binomial.

EXAMPLE 9   Mentally multiply $(x - y)(x^2 + xy + y^2)$.

Do the factors match a pattern? Yes. The role of a is assumed by x and the role of b is assumed by y. Cube both terms of the binomial. Since the binomial factor is a difference, insert a minus sign between the cubes.

$$(x \boxed{-} y)(x^2 + xy + y^2) = \boldsymbol{x^3} \boxed{\boldsymbol{-}} \boldsymbol{y^3}$$

EXAMPLE 10   Mentally multiply $(m + p)(m^2 - mp + p^2)$.

Do the factors match a pattern? Yes. The role of a is assumed by m and the role of b is assumed by p. Cube both terms of the binomial. Since the binomial factor is a sum, insert a plus sign between the cubes.

$$(m \boxed{+} p)(m^2 - mp + p^2) = \boldsymbol{m^3} \boxed{\boldsymbol{+}} \boldsymbol{p^3}$$

EXAMPLE 11   Mentally multiply $(4 - d)(16 + 4d + d^2)$.

Do the factors match a pattern? Yes. The role of a is assumed by 4 because $4^2 = 16$. The role of b is assumed by d. 16 is the same as 4^2, so this problem is worked like Example 9. Cube the 4 and the d; insert a minus sign between the cubes because the binomial is a difference.

$$(4 \boxed{-} d)(16 + 4d + d^2) = \boldsymbol{64} \boxed{\boldsymbol{-}} \boldsymbol{d^3} \qquad 4^3 = 64$$

SELF-STUDY EXERCISES 14–2

[1] Use the FOIL method to find the products. Practice combining the outer and inner products mentally.

1. $(a + 3)(a + 8)$ **2.** $(x - 4)(x + 5)$ **3.** $(y - 7)(y - 3)$
4. $(2a - 3)(a + 4)$ **5.** $(3a - 2b)(a - 2b)$ **6.** $(5x - y)(c - 5y)$
7. $(3x - 4)(2x - 3)$ **8.** $(a - b)(2a - 5b)$ **9.** $(7 - m)(3 - 7m)$
10. $(5 - 2x)(8 - x)$ **11.** $(x + 7)(x + 4)$ **12.** $(y - 7)(y - 5)$
13. $(m + 3)(m - 7)$ **14.** $(3b - 2)(x + 6)$ **15.** $(4r - 5)(3r + 2)$
16. $(5 - x)(7 - 3x)$ **17.** $(4 - 2m)(1 - 3m)$ **18.** $(2 + 3x)(3 + 2x)$
19. $(x + 3)(2x - 5)$ **20.** $(5x - 7y)(4x + 3y)$ **21.** $(2a + 3b)(7a - b)$
22. $(5a + 2b)(6a - 5b)$ **23.** $(9x - 2y)(3x + 4y)$ **24.** $(5x - 8y)(4x - 3y)$
25. $(7m - 2n)(3m + 5n)$

[2] Find the special products mentally.

26. $(a + 3)(a - 3)$ **27.** $(2x + 3)(2x - 3)$ **28.** $(a - y)(a + y)$
29. $(4r + 5)(4r - 5)$ **30.** $(5x + 2)(5x - 2)$ **31.** $(7 + m)(7 - m)$
32. $(2 - 3x)^2$ **33.** $(3x + 4)^2$ **34.** $(Q + L)^2$
35. $(a^2 + 1)^2$ **36.** $(2d - 5)^2$ **37.** $(3a + 2x)^2$
38. $(3x - 7)(3x + 7)$ **39.** $(6 + Q)^2$ **40.** $(y - 5x)(y + 5x)$
41. $(4 - 3j)^2$ **42.** $(3m - 2p)(3m + 2p)$ **43.** $(m^2 + p^2)^2$
44. $(2a - 7c)(2a + 7c)$ **45.** $(9 - 13a)^2$

Use the extension of the FOIL method to find the products. Perform the work mentally.

46. $(x + p)(x^2 - xp + p^2)$ **47.** $(Q - L)(Q^2 + QL + L^2)$ **48.** $(3 + a)(9 - 3a + a^2)$
49. $(2x - 4p)(4x^2 + 8xp + 16p^2)$ **50.** $(3m - 2)(9m^2 + 6m + 4)$ **51.** $(6 + p)(36 - 6p + p^2)$
52. $(5y - p)(25y^2 + 5yp + p^2)$ **53.** $(x + 2y)(x^2 - 2xy + 4y^2)$ **54.** $(Q - 6)(Q^2 + 6Q + 36)$
55. $(3T + 2)(9T^2 - 6T + 4)$

Learning Objectives

1. Recognize and factor the difference of two perfect squares.
2. Recognize and factor a perfect-square trinomial.
3. Recognize and factor the sum or difference of two perfect cubes.

To factor any of the special products of the previous section requires us to apply the inverse of the process. Instead of multiplying a pair of factors to obtain the difference of two perfect squares or a perfect square trinomial or the sum or difference of two perfect cubes, we start with the product and "work back" to the factors that produce these special products.

To rewrite a special product in factored form requires us to recognize the product as a pattern. Once we identify the special product, then we must know the pattern of the product in factored form.

1   Recognize and Factor the Difference of Two Perfect Squares.

The difference of two perfect squares, such as $x^2 - a^2$, factors into the product of two binomials, $(x + a)(x - a)$. These factors are the sum and difference of the same two terms. The terms of each binomial are the square roots of the terms of the expression to be factored. Before we can factor such a special product, we must be able to recognize an expression as the difference of two perfect squares.

EXAMPLE 1   Identify the special products that are the difference of two perfect squares in the following examples.

    (a) $x^2 - 9$ (b) $a^2 + 49$ (c) $m^2 - 27$ (d) $3y^2 - 25$
    (e) $9x^2 - 4$ (f) $4x^2 - 4x + 1$

    (a) Difference of two perfect squares.
    (b) Not the difference of two perfect squares; this is a *sum*, not a difference.
    (c) Not the difference of two perfect squares; 27 is not a perfect square.
    (d) Not the difference of two perfect squares; in $3y^2$, 3 is not a perfect square.
    (e) Difference of two perfect squares.
    (f) Not the difference of two perfect squares; this is not a binomial.

Rule 14–6   *To factor the difference of two perfect squares:*

1. Take the square root of the first term.
2. Take the square root of the second term.
3. Write one factor as the *sum* of the square roots found in Steps 1 and 2, and write the other as the *difference* of the square roots found in Steps 1 and 2.

EXAMPLE 2   Factor the following special products, which are the differences of two perfect squares.

    (a) $a^2 - 9$ (b) $x^2 - 36$ (c) $4x^2 - 1$ (d) $16m^2 - 49$

    (a) $a^2 - 9 = (a + 3)(a - 3)$ (b) $x^2 - 36 = (x + 6)(x - 6)$
    (c) $4x^2 - 1 = (2x + 1)(2x - 1)$ (d) $16m^2 - 49 = (4m + 7)(4m - 7)$

Order of Factors:
Because multiplication is commutative, the factors given as answers for Example 2 may be expressed in any order, such as $(a + 3)(a - 3)$ or $(a - 3)$ $(a + 3)$, $(x + 6)(x - 6)$ or $(x - 6)(x + 6)$, and so on.

2 Recognize and Factor a Perfect-Square Trinomial.

A trinomial is a *perfect-square trinomial* if the first and last terms are positive perfect squares and the absolute value of the middle term is *twice* the product of the square roots of the first and last terms. Again, we need to be able to distinguish these special products from other expressions before we can factor them.

EXAMPLE 3 Tell why the following trinomials are *not* perfect square trinomials.

(a) $x^2 + 2x - 1$ (b) $4x^2 + 6x + 9$
(c) $x^2 - 5x + 4$ (d) $-4x^2 - 4x + 1$

(a) The last term, -1, is negative. This term must be positive in a perfect square trinomial.
(b) The middle term, $6x$, is not twice the product of $2x$ and 3.
(c) The middle term, $-5x$, is not twice the product of x and 2.
(d) The first term, $-4x^2$, is negative. It should be positive.

EXAMPLE 4 Verify that the following trinomials are perfect square trinomials.

(a) $x^2 + 14x + 49$ (b) $4m^2 - 12m + 9$ (c) $9x^2 + 24xy + 16y^2$

(a) The first and last terms, x^2 and 49, are positive perfect squares. The middle term, $14x$, has an absolute value of twice the product of the square roots of x^2 and 49. $2(7x) = 14x$.
(b) The first and last terms, $4m^2$ and 9, are positive perfect squares. The middle term, $-12m$, has an absolute value of twice the product of the square roots of $4m^2$ and 9. $2(2m \cdot 3) = 12m$.
(c) The first and last terms, $9x^2$ and $16y^2$, are positive perfect squares. The middle term, $24xy$, has an absolute value of twice the product of the square roots of $9x^2$ and $16y^2$. $2(3x \cdot 4y) = 24xy$.

Rule 14–7 *To factor a perfect-square trinomial:*

1. Write the square root of the first term.
2. Write the sign of the middle term.
3. Write the square root of the last term.
4. Indicate the square of this binomial quantity.

EXAMPLE 5 Factor the perfect-square trinomials in Example 4.

(a) $x^2 + 14x + 49$

Square root of *first* term	Sign of *middle* term	Square root of *last* term	Indicate *square* of this quantity
x	$+$	7	$(x + 7)^2$

(b) $4m^2 - 12m + 9$ (c) $9x^2 + 24xy + 16y^2$

$(2m - \qquad 3)^2$ $(3x + \qquad 4y)^2$

3 Recognize and Factor the Sum or Difference of Two Perfect Cubes.

The sum or difference of two *perfect cubes*, such as $x^3 + y^3$ or $a^3 - b^3$, factors into the product of a binomial and a trinomial, $(x + y)(x^2 - xy + y^2)$ or $(a - b)(a^2 + ab + b^2)$. Before we can factor these special products, we must be able to recognize an expression as the sum or difference of two perfect cubes.

EXAMPLE 6 Identify the special products that are the sum or difference of two perfect cubes in the following examples.

(a) $a^3 - 27$ (b) $8x^3 + 9$ (c) $8y^3 - 125$ (d) $64 - 8a^3$
(e) $a^3 + 9b^3$ (f) $c^3 + y^3 - 27$ (g) $27x^3 + 8$ (h) $3a^3 - 64$

(a) Difference of two perfect cubes.
(b) Not the sum or difference of two perfect cubes; 9 is not a perfect cube.
(c) Difference of two perfect cubes.
(d) Difference of two perfect cubes.
(e) Not the sum or difference of two perfect cubes; $9b^3$ is not a perfect cube.
(f) Not the sum or difference of two perfect cubes; this is a trinomial.
(g) Sum of two perfect cubes.
(h) Not the sum or difference of two perfect cubes; $3a^3$ is not a perfect cube.

Rule 14–8 *To factor the sum of two perfect cubes:*

1. Write the binomial factor as the *sum* of the cube roots of the two terms.
2. Write the trinomial factor as the square of the first term from Step 1, *minus* the product of the two terms from Step 1, plus the square of the second term from Step 1.

$$a^3 + b^3 = (a + b)(a^2 - ab + b^2)$$

Rule 14–9 *To factor the difference of two perfect cubes:*

1. Write the binomial factor as the *difference* of the cube roots of the two terms.
2. Write the trinomial factor as the square of the first term from Step 1, *plus* the product of the two terms from Step 1, plus the square of the second term from Step 1.

$$a^3 - b^3 = (a - b)(a^2 + ab + b^2)$$

EXAMPLE 7 Factor the following special products, which are the sum or difference of two perfect cubes.

(a) $27a^3 - 8$ (b) $m^3 + 125n^3$

　　　　　　　　　　Minus sign　　　　Plus sign

(a) $27a^3 - 8 = \qquad (3a - 2)(9a^2 + 6a + 4)$

　　　　　　　Cube roots　　Square　　Product　　Square
　　　　　　of $27a^3$ and 8　　of $3a$　　of $(3a)(2)$　　of 2

　　　　　　　　　　Plus sign　　　　Minus sign

(b) $m^3 + 125n^3 = \qquad (m + 5n)(m^2 - 5mn + 25n^2)$

　　　　　　　Cube roots　　Square　　Product of　　Square
　　　　　　of m^3 and $125n^3$　　of m　　$m(5n)$　　of $5n$

1   Identify the special products that are the difference of two perfect squares.

1. $r^2 - s^2$ **2.** $d^2 - 4d + 10$ **3.** $4y^2 - 16$
4. $36m^2 - 9n^2$ **5.** $9 + 4a^2$ **6.** $64p^2 - q^2$

Factor the following special products.

7. $y^2 - 49$ **8.** $16x^2 - 1$ **9.** $9a^2 - 100$
10. $4m^2 - 81n^2$ **11.** $9x^2 - 64y^2$ **12.** $25x^2 - 64$
13. $100 - 49x^2$ **14.** $4x^2 - 49y^2$ **15.** $121m^2 - 49n^2$
16. $81x^2 - 169$

2   Verify which of the following trinomials are perfect-square trinomials.

17. $4y^2 + 2y + 16$ **18.** $9m^2 - 24mn + 16n^2$ **19.** $16a^2 + 8a - 1$
20. $-9r^2 + 12r + 4$ **21.** $y^2 - 14y + 49$ **22.** $p^2 + 10p + 25$

Factor the following special products.

23. $x^2 + 6x + 9$ **24.** $x^2 + 14x + 49$ **25.** $x^2 - 12x + 36$
26. $x^2 - 16x + 64$ **27.** $4a^2 + 4a + 1$ **28.** $25x^2 - 10x + 1$
29. $9m^2 - 48m + 64$ **30.** $4x^2 - 36x + 81$ **31.** $x^2 - 12xy + 36y^2$
32. $4a^2 - 20ab + 25b^2$

3   Identify the special products that are the sum or difference of two perfect cubes.

33. $8b^3 - 125$ **34.** $y^2 - 14y + 49$ **35.** $T^3 + 27$
36. $c^3 + 16$ **37.** $125x^3 - 8y^3$ **38.** $64 + a^3$

Factor the following special products, which are the sum or difference of two perfect cubes.

39. $8m^3 - 343$ **40.** $y^3 - 125$ **41.** $Q^3 + 27$
42. $c^3 + 1$ **43.** $125d^3 - 8p^3$ **44.** $a^3 + 64$
45. $216a^3 - b^3$ **46.** $x^3 + Q^3$ **47.** $8p^3 - 125$
48. $27 - 8y^3$

14–4   FACTORING GENERAL TRINOMIALS

Learning Objectives

1   Factor general trinomials whose squared term has a coefficient of 1.

2   Remove common factors after grouping an expression.

3   Factor a general trinomial by grouping.

4   Factor a general trinomial by trial and error.

5   Factor any binomial or trinomial that is not prime.

We often have expressions that do not fit the special cases described in Section 14–3. We will use factoring as one method for solving quadratic equations (Chapter 15), for simplifying algebraic fractions (Chapter 16), and for solving quadratic inequalities (Chapter 17). Trinomials that are not perfect-square trinomials are called *general trinomials*.

1   Factor General Trinomials Whose Squared Term has a Coefficient of 1.

Recall the FOIL method of multiplying two binomials, $(x + 3)(x + 2)$. x^2 is the product of the <u>F</u>irst terms of each binomial. $2x + 3x = 5x$ is the sum of the products of the <u>O</u>uter and <u>I</u>nner terms. $+6$ is the product of the <u>L</u>ast terms of each binomial.

The trinomial is $x^2 + 5x + 6$. If we wish to factor this trinomial, we begin by indicating that it factors into two binomials.

$$(\quad)(\quad)$$

If the sign of the *third* term is positive, the signs of the two binomial factors will be alike. The sign of the middle term will tell us which sign to use. In the trinomial $x^2 + 5x + 6$, the sign of the 6 is positive, so the signs will be alike. Because the sign of the $5x$ is positive, the signs of both binomial factors will be positive.

$$(\quad+\quad)(\quad+\quad)$$

Next, we write the factors of the *first* term in the *first* position of each binomial.

$$(x+\quad)(x+\quad)$$

The factors of x^2 are x and x.

Then we write the factors of the *last* term in the *last* position of each binomial. Although we have several pairs whose product is $+6$ ($+3$ and $+2$; -3 and -2; $+6$ and $+1$; -6 and -1), *we choose the pair that will give us the correct middle term, including its sign when we multiply the two binomials together*. The positive sign on the third term indicates that we need two factors whose *sum* is $+5$. So we choose $+3$ and $+2$.

$(x + 3)(x + 2) = x^2 + 5x + 6$ Only this choice gives the correct middle term, $+5x$.

$$\left.\begin{array}{l} (x - 3)(x - 2) = x^2 - 5x + 6 \\ (x + 6)(x + 1) = x^2 + 7x + 6 \\ (x - 6)(x - 1) = x^2 - 7x + 6 \end{array}\right\}$$ These choices give the wrong sign and/or the wrong numerical value for the middle term.

Therefore, the factors of the trinomial $x^2 + 5x + 6$ are $(x + 3)(x + 2)$. Because multiplication is commutative, the two factors may be written in the opposite order, $(x + 2)(x + 3)$.

Tips and Traps   *Start with the Signs:*
In the examples that follow, pay particular attention to the signs of the middle and last terms of the trinomials and their effect on the signs of the binomial factors.

One procedure for determining the factors that will produce a trinomial product is to *factor by inspection*. This means that after examining the problem, we logically deduce the appropriate solution.

EXAMPLE 1   Factor $x^2 + 6x + 5$ by inspection.

$(\quad)(\quad)$ Write parentheses for two binomial factors.

$(x+\quad)(x+\quad)$ The positive sign on $+5$ indicates like signs, and the positive sign on the $+6x$ indicates both signs are positive. Write the factors of x^2 as the first term of each binomial.

$(x + 5)(x + 1)$ Write the factors of $+5$, ($+5$ and $+1$), as the last term of each binomial. Note: $+5$ of the trinomial has two sets of factors: $+5$, $+1$ and -5, -1. We choose $+5$ and $+1$ because the sign of the middle term is positive.

$(x + 5)(x + 1) = x^2 + 6x + 5$ Multiply the binomials to check factoring.

The correct factoring of $x^2 + 6x + 5$ is $(x + 5)(x + 1)$.

EXAMPLE 2   Factor $x^2 - 4x + 3$ by inspection.

( )( )   Indicate binomial factors.

$(x - \quad)(x - \quad)$ The positive sign on +3 indicates that both signs are the same; the negative sign on $-4x$ indicates that we should use negative signs. Write the factors of x^2 as the first term of each binomial.

$(x - 1)(x - 3)$ Write the factors of +3, (−1 and −3), as the last terms. Note: +3 of the trinomial has two sets of factors: −1, −3 and +1, +3. We chose −1 and −3 because the sign of the middle term is negative.

$(x - 1)(x - 3) = x^2 - 4x + 3$ Check by multiplying.

The correct factoring of $x^2 - 4x + 3$ is $(x - 1)(x - 3)$.

EXAMPLE 3   Factor $x^2 - 2x - 8$ by inspection.

( )( )   Indicate binomial factors.

$(x - \quad)(x + \quad)$ Write the factors of x^2. The negative sign of −8 indicates that the signs will be different.

$(x - 2)(x + 4)$
$(x - 4)(x + 2)$ To find the factors of −8 that will give a middle term of $-2x$, find the two factors of 8 that have a difference of 2. Then use the negative sign with the larger factor.
$(x - 8)(x + 1)$
$(x - 1)(x + 8)$

$(x - 4)(x + 2) = x^2 - 2x - 8$

The correct factoring of $x^2 - 2x - 8$ is $(x - 4)(x + 2)$.

When trinomials have terms whose coefficients have several pairs of factors, such as in the trinomial $12a^2 + 28a + 15$, more cases have to be considered. Two methods of factoring general trinomials are common. The one method is systematic, while the other requires inspection or trial and error. Before we use the systematic method, we need to learn to factor by grouping.

2 Remove Common Factors After Grouping an Expression.

Some algebraic expressions that have no factors common to every term in the expression may have common factors for some groups of terms. Some of these cases will allow the expression to be written in factored form. In the expression $2x^2 - 2xb + ax - ab$, there are four terms and there is no factor common to each term. To factor, write the expression as two terms by *grouping* the first two terms and the last two terms; then look for common factors in each grouping.

$$(2x^2 - 2xb) + (ax - ab)$$

The first grouping $(2x^2 - 2xb)$ has a common factor of $2x$ and can be written in factored form as $2x(x - b)$. The second grouping $(ax - ab)$ has a common factor of a and can be written in factored form as $a(x - b)$. If we look at the entire expression $2x(x - b) + a(x - b)$, we see the common binomial factor $(x - b)$. When we factor this common factor from both terms, we have $(x - b)(2x + a)$. The expression is now written as a product of two factors. We can check the factoring by using the distributive property or the FOIL method of multiplying.

$$(x - b)(2x + a) = 2x^2 + ax - 2xb - ab$$

This result is the same as the original expression.

EXAMPLE 4 Write the following expressions in factored form by using grouping.

(a) $mx + 2m - 4x - 8$ (b) $y^2 + 2xy + 3y + 6x$
(c) $3x^2 - 9x - 7x + 21$ (d) $2x^2 + 8x + 5y - 15$

(a) $mx + 2m - 4x - 8 =$ | Group the four-termed expression into two terms.

$(mx + 2m) + (-4x - 8)$ | Factor out any common factor in each of the two terms.

$m(x + 2)\ +\ -4(x + 2) =$ | Convert double signs to an equivalent single sign.

$m\,(x + 2)\ -\ 4\,(x + 2)$ | Factor out the common factor $(x + 2)$.

$(x + 2)\,(m - 4)$ | Check the result by using the FOIL method.

$(x + 2)(m - 4) = xm - 4x + 2m - 8$
$\qquad\qquad\quad = mx + 2m - 4x - 8$ | Rearrange terms and factors because of the commutative properties of multiplication and addition.

(b) $\qquad y^2 + 2xy + 3y + 6x =$

$(y^2 + 2xy) + (3y + 6x) =$ | Group into two terms.

$y\,(y + 2x)\ +\ 3\,(y + 2x) =$ | Factor out the common factor in each term.

$(y + 2x)\,(y + 3)$ | Factor into one term by factoring out the common binomial factor.

(c) $\quad 3x^2 - 9x - 7x + 21$

$(3x^2 - 9x) + (-7x + 21)$ | Group into two terms.

$3x(x - 3)\ +\ -7(x - 3)$ | Factor each term.

$3x\,(x - 3)\ -\ 7\,(x - 3)$ | Convert double signs to a single sign.

$(x - 3)\,(3x - 7)$ | Factor into one term by factoring out the common binomial factor.

(d) $2x^2 + 8x + 5y - 15$

$(2x^2 + 8x) + (5y - 15)$ | Group into two terms.

$2x(x + 4) + 5(y - 3)$ | Factor each term.

In this example the two terms do not have a common factor. Thus, the expression cannot be factored into one term. Even if we rearrange the terms, we will not be able to write the expression as a single term.

$$2x^2 + 8x + 5y - 15 \text{ is prime.}$$

Tips and Traps *Make Leading Coefficient of Binomial Positive:*
When factoring by grouping, manipulate the signs of the common factor so that the leading coefficient of the binomial is positive.

$$-3x + 9$$

Factor as $-3(x - 3)$, *not* $3(-x + 3)$.

3 **Factor a General Trinomial by Grouping.**

We can now use a systematic method of factoring by grouping to factor general trinomials. First, we will factor trinomials of the type $ax^2 + bx + c$.

> **Rule 14–10** *To factor a general trinomial by grouping:*
>
> 1. Multiply the coefficient of the first term by the coefficient of the last term.
> 2. Factor the product from Step 1 into two factors:
>   (a) whose *sum* is the coefficient of the middle term if the sign of the last term is positive, or
>   (b) whose *difference* is the coefficient of the middle term if the sign of the last term is negative.
> 3. Rewrite the trinomial so that the middle term is the sum or difference from Step 2.
> 4. Divide the trinomial from Step 3 into two groups with a common factor in each.
> 5. Factor out the common factor.

We learned earlier that trinomials with a positive third term will factor into two binomials whose signs are alike.

EXAMPLE 5   Factor $6x^2 + 19x + 10$ by grouping.

$6(10) = 60$ Multiply the coefficients of the first and third terms.

$60 = (1)60$ List all factor pairs of 60.

     $(2)30$

     $(3)20$

     $(4)15$ Identify the pair that adds to 19, since the sign of the last term is positive.

     $(5)12$

     $(6)10$

$6x^2 + 19x + 10$ Separate $+19x$ into two terms using the coefficients 4 and 15.

$6x^2 + 4x + 15x + 10$ Group into two terms.

$(6x^2 + 4x) + (15x + 10)$ Factor each term.

$2x(3x + 2) + 5(3x + 2)$ Factor common binomial.

$(3x + 2)(2x + 5)$ Check using the FOIL method.

EXAMPLE 6   Factor $20x^2 - 23x + 6$ by grouping.

$20(6) = 120$ Find the product of 20 and 6.

List the factor pairs of 120 and select the pair that has a sum of 23.

$120 = (1)120$ List all factor pairs of 120.

     $(2)60$

     $(3)40$

     $(4)30$

     $(5)24$

     $(6)20$

     $(8)15$ Identify the pair that adds to 23, since the sign of the last term is positive.

     $(10)12$

$$20x^2 - 23x + 6$$

Separate $-23x$ into two terms using the coefficients of -8 and -15.

$$20x^2 - 8x - 15x + 6$$

Group into two terms.

$$(20x^2 - 8x) + (-15x + 6)$$

Factor each term.

$$4x(5x - 2) - 3(5x - 2)$$

Factor the common binomial.

$$(5x - 2)(4x - 3)$$

Check using the FOIL method.

If the third term of the trinomial has a negative sign, we use the same procedure but look for two factors whose *difference* is the coefficient of the middle term.

EXAMPLE 7   Factor $10x^2 + 19x - 15$ by grouping.

$$10(15) = 150$$ Find the product of 10 and 15.

Systematically factor 150 to find two factors whose difference is 19.

$150 = (1)150$ List all factor pairs of 150.

$(2)75$

$(3)50$

$(5)30$

$(6)25$ Identify the pair that subtracts to 19, since the last term is negative.

$(10)15$

The factors 25 and 6 have a difference of 19. When we rewrite the trinomial, we rewrite $19x$ as $25x - 6x$.

$$10x^2 + 19x - 15$$

Separate $+19x$ into two terms using the coefficients -6 and 25. The signs will be different and the larger factor will be positive.

$$10x^2 + 25x - 6x - 15$$

Group into two terms.

$$(10x^2 + 25x) + (-6x - 15)$$

Factor each term.

$$5x(2x + 5) - 3(2x + 5)$$

Factor the common binomial.

$$(2x + 5)(5x - 3)$$

Check using the FOIL method.

Now, let's factor by grouping a trinomial whose middle term and end term are negative. Because the last term is negative, we will look for two factors whose *difference* is the coefficient of the middle term.

EXAMPLE 8   Factor $x^2 - 2x - 8$ by grouping.

$$1(8) = 8$$ Find the product of 1 and 8.

Factor 8 so that the difference of the absolute values of the factors is 2.

$8 = (1)8$ List all factor pairs of 8.

$(2)4$ Identify the pair that subtracts to 2.

$$x^2 - 2x - 8$$

Separate $-2x$ into two terms using the coefficients 2 and 4. The signs will be different. The larger factor will be negative.

$$x^2 + 2x - 4x - 8$$

Group into two terms.

$$(x^2 - 4x) + (2x - 8)$$

Factor each term.

$$x(x - 4) + 2(x - 4)$$

Factor the common binomial.

$$(x - 4)(x + 2)$$

Check using the FOIL method.

Tips and Traps *Factor by Grouping:*

The following is a variation of the factor-by-grouping method. The variation is mathematically sound and employs a strategy using the property of 1 that is often overlooked.

Factor $6x^2 + 7x - 20$.

$6x^2 + 7x - 20$ Multiply the coefficients of the first and third terms: 6(−20) = −120.

$120 =$	1	120	List all factor pairs of 120.
	2	60	
	3	40	
	4	30	
	5	24	
	6	20	
	8	15	Identify the factor pair whose algebraic sum is +7. The larger factor will be positive. −8 + 15 = 7
	10	12	

Variation in procedure begins here.

$$\frac{(6x\quad)(6x\quad)}{6}$$ Use the coefficient of the first term as the coefficient of the first term in each binomial. This gives us an extra factor of 6, and we will compensate by dividing the expression by 6.

$$\frac{(6x - 8)(6x + 15)}{6}$$ Use the factors of −120 whose algebraic sum is 7 as the second term of each binomial.

$$\frac{2(3x - 4)(3)(2x + 5)}{6}$$ Factor the common factors from each binomial.

$$\frac{6(3x - 4)(2x + 5)}{6}$$ The factors of the trinomial will be the two binomial factors. The factor 6 reduces.

$(3x - 4)(2x + 5)$ Check using the FOIL method.

Why does this variation work? The coefficient of the first term in the original trinomial, 6, was used in each binomial. Thus, the resulting product would have an extra factor of 6. This extra factor of 6 must be removed to obtain the correct factors. Common factors of the original trinomial will be part of the final answer.

Factor $30x^2 + 8 - 32x$.

$30x^2 - 32x + 8$ Arrange in descending powers of x.

$2(15x^2 - 16x + 4)$ Factor any common factors.

$60 =$	1	60	Find the factors of 60 whose algebraic sum is −16.
	2	30	
	3	20	
	4	15	
	5	12	
	6	10	−6 + (−10) = −16

$$\dfrac{2(\,15\,x\qquad)(\,15\,x\qquad)}{15}$$	Keep the common factor, 2, as a factor. Use the coefficient of the first term as the coefficient of each binomial. An extra factor of 15 is produced, and we will compensate by dividing by 15.
$$\dfrac{2(\,15\,x\ -\ 6\,)(\,15\,x\ -\ 10\,)}{15}$$	Use the factors of 60 whose algebraic sum is -16 as the second term in each binomial. Remember, at this point an extra factor of 15 is still present.
$$\dfrac{2(\,\cancel{3}\,)(5x\ -\ 2)(\,\cancel{5}\,)(3x\ -\ 2)}{\cancel{15}}$$	Factor the common factors from each binomial.
$$2(5x\ -\ 2)(3x\ -\ 2)$$	Reduce extra factor of 15: 3(5). Common factor of 2 remains in answer.

4 Factor a General Trinomial by Trial and Error.

Another method of factoring is commonly referred to as *trial and error*. The name of this method might be misleading. The procedure involves mentally using the FOIL method for all possible binomial pairs that give the correct first and last terms until you find the one that gives the correct middle term. The binomial pairs that do not produce the correct middle term are not actually errors. They are logical choices that did not produce the desired result. With experience using this method, the correct choice is often found by testing other choices mentally and only writing the correct choice.

Rule 14–11 *To factor any general trinomial by trial and error:*

1. Indicate binomial factors with double parentheses ( )( ).
2. List all possible factors of the first and last terms of the trinomial.
3. List all possible binomial factors using factors from Step 2.
4. Multiply binomial factors from Step 3 to find the pair that gives the correct middle term of the trinomial.
5. Check all work to be sure that the signs are correct.

To factor the trinomial $2a^2 + 5a + 3$ by trial and error, we place the factors of the first term, $2a^2$, in the first position of each binomial factor.

$$(2a\qquad)(a\qquad)$$

We place the factors of the last term, 3, in the second position of each binomial factor.

$$(2a\qquad 3)(a\qquad 1)$$

$$\text{or}$$

$$(2a\qquad 1)(a\qquad 3)$$

We have to consider *both* possible arrangements of the factors. The signs of 3 and 1 are both positive because all signs in the trinomial are positive.

$$(2a\ +\ 3)(a\ +\ 1)$$

$$\text{or}$$

$$(2a\ +\ 1)(a\ +\ 3)$$

To determine the correct factoring, we multiply each product to see which yields $2a^2 + 5a + 3$.

$$(2a\ +\ 3)(a\ +\ 1) = 2a^2\ +\ 2a\ +\ 3a\ +\ 3$$

$$= 2a^2\ +\ 5a\ +\ 3$$

$$(2a + 1)(a + 3) = 2a^2 + \boxed{6a + a} + 3$$
$$= 2a^2 + \boxed{7a} + 3$$

The first factoring, $(2a + 3)(a + 1)$, is correct.

Tips and Traps   ***Shortening the Trial and Error Method:***
To make our work easier and less time consuming, we should learn to obtain the outer and inner products and add them *mentally*.

EXAMPLE 9   Factor $6x^2 + 7x - 5$ by the trial and error method.

$6x$ and x; $3x$ and $2x$ Choices for the first terms of the binomials.

-5 and 1; 5 and -1 Choices for the second terms of the binomials.

Possible binomial factors	(Middle term)	Observations
$(6x + 5)(x - 1)$	$-x$	Since the sign of the last term, -5, is negative, we must use unlike signs for the factors.
$(6x - 5)(x + 1)$	$+x$	
$(6x + 1)(x - 5)$	$-29x$	
$(6x - 1)(x + 5)$	$+29x$	
$(3x + 5)(2x - 1)$	$+7x$	
$(3x - 5)(2x + 1)$	$-7x$	
$(3x + 1)(2x - 5)$	$-13x$	
$(3x - 1)(2x + 5)$	$+13x$	

The fifth pair of factors, $(3x + 5)(2x - 1)$, gives the correct middle term, $+7x$. **Thus, $(3x + 5)(2x - 1)$ is the correct factoring of $6x^2 + 7x - 5$.**

5   Factor any Binomial or Trinomial That is Not Prime.

Now we will develop an organized strategy for factoring any binomial or trinomial that can be factored. This strategy will also allow us to say with confidence that a particular binomial or trinomial will not factor or is prime. We used the word *prime* to identify positive numbers that have no factors except themselves and 1. When the word *prime* refers to algebraic expressions, it describes algebraic expressions that have only themselves and 1 as factors.

Rule 14–12   *To factor any binomial or trinomial:*

Perform the following steps in order.

1. First, factor out the greatest common factor (if any) using Rule 14–1.
2. Check the binomial or trinomial to see if it is a special product.
   (a) If it is the difference of two perfect squares, use Rule 14–6.
   (b) If it is a perfect-square trinomial, use Rule 14–7.
   (c) If it is the sum or difference of two perfect cubes, use Rule 14–8 or Rule 14–9.
3. If there is no special product, use Rule 14–10 to factor by grouping or use Rule 14–11 for "trial-and-error" factoring of a general trinomial.
4. Examine each factor to see that it cannot be factored further.
5. Check factoring by multiplying.

EXAMPLE 10   Completely factor $4x^3 - 2x^2 - 6x$.

$2x(2x^2 - x - 3)$ Look for a common factor: 2x is the common factor.

$2x(2x - 3)(x + 1)$ Factor the trinomial but *keep* the 2x factor also. Check by multiplying.

EXAMPLE 11   Completely factor $12x^2 - 27$.

$3(4x^2 - 9)$ Look for a common factor: 3 is the common factor.

$3(2x - 3)(2x + 3)$ Factor the difference of two perfect squares. Keep the factor of 3. Check.

EXAMPLE 12   Completely factor $-18x^2 + 24x^2 - 8x$.

$-2x(9x^2 - 12x + 4)$ Look for a common factor: $-2x$. We often factor a negative when there are more negative terms than positive terms or when the term with the highest degree is negative. It is preferred that the leading coefficient in the binomial or trinomial be positive.

$-2x(3x - 2)^2$ Factor the perfect-square trinomial.

Tips and Traps   ***Why Do We Factor?***

There is general disagreement on the importance of factoring polynomials.

For factoring:

• Many problem-solving techniques that use factoring can be done more quickly than with the formula-based or systematic technique.
• Properties and patterns are generally easier to recognize from factored form.
• Computers (and even humans) can often evaluate factored expressions more quickly.

Evaluate the following when $x = 4$ and $y = 3$.

Factoring technique:	Formula-based technique:
$4x(2x + y)(x - 2y)$	$8x^3 - 12x^2y - 8xy^2$
$4 \cdot 4(2 \cdot 4 + 3)(4 - 2 \cdot 3)$	$8 \cdot 4^3 - 12 \cdot 4^2 \cdot 3 - 8 \cdot 4 \cdot 3^2$
$16(11)(-2)$	$8 \cdot 64 - 12 \cdot 16 \cdot 3 - 8 \cdot 4 \cdot 9$
-352	$512 - 576 - 288$
	-352

Against factoring:

• Real-world applications rarely have numbers and expressions that will factor.
• Factoring techniques apply to special cases. Formula-based techniques apply in general.
• With the accessibility of programmable calculators and inexpensive computer software, formula-based techniques are more practical.

Information Systems: Security Coding

One of the most important areas of information systems is security for computer programs, files, and data bases. Governments, companies, and individuals need sophisticated methods of coding passwords and log-on procedures. One type of advanced code uses a branch of mathematics called matrix theory to encode and decode. A rectangular array of numbers, called a matrix, is multiplied by the password matrix after the letters of the password have been assigned numeric values (A=1, B=2, for example). The inverse of the encoding matrix is then used to decode the password. It is desirable to choose an encoding matrix whose determinant is not a prime number because prime numbers make it much easier for thieves to learn the encoding matrix (and hence the password).

About 250 B.C. a Greek scholar named Erastosthenes devised a method, called a sieve, for finding all prime numbers up to any given number N. Use his method to find all prime numbers up to 200 (which would be eliminated as choices for encoding matrix determinants). First, write the numbers 1 to 200 in rows, putting ten numbers (1–10, 11–20, etc.) in each row. Cross out all the multiples of 2 beginning with 4. Now cross out all the multiples of 3 beginning with 6 (which was already crossed out). Cross out all the multiples of 5 beginning with 10. Continue to cross out all the multiples of the prime numbers 2, 3, 5, 7, 11, 13, 17, 19, ... Eventually you will not find any multiples to cross out. When this happens the numbers remaining are prime. The numbers that were crossed out are called composite numbers. The best choices for encoding matrix determinants are composite numbers with numerous factors. These numbers would have been crossed out repeatedly.

Exercises

Use the sieve that you constructed above to answer the following.

1. Count and list the prime numbers less than 200. Remember that 1 is neither prime nor composite because it has exactly one factor; the number 1. (Prime numbers have exactly 2 factors and composite numbers have more than two factors.)
2. What was the first prime number to have no multiples crossed out? Erastosthenes proved that it is never necessary to go higher than the square root of N. Did your sieve comply with this rule?
3. Select five good choices for encoding matrix determinants, which would be composite numbers less than or equal to 200 with many factors.
4. If possible, write a computer program to construct the same sieve, and compare results between the computer method and the manual method. If they differ, explain why.
5. Use the same program to find the primes less than 400. List and count them. Are there double the number of primes less than 200? Why do you think this is so?
6. If possible, alter your program to list the composites with 10 or more factors.

Answers

1. 46 primes: 2, 3, 5, 7, 11, 13, 17, 19, 23, 29, 31, 37, 41, 43, 47, 53, 59, 61, 67, 71, 73, 79, 83, 89, 97, 101, 103, 107, 109, 113, 127, 131, 137, 139, 149, 151, 157, 163, 167, 173, 179, 181, 191, 193, 197, and 199.
2. 17 was the first prime number to have no multiples crossed out. Note that the square root of 200 is about 14.2, and the next largest prime is 17.
3. Good choices: 48, 60, 72, 90, 96, 120, 144, 168, 180, and 192.

4. The results should be the same, but a common error is to forget to eliminate the number 1 from the list of primes.

5. 78 primes: 2, 3, 5, 7, 11, 13, 17, 19, 23, 29, 31, 37, 41, 43, 47, 53, 59, 61, 67, 71, 73, 79, 83, 89, 97, 101, 103, 107, 109, 113, 127, 131, 137, 139, 149, 151, 157, 163, 167, 173, 179, 181, 191, 193, 197, 199, 211, 223, 227, 229, 233, 239, 241, 251, 257, 263, 269, 271, 277, 281, 283, 293, 307, 311, 313, 317, 331, 337, 347, 349, 353, 359, 367, 373, 379, 383, 389, and 397. There are fewer and fewer primes as the numbers increase because larger numbers are more likely than smaller numbers to have additional factors. An interesting problem in mathematics is whether primes continue into infinity or stop at a certain point.

6. Good choices: 240, 252, 264, 270, 288, 300, 336, 360, 384, and 400.

SELF-STUDY EXERCISES 14–4

1 Factor.

1. $x^2 + 5x + 6$ **2.** $x^2 - 11x + 28$ **3.** $x^2 + 8x + 12$
4. $x^2 - 4x + 3$ **5.** $x^2 + 8x + 7$ **6.** $x^2 + 7x + 10$
7. $x^2 - x - 12$ **8.** $y^2 - 3y - 10$ **9.** $y^2 - y - 6$
10. $a^2 + a - 12$ **11.** $b^2 + 2b - 3$ **12.** $14 - 5b - b^2$
13. $x^2 - 7x + 12$ **14.** $x^2 - x - 30$ **15.** $x^2 + 11x + 18$
16. $x^2 - 9x + 18$ **17.** $x^2 - 7x - 18$ **18.** $x^2 + 17x - 18$
19. $x^2 + 9x + 20$ **20.** $x^2 - 12x + 20$ **21.** $x^2 - 10x + 16$
22. $x^2 - 17x + 16$ **23.** $x^2 - 13x - 14$ **24.** $x^2 - 5x - 14$

2 Factor the following polynomials by removing the common factors after grouping.

25. $x^2 + xy + 4x + 4y$ **26.** $6x^2 + 4x - 3xy - 2y$ **27.** $3mx + 5m - 6nx - 10n$
28. $30xy - 35y - 36x + 42$ **29.** $x^2 - 2x + 8x - 16$ **30.** $6x^2 - 2x - 21x + 7$
31. $x^2 - 4x + x - 4$ **32.** $8x^2 - 4x + 6x - 3$ **33.** $x^2 - 5x + 4x - 20$
34. $3x^2 - 6x + 5x - 10$

3 Factor the following trinomials by grouping.

35. $3x^2 + 7x + 2$ **36.** $3x^2 + 14x + 8$ **37.** $6x^2 + 13x + 6$
38. $x^2 - 18x + 32$ **39.** $6x^2 - 17x + 12$ **40.** $2x^2 - 9x + 10$
41. $6x^2 - 13x + 5$ **42.** $p^2 + 5p - 36$ **43.** $6x^2 - 11x + 5$
44. $8x^2 + 26x + 15$ **45.** $15x^2 - 22x - 5$ **46.** $y^2 - 15y + 36$
47. $2x^2 - 5x - 7$ **48.** $12x^2 + 8x - 15$ **49.** $10x^2 + x - 3$
50. $Q^2 - 7Q - 44$

4 Factor the following trinomials by trial and error.

51. $6x^2 + 7xy - 10y^2$ **52.** $6a^2 - 17ab - 14b^2$ **53.** $18x^2 - 3x - 10$
54. $20x^2 - xy - 12y^2$ **55.** $x^2 + x - 56$ **56.** $a^2 + 21a + 38$

5 Factor the following polynomials completely. Be sure to look for the common factors first and then look for special cases.

57. $4x - 4$ **58.** $x^2 + x - 6$ **59.** $2x^2 + x - 3$
60. $x^2 - 9$ **61.** $4x^2 - 16$ **62.** $m^2 + 2m - 15$
63. $2a^2 + 6a + 4$ **64.** $b^2 + 6b + 9$ **65.** $16m^2 - 8m + 1$
66. $x^2 + 8x + 7$ **67.** $2m^2 + 5m + 2$ **68.** $2m^2 - 5m - 3$
69. $2a^2 - 3a - 5$ **70.** $3x^2 + 10x - 8$ **71.** $6x^2 + x - 15$
72. $8x^2 + 10x - 3$

Section 14–1
Factor completely. Check.

1. $5x + 5y$

2. $2x + 5x^2$

3. $12m^2 - 8n^2$

4. $25x^2y - 10xy^3 + 5xy$

5. $2a^3 - 14a^2 - 2a$

6. $30a^3 - 18a^2 - 12a$

7. $15x^3 - 5x^2 - 20x$

8. $9a^2b^3 - 6a^3b^2$

9. $18a^3 + 12a^2$

10. $a^2b + ab^2$

Section 14–2
Use the FOIL method to find the products. Combine the outer and inner products mentally.

11. $(x + 7)(x + 4)$

12. $(y - 7)(y - 5)$

13. $(m + 3)(m - 7)$

14. $(3b - 2)(x + 6)$

15. $(4r - 5)(3r + 2)$

16. $(5 - x)(7 - 3x)$

17. $(4 - 2m)(1 - 3m)$

18. $(2 + 3x)(3 + 2x)$

19. $(x + 3)(2x - 5)$

20. $(5x - 7y)(4x + 3y)$

21. $(2a + 3b)(7a - b)$

22. $(5a + 2b)(6a - 5b)$

23. $(9x - 2y)(3x + 4y)$

24. $(5x - 8y)(4x - 3y)$

25. $(7m - 2n)(3m + 5n)$

Find the following special products mentally.

26. $(y - 4)(y + 4)$

27. $(6x - 5)(6x + 5)$

28. $(3m + 4)(3m - 4)$

29. $(7y + 11)(7y - 11)$

30. $(5x - 2y)(5x + 2y)$

31. $(8a - 5b)(8a + 5b)$

32. $(12r - 7s)(12r + 7s)$

33. $(8x + y)(8x - y)$

34. $(5m + 11n)(5m - 11n)$

35. $(3y - 4z)(3y + 4z)$

36. $(x + 8)^2$

37. $(x + 9)^2$

38. $(x - 7)^2$

39. $(x - 3)^2$

40. $(2x - 3)^2$

41. $(4x - 15)^2$

42. $(5 + 3m)^2$

43. $(8 + 7m)^2$

44. $(5x - 13)^2$

45. $(4x - 11)^2$

Use the extension of the FOIL method to find the products.

46. $(K + L)(K^2 - KL + L^2)$

47. $(g - h)(g^2 + gh + h^2)$

48. $(4 + a)(16 - 4a + a^2)$

49. $(2H - 3T)(4H^2 + 6HT + 9T^2)$

50. $(3a - 5)(9a^2 + 15a + 25)$

51. $(6 + i)(36 - 6i + i^2)$

52. $(9y - p)(81y^2 + 9yp + p^2)$

53. $(z + 2t)(z^2 - 2zt + 4t^2)$

54. $(g - 2)(g^2 + 2g + 4)$

55. $(7T + 2)(49T^2 - 14T + 4)$

Section 14–3
Identify the special products that are the difference of two perfect squares.

56. $25m^2 - 4n^2$

57. $64 + 4a^2$

58. $4f^2 - 9g^2$

59. $H^2 - G^2$

60. $s^2 - 4s + 10$

61. $64b^2 - 49$

Verify which of the following trinomials are perfect-square trinomials.

62. $16c^2 + 8c - 1$

63. $-9x^2 + 12x + 4$

64. $4d^2 + 2d + 16$

65. $9t^2 - 24tp + 16p^2$

66. $a^2 - 14a + 49$

67. $j^2 + 10j + 25$

Identify the special products that are the sum or difference of two perfect cubes.

68. $R^3 + 81$

69. $125a^3 - 8b^3$

70. $64 + m^3$

71. $8z^3 - 125$

72. $d^3 - 12d + 36$

73. $64W^3 + 27$

Factor the expressions that are special products. Explain why the other expressions are not special products.

74. $x^2 - 81$

75. $25y^2 - 4$

76. $100a^2 - 8ab^2$

77. $a^2b^2 + 49$

78. $121 - 9m^2$

79. $a^2 + 2a + 1$

80. $4x^2 + 12x + 9$

81. $16c^2 - 24bc + 9b^2$

82. $9y^2 - 12y - 4$

83. $n^2 + 169 - 26n$
(*Hint:* Rearrange.)

84. $16d^2 - 20d + 25$

85. $36a^2 + 84ab + 49b^2$

86. $4x^2 - 25y^2$

87. $49 - 14x + x^2$

88. $9x^2y^2 - 49z^2$

89. $64 + 25x^2$

90. $4x^2 + 12xy + 9y^2$

91. $16x^2 + 24x + 9y^2$

92. $36 - x^2$

93. $49 - 81y^2$

94. $64x^2 - 25y^2$

95. $9x^2 - 100y^2$

96. $a^2 - 10a + 25$

97. $9x^2 - 6xy + y^2$

98. $25 - 16a^2b^2$

99. $9x^2y^2 - z^2$

100. $x^2 - y^2$

101. $x^2 + 4x + 4$

102. $\frac{1}{4}x^2 - \frac{1}{9}y^2$

103. $\frac{4}{25}x^2 - \frac{1}{16}y^2$

104. $27v^3 - 8$

105. $T^3 - 8$

106. $8r^3 + 27$

107. $d^3 + 729$

108. $125c^3 - 216d^3$

109. $27K^3 + 64$

Section 14–4

Factor the following trinomials.

110. $x^2 + 10x + 21$

111. $x^2 + 11x + 24$

112. $x^2 + 29x + 28$

113. $x^2 + 13x + 30$

114. $x^2 - 13x + 40$

115. $x^2 - 9x + 8$

116. $x^2 - 17x - 18$

117. $x^2 - 11x - 26$

118. $x^2 + x - 30$

119. $x^2 + 5x - 24$

120. $6x^2 + 25x + 14$

121. $6x^2 + 25x + 4$

122. $4x^2 - 23x + 15$

123. $5x^2 - 34x + 24$

124. $3x^2 - x - 14$

125. $6x^2 - x - 35$

126. $3x^2 + 11x - 4$

127. $7x^2 - 13x - 24$

Factor the following polynomials. Look for common factors and special cases.

128. $5mn - 25m$

129. $9a^2 - 100$

130. $a^2 - b^2$

131. $2x^2 - 3x - 2$

132. $b^3 - 8b^2 - b$

133. $a^2 - 81$

134. $x^2 - 14x + 13$

135. $y^2 - 14y + 49$

136. $m^2 - 3m + 2$

137. $b^2 + 8b + 15$

138. $x^2 - 13x + 30$

139. $169 - m^2$

140. $5x^2 + 13x + 6$

141. $x^2 - 4x - 32$

142. $x^2 + 9x + 14$

143. $x^2 + 19x - 20$

144. $x^2 + 8x - 20$

145. $2x^2 - 4x - 16$

146. $5x^2 - 20$

147. $2x^3 - 10x^2 - 12x$

CHALLENGE PROBLEMS

148. Explain the value of factoring a general trinomial by the grouping method as opposed to the trial-and-error method.

150. A standard-sized rectangular swimming pool is 25 ft long and 15 ft wide. This gives a water-surface area of 375 ft². A customer wants to examine some options for varying the size of the pool. If x is the amount of adjustment to the length and y is the amount of adjustment to the width, find the following.

  (a) Write a formula in both factored and expanded form for finding the water-surface area of a pool that is adjusted x feet in length and y feet in width.

  (b) Write a formula using one variable in both factored and expanded form for the water-surface area of a pool when the length and width are increased the same amount.

  (c) Will the formulas in parts (a) and (b) work for both increasing and decreasing the size of the pool?

  (d) Illustrate your answer in part (c) with numerical examples.

149. What is the value of being able to recognize the special products when we factor algebraic expressions?

CONCEPTS ANALYSIS

1. Explain how the FOIL process for multiplying two binomials is an application of the distributive property.

2. (a) How is the distributive property used to multiply a binomial and a trinomial?

  (b) How is the distributive property used to multiply two trinomials?

3. Give an example of the product of a binomial and a trinomial to illustrate the procedure given in Question 2a.

5. List the properties of a binomial that is the difference of two perfect squares.

7. Explain in sentence form how the sign of the third term of a general trinomial affects the signs between the terms of its binomial factors.

9. Write a brief comment explaining each lettered step of the following example.

Factor $10x^2 - 19x - 12$ using grouping.

(a)
$$\underline{120}$$
1, 120
2, 60
3, 40
4, 30
5, 24*   Factors whose
                difference = 19
6, 20
8, 15
10, 12
$$\underline{10x^2 - 19x - 12}$$

(b) $10x^2 + 5x - 24x - 12$

(c) $5x(2x + 1) - 12(2x + 1)$

(d) $(2x + 1)(5x - 12)$

4. Give an example of the product of two trinomials to illustrate the procedure given in Question 2b.

6. List the properties of a perfect-square trinomial.

8. What do we mean if we say that a polynomial is prime?

10. Find a mistake in each of the following. Briefly explain the mistake, then work the problem correctly.
   (a) $(3x - 2y)^2 = 9x^2 + 4y^2$
   (b) $5xy^2 - 45x = 5x(y^2 - 9)$
   (c) $2x^2 - 28x + 96 = (2x - 12)(x - 8)$
   (d) $x^2 - 5x - 6 = (x - 2)(x - 3)$

CHAPTER SUMMARY

| Objectives | What to Remember with Examples |

Section 14–1

1 Factor an expression containing a common factor.

To factor an expression containing a common factor:
1. Find the largest factor common to each term of the expression.
2. Rewrite the expression as the indicated product of the largest common factor and the remaining quantity.

> Factor $3ab + 9b^2$
>
> $3b(a + 3b)$

Section 14–2

1 Use the FOIL method to multiply two binomials.

To multiply two binomials by the FOIL method: Multiply First terms, Outer terms, Inner terms, and Last terms. Combine like terms.

> Multiply $(3a - 2)(a + 3)$.
>
> F   O   I   L
> $3a^2 + 9a - 2a - 6$
> $3a^2 + 7a - 6$

2 Multiply polynomials to obtain three special products.

To multiply the sum and difference of the same two terms:
1. Square the first term.
2. Insert minus sign.
3. Square the second term.

Multiply $(3b - 2)(3b + 2)$.

$9b^2 - 4$

To square a binomial:
1. Square the first term.
2. Double the product of the two terms.
3. Square the second term.

Multiply $(2a - 3)^2$.

$4a^2 - 12a + 9$

To multiply a binomial and a trinomial of the types $(a + b)(a^2 - ab + b^2)$ and $(a - b)(a^2 + ab + b^2)$:
1. Cube the first term of the binomial.
2. If the binomial is a sum, insert a plus sign; if the binomial is a difference, insert a minus sign.
3. Cube the second term of the binomial.

Multiply

$(8b - 2)(64b^2 + 16b + 4)$.
$512b^3 - 8$

Multiply

$(2x + 5)(4x^2 - 10x + 25)$.
$8x^3 + 125$

Section 14–3

[1] Recognize and factor the difference of two perfect squares.

The difference of two perfect squares, $4z^2 - 9$, factors into two binomials, $(2z - 3)(2z + 3)$. To factor the difference of two perfect squares:
1. Take the square root of the first term.
2. Take the square root of the second term.
3. Write one factor as the sum and one factor as the difference of the square roots from Steps 1 and 2.

Identify which expression is the special product, the difference of two squares.
(a) $36 - 27A^2$: No, $27A^2$ is not a perfect square.
(b) $9c^2 - 4$: Yes, both terms are perfect squares.
Factor $9c^2 - 4$.

$(3c + 2)(3c - 2)$

[2] Recognize and factor a perfect-square trinomial.

In a perfect-square trinomial, the first and last terms are positive perfect squares and the absolute value of the middle term is twice the product of the square roots of the first and last terms, such as $a^2 + 2ab + b^2$.

Identify which expression is a perfect-square trinomial.
(a) $4x^2 + 18x + 9$: Not a perfect square trinomial, the middle term is not twice the product of the square roots of the first and last terms.
(b) $9a^2 + 6a + 4$: The first and last terms are perfect squares. The middle term is twice the product of the square roots of the first and last terms, so we factor.

To factor a perfect-square trinomial:
1. Write the square root of the first term.
2. Write the sign of the middle term.
3. Write the square root of the last term.
4. Indicate the square of the quantity.

> Factor $9a^2 + 6a + 4$.
>
> $(3a + 2)^2$

3 Recognize and factor the sum or difference of two perfect cubes.

The sum of two perfect cubes is two perfect cubes with a plus sign between them. The difference of two perfect cubes is two perfect cubes with a minus sign between them. Each factors into a binomial and a trinomial.

> Identify the sum or difference of two perfect cubes:
> (a) $b^3 - 27$: Yes, both terms are perfect cubes; difference of two cubes.
> (b) $X^3 - 6$: No, 6 is not a perfect cube.
> (c) $8 + y^3$: Yes, both terms are perfect cubes; sum of two cubes.

To factor the sum (difference) of two perfect cubes:
1. Write the binomial factor as the sum (difference) of the cube roots of the two terms.
2. Write the trinomial factor as the square of the first term from Step 1, insert a minus sign if factoring a sum (a plus sign if factoring a difference). Write the product of the two terms from Step 1, plus the square of the second term from Step 1.

> Factor $8 + y^3$.
>
> $(2 + y)(4 - 2y + y^2)$
>
> Factor $b^3 - 27$.
>
> $(b - 3)(b^2 + 3b + 9)$

Section 14–4

1 Factor general trinomials whose squared term has a coefficient of 1.

To factor a general trinomial whose squared term has a coefficient of 1: Factor using the FOIL method in reverse. Use the factors of the third term that will give the desired middle term and sign.

> Factor $a^2 - 5a + 6$.
>
> ( )( )
> $(a \quad 3)(a \quad 2)$
> $(a - 3)(a - 2)$
>
> These factors of 6, -3 and -2, give $-5a$ as the middle term.

2 Remove common factors after grouping an expression.

Arrange terms of expression into groups whose terms each have a common factor. Rewrite the expression as an indicated product of the common factor and the remaining quantity. Factor the common binomial factor.

> Factor $2x^2 - 4x + xy - 2y$ completely:
>
> $(2x^2 - 4x) + (xy - 2y)$
> $2x(x - 2) + y(x - 2)$
> $(x - 2)(2x + y)$

3 Factor a general trinomial by grouping.	To factor a general trinomial by grouping:

3 Factor a general trinomial by grouping.

To factor a general trinomial by grouping:
1. Multiply the coefficients of the first term and last term.
2. Factor the product from Step 1 into two factors (a) whose sum is the coefficient of the middle term if the last term is positive, or (b) whose difference is the coefficient of the middle term if the last term is negative.
3. Rewrite the trinomial so the middle term is the sum or difference from Step 2.
4. Write the trinomial from Step 3 in two groups with a common factor in each.
5. Factor out the common factor.
6. Factor the common binomial factor.

Factor $6x^2 - 17x + 12$.

$6 \cdot 12 = 72$
$1 \cdot 72$
$2 \cdot 36$
$3 \cdot 24$
$4 \cdot 18$
$6 \cdot 12$
$8 \cdot 9*$

$6x^2 - 8x - 9x + 12$

$(6x^2 - 8x) + (-9x + 12)$

$2x(3x - 4) - 3(3x - 4)$
$(3x - 4)(2x - 3)$

4 Factor a general trinomial by trial and error.

To factor a general trinomial by trial and error:
1. Indicate the binomial factors by ( )( ).
2. List all possible factors of the first and last terms.
3. List all possible binomial factors.
4. Multiply the binomial factors to find the product that is the original trinomial.
5. Check to be sure the signs are correct.

Factor $3x^2 - 13x - 10$.

( )( )

Factors of $3x^2$: $3x(x)$

Factors of 10: $1(10)$
                   $2(5)$

$3x^2 + 29x + 10 = (3x - 1)(x + 10)$	$(3x - 10)(x + 1) = 3x^2 - 7x - 10$
$3x^2 - 29x - 10 = (3x + 1)(x - 10)$	$(3x + 10)(x - 1) = 3x^2 + 7x - 10$
$3x^2 + x - 10 = (3x - 5)(x + 2)$	$(3x - 2)(x + 5) = 3x^2$
$3x^2 - x - 10 = (3x + 5)(x - 2)$	$(3x + 2)(x - 5) = 3x^2 - 13x - 10$

The factorization of $3x^2 - 13x - 10$ is $(3x + 2)(x - 5)$.

5 Factor any binomial or trinomial that is not prime.

To factor any polynomial:
1. Factor out the greatest common factor (if any).
2. Check for any special products.
3. If there is no special product, factor by grouping or trial and error.
4. Examine each factor to see that it cannot be factored further.
5. Check factoring by multiplying.

Factor $12x^3 + 6x^2 - 18x$ completely.

$6x(2x^2 + x - 3)$
$6x(2x + 3)(x - 1)$

(After removing the common factor, grouping or trial and error can be used to factor the trinomial.)

factor (p. 538)
common factor (p. 538)
polynomial (p. 540)
monomial (p. 540)
binomial (p. 540)
trinomial (p. 540)

FOIL (p. 540)
perfect squares (p. 541)
conjugate pairs (p. 541)
binomial square (p. 541)
perfect square trinomial
  (p. 542, 546)
perfect cubes (p. 547)

general trinomial (p. 548)
factor by inspection (p. 549)
grouping (p. 550)
trial and error (p. 555)
prime (p. 556)

CHAPTER TRIAL TEST

Find the following products.

1. $3(x + 2y)$
2. $7x(x - 5)$
3. $7x(2x^2 + 3x - 5)$
4. $2x^2(2x + 3)$
5. $(m - 7)(m + 7)$
6. $(3x - 2)(3x + 2)$
7. $(a + 3)^2$
8. $(2x - 7)^2$
9. $(x - 3)(2x - 5)$
10. $(7x - 3)(2x + 1)$
11. $(x - 2)(x^2 + 2x + 4)$
12. $(3y + 4)(9y^2 - 12y + 16)$
13. $(5a - 3)(25a^2 + 15a + 9)$
14. $(a + 6)(a^2 - 6a + 36)$

Factor completely.

15. $7x^2 + 8x$
16. $6ax + 15bx$
17. $7a^2b - 14ab$
18. $9x^2 - 42x + 49$
19. $x^2 + 9x + 8$
20. $x^3 + 8x$
21. $x^2 - 5x - 36$
22. $x^2 + 5x - 24$
23. $6x^2 - 5x - 6$
24. $x^3 - 27$
25. $3x^2 + 23x + 30$
26. $12y^2 - 27$
27. $a^2 + 16ab + 64b^2$
28. $y^2 - y - 6$
29. $b^2 - 3b - 10$
30. $x^2 - 7x + 10$
31. $3x^2 + x - 4$
32. $a^2 - 4a + 3$
33. $3m^2 - 5m + 2$
34. $4c^2 - 28c + 49$
35. $27a^3 - 8$
36. $3x^2 + 11xy + 3y^2$

566   CHAPTER 14 Products and Factors

CHAPTER

15

Solving Quadratic and Higher-Degree Equations

GOOD DECISIONS THROUGH TEAMWORK

Today's law enforcement officers often act as Sherlock Holmes when investigating motor vehicle accidents because drivers may die in the accident or lie about their driving speed and the road conditions. In addition, eyewitnesses may not be available or they may give conflicting reports. Therefore, physical evidence in the form of skid marks, distance, and the percentage of braking before stop-

ping, as well as the computed drag factor, are used in mathematical equations to estimate the speed of the car. Officers can then state with confidence that the driver was driving at a speed of at least a certain number of miles per hour.

The drag factor, or coefficient of friction, is a number between 0.03 (poor stopping conditions) and 1.3 (ideal stopping conditions). The drag factor varies according to road surface type (asphalt, concrete, gravel, dirt, or grass), age of road, weather conditions (precipitation, wind, air temperature, air pressure, relative humidity), road surface conditions (wet, dry, icy, oily, sandy, covered with dirt, gravel, or debris), grade of road, etc. Officers can conduct drag factor tests under similar conditions to determine the drag factor.

Each test consists of braking to a stop from a specified speed (usually 25 to 35 mph), denoted by s, then measuring the distance of the skid marks (D). At least two tests are conducted, using the longest skid distance (if they are within 5% of each other) in the following formula to calculate the drag factor (f):

$$f = \frac{s^2}{30D}$$

These tests are especially important if the accident resulted in a fatality. Usually the weight of the vehicle is negligible in determining speed, but weight can be a factor under extremely poor stopping conditions such as black ice, oil slick, or oil mixed with water (which happens at the beginning of a rain occurring after a long dry period).

Under most stopping conditions, the *distance* of the skid marks, *D*, is related to the *speed* of the car, *s*; the *drag factor, f*; and the *percentage* of braking before stopping, *n* (written as a decimal) by the formula:

$$D = \frac{s^2}{30fn}$$

Divide responsibilities among team members to gather additional information. Contact the police departments of a large metropolitan city (population of 1 million or more) and a small town (population of less than 25,000). Ask each to describe the process they use to determine a car's speed when given the length of the skid marks and the road conditions. Make sure to specify how the calculations are done (by hand, with a calculator, with tables or charts, or with computer software). Describe the advantages and disadvantages of each method.

Ask a law enforcement officer to describe how accident conditions are simulated accurately during drag factor tests. What happens if their calculated speed is disputed?

Most of the equations we have been solving so far are called linear equations. In such equations, the variable appears only to the first power, such as *x* or *y*. In Chapter 12 we looked briefly at one type of quadratic equation. In this chapter we will continue to examine quadratic equations.

15-1 QUADRATIC EQUATIONS

Learning Objectives

1. Write quadratic equations in standard form.
2. Identify the coefficients of the quadratic, linear, and constant terms of a quadratic equation.

1 Write Quadratic Equations in Standard Form.

There are several methods for solving quadratic equations. Each method will have some strengths and weaknesses based on the characteristics of the equation being solved. First, we'll examine the basic characteristics of quadratic equations, starting with quadratic equations having only one variable. Then, we will look at quadratic equations written as functions, which involve both independent and dependent variables.

■ **DEFINITION 15–1: Quadratic Equation.** A *quadratic equation* is an equation in which at least one letter term is raised to the second power and no letter terms have a power higher than 2 or less than 0. The *standard form* for a quadratic equation is $ax^2 + bx + c = 0$, where *a*, *b*, and *c* are real numbers and $a > 0$.

EXAMPLE 1 Arrange the following quadratic equations in standard form.

(a) $3x^2 - 3 + 5x = 0$ (b) $7x - 2x^2 - 8 = 0$
(c) $12 = 7x^2 - 4x$ (d) $7x^2 = 3$

(a) $3x^2 - 3 + 5x = 0$
  $\mathbf{3x^2 + 5x - 3 = 0}$ Arrange the terms in descending powers of *x*.

(b) $7x - 2x^2 - 8 = 0$
        $-2x^2 + 7x - 8 = 0$ Arrange the terms in descending powers of *x*.
  $-1(-2x^2 + 7x - 8) = -1(0)$ Multiply both sides by −1 to make the coefficient of x^2 term positive.
      $\mathbf{2x^2 - 7x + 8 = 0}$ Standard form

(c) $$12 = 7x^2 - 4x$$

Rearrange the terms so that one side of the equation equals zero.

$$0 = 7x^2 - 4x - 12$$

$$\mathbf{7x^2 - 4x - 12 = 0}$$

Interchange sides of equation if desired.

(d) $$7x^2 = 3$$

$$\mathbf{7x^2 - 3 = 0}$$ Rearrange so that one side equals zero.

[2] Identify the Coefficients of the Quadratic, Linear, and Constant Terms of a Quadratic Equation.

Some methods for solving quadratic equations can be used for any type of quadratic equation, but are very time consuming. Other methods are quicker, but apply only to certain types of quadratic equations. In choosing an appropriate method, it is helpful to be able to recognize similar and different characteristics of quadratic equations.

In the standard form of quadratic equations, $ax^2 + bx + c = 0$, we have three types of terms.

1. ax^2 is a *quadratic term*; that is, the degree of the term is 2. In standard form, this term is the leading term and a is the leading coefficient.
2. bx is a *linear term*; that is, the degree of the term is 1, the coefficient b.
3. c is a *number* term or *constant term*; that is, the degree of the term is 0. The coefficient is c ($c = cx^0$).

A quadratic equation that has all three types of terms is sometimes referred to as a *complete quadratic equation*.

$ax^2 + bx + c = 0$ Complete quadratic equation

A quadratic equation that has a quadratic term (ax^2) and a linear term (bx) but no constant (c) is sometimes referred to as an *incomplete quadratic equation*.

$ax^2 + bx = 0$ Incomplete quadratic equation

A quadratic equation that has a quadratic term (ax^2) and a constant (c) but no linear term (bx) is sometimes referred to as a *pure quadratic equation,* the type that we worked with in Chapter 12.

$ax^2 + c = 0$ Pure quadratic equation

It will be helpful in planning our strategy for solving a quadratic equation to be able to identify the types of terms in a quadratic equation and to identify the coefficients a, b, and c.

EXAMPLE 2   Write each equation in standard form and identify a, b, and c.

(a) $3x^2 + 5x - 2 = 0$ (b) $5x^2 = 2x - 3$
(c) $3x = 5x^2$ (d) $-6x^2 + 2x = 0$
(e) $5x^2 - 4 = 0$ (f) $x^2 = 9$
(g) $x^2 + 3x = 5x$ (h) $7 - x^2 = 3 + x$

(a) $3x^2 + 5x - 2 = 0$ In standard form
$\mathbf{a = 3, \quad b = 5, \quad c = -2}$

(b) $5x^2 = 2x - 3$
$5x^2 - 2x + 3 = 0$ Write in standard form.
$\mathbf{a = 5, \quad b = -2, \quad c = 3}$

(c) $3x = 5x^2$
$0 = 5x^2 - 3x$ Write in standard form.
$5x^2 - 3x = 0$ Interchange sides of equation.
$\mathbf{a = 5, \quad b = -3, \quad c = 0}$

(d) $-6x^2 + 2x = 0$
$-1(-6x^2 + 2x = 0)$ Multiply by -1.
$6x^2 - 2x = 0$ Standard form
$a = 6, \quad b = -2, \quad c = 0$

(e) $5x^2 - 4 = 0$ Standard form
$a = 5, \quad b = 0, \quad c = -4$

(f) $x^2 = 9$
$x^2 - 9 = 0$ Write in standard form.
$a = 1, \quad b = 0, \quad c = -9$

(g) $x^2 + 3x = 5x$
$x^2 + 3x - 5x = 0$ Rearrange.
$x^2 - 2x = 0$ Combine like terms.
$a = 1, \quad b = -2, \quad c = 0$

(h) $7 - x^2 = 3 + x$
$-x^2 - x + 7 - 3 = 0$ Rearrange.
$-x^2 - x + 4 = 0$ Combine like terms.
$-1(-x^2 - x + 4 = 0)$ Multiply by -1.
$x^2 + x - 4 = 0$ Standard form
$a = 1, \quad b = 1, \quad c = -4$

SELF-STUDY EXERCISES 15-1

1   Arrange the following quadratic equations in standard form.

1. $5 - 4x + 7x^2 = 0$
2. $8x^2 - 3 = 6x$
3. $7x^2 = 5$
4. $x^2 = 6x - 8$
5. $x^2 - 3x = 6x - 8$
6. $8 - x^2 + 6x = 2x$
7. $3x^2 + 5 - 6x = 0$
8. $5 = x^2 - 6x$
9. $x^2 - 16 = 0$
10. $8x^2 - 7x = 8$
11. $8x^2 + 8x - 2 = 8$
12. $0.3x^2 - 0.4x = 3$

2   Write each equation in standard form and identify a, b, and c.

13. $5x = x^2$
14. $7x - 3x^2 = 5$
15. $7x^2 - 4x = 0$
16. $8 + 3x^2 = 5x$
17. $5x - 6 = x^2$
18. $11x^2 = 8x$
19. $x = x^2$
20. $9x^2 - 7x = 12$
21. $5 = x^2$
22. $-x^2 - 6x + 3 = 0$
23. $0.2x - 5x^2 = 1.4$
24. $\dfrac{2}{3}x^2 - \dfrac{5}{6}x = \dfrac{1}{2}$
25. $1.3x^2 - 8 = 0$
26. $\sqrt{3}x^2 + \sqrt{5}x - 2 = 0$

Write equations in standard form for the following.

27. Coefficient of the quadratic term is 8; coefficient of the linear term is -2; the constant term is -3.

28. The constant is zero, the coefficient of the linear term is 3, and the coefficient of the quadratic term is 1.

29. $a = 5, \quad b = 2, \quad c = -7$

30. $a = 2.5, \quad c = 0.8$

15-2   SOLVING QUADRATIC EQUATIONS USING THE QUADRATIC FORMULA

Learning Objectives

1   Solve quadratic equations using the quadratic formula.

2   Solve applied problems that use quadratic equations.

The first method for solving quadratic equations that we will examine is a method that works for all types of quadratic equations. It is time-consuming and requires some cumbersome calculations. Using a programmable calculator or a computer program makes this a popular method for problem solving.

⬜1 Solve Quadratic Equations Using the Quadratic Formula.

One method for solving quadratic equations is to use the *quadratic formula*. This formula results from solving the standard quadratic equation, $ax^2 + bx + c = 0$, by a method called *completing the square*.

Formula 15-1 *Quadratic formula:*

$$x = \frac{-b \pm \sqrt{b^2 - 4ac}}{2a}$$

a, b, and c are coefficients of a quadratic equation in the form $ax^2 + bx + c = 0$.

Because the formula is derived from the standard quadratic equation $ax^2 + bx + c = 0$, the a, b, and c in the formula are the same a, b, and c in the standard quadratic equation. To evaluate the formula, we first identify a, b, and c. Then we evaluate the formula to solve equations. Even though we have one formula, we will have *two* values for x because the formula requires us to *add* or *subtract* the radical.

$$x = \frac{-b \pm \sqrt{b^2 - 4ac}}{2a}$$

Tips and Traps   *Common Cause for Errors:*
When using the formula to solve problems, we should begin by writing the formula to help us remember it. When writing the formula, be sure to extend the fraction bar beneath the *entire* numerator. This omission is a common cause for errors.

EXAMPLE 1   Use the quadratic formula to solve $x^2 + 5x + 6 = 0$ for x.

$a = 1, \qquad b = 5, \qquad c = 6$ Identify a, b, and c.

$$x = \frac{-b \pm \sqrt{b^2 - 4ac}}{2a}$$ Quadratic formula. Substitute for a, b, and c.

$$x = \frac{-5 \pm \sqrt{5^2 - 4 \cdot 1 \cdot 6}}{2 \cdot 1}$$ Multiply in groupings.

$$x = \frac{-5 \pm \sqrt{25 - 24}}{2}$$ Combine terms under radical.

$$x = \frac{-5 \pm \sqrt{1}}{2}$$ Evaluate radical.

$$x = \frac{-5 \pm 1}{2}$$

At this point we separate the formula into two parts, one using the $+1$ and the other using the -1.

$$x = \frac{-5 + 1}{2} \qquad\qquad\qquad x = \frac{-5 - 1}{2}$$

$$x = \frac{-4}{2} \qquad\qquad\qquad\qquad x = \frac{-6}{2}$$

$$x = -2 \qquad\qquad\qquad x = -3$$

Check $x = \boxed{-2}$ or   Check $x = \boxed{-3}$

$$\boxed{x}^2 + 5\boxed{x} + 6 = 0 \qquad\qquad \boxed{x}^2 + 5\boxed{x} + 6 = 0$$

$$(\boxed{-2})^2 + 5(\boxed{-2}) + 6 = 0 \qquad\qquad (\boxed{-3})^2 + 5(\boxed{-3}) + 6 = 0$$

$$4 - 10 + 6 = 0 \qquad\qquad 9 - 15 + 6 = 0$$

$$-6 + 6 = 0 \qquad\qquad -6 + 6 = 0$$

$$0 = 0 \qquad\qquad 0 = 0$$

Tips and Traps   ***Completing the Square and the Quadratic Formula:***
The completing-the-square method for solving quadratic equations is another method that works for all types of quadratic equations. It also requires several tedious manipulations. We will use this method to show how the quadratic formula was derived, but we will not use this method in the examples in this chapter. In Chapter 12, we saw that we could take the square root of both sides of an equation and maintain equality. That principle is applied in this method.

$ax^2 + bx + c = 0 \qquad$ a, b, and c are real numbers and $a > 0$.

$ax^2 + bx = -c \qquad$ Isolate the terms containing variables.

$x^2 + \dfrac{bx}{a} = -\dfrac{c}{a} \qquad$ Divide the entire equation by a so that the coefficient of x^2 is 1.

$x^2 + \dfrac{bx}{a} + \dfrac{b^2}{4a^2} = \dfrac{b^2}{4a^2} - \dfrac{c}{a} \qquad$ Add a constant to each side so that one side becomes a perfect-square trinomial. The appropriate constant will be $\left(\dfrac{1}{2}\dfrac{b}{a}\right)^2$ or $\dfrac{1}{4}\dfrac{b^2}{a^2}$ or $\dfrac{b^2}{4a^2}$.

$\left(x + \dfrac{b}{2a}\right)^2 = \dfrac{b^2}{4a^2} - \dfrac{c}{a} \qquad$ Factor the left side.

$\sqrt{\left(x + \dfrac{b}{2a}\right)^2} = \sqrt{\dfrac{b^2}{4a^2} - \dfrac{c}{a}} \qquad$ Take the square root of both sides.

$x + \dfrac{b}{2a} = \pm\sqrt{\dfrac{b^2}{4a^2} - \dfrac{c}{a}} \qquad$ Get a common denominator for the terms under the radical.

$x + \dfrac{b}{2a} = \pm\sqrt{\dfrac{b^2}{4a^2} - \dfrac{4ac}{4a^2}} \qquad$ Simplify the radicand.

$x + \dfrac{b}{2a} = \dfrac{\pm\sqrt{b^2 - 4ac}}{2a} \qquad$ Solve for x.

$x = -\dfrac{b}{2a} \dfrac{\pm\sqrt{b^2 - 4ac}}{2a} \qquad$ Combine like fractions.

$x = \dfrac{-b \pm \sqrt{b^2 - 4ac}}{2a} \qquad$ Quadratic formula.

EXAMPLE 2   Use the quadratic formula to solve $3x^2 - 3x - 7 = 0$ for x. Round answers to the nearest hundredth.

$a = 3, \qquad b = -3, \qquad c = -7 \qquad$ Identify a, b, and c.

$$x = \frac{-b \pm \sqrt{b^2 - 4ac}}{2a}$$

Quadratic formula

$$x = \frac{+3 \pm \boxed{\sqrt{(-3)^2 - 4(3)(-7)}}}{2 \cdot 3}$$

Substitute $-(b) = -(-3) = +3$.

$$x = \frac{+3 \pm \boxed{\sqrt{9 + 84}}}{6}$$

Perform calculations under the radical.

$$x = \frac{+3 \pm \boxed{\sqrt{93}}}{6}$$

Evaluate the radical.

$$x = \frac{+3 \pm \boxed{9.643650761}}{6}$$

Separate two solutions.

$$x = \frac{3 \boxed{+} 9.643650761}{6} \qquad x = \frac{3 \boxed{-} 9.643650761}{6}$$

$$x = \frac{12.643650761}{6} \qquad x = \frac{-6.643650761}{6}$$

$$x = \mathbf{2.11} \quad \text{Rounded} \qquad x = \mathbf{-1.11} \quad \text{Rounded}$$

Notice that rounding to hundredths takes place *after* the final calculation has been made.

Check $x = \boxed{2.11}$ Check $x = \boxed{-1.11}$

$$3x^2 - 3x - 7 = 0 \qquad\qquad 3x^2 - 3x - 7 = 0$$
$$3(\boxed{2.11})^2 - 3(\boxed{2.11}) - 7 = 0 \qquad 3(\boxed{-1.11})^2 - 3(\boxed{-1.11}) - 7 = 0$$
$$3(4.4521) - 6.33 - 7 = 0 \qquad 3(1.2321) + 3.33 - 7 = 0$$
$$13.3563 - 6.33 - 7 = 0 \qquad 3.6963 + 3.33 - 7 = 0$$
$$0.0263 \doteq 0 \qquad\qquad 0.0263 \doteq 0$$

Tips and Traps   *Rounding Discrepancies:*
The symbol $\doteq$ means "is approximately equal to."
If we had checked *before* rounding, the check would have been "closer."

■ General Tips for Using the Calculator

Use the full calculator value to check.
This can be done efficiently using a calculator. Calculate the first root.

$$3 \boxed{+} \boxed{\sqrt{}} \ 93 \ \boxed{\text{EXE}} \ \boxed{\div} \ 6 \ \boxed{\text{EXE}} \Rightarrow 2.107275127$$

Check:

$$3 \ \boxed{\text{ANS}} \ \boxed{x^2} \ \boxed{-} \ 3 \ \boxed{\text{ANS}} \ \boxed{-} \ 7 \ \boxed{\text{EXE}} \Rightarrow -1.6\text{E} - 09 \text{ or } -0.0000000016$$

Calculate the second root.

$$3 \boxed{-} \boxed{\sqrt{}} \ 93 \ \boxed{\text{EXE}} \ \boxed{\div} \ 6 \ \boxed{\text{EXE}} \Rightarrow -1.107275127$$

Check:

$$3 \ \boxed{\text{ANS}} \ \boxed{x^2} \ \boxed{-} \ 3 \ \boxed{\text{ANS}} \ \boxed{-} \ 7 \ \boxed{\text{EXE}} \Rightarrow 8.\text{E} - 12 \text{ or } 0.000000000008$$

In the next example, the equation contains no squared term at first, but the squared term appears *after* the multiplication to eliminate the denominator.

EXAMPLE 3 Solve $\frac{3}{5-x} = 2x$ for x. Round answers to the nearest hundredth.

$$\frac{3}{5-x} = 2x$$

Eliminate the denominator and then write the equation in standard form.

$$(5-x)\frac{3}{5-x} = 2x(5-x)$$

Multiply both sides by the denominator to eliminate it.

$$3 = 2x(5-x)$$

Distribute.

$$3 = 10x - 2x^2$$

$$2x^2 - 10x + 3 = 0$$

Write the equation in standard form.

$$a = 2, \qquad b = -10, \qquad c = 3$$

$$x = \frac{-b \pm \sqrt{b^2 - 4ac}}{2a}$$

Write the formula.

$$x = \frac{+10 \pm \sqrt{(-10)^2 - 4(2)(3)}}{2(2)}$$

Substitute.

$$x = \frac{10 \pm \sqrt{100 - 24}}{4}$$

$$x = \frac{10 \pm \sqrt{76}}{4}$$

$$x = \frac{10 \pm 8.717797887}{4}$$

Separate into two solutions.

$$x = \frac{10 + 8.717797887}{4} \qquad\qquad x = \frac{10 - 8.717797887}{4}$$

$$x = \frac{18.717797887}{4} \qquad\qquad x = \frac{1.282202113}{4}$$

$x = \mathbf{4.68}$ $\qquad\qquad$ $x = \mathbf{0.32}$ $\quad$ Round after final calculation.

Check $x = 4.68$ $\qquad\qquad$ Check $x = 0.32$

$$\frac{3}{5-x} = 2x \qquad\qquad \frac{3}{5-x} = 2x$$

$$\frac{3}{5 - (4.68)} = 2(4.68) \qquad \frac{3}{5 - (0.32)} = 2(0.32)$$

$$\frac{3}{0.32} = 9.36 \qquad\qquad \frac{3}{4.68} = 0.64$$

$$9.375 \doteq 9.36 \qquad\qquad 0.641025641 \doteq 0.64$$

Tips and Traps **Extraneous Roots:**
When an equation is multiplied by a factor that contains a variable, as in the previous example, an *extraneous root* can be introduced. An extraneous root is a root that will *not* check when substituted into the original equation. Thus, it is very important to check each root.

▣ 2 Solve Applied Problems that Use Quadratic Equations.

Many applications result in quadratic equations. Even though quadratic equations may have two roots, a root may be disregarded in an applied problem because it is not appropriate within the context of the problem.

EXAMPLE 4 Find the length and width of a rectangular table if the length is 8 inches more than the width and the area is 260 in.²

Unknown facts Length and width of rectangle

Known facts Area = 260 in.², length = 8 in. more than width

Relationships Let x = the number of inches in the width

$x + 8$ = the number of inches in the length

Area = length times width, or $A = lw$

Estimation The square root of 260 is between 16 and 17; width should be less than 16 and length more than 16.

Solution

$260 = (x + 8)(x)$ Substitute into area formula.

$260 = x^2 + 8x$ Distribute.

$x^2 + 8x - 260 = 0$ Write in standard form.

$a = 1, \qquad b = 8, \qquad c = -260$

$x = \dfrac{-b \pm \sqrt{b^2 - 4ac}}{2a}$ Quadratic formula

$x = \dfrac{-8 \pm \sqrt{(8)^2 - 4(1)(-260)}}{2(1)}$ Substitute.

$x = \dfrac{-8 \pm \sqrt{64 + 1040}}{2}$

$x = \dfrac{-8 \pm \sqrt{1104}}{2}$

$x = \dfrac{-8 \pm 33.22649545}{2}$ Separate into two solutions.

$x = \dfrac{-8 + 33.22649545}{2} \qquad\qquad x = \dfrac{-8 - 33.22649545}{2}$

$x = \dfrac{25.22649545}{2} \qquad\qquad x = \dfrac{-41.22649545}{2}$

$x = 12.6 \qquad\qquad\qquad x = -20.6$

$x + 8 = 20.6$

Interpretation Measurements are positive, so disregard negative solution. **Thus, the width is 12.6 in. and the length is 20.6 in.**

SELF-STUDY EXERCISES 15–2 ─────────────

▣ 1 Solve the following quadratic equations by using the quadratic formula.

1. $3x^2 - 7x - 6 = 0$ **2.** $x^2 + x - 12 = 0$ **3.** $5x^2 - 6x = 11$

4. $x^2 - 6x + 9 = 0$ **5.** $x^2 - x - 6 = 0$ **6.** $8x^2 - 2x = 3$

Solve the following quadratic equations by using the quadratic formula. Round each final answer to the nearest hundredth.

7. $x^2 = -9x + 20$

8. $3x^2 + 6x + 1 = 0$

9. $\dfrac{2}{x} + \dfrac{5}{2} = x$

10. $\dfrac{2}{x} + 3x = 8$

11. $2x^2 + 3x + 3 = 0$

12. $x^2 - x + 2 = 0$

---

2 Solve the following applied problems.

13. A rectangular table top is 3 cm longer than it is wide. Find the length and width to the nearest hundredth if the area is 47.5 cm². (Area = length × width, or $A = lw$.)

14. A rectangular instrument case has an area of 40 in.². If the length is 6 in. more than the width, find the dimensions (length and width) of the instrument case.

15. A bricklayer plans to build an arch with a span (*s*) of 8 m and a radius (*r*) of 4 m. How high (*h*) is the arch? (Use the formula $h^2 - 2hr + \dfrac{s^2}{4} = 0$.)

16. A rectangular patio slab is 180 ft². If the length is 1.5 times the width, find the width and length of the slab to the nearest whole number.

17. A farmer normally plants a field 80 ft by 120 ft. This year the government requires the planting area to be decreased by 20%, and the farmer chooses to decrease the width and length of the field by an equal amount. Show an illustration of the original field and the reduced-size field. If *x* is the amount by which the length and width are decreased, find the length and width of the new field.

18. The recommended dosage of a certain type of medicine is determined by the patient's weight. The formula to determine the dosage is given by $D = 0.1w^2 + 5w$, where *D* is the dosage in milligrams and *w* is the patient's body weight in kilograms. A doctor ordered a dosage of 1800 milligrams. This dosage is to be administered to what weight patient?

15–3   SOLVING PURE QUADRATIC EQUATIONS: $ax^2 + c = 0$

Learning Objective

1 Solve pure quadratic equations by the square-root method ($ax^2 + c = 0$).

The quadratic formula can be used to solve any quadratic equation; however, other methods are less cumbersome for equations with certain characteristics. We will examine alternative methods for pure, incomplete, and complete quadratic equations.

1 Solve Pure Quadratic Equations by the Square-Root Method ($ax^2 + c = 0$).

To solve pure quadratic equations, we solve for the squared letter, then take the square root of both sides of the equation. This procedure is the same one we used in Chapter 12. We will continue to check all roots or solutions.

EXAMPLE 1   Solve $3y^2 = 27$.

$3y^2 = 27$ Divide both sides by the coefficient of the quadratic term.

$y^2 = 9$ Take the square root of both sides.

$y = \pm 3$

Check $y = +3$ Check $y = -3$

$3y^2 = 27$ $3y^2 = 27$

$3(3)^2 = 27$ $3(-3)^2 = 27$

$$3(9) = 27 \qquad 3(9) = 27$$
$$27 = 27 \qquad 27 = 27$$

Rule 15–1   *To solve a pure quadratic equation ($ax^2 + c = 0$):*

1. Rearrange the equation, if necessary, so that quadratic or squared-letter terms are on one side and constants or number terms are on the other side of the equation.
2. Combine like terms, if appropriate.
3. Solve for the squared letter so that the coefficient is $+1$.
4. Apply the square-root principle of equality by taking the square root of both sides.
5. Check both answers.

EXAMPLE 2   Solve $4x^2 - 9 = 0$.

$$4x^2 - 9 = 0$$

$$4x^2 = 9 \qquad \text{Transpose.}$$

$$x^2 = \frac{9}{4} \qquad \text{Solve for the squared letter.}$$

$$x = \pm\frac{3}{2} \qquad \text{Take the square root of the numerator and denominator if they are perfect square whole numbers.}$$

$$x = \pm\mathbf{1.5} \qquad \text{Divide if a decimal answer is desired.}$$

$$\text{Check } x = \boxed{+1.5} \qquad\qquad \text{Check } x = \boxed{-1.5}$$

$$4x^2 - 9 = 0 \qquad\qquad\qquad 4x^2 - 9 = 0$$

$$4(\,1.5\,)^2 - 9 = 0 \qquad\qquad 4(\,-1.5\,)^2 - 9 = 0$$

$$4(2.25) - 9 = 0 \qquad\qquad 4(2.25) - 9 = 0$$

$$9 - 9 = 0 \qquad\qquad\qquad 9 - 9 = 0$$

$$0 = 0 \qquad\qquad\qquad\qquad 0 = 0$$

In some applications the fractional answer may be more convenient. In other applications the decimal answer may be more convenient.

Not all pure quadratic equations have *real solutions*. The next example illustrates such an equation.

EXAMPLE 3   Solve $3m^2 + 48 = 0$

$$3m^2 + 48 = 0$$

$$3m^2 = -48 \qquad \text{Isolate the variable.}$$

$$m^2 = \frac{-48}{3} \qquad \text{Solve for } m^2.$$

$$m^2 = -16$$

$$m = \pm\sqrt{-16} \qquad \text{Apply the square root principle.}$$

There are *no real solutions* to this equation. **If imaginary roots are acceptable, the roots are $\pm 4i$.**

SELF-STUDY EXERCISES 15–3

1 Solve the following equations. Round to thousandths when necessary.

1. $x^2 = 9$ **2.** $x^2 - 49 = 0$ **3.** $9x^2 = 64$
4. $16x^2 - 49 = 0$ **5.** $0.09y^2 = 0.81$ **6.** $0.04x^2 = 0.81$
7. $2x^2 = 10$ **8.** $3x^2 + 4 = 7$ **9.** $5x^2 - 8 = 12$
10. $7x^2 - 2 = 19$

11. The label on a new tarpaulin shows it is a square and contains 529 ft². What is the length of each side of the tarpaulin?

12. A farmer has a field designed in the shape of a circle for efficiency in irrigation. The area of the field is known to be 21,000 yd². What length of irrigation line is required to provide water to the entire field? (*Note:* The line moves from the center of the field in a circle around the field.)

13. Describe the process for solving a pure quadratic equation.

14. What is the relationship between the roots of a pure quadratic equation?

15-4 **SOLVING INCOMPLETE QUADRATIC EQUATIONS: $ax^2 + bx = 0$**

Learning Objective

1 Solve incomplete quadratic equations by factoring ($ax^2 + bx = 0$).

Incomplete quadratic equations have no number terms or constants. Thus, in standard form, $c = 0$. The two remaining terms both have variable factors. This characteristic leads to another method for solving quadratic equations.

1 **Solve Incomplete Quadratic Equations by Factoring ($ax^2 + bx = 0$).**

Incomplete quadratic equations like $x^2 + 2x = 0$ can be solved by factoring a common factor from $x^2 + 2x$. Because an incomplete quadratic equation will always have a variable, we can factor the expression to $x(x + 2) = 0$. We now have two factors whose product is 0. If we multiply two factors and get a product of 0, one or both factors *must* be 0 because $0 \cdot 0 = 0$ as well as $0 \cdot a = 0$, where a is any number. This is called the *zero-product property*.

> **Property 15–1** *Zero-product property:*
>
> If $ab = 0$, then a or b or both equal zero.

If $x = 0$, we can check by substituting 0 for x wherever x occurs in the original equation.

For $x = \boxed{0}$, $x^2 + 2x = 0$

$$\boxed{0}^2 + 2(\boxed{0}) = 0$$

$$0 + 0 = 0$$

$$0 = 0$$

If $x + 2 = 0$, we first solve for x.

$$x + 2 = 0$$

$$x = -2$$

Because $x = -2$, we can check by substituting -2 for x wherever x occurs in the original equation.

For $x = \boxed{-2}$, $x^2 + 2x = 0$

$$(\boxed{-2})^2 + 2(\boxed{-2}) = 0 \qquad {\scriptstyle (-2)^2\,=\,+4}$$

$$4 + (-4) = 0$$

$$0 = 0$$

Therefore, we find that the original equation checks with either value of x. This means that x may be 0 or -2 in this incomplete quadratic equation.

> **Rule 15–2** *To solve an incomplete quadratic equation $(ax^2 + bx = 0)$ by factoring:*
>
> 1. If necessary, transpose so that all terms are on one side of the equation in standard form, with zero on the other side.
> 2. Factor the expression that always has a common factor and set each factor equal to zero.
> 3. Solve for the variable in each equation formed in Step 2.

EXAMPLE 1   Solve $2x^2 = 5x$ for x.

$$2x^2 = 5x \qquad \text{Put in standard form.}$$

$$2x^2 - 5x = 0$$

$$x\,\boxed{(2x - 5)} = 0 \qquad \text{Factor common factor.}$$

$$\boxed{x = 0} \quad \boxed{2x - 5 = 0} \qquad \text{Set each factor equal to zero.}$$

$$\mathbf{x = 0} \qquad 2x = 5 \qquad \text{Solve each equation.}$$

$$x = \frac{5}{2}$$

Check $x = \boxed{0}$ $\qquad$ Check $x = \boxed{\dfrac{5}{2}}$

$$2x^2 = 5x \qquad\qquad 2x^2 = 5x$$

$$2(\boxed{0})^2 = 5(\boxed{0}) \qquad 2\left(\frac{\boxed{5}}{\boxed{2}}\right)^2 = 5\left(\frac{\boxed{5}}{\boxed{2}}\right)$$

$$2(0) = 5(0) \qquad \overset{1}{2}\left(\frac{25}{\underset{2}{4}}\right) = \frac{25}{2}$$

$$0 = 0 \qquad \frac{25}{2} = \frac{25}{2}$$

EXAMPLE 2 Solve $4x^2 - 8x = 0$ for x.

$$4x^2 - 8x = 0 \qquad \text{Standard form.}$$

$$\boxed{4x}\,\boxed{(x - 2)} = 0 \qquad \text{Factor common factors.}$$

$$\boxed{4x = 0} \quad \boxed{x - 2 = 0} \qquad \text{Set each factor equal to zero.}$$

$$x = \frac{0}{4} \qquad x = 2 \qquad \text{Solve each equation.}$$

$$x = 0$$

$$\text{Check } x = \boxed{0} \qquad\qquad \text{Check } x = \boxed{2}$$

$$4(\boxed{0})^2 - 8(\boxed{0}) = 0 \qquad 4(\boxed{2})^2 - 8(\boxed{2}) = 0$$

$$4(0) - 8(0) = 0 \qquad 4(4) - 8(2) = 0$$

$$0 - 0 = 0 \qquad 16 - 16 = 0$$

$$0 = 0 \qquad\qquad 0 = 0$$

Tips and Traps *Roots of Incomplete Quadratic Equations:*

In an incomplete quadratic equation, the value of $c = 0$: $ax^2 + bx + 0 = 0$.

$$x = \frac{-b \pm \sqrt{b^2 - 4a(c)}}{2a}$$

$$x = \frac{-b \pm \sqrt{b^2}}{2a}$$

$$x = \frac{-b + b}{2a} \qquad x = \frac{-b - b}{2a}$$

$$x = \frac{0}{2a} \qquad x = \frac{-2b}{2a}$$

$$x = 0 \qquad x = \frac{-b}{a}$$

- One root will always be zero.
- The other root will be $\frac{-b}{a}$.

SELF-STUDY EXERCISES 15–4

1 Solve the following equations.

1. $x^2 - 3x = 0$

2. $3x^2 = 7x$

3. $5x^2 - 10x = 0$

4. $2x^2 + x = 0$

5. $8x^2 - 4x = 0$

6. $5x^2 = 15x$

7. $x^2 - 6x = 0$

8. $y^2 = -4y$

9. $3x^2 + 2x = 0$

10. $9x^2 - 12x = 0$

11. The square of a number is 8 times the number. Find the number.

12. A square rug has an area that is 15 times the length of one of the sides. What is the length of a side?

13. How does an incomplete quadratic equation differ from a pure quadratic equation?

14. Will one of the two roots of an incomplete quadratic equation always be zero? Justify your answer.

15-5 SOLVING COMPLETE QUADRATIC EQUATIONS BY FACTORING

Learning Objective

1 Solve complete quadratic equations by factoring ($ax^2 + bx + c = 0$).

Complete quadratic equations will have all three types of terms. If the expression on the left will factor into two binomials, we can apply the zero-product property.

1 Solve Complete Quadratic Equations by Factoring ($ax^2 + bx + c = 0$).

In this section, we use knowledge of factoring trinomials to solve quadratic equations by factoring. If the trinomial will not factor, another method for solving quadratic equations will have to be used.

> **Rule 15-3** *To solve a complete quadratic equation by factoring:*
>
> **1.** If necessary, transpose terms so that all terms are on one side of the equation in standard form, with zero on the other side.
> **2.** Factor the trinomial, if possible, into the product of two binomials and set each factor equal to zero.
> **3.** Solve for the variable in each equation formed in Step 2.

EXAMPLE 1 Solve $x^2 + 6x + 5 = 0$ for x.

$$x^2 + 6x + 5 = 0$$
$$(x + 5)(x + 1) = 0 \quad \text{Factor into the product of two binomials.}$$
$$x + 5 = 0 \quad x + 1 = 0 \quad \text{Set each factor equal to 0.}$$
$$x = -5 \quad x = -1 \quad \text{Solve for } x \text{ in each equation.}$$

Check $x = -5$	Check $x = -1$
$x^2 + 6x + 5 = 0$	$x^2 + 6x + 5 = 0$
$(-5)^2 + 6(-5) + 5 = 0$	$(-1)^2 + 6(-1) + 5 = 0$
$25 + (-30) + 5 = 0$	$1 + (-6) + 5 = 0$
$-5 + 5 = 0$	$-5 + 5 = 0$
$0 = 0$	$0 = 0$

EXAMPLE 2 Solve $6x^2 + 4 = 11x$ for x.

$$6x^2 + 4 = 11x$$
$$6x^2 - 11x + 4 = 0 \quad \text{Transpose into standard form.}$$
$$(3x - 4)(2x - 1) = 0 \quad \text{Factor into the product of two binomials.}$$
$$3x - 4 = 0 \quad 2x - 1 = 0 \quad \text{Set each factor equal to 0.}$$

$$3x = 4 \qquad 2x = 1 \qquad \text{Solve each equation.}$$

$$x = \frac{4}{3} \qquad x = \frac{1}{2}$$

$$\text{Check } x = \frac{4}{3} \qquad\qquad \text{Check } x = \frac{1}{2}$$

$$6x^2 + 4 = 11x \qquad\qquad 6x^2 + 4 = 11x$$

$$6\left(\frac{4}{3}\right)^2 + 4 = 11\left(\frac{4}{3}\right) \qquad 6\left(\frac{1}{2}\right)^2 + 4 = 11\left(\frac{1}{2}\right)$$

$$\overset{2}{6}\left(\frac{16}{\underset{3}{9}}\right) + 4 = \frac{44}{3} \qquad\qquad \overset{3}{6}\left(\frac{1}{\underset{2}{4}}\right) + 4 = \frac{11}{2}$$

$$\left(4 = \frac{12}{3}\right) \quad \frac{32}{3} + \frac{12}{3} = \frac{44}{3} \qquad\qquad \frac{3}{2} + \frac{8}{2} = \frac{11}{2} \quad \left(4 = \frac{8}{2}\right)$$

$$\frac{44}{3} = \frac{44}{3} \qquad\qquad \frac{11}{2} = \frac{11}{2}$$

Tips and Traps **Roots of Complete Quadratic Equations:**
Because all three types of terms are included in complete quadratic equations, we see that the roots will *not* have the same characteristics as the roots of pure or incomplete quadratic equations.

- Zero will not be a root.
- The two roots will not have the same absolute value.

SELF-STUDY EXERCISES 15–5

[1] Solve the following equations by factoring.

1. $x^2 + 5x + 6 = 0$ **2.** $x^2 - 6x + 9 = 0$ **3.** $x^2 - 5x - 14 = 0$

4. $x^2 + 3x - 18 = 0$ **5.** $x^2 + 7x + 12 = 0$ **6.** $y^2 - 8y = -15$

7. $a^2 - 13a - 14 = 0$ **8.** $b^2 - 9b = -18$ **9.** $2x^2 - 7x + 3 = 0$

10. $3x^2 + 13x + 4 = 0$ **11.** $10x^2 - x = 3$ **12.** $6x^2 + 11x + 3 = 0$

13. $2x^2 + 13x + 15 = 0$ **14.** $3x^2 - 10x + 8 = 0$ **15.** $6x^2 + 17x = 3$

16. A rectangular hallway is 6 ft longer than its width. The area is 55 ft^2. What are the length and width of the hallway?

17. A rectangular metal plate covering a spare tire well that has an opening of 378 in.^{2} is broken and must be reconstructed. The width is 3 in. less than the length. What dimensions should the metalsmith use when making the replacement part?

15–6 **SELECTING AN APPROPRIATE METHOD FOR SOLVING QUADRATIC EQUATIONS**

Learning Objective [1] Determine the nature of the roots of a quadratic equation by examining the discriminant.

In the previous sections, we have made various observations about the roots of the different types of quadratic equations. Now, we will make some additional observations about the roots of quadratic equations.

1 Determine the Nature of the Roots of a Quadratic Equation by Examining the Discriminant.

Which method for solving quadratic equations is best? Since any methods that are mathematically sound and produce consistently correct solutions are good, a "best" method may be evaluated differently. If you have mastered factoring, the factoring methods are generally quicker, especially when a calculator or computer is unavailable.

In general, follow these suggestions for solving quadratic equations.

1. Write the equation in standard form: $ax^2 + bx + c = 0$.
2. Identify the type of quadratic equation.
3. Solve pure quadratic equations by the square-root method.
4. Solve incomplete quadratic equations by finding common factors.
5. Solve complete quadratic equations by factoring into two binomials, if possible.
6. If factoring is not possible or is difficult, use the quadratic formula to solve complete quadratic equations.

All types of quadratic equations can be solved using the completing-the-square method (not presented in this text) or the quadratic formula. Only certain types of quadratic equations can be solved by factoring or applying the square root principle. The radicand of the radical portion of the quadratic formula, $b^2 - 4ac$, is called the *discriminant*. The general characteristics of the quadratic equation can be determined by examining the discriminant.

Property 15–2 *Properties of the discriminant, $b^2 - 4ac$:*

1. If $b^2 - 4ac \geq 0$, the equation has real-number roots.
   a. If $b^2 - 4ac$ is a perfect square, there are two rational roots.
   b. If $b^2 - 4ac = 0$, there is one rational root.
   c. If $b^2 - 4ac$ is not a perfect square, there are two irrational roots.
2. If $b^2 - 4ac < 0$, the equation has no real-number roots. The roots are imaginary or complex.

EXAMPLE 1   Examine the discriminant of each equation and determine the nature of the roots. Then solve the equation.

(a) $5x^2 + 3x - 1 = 0$ (b) $3x^2 + 5x = 2$ (c) $4x^2 + 2x + 3 = 0$

(a) $5x^2 + 3x - 1 = 0$ $a = 5, b = 3, c = -1$
$\quad b^2 - 4ac = 3^2 - 4(5)(-1)$ Examine the discriminant.
$\qquad\qquad = 9 + 20$
$\qquad\qquad = \boxed{29}$ There will be two irrational roots.

$x = \dfrac{-b \pm \sqrt{b^2 - 4ac}}{2a}$ Use the quadratic formula.

$x = \dfrac{-3 \pm \sqrt{29}}{2(5)}$ Substitute.

$\boldsymbol{x = \dfrac{-3 + \sqrt{29}}{10}}$ or $\boldsymbol{x = \dfrac{-3 - \sqrt{29}}{10}}$ Exact roots

(b) $\quad 3x^2 + 5x = 2$
$\quad 3x^2 + 5x - 2 = 0$ Standard form: $a = 3, b = 5, c = -2$
$\quad b^2 - 4ac = 5^2 - 4(3)(-2)$ Examine the discriminant.
$\qquad\qquad = 25 + 24$

$$= 49 \qquad \text{There are two rational roots.}$$
$$(3x - 1)(x + 2) = 0 \qquad \text{Factor.}$$
$$3x - 1 = 0 \quad x + 2 = 0$$
$$3x = 1$$
$$\boldsymbol{x = \frac{1}{3}} \qquad \boldsymbol{x = -2} \qquad \text{Exact roots}$$

(c) $4x^2 + 2x + 3 = 0$ $\qquad a = 4,\ b = 2,\ c = 3$
$$b^2 - 4ac = 2^2 - 4(4)(3) \qquad \text{Examine the discriminant.}$$
$$= 4 - 48$$
$$= \boxed{-44} \qquad \text{There are no real roots.}$$
$$x = \frac{-2 \pm \sqrt{-44}}{2 \cdot 4} \qquad \text{Quadratic equation}$$
$$x = \frac{-2 \pm 2i\sqrt{11}}{8} \qquad \text{Simplify the radical and reduce.}$$
$$x = \frac{2(-1 \pm i\sqrt{11})}{8} \qquad \text{Factor and simplify the imaginary number.}$$
$$\boldsymbol{x = \frac{-1 \pm i\sqrt{11}}{4}} \qquad \text{Reduce.}$$

SELF-STUDY EXERCISES 15–6

1 Examine the discriminant of each equation and find the roots. Round to hundredths if necessary.

1. $3x^2 + x - 2 = 0$ **2.** $x^2 - 3x = -1$ **3.** $2x^2 + x = 2$
4. $3x^2 - 2x + 1 = 0$ **5.** $x^2 - 3x - 7 = 0$ **6.** $3x^2 + 5x - 6 = 0$
7. Use the discriminant to write a quadratic equation that has real roots.   **8.** Use the discriminant to write a quadratic equation that has roots that are real, rational, and unequal.

15–7 SOLVING HIGHER-DEGREE EQUATIONS BY FACTORING

Learning Objectives

1 Identify the degree of an equation.
2 Solve higher-degree equations by factoring.

Some equations contain terms that have a higher degree than two. We will look at a few types of equations of higher degree that can be written in factored form.

1 Identify the Degree of an Equation.

In Section 15–1 we defined a *quadratic* equation as an equation that has at least one letter term raised to the second power, and no letter terms have a power higher than 2. Similarly, a *cubic equation* is an equation that has at least one letter term raised to the third power, and no letter terms have a power higher than 3. A quadratic equation can also be referred to as a *second-degree* equation and a cubic equation as a *third-degree* equation.

■ **DEFINITION 15–2: Degree of an Equation.** The *degree of an equation* in one variable is the highest power of any letter term that appears in the equation.

When we solved linear or first-degree equations, we obtained at most one solution or root for the equation. When we solved quadratic or second-degree

584 CHAPTER 15 Solving Quadratic and Higher-Degree Equations

equations, we obtained at most two solutions or roots. Similarly, a cubic or third-degree equation will have at most three solutions or roots, and fourth-degree equations will have at most four solutions or roots.

EXAMPLE 1   State the degree of each of the following equations.

(a) $x^2 + 3x + 4 = 0$ (b) $x^3 = 27$ (c) $x^4 + 3x^3 + 2x^2 + x + 4 = 0$
(d) $x^8 = 256$ (e) $3x + 7 = 5x - 3$

(a) $x^2 + 3x + 4 = 0$ (b) $x^3 = 27$
    Quadratic or second-degree equation   Cubic or third-degree equation
(c) $x^4 + 3x^3 + 2x^2 + x + 4 = 0$ (d) $x^8 = 256$
    Fourth-degree equation   Eighth-degree equation
(e) $3x + 7 = 5x - 3$
    Linear or first-degree equation

2 Solve Higher-Degree Equations by Factoring.

In our discussion, we will limit our study to a few basic types of *higher-degree equations*. A more thorough study is reserved for advanced mathematics courses. We will look only at higher-degree equations that are written in factored form or can be written easily in factored form. We will solve these equations by using the zero-product property; that is, if the product of the factors equals 0, then all factors containing a variable may be equal to 0. (See Section 15–4.)

EXAMPLE 2   Solve the following equations by using the zero-product property.

(a) $x(x + 4)(x - 3) = 0$ (b) $5x(x - 7)(2x - 1) = 0$
(c) $2x^3 - 14x^2 + 20x = 0$ (d) $3x^3 - 15x = 0$

(a) $x(x + 4)(x - 3) = 0$ Already factored

$x = 0$ $x + 4 = 0$ $x - 3 = 0$ Set each factor equal to zero.
$\mathbf{x = 0}$ $\mathbf{x = -4}$ $\mathbf{x = 3}$ Solve each equation.

(b) $5x(x - 7)(2x - 1) = 0$ Already factored

$5x = 0$ $x - 7 = 0$ $2x - 1 = 0$ Set each factor to zero.
$\dfrac{5x}{5} = \dfrac{0}{5}$ $\mathbf{x = 7}$ $2x = 1$ Solve each equation.
$\mathbf{x = 0}$ $\dfrac{2x}{2} = \dfrac{1}{2}$
     $\mathbf{x = \dfrac{1}{2}}$

(c) $2x^3 - 14x^2 + 20x = 0$
   $2x(x^2 - 7x + 10) = 0$ Factor out the common factors.
  $2x(x - 5)(x - 2) = 0$ Factor the trinomial.

$2x = 0$ $x - 5 = 0$ $x - 2 = 0$ Set each factor equal to zero.
$\dfrac{2x}{2} = \dfrac{0}{2}$ $\mathbf{x = 5}$ $\mathbf{x = 2}$ Solve each equation.
$\mathbf{x = 0}$

(d) $3x^3 - 15x = 0$ Factor out the common factors.
         $3x(x^2 - 5) = 0$ Factor the difference of two squares,
     $3x(x^2 - 5) = 0$ if possible.

$3x = 0$ $x^2 - 5 = 0$ Set each factor equal to zero.
$\dfrac{3x}{3} = \dfrac{0}{3}$ $x^2 = 5$ Solve each equation.
$\mathbf{x = 0}$ $\mathbf{x = \pm\sqrt{5}}$

1 State the degree of each of the following equations.

1. $x^2 + 2x - 3 = 0$ **2.** $3x + 2x = 5$ **3.** $x^4 = 42$

4. $2x - 7 - 4x = 3$ **5.** $3x^3 + 2x^4 + 3 = x^2$ **6.** $x^7 = 128$

2 Find the roots of the following equations. Factor if necessary.

7. $x(x - 2)(x + 3) = 0$ **8.** $2x(2x - 1)(x + 3) = 0$ **9.** $3x(2x - 5)(3x - 2) = 0$

10. $x^3 - 7x^2 + 10x = 0$ **11.** $3x^3 - 3x^2 = 18x$ **12.** $4x^3 + 10x^2 + 4x = 0$

13. $2x^3 - 18x = 0$ **14.** $12x^3 = 3x$ **15.** $16x^3 = 9x$

16. Thirty-two times a number is subtracted from twice the cube of the number and the result is zero. How many numbers meet these conditions? What are they?

17. Find all the numbers that satisfy the following conditions: A number cubed is increased by 5 times the number and the result is zero.

18. A shipping container is a large cardboard box. The volume is found by multiplying the length times the width times the height. The length of the box is 3 feet more than the width and the height is 7 feet more than the width. The volume is 421 ft^3. Write an equation to find the length, width, and height. Can this problem be solved by the methods found in Section 15–7? If so, solve it. If not, explain why it can't be solved.

CAREER APPLICATION

Civil Engineering: Freeway Supports

Safe, dependable, durable freeway support design is extremely important because highways carry an ever-increasing number of heavy vehicles (such as tandem tractor-trailer trucks) and heavier weight loads. Civil engineers try to design supports that hold up during temperature extremes, earthquakes, and high wind conditions (for example, tornadoes and hurricanes). When designing a rectangular box support, researchers found that support is optimal when the depth is at least one half of the width, and the volume is at least 10 ft^3 per ft of support height.

Civil engineers lobby a state government to allot construction money for a freeway passing through an earthquake fault zone at a higher, safer volume rate of 12.5 ft^3 per ft of support height. At intersections, rectangular box supports 16 ft high are required.

Exercises

Use the above information to answer the following. Round answers to the nearest tenth.

1. Draw a diagram of a rectangular box support with a 16-ft height. Label the sides W for width, and D for depth.

2. What is the state-allotted volume of this 16-ft-high support?

3. Use W to write an expression for the depth if it is to be $\frac{1}{2}$ the width.

4. Use the volume formula $V = LWD$ to write an equation to find the width and depth of the support. Solve the equation, and state your answers rounded to the nearest tenth of a ft. What is the volume of this support?

5. If a support has a depth $\frac{3}{4}$ the size of the width, find its dimensions and volume.

6. This freeway will be in an earthquake fault zone, so state civil engineers decide to make the depth equal to the width of the support to withstand maximum vibration stress. Find the dimensions and volume of each support.

7. Why did the dimensions of Exercise 4 have a volume of exactly 200 ft^3, while the dimensions of Exercises 5 and 6 had volumes over and under the prescribed 200 ft^3?

8. A recent computer modeling simulation found that supports with the smallest cross-sectional perimeter per volume are best able to endure vibration stress, which means that supports in earthquake zones should be in the shape of a cylinder (a cylinder has a circular cross-section perimeter). Use $V = \pi r^2 h$ to find the radius of the support with a height of 16 ft and a volume of 200 ft^3.

9. Find the cross-sectional perimeters of the supports designed in Exercises 4, 5, 6, and 8. Does the circular cross-section have the smallest perimeter?

Answers

1. Support diagram:

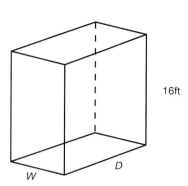

2. At 12.5 ft^3 per ft of height, 200 ft^3 is allotted.

3. $D = (\frac{1}{2})W$ or $D = \frac{W}{2}$ or $D = 0.5W$.

4. The width should be 5.0 ft and the depth 2.5 ft, with a volume of 200.0 ft^3.

5. From the equation $0.75W(W) = 200$, the width should be 4.1 ft, the depth should be 3.1 ft, and the height should be 16 ft. These dimensions would have a volume of 203.4 ft^3.

6. With $D = W$, both the width and depth should be 3.54 ft, and the height should be 16 ft. These dimensions would have a volume of 200.5 ft^3.

7. When solving the pure quadratic equation of Exercise 4 by the square root method, an exact square root was found. In Exercises 5 and 6, square roots were approximated to the nearest tenth of a foot, which resulted in round-off error propagation in the volume formula.

8. The radius should be 2.0 ft, which produces a volume of 201.1 ft^3.

9. The perimeter of the rectangular cross-sections is 15 ft in Exercise 4 and 14.4 ft in Exercise 5. The square cross-section of Exercise 6 is 14.2 ft, while the perimeter of the circular cross-section of Exercise 8 is 12.6 ft. The circular cross-section does have the smallest perimeter per volume.

ASSIGNMENT EXERCISES

Section 15–1
Identify the following quadratic equations as pure, incomplete, or complete.

1. $x^2 = 49$
2. $x^2 - 5x = 0$
3. $5x^2 - 45 = 0$
4. $3x^2 + 2x - 1 = 0$
5. $8x^2 + 6x = 0$
6. $5x^2 + 2x + 1 = 0$
7. $x^2 - 32 = 0$
8. $x^2 + x = 0$
9. $3x^2 + 6x + 1 = 0$

Section 15–2
Indicate the values for a, b, and c in the following quadratic equations.

10. $5x^2 + x + 6 = 0$
11. $x^2 - 2x = 8$
12. $x^2 - 7x + 12 = 0$
13. $x^2 + 3x = 4$
14. $3x^2 = 2x + 7$
15. $x^2 - 3x = -2$

Solve the following quadratic equations by using the quadratic formula.

16. $x^2 - 9x + 20 = 0$
17. $x^2 - 8x - 9 = 0$
18. $x^2 - 5x = -6$
19. $x^2 + 2x = 8$
20. $x^2 - x - 12 = 0$
21. $2x^2 - 3x - 2 = 0$

Solve the following quadratic equations by using the quadratic formula. Round each final answer to the nearest hundredth.

22. $3x^2 + 6x + 2 = 0$

23. $2x^2 - 3x - 1 = 0$

24. $5x^2 + 4x - 8 = 0$

25. $3x^2 + 5x + 1 = 0$

26. A bricklayer plans to build an arch with a span (s) of 10 m and a radius (r) of 5 m. How high (h) is the arch? (Use the formula $h^2 - 2hr + \frac{s^2}{4} = 0$.)

27. A rectangular kitchen contains 240 ft^2. If the length is two times the width, find the length and width of the room to the nearest whole number. (Area = length × width, or $A = lw$.)

28. What are the dimensions of a rectangular tool storage room if the area is 45.5 m^2 and the room is 0.5 m longer than it is wide? Round to the nearest tenth.

29. Find the length and width of a piece of fiberglass if its length is three times the width and the area is 591 in.^{2}. Round to the nearest inch.

Section 15–3

Solve the following equations. Round to thousandths when necessary.

30. $x^2 = 121$

31. $x^2 = 100$

32. $x^2 - 64 = 0$

33. $4x^2 = 9$

34. $64x^2 - 49 = 0$

35. $0.36y^2 = 1.09$

36. $0.16x^2 = 0.64$

37. $5x^2 = 40$

38. $2x^2 - 5 = 3$

39. $6x^2 + 4 = 34$

40. $5x^2 - 6 = 19$

41. $3x^2 = 12$

42. $3x^2 - 4 = 8$

43. $2x^2 = 34$

44. $5x^2 - 9 = 30$

45. $3y^2 - 36 = -8$

46. $2x^2 + 3 = 51$

47. $\frac{1}{2}x^2 = 8$

48. $\frac{2}{3}x^2 = 24$

49. $\frac{1}{4}x^2 - 1 = 15$

50. $\frac{2}{5}x^2 + 2 = 8$

51. A circle has an area of 845 cm^2. What is the radius of the circle?

52. An oil painting that is a square is known to be 9072 cm^2. What are the inside dimensions of its picture frame?

Section 15–4

Solve the following equations.

53. $x^2 - 5x = 0$

54. $4x^2 = 8x$

55. $6x^2 - 12x = 0$

56. $3x^2 + x = 0$

57. $10x^2 + 5x = 0$

58. $3y^2 = 12y$

59. $y^2 - 5y = 0$

60. $x^2 = 16x$

61. $12x^2 + 8x = 0$

62. $8x^2 - 12x = 0$

63. $x^2 + 3x = 0$

64. $4x^2 - 28x = 0$

65. $5x^2 = 45x$

66. $7x^2 = 28x$

67. $y^2 + 8y = 0$

68. $z^2 - 6z = 0$

69. $3m^2 - 5m = 0$

70. $4n^2 - 3n = 0$

71. $2x^2 = x$

72. $5y^2 = y$

73. Three times the square of a number is the same as 12 times the number. What is the number?

74. Describe the steps for solving an incomplete quadratic equation. Include a clear description of the nature of the roots.

Section 15–5

Solve the following equations by factoring.

75. $x^2 - 4x + 3 = 0$

76. $x^2 + 7x + 12 = 0$

77. $x^2 + 3x = 10$

78. $x^2 - 7x + 12 = 0$

79. $x^2 + 7x = -6$

80. $x^2 + 3 = -4x$

81. $x^2 - 6x + 8 = 0$

82. $6y + 7 = y^2$

83. $6y^2 - 5y - 6 = 0$

84. $5y^2 + 23y = 10$

85. $10y^2 - 21y - 10 = 0$

86. $6x^2 - 16x + 8 = 0$

87. $4x^2 + 7x + 3 = 0$

88. $3x^2 = -7x + 6$

89. $12y^2 - 5y - 3 = 0$

90. $x^2 - 3x = 18$

91. $x^2 + 19x = 42$

92. $3x^2 + x - 2 = 0$

93. $3y^2 + y - 2 = 0$

94. $2x^2 - 4x - 6 = 0$

95. $2x^2 - 10x + 12 = 0$

96. $y^2 + 18y + 45 = 0$

97. $x^2 - 3x - 18 = 0$

98. $3x^2 - 9x - 30 = 0$

99. $2y^2 + 22y + 60 = 0$

100. An office building in the shape of a rectangle is known to have 47,500 ft^2 of space on the ground floor. The tenant wants to landscape the two longer sides of the building. The tenant also knows that the building is about 60 ft longer than it is wide. How many feet of land along the building will need to be landscaped?

101. Jerri Amour is an architect and is designing a hospital. She knows that a kidney dialysis machine will need a space that is 7 ft longer than it is wide. The total area needed is 228 ft^2 of space. Advise Jerri about the number of feet of space she should include in the length and width of the planned space.

Section 15–6

Use the discriminant of the quadratic formula to describe the roots of the following equations.

102. $x^2 - 3x + 2 = 0$
103. $x^2 + 8x + 16 = 0$
104. $2x^2 - 3x - 5 = 0$
105. $5x^2 - 100 = 0$
106. $3x^2 - 2x + 4 = 0$
107. $2x = 5x^2 - 3$

Section 15–7

State the degree of each of the following equations.

108. $2x + 5x = 15$
109. $3x - 2x^3 + 8 = 0$
110. $16 = x^4$
111. $6 - 3x - 3 = 2x + 4$
112. $y^6 = 729$
113. $5y^8 + 2y^3 - 6 = y^2$

Find the roots of the following equations. Factor if necessary.

114. $x(x + 2)(x - 3) = 0$
115. $2x(3x - 2)(x - 2) = 0$
116. $3x(2x + 1)(x + 4) = 0$
117. $2x^3 + 10x^2 + 12x = 0$
118. $x^3 = 2x^2$
119. $2x^3 + 9x^2 = 5x$
120. $6x^3 + 3x^2 - 18x = 0$
121. $3x^3 - 6x^2 = 0$
122. $3x^3 - x^2 - 2x = 0$
123. $x^3 + 6x^2 + 8x = 0$
124. $x^3 - 8x^2 + 15x = 0$
125. $x^3 - x^2 - 20x = 0$
126. $x^3 + x^2 - 20x = 0$
127. $y^3 - 6y^2 + 7y = 0$
128. $y^3 + 2y^2 + 5y = 0$
129. $x^3 - 3x^2 - 4x = 0$

130. A number raised to the third power is the same as 121 times the number. What is the number? Give all possibilities.

131. A number multiplied by 51 is the same as three times the cube of the number. What is the number?

CHALLENGE PROBLEM

132. Many objects are designed with dimensions according to the ratio of the *Golden Ratio.* Objects that have measurements according to this ratio are said to be most pleasing to the eye. The *Golden Rectangle* has dimensions of length (*l*) and height (*h*) that satisfy the following formula.

$$\frac{l + h}{l} = \frac{l}{h}$$

You can cross-multiply to obtain a quadratic equation for the Golden Rectangle.

You have been commissioned to construct a wall hanging for the lobby of a new office building. The wall is 32 ft long and the ceiling is 20 ft high. The owners want the wall hanging to be at least 2 ft from the ceiling, the floor, and each of the side corners. They also want the wall hanging to have dimensions according to the Golden Ratio.

Using the given information, determine the largest-size wall hanging that can be placed in the lobby that also has the dimensions of the Golden Rectangle.

CONCEPTS ANALYSIS

1. State the zero-product property and explain how it applies to solving quadratic equations.

2. What is an extraneous root?

3. What technique for solving equations can produce an extraneous root?

4. When will a quadratic equation have no real solutions?

5. When will a quadratic equation have irrational roots?

6. When will a quadratic equation have rational roots?

Find a mistake in each of the following. Correct and briefly explain the mistake.

7. $2x^2 - 2x - 12 = 0$

$2(x^2 - x - 6) = 0$

$2(x + 2)(x - 3) = 0$

$2 = 0, \quad x + 2 = \quad 0, \quad x - 3 = 0$

$\qquad\qquad x = -2, \qquad\qquad x = 3$ The roots are 0, -2, and 3.

8. $2x^3 + 5x^2 + 2x = 0$

$\dfrac{x(2x^2 + 5x + 2)}{x} = \dfrac{0}{x}$

$2x^2 + 5x + 2 = 0$

$(2x + 1)(x + 2) = 0$

$2x + 1 = \quad 0 \qquad x + 2 = \quad 0$

$2x = -1 \qquad\qquad x = -2$

$x = -\dfrac{1}{2}$ The roots are $-\frac{1}{2}$ and -2.

9. Why are you not allowed to divide both sides of an equation by a variable?

10. What is the maximum number of roots that the equation $x^4 = 16$ *could* have? How many real roots does the equation have?

CHAPTER SUMMARY

Objectives	What to Remember with Examples

Section 15–1

1 Write quadratic equations in standard form.

Write the equation with all terms on the left side of the equation and zero on the right and arrange with the terms in descending powers. The leading coefficient should be positive.

Write the equation $5x + 3 = 4x^2$ in standard form.

$-4x^2 + 5x + 3 = 0$

$-1(-4x^2 + 5x + 3) = -1(0)$

$4x^2 - 5x - 3 = 0$

2 Identify the coefficients of the quadratic, linear, and constant terms of a quadratic equation.

The coefficient of the quadratic term is the coefficient of the squared variable; the coefficient of the linear term is the coefficient of the first-power variable and the constant term is its own coefficient.

Identify the coefficient of the quadratic and linear terms and identify the constant term in the following:

$5x = 4x^2 - 3$

First, write the equation in standard form:

$4x^2 - 5x - 3 = 0.$

The coefficient of the quadratic term is 4. The coefficient of the linear term is -5 and the constant term is -3.

Section 15–2

1 Solve quadratic equations using the quadratic formula.

In the quadratic formula

$$x = \frac{-b \pm \sqrt{b^2 - 4ac}}{2a},$$

a is the coefficient of the quadratic term, b is the coefficient of the linear term, and c is the constant when the equation is written in standard form. The variable is x.

The characteristics of the roots are determined by examining the discriminant (Section 15–6).

Solve using the quadratic formula:

$5x^2 + 7x - 6 = 0$
$a = 5, b = 7, c = -6$

$$x = \frac{-7 \pm \sqrt{7^2 - 4(5)(-6)}}{2(5)}$$

$$x = \frac{-7 \pm \sqrt{49 + 120}}{10}$$

$$x = \frac{-7 \pm \sqrt{169}}{10}$$

$$x = \frac{-7 \pm 13}{10}$$

$$x = \frac{6}{10} = \frac{3}{5}$$

$$x = \frac{-20}{10} = -2$$

2 Solve applied problems that use quadratic equations.

Solving applied problems often requires knowledge of other mathematical formulas, for example, $A = lw$ (area of a rectangle).

Find the length and width of a rectangular parking lot if the length is to be 12 m longer than the width and the area is to be 6205 m^2.

Let x = number of m in width
$x + 12$ = number of m in length

$$lw = A$$
$$x(x + 12) = 6205$$
$$x^2 + 12x = 6205$$
$$x^2 + 12x - 6205 = 0$$

Use a calculator with the formula.

$$x = \frac{-12 \pm \sqrt{12^2 - 4(1)(-6205)}}{2(1)}$$

$$x = \frac{-12 \pm \sqrt{144 + 24820}}{2}$$

$$x = \frac{-12 \pm \sqrt{24964}}{2}$$

$$x = \frac{-12 \pm 158}{2}$$

$$x = \frac{-12 + 158}{2}, x = \frac{-12 - 158}{2}$$

$$x = \frac{146}{2}$$

$x = 73$ width   Disregard the negative root.
$x + 12 = 85$ length

Section 15–3

1 Solve pure quadratic equations by the square-root method ($ax^2 + c = 0$).

To solve pure quadratic equations, solve for the squared variable; then take the square root of both sides of the equation. The two roots have the same absolute value.

> Solve $2x^2 - 72 = 0$.
>
> $2x^2 = 72$
> $x^2 = 36$
> $x = \pm 6$

Section 15–4

1 Solve incomplete quadratic equations by factoring ($ax^2 + bx = 0$).

Arrange the equation in standard form. Factor the common factor. Then set each of the two factors equal to zero and solve for the variable. Both roots are rational and one root is always zero.

> Solve $5x^2 - 15x = 0$.
>
> $5x(x - 3) = 0$
> $5x = 0 \qquad x - 3 = 0$
> $x = 0 \qquad x = 3$

Section 15–5

1 Solve complete quadratic equations by factoring ($ax^2 + bx + c = 0$).

Arrange the quadratic equation in standard form and factor the trinomial. Then set each factor equal to zero and solve for the variable.

If the expression factors, the roots are rational.

> Solve $2x^2 - 5x - 3 = 0$.
>
> $(x - 3)(2x + 1) = 0$
> $x - 3 = 0 \qquad 2x + 1 = 0$
> $x = 3 \qquad 2x = -1$
> $\qquad\qquad\qquad\qquad x = -\dfrac{1}{2}$

Section 15–6

1 Determine the nature of the roots of a quadratic equation by examining the discriminant.

The radicand of the quadratic formula, $b^2 - 4ac$, is the discriminant of the quadratic equation.
1. If $b^2 - 4ac \geq 0$, the equation has real-number solutions.
   a. If $b^2 - 4ac$ is a perfect square, the two solutions are rational.
   b. If $b^2 - 4ac = 0$, there is one rational solution.
   c. If $b^2 - 4ac$ is not a perfect square, the two roots are irrational.
2. If $b^2 - 4ac < 0$, the equation has no real-number solutions.

> Use the discriminant to determine the characteristics of the roots of the equation $3x^2 - 5x + 7 = 0$.
>
> $(-5)^2 - 4(3)(7) = 25 - 84$
> $\qquad\qquad\qquad\quad = -59$
>
> The roots are not real.

Section 15–7

1 Identify the degree of an equation.

The degree of an equation in one variable is the highest power of any term that appears in the equation.

> State the degree: $x^3 + 4x = 0$ is a cubic or third-degree equation.
> $x^4 = 81$ is a fourth-degree equation.

2 Solve higher-degree equations by factoring.

The higher-degree equations discussed in this section will have a common variable factor and can be solved by factoring.

Solve. $x^3 + 2x^2 - 3x = 0$.

$x(x^2 + 2x - 3) = 0$
$x(x + 3)(x - 1) = 0$
$x = 0 \qquad x + 3 = 0 \qquad x - 1 = 0$
$\qquad\qquad\qquad x = -3 \qquad\quad x = 1$

WORDS TO KNOW

quadratic equation (p. 568)
quadratic term (p. 569)
linear term (p. 569)
constant term (p. 569)
complete quadratic equation
    (p. 569)

incomplete quadratic equation
    (p. 569)
pure quadratic equation
    (p. 569)
quadratic formula (p. 571)
real solutions (p. 577)

zero-product property (p. 578–579)
discriminant (p. 583)
cubic equation (p. 584)
degree of an equation (p. 584)
higher-degree equations
    (p. 585)

CHAPTER TRIAL TEST

Identify the following quadratic equations as pure, incomplete, or complete.

1. $3x^2 = 42$

2. $7x^2 - 3x + 2 = 0$

3. $5x^2 = 7x$

4. $4x^2 - 1 = 0$

Solve the following quadratic equations. Find the exact solutions. Also, give approximate solutions to the nearest hundredth when appropriate.

5. $x^2 = 81$

6. $x^2 - 32 = 0$

7. $9x^2 = 16$

8. $81x^2 - 64 = 0$

9. $0.09x^2 = 0.49$

10. $2x^2 + 3x + 1 = 0$

11. $3x^2 - 6x = 0$

12. $3x^2 - 6x - 1 = 0$

13. $a^2 - 5a + 6 = 0$

14. $x^2 - 3x - 4 = 0$

15. $2x^2 + 12 = 11x$

16. $3x^2 - 5x + 4 = 0$

17. Find the diameter (d) in mils to the nearest hundredth of a copper wire conductor whose resistance (R) is 1.314 ohms and whose length (L) is 3642.5 ft. (Formula: $R = \frac{KL}{d^2}$, where K is 10.4 for copper wire.)

18. Find the radius (r) of a circle whose area (A) is 35.15 cm^2. Round the answer to the nearest hundredth centimeter. (Formula: $A = \pi r^2$.)

19. What is the current in amps (I) to the nearest hundredth if the resistance (R) of the circuit is 52.29 ohms and the watts (W) used are 205? (Formula: $R = \frac{W}{I^2}$.)

20. A square parcel of land is 156.25 m^2 in area. What is the length of a side to the nearest hundredth? (Use the formula $A = s^2$, where A is the area and s is the length of a side.)

21. In the formula $E = 0.5 \, mv^2$, solve for v if $E = 180$ and $m = 10$.

Find the roots of the following equations.

22. $x(2x - 5)(x - 3) = 0$

23. $6x^3 + 21x^2 = 45x$

24. $2x^3 - x^2 - 6x = 0$

25. $6x^3 - 18x^2 = 0$

16

Rational Expressions

GOOD DECISIONS THROUGH TEAMWORK

Accurate, up-to-date, and complete patient data has always been an important part of medical treatment, whether in a doctor's office or a hospital. It is vital in the intensive care unit (ICU) of a hospital, where a patient's life may hang by a thread.

The type, amount, and frequency of hospital patient data vary tremendously according to many factors, such as diagnosis, age, condition, and nurse-to-patient ratio. Typical hospital data may include vital signs (temperature, blood pressure, pulse), general patient appearance (for example, pallor, pupil dilation, speech, muscle tone), medicine name/dosage, medicinal side effects, description of procedures performed (for example, redressing incision, turning patient onto left side, enema given), tolerance level for each procedure, acuity chart level (which determines the frequency of nursing visits according to various factors), patient or family instruction and education given by the nurse, list of family visitors (for ICU), and charges for products purchased during the visit (such as Fleet enemas, extra pillows, bandages, etc.).

Although nurses generally enjoy gathering patient data during visits, most nurses dislike the tedious task of "charting," transcribing their handwritten data into the patient's hospital file. Most nurses dislike charting because it is time-consuming and boring, and it pulls them away from delivering patient care. Only the nurse who made the visit can chart the data because of possible errors and lawsuit liability. Some hospitals have switched to computerized charting, in which each patient has a computer mounted just outside his or her room.

Research a large hospital (more than 300 beds) and a small hospital (less than 300 beds) to find out the following information. How is patient data charted (manually or computerized)? If computerized charting is used, ask for two advantages and two disadvantages of this method. What percent of the nurse's time is spent charting patient records by computer? What is the name of the company that developed the software? If manual charting is used, ask for two advantages and two disadvantages of this method. What percent of the nurse's time is spent charting patient records manually?

Use the Internet to communicate with at least two computer software development companies specializing in medical services. Find out if there is a laptop computer on the market with appropriate software that is light enough to be carried by nurses as they do their rounds, enabling them to enter data directly into each patient's file. Ask the software representatives to predict what will be the next generation of patient records software.

Once we become proficient in factoring algebraic expressions, we can use this skill to simplify *rational expressions*. Factoring was used in solving quadratic and higher-degree equations. Now we will see how factoring helps us in simplifying rational expressions, also called algebraic fractions.

16–1 SIMPLIFYING RATIONAL EXPRESSIONS

Learning Objective

1. Simplify, or reduce, rational expressions.

1. Simplify, or Reduce, Rational Expressions.

In arithmetic we learned to reduce fractions by removing any *factors* that were common to both the numerator and denominator. This process of removing common factors is also referred to as *reducing* or *dividing out common factors*. In our study of the laws of exponents, we also *reduced* or *simplified* algebraic fractions by reducing common factors. This process is accomplished whenever the numerator and denominator of the fraction are written in factored form. In the following example, we will review the process with numerical and algebraic fractions. To draw attention to the process, we will write steps that are often done mentally.

EXAMPLE 1   Simplify or reduce the following fractions.

(a) $\dfrac{12}{15}$ (b) $\dfrac{24}{36}$ (c) $\dfrac{2x^2yz^3}{4xy^2z}$ (d) $\dfrac{9a^2b^3}{3ab}$

(a) $\dfrac{12}{15} = \dfrac{(2)(2)(3)}{(3)(5)}$ Write in factored form using prime factorization.

$\dfrac{(2)(2)(\cancel{3})}{(\cancel{3})(5)} = \dfrac{4}{5}$ Reduce the common factors and multiply the remaining factors.

(b) $\dfrac{24}{36} = \dfrac{(2)(2)(2)(3)}{(2)(2)(3)(3)}$ Write in factored form using prime factorization.

$\dfrac{(\cancel{2})(\cancel{2})(2)(\cancel{3})}{(\cancel{2})(\cancel{2})(\cancel{3})(3)} = \dfrac{2}{3}$ Reduce the common factors.

(c) $\dfrac{2x^2yz^3}{4xy^2z} = \dfrac{2x^{2-1}y^{1-2}z^{3-1}}{(2)(2)}$ Write numbers in factored form and reduce the coefficients. Apply the laws of exponents to literal factors with like bases.

$\dfrac{xy^{-1}z^2}{2} = \dfrac{xz^2}{2y}$ Express all literal factors with positive exponents.

(d) $\dfrac{9a^2b^3}{3ab} = \dfrac{(3)(\cancel{3})a^{2-1}b^{3-1}}{\cancel{3}}$ Write numbers in factored form and reduce. Apply the laws of exponents to literal factors with like bases.

$= 3ab^2$

Before continuing, we must understand clearly that we are reducing common *factors*. This procedure *will not* apply to *addends* or *terms*. Look at the next Tips and Traps.

Tips and Traps   *Can Common Addends Be Reduced?*

Does $\frac{5}{10}$ reduce to $\frac{3}{8}$?
We know from previous experience that $\frac{5}{10}$ reduces to $\frac{1}{2}$ or 0.5. Then what is wrong with the following argument?
  Both the numerator and the denominator of the fraction $\frac{5}{10}$ can be rewritten as addends or terms:

$$\frac{5}{10} = \frac{\cancel{2} + 3}{\cancel{2} + 8} = \frac{3}{8} \qquad \text{When common addends are reduced.}$$

Therefore,

$$\frac{5}{10} = \frac{3}{8}$$

Because we can verify with our calculator that this is an incorrect statement, then the *process must be incorrect. Common addends cannot be reduced.* To correct the process, we must reduce *only factors.* $\frac{5}{10}$ can be rewritten as factors:

$$\frac{5}{10} = \frac{(1)(\cancel{5})}{(2)(\cancel{5})} = \frac{1}{2}$$

If common factors are reduced,

$$\frac{5}{10} = \frac{1}{2}$$

Now we will use our knowledge of factoring *polynomials* to simplify rational expressions.

■ **DEFINITION 16–1: Rational Expression.** A *rational expression* is an algebraic fraction in which the numerator or denominator or both are polynomials.

First, we will look at some examples that are written in factored form.

EXAMPLE 2   Simplify the rational expressions that are already in factored form.

(a) $\dfrac{x(x + 2)}{(x + 2)(x + 3)}$ (b) $\dfrac{3ab(a + 4)}{6ab(a + 3)}$

(c) $\dfrac{2x^2(2x - 1)}{x^3(x - 1)}$ (d) $\dfrac{(x + 3)(x - 4)}{(x - 3)(x + 4)}$

(a) $\dfrac{x(x + 2)}{(x + 2)(x + 3)} = \dfrac{x(\cancel{x + 2})}{(\cancel{x + 2})(x + 3)}$ Reduce common binomial factors.

$= \dfrac{x}{x + 3}$

The remaining x's cannot be reduced. The x in the numerator is a factor $(x = 1 \cdot x)$; however, the x in the denominator is an *addend* or *term*.

(b) $\dfrac{3ab(a + 4)}{6ab(a + 3)} = \dfrac{\overset{1}{\cancel{3ab}}(a + 4)}{\underset{2}{\cancel{6ab}}(a + 3)}$ $1(a + 4) = a + 4$

$= \dfrac{a + 4}{2(a + 3)}$ or $\dfrac{a + 4}{2a + 6}$

(c) $\dfrac{2x^2(2x - 1)}{x^3(x - 1)}$ $\dfrac{x^2}{x^3} = x^{2-3} = x^{-1} = \dfrac{1}{x}$

$= \dfrac{2(2x - 1)}{x(x - 1)}$ or $\dfrac{4x - 2}{x^2 - x}$ Reduced rational expressions can be written in *either* factored or expanded form.

(d) $\dfrac{(x + 3)(x - 4)}{(x - 3)(x + 4)} = \dfrac{x^2 - x - 12}{x^2 + x - 12}$

There are no common factors; therefore, the fraction is already in lowest terms. However, rational expressions can also be written in expanded form.

Rule 16–1   *To simplify rational expressions:*

1. Factor *completely* both the numerator and denominator.
2. Reduce factors common to both the numerator and denominator.
3. Write the simplified expression in either factored or expanded form.

EXAMPLE 3   Reduce each rational expression to its simplest form.

(a) $\dfrac{x + y}{4x^2 + 4xy}$ (b) $\dfrac{a^2 + b^2}{a^2 - b^2}$

(c) $\dfrac{x^2 - 6x + 9}{x^2 - 9}$ (d) $\dfrac{3x^2 - 12}{6x + 12}$ (e) $\dfrac{a - b}{b - a}$

(a) $\dfrac{x + y}{4x^2 + 4xy} = \dfrac{(x + y)}{4x(x + y)}$ Factor out the common factor in the denominator and recall that the numerator is a grouping.

$\dfrac{\cancel{(x + y)}}{4x\cancel{(x + y)}} = \dfrac{1}{4x}$ Reduce. $x + y = 1(x + y)$

(b) $\dfrac{a^2 + b^2}{a^2 - b^2} = \dfrac{a^2 + b^2}{(a + b)(a - b)}$ Factor the difference of the squares in the denominator.

The numerator, which is the *sum* of the squares, will not factor. There are no factors common to both the numerator and denominator. Thus, the fraction is in simplest form:

$\dfrac{a^2 + b^2}{(a + b)(a - b)}$ or $\dfrac{a^2 + b^2}{a^2 - b^2}$

(c) $\dfrac{x^2 - 6x + 9}{x^2 - 9} = \dfrac{(x - 3)(x - 3)}{(x + 3)(x - 3)}$ Write both the numerator and the denominator in factored form.

$\dfrac{(x - 3)\cancel{(x - 3)}}{(x + 3)\cancel{(x - 3)}} = \dfrac{x - 3}{x + 3}$ Reduce.

(d) $\dfrac{3x^2 - 12}{6x + 12} = \dfrac{3(x^2 - 4)}{6(x + 2)} = \dfrac{3(x + 2)(x - 2)}{(3)(2)(x + 2)}$ Factor the numerator and the denominator completely.

$= \dfrac{\cancel{3}(\cancel{x + 2})(x - 2)}{(\cancel{3})(2)(\cancel{x + 2})} = \dfrac{x - 2}{2}$ Reduce.

(e) $\dfrac{a - b}{b - a} =$ Write terms in numerator and denominator in same order.

$\dfrac{a - b}{-a + b} =$ Factor the denominator so that the leading coefficient of the binomial is positive.

$\dfrac{a - b}{-1(a - b)} =$ Reduce.

$\dfrac{1}{-1} = -1$

Tips and Traps   *Don't forget 1 or −1:*
In Example 3, parts (a) and (e), we have some very important observations to make.

- When all factors of a numerator or denominator reduce out, a factor of 1 remains.
- Polynomial factors are opposites when every term of one grouping is the opposite of every term of the other grouping.

$(a - b)$ and $(b - a)$ are opposites.
$(2m - 5)$ and $(5 - 2m)$ are opposites.

- Opposites differ by a factor of minus one (-1).

$b - a = -1(a - b)$
$5 - 2m = -1(2m - 5)$

- When a numerator and denominator are opposites, the fraction reduces to -1.

$\dfrac{a - b}{b - a} = \dfrac{a - b}{-1(a - b)} = -1 \qquad \dfrac{2m - 5}{5 - 2m} = \dfrac{2m - 5}{-1(2m - 5)} = -1$

- It is helpful in recognizing common factors to rearrange or factor a -1 so that lead coefficients within a grouping are positive: $(-x + 5) = (5 - x)$ or $-1(x - 5)$.

SELF-STUDY EXERCISES 16–1

1 Simplify the following rational expressions.

1. $\dfrac{8}{18}$

2. $\dfrac{9}{24}$

3. $\dfrac{4a^2b^3}{2ab}$

4. $\dfrac{x(x + 3)}{(x + 3)(x + 2)}$

5. $\dfrac{x + y}{2x + 2y}$

6. $\dfrac{3x^2(3x + 2)}{x^3(2x - 1)}$

7. $\dfrac{a + b}{a^2 - b^2}$

8. $\dfrac{(x + 2)(x - 5)}{(x + 5)(x - 2)}$

9. $\dfrac{x^2 - 4x + 4}{x^2 - 4}$

10. $\dfrac{4x^2 - 16}{6x + 12}$

11. $\dfrac{3m - 11}{11 - 3m}$

12. $\dfrac{x^2 - x - 6}{3 - x}$

16-2 MULTIPLYING AND DIVIDING RATIONAL EXPRESSIONS

Learning Objectives
1. Multiply and divide rational expressions.
2. Simplify complex rational expressions.

1 Multiply and Divide Rational Expressions.

Again, we will connect our knowledge of arithmetic of fractions and the laws of exponents to expand our mathematical experience. As usual, we express results in lowest terms or in simplified form.

EXAMPLE 1 Multiply or divide as indicated. Express answers in simplest form.

(a) $\dfrac{5}{8} \cdot \dfrac{16}{25}$ (b) $\dfrac{\frac{7}{16}}{2}$ (c) $\dfrac{15xy^2}{xy^3} \cdot \dfrac{(xy)^3}{5x}$

(a) $\dfrac{5}{8} \cdot \dfrac{16}{25} = \dfrac{\overset{1}{5}}{\underset{1}{8}} \cdot \dfrac{\overset{2}{16}}{\underset{5}{25}}$ Reduce factors common to a numerator and denominator.

$= \dfrac{2}{5}$

If factors that are common to a numerator and denominator are reduced or canceled *before* multiplying, the product will be in lowest terms.

(b) $\dfrac{\frac{7}{16}}{2}$ Rewrite complex fraction as a division.

$\dfrac{7}{16} \div 2 = \dfrac{7}{16} \div \dfrac{2}{1}$ Rewrite division as multiplication.

$= \dfrac{7}{16} \cdot \dfrac{1}{2} = \dfrac{7}{32}$

(c) $\dfrac{15xy^2}{xy^3} \cdot \dfrac{(xy)^3}{5x}$

$= \dfrac{15xy^2}{xy^3} \cdot \dfrac{x^3y^3}{5x} = \dfrac{15x^4y^5}{5x^2y^3}$ Apply the laws of exponents.

$= \dfrac{\overset{3}{\cancel{15}}\overset{x^2}{x^4}\overset{y^2}{y^5}}{\underset{11}{\cancel{5}}\underset{1}{x^2}y^3}$ Reduce.

$= 3x^2y^2$

When applying the process of multiplying or dividing fractions to rational expressions, the key fact to remember is that the numerators and denominators should be written in *factored* form whenever possible.

> **Rule 16–2** *To multiply or divide rational expressions:*
> 1. Convert any division to an equivalent multiplication.
> 2. Factor completely every numerator and denominator.

3. Reduce factors as much as possible.
4. Multiply.
5. The result can be written in factored or expanded form.

EXAMPLE 2 Perform the operation and simplify.

(a) $\dfrac{x^2 - 4x - 12}{2x - 12} \cdot \dfrac{x - 4}{x^2 + 4x + 4}$ (b) $\dfrac{4y^2 - 9}{2y^2} \div \dfrac{y^2 - 2y - 15}{4y^2 + 12y}$

(c) $\dfrac{2x - 6}{1 - x} \div \dfrac{x^2 - 2x - 3}{x^2 - 1}$

(a) $\dfrac{x^2 - 4x - 12}{2x - 12} \cdot \dfrac{x - 4}{x^2 + 4x + 4}$ Factor.

$= \dfrac{(x + 2)(x - 6)}{2(x - 6)} \cdot \dfrac{x - 4}{(x + 2)(x + 2)}$ Reduce.

$= \dfrac{(x + 2)(x - 6)}{2(x - 6)} \cdot \dfrac{x - 4}{(x + 2)(x + 2)}$ Multiply.

$= \dfrac{x - 4}{2(x + 2)}$ or $\dfrac{x - 4}{2x + 4}$

(b) $\dfrac{4y^2 - 9}{2y^2} \div \dfrac{y^2 - 2y - 15}{4y^2 + 12y}$ Convert to multiplication.

$= \dfrac{4y^2 - 9}{2y^2} \cdot \dfrac{4y^2 + 12y}{y^2 - 2y - 15}$ Factor.

$= \dfrac{(2y + 3)(2y - 3)}{2y^2} \cdot \dfrac{4y(y + 3)}{(y + 3)(y - 5)}$ Reduce.

$= \dfrac{(2y + 3)(2y - 3)}{2y^2} \cdot \dfrac{\overset{2}{4}y(y + 3)}{(y + 3)(y - 5)}$ Multiply.
$\phantom{= \dfrac{(2y + 3)(2y - 3)}{2y^2}}{\scriptstyle y}$

$= \dfrac{2(2y + 3)(2y - 3)}{y(y - 5)}$ or $\dfrac{2(4y^2 - 9)}{y^2 - 5y}$ or $\dfrac{8y^2 - 18}{y^2 - 5y}$

(c) $\dfrac{2x - 6}{1 - x} \div \dfrac{x^2 - 2x - 3}{x^2 - 1}$ Convert to multiplication.

$= \dfrac{2x - 6}{1 - x} \cdot \dfrac{x^2 - 1}{x^2 - 2x - 3}$ Factor. Note the effect of factoring out -1 in the denominator of the first fraction so the $x - 1$ will reduce with $x - 1$ in the numerator of the second fraction.

$= \dfrac{2(x - 3)}{-1(x - 1)} \cdot \dfrac{(x + 1)(x - 1)}{(x + 1)(x - 3)}$ Reduce.

$= \dfrac{2(x - 3)}{-1(x - 1)} \cdot \dfrac{(x + 1)(x - 1)}{(x + 1)(x - 3)}$ Multiply.

$= \dfrac{2}{-1} = -2$

2 Simplify Complex Rational Expressions.

A *complex rational expression* is a rational expression that has a rational expression in its numerator or its denominator or both. Some examples of complex rational expressions are:

$$\frac{4xy}{\frac{2x}{5}} \qquad \frac{\frac{x^2 - y^2}{2x}}{\frac{x - y}{3x^2}} \qquad \frac{1 + \frac{1}{x}}{\frac{2}{3}} \qquad \frac{\frac{2}{x} - \frac{5}{2x}}{\frac{3}{4x} + \frac{3}{x}}$$

> **Rule 16–3** *To simplify a complex rational expression:*
>
> **1.** Rewrite the complex rational expression as a division of rational expressions.
> **2.** Proceed using Rule 16–2. (Multiply or divide rational expressions.)

Before we look at examples involving complex rational expressions, the following Tips and Traps will eliminate some of our written steps.

Tips and Traps   *Mentally Converting from Complex Form to Division and to Multiplication:* Examine the symbolic representation for converting a complex expression to multiplication. The numerator is multiplied by the reciprocal of the denominator.

$$\frac{\frac{a}{b}}{\frac{c}{d}} = \frac{a}{b} \div \frac{c}{d} = \frac{a}{b} \cdot \frac{d}{c} \qquad b, c, \text{ and } d \neq 0.$$

When simplifying a complex rational expression, we can eliminate some written steps by making this conversion mentally; for example, $\frac{4xy}{\frac{2x}{5}}$ becomes $\frac{4xy}{1} \cdot \frac{5}{2x}$.

EXAMPLE 3   Simplify.

(a) $\dfrac{4xy}{\frac{2x}{5}}$ (b) $\dfrac{\frac{x^2 - y^2}{2x}}{\frac{x - y}{3x^2}}$

(a) $\dfrac{4xy}{\frac{2x}{5}} = \dfrac{4xy}{1} \cdot \boxed{\dfrac{5}{2x}}$ Multiply the numerator times the reciprocal of the denominator.

$= \dfrac{\overset{2}{\cancel{4}}xy}{1} \cdot \dfrac{5}{\cancel{2}x}$ Reduce.

$= \dfrac{10y}{1} = \mathbf{10y}$ Simplify.

(b) $\dfrac{\dfrac{x^2 - y^2}{2x}}{\dfrac{x - y}{3x^2}} = \dfrac{x^2 - y^2}{2x} \cdot \dfrac{3x^2}{x - y}$ Multiply the numerator times the reciprocal of the denominator.

$= \dfrac{(x + y)(x - y)}{2x} \cdot \dfrac{3x^2}{x - y}$ Factor.

$= \dfrac{(x + y)(x - y)}{2x} \cdot \dfrac{3x^2}{x - y}$ Reduce.

$= \dfrac{3x(x + y)}{2}$ or $\dfrac{3x^2 + 3xy}{2}$ Multiply.

SELF-STUDY EXERCISES 16–2

1 Multiply or divide the following fractions. Reduce to simplest terms.

1. $\dfrac{5x^2}{3y} \cdot \dfrac{2x}{3y}$

2. $\dfrac{a^2 - b^2}{4} \cdot \dfrac{12}{a + b}$

3. $\dfrac{2u + 2v}{5} \cdot \dfrac{10}{u + v}$

4. $\dfrac{a^2 - 49}{b^2 - 25} \cdot \dfrac{b - 5}{a + 7}$

5. $\dfrac{x^2 + 2x + 1}{5x - 5} \cdot \dfrac{15}{x + 1}$

6. $\dfrac{13r^2}{20a^2} \div \dfrac{39r^2}{5a}$

7. $\dfrac{a - b}{4} \div \dfrac{a - b}{2}$

8. $\dfrac{x}{x^2 - 4x + 4} \div \dfrac{1}{x - 2}$

9. $\dfrac{5a^2 - 5b^2}{a^2 b^2} \div \dfrac{a + b}{10ab}$

10. $\dfrac{x^2 + 4x + 3}{x^2 - 4x - 5} \div \dfrac{x + 3}{x - 5}$

2 Simplify.

11. $\dfrac{\dfrac{5}{9}}{\dfrac{3}{5}}$

12. $\dfrac{\dfrac{7}{8}}{\dfrac{5}{6}}$

13. $\dfrac{\dfrac{x - 5}{4}}{3x}$

14. $\dfrac{\dfrac{6}{x - 2}}{9x}$

15. $\dfrac{\dfrac{4x}{8x}}{x + 3}$

16. $\dfrac{\dfrac{7x}{21x}}{x - 3}$

17. $\dfrac{\dfrac{x^2 - 36}{5x}}{\dfrac{x + 6}{15x}}$

18. $\dfrac{\dfrac{x^2 - 5x}{8}}{\dfrac{2x - 10}{12x}}$

16-3 ADDING AND SUBTRACTING RATIONAL EXPRESSIONS

Learning Objective

1 Add and subtract rational expressions.

1 Add and Subtract Rational Expressions.

In adding and subtracting fractions in arithmetic, the key is to remember that we can only add or subtract fractions with like denominators. Whenever we have unlike fractions, we must first find a common denominator for the fractions. We then convert each fraction to an equivalent fraction having the common denominator, and add or subtract the numerators. We will refresh our memory of the procedure for adding and subtracting fractions.

EXAMPLE 1   Add or subtract as indicated.

(a) $\dfrac{3}{8} + \dfrac{1}{8}$ (b) $\dfrac{7}{12} - \dfrac{1}{3}$ (c) $5 - \dfrac{2}{3}$ (d) $\dfrac{5}{2x} + \dfrac{3}{x} + \dfrac{9}{4}$

(a) $\dfrac{3}{8} + \dfrac{1}{8} = \dfrac{4}{8} = \mathbf{\dfrac{1}{2}}$

Add the numerators and *keep* the *like* denominator. Reduce.

(b) $\dfrac{7}{12} - \boxed{\dfrac{1}{3}} =$

Select a common denominator and change to equivalent fractions with the common denominator.

$\dfrac{7}{12} - \dfrac{\boxed{4}}{\boxed{12}} = \dfrac{3}{12} = \mathbf{\dfrac{1}{4}}$

Subtract and reduce.

(c) $\boxed{5} - \dfrac{2}{3} =$

Convert 5 to a fraction with a denominator of 3.

$\dfrac{\boxed{15}}{\boxed{3}} - \dfrac{2}{3} = \mathbf{\dfrac{13}{3}}$ or $\mathbf{4\dfrac{1}{3}}$

Subtract.

(d) $\dfrac{5}{2x} + \dfrac{3}{x} + \dfrac{9}{4} =$

Convert to equivalent fractions with a common denominator.

$\dfrac{5(\boxed{2})}{2x(\boxed{2})} + \dfrac{3(\boxed{4})}{x(\boxed{4})} + \dfrac{9(\boxed{x})}{4(\boxed{x})} =$

$\dfrac{10}{4x} + \dfrac{12}{4x} + \dfrac{9x}{4x} =$

Add numerators and *keep* the like (common) denominator.

$\mathbf{\dfrac{22 + 9x}{4x}}$

Combine like terms in the numerator.

Tips and Traps   *Improper Fraction Versus Mixed Number:*

In arithmetic, we often write an improper fraction as a mixed number. In algebra, we use the mixed number form only when the final result contains only numbers and when we are interpreting the result within the context of an applied problem. A rational expression like the solution in Example 1, part (d), can be written in an alternative form.

$$\dfrac{22 + 9x}{4x} = \dfrac{22}{4x} + \dfrac{9x}{4x} = \dfrac{22}{4x} + \dfrac{9}{4}$$

Unless there is a reason to do this within the context of the problem, however, we do not choose to do the extra work.

Rule 16–4   *To add or subtract rational expressions with unlike denominators:*

1. Find the *least common denominator (LCD)*.
2. Change *each* fraction to an equivalent fraction having the least common denominator.
3. Add or subtract numerators.
4. Keep the same denominator.
5. Reduce (or simplify) if possible.

EXAMPLE 2   Add $\dfrac{4}{x+3} + \dfrac{3}{x-3}$.

First, convert to equivalent expressions with a common denominator. The LCD is the product $(x+3)(x-3)$.

$$\frac{4}{x+3} = \frac{4(x-3)}{(x+3)(x-3)} = \frac{4x-12}{(x+3)(x-3)}$$

Multiply the numerator and denominator by $(x-3)$.

$$\frac{3}{(x-3)} = \frac{3(x+3)}{(x-3)(x+3)} = \frac{3x+9}{(x+3)(x-3)}$$

Multiply the numerator and denominator by $(x+3)$. Order does not matter in multiplication.

Next, use the equivalent expressions to proceed.

$$\frac{4x-12}{(x+3)(x-3)} + \frac{3x+9}{(x+3)(x-3)} =$$

Add numerators.

$$\frac{4x-12+3x+9}{(x+3)(x-3)} =$$

Combine like terms in the numerator.

$$\frac{7x-3}{(x+3)(x-3)} \quad \text{or} \quad \frac{7x-3}{x^2-9}$$

Factored or expanded form

Some of the steps in Example 2 might have been combined into one step, thus requiring fewer *written* steps, but the example illustrates each step that must be performed, either mentally or on paper, to add the rational expressions. Look at the next example.

EXAMPLE 3   Subtract $\dfrac{x}{x-2} - \dfrac{5}{x+4}$.

Change to equivalent expressions with a common denominator. The LCD is the product $(x-2)(x+4)$.

$$\frac{x}{x-2} = \frac{x(x+4)}{(x-2)(x+4)} = \frac{x^2+4x}{(x-2)(x+4)}$$

Multiply the numerator and denominator by $(x+4)$.

$$\frac{5}{(x+4)} = \frac{5(x-2)}{(x+4)(x-2)} = \frac{5x-10}{(x-2)(x+4)}$$

Multiply the numerator and denominator by $(x-2)$.

Use equivalent expressions and proceed.

$$\frac{x^2+4x}{(x-2)(x+4)} - \frac{5x-10}{(x-2)(x+4)} =$$

Add (or subtract) numerators.

$$\frac{x^2+4x - (5x-10)}{(x-2)(x+4)} =$$

Note: the *entire* numerator following the subtraction sign is subtracted.

$$\frac{x^2+4x - 5x+10}{(x-2)(x+4)} =$$

Be careful with the signs when the parentheses are removed.

$$\frac{x^2-x+10}{(x-2)(x+4)} \quad \text{or} \quad \frac{x^2-x+10}{x^2+2x-8}$$

Because x^2-x+10 will not factor, the fraction cannot be reduced.

Recall that a complex rational expression may have rational expressions in the numerator or denominator or both. Thus, there may be complex expressions that require the addition or subtraction of rational expressions before conversion to an equivalent multiplication.

EXAMPLE 4   Simplify.

(a) $\dfrac{1 + \dfrac{1}{x}}{\dfrac{2}{3}}$ (b) $\dfrac{\dfrac{2}{x} - \dfrac{5}{2x}}{\dfrac{3}{4x} + \dfrac{3}{x}}$

(a) $\dfrac{1 + \dfrac{1}{x}}{\dfrac{2}{3}} = \dfrac{\dfrac{x}{x} + \dfrac{1}{x}}{\dfrac{2}{3}} =$ Find the common denominator for numerator $1 = \frac{x}{x}$.

Add terms in the numerator.

$\dfrac{\dfrac{x + 1}{x}}{\dfrac{2}{3}} =$

$\dfrac{x + 1}{x} \cdot \dfrac{3}{2} =$ Multiply the numerator times the reciprocal of the denominator.

$\dfrac{3(x + 1)}{2x}$ or $\dfrac{3x + 3}{2x}$

(b) $\dfrac{\dfrac{2}{x} - \dfrac{5}{2x}}{\dfrac{3}{4x} + \dfrac{3}{x}} = \dfrac{\dfrac{4}{2x} - \dfrac{5}{2x}}{\dfrac{3}{4x} + \dfrac{12}{4x}} =$ Find the common denominator for the numerator, $2x$, and the denominator, $4x$, and convert to equivalent fractions.

$\dfrac{\dfrac{-1}{2x}}{\dfrac{15}{4x}} =$ Add or subtract fractions in the numerator and in the denominator.

$\dfrac{-1}{2x} \cdot \dfrac{4x}{15} =$ Multiply the numerator times the reciprocal of the denominator.

$\dfrac{-1}{2x} \cdot \dfrac{\overset{2}{4x}}{15} =$ Reduce.

$\dfrac{-2}{15} = -\dfrac{2}{15}$ Multiply.

SELF-STUDY EXERCISES 16-3

[1]   Add or subtract the following rational expressions. Reduce to simplest terms.

1. $\dfrac{3}{7} + \dfrac{2}{7}$

2. $\dfrac{3}{8} + \dfrac{7}{16}$

3. $\dfrac{2x}{3} + \dfrac{5x}{6}$

4. $\dfrac{3x}{8} + \dfrac{7x}{12}$

5. $\dfrac{2}{x} + \dfrac{3}{2}$

6. $\dfrac{5}{2x} + \dfrac{6}{x} + \dfrac{2}{3x}$

7. $\dfrac{5}{x + 1} + \dfrac{4}{x - 1}$

8. $\dfrac{6}{x + 3} + \dfrac{4}{x - 1}$

9. $\dfrac{5}{2x + 1} - \dfrac{2}{x - 1}$

10. $\dfrac{7}{x-6} - \dfrac{6}{x-5}$

11. $\dfrac{2 + \dfrac{1}{x}}{\dfrac{3}{5}}$

12. $\dfrac{3 + \dfrac{2}{x}}{\dfrac{5}{8}}$

13. $\dfrac{\dfrac{3}{x} + \dfrac{5}{3x}}{\dfrac{3}{6x} - \dfrac{2}{x}}$

14. $\dfrac{\dfrac{x}{6} - \dfrac{3x}{9}}{\dfrac{5x}{12} + \dfrac{3x}{4}}$

16-4 SOLVING EQUATIONS WITH RATIONAL EXPRESSIONS

Learning Objectives

1 Exclude certain values as solutions of rational equations.

2 Solve rational equations with variable denominators.

1 Exclude Certain Values as Solutions of Rational Equations.

A rational equation is an equation that contains one or more rational expressions. Some examples of rational equations are

$$\frac{1}{2} + \frac{1}{x} = \frac{1}{3}, \qquad x - \frac{3}{x} = 4, \qquad \frac{x}{x-2} = \frac{1}{x+3}$$

We solved such equations in Chapter 10 by first clearing the equation of all fractions. We cleared the fractions by multiplying both sides of the equation by the *least common multiple (LCM)* of all the denominators in the equation. Then we solved the equation as any other equation. However, when the variable is in the denominator of the equation, an important caution must be observed.

Tips and Traps *Excluded Values:*
Because division by zero is impossible, any value of the variable that makes the value of any denominator zero is excluded as a possible solution of an equation. Therefore, it is important (1) to determine which values are *excluded values* and cannot be used as possible solutions, and (2) to check each possible solution.

Rule 16-5 *To find excluded values of rational equations:*

1. Set each denominator containing a variable equal to zero and solve for the variable.
2. Each equation in Step 1 produces an excluded value; however, some values may be repeats.

EXAMPLE 1 Determine the value or values that must be excluded as possible solutions of the following.

(a) $\dfrac{1}{x} + \dfrac{1}{2} = 5$

(b) $\dfrac{1}{x} = \dfrac{1}{x-3}$

(c) $\dfrac{1}{x+2} + \dfrac{1}{x-2} = \dfrac{1}{x^2-4}$

(d) $\dfrac{3x}{x+2} = 3 - \dfrac{5}{2x}$

(a) $\dfrac{1}{x} + \dfrac{1}{2} = 5$ Set each denominator containing a variable equal to 0.

$x = 0$ Excluded value

(b) $\dfrac{1}{x} = \dfrac{1}{x - 3}$ Set each denominator containing a variable equal to 0.

$x = 0, \qquad x - 3 = 0$ Solve each equation.

$x = 3$

Excluded values are 0 and 3.

(c) $\dfrac{1}{x + 2} + \dfrac{1}{x - 2} = \dfrac{1}{x^2 - 4}$ Set each denominator containing a variable equal to 0.

$x + 2 = 0 \qquad x - 2 = 0 \qquad\qquad x^2 - 4 = 0$ Solve each equation.

$x = -2 \qquad\quad x = 2 \qquad\qquad (x + 2)(x - 2) = 0$

$x + 2 = 0 \qquad x - 2 = 0$

$x = -2 \qquad\quad x = 2$

Excluded values are −2 and 2.

(d) $\dfrac{3x}{x + 2} = 3 - \dfrac{5}{2x}$ Set each denominator containing a variable equal to 0.

$x + 2 = 0 \qquad 2x = 0$ Solve each equation.

$x = -2 \qquad x = 0$

Excluded values are −2 and 0.

2   Solve Rational Equations with Variable Denominators.

When solving rational equations with variable denominators, find any values that must be excluded as possible solutions by setting each variable denominator equal to zero and solving the resulting equations. Finding excluded values is another way to check whether roots are *extraneous roots* (Chapter 10). Even though we have found the excluded values, it is a good idea to check the solution to make sure the root makes a true statement in the original equation.

Rule 16–6   *To solve rational equations:*

1. Determine the excluded values.
2. Clear the equation of all denominators by multiplying the entire equation by the least common multiple of the denominators.
3. Complete the solution using previously learned strategies.
4. Eliminate any excluded values as solutions.
5. Check the solutions.

EXAMPLE 2   Solve the following rational equations. Check.

(a) $\dfrac{2}{y} + \dfrac{1}{3} = 1$ (b) $\dfrac{2}{y - 2} = \dfrac{4}{y + 1}$ (c) $\dfrac{5}{x - 2} - \dfrac{3x}{x - 2} = -\dfrac{1}{x - 2}$

(a) $\dfrac{2}{y} + \dfrac{1}{3} = 1$ Excluded value: $y = 0$.

$$3y\left(\frac{2}{y}\right) + 3y\left(\frac{1}{3}\right) = 3y(1) \qquad \text{Multiply by LCM, } 3y.$$

$$(3\cancel{y})\left(\frac{2}{\cancel{y}}\right) + (\cancel{3}y)\left(\frac{1}{\cancel{3}}\right) = 3y(1) \qquad \text{Cancel.}$$

$$6 + y = 3y \qquad \text{Solve for } y.$$

$$6 = 3y - y$$

$$6 = 2y$$

$$\frac{6}{2} = \frac{2y}{2}$$

$$\mathbf{3 = y} \qquad \text{Check because 3 is not an excluded value.}$$

Check: $$\frac{2}{y} + \frac{1}{3} = 1$$

$$\frac{2}{3} + \frac{1}{3} = 1 \qquad \text{Substitute 3 for } y.$$

$$\frac{3}{3} = 1$$

$$1 = 1 \qquad \text{Solution checks.}$$

(b) $$\frac{2}{y - 2} = \frac{4}{y + 1} \qquad \text{Excluded values:} \quad y - 2 = 0 \qquad y + 1 = 0$$
$$y = 2 \qquad\qquad y = -1$$

$$(y - 2)(y + 1)\,\frac{2}{y - 2} = (y - 2)(y + 1)\,\frac{4}{y + 1} \qquad \begin{array}{l}\text{Multiply by LCM,}\\ (y - 2)(y + 1).\end{array}$$

$$(\cancel{y - 2})(y + 1)\,\frac{2}{\cancel{y - 2}} = (y - 2)(\cancel{y + 1})\,\frac{4}{\cancel{y + 1}} \qquad \text{Cancel.}$$

$$2(y + 1) = 4(y - 2) \qquad \text{Solve for } y.$$

$$2y + 2 = 4y - 8$$

$$2y - 4y = -8 - 2$$

$$-2y = -10$$

$$\frac{-2y}{-2} = \frac{-10}{-2}$$

$$\mathbf{y = 5} \qquad \begin{array}{l}\text{Check because 5 is not}\\ \text{an excluded value.}\end{array}$$

Check: $$\frac{2}{y - 2} = \frac{4}{y + 1}$$

$$\frac{2}{5 - 2} = \frac{4}{5 + 1} \qquad \text{Substitute 5 for } y.$$

$$\frac{2}{3} = \frac{4}{6} \qquad \text{Reduce } \tfrac{4}{6}.$$

$$\frac{2}{3} = \frac{2}{3} \qquad \text{Solution checks.}$$

(c) $\dfrac{5}{x-2} - \dfrac{3x}{x-2} = -\dfrac{1}{x-2}$ Excluded value: $x - 2 = 0$

$x = 2$

$$(x-2)\dfrac{5}{x-2} - (x-2)\dfrac{3x}{x-2} = (x-2)\left(-\dfrac{1}{(x-2)}\right)$$

Multiply by LCM, $x - 2$, to clear the common denominator.

$$(\cancel{x-2})\dfrac{5}{\cancel{x-2}} - (\cancel{x-2})\dfrac{3x}{\cancel{x-2}} = (\cancel{x-2})\left(-\dfrac{1}{(\cancel{x-2})}\right)$$

Cancel.

$$5 - 3x = -1$$

Solve for x.

$$-3x = -1 - 5$$

$$-3x = -6$$

$$\dfrac{-3x}{-3} = \dfrac{-6}{-3}$$

$$x = 2$$

This is an excluded value. The root will not check.

There is no solution.

Check: $\dfrac{5}{x-2} - \dfrac{3x}{x-2} = -\dfrac{1}{x-2}$

$$\dfrac{5}{2-2} - \dfrac{3x}{2-2} = -\dfrac{1}{2-2}$$

Substitute 2 for x.

$$\dfrac{5}{0} - \dfrac{3(2)}{0} = -\dfrac{1}{0}$$

Division by zero is impossible. The equation has no solution.

A number of application problems can be solved by using rational equations. Let's look at an example.

EXAMPLE 3   At the January office products sale, Yan Yu purchased a package of 3.5-in. computer disks for $16 and a package of writing pens for $6. He paid 40 cents more per pen than he paid for each computer disk, and the computer disk package contained 15 more disks than the pen package. How many of the disks and how many pens did he purchase? Give the price of the disks and the pens.

Unknown facts   Let n = the number of pens. The number of pens, the number of disks, the cost per pen, and the cost per disk are all unknown.

Known facts   Disks cost $16 per package. Pens cost $6 per package. There are 15 more disks than pens. Each pen costs 40¢ more than each disk.

Relationships   Number of disks = $n + 15$

Cost ÷ number of items = cost per item

Cost per pen = $\dfrac{6}{n}$

Cost per disk = $\dfrac{16}{n+15}$

Cost per pen = cost per disk plus 40¢. $40¢ = \frac{40}{100}$ dollars.

$$\frac{6}{n} = \frac{16}{n+15} + \frac{40}{100}$$

Estimation

Pens cost more than disks and only \$6 was spent for pens, so there will be only a small number of pens.

$$\frac{6}{n} = \frac{16}{n+15} + \frac{40}{100}$$

$$\frac{6}{n} = \frac{16}{n+15} + \frac{2}{5} \qquad \frac{40}{100} = \frac{2}{5}$$

Solution

Excluded values: $n = 0$ and $n + 15 = 0$ or $n = -15$. These excluded values would be eliminated because they are inappropriate within the context of the problem.

$$(n)(n+15)(5)\frac{6}{n} = (n)(n+15)(5)\frac{16}{n+15} + (n)(n+15)(5)\frac{2}{5} \qquad \text{Multiply by LCM, } (n)(n+15)(5).$$

$$(\cancel{n})(n+15)(5)\frac{6}{\cancel{n}} = (n)(\cancel{n+15})(5)\frac{16}{\cancel{n+15}} + (n)(n+15)(\cancel{5})\frac{2}{\cancel{5}} \qquad \text{Cancel.}$$

$$(n+15)(5)(6) = (n)(5)(16) + (n)(n+15)(2)$$

$$30(n+15) = 80n + 2n(n+15) \qquad \text{Distribute.}$$

$$30n + 450 = 80n + 2n^2 + 30n$$

$$0 = 80n + 2n^2 + 30n - 30n - 450$$

$$0 = 2n^2 + 80n - 450$$

$$0 = 2(n^2 + 40n - 225)$$

$$0 = 2(n-5)(n+45)$$

$$n - 5 = 0 \qquad\qquad n + 45 = 0$$

$$\boldsymbol{n = 5 \text{ pens}} \qquad\qquad n = -45 \qquad \text{Disregard the negative value.}$$

Interpretation

There are 5 pens, and that fact can be used to find the other missing facts.

$$n + 15 = 5 + 15 = \textbf{20 disks}$$

$$\frac{6}{n} = \frac{\$6}{5} = \textbf{\$1.20 per pen}$$

$$\frac{16}{n+15} = \frac{\$16}{20} = \textbf{\$0.80 per disk}$$

Check: $\quad \dfrac{6}{n} = \dfrac{16}{n+15} + \dfrac{2}{5}$

$$\frac{6}{5} = \frac{16}{5+15} + \frac{2}{5} \qquad \text{Substitute 5 for } n.$$

$$\frac{6}{5} = \frac{16}{20} + \frac{2}{5}$$

$$\frac{6}{5} = \frac{16}{20} + \frac{8}{20}$$

$$\frac{6}{5} = \frac{24}{20}$$ $$\frac{24}{20} = \frac{6}{5}$$

$$\frac{6}{5} = \frac{6}{5}$$ The solution checks.

SELF-STUDY EXERCISES 16–4

1 Determine the value or values that must be excluded as possible solutions of the following.

1. $\frac{5}{x} + \frac{3}{5} = 7$

2. $\frac{6}{x} + 5 = \frac{5}{12}$

3. $\frac{9}{x-5} = \frac{7}{x}$

4. $\frac{15}{2x} = \frac{6}{x+9}$

5. $\frac{7}{x+8} + \frac{2}{x-8} = \frac{5}{x^2-64}$

6. $\frac{9}{x-2} + \frac{5}{x+2} = \frac{1}{x^2-4}$

7. $\frac{2x}{4x-3} - 4 = \frac{5}{6x}$

8. $\frac{1}{9x} + \frac{8x}{5x+15} = 11$

2 Solve the following equations. Check for extraneous roots.

9. $\frac{6}{7} + \frac{5}{x} = 1$

10. $\frac{4}{x} - \frac{3}{4} = \frac{1}{20}$

11. $\frac{2}{x} + \frac{5}{2x} - \frac{1}{2} = 4$

12. $\frac{5}{2x} - \frac{1}{x} = 6$

13. $\frac{2}{x-3} + \frac{7}{x+3} = \frac{21}{x^2-9}$

14. $\frac{1}{x^2-25} = \frac{3}{x-5} + \frac{2}{x+5}$

Use rational equations to solve the following problems. Check for extraneous roots.

15. Shawna can complete an electronic lab project in 3 days. Tom can complete the same project in 4 days. How many days will it take to complete the project if Shawna and Tom work together? (*Hint:* Rate × time = amount of work.)

16. A group of investors purchase land in Maine for $12,000. When four more investors joined the group, the cost dropped $500 per person in the original group. How many investors were in the original group?

17. Gomez commutes 50 miles on a motorbike to a technical college, but on the return trip he travels 15 miles per hour faster and makes the trip in $\frac{3}{4}$ hour less time. Find Gomez's speed to and from school. (*Hint:* Time = distance ÷ rate.)

18. Pipe 1 can fill a tank in 2 minutes and pipe 2 can empty the tank in 6 minutes. How long would it take to fill the tank if both pipes were working? (*Hint:* Rate × time = amount of work.)

19. Kim bought $150 worth of medium-roast coffee and, at the same cost per pound, $100 worth of dark-roast coffee. If she bought 25 pounds more of the medium-roast coffee than the dark-roast coffee, how many pounds of each did she buy and what was the cost per pound?

20. Find resistance$_2$ if resistance$_1$ is 10 ohms and resistance$_t$ is 3.75 ohms, using the formula $$R_t = \frac{R_1 R_2}{R_1 + R_2}.$$

CAREER APPLICATION

Consumer Interest: 2000 Olympic Games Deal

In preparation for the 2000 Summer Olympic Games in Sydney, Australia, several people decide to pool their money and share equally the $12,000 expense of renting a four-bedroom house in Sidney for two weeks. The original number of people who agreed to share the house changed after two people dropped out of the deal because they thought the house was too small for the number of people involved. Those left in the deal must now pay an additional \$300 each for the rental.

Exercises

Use the above information to answer the following.
1. Let x represent the original number of people in the deal, and write a rational expression for the price per person in the original group.
2. Write a rational expression for the price per person in the new, smaller group.
3. Use your expressions from Exercises 1 and 2 to write a rational equation showing how the cost per person increased as the size of the group decreased.
4. Solve the equation to find the original number of people and their share of the cost, as well as the new, smaller number of people and their higher share of the cost. Don't forget to list and consider the excluded values.
5. Why would the two excluded values also be rejected within the context of the problem?
6. If a group of 10 people decreased to 8 and the cost per person increased \$450, write an equation to find the total cost of the house rental. What type of equation is this?
7. Solve the equation above to find the total house rental cost, the original cost per person, and the new cost per person in the smaller group.

Answers

1. $\dfrac{12{,}000}{x}$ = original cost per person

2. $\dfrac{12{,}000}{(x-2)}$ = new cost per person

3. $\dfrac{12{,}000}{x} + 300 = \dfrac{12{,}000}{(x-2)}$

4. The original group of 10 people would have paid \$1200 each. The new, smaller group of 8 people will pay \$1500 each.
5. The excluded values of 0 and 2 do not make sense in the real world because the number of original people and the number of people in the smaller group cannot be zero.

6. Letting x = the total rental cost, the linear equation is $\dfrac{x}{10} + 450 = \dfrac{x}{8}$. It is not a rational equation because a variable does not appear in any denominator.
7. The total rental cost is \$18,000. The original cost per person was \$1800 and the new cost per person is \$2250.

ASSIGNMENT EXERCISES

Section 16–1
Simplify the following fractions.

1. $\dfrac{18}{24}$

2. $\dfrac{24}{42}$

3. $\dfrac{5a^2b^3c}{10a^3bc^2}$

4. $\dfrac{3x^2y}{9x^3y^3}$

5. $\dfrac{4xy(x-3)}{8xy(x+3)}$

6. $\dfrac{(x+7)(x-3)}{(x-3)(x-7)}$

7. $\dfrac{(x-4)(x+2)}{(x+2)(x-4)}$

8. $\dfrac{x+1}{3x+3}$

9. $\dfrac{m^2-n^2}{m^2+n^2}$

10. $\dfrac{3x^2+8x-3}{2x^2+5x-3}$

11. $\dfrac{x}{x+xy}$

12. $\dfrac{2x}{4x+6}$

13. $\dfrac{5x + 15}{x + 3}$

14. $\dfrac{x^3 + 2x^2 - 3x}{3x}$

15. $\dfrac{y^2 + 2y + 1}{y + 1}$

16. $\dfrac{y - 1}{y^2 - 2y + 1}$

17. $\dfrac{2x - 6}{x^2 + 3x - 18}$

18. $\dfrac{x^2 - 4x + 4}{x^2 - 2x}$

19. $\dfrac{3x - 9}{x - 3}$

20. $\dfrac{4x - 12}{x - 3}$

Section 16–2

Multiply or divide the following fractions. Reduce to simplest terms.

21. $\dfrac{3x^2}{2y} \cdot \dfrac{5x}{6y}$

22. $\dfrac{x^2 - y^2}{6} \cdot \dfrac{18}{x - y}$

23. $\dfrac{9}{x + b} \cdot \dfrac{5x + 5b}{3}$

24. $\dfrac{81 - x^2}{16 - f^2} \cdot \dfrac{4 - f}{9 + x}$

25. $\dfrac{4y^2 - 4y + 1}{6y - 6} \cdot \dfrac{24}{2y - 1}$

26. $\dfrac{x - 3}{x + 5} \cdot \dfrac{2x^2 + 10x}{2x - 6}$

27. $\dfrac{5 - x}{x - 5} \cdot \dfrac{x - 1}{1 - x}$

28. $\dfrac{3 - x}{x - 2} \cdot \dfrac{2x - 4}{x - 3}$

29. $\dfrac{x^2 + 6x + 9}{x^2 - 4} \cdot \dfrac{x - 2}{x + 3}$

30. $\dfrac{x^2}{x^2 - 9} \cdot \dfrac{x^2 - 5x + 6}{x^2 - 2x}$

31. $\dfrac{2a + b}{8} \div \dfrac{2a + b}{2}$

32. $\dfrac{17a^2}{21y^2} \div \dfrac{34a^2}{68}$

33. $\dfrac{y^2 - 2y + 1}{y} \div \dfrac{1}{y - 1}$

34. $\dfrac{x^2 y^2}{3x^2 - 3y^2} \div \dfrac{8xy}{x - y}$

35. $\dfrac{y^2 + 6y + 9}{y^2 + 4y + 4} \div \dfrac{y + 3}{y + 2}$

36. $\dfrac{2x + 2y}{3} \div \dfrac{x^2 - y^2}{y - x}$

37. $\dfrac{3x^2 + 6x}{x} \div \dfrac{2x + 4}{x^2}$

38. $\dfrac{x^2 - 7x}{x^2 - 3x - 28} \div \dfrac{1}{-x - 4}$

39. $\dfrac{y^2 - 16}{y + 3} \div \dfrac{y - 4}{y^2 - 9}$

40. $\dfrac{12x + 24}{36x - 36} \div \dfrac{6x + 12}{8x - 8}$

Simplify.

41. $\dfrac{\dfrac{5}{x - 3}}{4}$

42. $\dfrac{6ab}{\dfrac{3a}{4}}$

43. $\dfrac{\dfrac{x^2 - 4x}{6x}}{\dfrac{x - 4}{8x^2}}$

44. $\dfrac{2 + \dfrac{3}{x}}{\dfrac{2}{x}}$

45. $\dfrac{\dfrac{5}{x} - \dfrac{3}{4x}}{\dfrac{1}{3x} + \dfrac{2}{x}}$

46. $\dfrac{5 - \dfrac{x - 2}{x}}{\dfrac{x - 4}{2x} - 2}$

47. $\dfrac{\dfrac{3x}{6} - \dfrac{5}{x}}{\dfrac{x}{3} + \dfrac{4}{2x}}$

48. $\dfrac{\dfrac{2}{7}}{\dfrac{3}{4}}$

Section 16–3

Add or subtract the following fractions. Reduce to simplest terms.

49. $\dfrac{2}{9} + \dfrac{4}{9}$

50. $\dfrac{2}{3} + \dfrac{5}{12}$

51. $\dfrac{3x}{7} + \dfrac{2x}{14}$

52. $\dfrac{5x}{3} - \dfrac{2x}{4}$

53. $\dfrac{3x}{4} + \dfrac{5x}{6}$

54. $\dfrac{3}{4} + \dfrac{7}{x}$

55. $\dfrac{5}{x} - \dfrac{7}{3}$

56. $\dfrac{3}{x} + \dfrac{5}{7}$

57. $\dfrac{3}{4x} + \dfrac{2}{x} + \dfrac{3}{6x}$

58. $\dfrac{6}{2x + 3} + \dfrac{2}{2x - 3}$

59. $\dfrac{7}{x - 3} + \dfrac{3}{x + 2}$

60. $\dfrac{4}{3x + 2} - \dfrac{3}{x - 2}$

61. $\dfrac{8}{x + 3} - \dfrac{2}{x - 4}$

62. $\dfrac{3}{x - 2} + \dfrac{5}{2 - x}$

63. $\dfrac{x}{x - 5} - \dfrac{3}{5 - x}$

Determine the value or values that must be excluded as possible solutions of the following.

64. $\dfrac{3}{x} - \dfrac{4}{5} = 2$

65. $\dfrac{4}{x} = \dfrac{3}{x-2}$

66. $\dfrac{3}{x-5} - \dfrac{4}{x+5} = \dfrac{1}{x^2-25}$

67. $\dfrac{5x}{2x-1} - 6 = \dfrac{4}{3x}$

Solve the following equations. Check for extraneous roots.

68. $\dfrac{3}{x} + \dfrac{2}{3} = 1$

69. $\dfrac{4}{x} = \dfrac{1}{x+5}$

70. $\dfrac{3}{x-4} + \dfrac{1}{x+4} = \dfrac{1}{x^2-16}$

71. $-\dfrac{4x}{x+1} = 3 - \dfrac{4}{x+1}$

Use rational equations to solve the following problems. Check for extraneous or inappropriate roots.

72. Fugita can complete the assembly of 50 widgets in 6 days. Ohn can complete the same job in 8 days. How many days will it take to complete the project if Fugita and Ohn work together? (*Hint:* Rate × time = amount of work.)

73. Several students chipped in a total of $120 to buy a small refrigerator to use in the dorm. Later, another student joined the group, and the cost to the original group dropped $10 per person. How many students were in the original group?

CHALLENGE PROBLEMS

74. Pipe 1 can fill a tank in 4 minutes and pipe 2 can fill the same tank in 3 minutes. What proportion of the tank will be filled if both pipes were working for 1 minute? (*Hint:* Rate × time = amount of work.)

75. One machine can do a job in 5 hours alone. How many hours would it take a second machine to complete the job alone if both machines together can do the job in 3 hours?

CONCEPTS ANALYSIS

1. How are the properties $\dfrac{n}{n} = 1$ and $1 \times n = n$ applied in the following example: $\dfrac{4}{6} = \dfrac{2}{3}$?

2. Why is the following problem incorrect?

$$\frac{5}{10} = \frac{\cancel{2}+3}{\cancel{2}+8} = \frac{3}{8}$$

3. Write a brief comment for each step of the following example.

$$\frac{x^2+3x+2}{x^2-9} \div \frac{2x^2+3x-2}{2x^2-7x+3} =$$

$$\frac{x^2+3x+2}{x^2-9} \cdot \frac{2x^2-7x+3}{2x^2+3x-2} =$$

$$\frac{(x+1)(\cancel{x+2})}{(x+3)(\cancel{x-3})} \cdot \frac{(\cancel{2x-1})(\cancel{x-3})}{(\cancel{2x-1})(\cancel{x+2})} = \frac{x+1}{x+3}$$

Find the first mistake in each of the following and briefly explain the mistake. Then rework the problem correctly.

4. $\dfrac{x}{x+3} = \dfrac{\cancel{x}}{\cancel{x}+3} = \dfrac{1}{3}$

5. $\dfrac{x^2-4}{x^2+4x+4} \div x + 2 =$

$\dfrac{(\cancel{x+2})(x-2)}{(\cancel{x+2})(\cancel{x+2})} \cdot \dfrac{\cancel{x+2}}{1} = x - 2$

6. $\dfrac{5}{x+2} + \dfrac{3}{x-3} = \dfrac{8}{(x+2)(x-3)}$

7. $\dfrac{3}{x-1} - \dfrac{5}{x+1} =$

$\dfrac{3(x+1)}{(x-1)(x+1)} - \dfrac{5(x-1)}{(x-1)(x+1)} =$

$\dfrac{3x+3-5x-5}{(x+1)(x-1)} =$

$\dfrac{-2x-2}{(x+1)(x-1)} =$

$\dfrac{-2(x+1)}{(x+1)(x-1)} = \dfrac{-2}{x-1}$

8. $\dfrac{3x+2}{x-5} = \dfrac{x-4}{3x+2}$

$\dfrac{\cancel{3x+2}}{x-5} = \dfrac{x-4}{\cancel{3x+2}}$

$\dfrac{1}{x-5} = \dfrac{x-4}{1}$

$\dfrac{x-4}{x-5}$

9. How is a division of rational expressions related to a multiplication of rational expressions?

10. In your own words, write a rule for multiplying rational expressions.

CHAPTER SUMMARY

Objectives

What to Remember with Examples

Section 16–1

1 Simplify, or reduce, rational expressions.

If the rational expression is not in factored form, factor completely the numerator and the denominator. Then reduce factors common to both the numerator and denominator.

Simplify $\dfrac{a+b}{16a^2+16ab}$.

$\dfrac{a+b}{16a(a+b)}$ Factored form

$\dfrac{\cancel{a+b}}{16a(\cancel{a+b})} = \dfrac{1}{16a}$

Section 16–2

1 Multiply and divide rational expressions.

Write the numerator and denominator in factored form whenever possible. For multiplication, reduce factors common to both the numerator and denominator.

Multiply $\dfrac{x}{2x+8} \cdot \dfrac{2}{x+1}$.

$\dfrac{x}{2(x+4)} \cdot \dfrac{2}{x+1} =$ Factored form

$\dfrac{(2)(x)}{2(x+4)(x+1)} =$ Multiply and cancel.

$\dfrac{x}{(x+4)(x+1)}$ or $\dfrac{x}{x^2+5x+4}$

For division, multiply by the reciprocal of the second fraction; factor, then reduce factors common to both the numerator and denominator.

Divide $\dfrac{x}{x+2} \div \dfrac{2}{x+2}$.

$\dfrac{x}{\cancel{x+2}} \cdot \dfrac{\cancel{x+2}}{2} = \dfrac{x}{2}$ Multiply by reciprocal. Cancel.

2 Simplify complex rational expressions.

To simplify complex rational expressions, add or subtract any rational expressions in the numerator and/or denominator. Then rewrite the expression as the division of two rational expressions. Next, rewrite the division as a multiplication of the numerator by the reciprocal of the divisor. Factor and perform the indicated multiplication.

Simplify.

$$\dfrac{\dfrac{x-3}{5x}}{\dfrac{4x-12}{15x^2}}$$

Write as division. Factor.

$$\dfrac{x-3}{5x} \div \dfrac{4x-12}{15x^2} =$$

$$\dfrac{\cancel{x-3}}{\cancel{5}x} \cdot \dfrac{\cancel{15x^2}^{3x}}{4(\cancel{x-3})} =$$

Multiply by reciprocal.

$$\dfrac{3x}{4}$$

Section 16-3

1 Add and subtract rational expressions.

Add (or subtract) only rational expressions with the same denominator. For rational expressions with unlike denominators:
1. Find the LCD.
2. Change each fraction to an equivalent one having the LCD.
3. Add (or subtract) the numerators.
4. Keep same LCD.
5. Reduce if possible.

Perform the following operations:

(a) $\dfrac{x}{x+4} + \dfrac{2x}{x+4}$

$\dfrac{x+2x}{x+4}$ Add numerators.

$= \dfrac{3x}{x+4}$

(b) $\dfrac{3}{x+2} - \dfrac{2}{x+1}$

$$\text{LCD} = (x+2)(x+1)$$

$$\dfrac{3}{x+2} = \dfrac{3(x+1)}{(x+2)(x+1)}$$

$$= \dfrac{3x+3}{(x+2)(x+1)}$$

$$\dfrac{2}{x+1} = \dfrac{2(x+2)}{(x+1)(x+2)}$$

$$= \dfrac{2x+4}{(x+1)(x+2)}$$

$$\dfrac{3x+3-(2x+4)}{(x+1)(x+2)} = \dfrac{3x+3-2x-4}{(x+1)(x+2)}$$

$$= \dfrac{x-1}{(x+1)(x+2)}$$

Section 16-4

1 Exclude certain values as solutions of rational equations.

To find the values that make a denominator with a variable equal to zero, set each denominator with a variable equal to zero. Then solve the resulting equations. The excluded values obtained cannot be solutions of the rational equation.

Find the excluded values for the following:

(a) $4 = \dfrac{2}{3 + y}$

$$3 + y = 0$$
$$y = -3 \qquad \text{Excluded value}$$

(b) $\dfrac{2x}{x^2 - x - 6} = 8$

$$x^2 - x - 6 = 0$$
$$(x + 2)(x - 3) = 0$$
$$x + 2 = 0 \qquad x - 3 = 0$$
$$x = -2 \qquad x = 3$$
$$-2, 3 \qquad \text{Excluded values}$$

2 Solve rational equations with variable denominators.

To solve rational equations with variable denominators, multiply each term of the equation by the LCM of all denominators in the equation. Solve the resulting equation. Check all solutions to determine which are true roots and which are extraneous roots.

Solve $\dfrac{2}{x - 3} - \dfrac{1}{x + 3} = \dfrac{1}{x^2 - 9}$.

LCM is $x^2 - 9$ or $(x - 3)(x + 3)$. Excluded values are 3 and -3.

$$2(x + 3) - 1(x - 3) = 1$$
$$2x + 6 - x + 3 = 1$$
$$x + 9 = 1$$
$$x = -8$$

Check:

$$\frac{2}{-8 - 3} - \frac{1}{-8 + 3} = \frac{1}{(-8)^2 - 9}$$
$$\frac{2}{-11} - \frac{1}{-5} = \frac{1}{64 - 9}$$
$$\frac{-2}{11} + \frac{1}{5} = \frac{1}{55}$$
$$\frac{-10}{55} + \frac{11}{55} = \frac{1}{55}$$
$$\frac{1}{55} = \frac{1}{55}$$

WORDS TO KNOW

simplify (p. 595)
reduce (p. 595)
rational expression (p. 596)

addends (p. 596)
terms (p. 596)
complex rational expression (p. 601)

excluded values (p. 606)
extraneous roots (p. 607)

Simplify the following expressions.

1. $\dfrac{x-3}{2x-6}$

2. $\dfrac{x^2-16}{x-4}$

3. $\dfrac{6x^2-11x+4}{2x^2+5x-3}$

4. $\dfrac{x^2-6x+8}{2-x}$

5. $\dfrac{(x-2)(x-4)}{(4-x)(x+2)}$

6. $\dfrac{y^2+x^2}{y^2-x^2}$

7. $\dfrac{6xy}{ab}\cdot\dfrac{a^2b}{2xy^2}$

8. $\dfrac{x^2-y^2}{2x+y}\cdot\dfrac{4x^2+2xy}{y-x}$

9. $\dfrac{x-2y}{x^3-3x^2y}\div\dfrac{x^2-4y^2}{x-3y}$

10. $\dfrac{2a^2-ab-b^2}{6x^2+x-1}\div\dfrac{a^2-b^2}{8x+4}$

11. $\dfrac{2x^2+3x+1}{x}\div\dfrac{x+1}{1}$

12. $\dfrac{4y^2}{2x}\cdot\dfrac{x}{8x}$

13. $\dfrac{1}{x+2}-\dfrac{1}{x-3}$

14. $\dfrac{2}{x+2}+\dfrac{3}{x-1}$

15. $\dfrac{3}{x}+\dfrac{1}{4}$

16. $\dfrac{2}{3y}-\dfrac{7}{y}$

17. $\dfrac{5}{3x-2}+\dfrac{7}{2-3x}$

18. $\dfrac{2x}{3}-\dfrac{5x}{2}$

19. $\dfrac{x-2y}{x^2-4y^2}$

20. $\dfrac{3}{2y}+\dfrac{2}{y}+\dfrac{1}{5y}$

21. $\dfrac{2x}{1-\dfrac{3}{x}}$

22. $\dfrac{\dfrac{1}{a}+\dfrac{1}{b}}{ab}$

Determine the excluded values of x.

23. $\dfrac{5}{x}=\dfrac{2}{x+3}$

24. $\dfrac{3}{x-4}+\dfrac{5}{x+4}=6$

Solve the following rational equations. Check for extraneous roots.

25. $\dfrac{3x}{x-2}+4=\dfrac{3}{x-2}$

26. $\dfrac{x}{x+3}=5$

27. $\dfrac{x}{x-2}=\dfrac{-4}{x+1}$

Use rational equations to solve the following problems. Check for extraneous roots.

28. Henry can assemble four computers in 3 hours. Lester can assemble four computers in 4 hours. How long will it take to assemble four computers if both work together? (*Hint:* Rate × time = amount of work.)

29. A medical group purchases land in Colorado for $100,000. When five more persons joined the group, the cost dropped $10,000 per person in the original group. How many persons were in the original group?

30. Cedric Partee drove 300 miles in one day while on a vacation to the mountains, but on the return trip by the same route he drove 10 miles per hour less and the return trip took 1 hour more time. Find Partee's speed to and from the mountains. (*Hint:* Time = distance ÷ rate.)

17

Inequalities and Absolute Values

GOOD DECISIONS THROUGH TEAMWORK

it is difficult to pinpoint with absolute accuracy, your actual IQ may vary five points either way from your test score. In addition, there are many factors that can affect your score: you may be tired, ill, nervous, distracted, hungry, overcaffeinated, etc., during testing. There are many abilities not measured by IQ testing, such as musical, artistic, mechanical, or technological talent; manual dexterity; ability to read emotions; and the ability to communicate with others. IQ testing can give you an indication of your ability only to think logically, reason, and solve problems within the field of topics tested.

Mathematically speaking, an IQ score is a *quotient* or *ratio* of two numbers:

$$IQ = \frac{MA}{CA} \cdot 100$$

where MA is the mental age of the person, and CA is the chronological age (both measured in years).

MENSA is an international organization whose mission is to offer information, support, and fraternity to intellectually gifted people. Persons of all ages who score in the top 2% on internationally recognized intelligence tests are eligible for membership. The average minimum IQ score needed is 132, but it ranges from 130 to 135 because IQ tests differ slightly in scoring.

IQ, or intelligence quotient, is a general assessment of your ability to think and reason. An IQ score is actually an indication of how you compare in reasoning ability with the majority of people in your age group. For example, a rating of 100 means that, compared to the majority of other people in your age group, you have a normal rate of intelligence. Psychologists consider scores in the range of 95 to 105 as indicating a normal or average IQ. Because

Research the highest IQ measured, and then calculate the person's mental and chronological ages. Because the chronological age is known, give the range of mental ages possible (as an inequality) to be within the five-point tolerance. You may obtain the information from MENSA, the Guiness Book of World Records, the American Psychological Association (APA), or the Internet.

After acquiring a basic understanding of signed numbers, solving equations, and the laws of exponents, we can apply these concepts to another type of mathematical statement, inequalities. In this chapter we will examine inequalities while reviewing the basic procedures for solving equations.

17-1 INEQUALITIES AND SETS

Learning Objectives

1. Use set terminology.
2. Illustrate sets of numbers that are inequalities on a number line and write inequalities in interval notation.

In many mathematical applications, you will need to deal with statements of inequality. An *inequality* is a mathematical statement showing quantities that *are not equal*. The symbol $\neq$ is read "is not equal to."

$$5 \neq 7 \qquad \text{Five } \textit{is not equal to} \text{ seven.}$$

In most cases, to show that quantities are not equal to each other does not give us enough information. We may need to know which quantity is larger (or smaller). When we want specific information, such as 5 *is less than* 7, or 7 *is greater than* 5, we use specific inequality symbols, $<$ (is less than) and $>$ (is greater than). The symbol $\leq$ indicates *is less than or equal to,* and the symbol $\geq$ indicates *is greater than or equal to.*

$$5 < 7 \qquad \text{Five } \textit{is less than} \text{ seven.}$$

$$7 > 5 \qquad \text{Seven } \textit{is greater than} \text{ five.}$$

$$5 \leq 7 \qquad \text{Five } \textit{is less than or equal to} \text{ seven.}$$

$$7 \geq 5 \qquad \text{Seven } \textit{is greater than or equal to} \text{ five.}$$

To help distinguish the symbols, we can think of them as arrowheads. The arrowhead that resembles the left end of a number line is the less than symbol. The arrowhead that resembles the right end of a number line is the greater than symbol.

As in equations, inequalities can have missing amounts that are represented by letters. To *solve* an inequality is to find the value or set of values of the unknown quantity that makes the statement true.

1 Use Set Terminology.

Before solving inequalities, we will introduce some terminology and notations used with sets of numbers. A *set* is a group or collection of items. For example, a set of days of the week that begin with the letter T would include Tuesday and Thursday. However, in this chapter we will examine sets of numbers. The items or numbers that belong to a set are called *members* or *elements of a set*. The description of a set should distinguish clearly between the elements that belong to the set and those that do not belong. This description can be given in words or by using various types of *set notation*.

To illustrate the various types of set notation, we will examine the set of whole numbers between 1 and 8. One notation is to make a list or *roster* of the elements of a set. These elements are enclosed in braces and separated with commas.

Set of whole numbers between 1 and 8 = {2, 3, 4, 5, 6, 7}

Another method for illustrating a set is with *set-builder notation*. The elements of the set are written in the form of an inequality using a variable to represent all the elements of the set. The types of numbers that are to be included in a set will be denoted by the following guide.

Tips and Traps *Common Symbols for Sets:*
The following capital letters are used to denote the indicated set of numbers:

N = natural numbers W = whole numbers

Z = integers Q = rational numbers

I = irrational numbers R = real numbers

M = imaginary numbers C = complex numbers

Symbols that substitute for phrases and words that are often used in describing sets are:

| is read "such that."
∈ is read "is an element of."

Now, we can "build" the set of whole numbers between 1 and 8 with the following symbolic statement.

Set of whole numbers between 1 and 8 = $\{x|x \in W \text{ and } 1 < x < 8\}$

This statement is read "the set of values of x such that each x is an element of the set of whole numbers and x is between 1 and 8."

A special set is the empty set. The *empty set* is a set containing no elements. Symbolically, the empty set is identified as { } or ϕ. The symbol ϕ is the Greek letter phi, (pronounced "fee"). An example of an empty set would be the set of whole numbers between 1 and 2. The set of rational numbers between 1 and 2 would include numbers like $1\frac{1}{2}$ and 1.3, but there are no whole numbers between 1 and 2. Thus the set of whole numbers between 1 and 2 is an empty set.

EXAMPLE 1   Answer the following statements as true or false.

(a) 5 is an element of the set of whole numbers.
(b) $\frac{3}{4}$ is an element of the set of Z.
(c) $-8 \in$ the set of real numbers with the property $\{x|x < -5\}$.
(d) 3.7 is an element of the set of rational numbers with the property $\{x|x > 3.7\}$.
(e) The set of prime numbers that are evenly divisible by 2 is an empty set.

(a) 5 is an element of the set of whole numbers. **True.**
(b) $\frac{3}{4}$ is an element of the set of integers. **False.** Integers include only whole numbers and their opposites.
(c) -8 is an element of the set of real numbers with the property $\{x|x < -5\}$. **True.** -8 is a real number and it is less than -5. Both conditions are satisfied.
(d) 3.7 is an element of the set of rational numbers with the property $\{x|x > 3.7\}$. **False.** 3.7 is a rational number. A number can equal itself, but a number cannot be greater than itself.
(e) The set of prime numbers that are evenly divisible by 2 is the empty set. **False.** 2 is a prime number, and 2 is divisible by 2. Therefore, the set contains the element 2.

2 Illustrate Sets of Numbers that are Inequalities on a Number Line and Write Inequalities in Interval Notation.

Another type of notation used to represent inequalities is *interval notation*. The two *boundaries* are separated by a comma and enclosed with a symbol that indicates whether the boundary is included or not. If there is no boundary in one or both directions, an infinity symbol, ∞, is used. We can say the set has an infinite number of elements and is *unbounded*.

> **Rule 17–1** *To write an inequality in interval notation:*
>
> 1. Determine the boundaries of the interval, if they exist.
> 2. Write the left boundary or $-\infty$ first and write the right boundary or $+\infty$ second. Separate the two with a comma.
> 3. Enclose the boundaries with the appropriate grouping symbols. A parenthesis indicates that the boundary is *not* included. A bracket indicates that a boundary *is* included.

EXAMPLE 2 Represent the following sets of numbers on the number line and by using interval notation.

(a) $1 < x < 8$ (b) $1 \leq x \leq 8$ (c) $x < 3$
(d) $x \geq 3$ (e) all real numbers

Figure 17–1   Figure 17–2

(a) **$1 < x < 8$ is represented by the number line in Fig. 17–1 and by the interval notation (1, 8).** x is greater than 1, so 1 is not included. x is also less than 8, so 8 is not included. They are indicated by unshaded circles and the solution can be described as the values of x between 1 and 8.

(b) **$1 \leq x \leq 8$ is represented by the number line in Fig. 17–2 and by the interval notation [1, 8].** x is greater than or equal to 1, so 1 is included. x is also less than or equal to 8, so 8 is included. This can also be described as the values between 1 and 8, inclusive.

Figure 17–3   Figure 17–4

(c) **$x < 3$ is represented by the number line in Fig. 17–3 and by the interval notation $(-\infty, 3)$.** x is less than 3, so 3 is not included. The values in the negative direction beyond 3 have no boundary.

(d) **$x \geq 3$ is represented by the number line in Fig. 17–4 and by the interval notation $[3, +\infty)$ or $[3, \infty)$.** x is greater than or equal to 3, so 3 is included. In the positive direction beyond 3 there is no boundary.

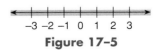

Figure 17–5

(e) **All real numbers are represented by the number line in Fig. 17–5 and by the interval notation $(-\infty, +\infty)$ or $(-\infty, \infty)$.** Real numbers extend in both directions without boundaries.

SELF-STUDY EXERCISES 17–1

1 Answer the following statements as true or false.

1. 0 is an element of the natural numbers.

2. $\frac{5}{8} \in N$.

3. $-6 \in Z$.

4. 5.3 is an element of the set of rational numbers with the property $\{x \mid x \leq 5.3\}$.

2 Represent the following sets on the number line and by using interval notation.

5. $8 \leq x \leq 12$

6. $x > 5$

7. $x \leq -2$

8. All real numbers greater than -6.

9. University Trailer Company had sales of $843,000 for the previous year. The projected sales for the current year are more than the previous year, but less than $1,000,000, which is projected for the next year. Express sales for the current year using interval notation.

10. All real numbers with a minimum of 4 and a maximum of 18.

11. College classes generally must have a minimum number of students to avoid cancellation. If that number is 12, write an inequality to represent the number of students that are in a cancelled class.

17–2 SOLVING SIMPLE LINEAR INEQUALITIES

Learning Objective

1 Solve a simple linear inequality.

1 Solve a Simple Linear Inequality.

The procedures for solving simple inequalities are similar to the procedures for solving equations. We will begin by examining these similarities.

EXAMPLE 1 Find the set of numbers that makes the statements $x < 5 + 4$ true.

Graph the solution (see Fig. 17–6) and express the set as an inequality and in interval notation.

$x < 5 + 4$ Combine number terms.

$x < 9$ or $(-\infty, 9)$

Figure 17–6

This means that the missing value, x, can be any amount, fraction, whole number, irrational number, etc., that *is less than* 9. The solution to an inequality is a *set* of numbers that satisfies the conditions of the statement. The number 9 in Example 1 represents the upper limit or boundary. In this case the *solution set* includes all numbers less than *but not equal to* 9.

Another pair of symbols used to show inequalities is $\leq$ and $\geq$. The symbol $\leq$ is read *"is less than or equal to."* The symbol $\geq$ is read *"is greater than or equal to."* The statement $x \leq 9$ means that the solution set is 9 or any number less than 9.

Compare Figs. 17–7 and 17–8. In Fig. 17–7, 9 is *not* part of the solution set; thus, 9 is represented on the number line with an open circle. In Fig. 17–8, 9 *is* part of the solution set; thus, 9 is represented on the number line with a dot or darkened circle. The statement $x \leq 9$ is written in interval notation as $(-\infty, 9]$.

Figure 17–7

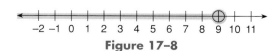

Figure 17–8

In the next example we continue to illustrate the similarities between solving equations and solving inequalities. To understand the interpretation of the solution of an inequality, we will represent the solution set on a number line and write the solution in interval notation.

EXAMPLE 2   Solve the equation, $4x + 6 = 2x - 2$, and the inequality, $4x + 6 > 2x - 2$.

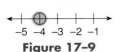

Figure 17–9

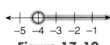

Figure 17–10

$4x + 6 = 2x - 2$	$4x + 6 > 2x - 2$	
$4x - 2x = -6 - 2$	$4x - 2x > -6 - 2$	Transpose.
$2x = -8$	$2x > -8$	Combine.
$\dfrac{2x}{2} = \dfrac{-8}{2}$	$\dfrac{2x}{2} > \dfrac{-8}{2}$	Divide.
$x = -4$	$x > -4$ or $(-4, \infty)$	

The solution for the equation is -4 (Fig. 17–9).

The solution set for the inequality is any number greater than, but not including, -4 (Fig. 17–10).

EXAMPLE 3   Solve the equation, $3x + 5 = 7x + 13$, and the inequality, $3x + 5 < 7x + 13$.

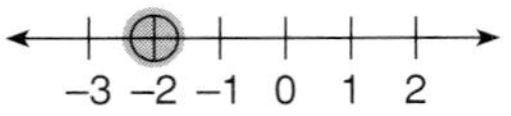

$3x + 5 = 7x + 13$	$3x + 5 < 7x + 13$	
$5 - 13 = 7x - 3x$	$5 - 13 < 7x - 3x$	Transpose.
$-8 = 4x$	$-8 < 4x$	Combine.
$\dfrac{-8}{4} = \dfrac{4x}{4}$	$\dfrac{-8}{4} < \dfrac{4x}{4}$	Divide.
$-2 = x$	$-2 < x$	

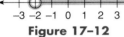

Figure 17–11

Figure 17–12

Recall, in equations we can rewrite $-2 = x$ as $x = -2$. (See Fig. 17–11.)
$x = -2$

In equalities, to say -2 *is less than x* is the same as saying *x is greater than* -2. (See Fig. 17–12.)
$x > -2$ or $(-2, \infty)$

We prefer writing the variable on the left.

Example 3 illustrates a very important difference in solving equations and solving inequalities. The sides of an equation can be interchanged without making any changes in the equal sign. This interchangeability is due to the *symmetric* property of equality; that is, if $a = b$, then $b = a$.

The symmetric property does *not* apply to inequalities. If *a is less than b*, then *b is not less than a*. In fact, the opposite is true. If *a is less than b*, then *b is greater than a*. In symbols: If $a < b$, then $b > a$.

Rule 17–2   *Interchanging the sides of an inequality:*

When the sides of an inequality are interchanged, the sense of the inequality is reversed.

If $a < b$, then $b > a$.   If $a > b$, then $b < a$.

If $a \leq b$, then $b \geq a$.   If $a \geq b$, then $b \leq a$.

The *sense of an inequality* is the appropriate comparison symbol: less than, greater than, less than or equal to, and greater than or equal to.

Let's now look at Example 3 again. This time we collect letter terms on the left in solving both the equation and the inequality.

EXAMPLE 4 Solve the equation, $3x + 5 = 7x + 13$, and the inequality, $3x + 5 < 7x + 13$.

$$3x + 5 = 7x + 13 \qquad\qquad 3x + 5 < 7x + 13$$

$$3x - 7x = 13 - 5 \qquad\quad 3x - 7x < 13 - 5 \qquad \text{Transpose.}$$

$$-4x = 8 \qquad\qquad\quad -4x < 8 \qquad\qquad \text{Combine.}$$

$$\frac{-4x}{-4} = \frac{8}{-4} \qquad\qquad \frac{-4x}{-4} \;?\; \frac{8}{-4} \qquad\qquad \text{Divide.}$$

$$x = -2 \qquad\qquad\qquad x \;?\; -2$$

Is the solution set for the inequality $x < -2$ or $x > -2$? To illustrate which solution set is appropriate, we will substitute a value that is less than -2 to see if the statement is true. Then we will substitute a value that is greater than -2 to see if the statement is true.

First, we try $x = -3$, which is less than -2.

$$3x + 5 < 7x + 13$$

$$3(-3) + 5 \stackrel{?}{<} 7(-3) + 13$$

$$-9 + 5 \stackrel{?}{<} -21 + 13$$

$$-4 \stackrel{?}{<} -8 \qquad \text{False}$$

This statement is false. -4 *is not less than* -8. Thus $x < -2$ is *not* the solution set.

Next, we try $x = -1$, which is greater than -2.

$$3x + 5 < 7x + 13$$

$$3(-1) + 5 \stackrel{?}{<} 7(-1) + 13$$

$$-3 + 5 \stackrel{?}{<} -7 + 13$$

$$2 \stackrel{?}{<} 6 \qquad \text{True}$$

This statement is true. 2 *is less than* 6. **Thus, $x > -2$ is the solution set.**

In Example 4, we logically tested possible solutions to decide which relationship, $<$ or $>$, was appropriate. Another way to illustrate this property of inequalities is to look at specific numbers and the number line. We will start with a true statement and its representation on the number line.

$$2 < 5 \qquad \text{2 is to the left of 5.}$$

Then, multiply each side of the inequality by the negative number -1.

$$-1(2) < -1(5)$$

$$-2 < -5 \qquad \text{This is a false statement because } -2 \text{ is to the right of } -5.$$

To make the statement true, the sense of the inequality must be reversed.

$$-2 > -5$$

Rule 17-3 *Multiplying or dividing an inequality by a negative number:*

If both sides of an inequality are multiplied or divided by a negative number, the sense of the inequality is reversed.

$$\text{If } a < b, \text{ then } -a > -b. \qquad \text{If } a > b, \text{ then } -a < -b.$$
$$\text{If } a \leq b, \text{ then } -a \geq -b. \qquad \text{If } a \geq b, \text{ then } -a \leq -b.$$

Inequalities are solved by performing similar procedures for solving equations *except* when the sides of the inequality are interchanged or when both sides of the inequality are multiplied or divided by a negative number.

Rule 17-4 *To solve inequalities:*

1. Follow the same sequence of steps that would normally be used to solve a similar equation.
2. The sense of the inequality remains the same unless the following situations occur:
   a. The sides of the inequality are interchanged.
   b. The steps used in solving the inequality require that the entire inequality (both sides) be multiplied or divided by a negative number.
3. If either of the situations a or b in Step 2 occurs in solving an inequality, *reverse* the sense of the inequality; that is, less than ($<$) becomes greater than ($>$), and vice versa.

EXAMPLE 5   Solve the inequality, $4x - 2 \leq 3(25 - x)$.

$$4x - 2 \leq 3(25 - x)$$
$$4x - 2 \leq 75 - 3x \qquad \text{Remove the parentheses.}$$
$$4x + 3x \leq 75 + 2 \qquad \text{Collect letter terms on the left and number terms on the right.}$$
$$7x \leq 77 \qquad \text{Combine like terms.}$$
$$\frac{7x}{7} \leq \frac{77}{7} \qquad \text{Divide by the coefficient of the letter.}$$
$$x \leq 11 \qquad \text{or} \qquad (-\infty, 11] \qquad \text{Solution set}$$

EXAMPLE 6   Solve the inequality, $2x - 3(x + 2) > 5x - (x - 5)$.

$$2x - 3(x + 2) > 5x - (x - 5)$$
$$2x - 3x - 6 > 5x - x + 5 \qquad \text{Remove the parentheses.}$$
$$-x - 6 > 4x + 5 \qquad \text{Combine like terms.}$$
$$-x - 4x > 5 + 6 \qquad \text{Transpose.}$$
$$-5x > 11 \qquad \text{Combine like terms.}$$
$$\frac{-5x}{-5} < \frac{11}{-5} \qquad \text{Divide by the coefficient of the letter.}$$
$$x < -\frac{11}{5} \qquad \text{or} \qquad \left(-\infty, -\frac{11}{5}\right) \qquad \text{Because both sides have been divided by a negative number, reverse the sense of the inequality.}$$

EXAMPLE 7 An electrically controlled thermostat is set so that the heating unit automatically comes on and continues to run when the temperature is equal to or below 72°F. At what Celsius temperatures will the heating unit come on? One formula relating Celsius and Fahrenheit temperatures is $°F = \frac{9}{5}°C + 32$.

Using the formula $°F = \frac{9}{5}°C + 32$, the heating unit will operate when the expression $\frac{9}{5}°C + 32$ is less than or equal to 72.

$$\frac{9}{5}C + 32 \le 72$$

$$\frac{9}{5}C \le 72 - 32 \qquad \text{Transpose.}$$

$$\frac{9}{5}C \le 40 \qquad \text{Combine like terms.}$$

$$\frac{5}{9}\left(\frac{9}{5}\right)C \le \left(\frac{5}{9}\right)(40) \qquad \begin{array}{l}\text{Multiply by the reciprocal of the}\\ \text{coefficient of the letter.}\end{array}$$

$$C \le \frac{200}{9}$$

$$C \le 22.22° \qquad \text{or} \qquad (-\infty, 22.22] \qquad \text{To nearest hundredth}$$

The heating unit will come on and continue to run if the temperature is less than or equal to 22.22°C.

SELF-STUDY EXERCISES 17–2

1 Solve the following inequalities. Show the solution set on a number line and by using interval notation.

1. $3x + 7x < 60$
2. $5a - 6a \le 3$
3. $y + 3y \le 32$
4. $b + 6 > 5$
5. $5t - 18 < 12$
6. $2y + 7 < 17$
7. $3a - 8 \ge 7a$
8. $4x + 7 \le 8$
9. $2t + 6 \le t + 13$
10. $4y - 8 > 2y + 14$
11. $3(7 + x) \ge 30$
12. $6(3x - 1) < 12$
13. $2x > 7 - (x + 6)$
14. $4a + 5 \le 3(2 + a) - 4$
15. $15 - 3(2x + 2) > 6$
16. Kevin Presley sold $196 more than twice as much merchandise as Robyn Presley. If Kevin sold at least $52,800, how much did Robyn sell? Write an inequality to represent the facts and solve.

17–3 COMPOUND INEQUALITIES

Learning Objectives

1 Identify subsets of sets and perform set operations.
2 Solve compound inequalities with the conjunction condition.
3 Solve compound inequalities with the disjunction condition.

In many applications of inequalities, more than one condition is placed on the solution. As an example, when measurements are made, a range of acceptable values is specified. All measurements are approximations, and the range of acceptable values is generally stated as a tolerance. If an acceptable measurement is specified to be within ±0.005 inch, then the range of acceptable values can be from 0.005 inch *less than* the ideal measurement to 0.005 inch *more than* the ideal measurement. This range can be stated symbolically as a *compound inequality*.

■ **DEFINITION 17–1: Compound Inequality.** A *compound inequality* is a mathematical statement that combines two statements of inequality.

The conditions placed on a compound inequality may use the connective *and* to indicate that both conditions must be met simultaneously. Such compound inequalities may be written as a continuous statement. The conditions placed on a compound inequality may use the connective *or* to indicate that either condition may be met. Such compound inequalities **must** be written as two separate statements using the connective *or*.

If an ideal measurement is 3 in. with a tolerance of ± 0.005 in., the range of acceptable values would be

$$3 - 0.005 \leq x \leq 3 + 0.005$$
$$2.995 \leq x \leq 3.005$$

① Identify Subsets of Sets and Perform Set Operations.

Before solving compound inequalities, we will look at some additional properties of sets. Another concept that is common when working with sets is the concept of subsets. The symbol $\subset$ is read "is a subset of." One set of items can be a subset of another set.

■ **DEFINITION 17–2: Subset.** A set is a *subset* of a second set if every element of the first set is also an element of the second set.

If $A = \{1, 2, 3\}$ and $B = \{2\}$, then B is a subset of A. This is written in symbols as $B \subset A$ and is illustrated in Fig. 17–13.

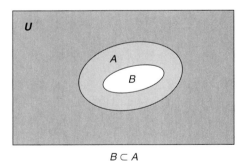

$B \subset A$

Figure 17–13

Two special sets are the universal set and the empty set.

■ **DEFINITION 17–3: Universal Set.** The *universal set* includes all the elements for a given description. The universal set is sometimes identified as U.

■ **DEFINITION 17–4: Empty Set.** The *empty set* is a set containing no elements. The empty set is identified as $\{\ \ \}$ or ϕ.

Property 17–1 *Special set relationships:*

Every set is a subset of itself: $A \subset A$

The empty set is a subset of every set: $\phi \subset A$

EXAMPLE 1   Using the given set definitions, answer the statements (a) through (h) as true or false.

$$U = \{0, 1, 2, 3, 4, 5, 6, 7, 8, 9\}; \ A = \{1, 3, 5, 7, 9\};$$

$$B = \{0, 2, 4, 6, 8\}; \ C = \{1, 2, 3\}; \ D = \{1\}; \ E = \{ \quad \}$$

(a) $A \subset U$ (b) $B \subset U$ (c) $A \subset B$ (d) $C \subset A$
(e) $C \subset U$ (f) $D \subset A$ (g) $E \subset U$ (h) $E \subset B$

(a) $A \subset U$ **True;** every element of A is also an element of U.
(b) $B \subset U$ **True;** every element of B is also an element of U.
(c) $A \subset B$ **False;** 1, 3, 5, 7, and 9 are not elements of B.
(d) $C \subset A$ **False;** 2 is not an element of A.
(e) $C \subset U$ **True;** every element of C is also an element of U.
(f) $D \subset A$ **True;** 1 is an element of A.
(g) $E \subset U$ **True;** the empty set is a subset of every set.
(h) $E \subset B$ **True;** the empty set is a subset of every set.

Two common set operations are union and intersection.

■ **DEFINITION 17–5:   Union of Two Sets.**   The *union* of two sets is a set that includes all elements that appear in *either* of the two sets.

Union is generally associated with the condition "or." The symbol for union is $\cup$. The blue shaded portion in Fig. 17–14 represents $A \cup B$.

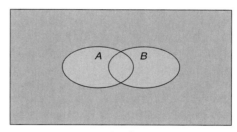

$A \cup B$

Figure 17–14

■ **DEFINITION 17–6:   Intersection of Sets.**   The *intersection* of two sets is a set that includes all elements that appear in *both* of the two sets.

Intersection is generally associated with the condition "and." The symbol for intersection is $\cap$. The blue shaded portion of Fig. 17–15 represents $A \cap B$.

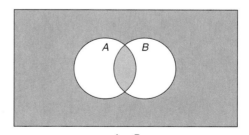

$A \cap B$

Figure 17–15

EXAMPLE 2   If $A = \{1, 2, 3, 4, 5\}$, $B = \{1, 3, 5, 7, 9\}$, and $C = \{2, 4, 6, 8, 10\}$, list the elements in the following sets.

(a) $A \cup B$ (b) $A \cup C$ (c) $B \cup C$
(d) $A \cap B$ (e) $A \cap C$ (f) $B \cap C$

(a) $A \cup B = \{1, 2, 3, 4, 5, 7, 9\}$ All elements in either A or B.
(b) $A \cup C = \{1, 2, 3, 4, 5, 6, 8, 10\}$ All elements in either A or C.
(c) $B \cup C = \{1, 2, 3, 4, 5, 6, 7, 8, 9, 10\}$ All elements in either B or C.
(d) $A \cap B = \{1, 3, 5\}$ Elements common to both A and B.
(e) $A \cap C = \{2, 4\}$ Elements common to both A and C.
(f) $B \cap C = \{\ \ \}$ or ϕ Elements common to both B and C.

Another set operation is complement.

■ **DEFINITION 17–7: Complement of a Set.** The *complement* of a set is a set that includes every element of the universal set that is *not* an element of the given set.

The symbol for the complement of a set is $'$ and is read "prime." A$'$ is represented by the blue shaded portion of Fig. 17–16.

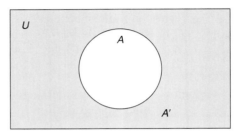

Figure 17–16

EXAMPLE 3   If $U = \{0, 1, 2, 3, 4, 5\}$, $A = \{1, 3, 5\}$, $B = \{0, 2, 4\}$, and $C = \{1, 2, 3\}$, list the elements in the following sets.

(a) A' (b) B' (c) C'
(d) $(A \cup B)'$ (e) $(A \cap C)'$

(a) $A' = \{0, 2, 4\}$ Elements of U not in A.
(b) $B' = \{1, 3, 5\}$ Elements of U not in B.
(c) $C' = \{0, 4, 5\}$ Elements of U not in C.
(d) $(A \cup B)'$
$\quad A \cup B = \{0, 1, 2, 3, 4, 5\}$ Elements in either A or B.
$\quad (A \cup B)' = \{\ \ \}$ or ϕ Elements in U not in either A or B.
(e) $(A \cap C)'$
$\quad A \cap C = \{1, 3\}$ Elements common to both A and C.
$\quad (A \cap C)' = \{0, 2, 4, 5\}$ Elements in U not common to both
$\qquad\qquad\qquad\qquad\qquad A$ and C.

2   **Solve Compound Inequalities with the Conjunction Condition.**

A compound inequality is a statement that places more than one condition on the variable of the inequality. The solution set is the set of values for the variable that meets all conditions of the problem. One type of compound inequality is *conjunction*.

■ **DEFINITION 17–8: Conjunction.** A *conjunction* is an intersection or "and" set relationship.

Both conditions must be met simultaneously in a conjunction. A conjunction can be written as a continuous statement.

Property 17–2 *Conjunctions:*

If $a < x$ and $x < b$, then $a < x < b$.

If $a > x$ and $x > b$, then $a > x > b$.

Similar compound inequalities may also use $\leq$ and $\geq$.

If $a \leq x$ and $x \leq b$, then $a \leq x \leq b$.

If $a \geq x$ and $x \geq b$, then $a \geq x \geq b$.

Rule 17–5: *To solve a compound inequality that is a conjunction:*

1. Separate the compound inequality into two simple inequalities using the conditions of the conjunction.
2. Solve each simple inequality.
3. Determine the solution set that includes the *intersection* of the solution sets of the two simple inequalities.

EXAMPLE 4 Find the solution set for each compound inequality. Graph the solution set on a number line.

(a) $5 < x + 3 < 12$ (b) $-7 \leq x - 7 \leq 1$
(c) $-3 > 3x > -6$ (d) $32 \geq 5x + 7 \geq 17$

(a) $5 < x + 3 < 12$ Separate into two simple inequalities.

$\quad 5 < x + 3 \qquad\qquad x + 3 < 12$

$5 - 3 < x \qquad\qquad\quad x < 12 - 3$

$\quad \mathbf{2 < x}$ or $\mathbf{x > 2} \qquad x < 9$

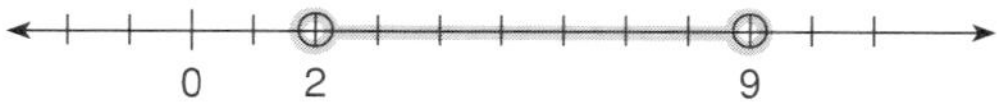

Figure 17–17

Figure 17–17 shows the solution set graphically. The solution set as a continuous statement is **$2 < x < 9$ and in interval notation, (2, 9).**

(b) $-7 \leq x - 7 \leq 1$ Separate into two simple inequalities.

$\quad -7 \leq x - 7 \qquad\qquad x - 7 \leq 1$

$-7 + 7 \leq x \qquad\qquad\quad x \leq 1 + 7$

$\quad \mathbf{0 \leq x}$ or $\mathbf{x \geq 0} \qquad x \leq 8$

Figure 17–18

Figure 17–18 shows the solution set graphically. The solution set as a continuous statement is **$0 \leq x \leq 8$ and in interval notation, [0, 8].**

(c) $-3 > 3x > -6$ Separate into two simple inequalities.

$\quad -3 > 3x \qquad\qquad\qquad 3x > -6$

$\quad \dfrac{-3}{3} > \dfrac{3x}{3} \qquad\qquad\quad \dfrac{3x}{3} > \dfrac{-6}{3}$

$\quad \mathbf{-1 > x}$ or $\mathbf{x < -1} \qquad x > -2$

Figure 17-19

Figure 17–19 shows the solution set graphically. The solution set is $-1 > x > -2$ **or** $-2 < x < -1$ and in interval notation is $(-2, -1)$.

(d) $32 \geq 5x + 7 \geq 17$ Separate into two simple inequalities.

$$32 \geq 5x + 7 \qquad\qquad 5x + 7 \geq 17$$
$$32 - 7 \geq 5x \qquad\qquad 5x \geq 17 - 7$$
$$25 \geq 5x \qquad\qquad 5x \geq 10$$
$$\mathbf{5 \geq x} \quad \text{or} \quad \mathbf{x \leq 5} \qquad\qquad \mathbf{x \geq 2}$$

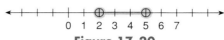

Figure 17-20

Figure 17–20 shows the solution set graphically. The solution set as a continuous statement is $\mathbf{5 \geq x \geq 2}$ **or** $\mathbf{2 \leq x \leq 5}$ and in interval notation, $\mathbf{[2, 5]}$.

Tips and Traps **_Greater Than Versus Less Than:_**
Even though continuous compound inequalities can be written using greater than or less than, using less than follows the natural positions of the boundaries on the number line. If $a > b > c$, then $c < b < a$. For instance, if $3 > 0 > -2$, then $-2 < 0 < 3$.

As with equations and simple inequalities, a compound inequality may not have a solution.

EXAMPLE 5 Find the solution set for the compound inequality, $x + 1 < 1 < 3x - 2$.

Separate into two simple inequalities.

$$x + 1 < 1 \qquad\qquad 1 < 3x - 2$$
$$x < 1 - 1 \qquad 2 + 1 < 3x$$
$$\mathbf{x < 0} \qquad\qquad 3 < 3x$$
$$\qquad\qquad \mathbf{1 < x} \quad \text{or} \quad \mathbf{x > 1}$$

Figure 17-21

As shown in Fig. 17–21, it is impossible for both conditions to be met at the same time. **Thus, the compound inequality has no solution.**

[3] Solve Compound Inequalities with the Disjunction Condition.

Another type of compound inequality is *disjunction.*

■ **DEFINITION 17–9: Disjunction.** A *disjunction* is a union or "or" set relationship. Either condition can be met in a disjunction. A disjunction *cannot* be written as a continuous statement.

Rule 17–6 *To solve a compound inequality that is a disjunction:*

1. Solve each simple inequality.
2. Determine the solution set that includes the union of the solution sets of the two simple inequalities.

EXAMPLE 6   Find the solution set for each compound inequality. Graph the solution set on a number line.

(a) $x + 3 < -2$ or $x + 3 > 2$ (b) $x - 5 \leq -3$ or $x - 5 \geq 3$

(a) $x + 3 < -2$ or $x + 3 > 2$ Solve each simple inequality.

$\quad\quad x < -2 - 3 \quad\quad\quad\quad x > 2 - 3$

$\quad\quad x < -5 \quad$ or $\quad\quad x > -1$ Solution set. See Fig. 17–22.

$\quad\quad \mathbf{(-\infty, -5)} \quad$ or $\quad\quad \mathbf{(-1, \infty)}$

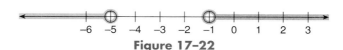

Figure 17–22

(b) $x - 5 \leq -3$ or $x - 5 \geq 3$ Solve each simple inequality.

$\quad\quad x \leq -3 + 5 \quad\quad\quad\quad x \geq 3 + 5$

$\quad\quad x \leq 2 \quad$ or $\quad\quad x \geq 8$ Solution set. See Fig. 17–23.

$\quad\quad \mathbf{(-\infty, 2]} \quad$ or $\quad\quad \mathbf{[8, \infty)}$

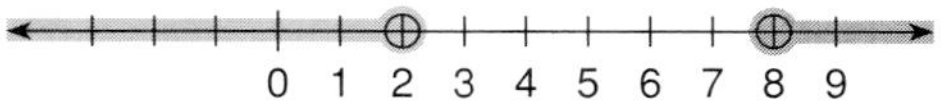

Figure 17–23

SELF-STUDY EXERCISES 17–3

Use the following sets for problems 1–10.

$$U = \{-5, -4, -3, -2, -1, 0, 1, 2, 3, 4, 5\}$$
$$A = \{-4, -2, 0\}, \quad B = \{1, 2, 3, 4, 5\}$$
$$C = \{2, -2\}, \quad D = \{0\}, \quad E = \{\ \}$$

1 Answer the following statements as true or false and justify your answers.

1. $B \subset U$ **2.** $C \subset A$ **3.** $E \subset B$

4. $A \subseteq B$

List the elements in the following sets.

5. $A \cap B$ **6.** $C \cup D$ **7.** $A \cup B$

8. A' **9.** $(A \cap D)'$ **10.** $(B \cup C)'$

2 Solve the following compound inequalities. Show the solution set on a number line and by using interval notation.

11. $x - 1 < 5 < 2x + 1$ **12.** $2x - 7 < x < 3x - 4$ **13.** $x + 3 < 8 < 2x - 12$

14. $x + 2 \leq 7 \leq 2x - 3$ **15.** $2x + 3 < 15 < 3x - 9$ **16.** $1 < x - 2 < 5$

17. $0 < 5x < 15$ **18.** $4 \leq 6 - x \leq 8$ **19.** $6 < -2x < 12$

20. $1 < 6 - x < 3$

3

21. $x + 3 < 5$ or $x - 7 > 2$ **22.** $x - 1 \leq 2$ or $x \geq 7$

23. $2x - 1 \leq 7$ or $3x \geq 15$ **24.** $5 - x \leq 2$ or $x + 1 \leq 2$

25. $5x - 2 \leq 3x + 1$ or $2x \geq 7$

26. The blueprint specifications for a part show it should have a measure of 5.27 cm with a tolerance of ± 0.05 cm. Express the limit dimensions of the part with a compound inequality. If x represents the measure of a different part on the blueprint with the same tolerance, express the limit dimensions with inequalities.

17-4 SOLVING QUADRATIC AND RATIONAL INEQUALITIES

Learning Objectives

1. Solve quadratic inequalities.
2. Solve rational inequalities.

Solving *quadratic* and *rational inequalities* has some basic similarities to solving quadratic and rational equations. The solution to these inequalities, like the solution to linear inequalities, is a *set* of numbers with specific boundaries.

1 Solve Quadratic Inequalities.

Quadratic inequalities can be solved by all of the same methods that were used to solve quadratic equations. In fact, you first treat the inequality as an equation to find the critical values or boundaries.

To solve quadratic inequalities by factoring, we will begin as we did when solving quadratic equations. Rearrange the terms of the inequality so that the inequality is in a form similar to the standard form of a quadratic equation. Then write the terms on the left side in factored form. For example, in the inequality $x^2 < 5x$, we have $x^2 - 5x < 0$, which is $x(x - 5) < 0$.

We determine the boundaries or *critical values* of the solution set by determining the values of x that make each factor *equal* zero.

$$x = 0 \qquad x - 5 = 0$$
$$x = 5$$

The *critical values* are 0 and 5. These critical values are plotted on a number line (see Fig. 17–24).

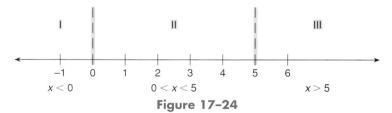

Figure 17–24

These values divide the number line into three regions. Region I indicates all values *less than* the leftmost critical value ($x < 0$). Region II indicates values between the two critical values ($0 < x < 5$). Region III indicates values *greater than* the rightmost critical value ($x > 5$).

The solution set for the inequality will include one or more of these regions. To determine which regions are in the solution set, select *any* point in each region, substitute that value into the inequality, and determine if the resulting inequality is a true or false statement. Regions generating a true statement are in the solution set.

Region I: $x < 0$

Region I test point: $x = -1$.

$$x(x - 5) < 0$$
$$-1(-1 - 5) < 0$$
$$-1(-6) < 0$$
$$+6 < 0$$

The inequality is false.

Therefore, Region I is not included in the solution set.

Region II: $0 < x < 5$

Region II test point: $x = +1$.

$$x(x - 5) < 0$$
$$1(1 - 5) < 0$$
$$1(-4) < 0$$
$$-4 < 0$$

The inequality is true.

Therefore, Region II, $0 < x < 5$, is in the solution set.

Region III: $x > 5$

Region III test point: $x = 6$.

$$x(x - 5) < 0$$
$$6(6 - 5) < 0$$
$$6(1) < 0$$
$$6 < 0$$

The inequality is false.

Therefore, Region III is not included in the solution set.

The solution set includes only Region II, $0 < x < 5$. Figure 17–25 shows the solution graphically.

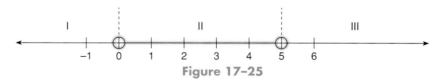

Figure 17–25

We summarize this procedure in Rule 17–7.

Rule 17–7 *To solve quadratic inequalities by factoring:*

1. Rearrange the inequality in standard form so that the right side of the inequality is zero (0).
2. Write the left side of the inequality in factored form.
3. Determine the critical values by finding the values that make each factor equal to zero (0).
4. Plot the critical values on a number line to form the various regions.
5. Test each region of values by selecting any point within a region, substituting that value into the inequality, solving the inequality, and deciding if the resulting inequality is a *true* statement.
6. The solution set for the quadratic inequality will be the region or regions that produce a *true* statement in Step 5.

EXAMPLE 1 Solve the inequality $x^2 + 5x + 6 \leq 0$.

$(x + 3)(x + 2) \leq 0$ Write in factored form.

$x + 3 = 0 \qquad x + 2 = 0$ Determine the critical values.

$\qquad x = -3 \qquad\quad x = -2$

Plot the critical values and label the corresponding regions (Fig. 17–26).

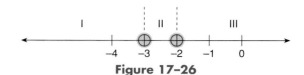

Figure 17–26

Region I: $x \leq -3$ To the left of the smallest critical value, -3.

Region II: $-3 \leq x \leq -2$ Between the critical values, -3 and -2.

Region III: $x \geq -2$ To the right of the largest critical value.

Test each region.

Region I: $x \leq -3$

Region I test point: $x = -4$.

$$(x + 3)(x + 2) \leq 0$$
$$(-4 + 3)(-4 + 2) \leq 0$$
$$(-1)(-2) \leq 0$$
$$2 \leq 0$$

The inequality is false, so Region I is not in the solution set.

Region II: $-3 \leq x \leq -2$

Region II test point: $x = -2.5$.

$$(-2.5 + 3)(-2.5 + 2) \leq 0$$
$$(0.5)(-0.5) \leq 0$$
$$-0.25 \leq 0$$

The inequality is true, so Region II is in the solution set.

Region III: $x \geq -2$

Region III test point: $x = -1$.

$$(-1 + 3)(-1 + 2) \leq 0$$
$$(2)(1) \leq 0$$
$$2 \leq 0$$

The inequality is false, so Region III is not in the solution set.

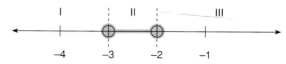

Figure 17–27

The solution set is $-3 \leq x \leq -2$ or $[-3, -2]$. (Fig. 17–27).

EXAMPLE 2 Solve the inequality $2x^2 + x - 6 > 0$.

$\quad (2x - 3)(x + 2) > 0 \qquad$ Write in factored form.

$\quad 2x - 3 = 0 \qquad x + 2 = 0 \qquad$ Determine the critical values.

$\quad\quad 2x = 3 \qquad\qquad x = -2$

$\quad\quad\quad x = \dfrac{3}{2}$

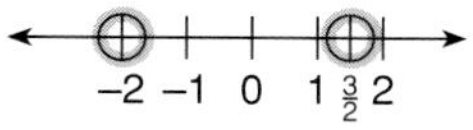

Figure 17–28

Plot the critical values and label the corresponding regions (see Fig. 17–28).

Region I: $x < -2$ $\qquad$ Region II: $-2 < x < \frac{3}{2}$ $\qquad$ Region III: $x > \frac{3}{2}$

Test each region.

> **Region I: $x < -2$**

Region I test point: $x = -3$.

$$(2x - 3)(x + 2) > 0$$
$$[2(-3) - 3][-3 + 2] > 0$$
$$(-6 - 3)(-1) > 0$$
$$(-9)(-1) > 0$$
$$9 > 0$$

The inequality is true.

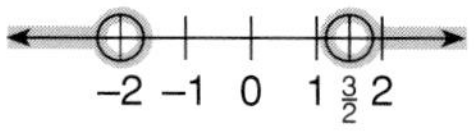

> **Region II: $-2 < x < \dfrac{3}{2}$**

Region II test point: $x = 0$.

$$(2x - 3)(x + 2) > 0$$
$$[2(0) - 3][0 + 2] > 0$$
$$(0 - 3)(2) > 0$$
$$(-3)(2) > 0$$
$$-6 > 0$$

The inequality is false.

> **Region III: $x > \dfrac{3}{2}$**

Region III test point: $x = 2$.

$$(2x - 3)(x + 2) > 0$$
$$[2(2) - 3][2 + 2] > 0$$
$$(4 - 3)(4) > 0$$
$$(1)(4) > 0$$
$$4 > 0$$

Figure 17–29

The inequality is true.

The solution set is $x < -2$ or $x > \frac{3}{2}$. In interval notation the solution is $(-\infty, -2)$ or $(\frac{3}{2}, \infty)$. See Fig. 17–29.

There are other ways of determining the solution set of a quadratic inequality. One way that is similar to finding a test point in each region is to determine the sign of each factor both to the left and right of each critical value. Then using the rules for signed numbers, you can determine which region or regions match the conditions of the inequality. Let's examine the inequalities in Examples 1 and 2 again.

EXAMPLE 3   Determine the solution set of the inequalities in Examples 1 and 2 by examining the signs of each factor in each region.

(a) $x^2 + 5x + 6 \leq 0$ (b) $2x^2 + x - 6 > 0$

(a) $x^2 + 5x + 6 \leq 0$ Original inequality

   $(x + 3)(x + 2) \leq 0$ Factored form of inequality

Critical values: $x = -3$ and $x = -2$

The critical value of a factor is the point on the number line where the factor equals zero. Negative values of the factor are to the left of the critical value and positive values are to the right (see Fig. 17–30.)

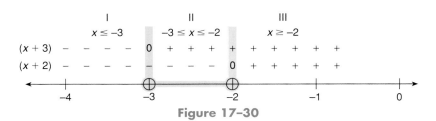

Figure 17–30

The product $(x + 3)(x + 2)$ is positive where the signs of each factor are alike, regions I and III.

The product $(x + 3)(x + 2)$ is negative where the signs of each factor are unlike, region II.

The factored form of the inequality is "less than or equal to" zero. Therefore, the solution set includes the region or regions where the product is negative (see Fig. 17–31).

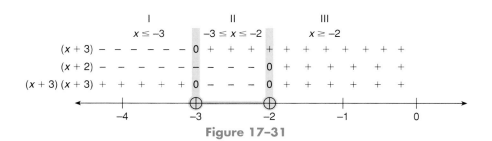

Figure 17–31

The solution set is $-3 \leq x \leq -2$.

(b) $2x^2 + x - 6 > 0$

   $(2x - 3)(x + 2) > 0$

Critical values: $x = \dfrac{3}{2}$ and $x = -2$

The solution set will be in the region or regions where the product is positive, or Regions I and III.

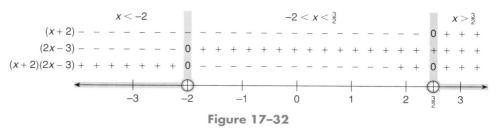

Figure 17–32

The solution set is $x < -2$ or $x > \frac{3}{2}$ (see Fig. 17–32).

After examining several problems, you may notice a pattern for determining the solution set by inspection once the critical values are known. This pattern or generalization can then be applied to quadratic inequalities that do not factor.

Tips and Traps

Finding Solution Sets by Inspection for Quadratic Inequalities:
Once you understand how to identify the appropriate region or regions for the solution set of a quadratic inequality, you may observe a pattern that will allow you to select appropriate regions for such inequalities by inspection. Patterns and generalizations are appropriate only under special conditions. Here, the inequality **must** be written in standard form ($ax^2 + bx + c < 0$ or $ax^2 + bx + c > 0$) where $a > 0$ (a is positive). The critical values will be symbolically represented as s_1 and s_2 where $s_1 < s_2$. Figure 17–33 illustrates the critical values and signs of the two factors and the product.

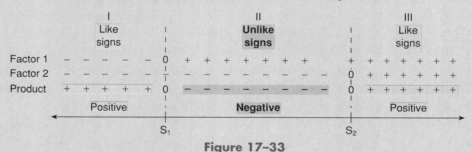

Figure 17–33

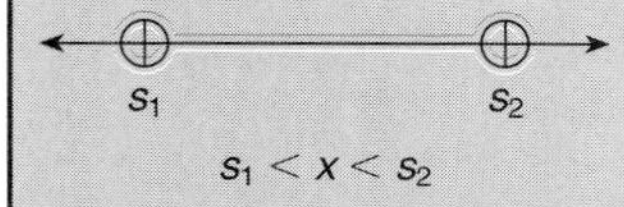

When the inequality $ax^2 + bx + c < 0$ is written in factored form, the two factors must have opposite signs to make a product that is less than zero (negative). From Figure 17–33, we see that the solution set will be in Region II and is written symbolically as $s_1 < x < s_2$. The graphical representation of this solution appears in Fig. 17–33.

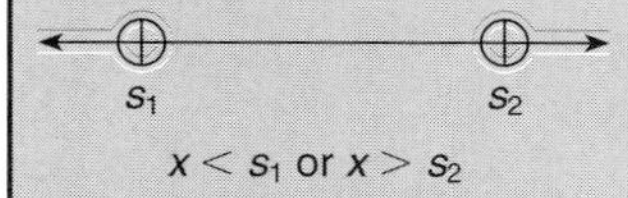

When the inequality $ax^2 + bx + c > 0$ is written in factored form, the two factors must have the same signs to make a product that is greater than zero (positive). From Figure 17–34, we see that the solution set will be in Regions I and III and is written symbolically as $x < s_1$ or $x > s_2$. The graphical representation of this solution appears in Fig. 17–35.

Now, if you solve a quadratic equation by *any* method, including using the quadratic formula, you can determine the solution set of the associated quadratic inequality by inspection after you have found the critical values. The solution can be checked by testing one point in the projected solution set. Similar generalizations can be made for $ax^2 + bx + c \le 0$ and $ax^2 + bx + c \ge 0$.

$s_1 < x < s_2$

Figure 17–34

$x < s_1$ or $x > s_2$

Figure 17–35

2 **Solve Rational Inequalities.**

When is a rational expression greater than zero? For solving a rational inequality, we can use strategies similar to the ones we used in solving quadratic inequalities.

To solve a *rational inequality* like $\frac{x+2}{x-5} > 0$, we will find the critical values by setting each factor equal to zero.

Rule 17–8 *To solve a rational inequality like $\frac{x+a}{x+b} < 0$ or $\frac{x+a}{x+b} > 0$:*

1. Find critical values by setting the numerator and denominator equal to zero.
2. Solve each equation and use the solutions (critical values) to divide the number line into three regions.
3. Regions that have like signs for both the numerator and denominator are in the solution set of $\frac{x+a}{x+b} > 0$ for $x \neq -b$.
4. Regions that have unlike signs for the numerator and denominator are in the solution set of $\frac{x+a}{x+b} < 0$ for $x \neq -b$.

EXAMPLE 4 Solve the following inequalities.

(a) $\dfrac{x+2}{x-5} < 0$ (b) $\dfrac{x+2}{x-5} > 0$

$x + 2 = 0$ $x - 5 = 0$ Critical values for inequalities in both *a* and *b*

$x = -2$ $x = 5$

See Fig. 17–36. The quotient in Region I is positive (like signs). The quotient in Region II is negative (unlike signs). The quotient in Region III is positive (like signs).

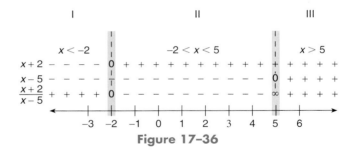

Figure 17–36

(a) Solution set (see Fig. 17–37):   (b) Solution set (see Fig. 17–38):

For $\dfrac{x+2}{x-5} < 0$,   For $\dfrac{x+2}{x-5} > 0$,

$-2 < x < 5$ $x < -2$ or $x > 5$

or   or

$(-2, 5)$ $(-\infty, -2)$ or $(5, \infty)$

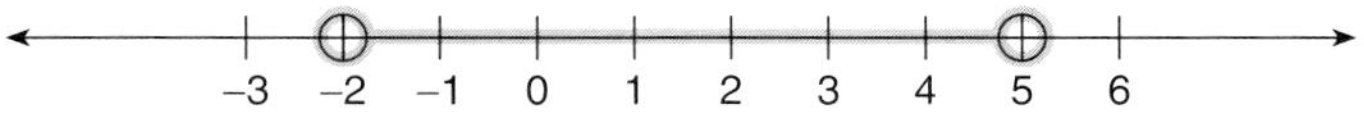

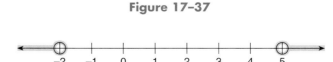

Figure 17–37

Figure 17–38

Tips and Traps	**Solution Patterns for Quadratic Inequalities and Rational Inequalities:**

Solution Patterns for Quadratic Inequalities and Rational Inequalities:
If we are careful to have the leading coefficients positive in each factor of a quadratic inequality or in the numerator and denominator of a rational inequality, we can select the appropriate solution set by inspection. If the critical values for the quadratic inequality $(ax + b)(cx + d) < 0$ are s_1 and s_2, $a > 0$ and $c > 0$, then the solution set is $s_1 < x < s_2$ or (s_1, s_2). See Fig. 17–39.

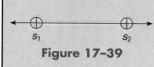

Figure 17–39

If the critical values for the rational inequality $\frac{ax + b}{cx + d} < 0$ are s_1 and s_2, $a > 0$ and $c > 0$, then the solution set is $s_1 < x < s_2$ or (s_1, s_2). See Fig. 17–39.

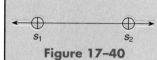

Figure 17–40

If the critical values for the quadratic inequality $(ax + b)(cx + d) > 0$ are s_1 and s_2, $a > 0$, and $c > 0$, then the solution set is $x < s_1$ or $x > s_2$. The solution is also written as $(-\infty, s_1)$ or (s_2, ∞). See Fig. 17–40.

If the critical values for the rational inequality $\frac{ax + b}{cx + d} > 0$ are s_1 and s_2, $a > 0$, and $c > 0$, then the solution set is $x < s_1$ or $x > s_2$. The solution is also written as $(-\infty, s_1)$ or (s_2, ∞). See Fig. 17–40.

Similar properties are true for $\leq$ and $\geq$ relationships.

SELF-STUDY EXERCISES 17–4

1 Solve the following quadratic inequalities and graph the solution on a number line.

1. $(x - 2)(x + 3) < 0$
2. $(2x - 1)(x + 3) > 0$
3. $(2x - 5)(3x - 2) < 0$
4. $(y + 6)(y - 2) \geq 0$
5. $(a + 4)(a - 6) < 0$
6. $x^2 - 7x + 10 \leq 0$
7. $x^2 - x - 6 \geq 0$
8. $x^2 - 4x + 3 \leq 0$
9. $4x^2 - 21x \leq -5$
10. $6x^2 + x > 1$

2 Solve the following rational inequalities and graph the solution on a number line.

11. $\dfrac{x - 3}{x + 7} < 0$
12. $\dfrac{x + 8}{x - 3} > 0$
13. $\dfrac{2x - 6}{3x - 1} > 0$
14. $\dfrac{x - 7}{x + 7} < 0$
15. $\dfrac{x}{x - 1} < 0$

17–5 EQUATIONS CONTAINING ONE ABSOLUTE-VALUE TERM

Learning Objective

1 Solve equations containing one absolute-value term.

1 Solve Equations Containing One Absolute-Value Term.

Earlier, absolute value was defined as the number of units from zero; that is, $|+7|$ or $|-7|$ is 7. If the *absolute value* of a term is indicated in an equation, then both possible values of the term must be considered in solving the equation. Look at the following example.

EXAMPLE 1   Solve the equation $|x| = 5$.

By definition of absolute value, the solutions are $x = +5$ or $x = -5$, which can be written symbolically as $x = \pm 5$.

Notice that the equation $|x| = 5$ has two roots, $+5$ and -5. To solve an equation that contains an absolute value, we must examine each of two cases. One case is when the expression within the absolute-value symbols is positive. The other case is when the expression within the absolute-value symbols is negative. Look at the next example. Recall that the symbolic definition of the absolute value of a number is: $|a| = a$ for $a \geq 0$, and $|a| = -a$ for $a < 0$, or $-|a| = a$ for $a < 0$. This gives us two ways of writing the second case.

EXAMPLE 2 Find the roots for the equation $|x - 4| = 7$.

By definition of absolute value, we have two cases to be examined. We solve the equation given in each case to find the *two* roots.

Case 1: $x - 4 = 7$ Case 2: $-(x - 4) = 7$ or $(x - 4) = -7$

$x - 4 = 7$ $-(x - 4) = 7$ or $(x - 4) = -7$

$\quad x = 7 + 4$ $-x + 4 = 7$ $x - 4 = -7$

$\quad x = 11$ $-x = 7 - 4$ $x = -7 + 4$

$\qquad\qquad\qquad -x = 3$ $x = -3$

$\qquad\qquad\qquad\quad x = -3$

Thus, the roots of the equation are 11 and −3. Each root will check to be a correct root. To check each root, substitute each root separately into the original equation.

If $x = 11$,   If $x = -3$,

$|x - 4| = 7$ $|x - 4| = 7$

$|11 - 4| = 7$ $|-3 - 4| = 7$

$|7| = 7$ $|-7| = 7$

$7 = 7$ $7 = 7$

Rule 17–9 *To solve an equation containing one absolute-value term:*

1. Isolate the absolute-value term on one side of the equation.
2. Separate the equation into two cases. One case considers the expression within the absolute-value symbols to be positive. The other case considers it to be negative.
3. Solve each case to obtain the *two* roots of the equation.
4. $|x| = b$ has no solution if b is negative ($b < 0$).
5. $|x| = b$ has one root when $b = 0$.

EXAMPLE 3 Find the roots of the equation $|y + 3| - 5 = 6$.

Before we can separate the equation into two cases, we should *isolate* the absolute-value term; that is, we rearrange the terms of the equation so that the absolute-value term is alone on one side of the equation.

$$|y + 3| - 5 = 6$$

$$|y + 3| = 6 + 5$$

$$|y + 3| = 11$$

Now, separate into two cases.

Case 1: $y + 3 = 11$

$$y = 11 - 3$$

$$\boxed{y = 8}$$

Case 2: $-(y + 3) = 11$ or $(y + 3) = -11$

$$-y - 3 = 11 \qquad\qquad y + 3 = -11$$

$$-y = 11 + 3 \qquad\qquad y = -11 - 3$$

$$-y = 14 \qquad\qquad\qquad \boxed{y = -14}$$

$$\boxed{y = -14}$$

Thus, the roots of the equation are 8 and −14. Each root will check to be a correct root.

EXAMPLE 4   Solve the equation $|2x - 5| + 14 = 7$.

$$|2x - 5| + 14 = 7$$

$$|2x - 5| = 7 - 14 \qquad \text{Isolate the absolute-value term.}$$

$$|2x - 5| = -7 \qquad \text{Cannot continue because the absolute value must be positive.}$$

Thus, $|2x - 5| + 14 = 7$ has no solution.

SELF-STUDY EXERCISES 17–5

1 Solve the following equations containing absolute values.

1. $|x| = 8$
2. $|x - 3| = 5$
3. $|x - 7| = 2$
4. $|2x - 3| = 9$
5. $|3x - 7| = 2$
6. $|x| - 4 = 7$
7. $|x - 2| + 5 = 3$
8. $-5 + |x - 4| = -2$
9. $|2x - 1| - 3 = 4$
10. $|3x - 2| + 1 = -6$

17–6   ABSOLUTE-VALUE INEQUALITIES

Learning Objectives

1 Solve absolute-value inequalities using the "less than" relationship.

2 Solve absolute-value inequalities using the "greater than" relationship.

1 Solve Absolute-Value Inequalities Using the "Less Than" Relationship.

An inequality that contains an absolute-value term may have a set of values that are solutions to the inequalities. If the inequality is a "less than" or "less than or equal to" relationship, then the following property is used.

Property 17-3 *"Less Than" Relationships for Absolute-Value Inequalities:*

If $|x| < b$ and $b > 0$, then $-b < x < b$.

or

If $|x| \leq b$ and $b > 0$, then $-b \leq x \leq b$.

On a number line (Fig. 17–41), the solution set is a continuous set of values. The values in the solution set of $|x| < b$ are between b and $-b$ when b is positive. Why are zero and negative values of b excluded in Property 17–3? Try some test points, $b = 0$ and $b = -1$. If $b = 0$, $|x| = 0$. Since zero is unsigned, we will not create an interval from -0 to 0. The solution is then a single value, 0. If $b = -1$, $|x| < -1$. A negative value will not check.

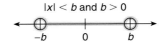

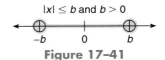

Figure 17–41

EXAMPLE 1   Find the solution set for each of the following inequalities.

(a) $|x| < 3$ (b) $|x + 5| < 8$ (c) $|x - 1| \leq -5$
(d) $|3x| + 2 \leq 14$

(a) $|x| < 3$ Apply the property of absolute-value inequalities having $<$ relationship.

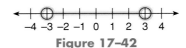

Figure 17–42

Thus, $-3 < x < 3$ or $(-3, 3)$. See Fig. 17–42.

(b) $|x + 5| < 8$ Apply the property of absolute-value inequalities having $<$ relationship.

$$-8 < \boxed{x + 5} < 8$$ Separate into two simple inequalities.

$$-8 < \boxed{x + 5} \qquad\qquad \boxed{x + 5} < 8$$ Solve each inequality.

$$-8 - 5 < x \qquad\qquad\qquad x < 8 - 5$$

$$-13 < x \quad \text{or} \quad x > -13 \qquad x < 3$$

Figure 17–43

Thus, $-13 < x < 3$ or $(-13, 3)$. See Fig. 17–43.

(c) $|x - 1| \leq -5$ Apply the property of absolute-value inequalities having $\leq$ relationship.

An absolute value term cannot be negative. **Thus, $|x - 1| \leq -5$ has no solution.**

(d) $|3x| + 2 \leq 14$ Isolate the absolute-value term.

$$|3x| \leq 14 - 2$$

$$|3x| \leq 12$$ Apply the property of absolute-value inequalities having $\leq$ relationship.

$$-12 \leq \boxed{3x} \leq 12$$ Separate into two simple inequalities.

$$\frac{-12}{3} \leq \frac{\boxed{3x}}{3} \qquad\qquad \boxed{3x} \leq 12$$

$$\qquad\qquad\qquad\qquad x \leq 4$$

$$-4 \leq x \text{ or } x \geq -4$$

Figure 17–44

Thus, $-4 \le x \le 4$ or $[-4, 4]$ **is the solution.** See Fig. 17–44.

2 Solve Absolute-Value Inequalities Using the "Greater Than" Relationship.

An absolute-value inequality having a "greater than" or "greater than or equal to" relationship will have a solution set that is represented by the extreme values on the number line.

Property 17–4 *"Greater Than" Relationships for Absolute-Value Inequalities:*

If $|x| > b$ and $b > 0$, then $x < -b$ or $x > b$.

or

If $|x| \ge b$ and $b > 0$, then $x \le -b$ or $x \ge b$.

The solution set for an absolute-value inequality having a $>$ or $\ge$ relationship is represented on the number lines in Figs. 17–45 and 17–46.

Figure 17–45   **Figure 17–46**

EXAMPLE 2   Find the solution set for each of the following inequalities.

(a) $|x| > 5$ (b) $|x - 2| < -3$ (c) $|x| + 1 \ge 4$
(d) $|2x - 2| \ge 7$

(a) $|x| > 5$ Apply the property of absolute-value inequalities having $>$ relationship.

Thus, $x > -5$ or $x > 5$; $(-\infty, -5)$ or $(5, \infty)$. See Fig. 17–47.

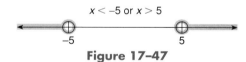

Figure 17–47

(b) $|x - 2| < -3$ Absolute-value terms cannot be negative.

Thus, $|x - 2| < -3$ has no solution.

(c) $|x| + 1 \ge 4$ Isolate the absolute value term.

$\qquad |x| \ge 4 - 1$

$\qquad |x| \ge 3$ Apply the property of inequalities having $\ge$ relationship.

$\qquad |x| \le -3$ or $|x| \ge 3$

Thus, $x \le -3$ or $x \ge 3$; $(-\infty, -3]$ or $[\,3, \infty)$. See Fig. 17–48.

Figure 17–48

(d) $|2x - 2| \geq 7$ Apply the property of absolute-value inequalities having $\geq$ relationship.

$$2x - 2 \leq -7 \quad\quad\text{or}\quad\quad 2x - 2 \geq 7$$
$$2x \leq -7 + 2 \quad\quad\quad\quad\quad 2x \geq 7 + 2$$
$$2x \leq -5 \quad\quad\quad\quad\quad\quad 2x \geq 9$$
$$x \leq -\frac{5}{2} \quad\quad\quad\quad\quad\quad x \geq \frac{9}{2}$$

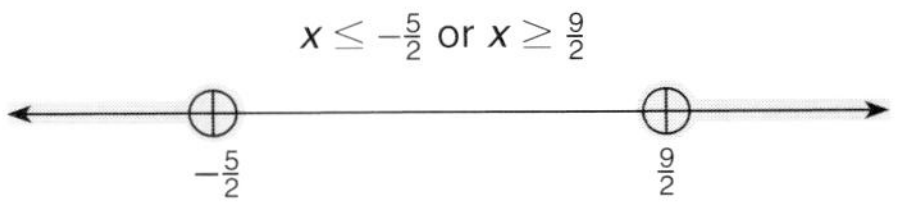

$$x \leq -\tfrac{5}{2} \text{ or } x \geq \tfrac{9}{2}$$

Figure 17–49

Thus, $x \leq -\frac{5}{2}$ **or** $x \geq \frac{9}{2}$**,** $(-\infty, -\frac{5}{2}]$ **or** $[\frac{9}{2}, \infty)$**. See Fig. 17–49.**

Tips and Traps   *The Importance of Isolating the Absolute-Value Term:*
In finding the root for the case when the absolute value term is negative (case 2), we were very careful in Example 2 to isolate the absolute-value term first. Let's examine what happens if we do not isolate the absolute-value term first. We will look at a correct and an incorrect way to proceed.

Find the root of $|y + 3| - 5 = 6$ when $(y + 3)$ is negative (case 2). Do not isolate the absolute-value term first.
Case 2:

Correct Procedure	Incorrect Procedure
$-(y + 3) - 5 = 6$	$y + 3 - 5 = -6$
$-y - 3 - 5 = 6$	$y - 2 = -6$
$-y - 8 = 6$	$y = -6 + 2$
$-y = 6 + 8$	$y = -4$
$-y = 14$	
$y = -14$	
Check: $y = -14$	Check: $y = -4$
$|y + 3| - 5 = 6$	$|y + 3| - 5 = 6$
$|-14 + 3| - 5 = 6$	$|-4 + 3| - 5 = 6$
$|-11| - 5 = 6$	$|-1| - 5 = 6$
$11 - 5 = 6$	$1 - 5 = 6$
$6 = 6$	$-4 = 6$ Wrong
	$-4 \neq 6$

Therefore, we must be very careful when applying the absolute-value properties for case 2 (when $a < 0$). If we choose *not* to isolate the absolute-value term first, then we must take the opposite of the absolute-value grouping rather than the value on the other side of the equation.

    CHAPTER 17   Inequalities and Absolute Values

1 Find the solution set for each of the following inequalities. Graph the solution set on a number line.

1. $|x| < 5$ **2.** $|x| < 1$ **3.** $|x| < 2$
4. $|x| \leq 7$ **5.** $|x + 3| < 2$ **6.** $|x + 4| < 3$
7. $|x - 3| < 4$ **8.** $|x - 2| < 3$ **9.** $|x - 5| \leq 6$
10. $|x - 1| \leq 7$ **11.** $|3x - 5| < 7$ **12.** $|2x - 4| < 3$
13. $|5x + 2| \leq 3$ **14.** $|4x + 7| \leq 5$

2 Find the solution set for each of the following inequalities. Graph the solution set on a number line.

15. $|x| > 2$ **16.** $|x| > 3$ **17.** $|x| \geq 5$
18. $|x| \geq 6$ **19.** $|x - 5| > 2$ **20.** $|x - 4| > 6$
21. $|x + 3| \geq 5$ **22.** $|x + 4| \geq 6$ **23.** $|2x - 5| \geq 0$
24. $|3x - 2| \geq 4$ **25.** $|5x + 8| \geq 2$ **26.** $|4x - 3| \geq 9$

Write the solution set for each of the following inequalities in set-builder notation and graph the solution set.

27. $|3x - 2| + 4 \leq 5$ **28.** $|2x - 5| - 3 \leq 2$ **29.** $|5x - 3| - 6 \geq 1$
30. $|4x - 6| + 4 \geq 5$ **31.** $|3x - 8| - 2 \leq -6$ **32.** $|5x + 7| + 3 \geq -6$

Find the solution set for each of the following inequalities.

33. $|x - 3| \geq 3$ **34.** $|4x + 1| > 1$ **35.** $|x - 4| \leq 3$
36. $|2x - 1| < 0$ **37.** $|x| < 7$ **38.** $|x + 8| < 3$
39. $|x| > 12$ **40.** $|x + 8| > 2$

CAREER APPLICATION

Chemical Engineering: Antifreeze Levels

In many industrial plants manufacturing products such as farm equipment and auto parts, noncontact cooling water is used in heat-treating processes. Water is circulated in pipes and coils around the outside of equipment, such as furnaces, to cool it in much the same way as a car's radiator cools its engine. In climates with below-freezing temperatures, this water is stored in an outside storage tank. If it is not kept moving, it can freeze. A common solution is to add antifreeze to the water to lower the freezing point. (Antifreeze also raises the boiling point of water, which is why cars use coolant during the summer.)

Because cylindrical storage tanks are usually quite large (20,000+ gal), it's much harder for the water there to freeze than the water in a car's radiator (3 to 4 gal). This is due not only to the storage tank's greater volume but also to its shape. A cylinder has a much lower ratio of surface area to volume than the thin, rectangular box shape of car radiators. Less surface area means less exposure to cold air, hence warmer water. Although a 33% antifreeze solution (67% water) is needed to lower a car's radiator freezing point to 0°F, only a 10% antifreeze solution is required to lower a water storage tank's freezing point to −4°F.

Exercises

Use the above information to answer the following. Round answers to the nearest unit.

1. Let T = temperature in °F, and write two simple inequalities to show the minimum temperature protection described above.
2. Use your radiator inequality from Exercise 1 to determine whether a 10% antifreeze solution will protect a car's radiator from freezing at (a) −6°F (b) 6°F (c) 0°F (d) −2°F

3. Use your storage tank inequality from Exercise 1 to determine whether a 10% antifreeze solution will protect a water storage tank from freezing at (a) −6°F (b) 6°F (c) 0°F (d) −2°F
4. How much water from a 12 L radiator should be drained and replaced with antifreeze to have a 33% or higher antifreeze solution? Why couldn't you simply add 4 L of antifreeze to the existing 12 L of water?
5. How much water from a 12 L radiator should be drained and replaced with antifreeze to have a 70% or higher antifreeze solution, which is needed for temperatures as low as −60°F?
6. How much water from a full 20,000-gal storage tank should be drained and replaced with antifreeze to have a 10% or higher antifreeze solution?
7. Describe three methods of obtaining a 10% antifreeze solution in a half-full, 20,000-gal water storage tank.
8. Which of the above methods seems best and why?

Answers

1. 33% radiator antifreeze solution: T ≥ 0°F
   10% storage tank antifreeze solution: T ≥ −4°F
2. (a) No   (b) Yes   (c) Yes   (d) No
3. (a) No   (b) Yes   (c) Yes   (d) Yes
4. At least 4 L. A car's radiator is a closed, full system. If you add antifreeze without draining some water first, the antifreeze will pour out. And even if it would hold 4 L, 4 L in a 16 L volume would be a 25% antifreeze solution.
5. At least 8.4 L
6. At least 2000 gal
7. Method 1: Drain 1000 gal of water and replace it with 1000 gal of antifreeze.
   Method 2: Add an appropriate amount of antifreeze to the 10,000 gal of water to make a 10% solution. Using the equation $0.1(10,000 + x) = x$, where x represents the amount of antifreeze in gallons, 1112 gallons (rounded up) of antifreeze should be added.
   Method 3: Add 2000 gal of antifreeze and 8000 gal of water to fill the 20,000-gal storage tank with a 10% antifreeze solution.
8. Method 2 is better than Method 1 because less labor and time are needed, no water is wasted, and 11,112 gal of cooling water are available instead of 10,000 gal. However, Method 3 may be the best choice because the maximum amount of 20,000 gal of cooling water is made available and no draining or water waste occurs. Both Methods 2 and 3 assume that buying the additional antifreeze and water is not a problem—but if money is tight, Method 1 requires the least money.

ASSIGNMENT EXERCISES

Section 17–1
1. Describe the empty set.

2. Write the symbols used to indicate the empty set.

3. Use symbols to write the following: Five is an element of the set of whole numbers.

Represent the following sets of numbers on the number line and using interval notation.

4. $x \leq 3$

5. $x > -7$

6. $-3 < x < 5$

7. $-4 \leq x < 2$

8. All real numbers

9. $-2 < x$

10. $-5 > x$

Section 17–2

Solve the following inequalities. Show the solution set on a number line.

11. $42 > 8m - 2m$

12. $3m - 2m < 3$

13. $0 < 2x - x$

14. $1 \le x - 7$

15. $10 - 2x \ge 4$

16. $3x + 4 > x$

17. $10x + 18 > 8x$

18. $4x \le 5x + 8$

19. $12 + 5x > 6 - x$

20. $8 - 7y < y + 24$

21. $15 \ge 5(2 - y)$

22. $4t > 2(7 + 3t)$

23. $6x - 2(x - 3) \le 30$

24. $8x - (3x - 2) > 12$

25. $2(b + 1) < 5(b - 4) + 1$

26. A shirt costs $3 less than a certain tie. If the total cost for six ties and two shirts must be less than $130, what is the most each shirt and tie can cost?

Use sets U, A, B, C, and D to give the elements in the following sets.

$$U = \{-2, -1, 0, 1, 2, 3, 4, 5\} \qquad A = \{-2, -1\}, \qquad B = \{0\},$$

$$C = \{0, 1, 2, 3, 4\}, \qquad D = \{-1, 0, 1\}$$

27. $A \cap B$

28. $B \cup D$

29. $A \cap D$

30. A'

31. $(A \cup C)'$

Solve. Show the solution on the number line and by using interval notation.

32. $x - 7 < 2$

33. $x + 4 > 2$

34. $3x - 1 \le 8$

35. $3x - 2 \le 4x + 1$

36. $5 < x - 3 < 8$

37. $-3 < 2x - 4 < 5$

38. $-7 < x - 5 < 7$

Section 17–3

Solve the following compound inequalities. Show the solution set on a number line and by using interval notation.

39. $x + 1 < 5 < 2x + 1$

40. $3x - 8 < x < 5x + 24$

41. $x + 2 < 7 < 2x - 15$

42. $x + 3 \le 7 \le 2x - 1$

43. $2x + 3 < 15 < 3x + 9$

44. $2 < 3x - 4 < 8$

45. $-5 \le -3x + 1 < 10$

46. $2x + 3 \le 5x + 6 < -3x - 7$

47. $-3 \le 4x + 5 \le 2$

48. $4x - 2 < x + 8 < 9x + 1$

49. $x + 3 < 5$ or $x > 8$

50. $x - 7 < 4$ or $x + 3 > 8$

51. $x - 3 < -12$ or $x + 1 > 9$

52. $3x < -4$ or $2x - 1 > 9$

Section 17–4

Solve the following inequalities and graph the solution on a number line.

53. $(x - 5)(x - 2) > 0$

54. $(3x + 2)(x - 2) < 0$

55. $(3x + 1)(2x - 3) < 0$

56. $(5x - 6)(x + 1) \ge 0$

57. $(x + 1)(x - 2) \le 0$

58. $x^2 + x - 12 < 0$

59. $2x^2 \le 5x + 3$

60. $2x^2 - 3x \ge -1$

61. $2x^2 + 7x - 15 < 0$

62. $x^2 - 2x - 8 \ge 0$

63. $\dfrac{x - 7}{x + 1} < 0$

64. $\dfrac{x + 1}{x - 3} > 0$

65. $\dfrac{x}{x + 8} > 0$

66. $\dfrac{x - 8}{x + 1} < 0$

Section 17–5

Solve the following equations containing absolute values.

67. $|x| = 12$

68. $|x - 9| = 2$

69. $|x + 3| = 7$

70. $|x + 4| = 11$

71. $|x - 8| = 12$

72. $|x + 7| = 3$

73. $|4x - 7| = 17$

74. $|2x + 3| = 5$

75. $|7x + 8| = 15$

76. $|4x + 1| = 9$

77. $|7x - 4| = 17$

78. $|6x - 2| = 3$

79. $|3x - 9| = 2$

80. $|x| + 8 = 10$

81. $|x| + 12 = 19$

82. $|2x| - 1 = 9$

83. $|x| - 9 = 7$

84. $|x - 4| - 10 = 6$

85. $-5 + |x - 3| = 2$

86. $|4x - 2| + 1 = 5$

87. $|4x - 3| - 12 = -7$

88. $|3x + 4| - 5 = 17$

89. $|x + 2| + 8 = 9$

90. $|5x + 2| + 4 = 21$

Find and graph the solution set for each of the following inequalities.

91. $|x - 3| < 4$ **92.** $|x + 7| > 3$ **93.** $|x - 4| - 3 < 5$

94. $|x - 1| < 9$ **95.** $|x - 3| < -4$

Write inequality statements for the following.

96. You earn more than $35,000 annually.

97. The gross income for Riddle's market exceeds that of Smith's market and falls short of the income of Duke's market. Smith's gross income is $108,000, and Duke's gross income is $250,000. Write an inequality expressing these relationships.

■ CHALLENGE PROBLEM ■

98. You are a travel agent and have been asked to plan a trip for the local math group. The math group has raised $8700 to spend on the trip. Your agency charges a one-time setup fee of $300. The cost per person will be $620.

   (a) Write an inequality that shows the relationship of the maximum cost of the trip and the costs charged by the travel agency for x persons.

   (b) If the total cost of the trip is $8700, solve the inequality to determine how many math students can go on the trip.

■ CONCEPTS ANALYSIS ■

1. The symmetric property of equations states that if $a = b$ then $b = a$. Is an inequality symmetric? Illustrate your answer with an example.

2. In equations, if $a = b$, then $ac = bc$, if c equals any real number. Is the statement, if $a < b$ then $ac < bc$, true if c is any real number? For what values of c is the statement true? For what values of c is the statement false?

3. Write in your own words two differences in the procedures for solving equations and for solving simple inequalities.

4. Explain the difference between the statements $x < 2$ and $x \leq 2$.

5. If $x < 2$ or $x > 10$, is it correct to write the statement as $2 > x > 10$? Why or why not? Explain your answer.

6. Explain the difference between an *and* relationship and an *or* relationship in inequalities.

7. In your own words, write a procedure for solving a compound inequality such as $5 < x + 2 < 9$.

8. Find the mistake in the following. Correct and briefly explain the mistake.

$$3x + 5 < 5x - 7$$
$$3x - 5x < -7 - 5$$
$$-2x < -12$$
$$x < 6$$

9. In the inequality $x^2 + 6x + 5 < 0$, what roles do the numbers -5 and -1 play? Explain the solution for the inequality in words.

10. If p_1 and p_2 are real numbers that solve the equation $ax^2 + bx + c = 0$ and $p_1 < p_2$, then when will the solution of $ax^2 + bx + c < 0$ be $p_1 < x < p_2$? When will the solution be $x < p_1$ or $x > p_2$? Give an example to illustrate each answer.

Objectives

What to Remember with Examples

Section 17–1

1 Use set terminology.

Set-builder notation is sometimes used to show sets.

> Use set-builder notation to show the following set: the set of integers between -2 and 3, including 3.
>
> $\{x \mid x \in Z \text{ and } -2 < x \leq 3\}$

2 Illustrate sets of numbers that are inequalities on a number line and write inequalities in interval notation.

Solutions to inequalities can be shown on the number line or by using interval notation.

> Show the solution of $-2 < x \leq 3$ on the number line and by using interval notation. $(-2, 3]$
>
>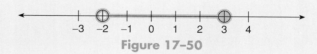
>
> **Figure 17–50**

Section 17–2

1 Solve a simple linear inequality.

To solve a simple linear inequality, isolate the letter as in solving equations with two exceptions. When reversing the sides of an inequality, reverse the sense of the inequality, and when multiplying or dividing by a negative number, reverse the sense of the inequality. Show both the algebraic and graphic solutions of the inequality.

> Solve.
>
>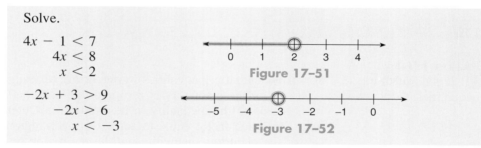
>
> $4x - 1 < 7$
> $4x < 8$
> $x < 2$
>
> **Figure 17–51**
>
> $-2x + 3 > 9$
> $-2x > 6$
> $x < -3$
>
> **Figure 17–52**

Section 17–3

1 Identify subsets of sets and perform set operations.

Every set has the set itself and the empty set as subsets. Set operations include union ($\cup$), intersection ($\cap$), and complements (A').

> List the subsets of the set $\{3, 5, 8\}$:
> ϕ, $\{3, 5, 8\}$ (the set itself), $\{3\}$, $\{5\}$, $\{8\}$, $\{3, 5\}$, $\{3, 8\}$, $\{5, 8\}$.
>
> If $U = \{1, 2, 3, 4, 5\}$, $A = \{1, 3\}$, and $B = \{2, 4\}$, find $A \cup B$, $A \cap B$, and A'.
>
> $A \cup B = \{1, 2, 3, 4\}$
> $A \cap B = \phi$
> $A' = \{2, 4, 5\}$

2 Solve compound inequalities with the conjunction condition.

To solve a compound inequality that is a conjunction, separate the compound inequality into two simple inequalities using the conditions of the conjunction. Solve each simple inequality. Determine the solution set that includes the *intersection* of the solution sets of the two simple inequalities. *Note:* If there is no overlap in the sets, the solution is the empty set.

Indicate the solution using a number line and by using interval notation.

$$-3 < x + 2 \le 2$$
$$-3 < x + 2 \quad \text{and} \quad x + 2 \le 2$$
$$-3 - 2 < x, \qquad\qquad x \le 2 - 2$$
$$-5 < x \text{ or } x > -5 \qquad x \le 0$$
$$-5 < x \le 0 \text{ or } (-5, 0]$$

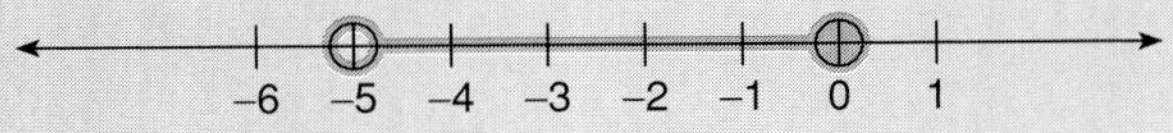

Figure 17-53

3 Solve compound inequalities with the disjunction condition.

To solve a compound inequality that is a disjunction: Solve each simple inequality. Determine the solution set that includes the union of the solution sets of the two simple inequalities.

Solve and indicate the solution set on the number line and by using interval notation.

$$x - 1 < 3 \qquad \text{or} \quad x + 2 > 8$$
$$x < 3 + 1 \qquad\qquad x > 8 - 2$$
$$x < 4 \qquad \text{or} \qquad x > 6$$

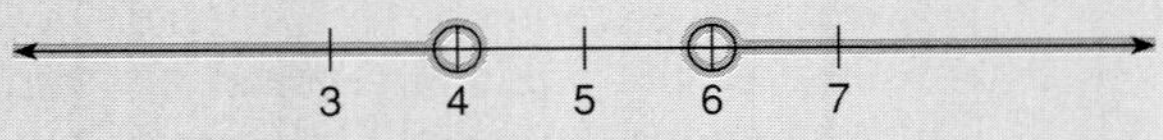

Figure 17-54

$(-\infty, 4)$ or $(6, \infty)$

Section 17-4
1 Solve quadratic inequalities.

To solve quadratic inequalities by factoring, rearrange the inequality so that the right side of the inequality is zero and the lead coefficient is positive. Write the left side of the inequality in factored form. Determine the critical values that make each factor equal to zero. Test each region of values. The solution set for the quadratic inequality will be the region or regions that produce a true statement.

Solve $x^2 + 4x - 21 < 0$.

$$(x + 7)(x - 3) = 0$$
$$x + 7 = 0 \qquad x - 3 = 0$$
$$x = -7 \qquad\quad x = 3 \qquad\qquad \text{Critical Values}$$

Test Region I. Let $x = -8$.
$(-8)^2 + 4(-8) - 21 < 0$?
$64 - 32 - 21 < 0$?
$11 < 0$. False.

Test Region II. Let $x = 0$.
$0^2 + 4(0) - 21 < 0$?
$0 + 0 - 21 < 0$?
$-21 < 0$. True.

Test Region III. Let $x = 4$.
$4^2 + 4(4) - 21 < 0$?
$16 + 16 - 21 < 0$?
$11 < 0$. False.

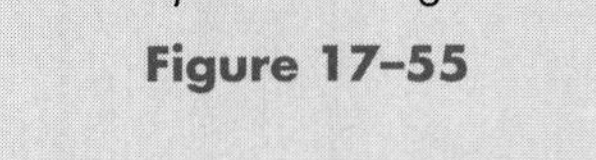

Figure 17-55

The solution set is Region II.
$-7 < x < 3$ or $(-7, 3)$.

2 Solve rational inequalities.

To solve rational inequalities, set both the numerator and denominator equal to zero and solve for the variable. These solutions are the boundaries for the solution regions of the inequality solution. The region or regions that have like signs for the numerator and denominator are solutions for the $>$ and $\geq$ inequalities. The region or regions that have unlike signs for the numerator and denominator are solutions for the $<$ or $\leq$ inequalities.

Solve the rational inequality. $\dfrac{x + 3}{x - 2} \leq 0$.

$x = 2$ is an excluded value.
$$x + 3 = 0 \qquad x - 2 = 0$$
$$x = -3 \qquad\qquad x = 2$$

Region I has like signs $\left(\dfrac{-}{-}\right)$.

Region II has unlike signs $\left(\dfrac{+}{-}\right)$.

Region III has like signs $\left(\dfrac{+}{+}\right)$.

Unlike signs indicate < 0.

Critical Values

Figure 17–56

The solution set is Region II.
$-3 \leq x < 2.\ [-3, 2)$

Section 17–5

1 Solve equations containing one absolute-value term.

To solve absolute-value equations, isolate the absolute-value expression, then form two equations. One equation considers the expression within the absolute-value symbol to be positive; the other equation considers the expression to be negative. Solve each case to obtain the two roots of the equation.

Solve $|x - 3| + 1 = 5$.

$$|x - 3| + 1 = 5$$
$$|x - 3| = 4$$

Case 1:
$$x - 3 = 4$$
$$x = 7$$

Case 2:
$$-(x - 3) = 4$$
$$-x + 3 = 4$$
$$-x = 1$$
$$x = -1$$

Section 17-6

1 Solve absolute-value inequalities using the "less than" relationship.

To solve absolute-value inequalities using the "less than" relationship, use the property, if $|x| < b$ and $b > 0$, then $-b < x < b$.

Solve $|x - 3| < 4$.

$$-4 < x - 3 < 4$$
$$-4 < x - 3 \qquad\qquad x - 3 < 4$$
$$-1 < x \text{ or } x > -1 \qquad x < 7$$
$$-1 < x < 7;\ (-1, 7)$$

2 Solve absolute-value inequalities using the "greater than" relationship.

To solve absolute-value inequalities using the "greater than" relationship, use the property, if $|x| > b$ and $b > 0$, then $x < -b$ or $x > b$.

Solve $|x + 7| > 1$.

$$x + 7 < -1 \qquad \text{or} \qquad x + 7 > 1$$
$$x < -8 \qquad\qquad\qquad x > -6;\ (-\infty, -8) \text{ or } (-6, \infty)$$

inequality, (p. 620)
set, (p. 620)
member or element of a set,
    (p. 620)
roster, (p. 620)
set notation, (p. 620)
set-builder notation, (p. 621)
empty set, (p. 621, 628)
interval notation, (p. 622)
boundary, (p. 622)
sense of an inequality, (p. 624)

compound inequality, (p. 628)
subset, (p. 628)
universal set (p. 628)
union of sets, (p. 629)
intersection of sets, (p. 629)
complement of a set, (p. 630)
conjunction, (p. 630)
disjunction, (p. 632)
quadratic inequality, (p. 634)
critical values, (p. 634)

rational inequality, (p. 634,
    640)
absolute value terms in
    equations, (p. 641)
property of absolute-value
    inequalities using $<$ or $\leq$,
    (p. 643)
property of absolute-value
    inequalities using $>$ or $\geq$,
    (p. 645)

CHAPTER TRIAL TEST

Represent the following sets of numbers on the number line and by using interval notation.

1. $x \geq -12$

2. $-3 < x \leq -2$

Solve the following inequalities. Graph the solution on a number line and write the solution using interval notation.

3. $3x - 1 > 8$

4. $2 - 3x \geq 14$

5. $10 < 2 + 4x$

6. $-5b > -30$

7. $\frac{1}{3}x + 5 \leq 3$

8. $2(1 - y) + 3(2y - 2) \geq 12$

9. $5 - 3x < 3 - (2x - 4)$

10. $7 + 4x \leq 2x - 1$

11. $-5 < x + 3 < 7$

12. $\frac{1}{4}x + 2 < 3 < \frac{1}{3}x + 9$

13. $3x - 1 \leq 5 \leq x - 5$

14. $(x + 4)(x - 2) < 0$

15. $(2x + 3)(x - 1) > 0$

16. $2y^2 + y < 15$

17. $2x - 3 < 1$ or $x + 1 > 7$

18. $2(x - 1) < 3$ or $x + 7 > 15$

Use the following sets to give the elements in the indicated sets.

$U = \{1, 2, 3, 4, 5, 6, 7, 8, 9\}$ $A = \{5, 8, 9\}$, $B = \{1, 2, 3, 4, 5, 6, 7, 8\}$

19. $A \cup B$

20. $A \cap B$

21. $A \cap B'$

22. List all the subsets of set A.

Solve. Graph the solution set on a number line and write the solution set in interval notation.

23. $\dfrac{x - 2}{x + 5} < 0$

24. $\dfrac{x - 3}{x} > 0$

Solve.

25. $|x| = 15$

26. $|x + 8| = 7$

27. $|x| + 8 = 10$

Solve and graph the solution set on a number line and write the solution set in interval notation.

28. $|4x - 7| < 17$

29. $|x| + 8 > 10$

30. $|x + 1| - 3 < 2$

18

Graphical Representation

_____ GOOD DECISIONS THROUGH TEAMWORK _____

Today's investors have various investing options. One field of investment that can be very profitable in a short time is the worldwide commodities market. Commodities are tangible items such as grains (wheat, corn, rice), beef, hogs, pork bellies, metals (gold, silver, copper), or agricultural products (soybeans, cotton, orange juice).

The commodities market is volatile: money can be made or lost quickly depending on unpredictable factors such as drought, flood, late or early frosts, hurricanes, workers' strikes (mining, dock, or transportation), fuel price increases, fuel shortages, or political unrest. Grain companies, ranchers, and meat-packing houses invest in specific products to use in their businesses to "lock in" the price ahead of time. For example, in February a rancher may invest in 10,000 bushels of May corn to feed the cattle all summer. The February price might be low because numerous acres of May corn may have been planted worldwide. By May the price may have increased dramatically due to floods wiping out the U.S. corn crop. The rancher would have saved money by purchasing the needed corn at the lower price.

Other commodities investors, like you and your team members who may have no personal knowledge of the commodities, typically invest in futures. A futures investor buys the rights to purchase products that have not yet been produced, at a specific price, with the hope that the price will increase before the investment is sold. Because commodity futures are typically held less than a year before selling, a commodities market investor must be prepared to lose the entire investment in a short time, usually with no warning.

Suppose that each member of your team invests $10,000 in an agricultural crop commodity today. Each team member must invest in a different crop. Use the Internet to check today's prices as listed in the U.S. Department of Agriculture's crop reports at their worldwide web site: www.usda.gov. If you do not have access to the Internet, use the prices quoted in today's newspaper.

One month later, check the prices at the same site. List each person's profit or loss and the annual interest rate. For each profit or loss, write a sentence offering a possible explanation for the profit or loss. Assume that these investments experienced linear growth (or loss). Have each member graph his or her investment for $t = 0$ to $t = 1$ month, and shade the profit or loss region.

In several previous chapters, we have solved various types of equations. The solution was usually one or two values that would make a true statement when substituted into the equation. In this chapter, we will look at the graphical representation of some of these types of equations.

18-1 GRAPHICAL REPRESENTATION OF EQUATIONS

Learning Objectives

1. Represent the solutions of a function in a table of values and on a graph.
2. Write an equation in one variable as a function and find the solution graphically.

Sometimes the solution of an equation answers a specific question. How much can we pay for a new copier with a 3-year service contract, if the service contract is $100 and the total amount we are budgeted to spend is $900? We can represent this situation with the equation $x + \$100 = \$900$, where x represents the cost of the copier.

In other situations an equation serves as a model to show the effect of change. These models can be written as functions and can be represented graphically.

1 Represent the Solutions of a Function in a Table of Values and on a Graph.

The monthly cost of a mobile phone is $15 plus $0.25 for each minute of use. An equation to represent this situation must have two variables. One variable will represent the number of minutes of phone use (m), and the other variable will represent the monthly cost (c). One variable is dependent on the other. For instance, the monthly cost is dependent on the number of minutes of use: $c = \$15 + \$0.25m$. We can also write equations with two variables in function notation: $C(m) = \$15 + \$0.25m$ or $f(x) = 15 + 0.25m$.

EXAMPLE 1 Make a table of values and a graph of the function $f(x) = 15 + 0.25m$ for five values of x in 50-minute intervals.

x	$f(x)$
0	$15.00
50	$27.50
100	$40.00
150	$52.50
200	$65.00

$f(0) = 15 + 0.25(0)$
$f(0) = 15 + 0$
$f(0) = 15$, or $\$15.00$

$f(50) = 15 + 0.25(50)$
$f(50) = 15 + 12.5$
$f(50) = 27.5$, or $\$27.50$

$f(100) = 15 + 0.25(100)$
$f(100) = 15 + 25$
$f(100) = 40$, or $\$40.00$

$f(150) = 15 + 0.25(150)$
$f(150) = 15 + 37.5$
$f(150) = 52.5$, or $\$52.50$

$f(200) = 15 + 0.25(200)$
$f(200) = 15 + 50$
$f(200) = 65$, or $\$65.00$

To represent this model graphically, use the x- or horizontal axis to represent the values of the independent variable. Use the y- or vertical axis to represent the values of the dependent variable (Fig. 18–1).

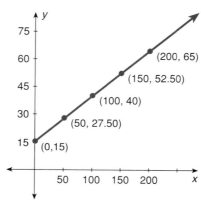

Figure 18-1

Let's examine the graphical solution. What does the point (0, $15) represent? This is a fixed cost. Even if no minutes are used, the monthly cost will be $15. What if the monthly phone use is more than 200 minutes? When the table of values was represented on the graph, a pattern was established. We can use this pattern or model to project other values of x.

EXAMPLE 2 Find the monthly cost from the graph of $f(x) = 15 + 0.25x$ (Fig. 18–2) for 250 minutes and 300 minutes.

From the graph find the corresponding values for $f(x)$ when $x = 250$ and when $x = 300$.

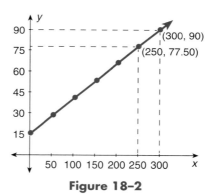

Figure 18-2

For $x = 250$, $f(x) = \$77.50$

For $x = 300$, $f(x) = \$90.00$

In Examples 1 and 2, we were looking at a solution that restricted the values of x and $f(x)$ to positive values. In many cases we want to view the pattern established by considering all types of numbers. After viewing all values, some of these values may still be disregarded because they are impractical or impossible within the context of the situation.

Rule 18-1 *To graph a function using a table of values:*

1. Prepare a table of values by evaluating the function at different values (at least three) of the independent variable.
2. Plot the points in the table of values on the rectangular coordinate system, with the independent variable on the x-axis and the dependent variable on the y-axis.

EXAMPLE 3 Prepare a table of values and graph the function $f(x) = x^2 + x - 6$ for integral values of x between -3 and 3, inclusive.

x	$f(x)$
-3	0
-2	-4
-1	-6
0	-6
1	-4
2	0
3	6

$f(-3) = (-3)^2 + (-3) - 6$
$f(-3) = 9 - 3 - 6$
$f(-3) = 0$

$f(-2) = (-2)^2 + (-2) - 6$
$f(-2) = 4 - 2 - 6$
$f(-2) = -4$

$f(-1) = (-1)^2 + (-1) - 6$
$f(-1) = 1 - 1 - 6$
$f(-1) = -6$

$f(0) = 0^2 + 0 - 6$
$f(0) = 0 + 0 - 6$
$f(0) = -6$

$f(1) = 1^2 + 1 - 6$
$f(1) = 1 + 1 - 6$
$f(1) = -4$

$f(2) = 2^2 + 2 - 6$
$f(2) = 4 + 2 - 6$
$f(2) = 0$

$f(3) = 3^2 + 3 - 6$
$f(3) = 9 + 3 - 6$
$f(3) = 6$

Next, plot the points indicated in the table of values and connect the points with a smooth, continuous curve (Fig. 18–3).

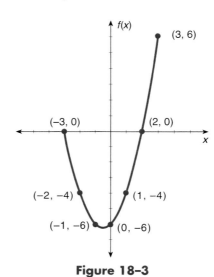

Figure 18-3

EXAMPLE 4 Prepare a table of values and graph the function $f(x) = 2^x$ for integral values of x between -3 and 3, inclusive.

x	$f(x)$
-3	0.125
-2	0.25
-1	0.5
0	1
1	2
2	4
3	8

$f(-3) = 2^{-3}$
$f(-3) = \frac{1}{2^3}$
$f(-3) = \frac{1}{8}$, or 0.125

$f(-2) = 2^{-2}$
$f(-2) = \frac{1}{2^2}$
$f(-2) = \frac{1}{4}$, or 0.25

$f(-1) = 2^{-1}$
$f(-1) = \frac{1}{2^1}$
$f(-1) = \frac{1}{2}$, or 0.5

$f(0) = 2^0$
$f(0) = 1$

$f(1) = 2^1$
$f(1) = 2$

$f(2) = 2^2$
$f(2) = 4$

$f(3) = 2^3$
$f(3) = 8$

Next, plot the points indicated in the table of values and connect the points with a smooth, continuous curve (Fig. 18–4).

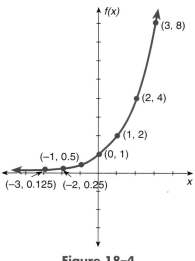

Figure 18–4

EXAMPLE 5 Prepare a table of values and graph the function $f(x) = \ln 4x$ for the following values of x: 0.125, 0.25, 0.5, 1, 2, 3, 4.

Using a calculator:

x	$f(x)$
0.125	−0.7
0.25	0
0.5	0.7
1	1.4
2	2.1
3	2.5
4	2.8

$f(0.125) = \ln 4(0.125)$ $f(0.25) = \ln 4(0.25)$ $f(0.5) = \ln 4(0.5)$

$f(0.125) = \ln 0.5$ $f(0.25) = \ln 1$ $f(0.5) = \ln 2$

$\mathbf{f(0.125) = -0.6931471806}$ $\mathbf{f(0.25) = 0}$ $\mathbf{f(0.5) = 0.6931471806}$

$f(1) = \ln 4(1)$ $f(2) = \ln 4(2)$ $f(3) = \ln 4(3)$

$f(1) = \ln 4$ $f(2) = \ln 8$ $f(3) = \ln 12$

$\mathbf{f(1) = 1.386294361}$ $\mathbf{f(2) = 2.079441542}$ $\mathbf{f(3) = 2.48490665}$

$f(4) = \ln 4(4)$

$f(4) = \ln 16$

$\mathbf{f(4) = 2.772588722}$

Next, plot the points indicated in the table of values and connect the points with a smooth, continuous curve (Fig. 18–5).

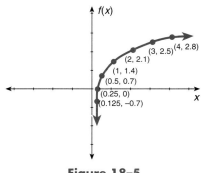

Figure 18–5

As you can see from these examples, all functions can be expressed graphically by using a table of values.

Write an Equation in One Variable as a Function and Find the Solution Graphically.

An equation in one variable, such as $2x + 5 = 3$, can be written in the form of a function by first rewriting all terms on either side of the equation.

$$2x + 5 = 3 \quad \text{or} \quad 2x + 5 = 3$$
$$2x + 5 - 3 = 0 \qquad\qquad 0 = 3 - 2x - 5$$
$$2x + 2 = 0 \qquad\qquad 0 = -2x - 2$$

Then, write the nonzero side of the equation as a function.

$$f(x) = 2x + 2 \quad \text{or} \quad f(x) = -2x - 2$$

The solution of the equation is the value of x that makes the value of the function zero.

$$2x + 5 = 3 \qquad f(-1) = 2(-1) + 2 \quad \text{or} \quad f(-1) = -2(-1) - 2$$
$$2x = 3 - 5 \qquad f(-1) = -2 + 2 \qquad\qquad f(-1) = 2 - 2$$
$$2x = -2 \qquad f(-1) = 0 \qquad\qquad f(-1) = 0$$
$$x = -1$$

Other values of the function can be found and used to graph the function. The graphs of $f(x) = 2x + 2$ and $f(x) = -2x - 2$ are shown in Fig. 18–6.

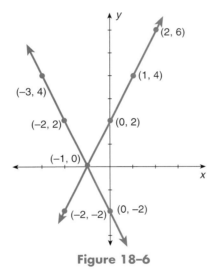

x	$f(x)$
-2	-2
-1	0
0	2
1	4
2	6

$$f(-2) = 2(-2) + 2 = -2$$
$$f(-1) = 2(-1) + 2 = 0$$
$$f(0) = 2(0) + 2 = 2$$
$$f(1) = 2(1) + 2 = 4$$
$$f(2) = 2(2) + 2 = 6$$

x	$f(x)$
-3	4
-2	2
-1	0
0	-2

$$f(-3) = -2(-3) - 2 = 4$$
$$f(-2) = -2(-2) - 2 = 2$$
$$f(-1) = -2(-1) - 2 = 0$$
$$f(0) = -2(0) - 2 = -2$$

Figure 18–6

If an equation in one variable is written in function notation, the solution of the equation is represented on the graph at the point where the graph crosses the x-axis. In Fig. 18–6, the solution is $x = -1$.

> **Rule 18–2** *To write an equation in one variable as a function:*
>
> 1. Rewrite the equation so that all terms are on one side of the equation and the leading coefficient is positive.
> 2. Write the equation in function notation using $f(x) =$ the nonzero side of the equation from Step 1.

EXAMPLE 6 Write $3x + 7 = x - 3$ as a function, graph the function, and find the solution from the graph.

$$3x + 7 = x - 3 \qquad \text{Write all terms on the left side of the equation.}$$

$$3x + 7 - x + 3 = 0 \qquad \text{Combine like terms.}$$
$$2x + 10 = 0 \qquad \text{Write as a function.}$$
$$f(x) = 2x + 10$$

Evaluate the function at -2, 0, and 2.

x	$f(x)$
-2	6
0	10
2	14

$$f(-2) = 2(-2) + 10 \qquad f(0) = 2(0) + 10 \qquad f(2) = 2(2) + 10$$
$$f(-2) = -4 + 10 \qquad f(0) = 0 + 10 \qquad f(2) = 4 + 10$$
$$f(-2) = 6 \qquad f(0) = 10 \qquad f(2) = 14$$

Graph the points and extend the graph to cross the x-axis (see Fig. 18–7). Read the solution from the graph where the graph crosses the x-axis.
The solution is -5.
Check:

$$f(-5) = 2(-5) + 10$$
$$f(-5) = -10 + 10$$
$$f(-5) = 0$$

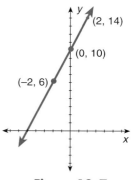

Figure 18–7

The table of values method of graphing and the graphing method of solving an equation are tedious. These methods are used when a computer algebra system and/or a graphics calculator is available. It is helpful when you can visualize a problem and solution. However, there are other methods for graphing that are less time consuming because they are based on the properties of various types of equations. In the sections that follow, we will examine several different types of equations and observe properties that will allow us to sketch a graph before using modern technology. Sketching is a form of estimating.

SELF-STUDY EXERCISES 18–1

1. Make a table of values for the function $f(x) = 3x - 7$ for 5 values of x and graph the function.

2. Rental cars cost $39 for a day plus $0.10 per mile. Write a function that represents the cost of the car for a single day if x represents the number of miles and $f(x)$ represents the car rental cost.

3. Make a table of values for the function in Exercise 2. Graph the function.

4. From the graph prepared in Exercise 3, determine the cost of driving the rental car for 400 miles in a single day.

Prepare a table of values and graph each of the following functions.

5. $f(x) = x^2 - 4x + 5$

6. $f(x) = 3^x$

7. $f(x) = \ln 3x$

2 Write each of the following equations as a function and find the solution graphically.

8. $3x + 2 = 7$

9. $5x - 4 = 2$

10. $4 = 6x - 1$

11. $3x + 7 = 4x + 6$

12. $3x - 1 = 4(x + 2)$

Learning Objectives

1. Check the solution of an equation with two variables.
2. Find a specific solution of an equation with two variables when given the value of one variable.

Most of the equations we have studied in this text have had only one variable, such as x or y, and took the form $3x - 2 = 7$ or $5y + 3 = 9$. However, equations may have two variables, both x *and* y, such as $2x + 3y = 12$. Both types of equations are *linear equations*.

■ **DEFINITION 18–1: Linear Equation.** A *linear equation* is one whose graph is a straight line. It can be written in the standard form

$$ax + by = c$$

where a, b, and c are real numbers and a and b are not both zero.

Before we begin examining properties of linear equations, we will be sure we can verify that a possible solution is correct, or that we can find a specific solution if we know one value of the solution.

1 Check the Solution of an Equation with Two Variables.

The solutions to a linear equation in two variables are *ordered pairs* of numbers. There is one number for each variable. The numbers are ordered in the alphabetical order of the two variables rather than the order the letters appear in the equation. Thus, in the equation $2y + x = 6$, a solution of $(4, 2)$ indicates that $x = 4$ and $y = 2$. An ordered pair is enclosed in parentheses and the numbers are separated by a comma.

Solutions to equations in two variables are ordered pairs of numbers that make a true statement when used to evaluate the original equation. An equation in two variables will have many solutions. We can check each solution to an equation in two variables by substituting the numerical values for the respective variables and performing the operations indicated.

EXAMPLE 1 Check the solution $(-2, 16)$ for the equation $2x + y = 12$.

$$2x + y = 12$$
$$2(-2) + 16 = 12 \quad \text{Substitute } -2 \text{ for } x \text{ and } 16 \text{ for } y.$$
$$-4 + 16 = 12$$
$$12 = 12$$

The ordered pair makes the equation true. So $(-2, 16)$ is a solution.

EXAMPLE 2 Check the ordered pair $(3, -2)$ for the equation $y = 2x - 5$.

$$y = 2x - 5$$
$$-2 = 2(3) - 5 \quad \text{Substitute 3 for } x \text{ and } -2 \text{ for } y.$$
$$-2 = 6 - 5$$
$$-2 = 1$$

The ordered pair does not make the equation true. So $(3, -2)$ is not a solution.

2   Find a Specific Solution of an Equation with Two Variables When Given the Value of One Variable.

If we are given the value of one variable for an equation with two variables, we can substitute the given value in the equation and solve the equation for the other variable.

EXAMPLE 3 Solve the equation $2x - 4y = 14$, if $x = 3$.

$$2x - 4y = 14$$

$$2(3) - 4y = 14 \qquad \text{Substitute 3 for } x.$$

$$6 - 4y = 14$$

$$-4y = 14 - 6 \qquad \text{Subtract 6 from both sides or transpose.}$$

$$-4y = 8$$

$$\frac{-4y}{-4} = \frac{8}{-4} \qquad \text{Divide both sides by } -4.$$

$$y = -2$$

The solution is $(3, -2)$; $x = 3$ and $y = -2$.

The standard form of an equation in two variables, $ax + by = c$, is such that both a and b are not zero. However, either a or b may be zero.

When a or b is zero, the variable for which zero is the coefficient is eliminated from the equation. Let's take the equation $5x + 3y = 15$ and replace the coefficient 3 with zero:

$$5x + 3y = 15$$

$$5x + 0(y) = 15 \qquad \text{Replace 3 with 0.}$$

$$5x + 0 = 15 \qquad 0(y) = 0 \text{ and 0 times any number is 0.}$$

$$5x = 15$$

$$\frac{5x}{5} = \frac{15}{5} \qquad \text{Divide both sides by 5.}$$

$$x = 3$$

Linear equations may also be equations in one variable. Thus, an equation like $x = 3$ is a linear equation and we think of the coefficient of the missing variable term as zero:

$$x = 3 \qquad \text{is considered as} \qquad x + 0y = 3$$

This means that, for any value of y, x is 3.

EXAMPLE 4 For the equation $x = 6$, complete the given ordered pairs: $(\ \ , 2)$; $(\ \ , -3)$; $(\ \ , 1)$.

Equation   Ordered Pairs

$x = 6 \qquad (\ \ , 2) \qquad (\ \ , -3) \qquad (\ \ , 1)$

$\qquad\qquad \mathbf{(6, 2)} \qquad \mathbf{(6, -3)} \qquad \mathbf{(6, 1)} \qquad x \text{ is 6 for any value of } y.$

Numerous application problems can be solved with equations in two variables.

EXAMPLE 5   A small business photocopied a report that included 12 black-and-white pages and 25 color pages. The cost was $23.70. Letting b = the cost for a black-and-white page and c = the cost for a color page, set up an equation with two variables.

Cost of black-and-white copies: $12b$ (12 times cost per page)

Cost of color copies: $25c$ (25 times cost per page)

Total cost: black-and-white plus color = $23.70

Equation: $12b + 25c = 23.70$

An equation containing two variables shows the relationships between the two unknown amounts. If a value is known for either amount, the other amount can be found by solving the equation for the missing amount.

EXAMPLE 6   Solve the equation from Example 5 to find the cost for each black-and-white copy if color copies cost $0.90 each.

$$12b + 25c = 23.70$$

$$12b + 25(0.90) = 23.70 \qquad \text{Substitute 0.10 for } c.$$

$$12b + 22.50 = 23.70$$

$$12b = 23.70 - 22.50 \qquad \text{Transpose.}$$

$$12b = 1.20$$

$$\frac{12b}{12} = \frac{1.20}{12} \qquad \text{Divide both sides by 12.}$$

$$b = 0.10$$

Each black-and-white copy costs $0.10.

SELF-STUDY EXERCISES 18–2

1 Determine which of the following ordered pairs are solutions for the equation $2x - 5y = 9$.

1. (4, 5)   **2.** (17, 5)   **3.** (12, 3)
4. (6, 1)   **5.** (7, 1)   **6.** (4.5, 0)

Determine which of the following ordered pairs are solutions for the equation $3y = 2 - 4x$.

7. (2, −2)   **8.** (−2, 2)   **9.** (5, −6)

10. (−6, 5)   **11.** $\left(0, \dfrac{2}{3}\right)$ **12.** $\left(6, -7\dfrac{1}{3}\right)$

Determine which of the following ordered pairs are solutions for the equation $4y - x = 7$.

13. (1, 2)   **14.** (−2, −12)   **15.** (−2, −15)
16. (3, 5)   **17.** (5, 3)   **18.** (7, 0)

2 Write the specific solution for each equation in ordered pair form.

19. $x - y = 3$, if $x = 8$ **20.** $x + y = 7$, if $x = 2$
21. $x + 2y = -4$, if $y = 1$ **22.** $x - 3y = 12$, if $y = 3$
23. $5x - y = 10$, if $x = 5$ **24.** $4x + y = 8$, if $x = 2$
25. $3x + 4y = 16$, if $y = -2$ **26.** $5x + 2y = 9$, if $y = -3$

27. $3x - y = 16$, if $x = 6$ **28.** $x + 3y = 12$, if $x = 9$ **29.** $\dfrac{2}{3}x - y = 6$, if $y = 2$

30. $x - \dfrac{3}{4}y = 9$, if $x = 0$ **31.** $6 = x + 3y$, if $y = -4$ **32.** $8 = 2x + y$, if $x = -2$

Complete the ordered pairs for each equation.

33. $x = 1$ (, 3); (, -2); (, 0); (, 1)   **34.** $y = 4$ (-1,); (3,); (0,); (2,)

35. $y = -7$ (2,); (-3,); (0,); (5,)   **36.** $x = 9$ (, -2); (, 3); (, 0); (, 2)

Solve the following using equations with two variables.

37. Diane purchased three shirts at one price and five belts at another price. Let s = the price of each shirt and b = the price of each belt. Each shirt cost \$22. How much did each belt cost if the total purchase price was \$126?

38. Paulette and Terry Fink paid \$13 for four spark plugs and five quarts of motor oil. Let p = the cost of each plug and q = the cost of each quart of oil. If the oil cost \$1 per quart, how much did each spark plug cost?

39. Paul and Donna were charged \$160 for four pairs of pants and two shirts. If the shirts cost \$12 each, how much did all four pairs of pants cost? Let p = the cost of each pair of pants and s = the cost of each shirt.

40. David paid \$15,000 for an automobile. The purchase price included three extended warranty payments of \$400 each. How much did the car cost without the extended warranty? Let c = the cost of the car and w = the cost of each warranty payment.

18-3   GRAPHING LINEAR EQUATIONS WITH TWO VARIABLES

Learning Objectives

1   Graph linear equations using intercepts.

2   Graph linear equations using the slope–intercept method.

3   Graph linear equations using a graphics calculator.

We have already graphed linear equations using a table of values. We will now examine other procedures and specific properties of linear equations.

1   Graph Linear Equations Using Intercepts.

■ **DEFINITION 18–2: x-intercept.** The *x-intercept* is the point on the x-axis through which the line of the equation passes; that is, the y-value is zero $(x, 0)$.

■ **DEFINITION 18–3: y-intercept.** The *y-intercept* is the point on the y-axis through which the line of the equation passes; that is, the x-value is zero $(0, y)$.

The following rule may help.

Rule 18–3   *To find the intercepts of the axes of a linear equation:*

1. Find the x-intercept by letting $y = 0$ and solving for x.
2. Find the y-intercept by letting $x = 0$ and solving for y.

EXAMPLE 1   Find the intercepts of the equation $3x - y = 5$.

$$3x - y = 5$$ For the x-intercept, let $y = 0$.

$$3x - 0 = 5$$

$$3x = 5$$

$$x = \frac{5}{3}$$ A mixed number or decimal is easier to plot than an improper fraction.

$$x = 1\frac{2}{3}$$

Thus, the *x*-intercept is $(1\frac{2}{3}, 0)$.

$$3x - y = 5$$

$$3(0) - y = 5 \qquad \text{For the } y\text{-intercept, let } x = 0.$$

$$-y = 5$$

$$y = -5$$

Thus, the *y*-intercept is $(0, -5)$.

Rule 18-4 *To graph linear equations by the intercepts method:*

1. Find the *x*- and *y*-intercepts.
2. Plot the intercepts on a rectangular coordinate system.
3. Draw the line through the two points and extend it beyond each point.

EXAMPLE 2 Graph the equation $3x - y = 5$ by using the intercepts of each axis.

Plot the two intercepts found in Example 1, $(1\frac{2}{3}, 0)$ and $(0, -5)$. Draw the graph connecting these points and extending beyond. See Fig. 18–8. We can check by finding one other point in the usual way and plotting it. If it is on the line, the graph is correct. Check for $x = 2$:

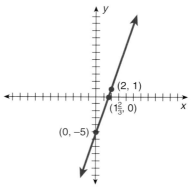

Figure 18–8

$$3x - y = 5$$

$$3(2) - y = 5 \qquad \text{Substitute 2 for } x.$$

$$6 - 5 = y \qquad \text{Solve for } y.$$

$$1 = y \qquad \text{The point (2, 1) is on the graph.}$$

Tips and Traps *Graphing Equations When Both Intercepts Are (0, 0):*
If both intercepts are (0, 0), they coincide on the origin, forming only *one point*. An additional point must be found by the table-of-values method so that you will have two distinct points for drawing the line. A third point is still useful to check your work.

$$y = 7x$$

If $x = 0$, then $y = 7(0)$ or $y = 0$. Both the *x*- and *y*-intercepts are (0, 0). So let $x = 1$, then $y = 7(1)$ or $y = 7$. Plot the points (1, 7) and (0, 0) and draw the graph.

EXAMPLE 3   Graph $y = 7x$ using the intercepts method.

Plot the points (1, 7) and (0, 0) as shown in the Tips and Traps feature. Draw the line through the two points and extend it beyond each point, as shown in Fig. 18–9.

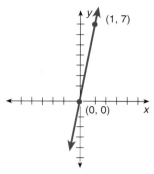

Figure 18–9

[2] Graph Linear Equations Using the Slope–Intercept Method.

One method for graphing linear equations allows us to identify characteristics of the graph of this equation by inspection. First, let's examine the characteristics identified by the constant. We will let the coefficient of x equal 1 and look at the graphs of the equations in Fig. 18–10.

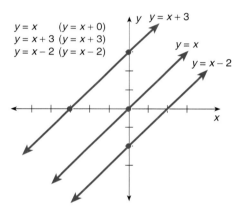

Figure 18–10

$y = x$			$y = x + 3$			$y = x - 2$	
x	y		x	y		x	y
0	0		0	3		0	-2
1	1		-3	0		2	0

The three graphs have the same slope (steepness or slant), but they have different y-intercepts.

$$y = x + 0 \quad \text{crosses } y \text{ at} \quad 0; \ y\text{-intercept} = (0, 0)$$

$$y = x + 3 \quad \text{crosses } y \text{ at} +3; \ y\text{-intercept} = (0, 3)$$

$$y = x - 2 \quad \text{crosses } y \text{ at} -2; \ y\text{-intercept} = (0, -2)$$

Notice that the point where the graph crosses the y-axis is the same as the constant in the equation. The constant (b) in an equation in the form $y = mx + b$ is the y-coordinate of the y-intercept.

Now, let's examine equations that have graphs with a different slant or steepness. The x- and y-intercepts in each of the three equations is $(0, 0)$. Thus, we will find one additional point to graph (see Fig. 18–11).

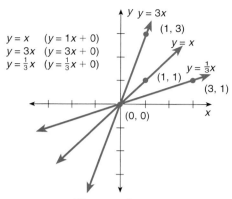

$$y = x \quad (y = 1x + 0)$$
$$y = 3x \quad (y = 3x + 0)$$
$$y = \tfrac{1}{3}x \quad (y = \tfrac{1}{3}x + 0)$$

Figure 18–11

$$y = x \qquad y = 3x \qquad y = \frac{1}{3}x$$

x	y
0	0
1	1

x	y
0	0
1	3

x	y
0	0
3	1

The three graphs have the same y-intercepts, but they have different slopes (steepnesses or slants). We will define this slope or steepness so that a numerical value will identify it. The *slope* of a line is the ratio of the vertical change from one point to another, to the horizontal change from the first point to the second.

A slope of 1 (written as $\tfrac{1}{1}$) means a change of 1 vertical unit for every 1 horizontal unit of change. A slope of 3 (written as the ratio $\tfrac{3}{1}$) means a change of 3 vertical units for every 1 horizontal unit. A slope of $\tfrac{1}{3}$ means a change of 1 vertical unit for every 3 horizontal units (see Fig. 18–12). This slope can be determined from any two points on the graph.

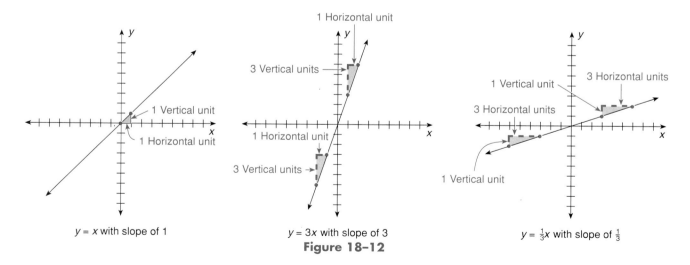

$y = x$ with slope of 1 $y = 3x$ with slope of 3 $y = \tfrac{1}{3}x$ with slope of $\tfrac{1}{3}$

Figure 18–12

Notice that the slope of the line is the same as the coefficient of the x-term in the equation. We must make one final observation. Look again at the equations in Figs. 18–10 and 18–11.

$$y = x \qquad\qquad y = x$$
$$y = x + 3 \qquad y = 3x$$
$$y = x - 2 \qquad y = \frac{1}{3}x$$

In each case, the equation is solved for y. This form of equation is called the *slope–intercept form* of a linear equation, or $y = mx + b$, where m is the slope and b is the y-coordinate of the y-intercept. When equations are written in this form, the slope and y-intercept can be identified by inspection.

EXAMPLE 4 Write the equations in slope–intercept form and identify the slope and y-intercept.

(a) $2x + y = 4$ (b) $5x - y = -2$ (c) $3x + 4y = -12$

(a) $2x + y = 4$ Solve for y.
  $\mathbf{y = -2x + 4}$
  slope $= -2$ or $-\dfrac{2}{1}$ or $\dfrac{2}{-1}$
  y-**intercept** $= 4$

(b) $5x - y = -2$ Solve for y.
  $\dfrac{-y}{-1} = \dfrac{-5x - 2}{-1}$ Divide by -1.
  $\mathbf{y = 5x + 2}$
  slope $= 5$ or $\dfrac{5}{1}$
  y-**intercept** $= 2$

(c) $3x + 4y = -12$ Solve for y.
  $\dfrac{4y}{4} = \dfrac{-3x - 12}{4}$ Divide by 4.
  $\mathbf{y = -\dfrac{3}{4}x - 3}$
  slope $= -\dfrac{3}{4}$
  y-**intercept** $= -3$

An equation in the slope–intercept form can be graphed by using just the slope and y-intercept.

Rule 18–5 *To graph a linear equation in the form $y = mx + b$ by using the slope and y-intercept:*

1. Locate the y-intercept on the y-axis.
2. Using the slope, determine the amount of vertical and horizontal movement indicated.
3. From the y-intercept, locate additional points on the graph of the equation by counting the indicated vertical and horizontal movement.
4. Draw the line graph connecting the points and extending beyond the points.

EXAMPLE 5   Graph the following equations using the slope and *y*-intercept.

(a) $2x + y = 4$ (b) $5x - y = -2$ (c) $3x + 4y = -12$

(a) $2x + y = 4$

$y = -2x + 4$ Solve for *y*.

y-intercept $= (0, 4)$ Locate this point on the *y*-axis.

Slope $= \dfrac{-2}{1}$ or $\dfrac{2}{-1}$

$\frac{-2}{1}$ indicates vertical movement of -2 and horizontal movement of $+1$ from the *y*-intercept. $\frac{2}{-1}$ indicates vertical movement of $+2$ and horizontal movement of -1. See Fig. 18–13.

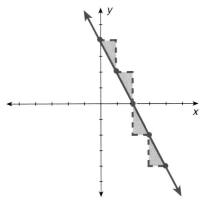

Figure 18–13

(b) $5x - y = -2$ Solve for *y*.

$y = 5x + 2$

y-intercept $= (0, 2)$.   Locate this point on the *y*-axis.

Slope $= \dfrac{5}{1}$ or $\dfrac{-5}{-1}$.

$\frac{5}{1}$ indicates vertical movement of $+5$ and horizontal movement of $+1$ from the *y*-intercept. $\frac{-5}{-1}$ indicates vertical movement of -5 and horizontal movement of -1. See Fig. 18–14.

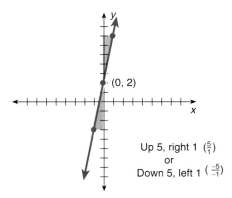

Figure 18–14

(c) $3x + 4y = -12$ Solve for *y*.

$y = -\dfrac{3}{4}x - \dfrac{12}{4}$

y-intercept $= (0, -3)$.   Locate this point on the *y*-axis.

Slope $= \dfrac{-3}{4}$ or $\dfrac{3}{-4}$.

$\frac{-3}{4}$ indicates vertical movement of -3 and horizontal movement of $+4$ from the y-intercept. $\frac{3}{-4}$ indicates vertical movement of $+3$ and horizontal movement of -4. See Fig. 18–15.

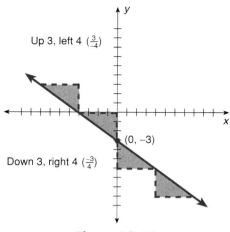

Up 3, left 4 $(\frac{3}{-4})$

(0, –3)

Down 3, right 4 $(\frac{-3}{4})$

Figure 18–15

3 Graph Linear Equations Using a Graphics Calculator.

A graphics calculator can be used to show the graphs of linear equations that can be written in the form $y = mx + b$. Also, computer algebra software can be used to show the graphs of linear equations. These calculators and computer software are tools that free us from the tedious task of graphing equations and allow us to focus on the patterns and properties of graphs.

Every model of graphics calculator and every brand of computer algebra software will have a different set of key strokes for graphing equations. You will need to refer to the owner's manual to adapt to a particular model or brand. However, we will illustrate the usefulness of these tools by showing a few features common to most calculators.

EXAMPLE 6   Graph $y = 5x + 2$ using a graphics calculator.

Casio *fx*-7700:

Range INIT Range Range   To set range of x and y to initial settings.

Graph 5 X, θ, T + 2 EXE $y =$ appears. X, θ, T enters the variable x.

G↔T   Toggles between the graph screen and the text screen.

To see a closer view we will define a box and view the box. Select the Zoom feature. Select the Box feature. The cursor (blinking dot) is at the origin. Use the left arrow key to move the cursor to a corner of the desired viewing box ($x = -1$, $y = -0.5$). Notice that the coordinates of x and y change as you move the cursor. When the cursor is in the desired place, press EXE to "nail down" the corner.

Next, use an arrow key to move the cursor to an adjacent corner of the desired viewing box ($x = -1$, $y = 3$). A line connects the first corner and the second. Press EXE, then use an arrow key to move the cursor to the next adjacent corner ($x = 1$, $y = 3$). A box is forming when you move the arrow keys. Once the box is defined, press EXE to view the graph in the box. Be sure to include critical values like the x- and y-intercepts in the box. To see the approximate values of the x- and y-intercepts, use the Trace function on the calculator. Press Trace. The coordinates of the leftmost point on the graph appear at the bottom of the screen. As you move the right arrow key, the cursor moves along the graph, and the coordinates change accordingly. Move the cursor to the value of y that is closest to 0 to determine the coordinates of the x-intercept.

$$y = -0.021276 \qquad \text{Closer to 0.}$$

$$y = 0.0851063$$

The approximate value of x when $y = 0$ is -0.4.

Next, find the value of x that is closest to 0 to determine the coordinates of the y-intercept.

$$x = -4.E\text{-}14 \quad (-0.00000000000004) \qquad \text{Closer to 0.}$$

$$x = 0.0212765$$

The approximate value of y is 2 when $x = 0$. If the desired accuracy is not achieved, you can define a viewing box just around the x-intercept or y-intercept for closer views. There are other ways to "zoom" critical values. Refer to your calculator owner's manual for more details.

To return to the initial range settings, press $\boxed{\text{Range}}$ $\boxed{\text{INIT}}$ $\boxed{\text{Range}}$ $\boxed{\text{Range}}$.

To clear the graph from the calculator, press $\boxed{\text{Cls}}$ $\boxed{\text{EXE}}$ $\boxed{\text{AC}}$.

TI-81:

The keyboard layout of the TI-81 is different from that of the Casio fx-7700, but the procedures for graphing are similar. To enter the equation to be graphed, use the $\boxed{y=}$ key. The variable key is labeled $\boxed{\text{X}|\text{T}}$.

$$\boxed{y=} \;\; 5 \;\; \boxed{\text{X}|\text{T}} \;\; \boxed{+} \;\; 2 \;\; \boxed{\text{Graph}}$$

For a closer view, use the $\boxed{\text{Zoom}}$ key. Select option 1, box, by pressing 1. The cursor begins at the origin rather than on the graph. Use the right or left arrow to move the cursor to the right or left and the up or down arrow key to move the cursor up or down. Move the cursor to any corner of the box and press $\boxed{\text{ENTER}}$. Move the cursor to the diagonal corner of your desired box and press $\boxed{\text{ENTER}}$. The portion of the graph inside the box is displayed. To find the approximate values of the x- and y-intercepts, use the $\boxed{\text{Trace}}$ key and the left or right arrow keys. For more details, refer to your calculator owner's manual.

To clear the graph, return to the text screen using the $\boxed{y=}$ key. Use the delete key, $\boxed{\text{Del}}$, to delete the equation. The range can be changed using the $\boxed{\text{Range}}$ key or the $\boxed{\text{Zoom}}$ key.

To return to the default range, press $\boxed{\text{Zoom}}$ and select the standard, option 6. To display a graph with equal intervals on x and y, select option 5, square, from the zoom menu.

SELF-STUDY EXERCISES 18–3

1 Graph the following equations using the intercepts procedure.

1. $x + y = 5$
2. $x + 3y = 5$
3. $\frac{1}{2}y = 4 + x$

4. $y = 3x - 1$
5. $5x = y + 2$
6. $x = -2y + 3$

2 Graph the following equations using the slope–intercept procedure.

7. $y = 2x - 3$
8. $y = -\frac{1}{2}x - 2$
9. $y = -\frac{3}{5}x$

10. $x - 2y = 3$
11. $2x + y = 1$

3 Verify the graphs for Exercises 1 through 11 using a graphics calculator.

GRAPHING LINEAR INEQUALITIES WITH TWO VARIABLES

Learning Objectives

1. Graph linear inequalities with two variables using test points.
2. Graph linear inequalities with two variables using a graphics calculator.

We saw when we solved inequalities in one variable that the solution is a set of numbers that we represent graphically. The solution of linear inequalities with two variables is also a set of points.

1 Graph Linear Inequalities with Two Variables Using Test Points.

To graph a linear inequality, first find the boundary of the inequality by graphing a similar mathematical statement that substitutes an equals sign for the inequality symbol. The graphical solution for a linear inequality, $<$ or $>$, includes all points on one side of the line representing the graph of the similar equation. If either the "less than or equal" ($\leq$) or "greater than or equal" ($\geq$) symbol is used, the points on the boundary line are also included.

Rule 18-6 *To graph a linear inequality with two variables:*

1. Find the boundary by graphing an equation that substitutes an equals sign for the inequality symbol. Make a solid line if the boundary is included or a dashed line if the boundary is not included in the solution set.
2. Test any point that is not on the boundary line.
3. If the test point makes a true statement with the original inequality, shade the side containing the test point.
4. If the test point makes a false statement with the original inequality, shade the side opposite the side containing the test point.

EXAMPLE 1 Graph the inequality $2x + y < 3$.

First graph the equation $2x + y = 3$ (Fig. 18–16).

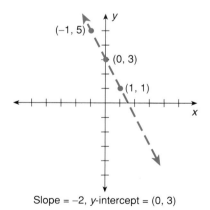

Slope = −2, *y*-intercept = (0, 3)

Figure 18–16

$y = -2x + 3$ Solve the equation for *y*.

y-intercept $= 3$

Slope $= -2$ or $\dfrac{-2}{1}$ or $\dfrac{2}{-1}$

To show that the boundary is not included in the solution set, a dashed or broken line is used. The graph of the equation $y = -2x + 3$ divides the surface into two parts and it is the *boundary* separating the parts. In graphing inequalities, one of the parts is included in the solution. To represent this solution, we will *shade* the appropriate part.

Now, select one point, on either side of the boundary, to determine which side of the line contains the set of points that solves the inequality. Suppose that we choose $(2, 2)$ as the point above the boundary.

$$\text{Is } 2x + y < 3 \quad \text{when} \quad x = 2 \text{ and } y = 2?$$

$$2(2) + 2 < 3$$

$$4 + 2 < 3$$

$$6 < 3 \qquad \text{False statement}$$

Thus, the side of the line including the point $(2, 2)$ is not in the solution set. Shade the opposite side. See Fig. 18–17.

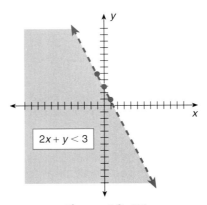

Figure 18–17

Tips and Traps   *Solid or Dashed Boundary Lines:*
In inequalities with one variable, we used open circles and solid circles to show that boundaries are excluded or included. In inequalities with two variables, we will represent boundaries that are included in the solution with a solid line and boundaries that are excluded from the solution with a dashed line. Figure 18–18 shows the graph of $2x + y \leq 3$.

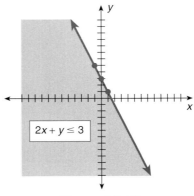

Figure 18–18

Selecting Test Points:
Choose numbers that are easy to work with when selecting a test point for finding the area to shade. If the boundary line does not pass through the origin, a good point to use is (0, 0).

EXAMPLE 2   Graph the inequality $y \geq 3x + 1$.

Graph the line $y = 3x + 1$. See Fig. 18–19.

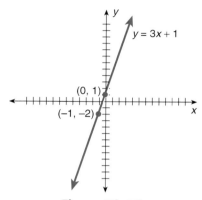

Figure 18–19

y-intercept $= (0, 1)$

Slope $= 3$ or $\dfrac{3}{1}$ or $\dfrac{-3}{-1}$

The boundary line is shown in Fig. 18–19.

Because the inequality $y \geq 3x + 1$ *does* include the line $y = 3x + 1$, we will represent the line with an unbroken or solid line (see Fig. 18–20).

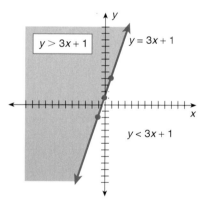

Figure 18–20

The point (0, 0) is not on the boundary line and it is on the right side of the boundary. We will use (0, 0) as our test point.

$$\text{Is } y \geq 3x + 1 \quad \text{when} \quad x = 0 \text{ and } y = 0?$$

$$0 \geq 3(0) + 1$$

$$0 \geq 0 + 1$$

$$0 \geq 1 \qquad \text{False statement}$$

Thus, the boundary and the side opposite to the side of the boundary including the point (0, 0) represents

$$y \geq 3x + 1$$

To show the solution set, we shade the portion of the graph that represents $y \geq 3x + 1$, the given inequality (see Fig. 18–20). The shaded portion along with the boundary line is the graph of $y \geq 3x + 1$.

<table>
<tr><td>**Tips and Traps**</td><td>

Two Observations About Inequalities:

Do two points have to be tested to determine which region to shade? Only one test point is necessary; however, a second test point can be used to check your work.

If the first test point makes a true statement, the side of the boundary that includes the test point is shaded. If the first test point makes a false statement, the side of the boundary that does not include the test point is shaded.

Other properties of inequalities can be applied when the inequality is in the form $y < mx + b$, $y \leq mx + b$, $y > mx + b$, or $y \geq mx + b$.

In general, when an inequality is in the form $y < mx + b$ or $y \leq mx + b$, the solution set includes all points *below* the boundary line. When an inequality is in the form $y > mx + b$ or $y \geq mx + b$, the solution set includes all points *above* the boundary line.
</td></tr>
</table>

2 Graph Linear Inequalities with Two Variables Using a Graphics Calculator.

A graphics calculator can be used to show the graph of the boundary of an inequality solution set. Some graphics calculators will also show the shaded region. Again, the owner's manual should be consulted to determine the capabilities of a given model or brand.

Using the Casio *fx*-7700, inequalities can be selected by accessing the menu system.

$\boxed{\text{Mode}}$ $\boxed{\text{Shift}}$ $\boxed{\div}$ accesses inequalities mode.

The appropriate inequality is selected by the function keys.

$$
\begin{array}{cccc}
y > & y < & y \geq & y \leq \\
\boxed{\text{F1}} & \boxed{\text{F2}} & \boxed{\text{F3}} & \boxed{\text{F4}}
\end{array}
$$

The right side of the inequality is entered as before.

EXAMPLE 3   Use a graphics calculator to graph $2x + y < 3$.

$y < -2x + 3$ Solve for *y* and write in the form $y < mx + b$.

$\boxed{\text{Cls}}$ $\boxed{\text{EXE}}$ Clear previous graphs from the calculator.

$\boxed{\text{Mode}}$ $\boxed{\text{Shift}}$ $\boxed{\div}$ Put the calculator in the inequality mode.

$\boxed{\text{Graph}}$ $\boxed{y<}$ $\boxed{-}$ 2 $\boxed{\text{X, θ, T}}$ $\boxed{+}$ 3 $\boxed{\text{EXE}}$ Enter the inequality. (See Fig. 18–21.)

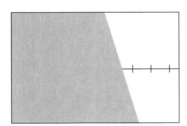

Figure 18–21

The calculator graph does not distinguish between a dashed and a solid boundary line.

To put the calculator back in the equation mode, press

$\boxed{\text{Mode}}$ $\boxed{\text{Shift}}$ $\boxed{+}$.

SELF-STUDY EXERCISES 18–4

1 Graph the following linear inequalities using test points.

1. $x - y < 8$
2. $x + y > 4$
3. $3x + y < 2$
4. $2x + y \leq 1$
5. $x + 2y < 3$
6. $2x + 2y \geq 3$
7. $y \geq 2x - 3$
8. $y \geq -3x + 1$
9. $y > \frac{1}{2}x - 3$

10. $y > -\frac{3}{2}x + \frac{1}{2}$

2 Verify the graphs in Problems 1 through 10 using a graphics calculator.

18–5 GRAPHING QUADRATIC EQUATIONS AND INEQUALITIES

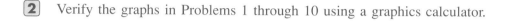

Learning Objectives

1 Identify nonlinear equations.

2 Graph quadratic equations using the table-of-solutions method or by examining properties.

3 Graph quadratic equations and inequalities using a graphics calculator.

1 **Identify Nonlinear Equations.**

In the previous sections, we discussed graphing of linear equations and inequalities. Linear equations have two basic characteristics.

1. No letter term will have a letter with an exponent greater than 1.
2. No letter term will have two or more different letter factors.

Quadratic equations also have two basic characteristics.

1. No letter term will have a degree greater than 2. Quadratic equations may have terms with two different letter factors as long as the degree of the term does not exceed 2.
2. At least one term of the equation must have degree 2.

EXAMPLE 1 Identify the following equations as linear, quadratic, or other nonlinear equations.

(a) $3x + 4y = 12$
(b) $3x + 4 = 8$
(c) $3x + 4xy + 2y = 0$
(d) $y = x^2 + 4$
(e) $y = 3x^3 - 2x^2 + x$

(a) $3x + 4y = 12$
This is a linear equation. The letter terms have neither an exponent greater than 1 nor two different letter factors in the same term.
(b) $3x + 4 = 8$
This is a linear equation. The only letter term does not have an exponent greater than 1 nor two different letter factors.
(c) $3x + 4xy + 2y = 0$
This is *not* a linear equation. The term $4xy$ has two different letter factors and a degree of 2. **This is a quadratic equation.**

(d) $y = x^2 + 4$

This is *not* a linear equation. The term x^2 has a degree of 2. **The equation is a quadratic equation.**

(e) $y = 3x^3 - 2x^2 + x$

This equation is neither linear nor quadratic. It has a term that has degree 3. **The equation is called a *cubic equation*.**

2 Graph Quadratic Equations Using the Table-of-Solutions Method or by Examining Properties.

Linear equations graph into *straight* lines. We will examine the graphs of some equations that have a degree higher than 1.

The graph of an equation of a degree higher than 1 will be a *curved* line. The curved line can be a parabola, hyperbola, circle, ellipse, or an irregular curved line. We will not define these terms at this time, but Fig. 18–22 illustrates them. The equation for a parabola that opens up or down is distinguished from other quadratic equations. The y variable has degree 1. The x variable must have one term with degree 2. Such a quadratic equation can be written in function notation: $f(x) = ax^2 + bx + c$.

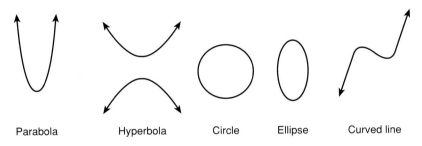

| Parabola | Hyperbola | Circle | Ellipse | Curved line |

Figure 18–22

One method of graphing nonlinear equations is to form a table of solutions and plot the points like we did in Section 18–1. This method will require *more* points than we customarily use in graphing linear equations or inequalities.

A parabola is symmetrical; that is, it can be folded in half and the two halves match. The fold line is called the *axis of symmetry*. For a parabola in the form $y = ax^2 + bx + c$, the equation of the axis of symmetry is $x = -\frac{b}{2a}$.

The point of the graph that crosses the axis of symmetry is the *vertex* of the parabola. Thus, the x-coordinate of the vertex of the parabola is $-\frac{b}{2a}$.

Rule 18–7 *To graph quadratic equations:*

1. Find the axis of symmetry: $x = -\frac{b}{2a}$.
2. Find the vertex: x-coordinate of vertex $= -\frac{b}{2a}$. To find the y-coordinate of the vertex, substitute the x-coordinate into an equation that is similar to the original inequality.
3. Find one or two additional points that are to the right of the axis of symmetry.
4. Apply the property of symmetry to find additional points to the left of the line of symmetry.
5. Connect the plotted points with a smooth, continuous curved line.

EXAMPLE 2 Graph the equation $y = x^2$ by finding the vertex, the axis of symmetry, and some additional points.

$$y = x^2$$

$$y = x^2 + 0x + 0 \qquad \text{Standard form: } a = 1, b = 0, c = 0.$$

$$-\frac{b}{2a} = -\frac{0}{2(1)}$$

$$-\frac{b}{2a} = 0 \qquad \text{x-coordinate of vertex}$$

$$y = 0^2$$

$$y = 0 \qquad \text{y-coordinate of vertex}$$

Vertex: (0, 0)
Axis of symmetry: $x = 0$

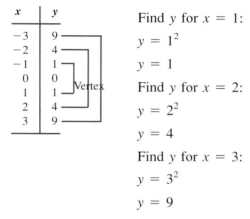

Find y for $x = 1$:

$$y = 1^2$$

$$y = 1$$

Find y for $x = 2$:

$$y = 2^2$$

$$y = 4$$

Find y for $x = 3$:

$$y = 3^2$$

$$y = 9$$

Apply the principle of symmetry to complete the table. Plot the points and connect them with a smooth, continuous curve (Fig. 18–23).

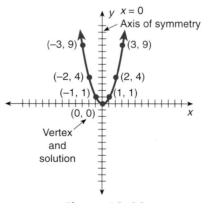

Figure 18–23

If the equation $y = x^2$ is written in function notation as $f(x) = x^2$, the solutions of the equation are the value or values of x that make the function zero. If $0 = x^2$ or $x^2 = 0$, then $x = 0$ is the only solution of the equation. Again, the solutions of the equation are found where the graph crosses or meets the x-axis. When we graph a quadratic equation, we can test the solutions as points on the graph. If the graph does *not* cross or meet the x-axis, the equation does *not* have any real-number solutions.

EXAMPLE 3 Graph the equation $y = x^2 + x - 6$ by using the axis of symmetry, the vertex, and the solutions.

For the parabola $y = x^2 + x - 6$, the equation of the axis of symmetry is

Axis of symmetry: $\quad x = -\dfrac{b}{2a}$

$$x = -\dfrac{1}{2(1)} \qquad (a = 1, b = 1)$$

$$x = -\dfrac{1}{2}$$

Vertex: $\quad \left(-\dfrac{1}{2}, y \right)$

$$y = x^2 + x - 6 \qquad \text{for} \qquad x = -\dfrac{1}{2}$$

$$y = \left(-\dfrac{1}{2} \right)^2 + \left(-\dfrac{1}{2} \right) - 6$$

$$y = \dfrac{1}{4} - \dfrac{1}{2} - 6$$

$$y = \dfrac{1}{4} - \dfrac{2}{4} - \dfrac{24}{4}$$

$$y = -\dfrac{25}{4} \text{ or } -6\dfrac{1}{4} \quad \left(-\dfrac{1}{2}, -6\dfrac{1}{4} \right)$$

Solutions: $\quad 0 = x^2 + x - 6$

$$0 = (x + 3)(x - 2)$$

$$x + 3 = 0 \qquad x - 2 = 0$$

$$x = -3 \qquad\quad x = 2$$

$$\mathbf{(-3, 0); (2, 0)}$$

By using the axis of symmetry, vertex, and two solutions, we can get a general idea of the shape of the graph (see Fig. 18–24).

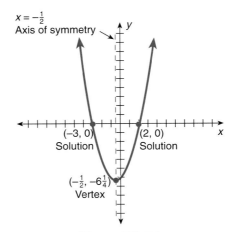

Figure 18-24

EXAMPLE 4 Graph the equation $y = -x^2 + 3$ by using the axis of symmetry, vertex, and the solutions.

Axis of symmetry: $x = -\dfrac{b}{2a}$ $y = -x^2 + 0x + 3; \ a = -1, b = 0$

$$x = -\dfrac{0}{2(-1)}$$

$$\mathbf{x = 0}$$

Vertex: $(0, y)$

$$y = -0^2 + 0 + 3$$

$$y = 3$$

$$\mathbf{(0, 3)}$$

Solutions: $0 = -x^2 + 3$

$$x^2 = 3$$

$$x = \pm\sqrt{3}$$

$$x \approx \pm 1.7$$

$$\mathbf{(1.7, 0); \ (-1.7, 0)}$$

Plot the points and connect them with a smooth, continuous curve. Two additional points are plotted to give a more complete view of the parabola. See Fig. 18–25.

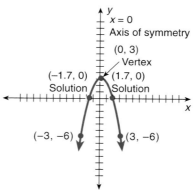

Figure 18–25

Tips and Traps   *Tips for Sketching Curves:*
When the coefficient of the squared letter term is negative as in Example 4, the graph of the parabola opens downward. When the coefficient of the squared letter term is positive, the graph of the parabola opens upward as in Example 3. Think of the parabola as a cup or glass: positive holds water; negative spills water.

3 Graph Quadratic Equations and Inequalities Using a Graphics Calculator.

Quadratic equations and inequalities are graphed on a graphics calculator using the same procedures as for linear equations and inequalities. The equation or inequality must first be solved for y.

EXAMPLE 5 Graph the following using a graphics calculator.

(a) $y = 2x^2 - 5x - 3$ (b) $y < 2x^2 - 5x - 3$ (c) $y \geq 2x^2 - 5x - 3$

(a) $y = 2x^2 - 5x - 3$

Casio *fx*-7700:

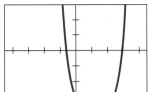

Figure 18–26

Cls EXE   Clear previous graphs.

Range INIT Range Range   Initiate range to preset range.

Graph 2 X, θ, T x^2 − 5 X, θ, T − 3 EXE   See Fig. 18–26.

To view the portion of the parabola containing the vertex, change the range to show the *x*-values from -2 to $+4$ and the *y*-values from -10 to 5. Leave the scale at 1.

Range − 2 EXE 4 EXE EXE − 10 EXE 5 EXE EXE Range

Graph again by pressing the left arrow key to remove the "done" indication. Then press EXE. See Fig. 18–27.

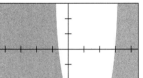

X = 1.2553191
Y = −6.124943

Figure 18–27

Different views can be obtained by changing the range. More accurate values for *x* and *y* at critical points such as *x*- and *y*-intercepts and the vertex can be displayed by using the trace function with closer views of particular points.

(b) $y < 2x^2 - 5x - 3$

Cls EXE Range INIT Range Range   To clear the previous graph and initialize range settings

Mode Shift ÷   To select the inequality mode

Graph *y<* 2 X, θ, T x^2 − 5 X, θ, T − 3 EXE   See Fig. 18–28.

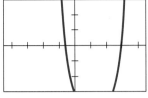

Figure 18–28

To see a different view, change the range settings to the settings used in part (a). To draw the graph again, go to the text screen G↔T, press the left arrow key to remove the "done" indication, and press EXE. See Fig. 18–29.

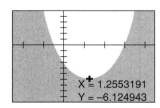

X = 1.2553191
Y = −6.124943

Figure 18–29

Tips and Traps   *Features of the Graphics Calculator:*
Other features of the calculator, such as zoom and box, allow you to view different portions of the graph. A smaller range allows a closer examination of critical points. A larger range allows a more complete overview of the graph.

On the Casio *fx*-7700, the initial settings for the range that are programmed into the calculator give a view with *x* and *y* intervals equally spaced, thus yielding a minimum of distortion. Multiplying (or dividing) the *x* and *y* values by the same amount will allow different views with a minimum of distortion.

(c) $y \geq 2x^2 - 5x - 3$

| CLS | EXE | Range | INIT | Range | Range | To clear previous graph and initialize
                                                range settings

| Mode | Shift | ÷ | To select the inequality mode

| Graph | F3 | 2 | X, θ, T | x^2 | − | 5 | X, θ, T | − | 3 | EXE | See Fig. 18–30.

Change the range settings to -2 to 4 for x and -10 to 5 for y.

To draw the graph again, go to text screen | G↔T |, press the left arrow key to remove the "done" indication, and press | EXE |. See Fig. 18–31.

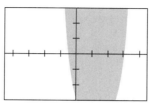

Figure 18–30

Figure 18–31

SELF-STUDY EXERCISES 18-5

[1] Identify the following equations as linear, quadratic, or other.

1. $y = 3x - 7$

2. $y = x^2 - 4x + 2$

3. $x^2 + y = 5x$

4. $x^2 + y^2 - 3 = 2x$

5. $2x + 3 = 8$

6. $\dfrac{x}{y} = 3$

[2] Use a table of solutions to graph the following quadratic equations.

7. $y = x^2$

8. $y = 3x^2$

9. $y = \dfrac{1}{3}x^2$

10. $y = -4x^2$

11. $y = -\dfrac{1}{4}x^2$

12. $y = x^2 - 4$

13. $y = x^2 + 4$

14. $y = x^2 - 6x + 9$

15. $y = -x^2 + 6x - 9$

[3] Graph the following using a graphics calculator. Reset the range if necessary to show the vertex of the parabola.

16. $y = 3x^2 + 5x - 2$

17. $y = (2x - 3)(x - 1)$

18. $y > 2x^2 - 9x - 5$

19. $y \geq -2x^2 + 9x + 5$

CAREER APPLICATION

Statistics: Predicting College Class Size

A common method of organizing undergraduate mathematics classes at large universities is to have a professor lecture to a large class of 200 or more students on Monday, Wednesday, and Friday, and then have graduate students in mathematics, engineering, computer science, etc., lead recitation sessions for 30 to 40 students on Tuesday and Thursday. In recitation sessions, students can ask questions about the lecture, homework assignments, or exam errors.

A statistics professor at a major university noticed a relationship between the number of hours of homework assigned and the number of students continuing in the class. She found that as the number of homework hours increased in her algebra class, the number of students decreased. In a class beginning with 225 students, she started with no required homework and then systematically increased the homework requirement three times during the semester, recording the class enrollment one week after each increase. She used this data to derive the following function:

$$f(x) = -6x + 225$$

where x represents the approximate number of hours of homework required weekly, and $f(x)$ represents the approximate class enrollment.

Exercises

Use the above function to answer the following in complete sentences. Show all computations. Round to the nearest number of people or hours.

1. Approximate the algebra class enrollment when the following number of hours of weekly homework was required: (a) none (b) 5 (c) 10 (d) 15
2. Approximate the weekly homework requirement if the enrollment was the following: (a) 201 students (b) 177 students (c) 190 students (d) 100 students
3. Graph the function and explain why only the first quadrant is of any interest. Increment by 1's along the x axis, and by 25's along the $f(x)$ axis. Recall that $f(x) = y$ for graphing purposes.
4. Explain the real-life meaning of the graph's slope: $m = -\frac{6}{1}$.
5. Use the graph to estimate the homework required for all students to drop the class. Now find this answer exactly by using the function definition. Explain any difference in these two answers.
6. From prior experience the professor knows that students learn algebra much more efficiently when the class size is no more than 150 students. How much homework should be assigned to achieve this number of students?
7. Do you think this function is valid for all non-negative values of x? If you do not, state the highest value of x you think is appropriate, and give your reason(s).
8. Do you think a homework requirement is the only factor influencing class size? If not, give other appropriate factors.
9. Which factor do you think most influences the class retention rate, and why?
10. What would you say to this professor regarding her use of this function to predict class size?

Answers

1. (a) 225 (b) 195 (c) 165 (d) 135
2. (a) 4 (b) 8 (c) 6 (d) 21

3.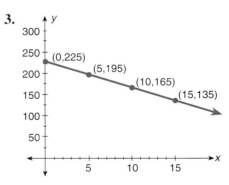

4. The slope of the graph in this application means that, for every hour of homework required, 6 students withdraw from the class.
5. Graph estimate: ~35 hr
   Exact answer: 37.5, or ~38, hr
6. No more than 12.5 or ~12 hr should be assigned.
7. Clearly x cannot be greater than 37.5 hr because a negative number of students in the class would result! Also, because the rule for a weekly home-

work requirement is twice the number of class hours per week (and this class meets 5 hr per week), the professor should not require much more than 10 hr per week. If she does, she will receive poor student evaluations and low initial class enrollments (because word spreads about these professors), and she will probably be reprimanded by her department chair and/or the dean of the college.

8. Other factors that can influence class size include the instructor's teaching ability, length and difficulty of exams, changes in students' work schedules, students' prerequisite mathematical skills, students' efforts in the class, and amount of time available in students' lives to devote to the class.

9. Answers vary.

10. Answers vary.

ASSIGNMENT EXERCISES

Section 18–1

Represent the solutions of the following functions in a table of values and on a graph.

1. $f(x) = 2x - 3$
2. $f(x) = -4x + 1$
3. $f(x) = 3x$
4. $f(x) = 4x$
5. $f(x) = -3x$
6. $f(x) = -4x$
7. $f(x) = 2x + 1$
8. $f(x) = 2x + 5$
9. $f(x) = x^2 + 2x - 8$
10. $f(x) = 4^x$
11. $f(x) = \ln x$
12. $f(x) = x^3$

Write each of the following equations as a function and find the solution graphically.

13. $x + 2 = 8$
14. $3x - 7 = 1$
15. $2x + 1 = 5x + 7$
16. $-14 = 2(x - 7)$
17. $5(x + 2) = 3(x + 4)$
18. $2x + 1 = 7x + 2 - 8$

Section 18–2

Which of the following coordinate pairs are solutions for the equation $2x - 3y = 12$?

19. $(-2, -3)$
20. $(1, -3)$
21. $(3, -2)$
22. $(0, 4)$
23. $(6, 0)$
24. In the equation $2x + y = 8$, find x when $y = 10$.
25. Find y in the equation $x - 3y = 5$ when $x = 8$.
26. Find y if $x = 7$ in the equation $3x - 4 = -2$.

Section 18–3

Graph the following equations using the intercepts procedure.

27. $x = -4y - 1$
28. $x + y = -4$
29. $3x - y = 1$
30. $x = -4y$

Graph the following using the slope–intercept procedure.

31. $y = 5x - 2$
32. $y = -x$
33. $y = -3x - 1$
34. $y = \frac{1}{2}x + 3$
35. $x - y = 4$
36. $2y + 4 = -3$
37. $x - 2y = -1$
38. Verify Exercises 27 through 37 with a graphics calculator.

Section 18–4

Graph the following linear inequalities using test points, and verify with a graphics calculator.

39. $4x + y < 2$
40. $x + y > 6$
41. $3x + y \leq 2$
42. $x + 3y > 4$
43. $x - 2y < 8$
44. $3x + 2y \geq 4$
45. $y \geq 3x - 2$
46. $y \leq -2x + 1$
47. $y > \frac{2}{3}x - 2$

Identify the equations as linear, quadratic, or other nonlinear equations.

48. $4x - 7y = 8$ **49.** $y = 5x + 3x^2$ **50.** $y = 2^x$

51. $y = \ln 5x$ **52.** $x^2 + y = x + 2$ **53.** $xy = 5x - 5y$

Graph the quadratic equations by examining their properties, and verify with a graphics calculator.

54. $y = x^2 - 1$ **55.** $y = -x^2 - 1$ **56.** $y = x^2 + 3x - 10$

57. $y = x^2 - 6x + 8$ **58.** $y = x^2 - 2x + 1$ **59.** $y = -x^2 + 2x - 1$

60. $y = x^2 - 4x + 4$ **61.** $y = -x^2 + 4x - 4$

Use a calculator to graph the following. Set the range to display the vertex.

62. $y < x^2 - 2x + 1$ **63.** $y \geq -2x^2$ **64.** $y \leq \frac{1}{2}x^2 - 3$

CHALLENGE PROBLEMS

65. A 1-gallon can of indoor house paint is advertised to cover 400 square feet of wall surface.
 (a) Make a table to show the amount of wall surface area that can be covered by 1, 2, 3, . . . , or 10 gallons of paint.
 (b) Draw a graph of these data.
 (c) Use the graph to decide how many 1-gallon cans of paint it would take to cover 5500 square feet of wall surface.
 (d) The paint being used for this job can be purchased for $19.95 a gallon. Find the cost of the paint for the 5500 square feet of wall surface.
 (e) The sales tax rate is 8.25%. Calculate the total cost of the paint.

66. Using column 1 of the table below for the horizontal scale and one of the other columns for the vertical scale, select the column illustrated in Fig. 18–32, graph A, graph B, and graph C.

Payment Number	Payment Amount	Applied to Interest	Applied to Principal	Balance Owed	Total Interest
1	611.09	580.00	31.09	69,568.91	580.00
2	611.09	579.74	31.35	69,537.56	1,159.74
3	611.09	579.48	31.61	69,505.95	1,739.22
4	611.09	579.22	31.87	69,474.08	2,318.44
5	611.09	578.95	32.14	69,441.94	2,897.39
6	611.09	578.68	32.41	69,409.53	3,476.07
7	611.09	578.41	32.68	69,376.85	4,054.48
8	611.09	578.14	32.95	69,343.90	4,632.62
9	611.09	577.87	33.22	69,310.68	5,210.49
10	611.09	577.59	33.50	69,277.18	5,788.08

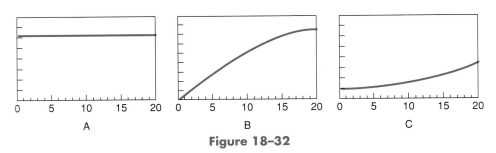

Figure 18–32

67. Sketch graphs to describe the pattern shown by the numbers in the two columns not illustrated by graphs A, B, and C.

1. What do we mean when we say a point is represented by an ordered pair of numbers?

2. What does the graph of an equation represent?

3. Explain the procedure for plotting the point $(5, -2)$ on a grid representing the rectangular coordinate system.

4. How can you tell from an equation if the graph will be a straight line?

5. Discuss the similarities and differences in the table-of-solutions method, the intercepts method, and the slope–intercept method for graphing a linear equation.

6. Why is it generally helpful to solve an equation for y before using the table-of-solutions method for graphing?

7. How is the graph of a linear inequality different from the graph of a linear equation?

8. When graphing linear inequalities, how do you determine which portion of the graph to shade?

9. How is the graph of a quadratic equation different from the graph of a linear equation?

10. What do *axis of symmetry* and *vertex* refer to on the graph of a quadratic equation that represents a parabola?

CHAPTER SUMMARY

Objectives

What to Remember with Examples

Section 18–1

1 Represent the solutions of a function in a table of values and on a graph.

To graph a function using a table of values, 1) prepare a table of values, 2) plot the points on a rectangular coordinate system, and 3) connect the points with a straight line or a smooth, continuous curve as appropriate.

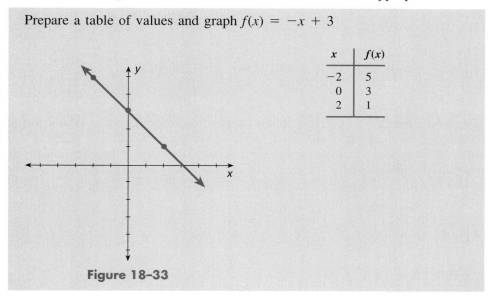

Prepare a table of values and graph $f(x) = -x + 3$

x	$f(x)$
-2	5
0	3
2	1

Figure 18–33

2 Write an equation in one variable as a function and find the solution graphically.

To write an equation in one variable as a function, 1) rewrite the equation so that all terms are on one side of the equation, and 2) write the equation in function notation using $f(x) =$ the nonzero side of the equation. Then, evaluate the function for at least 2 values and graph. The solution is the x-coordinate where the line crosses the x-axis.

Write $2x + 3 = 6$ as a function, graph the function, and find the solution from the graph.

$2x + 3 = 6$
$2x - 3 = 0$
$\quad f(x) = 2x - 3$
$\quad f(0) = 2(0) - 3$
$\quad f(0) = -3$
$\quad f(1) = 2(1) - 3$
$\quad f(1) = -1$

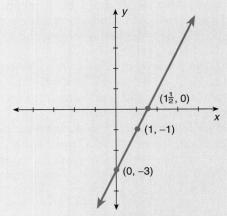

Figure 18–34

Solution: $x = \dfrac{3}{2}$ The x coordinate of the point where the line crosses the x-axis.

Section 18–2

1 Check the solution of an equation with two variables.

To check the solution of an equation in two variables, substitute the numerical value for x in each place it occurs and substitute the numerical value for y in each place it occurs. Simplify each side of the equation using the order of operations. The solution will make a true statement; that is, both sides of the equation will be equal to the same number.

A solution for the equation $3x - y = 1$ is (1, 2). Check to verify the solution.

$3(1) - 2 = 1$
$\quad 3 - 2 = 1$
$\quad\quad 1 = 1$ True

2 Find a specific solution of an equation with two variables when given the value of one variable.

To evaluate an equation when the value of one variable is given, substitute the numerical value of the given variable in place of that variable each place it occurs in the equation; then solve the equation for the other variable.

Evaluate the equation $4x - y = 6$ for x, if $y = 14$.

$4x - 14 = 6$
$\quad 4x = 20$
$\quad\quad x = 5$

Section 18–3

1 Graph linear equations using intercepts.

Find the x-intercept by letting $y = 0$ and solving for x.
Find the y-intercept by letting $x = 0$ and solving for y.
Plot the two points and draw the graph of the equation.

Graph the equation $y = 2x - 1$ by using the intercepts method.

x-intercept: $0 = 2x - 1$
$\quad\quad\quad\quad\quad 1 = 2x$
$\quad\quad\quad\quad\quad \dfrac{1}{2} = x$

$$\left(\frac{1}{2}, 0\right)$$

y-intercept: $y = 2(0) - 1$
$\qquad\qquad\quad y = 0 - 1$
$\qquad\qquad\quad y = -1$

$(0, -1)$

Plot the two points; then draw the graph.

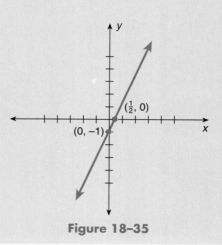

Figure 18–35

2 Graph linear equations using the slope–intercept method.

Write the equation in the form $y = mx + b$. The slope is m. In fraction form the numerator of m is the vertical movement and the denominator of m is the horizontal movement. The y-coordinate of the y-intercept is b. To graph using the slope–intercept method, first locate the y-intercept or b by counting vertically along the y-axis. From this point, count the slope (vertical, then horizontal) and locate the second point. Draw the graph through the two points.

Use the slope–intercept method to graph $y = 2x - 1$. The y-intercept is -1, so count down 1 from the origin. The coordinates of this point are $(0, -1)$. From this point, move $+2$ vertically and $+1$ horizontally. The coordinates of the second point are $(1, 1)$. Connect the two points.

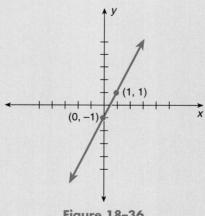

Figure 18–36

3 Graph linear equations using a graphics calculator.

First, solve the equation for *y*. Press the GRAPH key; then enter the right side of the equation. Next, press EXE to show the graph on the screen.

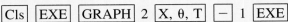

Graph the equation $y = 2x - 1$. Steps for the Casio *fx*-7700 are:

Cls | EXE | GRAPH | 2 | X, θ, T | – | 1 | EXE

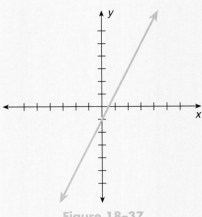

Figure 18–37

Section 18–4

1 Graph linear inequalities with two variables using test points.

Graph a linear inequality by first graphing the corresponding equation. Select a point on one side of the boundary line and substitute each coordinate in place of the appropriate letter. If the statement is true, shade the side of the boundary that contains the point. If false, shade the other side of the boundary. For $>$ or $<$ inequalities, make the boundary line dashed. For $\geq$ or $\leq$ inequalities, make the boundary line solid.

Graph $y \geq 2x - 1$.
Graph $y = 2x - 1$.
Make boundary line solid.
Test (1, 3):

$$3 \geq 2(1) - 1$$
$$3 \geq 2 - 1$$
$$3 \geq 1 \qquad \text{True}$$

Shade the side of the boundary line that contains the point since the coordinates of the test point made a true statement.

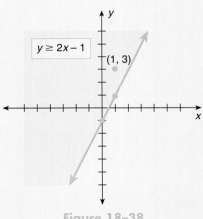

Figure 18–38

2 Graph linear inequalities with two variables using a graphics calculator.

Set the initial settings on the graphics calculator to graph inequalities.

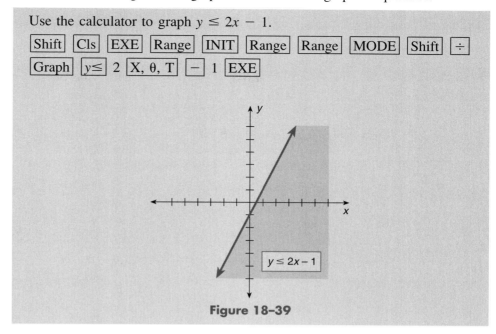

Use the calculator to graph $y \leq 2x - 1$.

$\boxed{\text{Shift}}$ $\boxed{\text{Cls}}$ $\boxed{\text{EXE}}$ $\boxed{\text{Range}}$ $\boxed{\text{INIT}}$ $\boxed{\text{Range}}$ $\boxed{\text{Range}}$ $\boxed{\text{MODE}}$ $\boxed{\text{Shift}}$ $\boxed{\div}$
$\boxed{\text{Graph}}$ $\boxed{y\leq}$ 2 $\boxed{\text{X, θ, T}}$ $\boxed{-}$ 1 $\boxed{\text{EXE}}$

$y \leq 2x - 1$

Figure 18–39

Section 18–5

1 Identify nonlinear equations.

Linear equations can have letter terms with an exponent of *one* only. And no term can have two letter factors. Quadratic equations must have at least one term with degree 2 and no terms with a degree greater than 2.

Identify the linear and quadratic equations from the following.

(a) $2x - 3 = y$
(b) $x^2 - y = 8$
(c) $xy - 5x = 2$
(d) $y = x^3 + 5$

(a) is a linear equation. (b) and (c) are quadratic equations. (d) is a cubic equation.

2 Graph quadratic equations using the table-of-solutions method or by examining properties.

Before graphing quadratic equations, determine the x-coordinate of the vertex of the parabola by finding the value of $-\frac{b}{2a}$. Use this value, k, to write an equation $x = k$, which is the axis of symmetry. Then substitute the value as the x-coordinate into the equation to find the corresponding y-value. Replace y with zero and solve the resulting equation for x. Plot these points if any, which are the solutions for the equation. Connect the plotted points with a smooth, continuous curved line.

Graph $y = x^2 - 6x + 8$.
$$-\frac{b}{2a} = -\frac{-6}{2(1)} = \frac{6}{2} = 3.$$

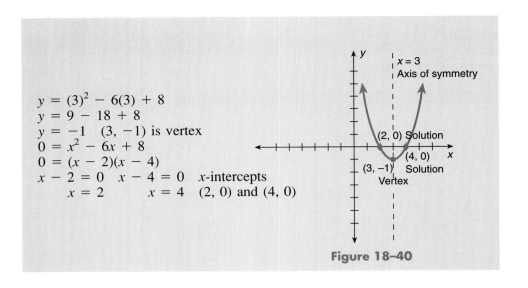

$y = (3)^2 - 6(3) + 8$
$y = 9 - 18 + 8$
$y = -1 \quad (3, -1)$ is vertex
$0 = x^2 - 6x + 8$
$0 = (x - 2)(x - 4)$
$x - 2 = 0 \quad x - 4 = 0 \quad x$-intercepts
$\quad\quad x = 2 \quad\quad\quad x = 4 \quad (2, 0)$ and $(4, 0)$

Figure 18–40

3 Graph quadratic equations and inequalities using a graphics calculator.

To graph quadratic equations using a graphics calculator, clear and initialize the calculator.

$\boxed{\text{Cls}}$ $\boxed{\text{EXE}}$ $\boxed{\text{Range}}$ $\boxed{\text{F1}}$ $\boxed{\text{Range}}$ $\boxed{\text{Range}}$

Be sure that the calculator is in equation mode.

$\boxed{\text{Mode}}$ $\boxed{\text{Shift}}$ $\boxed{+}$

To graph quadratic inequalities, clear and initialize the calculator. Put the calculator in inequality mode.

Graph $y = x^2 - 6x + 8$.
Clear and initialize the calculator.

$\boxed{\text{Graph}}$ $\boxed{\text{X, θ, T}}$ $\boxed{x^2}$ $\boxed{-}$ 6 $\boxed{\text{X, θ, T}}$ $\boxed{+}$ 8 $\boxed{\text{EXE}}$ (See Fig. 18–40.)

Graph $y \le x^2 - 6x + 8$.
Clear and initialize the calculator.

$\boxed{\text{Cls}}$ $\boxed{\text{EXE}}$ $\boxed{\text{Range}}$ $\boxed{\text{INIT}}$ $\boxed{\text{Range}}$ $\boxed{\text{Range}}$ $\boxed{\text{Mode}}$ $\boxed{\text{Shift}}$ $\boxed{÷}$ $\boxed{\text{Graph}}$
$\boxed{y\le}$ $\boxed{\text{X, θ, T}}$ $\boxed{x^2}$ $\boxed{-}$ 6 $\boxed{\text{X, θ, T}}$ $\boxed{+}$ 8 $\boxed{\text{EXE}}$

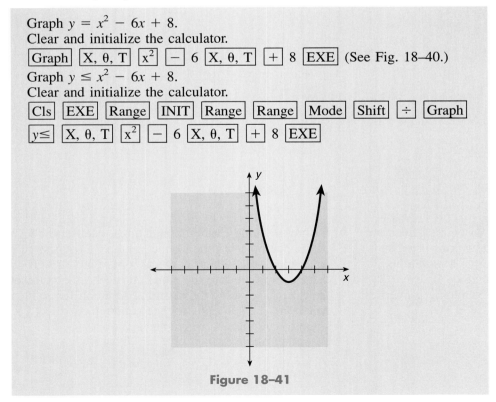

Figure 18–41

CHAPTER TRIAL TEST

Make a table of values to represent the solutions to the following equations and show the solutions on a graph.

1. $f(x) = \dfrac{1}{2}x$

2. $f(x) = \dfrac{1}{2}x + 1$

3. $f(x) = 2x - 4$

4. $f(x) = 5 - x$

Write as functions and find the solution graphically.

5. $3x + 2 = 5$

6. $2(x + 1) = 3x$

Write the specific solution for each equation in ordered pair form.

7. $x + y = 7$, if $x = 2$

8. $2x - y = 1$, if $y = -3$

9. Find the x- and y-intercepts of the equation $x - 4y = 2$.

10. Find the x- and y-intercepts of the equation $y = -8 + x$.

11. Plot the graph for the relationship between horsepower (hp) and revolutions per minute (rpm) in a test engine. Use the same intervals as those in the following data table.

x-axis (rpm)	y-axis (hp)
500	30
1000	45
1500	60
2000	75

Graph using the intercepts method.

12. $2x - 3y = 6$

13. $x + 2y = 8$

Graph using the slope–intercept method.

14. $y = -3x + 1$

15. $2x + y = -3$

16. $x + 2y = 1$

17. $y = x - 5$

Graph the following linear inequalities.

18. $x + 2y < 4$

19. $2x - y \le 2$

20. $y \le 2x + 2$

21. $x + y < 1$

Graph the following quadratic equations by examining properties. Show the axis of symmetry, vertex, and solutions.

22. $y = x^2 + 2$

23. $y = x^2 + 2x + 1$

24. $y = x^2 + 4x + 4$

Graph the following quadratic inequalities with a graphics calculator.

25. $y \le x^2 - 6x + 8$

26. $y > 2x^2$

27. $y < -\dfrac{1}{2}x^2$

GOOD DECISIONS THROUGH TEAMWORK

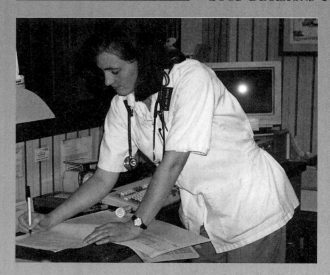

A technical college graduate with a degree in computer engineering started a company called Health Sciences Analysis, Inc., which specializes in designing software to allow physicians to enter hospital chart data electronically at each patient's bedside so that nurses don't have to spend hours transcribing this same data. Her first year of oper-

ation in 1995 yielded a gross income (revenue before expenses) of about $185,000. In 1998, her gross income was $326,250.

You and your team have been hired as financial analysts to track company growth thus far for this company, and also to derive an equation to predict future growth, if we assume growth will follow the same linear trend. Your team will use the 1995 and 1998 gross income figures to form ordered pairs for writing a linear equation relating the year to gross income. Let $x = 1$ represent 1995 and $x = 4$ represent 1998. Let y represent the corresponding yearly gross income. Find the y-intercept and give its real-life meaning, and draw a graph that represents the given information.

From the graph, predict gross income for the years 1999, 2000, and 2002. If this trend continues, in what year will gross income top $500,000? Verify your predictions using the equation of the graph.

As her financial analysts and counselors, you and your team members must explain how you arrived at your model and what assumptions you used. But you must also discuss any limitations of your model. Prepare a presentation for the company owner.

The concept of slope is becoming increasingly important in real-life applications. Not only do we use slope when we consider slant, like the slant of a roof or the grade of a roadway, but we also use slope in applications dealing with change, like a change in temperature, a change in the quality of a product, or a change in sales.

19-1 SLOPE

Learning Objectives

1. Calculate the slope of a line, given two points on the line.
2. Determine the slope of a horizontal or vertical line.

In Chapter 18 we learned to graph linear equations using slope. Now we will extend our knowledge of slope and examine procedures for calculating it.

1 Calculate the Slope of a Line, Given Two Points on the Line.

When we know the model or the equation of a line, we can find the slope by rewriting the equation in the form $y = mx + b$. If we do not know the equation of a line, we can calculate the slope using a formula.

Figure 19-1

DEFINITION 19-1: Slope.   The *slope* of a line is the ratio of the vertical rise of a line to the horizontal run of a line (see Fig. 19-1).

The definition of slope may be expressed as a formula.

Formula 19-1   *Find the slope from two given points:*

$$\text{Slope} = \frac{\text{rise}}{\text{run}}$$

$$= \frac{\Delta y}{\Delta x} = \frac{y_2 - y_1}{x_2 - x_1}$$

where Δy is vertical change and Δx is horizontal change.

$$P_1 = (x_1, y_1), \qquad P_2 = (x_2, y_2)$$

P_1 and P_2 are any two points on the line.

The formula outlines a procedure for finding the slope from the coordinates of two points on a line. We calculate the change (*difference*) in the y-coordinates to find the vertical *rise*. We calculate the change (*difference*) in the x-coordinates to find the horizontal *run*. Then we make a ratio of the rise to the run and reduce the ratio to lowest terms.

EXAMPLE 1   Find the slope of a line if the points $(2, -1)$ and $(5, 3)$ are on the line (see Fig. 19-2).

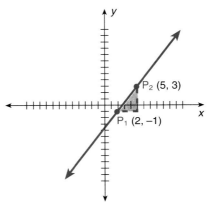

Figure 19-2

We begin by labeling $(2, -1)$ as point 1 (P_1) and $(5, 3)$ as point 2 (P_2). The coordinates of P_1 are referred to as (x_1, y_1) and the coordinates of P_2 are referred to as (x_2, y_2). We will use the Greek capital letter delta (Δ) to indicate a change and write the slope definition symbolically.

$$\text{Change in } y = \Delta y = \text{rise} = y_2 - y_1 = 3 - (-1) = 4$$

$$\text{Change in } x = \Delta x = \text{run} = x_2 - x_1 = 5 - 2 = 3$$

$$\text{Slope} = \frac{\Delta y}{\Delta x} = \frac{\text{rise}}{\text{run}} = \frac{y_2 - y_1}{x_2 - x_1} = \frac{3 - (-1)}{5 - 2} = \frac{4}{3}$$

Thus, the slope of the line through points $(2, -1)$ and $(5, 3)$ is $\frac{4}{3}$.

Tips and Traps   ***What Points on a Line Are Used? In What Order Are They Used?***
Any line has an infinite number of points. Any two points can be used to find the slope. Also, the designation of P_1 and P_2 is not critical. Let's find the slope of the same line in Example 1 by designating P_1 as $(5, 3)$ and P_2 as $(2, -1)$.

$$\Delta y = \text{rise} = y_2 - y_1 = -1 - 3 = -4$$

$$\Delta x = \text{run} = x_2 - x_1 = 2 - 5 = -3$$

$$\frac{\Delta y}{\Delta x} = \frac{\text{rise}}{\text{run}} = \frac{-4}{-3} = \frac{4}{3}$$

Notice in Example 1 that, with either point designated as point 1, the slope is positive. When the slope of a line is positive, the line *rises* from left to right. When the slope of a line is negative, the line *falls* from left to right. Figure 19-3 shows a line with a slope of $\frac{4}{3}$ and a line with a slope of $-\frac{4}{3}$.

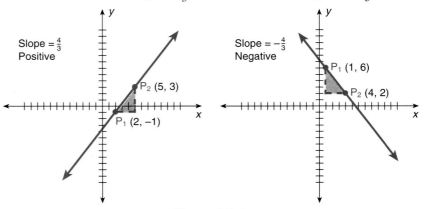

Figure 19-3

EXAMPLE 2 Find the slope of the line passing through the following points.

(a) $(1, 6)$ and $(4, 2)$ (b) $(-5, -1)$ and $(3, -2)$ (c) $(0, 0)$ and $(-5, 3)$

(a) Either point can be designated as P_1. The other point will be P_2. If $P_1 = (x_1, y_1)$ and $P_2 = (x_2, y_2)$, then the slope $= \frac{\Delta y}{\Delta x} = \frac{y_2 - y_1}{x_2 - x_1}$. Let $P_1 = (1, 6)$ and $P_2 = (4, 2)$. Then

$$\frac{\Delta y}{\Delta x} = \frac{2 - 6}{4 - 1} = \frac{-4}{3} = -\frac{4}{3}$$

The slope is $-\frac{4}{3}$.

(b) Let $P_1 = (-5, -1)$ and $P_2 = (3, -2)$.

$$\text{Slope} = \frac{\Delta y}{\Delta x} = \frac{-2 - (-1)}{3 - (-5)} = \frac{-2 + 1}{3 + 5} = \frac{-1}{8} = -\frac{1}{8}$$

The slope is $-\frac{1}{8}$.

(c) Let $P_1 = (0, 0)$ and $P_2 = (-5, 3)$.

$$\text{Slope} = \frac{\Delta y}{\Delta x} = \frac{3 - 0}{-5 - 0} = \frac{3}{-5} = -\frac{3}{5}$$

The slope is $-\frac{3}{5}$.

---

2 Determine the Slope of a Horizontal or Vertical Line.

There are two special lines whose slopes we want to examine, the horizontal line and the vertical line. See Figs. 19–4 and 19–5. Let's first look at the horizontal line that passes through the points $(3, 2)$ and $(-3, 2)$.

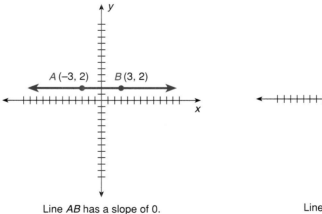

Line *AB* has a slope of 0.

Figure 19–4

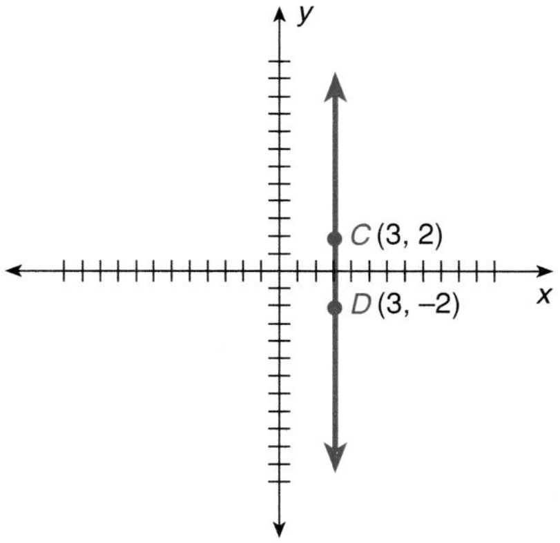

Line *CD* has no slope.

Figure 19–5

EXAMPLE 3 Find the slope of the horizontal line that passes through $(3, 2)$ and $(-3, 2)$.

We will let $B = (3, 2)$ and $A = (-3, 2)$.

$$\text{Slope} = \frac{\Delta y}{\Delta x} = \frac{y_2 - y_1}{x_2 - x_1} = \frac{2 - 2}{-3 - 3} = \frac{0}{-6} = 0$$

Since any horizontal line has no change in *y* values or no rise, *the slope of any horizontal line is zero.*

Next, let's look at the vertical line that passes through the points (3, 2) and (3, −2).

EXAMPLE 4 Find the slope of the vertical line that passes through the points (3, 2) and (3, −2).

We will let $C = (3, 2)$ and $D = (3, -2)$.

$$\text{Slope} = \frac{\Delta y}{\Delta x} = \frac{y_2 - y_1}{x_2 - x_1} = \frac{-2 - 2}{3 - 3} = \frac{-4}{0}$$

Remember, division by zero is impossible; **therefore, the slope of the vertical line passing through (3, 2) and (3, −2) is undefined.**

Since any vertical line has no change in x values or no run, *the slope of any vertical line is undefined.* We can say that a vertical line has *no slope.*

Tips and Traps *Slope of Zero Versus No Slope:*
A slope of zero and no slope are *not* the same. Zero is a real number. A slope of zero is real. When zero is the denominator of a fraction, the number is undefined. No slope is the same as an undefined slope. As a memory aid, think of the number zero as a ball. It will balance on a horizontal line. It will not balance on a vertical line.

SELF-STUDY EXERCISES 19–1

1 Find the slope of the line passing through the following pairs of points.

1. (−3, 3) and (1, 5)
2. (4, −1) and (1, 4)
3. (3, 1) and (5, 7)
4. (−1, −1) and (3, 3)
5. (4, 5) and (−4, −1)
6. (7, 2) and (−3, 2)
7. (4, −5) and (0, 0)
8. (2, −2) and (5, −5)
9. (−5, 2) and (−5, −4)
10. (4, −4) and (1, 5)

2 **11.** Identify the horizontal and vertical lines in Exercises 1 through 10 and explain why you identified each as horizontal or vertical.

Learning Objectives

[1] Find the equation of a line, given the slope and one point.

[2] Find the equation of a line, given two points on the line.

The slope of a line can be determined from any two points on that line. However, it is sometimes desirable to know the equation of the line. The equation of the line can serve as a model for solving various applications.

[1] **Find the Equation of a Line, Given the Slope and One Point.**

To write the equation of a line, there are certain facts about the line that we need to know. From the basic definition of slope, we can determine the minimum amount of information needed. To help us in writing equations, we rewrite the slope definition using the letter m to represent slope.

$$\text{Slope} = \frac{y_2 - y_1}{x_2 - x_1} \qquad m = \frac{y_2 - y_1}{x_2 - x_1}$$

We used this definition in the last section to find the slope when two points were known. Now we will use the definition to find the equation of a line when we know the slope and one point.

Formula 19-2 *Point–slope form of an equation of a straight line:*

$$y - y_1 = m(x - x_1) \qquad \text{or} \qquad m = \frac{y - y_1}{x - x_1}$$

where y_1 and x_1 are coordinates of the given point and m is the slope of the line passing through that point. The x and y represent the variables of the equation being sought.

EXAMPLE 1 Find the equation of the line passing through the point $(3, -2)$ with a slope of $\frac{2}{3}$.

$$y - y_1 = m(x - x_1) \qquad \text{Substitute into the formula.}$$

$$y - (-2) = \frac{2}{3}(x - 3) \qquad m = \frac{2}{3}; \ x_1 = 3; \ y_1 = -2$$

$$y + 2 = \frac{2}{3}(x - 3) \qquad \text{Distribute. } \frac{2}{\cancel{3}} \cdot \frac{\cancel{-3}^{-1}}{1} = -2$$

$$y + 2 = \frac{2}{3}x - 2 \qquad \text{Transpose.}$$

$$y = \frac{2}{3}x - 2 - 2 \qquad \text{Combine like terms.}$$

$$y = \frac{2}{3}x - 4 \qquad \text{Equation in slope–intercept form}$$

Another method for finding the equation is based on a variation of the slope formula that is similar to the definition of slope. (x_1, y_1) is the known point. (x, y) is the unknown point.

$$m = \frac{y - y_1}{x - x_1}$$
Substitute.

$$\frac{2}{3} = \frac{y - (-2)}{x - 3}$$
$m = \frac{2}{3}$; $x_1 = 3$; $y_1 = -2$

$$\frac{2}{3} = \frac{y + 2}{x - 3}$$
Solve the equation for y.

$$3(y + 2) = 2(x - 3)$$
Cross multiply.

$$3y + 6 = 2x - 6$$
Distribute.

$$3y = 2x - 6 - 6$$
Transpose.

$$3y = 2x - 12$$

$$\frac{3y}{3} = \frac{2x - 12}{3}$$

$$y = \frac{2x - 12}{3} \quad \text{or} \quad y = \frac{2x}{3} - \frac{12}{3}$$

$$y = \frac{2x}{3} - 4 \quad \text{or} \quad y = \frac{2}{3}x - 4$$

Tips and Traps   ***Which Point–Slope Formula Is Better?***
If you surveyed numerous math students and teachers to determine the preferred version of the point–slope formula, you would probably get an inconclusive result. It is strictly a matter of personal preference. Some may argue that they prefer one version if the slope is a fraction and another if the slope is an integer.

2 Find the Equation of a Line, Given Two Points on the Line.

Many times the facts that we are given do not match the facts that are required in a particular formula. It might be necessary to use more than one formula. To find the equation of a line when two points on the line are known, we will apply Formulas 19–1 and 19–2.

EXAMPLE 2   Find the equation of the line that passes through the points (0, 8) and (5, 0).

We first need to find the slope of the line passing through the points (0, 8) and (5, 0). Let $P_1 = (0, \boxed{8})$ and $P_2 = (5, \boxed{0})$.

$$\text{Slope} = \frac{\Delta y}{\Delta x} = \frac{y_2 - y_1}{x_2 - x_1} = \frac{0 - 8}{5 - 0} = -\frac{8}{5} \qquad \text{Formula 19–1}$$

Now we can use the slope and one of the points and the point–slope form of an equation. Using P_1 gives us

$$y - y_1 = m(x - x_1) \qquad \text{Formula 19–2}$$

$$y - 8 = -\frac{8}{5}(x - 0) \qquad m = -\frac{8}{5}; \; x_1 = 0; \; y_1 = 8$$

$$y - 8 = -\frac{8}{5}x \qquad \text{Solve for } y.$$

$$y = -\frac{8}{5}x + 8$$

Suppose that we had used P_2 (5, 0) instead of P_1 in the point–slope form of the equation. Because there is only one line that passes through (0, 8) and (5, 0), we must have the same equation, no matter which point is used. Let's find the equation of the line passing through (0, 8) and (5, 0) using the point (5, 0) and the slope $-\frac{8}{5}$.

$$y - y_1 = m(x - x_1)$$

$$y - 0 = -\frac{8}{5}(x - 5) \qquad m = -\frac{8}{5}; \ x_1 = 5; \ y_1 = 0$$

$$y = -\frac{8}{5}x - \frac{8}{5}(-5) \qquad \text{Distribute.}$$

$$y = -\frac{8}{5}x + 8 \qquad \text{Same equation as before}$$

EXAMPLE 3 Find the equation of the line that passes through the points $(-3, 4)$ and $(7, 4)$.

We will let $P_1 = (-3, \boxed{4})$ and $P_2 = (7, \boxed{4})$.

$$\text{Slope} = \frac{\Delta y}{\Delta x} = \frac{\boxed{y_2} - \boxed{y_1}}{x_2 - x_1} = \frac{\boxed{4} - \boxed{4}}{7 - (-3)} = \frac{0}{7 + 3} = \frac{0}{10} = 0$$

Using the point–slope form of an equation and P_1, we have

$$y - y_1 = m(x - x_1)$$

$$y - 4 = 0[x - (-3)]$$

$$y - 4 = 0$$

$$y = 4$$

The situation in Example 3 can be generalized in the property of the equation of a horizontal line.

Property 19-3 *Equation of a horizontal line:*

The equation of a line whose slope is zero (a horizontal line) is

$$y = k$$

where k is the common y-coordinate for all points on the line.

There is one special situation when the point–slope form of an equation *cannot* be used to determine the equation of a line. This is when the slope is undefined. Remember, *any vertical line has an undefined slope.* We can readily identify such a line if the coordinates of the two points have the same x-coordinate. The equation of the line will be $x = k$, where k is the common x-coordinate.

Property 19-4 *Equation of a vertical line:*

The equation of a line whose slope is undefined (a vertical line) is

$$x = k$$

where k is the common x-coordinate for all points on the line.

EXAMPLE 4   Find the equation of the line that passes through the points (5, 3) and (5, 7).

Since the *x*-coordinate is the same in both points, we know the slope of the line is undefined and the equation of the line is $x = k$, where k is the common *x*-coordinate. In this example, $k = 5$; **thus, the equation of the line through (5, 3) and (5, 7) is $x = 5$.**

SELF-STUDY EXERCISES 19–2

1 Find the equation of a line passing through the given point with the given slope. Solve the equation for *y* when necessary.

1. $(-8, 3)$, $m = \dfrac{2}{3}$
2. $(4, 1)$, $m = -\dfrac{1}{2}$
3. $(-3, -5)$, $m = 2$
4. $(0, -1)$, $m = 1$

2 Find the equation of a line passing through the given pairs of points. Solve the equation for *y*.

5. $(4, 6)$ and $(7, 1)$
6. $(-1, 6)$ and $(-1, 4)$
7. $(-1, -3)$ and $(3, -3)$
8. $(-4, 4)$ and $(-4, -2)$
9. $(-2, -4)$ and $(5, 10)$
10. $(-4, 0)$ and $(6, 0)$

19–3   SLOPE–INTERCEPT FORM OF AN EQUATION

Learning Objectives

1 Find the slope and *y*-intercept of a line, given the equation of the line.
2 Find the equation of a line, given the slope and *y*-intercept.

1   Find the Slope and *y*-Intercept of a Line, Given the Equation of the Line.

So far in our discussion of equations of straight lines, we have always preferred that the final form of the equation be solved for *y*. When a linear equation is solved for *y*, we say that it is the *slope–intercept form* of an equation of a straight line.

> **Formula 19–3**   *Slope–intercept form of an equation of a straight line:*
> $$y = mx + b$$
> where *m* is the slope and *b* is the *y*-coordinate of the *y*-intercept, (0, *b*).

When an equation is in the form $y = mx + b$, we can determine the slope of the line and the *y*-intercept by inspection.

EXAMPLE 1   Find the slope and *y*-intercept of the lines with the following equations.

(a) $y = 4x + 3$ (b) $y = -5x - 3$
(c) $y = -\dfrac{3}{5}x + \dfrac{1}{2}$ (d) $y = 2.3x - 4.7$

(a) $y = 4x + 3$
   Slope, $m = 4$ (coefficient of *x*); $b = 3$, *y*-intercept = (0, 3)

(b) $y = -5x - 3$
   Slope, $m = -5$; $b = -3$, *y*-intercept = (0, -3)

(c) $y = -\dfrac{3}{5}x + \dfrac{1}{2}$

$$\text{Slope, } m = -\frac{3}{5}\,; \; b = \frac{1}{2}\,, \; y\text{-intercept} = \left(0, \; \frac{1}{2}\right)$$

(d) $y = 2.3x - 4.7$
$$\textbf{Slope, } \boldsymbol{m = 2.3}\,; \; \boldsymbol{b = -4.7}\,, \; \boldsymbol{y}\textbf{-intercept} = (0, \; \boldsymbol{-4.7})$$

Tips and Traps   ***Why Do We Say That b = the y-Intercept?***
Even though b is the y-coordinate of the y-intercept, it is common to refer to b as the y-intercept. This has become an acceptable shortcut.

EXAMPLE 2   Rewrite the following equations in the slope–intercept form. Then find the slope and y-intercept.

(a) $x = y + 5$ (b) $2x - 3y = 9$ (c) $5y + 2x = 0$
(d) $1.5x - 3y = 4.5$ (e) $2y = 4$

To rewrite an equation in slope–intercept form, we solve the equation for y. Then the coefficient of x is the slope and the number term or constant is the y-intercept.

(a) $x = y + 5$ Transpose.
$\quad\; x - 5 = y$ Interchange sides.
$\quad\; \boldsymbol{y = x - 5}$
$\quad\; \textbf{Slope, } \boldsymbol{m = 1}\,; \; \boldsymbol{b = -5}$

(b) $2x - 3y = 9$ Transpose.
$\quad\;\; 2x - 9 = 3y$ Interchange sides.

$\quad\; 3y = 2x - 9$ Divide.
$\quad\; \dfrac{3y}{3} = \dfrac{2x - 9}{3}$

$\quad\; y = \dfrac{2x}{3} - \dfrac{9}{3}$ Reduce.

$\quad\; \boldsymbol{y = \dfrac{2}{3}x - 3}$

$\quad\; \textbf{Slope, } \boldsymbol{m = \frac{2}{3}}\,; \; \boldsymbol{y}\textbf{-intercept, } \boldsymbol{b = -3}$

(c) $5y + 2x = 0$ Transpose.
$\quad\;\;\;\; 5y = -2x$ Divide.
$\quad\;\;\; \dfrac{5y}{5} = -\dfrac{2x}{5}$

$\quad\;\;\;\;\; y = -\dfrac{2}{5}x$ Because there is no number term, the value of b is zero.

$\quad\; \textbf{Slope, } \boldsymbol{m = -\frac{2}{5}}\,; \; \boldsymbol{y}\textbf{-intercept, } \boldsymbol{b = 0}$

(d) $1.5x - 3y = 4.5$ Transpose.
$\quad\;\; 1.5x - 4.5 = 3y$ Interchange sides.

$\quad\; 3y = 1.5x - 4.5$ Divide.
$\quad\; \dfrac{3y}{3} = \dfrac{1.5x - 4.5}{3}$

$\quad\; \boldsymbol{y = 0.5x - 1.5}$
$\quad\; \textbf{Slope, } \boldsymbol{m = 0.5}\,; \; \boldsymbol{y}\textbf{-intercept, } \boldsymbol{b = -1.5}$

(e) $2y = 4$ Divide.
$\quad\; \dfrac{2y}{2} = \dfrac{4}{2}$

$$y = 2$$

Because there is no *x*-term, the coefficient of *x* or the slope is zero.

Slope, *m* = 0; *y*-intercept, *b* = 2

Tips and Traps   *Vertical Lines and the Slope–Intercept Form:*
Any vertical line will be an exception in using the slope–intercept form of an equation. In the equation of a vertical line such as $x = 4$, there is no *y*-term; so the equation cannot be solved for *y*. We recall that the graph is a vertical line that crosses the *x*-axis at (4, 0). However, it does not cross the *y*-axis, so it does not have a *y*-intercept. We also recall that the slope of a vertical line is undefined. We say that it has no slope.

---

2 Find the Equation of a Line, Given the Slope and *y*-Intercept.

So far we have written the equation of a line if we knew either two points on the line or the slope and at least one point on the line. If the one point is the *y*-intercept (0, *b*), then we can use the slope–intercept form of an equation, $y = mx + b$. Thus, the equation can be written by inspection.

EXAMPLE 3   Write the equation for a line with a slope of -3 and a *y*-intercept of 5.

Slope $= m = -3$;   *y*-intercept $= b = 5$

$y = mx + b$ Substitute values.

$y = -3x + 5$

The necessary facts for writing the equation of a line are often obtained from a graph.

EXAMPLE 4   Figure 19–6 shows the cost of producing picture frames.

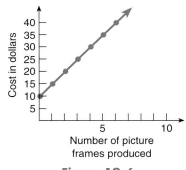

Figure 19–6

(a) Write an equation that represents the graph.
(b) Using the equation, find the cost of producing 20 picture frames.

(a) The *y*-intercept represents the fixed cost of producing picture frames. An example of a fixed cost would be the cost of the necessary tools. From the graph, $b = \$10$. The slope *m* is 5 units vertical change for every 1 unit horizontal change, which is $\frac{5}{1} = 5$.

$y = 5x + 10$ Equation of a line

This type of equation is typically referred to in the business world as a *cost function*. The notation used is $C(x) = 5x + 10$.

(b) Use the equation of the line found in part (a) to find y when $x = 20$.

$$y = 5x + 10 \qquad \text{or} \qquad C(x) = 5x + 10$$
$$y = 5(20) + 10 \qquad\qquad C(20) = 5(20) + 10$$
$$y = 100 + 10 \qquad\qquad C(20) = 100 + 10$$
$$y = \$110 \qquad\qquad C(20) = \$110$$

The cost of producing 20 frames is $110.

SELF-STUDY EXERCISES 19–3

1 Determine the slope and y-intercept of the following equations by inspection.

1. $y = 4x + 3$

2. $y = -5x + 6$

3. $y = -\dfrac{7}{8}x - 3$

4. $y = 3$

Rewrite the following equations in slope–intercept form and determine the slope and y-intercept.

5. $4x - 2y = 10$

6. $2y = 5$

7. $\dfrac{1}{2}y + x = 3$

8. $2.1y - 4.2x = 10.5$

9. $2x = 8$

10. $x - 5 = 4$

2 Write the equations of lines with the following slopes and y-intercepts.

11. $m = \dfrac{1}{4}, b = 7$

12. $m = -8, b = -4$

Write the equations of the lines that are graphed in Exercises 13–15.

13.

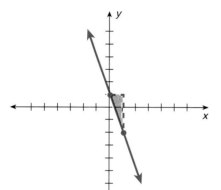

Figure 19–7

14.

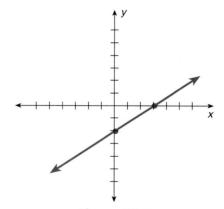

Figure 19–8

15.

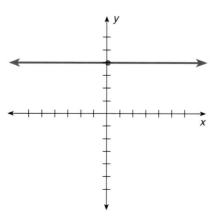

Figure 19–9

Learning Objective

1️⃣ Find the equation of a line, given a point on the line and the equation of a line parallel to a given line.

1️⃣ Find the Equation of a Line, Given a Point on the Line and the Equation of a Line Parallel to a Given Line.

Parallel lines are lines that are the same distance apart. No matter how far the lines are extended, the lines do not get any closer together or farther apart. Since slope represents the slant or steepness of a line, parallel lines have the same slope. Any line can have several lines that are parallel to it. Figure 19–10 illustrates two examples of lines that are parallel to the line $x + y = 5$.

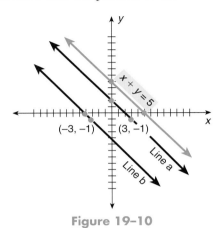

Figure 19–10

◼ **DEFINITION 19–2: Parallel Lines.** *Parallel lines* are two or more lines in the same plane that are the same distance apart everywhere. They have no points in common.

Property 19–5 *Slope of parallel lines:*

The slopes of parallel lines are equal.

If we write $x + y = 5$ in slope–intercept form, we see that $y = -x + 5$, and the slope is -1. Thus, *any* equation that has a slope of -1 is parallel to (or coincides with) the line whose equation is $x + y = 5$. *Coincides* means that one line lies on top of the other; that is, they are the same line.

We have examined the similarities of parallel lines, but what are the differences? Parallel lines will have different x- and y-intercepts.

EXAMPLE 1   Write the equations for lines a and b in Figure 19–10.

Line a has a y-intercept of 2. It is parallel to $x + y = 5$. Solved for y, $x + y = 5$ is $y = -x + 5$, with a slope of -1.

Line a: $y = -x + 2$ $m = -1, b = 2$

Line b: $y = -x - 4$ $m = -1, b = -4$

Another form of equations of lines is the *standard form*. The standard form of an equation has the variables on the left, with the x-term first, and the constant on the right. Thus, the equation of line a is $y = -1x + 2$, or $y = -x + 2$, or $x + y = 2$. The equation $x + y = 2$ is said to be in *standard form*.

Formula 19–4   *Standard form of an equation of a straight line:*

$$ax + by = c$$

where a, b, and c are integers and $a \geq 0$.

The equation of a parallel line can be found if any point, not just the y-intercept, is known. In line a of Figure 19–10, one point is $(3, -1)$. We can substitute $x = 3$, $y = -1$, and $m = -1$ into the point–slope form of an equation.

$$y - y_1 = m(x - x_1)$$
$$y - (-1) = -1(x - 3)$$
$$y + 1 = -x + 3$$
$$y = -x + 3 - 1$$
$$y = -x + 2$$

Any point we choose on line a will give the same equation, $y = -x + 2$ or $x + y = 2$.

Now, let's write the equation for line b using the slope and a known point other than the y-intercept. To do so, we will use the following procedure. Line b in Fig. 19–10 is parallel to the line $x + y = 5$ and passes through the point $(-3, -1)$.

Rule 19–1   *To find the equation of a line that passes through a given point and is parallel to a given line:*

1. Determine the slope (m) of the *given* line. The slope of the parallel line will be the same as the slope of the given line.
2. Write the equation for the parallel line by substituting the values for m, x_1, and y_1 into the point–slope form of the equation, $y - y_1 = m(x - x_1)$.
3. Write the equation in slope–intercept or standard form.

EXAMPLE 2   Find the equation of line b in Fig. 19–10.

The slope of the line $x + y = 5$ is -1 because $x + y = 5$ is equal to $y = -x + 5$. Substitute -1 for m, -3 for x_1, and -1 for y_1 into the point–slope form of the equation.

$$y - y_1 = m(x - x_1)$$
$$y - (-1) = -1[x - (-3)] \qquad m = -1; x_1 = -3; y_1 = -1$$
$$y + 1 = -1(x + 3) \qquad \text{Distribute.}$$
$$y + 1 = -x - 3 \qquad \text{Solve for } y.$$
$$y = -x - 3 - 1 \qquad \text{Combine like terms.}$$
$$\mathbf{y = -x - 4} \qquad \text{Slope–intercept form}$$
$$\mathbf{x + y = -4} \qquad \text{Standard form}$$

EXAMPLE 3 Find the equation of a line that is parallel to $2y = 3x + 8$ and passes through the point $(2, 1)$.

Find the slope of the given line, $2y = 3x + 8$.

$$2y = 3x + 8$$

$$\frac{2y}{2} = \frac{3x + 8}{2}$$

$$y = \frac{3}{2}x + 4$$

Thus, the slope of $2y = 3x + 8$ is $\frac{3}{2}$.

Next, find the equation of a line passing through the point $(2, 1)$ that is parallel to $2y = 3x + 8$.

$$y - y_1 = \boldsymbol{m}(x - x_1) \qquad \text{Substitute. The slope of a parallel line is equal to the slope of the given line.}$$

$$y - 1 = \frac{3}{2}(x - 2) \qquad m = \frac{3}{2}; \; x_1 = 2; \; y_1 = 1$$

$$y - 1 = \frac{3}{2}x - 3$$

$$y = \frac{3}{2}x - 3 + 1$$

$$\boldsymbol{y = \frac{3}{2}x - 2} \qquad \text{Slope–intercept form}$$

To write the equation in standard form, we start by clearing fractions.

$$2y = 2\left(\frac{3}{2}x - 2\right) \qquad \text{Clear fractions.}$$

$$2y = 3x - 4 \qquad \text{Transpose variable terms to the left.}$$

$$-3x + 2y = -4 \qquad \text{Each term is multiplied by } -1 \text{ so that the coefficient of } x \text{ is positive.}$$

$$\boldsymbol{3x - 2y = 4} \qquad \text{Standard form}$$

SELF-STUDY EXERCISES 19–4

1   Write equations for the following in standard form. Verify your answers using a graphics calculator.

1. Find the equation of the line that is parallel to the line $x + y = 6$ and passes through the point $(2, -3)$.

2. Find the equation of the line that is parallel to the line $2x + y = 5$ and passes through the point $(1, 7)$.

3. Find the equation of the line that is parallel to the line $3y = x - 2$ and passes through the point $(4, 0)$.

4. Find the equation of the line that is parallel to the line $3x - y = -2$ and passes through the point $(-3, -2)$.

5. Find the equation of the line that is parallel to the line $x + 3y = 7$ and passes through the point $(4, 1)$.

6. Find the equation of the line that is parallel to the line $3x - y = 4$ and passes through the point $(0, 3)$.

7. Find the equation of the line that is parallel to the line $2x + 3y = 5$ and passes through the point $(1, 1)$.

8. Find the equation of the line that is parallel to the line $3x + 2y = 1$ and passes through the point $(2, 0)$.

9. Find the equation of the line that is parallel to the line $2x - 5y = 0$ and passes through the point $(3, -1)$.

10. Find the equation of the line that is parallel to the line $-3x + 4y = -1$ and passes through the point $(\frac{1}{2}, 0)$.

19–5 PERPENDICULAR LINES

Learning Objective

1 Find the equation of a line, given a point on the line and the equation of a line perpendicular to a given line.

1 **Find the Equation of a Line, Given a Point on the Line and the Equation of a Line Perpendicular to a Given Line.**

Perpendicular lines are lines that meet and make a square corner. Any line can have several lines that are perpendicular to it, but there is only one line perpendicular to a *given* line that passes through a *given* point. Figure 19–11 illustrates a line that is perpendicular to the line $2x + y = 7$ and passes through the point $(6, 8)$. In some cases the given point may lie on the given line.

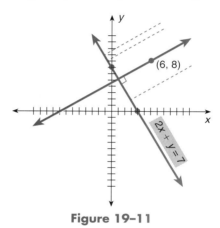

Figure 19–11

■ **DEFINITION 19–3: Perpendicular Lines.** *Perpendicular lines* are two lines that intersect to form right angles (90° angles). Another term for *perpendicular* is *normal*.

In Fig. 19–11, the perpendicular (or normal) line that passes through the given point is indicated by the symbol ⊾, which means a right or 90° angle is formed by the lines.

The slope of the equation $2x + y = 7$ is -2, which can be seen if the equation is written in the slope–intercept form as $y = -2x + 7$. We will determine the equation of a line that passes through $(6, 8)$ and is perpendicular to $2x + y = 7$ by using an important property of perpendicular lines.

> **Property 19–6** *Slope of perpendicular lines:*
>
> The slope of *any* line perpendicular to a given line is the *negative reciprocal* of the slope of the given line.

Using Property 19–6, if the slope of $2x + y = 7$ or $y = -2x + 7$ is -2, then the slope of *any* line perpendicular to that line is $+\frac{1}{2}$(the negative reciprocal of -2, or $-\frac{2}{1}$).

Now, we know the slope $(+\frac{1}{2})$ and one point $(6, 8)$ through which the perpendicular line passes. We can find the equation of the perpendicular line by using the point–slope form of the equation.

$$y - y_1 = m(x - x_1) \qquad m = \tfrac{1}{2}, \ P_1 = (6, 8)$$

$$y - 8 = \frac{1}{2}(x - 6)$$

$$y - 8 = \frac{1}{2}x - 3$$

$$y = \frac{1}{2}x - 3 + 8$$

$$y = \frac{1}{2}x + 5$$

Then the equation of the perpendicular line is

$$y = \frac{1}{2}x + 5 \qquad \text{Slope–intercept form}$$

or

$$x - 2y = -10 \qquad \text{Standard form}$$

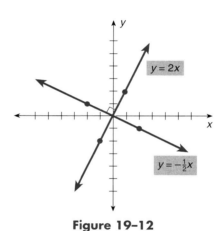

Figure 19–12

Tips and Traps   *Negative Reciprocals:*
The negative reciprocal of a number is not necessarily a negative value. It will have the opposite sign.
  To find the negative reciprocal:

1. Interchange the numerator and denominator.
2. Give the reciprocal the opposite sign.

The negative reciprocal of -5 is $+\frac{1}{5}$. The negative reciprocal of $-\frac{3}{4}$ is $+\frac{4}{3}$.
The negative reciprocal of $\frac{4}{5}$ is $-\frac{5}{4}$. The negative reciprocal of 3 is $-\frac{1}{3}$.

To further illustrate the relationship of the slopes of perpendicular lines, examine Figs. 19–12 and 19–13.

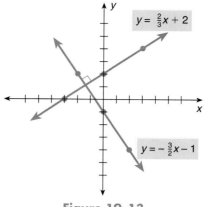

$y = \frac{2}{3}x + 2$

$y = -\frac{3}{2}x - 1$

Figure 19–13

We summarize the procedures for finding the equation of a line perpendicular to a given line in Rule 19–2.

Rule 19–2 *To find the equation of a line that is perpendicular (normal) to a given line at a given point on the line or through a given point not on the line:*

1. Determine the slope of the *given* line.
2. Find the *negative reciprocal* of this slope. The negative reciprocal is the slope of the perpendicular line.
3. Write the equation for the perpendicular line by substituting the coordinates of the *given* point for x_1 and y_1 and the slope of the *perpendicular* line for m into the point–slope form of the equation $y - y_1 = m(x - x_1)$.
4. Write the equation in slope–intercept or standard form.

EXAMPLE 1 Find the equation of a line that is perpendicular to the line $y = \frac{2}{3}x + 8$ and passes through the point (3, 10).

First, find the slope of the given line, $y = \frac{2}{3}x + 8$. The negative reciprocal of $\frac{2}{3}$ is $-\frac{3}{2}$.

Write the equation.

$$y - y_1 = m(x - x_1)$$ Point–slope form

$$y - 10 = -\frac{3}{2}(x - 3)$$ Substitute values: $m = -\frac{3}{2}$; $x_1 = 3$; $y_1 = 10$.

$$y - 10 = -\frac{3}{2}x + \frac{9}{2}$$ Distribute.

$$y = -\frac{3}{2}x + \frac{9}{2} + 10$$ Solve for y.

$$y = -\frac{3}{2}x + \frac{9}{2} + \frac{20}{2}$$ Combine like terms.

$$y = -\frac{3}{2}x + \frac{29}{2}$$ Slope–intercept form

or

$$2y = -3x + 29$$ Clear fractions.

$$3x + 2y = 29$$ Standard form

EXAMPLE 2   Find the equation of the line normal to $4x + y = -3$ and passing through $(0, -3)$.

$$4x + y = -3 \qquad \text{Given equation}$$

$$y = -4x - 3 \qquad \text{Solve for } y.$$

$$\text{Slope}_{given} = -4$$

$$\text{Slope}_{perpendicular} = +\frac{1}{4} \qquad \text{Negative reciprocal of } -4$$

$$y - y_1 = m(x - x_1)$$

$$y - (-3) = \frac{1}{4}(x - 0) \qquad m = \tfrac{1}{4}, \ p_1 = (0, -3)$$

$$y + 3 = \frac{1}{4}x$$

$$\mathbf{y = \frac{1}{4}x - 3} \qquad \text{Equation of the normal in slope–intercept form}$$

or

$$4y = x - 12 \qquad \text{Clear fractions.}$$

$$\mathbf{x - 4y = 12} \qquad \text{Equation of the normal in standard form}$$

EXAMPLE 3   Which of the following equations represents a normal of the line $2x - 3y = 1$ passing through $(2, -1)$?

(a) $y = \dfrac{2}{3}x - \dfrac{1}{3}$ (b) $y = \dfrac{2}{3}x - \dfrac{7}{3}$

(c) $y = -\dfrac{3}{2}x - \dfrac{1}{3}$ (d) $y = -\dfrac{3}{2}x + 2$

$$2x - 3y = 1 \qquad \text{Rewrite in slope–intercept form.}$$

$$-3y = -2x + 1$$

$$\frac{-3y}{-3} = \frac{-2x}{-3} + \frac{1}{-3}$$

$$y = \frac{2}{3}x - \frac{1}{3} \qquad \text{Slope–intercept form}$$

$$\text{Slope}_{given} = \frac{2}{3} \qquad \text{From } y = \tfrac{2}{3}x - \tfrac{1}{3}$$

$$\text{Slope}_{\text{perpendicular}} = -\frac{3}{2} \qquad \text{Negative reciprocal}$$

$$y - y_1 = m(x - x_1) \qquad \text{Substitute } m = -\tfrac{3}{2},\ x_1 = 2,\ y_1 = -1.$$

$$y - (-1) = -\frac{3}{2}(x - 2)$$

$$y + 1 = -\frac{3}{2}x + 3$$

$$y = -\frac{3}{2}x + 3 - 1$$

$$y = -\frac{3}{2}x + 2$$

Thus, the correct equation is $y = -\frac{3}{2}x + 2$, or choice (d).

SELF-STUDY EXERCISES 19–5

1 Write the equations for the following exercises in standard form. Verify your results using a graphics calculator. Be sure the range is set so that the vertical and horizontal increments are equal.

1. Find the equation of the line that is perpendicular to the line $x + y = 6$ and passes through the point $(2, 3)$.

2. Find the equation of the line that is normal to the line $2x + y = 5$ and passes through the point $(1, 7)$.

3. Find the equation of the line that is perpendicular to the line $3y = x - 2$ and passes through the point $(4, 0)$.

4. Find the equation of the line that is normal to the line $3x - y = 2$ and passes through the point $(-3, -2)$.

5. Find the equation of the line that is normal to the line $x + 3y = 7$ and passes through the point $(4, 1)$.

6. Find the equation of the line that is perpendicular to the line $x + 2y = 7$ and passes through the point $(-2, 3)$.

7. Find the equation of the line that is perpendicular to the line $2x + 3y = 4$ and passes through the point $(3, -1)$.

8. Find the equation of the line that is normal to the line $4x + y = 1$ and passes through the point $(0, 0)$.

9. Find the equation of the line that is perpendicular to the line $2x + 2y = 3$ and passes through the point $(\frac{1}{2}, 2)$.

10. Find the equation of the line that is perpendicular to the line $5x - y = 6$ and passes through the point $(5, -\frac{1}{5})$.

19-6 DISTANCE AND MIDPOINTS

Learning Objectives

1 Find the distance between two points.

2 Find the midpoint between two points.

1 Find the Distance Between Two Points.

Many times it is desirable to know the straight-line distance between two points. If we consider two points such as $P_1 = (4, 2)$ and $P_2 = (7, 6)$, we can visualize the straight-line distance between these two points on a pair of coordinate axes as in Fig. 19–14. To find the distance, we extend a vertical line down from P_2 and a horizontal line across from P_1 to form a triangle.

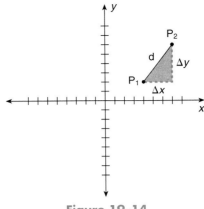

Figure 19–14

We have labeled the straight-line distance from P_1 to P_2 as d, the change in horizontal movement as Δx, and the change in vertical movement as Δy. The triangle formed is a *right triangle*. A horizontal and a vertical line meet at a right angle. The *distance formula* is based on the special property of right triangles, the *Pythagorean theorem*.

The distance from P_1 to P_2 is the principal square root of the sum of the squares of the change in horizontal movement and the change in vertical movement; that is, $d^2 = (\Delta x)^2 + (\Delta y)^2$. Solving for d by taking the square root of each side of the equation, we obtain

$$d = \sqrt{(\Delta x)^2 + (\Delta y)^2}$$

If $P_1 = (x_1, y_1)$ and $P_2 = (x_2, y_2)$, the distance formula can be stated using the coordinates of the given points.

Formula 19–5 *Distance between two points:*

Distance: $d = \sqrt{(x_2 - x_1)^2 + (y_2 - y_1)^2}$, where P_1 and P_2 are two points with coordinates (x_1, y_1) and (x_2, y_2), respectively.

EXAMPLE 1 Find the distance from (4, 2) to (7, 6).

These points are illustrated in Fig. 19–14. Point (4, 2) is labeled P_1, and point (7, 6) is labeled P_2.

$d = \sqrt{(x_2 - x_1)^2 + (y_2 - y_1)^2}$ Substitute: $x_1 = 4$, $x_2 = 7$, $y_1 = 2$, and $y_2 = 6$.

$d = \sqrt{(7 - 4)^2 + (6 - 2)^2}$ Combine in each grouping.

$d = \sqrt{3^2 + 4^2}$ Square each term in the radicand.

$d = \sqrt{9 + 16}$ Add terms in the radicand.

$d = \sqrt{25}$ Find the principal square root.

$d = 5$ Distance

Thus, the distance from (4, 2) to (7, 6) is 5 units.

Point Selection Does Not Matter:

The distance formula can be used to calculate the distance between two points, no matter which point we designate as P_1 and which as P_2. Let's rework Example 1, letting $P_1 = (7, 6)$ and $P_2 = (4, 2)$. Thus, $x_1 = 7$, $x_2 = 4$, $y_1 = 6$, and $y_2 = 2$.

$d = \sqrt{(4 - 7)^2 + (2 - 6)^2}$ Substitute.

$d = \sqrt{(-3)^2 + (-4)^2}$ $(-3)^2 = +9; (-4)^2 = +16$

$d = \sqrt{9 + 16}$

$d = \sqrt{25}$

$d = 5$

The change in vertical movement has the same *absolute value*, no matter which point is used first. The result after squaring will be positive whether the difference is positive or negative. For similar reasons, the change in horizontal movement squared will be the same, no matter which point is used first.

EXAMPLE 2   Find the distance from $(5, -2)$ to $(-3, -4)$.

Let $(5, -2)$ be P_1 and $(-3, -4)$ be P_2. [We could have designated $(-3, -4)$ as P_1 and $(5, -2)$ as P_2.]

$d = \sqrt{(x_2 - x_1)^2 + (y_2 - y_1)^2}$ Substitute: $x_1 = 5$, $x_2 = -3$, $y_1 = -2$, and $y_2 = -4$.

$d = \sqrt{(-3 - 5)^2 + (-4 - (-2))^2}$ $(-4 + 2) = -2$

$d = \sqrt{(-8)^2 + (-2)^2}$

$d = \sqrt{64 + 4}$

$d = \sqrt{68}$

$d = 8.246$ To the nearest thousandth

The distance from $(5, -2)$ to $(-3, -4)$ is 8.246 units.

2   **Find the Midpoint Between Two Points.**

Two points determine a line that is infinite in length. However, the two points also determine a *line* segment that has a certain length. The two points are called the *end points* of the line segment.

On a number line or measuring device, the *coordinate of the midpoint between two points* is the average of the coordinates of the points.

Formula 19–6   *Midpoint of two points on a line:*

To find the coordinate of the midpoint between two points *on the number line* or linear measuring device, average the end points.

$$\text{Midpoint} = \frac{P_1 + P_2}{2}$$

where P_1 and P_2 are points on the number line or measuring device.

EXAMPLE 3 Find the midpoint between two points on a metric rule at 2.8 and 5.6.

$$\text{Midpoint} = \frac{P_1 + P_2}{2}$$

$$\text{Midpoint} = \frac{2.8 + 5.6}{2}$$

$$\text{Midpoint} = \frac{8.4}{2}$$

$$\text{Midpoint} = 4.2$$

The midpoint of the points 2.8 and 5.6 is 4.2.

The coordinates of the midpoint between two points on a coordinate system can be found by averaging the respective coordinates.

Formula 19-7 *Midpoint of two points on a coordinate system:*

To find the coordinates of the midpoint of a line segment *on a coordinate system*, average the respective coordinates of the end points of the segment.

$$\text{Midpoint} = \left(\frac{x_1 + x_2}{2}, \frac{y_1 + y_2}{2} \right)$$

where P_1 and P_2 are end points of the segment and $P_1 = (x_1, y_1)$ and $P_2 = (x_2, y_2)$.

EXAMPLE 4 Find the midpoint of each segment with the given end points.

(a) (2, 4) and (6, 10)   (b) (−1, 4) and (2, −3)
(c) (3, −5) and origin

(a) (2, 4) and (6, 10). See Fig. 19–15.

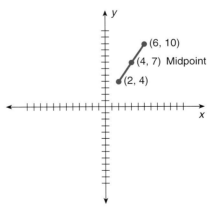

Figure 19–15

$$\text{Midpoint} = \left(\frac{x_1 + x_2}{2}, \frac{y_1 + y_2}{2} \right)$$ Substitute values: $x_1 = 2$, $y_1 = 4$, $x_2 = 6$, $y_2 = 10$.

$$\text{Midpoint} = \left(\frac{2 + 6}{2}, \frac{4 + 10}{2} \right)$$

$$\text{Midpoint} = \left(\frac{8}{2}, \frac{14}{2}\right)$$

Midpoint = (4, 7)

(b) $(-1, 4)$ and $(2, -3)$. See Fig. 19–16.

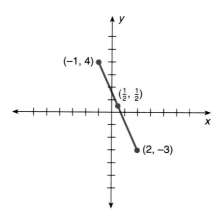

Figure 19–16

$$\text{Midpoint} = \left(\frac{x_1 + x_2}{2}, \frac{y_1 + y_2}{2}\right)$$ Substitute values: $x_1 = -1$, $y_1 = 4$, $x_2 = 2$, $y_2 = -3$.

$$\text{Midpoint} = \left(\frac{-1 + 2}{2}, \frac{4 - 3}{2}\right)$$ $4 + (-3) = 4 - 3 = 1$

$$\textbf{Midpoint} = \left(\frac{1}{2}, \frac{1}{2}\right)$$

(c) $(3, -5)$ and origin. See Fig. 19–17.

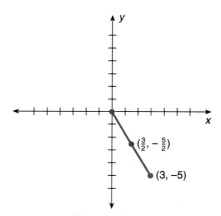

Figure 19–17

$$\text{Midpoint} = \left(\frac{x_1 + x_2}{2}, \frac{y_1 + y_2}{2}\right)$$ Substitute values: $x_1 = 3$, $y_1 = -5$, $x_2 = 0$, $y_2 = 0$.

$$\text{Midpoint} = \left(\frac{3 + 0}{2}, \frac{-5 + 0}{2}\right)$$

$$\textbf{Midpoint} = \left(\frac{3}{2}, \frac{-5}{2}\right) \textbf{ or } \left(1\frac{1}{2}, -2\frac{1}{2}\right)$$

SELF-STUDY EXERCISES 19–6

1 Graph the line segment determined by the given end points and find the distance between each of the following pairs of points. Express the answer to the nearest thousandth when necessary.

1. (7, 10) and (1, 2)   **2.** (7, −7) and (2, 5)   **3.** (5, 0) and (0, 5)
4. (−2, −2) and (3, −4)   **5.** (5, 7) and (0, −3)   **6.** (8, −2) and (−4, 3)
7. (−4, 6) and (2, −2)   **8.** (3, 5) and (0, 1)   **9.** (5, 4) and (−7, −5)
10. (7, 2) and (−2, −3)

2 11–20. Find the coordinates of the midpoints of the segments determined in Exercises 1 through 10.

CAREER APPLICATION

Electronics: Using the Slope to Find Resistance

To find the slope of the line that goes through the two points (x_1, y_1) and (x_2, y_2), the formulas are

$$\text{Slope} = \frac{\text{rise}}{\text{run}} = \frac{y_2 - y_1}{x_2 - x_1} \quad \text{or} \quad \text{slope} = \frac{y_1 - y_2}{x_1 - x_2}$$

Assume that you have a graph with current on the horizontal axis and voltage on the vertical axis. See Fig. 19–18. Find the slope of the line through the points (1 mA, 3 V) and (4 mA, 6 V).

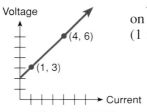

Figure 19–18

$$\text{Slope} = \frac{3 \text{ V} - 6 \text{ V}}{1 \text{ mA} - 4 \text{ mA}} = \frac{-3 \text{ V}}{-3 \text{ mA}} = 1 \text{ k}\Omega \quad \text{or}$$

$$\text{Slope} = \frac{6 \text{ V} - 3 \text{ V}}{4 \text{ mA} - 1 \text{ mA}} = \frac{3 \text{ V}}{3 \text{ mA}} = 1 \text{ k}\Omega$$

To analyze the dimensions, use the following relationships. Volts divided by amperes gives ohms, $\frac{V}{A} = \Omega$. $m = \frac{1}{1000}$; $1 \div \frac{1}{1000} = 1000$; $1000 = k$; that is, the reciprocal of the unit m is the unit k.

Find the slope of the line through (4 V, 5 mA) and (6 V, 4 mA).

$$\text{Slope} = \frac{5 \text{ mA} - 4 \text{ mA}}{4 \text{ V} - 6 \text{ V}} = \frac{1 \text{ mA}}{-2 \text{ V}} = -0.5 \text{ mS}$$

To analyze the dimensions, amperes divided by volts gives siemens, $\frac{A}{V} = S$.

A graph like the one in Fig. 19–19 that compares current and voltage for a **series circuit** is usually drawn with the current or the independent variable on the horizontal axis because the current is the reference. The voltage or dependent variable is on the vertical axis. A line drawn on this graph that goes through the origin (vertical or *y*-intercept = 0) has

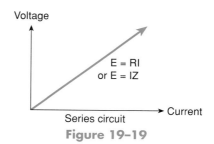

Figure 19–19

$$\text{Slope} = \frac{\text{rise}}{\text{run}} = \frac{\Delta \text{ voltage } (E)}{\Delta \text{ current } (I)} = \text{resistance } (R \text{ or } Z)$$

The equation of the line for a **series circuit** is

$$E = RI \text{ (for dc circuits)} \qquad \text{or} \qquad E = IZ \text{ (for ac circuits)}$$

A graph like the one in Fig. 19–20 that compares current and voltage for a **parallel circuit** is usually drawn with the voltage on the horizontal axis because the voltage is the reference or independent variable, and the current is on the vertical axis as the dependent variable. A line drawn on this graph that passes through the origin has

$$\text{Slope} = \frac{\text{rise}}{\text{run}} = \frac{\Delta \text{ current } (I \text{ or } V)}{\Delta \text{ voltage } (E)} = \text{conductance } (G \text{ or } Y)$$

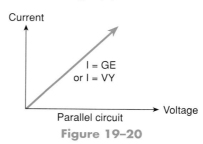

Figure 19–20

The equation of the line is

$$I = GE \text{ (for dc circuits)} \qquad \text{or} \qquad I = VY \text{ (for ac circuits)}$$

Exercises

Find the slope of the line that passes through each of the following pairs of points. Be sure to include sign, prefix, and units as well as the number in the answer.

1. (4 mA, 12 V) and (7 mA, 15 V)
2. (4 mA, 12 V) and (9 mA, 8 V)
3. (15 V, 7 mA) and (20 V, 9 mA)
4. (25 V, 4 mA) and (37 V, 1 mA)
5. (0.65 V, 1 mA) and (0.75 V, 2 mA)
6. (9 μS, 11 V) and (12 μS, 17 V)
7. (9 μS, 11 V) and (12 μS, 5 V)
8. (4 hours, 200 miles) and (5 hours, 250 miles)
9. (7 hours, 700 kilometers) and (9 hours, 900 kilometers)
10. (2 class hours, 4 hours of homework) and (3 class hours, 6 hours of homework)

Answers for Exercises

Find the slope of the line that passes through each of the following pairs of points. Be sure to include sign, prefix, and units as well as the number in the answer.

1. $\dfrac{(12\text{ V} - 15\text{ V})}{(4\text{ mA} - 7\text{ mA})} = \dfrac{-3\text{ V}}{(-3\text{ mA})} = +1\text{ k}\Omega$

2. $\dfrac{(12\text{ V} - 8\text{ V})}{(4\text{ mA} - 9\text{ mA})} = \dfrac{4\text{ V}}{(-5\text{ mA})} = -0.8\text{ k}\Omega \text{ or } -800\ \Omega$

3. $\dfrac{(7\text{ mA} - 9\text{ mA})}{(15\text{ V} - 20\text{ V})} = \dfrac{-2\text{ mA}}{(-5\text{ V})} = 0.4\text{ mS or } 400\ \mu\text{S}$

4. $\dfrac{(4\text{ mA} - 1\text{ mA})}{(25\text{ V} - 37\text{ V})} = \dfrac{3\text{ mA}}{(-12\text{ V})} = -0.25\text{ mS or } -250\ \mu\text{S}$

5. $\dfrac{(1\text{ mA} - 2\text{ mA})}{(0.65\text{ V} - 0.75\text{ V})} = \dfrac{-1\text{ mA}}{(-0.10\text{ V})} = 10\text{ mS}$

6. $\dfrac{(11\text{ V} - 17\text{ V})}{(9\ \mu\text{S} - 12\ \mu\text{S})} = \dfrac{-6\text{ V}}{(-3\ \mu\text{S})} = 2\text{ M}\Omega \text{ or } 2{,}000{,}000\ \Omega$

7. $\dfrac{(11\text{ V} - 5\text{ V})}{(9\ \mu\text{S} - 12\ \mu\text{S})} = \dfrac{+6\text{ V}}{(-3\ \mu\text{S})} = -2\text{ M}\Omega \text{ or } -2{,}000{,}000\ \Omega$

8. $\dfrac{(200\text{ miles} - 250\text{ miles})}{(4\text{ hours} - 5\text{ hours})} = 50\text{ miles per hour}$

9. $\dfrac{(700\text{ kilometers} - 900\text{ kilometers})}{(7\text{ hours} - 9\text{ hours})} = 100\text{ kilometers per hour}$

10. $\dfrac{(4\text{ hours of homework} - 6\text{ hours of homework})}{(2\text{ class hours} - 3\text{ class hours})} = 2\text{ hours of homework per class hour}$

ASSIGNMENT EXERCISES

Section 19–1

Find the slope of the line passing through the following pairs of points.

1. $(-2, 2)$ and $(1, 3)$ **2.** $(3, -1)$ and $(1, 3)$ **3.** $(3, 2)$ and $(5, 6)$

4. $(-1, -1)$ and $(2, 2)$ **5.** $(4, 3)$ and $(-4, -2)$ **6.** $(6, 2)$ and $(-3, 2)$

7. $(3, -4)$ and $(0, 0)$ **8.** $(1, -1)$ and $(5, -5)$ **9.** $(-4, 1)$ and $(-4, 3)$

10. $(4, -4)$ and $(1, 3)$ **11.** $(5, 0)$ and $(-2, 4)$ **12.** $(-2, 1)$ and $(0, 3)$

13. $(-4, -8)$ and $(-2, -1)$ **14.** $(3, 3)$ and $(3, 0)$ **15.** $(5, -3)$ and $(-1, -3)$

16. $(-5, -1)$ and $(-7, -3)$ **17.** $(-7, 0)$ and $(-7, 5)$ **18.** $(3, 5)$ and $(2, 5)$

19. $(5, 9)$ and $(7, 11)$ **20.** $(3, 5)$ and $(5, 3)$

21. Write the coordinates of two points that lie on the same horizontal line.

22. Write the coordinates of two points that lie on the same vertical line.

Section 19–2

Find the equation of a line passing through the given point with the given slope.
Solve the equation for y if necessary.

23. $(-6, 2)$, $m = \dfrac{1}{3}$ **24.** $(3, 2)$, $m = -\dfrac{2}{5}$ **25.** $(4, 0)$, $m = \dfrac{3}{4}$

26. $(0, -2)$, $m = 2$ **27.** $(2, 3)$, $m = 4$ **28.** $(6, 0)$, $m = -1$

29. $(5, -4)$, $m = -\dfrac{2}{3}$ **30.** $(-1, -5)$, $m = -3$

Find the equation of a line passing through the given pairs of points. Solve the equation for y if necessary.

31. $(-5, 2)$ and $(6, 1)$ **32.** $(1, 4)$ and $(-1, 3)$ **33.** $(-1, -3)$ and $(3, 4)$

34. $(-3, 0)$ and $(4, 0)$ **35.** $(-2, -3)$ and $(3, 6)$ **36.** $(2, -4)$ and $(3, -4)$

37. $(5, 2)$ and $(6, 3)$ **38.** $(4, 6)$ and $(1, -1)$ **39.** $(-1, -2)$ and $(-3, -4)$

40. $(4, 0)$ and $(4, -3)$ **41.** $(5, -2)$ and $(3, -2)$ **42.** $(5, 4)$ and $(0, 4)$

Section 19–3

Determine the slope and y-intercept of the following equations by inspection.

43. $y = 3x + \dfrac{1}{4}$
44. $y = \dfrac{2}{3}x - \dfrac{3}{5}$
45. $y = -5x + 4$

46. $y = 7$
47. $x = 8$
48. $y = \dfrac{1}{3}x - \dfrac{5}{8}$

49. $y = \dfrac{x}{8} - 5$
50. $y = -\dfrac{x}{5} + 2$

Rewrite the following equations in slope–intercept form and determine the slope and y-intercept.

51. $2x + y = 8$
52. $4x + y = 5$
53. $3x - 2y = 6$

54. $5x - 3y = 15$
55. $\dfrac{3}{5}x - y = 4$
56. $2.2y - 6.6x = 4.4$

57. $3y = 5$
58. $3x - 6y = 12$

Write the equations using the given slope and y-intercept.

59. $m = 3, b = -2$
60. Slope $= \dfrac{3}{5}$; y-intercept $= -7$

Write the equations using information from the graphs.

61.

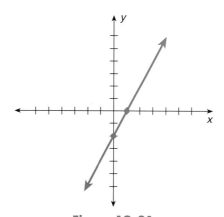

Figure 19–21

62.

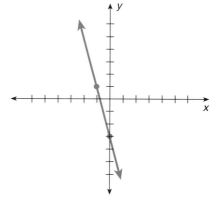

Figure 19–22

Section 19–4

Write equations for the following exercises in standard form. Graph both equations for each exercise on a graphics calculator to verify parallelism.

63. Find the equation of the line that is parallel to the line $x + y = 4$ and passes through the point $(2, 5)$.

64. Find the equation of the line that is parallel to the line $3x + y = 6$ and passes through the point $(1, 0)$.

65. Find the equation of the line that is parallel to the line $2y = x - 3$ and passes through the point $(2, -3)$.

66. Find the equation of the line that is parallel to the line $4x - y = -1$ and passes through the point $(0, -2)$.

67. Find the equation of the line that is parallel to the line $x - 3y = 5$ and passes through the point $(5, -5)$.

68. Find the equation of the line that is parallel to the line $3x - 2y = 2$ and passes through the point $(0, -3)$.

69. Find the equation of the line that is parallel to the line $x + 3y = 6$ and passes through the point $(-4, -2)$.

70. Find the equation of the line that is parallel to the line $4x + 3y = 1$ and passes through the point $(3, -\frac{1}{2})$.

71. Find the equation of the line that is parallel to the line $3x - 4y = 0$ and passes through the point $(\frac{1}{3}, 2)$.

72. Find the equation of the line that is parallel to the line $-2x + 3y = 2$ and passes through the point $(-1, -1)$.

Section 19–5

Write the equations for the following exercises in standard form. Use a graphics calculator to verify that the two lines in each exercise are perpendicular.

73. Find the equation of the line that is perpendicular to the line $x + y = 4$ and passes through the point $(-3, 1)$.

74. Find the equation of the line that is normal to the line $3x + y = 6$ and passes through the point $(1, 4)$.

75. Find the equation of the line that is perpendicular to the line $x + 2y = 5$ and passes through the point $(-2, 0)$.

76. Find the equation of the line that is normal to the line $2x + 2y = 4$ and passes through the point $(0, 0)$.

77. Find the equation of the line that is normal to the line $5x + y = 8$ and passes through the point $(-1, 2)$.

78. Find the equation of the line that is perpendicular to the line $3y = x - 4$ and passes through the point $(2, 3)$.

79. Find the equation of the line that is perpendicular to the line $5x - y = 10$ and passes through the point $(\frac{1}{2}, 3)$.

80. Find the equation of the line that is normal to the line $x - 3y = 6$ and passes through the point $(-2, 4)$.

81. Find the equation of the line that is perpendicular to the line $4x - y = 8$ and passes through the point $(4, -\frac{1}{2})$.

82. Find the equation of the line that is perpendicular to the line $4y = 2x + 1$ and passes through the point $(3, -1)$.

Section 19–6

Find the distance between each of the following pairs of points. Express the answer to the nearest thousandth when necessary.

83. $(3, 6)$ and $(-1, 4)$
84. $(4, 0)$ and $(0, -1)$
85. $(3, -3)$ and $(0, 7)$
86. $(2, 7)$ and $(-3, -2)$
87. $(0, 0)$ and $(-3, 5)$
88. $(-5, -5)$ and $(4, 0)$
89. $(5, 2)$ and $(-3, -3)$
90. $(1, -5)$ and $(-2, -5)$
91. $(-5, -4)$ and $(2, -2)$
92. $(8, 3)$ and $(3, -4)$
93–102. Calculate the coordinates for the midpoint of each segment whose end points are given in Exercises 83 through 92.

■ CHALLENGE PROBLEM ■

103. Christy Hodge is introducing a new lipstick in Brightglow's product line. As marketing manager, she estimates the cost of the new lipstick will be $4.53 per item, plus an additional cost of $5000.00.

(a) Make a table of the estimated cost for producing 0 lipsticks, 100 lipsticks, 1000 lipsticks, and 2000 lipsticks.

(b) Represent these costs as ordered pairs (number of lipsticks, cost).

(c) Graph the ordered pairs and sketch a line that fits the ordered pairs.

(d) Write a formula (using function notation) that represents the cost of the new lipstick as a function of the number of lipsticks produced.

(e) Christy Hodge projects the selling price of the new lipstick to be $8.99. How many lipsticks must be sold for the company to recover its cost of producing the new item?

1. (a) Describe the graph of a line whose slope is positive.
   (b) Describe the graph of a line whose slope is negative.
   (c) Describe the graph of a line whose slope is a fraction between 0 and 1.
   (d) Describe the graph of a line whose slope is a number > 1.
   (e) Describe the graph of a line whose slope is a fraction between -1 and 0.
   (f) Describe the graph of a line whose slope is a number < -1.

3. What is the slope of a vertical line? Why?

5. If the standard form of an equation of a vertical line is $x = k$, what does k represent?

7. How do the slopes of parallel lines compare?

9. How are the distance formula and the Pythagorean theorem related?

2. What is the slope of a horizontal line? Why?

4. If the standard form of an equation of a horizontal line is $y = k$, what does k represent?

6. If an equation is solved for y, what information about the graph can we read from the equation?

8. How do the slopes of perpendicular lines compare?

10. List the steps necessary to find the x-intercept of a line if you are given two points on the line.

CHAPTER SUMMARY

Objectives	What to Remember with Examples

Section 19–1

1 Calculate the slope of a line, given two points on the line.

The slope of a line joining two points is the difference of the y-coordinates divided by the difference of the x-coordinates, that is, $\frac{\text{rise}}{\text{run}}$.

$$m = \frac{\text{rise}}{\text{run}} = \frac{\Delta y}{\Delta x} = \frac{y_2 - y_1}{x_2 - x_1}$$

Find the slope of the line passing through the points $(3, 1)$ and $(-2, 5)$.

$$\frac{5 - 1}{-2 - 3} = \frac{4}{-5} \quad \text{or} \quad -\frac{4}{5}$$

2 Determine the slope of a horizontal or vertical line.

The slope of a horizontal line is zero, and points on the same horizontal line have the same y-coordinate. The slope of a vertical line is not defined, and points on the same vertical line have the same x-coordinate.

Examine the following coordinate pairs and indicate which points lie on the same horizontal line and which lie on the same vertical line.
$A\ (4, -3)$, $B\ (2, 5)$, $C\ (4, 5)$, $D\ (2, -3)$. Points A and D and points B and C lie on the same horizontal lines, respectively. Points A and C and points B and D lie on the same vertical lines, respectively.

Section 19–2

1 Find the equation of a line, given the slope and one point.

Use the point–slope form of an equation to find the equation when given the slope and coordinates of a point on a line.

$$y - y_1 = m(x - x_1)$$

A line has slope 2 and passes through the point (3, 4). Find the equation of the line.

$$y - y_1 = m(x - x_1)$$
$$y - 4 = 2(x - 3)$$
$$y - 4 = 2x - 6$$
$$y = 2x - 2$$

2 Find the equation of a line, given two points on the line.

Find the slope, then use the point–slope form of the equation to find the equation when given coordinates of two points on a line.

Find the equation of a line that passes through the two points (3, 1), (−3, 2). First, find the slope.

$$m = \frac{2 - 1}{-3 - 3} = \frac{1}{-6} \qquad \text{or} \qquad -\frac{1}{6}$$
$$y - y_1 = m(x - x_1)$$
$$y - 1 = -\frac{1}{6}(x - 3)$$
$$y - 1 = -\frac{1}{6}x + \frac{1}{2}$$
$$y = -\frac{1}{6}x + \frac{3}{2} \qquad \text{Slope-intercept form}$$

or

$$6y = -1x + 9$$
$$x + 6y = 9 \qquad \text{Standard form}$$

Section 19–3

1 Find the slope and *y*-intercept of a line, given the equation of the line.

Write the equation in slope–intercept form (solve for *y*).

$$y = mx + b$$

The coefficient of *x* is the slope and the constant is the *y*-coordinate of the *y*-intercept.

Find the slope and *y*-intercept for the equation $2x - 4y = 8$.

$$-4y = -2x + 8 \qquad \text{Solve for } y.$$
$$y = \frac{1}{2}x - 2$$

Slope $= \frac{1}{2}$.   *y*-intercept $= (0, -2)$.

2 Find the equation of a line, given the slope and *y*-intercept.

Use the slope–intercept form of the equation, $y = mx + b$, and substitute values for *m* and *b*, the slope and *y*-intercept, respectively to find the equation of a line.

Write the equation of a line that has slope $\frac{2}{3}$ and the *y*-intercept $(0, -1)$. Use the slope–intercept form of the equation:

$$y = mx + b$$
$$y = \frac{2}{3}x - 1$$

CHAPTER 19 Slope and Distance

Section 19–4

1 Find the equation of a line, given a point on the line and the equation of a line parallel to a given line.

The slopes of parallel lines are equal; that is, lines that have the same slope are parallel lines. Equations that have equal slopes will have graphs that are parallel lines.

Determine which two of the three given equations have graphs that are parallel lines:

(a) $y = 3x - 5$;   (b) $3x - 2y = 10$;   (c) $6x - 2y = 8$.

Write each of the three equations in slope–intercept form and compare the slopes (coefficients of x).

(a) $y = 3x - 5$ (b) $y = \frac{3}{2}x - 5$ (c) $y = 3x - 4$

(a) and (c) have the same slope and thus their graphs are parallel.

To find the equation of a line parallel to a given line and passing through a given point, use the point–slope form of the linear equation. Substitute the value of the slope of the given equation and the coordinates of the given point into the point–slope form of an equation.

Find the equation of a line that passes through the point (2, 3) and is parallel to the line whose equation is $y = 4x - 1$. Write the new equation in slope–intercept form.

Use the slope of the given equation, 4, for the slope of the new equation.

$y - y_1 = m(x - x_1)$
$y - 3 = 4(x - 2)$
$y - 3 = 4x - 8$
$y = 4x - 5$ Slope-intercept form

Section 19–5

1 Find the equation of a line, given a point on the line and the equation of a line perpendicular to a given line.

The slopes of perpendicular lines are negative reciprocals; that is, lines that have slopes that are negative reciprocals are perpendicular lines. Equations with slopes that are negative reciprocals have graphs that are perpendicular lines.

Determine which two of the three given equations have graphs that are perpendicular lines:

(a) $y = 3x - 5$;   (b) $3x + 9y = 10$;   (c) $9x + 3y = 12$

Write each of the three equations in slope–intercept form and compare the slopes (coefficients of x).

(a) $y = 3x - 5$ (b) $y = -\frac{1}{3}x + \frac{10}{9}$ (c) $y = -3x + 4$

(a) and (b) have the slopes that are negative reciprocals and thus their graphs are perpendicular. Note that (a) and (c) are not perpendicular. Their slopes are opposites but *not* reciprocals.

To write the equation of a line perpendicular to a given line, substitute the values of the negative reciprocal of the slope of the given equation and the coordinates of the given point into the point–slope form of an equation.

Find the equation of a line that passes through the point $(-1, 2)$ and is perpendicular to the line whose equation is $y = \frac{1}{3}x - 5$. Write the equation in standard form. The slope for the new equation is the negative reciprocal of $\frac{1}{3}$, which is -3.

$$y - y_1 = m(x - x_1)$$
$$y - 2 = -3(x - (-1))$$
$$y - 2 = -3x - 3$$
$$3x + y = -3 + 2$$
$$3x + y = -1 \qquad \text{Standard form}$$

Section 19–6

1 Find the distance between two points.

Use the distance formula: $\quad d = \sqrt{(x_2 - x_1)^2 + (y_2 - y_1)^2}$

Find the distance between the points $(3, 2)$ and $(-1, 5)$.

$$d = \sqrt{(x_2 - x_1)^2 + (y_2 - y_1)^2}$$
$$d = \sqrt{(-1 - 3)^2 + (5 - 2)^2}$$
$$d = \sqrt{(-4)^2 + 3^2}$$
$$d = \sqrt{16 + 9}$$
$$d = \sqrt{25}$$
$$d = 5$$

2 Find the midpoint between two points.

Use the midpoint formula:

$$\left(\frac{(x_1 + x_2)}{2}, \frac{(y_1 + y_2)}{2} \right)$$

Find the midpoint of the segment joining the points $(3, 2)$ and $(-1, 5)$.

$$\left(\frac{(x_1 + x_2)}{2}, \frac{(y_1 + y_2)}{2} \right)$$
$$\left(\frac{3 + (-1)}{2}, \frac{2 + 5}{2} \right)$$
$$\left(\frac{2}{2}, \frac{7}{2} \right)$$
$$\left(1, \frac{7}{2} \right) \text{ or } (1, 3.5)$$

WORDS TO KNOW

slope (p. 695)
rise (p. 695)
run (p. 695)
horizontal line (p. 697)
vertical line (p. 697)
zero slope (p. 697–698)
undefined slope (p. 698)
point–slope form of an
   equation (p. 699)

slope–intercept form of an
   equation (p. 702)
cost function (p. 704)
parallel lines (p. 706)
standard form of equation
   (p. 706–707)
slope of parallel lines (p. 706)
perpendicular lines (p. 709)

normal (p. 709)
slope of perpendicular lines
   (p. 709)
negative reciprocal (p. 709)
distance formula (p. 714)
Pythagorean theorem (p. 714)
coordinates of midpoint of line
   segment (p. 715)

Find the slope of the line passing through the following pairs of points.

1. $(-3, 6)$ and $(3, 2)$
2. $(0, 4)$ and $(-1, 6)$
3. $(1, -5)$ and $(3, 0)$
4. $(5, 3)$ and $(-2, 3)$
5. $(-1, -1)$ and $(2, 2)$
6. $(-1, 5)$ and $(-1, 7)$

Find the equation of the line passing through the given point with the given slope. Solve the equation for y.

7. $(3, -5)$, $m = \dfrac{2}{3}$
8. $(5, 1)$, $m = -2$

Find the equation of the line passing through the following pairs of points. Solve the equation for y.

9. $(1, 3)$ and $(4, 5)$
10. $(-1, 1)$ and $(4, -4)$
11. $(5, 2)$ and $(-1, 2)$
12. $(7, 4)$ and $(-3, -1)$
13. What is the slope and y-intercept of a line whose equation is $y = 3x - 22$?

Rewrite the following equations in slope–intercept form and determine the slope and y-intercept.

14. $x - y = 4$
15. $x = 4y$
16. $2y - x = 3$
17. $\dfrac{1}{3}y + 2x = 1$

Write equations in slope–intercept form using the given slope and y-intercept or the graph in Fig. 19–23.

18. slope $= -2$, y-intercept $= -3$
19.

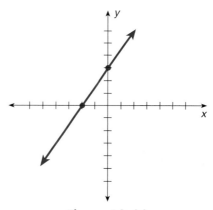

Figure 19–23

Write the equations for the following in standard form.

20. Find the equation of the line that is parallel to $y - 2x = 3$ and passes through the point $(2, 5)$.

21. Find the equation of the line that is parallel to $2x + y = 4$ and passes through the point $(4, -3)$.

22. Find the equation of the line perpendicular to $y - 2x = 3$ and passing through the point $(2, 5)$.

23. Find the equation of the line perpendicular to $2x + y = 4$ and passing through the point $(4, -3)$.

Find the distance between the following points and find the coordinates of the midpoint of the line segment made by the following pairs of points.

24. $(5, 2)$ and $(-7, -3)$
25. $(-2, 1)$ and $(3, 3)$

CHAPTER
20

Systems of Equations and Inequalities

GOOD DECISIONS THROUGH TEAMWORK

While going to school at night to get a degree in horticulture, a college student decides to give up his full-time, minimum-wage job selling clothes for a job that would require less time, pay the same or more money, be in the field of horticulture, and allow for flexible working hours. He decides to capitalize on a hobby he has had for several

years—making holiday table centerpieces out of greenery, pine cones, dried flowers, and candles. He already has a large stockpile of greenery, cones, and flowers he collected or grew and dried himself. He would have to buy candles, ribbon, and foam bases, but he can buy these in bulk at low prices. He already has a workshop set up in his garage, so he can work at home.

He knows from experience that most customers want either a 12-in. or a 16-in. base, and that he can sell these to local florist shops for $20 and $25, respectively. (The florist shops resell them for $40 and $50). The 12-in. base requires a half hour of production time and costs $3 for materials. The 16-in. base requires the same amount of production time but costs $4.50 for materials.

You and your team decide to help your fellow student with his economic forecasting. Let x equal the number of 12-in. centerpieces and y equal the number of 16-in. centerpieces he can make per week in the home business while attending college full-time. Derive the following economic forecasting formulas appropriate for different production levels. Write a linear equation in x and y to show the revenue received for any number of 12-in. and 16-in. centerpieces. Write a linear equation in x and y to show the cost of materials for making any number of 12-in. and 16-in. centerpieces. Finally, write a linear equation in x and y to show the profit made on any number of 12-in. and 16-in. centerpieces.

Graph your cost and revenue equations. Shade the area representing profit and the area repre-

senting loss, and identify the breakeven point. Determine how many hours per week he must work to make more than minimum wage. In your report, offer your advice to him in this business venture.

Many real-life situations involve problems that have several conditions or constraints that have to be considered. These conditions can be written in separate equations or inequalities forming a system of equations or inequalities. The solution of the system will be the value or values that satisfy all conditions.

20-1   SOLVING SYSTEMS OF EQUATIONS AND INEQUALITIES GRAPHICALLY

Learning Objectives

1. Solve a system of equations by graphing.
2. Solve a system of inequalities by graphing.

In Chapter 18 we learned to graph equations and inequalities with two variables. Because an equation with two variables has many ordered pairs of solutions, a graph gives an overall pictorial view of these solutions. In a similar way, since an inequality with two variables has many values that satisfy the conditions of the inequality, a graph gives a pictorial view of these values as well. In this section we want to look at solutions that two or more equations or inequalities have in common.

1   Solve a System of Equations by Graphing.

A system of two linear equations, each having two variables, is solved when we find the one ordered pair of solutions that satisfies *both* equations. One method of solving systems of two equations is to graph the ordered pairs of solutions of each equation and find the intersection of these graphs. The point where the two graphs intersect represents the ordered pair of solutions that the two graphs have in common.

First, let's look at an example that could be described as a system of two equations with two variables.

Rule 20-1   *To solve a system of two equations with two variables by graphing:*

1. Graph each equation on the same pair of axes.
2. The solution will be the common point or points.

EXAMPLE 1   A board is 20 ft long. It needs to be cut so that one piece is 2 ft longer than the other. What should be the length of each piece?

First, we write equations to describe all the conditions of the problem. Since the board is not cut into equal pieces, we let the letter l represent the *longer* piece and the letter s represent the *shorter* piece.

The first condition of the problem states that the total length of the board is 20 ft. Thus, the two pieces (l and s) total 20 ft: $l + s = 20$ (condition 1).

The second condition of the problem is that one piece is 2 ft longer than the other. Thus, the shorter piece plus 2 ft equals the longer piece: $s + 2 = l$ (condition 2).

The two equations become a *system of equations.*

$l + s = 20$ Condition 1

$s + 2 = l$ Condition 2

Graph each equation on the same set of axes and examine the intersection of the graphs. To do this, we can make a table of solutions for each equation.

Condition 1		Condition 2	
$l + s = 20$		$s + 2 = l$	
or		or	
$l = 20 - s$		$l = s + 2$	
s	l	s	l
9	11	9	11
10	10	10	12
11	9	11	13

Values for s like 1 and 2 would require 19 and 18 units for l on the graph. Therefore, to fit the values easily on a smaller graph, we selected s values near 10. Now we graph the s values on the x-axis and the l values on the y-axis. We then write each equation along its line graph for easy identification (see Fig. 20–1).

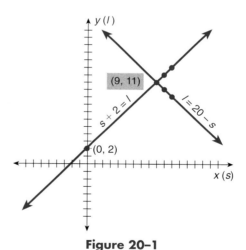

Figure 20–1

The point of intersection is (9, 11), which means that $s = 9$, $l = 11$. **The shorter length is 9 ft and the longer length is 11 ft.**

We can check to see if the solution ($s = 9$, $l = 11$) satisfies both equations.

$l + s = 20$ $s + 2 = l$

$11 + 9 = 20$ $9 + 2 = 11$

$20 = 20$ $11 = 11$

The ordered pair checks in both equations.

CHAPTER 20   Systems of Equations and Inequalities

Property 20–1 *Intersecting lines:*

When graphing two straight lines, three possibilities can occur.

1. The two lines can intersect in **just one point.** This means the system is *independent.*
2. The two lines **will not intersect** at all. This means that the system is *inconsistent* and there is no pair of values that will satisfy both equations.
3. The two lines **will coincide** or fall exactly in the same place. This means that the system is *dependent* and the equations are identical or are multiples, and any pair of values that satisfy one equation will satisfy both equations.

The last two possibilities rarely occur in practical applications.

■ **General Tips for Using the Calculator**

Systems of linear equations can be graphed on a graphics calculator or computer.

1. Graph the first equation.
2. Graph the second equation without clearing the graph screen.
3. Use the $\boxed{\text{Trace}}$ feature to determine the approximate coordinates of the intersection of the graphs.
4. Use the $\boxed{\text{Zoom}}$ or $\boxed{\text{Box}}$ feature to get a closer view and thus a more accurate approximation of the intersection of the graphs.

$\boxed{2}$ **Solve a System of Inequalities by Graphing.**

In some instances it is desirable to graph two inequalities and find the overlapping or common portion of the two graphs.

Rule 20–2 *To solve a system of two inequalities with two variables by graphing:*

1. Graph each inequality on the same pair of axes.
2. The solution set of the system will be the overlapping portion of the solution sets of the two inequalities.

EXAMPLE 2   Shade the portion on the graph that is represented by the following conditions: $y \leq 3x + 5$ and $x + y > 7$.

Graph the inequality $y \leq 3x + 5$. The boundary $y = 3x + 5$ has a slope of 3 and a y-intercept of 5. The boundary will be included in the solution set.

Because $(0, 0)$ does not fall on the graph of $y = 3x + 5$, we will use it as our test point in the original inequality, $y \leq 3x + 5$.

For $(0, 0)$

$$y \leq 3x + 5$$

$$0 \leq 3(0) + 5$$

$$0 \leq 0 + 5$$

$$0 \leq 5 \qquad \text{True}$$

Thus, the point (0, 0) is included in the solution set (see Fig. 20–2).

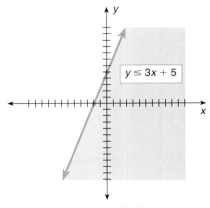

Figure 20–2

Graph the inequality, $x + y > 7$. The boundary $x + y = 7$ in slope–intercept form is $y = -x + 7$. The slope is -1 and the y-intercept is 7. ($x + y = 7$ will not be included.)

Again, the point (0, 0) does not fall on the graph of $x + y = 7$. We will use it as our test point in the original inequality $x + y > 7$.

$$\text{For } (0, 0)$$
$$x + y > 7$$
$$0 + 0 > 7$$
$$0 > 7 \qquad \text{False}$$

Thus, the point (0, 0) is not included in the solution set. The solution set will be the opposite side of the boundary (see Fig. 20–3).

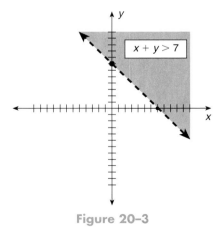

Figure 20–3

Find the solution set common to both inequalities. Visualize both graphs on the same axes. The portion that has overlapping shading represents the points that satisfy *both* conditions and forms the solution set for the system of inequalities (see Fig. 20–4).

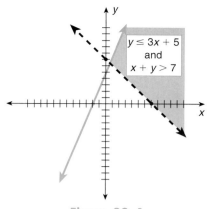

$y \leq 3x + 5$
and
$x + y > 7$

Figure 20–4

■ **General Tips for Using the Calculator**

Not all graphics calculators or computer graphing software will shade the intersecting region for the solution of a system of inequalities. However, the calculator can graph the boundaries. Through practice and watching for patterns, you will be able to shade the appropriate region from a mental analysis of the conditions of the problem.

SELF-STUDY EXERCISES 20–1

1 Solve the following systems of equations by graphing.

1. $x + y = 12$
   $x - y = 2$
2. $2x + y = 9$
   $3x - y = 6$
3. $2x - y = 5$
   $4x - 2y = 8$
4. $x - y = 9$
   $3x - 3y = 27$
5. $3x + 2y = 8$
   $x + y = 2$

2 Graph each inequality and shade the portion on the graph represented by the following sets of conditions.

6. $y < 2x + 1$ and $x + y > 5$
7. $x + y \geq 3$ and $x - y \geq 2$
8. $x - 2y \leq -1$ and $x + 2y \geq 3$
9. $x + y > 4$ and $x - y > -3$
10. $3x - 2y < 8$ and $2x + y \leq -4$

20–2 SOLVING SYSTEMS OF EQUATIONS USING THE ADDITION METHOD

Learning Objectives

1 Solve a system of equations containing opposite variable terms by the addition method.

2 Solve a system of equations not containing opposite variable terms by the addition method.

3 Apply the addition method to a system of equations with no solution or with many solutions.

We can see that solving systems of equations graphically is tedious and time consuming. Also, many solutions to systems of equations are not whole numbers. Graphically, it is difficult to plot fractions or to read fractional intersection points. Therefore, we need a convenient algebraic procedure for solving systems of equations. We will introduce only the two methods most commonly used to solve a system of equations with two variables. The first method is the *addition method.*

1 Solve a System of Equations Containing Opposite Variable Terms by the Addition Method.

The addition method incorporates the concept that equals added to equals give equals. Thus, when we add two equations, the result is still an equation. However, the addition method is effective *only* if one of the variables is *eliminated* in the addition process. The following examples contain two opposite terms that add to zero when the equations are added.

EXAMPLE 1 Add the following pairs of equations.

$$\text{(a)} \quad \begin{aligned} x + y &= 7 \\ x - y &= 5 \end{aligned} \qquad \text{(b)} \quad \begin{aligned} 3x + 8y &= 7 \\ -3x - 3y &= 8 \end{aligned} \qquad \text{(c)} \quad \begin{aligned} x - 2y &= 4 \\ 3x + 2y &= 8 \end{aligned}$$

$$\text{(a)} \quad \begin{aligned} x + y &= 7 \\ \underline{x - y} &= \underline{5} \\ 2x + 0y &= 12 \\ \mathbf{2x} \phantom{{}+ 0y} &= \mathbf{12} \end{aligned} \qquad \text{(b)} \quad \begin{aligned} 3x + 8y &= 7 \\ \underline{-3x - 3y} &= \underline{8} \\ 0x + 5y &= 15 \\ \mathbf{5y} &= \mathbf{15} \end{aligned} \qquad \text{(c)} \quad \begin{aligned} x - 2y &= 4 \\ \underline{3x + 2y} &= \underline{8} \\ 4x + 0y &= 12 \\ \mathbf{4x} \phantom{{}+ 0y} &= \mathbf{12} \end{aligned}$$

In each problem in Example 1, one unknown was eliminated by the addition method. Now the remaining equation that contains only one variable can be solved to find the value of that variable in the system. Then the other variable can be found by substitution. Let's finish solving each problem in Example 1.

EXAMPLE 2 Solve the systems of equations in Example 1 by the addition method.

$$\text{(a)} \quad \begin{aligned} x + y &= 7 \qquad &&\text{Equation 1} \\ \underline{x - y} &= \underline{5} \qquad &&\text{Equation 2} \\ 2x \phantom{{}+ y} &= 12 \end{aligned}$$

$$\qquad\qquad x = 6 \qquad \text{Substitute 6 in place of } x \text{ in } \textit{either} \text{ of the given equations.}$$

$$\begin{aligned} x + y &= 7 \qquad \text{Equation 1} \\ 6 + y &= 7 \\ y &= 7 - 6 \\ y &= 1 \end{aligned}$$

The solution is $x = 6$ and $y = 1$, or (6, 1).
Check in the *other* equation:

$$\begin{aligned} x - y &= 5 \qquad \text{Equation 2} \\ 6 - 1 &= 5 \\ 5 &= 5 \end{aligned}$$

Checking in the *other* equation helps ensure that no errors have been made in the addition process or in the substitution process.

$$\text{(b)} \quad \begin{aligned} 3x + 8y &= 7 \qquad &&\text{Equation 1} \\ \underline{-3x - 3y} &= \underline{8} \qquad &&\text{Equation 2} \\ 5y &= 15 \end{aligned}$$

$$\qquad\qquad y = 3 \qquad \text{Substitute 3 for } y \text{ in the first equation.}$$

$$3x + 8y = 7 \qquad \text{Equation 1}$$
$$3x + 8(3) = 7$$
$$3x + 24 = 7$$
$$3x = 7 - 24$$
$$3x = -17$$
$$x = \frac{-17}{3}$$

The solution is $(-\frac{17}{3}, 3)$.

Check in the other equation:

$$-3x - 3y = 8 \qquad \text{Equation 2}$$
$$-3\left(\frac{-17}{3}\right) - 3(3) = 8$$
$$17 - 9 = 8$$
$$8 = 8$$

(c) $\quad x - 2y = 4 \qquad \text{Equation 1}$

$\quad \underline{3x + 2y = 8} \qquad \text{Equation 2}$

$\quad 4x \qquad\quad = 12$

$\qquad\qquad x = 3 \qquad \text{Substitute 3 for } x \text{ in the first equation.}$

$$x - 2y = 4 \qquad \text{Equation 1}$$
$$3 - 2y = 4$$
$$-2y = 4 - 3$$
$$-2y = 1$$
$$y = -\frac{1}{2}$$

The solution is $(3, -\frac{1}{2})$.

Check:

$$3x + 2y = 8 \qquad \text{Equation 2}$$
$$3(3) + 2\left(-\frac{1}{2}\right) = 8$$
$$9 + -1 = 8$$
$$8 = 8$$

[2] Solve a System of Equations Not Containing Opposite Variable Terms by the Addition Method.

Remember, in the addition method one of the unknowns must be eliminated; that is, the terms must add to 0. We can also use the property that both sides of an equation may be multiplied by the same number so that the equality is preserved and the value of the variable remains the same. Thus, if neither pair of variable terms in a system of equations is opposite terms that will add to 0, we multiply one or both of the equations by numbers that will *cause* the terms of one of the variables to add to 0.

> **Rule 20-3** *To solve a system of equations using the addition method (elimination):*
>
> 1. If necessary, multiply one or both equations by numbers that will cause the terms of one of the variables to add to zero.
> 2. Add the two equations to eliminate a variable.
> 3. Solve the equation from Step 2.
> 4. Substitute the solution from Step 3 in either equation and solve for the remaining variable.
> 5. Check in both equations.

EXAMPLE 3 Solve the system of equations: $2x + y = 7$ and $x + y = 3$.

In this system, neither of the like variable terms will immediately add to 0. Several different strategies can be used. If the x in the second equation had a coefficient of -2, the x terms would add to 0. Or if the y term in either the first or second equation had a coefficient of -1, the y terms would add to 0. Thus, we have at least three choices: multiply the first equation by -1, multiply the second equation by -1, or multiply the second equation by -2.

Choice 1.
$$-2x - y = -7 \qquad \text{Multiply the first equation by } -1.$$
$$\underline{x + y = 3} \qquad \text{Add the equations.}$$
$$-x = -4$$
$$x = 4$$

Choice 2.
$$2x + y = 7$$
$$\underline{-x - y = -3} \qquad \text{Multiply the second equation by } -1 \text{ and add the equations.}$$
$$x = 4$$

Choice 3.
$$2x + y = 7$$
$$\underline{-2x - 2y = -6} \qquad \text{Multiply the second equation by } -2 \text{ and add the equations.}$$
$$-y = 1$$
$$y = -1$$

In choices 1 and 2, substitute $x = 4$.
$$-2x - y = -7$$
$$-2(4) - y = -7$$
$$-8 - y = -7$$
$$-y = -7 + 8$$
$$-y = 1$$
$$y = -1$$

In choice 3, substitute $y = -1$.
$$2x + y = 7$$
$$2x + (-1) = 7$$
$$2x - 1 = 7$$
$$2x = 7 + 1$$
$$2x = 8$$
$$x = 4$$

The solution is $(4, -1)$.

Check the solution in the first original equation.
$$2x + y = 7$$
$$2(4) + (-1) = 7$$

Check the solution in the second original equation.
$$x + y = 3$$
$$4 + (-1) = 3$$

$$8 + (-1) = 7 \qquad\qquad 3 = 3$$
$$7 = 7$$

Tips and Traps *Importance of Checking Solutions:*
When an original equation is altered, it is very important to check your solution in *both* original equations. This will enable you to identify mistakes such as forgetting to multiply *each* term in the equation by a number.

EXAMPLE 4 Solve the system of equations: $x = 3y + 7$ and $2x + 3y = 2$.

In this system the first equation does not have both letter terms on one side of the equation. Our standard form for solving systems of equations using the addition method requires that the letter terms be on one side and the number term on the other. Also, the letter terms in both equations need to be in the same order so that like terms will be in columns. Thus, we will rearrange the first equation.

$$x = 3y + 7 \rightarrow x - 3y = 7$$

Now, we will continue to solve the system.
  Add to eliminate the y terms.

$$
\begin{aligned}
x - 3y &= 7 \qquad \text{Equation 1}\\
\underline{2x + 3y} &= \underline{2} \qquad \text{Equation 2}\\
3x \phantom{{}+ 3y} &= 9\\
x &= 3
\end{aligned}
$$

$$
\begin{aligned}
x - 3y &= 7 \qquad \text{Substitute } x = 3.\\
3 - 3y &= 7\\
-3y &= 7 - 3\\
-3y &= 4\\
y &= -\frac{4}{3}
\end{aligned}
$$

The solution is $(3, -\frac{4}{3})$.

Check the solution in the original equations:

Equation 1:

$$x = 3y + 7$$
$$3 = 3\left(-\frac{4}{3}\right) + 7$$
$$3 = -4 + 7$$
$$3 = 3$$

Equation 2:

$$2x + 3y = 2$$
$$2(3) + 3\left(-\frac{4}{3}\right) = 2$$
$$6 + (-4) = 2$$
$$2 = 2$$

EXAMPLE 5 Solve the system of equations: $2x + 3y = 1$ and $3x + 4y = 2$.

In this system there is no number that we can multiply just one equation by so that a letter will be eliminated. Therefore, we need to multiply each equation by some number. There are several possibilities; we will look at just two.

Choice 1.   We multiply the first equation by -3 so that the coefficient of x will be -6. Then we multiply the second equation by $+2$ so that the coefficient of x will be $+6$. We then have

$$
\begin{array}{ll}
-6x - 9y = -3 & \text{-3 (Equation 1)} \\
\underline{6x + 8y = 4} & \text{2 (Equation 2)} \\
-y = 1 & \\
\ \boldsymbol{y = -1} &
\end{array}
$$

$$
\begin{array}{ll}
2x + 3y = 1 & \text{Substitute } y = -1. \\
2x + 3(-1) = 1 & \\
2x - 3 = 1 & \\
2x = 1 + 3 & \\
2x = 4 & \\
\boldsymbol{x = 2} &
\end{array}
$$

Choice 2.   To eliminate y, we could multiply the first equation by 4 and the second equation by -3. There is no real preference between choices 1 and 2, so we will complete choice 2.

$$
\begin{array}{ll}
8x + 12y = 4 & \text{4 (Equation 1)} \\
\underline{-9x - 12y = -6} & \text{-3 (Equation 2)} \\
-\ x = -2 & \\
\boldsymbol{x = 2} &
\end{array}
$$

$$
\begin{array}{ll}
2x + 3y = 1 & \text{Substitute } x = 2. \\
2(2) + 3y = 1 & \\
3y = 1 - 4 & \\
3y = -3 & \\
\boldsymbol{y = -1} &
\end{array}
$$

The solution is $(2, -1)$.

Check the solution in the original equations:

$$
\begin{array}{ll}
2x + 3y = 1 & 3x + 4y = 2 \\
2(2) + 3(-1) = 1 & 3(2) + 4(-1) = 2 \\
4 + (-3) = 1 & 6 + (-4) = 2 \\
1 = 1 & 2 = 2
\end{array}
$$

3 Apply the Addition Method to a System of Equations with no Solution or with Many Solutions.

In Section 20–1, we saw that there are situations when there is no solution to a system of equations. In one situation, the graphs of the two equations did not intersect and, in the other, the graphs of the two equations coincided. We will look at examples of these situations again and attempt to solve them using the addition method.

EXAMPLE 6   Solve the system $x + y = 7$ and $x + y = 5$.

To eliminate the x terms, let's multiply the second equation by -1 and add the equations.

$$x + y = 7 \qquad \text{Equation 1}$$
$$\underline{-x - y = -5} \qquad \text{-1 (Equation 2)}$$
$$0 = 2$$

Notice, both variables are eliminated and the resulting equation, $0 = 2$, is *false*. **Thus, there are no solutions to this system.**

Tips and Traps   *Inconsistent Equations—No Solution:*
When solving a system of equations, if both variables are eliminated and the resulting statement is false, then the equations are *inconsistent* and have no solution.

EXAMPLE 7   Solve the system $2x - y = 7$ and $4x - 2y = 14$.

The y's can be eliminated by multiplying the first equation by -2 and adding the equations.

$$-4x + 2y = -14 \qquad \text{-2 (Equation 1)}$$
$$\underline{4x - 2y = 14} \qquad \text{Equation 2}$$
$$0 = 0$$

Again, both variables are eliminated; however, this time the result is a *true* statement ($0 = 0$). In this situation, **all solutions of one equation are also solutions of the other equation.**

Tips and Traps   *Dependent Equations—All Solutions in Common:*
When solving a system of equations, if both variables are eliminated and the resulting statement is true, then the equations are *dependent* and have many solutions. The equations are equivalent.

SELF-STUDY EXERCISES 20-2

1 Solve the following systems of equations using the addition method.

1. $a - 2b = 7$
$3a + 2b = 13$

2. $3m + 4n = 8$
$2m - 4n = 12$

3. $x - 4y = 5$
$-x - 3y = 2$

4. $a - b = 6$
$2a + b = 3$

5. $x + 2y = 5$
$3x - 2y = 3$

2 Solve the following systems of equations using the addition method.

6. $3x + y = 9$
$x + y = 3$

7. $7x + 2y = 17$
$y = 3x + 2$

8. $a + 6b = 18$
$4a - 3b = 0$

9. $3x + y = -1$
$4x - 2y = -8$

10. $a = 6y$
$2a - y = 11$

③ Solve the following systems of equations using the addition method.

11. $x + y = 8$
   $x + y = 3$

12. $x + 2y = 9$
   $x + y = 3$

13. $3x - 2y = 6$
   $9x - 6y = 18$

14. $2a + 4b = 10$
   $a + 2b = 5$

15. $3a - b = 14$
   $a - 3b = 2$

20-3   SOLVING SYSTEMS OF EQUATIONS BY THE SUBSTITUTION METHOD

Learning Objective

1   Solve a system of equations by the substitution method.

1   Solve a System of Equations by the Substitution Method.

Another method for solving systems of equations is by substitution. Recall that in formula rearrangement (Section 13–2), whenever more than one variable is used in an equation or formula, we can rearrange the equation or solve for a particular variable. In the *substitution method* for solving systems of equations, we solve one equation for one of the variables and then substitute the equivalent expression in place of the variable in the other equation.

> **Rule 20–4**   *To solve a system of equations by substitution:*
>
> 1. Rearrange either equation to isolate one of the variables.
> 2. Substitute the equivalent expression from Step 1 into the *other* equation and solve for the variable.
> 3. Substitute the solution from Step 2 into the equation from Step 1 to find the value of the other variable.
> 4. Check both original equations.

EXAMPLE 1   Solve the following system of equations using the substitution method.

$$x + y = 15$$
$$y = 2x$$

$x + y = 15$

$\boxed{y = 2x}$ The second equation is already solved for *y* (Step 1).

$x + \boxed{y} = 15$ Substitute 2*x* for *y* in the first equation and solve (Step 2).

$x + \boxed{2x} = 15$

$3x = 15$

$\boxed{x = 5}$

$y = 2\boxed{x}$ Substitute the solution for *x* in the second equation to find *y* (Step 3).

$y = 2(\boxed{5})$

$\boxed{y = 10}$

The solution is (5, 10).
   Check:

$\boxed{x} + \boxed{y} = 15$ $\boxed{y} = 2\boxed{x}$

$\boxed{5} + \boxed{10} = 15$ $\boxed{10} = 2(\boxed{5})$

$15 = 15$ $10 = 10$ The solution checks in both equations.

EXAMPLE 2 Solve the following system of equations using the substitution method.

$$2x - 3y = -14$$
$$x + 5y = \ \ \ 19$$

When using the substitution method, either equation can be solved for either unknown. In this example, the x term in the second equation has a coefficient of 1, so the simplest choice would be to solve the second equation for x.

Step 1	Step 2	Step 3
$x + 5y = 19$	$2x - 3y = -14$	$x = 19 - 5y$
$x = 19 - 5y$	$2(19 - 5y) - 3y = -14$	$x = 19 - 5(4)$
	$38 - 10y - 3y = -14$	$x = 19 - 20$
	$38 - 13y = -14$	$x = -1$
	$-13y = -14 - 38$	
	$-13y = -52$	
	$\dfrac{-13y}{-13} = \dfrac{-52}{-13}$	
	$y = 4$	

The solution is $(-1, 4)$.
Check the roots $x = -1$, $y = 4$ in both original equations.

$$2x - 3y = -14 \qquad x + 5y = 19$$
$$2(-1) - 3(4) = -14 \qquad -1 + 5(4) = 19$$
$$-2 - 12 = -14 \qquad -1 + 20 = 19$$
$$-14 = -14 \qquad 19 = 19$$

SELF-STUDY EXERCISES 20–3

1 Solve the following systems of equations using the substitution method.

1. $2a + 2b = 60$
$a = 10 + b$

2. $7r + c = 42$
$3r - 8 = c$

3. $x - 35 = -2y$
$3x - 2y = 17$

4. $x + y = 12$
$x = 2 + y$

5. $2p + 3k = 2$
$2p - 3k = 0$

6. $x + 2y = 5$
$x = 3y$

MATHEMATICS IN THE WORKPLACE

Business Management: Supply and Demand Equations

An application of systems of equations important to business management concerns supply and demand. Supply is the quantity of goods manufacturers will produce at a specific price. As the price increases, supply usually increases. Demand is the quantity of goods consumers will buy at a specific price. Typically, as the price increases, demand decreases. By studying past records of a company's supply and demand amounts at various prices, financial analysts can construct an equation in two variables that is an approximate model of the supply or demand for a given product. These equations may be linear or nonlinear, but the graph of the supply equation is usually increasing (uphill) and the graph of the demand equation is usually decreasing (downhill). The point at which the graphs intersect is called the equilibrium point, and reaching that point is the goal of business management. The equilibrium point is the price at which manufacturers produce almost exactly the same quantity of goods as the public is willing to buy at that price.

Finding the equilibrium point is done by solving the supply and demand system of equations. To the left of the equilibrium point on the graph, demand is higher than supply and shortages result. Remember the "Tickle Me Elmo Doll" shortage of Christmas 1996? To the right of the equilibrium point on the graph, supply is higher than demand, which results in too much inventory for both manufacturers and retail stores. To clear inventory space, prices will be reduced, which results in lower profits for all. Both sides of the equilibrium point are to be avoided by business management.

An increase in price most often produces an increase in the supply of an item. There are a few cases, however, in which an increase in price will produce a decrease in supply. An example is labor. In developing countries with few consumer goods to purchase, an increase in labor wage rates can lead to workers putting in fewer hours (while earning the same amount of money). Other examples are diamonds, wine, perfume, gold, and oil production. When the worldwide market price of, say, diamonds increases, some countries with limited natural resources restrict diamond mining to conserve their scarce resources and to keep the price of diamonds high. Due to increased market prices, diamond mining companies make the same amount of profit, shelter diamond resources, maintain high prices, and give their workers more leisure time. These situations result in backward-bending supply graphs that start out to be in-

creasing but then turn left and bend back almost like an arrow point: ⟩.

Demand graphs can also be backward bending: ⟩. Although usually an increase in price results in lower demand, occasionally a price increase causes greater demand. For example, when a certain line of collectible dolls was first introduced at a midrange price, demand was lukewarm. The manufacturer pulled the dolls from the market and reissued them two years later at more than double the price. Only a few upscale stores carried the line at the higher price. But within six months the popularity of these now prestigious dolls drove the demand way up.

In the opposite scenario, sometimes a significant decrease in product price *decreases* demand instead of increasing demand, as expected. "Everything under $1" stores are an example—many of their products are half the price of the *identical item* at another store, but many people don't want to shop at a dollar store. They sometimes believe that something so inexpensive isn't worth buying because they deserve and/or can afford better, there must be something wrong with it, or everybody will have one and they want something different.

Finding the equilibrium point with backward-bending supply or demand graphs requires graphing because these graphs typically come from piecewise (multipart) equations. The intersection of one or two backward-bending graphs can result in more than one equilibrium point.

20–4 PROBLEM SOLVING USING SYSTEMS OF EQUATIONS

Learning Objective

1 Use a system of equations to solve application problems.

1 Use a System of Equations to Solve Application Problems.

Many job-related problems can be solved by setting up and solving systems of equations such as in Example 1. We will look at some additional situations.

EXAMPLE 1 A radio–television repair person purchased 10 pairs of rabbit-ear antennas and four power supply lines for $48. Soon after, the repair person purchased another three pairs of rabbit-ear antennas and five power supply lines for $22. How much did each antenna and each power supply line cost?

Known facts 10 pairs of antennas and 4 power supply lines cost $48
3 antennas and 5 power supply lines cost $22

Unknown facts Cost of one antenna (a)
Cost of one power supply line (l)

Relationships Total cost = number of items × cost of each item

$10a + 4l = \$48$ Equation 1

$3a + 5l = \$22$ Equation 2

In the first pair, 14 items cost $48, which averages to less than $4 per item. In the second pair, 8 items cost $22, which also averages to less than $4 per item.

Solution

$$10a + 4l = \$48$$

Eliminate a by multiplying Equation 1 by 3 and equation 2 by -10.

$$3a + 5l = \$22$$

$$30a + 12l = 144$$

Add equations and solve.

$$\underline{-30a - 50l = -220}$$

$$-38l = -76$$

$$l = \frac{-76}{-38}$$

$$\boxed{l = 2}$$

A power supply line costs $2.00. Substituting into equation 1, we get

$$10a + 4l = 48$$

$$10a + 4(2) = 48$$

$$10a + 8 = 48$$

$$10a = 48 - 8$$

$$10a = 40$$

$$a = \frac{40}{10}$$

$$\boxed{a = 4}$$

Interpretation

Rabbit-ear antennas cost $4.00 each and a supply line costs $2.00.

EXAMPLE 2

A restaurant ordered three hampers of blue crabs and one hamper of shrimp that together weighed 89 lb. In another order, two hampers of shrimp and five hampers of blue crabs weighed 160 lb. How much did one hamper of crabs weigh and how much did one hamper of shrimp weigh?

Known facts

3 hampers of blue crabs and 1 hamper of shrimp weigh 89 lb
5 hampers of blue crabs and 2 hampers of shrimp weigh 160 lb

Unknown facts

What is the weight of one hamper of blue crabs (c)?
What is the weight of one hamper of shrimp (s)?

Relationships

Total weight = number of containers $\times$ weight of each container

$$3c + 1s = 89 \qquad \text{Equation 1}$$

$$5c + 2s = 160 \qquad \text{Equation 2}$$

Estimation

In the first pair, 4 hampers weigh 89 lbs, which averages to more than 20 lb per hamper.
In the second pair, 7 hampers weigh 160 lb, which also averages to more than 20 lb per hamper.

Solution

$$3c + 1s = 89 \qquad \text{Eliminate } s \text{ by multiplying Equation 1 by } -2.$$

$$5c + 2s = 160$$

$$-6c - 2s = -178$$

$$\underline{5c + 2s = 160}$$

$$-c = -18$$

$$c = 18$$

$$5c + 2s = 160 \qquad \text{Substitute for } c \text{ into Equation 1.}$$

$$5(18) + 2s = 160$$

$$90 + 2s = 160$$

$$2s = 160 - 90$$

$$2s = 70$$

$$s = \frac{70}{2}$$

$$s = 35$$

Interpretation

One hamper of blue crabs weighed 18 lb, and one hamper of shrimp weighed 35 lb.

EXAMPLE 3 Two dry cells connected in series have a total internal resistance of 0.09 ohm. The difference between the internal resistance of each dry cell is 0.03 ohm. How much is each internal resistance?

Known facts   The total internal resistance of the two dry cells is 0.09 ohm
The difference in internal resistance of the two dry cells is 0.03 ohm

Unknown facts   What is the amount of resistance of dry cell 1 (r_1)?
What is the amount of resistance of dry cell 2 (r_2)?

Relationships $r_1 + r_2 = 0.09 \qquad$ Equation 1

$r_1 - r_2 = 0.03 \qquad$ Equation 2

Estimation   If resistances were the same, they would each be 0.045 ohm. Because they are not the same, one will be more than 0.045 and one will be less than 0.045.

Solution

$$r_1 + r_2 = 0.09$$

$$\underline{r_1 - r_2 = 0.03} \qquad \text{Add and solve.}$$

$$2r_1 \quad = 0.12$$

$$r_1 = \frac{0.12}{2}$$

$$r_1 = 0.06$$

$$r_1 + r_2 = 0.09 \qquad \text{Substitute for } r_1 \text{ into Equation 1.}$$

$$0.06 + r_2 = 0.09$$

$$r_2 = 0.09 - 0.06$$

$$r_2 = 0.03$$

Interpretation

Thus, the larger internal resistance is 0.06 ohm and the smaller internal resistance is 0.03 ohm.

EXAMPLE 4 A tank holds a solution that is 10% herbicide. Another tank holds a solution that is 50% herbicide. If a farmer wants to mix the two solutions to get 200 gallons of a solution that is 25% herbicide, how many gallons of each solution should be mixed?

Known facts   There are two strengths of herbicide, 10% and 50%.
200 gallons of 25% herbicide are needed.

Unknown facts   How many gallons of 10% herbicide (h) are needed?
How many gallons of 50% herbicide (H) are needed?

Relationships

200 gallons of the new herbicide are needed: $h + H = 200$
Amount of pure herbicide in h gallons of 10% herbicide: $0.10h$
Amount of pure herbicide in H gallons of 50% herbicide: $0.50H$
Amount of pure herbicide in 200 gallons of 25% herbicide: $0.25(200)$

$$h + H = 200 \qquad \text{Equation 1 (total gallons)}$$

$$0.10h + 0.50H = 0.25(200) \qquad \text{Equation 2 (gallons of pure herbicide)}$$

Estimation

If equal amounts of herbicide were needed, we would need 100 gallons of each solution. However, since the desired solution strength is not exactly halfway between the two original herbicide strengths, we will need unequal amounts of herbicide. One amount will be less than 100 gallons and the other will be more than 100 gallons.

Solution

Solve by the substitution method.

$$h + H = 200 \qquad \text{Solve Equation 1 for } h.$$

$$h = 200 - H \qquad \text{Substitute into Equation 2.}$$

$$0.10h + 0.50H = 0.25(200)$$

$$0.10(200 - H) + 0.50H = 0.25(200) \qquad \text{Substitute into Equation 2: } h = 200 - H.$$

$$20 - 0.10H + 0.50H = 50$$

$$20 + 0.40H = 50$$

$$0.40H = 50 - 20$$

$$0.40H = 30$$

$$H = \frac{30}{0.40}$$

$$H = 75 \text{ gal}$$

$$h + H = 200 \qquad \text{Substitute 75 for } H \text{ in Equation 1.}$$

$$h + 75 = 200$$

$$h = 200 - 75$$

$$h = 125 \text{ gal}$$

Interpretation

The farmer must mix 75 gallons of the 50% solution and 125 gallons of the 10% solution to make 200 gallons of a 25% solution.

EXAMPLE 5

Rosita has $5500 to invest and for tax purposes wants to earn exactly $500 interest for 1 year. She wants to invest part at 10% and the remainder at 5%. How much must she invest at each interest rate to earn exactly $500 interest in 1 year?

Let x = the amount invested at 10%. Let y = the amount invested at 5%. Interest for one year = rate × amount invested. Remember to convert percents to decimals. Using these relationships, we derive a system of equations.

Known facts

Total of $5500 to be invested
$500 interest to be earned

Unknown facts

How much should be invested at 10%?
How much should be invested at 5%?

Relationships

Amount invested at 10%: x
Interest earned at 10%: $0.1x$
Amount invested at 5%: y
Interest earned at 5%: $0.05y$

$$x + y = 5500 \qquad \text{Equation 1 (total investment)}$$
$$0.1x + 0.05y = 500 \qquad \text{Equation 2 (total interest)}$$

Estimation If the total amount were invested at 10%, the interest would be $550 (0.1 × $5500). Since we want $500 in interest, most of the money will need to be invested at 10%.

Solution Solve by the substitution method.

$$x + y = 5500 \qquad \text{Solve Equation 1 for } x.$$

$$x = 5500 - y \qquad \text{Substitute into Equation 2.}$$

$$0.1x + 0.05y = 500$$

$$0.1(5500 - y) + 0.05y = 500 \qquad \text{Substitute } 5500 - y \text{ for } x.$$

$$550 - 0.1y + 0.05y = 500$$

$$550 - 0.05y = 500 \qquad -0.10 + 0.05 = -0.05$$

$$-0.05y = 500 - 550$$

$$-0.05y = -50$$

$$y = \frac{-50}{-0.05}$$

$$y = \$1000 \text{ at } 5\%$$

$$x + y = \$1000$$

$$x + 1000 = 5500 \qquad \text{Substitute } \$1000 \text{ for } y \text{ in Equation 1.}$$

$$x = 5500 - 1000$$

$$x = \$4500 \text{ at } 10\%$$

Interpretation **Rosita must invest $4500 at 10% and $1000 at 5% for 1 year to earn $500 interest.**

SELF-STUDY EXERCISES 20-4

1 Solve the following word problems using systems of equations with two unknowns.

1. Two boards together are 48 in. If one board is 17 in. shorter than the other, find the length of each board.

2. A broker invested $35,000 in two different stocks. One earned dividends at 4% and the other at 5%. If a $1570 dividend was earned on both stocks together, how much was invested in each? (*Reminder:* Change 4% to 0.04 and 5% to 0.05.)

3. A department store buyer ordered 12 shirts and 8 hats for $180 one month and 24 shirts and 10 hats for $324 the following month. What was the cost of each shirt and each hat?

4. A mechanic makes $15 on each 8-cylinder-engine tune-up and $10 on each 4-cylinder-engine tune-up. If the mechanic did 10 tune-ups and made a total of $135, how many 8-cylinder jobs and how many 4-cylinder jobs were completed?

5. Thirty resistors and 15 capacitors cost $12. Ten resistors and 20 capacitors cost $8.50. How much does each capacitor and resistor cost?

6. A private airplane flew 420 miles in 3 hours with the wind. The return trip took 3.5 hours. Find the rate of the plane in calm air and the rate of the wind.

7. In 1 year, Doug Call earned $660 in interest on two investments totaling $8000. If he received 7% and 9% rates of return, how much did he invest at each rate?

8. A motorboat went 40 miles with the current in 3 hours. The return trip against the current took 4 hours. How fast was the current? What would have been the speed of the boat in calm water?

9. A visitor to south Louisiana purchased 3 lb of dark-roast pure coffee and 4 lb of coffee with chicory for $15.10 in a local supermarket. Another visitor at the same store purchased 2 lb of coffee with chicory and 5 lb of dark-roast pure coffee for $16.30. How much did each coffee cost per pound?

10. A lawn-care technician wants to spread a 200-lb seed mixture that is 50% bluegrass. If the technician has on hand a mixture that is 75% bluegrass and a mixture that is 10% bluegrass, how many pounds of each mixture are needed to make 200 lb of the 50% mixture? Round to the nearest whole lb.

11. A college bookstore received a partial shipment of 50 scientific calculators and 25 graphics calculators at a total cost of $1300. Later the bookstore received the balance of the calculators: 25 scientific and 50 graphics at a cost of $1775. Find the cost of each calculator.

12. A consumer received two 1-yr loans totaling $10,000 at interest rates of 10% and 15%. If the consumer paid $1300 interest, how much money was borrowed at each rate?

13. For the first performance at the Overton Park Shell, 40 reserved seats and 80 general admission seats were sold for $2000. For the second performance, 50 reserved seats and 90 general admission seats were sold for $2350. What was the cost for a reserved seat and for a general admission seat?

14. A photographer has a container with a solution of 75% developer and a container with a solution of 25% developer. If she wants to mix the solutions to get 8 pt of solution with 50% developer, how many pints of each solution does she need to mix?

CAREER APPLICATION

Electronics: Mesh Currents

Circuits are frequently solved using mesh currents to determine individual currents in a circuit. For instance, the analysis of a circuit with three currents might yield the following system of equations for I_1, I_2, and I_3.

$$I_1 - I_2 - I_3 = 0 \qquad \text{Equation 1}$$

$$6I_1 + 4I_3 = 12 \text{ A} \qquad \text{Equation 2}$$

$$3I_2 - 4I_3 = -3 \text{ A} \qquad \text{Equation 3}$$

There are several ways to solve this system. One way is to solve the first equation for I_1 in terms of I_2 and I_3 and substitute that into the second equation. Then the second and third equations will form a pair of equations in two unknowns, which are easy to solve using many methods. The proof is to put all the final values back into the original equation to see if they work.

$$I_1 - I_2 - I_3 = 0 \qquad \text{Rearrange Equation 1 to give } I_1 = I_2 + I_3$$

Substituting that expression for I_1 into the second equation gives

$$6I_1 + 4I_3 = 12 \qquad \text{Equation 2}$$

$$6(I_2 + I_3) + 4I_3 = 12$$

$$6I_2 + 10I_3 = 12$$

$$3I_2 + 5I_3 = 6 \qquad \text{Divide each term by 2.}$$

This gives the following pair from the modified Equation 2 and the original Equation 3. (Illustrated is the addition method of solving a pair of equations.)

New Equation 2: $3I_2 + 5I_3 = 6$ A $\qquad$ $3I_2 + 5I_3 = 6$ $\qquad$ Substituting into Equation 3:

Equation 3: $3I_2 - 4I_3 = -3$ A $\qquad$ $\underline{-3I_2 + 4I_3 = 3}$ $\qquad$ $3I_2 - 4I_3 = -3$ A

$$9I_3 = 9$$

$$3I_2 - 4(1\ \text{A}) = -3\ \text{A}$$

$$\boxed{I_3 = 1\ \text{A}}$$

$$3I_2 = -3\ \text{A} + 4\ \text{A}$$

$$3I_2 = 1\ \text{A}$$

$$I_2 = 0.333333\ \text{A}$$

Then go back to rearranged Equation 1 and substitute values for I_2 and I_3.

$$I_1 = I_2 + I_3 = 0.333333\ \text{A} + 1\ \text{A} = +1.333333\ \text{A}$$

Thus, the solution is $I_1 = 1.333333$ A, $I_2 = 0.333333$ A, and $I_3 = 1$ A.

Proof:

Equation 1

$$I_1 - I_2 - I_3 = 0$$

$$1.333333\ \text{A} - 0.333333\ \text{A} - 1\ \text{A} = 0$$

$$0 = 0$$

Equation 2

$$6I_1 + 4I_3 = 12\ \text{A}$$

$$6(1.333333\ \text{A}) + 4(1\ \text{A}) = 12\ \text{A}$$

$$12\ \text{A} = 12\ \text{A}$$

Equation 3

$$3I_2 - 4I_3 = -3\ \text{A}$$

$$3(0.333333\ \text{A}) - 4(1\ \text{A}) = -3\ \text{A}$$

$$-3\ \text{A} = -3\ \text{A}$$

Exercises

Solve each of the following systems of equations, which are derived from actual circuits. Prove your answers by going back to the original equations.

1. Set A: $\quad I_1 - 2I_2 + 3I_3 = 4$ A
   $\qquad 2I_1 + I_2 - 4I_3 = 3$ A
   $\qquad\qquad\quad I_1 + 2I_3 = 8$ A

2. Set B: $\quad I_1 + I_2 + I_3 = -4$ A
   $\qquad 2I_1 + 3I_2 + 4I_3 = 0$
   $\qquad -I_1 - I_2 + 2I_3 = 8$ A

3. Set C: $\quad I_1 + I_2 + I_3 = +4$ A
   $\qquad 2I_1 + 3I_2 + 4I_3 = 0$
   $\qquad -I_1 - I_2 + 2I_3 = 8$ A

4. Set D: $\quad 3I_1 + 2I_2 + 2I_3 = 3$ A
   $\qquad 2I_1 + 6I_2 + 3I_3 = 0$
   $\qquad\quad I_1 + 2I_2 + I_3 = 1$ A

5. Set E: $\quad I_1 - I_2 + 3I_3 = 2$ A
   $\qquad\quad -I_1 + I_2 = 7$ A
   $\qquad -I_1 + 2I_2 + 6I_3 = 4$ A

Answers for Exercises

Set	Solutions	Proof
1. Set A: $I_1 - 2I_2 + 3I_3 = 4$ A $2I_1 + I_2 - 4I_3 = 3$ A $I_1 + 2I_3 = 8$ A	$I_1 = 4$ A $I_2 = 3$ A $I_3 = 2$ A	$4 - 6 + 6 = 4$ $8 + 3 - 8 = 3$ $4 + 4 = 8$
2. Set B: $I_1 + I_2 + I_3 = -4$ A $2I_1 + 3I_2 + 4I_3 = 0$ $-I_1 - I_2 + 2I_3 = 8$ A	$I_1 = -10.667$ A $I_2 = 5.333$ A $I_3 = 1.333$ A	$-10.667 + 5.333 + 1.333 = -4$ $-21.333 + 16 + 5.333 = 0$ $10.667 - 5.333 + 2.667 = 8$ A
3. Set C: $I_1 + I_2 + I_3 = +4$ A $2I_1 + 3I_2 + 4I_3 = 0$ $-I_1 - I_2 + 2I_3 = 8$ A	$I_1 = 16$ A $I_2 = -16$ A $I_3 = 4$ A	$16 - 16 + 4 = 4$ $32 - 48 + 16 = 0$ $-16 + 16 + 8 = 8$
4. Set D: $3I_1 + 2I_2 + 2I_3 = 3$ A $2I_1 + 6I_2 + 3I_3 = 0$ $I_1 + 2I_2 + I_3 = 1$ A	$I_1 = 3$ A $I_2 = 1$ A $I_3 = -4$ A	$9 + 2 - 8 = 3$ $6 + 6 - 12 = 0$ $3 + 2 - 4 = 1$
5. Set E: $I_1 - I_2 + 3I_3 = 2$ A $-I_1 + I_2 = 7$ A $-I_1 + 2I_2 + 6I_3 = 4$ A	$I_1 = -28$ A $I_2 = -21$ A $I_3 = 3$ A	$-28 + 21 + 9 = 2$ $28 - 21 = 7$ $28 - 42 + 18 = 4$

ASSIGNMENT EXERCISES

Section 20–1

Solve the following systems of equations by graphing.

1. $x + y = 8$
$x - y = 2$

2. $3x + 2y = 13$
$x - 2y = 7$

3. $2x + 2y = 10$
$3x + 3y = 15$

4. $x + y = 1$
$3x - 4y = 10$

5. $2x - y = 5$
$4x - 2y = 2$

Shade the portion of the graph represented by the following sets of conditions.

6. $y \geq x + 3$ and $x + y < 4$

7. $2x + y < 6$ and $x - y < 1$

8. $x + 2y < -1$ and $x + 2y > 3$

9. $2x + y > 3$ and $x - y \leq 1$

10. $x + 2y > -4$ and $x - 2y > -1$

Section 20–2

Solve the following systems of equations using the addition method.

11. $3x + y = 9$
$2x - y = 6$

12. $2a + 3b = 8$
$a - b = 4$

13. $Q = 2P + 8$
$2Q + 3P = 2$

14. $4j + k = 3$
$8j + 2k = 6$

15. $r = 2y + 6$
$2r + y = 2$

16. $3a + 3b = 3$
$2a - 2b = 6$

17. $c = 2y$
$2c + 3y = 21$

18. $2x + 4y = 9$
$x + 2y = 3$

19. $3R - 2S = 7$
$-14 = -6R + 4S$

20. $x - 3 = -y$
$2y = 9 - x$

21. $c = 2 + 3d$
$3c - 14 = d$

22. $Q - 10 = T$
$T = 2 - 2Q$

23. $x - 18 = -6y$
$4x - 0 = 3y$

24. $R + S = 3$
$S - 9 = -3R$

25. $3a - 2b = 6$
$6a - 12 = b$

26. $a - b = 2$
$a + b = 12$

27. $x + 2y = 7$
$x - y = 1$

28. $2c + 3b = 2$
$2c - 3b = 0$

29. $x + 2r = 5.5$
$2x = 1.5r$

30. $7x + 2y = 6$
$4y = 12 - x$

Section 20-3

Solve the following systems of equations using either the addition or the substitution method.

31. $a + 7b = 32$
$\quad 3a - b = 8$

32. $x + y = 1$
$\quad 4x + 3y = 0$

33. $c - d = 2$
$\quad c = 12 - d$

34. $3a + 4b = 0$
$\quad a + 3b = 5$

35. $7x - 4 = -4y$
$\quad 3x + y = 6$

36. $5Q - 4R = -1$
$\quad R + 3Q = -38$

37. $a = 2b + 11$
$\quad 3a + 11 = -5b$

38. $y = 5 - 2x$
$\quad 3x - 2y = 4$

39. $c = 2q$
$\quad 2c + q = 2$

40. $3x + 2y = 10$
$\quad y = 6 - x$

41. $4x - 2.5y = 2$
$\quad 2x - 1.5y = -10$

42. $2a - c = 4$
$\quad a = 2 + c$

43. $4d - 7 = -c$
$\quad 3c - 6 = -6d$

44. $x = 10 - y$
$\quad 5x + 2y = 11$

45. $3.5a + 2b = 2$
$\quad 0.5b = 3 - 1.5a$

46. $c + d = 12$
$\quad c - d = 2$

47. $x + 4y = 20$
$\quad 4x + 5y = 58$

48. $a + 5y = 7$
$\quad a + 4y = 8$

49. $3a + 1 = -2b$
$\quad 4b + 23 = 15a$

50. $6y + 0 = -5p$
$\quad 4y - 3p = 38$

Section 20-4

Solve the following word problems using systems of equations with two unknowns.

51. Three electricians and four apprentices earned a total of $365 on one job. At the same rate of pay, one electrician and two apprentices earned a total of $145. How much pay did each apprentice and electrician receive?

52. Six bushels of bran and 2 bushels of corn weigh 182 lb. If 2 bushels of bran and 4 bushels of corn weigh 154 lb, how much does 1 bushel of bran and 1 bushel of corn each weigh?

53. A painter paid $22.50 for 2 qt of white shellac and 5 qt of thinner. If 3 qt of shellac and 2 qt of thinner cost the painter's helper $14.50, what is the cost of each qt of shellac and thinner?

54. A main current of electricity is the sum of two smaller currents whose difference is 0.8 A. What are the two smaller currents if the main current is 10 A?

55. The sum of two angles is 175°. Their difference is 63°. What is the measure of each angle?

56. A jonboat traveled 20 m in 2 hr with the current. The return trip took 3 hr. Find the rate of the boat in calm water and the rate of the current.

57. In 1 year, Sholanda Brown earned $560 on two investments totaling $5000. If she received 10% and 12% rates of return, how much did she invest at each rate?

58. A plane flew 300 m against the wind in 4 hr. The return trip with the wind took 3 hr. How fast was the wind? What would have been the speed of the plane in calm air?

59. A restaurant purchased 30 lb of Columbian coffee and 10 lb of blended coffee for $190. In a second purchase, the same restaurant paid $120 for 20 lb of Columbian coffee and 5 lb of blended coffee. How much did each coffee cost per lb?

60. A rancher wants to spread a 300-lb grass seed mixture that is 50% tall fescue. If the rancher has on hand a seed mixture that is 80% tall fescue and a mixture that is 20% tall fescue, how many pounds of each mixture are needed to make 300 lb of the 50% mixture?

61. An automotive service station purchased 25 maps of Ohio and 8 maps of Alaska at a total cost of $65.55. Later the station purchased 20 maps of Ohio and 5 maps of Alaska for $49.50. Find the cost of each map.

62. Bev Witonski made two 1-yr investments totaling $7000 at interest rates of 8% and 12%. If she received $760 in return, how much money was invested at each rate?

63. At the first of the month, store buyer Kathy Miller placed a $12,525 order for 20 name-brand suits and 35 suits with generic labels. At the end of the month, she placed a $15,725 order for 30 name-brand suits and 35 generic-label suits. How much did she pay for each type of suit?

64. A taxidermist has a container with a solution of 10% tanning chemical and a container with a solution of 50% tanning chemical. If the taxidermist wants to mix the solutions to get 10 gal of solution with 25% tanning chemical, how many gal of each solution should be mixed?

65. Jorge makes 5% commission on telephone sales and 6% commission on showroom sales. If his sales totaled $40,000 and his commission was $2250, how much did he sell by telephone? How much did he sell on the showroom floor?

66. Sing-Fong has 60 coins in nickels and quarters. The total value of the coins is $12. How many coins of each type does she have?

67. How many gal of 75% fertilizer and 25% fertilizer are needed to make a mixture of 8 gal of 50% fertilizer?

Examine the following problems to see whether they are worked correctly. If there are errors, make corrections.

68. Solve by addition: $a - b = 6$ and $a + b = 2$.

$$\begin{array}{r} a - b = 6 \\ a + b = 2 \\ \hline 2a \quad\;\; = 8 \\ a = 4 \end{array}$$

$$\begin{aligned} 4 - b &= 6 \\ -b &= 6 - 4 \\ -b &= 2 \\ b &= -2 \end{aligned}$$

Check:
$$\begin{aligned} a - b &= 6 \\ 4 - (-2) &= 6 \\ 4 + 2 &= 6 \\ 6 &= 6 \end{aligned}$$

69. Solve by addition: $2c + 3d = 9$ and $3c + d = 10$.

$$\begin{aligned} 2c + 3d &= 9 \\ 3c + d &= 10 \qquad \text{Multiply by } -3. \\ 2c + 3d &= 9 \\ -6c - 3d &= -30 \\ \hline -4c \qquad\;\; &= -21 \\ \frac{-4c}{-4} &= \frac{-21}{-4} \\ c &= 5.25 \end{aligned}$$

$$\begin{aligned} 2c + 3d &= 9 \\ 2(5.25) + 3d &= 9 \\ 10.5 + 3d &= 9 \\ 3d &= 9 - 10.5 \\ 3d &= -1.5 \\ \frac{3d}{3} &= -\frac{1.5}{3} \\ d &= -0.5 \end{aligned}$$

CHALLENGE PROBLEMS

70. Write an applied mixture problem that can be solved with the given system of equations with two unknowns. Solve the system and check the results.

$$0.5x + 0.3y = 8$$
$$x + y = 20$$

71. Write an applied problem that can be solved with the given system of equations with two unknowns. Solve the system and check the results.

$$3x + 5y = 28$$
$$x - y = 4$$

72. Solve Problem 55 by writing one equation with one variable.

CONCEPTS ANALYSIS

1. What does the graphical solution of a system of two equations with two unknowns represent?

2. What can be said about the solution of a system of two equations if the graphs of the two equations are parallel?

3. What can be said about the solution of a system of two equations if the graphs of the two coincide?

4. What can be said about the solution of a system of two equations if the graphs of the two intersect in exactly one point?

5. How do you determine the solution to a system of inequalities graphically? Can a system of inequalities be solved algebraically?

6. Explain the addition or elimination method of solving a system of two equations with two unknowns.

7. Explain the substitution method of solving a system of two equations with two unknowns.

8. What does it mean when solving a system of two equations with two unknowns if the system produces a statement like $0 = 0$?

9. What does it mean when solving a system of two equations with two unknowns if the system produces a statement like $3 = 0$?

10. Can a system of two linear equations in two unknowns have exactly two ordered pairs as solutions? Explain your answer.

CHAPTER SUMMARY

| Objectives | What to Remember with Examples |

Section 20–1

1 Solve a system of equations by graphing.

Graph each equation. The intersection of the two lines is the solution to the system. (The table-of-solutions, intercepts, or slope–intercept methods may also be used to graph each equation.)

Graph $x + y = 2$ and $x - y = 4$.

$$x + y = 2 \qquad x - y = 4$$
$$y = -x + 2 \qquad y = x - 4$$

x	y	x	y
0	2	0	-4
1	1	1	-3
2	0	2	-2

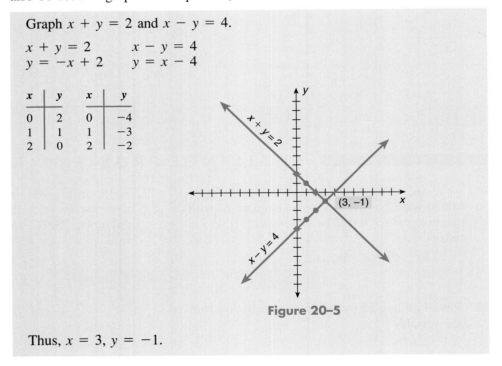

Figure 20–5

Thus, $x = 3$, $y = -1$.

2 Solve a system of inequalities by graphing.

Rewrite each inequality as an equation. Graph each equation and shade the area that satisfies each inequality. Where shaded areas overlap is the set of solutions to the system of inequalities.

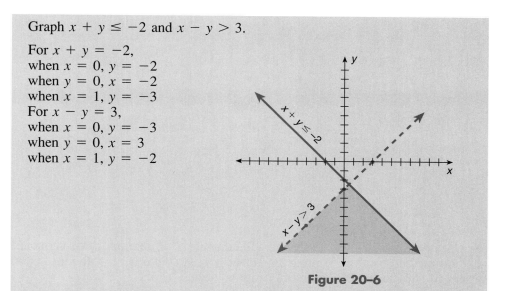

Graph $x + y \leq -2$ and $x - y > 3$.

For $x + y = -2$,
when $x = 0$, $y = -2$
when $y = 0$, $x = -2$
when $x = 1$, $y = -3$
For $x - y = 3$,
when $x = 0$, $y = -3$
when $y = 0$, $x = 3$
when $x = 1$, $y = -2$

Figure 20–6

The overlapping shaded area meets both conditions of the system of inequalities.

Section 20–2

1 Solve a system of equations containing opposite variable terms by the addition method.

To solve a system of equations with two opposite variable terms, add the equations so that the opposite terms add to zero. Solve the new equation and substitute the root in an original equation.

Solve $x - y = 6$ and $x + y = 4$ ($-y$ and $+y$ are opposites).

$$\begin{array}{r} x - y = 6 \\ x + y = 4 \\ \hline 2x = 10 \end{array}$$

$$\frac{2x}{2} = \frac{10}{2}$$

$$x = 5$$

$$x - y = 6$$
$$5 - y = 6$$
$$-y = 6 - 5$$
$$-y = 1$$
$$y = -1$$

The solution is $(5, -1)$.

2 Solve a system of equations not containing opposite variable terms by the addition method.

To solve a system of equations not containing opposite variables that will add to zero:
1. Multiply one or both equations by signed numbers that will cause the terms of one of the variables to add to zero.
2. Add the equations.
3. Solve the equation from Step 2.
4. Substitute the root in an original equation and solve for the remaining variable.
5. Check solutions.

Solve $2x - y = 6$ and $4x + 2y = 4$.

$$\begin{array}{r} 4x - 2y = 12 \\ 4x + 2y = 4 \\ \hline 8x = 16 \end{array}$$

Multiply first equation by 2.

$$\frac{8x}{8} = \frac{16}{8}$$

$$x = 2$$

$$4x + 2y = 4$$

$$4(2) + 2y = 4$$

$$8 + 2y = 4$$

$$2y = 4 - 8$$

$$2y = -4$$

$$\frac{2y}{2} = \frac{-4}{2}$$

$$y = -2$$

Solution: $(2, -2)$.

Check: $4x - 2y = 12$

$$4(2) - 2(-2) = 12$$

$$8 + 4 = 12$$

$$12 = 12$$

$$2x - y = 6$$

$$2(2) - (-2) = 6$$

$$4 + 2 = 6$$

$$6 = 6$$

3 Apply the addition method to a system of equations with no solution or with many solutions.

Apply the addition rule. But note that *both* variables are eliminated. If the resulting statement is false, there is no solution. If the resulting statement is true, there are many solutions.

Solve $a + b = 2$ and $a + b = 4$.

$$-a - b = -2 \qquad \text{Multiply first equation by } -1.$$
$$\underline{a + b = 4}$$
$$0 = 2 \qquad \text{No solutions}$$

Section 20–3

1 Solve a system of equations by the substitution method.

To solve a system of equations by substitution:
1. Rearrange either equation to isolate one of the variables.
2. Substitute the expression from Step 1 in the *other* equation and solve.
3. Substitute the root from Step 2 in the equation from Step 1 and solve for the other variable.
4. Check.

Solve $2x + y = 6$ and $2x - 2y = 8$.

$$y = 6 - 2x \qquad \text{Isolate } y \text{ in first equation.}$$

$$2x - 2(6 - 2x) = 8 \qquad \text{Substitute } 6 - 2x \text{ for } y \text{ in second equation.}$$

$$2x - 12 + 4x = 8$$

$$6x - 12 = 8$$

$$6x = 8 + 12$$

$$6x = 20$$

$$\frac{6x}{6} = \frac{20}{6}$$

$$x = \frac{10}{3}$$

$$y = 6 - 2\left(\frac{10}{3}\right)$$

$$y = \frac{18}{3} - \frac{20}{3}$$

$$y = -\frac{2}{3}$$

Check: $2x - 2y = 8$

$$2\left(\frac{10}{3}\right) - 2\left(-\frac{2}{3}\right) = 8$$

$$\frac{20}{3} + \frac{4}{3} = 8$$

$$\frac{24}{3} = 8$$

$$8 = 8$$

$2x + 6y = 6$

$$2\left(\frac{10}{3}\right) + \left(\frac{-2}{3}\right) = 6$$

$$\frac{20}{3} + \frac{-2}{3} = 6$$

$$\frac{18}{3} = 6$$

$$6 = 6$$

Solution: $\left(\frac{10}{3}, -\frac{2}{3}\right)$

1 Use a system of equations to solve application problems.

Let two variables represent the unknown amounts. Use numbers and the variables to represent the conditions of the problem. Set up a system of equations and solve.

A taxidermist bought two pairs of glass deer eyes and five pairs of glass duck eyes for $15. She later purchased three pairs of deer eyes and five pairs of duck eyes for $20. Find the price of each type of glass eye.

Let x = the price of a pair of deer eyes. Let y = the price of a pair of duck eyes. Price = the number of eyes times price of eye type:

$2x + 5y = 15$ Multiply first equation by −1.
$3x + 5y = 20$

$-2x - 5y = -15$
$\underline{3x + 5y = 20}$
$x = 5$
$2x + 5y = 15$
$2(5) + 5y = 15$
$10 + 5y = 15$
$5y = 15 - 10$
$5y = 5$
$\dfrac{5y}{5} = \dfrac{5}{5}$
$y = 1$

The deer eyes cost $5 a pair and the duck eyes cost $1 a pair.

WORDS TO KNOW

system of equations (p. 729–730)
system of inequalities (p. 731)

inconsistent system (p. 731)
independent system (p. 731)
dependent system (p. 731)

addition method (p. 734)
substitution method (p. 740)

CHAPTER TRIAL TEST

Solve the following systems of equations graphically.

1. $2a + b = 10$
$a - b = 5$

2. $x + y = 8$
$3x + 2y = 12$

3. $3x + 4y = 6$
$x + y = 5$

4. $a - 3b = 7$
$a - 5 = b$

5. $2c - 3d = 6$
$c - 12 = 3d$

Shade the portion on the graph represented by the conditions in the following problems.

6. $y \le 2x + 1$ and $x + y \ge 5$

7. $x + y < 4$ and $y > 3x + 2$

Solve the following systems of equations using the addition method.

8. $x + y = 6$
$x - y = 2$

9. $p + 2m = 0$
$2p = -m$

10. $6p + 5t = -16$
$3p - 3 = 3t$

11. $3x + y = 5$
$2x - y = 0$

12. $7c - 2b = -2$
$c - 4b = -4$

Solve the following systems of equations with two unknowns using the substitution method.

13. $4x + 3y = 14$
$x - y = 0$

14. $a + 2y = 6$
$a + 3y = 3$

15. $7p + r = -6$
$3p + r = 6$

Solve the following systems of equations with two unknowns using either the addition or the substitution method.

16. $4x + 4 = -4y$
$6 + y = -6x$

17. $38 + d = -3a$
$5a + 1 = 4d$

Solve the following word problems using systems of equations with two unknowns.

18. Two lengths of stereo speaker wire total 32.5 ft. One length is 2.9 ft longer than the other. How long is each length of speaker wire?

19. Two currents add up to 35 A and their difference is 5 A. How many amperes are in each current?

20. Six packages of common nails and four packages of finishing nails weigh 6.5 lb. If two packages of common nails and three packages of finishing nails weigh 3.0 lb, how much does one package of each kind of nail weigh?

21. The length of a piece of sheet metal is $1\frac{1}{2}$ times the width. The difference between the length and the width is 17 in. Find the length and the width.

22. A mixture of fieldstone is needed for a construction job and will cost $378 for the 27 tons of stone. The stone is of two types, one costing $18 per ton and one costing $12 per ton. How many tons of each are required?

23. A broker invested $25,000 in two different stocks. One stock earned dividends at 11% and the other at 12.5%. If a dividend of $3050 was earned on both stocks together, how much was invested in each stock?

24. A mason purchased 2-in. cold-rolled channels and $\frac{3}{4}$-in. cold-rolled channels whose total weight was 820 lb. The difference in weight between the heavier 2-in. and lighter $\frac{3}{4}$-in. channels was 280 lb. How many pounds of each type of channel did the mason purchase?

25. A total capacitance in parallel is the sum of two capacitances. If the capacitance totals 0.00027 farad (F) and the difference between the two capacitances is 0.00016 F, what is the value of each capacitance in the system?

CHAPTER

21

Selected Concepts of Geometry

GOOD DECISIONS THROUGH TEAMWORK

What is the best pizza deal in town? How do you conduct comparative shopping when different pizza stores have different size pans? Are prices for some large pizzas for a particular store proportional to the amount of pizza for each size? Does any com-

bination of two pan sizes give a better buy than a larger pan size? Is pizza available in any shapes besides circles? If so, what are the advantages and disadvantages of these shapes from a marketing viewpoint? From a consumer viewpoint?

To answer these and other questions, your team will collect statistics on pan sizes, prices, topping options, and other information affecting a decision to purchase pizza. As a class, before you begin your research, rate pizza available in your area based on quality or taste and on your initial impressions of which pizza store has the best deal in town.

To begin your research, compare cost per square inch of pizza for each pan size offered by a restaurant. Then compare cost per square inch of pizza with customer-chosen toppings versus store-established combinations of toppings. Compare cost per square inch of pizza for small, medium, and large pizzas among various restaurants.

After the information is collected, discuss your findings among your team members to determine how it will be organized for reporting team results. Charts, tables, graphs, lists, and narrative are some choices you may want to consider for a presentation.

Geometry is one of the oldest and most useful of the mathematical sciences. *Geometry* involves the study and measurement of shapes according to their sizes, volumes, and positions. A knowledge of geometry is necessary in many careers. In the following section, notice the precise meanings given to some ordinary words when they are used in geometry.

21-1 BASIC TERMINOLOGY AND NOTATION

Learning Objectives

1. Use various notations to represent points, lines, line segments, rays, and planes.
2. Distinguish among lines that intersect, that coincide, and that are parallel.
3. Use various notations to represent angles.
4. Classify angles according to size.

1 Use Various Notations to Represent Points, Lines, Line Segments, Rays, and Planes.

Geometry can be described as the study of size, shape, position, and other properties of the objects around us. The basic terms used in geometry are *point, line,* and *plane.* Generally, these terms are not defined. Instead, they are usually merely described. Once described, they may be used in definitions of other terms and concepts.

A *point* is a location or position that has no size or dimension. A dot will be used to represent a point, and a capital letter will usually be used to label the point (Fig. 21–1).

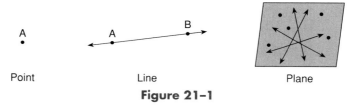

Point   Line   Plane

Figure 21-1

A *line* extends indefinitely in both directions. It has length but no width. In our discussions the word *line* will always refer to a straight line unless otherwise specified. In Fig. 21–1, the line is identified by naming any two points on the line (such as *A* and *B*).

A *plane* is a flat, smooth surface that extends indefinitely. A plane contains an infinite number of points and lines (Fig. 21–1).

Since a line extends indefinitely in both directions, most geometric applications deal with parts of lines. A part of a line is called a *line segment* or *segment.* A line segment starts and stops at distinct points that we call *end points.* A line segment is defined as follows:

Figure 21-2

■ **DEFINITION 21–1: Line Segment** or **Segment.** A *line segment,* or *segment,* consists of all points on the line between and including two points that are called end points (Fig. 21–2).

The notation for a line that extends through points *A* and *B* is $\overleftrightarrow{AB}$ (read "line *AB*"). The notation for the line segment including points *A* and *B* and all the points between is $\overline{AB}$ (read "line segment *AB*").

Another term used in connection with parts of a line is *ray.* Before the definition of a ray is given, consider the beam of light from a flashlight. The beam is like a ray. It seems to continue indefinitely in one direction.

■ **DEFINITION 21–2: Ray.** A *ray* consists of a point on a line and all points of the line on one side of the point (Fig. 21–3).

Figure 21–3

The point from which the ray originates is called the *end point,* and all other points on the ray are called *interior points* of the ray. A ray is named by its end point and any interior point on the ray. In Fig. 21–3, we use the notation $\overrightarrow{RS}$ to denote the ray whose end point is R and that passes through S.

To see the contrast in the notation used for a line, line segment, and ray, look at Fig. 21–4.

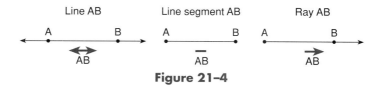

Figure 21–4

To further understand the differences among lines, segments, and rays, consider a line with several points designated on the line (Fig. 21–5).

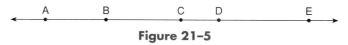

Figure 21–5

Any two points can be used to name the line in Fig. 21–5. For example, $\overleftrightarrow{AB}$, $\overleftrightarrow{AC}$, $\overleftrightarrow{CE}$, $\overleftrightarrow{BD}$, and $\overleftrightarrow{BC}$ are some of the possible ways to name the line. However, a segment is named *only* by its end points. Thus, in Fig. 21–5, $\overline{AB}$ is not the same segment as $\overline{AC}$, but $\overleftrightarrow{AB}$ and $\overleftrightarrow{AC}$ represent the same line.

Again in Fig. 21–5, $\overrightarrow{BC}$ and $\overrightarrow{BD}$ represent the same ray, but $\overline{BC}$ and $\overline{BD}$ do not represent the same segment.

Notation is important in geometry because it allows us to shorten the amount of writing needed to describe a situation, and it allows us to recognize or recall certain properties at a glance.

2 Distinguish Among Lines that Intersect, that Coincide, and that are Parallel.

A line can be extended indefinitely in either direction. If two lines are drawn in the same plane, there are three things that can possibly happen:

1. The two lines will *intersect* in *one and only one point.* In Fig. 21–6, $\overleftrightarrow{AB}$ and $\overleftrightarrow{CD}$ intersect at point E.

Figure 21–6

2. The two lines will *coincide;* that is, one line will fit exactly on the other. In Fig. 21–6, $\overleftrightarrow{EF}$ and $\overleftrightarrow{GH}$ coincide.
3. The two lines will never intersect. In Fig. 21–6, $\overleftrightarrow{IJ}$ and $\overleftrightarrow{KL}$ are the same distance from each other along their entire lengths and so never touch.

The relationship described in the third situation has a special name, *parallel lines.* The symbol ∥ is used for parallel lines. (See Section 19–4.)

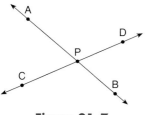

Figure 21-7

This objective deals with a special type of intersection of straight lines—the angle. Most of us are familiar with angles in everyday life. The corners of a soccer field are angles. The lines between bricks and tiles form angles, and so on. This section and the ones that follow will introduce us formally to the study of angles for technical applications.

When two lines intersect in a point, *angles* are formed, as shown in Fig. 21–7.

■ **DEFINITION 21–3: Angle.** An *angle* is a geometric figure formed by two rays that intersect in a point, and the point of intersection is the end point of each ray.

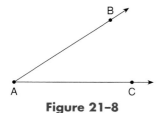

Figure 21-8

In Fig. 21–8, rays $\overrightarrow{AB}$ and $\overrightarrow{AC}$ intersect at point *A*. Point *A* is the end point of $\overrightarrow{AB}$ and $\overrightarrow{AC}$. $\overrightarrow{AB}$ and $\overrightarrow{AC}$ are called the *sides* of the angle. Point *A* is called the *vertex* of the angle.

There are several ways to name an angle. An angle can be named using a number or lowercase letter, or it can be named by the capital letter that names the vertex point, or it can be named by using three capital letters. If three capital letters are used, two of the letters will name interior points of each of the two rays, and the middle letter will name the vertex point of the angle. Using the symbol ∠ for angle, the angle in Fig. 21–9 can be named ∠1, ∠*KLM*, ∠*MLK*, or ∠*L*. If one capital letter is used to name an angle, this letter is always the vertex letter. One capital letter is used only when it is perfectly clear which angle is designated by this letter. If three letters are used to name an angle, the vertex letter will be the center letter. To name the angle in Fig. 21–10 with three letters, we write ∠*XZY* or ∠*YZX*. This angle can also be named ∠*a* or ∠*Z*.

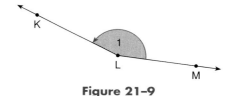

Figure 21-9

Figure 21–11 illustrates how the intersection of two rays actually forms two angles. Throughout this textbook, we will refer to the smaller of the two angles formed by two rays unless the other angle is specifically indicated.

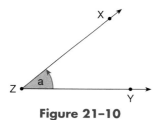

Figure 21-10

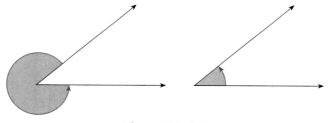

Figure 21-11

4 **Classify Angles According to Size.**

The measure of an angle is determined by the amount of opening between the two sides of the angle. The length of the sides does not affect the angle measure. Two measuring units are commonly used to measure angles, *degrees* and *radians*. In this chapter we will use only degrees to measure angles. Radian measures of angles will be discussed in Chapter 22 on trigonometry.

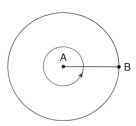

Figure 21–12

Consider the hands of a clock to be the sides of an angle. When the two hands of the clock are both pointing to the same number, the measure of the angle formed is 0 degrees (0°). An angle of 0 degrees is used in trigonometry but is seldom used in geometric applications. During 1 hour the minute hand makes one complete revolution. Ignoring the movement of the hour hand, this revolution of the minute hand contains 360 degrees (360°). Figure 21–12 illustrates a revolution or rotation of 360°. This rotation can be either clockwise or counterclockwise. A complete rotation counterclockwise is +360°. A complete rotation clockwise is −360°.

■ **DEFINITION 21–4: Degree.** A *degree* is a unit for measuring angles. It represents $\frac{1}{360}$ of a complete rotation about the vertex.

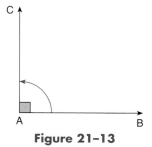

Figure 21–13

Suppose, in Fig. 21–13, that $\overrightarrow{AC}$ rotates from $\overrightarrow{AB}$ through one-fourth of a circle. Then $\overrightarrow{AC}$ and $\overrightarrow{AB}$ form a 90° angle ($\frac{1}{4}$ of 360 = 90). This angle is called a *right angle*. The symbol for a right angle is ∟.

If two lines intersect so that right angles are formed (90° angles), the lines are *perpendicular* to each other. (See Section 19–5.) The symbol for "perpendicular" is ⊥. However, the right-angle symbol implies that the lines forming the angle are perpendicular.

If a string is suspended at one end and weighted at the other, the line it forms is a *vertical line*. A line that is perpendicular to the vertical line is a *horizontal line*. In Fig. 21–14, $\overleftrightarrow{AB}$ is a vertical line, and $\overleftrightarrow{AB}$ and $\overleftrightarrow{CD}$ form right angles. Thus, $\overleftrightarrow{AB} \perp \overleftrightarrow{CD}$, and $\overleftrightarrow{CD}$ is a horizontal line.

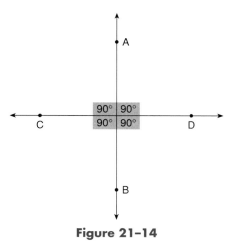

Figure 21–14

Look now at Fig. 21–15. When $\overrightarrow{AC}$ rotates one-half a circle from $\overrightarrow{AB}$, an angle of 180° is formed ($\frac{1}{2}$ of 360 = 180). This angle is called a *straight angle*.

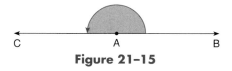

Figure 21–15

These two angles, the right angle (90°) and the straight angle (180°), are used to define two sets of angles that are used often in geometry. An angle that is less than 90° but more than 0° is called an *acute angle*. An angle that is more than 90° but less than 180° is called an *obtuse angle* (see Fig. 21–16).

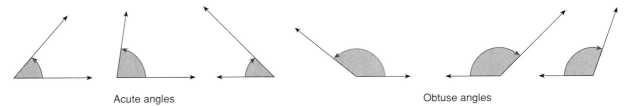

Acute angles   Obtuse angles

Figure 21–16

The following definitions will help us by summarizing the classification of angles.

■ **DEFINITION 21–5: Right Angle.** A *right angle* (90°) represents one-fourth of a circle or one-fourth of a complete rotation.

■ **DEFINITION 21–6: Straight Angle.** A *straight angle* (180°) represents one-half of a circle or one-half of a complete rotation.

■ **DEFINITION 21–7: Acute angle.** An *acute angle* is an angle less than 90° but more than 0°.

■ **DEFINITION 21–8: Obtuse Angle.** An *obtuse angle* is an angle more than 90° but less than 180°.

Two other useful terms involving angles are *complementary* and *supplementary* angles.

■ **DEFINITION 21–9: Complementary Angles.** *Complementary angles* are two angles whose sum is one right angle or 90° (Fig. 21–17).

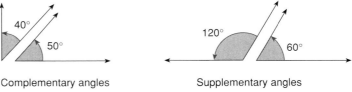

Complementary angles   Supplementary angles

Figure 21–17

■ **DEFINITION 21–10: Supplementary Angles.** *Supplementary angles* are two angles whose sum is one straight angle or 180° (Fig. 21–17).

Tips and Traps

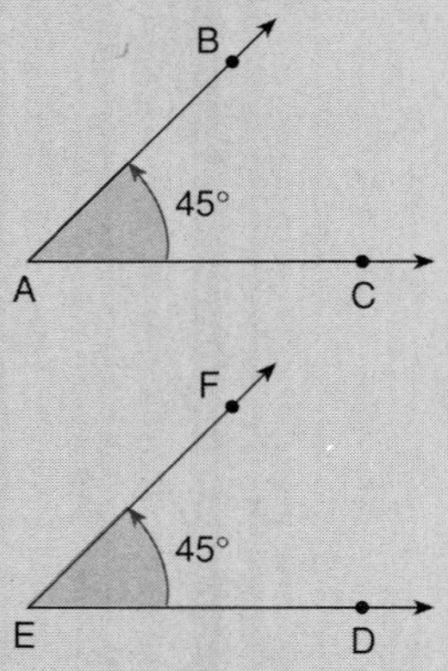

Equal Versus Congruent:
When the word *equal* is used to describe the relationship between two angles, we are implying that the measures of the angles are equal. Another word that is often used in geometry is *congruent*. When geometric figures are congruent, one figure can be placed on top of the other, and the two figures will match perfectly. If two angles are congruent, the measures of the angles are equal. Also, if the measures of two angles are equal, the angles are congruent. The symbol for congruence is ≅. For instance, in Fig. 21–18, the two angles have equal measures, both 45°. Because they have the same measures, they are said to be congruent; that is, ∠*BAC* ≅ ∠*FED*.

A traditional notation to indicate the measures of angles is to precede the angle symbol with the letter *m*. Thus *m*∠*BAC* = *m*∠*FED* is another way to express that the measures of angles *BAC* and *FED* are equal. In this text, we will not use this notation.

Figure 21–18

1   Use Fig. 21–19 for the following exercises.

1. Name the line in three different ways.
2. Name in two different ways the ray whose end point is P and whose interior points are Q and R.
3. Name the segment whose end points are Q and R.

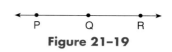

Figure 21–19

Use Fig. 21–20 for the following exercises.

4. Does $\overleftrightarrow{XY}$ represent the same line as $\overleftrightarrow{YZ}$?
5. Does $\overrightarrow{XY}$ represent the same ray as $\overrightarrow{YZ}$?
6. Does $\overline{XY}$ represent the same segment as $\overline{YZ}$?
7. Is $\overleftrightarrow{WX}$ the same as $\overleftrightarrow{WY}$?
8. Is $\overrightarrow{XY}$ the same as $\overrightarrow{XZ}$?
9. Is $\overline{XY}$ the same as $\overline{YX}$?
10. Is $\overrightarrow{XW}$ the same as XY?

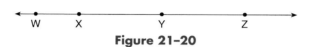

Figure 21–20

2   Use Fig. 21–21 for the following exercises.

11. $\overleftrightarrow{AB}$ and $\overleftrightarrow{CD}$ are _____ lines.
12. $\overleftrightarrow{EF}$ and $\overleftrightarrow{GH}$ _____ at point O.
13. $\overleftrightarrow{IJ}$ and $\overleftrightarrow{KL}$ _____.
14.–15. $\overleftrightarrow{AB}$ and $\overleftrightarrow{CD}$ will never _____ or _____.

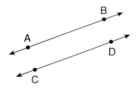 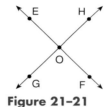 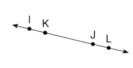

Figure 21–21

3   Use Fig. 21–22 for the following exercises.

16. Name the angle in two different ways using three capital letters.
17. Name the angle using one capital letter.
18. Name the angle using a number.

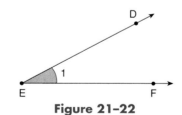

Figure 21–22

Use Fig. 21–23 for the following exercises.

19. Name the angle using one lowercase letter.
20. Name the angle using three capital letters.
21. Name the angle using one capital letter.

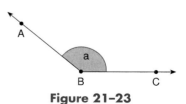

Figure 21–23

4   Fill in the blanks.

22. One complete rotation is _____°.
23. One-half a complete rotation is _____°.
24. One-fourth a complete rotation is _____°.

Classify the following angle measures using the terms *right, straight, acute,* or *obtuse.*

25. 38°   **26.** 95°   **27.** 90°   **28.** 153°

29. 10°   **30.** 180°   **31.** 60°   **32.** 120°

Tell whether the following angle pairs are complementary.

33. 17°, 73°   **34.** 38°, 142°   **35.** 52°, 48°

Tell whether the following angle pairs are supplementary.

36. 60°, 30°   **37.** 110°, 70°   **38.** 42°, 138°   **39.** 23°, 67°

40. Two perpendicular lines form a _____° angle.   **41.** Two 35° angles are said to be _____ angles.

21-2   ANGLE CALCULATIONS

Learning Objectives

1. Add and subtract angle measures.
2. Change minutes and seconds to a decimal part of a degree.
3. Change the decimal part of a degree to minutes and seconds.
4. Multiply and divide angle measures.

A device used to measure angles is called a *protractor.* The most common type of protractor is a semicircle with two scales from 0° to 180°. An *index mark* is in the middle of the straight edge of the protractor (Fig. 21–24).

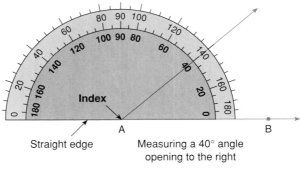

Measuring a 40° angle
opening to the right

Figure 21–24

To measure an angle that opens to the right, the degree measure is read from the lower scale. Notice in Fig. 21–24 that the lower scale starts with 0 at the right. The index of the protractor is aligned with the vertex of the angle, and the straight edge lies along the lower side of the angle.

To measure an angle that opens to the left, the degree measure is read from the upper scale. Notice in Fig. 21–25 that the upper scale starts with 0 at the left. The index of the protractor is aligned with the vertex of the angle and lies along the lower side of the angle.

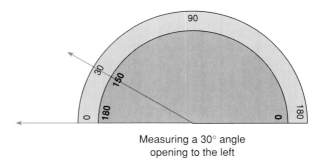

Measuring a 30° angle
opening to the left

Figure 21–25

Degrees are divided into 60 equal parts. Each of the equal parts is called a *minute*.

$$1 \text{ degree } (1°) = 60 \text{ minutes } (60')$$

The symbol $'$ is used for minutes. Similarly, the minute is divided into 60 equal parts called *seconds*.

$$1 \text{ minute } (1') = 60 \text{ seconds } (60'')$$

The symbol $''$ is used for seconds. Thus,

$$1° = 60' = 3600''$$

① Add and Subtract Angle Measures.

Angle measures can be added or subtracted. Keep in mind that only like measures can be added or subtracted.

EXAMPLE 1 Add $12°15'54''$ and $82°28'19''$.

$$12°\ 15'\ 54''$$
$$+\ 82°\ 28'\ 19''$$
$$94°\ 43'\ 73''$$

Arrange the measures in columns of like measures and add.

Because $73'' = 1'13''$, simplify by adding $1'$ to the minutes column and subtracting $60''$ from the seconds column. **Thus, $94°43'73'' = 94°44'13''$.**

EXAMPLE 2 Add $71°14'$ and $82°12''$.

$$71°\ 14'$$
$$+\ 82°\qquad 12''$$
$$\mathbf{153°14'12''}$$

Arrange in columns of like measures and add.

EXAMPLE 3 Subtract $15°32'$ from $37°15'$.

$$37°15' = 36°75'$$
$$-\ 15°32' = 15°32'$$
$$\mathbf{21°43'}$$

Borrow $1°$ from $37°$. $1° = 60'$ and $60' + 15' = 75'$.

EXAMPLE 4 Subtract $3°12'30''$ from $15°$.

$$\cancel{15°} = 14°59'60''$$
$$-\ 3°12'30'' = 3°12'30''$$
$$\mathbf{11°47'30''}$$

Borrow $1°$ from $15°$. $1° = 60'$. Borrow $1'$ from $60'$. $1' = 60''$. Then subtract.

EXAMPLE 5 Find the complement of an angle of $35°25'40''$.

To find the complement of an angle, we subtract the given angle measure from $90°$.

$$\cancel{90°} = 89°59'60''$$
$$-\ 35°25'40'' = 35°25'40''$$
$$\mathbf{54°34'20''}$$

2 Change Minutes and Seconds to a Decimal Part of a Degree.

With the increased popularity of the calculator, it is sometimes necessary to change minutes or seconds to decimal equivalents. Minutes or seconds are first changed to their fractional part of a degree. Then the fraction is changed to its decimal equivalent by dividing the numerator by the denominator. Remember, $1' = \frac{1}{60}$ of a degree, and $1'' = \frac{1}{3600}$ of a degree.

> **Rule 21-1** *To change minutes to a decimal part of a degree:*
> Divide minutes by 60.

> **Rule 21-2** *To change seconds to a decimal part of a degree:*
> Divide seconds by 3600.

EXAMPLE 6   Change $15'$ to its decimal degree equivalent.

$$\frac{15}{60} = \mathbf{0.25°}$$

EXAMPLE 7   Change $37''$ to its decimal degree equivalent.

$$\frac{37}{3600} = \mathbf{0.0103°} \qquad \text{To the nearest ten-thousandth}$$

EXAMPLE 8   Change $22'35''$ to its decimal degree equivalent.

$$22' = \frac{22}{60} = 0.3667° \qquad \text{To the nearest ten-thousandth}$$

$$35'' = \frac{35}{3600} = 0.0097° \qquad \text{To the nearest ten-thousandth}$$

$$0.3667° + 0.0097° = 0.3764°$$

■ General Tips for Using the Calculator

Many calculators have a key that will automatically convert between degrees, minutes, and seconds and decimal degrees. The key is often labeled $\boxed{° \ ' \ ''}$. This function key is also called a sexagesimal function key because of its relationship with the number 60. The same key is also used for hours, minutes, and seconds.

Scientific Calculator
To enter a degree measure in degrees, minutes, and seconds notation:

1. Enter the number of degrees, followed by the $\boxed{° \ ' \ ''}$ key.

2. Enter the number of minutes, followed by the $\boxed{° \ ' \ ''}$ key.

3. Enter the number of seconds, followed by the $\boxed{° \ ' \ ''}$ key. The display on many calculators will automatically show the equivalent measure in decimal degrees. To see the degree, minutes, and seconds notation, press the shift, inverse, or 2nd function key, followed by the $\boxed{° \ ' \ ''}$ key.

Calculations can be performed in degrees, minutes, and seconds and the results displayed in either notation.

To enter a degree measure in decimal notation:

1. Enter the measure in decimal degrees.

2. To see the equivalent degrees, minutes, and seconds notation, press the shift, inverse, or 2nd function key, followed by the $\boxed{° \ ' \ ''}$ key.

Here are some previously worked examples, performed on the calculator.

- Example 5

$$90 \ \boxed{-} \ 35 \ \boxed{° \ ' \ ''} \ 25 \ \boxed{° \ ' \ ''} \ 40 \ \boxed{° \ ' \ ''} \ \boxed{=} \ \Rightarrow 54.57222222$$
$$\boxed{\text{INV}} \ \boxed{° \ ' \ ''} \ \Rightarrow 54°34°20.$$

Some calculator displays separate the degrees and minutes and the minutes and seconds with the same notation. The traditional ° ' " symbols should be used in handwritten displays of the results.

- Example 10

$$.43 \ \boxed{\text{INV}} \ \boxed{° \ ' \ ''} \ \Rightarrow 0°25°48 \text{ is written as } 0°25'48''.$$

Graphics Calculator

On some graphics calculators, the degrees, minutes, and seconds function is found on the $\boxed{\text{MATH}}$ menu, accessed by pressing $\boxed{\text{SHIFT}}$ $\boxed{\text{Graph}}$. Then select $\boxed{\text{DMS}}$ (F4). Enter the angle measure using $\boxed{\text{F1}}$ for degrees, minutes, and seconds. Press $\boxed{\text{EXE}}$ to display the decimal degrees equivalent.

An angle entered as decimal degrees can be displayed as degrees, minutes, and seconds by pressing the $\boxed{\text{EXE}}$ key, followed by $\boxed{\text{F2}}$.

---

3 ## Change the Decimal Part of a Degree to Minutes and Seconds.

A decimal part of a degree can be changed to minutes and seconds by reversing the procedure for changing minutes and seconds to degrees. To change a decimal part of a degree to minutes, multiply by 60. Similarly, to change the decimal part of a minute to seconds, multiply by 60.

Rule 21–3 *To change a decimal part of a degree to minutes:*

Multiply the decimal part of a degree by 60.

Rule 21–4 *To change a decimal part of a minute to seconds:*

Multiply the decimal part of a minute by 60.

If we are changing a decimal part of a *degree* to *seconds,* we can multiply by 3600.

EXAMPLE 9   Change 0.75° to minutes.

$$\mathbf{0.75 \times 60 = 45'}$$

EXAMPLE 10   Change 0.43° to minutes and seconds.

$0.43 \times 60 = 25.8'$ Degrees to minutes

$0.8 \times 60 = 48''$ Decimal part of minute to seconds

Thus, 0.43° = 25′48″.

4 Multiply and Divide Angle Measures.

Angle calculations involving multiplication and division are sometimes necessary. We may multiply or divide angle measures by a number as we do with U.S. customary measures and then simplify as in Example 1.

EXAMPLE 11   An angle of 42°27′32″ needs to be increased to four times its size. Find the measure of the new angle in degrees, minutes, and seconds.

$$42°27'32''$$ Multiply each part of the measure by 4.

$$\underline{\times \qquad\qquad 4}$$

$$168°108'128'' = \mathbf{169°50'8''}$$ Simplify result.

EXAMPLE 12   An angle of 70°15′16″ is divided into three equal angles. Find the measure of each angle in degrees, minutes, and seconds.

$$
\begin{array}{r}
23° \quad\; 25'5\frac{1}{3}'' \text{ or } \mathbf{23°25'5''} \\
3\overline{)70° \quad 15'16''} \\
\underline{69} \\
1° = 60' \\
\overline{75'} \\
\underline{75'} \\
16'' \\
\underline{15''} \\
1''
\end{array}
$$

To nearest second

EXAMPLE 13   An angle of 57°42′17″ is divided into four equal parts. Change the measure to its decimal equivalent and find the measure of each part in decimal degrees.

$$57°42'17'' = 57 + \frac{42}{60} + \frac{17}{3600}$$

$$= 57 + 0.7 + 0.0047 = 57.7047°$$ Rounded

$$\frac{57.7047}{4} = 14.4262°$$ Rounded

SELF-STUDY EXERCISES 21-2 _____

1   Add or subtract as indicated. Simplify the answers.

1. $15°47'18''$
$+\;38°12'42''$

2. $83°19'54''$
$-\;37°11'36''$

3. $152°28'19''$
$-\;114°35'23''$

4. $45°15'38''$
$+\;28°47'34''$

5. $47°30'$
$-\;30°30'15''$

6. $90°$
$-\;35°15'48''$

7. Find the supplement of an angle whose measure is 115°35′14″.

Change the following to decimal degree equivalents. Express the decimals to the nearest ten-thousandth.

8. 47′   **9.** 36″   **10.** 5′14″   **11.** 10′15″

12. An angle of 59°24′ is divided into two equal angles. Change the measure to its decimal equivalent and express the measure of the two equal angles in decimal degrees.

3 Change the following to equivalent minutes and seconds. Round to the nearest second when necessary.

13. 0.35°   **14.** 0.20°   **15.** 0.12°   **16.** 0.213°   **17.** 0.3149°

4 In the following exercises, round angle measures to the nearest second when necessary.

18. An angle of 90° is divided into 12 equal parts. Find the measure of each angle in degrees and minutes.

19. An angle of 80° is divided into seven equal parts. Find the measure of each angle in degrees, minutes, and seconds.

20. An angle of 12°43′49″ needs to be three times as large. Find the measure of the new angle in degrees to the nearest ten-thousandth.

21. An angle of 25°17′39″ is divided into two equal angles. Find the measure of each angle in degrees, minutes, and seconds.

21-3   TRIANGLES

Learning Objectives

1. Classify triangles by sides.
2. Relate the sides and angles of a triangle.
3. Determine whether two triangles are congruent using inductive and deductive reasoning.
4. Use the properties of a 45°, 45°, 90° triangle to find missing parts and to solve applied problems.
5. Use the properties of a 30°, 60°, 90° triangle to find missing parts and to solve applied problems.

We have examined various relationships among lines and angles in the previous sections. In this section we study one of the most useful mathematical figures for any technician who uses geometry—the *triangle*. The relationships among the sides and angles in the triangle enable us to obtain much information that is implied but not always expressed in certain applications.

Triangles can be classified according to the relationship of their sides or their angles. We classified triangles by angles in Chapter 13. In this section, we will classify triangles by their sides.

1   Classify Triangles by Sides.

There are three possible relationships among the three sides of a triangle, and these relationships result in a special name for each type of triangle.

■ **DEFINITION 21–11: Equilateral Triangle.** An *equilateral triangle* is a triangle with three equal sides. The three angles of an equilateral triangle are also equal. Each angle measures 60° (Fig. 21–26).

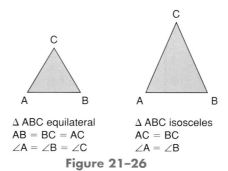

△ ABC equilateral
AB = BC = AC
∠A = ∠B = ∠C

△ ABC isosceles
AC = BC
∠A = ∠B

Figure 21–26

■ **DEFINITION 21–12: Isosceles Triangle.** An *isosceles triangle* is a triangle with *exactly* two equal sides. The angles opposite these equal sides are also equal (Fig. 21–26).

■ **DEFINITION 21–13: Scalene Triangle.** A *scalene triangle* is a triangle with *all* three sides unequal (Fig. 21–27).

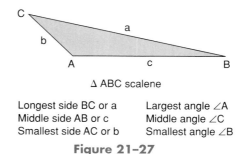

△ ABC scalene

Longest side BC or a   Largest angle ∠A
Middle side AB or c   Middle angle ∠C
Smallest side AC or b   Smallest angle ∠B

Figure 21–27

2 Relate the Sides and Angles of a Triangle.

An equilateral (three sides equal) triangle also has three equal angles. The two equal sides of an isosceles triangle have opposite angles that are equal (Fig. 21–26).

In a scalene triangle, where no sides are equal, we can also state an important relationship between the sides and their opposite angles (Fig. 21–27).

Rule 21–5   *To determine the longest and shortest sides of a triangle:*

If the three sides of a triangle are unequal, the *largest* angle will be opposite the *longest* side and the *smallest* angle will be opposite the *shortest* side.

EXAMPLE 1   Use Rule 21–5 to identify the longest and shortest sides of the triangle in Fig. 21–28.

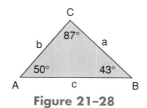

Figure 21–28

The longest side is *AB* or *c* (87° is the largest angle). The shortest side is *AC* or *b* (43° is the smallest angle).

EXAMPLE 2   Use Rule 21–5 to identify the largest and smallest angles in the triangle in Fig. 21–29.

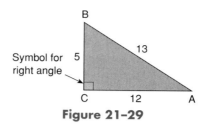

Figure 21–29

The largest angle is ∠C (13 is the longest side). The smallest angle is ∠A (5 is the shortest side).

When applying Rule 21–5 to a right triangle, we see that the hypotenuse (side opposite the 90° angle) is always the longest side. Figure 21–29 shows the hypotenuse (side *AB*) to be the longest side.

3 Determine Whether Two Triangles are Congruent Using Inductive and Deductive Reasoning.

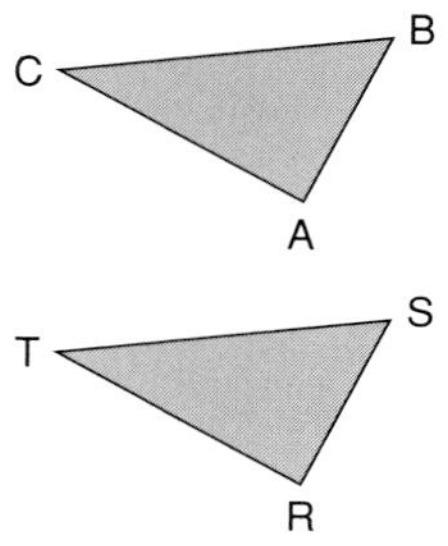

When we work with triangles that have the same size and shape, we are working with *congruent* triangles. These triangles can fit exactly on top of each other.

In Fig. 21–30, △*ABC* will fit exactly over △*RST*. They are congruent triangles. The symbol ≅ means congruent. That is, △*ABC* ≅ △*RST*. Each angle in △*ABC* has an angle in △*RST* that is its equal. We say that these pairs of angles *correspond*. In Fig. 21–30, ∠*A* corresponds to ∠*R* because they are equal. Also, ∠*C* corresponds to ∠*T*, and ∠*B* corresponds to ∠*S*. The equal sides also correspond. $\overline{AB}$ corresponds to $\overline{RS}$, $\overline{CB}$ corresponds to $\overline{TS}$, and $\overline{AC}$ corresponds to $\overline{RT}$.

Figure 21–30

■ **DEFINITION 21–14: Congruent Triangles.** *Congruent triangles* are triangles in which the corresponding sides and angles are equal.

We can establish congruent triangles even when we know that only certain angles and sides are equal.

> **Rule 21–6** *Use three sides to determine congruent triangles:*
>
> If the three sides of one triangle are equal to the corresponding three sides of another triangle, the triangles are congruent. (SSS)

EXAMPLE 3   Write the corresponding sides of the two triangles in Fig. 21–31.

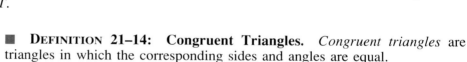

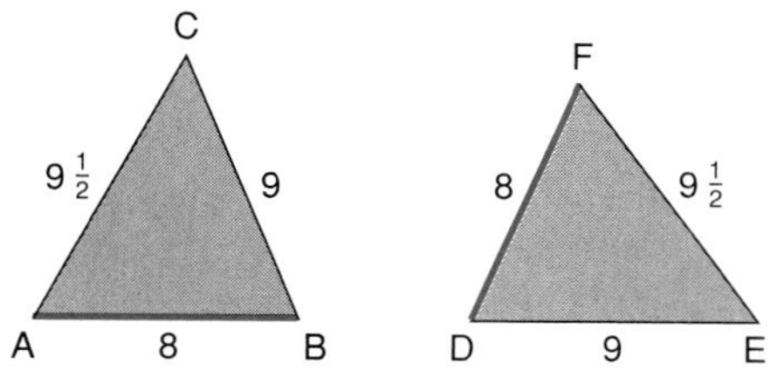

△*ABC* ≅ △*FDE*

$\overline{AB}$ corresponds to $\overline{FD}$

$\overline{BC}$ corresponds to $\overline{DE}$

$\overline{AC}$ corresponds to $\overline{FE}$

Figure 21–31

Rule 21–7 *Use two sides and the included angle to determine congruent triangles:*

If two sides and the *included angle* of one triangle are equal to two sides and the included angle of another triangle, the triangles are congruent. (SAS)

EXAMPLE 4   List the corresponding sides and angles of the two triangles in Fig. 21–32.

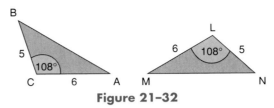

Figure 21–32

$\overline{BC}$ corresponds to $\overline{NL}$.

$\overline{AC}$ corresponds to $\overline{ML}$.

$\angle C$ corresponds to $\angle L$.

$\overline{AB}$ corresponds to $\overline{MN}$.   They are opposite equal angles.

$\angle A$ corresponds to $\angle M$.   They are opposite equal sides.

$\angle B$ corresponds to $\angle N$.   They are opposite equal sides.

Rule 21–8 *Use two angles and the included side to determine congruent triangles:*

If two angles and the common side of one triangle are equal to two angles and the common side of another triangle, the triangles are congruent. (ASA)

EXAMPLE 5   List the corresponding angles and sides of the two triangles in Fig. 21–33.

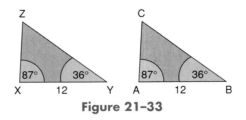

Figure 21–33

$\angle X$ corresponds to $\angle A$. $\overline{XY}$ corresponds to $\overline{AB}$.

$\angle Y$ corresponds to $\angle B$. $\overline{XZ}$ corresponds to $\overline{AC}$.

$\angle Z$ corresponds to $\angle C$. $\overline{YZ}$ corresponds to $\overline{BC}$.

The study of geometry applies two basic types of reasoning, *inductive* and *deductive* reasoning. *Inductive reasoning* starts with investigation and experimentation. Early mathematicians discovered the properties of geometry through inductive reasoning. After extensive investigation, general conclusions are drawn from the results of specific cases.

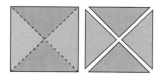

Figure 21-34

An example of inductive reasoning would be to examine the results of cutting a square into four parts by cutting along the two diagonals (Fig. 21–34). Compare the resulting four triangles. Are they congruent triangles? Make a square of a different size and repeat the exercise. Continued repetitions will lead you to conclude that the diagonals of a square divide the square into four congruent triangles.

Now, cut a rectangle along the two diagonals (Fig. 21–35). Are the resulting four triangles congruent? No. What conclusions appear to result from this experiment? Repeat the experiment with different sizes of rectangles. What properties seem apparent? There are two pairs of congruent triangles.

Figure 21-35

Deductive reasoning starts with accepted principles and additional properties are concluded from these accepted principles. Accepting the congruent triangle properties in Rules 21–6, 21–7, and 21–8 and an additional property that the diagonals of a square bisect (cut in half) the angles of the square, we will use deductive reasoning to show that the two diagonals of a square form four congruent triangles.

Square $ABCD$ forms the four triangles $\triangle AED$, $\triangle AEB$, $\triangle BEC$, and $\triangle DEC$ (Fig. 21–36). The four sides of a square are equal in measure, so sides AD and AB are equal in measure. Because the diagonals of a square bisect the angles, $\angle DAE$, $\angle EAB$, $\angle ABE$, $\angle EBC$, $\angle BCE$, $\angle ECD$, $\angle CDE$, and $\angle EDA$ are all 45° angles. Then, in $\triangle AED$ and $\triangle AEB$ we have two angles and a common side of one triangle equal to two angles and a common side of the other. By applying Rule 21–8, we can say that $\triangle AED$ and $\triangle AEB$ are congruent. Similarly, we could continue the argument to show that all four triangles are congruent.

In a systematic or axiomatic study of geometric concepts, all properties are developed or proved from a limited number of basic principles. In our study of geometric concepts, we will use an informal approach. While we will apply both inductive and deductive reasoning in our arguments for solving problems, we will not introduce all the concepts necessary to make formal proofs of geometric properties.

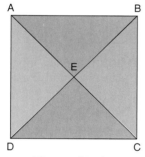

Figure 21-36

4 Use the Properties of a 45°, 45°, 90° Triangle to Find Missing Parts and to Solve Applied Problems.

We saw that an isosceles triangle is a triangle with two equal sides. An isosceles right triangle is a right triangle whose legs are equal. The angles opposite the equal legs are *base angles.* Then the two base angles are also equal because they are the angles opposite the equal sides (Fig. 21–37).

Because the sum of the angles of the triangle is 180°, the sum of the two base angles of a right triangle is 180° − 90° or 90°. If both angles are equal, as in the isosceles right triangle, then each angle is $\frac{1}{2}(90°)$ or 45°. Thus, we get the name 45°, 45°, 90° triangle.

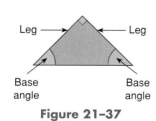

Figure 21-37

■ **DEFINITION 21–15: 45°, 45°, 90° Triangle.** A *45°, 45°, 90° triangle* is an isosceles right triangle in which the two base angles are 45° each.

This triangle is frequently used in applications of the Pythagorean theorem.

EXAMPLE 6 Find the hypotenuse of an isosceles right triangle, each of whose equal sides is 2 m (Fig. 21–38).

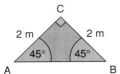

Figure 21–38

Using the Pythagorean theorem, we have

$$2^2 + 2^2 = (AB)^2$$
$$4 + 4 = (AB)^2$$
$$8 = (AB)^2$$
$$\sqrt{8} = AB$$
$$2.828 = AB \qquad \text{Rounded}$$

Thus, the hypotenuse is 2.828 m.
Notice that we can also calculate $AB = 2.828$ by simplifying $\sqrt{8}$.

$$AB = \sqrt{8} = \sqrt{4 \cdot 2} = 2\sqrt{2}$$
$$AB = 2(1.414213562)$$
$$AB = 2.828 \qquad \text{Rounded}$$

The hypotenuse, then, is the product of a leg such as AC and $\sqrt{2}$.

Rule 21–9 *To find the hypotenuse of a 45°, 45°, 90° triangle:*
Multiply the measure of a leg by $\sqrt{2}$ or 1.414213562 (Fig. 21–40); that is,
$$\text{hypotenuse} = \text{leg}\sqrt{2}$$

If this relationship is remembered, it can save computational time on the job.
If we need to find a leg of a 45°, 45°, 90° triangle, we can use another rule based on Rule 21–9.

EXAMPLE 7 Find AC and BC if $AB = 5$ cm (Fig. 21–39).

Figure 21–39

The hypotenuse is equal to the product of a leg and $\sqrt{2}$.

$$5 = AC\sqrt{2}$$
$$\frac{5}{\sqrt{2}} = \frac{AC\sqrt{2}}{\sqrt{2}} \qquad \text{Solve for } AC, \text{ a leg.}$$
$$\frac{5}{\sqrt{2}} = AC$$
$$\frac{5}{\sqrt{2}} \cdot \frac{\sqrt{2}}{\sqrt{2}} = AC \qquad \text{Rationalize.}$$
$$\frac{5\sqrt{2}}{2} = AC$$
$$3.536 \text{ cm} = AC \qquad \text{Rounded}$$

When using a calculator, we can find the decimal value of AC just as easily without rationalizing.

$$\frac{5}{\sqrt{2}} = AC$$

$$AC = 3.536 \text{ cm}, \quad \text{from } \boxed{5} \; \boxed{\div} \; \boxed{2} \; \boxed{\sqrt{}} \; \boxed{=} \Rightarrow 3.536 \qquad \text{Rounded}$$

Since $AC = BC$ then $BC = 3.536$ cm.

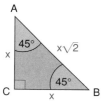

Figure 21–40

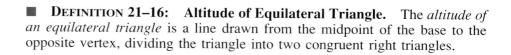

Rule 21–10 *To find a leg of a 45°, 45°, 90° triangle:*

Divide the product of the hypotenuse and $\sqrt{2}$ by 2 (Fig. 21–40); that is,

$$\text{leg} = \frac{\text{hypotenuse } \sqrt{2}}{2}$$

If the sides opposite the 45° angles are *x* units, the sides of a 45°, 45°, 90° triangle are *x*, *x*, $x\sqrt{2}$, respectively.

5 **Use the Properties of a 30°, 60°, 90° Triangle to Find Missing Parts and to Solve Applied Problems.**

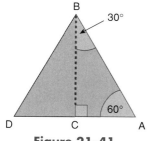

Figure 21–41

Another special case of the Pythagorean theorem is the *30°, 60°, 90° triangle,* which often arises in applications. If we draw the altitude of an *equilateral* triangle, we form two 30°, 60°, 90° triangles (Fig. 21–41).

■ **DEFINITION 21–16: Altitude of Equilateral Triangle.** The *altitude of an equilateral triangle* is a line drawn from the midpoint of the base to the opposite vertex, dividing the triangle into two congruent right triangles.

Because the altitude of an equilateral triangle bisects (divides in half) the vertex angle and the base, we know that two 30° angles are formed at *B* and that $AC = CD$. Since $AD = AB$ and $2AC = AD$, we know that $2AC = AB$, or $AC = \frac{1}{2}AB$, or the side opposite the 30° angle is one-half the hypotenuse.

EXAMPLE 8   If $AC = 5$ cm, find *AB* and *BC* (Fig. 21–42).

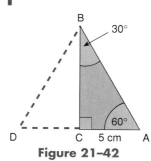

Figure 21–42

Side *AB* (the hypotenuse) is twice *AC* because the altitude divides *AD* in half, and in an equilateral triangle all sides are equal. Therefore, $AB = 2(5) = 10$ cm. Now that we know two sides of the right triangle, we can use the Pythagorean theorem to find the third side.

$$(AB)^2 = (AC)^2 + (BC)^2$$
$$10^2 = 5^2 + (BC)^2$$
$$100 = 25 + (BC)^2$$
$$75 = (BC)^2$$
$$\sqrt{75} = BC$$
$$8.660 = BC \qquad \text{Rounded}$$

Thus, *BC* is 8.660 cm.

Notice that we can also calculate $BC = 8.660$ by simplifying $\sqrt{75}$.

$$BC = \sqrt{75} = \sqrt{25 \cdot 3} = 5\sqrt{3}$$
$$BC = 5(1.732050808)$$
$$BC = 8.660 \qquad \text{Rounded}$$

Then *BC* (the side opposite the 60° angle) is the product of *AC* (the side opposite the 30° angle) and $\sqrt{3}$.

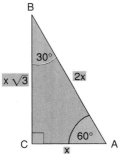

Figure 21–43

Once we have learned the relationships of the three sides, we can find two sides of any 30°, 60°, 90° triangle if we know only one side (Fig. 21–43).

> **Rule 21–11** *To find the hypotenuse of a 30°, 60°, 90° triangle:*
> Multiply the side opposite the 30° angle by 2.

> **Rule 21–12** *To find the side opposite the 30° angle in a 30°, 60°, 90° triangle:*
> Divide the hypotenuse by 2.

> **Rule 21–13** *To find the side opposite the 60° angle in a 30°, 60°, 90° triangle:*
> Multiply the side opposite the 30° angle by $\sqrt{3}$.

Tips and Traps *Summary of 30°, 60°, 90° Triangle:*
To summarize the relationships in Fig. 21–43:

$$AB = 2(AC) \text{ or } \frac{2\sqrt{3}BC}{3} \qquad AC = \frac{AB}{2} \text{ or } \frac{BC\sqrt{3}}{3} \qquad BC = AC\sqrt{3} \text{ or } \frac{AB\sqrt{3}}{2}$$

If the side opposite the 30° angle is x units, the sides of a 30°, 60°, 90° triangle are x, $x\sqrt{3}$, $2x$, respectively.

EXAMPLE 9 Find AC and AB if $BC = 8$ cm (Fig. 21–44).

$$BC = x\sqrt{3} = 8 \qquad x = AC$$

Solving for x, we have $\dfrac{x\sqrt{3}}{\sqrt{3}} = \dfrac{8}{\sqrt{3}}$. Rationalizing, we have

$$x = \frac{8}{\sqrt{3}} \cdot \frac{\sqrt{3}}{\sqrt{3}}$$

$$x = \frac{8\sqrt{3}}{3}$$

$$x = 4.619 \text{ cm} \qquad \text{Rounded}$$

$$AB = 2x = 2\left(\frac{8\sqrt{3}}{3}\right) = \frac{16\sqrt{3}}{3} = 9.238 \text{ cm} \qquad \text{Rounded}$$

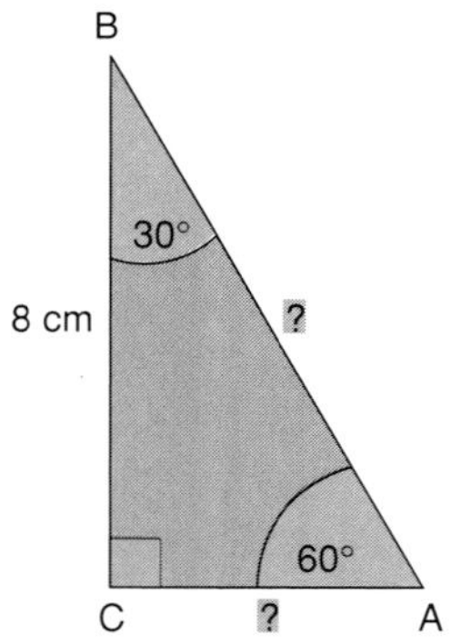

When using a scientific calculator, we can find the decimal value of AC without rationalizing.

$$AC = \frac{8}{\sqrt{3}} = 8 \div \sqrt{3}$$

$$AC = 4.619 \text{ cm}, \quad \text{from } \boxed{8} \ \boxed{\div} \ \boxed{\sqrt{\ }} \ \boxed{3} \ \boxed{=} \ \Rightarrow 4.618802154$$

Thus, $AB = 9.238$ cm and $AC = 4.619$ cm.

Figure 21–44

1 Fill in the blanks.

1. A triangle with no equal sides is called a(n) _____ triangle.

2. A triangle with three equal sides is called a(n) _____ triangle.

3. A triangle with only two equal sides is called a(n) _____ triangle.

2 Identify the longest and shortest sides in Figs. 21–45 and 21–46.

4.

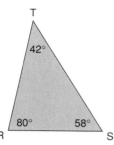

Figure 21–45

5.

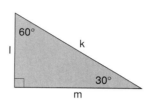

Figure 21–46

List the angles in order of size from largest to smallest in Figs. 21–47 and 21–48.

6.

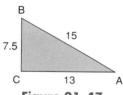

Figure 21–47

7.

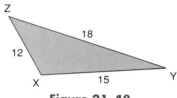

Figure 21–48

3 Write the corresponding parts not given for the congruent triangles in Figs. 21–49 to 21–51.

8.

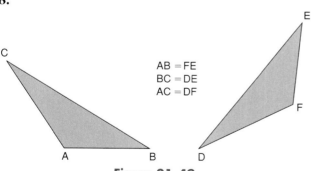

AB = FE
BC = DE
AC = DF

Figure 21–49

9.

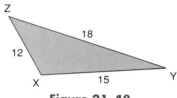

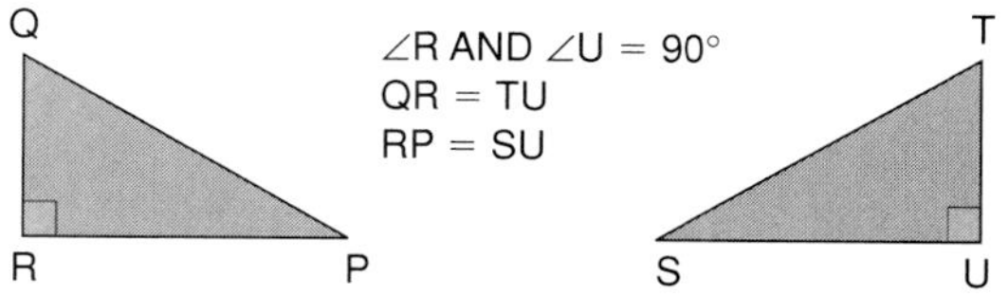

∠R AND ∠U = 90°
QR = TU
RP = SU

Figure 21–50

10.

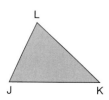

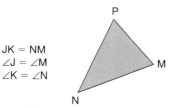

JK = NM
∠J = ∠M
∠K = ∠N

Figure 21–51

4 Use Fig. 21–52 to solve the following exercises. Round the final answers to the nearest thousandth if necessary.

11. $AC = 12$ cm; find BC and AB.
12. $AB = 10$ m; find AC and BC.
13. $BC = 7\sqrt{2}$ m; find AC and AB.
14. $AB = 8\sqrt{2}$ m; find AC and BC.
15. $AB = 12\sqrt{3}$ mm; find AC and BC.

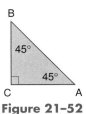

Figure 21–52

16. A rafter 19.5 ft long makes a 45° angle with a joist. If the rafter has an 18-in. overhang, find the length of the joist to the nearest inch (Fig. 21–53).

17. An elevated tank must be connected to a pipe on the ground. The connecting pipe is 53.5 ft long and will form a 45° angle where it connects to the tank (Fig. 21–54). At what horizontal distance from the tank should the ground pipe stop so that the connection can be made?

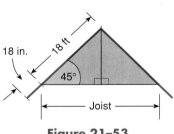

Figure 21–53

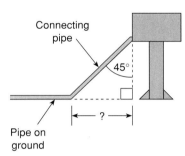

Figure 21–54

5 Use Fig. 21–55 to solve the following exercises. Round the final answers to the nearest thousandth if necessary.

18. $AC = 6$ cm; find AB and BC.
19. $AB = 18$ mm; find AC and BC.
20. $BC = 8$ inches; find AC and AB.
21. $BC = 7\sqrt{2}$ cm; find AC and AB.
22. $AC = 2$ ft 9 in.; find AB and BC to the nearest inch.

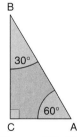

Figure 21–55

Solve the following problems. Round the final answers to the nearest thousandth if necessary.

23. Find the length of the conduit $ABCD$ (Fig. 21–56) if $AB = 18$ ft, $CK = 6$ ft, $CD = 5$ ft, and $\angle KCB = 60°$.

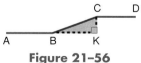

Figure 21–56

24. A rafter makes a 30° angle with the horizontal. If the rise is 9 ft, find the rafter length and the run to the nearest inch (Fig. 21–57).

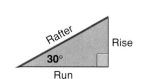

Figure 21–57

25. Find the depth of a V-slot in the form of an equilateral triangle if the cross-sectional opening is 5.2 cm across (Fig. 21–58).

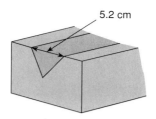

Figure 21–58

21–4 POLYGONS

Learning Objectives

1. Find the missing dimensions of composite shapes.
2. Find the perimeter and area of composite shapes.
3. Find the number of degrees in each angle of a regular polygon.

As we look about us, we see polygons that are not squares, rectangles, parallelograms, trapezoids, or triangles. Yet our work may require us to calculate the perimeter and area of these polygons. In such cases we can use what we already know to find both perimeter and area. Such figures can be considered as *composites*.

■ **DEFINITION 21–17: Composite.** A *composite* is a geometric figure made up of two or more geometric figures.

For instance, an L-shaped slab foundation for a building is really a composite of two rectangles. As suggested in Fig. 21–59, the two rectangles forming the composite figure may be considered in more than one way. Here layout I is partitioned vertically, whereas layout II is partitioned horizontally.

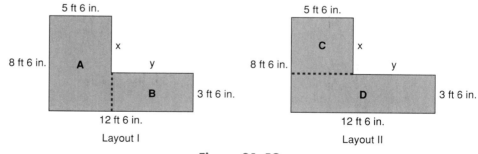

Figure 21–59

1 Find the Missing Dimensions of Composite Shapes.

Observe also that such a layout may not always have all its dimensions indicated. However, the missing dimensions can be inferred or calculated from our knowledge of polygons and the dimensions indicated on the layout itself.

EXAMPLE 1 Find the missing dimensions x and y on the slab layout in Fig. 21–59.

In layout I, the side of B opposite its 3′6″ side is also 3′6″ because opposite sides of a rectangle are equal. The side of A opposite its 8′6″ side is, for the same reason, 8′6″. Dimension x must therefore be the difference between 8′6″ and 3′6″.

$$x = 8'6'' - 3'6''$$

$$x = 5'$$

If we think of layout II as two horizontal rectangles, we can find dimension *y*. The side opposite the 5′6″ side of rectangle *C* must be 5′6″. The side of *D* opposite the 12′6″ side must also be 12′6″. Dimension *y* must therefore be the difference between 12′6″ and 5′6″.

$$y = 12'6'' - 5'6''$$

$$y = 7'$$

The missing dimensions are therefore $x = 5'$ and $y = 7'$.

2 Find the Perimeter and Area of Composite Shapes.

Perimeter is the sum of the lengths of the sides of a figure or layout. The number and the length of the sides will vary from one composite shape to the next, so no specific formula can be given to cover the variety of composite shapes that can be found. We can use a general formula.

Formula 21–1 *Perimeter of a composite:*

$$P = a + b + c, \text{etc.}$$

EXAMPLE 2   Find the number of feet of 4-in. stock needed for the base plates of a room that has the layout shown in Fig. 21–60. Make no allowances for openings when estimating the linear footage of the base plates that form the perimeter.

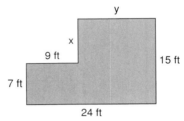

Figure 21–60

1. Find the missing dimensions.

$$x = 15 \text{ ft} - 7 \text{ ft} \qquad y = 24 \text{ ft} - 9 \text{ ft}$$

$$x = 8 \text{ ft} \qquad\qquad y = 15 \text{ ft}$$

2. Apply the general formula for the perimeter of a polygon.

$$P = a + b + c, \text{etc.}$$

$$P = 7 \text{ ft} + 24 \text{ ft} + 15 \text{ ft} + 15 \text{ ft} + 8 \text{ ft} + 9 \text{ ft} = 78 \text{ ft}$$

3. Count the number of sides on the layout to make sure that each has been substituted into the formula.

The room needs 78 ft of 4-in. stock for the base plates.

EXAMPLE 3   Find the number of square yards of carpeting required for the room in Example 2 ($9 \text{ ft}^2 = 1 \text{ yd}^2$).

1. Divide the composite into two polygons whose areas we can compute (Fig. 21–61). In this case, *A* is a rectangle and *B* is a square.

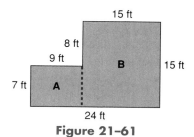

Figure 21–61

2. Find the areas of each smaller polygon and add them together.

Rectangle *A*	Square *B*
$A_1 = lw$	$A_2 = s^2$
$A_1 = 9 \times 7$	$A_2 = 15^2$
$A_1 = 63 \text{ ft}^2$	$A_2 = 225 \text{ ft}^2$

$$A_1 + A_2 = \text{total area}$$

$$63 + 225 = 288 \text{ ft}^2$$

3. Convert square feet to square yards using a unity ratio.

$$\frac{\overset{32}{\cancel{288}} \text{ ft}^2}{1} \times \frac{1 \text{ yd}^2}{\underset{1}{\cancel{9}} \text{ ft}^2} = 32 \text{ yd}^2$$

The room requires 32 yd² of carpeting.

EXAMPLE 4   Find the area of a gable end of a gambrel roof whose dimensions are shown in Fig. 21–62.

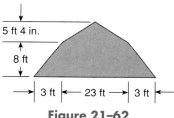

Figure 21–62

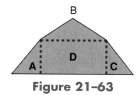

Figure 21–63

1. Divide the gable end of the gambrel roof into polygons whose areas can be calculated (Fig. 21–63).
2. Calculate the areas of the three triangles and the rectangle. Find the sum of the areas.

Triangle *B*	Triangle *A* or *C*	Rectangle *D*
Base = 23′	Base = 3′	Length = 23′
Height = 5′4″ = $5\frac{1}{3}'$	Height = 8′	Width = 8′
$A_1 = \frac{1}{2} bh$	$A_2 = \frac{1}{2} bh$	$A_3 = lw$
$A_1 = \frac{1}{2}(23)\left(5\frac{1}{3}\right)$	$A_2 = \frac{1}{\cancel{2}}(3)(\overset{4}{\cancel{8}})$	$A_3 = 23(8)$

$$A_1 = \frac{1}{2}(23)\left(\frac{\overset{8}{\cancel{16}}}{3}\right) \qquad\qquad A_2 = 12 \text{ ft}^2 \qquad\qquad A_3 = 184 \text{ ft}^2$$

$$A_1 = \frac{184}{3} = 61\frac{1}{3} \text{ ft}^2$$

doubled to include $\triangle A$ and $\triangle C$

$$A_1 + 2(A_2) + A_3 = \text{area of gable end}$$

$$61\frac{1}{3} + 2(12) + 184 = 61\frac{1}{3} + 24 + 184 = 269\frac{1}{3} \text{ ft}^2$$

The area of the gable end of the gambrel roof is $269\frac{1}{3}$ ft².

Sometimes when we figure area we may find it more convenient to calculate an overall area and subtract a smaller area. We saw this for rectangles previously when we figured the area of a room and then subtracted the area of a fireplace hearth that projects into the room. Our purpose was to find the area to be covered with carpet, so we had to exclude the area of the fireplace.

The following example is a similar case that involves composite shapes.

EXAMPLE 5 Find the area of the flat metal piece shown in Fig. 21–64.

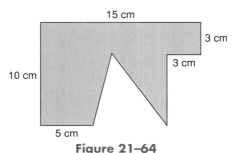

Figure 21–64

1. Divide the figure into polygons for which we can calculate areas (Fig. 21–65).

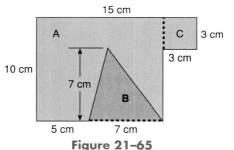

Figure 21–65

2. Find the missing dimensions.
3. Find the area of the square C (A_1), rectangle A (A_2), and triangle B (A_3). The area of the piece of metal is $A_1 + A_2 - A_3$.

Square	Rectangle	Triangle
$A_1 = s^2$	$A_2 = lw$	$A_3 = \frac{1}{2}bh$
$A_1 = 3^2$	$A_2 = 12(10)$	$A_3 = \frac{1}{2}(7)(7)$

$$A_1 = 9 \text{ cm}^2 \qquad A_2 = 120 \text{ cm}^2 \qquad A_3 = \frac{1}{2}(49)$$

$$A_3 = 24.5 \text{ cm}^2$$

Area of piece $= A_1 + A_2 - A_3 = 9 + 120 - 24.5 = 104.5 \text{ cm}^2$

The area of the flat piece of metal is 104.5 cm².

③ Find the Number of Degrees in Each Angle of a Regular Polygon.

Polygons with equal sides and angles like squares are called *regular polygons*. There are several other regular polygons with definite shapes and specific names that are treated as composite shapes for calculating their areas. Let's examine the regular polygons more closely.

■ **DEFINITION 21–18: Regular Polygon.** A *regular polygon* is a polygon with equal sides and equal angles.

Rule 21–14 *To find the number of degrees in each angle of a regular polygon:*

Multiply the number of sides less 2 (which is the number of triangles) by 180° and divide by the number of sides or angles.

$$\text{Degrees per angle} = \frac{180° \ (\text{number of sides} - 2)}{\text{number of sides}}$$

EXAMPLE 6   Find the number of degrees in each angle of the following regular polygons.

| Triangle | (3 sides) | Pentagon | (5 sides) | Octagon | (8 sides) |

| Quadrilateral | (4 sides) | Hexagon | (6 sides) |

Triangle: $\dfrac{180° \ (3 - 2)}{3} = \dfrac{180° \ (1)}{3} = 60°$

Quadrilateral: $\dfrac{180° \ (4 - 2)}{4} = \dfrac{180° \ (2)}{4} = \dfrac{360°}{4} = 90°$

Pentagon: $\dfrac{180° \ (5 - 2)}{5} = \dfrac{180° \ (3)}{5} = \dfrac{540°}{5} = 108°$

Hexagon: $\dfrac{180° \ (6 - 2)}{6} = \dfrac{180° \ (4)}{6} = \dfrac{720°}{6} = 120°$

Octagon: $\dfrac{180° \ (8 - 2)}{8} = \dfrac{180° \ (6)}{8} = \dfrac{1080°}{8} = 135°$

Rule 21–15 *To form congruent triangles in a regular polygon:*

If lines are drawn from the center of a regular polygon to each vertex, congruent triangles are formed.

Figure 21–66 shows how this property applies to several different regular polygons: the equilateral triangle, square, regular pentagon, and regular hexagon.

To inductively accept this property, construct various regular polygons, draw line segments from the vertex to the center, and compare the resulting triangles.

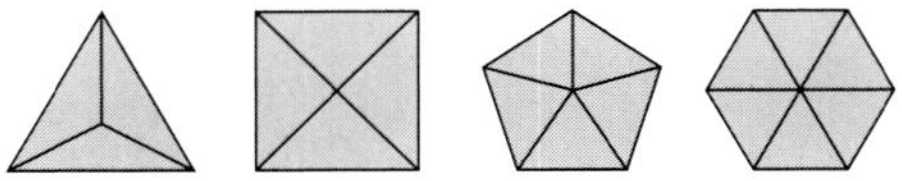

Figure 21–66

EXAMPLE 7   Find the floor area of a recreational building at a park if it forms a regular hexagon and each side is 20 ft long. The perpendicular distance from one side to the center of the building is 17.3 ft (see Fig. 21–67).

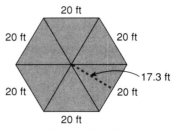

20 ft

20 ft   20 ft

17.3 ft

20 ft   20 ft

20 ft

Figure 21–67

Divide the regular hexagon into congruent triangles by connecting the center of the hexagon with each vertex of the hexagon.

$$A = \frac{1}{2}hb$$ Find the area of one triangle.

$$A = \frac{1}{\cancel{2}}(17.3)(\overset{10}{\cancel{20}})$$ We use the known distance from the side of the hexagon to the center for the height of the triangle.

$$A = 173 \text{ ft}^2$$

$$173 \times 6 = 1038 \text{ ft}^2$$ Multiply the area of the one triangle by 6 because the hexagon has been divided into six congruent triangles.

The floor area of the recreational building is therefore 1038 ft².

SELF-STUDY EXERCISES 21–4

1   Find the missing dimensions x and y of Figs. 21–68 and 21–69.

1.

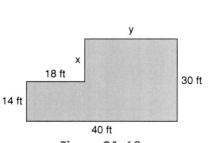

y

x

18 ft

14 ft   30 ft

40 ft

Figure 21–68

2.

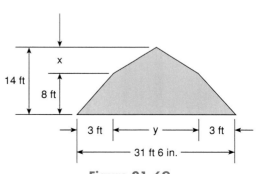

x

14 ft

8 ft

3 ft   y   3 ft

31 ft 6 in.

Figure 21–69

2 Find the perimeter and the area of Figs. 21–70 and 21–71.

3.

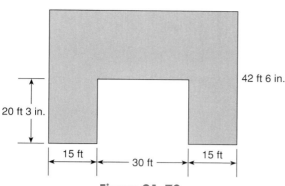

Figure 21-70

4.

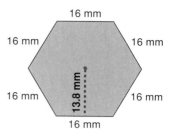

Figure 21-71

Solve the following problems involving perimeter and area of composite figures.

5. A house features a den with the layout shown in Fig. 21–72. Find the number of feet of 4-in. stock needed for the base plates around the perimeter. Make no allowances for doorways or other openings.

6. If the den in Exercise 5 is to be covered with roll vinyl that costs \$16.50 per yd^2 for materials and labor, how much would it cost to install the vinyl? (*Note:* Round to the next whole yd^2 before figuring cost.)

7. The metal piece in Fig. 21–73 has the dimensions as indicated. If the 18-gauge steel weighs 2 lb per ft^2, how much does the piece weigh to the nearest pound? ($144 in.^2 = 1 ft^2$)

8. A swimming pool is in the form of an octagon 15 ft on a side. If the distance from the center of the pool to the midpoint of a side is 18.1 ft, what is the area of the pool? If the area is to be protected by an octagonal cover, what is the measure of each angle of the cover?

9. A no. 5 soccer ball covering is made of 12 colored pentagons 1.75 in. on a side and 20 white hexagons 3.19 in. on a side. If the distance from the midpoint of a side of a colored pentagon to the center of the pentagon is 1.20 in., how many $in.^2$ are made of the colored covering?

10. Find the area of the layout shown in Fig. 21–74.

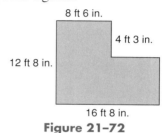

Figure 21-72

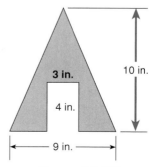

Figure 21-73

11. Find the area of the layout shown in Fig. 21–75.

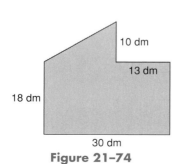

Figure 21-74

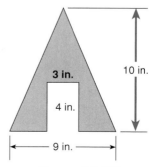

Figure 21-75

3 Give the specific name of each of the shapes in Figs. 21–76 to 21–78 and the number of degrees in each angle.

12.

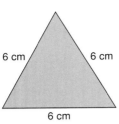

6 cm   6 cm

6 cm

Figure 21–76

13.

10 ft

10 ft   10 ft

10 ft

Figure 21–77

14.

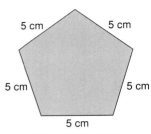

5 cm   5 cm

5 cm   5 cm

5 cm

Figure 21–78

15. Identify the figure in Exercise 4 by giving its specific name and the number of degrees in each angle.

21–5   SECTORS AND SEGMENTS OF A CIRCLE

Learning Objectives

1 Find the area of a sector.
2 Find the arc length of a sector.
3 Find the area of a segment.

We have already seen that we sometimes work with figures that are less than a whole circle. We saw previously, for example, the semicircle in several composite figures. In addition to the semicircle, which is half a circle, there are *sectors* and *segments* of a circle.

1 **Find the Area of a Sector.**

■ **DEFINITION 21–19: Sector.** A *sector* of a circle is the portion of the area of a circle cut off by two radii.

To find the area of a sector, we need to calculate how much of the circle is taken up by the sector. In Fig. 21–79, the sector takes up 45° out of the 360° of the whole circle. We can represent this fractional part of the circle as $\frac{45}{360}$. The area of this sector, then, is $\frac{45}{360}$ of the area of the circle.

If we use the Greek letter *theta* (θ) to represent the number of degrees in the angle of the sector, we can express the area of a sector as shown in Formula 21–2.

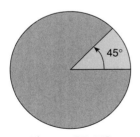

45°

Figure 21–79

Formula 21–2 *Area of a sector:* $A = \frac{\theta}{360}\,\pi r^2$

$\frac{\theta}{360}$ = fractional part of circle πr^2 = area of circle

θ is a central angle measured in degrees.

EXAMPLE 1 Find the area of a sector with a central angle of 45° in a circle with a radius of 10 in. Round to hundredths.

$$A = \frac{\theta}{360}\,\pi r^2$$

$$A = \frac{45}{360}(\pi)(10)^2 \qquad \text{Substitute for } \theta \text{ and } r.$$

$$A = 0.125\,(\pi)(100) \qquad \text{Perform the indicated operations.}$$

$$A = 39.26990817 \text{ in.}^2$$

The area of the sector is 39.27 in.²

EXAMPLE 2 A cone is to be made from sheet metal. To form a cone, a sector with a central angle of 40°20′ is cut from a metal circle whose diameter is 20 in. Find the area of the stretchout (portion of the circle) to be formed into the cone to the nearest hundredth (Fig. 21–80).

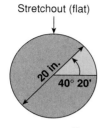

Stretchout (flat)

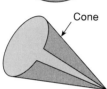

Cone

Figure 21–80

Area used for cone = area of circle − area of sector

Circle	Sector

$$A_1 = \pi r^2 \qquad\qquad A_2 = \frac{\theta}{360}\pi r^2$$

$$A_1 = \pi(10)^2 \qquad\qquad A_2 = \frac{40.3\overline{3}}{360}(314.1592654)$$

$$A_1 = \pi(100) \qquad\qquad A_2 = 0.112037037(314.1592654)$$

$$A_1 = 314.1592654 \text{ in.}^2 \qquad A_2 = 35.19747324 \text{ in.}^2$$

Convert 40°20′ to 40.333333°. πr^2 has already been figured as 314.1592654 in.²

Area used for cone = $A_1 - A_2$

$$A_3 = 314.1592654 - 35.19747324$$

$$A_3 = 278.96 \text{ in.}^2 \qquad \text{Rounded}$$

The area of the metal sector used to form the cone is 278.96 in.²

2 Find the Arc Length of a Sector.

Besides finding the area of a sector, we sometimes need to find the arc length of a sector. The *arc length* is the portion of the circumference intercepted by the sides of the sector (see Fig. 21–81). A formal definition of *arc* is given in Definition 21–21.

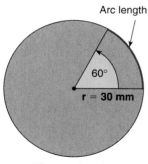

Arc length

60°

r = 30 mm

Figure 21–81

Formula 21–3 *Arc length:*

$$s = \frac{\theta}{360}(2\pi r) \qquad \text{or} \qquad s = \frac{\theta}{360}(\pi d)$$

θ is a central angle measured in degrees.

EXAMPLE 3 Find the arc length of the sector formed by a 60° central angle if the radius is 30 mm (Fig. 21–81).

$$s = \frac{\theta}{360}(2\pi r)$$

$$s = \frac{60}{360}(2)(\pi)(30) \qquad \text{Substitute values.}$$

$$s = 31.41592654$$

$$s = 31.42 \text{ mm} \qquad \text{Rounded}$$

3 Find the Area of a Segment.

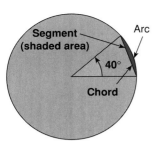

Figure 21-82

If a line segment (called a *chord*) joins the end points of the radii that form a sector, the sector is divided into two figures, a triangle and a *segment* (Fig. 21–82).

■ **DEFINITION 21–20: Chord.** A *chord* is a line segment joining two points on the circumference of a circle.

■ **DEFINITION 21–21: Arc.** The portion of the circumference cut off by a chord is an *arc*.

■ **DEFINITION 21–22: Segment.** A *segment* is the portion of the area of a circle bounded by a chord and an arc.

Because the chord divides the sector into a triangle and a segment, we can calculate the area of the segment by subtracting the area of the triangle from the area of the sector. Expressed symbolically, we have the formula for the area of a segment.

Formula 21–4 *Area of a segment:* $A = \frac{\theta}{360} \pi r^2 - \frac{1}{2} bh$

$\frac{\theta}{360} \pi r^2$ = area of sector $\quad \frac{1}{2} bh$ = area of triangle

---

EXAMPLE 4 Find the area of the segment in circle A of Fig. 21–83 to the nearest hundredth.

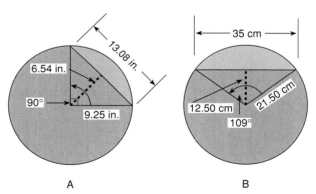

Figure 21–83

$A = \frac{\theta}{360} \pi r^2 - \frac{1}{2} bh$

$A = \frac{90}{360} (\pi)(9.25)^2 - \frac{1}{2} (13.08)(6.54)$ Substitute in formula.

$A = 67.20063036 - 42.7716$

$A = 24.43$ in.^{2} Rounded

The area of the segment is 24.43 in.^{2}.

EXAMPLE 5 A segment of circle B in Fig. 21–83 is removed so that a template for a cam can be made from the rest of the circle. What is the area of the template? Give your answer to the nearest hundredth.

$$\boxed{\text{Area of template } = \text{ area of circle } - \text{ area of segment}}$$

Circle   Segment

$$A_1 = \pi r^2 \qquad\qquad A_2 = \frac{\theta}{360}\,\pi r^2 - \frac{1}{2}\,bh$$

$$A_1 = \pi(21.50)^2 \qquad A_2 = \frac{109}{360}\,(1452.201204) - \frac{1}{2}(35)(12.50)$$

$$A_1 = 1452.201204 \text{ cm}^2 \qquad A_2 = 220.9442535 \text{ cm}^2$$

$$\boxed{\text{Area of template } = A_1 - A_2}$$

$$A_3 = 1452.201204 - 220.9442535$$

$$A_3 = 1231.26 \text{ cm}^2 \qquad \text{Rounded}$$

The area of the template is 1231.26 cm².

SELF-STUDY EXERCISES 21–5

1 Find the area of the sectors of a circle using Fig. 21–84. Round to hundredths.

1. $\angle = 54°$
$r = 16$ cm

2. $\angle = 25°16'$
$r = 30$ mm

3. $\angle = 120°30'$
$r = 1.52$ ft

4. $\angle = 65°$
$r = 5$ in.

5. $\angle = 150°$
$r = 1.45$ m

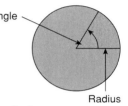

$\angle$ = Angle
r = Radius

Figure 21–84

6. The library of a contemporary elementary school is circular (Fig. 21–85). The floor plan includes pie-shaped areas reserved for science materials, literary materials, reference materials, and so on. Find the area of the reference section excluding its storage area.

7. A mason lays a tile mosaic featuring a four-sector design (Fig. 21–86). What is the area of the design to the nearest hundredth?

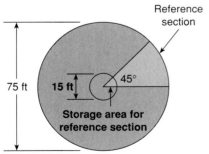

Figure 21–85

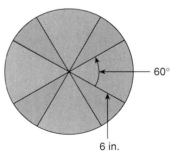

Figure 21–86

8. If the angle formed by a pendulum swing is $19°30'$ and the pendulum is 10 in. long (Fig. 21–87), what area does the pendulum cut as it swings? Express the answer to the nearest tenth. $(60' = 1°.)$

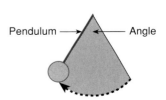

Figure 21–87

9. ∠ = 54°
    r = 30 mm

10. ∠ = 150°
    r = 5 in.

11. ∠ = 120°30′
    r = 1.52 ft

12. ∠ = 25°16′
    r = 16 cm

13. ∠ = 65°
    r = 1.45 m

14. ∠ = 90°
    r = 10 in.

3 Find the area of the segments of a circle using Fig. 21–88. Round to hundredths.

15. ∠ = 60°
    r = 13.3 cm
    h = 11.52 cm
    b = 13.3 cm

16. ∠ = 110°
    r = 10 in.
    h = 5.74 in.
    b = 16.38 in.

17. ∠ = 105°
    r = 11.25 in.
    h = 6.85 in.
    b = 17.85 in.

18. ∠ = 60°
    r = 24 cm
    h = 20.8 cm
    b = 24 cm

19. ∠ = 108°
    r = 14 in.
    h = 8.25 in.
    b = 22.65 in.

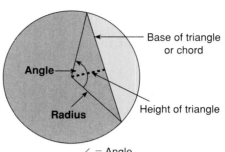

∠ = Angle
r = Radius
h = Height
b = Base

Figure 21–88

20. A motor shaft has milled on it a flat for a set-screw to rest so that it can hold a pulley on the shaft (Fig. 21–89). What is the cross-sectional area of the shaft after being milled? Round to hundredths.

21. A contractor pours a concrete patio in the shape of a circle except where the patio touches the exterior wall of the house (Fig. 21–90). What is the area of the patio in square feet? Round to hundredths.

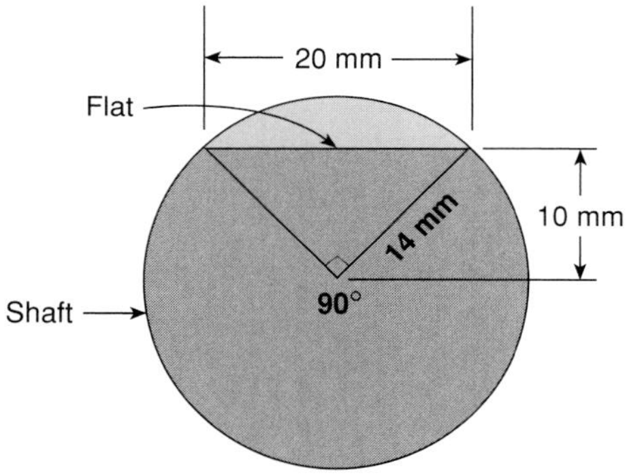

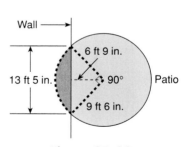

Figure 21–89

Figure 21–90

21-6 **INSCRIBED AND CIRCUMSCRIBED REGULAR POLYGONS AND CIRCLES**

Learning Objectives

1 Use the properties of inscribed and circumscribed equilateral triangles to find missing amounts and to solve applied problems.

2 Use the properties of inscribed and circumscribed squares to find missing amounts and to solve applied problems.

3 Use the properties of inscribed and circumscribed regular hexagons to find missing amounts and to solve applied problems.

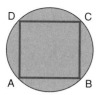

Vertices lie on circle
at points A, B, C, and D

Inscribed square

Sides of square are tangent
at points A, B, C, and D

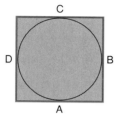

Circumscribed square

Figure 21–91

Previously, we studied polygons and their areas and perimeters. We also briefly looked at regular polygons, that is, polygons whose sides and angles are equal. In this section we study regular polygons *inscribed* in a circle and *circumscribed* about a circle.

■ **DEFINITION 21–23: Inscribed Polygon.** A polygon is *inscribed in a circle* when it is inside the circle and all its *vertices* (points where sides of each angle meet) are on the circle. The circle is said to be *circumscribed about the polygon* (Fig. 21–91).

■ **DEFINITION 21–24: Circumscribed Polygon.** A polygon is *circumscribed about a circle* when it is outside the circle and all its sides are *tangent* to (intersecting in exactly one point) the circumference. The circle is said to be *inscribed in the polygon* (Fig. 21–91).

In this section we examine the three most common polygons inscribed in or circumscribed about a circle: the triangle, the square, and the hexagon.

☐ **Use the Properties of Inscribed and Circumscribed Equilateral Triangles to Find Missing Amounts and to Solve Applied Problems.**

An equilateral triangle can be inscribed in a circle or circumscribed about a circle. If the height (or altitude) is drawn, two congruent right triangles are formed, as shown in Fig. 21–92. Each congruent triangle formed in this way is a 30°, 60°, 90° triangle in which the hypotenuse is twice the shortest side. The special 30°, 60°, 90° triangle and its properties may be reviewed in Section 21–3.

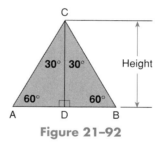

Figure 21–92

Radius = $\frac{2}{3}$ height

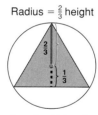

Inscribed triangle

Radius = $\frac{1}{3}$ height

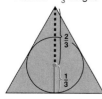

Circumscribed triangle

Figure 21–93

Examine Fig. 21–93. When the equilateral triangle is inscribed in a circle, the radius (from the vertex of the triangle to the center of the circle) is two-thirds the height of the triangle. When the equilateral triangle is circumscribed about a circle, the radius (from the center of the circle to the base of the triangle) is one-third the height of the triangle.

> **Rule 21–16** *Height of an equilateral triangle:*
>
> The height of an equilateral triangle forms two congruent 30°, 60°, 90° right triangles.

> **Rule 21–17** *Radius of a circumscribed circle:*
>
> The radius of a circle circumscribed about an equilateral triangle is two-thirds the height of the triangle or twice the distance from the center of the circle to the base of the triangle.

> **Rule 21–18** *Radius of an inscribed circle:*
>
> The radius of a circle inscribed in an equilateral triangle is one-third the height of the triangle or one-half the distance from the vertex of the triangle to the center of the circle.

EXAMPLE 1 Find the following dimensions for the inscribed equilateral triangle in Fig. 21–94, where $CD = 13$ mm and $AO = 15$ mm.

(a) Height of $\triangle ABC$ (b) BC (c) BD (d) $\angle CAD$
(e) $\angle ACD$ (f) Area of $\triangle ABC$

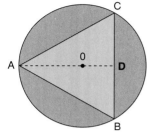

Figure 21–94

(a) $$AO = \frac{2}{3}AD \qquad \text{\small AO = two thirds of height}$$

$$15 = \frac{2}{3}AD \qquad \text{\small Substitute.}$$

$$\frac{3}{2}(15) = \left(\frac{3}{2}\right)\frac{2}{3}AD \qquad \text{\small Solve for AD.}$$

$$\frac{45}{2} = AD$$

22.5 cm = AD

Or use the formula for finding a leg of a 30°, 60°, 90° triangle (Section 21–3).

$$AD = CD\sqrt{3}$$
$$AD = 13\sqrt{3}$$
$AD = 22.5$ mm $\text{\small Rounded}$

(b) $BC = 2CD$ $\text{\small 30°, 60°, 90° }\triangle$
$BC = 2(13)$
$BC = 26$ mm

(c) **$BD = 13$ mm** $\text{\small }\triangle ADC \cong \triangle ADB \quad BD = CD$

(d) **$\angle CAD = 30°$**

(e) **$\angle ACD = 60°$**

(f) $A = \frac{1}{2}bh$

$$A = \frac{1}{2}(BC)(AD)$$

$$A = \frac{1}{2}(26)(22.5)$$

$A = 292.5$ mm^2

EXAMPLE 2 One end of a shaft 35 mm in diameter is milled as shown in Fig. 21–95. Find the radius of the smaller circle.

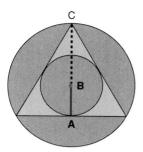

Figure 21–95

$$BC = \frac{1}{2} \text{ diameter of large circle}$$

$$BC = \frac{1}{2}(35)$$

$$BC = 17.5 \text{ mm}$$

$$AB = \frac{1}{2}BC$$

$$AB = \frac{1}{2}(17.5)$$

$$AB = 8.75 \text{ mm}$$

Thus, the radius of the smaller circle is 8.75 mm.

EXAMPLE 3   Find the area of the larger circle in Fig. 21–95. Round to hundredths.

$$A = \pi r^2 = \pi(17.5)^2 = 962.11 \text{ mm}^2 \qquad \text{Rounded}$$

Therefore, the area of the larger circle is 962.11 mm².

2   Use the Properties of Inscribed and Circumscribed Squares to Find Missing Amounts and to Solve Applied Problems.

A square has four equal sides and four equal angles, each 90°. If a diagonal divides the square, the result is two congruent right triangles whose angles are 45°, 45°, and 90°. A second diagonal results in a division of the square into four 45°, 45°, 90° congruent right triangles, as shown in Fig. 21–96. The special 45°, 45°, 90° triangle and its properties may be reviewed in Section 21–3. Diagonals *AC* and *BD* are equal in length and bisect each other, that is, divide each other equally. Point *O*, where the diagonals intersect, is the center of the square and the center of any inscribed or circumscribed circle (Fig. 21–97).

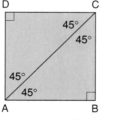  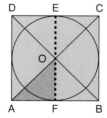 

Figure 21–96   **Figure 21–97**

When the circle is inscribed in the square, the diameter (such as *EF*) equals the length of a side. The radius (such as *FO*) equals the height of a 45°, 45°, 90° triangle formed (such as △*AOB*) by the two diagonals.

When the circle is circumscribed about the square, the diameter (such as *LJ*) equals the diagonal of the square. The radius (such as *KO*) is the height of a 45°, 45°, 90° right triangle (such as △*JKL*) formed by one diagonal.

Rule 21–19   *Diagonals of a square:*

The diagonals of a square form congruent 45°, 45°, 90° triangles.

> **Rule 21-20** *Diameter of a circle inscribed in a square:*
>
> The diameter of a circle inscribed in a square equals a side of the square.

> **Rule 21-21** *Diameter of a circle circumscribed about a square:*
>
> The diameter of a circle circumscribed about a square equals a diagonal of the square.

> **Rule 21-22** *Radius of a circle inscribed in a square:*
>
> The radius of a circle inscribed in a square equals the height of a 45°, 45°, 90° triangle formed by two diagonals.

> **Rule 21-23** *Radius of a circle circumscribed about a square:*
>
> The radius of a circle circumscribed about a square equals the height of a 45°, 45°, 90° triangle formed by one diagonal.

EXAMPLE 4   To mill a square bolt head $1\frac{1}{2}$ in. across flats, round stock of what diameter would be needed? Answer to the nearest hundredth (see Fig. 21–98).

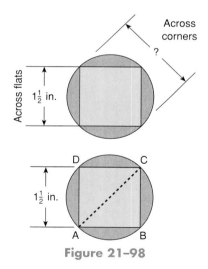

Figure 21–98

The distance across corners is the diameter of the circle as well as the diagonal of the square. It is also the hypotenuse of the two 45°, 45°, 90° triangles formed by one diagonal. Use the property of 45°, 45°, 90° triangles to find the hypotenuse (Rule 21–9).

Hypotenuse $= \text{leg } \sqrt{2}$

Hypotenuse $= 1.5\sqrt{2}$ $1\frac{1}{2}$ in. = 1.5 in.

Hypotenuse $= 2.12$ in.   Rounded

Therefore, the diameter of the round stock is 2.12 in.

EXAMPLE 5 Find the diagonal of a square if the area of the circumscribed circle is 22 in.² . Round to hundredths (Fig. 21–99).

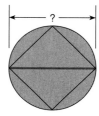

Figure 21–99

$$A = \pi r^2$$

$$22 = \pi r^2 \quad \text{Substitute.}$$

$$\frac{22}{\pi} = r^2$$

$$\sqrt{\frac{22}{\pi}} = r$$

$$2.646283714 = r$$

Diagonal = diameter = 2 × radius = 2(2.646283714) = 5.29 in. Rounded

The diagonal of the inscribed square is 5.29 in.

EXAMPLE 6 What is the minimum depth of cut required to mill a circle on the end of a square piece of stock whose side is 4 cm? Answer to hundredths (Fig. 21–100).

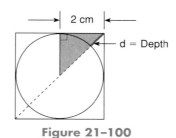

Figure 21–100

The side of the square is 4 cm, so the diameter of the circle is 4 cm and the radius is 2 cm. Also, if the radius is drawn perpendicular to the top side, a right triangle is formed as indicated.

Let d = depth of milling, which equals the hypotenuse of the small triangle minus the radius of the circle. Using the Pythagorean theorem, let c = hypotenuse.

$$c^2 = a^2 + b^2$$

$$c^2 = 2^2 + 2^2$$

$$c^2 = 4 + 4$$

$$c^2 = 8$$

$$c = \sqrt{8}$$

$$c = 2.828427125 \text{ cm}$$

Hypotenuse − radius = depth of milling.

2.828427125 − 2 = 0.83 cm Rounded

The minimum depth of milling should be 0.83 cm.

3 Use the Properties of Inscribed and Circumscribed Hexagons to Find Missing Amounts and to Solve Applied Problems.

A regular hexagon is a figure with six equal sides and six equal angles of 120° each. If three diagonals joining pairs of opposite vertices are drawn, they divide

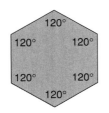

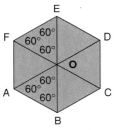

Figure 21–101

the hexagon into six congruent equilateral triangles (all angles 60°) (see Fig. 21–101).

In Fig. 21–102, for the inscribed circle, the height of an equilateral triangle (such as *GO*) is the radius, and for the circumscribed circle, any side of the equilateral triangles (such as *AO*) equals the radius. The height divides any of the six triangles into two congruent 30°, 60°, 90° right triangles.

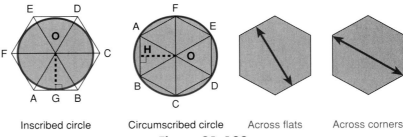

| Inscribed circle | Circumscribed circle | Across flats | Across corners |

Figure 21–102

Rule 21–24   *Diagonals join opposite vertices of a regular hexagon:*

The three diagonals that join opposite vertices of a regular hexagon form six congruent equilateral triangles.

Rule 21–25   *Height of equilateral triangles:*

The height of any of the equilateral triangles of Rule 21–24 forms two congruent 30°, 60°, 90° right triangles.

Rule 21–26   *Distance across corners:*

The diameter of a circle circumscribed about a regular hexagon is a diagonal of the hexagon (*distance across corners*).

Rule 21–27   *Radius of a circumscribed circle:*

The radius of a circle circumscribed about a regular hexagon is a side of an equilateral triangle formed by the three diagonals or one-half the distance across corners.

Rule 21–28   *Radius of a circle inscribed in a regular hexagon:*

The radius of a circle inscribed in a regular hexagon is the height of an equilateral triangle formed by the three diagonals.

Rule 21–29   *Distance across flats:*

The diameter of a circle inscribed in a regular hexagon is the *distance across flats* or twice the height of an equilateral triangle formed by the three diagonals.

EXAMPLE 7 Find the indicated dimensions. When appropriate, round to hundredths. Use Fig. 21–103 for (a) through (f) and Fig. 21–104 for (g) through (j).

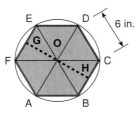

Figure 21–103

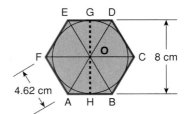

Figure 21–104

(a) $\angle EOD$ (b) $\angle COH$ (c) $\angle BHO$
(d) Radius   (e) Distance across corners   (f) FO
(g) Distance across flats   (h) EO (i) Diagonal   (j) $\angle FED$

(a) $\angle EOD = 60°$ This is an angle of an equilateral triangle.

(b) $\angle COH = 30°$ Height bisects the 60° angle to form a 30°, 60°, 90° triangle.

(c) $\angle BHO = 90°$ The height forms a right angle with the base.

(d) Radius = 6 in.   The radius is a side of the triangle.

(e) Distance across corners = 12 in.   The distance across corners is the diameter.

(f) $FO = 6$ in.   This is a side of an equilateral triangle.

(g) Distance across flats = 8 cm   This is the diameter of the inscribed circle.

(h) $EO = 4.62$ cm   An equilateral triangle has equal sides.

(i) Diagonal = 9.24 cm   The diagonal is twice the side of the equilateral triangle.

(j) $\angle FED = 120°$ This is the angle at the vertex of a hexagon.

EXAMPLE 8 If a regular hexagon is cut from a circle whose radius is 6.45 in., how much of the circle is not used? Round to hundredths (Fig. 21–105).

To solve, we find the area of the circle and the area of the hexagon and subtract to find the difference (portion not used).

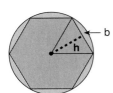

Figure 21–105

$A_1 = \pi r^2$ A_1 = area of circle.

$A_1 = \pi(6.45)^2$

$A_1 = \pi(41.6025)$

$A_1 = 130.6981084$ in.^{2}

$A_2 = 6\left(\dfrac{1}{2}\,bh\right)$ A_2 = area of six triangles.
Because the height of an equilateral triangle forms a 30°, 60°, 90° triangle, $h = \dfrac{b}{2}\sqrt{3} = 3.225\ \sqrt{3} = 5.585863854$ in.

$A_2 = 6\left[\dfrac{1}{2}(6.45)(5.585863854)\right]$

$A_2 = 108.0864656$ in.^{2}

Waste = $A_1 - A_2$

Waste = 22.61 in.^{2} Rounded

There would be 22.61 in.^{2} of waste from the circle.

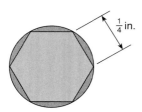

EXAMPLE 9 Use Fig. 21–106 to answer the following questions. Round answers to hundredths.

(a) What is the smallest-diameter round stock from which a hex-bolt head $\frac{1}{4}$ in. on a side can be milled?
(b) What is the distance across the corners of the hex-bolt head?
(c) What is the distance across the flats?

(a) 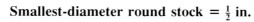 $2 \times \frac{1}{4} = \frac{1}{2}$.

Smallest-diameter round stock = $\frac{1}{2}$ in.

(b)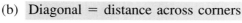

Distance across corners = $\frac{1}{2}$ in.

(c) Use the properties of a 30°, 60°, 90° right triangle.

$$a = b\sqrt{3}$$ a = height of 30°, 60°, 90° triangle;
b = base of 30°, 60°, 90° triangle.

$$a = \frac{1}{8}\sqrt{3}$$

$$a = 0.21650635 \text{ in.}$$

 To nearest hundredth

The distance across the flats is 0.43 in.

Figure 21–106

SELF-STUDY EXERCISES 21–6

1   Use Fig. 21–107 to find the following dimensions for an equilateral triangle inscribed in a circle. Round to hundredths.

1. $\angle CAB$ **2.** $\angle BCD$ **3.** $\angle ADC$ **4.** CB **5.** AC
6. Diameter   **7.** Area of $\triangle ABC$ **8.** AD **9.** Radius   **10.** Circumference

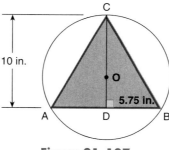

Figure 21–107

Solve the following problems involving inscribed or circumscribed equilateral triangles. Round to hundredths.

11. The area of an equilateral triangle with a base of 10 cm is 43.5 cm². The triangle is circumscribed about a circle. What is the radius of the inscribed circle?

12. One end of a circular steel rod is milled into an equilateral triangle whose height is $\frac{3}{4}$ in. What is the diameter of the rod if the smallest diameter possible for the job was used?

2 Find the measures of the inscribed square in Fig. 21–108.

13. ∠*DOC* **14.** ∠*BCO* **15.** *BO*
16. ∠*BOE* **17.** *CE*

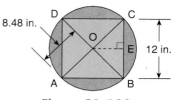

Figure 21–108

Find the measures of the circumscribed square in Fig. 21–109.

18. *DO* **19.** Radius of circle
20. ∠*AOB* **21.** *BE*

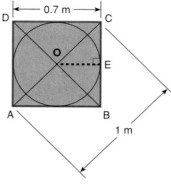

Figure 21–109

Solve the following problems involving squares and circles. Round the answers to hundredths.

22. To what depth must a 2-in. shaft (Fig. 21–110) be milled on an end to form a square 1.414 in. on a side?

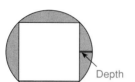

Figure 21–110

23. What is the smallest-diameter round stock (Fig. 21–111) needed to mill a square $\frac{3}{4}$ in. on a side on the end of the stock?

Figure 21–111

24. What is the area of the largest circle inscribed in a square with a perimeter of 36 dkm?

3 Find the following dimensions for Fig. 21–112 if *GO* = 17.32 mm and *FO* = 20 mm.

25. Distance across flats **26.** ∠*CDE*
27. ∠*EFO* **28.** Distance across corners
29. ∠*FGO* **30.** *GA*
31. *AB*

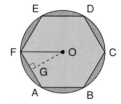

Figure 21–112

Use Fig. 21–113. Follow the preceding directions.

32. ∠*ODC*
33. ∠*EOG*
34. *OC*

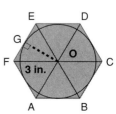

Figure 21–113

35. What is the distance across the flats of a hexagon cut from a circle blank with a circumference of 145 mm (Fig. 21–114)? Round to hundredths.

Figure 21–114

36. The end of a hexagonal rod is milled into the largest circle possible (Fig. 21–115). What is the distance across the corners of the hexagon if the area of the circle is 4.75 in.^{2}? Round to hundredths.

Figure 21–115

CAREER APPLICATION

Electronics: Phase Angle

The phase angle is very important in electronics. One application is comparing voltage and current.

An inductive circuit contains an inductor or coil, usually denoted by the symbol L, and the inductance across the coil is measured in henrys (H). If an alternating current is applied (ac), the voltage in that circuit will always lead the current. This can be measured easily on an oscilloscope. You can also say that the current lags the voltage.

Assume that the sine waves that represent the voltage and the current are shown on the oscilloscope. The two sine waves will be slightly offset from each other. The distance of separation will be constant and is called the *phase angle*.

Suppose one full cycle (from zero up to the maximum and then back to zero and to the negative maximum and then back to zero) measures 8 cm on the oscilloscope, and the amount of offset is 1 cm. Then the two sine waves are $\frac{1}{8}$ of a cycle apart. But one full cycle is considered to be 360°. Therefore, 1 cm of offset can be converted to degrees by taking $360° \times \frac{1}{8} = 45°$ = the phase angle. If the offset is 0.8 cm, then the phase angle $= 360° \times \frac{0.8}{8} = 36°$.

If the inductive circuit is a simple series circuit with only one inductor and only one resistor in it, and an alternating current is applied, then the voltage measured across the inductor (called V_L) will lead the voltage across the resistor (called V_R) by 90°. This always happens. We can also state this result by saying that the voltage across the resistor lags the voltage across the inductor by 90°. Again, this is easy to measure on an oscilloscope.

If the circuit is capacitive (rather than inductive) and is a simple series circuit with only one capacitor and only one resistor in it, and an alternating current is applied, then the voltage measured across the capacitor (called V_C) will lag the voltage across the resistor (called V_R) by 90°. This always happens. We can also state this result by saying that the voltage across the capacitor lags the voltage across the resistor by 90°. Again, this is easy to measure on an oscilloscope.

When measuring current and voltage for a capacitive circuit, the current will always lead the voltage, or the voltage will always lag the current. When measuring current and voltage for an inductive circuit, the current will always lag the voltage, or the voltage will always lead the current. We remember this by using

ELI the ICE man

where E = electromotive force (voltage), I = current, L = inductive circuit, and C = capacitive circuit.

Exercises

Fill in all the blanks in the following chart, which shows readings and interpretations from two sine waves on an oscilloscope. Also, for each pair of sine waves, is the circuit inductive or capacitive?

	Sine Wave 1	Sine Wave 2	Length of One Cycle (cm)	Amount of Offset (cm)	Phase Angle (°)	Inductive or Capacitive?
1.	E	I	8	3		
2.	E	I	6	2		
3.	E	I	5	1		
4.	I	E	4	0.5		
5.	I	E	8	0.5		
6.	I	E	6	1		
7.	V_L	V_R	5			
8.	V_R	V_L		1		
9.	V_C	V_R		2		
10.	V_R	V_C		1.5		

Answers for Exercises

	Sine Wave 1	Sine Wave 2	Length of One Cycle (cm)	Amount of Offset (cm)	Phase Angle (°)	Inductive or Capacitive?
1.	E	I	8	3	135	Inductive
2.	E	I	6	2	120	Inductive
3.	E	I	5	1	72	Inductive
4.	I	E	4	0.5	45	Capacitive
5.	I	E	8	0.5	22.5	Capacitive
6.	I	E	6	1	60	Capacitive
7.	V_L	V_R	5	1.25	90	Inductive
8.	V_R	V_L	4	1	90	Inductive
9.	V_C	V_R	8	2	90	Capacitive
10.	V_R	V_C	6	1.5	90	Capacitive

ASSIGNMENT EXERCISES

Section 21–1

Give the proper notation for the following.

1. Line AB

2. Segment AB

3. Ray AB

Use Fig. 21–116 for the following exercises.

4. Name the line in two different ways.

5. Name the ray with end point N and with an interior point O.

6. Name the ray with end point M and with an interior point L.

7. Is $\overline{MN}$ the same as $\overline{MO}$?

8. Is $\overleftrightarrow{NO}$ the same as $\overleftrightarrow{NM}$?

9. Is $\overrightarrow{NO}$ the same as $\overrightarrow{NM}$?

10. Is $\overrightarrow{MN}$ the same as $\overrightarrow{MO}$?

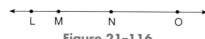

Figure 21–116

Use Fig. 21–117 for the following exercises.

11. Which lines are parallel?
12. Which lines coincide?
13. Which lines intersect?
14. When will $\overleftrightarrow{AB}$ and $\overleftrightarrow{CD}$ meet?
15. Is $\overleftrightarrow{IJ}$ the same as $\overleftrightarrow{KJ}$?

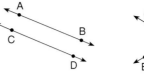 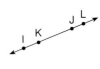

Figure 21–117

Use Fig. 21–118 for the following exercises.

16. Name $\angle a$ using three capital letters.
18. Name $\angle c$ using three capital letters.
20. Name $\angle b$ using one capital letter.

17. Name $\angle b$ using three capital letters.
19. Name $\angle a$ using one capital letter.
21. Name $\angle c$ using one capital letter.

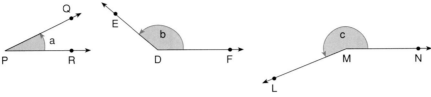

Figure 21–118

Classify the following angle measures using the terms *right, straight, acute,* or *obtuse.*

22. 50° **23.** 90° **24.** 120° **25.** 18° **26.** 75°
27. 180° **28.** 89° **29.** 179° **30.** 30°

State whether the following angle pairs are complementary, supplementary, or neither.

31. 63°, 37° **32.** 98°, 62° **33.** 135°, 45°
34. 45°, 35° **35.** 21°, 79° **36.** 90°, 90°
37. Congruent angles have _____ measures.
38.–39. Perpendicular lines are formed when a _____ line and a _____ line intersect.

Section 21–2

Add or subtract as indicated. Simplify the answers.

40. 34°28′41″
  + 18°37′50″

41. 115°34′29″
  − 84°26′18″

42. 34°29′35″
  + 19°30′25″

43. 64°15′37″
  − 29°37′41″

44. 80°
  − 28°14′28″

45. 74°
  − 13°19′42″

46. Find the complement of an angle whose measure is 35°29′14″.

Change to decimal degree equivalents. Express the decimals to the nearest ten-thousandth.

47. 29′ **48.** 47″ **49.** 7′34″

50. An angle of 34°36′48″ is divided into two equal angles. Express the measure of each of the equal angles in decimal degrees. Round to the nearest ten-thousandth.

Change to equivalent minutes and seconds. Round to the nearest second when necessary.

51. 0.75° **52.** 0.46° **53.** 0.2176°

In the following exercises, round angle measures to the nearest second when necessary.

54. A right angle is divided into eight equal parts. Find the measure of each angle in degrees and minutes.

55. An angle of 140° is divided into six equal parts. Find the measure of each angle in degrees and minutes.

56. An angle of 18°52′48″ needs to be three times as large. Find the measure of the new angle in decimal degrees.

57. An angle of 37°19′41″ is divided into two equal angles. Find the measure of each equal angle in degrees, minutes, and seconds.

Section 21–3
Classify the triangles in Figs. 21–119 to 21–121 according to their sides.

58.

Figure 21–119

59.

Figure 21–120

60.

Figure 21–121

Identify the longest and shortest sides in Figs. 21–122 and 21–123.

61.

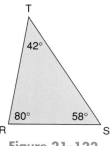

Figure 21–122

62.

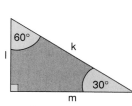

Figure 21–123

List the angles in order of size from largest to smallest in Figs. 21–124 and 21–125.

63.

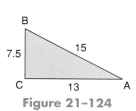

Figure 21–124

64.

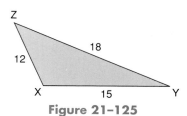

Figure 21–125

Write the corresponding parts not given for the congruent triangles in Figs. 21–126 to 21–128.

65.

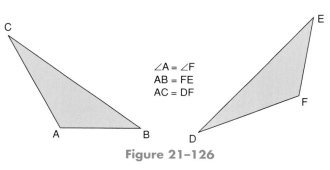

∠A = ∠F
AB = FE
AC = DF

Figure 21–126

66.

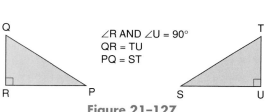

∠R AND ∠U = 90°
QR = TU
PQ = ST

Figure 21–127

67.

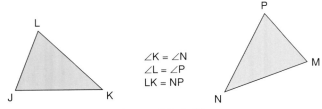

∠K = ∠N
∠L = ∠P
LK = NP

Figure 21-128

Use Fig. 21–129 to solve the following exercises. Round the final answers to the nearest thousandth if necessary.

68. $RS = 8$ mm; find ST and RT.
69. $RT = 15$ cm; find RS and ST.
70. $ST = 7.2$ ft; find RS and RT.
71. $RT = 9\sqrt{2}$ hm; find RS and ST.
72. $RT = 8\sqrt{5}$ dkm; find RS and ST.

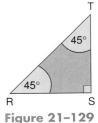

Figure 21-129

Use Fig. 21–130 to solve the following exercises. Round the final answers to the nearest thousandth if necessary.

73. $AC = 12$ dm; find AB and BC.
74. $AB = 15$ km; find AC and BC.
75. $BC = 10$ in.; find AC and AB.
76. $BC = 8\sqrt{2}$ hm; find AC and AB.
77. $AC = 40$ ft 7 in.; find AB and BC to the nearest inch.

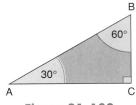

Figure 21-130

Solve the following problems. Round the final answers to the nearest thousandth if necessary.

78. The sides of an equilateral triangle are 4 dm in length. Find the altitude of the triangle. Round the answer to the nearest thousandth of a decimeter.

79. Find the total length to the nearest inch of the conduit $ABCD$ if $XY = 8$ ft, $CE = 14$ in., and angle $CBE = 30°$ (Fig. 21–131).

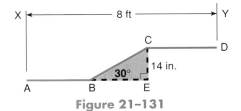

Figure 21-131

80. A piece of round steel is milled to a point at one end. The angle formed by the point is the angle of taper. If the steel has a 16-mm diameter and a 60° angle of taper, find the length c of the taper (Fig. 21–132).

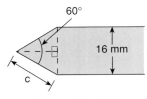

Figure 21-132

81. A V-slot forms a triangle whose angle at the vertex is 60°. The depth of the slot is 17 mm (see Fig. 21–133). Find the width of the V-slot.

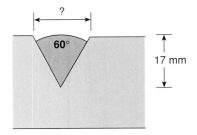

Figure 21-133

82. A manufacturer recommends attaching a guy wire to its 30-ft antenna at a 45° angle. If the antenna is installed on a flat surface, how long must the guy wire be (to the nearest foot) if it is attached to the antenna 4 ft from the top?

Section 21–4

Find the perimeter and the area of Figs. 21–134 and 21–135.

83.

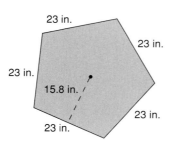

Figure 21–134

84.

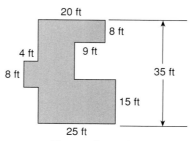

Figure 21–135

85. Identify the figure in Exercise 83 by giving its specific name and the number of degrees in each angle.

Solve the following problems involving the perimeter and area of composite figures.

86. How many sections of 2- × 4-ft ceiling tiles will be required for the layout in Fig. 21–136?

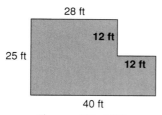

Figure 21–136

87. How many feet of baseboard will be needed to install around the room in Fig. 21–136 if we make no allowance for openings or doorways?

88. A gable end of a gambrel roof (Fig. 21–137) is to be covered with 12-in. bevel siding at an $8\frac{1}{2}$-in. exposure to the weather. Estimate the square footage of siding needed if 18% is allotted for lap and waste. Answer to the nearest foot.

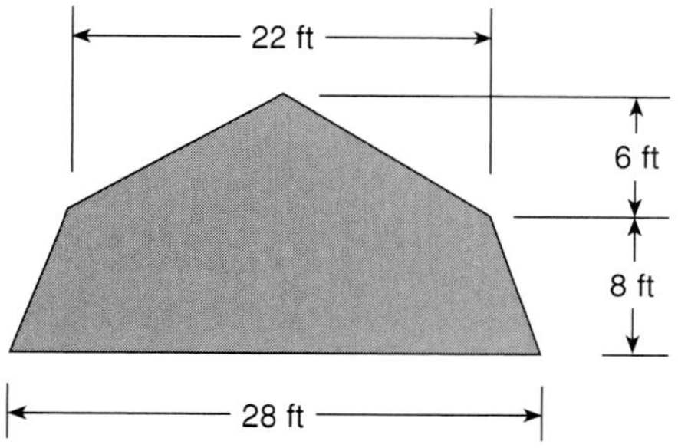

89. Find the square feet of floor space inside the walls of the plan of Fig. 21–138. Answer to the nearest square foot.

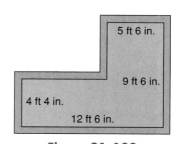

Figure 21–138

Figure 21–137

90. If one 2- × 4-in. stud is estimated for every linear foot of a wall when studs are 16 in. on center from each other, how many studs are needed for the outside walls of the floor plan in Fig. 21–139?

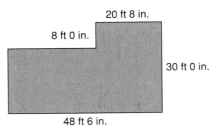

20 ft 8 in.

8 ft 0 in.

30 ft 0 in.

48 ft 6 in.

Figure 21–139

91. A home has concrete steps in the rear (Fig. 21–140). If the two sides are covered with a brick facing, how many bricks are needed if $\frac{1}{4}$-in. mortar joints are used? (Six bricks cover 1 square foot.) Round any part of a square foot to the next-highest square foot. (144 in.^{2} = 1 ft^2.)

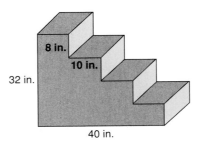

8 in.

10 in.

32 in.

40 in.

Figure 21–140

92. A mason will lay tile to cover an area in the form of a regular octagon whose eight sides are each 18 ft. If the distance from the midpoint of a side to the center of the octagon is 21.7 ft, how many tiles will the job require if $7\frac{1}{2}$ tiles will cover 1 square foot and 36 tiles are added for waste?

Section 21–5

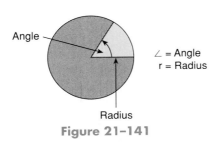

Angle

∠ = Angle
r = Radius

Radius

Figure 21–141

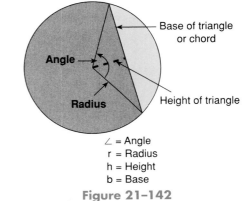

Base of triangle or chord

Angle

Radius

Height of triangle

∠ = Angle
r = Radius
h = Height
b = Base

Figure 21–142

Find the area of the sectors of a circle using Fig. 21–141. Round to hundredths.

93. ∠ = 45°9′
r = 2.58 cm

94. ∠ = 165°
r = 15 in.

95. ∠ = 15°15′
r = 110 mm

96. ∠ = 75°
r = 1 m

97. ∠ = 40°
$r = 2\frac{1}{2}$ ft

Find the area of the segments of a circle using Fig. 21–142. Round to hundredths.

98. ∠ = 52°30′
r = 11.25 in.
h = 10.09 in.
b = 10 in.

99. ∠ = 45°
r = 14 mm
h = 12.9 mm
b = 10.9 mm

100. ∠ = 55°
r = 10″
h = 8.9″
b = 9.2″

101. ∠ = 30°
r = 24 cm
h = 23.2 cm
b = 12.4 cm

102. ∠ = 60°
r = 13.3 cm
h = 11.5 cm
b = 13.3 cm

Solve the following problems involving sectors and segments of a circle.

103. What is the area (to the nearest tenth) of the flat top bifocal lens shown in Fig. 21–143?

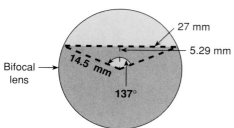

Figure 21–143

104. The minute hand on a grandfather clock cuts an angle of 30° as it moves from 12 to 1 on the clock face. If the minute hand is 27.5 cm long, what is the area of the clock face over which the minute hand moves from 12 to 1? Round to tenths.

105. What is the area of a patio (Fig. 21–144) that has one rounded corner? Round to hundredths.

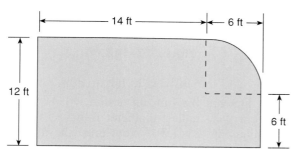

Figure 21–144

106. A machine cuts a 12-in.-diameter frozen pizza into slices whose sides form 72° angles at the center of the pizza. What is the surface area of each slice to the nearest square inch?

107. A drain pipe 20 in. in diameter has 4 in. of water in it (Fig. 21–145). What is the cross-sectional area of the water in the pipe to the nearest hundredth?

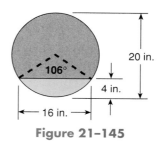

Figure 21–145

Find the arc length of the sectors of a circle using Fig. 21–141. Round to hundredths.

108. ∠ = 40°
r = 1.45 ft

109. ∠ = 180°
r = 10 in.

110. ∠ = 30°15′
r = 20 cm

111. ∠ = 70°10′
r = 30 mm

Section 21–6

For each figure, supply the missing dimensions. For an inscribed triangle, see Fig. 21–146.

112. If FB = 9, FO = _____
113. If AB = 10, AE = _____
114. ∠ABF = _____

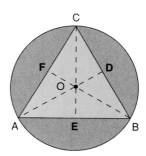

Figure 21–146

For a circumscribed square, see Fig. 21–147.

115. $\angle GJO =$ _____
116. If $GO = 7$, $HO =$ _____
117. If $KO = 10$, $IJ =$ _____

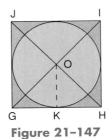

Figure 21–147

For an inscribed hexagon, see Fig. 21–148.

118. If $QO = 15$, $MN =$ _____
119. $\angle MOP =$ _____
120. $\angle MNO =$ _____

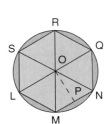

Figure 21–148

Solve the following problems involving polygons and circles. Round the answers to hundredths.

121. A shaft with a 20-mm diameter is milled at one end, as illustrated in Fig. 21–149. What is the radius of the inscribed circle?

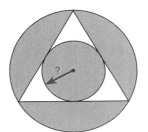

Figure 21–149

122. A hex nut 0.87 in. on a side will be milled from the smallest-diameter round stock possible (Fig. 21–150). What must the diameter of the stock be?

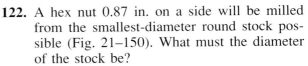
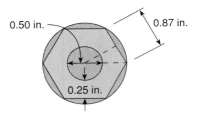

Figure 21–150

123. A square is milled on the end of a 5-cm shaft (Fig. 21–151). If the square is the largest that can be milled on the 5-cm shaft, what is the length of a side to hundredths?

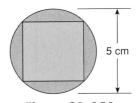

Figure 21–151

CHALLENGE PROBLEMS

124. Compare the area of a circle with a radius of 5 cm to the area of an inscribed equilateral triangle. Compare the area of a circle with a radius of 5 cm to the area of an inscribed square. Compare the area of a circle with a radius of 5 cm to the area of the inscribed polygon in Fig. 21–152.

125. Estimate the area of the inscribed regular pentagon in Fig. 21–153 by giving two values that the area will be between. Present a convincing argument for your estimate. Similarly, estimate the area of an inscribed regular octagon and give an argument for your estimate.

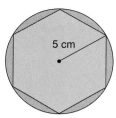

Figure 21-152

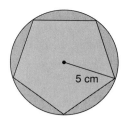

Figure 21-153

CONCEPTS ANALYSIS

1. Can the measures 8 cm, 12 cm, and 25 cm represent the sides of a triangle? Illustrate and explain your answer.

2. The Pythagorean theorem $a^2 + b^2 = c^2$ applies to right triangles where a and b are legs and c is the hypotenuse. For what type of triangle is the statement $a^2 + b^2 > c^2$ true? Illustrate your answer.

3. For what type of triangle is the statement $a^2 + b^2 < c^2$ true? Illustrate your answer.

4. A triangle has 6, 3, and $3\sqrt{3}$ as the measures of its sides. Sketch the triangle, place the measures on the appropriate sides, and show the measures of each of the three angles of the triangle.

5. Explain the process for finding the perimeter of a composite figure.

6. Explain the process for finding the area of a composite figure.

7. Draw a composite figure for which the area can be found by using either addition or subtraction and explain each approach.

8. Write in words the formula for finding the degrees in each angle of a regular polygon.

9. Describe a shortcut for finding the perimeter of an *L*-shaped figure. Illustrate your procedure with a problem.

10. Two pieces of property have the same area (acreage) and equal desirability for development. Both will require expensive fencing. One piece is a square 600 ft on each side. The other piece is a rectangle 400 ft by 900 ft. Which piece of property requires the least amount of fencing and is thus more desirable?

CHAPTER SUMMARY

Objectives

What to Remember with Examples

Section 21-1

1 Use various notations to represent points, lines, line segments, rays, and planes.

A dot represents a point and a capital letter names the point. A line extends in both directions and is named by any two points on the line (with $\leftrightarrow$ above the letters, such as $\overleftrightarrow{AB}$). A line segment has a beginning point and an ending point, which are named by letters (with a ‾ above the letters, such as $\overline{CD}$). A ray extends from a point on a line and includes all points on one side of the point and is named by the end point and any other point on the line forming the ray (with a → above the letters, such as $\overrightarrow{AC}$). A plane contains an infinite number of points and lines on a flat surface. (Line segments are often informally named just by letters, such as CD.)

Use proper notation for the following:

(a) Line *GH*   (b) Line segment *OP*   (c) Ray *ST*

(a) $\overleftrightarrow{GH}$ (b) $\overline{OP}$ (c) $\overrightarrow{ST}$

2 Distinguish among lines that intersect, that coincide, and that are parallel.

Lines that intersect meet at only one point. Lines that coincide will fit exactly on top of one another. Lines that are parallel are the same distance from one another and will never intersect.

Classify the pairs of lines in Fig. 21–154.

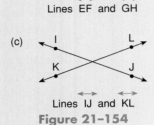

Figure 21–154

(a) Coinciding
(b) Parallel
(c) Intersecting

3 Use various notations to represent angles.

When two rays intersect in a point (end point of rays), an angle is formed. Angles may be named with three capital letters (end point in middle and one point on each ray), such as $\angle ABC$. They may be named by only the middle letter (vertex of the angle), such as $\angle B$. They may be assigned a number or lowercase letter placed within the vertex, such as $\angle 2$ or $\angle d$.

Name the angle in Fig. 21–155 three ways.

Figure 21–155

$\angle DEF$ or $\angle FED$, $\angle E$, $\angle a$

4 Classify angles according to size.

A right angle contains 90° or one-fourth a rotation. A straight angle contains 180° or one-half a rotation. An acute angle contains less than 90° but more than 0°. An obtuse angle contains more than 90° but less than 180°.

Identify the following angles:

(a) 30° (b) 100° (c) 90° (d) 180°

(a) acute (b) obtuse (c) right (d) straight

1 Add and subtract angle measures.

Only like measures can be added or subtracted. Arrange measures in columns of like measures. Add or subtract each column separately as for operations with U.S. customary measures (yards, feet, inches).

Perform the following:

(a) $35°15'10''$
    $+\ 10°55'\ 5''$
    $\overline{45°70'15''}\ =\ 46°10'15''$ Standard form

Borrowing and writing answers in standard form are also similar to operations with U.S. customary measures.

(b) Borrow.

$\overset{34}{\cancel{35°}}\overset{75}{\cancel{15'}}10''$ $1° = 60'.\ 60' + 15' = 75'.$
$-\ 10°55'\ 5''$
$\overline{24°20'\ 5''}$

2 Change minutes and seconds to a decimal part of a degree.

To change minutes to a decimal part of a degree: divide minutes by 60. To change seconds to a decimal part of a degree: divide seconds by 3600.

Change $15'36''$ to a decimal.

$15 \div 60 = 0.25$
$36 \div 3600 = 0.01$
$0.25 + 0.01 = 0.26°$

3 Change the decimal part of a degree to minutes and seconds.

To change a decimal part of a degree to minutes: multiply the decimal part of a degree by 60. To change a decimal part of a minute to seconds: multiply the decimal part of a minute by 60.

Change $0.48°$ to minutes and seconds.

$0.48 \times 60 = 28.8'$
$0.8 \times 60 = 48''$

Thus, $0.48° = 28'48''$.

4 Multiply and divide angle measures.

Multiply or divide angle measures as U.S. customary measures are multiplied or divided, and simplify the answers.

Perform the following:

(a) $25°\ 40'10''$
    $\times\ \qquad\ \ 3$
    $\overline{75°120'30''}$
    $=\ 77°\qquad 30''$

(b) $8°\qquad 33'\qquad 23\frac{1}{3}''$
    $3\overline{)25°\qquad 40'\qquad 10''}$
    $\ \ \underline{24}$
    $\ \ \ \overline{1°} =\ \ \underline{60'}$
    $\qquad\qquad 100'$
    $\qquad\qquad\ \ \underline{99}$
    $\qquad\qquad\ \ 1' =\ \underline{60''}$
    $\qquad\qquad\qquad\quad 70''$
    $\qquad\qquad\qquad\quad\underline{69}$
    $\qquad\qquad\qquad\qquad 1$

1 Classify triangles by sides.

A scalene triangle has all three sides unequal. An isosceles triangle has exactly two equal sides. An equilateral triangle has three equal sides.

> Identify the following triangles by the measures of their sides:
>
> (a) 6 cm, 6 cm, 6 cm (a) Equilateral
> (b) 12 in., 10 in., 12 in. (b) Isosceles
> (c) 10 cm, 12 cm, 15 cm (c) Scalene

2 Relate the sides and angles of a triangle.

An equilateral triangle has three equal angles. The equal sides of an isosceles triangle have opposite angles that are equal. If the three sides of a triangle are unequal, the largest angle will be opposite the longest side, and the smallest angle will be opposite the shortest side. In a right triangle, the hypotenuse is always the longest side.

> Identify the largest and the smallest angles in triangles with the following sides: 12 in., 10 in., 9 in.
>
> The largest angle is opposite the 12-in. side; the smallest angle is opposite the 9-in. side.

3 Determine whether two triangles are congruent using inductive and deductive reasoning.

Triangles are congruent (one can fit exactly over the other) (a) if the three sides of one equal the corresponding sides of the other, (b) if two sides and the included angle of one triangle are equal to two sides and the included angle of the other triangle, or (c) if two angles and the common side of one triangle are equal to two angles and the common side of the other triangle.

> Determine whether the pairs of triangles in Fig. 21–156 are congruent.
>
> (a) Congruent
> (b) Not congruent
> (c) Congruent
> (d) Congruent

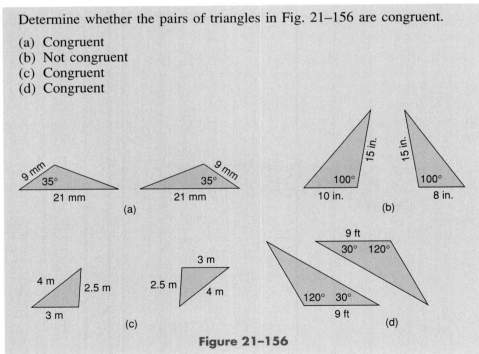

Figure 21–156

4 Use the properties of a 45°, 45°, 90° triangle to find missing parts and to solve applied problems.

To find the hypotenuse of a 45°, 45°, 90° triangle: multiply a leg by $\sqrt{2}$. To find a leg of a 45°, 45°, 90° triangle: divide the product of the hypotenuse and $\sqrt{2}$ by 2.

Given a 45°, 45°, 90° triangle, find:

(a) the hypotenuse if a leg is 10 cm
(b) a leg if the hypotenuse is 15 in.

(a) $10(\sqrt{2}) = 14.14213562$ or 14.1 cm   (rounded)

(b) $\dfrac{15(\sqrt{2})}{2} = 10.60660172$ or 10.6 in.   (rounded)

5 Use the properties of a 30°, 60°, 90° triangle to find missing parts and to solve applied problems.

To find the hypotenuse of a 30°, 60°, 90° triangle: multiply the side opposite the 30° angle by 2. To find the side opposite the 30° angle in a 30°, 60°, 90° triangle: divide the hypotenuse by 2. To find the side opposite the 60° angle of a 30°, 60°, 90° triangle: multiply the side opposite the 30° angle by $\sqrt{3}$.

Given a 30°, 60°, 90° triangle, find:

(a) the hypotenuse if the side opposite the 30° angle = 4.5 cm;
(b) the side opposite the 30° angle if the hypotenuse is 20 in.;
(c) the side opposite the 60° angle if the side opposite the 30° angle is 35 mm.

(a) $4.5(2) = 9$ cm   (b) $\dfrac{20}{2} = 10$ in.

(c) $35(\sqrt{3}) = 60.62177826$ or 60.6 in.   (rounded)

Section 21–4
1 Find the missing dimensions of composite shapes.

If a composite shape does not have all the dimensions marked, the missing dimensions can be inferred or calculated from our knowledge of polygons and the dimensions on the layout itself.

Find the dimensions x and y of the layout in Fig. 21–157. The layout is a square topped by an isosceles triangle.

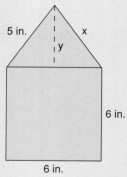

Figure 21–157

$x = 5$ in. because the legs of an isosceles triangle are equal. $y =$ the height of a triangle. From the apex of the triangle, drop a perpendicular line to the base, bisecting the base and forming two right triangles whose bases are each 3 in. (half of 6 in.). Use the Pythagorean theorem to find the altitude.

$$x^2 = y^2 + \left(\frac{1}{2} \text{ base}\right)^2$$
$$25 = y^2 + 3^2$$
$$25 = y^2 + 9$$
$$y^2 = 25 - 9$$
$$y^2 = 16$$
$$y = 4 \text{ in.}$$

2 Find the perimeter and area of composite shapes.

Perimeter of a composite: $P = a + b + c$, etc.

Area of a composite: use the area formulas for separate polygons.

Find the perimeter and area of the layout in Fig. 21–157.

$$P = 6 + 6 + 6 + 5 + 5$$
$$P = 28 \text{ in.}$$

$$A_{\text{square}} = 6^2 = 36 \text{ in.}^2$$

$$A_{\text{triangle}} = \frac{1}{2} bh$$

$$= \frac{1}{2}(6)(4)$$

$$= 12 \text{ in.}^2$$

$$A_{\text{layout}} = 36 + 12$$

$$= 48 \text{ in.}^2$$

3 Find the number of degrees in each angle of a regular polygon.

To find the number of degrees in each angle of a regular polygon: multiply the number of sides less 2 by 180° and divide the product by the number of sides.

Find the number of degrees in each angle of a pentagon.

A pentagon has five sides.

$$\text{Degrees per angle} = \frac{180(5 - 2)}{5}$$

$$= \frac{180(3)}{5}$$

$$= \frac{540}{5} = 108°$$

Section 21–5

1 Find the area of a sector.

A sector is a portion of a circle cut off by two radii.

$$\text{Area of a sector} = \frac{\theta}{360} \pi r^2$$

where θ is the degrees of the central angle formed by the radii.

CHAPTER 21 Selected Concepts of Geometry

Find the area of a sector whose central angle is 50° and whose radius is 20 cm (Fig. 21-158).

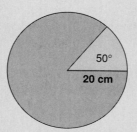

Figure 21-158

$$\text{Area} = \frac{\theta}{360}\,\pi r^2$$
$$= \frac{50}{360}\,(\pi)(20)^2$$
$$= 174.5329252 \quad \text{or} \quad 174.53\ \text{cm}^2 \quad \text{(rounded)}$$

2 Find the arc length of a sector.

Arc length is the portion of the circumference cut off by the sides of a sector.

$$\text{Arc length } (s) = \frac{\theta}{360}(2\pi r) \text{ or } (s) = \frac{\theta}{360}(\pi d)$$

Find the arc length of a sector formed by a 50° central angle if the radius is 20 cm (Fig. 21-159).

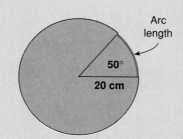

Figure 21-159

$$s = \frac{\theta}{360}(2\pi r)$$
$$s = \frac{50}{360}(2)(\pi)(20)$$
$$s = 17.45329252 \quad \text{or} \quad 17.45\ \text{cm}$$

3 Find the area of a segment.

A chord is a line segment joining two points on a circle. An arc is the portion of the circumference cut off by a chord. A segment is the portion of a circle bounded by a chord and an arc.

The area of a segment is the area of the sector less the area of the triangle formed by the chord and the central angle of the sector:

$$A = \frac{\theta}{360}\,\pi r^2 - \frac{1}{2}\,bh$$

Find the area of the segment formed by a 12-cm chord if the height of the triangle part of the sector is 6 cm. The central angle is 90° (Fig. 21–160).

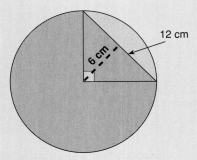

Figure 21–160

$$A = \frac{\theta}{360}\,\pi r^2 - \frac{1}{2}bh$$

$$A = \frac{90}{360}(\pi)(8.485281374)^2 - \frac{1}{2}(12)(6)$$

Use Pythagorean theorem to find radius = 8.485281374 cm.

$$A = 0.25(\pi)(72) - 0.5(72)$$
$$A = 20.54866776 \qquad \text{or} \qquad 20.48\ \text{cm}^2 \qquad \text{(rounded)}$$

Section 21–6

1 Use the properties of inscribed and circumscribed equilateral triangles to find missing amounts and to solve applied problems.

The height of an equilateral triangle forms congruent 30°, 60°, 90° triangles.

The radius of a circle circumscribed about an equilateral triangle is two-thirds the height of the triangle or twice the distance from the center of the circle to the base of the triangle.

The radius of a circle inscribed in an equilateral triangle is one-third the height of the triangle or one-half the distance from the vertex of the triangle to the center of the circle.

One end of a shaft 30 mm in diameter is milled as shown in Fig. 21–161. Find the diameter of the smaller circle.

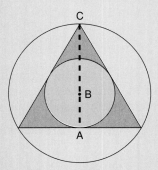

Figure 21–161

BC is half the diameter of the large circle.

$$BC = \frac{1}{2}(30)$$
$$BC = 15 \text{ mm}$$
$$AB = \frac{1}{2}BC$$
$$AB = \frac{1}{2}(15)$$
$$AB = 7.5 \text{ mm} \qquad \text{(radius of small circle)}$$

The diameter of the small circle = 2(7.5) = 15 mm.

2 Use the properties of inscribed and circumscribed squares to find missing amounts and to solve applied problems.

The diagonal of a square inscribed in or circumscribed about a circle forms congruent 45°, 45°, 90° triangles.

The diameter of a circle inscribed in a square equals a side of the square.

The diameter of a circle circumscribed about a square equals a diagonal of the square.

The radius of a circle inscribed in a square equals the height of a 45°, 45°, 90° triangle formed by two diagonals.

The radius of a circle circumscribed about a square equals the height of a 45°, 45°, 90° triangle formed by one diagonal.

The end of a 20-mm round rod is milled as shown in Fig. 21–162. Find the area of the square.

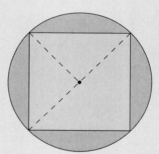

Figure 21–162

The radius equals the height of a right triangle formed by one diagonal. The diameter is the base. Find the area of the triangle and double it.

$$A_{\text{triangle}} = \frac{1}{2}bh$$
$$= \frac{1}{2}(20)(10)$$
$$= 100 \text{ mm}^2$$

Area of square end: $100 \times 2 = 200 \text{ mm}^2$.

3 Use the properties of inscribed and circumscribed regular hexagons to find missing amounts and to solve applied problems.

The three diagonals that join opposite vertices of a regular hexagon form six congruent equilateral triangles.

The height of any of the equilateral triangles described above forms two congruent 30°, 60°, 90° triangles. The diameter of a circle circumscribed about a regular hexagon is a diagonal of the hexagon (distance across corners).

The radius of a circle circumscribed about a regular hexagon is a side of an equilateral triangle formed by the three diagonals or one-half the distance across corners.

The radius of a circle inscribed in a regular hexagon is the height of an equilateral triangle formed by the three diagonals.

The diameter of a circle inscribed in a regular hexagon is the distance across flats or twice the height of an equilateral triangle formed by the three diagonals.

Find the following dimensions for Fig. 21–163 if the diameter is 10 cm.

(a) $\angle AOF$
(b) Radius
(c) Distance across corners

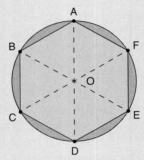

Figure 21–163

(a) $\angle AOF = 60°$ (angle of equilateral triangle)
(b) 5 cm (radius of a circumscribed circle is a side of the equilateral triangle)
(c) 10 cm (diameter of a circumscribed circle is the distance across corners.)

WORDS TO KNOW

geometry (p. 758)
point (p. 758)
line (p. 758)
plane (p. 758)
line segment (p. 758)
end point (p. 758)
ray (p. 758)
interior point (p. 759)
intersect (p. 759)
coincide (p. 759)
parallel (p. 759)
angle (p. 760)
side of an angle (p. 760)
vertex (p. 760)
degree (p. 761)
perpendicular (p. 761)
vertical (p. 761)
right angle (p. 762)

straight angle (p. 762)
acute angle (p. 762)
obtuse angle (p. 762)
complementary angles (p. 762)
supplementary angles (p. 762)
congruent (p. 762)
protractor (p. 764)
index mark on protractor
    (p. 764)
minute (p. 765)
second (p. 765)
equilateral triangle (p. 769)
scalene triangle (p. 770)
isosceles triangle (p. 770)
corresponding sides (p. 771)
congruent triangles (p. 771)
included angle (p. 772)
inductive reasoning (p. 772)

deductive reasoning (p. 773)
45°, 45°, 90° triangle (p. 773)
30°, 60°, 90° triangle (p. 775)
altitude (p. 775)
composite (p. 779)
perimeter (p. 780)
regular polygon (p. 783)
sector (p. 786)
arc length (p. 787)
chord (p. 788)
arc (p. 788)
segment (p. 788)
inscribed polygon (p. 791)
circumscribed polygon (p. 791)
tangent (p. 791)
distance across corners (p. 796)
distance across flats (p. 796)

1. Show the proper notation to name the line segment from *B* to *C* in Fig. 21–164.

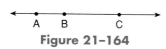

Figure 21-164

2. Name the angle in Fig. 21–165 in four different ways.

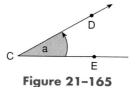

Figure 21-165

Classify the following angles as right, straight, acute, or obtuse.

3. 42°

4. 158°

Identify the lines in Figs. 21–166 to 21–168 as intersecting, parallel, or perpendicular.

5.

Figure 21-166

6.

Figure 21-167

7.

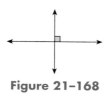

Figure 21-168

8. Subtract 40°37′26″ from 75°.

9. An angle of 47°16′28″ is divided into three equal angles. Find the measure of each angle in degrees, minutes, and seconds (to the nearest second).

10. Change 15′32″ to degrees to the nearest ten-thousandth.

11. Change 0.3125° to minutes and seconds.

12. Name the following triangles: (a) all sides unequal, (b) exactly two sides equal, and (c) all sides equal.

13. Identify the largest and the smallest angles in Fig. 21–169.

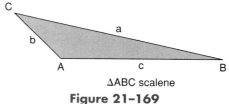

△ABC scalene

Figure 21-169

14. In Fig. 21–170, identify (a) the angle corresponding to ∠*DEF* and (b) the side corresponding to *HI*.

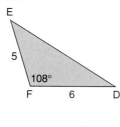

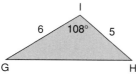

Figure 21-170

15. A conduit *ABCD* must be made so that ∠*CBE* is 45° (Fig. 21–171). If *BE* = 4 cm and *AK* = 12 cm, find the length of the conduit to the nearest thousandth centimeter.

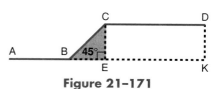

Figure 21-171

16. The rafters of a house make a 30° angle with the joists (Fig. 21–172). If the rafters have an 18-in. overhang and the center of the roof is 10 ft above the joists, how long must the rafters be cut?

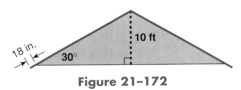

Figure 21–172

18. A section of a hip roof is a trapezoid measuring 35 ft at the bottom, 15 ft at the top, and 10 ft high. Find the area of this section of the roof in square feet.

20. Find the area of the sector in Fig. 21–174 if the central angle is 55° and the radius is 45 mm. Round to hundredths.

21. Find the arc length cut off by the sector in Fig. 21–174. Round to hundredths.

22. A segment is removed from a flat metal circle so that the piece will rest on a horizontal base (Fig. 21–175). Find the area of the segment that was removed.

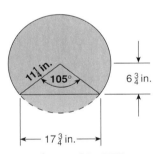

Figure 21–175

24. Taylor Fink milled a square metal rod so that a circle is formed at the end (see Fig. 21–177). If the square cross-section is 1.8 in. across the corners, what is the diameter of the circle?

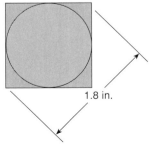

Figure 21–177

17. Find the perimeter of the composite in Fig. 21–173.

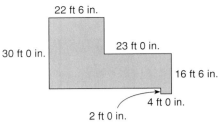

Figure 21–173

19. Find the number of degrees in each angle of a regular octagon (eight sides).

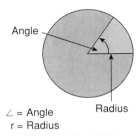

∠ = Angle
r = Radius

Figure 21–174

23. Find the circumference of a circle inscribed in an equilateral triangle if the height of the triangle is 12 in. (Fig. 21–176).

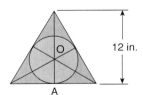

Figure 21–176

25. Haley Fink milled a circle on the cross-sectional end of a hexagonal rod (Fig. 21–178). If the distance across the flats of the cross-section is $\frac{1}{2}$ in., what is the circumference of the circle?

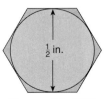

Figure 21–178

CHAPTER

22

Introduction to Trigonometry

GOOD DECISIONS THROUGH TEAMWORK

Trigonometry allows us to find lengths, angles, or other values, such as the length of the arc formed when a pendulum swings or the force of an electric current, that cannot be measured directly. Your team is assigned to find out what are some of the major practical uses of trigonometry, who uses trigonometry in the workplace, and what are the names of some of these workplaces.

To complete this project, your team members may investigate other mathematics books and library resources, and they can interview professionals like surveyors, manufacturers, architects, engineers, and college professors in technical and scientific fields. Team members may also want to visit workplaces suggested in the interviews, such as manufacturing plants, architectural firms that design buildings and roads, and airports.

Collect the information gathered by team members and organize it into categories, such as "Users," "Uses," and "Places." The team may want to organize the information on charts, overhead transparencies, or similar media for an oral presentation to the class.

One of the most important uses of *trigonometry* is to allow us to find by indirect measurement lengths or distances that are difficult or impossible to measure directly. Although trigonometry, which means "triangle measurement," involves more than a study of just the triangle, we will limit our study to the trigonometry of the triangle.

In our study of trigonometry, we will build on much of our knowledge of angles and triangles from geometry. We will also see that certain calculations are made more quickly and in fewer steps than they can be made using geometry alone. Also, we will see that certain calculations that cannot be made using geometry alone can be made using trigonometry.

22–1 RADIANS AND DEGREES

Learning Objectives

1. Find arc length, given a central angle in radian measure.
2. Find the area of a sector, given a central angle in radian measure.
3. Convert angles in degree measure to angles in radian measure.
4. Convert angles in radian measure to angles in degree measure.

In this section we study the angles formed by radii of a circle and how such angles are measured. We also use that measure to find additional information about the circle.

In geometry, we learned that angles can be measured in units called degrees. Another unit for measuring angles is the *radian*.

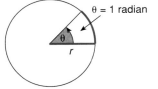

Figure 22–1

■ **DEFINITION 22–1: Radian.** A *radian* is the measure of a central angle of a circle whose intercepted arc is equal in length to the radius of the circle (Fig. 22–1). The abbreviation for radian is *rad*.

The circumference of a circle is related to the radius by the formula $C = 2\pi r$. Thus, the ratio of the circumference to the radius of any circle is $\frac{C}{r} = 2\pi$; that is, the length of the radius could be measured off 2π times (about 6.28 times) along the circumference. A complete rotation is 2π radians (see Fig. 22–2).

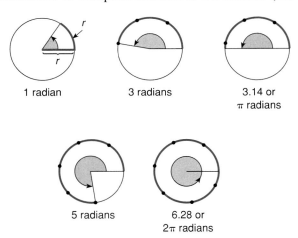

Figure 22–2

How are radians and degrees related? A central angle measuring 1 radian makes an arc length equal to the radius, so we will use the arc-length formula that uses degrees to find the equivalent angle measure in degrees for 1 radian.

$$\text{Arc length} = \frac{\theta}{360} 2\pi r$$

If the radius and arc length are equal to 1,

$$1 = \frac{\theta}{360} \, 2\pi \ (1)$$

$$360 = \theta \, (2\pi)$$ Multiply both sides by 360.

$$\frac{360}{2\pi} = \frac{\theta \, (2\pi)}{2\pi}$$ Divide both sides by 2π.

$$57.29577951 = \theta$$

One radian is approximately 57.3°.

Tips and Traps   **Degree and Radian Notation:**
A degree measure always requires the word *degree* or the degree symbol to be written. There is no comparable symbol for the radian measure. The abbreviation *rad,* or no indication at all, identifies radian angle measures.

---

1 Find Arc Length, Given a Central Angle in Radian Measure.

Radian measure is used very often in physics and engineering mechanics. One application of radian measure is in determining *arc length,* which is the length of the arc intercepted by a central angle.

The circumference of a circle is $2\pi r$, where r is the radius of the circle. 2π is the radian measure of a complete rotation. To find the arc length, s, of an arc, we can multiply the radian measure of the central angle formed by the end points of the arc and the center of the circle times the radius of the circle.

Arc length = (radian measure of angle)(radius)

We will use the Greek lowercase letter theta (θ) to represent the radian measure of the central angle in the formula for arc length.

Formula 22–1   *Arc length:*

$$s = \theta r$$

where θ is the central angle measured in radians.

EXAMPLE 1   Find the arc length intercepted on the circumference of the circle in Fig. 22–3 by a central angle of $\frac{\pi}{3}$ radians (rad) if the radius of the circle is 10 cm.

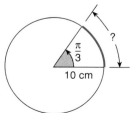

Figure 22–3

$s = \theta r$ Substitute $\frac{\pi}{3}$ for θ and 10 for r.

$s = \dfrac{\pi}{3}(10)$

$s = \dfrac{\pi(10)}{3}$

$s = 10.47$ cm   To the nearest hundredth

Thus, the arc length of the intercepted arc is 10.47 cm.

EXAMPLE 2 Find the radian measure of an angle at the center of a circle of radius 5 m. The angle intercepts an arc length of 12.5 m (see Fig. 22–4).

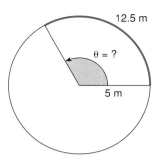

12.5 m

θ = ?

5 m

Figure 22–4

Because $s = \theta r$,

$\theta = \dfrac{s}{r}$ Formula is rearranged for θ.

$\theta = \dfrac{12.5}{5}$ Substitute 12.5 for *s* and 5 cm for *r*.

$\theta = 2.5$ rad

Thus, the angle is 2.5 rad.

EXAMPLE 3 Find the radius of an arc if the length of the arc is 8.22 cm and the intercepted central angle is 3 rad (see Fig. 22–5).

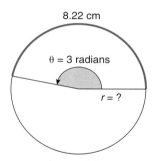

8.22 cm

θ = 3 radians

r = ?

Figure 22–5

Because $s = \theta r$,

$r = \dfrac{s}{\theta}$ Formula is rearranged for *r*.

$r = \dfrac{8.22}{3}$ Substitute 8.22 for *s* and 3 rad for θ.

$r = 2.74$ cm

Therefore, the radius of the arc is 2.74 cm.

2 Find the Area of a Sector, Given a Central Angle in Radian Measure.

The formula for the area of a circle is $A = \pi r^2$. A sector is a part of a circle. In this formula we will examine how the area of a complete rotation (circle) relates to the angle measure in radians of a complete rotation (2π). First, we rewrite the area formula so that we can see the factors 2π. Because $\frac{1}{2}(2) = 1$

and multiplying by 1 does not change the value of an expression, we can rewrite $A = \pi r^2$ as $A = \frac{1}{2}(2)\pi r^2$. Then, $A = \frac{1}{2}(2\pi)r^2$, or $A = \frac{1}{2}$(radian angle measure of a complete rotation)r^2. In Section 21–5, we defined a *sector* as the part of a circle determined by two radii of the circle and the included arc. The area of the sector was a fractional part of the area of a complete circle, $A = \frac{\theta}{360}\pi r^2$, where θ is the central angle measure of the sector in degrees. Similarly, the area of a sector can be written as $A = \frac{1}{2}\theta r^2$, where θ is the radian measure of the central angle of the sector in radians.

Formula 22–2 *Area of sector:*

$$A = \frac{1}{2}\theta r^2$$

where θ is the central angle measured in radians.

EXAMPLE 4   Find the area of a sector whose central angle is 5 rad and whose radius is 7.2 in.

$$A = \frac{1}{2}\theta r^2$$

$$A = \frac{1}{2}(5)(7.2)^2$$

$$A = 129.6 \text{ in.}^2$$

Thus, the area of the sector is 129.6 in.^{2}.

[3] Convert Angles in Degree Measure to Angles in Radian Measure.

Many job-related problems will require us to convert radians to degrees for certain applications and to convert degrees to radians for others. A complete rotation is 360°, or 2π rad. To convert from one type of angle measure to another, we multiply by a *unity ratio* that relates degrees and radians. Because 360° = 2π rad, we can simplify the relationship to 180° = π rad $\left(\frac{360°}{2} = \frac{2\pi}{2}\text{ rad}\right)$.

Rule 22–1 *To convert degrees to radians:*

Multiply degrees by the unity ratio $\frac{\pi \text{ rad}}{180°}$.

EXAMPLE 5   Convert the following degree measures to radians. Use the calculator value for π to change to a decimal equivalent rounded to the nearest hundredth.

(a) 20°   (b) 50°   (c) 175°   (d) 270°   (e) 67°

(a) Because 360° = 2π rad or 180° = π rad, we use this relationship in a unity ratio to convert degrees to radians. As in the procedure we used to change to different measuring units in Chapter 5, we arrange the measures in the unity ratio so the original units cancel.

$$20° \times \frac{\pi \text{ rad}}{180°} = \frac{20(\pi)}{180} = \mathbf{0.35 \text{ rad}}$$

(b) $50° \times \dfrac{\pi \text{ rad}}{180°} = \dfrac{50(\pi)}{180} = \mathbf{0.87 \text{ rad}}$

(c) $175° \times \dfrac{\pi \text{ rad}}{180°} = \dfrac{175(\pi)}{180} = \textbf{3.05 rad}$

(d) $270° \times \dfrac{\pi \text{ rad}}{180°} = \dfrac{270(\pi)}{180} = \textbf{4.71 rad}$

(e) $67° \times \dfrac{\pi \text{ rad}}{180°} = \dfrac{67(\pi)}{180} = \textbf{1.17 rad}$

When parts of a degree are expressed in decimals, we use the same procedure as before to convert to radians. When angle measures are expressed in degrees, minutes, and seconds, we must first convert the minutes and seconds to a decimal part of a degree before converting to radians. This procedure was discussed in detail in Section 21–2.

EXAMPLE 6 Convert the following measures to radians. Round the degree measure to the nearest ten-thousandth; then round the radian measure to the nearest hundredth.

(a) $15.25°$ (b) $25°20'$ (c) $110°25'30''$

(a) $15.25° \times \dfrac{\pi \text{ rad}}{180°} = \dfrac{15.25(\pi)}{180} = \textbf{0.27 rad}$ Rounded

(b) $25°20' = 25.3333°$ Convert minutes to a decimal part of a degree: $20' \times \dfrac{1°}{60'} = \dfrac{20°}{60} = 0.3333°$

$25.3333° \times \dfrac{\pi \text{ rad}}{180°} = \dfrac{25.3333(\pi)}{180} = \textbf{0.44 rad}$ Rounded

(c) $110°25'30'' = 110.425°$ Convert minutes and seconds to a decimal part of a degree: $110 + \dfrac{25}{60} + \dfrac{30}{3600} = 110.425$

$110.425° \times \dfrac{\pi \text{ rad}}{180°} = \dfrac{110.425(\pi)}{180} = \textbf{1.93 rad}$ Rounded

4 Convert Angles in Radian Measure to Angles in Degree Measure.

When converting from radians to degrees, we multiply by a unity ratio so that the radian measures cancel and are replaced by degrees.

> **Rule 22–2** *To convert radians to degrees:*
> Multiply radians by the unity ratio $\dfrac{180°}{\pi \text{ rad}}$.

EXAMPLE 7 Convert the following to degrees. Round to the nearest ten-thousandth of a degree.

(a) 2 rad   (b) $\dfrac{\pi}{2}$ rad   (c) 0.25 rad

(a) $2 \text{ rad} \times \dfrac{180°}{\pi \text{ rad}} = \dfrac{360}{\pi} = \textbf{114.5916°}$ Substitute calculator value of π and round.

(b) $\dfrac{\overset{1}{\pi}}{2} \text{ rad} \times \dfrac{180°}{\underset{1}{\pi} \text{ rad}} = \dfrac{180}{2} = \textbf{90°}$ Because π canceled, we did not have to substitute for π.

(c) $0.25 \text{ rad} \times \dfrac{180°}{\pi \text{ rad}} = \dfrac{(0.25)(180)}{\pi} = \textbf{14.3239°}$ Use calculator value for π and round.

EXAMPLE 8   Convert the following to degrees, minutes, and seconds. Round to the nearest second.

(a) 1 rad   (b) 3.2 rad

(a) $1 \text{ rad} \times \dfrac{180°}{\pi \text{ rad}} = \dfrac{180°}{\pi} = 57.29577951°$ Continue with the decimal part of the degree measure.

$0.29577951° \times \dfrac{60'}{1°} = 17.74677077'$ Continue with the decimal part of the minute measure.

$0.74677077' \times \dfrac{60''}{1'} = 45''$ Convert the decimal part of minute to seconds, to the nearest second.

Thus, 1 rad = 57°17′45″.

(b) $3.2 \text{ rad} \times \dfrac{180°}{\pi \text{ rad}} = \dfrac{3.2(180)}{\pi} = 183.3464944°$

$0.3464944° \times \dfrac{60'}{1°} = 20.7896664'$ Convert the decimal part of degree to minutes.

$0.7896664' \times \dfrac{60''}{1'} = 47'$ Convert the decimal part of minute to seconds, to the nearest second.

Thus, 3.2 rad = 183°20′47″.

■ General Tips for Using the Calculator

Using Only the Decimal Part of a Value

When a procedure requires that we continue a calculation with only the decimal part of a result, first subtract the whole number part. Look at the calculator sequence for Example 8, part (a):

$\boxed{180}$ $\boxed{\div}$ $\boxed{\pi}$ $\boxed{=}$ $\boxed{-}$ 57 $\boxed{=}$ $\boxed{\times}$ 60 $\boxed{=}$ $\Rightarrow$ 17.74677077

Whenever the central angle measure is expressed in degrees, we may use either Formula 21–2 or Formula 22–2. To use Formula 22–2, we must convert the central angle measure from degrees to radians. In the following example, the area is found using both formulas.

EXAMPLE 9   Find the area of a sector whose central angle is 135° and whose radius is 2.7 cm.

Using Formula 21–2, $A = \frac{\theta}{360}\pi r^2$, where θ is given in degrees:

$A = \left(\dfrac{135}{360}\right)(\pi)(2.7)^2 = 8.59 \text{ cm}^2$ To nearest hundredth

Using Formula 22–2, $(A = \frac{1}{2}\theta r^2$, where θ is given in radians, 135° must first be changed to radians.

$$135° = 135° \times \dfrac{\pi \text{ rad}}{180°} = \dfrac{135(\pi)}{180} = 2.35619449 \text{ rad}$$

$A = \dfrac{1}{2}(2.35619449)(2.7)^2 = 8.59 \text{ cm}^2$ To nearest hundredth

Therefore, the area of the sector is approximately 8.59 cm².

Solve the following problems. Round answers to hundredths if necessary.

1. Find the arc length intercepted on the circumference of a circle by a central angle of 2.15 rad if the radius of the circle is 3 in.

2. Find the arc length intercepted on the circumference of a circle by a central angle of 4 rad if the radius of the circle is 3.5 cm.

3. Find the radian measure of an angle at the center of a circle of radius 2 in. if the angle intercepts an arc length of 8.5 in.

4. Find the radian measure of an angle at the center of a circle of radius 4.3 cm if the angle intercepts an arc length of 15 cm.

5. Find the radius of an arc if the length of the arc is 14.7 cm and the intercepted central angle is 2.1 rad.

6. Find the radius of an arc if the length of the arc is 12.375 in. and the intercepted central angle is 2.75 rad.

7. Find the area of a sector whose central angle is 2.14 rad and whose radius is 4 in.

8. Find the area of a sector whose central angle is 6 rad and whose radius is 1.2 cm.

9. Find the radius of a sector if the area of the sector is 7.5 cm^2 and the central angle is 3 rad.

10. How many radians does the central angle of a sector measure if its area is 1.7 in.^{2} and its radius is 2 in.?

 Convert the following degree measures to radians rounded to the nearest hundredth.

11. 45°   **12.** 56°   **13.** 78°   **14.** 140°

Convert the following measures to degrees rounded to the nearest ten-thousandth. Then convert to radians to the nearest hundredth.

15. 21°45′   **16.** 177°33′   **17.** 44°54′12″   **18.** 10°31′15″

4 Convert the following radian measures to degrees. Round to the nearest ten-thousandth of a degree.

19. $\frac{\pi}{4}$ rad   **20.** $\frac{\pi}{6}$ rad   **21.** 2.5 rad   **22.** 1.4 rad

Convert the following radian measures to degrees, minutes, and seconds. Round to the nearest second.

23. 0.5 rad   **24.** $\frac{\pi}{8}$ rad   **25.** 0.75 rad   **26.** 1.1 rad

27. Find the arc length of an arc whose intercepted angle is 38° and whose radius is 2.3 cm. Round to hundredths.

28. Find the number of degrees in a central angle whose arc length is 3.2 in. and whose radius is 3 in. Round to the nearest whole degree.

29. Find the area of a sector whose central angle is 105° and whose radius is 7.2 cm. Round to hundredths.

30. Find the number of degrees to the nearest ten-thousandth of a central angle of a sector whose area is 5.6 cm^2 and whose radius is 4 cm.

22–2 TRIGONOMETRIC FUNCTIONS

Learning Objectives

1 Find the sine, cosine, and tangent of angles of right triangles, given the measures of at least two sides.

2 Find the cosecant, secant, and cotangent of angles of right triangles, given the measures of at least two sides.

1 Find the Sine, Cosine, and Tangent of Angles of Right Triangles, Given the Measures of at Least Two Sides.

In geometry, we studied some basic properties of similar triangles and right triangles that allowed us to find missing measures of the sides of the triangle. Using trigonometry, we can determine the measure of either acute angle of a right triangle if we know the measure of at least two sides of the right triangle. We will define several functions that are ratios of various sides of a triangle, and these ratios will be used later in determining the angles of a triangle.

Figure 22–6 shows a right triangle, *ABC*, with the sides of the triangle labeled according to their relationship to angle *A*. The *hypotenuse* is the side opposite the right angle of the triangle, and the hypotenuse forms one of the sides of angle *A*. The other side that forms angle *A* will be called the *adjacent side* of angle *A*. The third side of the triangle is called the *opposite side* of angle *A*.

In Fig. 22–7, the sides of the right triangle *ABC* are labeled according to their relationship to angle *B*. The hypotenuse forms one of the sides of angle *B*, the other side that forms angle *B* is the adjacent side of angle *B*, and the third side is the opposite side of angle *B*. We used in Section 13–3 the term *hypotenuse*, but the terms *adjacent side* and *opposite side* are also very important in understanding trigonometric functions.

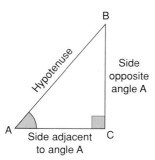

Figure 22–6

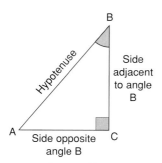

Figure 22–7

■ **DEFINITION 22–2: Adjacent Side.** The *adjacent side* of a given acute angle of a right triangle is the side that forms the angle with the hypotenuse.

■ **DEFINITION 22–3: Opposite Side.** The *opposite side* of a given acute angle of a right triangle is the side that does not form the given angle.

The three most commonly used trigonometric functions are the *sine, cosine,* and *tangent*. The sine, cosine, and tangent of angle *A* in Fig. 22–8 are defined to be the following ratios of sides of the right triangle.

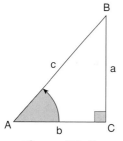

Figure 22–8

$$\text{Sine of angle } A = \frac{\text{side opposite angle } A}{\text{hypotenuse}}$$

$$\text{Cosine of angle } A = \frac{\text{side adjacent to angle } A}{\text{hypotenuse}}$$

$$\text{Tangent of angle } A = \frac{\text{side opposite angle } A}{\text{side adjacent to angle } A}$$

For convenience, we will abbreviate the sine, cosine, and tangent functions as *sin, cos,* and *tan,* respectively.

Furthermore, we will abbreviate the names of the ratios and use the letters that designate the sides of the *standard right triangle* in Fig. 22–8. In the standard right triangle, side *a* is opposite angle *A*, side *b* is opposite angle *B*, and side *c* (hypotenuse) is opposite angle *C* (right angle). Thus, we can identify the functions of angle *A* as follows.

Formula 22–3 *Trigonometric functions of angle A in a standard right triangle:*

$$\sin A = \frac{\text{side opposite } \angle A}{\text{hypotenuse}} = \frac{a}{c}$$

$$\cos A = \frac{\text{side adjacent to } \angle A}{\text{hypotenuse}} = \frac{b}{c}$$

$$\tan A = \frac{\text{side opposite } \angle A}{\text{side adjacent to } \angle A} = \frac{a}{b}$$

Similarly, we can identify the sine, cosine, and tangent of the other acute angle, angle B.

Formula 22–4   *Trigonometric functions of angle B in a standard right triangle*:

$$\sin B = \frac{\text{side opposite } \angle B}{\text{hypotenuse}} = \frac{b}{c}$$

$$\cos B = \frac{\text{side adjacent to } \angle B}{\text{hypotenuse}} = \frac{a}{c}$$

$$\tan B = \frac{\text{side opposite } \angle B}{\text{side adjacent to } \angle B} = \frac{b}{a}$$

We can now determine the sine, cosine, and tangent of a right triangle by writing the appropriate ratio of the sides of the triangle and expressing the ratio in lowest terms or as a decimal equivalent.

EXAMPLE 1   Find the sine, cosine, and tangent of angles A and B in Fig. 22–9. Leave the answers as fractions in lowest terms.

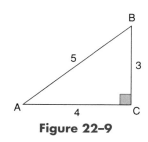

Figure 22–9

$$\sin A = \frac{\text{opposite}}{\text{hypotenuse}} = \frac{3}{5}$$

$$\cos A = \frac{\text{adjacent}}{\text{hypotenuse}} = \frac{4}{5}$$

$$\tan A = \frac{\text{opposite}}{\text{adjacent}} = \frac{3}{4}$$

$$\sin B = \frac{\text{opposite}}{\text{hypotenuse}} = \frac{4}{5}$$

$$\cos B = \frac{\text{adjacent}}{\text{hypotenuse}} = \frac{3}{5}$$

$$\tan B = \frac{\text{opposite}}{\text{adjacent}} = \frac{4}{3}$$

EXAMPLE 2   Find the sine, cosine, and tangent of angles A and B in Fig. 22–10. Write the answers as decimals. Round to the nearest ten-thousandth.

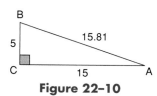

Figure 22–10

$$\sin A = \frac{\text{opposite}}{\text{hypotenuse}} = \frac{5}{15.81} = \mathbf{0.3163}$$

$$\cos A = \frac{\text{adjacent}}{\text{hypotenuse}} = \frac{15}{15.81} = \mathbf{0.9488}$$

$$\tan A = \frac{\text{opposite}}{\text{adjacent}} = \frac{5}{15} = \mathbf{0.3333}$$

$$\sin B = \frac{\text{opposite}}{\text{hypotenuse}} = \frac{15}{15.81} = \mathbf{0.9488}$$

$$\cos B = \frac{\text{adjacent}}{\text{hypotenuse}} = \frac{5}{15.81} = \mathbf{0.3163}$$

$$\tan B = \frac{\text{opposite}}{\text{adjacent}} = \frac{15}{5} = \mathbf{3}$$

EXAMPLE 3 Find the sine, cosine, and tangent of angles A and B in Fig. 22–11. Leave the
answers as fractions in lowest terms.

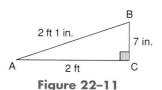

Figure 22–11

$$\sin A = \frac{\text{opposite}}{\text{hypotenuse}} = \frac{7 \text{ in.}}{2 \text{ ft 1 in.}} = \frac{7 \text{ in.}}{25 \text{ in.}} = \frac{7}{25}$$

$$\cos A = \frac{\text{adjacent}}{\text{hypotenuse}} = \frac{2 \text{ ft}}{2 \text{ ft 1 in.}} = \frac{24 \text{ in.}}{25 \text{ in.}} = \frac{24}{25}$$

$$\tan A = \frac{\text{opposite}}{\text{adjacent}} = \frac{7 \text{ in.}}{2 \text{ ft}} = \frac{7 \text{ in.}}{24 \text{ in.}} = \frac{7}{24}$$

$$\sin B = \frac{\text{opposite}}{\text{hypotenuse}} = \frac{2 \text{ ft}}{2 \text{ ft 1 in.}} = \frac{24 \text{ in.}}{25 \text{ in.}} = \frac{24}{25}$$

$$\cos B = \frac{\text{adjacent}}{\text{hypotenuse}} = \frac{7 \text{ in.}}{2 \text{ ft 1 in.}} = \frac{7 \text{ in.}}{25 \text{ in.}} = \frac{7}{25}$$

$$\tan B = \frac{\text{opposite}}{\text{adjacent}} = \frac{2 \text{ ft}}{7 \text{ in.}} = \frac{24 \text{ in.}}{7 \text{ in.}} = \frac{24}{7}$$

2 Find the Cosecant, Secant, and Cotangent of Angles of Right Triangles, Given the Measures of at Least Two Sides.

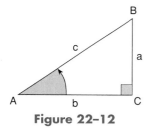

Figure 22–12

There are three other trigonometric functions. These functions are not used as
often as the sine, cosine, and tangent functions, but because of the relationships
between these new functions and the previously learned functions, these new
functions are useful to know.

Each of the three basic trigonometric functions (sine, cosine, and tangent) has
a *reciprocal function*. The *cosecant* (*csc*) function is the reciprocal of the sine
function, the *secant* (*sec*) function is the reciprocal of the cosine function, and
the *cotangent* (*cot*) function is the reciprocal of the tangent function.

Using Fig. 22–12, we can write the reciprocal trigonometric functions of
angle A.

Formula 22–5 *Reciprocal trigonometric functions of angle A in a
standard right triangle:*

$$\csc A = \frac{\text{hypotenuse}}{\text{opposite}} = \frac{c}{a}$$

$$\sec A = \frac{\text{hypotenuse}}{\text{adjacent}} = \frac{c}{b}$$

$$\cot A = \frac{\text{adjacent}}{\text{opposite}} = \frac{b}{a}$$

Similarly, we can write the reciprocal trigonometric functions of angle B in Fig. 22–12.

Formula 22–6   *Reciprocal trigonometric functions of angle B in a standard right triangle:*

$$\csc B = \frac{\text{hypotenuse}}{\text{opposite}} = \frac{c}{b}$$

$$\sec B = \frac{\text{hypotenuse}}{\text{adjacent}} = \frac{c}{a}$$

$$\cot B = \frac{\text{adjacent}}{\text{opposite}} = \frac{a}{b}$$

EXAMPLE 4   Write the six trigonometric ratios for angles A and B in Fig. 22–13.

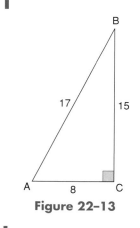

Figure 22–13

$$\sin A = \frac{\text{opp}}{\text{hyp}} = \frac{15}{17} \qquad \sin B = \frac{\text{opp}}{\text{hyp}} = \frac{8}{17}$$

$$\cos A = \frac{\text{adj}}{\text{hyp}} = \frac{8}{17} \qquad \cos B = \frac{\text{adj}}{\text{hyp}} = \frac{15}{17}$$

$$\tan A = \frac{\text{opp}}{\text{adj}} = \frac{15}{8} \qquad \tan B = \frac{\text{opp}}{\text{adj}} = \frac{8}{15}$$

$$\csc A = \frac{\text{hyp}}{\text{opp}} = \frac{17}{15} \qquad \csc B = \frac{\text{hyp}}{\text{opp}} = \frac{17}{8}$$

$$\sec A = \frac{\text{hyp}}{\text{adj}} = \frac{17}{8} \qquad \sec B = \frac{\text{hyp}}{\text{adj}} = \frac{17}{15}$$

$$\cot A = \frac{\text{adj}}{\text{opp}} \quad \frac{8}{15} \qquad \cot B = \frac{\text{adj}}{\text{opp}} \quad \frac{15}{8}$$

For convenience, *opposite*, *hypotenuse*, and *adjacent* are abbreviated here and elsewhere.

EXAMPLE 5   Find the six trigonometric functions for angles A and B in Fig. 22–14. Express ratios to the nearest ten-thousandth.

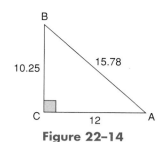

Figure 22–14

$$\sin A = \frac{10.25}{15.78} = \mathbf{0.6496} \qquad \sin B = \frac{12}{15.78} = \mathbf{0.7605}$$

$$\cos A = \frac{12}{15.78} = \mathbf{0.7605} \qquad \cos B = \frac{10.25}{15.78} = \mathbf{0.6496}$$

$$\tan A = \frac{10.25}{12} = \mathbf{0.8542} \qquad \tan B = \frac{12}{10.25} = \mathbf{1.1707}$$

$$\csc A = \frac{15.78}{10.25} = \mathbf{1.5395} \qquad \csc B = \frac{15.78}{12} = \mathbf{1.3150}$$

$$\sec A = \frac{15.78}{12} = \mathbf{1.3150} \qquad \sec B = \frac{15.78}{10.25} = \mathbf{1.5395}$$

$$\cot A = \frac{12}{10.25} = \mathbf{1.1707} \qquad \cot B = \frac{10.25}{12} = \mathbf{0.8542}$$

Tips and Traps   *Patterns and Relationships:*
Examine the color marking in Examples 4 and 5.

$$\sin A = \cos B \qquad \csc A = \sec B$$
$$\cos A = \sin B \qquad \sec A = \csc B$$

The unmarked functions are also related.

$$\tan A = \cot B \qquad \cot A = \tan B$$

SELF-STUDY EXERCISES 22–2

1   Using Fig. 22–15, find the indicated trigonometric ratios for the following exercises.

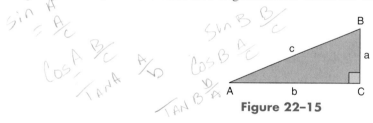

Figure 22–15

$a = 5$, $b = 12$, and $c = 13$. Express the ratios as fractions in lowest terms.

  1. $\sin A$ **2.** $\cos A$ **3.** $\tan A$ **4.** $\sin B$ **5.** $\cos B$ **6.** $\tan B$

$a = 9$, $b = 12$, and $c = 15$. Express the ratios as fractions in lowest terms.

  7. $\sin A$ **8.** $\cos A$ **9.** $\tan A$ **10.** $\sin B$ **11.** $\cos B$ **12.** $\tan B$

$a = 16$, $b = 30$, and $c = 34$. Express the ratios as fractions in lowest terms.

13. $\sin A$ **14.** $\cos A$ **15.** $\tan A$ **16.** $\sin B$ **17.** $\cos B$ **18.** $\tan B$

$a = 9$, $b = 14$, and $c = 16.64$. Express the ratios as decimals to the nearest ten-thousandth.

19. $\sin A$ **20.** $\cos A$ **21.** $\tan A$ **22.** $\sin B$ **23.** $\cos B$ **24.** $\tan B$

2   Determine the six trigonometric ratios for angles A and B in Fig. 22–15.

25. $a = 12$, $b = 16$, $c = 20$ **26.** $a = 8.15$, $b = 5.32$, $c = 9.73$. Express the ratios as decimals to the nearest ten-thousandth.

USING A CALCULATOR
TO FIND TRIGONOMETRIC VALUES

Learning Objectives

1. Find trigonometric values for sine, cosine, and tangent using a calculator.
2. Find the angle measure, given a trigonometric value.
3. Find the trigonometric values for cosecant, secant, and cotangent using the reciprocal relationship.
4. Find the trigonometric values for cosecant, secant, and cotangent using the cofunction relationship.

In studying similar triangles we found that corresponding angles of similar triangles are equal and that corresponding sides are proportional. These properties of similar triangles lead us to a very important property of trigonometric functions.

> **Property 22-1** *Values of trigonometric functions:*
>
> Every angle of a specified measure has a specific set of values for its trigonometric functions.

For many years, even centuries, these trigonometric values were recorded in tables and were readily available in mathematics textbooks to facilitate calculations with trigonometric functions. Today, scientific and graphics calculators and computer software have these values of the trigonometric functions already programmed into memory. These calculators and computer values are expressed to more decimal places than even the most accurate of written tables, thus producing more accurate results than can be obtained with tables.

1. ## Find Trigonometric Values for Sine, Cosine, and Tangent Using a Calculator.

It is recommended that a calculator be used to find trigonometric values. To find trigonometric values on a calculator, the keys *sin*, *cos*, and *tan* are used most often. The angle measure can be entered as a degree measure or a radian measure, depending on the selected mode.

■ **General Tips for Using the Calculator**

To find trigonometric values for sine, cosine, and tangent using most scientific and graphics calculators, follow this procedure:

1. Set the calculator to the desired mode of angle measure. Calculators generally have three angle modes: degrees, radians, and gradients.
2. Press the appropriate trigonometric function key (sin, cos, tan) and then enter the angle measure.
3. Display the result by pressing the $\boxed{=}$ or $\boxed{\text{EXE}}$ key.

To find trigonometric values for sine, cosine, and tangent using a few scientific calculators, the sequence is reversed. Enter the angle measure, then press the appropriate function key. In this case, the $\boxed{=}$, $\boxed{\text{Enter}}$, or $\boxed{\text{EXE}}$ key is normally not pressed. The result is displayed immediately after pressing the trigonometric function key.

EXAMPLE 1   Using a calculator, find the following trigonometric values.

(a) sin 27°   (b) cos 52°   (c) tan 85°   (d) sin 20°30′

(e) cos 1.34   (f) tan$\frac{\pi}{4}$

(a) Be sure the calculator is in degree mode.

$\boxed{\text{SIN}}$ 27 $\boxed{\text{EXE}}$ ⇒ **0.4539904997**   Key strokes may vary.

(b) Be sure the calculator is in degree mode.

$\boxed{\text{COS}}$ 52 $\boxed{\text{EXE}}$ ⇒ **0.6156614753**

(c) Be sure the calculator is in degree mode.

$\boxed{\text{TAN}}$ 85 $\boxed{\text{EXE}}$ ⇒ **11.4300523**

(d) Be sure the calculator is in degree mode.

$\boxed{\text{SIN}}$ 20.5 $\boxed{\text{EXE}}$ ⇒ **0.3502073813**

(Some scientific calculators allow an angle measure to be entered using degrees, minutes, and seconds *or* using the decimal degree equivalent.)

(e) Reset the calculator to radian mode.

$\boxed{\text{COS}}$ 1.34 $\boxed{\text{EXE}}$ ⇒ **0.2287528078**

(f) Be sure the calculator is in radian mode.

$\boxed{\text{TAN}}$ $\boxed{(}$ $\boxed{\pi}$ $\boxed{\div}$ 4 $\boxed{)}$ $\boxed{\text{EXE}}$ ⇒ **1**

2 Find the Angle Measure, Given a Trigonometric Value.

When the measures of two sides of a triangle are known, it is possible to determine an angle measure from a trigonometric value.

■ **General Tips for Using the Calculator**

Finding the angle measure of an angle of a right triangle when given the trigonometric value is an *inverse operation* to finding the trigonometric value when given the measure of the angle. The notation most commonly used is $\sin^{-1}$, $\cos^{-1}$, or $\tan^{-1}$. Another notation is arcsin, arccos, or arctan.

Most Scientific and Graphics Calculators

1. Set the calculator to the desired mode of angle measure (degrees or radians).
2. Enter the appropriate inverse trigonometric function key ($\sin^{-1}$, $\cos^{-1}$, or $\tan^{-1}$). Enter the trigonometric value and $\boxed{\text{EXE}}$ or $\boxed{\text{ENTER}}$.

Some calculators use an inverse key $\boxed{\text{INV}}$ or shift key to find the inverse of a function.

Other Calculators

1. Set the calculator to the desired mode of angle measure (degrees or radians.)
2. Enter the trigonometric value and press the *inverse* function key $\boxed{\text{INV}}$ or shift function key, then the appropriate trigonometric function key (sin, cos, tan). Some calculators may have $\sin^{-1}$, $\cos^{-1}$, and $\tan^{-1}$ as inverse function keys.

EXAMPLE 2 Determine the degree measure of an angle of the given trigonometric values in parts (a) and (b). θ represents the unknown angle measure. Round to the nearest tenth of a degree. For part (c), find the radian measure to the nearest thousandth.

(a) $\sin \theta = 0.6561$ (b) $\cos \theta = 0.4226$ (c) $\tan \theta = 2.825$

(a) Be sure the calculator is in degree mode.

   $\boxed{\text{SIN}^{-1}}$ .6561 $\boxed{\text{EXE}}$ $\Rightarrow$ 41.0031105

               $\approx$ **41.0°**   Rounded

(b) Be sure the calculator is in degree mode.

   $\boxed{\text{COS}^{-1}}$ .4226 $\boxed{\text{EXE}}$ $\Rightarrow$ 65.00115448

               $\approx$ **65.0°**   Rounded

(c) Be sure the calculator is in radian mode.

   $\boxed{\text{TAN}^{-1}}$ 2.825 $\boxed{\text{EXE}}$ $\Rightarrow$ 1.230578215

               $\approx$ **1.231 rad**   Rounded

3   Find the Trigonometric Values for Cosecant, Secant, and Cotangent Using the Reciprocal Relationship.

When using a calculator, the reciprocal key, $\boxed{x^{-1}}$ or $\boxed{1/x}$, can be used to find the value of *reciprocal trigonometric functions*. For instance, after the sine of an angle is found, the cosecant can be determined by pressing the reciprocal key.

The following identities indicate the reciprocal relationships of the tangent and cotangent, the sine and cosecant, and the cosine and secant functions.

$$\cot \theta = \frac{1}{\tan \theta} \qquad \csc \theta = \frac{1}{\sin \theta} \qquad \sec \theta = \frac{1}{\cos \theta}$$

EXAMPLE 3 Determine the value of the indicated trigonometric functions.

(a) $\csc 18.5°$ (b) $\sec 1.0821$

(a) $\csc 18.5° = \dfrac{1}{\sin 18.5°} = \dfrac{1}{0.317304656} = $ **3.1515**   To nearest ten-thousandth

     In degree mode:

     $\boxed{\text{SIN}}$ 18.5 $\boxed{\text{EXE}}$ $\boxed{x^{-1}}$ $\boxed{\text{EXE}}$ $\Rightarrow$ 3.151545305

(b) $\sec 1.0821 = \dfrac{1}{\cos 1.0821} = \dfrac{1}{0.469475214} = $ **2.1300**   To nearest ten-thousandth

     In radian mode:

     $\boxed{\text{COS}}$ 1.0821 $\boxed{\text{EXE}}$ $\boxed{x^{-1}}$ $\boxed{\text{EXE}}$ $\Rightarrow$ 2.130037898

EXAMPLE 4 Find the degree measure of an angle whose trigonometric value is given. Express the measure to the nearest tenth of a degree.

(a) $\sec \theta = 2.7320$ (b) $\csc \theta = 5.9137$

Find the radian measure to the nearest hundredth radian.

(c) $\cot \theta = 0.2167$

First, find the reciprocal of the given value. Then find the inverse of the reciprocal function.

(a) If sec θ = 2.7320, then cos θ = $\frac{1}{2.732}$. In degree mode:

2.732 $\boxed{x^{-1}}$ $\boxed{\text{EXE}}$ $\boxed{\text{COS}^{-1}}$ $\boxed{\text{Ans}}$ $\boxed{\text{EXE}}$

θ = 68.5°

(b) In degree mode:

5.9137 $\boxed{x^{-1}}$ $\boxed{\text{EXE}}$ $\boxed{\text{SIN}^{-1}}$ $\boxed{\text{Ans}}$ $\boxed{\text{EXE}}$

θ = 9.7°

(c) In radian mode:

.2167 $\boxed{x^{-1}}$ $\boxed{\text{EXE}}$ $\boxed{\text{TAN}^{-1}}$ $\boxed{\text{Ans}}$ $\boxed{\text{EXE}}$

θ = 1.36 rad

4 Find the Trigonometric Values for Cosecant, Secant, and Cotangent Using the Cofunction Relationship.

A calculator can be used to investigate relationships among trigonometric functions. Because the sum of all three angles of a triangle is 180° and the right angle in a right triangle is 90°, the other two angles in a right triangle are both acute angles (less than 90°) and their sum is 90°. That makes the two acute angles of any right triangle *complementary angles*. Let's examine the relationship between the sine of an angle and the cosine of its complement.

sin 40° = 0.6428	cos 50° = 0.6428
sin 30° = 0.5	cos 60° = 0.5
sin 80° = 0.9848	cos 10° = 0.9848
sin 90° = 1	cos 0° = 1

We can generalize our observations.

Property 22–2 *The relationship of the sine of an angle and the cosine of its complement:*

The sine of an angle equals the cosine of its complement.

Now, let's look at the relationship between the tangent of an angle and the cotangent of its complement.

tan 0° = 0	cot 90° = 0
tan 30° = 0.5774	cot 60° = 0.5774
tan 10° = 0.1763	cot 80° = 0.1763
tan 45° = 1	cot 45° = 1
tan 90° = error Undefined	cot 0° = error Undefined

Also compare the secant of an angle with the cosecant of its complement.

sec 0° = 1	csc 90° = 1
sec 30° = 1.1547	csc 60° = 1.1547
sec 10° = 1.0154	csc 80° = 1.0154
sec 45° = 1.4142	csc 45° = 1.4142
sec 90° = error	csc 0° = error

The pairs of functions given in properties 22–2, 22–3, and 22–4 are referred to as trigonometric *cofunctions*.

Tips and Traps *Cofunction Versus Reciprocal Function:*
Trigonometric functions can be paired as cofunctions or as reciprocal functions.

Cofunctions

sine θ and cosine θ

secant θ and cosecant θ

tangent θ and cotangent θ

The value of a function of an angle and the value of the cofunction of the complement of the angle are equal.

For degrees:	For radians:
$\sin \theta = \cos (90° - \theta)$	$\sin \theta = \cos \left(\dfrac{\pi}{2} - \theta \right)$
$\sec \theta = \csc (90° - \theta)$	$\sec \theta = \csc \left(\dfrac{\pi}{2} - \theta \right)$
$\tan \theta = \cot (90° - \theta)$	$\tan \theta = \cot \left(\dfrac{\pi}{2} - \theta \right)$

Reciprocal Functions

$$\text{sine } \theta = \frac{1}{\text{cosecant } \theta}$$

$$\text{cosine } \theta = \frac{1}{\text{secant } \theta}$$

$$\text{tangent } \theta = \frac{1}{\text{cotangent } \theta}$$

EXAMPLE 5   Find the values to the nearest ten-thousandth using the cofunction relationship.

(a) cot 25°   (b) cot 1.414

(a) $\cot 25° = \tan (90° - 25°) = \tan 65° = $ **2.1445**

$$\text{(b) } \cot 1.414 = \tan\left(\frac{\pi}{2} - 1.414\right) = \tan 0.1568 = \mathbf{0.1581}$$

$$\boxed{\tan}\ \boxed{(}\ \boxed{\pi}\ \boxed{\div}\ 2\ \boxed{-}\ 1.414\ \boxed{)}\ \boxed{=} \Rightarrow 0.15809404$$

EXAMPLE 6   Find the angle measure of (a) cot $\theta = 2.6$, in degrees (b) cot $\theta = 0.24$, in radians.

$$\text{(a)} \quad \cot \theta = \tan(90° - \theta) = 2.6 \qquad\qquad \text{Find } \boxed{\text{TAN}^{-1}} \text{ of 2.6.}$$

$$90° - \theta = 68.96248897° \qquad\qquad \text{Find the complement of the answer in degrees.}$$

$$\theta = 90° - 68.96248897°$$

$$\boldsymbol{\theta = 21.03751103°} \quad \textbf{or} \quad \textbf{21.0°}$$

$$\text{(b)} \quad \cot \theta = \tan\left(\frac{\pi}{2} - \theta\right) = 0.24 \qquad\qquad \text{Find } \boxed{\text{TAN}^{-1}} \text{ of 0.24}$$

$$\frac{\pi}{2} - \theta = 0.23554498 \qquad\qquad \text{Find } \theta.$$

$$\theta = \frac{\pi}{2} - 0.23554498$$

$$\boldsymbol{\theta = 1.335251346} \quad \textbf{or} \quad \textbf{1.3353 rad}$$

SELF-STUDY EXERCISES 22–3

1   Use a calculator to find the following trigonometric values. Express answers in ten-thousandths.

1. sin 21°	**2.** cos 3.5°	**3.** tan 47°
4. cos 52.5°	**5.** sin 0.5498	**6.** cos 21°30′
7. cos 1.1519	**8.** cos 0.3665	**9.** sin 53°30′
10. tan 42.5°	**11.** tan 47.7°	**12.** sin 62°10′
13. cos 12°40′	**14.** cos 1.0530	**15.** tan 73°14′
16. sin 1.2363	**17.** cos 46.8°	**18.** cos 0.3549
19. cos 1.1636	**20.** tan 12.4°	

2   Determine the degree measure of the angles of the following trigonometric values. θ represents the unknown angle measure. Express each answer to the nearest tenth of a degree.

21. sin θ = 0.3420	**22.** cos θ = 0.9239	**23.** tan θ = 2.356
24. cos θ = 0.4617	**25.** cos θ = 0.540	**26.** sin θ = 0.5712
27. tan θ = 1.265	**28.** cos θ = 0.137	**29.** sin θ = 0.6298
30. cos θ = 0.9325		

Determine the radian measure of the angles of the following trigonometric values.
Express each answer to the nearest ten-thousandth.

31. sin θ = 0.5299	**32.** tan θ = 0.8098	**33.** cos θ = 0.6947
34. cos θ = 0.3907	**35.** cos θ = 0.968	**36.** sin θ = 0.9959
37. tan θ = 0.3160	**38.** tan θ = 2.430	**39.** cos θ = 0.9610
40. cos θ = 0.4210		

3   Find the indicated trigonometric values using the reciprocal relationship. Round to the nearest ten-thousandth.

41. cot 24.5°	**42.** csc 42°	**43.** sec 0.2443
44. csc 1.3788	**45.** sec 1.2165	**46.** cot 28.6°
47. cot 87°	**48.** sec 42°30′	**49.** csc 0.2136
50. sec 1.0372		

Find θ to the nearest tenth of a degree.

51. $\cot \theta = 0.4238$ **52.** $\sec \theta = 1.8291$ **53.** $\csc \theta = 3.7129$
54. $\sec \theta = 8.2156$ **55.** $\cot \theta = 1.7318$

 Find the indicated trigonometric value using the cofunction relationship. Round to the nearest ten-thousandth.

56. $\sin 82°$ **57.** $\cos 15°$ **58.** $\tan 1.023$ **59.** $\cot 0.896$

Find θ to the nearest tenth of a degree or hundredth of a radian.

60. $\sin \theta = 0.8724$ **61.** $\cos \theta = 0.4687$ **62.** $\csc \theta = 0.1067$ **63.** $\cot \theta = 5.213$

CAREER APPLICATION

Consumer Interest: Tornado Chasing Expeditions

Since the 1996 movie hit *Twister,* travel agents in Oklahoma City have done a brisk business booking adventure expeditions with local storm-chasers. Most tornadoes occur during hot weather, so bookings have followed a periodic yearly pattern, as illustrated in the graph. This graph comes from the following formula:

$$B = 80 \left[\sin\left(\frac{\pi}{26} \right) w \right] + 80$$

where B is the number of individual bookings per week, and w is the number of weeks since the week of April 1 (the first week of tornado season).

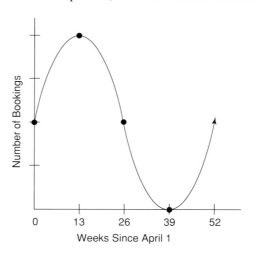

Exercises

Use the above information to answer the following. Round answers to the nearest unit.
1. The numbers on the vertical axis of the graph have been omitted. Use the formula and a calculator to fill in the numbers.
2. Under the five numbers on the horizontal axis, write the name of the corresponding month of the year.
3. Compare this graph with the standard graph of $y = \sin \theta$, for θ between 0 and 2π radians. Discuss amplitude (maximum high and low values), period (length of a complete cycle), and phase shift (shift vertically or horizontally).
4. Use the graph to estimate two dates at which the number of weekly bookings is about 50.

5. Now substitute your guesses from Exercise 4 into the formula to see how close you are.
6. Use the formula to find the number of bookings for the following weeks after the start of tornado season: (a) 3 (b) 14 (c) 20 (d) 45
7. Now use the graph to verify your answers in Exercise 6. If your answers differ, explain why.
8. During which months are bookings increasing? During which months are they decreasing?

Answers

1. The vertical axis is incremented by 40. The missing numbers are 40, 80, 120, and 160.
2. Because w = weeks since April 1, 0 = April, 13 = July, 26 = October, 39 = January, and 52 = April.
3. The graphs are very similar; both have the same shape, the same orientation, and the same period. In the bookings graph, the amplitude of the standard sine graph has been multiplied by 80, and the whole standard sine graph has been shifted vertically +80 units.
4. Around weeks 29 and 49.
5. Answers vary.
6. (a) About 108 (b) About 159 (c) About 133 (d) About 20
7. Answers vary, but they should be close.
8. Bookings increase from January to July, and they decrease from July to January.

ASSIGNMENT EXERCISES

Section 22–1

Convert the following degree measures to radians rounded to the nearest hundredth.

1. 60°
2. 212°
3. 300°

Convert the following measures to degrees rounded to the nearest ten-thousandth.
Then convert to radians to the nearest hundredth.

4. 25°30′
5. 99°45′
6. 120°20′40″

Convert the following radian measures to degrees. Round to the nearest ten-thousandth of a degree.

7. $\dfrac{5\pi}{6}$ rad
8. 2.4 rad
9. 1.7 rad

Convert the following radian measures to degrees, minutes, and seconds. Round to the nearest second.

10. 0.9 rad
11. $\dfrac{3\pi}{8}$ rad
12. 1.2 rad

Find the arc length, radius, or central angle (in radians) for each of the following.
Round to hundredths when necessary.

13. Find s if $\theta = 0.7$ rad and $r = 2.3$ cm.
14. Find θ if $s = 6.2$ cm and $r = 5$ cm.
15. Find r if $\theta = 2.1$ rad and $s = 3.6$ ft.

Find the area of the sector, the radius, or the central angle (in radians) of the sector for each of the following. Round to hundredths when necessary.

16. Find A if $\theta = 0.88$ rad and $r = 1.5$ m.
17. Find r if $\theta = 4.2$ rad and $A = 24$ in.^{2}.
18. Find θ if $r = 4$ cm and $A = 12$ cm^2.

19. A pendulum 6 in. long swings through an angle of 20°. Find the arc length the pendulum swings over from one extreme position to the other. Round to hundredths.

20. A movable part on a machine swings through an angle of 40° with an arc length of 8 in. What is the length of the part? Round to hundredths.

21. Find the area of a sector whose central angle is 85° and whose radius is 4.6 cm. ($A = \frac{1}{2}\theta r^2$, where θ is in radians.) Round to hundredths.

22. Find the number of degrees to the nearest ten-thousandth of a central angle of a sector whose area is 8.4 cm^2 and whose radius is 6 cm.

Section 22-2
Find the indicated trigonometric ratios using Fig. 22–16.

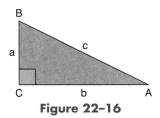

Figure 22-16

$a = 15$, $b = 20$, $c = 25$. Express the ratios as fractions in lowest terms.

23. sin A	**24.** cos A	**25.** tan A	**26.** csc A	**27.** sec A	**28.** cot A
29. sin B	**30.** cos B	**31.** tan B	**32.** csc B	**33.** sec B	**34.** cot B

$a = 2$ ft, $b = 10$ in., $c = 2$ ft 2 in. Express the ratios as fractions in lowest terms.

35. sin A	**36.** sec B	**37.** cot B	**38.** csc A	**39.** tan B	**40.** cos A

$a = 7$, $b = 10.5$, $c = 12.62$. Express the ratios in decimals to the nearest ten-thousandth.

41. cos B	**42.** csc A	**43.** tan A	**44.** sin B	**45.** sec A	**46.** sin A
47. cot A					

Section 22-3
Use a calculator to find the following trigonometric values. Round to the nearest ten-thousandth.

48. cos 42.5°	**49.** sin 0.4712	**50.** sin 65.5°	**51.** cot 73°
52. tan 1.0210	**53.** tan 47°	**54.** tan 15.6°	**55.** sin 0.8610
56. tan 25°40′	**57.** cos 32°50′	**58.** cot 0.7510	**59.** cos 80°10′

Determine the degree measure of an angle having the following trigonometric values. θ represents the unknown angle measure. Express answers to the nearest tenth of a degree.

60. sin $\theta = 0.5446$	**61.** cos $\theta = 0.6088$	**62.** tan $\theta = 0.8720$
63. cot $\theta = 0.9884$	**64.** cos $\theta = 0.8897$	**65.** cot $\theta = 3.340$

Determine the radian measure of an angle having the following trigonometric values. Round to the nearest ten-thousandth.

66. sin $\theta = 0.9205$	**67.** tan $\theta = 2.723$	**68.** cos $\theta = 0.9450$
69. cot $\theta = 0.3772$	**70.** sin $\theta = 0.2896$	**71.** tan $\theta = 0.3440$

Find the indicated trigonometric values. Round to the nearest ten-thousandth.

72. sec 15.5°	**73.** csc 71°	**74.** sec 0.4363
75. csc 1.0821	**76.** sec 1.2886	**77.** csc 0.4829

Use your knowledge of cofunctions to evaluate the following.

78. cot 45°	**79.** csc 82°20′	**80.** sec 38°
81. csc 1.235	**82.** cot 0.1823	**83.** sec 4.23

84. Is there an angle between 0° and 360° for which the sine and cosine of the angle are equal? Justify your answer by making a table of values for the sine and cosine of angles between 0° and 360° and by graphing sin x and cos x for values between 0° and 360°.

85. Investigate the relationship described in problem 84 for the sine and tangent functions and justify your answer.

CONCEPTS ANALYSIS

1. Explain how the sine function is a ratio.

2. Use a calculator to find the value of the sine of several angles in each of the four quadrants and make a general statement about the greatest and least values the sine function can have.

3. Draw three right triangles. The acute angles of the first triangle are 30° and 60°. The acute angles of the second triangle are 25° and 65°. The acute angles of the third triangle are 80° and 10°. For each triangle, find the sine of the first angle and the cosine of the second angle. Compare the sine and cosine values for each triangle. What generalization can you make from these three examples?

4. Draw three right triangles of your choice and verify the generalization you made in question 3.

5. Use a calculator to find the sine of the angles given below and make a table with the given angle in the first column and the sine of the angle in the second column. What is the measure of the angle at which the sine of the angle begins to repeat?
30°, 60°, 90°, 120°, 150°, 180°, 210°, 240°, 270°, 300°, 330°, 360°, 390°, 420°

6. Find the sine of several angles larger than those given in question 5 to determine where, if anywhere, the sine begins to repeat.

7. Make a general statement about the sine function as it repeats. Compare your general statement with the graph of the sine function.

8. Use a graphics calculator to graph the following functions on the same grid. sin x; 2 sin x; 3 sin x; $\frac{1}{2}$ sin x; $\frac{1}{3}$ sin x. Describe the impact the coefficient of the sine function has on the graph of the function. This coefficient is called the amplitude. Check a dictionary for the definition of the word *amplitude*. Does the dictionary definition make sense in view of your description? Set the calculator range for x as 0° to 360° and for y as −3 to +3. Calculator steps for first graph: | Graph | | Sin |
| X,θ,T | | EXE |

9. Graph sin x and sin $2x$ on the same graph. Compare these graphs. How do they differ and how are they alike?

10. Graph sin x and sin $(\frac{1}{2})x$ on the same graph. Compare these two graphs to the graphs for question 9. Describe the effect the coefficient of x has on the graph of the sine function. Graph other sine functions in which the coefficient of x is different to verify your description.

Objectives	What to Remember with Examples

Section 22–1

1 Find arc length, given a central angle in radian measure.

To find the length of an arc of a circle, multiply the radian measure of the central angle by the radius of the circle.

Use $s = \theta r$, where θ is given in radians.

> Find the arc length intercepted on the circumference of a circle by a central angle of $\frac{\pi}{4}$ if the radius of the circle is 24 meters.
>
> $s = \theta r$
>
> $s = \frac{\pi}{4}(24)$
>
> $s = \pi(6)$
>
> $s = 6\pi$
>
> $s = 18.850$ m

2 Find the area of a sector, given a central angle in radian measure.

The area of a sector is $\frac{1}{2}\theta r^2$, where θ is given in radians.

> Find the area of a sector that has a central angle of 1.4 rad and a radius of 5.2 ft.
>
> $A = \frac{1}{2}\theta r^2$
>
> $A = 0.5(1.4)(5.2)^2$
>
> $A = 18.928$ ft^2

3 Convert angles in degree measure to angles in radian measure.

Degree angle measures are converted to radian measures by multiplying by the unity ratio $\frac{\pi \text{ rad}}{180°}$.

> Write 82° in radian measure and round to the nearest thousandth.
>
> $82° \times \frac{\pi}{180°} = 1.431$ rad

4 Convert angles in radian measure to angles in degree measure.

Radian angle measures are converted to degree measures by multiplying by the unity ratio $\frac{180°}{\pi \text{ rad}}$.

> Write 0.45 radians in degree measure and round to the nearest tenth of a degree.
>
> $0.45 \text{ rad} \times \frac{180°}{\pi \text{ rad}} = 25.8°$

Section 22–2

1 Find the sine, cosine, and tangent of angles of right triangles, given the measures of at least two sides.

Use the ratios for each of the three trigonometric functions to calculate the value of the function when the measures of appropriate sides are given. The Pythagorean theorem may be needed to find the length of a third side in some instances.

$$\text{sine } A = \frac{\text{opposite}}{\text{hypotenuse}} \qquad \text{cosine } A = \frac{\text{adjacent}}{\text{hypotenuse}} \qquad \text{tangent } A = \frac{\text{opposite}}{\text{adjacent}}$$

Find the sine, cosine, and tangent of angle A in Fig. 22–17 and round to the nearest ten-thousandth.

$$\sin A = \frac{15}{17} = 0.8824$$

$$\cos A = \frac{8}{17} = 0.4706$$

$$\tan A = \frac{15}{8} = 1.875$$

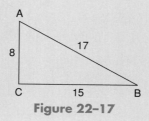

Figure 22–17

2 Find the cosecant, secant, and cotangent of angles of right triangles, given the measures of at least two sides.

The cosecant, secant, and cotangent are reciprocal functions of the sine, cosine, and tangent, respectively.

$$\text{cosecant } A = \frac{\text{hypotenuse}}{\text{opposite}}$$

$$\text{secant } A = \frac{\text{hypotenuse}}{\text{adjacent}}$$

$$\text{cotangent } A = \frac{\text{adjacent}}{\text{opposite}}$$

Find the cosecant, secant, and cotangent of angle A in Fig. 22–17 and round to the nearest ten-thousandth.

$$\csc A = \frac{17}{15} = 1.1333$$

$$\sec A = \frac{17}{8} = 2.125$$

$$\cot A = \frac{8}{15} = 0.5333$$

Section 22–3

1 Find trigonometric values for sine, cosine, and tangent using a calculator.

Use the $\boxed{\text{SIN}}$, $\boxed{\text{COS}}$, and $\boxed{\text{TAN}}$ keys to find the trigonometric value of a specified angle. Be sure the calculator is set to the appropriate mode, degree or radian.

Use a calculator to find the following: sin 35° cos 198° tan 125° sin 2.8 cos 2.98 tan 4.2

Most Calculators:
Be sure calculator is in degree mode.

$\boxed{\text{SIN}}$ 35 $\boxed{\text{EXE}}$ ⇒ 0.5735764364

$\boxed{\text{COS}}$ 198 $\boxed{\text{EXE}}$ ⇒ −0.9510565163

$\boxed{\text{TAN}}$ 125 $\boxed{\text{EXE}}$ ⇒ −1.428148007

Change to radian mode:

$\boxed{\text{SIN}}$ 2.8 $\boxed{\text{EXE}}$ ⟹ 0.3349881502

$\boxed{\text{COS}}$ 2.98 $\boxed{\text{EXE}}$ ⟹ −0.9869722927

$\boxed{\text{TAN}}$ 4.2 $\boxed{\text{EXE}}$ ⟹ 1.777779775

Other Calculators:
Be sure calculator is in degree mode.

35 $\boxed{\text{SIN}}$ ⟹ 0.573576436

198 $\boxed{\text{COS}}$ ⟹ −0.951056516

125 $\boxed{\text{TAN}}$ ⟹ −1.428148007

Change calculator to radian mode:

2.8 $\boxed{\text{SIN}}$ ⟹ 0.33498815

2.98 $\boxed{\text{COS}}$ ⟹ −0.986972292

4.2 $\boxed{\text{TAN}}$ ⟹ 1.777779774

---

2 Find the angle measure, given a trigonometric value.

To find the angle measure when the trigonometric value is given, use the inverse trigonometric function key, which is usually accessed by pressing the $\boxed{\text{Shift}}$ key.

Find the value of x in degrees: $0.906307787 = \cos x$ or $x = \cos^{-1} 0.906307787$.

$\boxed{\text{COS}^{-1}}$ 0.906307787 $\boxed{=}$ or 0.906307787 $\boxed{\text{COS}^{-1}}$

$x = 25°$

---

3 Find the trigonometric values for cosecant, secant, and cotangent using the reciprocal relationship.

Use the $\boxed{\text{SIN}}$, $\boxed{\text{COS}}$, and $\boxed{\text{TAN}}$ calculator keys and the inverse function key $\boxed{1/x}$ to find the values of the inverse functions.

Use a calculator to find the following: csc 25°, sec 1

Most calculators:
In degree mode:

$\boxed{\text{SIN}}$ 25 x^{-1} $\boxed{=}$ ⟹ 2.366201583

In radian mode:

$\boxed{\text{COS}}$ 1 x^{-1} $\boxed{=}$ ⟹ 1.850815718

Other calculators:
In degree mode:

25 $\boxed{\text{SIN}}$ $\boxed{1/x}$ ⟹ 2.366201583

In radian mode:

1 $\boxed{\text{COS}}$ $\boxed{1/x}$ ⟹ 2.366201583

4 Find the trigonometric values for cosecant, secant, and cotangent using the cofunction relationship.

To find the cotangent of an angle, find its complement by subtracting the angle from 90°. Then find the tangent of the complement. The tangent of the complement is the same as the cotangent of the angle.

> Find the cotangent of 35°.
>
> $$90° - 35° = 55°$$
> $$\tan 55° = 1.428148007$$
>
> Thus, $\cot 55° = 1.428148007$.

WORDS TO KNOW

radian (p. 822)
degree (p. 823)
arc length (p. 823)
sector (p. 825)
unity ratio (p. 825)
hypotenuse (p. 829)

adjacent side (p. 829)
opposite side (p. 829)
sine (p. 829)
cosine (p. 829)
tangent (p. 829)
standard right triangle (p. 829)

cosecant (p. 831)
secant (p. 831)
cotangent (p. 831)
reciprocal trigonometric
   functions (p. 831)
cofunction (p. 847)

CHAPTER TRIAL TEST

Convert the following degree measures to radians rounded to the nearest hundredth.

1. 35° **2.** 122° **3.** 315° **4.** 240°

Convert the following measures to degrees rounded to the nearest ten-thousandth.
Then convert to radians to the nearest hundredth.

5. 15°25′ **6.** 142°32′15″ **7.** 16°12′ **8.** 32°18′37″

Convert the following radian measures to degrees. Round to the nearest ten-thousandth of a degree.

9. $\dfrac{5\pi}{8}$ rad **10.** 3.1 rad

Convert the following radian measures to degrees, minutes, and seconds. Round to the nearest second.

11. 1.2 rad **12.** $\dfrac{\pi}{6}$ rad

Use the relationships of arc length, area, central angle, and radius to solve the following problems relating to sectors. Round to hundredths if necessary.

13. Find s if $\theta = 0.5$ and $r = 2$ in. **14.** Find θ if $s = 5.3$ m and $r = 7$ m.
15. Find r if $\theta = 1.7$ and $s = 2.9$ m. **16.** Find A if $\theta = 35°$ and $r = 7.3$ cm.

Write the ratios as fractions in lowest terms for the following trigonometric functions using triangle *ABC* in Fig. 22–18.
$a = 10$, $b = 24$, $c = 26$.

17. sin *A* **18.** tan *B* **19.** csc *A*

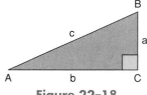

Figure 22–18

Express the following trigonometric values in decimals to the nearest ten-thousandth. Refer to Fig. 22–18.
$a = 5$, $b = 11.5$, $c = 12.54$

20. cot A **21.** cos A **22.** sin B

Use a calculator to find the following trigonometric values.

23. sin $53°$ **24.** sin $61°10'$ **25.** sin 1.1519 **26.** cos 1.0297

Determine the degree measure of an angle having the following trigonometric values.
θ represents the unknown angle measure. Express answers to the nearest tenth of a degree.

27. sin $\theta = 0.2756$ **28.** tan $\theta = 1.280$
29. cos $\theta = 0.9426$ **30.** cot $\theta = 1.540$

Determine the radian measure of an angle having the following trigonometric values. Round to the nearest ten-thousandth.

31. sin $\theta = 0.7660$ **32.** cos $\theta = 0.8387$ **33.** tan $\theta = 0.3259$

Find the indicated trigonometric values. Round to the nearest ten-thousandth.

34. sec $25.5°$ **35.** csc $47°$ **36.** cot 0.3316

23

Right-Triangle Trigonometry

Right triangles can be solved by means of the Pythagorean theorem or the principles of right-triangle trigonometry. Are both means equally useful for all situations in the workplace? Does one have advantages over the other? If there are situations where both methods can be used, which is preferable? How does one decide which method to use?

Investigate possible answers to the proposed questions or similar questions by interviewing professors of mathematics and science on your campus. Verify the findings by working problems using both methods and comparing and contrasting your experiences. Be sure to take notes to help you prepare for a debate about the two methods.

For your team's project, conduct a classroom debate that addresses these types of questions. Divide into two sides, one for the Pythagorean theorem and the other for the principles of right-triangle trigonometry, and debate in class.

Right triangles are used extensively in real-life applications. Thus, it is important for us to gain a working knowledge of solving right triangles, that is, finding the measures of all sides and angles. We also need to become skilled in determining how to use our knowledge to solve career-related applications of right triangles. By using the trigonometric relationships we learned in Chapter 22, we can find all the angles and sides of a right triangle if we know the measure of one side and any other part.

23-1 SINE, COSINE, AND TANGENT FUNCTIONS

Learning Objectives

1 Find the missing parts of a right triangle using the sine function.

2 Find the missing parts of a right triangle using the cosine function.

3 Find the missing parts of a right triangle using the tangent function.

The sine, cosine, and tangent functions can be used to find certain parts of a right triangle. Formula rearrangement and other algebraic principles may be needed also, depending on what information is given in the problem and what information needs to be found.

1 **Find the Missing Parts of a Right Triangle Using the Sine Function.**

We will abbreviate the formula of the sine function to read $\sin \theta = \frac{\text{opp}}{\text{hyp}}$. Using this formula, we can find parts of right triangles when we know any two of the parts that involve the sine function: one acute angle, the side opposite the known acute angle, and the hypotenuse.

> **Rule 23-1** *Use the sine function to find unknown measures of a right triangle:*
>
> 1. Two of these three parts must be known:
>   **a.** One acute angle
>   **b.** The side opposite the known acute angle
>   **c.** Hypotenuse
> 2. Substitute two known values in the formula $\sin \theta = \frac{\text{opp}}{\text{hyp}}$.
> 3. Solve for the missing part.

Tips and Traps *Does It Matter Which Acute Angle Is Known?*
Not really. The acute angles of a right triangle are complementary. Therefore, if we know either acute angle (a), we can find the other one ($90° - a$). Thus, we can actually use the sine function if we know *either* angle and any side.

EXAMPLE 1 Find angle A if $a = 7$ and $c = 21$ (see Fig. 23–1).

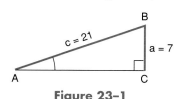

Figure 23–1

The two known values are the side opposite angle A and the hypotenuse, so we use the sine function for the acute angle A.

$$\sin A = \frac{\text{opp}}{\text{hyp}}$$

Substitute the known values: opp = 7, hyp = 21.

$$\sin A = \frac{7}{21}$$

Convert ratio to a decimal equivalent.

$$\sin A = 0.3333333333$$

Find $\boxed{\sin^{-1}}$ of 0.3333333333.

$$A = 19.47122063° \text{ or } 19.5°$$

Round to the nearest tenth of a degree.

Tips and Traps

How Do We Round Our Answers?

Rounding practices are generally dictated by the context of the problem or industry standards; however, **for consistency, we will round all trigonometric ratios to four significant digits and all angle values to the nearest 0.1° throughout chapters 23 and 24 unless otherwise indicated.**

The *significant digits* of a whole number or integer are the digits beginning with the first nonzero digit on the left and ending with the last nonzero digit on the right. The significant digits of a decimal number are the digits beginning with the first nonzero digit on the left and ending with the last digit on the *right of the decimal point.*

5000	1 significant digit	5 is first *and* last nonzero digit.
250	2 significant digits	2 is first and 5 is last nonzero digit.
205	3 significant digits	2 is first and 5 is last nonzero digit.
0.004	1 significant digit	4 is first nonzero *and* last digit.
3.05	3 significant digits	3 is first nonzero and 5 is last digit.
2.070	4 significant digits	2 is first nonzero digit and 0 is last digit.

To round a number to a certain number of significant digits:

1. From the left, count the number of significant digits desired. Notice the last of these significant digits.
2. If the next digit to the right is less than 5, do not change the last significant digit. If the next digit to the right is 5 or greater, add 1 to the last significant digit.
3. Then, if the last significant digit is to the *left* of a decimal point, replace all digits after the last significant digit up to the decimal point with zeros and drop all digits after the decimal point. If the last significant digit is to the *right* of a decimal point, drop all digits after the last significant digit.

In Example 1, the angle measure is rounded to the nearest tenth of a degree. In Example 2, the side measure is rounded to four significant digits.

EXAMPLE 2 Find side *a* in triangle *ABC* (see Fig. 23–2).

We are given $\angle A$ and the hypotenuse and need to find the side opposite $\angle A$, so again we use the sine function.

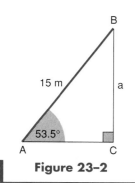

Figure 23–2

$$\sin A = \frac{\text{opp}}{\text{hyp}}$$

Substitute known values:
$\angle A = 53.5°$, hyp = 15 m.

$$\sin 53.5° = \frac{a}{15}$$

$\sin 53.5° = 0.80385686$

$$0.80385686 = \frac{a}{15}$$

$$15(0.80385686) = a$$

$$a = 12.05785291 \quad \text{or} \quad 12.06 \text{ m}$$

Rounded to four significant digits

Rearrange Formula Before Starting Calculations:
In Example 2 and the examples to follow, we can visualize a continuous sequence of calculator steps if we rearrange the formula for the missing part *before* we make any calculations.

$$\sin 53.5° = \frac{a}{15}$$

$$15(\sin 53.5°) = a$$

Then a continuous series of calculations can be made. In degree mode:

$$15 \boxed{\times} \boxed{\sin} 53.5 \boxed{=} \Rightarrow 12.05785291$$

EXAMPLE 3 Find the hypotenuse in triangle *RST* (see Fig. 23–3).

We are given an acute angle and the side opposite the angle and need to find the hypotenuse, so the sine function is used.

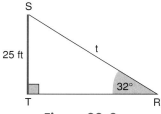

Figure 23–3

$$\sin R = \frac{\text{opp}}{\text{hyp}}$$ Substitute known values: ∠R = 32°, opp = 25 ft.

$$\sin 32° = \frac{25}{t}$$ Rearrange for *t*.

$$t(\sin 32°) = 25$$

$$t = \frac{25}{\sin 32°}$$

$$t = 47.17699787$$

t = 47.18 ft Rounded to four significant digits

Choose Given Values over Calculated Values When Possible:
In most problems where a triangle is being solved, there is a choice of which missing part can be found first. Because the rounded value for that missing part is sometimes used to find other missing parts, final answers may vary slightly due to rounding discrepancies. For instance, the angles of a triangle may add to as little as 179° or as much as 181°. Or the length of a side may be slightly different in the last significant digit. If the full calculator value is used in finding other missing parts, the rounding discrepancy will be reduced.

To *solve* a triangle means to find the measures of all sides and angles of a triangle. A right triangle can be solved if we know one side and any other part besides the right angle. In Example 4, we will find values of all sides and angles of the triangle using the sine function.

EXAMPLE 4 Solve triangle *DEF* (see Fig. 23–4). One side and one other part besides the right angle are known.

To use the sine function, we must have *two* of the following parts: an acute angle, its opposite side, or the hypotenuse. Because we have one acute angle and one side not opposite the known angle, we must find the other angle of the triangle. Angles *D* and *E* are complementary, so

$$∠E = 90° − 32.5° = 57.5°$$

Now we can use the sine function to find the hypotenuse because we have an acute angle, *E*, and its opposite side, *e*.

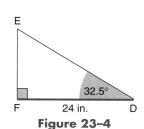

Figure 23–4

$$\sin E = \frac{\text{opp}}{\text{hyp}} \qquad \text{Substitute.}$$

$$\sin 57.5° = \frac{24}{f} \qquad \text{Solve for } f.$$

$$f(\sin 57.5°) = 24$$

$$f = \frac{24}{\sin 57.5°}$$

$$f = 28.45653714$$

$$\mathbf{f = 28.46 \text{ in.}} \qquad \text{Rounded to four significant digits}$$

To find side d, we have

$$\sin D = \frac{\text{opp}}{\text{hyp}} \qquad \text{Substitute; use full calculator value for } f \text{ to get the most accurate result.}$$

$$\sin 32.5° = \frac{d}{28.45653714} \qquad \text{Solve for } d.$$

$$28.45653714(\sin 32.5°) = d$$

$$15.28968626 = d$$

$$15.29 \text{ in.} = d \qquad \text{Rounded to four significant digits}$$

The solved triangle looks like Fig. 23–5.

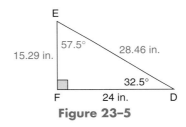

Figure 23–5

Tips and Traps   *Check Computations Using the Pythagorean Theorem:*
In this and other problems involving right triangles, our computations of the sides can be checked by using the Pythagorean theorem. We will check the solution to Example 4 as an illustration.

$$(\text{hyp})^2 = (\text{leg})^2 + (\text{leg})^2$$

$$(28.45653714)^2 = (15.28968626)^2 + (24)^2$$

$$809.774506 = 233.7745059 + 576$$

$$809.774506 = 809.7745059 \qquad \textbf{Difference is caused by rounding discrepancy.}$$

Rounding discrepancies are minimized when more significant digits are used.

---

② Find the Missing Parts of a Right Triangle Using the Cosine Function.

In some of the previous problems, when we had to find a side not opposite the given angle, we had to find the other angle first by subtracting the given acute angle from 90°. If we use the cosine function, however, we can find the same side by using the given angle, rather than its complement. The abbreviated cosine ratio is $\cos\theta = \frac{adj}{hyp}$.

Keep in mind that angle θ is made up of two sides of the right triangle. One of these sides is the hypotenuse and the other side is the side *adjacent* to angle θ.

Rule 23–2 *Use the cosine function to find unknown measures of a right triangle:*

1. Two of these three parts of a right triangle must be known:
  a. One acute angle
  b. The side adjacent to the known acute angle
  c. Hypotenuse
2. Substitute two known values in the formula $\cos\theta = \frac{adj}{hyp}$.
3. Solve for the missing part.

EXAMPLE 5   Find angle A of Fig. 23–6.

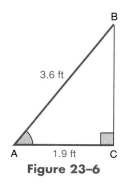

Figure 23–6

We are asked to find an angle and we are given the hypotenuse and the side adjacent to the angle, so the cosine function is used.

$$\cos A = \frac{adj}{hyp}$$ Substitute known values.

$$\cos A = \frac{1.9}{3.6}$$ Or $\cos^{-1}\frac{1.9}{3.6} = A$

$$A = 58.14456918$$

$$A = 58.1°$$ Rounded to nearest 0.1°

EXAMPLE 6   Find side b of Fig. 23–7.

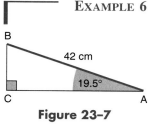

Figure 23–7

We can use either the sine or the cosine function because we are given the hypotenuse and an angle. However, we do not have to find the complement of the given angle if we use the cosine function.

$$\cos A = \frac{adj}{hyp}$$ Substitute.

$$\cos 19.5° = \frac{b}{42}$$ Solve for b.

$$42(\cos 19.5°) = b$$
$$39.59094263 = b$$
$$\mathbf{39.59\ cm = b}$$ Four significant digits

EXAMPLE 7 Find side t of Fig. 23–8.

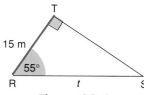

Figure 23–8

Again the cosine function is the most efficient function to use because we must find the hypotenuse and we are given an acute angle and its adjacent side.

$$\cos R = \frac{adj}{hyp}$$ Substitute known values.

$$\cos 55° = \frac{15}{t}$$ Solve for t.

$$t(\cos 55°) = 15$$

$$t = \frac{15}{\cos 55°}$$

$$t = 26.15170193$$

$$t = \mathbf{26.15\ m}$$ Four significant digits

3 Find the Missing Parts of a Right Triangle Using the Tangent Function.

If we have a right triangle in which we know only the length of the two legs, we cannot use the sine or cosine function to solve the triangle unless we use the Pythagorean theorem to find the length of the hypotenuse. However, we can use the tangent function directly: $\tan \theta = \frac{opp}{adj}$.

Rule 23–3 *Use the tangent function to find unknown measures of a right triangle:*

1. Two of the following three parts of a right triangle must be known:
   a. An acute angle
   b. The side opposite the acute angle
   c. The side adjacent to the acute angle
2. Substitute two known values in the formula $\tan \theta = \frac{opp}{adj}$.
3. Solve for the missing part.

Whenever we wish to check our calculation of the sides, we may use the Pythagorean theorem as we did in the Tips and Traps following Example 4.

EXAMPLE 8 Find angle A of Fig. 23–9.

Figure 23–9

We are looking for an angle and we are given the side opposite and the side adjacent to the angle, so we use the tangent function.

$$\tan A = \frac{opp}{adj}$$ Substitute known values.

$$\tan A = \frac{14}{16}$$ Or $\tan^{-1} \frac{14}{16} = A$

$$A = 41.18592517°$$

$$A = \mathbf{41.2°}$$ Rounded to nearest 0.1°

EXAMPLE 9   Use the tangent function to find a of Fig. 23–10.

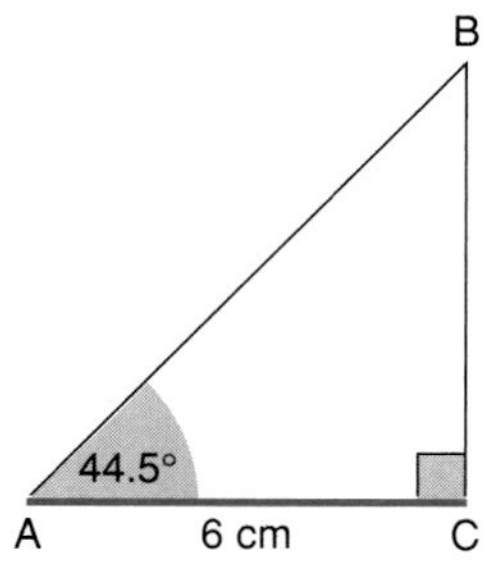

In this example, we are given an acute angle and the side adjacent to the acute angle. We are looking for the opposite side and thus use the tangent function.

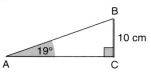

$$\tan A = \frac{\text{opp}}{\text{adj}} \qquad \text{Substitute known values.}$$

$$\tan 44.5° = \frac{a}{6} \qquad \text{Solve for } a.$$

$$6(\tan 44.5°) = a$$

$$5.896183579 = a$$

$$\textbf{5.896 cm} = \textbf{\textit{a}} \qquad \text{Four significant digits}$$

Figure 23–10

EXAMPLE 10   Find side b of Fig. 23–11.

Here we know an acute angle and its opposite side. We are looking for the side adjacent to the acute angle.

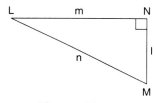

Figure 23–11

$$\tan A = \frac{\text{opp}}{\text{adj}} \qquad \text{Substitute known values.}$$

$$\tan 19° = \frac{10}{b} \qquad \text{Solve for } b.$$

$$b(\tan 19°) = 10$$

$$b = \frac{10}{\tan 19°}$$

$$b = 29.04210878$$

$$b = 29.04 \text{ cm} \qquad \text{Four significant digits}$$

SELF-STUDY EXERCISES 23-1

1 Use the sine function to find the indicated parts of the triangle LMN in Fig. 23–12. Round lengths of sides to four significant digits and angles to the nearest 0.1°.

1. Find M if $n = 15$ m and $m = 7$ m.
2. Find l if $n = 13$ in. and $L = 32°$.
3. Find m if $l = 15$ m and $L = 28°$.
4. Find n if $l = 12$ ft and $M = 42°$.
5. Find M if $m = 13$ cm and $n = 19$ cm.
6. Find M if $n = 3.7$ in. and $l = 2.4$ in.

Figure 23–12

Solve triangle STU of Fig. 23–13 using the sine function. Check the measures of the sides by using the Pythagorean theorem. Round as above.

7. Solve if $t = 18$ yd and $s = 14$ yd.
8. Solve if $U = 45°$ and $u = 4.7$ m.
9. Solve if $S = 34.5°$ and $t = 8.5$ mm.
10. Solve if $S = 16°$ and $s = 14$ m.

Figure 23–13

2 Use the cosine function to find the indicated parts of triangle *KLM* in Fig. 23–14. Round sides to four significant digits and angles to the nearest 0.1°.

11. Find *M* if *k* = 13 m and *l* = 16 m.
12. Find *k* if *l* = 11 cm and *M* = 24°.
13. Find *l* if *M* = 31° and *k* = 27 ft.
14. Find *l* if *m* = 15 dm and *M* = 25°.
15. Find *k* if *K* = 72° and *l* = 16.7 mm.
16. Find *l* if *K* = 67° and *k* = 13 yd.

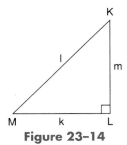

Figure 23–14

Use the sine *or* cosine function to solve triangle *QRS* of Fig. 23–15. Round as above.

17. Solve if *s* = 23 ft and *q* = 16 ft.
18. Solve if *s* = 17 cm and *R* = 46°.
19. Solve if *q* = 14 dkm and *Q* = 73.5°.
20. Solve if *R* = 59.5° and *q* = 8 m.

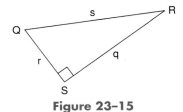

Figure 23–15

3 Use the tangent function to find the indicated parts of triangle *ABC* of Fig. 23–16. Round sides to four significant digits and angles to the nearest 0.1°.

21. Find *A* if *b* = 11 cm and *a* = 6 cm.
22. Find *b* if *a* = 1.9 m and *A* = 25°.
23. Find *a* if *A* = 40.5° and *b* = 7 ft.
24. Find *A* if *b* = 10.8 m and *a* = 4.7 m.
25. Find *a* if *A* = 43° and *b* = 0.05 cm.
26. Find *a* if *B* = 68° and *b* = 0.03 m.

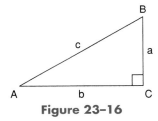

Figure 23–16

Solve triangle *DEF* of Fig. 23–17. Round sides to four significant digits and angles to the nearest 0.1°.

27. Solve if *e* = 4.6 m and *d* = 3.2 m.
28. Solve if *D* = 42° and *e* = 7 ft.
29. Solve if *E* = 73.5° and *e* = 20.13 in.
30. Solve if *d* = 11 ft and *e* = 8 ft.

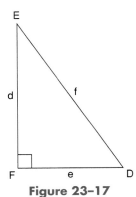

Figure 23–17

APPLIED PROBLEMS USING RIGHT-TRIANGLE TRIGONOMETRY

Learning Objectives

[1] Select the most direct method for solving right triangles.

[2] Solve applied problems using right-triangle trigonometry.

In this section, we focus on how to select the function (theorem, property, or the like) that minimizes the number of steps needed to solve a problem or that uses the fewest calculations. This economy improves both efficiency and accuracy.

[1] **Select the Most Direct Method for Solving Right Triangles.**

Our first task in solving any problem involving right triangles, especially problems on the job, is to *select the most convenient and efficient function* to use in each particular problem. Following are two basic guidelines to use.

> **Rule 23–4** *To select the most direct method:*
>
> 1. Where possible, choose the function that uses parts given in the problem rather than parts that are not given and must be calculated.
> 2. Where possible, choose the function that gives the desired part directly, that is, without having to find other parts first.

EXAMPLE 1 In $\triangle ABC$ of Fig. 23–18, find c.

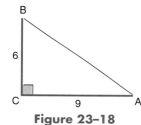

Figure 23–18

The most direct method of finding side c is the *Pythagorean theorem*. The two legs of a right triangle are given.

$c^2 = a^2 + b^2$ Substitute.

$c^2 = 6^2 + 9^2$

$c^2 = 36 + 81$

$c^2 = 117$

$c = 10.81665383$

$\boldsymbol{c = 10.82}$ Four significant digits

For illustrative purposes only, we will show how trigonometric functions can be used to find c.

$$\tan A = \frac{\text{opp}}{\text{adj}}$$ Legs or opposite and adjacent sides are given for this function.

$$\tan A = \frac{6}{9}$$ Solve for A.

$$A = \tan^{-1}\frac{6}{9}$$

$$A = 33.69006753$$

$$A = 33.7°$$ Rounded to nearest 0.1°

$$\sin A = \frac{\text{opp}}{\text{hyp}}$$ Values are now available for the sine function.

$$\sin 33.69006753 = \frac{6}{c}$$

$$c = \frac{6}{\sin 33.69006753}$$

$$c = 10.81665383$$

$$\boldsymbol{c = 10.82} \qquad \text{Four significant digits}$$

EXAMPLE 2   In $\triangle ABC$ of Fig. 23–19, find angle B.

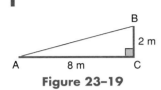

Figure 23–19

The Pythagorean theorem cannot be used to find any angle measure. Since we are given the sides adjacent to and opposite angle B, we should use the *tangent function* for a quick, direct solution.

$$\tan \theta = \frac{\text{opp}}{\text{adj}} \qquad \text{Substitute.}$$

$$\tan B = \frac{8}{2} \qquad \text{Reduce.}$$

$$\tan B = 4 \qquad \text{Or } \tan^{-1} 4 = B$$

$$B = 75.96375653$$

$$\boldsymbol{B = 76.0°} \qquad \text{Round to nearest } 0.1°$$

2   Solve Applied Problems Using Right-Triangle Trigonometry.

Many technical applications can be solved by the use of right triangles. The following examples show some career-related applications of right triangles. In solving technical problems, we should draw a diagram or picture to help us visualize the various relationships.

EXAMPLE 3   A jet takes off at a 30° angle (see Fig. 23–20). If the runway (from takeoff) is 875 ft long, find the altitude of the airplane as it flies over the end of the runway.

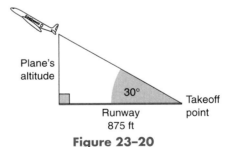

Figure 23–20

Known facts	One acute angle = 30°; adjacent side = 875 ft
Unknown fact	Plane's altitude or opposite side
Relationship	$\tan \theta = \dfrac{\text{opp}}{\text{adj}}$
Estimation	In a 45° 45° 90° right triangle, the legs are equal. Because 30° is less than 45°, the side opposite the 30° angle should be less than 875 feet.
Solution	$\tan \theta = \dfrac{\text{opp}}{\text{adj}}$

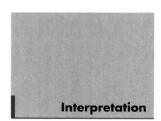

$$\tan 30° = \frac{a}{875}$$

$$875(\tan 30°) = a$$

$$505.1814855 = a$$

Interpretation **505.2 ft = plane's altitude** Four significant digits

Many right-triangle applications make use of the terminology *angle of elevation* and *angle of depression*. See Fig. 23–21. The angle of elevation is generally used when we are looking *up* at an object. The angle of depression is used to describe the location of an objective *below* eye level. *Both* angles are the angles formed by a line of sight and a horizontal line from the point of sight.

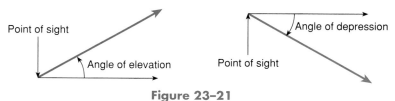

Figure 23–21

EXAMPLE 4   A stretch of roadway drops 30 ft for every 300 ft of road (see Fig. 23–22). Find the *angle of declination* of the road.

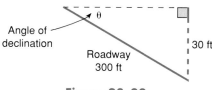

Figure 23–22

The *angle of declination* is the angle of depression. The opposite side and the hypotenuse are given.

$$\sin \theta = \frac{\text{opp}}{\text{hyp}}$$ Substitute.

$$\sin \theta = \frac{30}{300}$$

$$\sin \theta = 0.1$$ Or $\sin^{-1}0.1 = \theta$

$$\theta = 5.739170477°$$

The angle of declination of the road is 5.7° rounded to the nearest 0.1°.

EXAMPLE 5   A surveyor locates two points on a steel column so that it can be set plumb (perpendicular to the horizon). If the angle of elevation is 15° and the surveyor's transit is 175 ft from the column (see Fig. 23–23), find the distance from the transit to the upper point on the column. (A transit is a surveying instrument for measuring angles.)

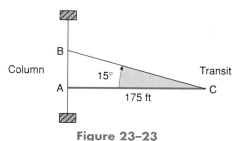

Figure 23–23

An acute angle and the adjacent side are given. To find the distance from the transit to point B on the column (hypotenuse) we use the *cosine function*.

$$\cos\ \theta = \frac{adj}{hyp} \qquad \text{Substitute.}$$

$$\cos\ 15° = \frac{175}{hyp} \qquad \text{Solve for hyp.}$$

$$hyp(\cos\ 15°) = 175$$

$$hyp = \frac{175}{\cos\ 15°}$$

$$hyp = 181.1733316$$

$$hyp = 181.2\ ft \qquad \text{Four significant digits}$$

Point B is 181.2 ft from the transit.

EXAMPLE 6 Find the angle a rafter makes with a joist of a house if the rise is 12 ft and the span is 30 ft (see Fig. 23–24). Find the angle that the rafter makes with the joist and the length of the rafter.

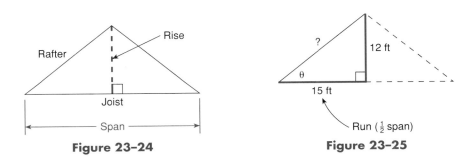

Figure 23–24   **Figure 23–25**

The span is twice the distance from the outside stud to the center point of the joist. Therefore, to solve the right triangle for the desired angle, we draw the triangle shown in Fig. 23–25. Because the legs of a right triangle are given, the tangent function is used.

$$\tan\ \theta = \frac{opp}{adj} \qquad \text{Substitute.}$$

$$\tan\ \theta = \frac{12}{15} \qquad \text{Tan}^{-1}\ \tfrac{12}{15} = \theta$$

$$\theta = 38.65980825$$

$$\theta = 38.7° \qquad \text{Rounded to nearest 0.1°}$$

We can find the length of the rafter directly by using the *Pythagorean theorem*.

$$(hyp)^2 = 12^2 + 15^2$$

$$(hyp)^2 = 144 + 225$$

$$(hyp)^2 = 369$$

$$hyp = 19.20937271$$

$$hyp = 19.21\ ft \qquad \text{Four significant digits}$$

The angle the rafter makes with the joist is 38.7° and the rafter is 19.21 ft.

EXAMPLE 7 Find the angle formed by the connecting rod in the mechanical assembly shown in Fig. 23–26.

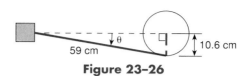

Figure 23–26

Given the hypotenuse and the side opposite the desired angle, we use the *sine function.*

$$\sin \theta = \frac{opp}{hyp}$$ Substitute.

$$\sin \theta = \frac{10.6}{59}$$ $\mathrm{Sin}^{-1} \frac{10.6}{59} = \theta$

$$\theta = 10.35001563$$

$$\theta = 10.4°$$ Rounded

The angle formed by the connecting rod is 10.4°.

EXAMPLE 8   Find the impedance Z of a circuit with 20 ohms of reactance X_L represented by the vector diagram in Fig. 23–27.

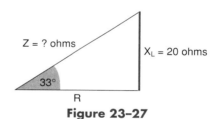

Figure 23–27

Because we are looking for the hypotenuse Z and are given an acute angle and the opposite side, we use the *sine function.*

$$\sin \theta = \frac{opp}{hyp}$$ Substitute.

$$\sin 33° = \frac{20}{Z}$$

$$Z(\sin 33°) = 20$$

$$Z = \frac{20}{\sin 33°}$$

$$Z = 36.72156918$$

$$Z = 36.72 \text{ ohms}$$

The impedance Z is 36.72 ohms.

1   Find the indicated part of the right triangles in Figs. 23–28 to 23–32 by the most direct method.

1.

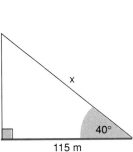

Figure 23–28

2.

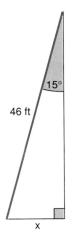

Figure 23–29

3.

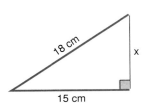

Figure 23–30

4.

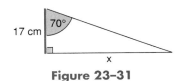

Figure 23–31

5.

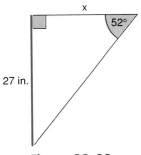

Figure 23–32

2   Use the trigonometric functions to solve the following technical problems. Round side lengths to four significant digits and angles to 0.1°.

6. A sign is attached to a building by a triangular brace. If the horizontal length of the brace is 48 in. and the angle at the sign is 25° (see Fig. 23–33), what is the length of the wall support piece?

7. A surveyor uses right triangles to measure inaccessible property lines. To measure a property line that crosses a pond, a surveyor sights to a point across the pond, then makes a right angle, measures 50 ft, and sights the point across the pond with a 47° angle (see Fig. 23–34). Find the distance across the pond from the initial point.

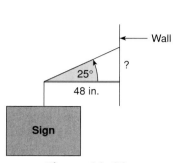

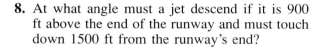

Figure 23–33

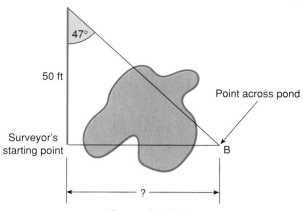

Figure 23–34

8. At what angle must a jet descend if it is 900 ft above the end of the runway and must touch down 1500 ft from the runway's end?

10. A roadway rises 4 ft for every 15 ft along the road. What is the angle of inclination of the roadway?

12. The vector diagram of the circuit in Fig. 23–35 has a known impedance Z. Find the reactance X_L. All units are in ohms.

13. Using Fig. 23–35, find resistance R.

9. A 50-ft wire is used to brace a utility pole. If the wire is attached 4 ft from the top of the 35-ft pole, how far from the base of the pole will the wire be attached to the ground?

11. A shadow cast by a tree is 32 ft long when the angle of inclination of the sun is 36°. How tall is the tree?

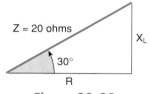

Figure 23–35

14. A piston assembly at the midpoint of its stroke forms a right triangle (see Fig. 23–36). Find the length of rod R.

15. Find the angle a rafter makes with a joist of a house if the rise is 18 ft and the span is 50 ft. Refer to Fig. 23–24 on page 861.

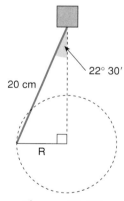

Figure 23–36

CAREER APPLICATION

Electronics: Series Circuits and Parallel Circuits

Technicians use many right triangles, but the two most common ones are the ohms triangle and the siemens triangle. Remember that ohms and siemens are reciprocals of each other. Series circuits use the ohms triangle (see Fig. 23–37) and parallel circuits use the siemens triangle (see Fig. 23–38).

For each triangle, the units are the same on all three legs. The hypotenuse and the angle of rotation combine to give the polar notation, and the horizontal side and the vertical side combine to give the coordinates of a point in rectangular notation. A polar number shows the location of a point differently from the rectangular coordinate system. The coordinates of a polar number are the angle of rotation and the distance of the hypotenuse.

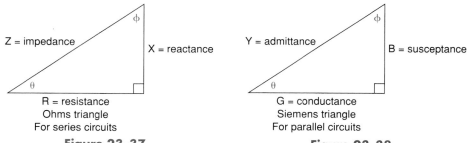

Ohms triangle
For series circuits

Figure 23–37

Siemens triangle
For parallel circuits

Figure 23–38

Most analysis is done with rectangular numbers. All measurements are done with polar numbers. It is important to understand the terminology of this application. All the triangles are handled in a similar way.

Some basic facts about circuits will help us in interpreting a problem.

- All three sides in the siemens triangle are measured in siemens, the reciprocal of ohms.
- If a component has a high resistance, it has a low conductance, and vice versa.
- R and G are used for both dc (direct current) and ac (alternating current) analysis.
- X, Z, B, and Y are used only for ac analysis.

The Pythagorean theorem applies to any right triangle. The formula can be rearranged to solve for any missing side if you know the other two sides. The following equations give a variation of the Pythagorean theorem for each side in both the ohms and siemens triangles.

Ohms Triangle for Series Circuits

$$Z = \sqrt{R^2 + X^2}$$
$$R = \sqrt{Z^2 - X^2}$$
$$X = \sqrt{Z^2 - R^2}$$

Siemens Triangle for Parallel Circuits

$$Y = \sqrt{G^2 + B^2}$$
$$G = \sqrt{Y^2 - B^2}$$
$$B = \sqrt{Y^2 - G^2}$$

Some additional basic facts in electronics are:

- The sides for R and G are always positive and rest on the horizontal axis.
- X and B are always perpendicular to the horizontal axis, and they may be positive or negative depending on whether the circuit is inductive or capacitive.
- The hypotenuse, Z or Y, is always considered positive because it is the square root of the sum of the squares.
- The angle θ is also called the *phase angle* and is either positive or negative, depending on the slope of the hypotenuse.
- The angle θ is the angle between the hypotenuse and the vertical side.

The angles θ and ϕ can be calculated in several different ways, depending on which sides or angles you were given originally. Another standard notation for $\sin^{-1}$, $\cos^{-1}$, and $\tan^{-1}$ is arcsin, arccos, and arctan, respectively.

Ohms

$$\theta = \arctan\left(\frac{\text{opp}}{\text{adj}}\right) = \arctan\left(\frac{X}{R}\right) = \tan^{-1}\left(\frac{X}{R}\right)$$

Siemens

$$\theta = \arctan\left(\frac{\text{opp}}{\text{adj}}\right) = \arctan\left(\frac{B}{G}\right) = \tan^{-1}\left(\frac{B}{G}\right)$$

$$\theta = \arcsin\left(\frac{\text{opp}}{\text{hyp}}\right) \qquad \theta = \arcsin\left(\frac{\text{opp}}{\text{hyp}}\right)$$

$$= \arcsin\left(\frac{X}{Z}\right) \qquad = \arcsin\left(\frac{B}{Y}\right)$$

$$= \sin^{-1}\left(\frac{X}{Z}\right) \qquad = \sin^{-1}\left(\frac{B}{Y}\right)$$

$$\theta = \arccos\left(\frac{\text{adj}}{\text{hyp}}\right) \qquad \theta = \arccos\left(\frac{\text{adj}}{\text{hyp}}\right)$$

$$= \arccos\left(\frac{R}{Z}\right) \qquad = \arccos\left(\frac{G}{Y}\right)$$

$$= \cos^{-1}\left(\frac{R}{Z}\right) \qquad = \cos^{-1}\left(\frac{G}{Y}\right)$$

Finally, θ and ϕ are complementary angles.

$$\phi = 90° - \theta \qquad \text{and} \qquad \theta = 90° - \phi$$

The calculator sequence for finding an arcfunction is the same as the inverse function.

Find all indicated missing sides and angles in Figs. 23–39 and 23–40.

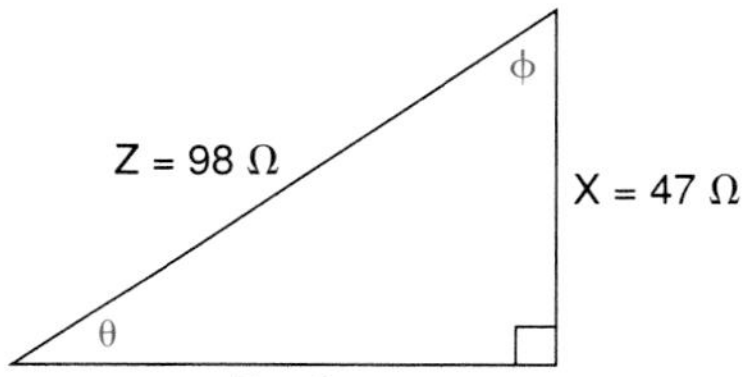

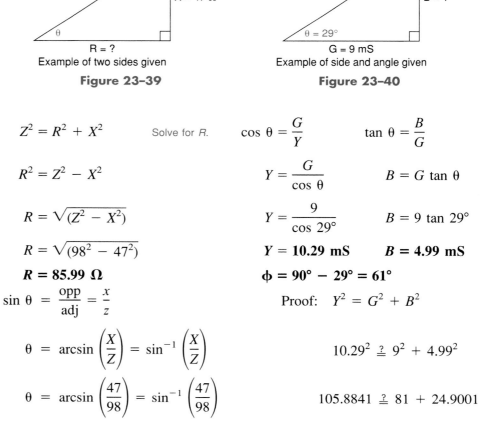

Example of two sides given

Figure 23–39

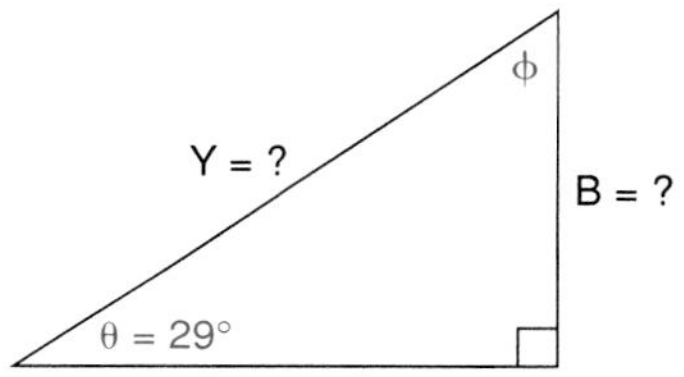

Example of side and angle given

Figure 23–40

$$Z^2 = R^2 + X^2 \qquad \text{Solve for } R. \qquad \cos\theta = \frac{G}{Y} \qquad \tan\theta = \frac{B}{G}$$

$$R^2 = Z^2 - X^2 \qquad\qquad Y = \frac{G}{\cos\theta} \qquad B = G\tan\theta$$

$$R = \sqrt{(Z^2 - X^2)} \qquad\qquad Y = \frac{9}{\cos 29°} \qquad B = 9\tan 29°$$

$$R = \sqrt{(98^2 - 47^2)} \qquad\qquad \mathbf{Y = 10.29\ mS} \qquad \mathbf{B = 4.99\ mS}$$

$$\mathbf{R = 85.99\ \Omega} \qquad\qquad \boldsymbol{\phi = 90° - 29° = 61°}$$

$$\sin\theta = \frac{\text{opp}}{\text{adj}} = \frac{x}{z} \qquad\qquad \text{Proof:} \quad Y^2 = G^2 + B^2$$

$$\theta = \arcsin\left(\frac{X}{Z}\right) = \sin^{-1}\left(\frac{X}{Z}\right) \qquad 10.29^2 \overset{?}{=} 9^2 + 4.99^2$$

$$\theta = \arcsin\left(\frac{47}{98}\right) = \sin^{-1}\left(\frac{47}{98}\right) \qquad 105.8841 \overset{?}{=} 81 + 24.9001$$

$$\theta = 28.7°$$

$$\phi = 90° - 28.7° = 61.3°$$

Proof: $\tan 28.7° \stackrel{?}{=} \dfrac{47}{85.99}$

$0.547484008 \doteq 0.546575183$ True, except for rounding discrepancy.

$105.8841 \stackrel{?}{=} 105.9001$

True, except for rounding discrepancy.

$$\phi + \theta = 90°$$

$61 + 29 \stackrel{?}{=} 90°$ True

Fill in all answers on the triangles in Figs. 23–41 and 23–42 with the correct units. Always give a proof.

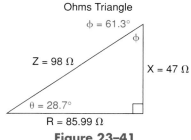

Ohms Triangle

Figure 23–41

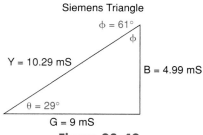

Siemens Triangle

Figure 23–42

Exercises

Redraw the triangles in Figs. 23–43 to 23–50. Fill in all missing sides and angles. Show your proof. Include correct units.

1.

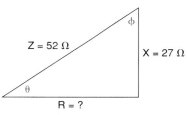

Figure 23–43

2.

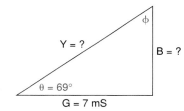

Figure 23–44

3.

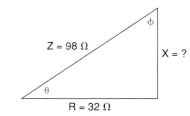

Figure 23–45

4.

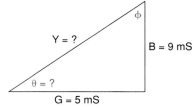

Figure 23–46

5.

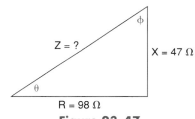

Figure 23–47

6.

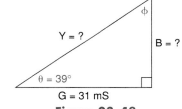

Figure 23–48

7.

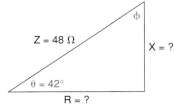

Figure 23–49

8.

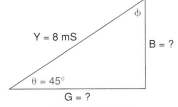

Figure 23–50

Answers for Exercises

Redraw the triangles in Figs. 23–51 to 23–58. Fill in all missing sides and angles. Include correct units.

1. $\phi = 58.7°$

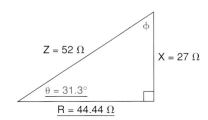

Figure 23–51

2. $\phi = 21°$

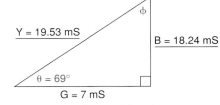

Figure 23–52

3. $\phi = 19.06°$

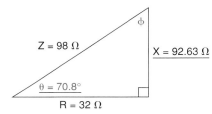

Figure 23–53

4. $\phi = 29.1°$

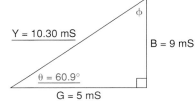

Figure 23–54

5. $\phi = 64.4°$

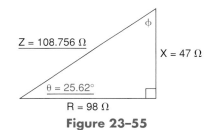

Figure 23–55

6. $\phi = 51°$

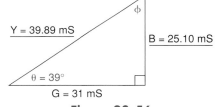

Figure 23–56

7. $\phi = 48°$

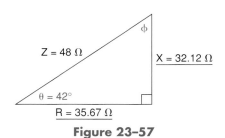

Figure 23–57

8. $\phi = 45°$

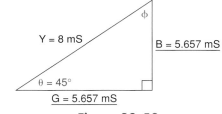

Figure 23–58

Section 23–1

Solve the triangles in Figs. 23–59 to 23–68. Round sides to four significant digits and round angles to the nearest 0.1°.

1.

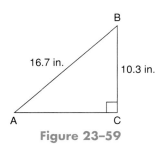

Figure 23–59

2.

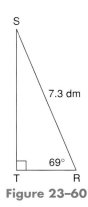

Figure 23–60

3.

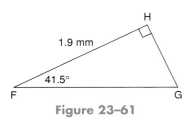

Figure 23–61

4.

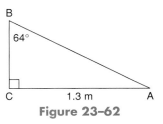

Figure 23–62

5.

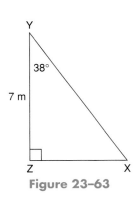

Figure 23–63

6.

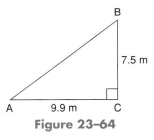

Figure 23–64

7.

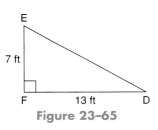

Figure 23–65

8.

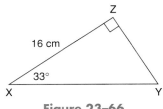

Figure 23–66

9.

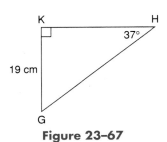

Figure 23–67

10.

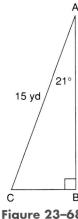

Figure 23–68

Section 23–2
Find the indicated parts of the triangles in Figs. 23–69 to 23–71. Round side lengths to four significant digits and angle measures to the nearest 0.1°.

11. Find *A*.

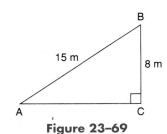

Figure 23–69

12. Find *A*.

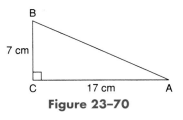

Figure 23–70

13. Find *a*.

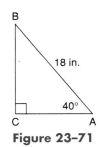

Figure 23–71

Solve the triangles in Figs. 23–72 and 23–73.

14.

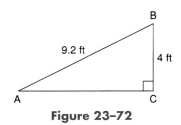

Figure 23–72

15.

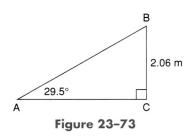

Figure 23–73

16. A railway inclines 14°. How many feet of track must be laid if the hill is 15 ft high?

17. A corner shelf is cut so that the sides placed on the wall are 37 in. and 42 in. What are the measures of the acute angles?

18. From a point 5 ft above the ground and 20 ft from the base of a building, a surveyor uses a transit to sight an angle of 38° to the top of the building. Find the height of the building.

19. A surveyor makes the measures indicated in Fig. 23–74. Solve the triangle. All parts of the triangle should be included in a report.

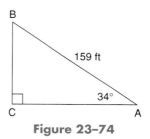

Figure 23–74

20. What length of rafter is needed for a roof if the rafters will form an angle of 35.5° with a joist and the rise is 8 ft?

━━━━━━━━━━━━━ **CHALLENGE PROBLEMS** ━━━━━━━━━━━━━

21. Draw triangle ABC so that angle C is 90° and angle A is 70°. Point E is on side $\overline{AB}$ and is 10 cm from A. Point D is on $\overline{BC}$ and $\overline{DE}$ is parallel to $\overline{CA}$. If $\overline{CA}$ is 24 cm, find the length of $\overline{BD}$.

22. You are designing a stairway for a home and you need to determine how many steps the stairway must have. If the stairway is to be 8 ft high and the floorspace allotted is 14 ft, determine how many steps are needed and determine the optimum measure (rise and run) for each step.

━━━━━━━━━━━━━ **CONCEPTS ANALYSIS** ━━━━━━━━━━━━━

1. Use a calculator to find sin 45° and cos 45°. Using your knowledge of geometry, explain the relationship between the sin 45° and cos 45°. Does it follow that the sine and cosine of the same angle are always equal? Are there other angles for which the two functions are equal?

3. You know the measure of both legs of a right triangle and the measure of one of the acute angles. Describe two different methods you could use to find the length of the hypotenuse.

5. Explain the differences and similarities between angle of inclination and angle of declination.

7. Devise a right triangle that has the measure of an angle (other than the right angle) and a side given. Then use trigonometric functions or other methods to find the measures of all the other sides and angles.

9. Use your calculator to discover which of the three trigonometric functions—sine, cosine, tangent—can have values greater than 1. For what range of angles is the value of this function greater than 1?

2. If you are given an angle and an adjacent side of a right triangle, explain how you would find the hypotenuse of the triangle.

4. What is the least number of measures of the parts of a triangle that must be known to find the measures of all the other parts of the triangle? What are the parts required?

6. Using a calculator, compare the sine of an acute angle with the sine of its complement for several angles. Compare the sine of an acute angle with the cosine of its complement for several angles. Generalize your findings.

8. Describe the calculator steps required to find the measure of an angle whose sine function is given.

10. Use your graphics calculator to graph the trigonometric functions sine and cosine. Tell how the graphs are similar and how they are different.

CHAPTER SUMMARY

Objectives	**What to Remember with Examples**

Section 23–1

1 Find the missing parts of a right triangle using the sine function.

$$\sin \theta = \frac{\text{opposite side}}{\text{hypotenuse}}$$

Use the sine function to find the part of the triangle indicated in Fig. 23–75.

$$\sin 25° = \frac{x}{12}$$
$$x = 12 \sin 25°$$
$$x = 5.071419141$$
$$x = 5.071 \text{ in.}$$

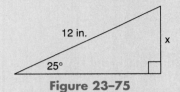

Figure 23–75

2 Find the missing parts of a right triangle using the cosine function.

$$\cos \theta = \frac{\text{adjacent side}}{\text{hypotenuse}}$$

Use the cosine function to find the part of the triangle indicated in Fig. 23–76.

$$\cos 37° = \frac{26}{x}$$
$$x = \frac{26}{\cos 37°}$$
$$x = 32.55552711$$
$$x = 32.56 \text{ m}$$

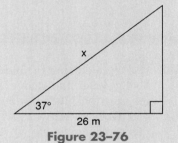

Figure 23–76

3 Find the missing parts of a right triangle using the tangent function.

$$\tan \theta = \frac{\text{opposite side}}{\text{adjacent side}}$$

Use the tangent function to find the part of the triangle indicated in Fig. 23–77.

$$\tan \theta = \frac{8}{11}$$
$$\tan \theta = 0.727272727$$
$$\theta = \tan^{-1} 0.727272727$$
$$\theta = 36.02737311$$
$$\theta = 36.0°$$

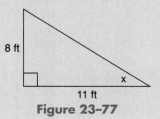

Figure 23–77

Section 23–2

1 Select the most direct method for solving right triangles.

Whenever possible, choose the function that uses parts given in the problem rather than parts that are not given and must be calculated.

Whenever possible, choose the function that gives the desired part directly, that is, without having to find other parts first.

Find the parts indicated in Fig. 23–78.

$$\cos 1.2 = \frac{x}{25}$$
$$x = 25(\cos 1.2)$$
$$x = 9.058943862$$
$$x = 9.059 \text{ m}$$

$$\sin 1.2 = \frac{y}{25}$$
$$y = 25(\sin 1.2)$$
$$y = 23.30097715$$
$$y = 23.30 \text{ m}$$

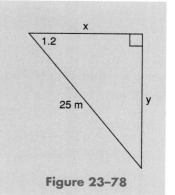

Figure 23–78

2 Solve applied problems using right-triangle trigonometry.

To solve a problem using the trigonometric functions, first determine the given parts and the missing part or parts; then identify the trigonometric function that relates the given and missing parts. Use this function and the given information to find the missing information.

A jet takes off at a 25° angle. Find the distance traveled by the plane from the takeoff point to the end of the runway if the runway is 950 feet long.

$$\cos 25° = \frac{950}{x}$$
$$x = \frac{950}{\cos 25°}$$
$$x = 1048.209023$$
$$x = 1048 \text{ ft}$$

WORDS TO KNOW

significant digits (p. 851)   angle of depression (p. 860)   angle of declination (p. 860)
angle of elevation (p. 860)

CHAPTER TRIAL TEST

The following measures are indicated for a standard right triangle, *ABC*, where *C* is the right angle. Draw the figures and use the sine, cosine, or tangent function to find the indicated parts of the triangle. Round side lengths to four significant digits and angles to the nearest 0.1°.

1. $a = 16$ m, $b = 14$ m, find A.
2. $a = 7$ in., $A = 33°$, find c.
3. $c = 17$ ft, $B = 25°$, find a.
4. $a = 21$ m, $A = 48.5°$, find b.
5. $a = 32$ cm, $c = 47$ cm, find A.
6. $c = 12$ m, $A = 35°$, find a.
7. $b = 21$ cm, $A = 17°$, find a.
8. $b = 1$ cm, $A = 87°$, find c.
9. $b = 3.1$ m, $c = 6.8$ m, find A.
10. $a = 0.15$ m, $c = 0.46$ m, find A.

11. Solve triangle *ABC* in Fig. 23–79. Round as above.

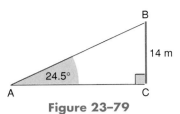

Figure 23–79

12. Solve triangle *DEF* in Fig. 23–80. Round as above.

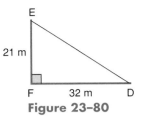

Figure 23–80

Solve the following problems. Round as above unless otherwise indicated.

13. A stair has a rise of 4 in. for every 5 in. of run. What is the angle of inclination of the stair?

14. A surveyor uses indirect measurement to find the width of a river at a certain point. The surveyor marks off 50 ft along the river bank at right angles with the river and then sights an angle of 43° to point *A* across the river (Fig. 23–81). Find the width of the river.

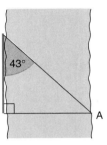

Figure 23–81

15. Steel girders are reinforced by placing steel supports between two runners so that right triangles are formed (see Fig. 23–82). If the runners are 24 in. apart and a 30° angle is desired between the support and a runner, find the length of the support that will be placed at a 30° angle.

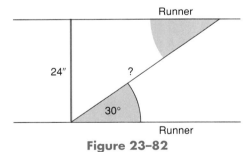

Figure 23–82

16. In the circuit represented by the diagram of Fig. 23–83, find the total current I_t. All units are in amps.

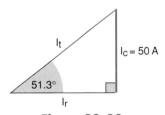

Figure 23–83

17. Find the current in the resistance branch I_r of the circuit in Exercise 16.

18. The minimum clearances for the installation of a metal chimney pipe are shown in Fig. 23–84. How far from the ridge should the hole be cut for the pipe to pass through the roof?

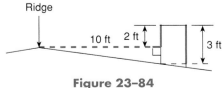

Figure 23–84

19. Refer again to Fig. 23–84. What angle is formed by the chimney pipe and the roof where the pipe passes through the roof?

20. Elbows are used to form bends in rigid pipe and are measured in degrees. What is the angle of bend of the elbow in the installation shown in Fig. 23–85 to the nearest whole degree?

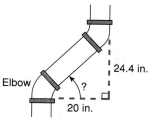

Figure 23–85

21. Find the angle formed by the connecting rod and the horizontal in the mechanical assembly shown in Fig. 23–86.

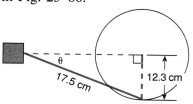

Figure 23–86

22. A utility pole is 40 ft above ground level. A guy wire must be attached to the pole 3 ft from the top to give it support. If the guy wire forms a 20° angle with the ground, how long must the guy wire be? Disregard the length needed for attaching the wire to the pole or ground. Round to hundredths.

23. Solve for reactance X_L and resistance R in Fig. 23–87. All units are in ohms. Round to hundredths.

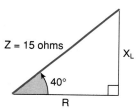

Figure 23–87

24

Oblique Triangles

GOOD DECISIONS THROUGH TEAMWORK

Survey reports describe property lines determined in part through the use of triangles, including oblique triangles, particularly for irregularly shaped properties. For this project, your team must diagram two irregularly shaped properties described in

survey reports. To begin, draw a scaled diagram for the property described below, including angle measures for each vertex. Calculate the acreage to verify that the acreage shown on the survey report is correct. The report reads: "Beginning at the northwest corner of said forty; thence east along the north line of said forty 600 feet; thence south 313 feet to the center of County Road as now located; thence in a northwesterly direction along center of said county road 613 feet to the point of beginning (3 acres)."

Next, your team should locate another survey report of an irregularly shaped property. You might check with real estate offices, abstract companies, or court house records for suitable survey reports. Then draw a diagram of the selected property as you did above.

Keep a record of the calculations your teammates had to make and any landmarks or other points used by the surveyors as reference points. After the diagrams are completed, report to your class on the tools you used to make the calculations and draw the diagrams, the reference points used in the survey reports, the calculations your team made, and the diagrams of the properties.

Although right triangles are perhaps the most frequently used triangles, we often work with other kinds of triangles. In this chapter, we will apply the laws of sines and cosines to oblique triangles, find the area of oblique triangles when only selected parts are given, and use *vectors* for solving triangles that have an angle greater than 90°.

24-1 VECTORS

Learning Objectives

1. Find the magnitude of a vector in standard position, given the coordinates of the end point.
2. Find the direction of a vector in standard position, given the coordinates of the end point.
3. Find the sum of vectors.

1 Find the Magnitude of a Vector in Standard Position, Given the Coordinates of the End Point.

Quantities that we have discussed so far in this text have been described by specifying their size or magnitude. Quantities such as area, volume, length, and temperature, which are characterized by magnitude only, are called *scalars*. There are many other quantities such as electrical current, force, velocity, and acceleration that are called *vectors*.

■ **DEFINITION 24-1: Vector.** A quantity described by magnitude (or length) and direction is a *vector*. A vector represents a shift from one point to another.

When we consider the speed of a plane to be 500 mph, we are considering a scalar quantity. However, when we consider the speed of a plane *traveling northeast from a given location,* we are concerned with both the distance traveled and the direction. This is a vector quantity. We have seen such quantities in the vector diagrams used to solve electronic problems by means of right triangles in Chapter 23 and elsewhere.

Vectors are represented by straight arrows. The length of the arrow represents the *magnitude* of the vector (see Fig. 24–1). The curved arrow shows the counterclockwise *direction* of the vector (see Fig. 24–2), with the end of the arrow being the beginning point and the arrowhead being the end point.

In relating trigonometric functions to vector quantities, we will place the vectors in *standard position*. Standard position places the beginning point of the vector at the origin of a rectangular coordinate system. *The direction of the vector is the counterclockwise angle measured from the positive x-axis (horizontal axis).*

In Fig. 24–2, vector P is a first quadrant vector that has a direction of θ_1 and an endpoint at the point P. The magnitude of a vector can be determined if the vector is in standard position and the x- and y-coordinates of the end point are known. Vector R is a third-quadrant vector that has a direction of θ_2 and an end point at point R. Its magnitude may be determined similarly.

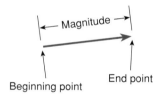

Figure 24–1

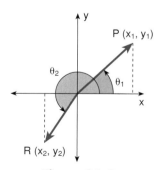

Figure 24–2

Property 24-1 *Magnitude of a vector in standard position:*

The *magnitude* of a vector in standard position is the length of the hypotenuse of the right triangle formed by the vector, the x-axis, and the vertical line from the endpoint of the vector to the x-axis.

> **Rule 24–1** *To find the magnitude of a vector in standard position:*
>
> 1. Substitute into the Pythagorean theorem the coordinates of the end points of the vector as the legs of a right triangle.
> 2. Solve for the hypotenuse.

EXAMPLE 1 Find the magnitude of vector P in Fig. 24–2 if the coordinates of P are (4, 3).

From P, we can draw a vertical line to the x-axis, forming a right triangle with the x-axis and the vector P. The length of the side of the triangle along the x-axis is the x-coordinate of the point P, or 4. The length of the vertical side of the triangle is the y-coordinate of the point P, or 3. The length of the hypotenuse of the triangle is the magnitude of the vector. We will use the Pythagorean theorem to determine the magnitude of vector P.

$p^2 = x^2 + y^2$ Substitute $x = 4$ and $y = 3$.

$p^2 = 4^2 + 3^2$ Solve for p.

$p^2 = 16 + 9$

$p^2 = 25$

$p = \pm\sqrt{25}$

$p = \pm 5$ Or +5 because length or magnitude is positive.

The magnitude of vector P is 5.

EXAMPLE 2 Find the magnitude of vector R in Fig. 24–2 if the coordinates of point R are $(-2, -5)$.

The magnitude of a vector is always positive; however, the x- and y-coordinates can be negative.

$p^2 = x^2 + y^2$ Substitute $x = -2$ and $y = -5$.

$p^2 = (-2)^2 + (-5)^2$ Solve for p.

$p^2 = 4 + 25$

$p^2 = 29$

$p = \sqrt{29}$

$p = 5.385$ Rounded

2 Find the Direction of a Vector in Standard Position, Given the Coordinates of the End Point.

If the vector in standard position falls in quadrant I, we can use right-triangle trigonometry to find the direction of the vector. For a vector in standard position with an end point in quadrant I, the tangent function can be used to find the angle or direction of the vector.

> **Rule 24–2** *To find the direction of a quadrant I vector in standard position:*
>
> 1. Identify the coordinates of the end point of the vector.
> 2. Substitute the coordinates of the end point into the tangent function:
> $\tan \theta = \dfrac{\text{opp}}{\text{adj}}$ or $\tan \theta = \dfrac{y}{x}$.
> 3. Solve for θ.

EXAMPLE 3 Find the direction of vector P in Fig. 24–2 if the coordinates of P are (4, 3).

The right triangle is formed by the vector, the x-axis, and a line from the end point of the vector to the x-axis. The angle at the origin has an opposite-side value of 3 and an adjacent-side value of 4.

$\tan \theta = \dfrac{y}{x}$ Substitute $x = 4$ and $y = 3$.

$\tan \theta = \dfrac{3}{4}$ Solve for θ.

$\tan \theta = 0.75$ Or $\tan^{-1} 0.75 = \theta$

$\theta = 36.9°$ Nearest 0.1°

The direction of vector P is 36.9°.

3 Find the Sum of Vectors.

Two vectors with the same direction can be added by aligning the beginning point of one vector with the end point of the other. The vector represented by the sum is called the *resultant* vector, or resultant. The resultant has the same direction as the vectors being added and a magnitude that is the sum of the two magnitudes.

> **Rule 24–3** *To add vectors of the same direction:*
> 1. Add the magnitudes of each vector.
> 2. The resultant vector will have the same direction as the original vectors.

EXAMPLE 4 Add two vectors with a direction of 35° if the magnitudes of the vectors are 4 and 5, respectively (see Fig. 24–3).

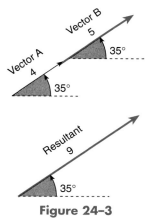

Figure 24–3

The resultant vector has a magnitude of 9 and a direction of 35°.

Two vectors have opposite directions if the directions of the vectors differ by 180°. Two vectors with opposite directions can be added by aligning the beginning point of the vector with the smaller magnitude, with the end point of the vector with the larger magnitude. The resultant has the same direction as the vector with the larger magnitude and a magnitude that is the difference of the two magnitudes.

EXAMPLE 5 Add a vector with a direction of 60° and a magnitude of 6 to a vector with a direction of 240° and a magnitude of 4 (see Fig. 24–4).

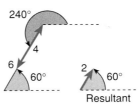

Figure 24–4

Subtract the magnitudes: $6 - 4 = 2$. The resultant has a direction of 60° and a magnitude of 2.

Adding any two vectors that have different directions is accomplished by performing the shifts of each vector in succession.

EXAMPLE 6 Show a graphical representation of the sum of a 45° vector with a magnitude of 5 (vector A) and a 60° vector with a magnitude of 6 (vector B) (see Fig. 24–5).

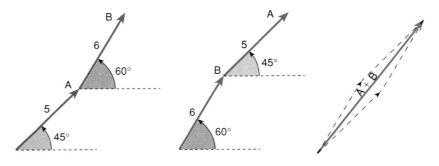

A + B or B + A = resultant A + B

Figure 24–5

Complex numbers are sometimes used to represent vectors. The *x*-coordinate of a vector in standard position is represented as the real component. The *y*-coordinate of a vector in standard position is represented as the imaginary component. Vector *A* could be written in complex notation by first finding the *x*- and *y*-coordinates of the end point.

EXAMPLE 7 Write the vectors A, B, and $A + B$ from Fig. 24–5 in complex notation.

Because the magnitude and direction are known for both vectors A and B, the x-coordinates can be found by using the cosine function, $\cos \theta = \frac{\text{adj}}{\text{hyp}} = \frac{x}{r}$, where r is the magnitude. The y-coordinates can be found by using the sine function, $\sin \theta = \frac{\text{opp}}{\text{hyp}} = \frac{y}{r}$.

Vector A: x-coordinate y-coordinate

$$\cos 45° = \frac{x}{5} \qquad \sin 45° = \frac{y}{5}$$

$$5(\cos 45°) = x \qquad 5(\sin 45°) = y$$

$$3.535533906 = x \qquad 3.535533906 = y$$

Vector A in complex form: 3.535533906 + 3.535533906i

Vector B: x-coordinate y-coordinate

$$\cos 60° = \frac{x}{6} \qquad \sin 60° = \frac{y}{6}$$

$$6(\cos 60°) = x \qquad 6(\sin 60°) = y$$

$$3 = x \qquad 5.196152423 = y$$

Vector B in complex form: 3 + 5.196152423i

Resultant $A + B = ($ 3.535533906 $+$ 3.535533906 $i) + ($ 3 $+$ 5.196152423 $i)$

Resultant $A + B$ = 6.535533906 + 8.731686329i or 6.536 + 8.732i

---

EXAMPLE 8 Find the magnitude and direction of the resultant $A + B$ in Example 7.

The resultant in complex form is $6.535533906 + 8.731686329i$; thus, the x-coordinate of the end point is 6.535533906 and the y-coordinate is 8.731686329.

$$\text{Magnitude} = \sqrt{(6.535533906)^2 + (8.731686329)^2}$$

$$\text{Magnitude} = \sqrt{118.9555496}$$

Magnitude = 10.91

$$\tan \theta = \frac{8.731686329}{6.535533906}$$

$$\tan \theta = 1.336032596 \qquad\qquad \text{Or } \tan^{-1} 1.336032596 = \theta.$$

$$\theta = 53.18570658$$

$$\theta = 53.2° \qquad\qquad \text{Nearest } 0.1°$$

Tips and Traps   ***Vector Notation in Electronics:***

Vectors written in complex form are often used in electronics. The imaginary part is referred to as the j-factor.

Resultant vector $A + B$ from Example 6 would be written as $6.536 + j8.732$. The customary notation is for j to be followed by the coefficient.

1   Find the magnitude of the vectors in standard position with end points at the following points. Round to the nearest thousandth.

1. (5, 12)   **2.** (−12, 9)   **3.** (2, −7)   **4.** (−8, −3)   **5.** (1.5, 2.3)

2   Find the direction of the vectors in standard position with end points at the following points. Round to the nearest hundredth.

6. (5, 12)   **7.** (6, 8)   **8.** (8, 3)   **9.** (2, 5)   **10.** (1, 4)

3

11. Find the resultant vector of two vectors that have a direction of 42° and magnitudes of 7 and 12, respectively.

12. Two vectors have a direction of 72°. Find the sum of the vectors if their magnitudes are 1 and 7, respectively.

13. Find the sum of two vectors if one has a direction of 45° and a magnitude of 7 and the other has a direction of 225° and a magnitude of 8.

14. Find the sum of two vectors if one has a direction of 75° and a magnitude of 15 and the other has a direction of 255° and a magnitude of 9.

Find the magnitude and direction of the resultant of the sum of the two given vectors in complex notation.

15. $5 + 3i$ and $7 + 2i$

16. $1 + 2i$ and $5 + 2i$

24–2   TRIGONOMETRIC FUNCTIONS FOR ANY ANGLE

Learning Objectives

1   Find related, acute angles for angles or vectors in quadrants II, III, and IV.

2   Determine the signs of trigonometric values of angles of more than 90°.

3   Find the trigonometric values of angles of more than 90° using a calculator.

1   **Find Related, Acute Angles for Angles or Vectors in Quadrants II, III, and IV.**

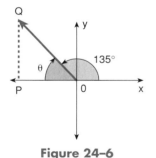

Figure 24–6

The direction of any vector in standard position can be determined, if the coordinates of the end point are known, by applying our knowledge of trigonometric functions. A right triangle that we will refer to as our *reference triangle* can be formed by drawing a vertical line from the end point of the vector to the *x*-axis. See the example of a reference triangle in Fig. 24–6. The angle θ will be called the *related angle*.

■   **DEFINITION 24–2:**   **Related Angle.**   The *related angle* is the acute angle formed by the *x*-axis and the vector.

In quadrant I, the related angle is the same as the direction of the vector. Therefore, the direction of the vector in quadrant I is always less than 90°. For vectors in quadrants II, III, and IV, see Figs. 24–6, 24–7, and 24–8.

In Fig. 24–6, $\overline{PQ}$, the *x*-axis, and vector *Q* form a right triangle. The direction of the vector is 135°; therefore, the related angle is 45° (180° − 135°). 135° is a second quadrant angle. Second quadrant angles are more than 90° and less than 180°, or more than $\frac{\pi}{2}$ radians (1.57) and less than π radians (3.14). The related angle for any second-quadrant vector can be found by subtracting the direction of the vector from 180° (or π radians).

Third-quadrant angles are more than 180° and less than 270° or more than π radians (3.14) and less than $\frac{3\pi}{2}$ radians (4.71). In Fig. 24–7, vector *S* is a third-

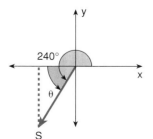

Figure 24–7

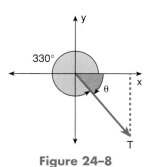

Figure 24-8

quadrant vector. The related angle θ is 60° (240° − 180°). The related angle for any third-quadrant vector is found by subtracting 180° (or π radians) from the direction of the vector.

Vector *T* in Fig. 24–8 is a fourth-quadrant vector. Fourth-quadrant angles are more than 270° and less than 360°, or more than $\frac{3\pi}{2}$ radians (4.71) and less than 2π radians (6.28). The related angle θ is 30° (360° − 330°). The related angle for any fourth-quadrant vector is found by subtracting the direction of the vector from 360° (or 2π radians).

Rule 24–6 *To find related angles for vectors that are more than 90°:*

The related angle for quadrant I angles and vectors is equal to the direction (angle) of the vector. The related angle for angles and vectors more than 90° can be found as follows:

Quadrant II angle: $180° − \theta_2$ or $\pi − \theta_2$

Quadrant III angle: $\theta_3 − 180°$ or $\theta_3 − \pi$

Quadrant IV angle: $360° − \theta_4$ or $2\pi − \theta_4$

Related angles will always be angles less than 90° or $\frac{\pi}{2}$ radians.

EXAMPLE 1 Find the related angle for the following angles.

(a) 210°   (b) 1.93 rad

(a) 210° is between 180° and 270° and is a quadrant III angle. Then 210° − 180° = 30°. **The related angle is 30°.**

(b) 1.93 rad is between π and $\frac{\pi}{2}$ rad and is a quadrant II angle. Then π − 1.93 = 1.211592654 rad. **The related angle is 1.21 rad to the nearest hundredth.**

---

2 **Determine the Signs of Trigonometric Values of Angles of More than 90°.**

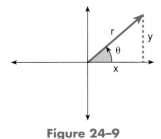

Figure 24-9

To determine the appropriate sign of trigonometric functions for angles more than 90°, we will examine the trigonometric functions in each quadrant. The sign of the function indicates the direction or quadrant of the vector.

The vector in Fig. 24–9 has a magnitude of *r* and direction of θ. To relate our trigonometric functions to vectors, the magnitude of the vector is the hypotenuse, the side opposite θ is *y*, and the side adjacent to θ is *x*. The magnitude of a vector (*r*) is always positive. In quadrant I, the *x*-value and the *y*-value are both positive.

■ **DEFINITION 24–3: Trigonometric Functions in Quadrant I:**

$$\sin \theta_1 = \frac{y}{r} \qquad \cos \theta_1 = \frac{x}{r} \qquad \tan \theta_1 = \frac{y}{x}$$

$$\csc \theta_1 = \frac{r}{y} \qquad \sec \theta_1 = \frac{r}{x} \qquad \cot \theta_1 = \frac{x}{y}$$

Figure 24-10

In quadrant I, the sign of all six trigonometric functions is positive.

For quadrant II vectors (Fig. 24–10), again the magnitude (*r*) is positive and the *y* value is positive, but the *x* value is negative.

■ **DEFINITION 24–4:** **Trigonometric Functions in Quadrant II:**

$$\sin \theta_2 = \frac{y}{r} \qquad \cos \theta_2 = \frac{-x}{r} = -\frac{x}{r} \qquad \tan \theta_2 = \frac{y}{-x} = -\frac{y}{x}$$

$$\csc \theta_2 = \frac{r}{y} \qquad \sec \theta_2 = -\frac{r}{x} \qquad \cot \theta_2 = -\frac{x}{y}$$

Thus, in quadrant II the sine and cosecant functions are positive, and the remaining functions are negative.

Quadrant III vectors (Fig. 24–11) have a positive magnitude (r), negative x value, and negative y value.

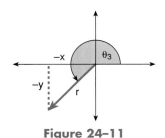

Figure 24–11

■ **DEFINITION 24–5:** **Trigonometric Functions in Quadrant III:**

$$\sin \theta_3 = \frac{-y}{r} = -\frac{y}{r} \qquad \cos \theta_3 = \frac{-x}{r} = -\frac{x}{r} \qquad \tan \theta_3 = \frac{-y}{-x} = \frac{y}{x}$$

$$\csc \theta_3 = -\frac{r}{y} \qquad \sec \theta_3 = -\frac{r}{x} \qquad \cot \theta_3 = \frac{x}{y}$$

Thus, in quadrant III the tangent and cotangent functions are positive and the remaining functions are negative.

Quadrant IV vectors (Fig. 24–12) have a positive magnitude (r), positive x value, and negative y value.

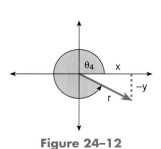

Figure 24–12

■ **DEFINITION 24–6:** **Trigonometric Functions in Quadrant IV:**

$$\sin \theta_4 = \frac{-y}{r} = -\frac{y}{r} \qquad \cos \theta_4 = \frac{x}{r} \qquad \tan \theta_4 = \frac{-y}{x} = -\frac{y}{x}$$

$$\csc \theta_4 = -\frac{r}{y} \qquad \sec \theta_4 = \frac{r}{x} \qquad \cot \theta_4 = -\frac{x}{y}$$

Thus, in quadrant IV the cosine and secant functions are positive and the remaining functions are negative.

Trigonometric functions, just like algebraic functions, can be represented graphically. This graphical representation will help you to visualize the sign patterns of the various functions and will draw attention to other properties of trigonometric functions. To graph these functions, we will make a table of values from 0° to 360°, or from 0 radians to 2π radians. See Figs. 24–13 to 24–15.

		Quadrant I				Quadrant II				Quadrant III				Quadrant IV			
$\sin \theta$	0	0.5	0.71	0.87	1	0.87	0.71	0.5	0	−0.5	−0.71	−0.87	−1	−0.87	−0.71	−0.5	0
θ	0	$\frac{\pi}{6}$	$\frac{\pi}{4}$	$\frac{\pi}{3}$	$\frac{\pi}{2}$	$\frac{2\pi}{3}$	$\frac{3\pi}{4}$	$\frac{5\pi}{6}$	π	$\frac{7\pi}{6}$	$\frac{5\pi}{4}$	$\frac{4\pi}{3}$	$\frac{3\pi}{2}$	$\frac{5\pi}{3}$	$\frac{7\pi}{4}$	$\frac{11\pi}{6}$	2π
$\theta°$	0	30	45	60	90	120	135	150	180	210	225	240	270	300	315	330	360

		Quadrant I				Quadrant II				Quadrant III				Quadrant IV			
$\cos \theta$	1	0.87	0.71	0.5	0	−0.5	−0.71	−0.87	−1	−0.87	−0.71	−0.5	0	0.5	0.71	0.87	1
θ	0	$\frac{\pi}{6}$	$\frac{\pi}{4}$	$\frac{\pi}{3}$	$\frac{\pi}{2}$	$\frac{2\pi}{3}$	$\frac{3\pi}{4}$	$\frac{5\pi}{6}$	π	$\frac{7\pi}{6}$	$\frac{5\pi}{4}$	$\frac{4\pi}{3}$	$\frac{3\pi}{2}$	$\frac{5\pi}{3}$	$\frac{7\pi}{4}$	$\frac{11\pi}{6}$	2π
$\theta°$	0	30	45	60	90	120	135	150	180	210	225	240	270	300	315	330	360

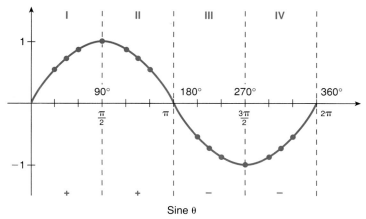

Sine θ

Figure 24–13

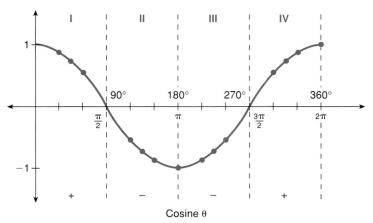

Cosine θ

Figure 24–14

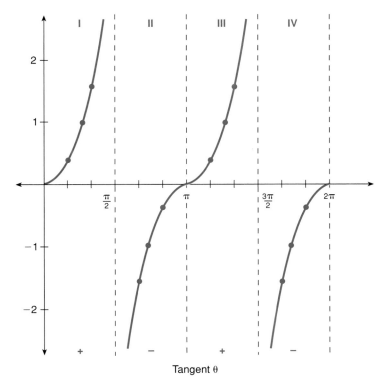

Tangent θ

Figure 24–15

		Quadrant I				Quadrant II				Quadrant III				Quadrant IV			
tan θ	0	0.58	1	1.7	∞	−1.7	−1	−0.58	0	0.58	1	1.7	∞	−1.7	−1	−0.58	0
θ	0	$\frac{\pi}{6}$	$\frac{\pi}{4}$	$\frac{\pi}{3}$	$\frac{\pi}{2}$	$\frac{2\pi}{3}$	$\frac{3\pi}{4}$	$\frac{5\pi}{6}$	π	$\frac{7\pi}{6}$	$\frac{5\pi}{4}$	$\frac{4\pi}{3}$	$\frac{3\pi}{2}$	$\frac{5\pi}{3}$	$\frac{7\pi}{4}$	$\frac{11\pi}{6}$	2π
θ°	0	30	45	60	90	120	135	150	180	210	225	240	270	300	315	330	360

Tips and Traps **Signs of Trigonometric Functions:**
To help remember the signs of the trigonometric functions in the various quadrants, we may find the following reminder helpful.

ALL-SIN-TAN-COS

This reminder gives the positive functions in the quadrants I to IV, respectively. Also, the functions cosecant, cotangent, and secant have the same signs as their respective reciprocal functions.

---

3 Find the Trigonometric Values of Angles of More Than 90° Using a Calculator.

■ General Tips for Using the Calculator

To find the value of the trigonometric function of an angle that is more than 90° using a calculator:

1. Set the calculator to degree or radian mode as desired.
2. Enter the trigonometric function.
3. Enter the degrees or radians.
4. Press $=$, ENTER, or EXE.

To find the related angle given the end point of a vector:

1. Set the calculator to degree or radian mode as desired.
2. Press the inverse tangent $\boxed{\tan^{-1}}$ or $\boxed{\arctan}$ function.
3. Enter the tangent ratio or the decimal equivalent.
4. Press $=$, ENTER, or EXE.

EXAMPLE 2   Using a calculator and the π key, find the values of the following trigonometric functions. Round the final answer to four significant digits.

(a) sin 155°   (b) cos 3   (c) tan 208°   (d) sin 4.2

(e) cos 304.5°   (f) $\tan\frac{5\pi}{3}$

(a) sin 155° = 0.4226   (b) cos 3 = −0.9900
(c) tan 208° = 0.5317   (d) sin 4.2 = −0.8716
(e) cos 304.5° = 0.5664   (f) $\tan\frac{5\pi}{3}$ = −1.732

EXAMPLE 3   Find the direction in degrees of a vector in standard position if the coordinates of its end point are (6, −8).

From the coordinates of the end point, we can determine that the vector is a quadrant IV vector. Also, from the coordinates of the end point we know that the *x*-coordinate is 6 and the *y*-coordinate is -8. Then $\tan \theta_4 = \frac{-8}{6} = -1.333333333$. From the calculator, we see that $\theta = -53.1°$ to the nearest $0.1°$. The related angle for a quadrant IV angle of $-53.1°$ is $53.1°$. If $360° - \theta_4 = 53.1°$, then $\theta_4 = 306.9°$.

Thus, the direction of the vector is 306.9°.

SELF-STUDY EXERCISES 24-2

1 Find the related angle for the following angles. (Use the calculator value for π and round to hundredths.)

1. 120°	**2.** 195°	**3.** 290°	**4.** 345°	**5.** 148°
6. 250°	**7.** 212°	**8.** 118°	**9.** 2.18 rad	**10.** 5.84 rad

2 Give the signs of all six trigonometric functions of vectors with the following end points.

11. (3, 5)	**12.** $(-2, 6)$	**13.** $(-4, -2)$	**14.** $(5, -3)$	**15.** (5, 0)

3 Using a calculator and the π key, find the value of the following functions. Round to four significant digits.

16. $\sin 210°$	**17.** $\tan 140°$	**18.** $\cos 2.5$	**19.** $\cos 4$	**20.** $\sin 300°$
21. $\tan 6$	**22.** $\cos 100°$	**23.** $\sin \dfrac{5\pi}{6}$		

24. Find the direction in degrees of a vector in standard position if the coordinates of its end point are $(-3, 2)$. Round to the nearest $0.1°$.

25. Find the direction in radians of a vector in standard position if the coordinates of its end point are $(-2, -1)$. Round to the nearest hundredth.

24-3 LAW OF SINES

Learning Objectives

1 Find the missing parts of an oblique triangle, given two angles and a side.

2 Find the missing parts of an oblique triangle, given two sides and an angle opposite one of them.

3 Solve applied problems using the law of sines.

In Chapter 21, we defined an oblique triangle as a triangle that does not contain a right angle. Because these triangles do not have right angles, we cannot use the trigonometric functions directly as we did previously to find sides and angles of right triangles. However, two formulas based on the trigonometric functions of right triangles can be used to solve oblique triangles. In this section we will study one of these formulas, the *law of sines*.

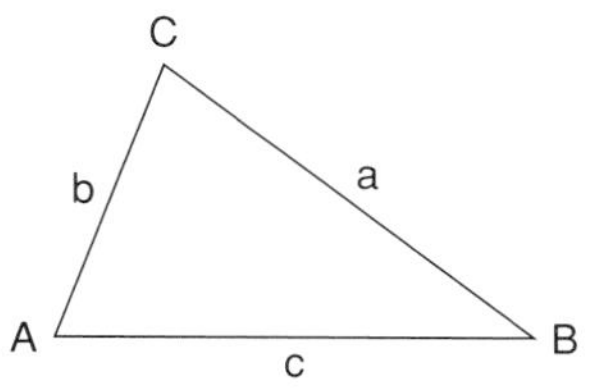

Figure 24-16

> **Property 24-2** *Law of sines:*
>
> The **law of sines** states that the ratios of the sides of a triangle to the sines of the angles opposite these respective sides are equal (see Fig. 24-16).
>
> $$\frac{a}{\sin A} = \frac{b}{\sin B} = \frac{c}{\sin C}$$

Tips and Traps ***Conditions for Using the Law of Sines:***
The law of sines is used to solve triangles when either of the following conditions exists:

1. Two angles and the side opposite one of them are known.
2. Two sides and the angle opposite one of them are known.

If we know two angles of a triangle, we can always find the third angle. Therefore, in Step 1 above, if we know two angles and any side, we can solve the triangle.

1 Find the Missing Parts of an Oblique Triangle, Given Two Angles and a Side.

Application problems often require solving triangles when we know two angles and a side of a triangle. In the examples that follow, all digits of the calculated value that show in the display will be given. Rounding should be done after the last calculation has been made.

EXAMPLE 1 Solve triangle ABC if $A = 50°$, $B = 75°$, and $b = 12$ ft.

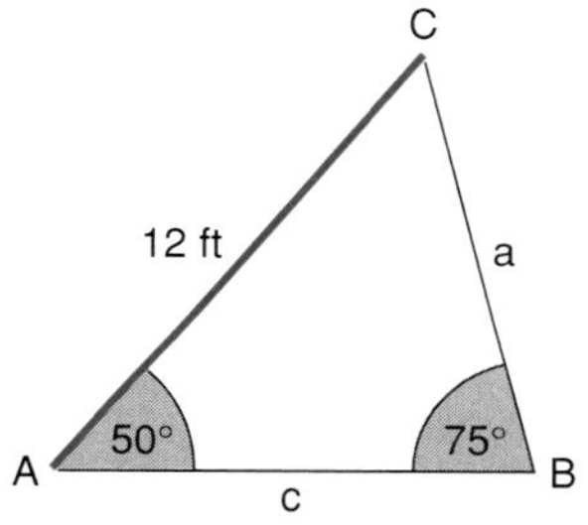

Figure 24–17

The first step in solving the triangle is to sketch it and label the parts, as shown in Fig. 24–17. To solve the triangle, we need to find angle C, side a, and side c. We find angle C by applying the property that the sum of the angles of a triangle equals 180°.

$$A + B + C = 180° \quad \text{Substitute } A = 50° \text{ and } B = 75°.$$

$$50° + 75° + C = 180° \quad \text{Solve for } C.$$

$$C = 180° - 50° - 75°$$

$$\mathbf{C = 55°}$$

To find a and c, we use the law of sines. First, we find a. To find a, we choose the proportion that contains a (the unknown quantity) and three known quantities.

$$\frac{a}{\sin A} = \frac{b}{\sin B} \quad \text{Substitute } b = 12,\ A = 50°,\ B = 75°.$$

$$\frac{a}{\sin 50°} = \frac{12}{\sin 75°} \quad \text{Solve for } a.$$

$$a \sin 75° = 12 \sin 50°$$

$$a = \frac{12 \sin 50°}{\sin 75°} \quad \text{Perform calculations.}$$

$$a = 9.516810781$$

$$\mathbf{a = 9.517} \quad \text{Four significant digits}$$

To find c, we choose a proportion that contains the unknown quantity c and three known quantities.

$$\frac{b}{\sin B} = \frac{c}{\sin C} \quad \text{Substitute } b = 12,\ B = 75°, \text{ and } C = 55°.$$

$$\frac{12}{\sin 75°} = \frac{c}{\sin 55°} \quad \text{Solve for } c.$$

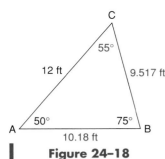

Figure 24–18

$12 \sin 55° = c \sin 75°$

$\dfrac{12 \sin 55°}{\sin 75°} = c$ Perform calculations.

$c = 10.1765832$

$c = \mathbf{10.18}$ Four significant digits

As a check for our work, the longest side should be opposite the largest angle, and the shortest side should be opposite the smallest angle (see Fig. 24–18).

EXAMPLE 2   Solve triangle ABC if $A = 35°$, $a = 7$ cm, and $B = 40°$.

We first sketch the triangle and label its parts. (See Fig. 24–19). To solve the triangle, we need to find angle C, side b, and side c.

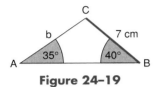

Figure 24–19

$A + B + C = 180°$ Substitute $A = 35°$ and $B = 40°$.

$35° + 40° + C = 180°$ Solve for C.

$C = 180° - 35° - 40°$

$C = \mathbf{105°}$

To find b and c, we use the law of sines. First, we find b.

$\dfrac{a}{\sin A} = \dfrac{b}{\sin B}$ Substitute $a = 7$, $A = 35°$, and $B = 40°$.

When choosing the proportion, such as the one above, be sure to choose one in which substitutes can be made for three terms.

$\dfrac{7}{\sin 35°} = \dfrac{b}{\sin 40°}$ Solve for b.

$7 \sin 40° = b \sin 35°$

$\dfrac{7 \sin 40°}{\sin 35°} = b$ Evaluate.

$b = 7.844661989$

$b = \mathbf{7.845}$ Four significant digits

To find c, we choose the proportion that allows substitution for as many of the *originally* given sides and angles as possible.

$\dfrac{a}{\sin A} = \dfrac{c}{\sin C}$ Substitute $a = 7$, $A = 35°$, and $C = 105°$.

$\dfrac{7}{\sin 35°} = \dfrac{c}{\sin 105°}$ Solve for c.

$7 \sin 105° = c \sin 35°$

$\dfrac{7 \sin 105°}{\sin 35°} = c$ Evaluate.

$c = 11.78828201$

$c = \mathbf{11.79}$ Four significant digits

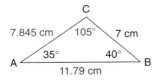

Figure 24–20

As a check for our work, the longest side should be opposite the largest angle, and the shortest side should be opposite the smallest angle (see Fig. 24–20).

EXAMPLE 3 Determine the unknown angles and sides of a piece of land described by the triangle in Fig. 24–21.

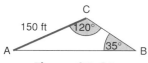

Figure 24-21

Find A.

$$A + B + C = 180°$$ Substitute $B = 35°$ and $C = 120°$.

$$A + 35° + 120° = 180°$$

$$A = 180° - 35° - 120°$$

$$\mathbf{A = 25°}$$

Next, we find side a.

$$\frac{b}{\sin B} = \frac{a}{\sin A}$$ Substitute $A = 25°$, $b = 150$, and $B = 35°$.

$$\frac{150}{\sin 35°} = \frac{a}{\sin 25°}$$ Solve for a.

$$150 \sin 25° = a \sin 35°$$

$$\frac{150 \sin 25°}{\sin 35°} = a$$ Evaluate.

$$a = 110.5218681$$

$$\mathbf{a = 110.5 \ ft}$$ Four significant digits

Next, we find c.

$$\frac{b}{\sin B} = \frac{c}{\sin C}$$ Substitute $B = 35°$, $b = 150$, and $C = 120°$.

$$\frac{150}{\sin 35°} = \frac{c}{\sin 120°}$$ Solve for c.

$$150 \sin 120° = c \sin 35°$$

$$\frac{150 \sin 120°}{\sin 35°} = c$$ Evaluate.

$$c = 226.4803823$$

$$\mathbf{c = 226.5 \ ft}$$ Four significant digits

[2] Find the Missing Parts of an Oblique Triangle, Given Two Sides and an Angle Opposite One of Them.

When two sides of a triangle and an angle opposite one of them are known, we do not always have a single triangle. If the given sides are a and b and the given angle is B, three possibilities may exist (see Fig. 24–22).

- If $b < a$ and $\angle A \neq 90°$, we have two possible solutions (case 1).
- If $b < a$ and $\angle A = 90°$, we have one solution (case 2).
- If $b \geq a$, and $\angle A \neq 90°$, we have one solution (case 3).

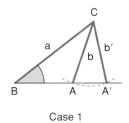

Case 1

Two solutions for A, C, and c since b can meet side c in two points, A and A′.

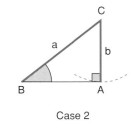

Case 2

One solution since b meets side c in exactly one point.

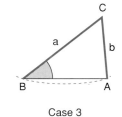

Case 3

One solution since b meets side c in only one point.

Figure 24–22

Note that side *b* above could be any side of the triangle that is opposite the given angle. Side *a* is then the other given side. Because of the lack of clarity when two sides and an angle opposite one of them are given, the situation is called the *ambiguous case*. (*Ambiguous* means that the given information can be interpreted in more than one way.)

EXAMPLE 4   Solve triangle *ABC* if $A = 35°$, $a = 8$, and $b = 11.426$ (see Fig. 24–23).

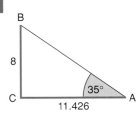

Figure 24–23

$$\frac{a}{\sin A} = \frac{b}{\sin B}$$ Substitute $A = 35°$, $a = 8$, and $b = 11.426$.

$$\frac{8}{\sin 35°} = \frac{11.426}{\sin B}$$ Solve for *B*.

$$8 \sin B = 11.426 \sin 35°$$

$$\sin B = \frac{11.426 \sin 35°}{8}$$ Evaluate.

$$\sin B = 0.8192105452$$ $\sin^{-1} 0.8192105452 = B$.

Possible Solution 1: $\sin^{-1} 0.8192105452 = 55.00584421°$

$$B = 55.0°$$ Nearest 0.1°

$$A + B + C = 180°$$

$$35° + 55° + C = 180°$$

$$C = 180° - 55° - 35°$$

$$C = 90°$$ *ABC* is a right triangle.

This is a right triangle, but we still have two possible solutions. The angle opposite the second given side is not the 90° angle. To find the third side of this right triangle, we can use the Pythagorean theorem or the law of sines.

Using the law of sines, we have

$$\frac{a}{\sin A} = \frac{c}{\sin C}$$ Substitute $a = 8$, $A = 35°$, and $C = 90°$.

$$\frac{8}{\sin 35°} = \frac{c}{\sin 90°}$$ Solve for *c*.

$$8 \sin 90° = c \sin 35°$$

$$\frac{8 \sin 90°}{\sin 35°} = c$$ Evaluate.

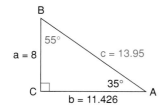

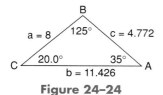

Figure 24–24

$$c = 13.94757436$$

$$\boldsymbol{c = 13.95}$$ Four significant digits

The solved triangle for this possibility is shown in Fig. 24–24.
The second possibility for angle B is

$$180° - 55.00584421° = 124.9941558°$$

To obtain angle C we have,

$$C = 180° - (35° + 124.9941558°) = 20.0058442°$$

$$\boldsymbol{C = 20.0°}$$

Side c can be found using the law of sines.

$$\frac{a}{\sin A} = \frac{c}{\sin C}$$

$$\frac{8}{\sin 35°} = \frac{c}{\sin 20.0058442°}$$

$$c \sin 35° = 8 \sin 20.0058442°$$

$$c = \frac{8 \sin 20.0058442°}{\sin 35°}$$

$$c = 4.771688222$$

or

$$\boldsymbol{c = 4.772}$$

EXAMPLE 5 Solve triangle ABC if $a = 12$, $b = 8$, and $B = 40°$ (see Fig. 24–25).

To find A, we have

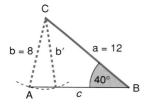

Figure 24–25

$$\frac{a}{\sin A} = \frac{b}{\sin B}$$ Substitute $a = 12$, $b = 8$, and $B = 40°$.

$$\frac{12}{\sin A} = \frac{8}{\sin 40°}$$ Solve for A.

$$12 \sin 40° = 8 \sin A$$

$$\frac{12 \sin 40°}{8} = \sin A$$ Evaluate.

$$\sin A = 0.9641814145$$ $\sin^{-1} 0.9641814145 = A$

$$\boldsymbol{A = 74.61856831°} \quad \textbf{or} \quad \boldsymbol{105.3814317°}$$

There are two angles less than 180° that have a sine of approximately 0.9641814145. These angles are 74.61856831° and 105.3814317°. The angle 74.61856831° is in quadrant I. The other angle is in quadrant II and has 74.61856831° as its reference angle. Therefore, $180° - 74.61856831° = 105.3814317°$ is the measure of the other angle. You can verify this by comparing the sine of the two angles. This situation occurs when the side opposite the given angle is less than the other given side. As a result, we have *two* possible solutions.

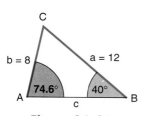

Figure 24–26

Possible solution 1: $A = 74.61856831°$ (see Fig. 24–26).

$$C = 180° - 74.61856831° - 40°$$

$$C = 65.38143169°$$

$C = 65.4°$ Nearest 0.1°

Now we find side c.

$$\frac{b}{\sin B} = \frac{c}{\sin C}$$

Substitute $b = 8$, $B = 40°$, $C = 65.38143169°$. Use full calculator value for C to maximize accuracy.

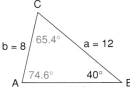

C
65.4°
b = 8   a = 12
74.6°   40°
A _____ B
c = 11.31

Figure 24–27

$$\frac{8}{\sin 40°} = \frac{c}{\sin 65.38143169°}$$

Solve for c.

$$8 \sin 65.38143169° = c \sin 40°$$

$$\frac{8 \sin 65.38143169°}{\sin 40°} = c$$ Evaluate.

$$c = 11.31448261$$

$$c = 11.31$$ Four significant digits

The solved triangle is shown in Fig. 24–27.

Possible solution 2: $A = 105.3814317°$ (see Fig. 24–28).

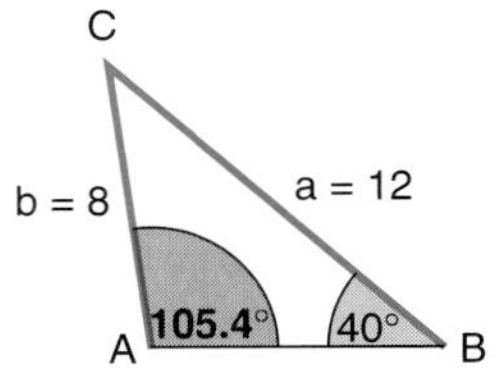

$$C = 180° - 105.3814317° - 40°$$

$$C = 34.6185683°$$

$$C = 34.6°$$ Nearest 0.1°

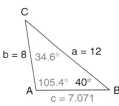

C
b = 8   a = 12
105.4°   40°
A   B

Figure 24–28

Now we find side c.

$$\frac{b}{\sin B} = \frac{c}{\sin C}$$

Substitute. Use full calculator value for C to maximize accuracy.

$$\frac{8}{\sin 40°} = \frac{c}{\sin 34.6185683°}$$

Solve for c.

$$8 \sin 34.6185683° = c \sin 40°$$

$$\frac{8 \sin 34.6185683°}{\sin 40°} = c$$

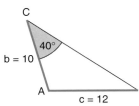

C
b = 8   34.6°   a = 12
105.4° 40°
A _____ B
c = 7.071

Figure 24–29

$$c = 7.070584025$$

$$c = 7.071$$ Four significant digits

The solved triangle is shown in Fig. 24–29.

EXAMPLE 6   Solve triangle ABC if $b = 10$, $c = 12$, and $C = 40°$ (Fig. 24–30).

Solve for B first. Because c, the side opposite the given angle C, is *longer* than the other side, b, we expect *one* solution.

C
40°
b = 10
A   B
c = 12

Figure 24–30

$$\frac{c}{\sin C} = \frac{b}{\sin B}$$ Substitute.

$$\frac{12}{\sin 40°} = \frac{10}{\sin B}$$ Solve for B.

$$12 \sin B = 10 \sin 40°$$

$$\sin B = \frac{10 \sin 40°}{12}$$ Evaluate.

$$\sin B = 0.5356563414$$ $\sin^{-1} 0.5356563414 = B$

$$B = 32.38843382°$$

$$B = 32.4°$$ Nearest 0.1°

Both 32.38843382° and 147.6115662° have a sine of approximately 0.5356563414; however, we cannot have an angle of 147.6115662° in this triangle because the triangle already has a 40° angle. 147.6115662° + 40° = 187.6115662°. The *three* angles of a triangle total only 180°. Therefore, we have only *one* solution. Finding *A*, we have

$$A = 180° - 40° - 32.38843382°$$ Use full calculator value for *B*.

$$A = 107.6115662°$$

$$\mathbf{A = 107.6°}$$

Next, we find *a*.

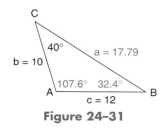

Figure 24–31

$$\frac{a}{\sin A} = \frac{c}{\sin C}$$ Substitute full calculator values.

$$\frac{a}{\sin 107.6115662°} = \frac{12}{\sin 40°}$$ Solve for *a*.

$$a \sin 40° = 12 \sin 107.6115662°$$

$$a = \frac{12 \sin 107.6115662°}{\sin 40°}$$ Evaluate.

$$a = 17.79367732$$

$$\mathbf{a = 17.79}$$ Four significant digits

The solved triangle is shown in Fig. 24–31.

3 Solve Applied Problems Using the Law of Sines.

Because real-world applications often do not involve right triangles, the law of sines is very useful in solving applied problems.

EXAMPLE 7   A technician checking a surveyor's report is given the information shown in Fig. 24–32. Calculate the missing information.

Find *B*.

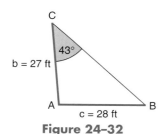

Figure 24–32

$$\frac{c}{\sin C} = \frac{b}{\sin B}$$ Substitute.

$$\frac{28}{\sin 43°} = \frac{27}{\sin B}$$ Solve for *B*.

$$28 \sin B = 27 \sin 43°$$

$$\sin B = \frac{27 \sin 43°}{28}$$

$$\sin B = 0.6576412758$$ $\sin^{-1} 0.6576412758 = B$

$$B = 41.12023021°$$

$$\mathbf{B = 41.1°}$$ Nearest 0.1°

There is only one case here because we are given the triangle. Also, the other angle whose sine is 0.6576412758 is 138.8797698°, which we exclude because 138.8797698° + 43° = 181.8797698°.

To find angle *A*, we have

$$A = 180° - (43° + 41.12023021°)$$

$$A = 95.87976979°$$

$$A = 95.9°$$

Nearest 0.1°

$$\frac{a}{\sin A} = \frac{c}{\sin C}$$

Substitute.

$$\frac{a}{\sin 95.87976979°} = \frac{28}{\sin 43°}$$

$$a \sin 43° = 28 \sin 95.87976979°$$

$$a = \frac{28 \sin 95.87976979°}{\sin 43°}$$

Evaluate.

$$a = 40.83982458$$

$$a = 40.84 \text{ ft}$$

Four significant digits

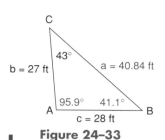

Figure 24–33

The completed survey should have the measures shown in Fig. 24–33.

SELF-STUDY EXERCISES 24-3

 Solve each of the following oblique triangles using the law of sines. Round the final answer for sides to four significant digits and angles to the nearest 0.1°.

1. $a = 46, A = 65°, B = 52°$
2. $b = 7.2, B = 58°, C = 72°$
3. $a = 65, B = 60°, A = 87°$
4. $c = 3.2, A = 120°, C = 30°$
5. $b = 12, A = 95°, B = 35°$

2 Use the law of sines to solve the following triangles. If a triangle has two possibilities, find both solutions. Round sides to four significant digits and angles to the nearest 0.1°.

6. $a = 42, b = 24, A = 40°$
7. $b = 15, c = 3, B = 70°$
8. $a = 18, c = 9, C = 20°$
9. $a = 8, b = 4, A = 30°$

3

10. Find the missing angle and sides of the plot of land described by Fig. 24–34.

11. Find the distance from A to B on the surveyed plot shown in Fig. 24–35.

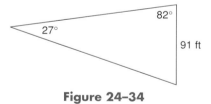

Figure 24–34

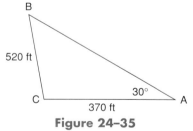

Figure 24–35

24-4 LAW OF COSINES

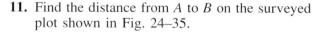

Learning Objectives

1 Find the missing parts of an oblique triangle, given three sides of the triangle.

2 Find the missing parts of an oblique triangle, given two sides and the included angle of the triangle.

3 Solve applied problems using the law of cosines and the law of sines.

In some cases, our given information will not allow us to use the law of sines. For example, if we know all three sides of a triangle, we cannot use the law of sines to find the angles. In a case such as this, however, we can use the *law of cosines*. This law is based on the trigonometric functions just as the law of sines is.

Property 24-3 *Law of cosines:*

The **law of cosines** states that the square of any side of a triangle equals the sum of the squares of the other sides minus twice the product of the other two sides and the cosine of the angle opposite the first side. For triangle *ABC*:

$$a^2 = b^2 + c^2 - 2bc \cos A$$
$$b^2 = a^2 + c^2 - 2ac \cos B$$
$$c^2 = a^2 + b^2 - 2ab \cos C$$

Tips and Traps   *When Do You Use the Law of Cosines?*
To use the law of cosines efficiently, we are given

1. Three sides of a triangle
   or
2. Two sides and the included angle of a triangle.

We can sometimes use *either* the law of sines or the law of cosines to solve certain triangles. However, whenever possible, the law of sines is generally preferred because it involves fewer calculations.

---

1   **Find the Missing Parts of an Oblique Triangle, Given Three Sides of the Triangle.**

EXAMPLE 1   Find the angles in triangle *ABC* (see Fig. 24–36).

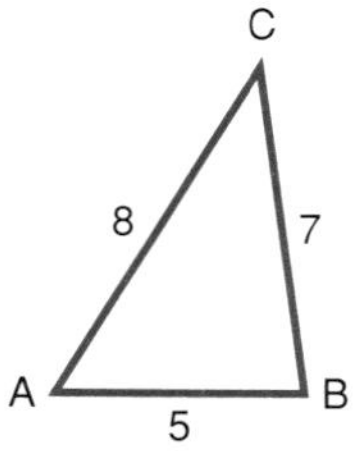

Figure 24–36

We may find any one of the angles first. If we choose to find angle *A* first, we must use the formula that contains cos *A*: $a^2 = b^2 + c^2 - 2bc \cos A$. We rearrange the formula to solve for cos *A*.

$$a^2 - b^2 - c^2 = -2bc \cos A$$

Now we multiply each term on both sides by -1 to reduce the number of negative signs.

$$-a^2 + b^2 + c^2 = 2bc \cos A \qquad \text{Solve for cos } A.$$

$$\frac{-a^2 + b^2 + c^2}{2bc} = \cos A \qquad \text{Substitute.}$$

$$\frac{-(7)^2 + 8^2 + 5^2}{2(8)(5)} = \cos A \qquad \text{Solve for } A.$$

$$\frac{-49 + 64 + 25}{80} = \cos A$$

$$\frac{40}{80} = \cos A$$

$$0.5 = \cos A \qquad \cos^{-1} 0.5 = A$$

$$\mathbf{60° = A}$$

To find angle B, we use the law of cosines so that we can use only given values. The law of sines could be used but would involve using a calculated value.

$$b^2 = a^2 + c^2 - 2ac \cos B \qquad \text{Solve for } \cos B.$$

$$2ac \cos B = a^2 + c^2 - b^2$$

$$\cos B = \frac{a^2 + c^2 - b^2}{2ac} \qquad \text{Substitute.}$$

$$\cos B = \frac{7^2 + 5^2 - (8)^2}{2(7)(5)}$$

$$\cos B = \frac{49 + 25 - 64}{70}$$

$$\cos B = \frac{10}{70}$$

$$\cos B = 0.1428571429 \qquad \cos^{-1} 0.1428571429 = B$$

$$B = 81.7867893°$$

$$\boldsymbol{B = 81.8°} \qquad \text{Nearest } 0.1°$$

The law of cosines or the law of sines can be used to find the third angle. However, the quickest way to find this angle is to subtract the sum of A and B from 180°.

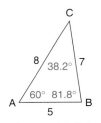

Figure 24–37

$$C = 180° - (A + B)$$

$$C = 180° - 60° - 81.7867893°$$

$$C = 180° - 141.7867893°$$

$$C = 38.2132107°$$

$$\boldsymbol{C = 38.2°} \qquad \text{Nearest } 0.1°$$

The solved triangle is shown in Fig. 24–37.

EXAMPLE 2 Find the angles in triangle ABC (see Fig. 24–38).

We will find A first.

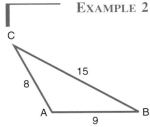

Figure 24–38

$$a^2 = b^2 + c^2 - 2bc \cos A \qquad \text{Solve for } \cos A.$$

$$\cos A = \frac{b^2 + c^2 - a^2}{2bc} \qquad \text{Substitute.}$$

$$\cos A = \frac{8^2 + 9^2 - 15^2}{2(8)(9)} \qquad \text{Evaluate.}$$

$$\cos A = \frac{64 + 81 - 225}{144}$$

$$\cos A = \frac{-80}{144}$$

$$\cos A = -0.5555555556 \qquad \cos^{-1} -0.5555555556 = A$$

$$A = 123.7489886°$$

$$\boldsymbol{A = 123.7°} \qquad \text{Nearest } 0.1°$$

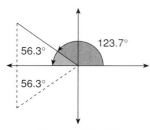

Figure 24–39

Recall from Section 24–2 that the cosine is *negative* in the second and third quadrants. Because A is either acute ($<90°$) or obtuse ($>90°$ but $<180°$) and because its cosine is negative (see Fig. 24–39), it must be in the second quadrant. If a calculator is used and -0.5555555556 is entered, $123.7489886°$ will appear in the display. We can use the law of sines to find the second angle, but keep in mind the risk involved in continuing with a calculated value instead of a given value.

$$\frac{a}{\sin A} = \frac{b}{\sin B} \qquad \text{Substitute.}$$

$$\frac{15}{\sin 123.7489886°} = \frac{8}{\sin B} \qquad \text{Solve for } B.$$

$$15 \sin B = 8 \sin 123.7489886°$$

$$\sin B = \frac{8 \sin 123.7489886°}{15}$$

$$\sin B = 0.4434556903$$

$$B = 26.32457654° \qquad \sin^{-1} 0.4434556903 = B$$

$$\mathbf{B = 26.3°} \qquad \text{Rounded to } 0.1°$$

$$C = 180° - A - B$$

$$C = 180° - 123.7489886° - 26.32457654°$$

$$\mathbf{C = 29.9°}$$

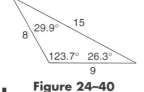

Figure 24–40

The solved triangle is given in Fig. 24–40.

2 Find the Missing Parts of an Oblique Triangle, Given Two Sides and the Included Angle of the Triangle.

The law of cosines is needed to solve oblique triangles if two sides and the included angle are given.

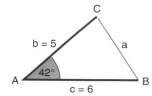

Figure 24–41

EXAMPLE 3 Solve triangle ABC in Fig. 24–41.

We can find a by using the law of cosines.

$$a^2 = b^2 + c^2 - 2bc \cos A \qquad \text{Substitute.}$$

$$a^2 = 5^2 + 6^2 - 2(5)(6) \cos 42°$$

$$a^2 = 25 + 36 - 60 \cos 42° \qquad \cos 42° = 0.7431448255$$

$$a^2 = 25 + 36 - 44.58868953$$

$$a^2 = 16.41131047$$

$$a = 4.051087566$$

$$\mathbf{a = 4.051} \qquad \text{Four significant digits}$$

Use the law of sines to find each of the two missing angles.

Find B.

$$\frac{a}{\sin A} = \frac{b}{\sin B} \qquad \text{Substitute.}$$

$$\frac{4.051087566}{\sin 42°} = \frac{5}{\sin B} \qquad \text{Solve for } B.$$

$$4.051087566 \sin B = 5 \sin 42°$$

$$\sin B = \frac{5 \sin 42°}{4.051087566}$$

$$\sin B = 0.8258653947$$

$$B = 55.67632739°$$

$$\boldsymbol{B = 55.7°}$$ Nearest 0.1°

Find C.

$$\frac{a}{\sin A} = \frac{c}{\sin C}$$ Substitute.

$$\frac{4.051087566}{\sin 42°} = \frac{6}{\sin C}$$ Solve for C.

$$4.051087566 \sin C = 6 \sin 42°$$

$$\sin C = \frac{6 \sin 42°}{4.051087566}$$

$$\sin C = 0.9910384737$$ $\sin^{-1} 0.9910384737 = C$

$$C = 82.32367267°$$

$$\boldsymbol{C = 82.3°}$$ Nearest 0.1°

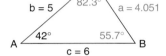

Figure 24–42

The solved triangle is given in Fig. 24–42.

To check, verify that the sum of the angles adds to 180°, the longest side is opposite the largest angle, and the shortest side is opposite the smallest angle.

3 **Solve Applied Problems Using the Law of Cosines and the Law of Sines.**

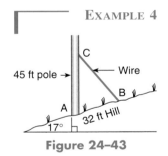

Figure 24–43

EXAMPLE 4 A vertical 45-ft pole is placed on a hill that is inclined 17° to the horizontal (see Fig. 24–43). How long a guy wire is needed if the guy wire is placed 5 ft from the top of the pole and attached to the ground at a point 32 ft uphill from the base of the pole?

Point A is where the pole enters the ground. Point B is where the wire is attached to the ground. Point C is where the wire is attached to the pole (5 ft from the top of the pole). Side b is the length of the pole from the wire to the ground ($45 - 5 = 40$ ft). The length of the wire is side a in the triangle. Because AC makes a 90° angle with the horizontal, we know that angle A is $90° - 17°$, or 73°. Using the law of cosines, we have

$$a^2 = b^2 + c^2 - 2bc \cos A$$ Substitute.

$$a^2 = 40^2 + 32^2 - 2(40)(32) \cos 73°$$

$$a^2 = 1600 + 1024 - 2560(0.2923717047)$$

$$a^2 = 1600 + 1024 - 748.471564$$

$$a^2 = 1875.528436$$

$$a = 43.30737161$$

$$a = 43.31 \text{ ft}$$ Four significant digits

Thus, the length of the wire is 43.31 ft.

1 Solve the triangles in Figs. 24–44 and 24–45. Round side lengths to four significant digits and angles to the nearest 0.1°.

1.

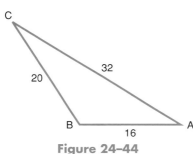

Figure 24–44

2.

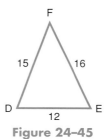

Figure 24–45

2 Solve the triangles in Figs. 24–46 and 24–47. Round as above.

3.

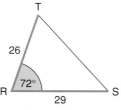

Figure 24–46

4.

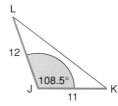

Figure 24–47

3 Solve the following problems. Round as above.

5. A hill is inclined 20° to the horizontal. A pole stands vertically on the side of the hill with 35 ft above the ground. How much wire will it take to reach from a point 2 ft from the top of the pole to a point on the ground 27 ft downhill from the base of the pole?

6. A triangular tabletop is to be 8.4 ft by 6.7 ft by 9.3 ft. What angles must be cut?

24–5 AREA OF TRIANGLES

Learning Objectives

1 Find the area of a triangle when the height is unknown and at least three parts of the triangle are known.

2 Find the area of a triangle using Heron's formula when the three sides of the triangle are known.

3 Solve applied problems involving the area of a triangle.

In Chapter 13, we learned how to find the area of right triangles and other triangles when the height was given. However, we can now use trigonometry to find the area of triangles even if we do not know the height. We can also find the area of a triangle if we know only the lengths of each of the three sides.

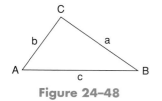

Figure 24–48

1 **Find the Area of a Triangle when the Height is Unknown and at Least Three Parts of the Triangle are Known.**

We can use trigonometric relationships to find the area of *any* triangle if we know the measure of any two sides and the included angle. In Fig. 24–48, the area of triangle *ABC* can be found by using one of three formulas.

Formula 24–1   *The area of a triangle:*

$$\text{Area} = \frac{1}{2}ab \sin C$$

$$\text{Area} = \frac{1}{2}ac \sin B$$

$$\text{Area} = \frac{1}{2}bc \sin A$$

These three formulas may be stated as a rule.

Rule 24–7   *To find the area of a triangle without the height:*

The area of a triangle equals one-half the product of two sides times the sine of the included angle.

EXAMPLE 1   Find the area of triangle *ABC* in Fig. 24–49.

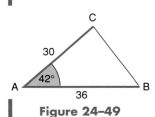

$$\text{Area} = \frac{1}{2}bc \sin A \qquad \text{Substitute.}$$

$$\text{Area} = \frac{1}{2}(30)(36)(\sin 42°)$$

$$\text{Area} = 361.3305274$$

Area = 361.3 square units   Four significant digits

Figure 24–49

EXAMPLE 2   Find the area of triangle *DEF* in Fig. 24–50.

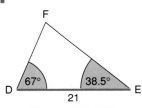

Figure 24–50

In this example we are not given two sides and the included angle. However, we can use the law of sines to find side *d* or side *e*. This would give us two sides and the included angle. To use the law of sines, we first need to find $\angle F$.

$$67° + 38.5° + F = 180° \qquad \text{Solve for } F.$$

$$F = 180° - 67° - 38.5°$$

$$F = 74.5°$$

We then find side *d*.

$$\frac{f}{\sin F} = \frac{d}{\sin D} \qquad \text{Substitute.}$$

$$\frac{21}{\sin 74.5°} = \frac{d}{\sin 67°} \qquad \text{Solve for } d.$$

$$21 \sin 67° = d \sin 74.5°$$

$$\frac{21 \sin 67°}{\sin 74.5°} = d$$

$$d = 20.06018164$$

$$d = 20.06 \qquad \text{Four significant digits}$$

Using the formula for area, we have

$$\text{Area} = \frac{1}{2}df \sin E \qquad \text{Substitute.}$$

$$\text{Area} = \frac{1}{2}(20.06018164)(21) \sin 38.5° \qquad \text{Angle } E \text{ is included between } d \text{ and } f.$$

$$\text{Area} = 131.1214452$$

Area = 131.1 square units   Four significant digits

EXAMPLE 3   Find the area of triangle ABC (Fig. 24–51).

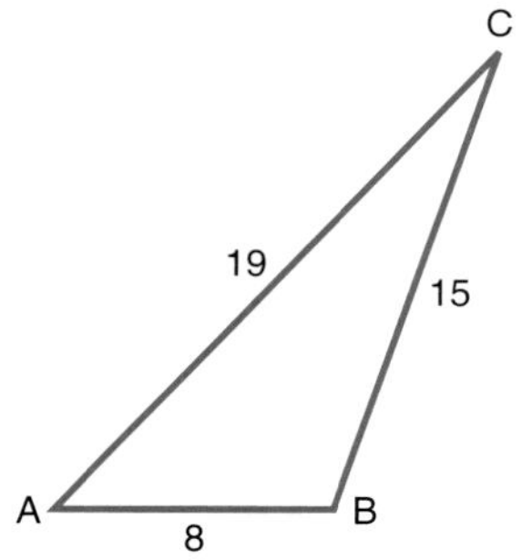

To find the area of $\triangle ABC$, we must find the measure of an angle. The law of cosines must be used here.

$$a^2 = b^2 + c^2 - 2bc \cos A \qquad \text{Solve for } \cos A.$$

$$\cos A = \frac{b^2 + c^2 - a^2}{2bc} \qquad \text{Substitute.}$$

$$\cos A = \frac{19^2 + 8^2 - 15^2}{2(19)(8)}$$

$$\cos A = 0.6578947368 \qquad \cos^{-1} 0.6578947368 = A$$

$$A = 48.86048959°$$

$$A = 48.9° \qquad \text{Nearest } 0.1°$$

Figure 24–51

Using the area formula, we have

$$\text{Area} = \frac{1}{2}(19)(8) \sin 48.86048959°$$

$$\text{Area} = 57.23635209$$

Area = 57.24 square units   Four significant digits

2   **Find the Area of a Triangle Using Heron's Formula When the Three Sides of the Triangle are Known.**

There is another way to calculate the area of a triangle when all three sides are known. The formula is known as *Heron's formula*.

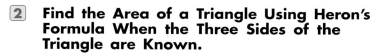

Formula 24–2   **Heron's formula *for the area of a triangle:***
$$\text{Area} = \sqrt{s(s - a)(s - b)(s - c)}$$
where $s = \frac{1}{2}(a + b + c)$ and a, b, and c are the lengths of the three sides.

EXAMPLE 4   Find the area of the triangle in Example 3 by using Heron's formula. Find s.

$$s = \frac{1}{2}(a + b + c) \qquad \text{Substitute.}$$

$$s = \frac{1}{2}(19 + 15 + 8)$$

$$s = 21$$

$$\text{Area} = \sqrt{s(s - a)(s - b)(s - c)} \qquad \text{Heron's formula}$$

$$\text{Area} = \sqrt{21(21 - 19)(21 - 15)(21 - 8)} \qquad \text{Substitute.}$$

$$\text{Area} = \sqrt{21(2)(6)(13)}$$

$$\text{Area} = \sqrt{3276}$$

$$\text{Area} = 57.23635209$$

Area = 57.24 square units   Four significant digits

The two calculated areas in Examples 3 and 4 are the same.

③ Solve Applied Problems Involving the Area of a Triangle.

EXAMPLE 5   A triangular piece of property has sides that measure 120 ft long, 150 ft long, and 100 ft long. Find the area of the lot.

Find s.

$$s = \frac{1}{2}(120 + 150 + 100) = \frac{370}{2} = 185$$

$$\text{Area} = \sqrt{s(s - a)(s - b)(s - c)}$$

$$\text{Area} = \sqrt{185(185 - 120)(185 - 150)(185 - 100)} \qquad \text{Heron's formula}$$

$$\text{Area} = \sqrt{185(65)(35)(85)}$$

$$\text{Area} = \sqrt{35,774,375}$$

$$\text{Area} = 5981.168364$$

Area = 5981 ft^2   Four significant digits

SELF-STUDY EXERCISES 24–5

①   Find the area of the triangles shown in Figs. 24–52 to 24–59. Use the law of sines or the law of cosines when necessary. Round answers to four significant digits.

1.

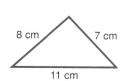

8 cm   7 cm

11 cm

Figure 24–52

2.

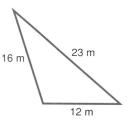

16 m   23 m

12 m

Figure 24–53

3.

15 ft

24°

24 ft

Figure 24–54

4.

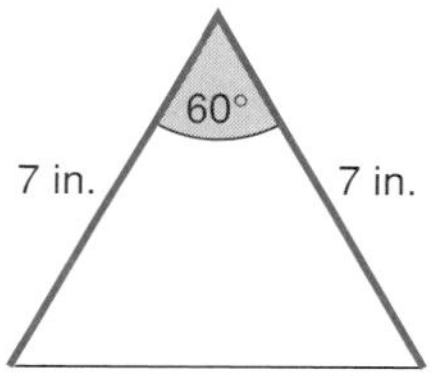

60°

7 in.   7 in.

Figure 24–55

5.

Figure 24–56

6.

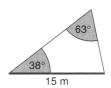

Figure 24–57

7.

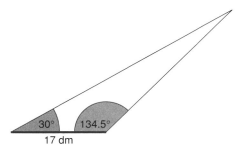

Figure 24–58

8.

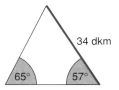

Figure 24–59

2 Find the area of the triangles shown in Figs. 24–60 to 24–63. Use Heron's formula. Round to four significant digits.

9.

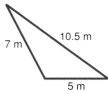

Figure 24–60

10.

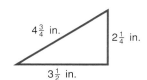

Figure 24–61

11.

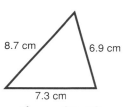

Figure 24–62

12.

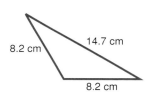

Figure 24–63

13. A tool and die maker for a medical implant manufacturer must calculate the area of a triangle whose sides are 2.06 cm, 3.27 cm, and 4.16 cm to make a die. What is the area of the triangle to three significant digits?

14. What is the area of a piece of carpeting that has sides measuring 14.3 ft, 12.8 ft, and 11.6 ft? Round to the nearest tenth of a square foot.

15. What is the area of the irregularly shaped property in Fig. 24–64? Round to the nearest square foot.

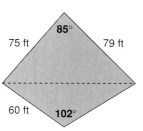

Figure 24–64

Law Enforcement: Vehicle Vaults

Today's traffic accident investigators often act as traffic reconstructionists because drivers may lie about their driving speed, be seriously injured (and so be unable to think clearly and answer questions), or die in the accident. When a vehicle becomes airborne with an uphill take-off angle of at least 6°, investigators use trigonometry to estimate the vehicle's speed at take-off with the following vault formula based on an oblique triangle:

$$S = \frac{2.73 \cdot D}{\sqrt{D \cdot \cos\theta \cdot \sin\theta \pm [H \cdot (\cos\theta)^2]}}$$

where S = speed in mph, D = horizontal distance from take-off point to first contact, H = vertical distance from take-off point to first contact, and θ = take-off angle in degrees. Use + in the formula when the vehicle lands at a point lower than take-off, and − when it lands at a point higher than take-off.

The take-off angle is usually the same as the road grade. The grade of a road is the slope of the road written as a percent, or

$$\text{Grade} = \frac{\text{vertical rise}}{\text{horizontal run}} \cdot 100$$

The take-off angle in degrees is $\tan^{-1}$ of the grade (in decimal form). In the case of a motorcycle driver vaulting off his vehicle, the take-off angle is assumed to be 45°, and his vertical drop on level ground is assumed to be 3 ft.

Exercises

Use the above information to estimate the speed of the vehicle in each of the following vault accidents. Round answers to the nearest unit.

1. The driver of an empty tractor-trailer truck falls asleep at the wheel, "straightens the curve" on an uphill 7% grade road, and vaults off a cliff. The truck's first contact point is 40 vertical ft and 55 horizontal ft below its take-off point. The driver of the car behind the truck said the truck was going between 20 and 25 mph when it left the road. Is this correct?
2. On a level street, a driver runs a red light at a T intersection and is hit by a motorcycle whose driver vaults over the top of the car. The motorcycle driver hits the street 57 horizontal ft from the take-off point. How fast was the motorcycle traveling when it hit the car?
3. A car traveling on a divided highway blows a tire, crosses the median ditch (which dips 5 vertical ft, then rises sharply), and vaults into the oncoming traffic lanes. It makes first contact at a point 12 ft higher than the take-off point and 162 ft horizontally from the take-off point. Combining the 6% uphill road grade with the ditch angle gives a take-off angle of 32°. Find the car's speed.
4. A drunk driver in a high-rise sport utility vehicle with big wheels drifts into the median strip of a divided highway at an overpass. He hits a large mound of dirt piled into the median and the truck is launched upward at a 50° angle. The truck vaults over the oncoming traffic lane, and falls 30 ft to the ground below. Find the take-off speed if the horizontal distance is 153 ft.
5. Do you think the take-off speed of the vehicle from Exercise 4 is the same speed the truck was going just before it hit the median strip? Explain.

Answers

1. The driver of the car estimated the truck's speed correctly at about 22 mph.
2. The motorcycle was traveling at least 25 mph.
3. The car's speed was at least 55 mph.

4. When the truck left the top of the dirt mound, its speed was at least 44 mph.

5. Climbing the dirt mound would have slowed the vehicle considerably, so the truck's speed when it left the road was greater.

ASSIGNMENT EXERCISES

Section 24–1

Find the direction and magnitude of vectors in standard position with the following end points. Round lengths to four significant digits and angles to the nearest tenth of a degree.

1. (4, 6) **2.** (2, 8) **3.** (6, 1)

Find the magnitude and direction of the resultant of the sum of the two given vectors in complex notation. Round as above.

4. $3 + 5i$ and $2 + 3i$ **5.** $1 + 4i$ and $6 + 8i$

Section 24–2

Find the related angle for the following angles.

6. 115° **7.** 3.04 rad **8.** 4.75 rad **9.** 221°
10. 305° **11.** 5.4 rad **12.** 138.5° **13.** 212°15′10″

Using the calculator, find the values of the following functions. Round to four significant digits.

14. cos 250° **15.** sin 2.1 **16.** tan 175° **17.** sin 340°

18. tan 4.5 **19.** cos 290° **20.** $\sin \dfrac{3\pi}{4}$ **21.** $\tan \dfrac{5\pi}{4}$

22. Find the magnitude and direction of a vector of an electrical current in standard position if the coordinates of the end point of the vector are (2, −3). Round the magnitude (in amps) to the nearest hundredth. Express the direction in degrees to the nearest 0.1°.

23. Find the direction and magnitude of a vector in standard position if the coordinates of the end point of the vector are (−2, 2). Express its direction in radians and round its direction and its magnitude to the nearest hundredth.

Section 24–3

Solve the following triangles. If a triangle has two possibilities, find both solutions. Round sides to tenths and angles to the nearest 0.1°.

24. $A = 60°$, $B = 40°$, $b = 20$
25. $B = 120°$, $C = 20°$, $a = 8$
26. $A = 60°$, $B = 60°$, $a = 10$
27. $a = 5$, $c = 7$, $C = 45°$
28. $b = 10$, $c = 8$, $B = 52°$
29. $a = 9.2$, $b = 6.8$, $B = 28°$
30. A surveyor needs the measure of JK in Fig. 24–65. Find JK to the nearest foot.
31. In Fig. 24–66, find RS.

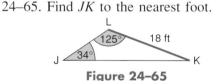

Figure 24–65

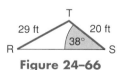

Figure 24–66

Section 24–4

Solve the triangles in Figs. 24–67 to 24–72. Round sides to four significant digits and angles to nearest 0.1°.

32.

Figure 24–67

33.

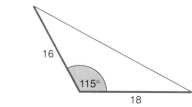

Figure 24–68

34.

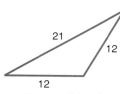

Figure 24–69

35.

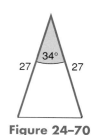

Figure 24–70

36.

Figure 24–71

37.

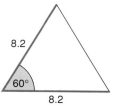

Figure 24–72

Solve the following problems. Round as in Problems 32 to 37.

38. A triangular lot has sides 180 ft long, 160 ft long, and 123.5 ft long. Find the angles of the lot.

39. A vertical 50-ft pole stands on top of a hill inclined 18° to the horizontal. What length of wire is needed to reach from a point 6 ft from the pole's top to a point 75 ft downhill from the base of the pole?

40. A ship sails from a harbor 35 nautical miles east, then 42 nautical miles in a direction 32° south of east (Fig. 24–73). How far is the ship from the harbor?

41. A hill with a 35° grade (inclined to the horizontal) is cut down for a roadbed to a 10° grade. If the distance from the base to the top of the original hill is 800 ft (Fig. 24–74), how many vertical feet will be removed from the top of the hill, and what is the distance from the bottom to the top of the hill for the roadbed?

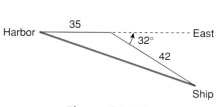

Figure 24–73

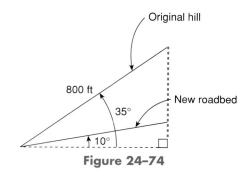

Figure 24–74

Section 24–5

Find the area of the triangles shown in Figs. 24–75 to 24–83. Use the law of sines, the law of cosines, or Heron's formula. Round the answers to four significant digits.

42.

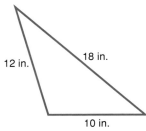

12 in. 18 in.
10 in.

Figure 24–75

43.

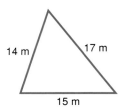

14 m 17 m
15 m

Figure 24–76

44.

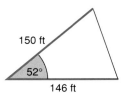

150 ft
52°
146 ft

Figure 24–77

45.

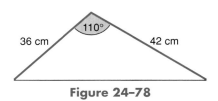

110°
36 cm 42 cm

Figure 24–78

46.

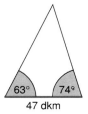

63° 74°
47 dkm

Figure 24–79

47.

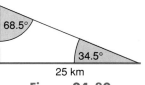

68.5°
34.5°
25 km

Figure 24–80

48.

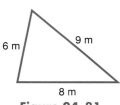

6 m 9 m
8 m

Figure 24–81

49.

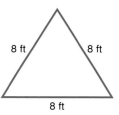

8 ft 8 ft
8 ft

Figure 24–82

50.

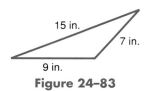

15 in.
7 in.
9 in.

Figure 24–83

51. Find the area of a triangular tabletop whose sides are 46 in., 37 in., and 40 in.

52. What is the area of a triangular plot of ground with sides that measure 146 ft, 85 ft, and 195 ft?

53. Find the area of the irregularly shaped property in Fig. 24–84.

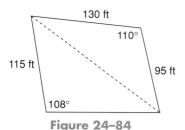

130 ft

110°

115 ft

95 ft

108°

Figure 24–84

━━━━━━━━ CHALLENGE PROBLEM ━━━━━━━━

54. Cy Pipkin needs to finish his surveying report by finding the area of the triangular plot of land he has surveyed. Read his survey report and finish it. Be sure to show all calculations so he can check them.

Survey report
   From a corner of the property, one property line measures 196 ft and the other property line measures 216 ft. The angle formed by these two property lines measures 42°.

━━━━━━━━ CONCEPTS ANALYSIS ━━━━━━━━

1. Speed is a scalar measure, whereas velocity is a vector measure. What are the differences and similarities in these two measures?

3. Locate a point in the second quadrant by a set of coordinates. Find the magnitude and direction of the vector determined by the point, and find the related angle for the vector.

5. Give the sign of the tangent function in each of the four quadrants. Use your calculator to find the value of an angle in each of the four quadrants to validate your answer.

7. Draw a triangle and assign values to three parts of the triangle, then use the law of sines and other methods to find the remaining angles and sides.

9. Write the law of sines in words.

2. Explain in your own words what a related angle is, and explain how to find the related angle for an angle located in the third quadrant.

4. Give the sign of the sine function in each of the four quadrants. Use your calculator to find the value of an angle in each of the four quadrants to validate your answer.

6. What parts of a triangle must be given to use the law of sines?

8. Write the law of cosines in words.

10. How would you know whether to select the law of sines or the law of cosines to find missing angles or sides of a triangle?

━━━━━━━━ CHAPTER SUMMARY ━━━━━━━━

Objectives	**What to Remember with Examples**

Section 24–1

1 Find the magnitude of a vector in standard position, given the coordinates of the end point.

Use the Pythagorean theorem.
$$P = \sqrt{x^2 + y^2}$$

Find the magnitude of a vector in standard position with the end point at (4, 2).
$$P = \sqrt{x^2 + y^2}$$
$$P = \sqrt{4^2 + 2^2}$$
$$P = \sqrt{16 + 4}$$
$$P = \sqrt{20}$$

Magnitude = 4.472   Rounded

2 Find the direction of a vector in standard position, given the coordinates of the end point.

Use the tangent function.

$$\tan \theta = \frac{y}{x}$$

> Find the direction of a vector in standard position with the end point at (2, 4).
>
> $$\tan \theta = \frac{4}{2}$$
> $$\tan \theta = 2$$
> $$\tan^{-1} 2 = 63.43494882° \quad \theta = 63.4° \quad \text{Rounded}$$
>
> **Direction = 63.4°**

3 Find the sum of vectors.

1. Represent the vectors in the form of complex numbers, where the *x*-coordinate of the end point of a vector in standard position is the real component and the *y*-coordinate of the end point of a vector is the imaginary component.
2. Add the like components.

> Add vectors *A* and *B*. See Fig. 24–85.
>
>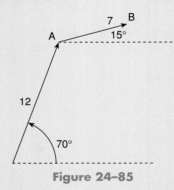
>
> **Figure 24–85**
>
> Vector *A*:
>   *x*-coordinate
>
> $$\cos 70° = \frac{x}{12}$$
> $$12(\cos 70°) = x$$
> $$4.10424172 = x$$
> $$x = 4.104 \quad \text{Rounded}$$
>
>   *y*-coordinate
>
> $$\sin 70° = \frac{y}{12}$$
> $$12(\sin 70°) = y$$
> $$11.27631145 = y$$
> $$y = 11.28 \quad \text{Rounded}$$
>
> Vector *A* = 4.104 + 11.28*i*
>
> Vector *B*:
>   *x*-coordinate
>
> $$\cos 15° = \frac{x}{7}$$
> $$7(\cos 15°) = x$$

$$6.761480784 = x$$
$$x = 6.761 \qquad \text{Rounded}$$

y-coordinate

$$\sin 15° = \frac{y}{7}$$
$$7(\sin 15°) = y$$
$$1.811733316 = y$$
$$y = 1.812 \qquad \text{Rounded}$$

Vector $B = 6.761 + 1.812i$
Vector $A + B = 10.87 + 13.09i$

Section 24–2

1 Find related, acute angles for angles or vectors in quadrants II, III, and IV.

To find a related angle for an angle in any quadrant, draw the angle and form the third side of a triangle by drawing a line perpendicular to the x-axis from the line forming the angle. The related angle is the angle formed by the ray forming the angle and the x-axis.

Quadrant II angle:

$$180° - \theta \qquad \text{or} \qquad \pi - \theta$$

Quadrant III angle:

$$\theta - 180° \qquad \text{or} \qquad \theta - \pi$$

Quadrant IV angle:

$$360° - \theta \qquad \text{or} \qquad 2\pi - \theta$$

Find the related angle for an angle of 156°. The angle is in quadrant II; thus, $180° - 156° = $ **24°.**

2 Determine the signs of trigonometric values of angles of more than 90°.

The memory device ALL-SIN-TAN-COS can be used to remember the signs of the trigonometric functions in the various quadrants. The signs of *all* trigonometric functions are positive in the first quadrant. The sign of the SIN function is positive in the second quadrant (COS and TAN are negative). The sign of the TAN function is positive in the third quadrant. The sign of the COS function is positive in the fourth quadrant.

What is the sign of cos 145°? 145° is a quadrant II angle and the cosine function is negative in quadrant II. **Thus, the sign of cos 145° is negative.**

3 Find the trigonometric values of angles of more than 90° using a calculator.

Most scientific calculators will now give the trigonometric value of an angle regardless of the quadrant in which it is found.

Use a scientific or graphics calculator to find the value of the following trigonometric functions. Round to four significant digits.

$$\sin 125° = \mathbf{0.8192}$$
$$\cos 215° = \mathbf{-0.8192}$$
$$\tan 335° = \mathbf{-0.4663}$$

Section 24–3

1 Find the missing parts of an oblique triangle, given two angles and a side.

Use the law of sines to find the missing part of an oblique triangle when two angles and a side opposite one of them are given.

$$\frac{a}{\sin A} = \frac{b}{\sin B} = \frac{c}{\sin C}$$

Find side *AB* in the triangle of Fig. 24–86.

$$\frac{a}{\sin A} = \frac{c}{\sin C}$$

$$\frac{13}{\sin 22°} = \frac{c}{\sin 97°}$$

$$c \sin 22° = 13 \sin 97°$$

$$c = \frac{13 \sin 97°}{\sin 22°}$$

$$c = 34.44440167 \text{ cm}$$

Rounded

$$\mathbf{AB = c = 34.44 \text{ cm}}$$

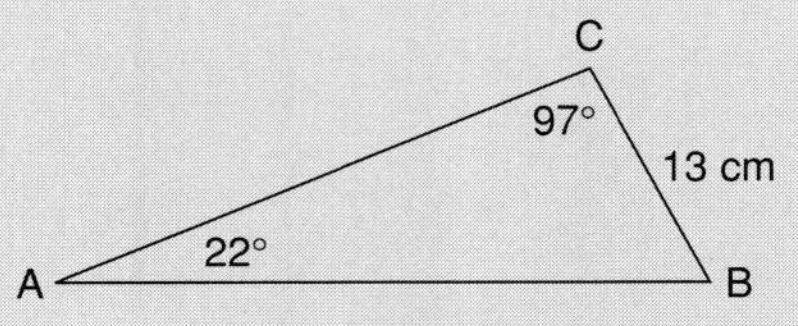

C
97°
13 cm
22°
A
B

Figure 24–86

2 Find the missing parts of an oblique triangle, given two sides and an angle opposite one of them.

Use the law of sines to find the missing part of an oblique triangle when two sides and an angle opposite one of them are given.

Find the measure of angle *A* in the triangle of Fig. 24–87.

$$\frac{a}{\sin A} = \frac{b}{\sin B}$$

$$\frac{18}{\sin A} = \frac{16}{\sin 32°}$$

$$16 \sin A = 18 \sin 32°$$

$$\sin A = \frac{18 \sin 32°}{16}$$

$$\sin A = 0.5961591723$$

$$\sin^{-1} 0.5961591723 = 36.59531105°$$

$$\mathbf{A = 36.6°}$$

Rounded

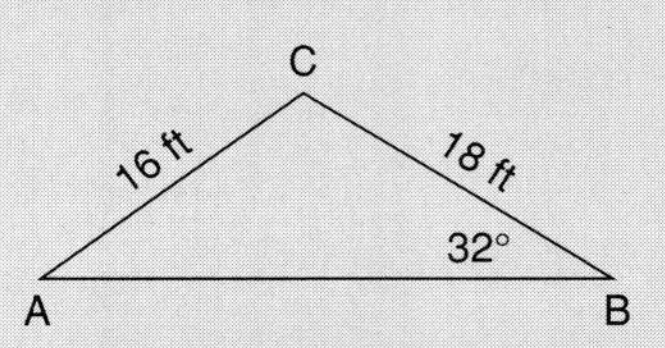

C
16 ft
18 ft
32°
A
B

Figure 24–87

Given $\triangle ABC$, if $b < a$ and $\angle A \neq 90°$, we have two possible solutions. If $b < a$ and $\angle A = 90°$, we have one solution. If $b > a$, we have one solution.

Determine how many solutions exist for the triangle in Fig. 24–88 based on the given facts. Explain your response.

C
8 cm
A
87°
15 cm
B

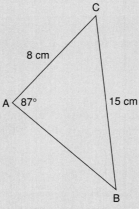

Figure 24–88

Because $b < a$ and $\angle A < 90°$, there are two possible solutions.

3 Solve applied problems using the law of sines.

Identify the given facts in the problem and represent these facts in graphic form. Use the law of sines to find the necessary parts of the problem.

A surveyor took the measurements shown in Fig. 24–89 for a triangular plot of land. Use the law of sines to find the missing angles and side of the plot.

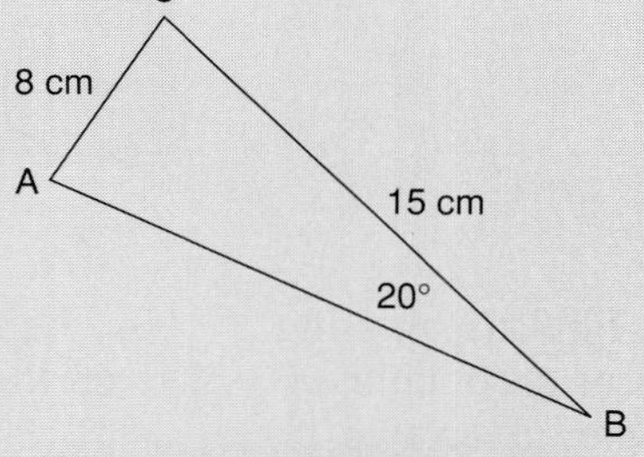

$$\frac{a}{\sin A} = \frac{b}{\sin B} = \frac{c}{\sin C}$$

Find $\angle A$.

$$\frac{15}{\sin A} = \frac{8}{\sin 20°}$$
$$8 \sin A = 15 \sin 20°$$
$$\sin A = \frac{15 \sin 20°}{8}$$
$$\sin A = 0.6412877688$$
$$\sin^{-1} 0.6412877688 = 39.88791233°$$
$$\mathbf{A = 39.9°} \quad \text{Rounded}$$

Figure 24–89

Find $\angle C$.

$$\angle C = 180° - 20° - 39.9°$$
$$\mathbf{\angle C = 120.1°}$$

Find AB.

$$\frac{c}{\sin C} = \frac{b}{\sin B}$$
$$\frac{c}{\sin 120.1°} = \frac{8}{\sin 20°}$$
$$c \sin 20° = 8 \sin 120.1°$$
$$c = \frac{8 \sin 120.1°}{\sin 20°}$$
$$c = 20.23626824$$
$$\mathbf{c = 20.24 \ cm} \quad \text{Rounded}$$

Section 24–4

1 Find the missing parts of an oblique triangle, given three sides of the triangle.

To find an angle when three sides are given, we need to use the law of cosines, but first we have to rearrange the formula to solve for cos A. The rearranged formula is:

$$\cos A = \frac{-a^2 + b^2 + c^2}{2bc}$$

Find the measure of angle A in the triangle of Fig. 24–90.

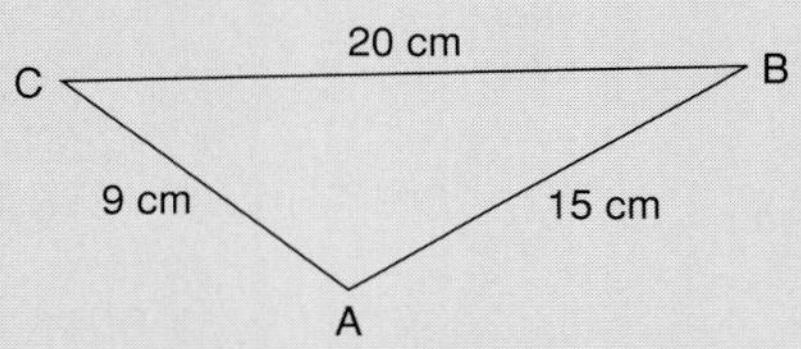

Figure 24–90

$$\cos A = \frac{-a^2 + b^2 + c^2}{2bc}$$

$$\cos A = \frac{-20^2 + 9^2 + 15^2}{2(9)(15)}$$

$$\cos A = \frac{-400 + 81 + 225}{270}$$

$$\cos A = -0.3481481481$$

$$\cos^{-1}(-0.3481481481) = 110.3740893$$

$$\boldsymbol{A = 110.4°} \quad \text{Rounded}$$

2 Find the missing parts of an oblique triangle, given two sides and the included angle of the triangle.

Use the law of cosines to find the missing parts of an oblique triangle when given three sides of the triangle.

$$a = \sqrt{b^2 + c^2 - 2bc \cos A}$$

In words, the side opposite the given angle equals the square root of the quantity found by the square of one given side plus the square of the other given side minus twice the product of the two given sides and the cosine of the given angle.

Find the third side of the triangle in Fig. 24–91.

$$a = \sqrt{b^2 + c^2 - 2bc \cos A}$$
$$a = \sqrt{17^2 + 23^2 - 2(17)(23) \cos 70°}$$
$$a = \sqrt{289 + 529 - 267.4597521}$$
$$a = \sqrt{550.5402479}$$
$$a = 23.4635941$$
$$\boldsymbol{a = 23.46 \text{ cm}} \quad \text{Rounded}$$

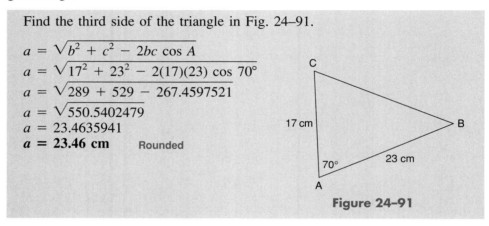

Figure 24–91

Find as many parts as possible without using information that you have calculated.

Find the measures of all the angles and sides of the triangle in Fig. 24–92.

$$\frac{a}{\sin A} = \frac{c}{\sin C}$$

$$\frac{6.8}{\sin 48°} = \frac{7.5}{\sin C}$$

$$\sin C = \frac{7.5 \sin 48°}{6.8}$$

$$\sin C = 0.8196450281$$

$$\sin^{-1} 0.8196450281 = 55.04927548°$$

$$\boldsymbol{C = 55.0°} \quad \text{Rounded}$$

$$\angle B = 180° - \angle A - \angle C$$
$$\angle B = 180° - 48° - 55.04927548°$$
$$\angle B = 76.95072452°$$
$$\boldsymbol{\angle B = 77.0°} \quad \text{Rounded}$$

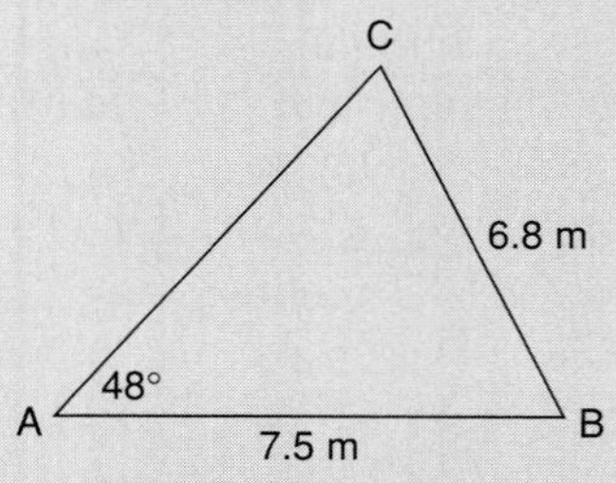

Figure 24–92

Use the law of sines or law of cosines to find b or AC.

$$\frac{a}{\sin A} = \frac{b}{\sin B}$$
$$b = \frac{a \sin B}{\sin A}$$
$$b = \frac{6.8 \sin 76.95072452°}{\sin 48°}$$
$$b = 8.914007364$$
$$\mathbf{b = 8.914 \ m}$$

3 Solve applied problems using the law of cosines and the law of sines.

Draw a sketch to represent the data given in the problem. Then identify which law is needed to solve the problem based on the given data.

The vertical utility pole in Fig. 24–93 is placed on a hill. The pole is 40 ft high and a 35-ft guy wire is placed at the top of the pole to be attached to a ground anchor. If the angle formed by the hill and the pole is 43°, how far up the hill should the guy wire be placed if no allowance is given for the amount of wire required to attach the guy wire to its anchor on the pole.

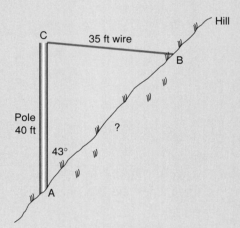

Figure 24–93

$$\frac{a}{\sin A} = \frac{b}{\sin B}$$
$$\frac{35}{\sin 43°} = \frac{40}{\sin B}$$
$$\sin B = \frac{40 \sin 43°}{35}$$
$$\sin B = 0.7794266971$$
$$\sin^{-1} 0.7794266971 = 51.20811426°$$
$$\angle C = 180° - 43° - 51.20811426°$$
$$\angle C = 85.79188573°$$
$$c = \sqrt{a^2 + b^2 - 2ab \cos C}$$
$$c = \sqrt{35^2 + 40^2 - 2(35)(40) \cos 85.79188573°}$$
$$c = \sqrt{1225 + 1600 - 205.462423}$$
$$c = \sqrt{2619.537577}$$
$$c = 51.18141828$$
$$\mathbf{c = 51.18 \ ft} \quad \text{Rounded}$$

Section 24–5

1 Find the area of a triangle when the height is unknown and at least three parts of the triangle are known.

Area $= \frac{1}{2}ab \sin C$, where a and b are two adjacent sides of a triangle and angle C is the angle between sides a and b.

Find the area of triangle ABC in Fig. 24–94.

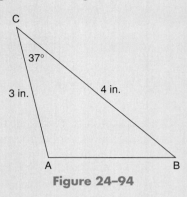

Figure 24–94

Area $= \dfrac{1}{2}ab \sin C$

Area $= 0.5(4)(3)(\sin 37°)$

Area $= 3.610890139$

Area = 3.611 in.^{2}

2 Find the area of a triangle using Heron's formula when the three sides of the triangle are known.

Area $= \sqrt{s(s - a)(s - b)(s - c)}$, where a, b, and c are the lengths of the three sides and s is one-half the perimeter of the triangle.

$$\left(s = \frac{1}{2}(a + b + c) \right)$$

Find the area of triangle ABC in Fig. 24–95.

$s = \dfrac{1}{2}(a + b + c)$

$s = \dfrac{1}{2}(10 + 8 + 12)$

$s = \dfrac{1}{2}(30)$

$s = 15$

Area $= \sqrt{s(s - a)(s - b)(s - c)}$

Area $= \sqrt{15(15 - 10)(15 - 8)(15 - 12)}$

Area $= \sqrt{15(5)(7)(3)}$

Area $= \sqrt{1575}$

Area $= 39.68626967$

Area = 39.69 ft^2

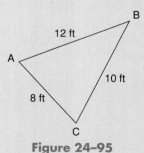

Figure 24–95

3 Solve applied problems involving the area of a triangle.

To solve applied problems, illustrate the data with a figure and use the formula appropriate for the given data.

Find the square footage in the triangular lot shown in Fig. 24–96 that measures 108 ft, 125 ft, and 97 ft on each of the sides.

$$s = \frac{1}{2}(a + b + c)$$

$$s = \frac{1}{2}(125 + 97 + 108)$$

$$s = \frac{1}{2}(330)$$

$$s = 165$$

$$A = \sqrt{s(s - a)(s - b)(s - c)}$$

$$A = \sqrt{165(165 - 125)(165 - 97)(165 - 108)}$$

$$A = \sqrt{165(40)(68)(57)}$$

$$A = \sqrt{25{,}581{,}600}$$

$$A = 5057.82562$$

$$\mathbf{A = 5058 \ ft^2}$$

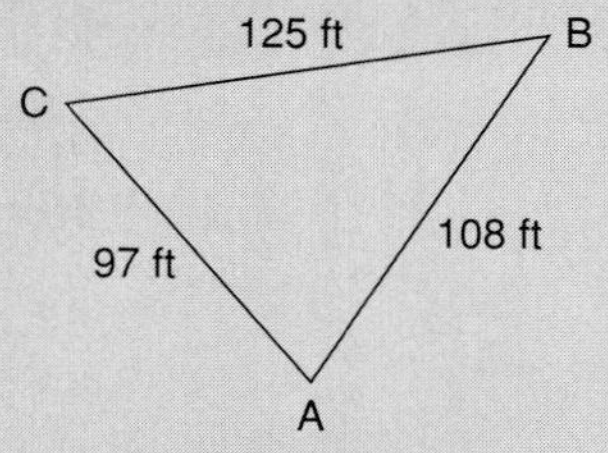

Figure 24–96

WORDS TO KNOW

vector (p. 877)
magnitude (p. 877)
direction (p. 877)
standard position (p. 882)
related angle (p. 882)

trigonometric functions in quadrant I (p. 883)
trigonometric functions in quadrant II (p. 884)
trigonometric functions in quadrant III (p. 884)

trigonometric functions in quadrant IV (p. 884)
law of sines (p. 887)
ambiguous case (p. 891)
law of cosines (p. 896)
Heron's formula (p. 902)

CHAPTER TRIAL TEST

Find the values of the following. Round to four significant digits.

1. sin 125°   **2.** tan 140°   **3.** cos 160°

4. Find the length of the vector whose end point coordinates are (8, 15).

5. Find the angle of the vector whose end point coordinates are (−8, 8).

Use the law of sines or the law of cosines to find the side or angle indicated in Figs. 24–97 to 24–104. Round sides to four significant digits and angles to the nearest 0.1°.

6.

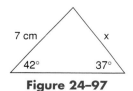

Figure 24–97

7.

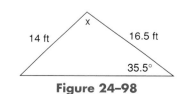

Figure 24–98

8.

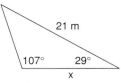

21 m

107°   29°

x

Figure 24–99

9.

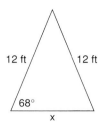

12 ft   12 ft

68°

x

Figure 24–100

10.

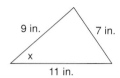

9 in.   7 in.

x

11 in.

Figure 24–101

11.

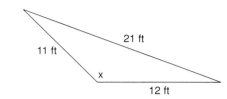

21 ft

11 ft

x

12 ft

Figure 24–102

12.

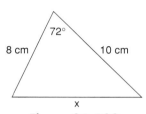

72°

8 cm   10 cm

x

Figure 24–103

13.

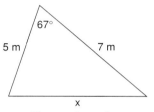

67°

5 m   7 m

x

Figure 24–104

14. Find the area of triangle *ABC* in Fig. 24–105. Round to four significant digits.

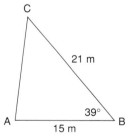

C

21 m

39°

A   B

15 m

Figure 24–105

15. Find the area of triangle *RST* in Fig. 24–106. Round to four significant digits.

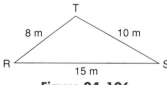

T

8 m   10 m

R   S

15 m

Figure 24–106

16. Find the direction in degrees (to the nearest 0.1°) of the vector I_t (total current) in Fig. 24–107.

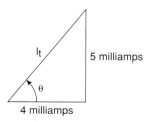

I_t 5 milliamps

θ

4 milliamps

Figure 24–107

17. A surveyor measures two sides of a triangular lot and the angle formed by these two sides. What is the length of the third side if the two measured sides are 76 ft and 110 ft and the angle formed by these two sides is 107°?

918   CHAPTER 24   Oblique Triangles

18. What is the area of a triangular cast if the sides measure 31 mm, 42 mm, and 27 mm?

19. Find the area of a triangular flower bed if two sides measure 12 ft and 15 ft and the included angle measures 48°.

20. A plane is flown from an airport due east for 32 mi, and then turns 15° north of east and travels 72 mi. How far is the plane from the airport?

21. Find the magnitude of the vector I_t in Problem 16 to the nearest tenth.

22. A connecting rod 30 cm long joins a crank 20 cm long to form a triangle with a third, imaginary line (Fig. 24–108). If the crank and the imaginary line form an angle of 150°, find the angle formed by the connecting rod and the imaginary line.

23. Find the magnitude of vector Z in Fig. 24–109 if vectors R and X_L form a right angle. All units are in ohms.

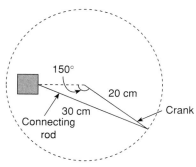

Figure 24–108

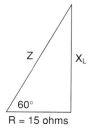

Figure 24–109

24. Find the area of the lot described in Problem 17. (Round to four significant digits.)

Selected Answers to Student Exercise Material

Answers to all Self-Study Exercises and answers to odd-numbered Assignment Exercises and odd-numbered problems in the Chapter Trial Tests are included here. Answers to all even-numbered exercises and problems may be found in the Instructor's Manual.

Self-Study Exercises 1–1

1. hundred millions   **2.** ten thousands   **3.** hundreds   **4.** ten millions
5. billions   **6.** hundred thousands   **7.** tens   **8.** ten millions
9. billions   **10.** ten thousands

11. six thousand, seven hundred, four   **12.** eighty-nine thousand, twenty-one
13. six hundred sixty-two million, nine hundred thousand, seven hundred
fourteen   **14.** three million, one hundred one
15. fifteen billion, four hundred seven million, two hundred ninety-four
thousand, three hundred seventy-six.   **16.** one hundred fifty
17. 7,000,000,400   **18.** 1,627,106   **19.** 58,201   **20.** $1006

21. thousandths   **22.** ones   **23.** hundred-thousandths   **24.** millionths
25. ten-thousandths   **26.** 4   **27.** 0   **28.** 6   **29.** 3   **30.** 7

31. twenty-one and three hundred eighty-seven thousandths
32. four hundred twenty and fifty-nine thousandths
33. eighty-nine hundredths   **34.** five hundred sixty-eight ten-thousandths
35. thirty and two thousand three hundred seventy-nine hundred-thousandths
36. twenty-one and two hundred five thousand eighty-five millionths
37. 3.42   **38.** 78.195   **39.** 500.0005   **40.** 0.75034

5

41. 0.5   **42.** 0.23   **43.** 0.07   **44.** 6.83   **45.** 0.079   **46.** 0.468
47. 5.87   **48.** 0.108   **49.** 6.03   **50.** 4.00

6

51. 3.72   **52.** 7.08   **53.** 0.3   **54.** 0.56   **55.** 2.75   **56.** 0.2
57. 8.88   **58.** 0.25   **59.** 0.913   **60.** 0.76   **61.** 5.983   **62.** 1.972
63. 0.179, 0.23, 0.314   **64.** 1.87, 1.9, 1.92   **65.** 72.07, 72.1, 73
66. 0.837 in.   **67.** the reading   **68.** yes   **69.** 0.04 in.
70. No. 10 wire

7

71. 500   **72.** 6200   **73.** 8000   **74.** 430,000   **75.** 40,000
76. 40,000   **77.** 285,000,000   **78.** 470   **79.** 83,000,000,000
80. 300,000,000   **81.** 43   **82.** 367   **83.** 8   **84.** 103   **85.** 3
86. 8.1   **87.** 12.9   **88.** 42.6   **89.** 83.2   **90.** 6.0   **91.** 7.04
92. 42.07   **93.** 0.79   **94.** 3.20   **95.** 7.77   **96.** 0.217   **97.** 0.020
98. 1.509   **99.** 4.238   **100.** 7.004   **101.** $219   **102.** $83
103. $507   **104.** $3   **105.** $6   **106.** $8.24   **107.** $0.29   **108.** $0.53
109. $5.80   **110.** $238.92   **111.** 0.784   **112.** 3.82 in.   **113.** 2.8 A

8

114. 500   **115.** 8   **116.** 60   **117.** 0.5   **118.** 0.009   **119.** 0.1
120. 3   **121.** 50   **122.** 80   **123.** 50   **124.** $3   **125.** 20
126. 0.4, 2, 0.5, 1, 0.07, 2   **127.** 0.03, 0.03, 0.02, 0.03, 0.04

Self-Study Exercises 1–2

1

1. 15   **2.** 27   **3.** 22   **4.** 36   **5.** 37 bolts   **6.** 9,192   **7.** 16,956
8. 106,285   **9.** 2310   **10.** 97,614   **11.** $456   **12.** 24,349 lb
13. 298 screws   **14.** 4553 bricks   **15.** 50 gal

2

16. 15.7   **17.** 34.18   **18.** 8.13   **19.** 87.4   **20.** 129.97   **21.** 26.03
22. 78.2   **23.** 45.7   **24.** 261.335   **25.** 356.612   **26.** 6.984
27. 9.525   **28.** 1126.6   **29.** 413.6   **30.** 18.1   **31.** 18.8   **32.** 0.81805
33. 1.17642   **34.** 55.513   **35.** 85.411   **36.** 69.987   **37.** 3.077 in.
38. 15.503 A   **39.** $181.25   **40.** 391.4 ft   **41.** $19.93   **42.** $5.04

3

43. 17,100; 17,018.21   **44.** 84,200; 84,213   **45.** 402,300; 402,199
46. $3440; $3443.60   **47.** $2300; $2260   **48.** 8900; 9239
49. 600; 624.18   **50.** 9000; 8449   **51.** 800; 801.39   **52.** 60,000; 52,801
53. 50,000; 49,241   **54.** 159 lb   **55.** Yes, total capacity available is
115 gal.   **56.** Yes, 479 pages needed   **57.** 350 ft

4

58. 190,000; 190,786   **59.** 1,900,000; 1,859,867   **60.** 50,000; 54,359.65
61. 51,000,000; 47,520,014   **62.** $17,039.04

Self-Study Exercises 1-3

1

1. 5 **2.** 1 **3.** 7 **4.** 2 **5.** 3 **6.** 7 **7.** 5 **8.** 2 **9.** 4 **10.** 1
11. 24 **12.** 401 **13.** 1020 **14.** 115 **15.** 53,036 **16.** 22 bags
17. 341 boxes **18.** 321 ft

2

19. 6.93 **20.** 15.834 **21.** 803.693 **22.** 56.14 **23.** 4.094 **24.** 3.9
25. 291.82 **26.** 7.5 **27.** 310.8 **28.** 4.4 **29.** 15.1 lb
30. 12.08 in., 12.10 in. **31.** 59.83 cm **32.** 4.189 in., 4.201 in.
33. 0.22 dm **34.** 11.55 A **35.** $8.75 **36.** $2.25 **37.** $4.75

3

38. 45 **39.** 608 **40.** 1573 **41.** 22,205 **42.** 100,000; 91,034
43. 2000; 2088.884 **44.** 0; 174 **45.** 200; 599.15 **46.** 70,000; 70,382
47. 1300; 1097.15

4

48. 4000; 4397 **49.** 1000; 1187.251 **50.** 500,000; 508,275 **51.** 17 L
52. 186 bricks **53.** 213.8 in. **54.** 125.5 in. **55.** 45 in.

Self-Study Exercises 1-4

1

1. (a) 15 (b) 56 (c) 63 (d) 24 **2.** $42
3. commutative property of multiplication
4. Numbers may be grouped in any way for multiplication.
$$3[(5)(2)] = [3(5)]2$$
$$3(10) = (15)2$$
$$30 = 30 \quad \text{Answers may vary.}$$
5. 0 **6.** 378 **7.** 84 **8.** 0 **9.** 224 **10.** 581 **11.** 630
12. 102,612 **13.** 864 pieces **14.** 4096 washers **15.** 84 books
16. 700 tickets **17.** $288 **18.** 249 students

2

19. 15.486 **20.** 56.55 **21.** 3.2445 **22.** 0.05805 **23.** 0.08672
24. 7.141 **25.** 0.0834 **26.** 170.12283 **27.** 0.38381871
28. 4.9386274 **29.** 30.66 **30.** 596.97 **31.** 50.7357 **32.** 38.6232
33. 339.04 **34.** 540.27 **35.** 254 **36.** 184.2 **37.** 0.9307
38. 0.7602 **39.** 0.58635 **40.** 0.73265 **41.** 9.21702 **42.** 11.42356
43. 0.0176 **44.** 0.027045 **45.** 0.915371 **46.** 0.390483 **47.** 0.00015
48. 0.00056 **49.** 3.957 in. **50.** 5.25 in. **51.** $27.48 **52.** 151.2 in.
53. $64.20 **54.** $10,728 **55.** 0.375 in. **56.** $1960.50 **57.** 15 A
58. $78

3

59. 55 **60.** 60 **61.** 196 **62.** 12 **63.** 15.6 **64.** 5.6
65. $P = 40$ ft **66.** $P = 61$ in. **67.** 56 ft **68.** 170 ft

4

69. $600; $768 **70.** 123.54 ft^2 **71.** 1164 cm^2 **72.** 1200 ft^2; 1374.75 ft^2
73. 96 ft^2

74. 3,578,040.664   **75.** 23,379,045   **76.** 561,500,160   **77.** 0
78. $36,180   **79.** $22,006.40   **80.** 147,000,000,000   **81.** 16,755,200,000
82. 420,000,000

Self-Study Exercises 1–5

1

1. $8 \div 4$; $4\overline{)8}$; $\dfrac{8}{4}$ **2.** $9 \div 3$; $3\overline{)9}$; $\dfrac{9}{3}$ **3.** $24 \div 6$; $6\overline{)24}$; $\dfrac{24}{6}$

4. $30 \div 7$; $7\overline{)30}$; $\dfrac{30}{7}$ **5.** $6 \div 2$; $2\overline{)6}$; $\dfrac{6}{2}$ **6.** $8 \div 4$; $4\overline{)8}$; $\dfrac{8}{4}$

2

7. 75   **8.** 23   **9.** 20   **10.** 3   **11.** 16   **12.** 12   **13.** 43   **14.** 47
15. 12   **16.** 23R1   **17.** 124R30   **18.** 56R80   **19.** 32 ft   **20.** $1245
21. 6 in.

3

22. 1.26   **23.** 3.09   **24.** 0.063   **25.** 285   **26.** 5.9   **27.** 45
28. 10.9   **29.** 0.19   **30.** 1.06   **31.** 0.33   **32.** 25   **33.** 20,700
34. 90,200   **35.** 10,700   **36.** 1.8   **37.** 6 lb   **38.** 23 rolls
39. 600 revolutions   **40.** 0.7 ft   **41.** 93.75 volts

42. A measure divided by a number equals a measure.   **43.** $19.84\dfrac{8}{13}$

44. $1.82\dfrac{1}{12}$ **45.** $2601.16\dfrac{2}{3}$ **46.** $1.98\dfrac{4}{7}$ **47.** $0.07\dfrac{27}{29}$ **48.** 169.3
49. 0.16   **50.** 9   **51.** $0.96   **52.** $1   **53.** $575   **54.** $1.98
55. 0.0812 in.   **56.** 0.2 in.

4

57. 100; 125.6   **58.** 4000; 3151R90   **59.** 4000; 3731R53   **60.** 8; 8.43
61. 128 loads   **62.** 127 cords   **63.** 80.4°   **64.** 85   **65.** 454 lb
66. $765.32   **67.** 1.69 in.   **68.** 3.5 A

6

69. 54   **70.** 301   **71.** 40   **72.** 89,500   **73.** 40,200   **74.** 6201
75. 1070 bricks   **76.** $10,500   **77.** $104.75   **78.** $1950   **79.** $2465
80. $4.45   **81.** 24 parts   **82.** 10 lengths   **83.** $78,505   **84.** $59.71
85. 0.7 in.

Self-Study Exercises 1–6

1

1. 4; 3   **2.** 9; 4   **3.** 2.7; 9   **4.** 3.375   **5.** 49   **6.** 1000   **7.** 16
8. 11.56   **9.** 15   **10.** 8   **11.** 1   **12.** 8^1 **13.** 14.5^1 **14.** 12^1
15. 23^1 **16.** leaves base unchanged   **17.** changes value to 1; 0^0 is
undefined

2

18. 64   **19.** 4   **20.** 12.25   **21.** 1.96   **22.** 169   **23.** 1   **24.** 10,000
25. 64   **26.** 81   **27.** 289   **28.** 324   **29.** 10,201   **30.** 484   **31.** 5

32. 7   **33.** 9   **34.** 14   **35.** Use it as a factor two times.
36. Find a number that is used as a factor two times to give the desired number.

③
37. $10^3 = 1000$ **38.** 32,000   **39.** 30,000   **40.** 20,000   **41.** 10,200
42. 22,000   **43.** 25   **44.** 21   **45.** 3   **46.** 9   **47.** 250   **48.** 1.2
49. Shift the decimal to the right in the number being multiplied by the power of ten as indicated by the exponent.
50. Shift the decimal to the left in the number being multiplied by the power of ten as indicated by the exponent.

④
51. 225   **52.** 343   **53.** 78,125   **54.** 18   **55.** 28   **56.** 33

Self-Study Exercises 1–7

①
1. 26   **2.** 18   **3.** 9   **4.** 12.5   **5.** 24   **6.** 32   **7.** 30   **8.** 15
9. 156   **10.** 109   **11.** 23   **12.** 145   **13.** 15.6   **14.** 25.04   **15.** 9
16. 5   **17.** 7   **18.** 15   **19.** 160   **20.** 1458

②
21. 483   **22.** 443   **23.** 70   **24.** 13   **25.** 11   **26.** 29   **27.** 5
28. 39   **29.** 7   **30.** 14   **31.** 20   **32.** 30   **33.** 20.52   **34.** undefined
35. 191.75   **36.** parentheses   **37.** addition or subtraction

Assignment Exercises, Chapter 1

1. (a) 0.3 (b) 0.15 (c) 0.04   **3.** thousandths   **5.** 7
7. (a) tens (b) ten-thousands (c) hundreds; millions
9. fifty six million, one hundred and nine thousand, one hundred ten
11. 1,265,401   **13.** six and eight hundred three thousandths   **15.** 0.625
17. (a) 40 (b) 70 (c) 24 (d) $43 (e) 80 (f) $8.94 (g) $1.00 (h) 0.0970
19. (a) 320 (b) 7000 (c) 500 (d) 50,000 (e) 27 b (f) 41.4 (g) 6.90
(h) 23.4610
21. 4.79   **23.** 0.02; 0.021; 0.0216   **25.** $\dfrac{7}{8}$
27. (a) 23 (b) 18 (c) 28 (d) 25   **29.** 34.9 kW
31. (a) 29,000; 29,092.09 (b) 36,000; 36,048
33. (a) 4.61 (b) 3.127 (c) 2 (d) 5 (e) 0 (f) 204.899 (g) 144 (h) 12,140
35. 8.291 in.; 8.301 in.   **37.** 200 miles, 190 miles   **39.** 0.430
41. 8.930 in.; 8.940 in.   **43.** 84   **45.** 1143   **47.** 13,725
49. 3,349,890   **51.** 394,254,080   **53.** $1407   **55.** $43,920   **57.** 20
59. 60   **61.** 80   **63.** $328
65. $14,800, $13,140   **67.** 1,140,000, 1,204,010.28   **69.** 0.12096 in.
71. $5 \div 3, 3\overline{)5}, \dfrac{5}{3}$ **73.** 1   **75.** not defined   **77.** 5.26875   **79.** 13
81. 2008.4   **83.** 23, 11R   **85.** 5.8375   **87.** 5; 5.52   **89.** $10; $10.63
91. 50; 48.79 ft   **93.** (a) 7, 3, 343 (b) 2.3, 4, 27.9841 (c) 8, 4, 4096
95. (a) 0.9 (b) 35 (c) 1   **97.** (a) 1 (b) 1 (c) 1 (d) 1
99. (a) 1 (b) 15.625 (c) 31.36 (d) 441   **101.** (a) 10^1 (b) 10^3 (c) 10^4 (d) 10^5
103. (a) 7 (b) 0.04056 (c) .605 (d) 2.3079 (e) 44.582   **105.** 75   **107.** 3
109. 13.6   **111.** 113.608   **113.** 11   **115.** 45×48
117. (a) 0.8 (b) 0.3 (c) 0.03 (d) 0.25 (e) 0.004 (f) 0.63245 . . .

Trial Test, Chapter 1

1. five million, thirty thousand, one hundred two   **3.** 7.027   **5.** 2.700
7. 5.09   **9.** 48.3   **11.** 1007   **13.** \$9,271,314   **15.** 134   **17.** 106
19. 0.0086   **21.** \$310, \$310   **23.** \$10, \$14.00
25. Commutative means we can add numbers in any order. $2 + 4 = 6$;
$4 + 2 = 6$. Associative means we can group numbers differently. $(2 + 1) +$
$3 = 6$; $2 + (1 + 3) = 6$.   **27.** 42,730   **29.** 11.6   **31.** 83   **33.** 0.6
35. 70 or between 70 and 80   **37.** \$17,500

Self-Study Exercises 2–1

1

1. right   **2.** left   **3.** infinity   **4.** 1   **5.** neither
6. $-3, -2, -1, 0, 1, 2, 3$ **7.** to the right   **8.** to the left

2

9. <   **10.** >   **11.** <   **12.** <   **13.** >   **14.** <   **15.** >   **16.** <
17. $x < y$ **18.** $a > b$ **19.** $3 + 5 > 6$ **20.** $9 < 18 - 6$ **21.** $k > t$
22. $r < s$

3

23. 23   **24.** 0   **25.** 10   **26.** -17 **27.** 13   **28.** 345   **29.** 67
30. 61

4

31. -7 0   7   **32.** -8 0   8
33. -4 0   4   **34.** -12 0   12
35. 0
36. a negative integer; a positive integer

5

37. $A = (4, 2)$ $B = (-3, 2)$ $C = (-2, -1)$ $D = (3, -2)$

38.–43.

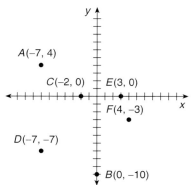

44. y-value = 0   **45.** x-value = 0   **46.** $(-x, +y)$ **47.** $(+x, -y)$

Self-Study Exercises 2–2

1

1. 17   **2.** -13 **3.** 99   **4.** -59 **5.** -48 **6.** -161

2

7. -6 8. 4 9. -2 10. 2 11. -10 12. -40 13. 7 14. 15
15. -8 16. -33 17. 9 18. -11 19. 13 20. 9 21. 3
22. 0 23. 0 24. 0 25. 0 26. 0 27. 9 28. -3 29. -7
30. 18 31. -8 32. -28 33. $5 + (-5) = 0$ or other example
34. 0 (zero)
35. Adding numbers of like signs increases the absolute value whether positive or negative.
36. $36   37. $66

Self-Study Exercises 2–3

1

1. -12 2. 6 3. -6 4. 4 5. -25 6. -3 7. 8 8. -3
9. -9 10. 13 11. -1 12. -8

2

13. 15 14. -8 15. -12 16. 8 17. 7 18. 10 19. 56
20. -92 21. 14 22. -36

3

23. -4 24. 1 25. 14 26. -3 27. -9 28. -5 29. 13
30. 2 31. -1 32. $107°$ 33. $107°$ 34. $115,054$
35. Subtracting zero from a number results in the same number with the same sign. Subtracting a number from zero results in the opposite of the number.

Self-Study Exercises 2–4

1

1. 40 2. 12 3. 35 4. 21 5. 24 6. 6 7. -15 8. -10
9. -32 10. -12 11. -56 12. -24 13. $4 \cdot (-28) = -$112$
14. $(7)(-40) = -$280$
15. When multiplying signed numbers, the absolute values are always multiplied. When numbers with like signs are multiplied, the result is positive. When numbers with unlike signs are multiplied, the result is negative.

2

16. -60 17. 36 18. 0 19. 90 20. 0 21. -42 22. 54
23. -210 24. 2268 25. -20160

3

26. 0 27. 0 28. 0 29. 0 30. 0 31. 0 32. 0 33. 0
34. 0 35. 0
36. The multiplicative inverse of a number is that number that, when multiplied by the original number, results in 1, the multiplicative identity.

4

37. 9 38. -8 39. 25 40. -25 41. -8 42. 1 43. 625
44. 81 45. 1 46. 1 47. $604,800$ 48. $1,209,600$
49. No, we don't normally use 000 as the first 3 digits.   50. $17,576,000$
51. $10,000$

926   Selected Answers to Student Exercise Material

Self-Study Exercises 2–5

1

1. 2   **2.** −3   **3.** −4   **4.** −4   **5.** −4   **6.** 8   **7.** 5   **8.** −6
9. 8   **10.** 5   **11.** $\dfrac{-1800}{6} = -\$300.00$

2

12. not defined   **13.** not defined   **14.** 0   **15.** 0   **16.** 0
17. not defined   **18.** not defined   **19.** not defined   **20.** 0
21. not defined   **22.** multiplication and division

Self-Study Exercises 2–6

1

1. −4   **2.** −30   **3.** −2   **4.** 17   **5.** −2   **6.** 24   **7.** 8   **8.** 24
9. −5   **10.** −7   **11.** −3   **12.** 14

2

13. −12   **14.** −1260   **15.** 27   **16.** −14   **17.** 28   **18.** −35
19. −29   **20.** −132
21. −E−; indicating that division by zero is not possible.
22. Numbers can be added in any order.
23. When the order of subtraction is changed, the result is not the same.
24. Numbers can be multiplied in any order.
25. The order of division of two numbers cannot be changed.

Assignment Exercises, Chapter 2

1. G   **3.** F   **5.** E   **7.** A   **9.** L   **11.** M   **13.** 5, 21, 987 (Answers
may vary.)   **15.** 0   **17.** 1   **19.** >   **21.** >   **23.** <   **25.** 5   **27.** 3
29. 7   **31.** 11   **33.** 12   **35.** −15   **37.** 2   **39.** 42   **41.** −87

43.–47. 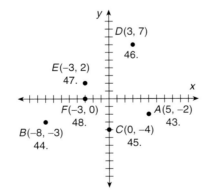   **49.** (0, 0)

51. $A(3, 0)$ $B(2, 2)$ $C(2, -5)$ $D(-4, -1)$ $E(-3, 1)$ **53.** 17   **55.** −7
57. −13   **59.** 0   **61.** −8   **63.** −25   **65.** run up the middle, 2 yd for
first down   **67.** \$569   **69.** The additive identity, 0, if added to any
number, will result in the same number. Numbers that are additive inverses,
when added, result in zero.   **71.** −13   **73.** 14   **75.** −15   **77.** −5
79. 2   **81.** 36   **83.** 70°   **85.** 40   **87.** −14   **89.** −12   **91.** 0
93. 0   **95.** −168   **97.** 343   **99.** −16   **101.** −10°   **103.** higher, 20,
$-4 \times 5 = -20$ **105.** 4   **107.** 4   **109.** −4   **111.** undefined
113. 56   **115.** 1   **117.** 36   **119.** 73   **121.** −2   **123.** 391, 464

125. 9:00 A.M.: $-1°C$ 10:00 A.M.: $0°C$
11:00 A.M.: $0°C$ 12:00 P.M.: $1°C$
1:00 P.M.: $1°C$ 2:00 P.M.: $4°C$
3:00 P.M.: $0°C$ 4:00 P.M.: $-7°C$
5:00 P.M.: $-15°C$ 6:00 P.M.: $-27°C$

Trial Test, Chapter 2

1. $<$ **3.** $<$ **5.** -8 **7.** 4   **9.** -48 **11.** 3   **13.** 8   **15.** 0
17. -7 **19.** not defined   **21.** -3 **23.** -2 **25.** 1   **27.** 33
29. 22   **31.** -5
33. If a number is multiplied by the multiplicative identity, or if a number is added to the additive identity, the result is the original number. $5 \times 1 = 5$ or $3 + 0 = 3$
35. $175°$

Self-Study Exercises 3–1

1

1. $\dfrac{5}{6}$ **2.** $\dfrac{3}{4}$ **3.** $\dfrac{3}{8}$ **4.** $\dfrac{7}{12}$ **5.** $\dfrac{1}{16}$ **6.** $\dfrac{3}{3}$ **7.** $\dfrac{4}{6}$
8. (a) 6 (b) 5 (c) 5 (d) 6 (e) 5; 6 (f) 6 (g) 5 (h) proper (i) 5; 6 (j) less than 1
9. (a) 8 (b) 8 (c) 8 (d) 8 (e) 8; 8 (f) 8 (g) 8 (h) improper (i) 8; 8 (j) equal to 1
10. (a) 5 (b) 11 (c) 11 (d) 5 (e) 11; 5 (f) 5 (g) 11 (h) improper (i) 11; 5 (j) greater than 1
11. (a) 3 (b) 12 (c) 12 (d) 3 (e) 12; 3 (f) 3 (g) 12 (h) improper (i) 12; 3 (j) greater than 1
12. b   **13.** a   **14.** f   **15.** b   **16.** d   **17.** d   **18.** a   **19.** e
20. c   **21.** c   **22.** c   **23.** $6\dfrac{27}{100}$ **24.** $\dfrac{3}{10}$ **25.** $3\dfrac{7}{10}$ **26.** $1\dfrac{53}{100}$

Self-Study Exercises 3–2

1

1. $5 = 5 \times 1, 10 = 5 \times 2, 15 = 5 \times 3, 20 = 5 \times 4, 25 = 5 \times 5, 30 = 5 \times 6$
2. $6 = 6 \times 1, 12 = 6 \times 2, 18 = 6 \times 3, 24 = 6 \times 4, 30 = 6 \times 5, 36 = 6 \times 6$
3. $8 = 8 \times 1, 16 = 8 \times 2, 24 = 8 \times 3, 32 = 8 \times 4, 40 = 8 \times 5, 48 = 8 \times 6$
4. $9 = 9 \times 1, 18 = 9 \times 2, 27 = 9 \times 3, 36 = 9 \times 4, 45 = 9 \times 5, 54 = 9 \times 6$
5. $10 = 10 \times 1, 20 = 10 \times 2, 30 = 10 \times 3, 40 = 10 \times 4, 50 = 10 \times 5, 60 = 10 \times 6$
6. $30 = 30 \times 1, 60 = 30 \times 2, 90 = 30 \times 3, 120 = 30 \times 4, 150 = 30 \times 5, 180 = 30 \times 6$
7. $5 \times 1 = 5, 5 \times 2 = 10, 5 \times 3 = 15, 5 \times 4 = 20, 5 \times 5 = 25$
8. $12 \times 1 = 12, 12 \times 2 = 24, 12 \times 3 = 36, 12 \times 4 = 48, 12 \times 5 = 60$
9. $7 \times 1 = 7, 7 \times 2 = 14, 7 \times 3 = 21, 7 \times 4 = 28, 7 \times 5 = 35$
10. $3 \times 1 = 3, 3 \times 2 = 6, 3 \times 3 = 9, 3 \times 4 = 12, 3 \times 5 = 15$
11. $50 \times 1 = 50, 50 \times 2 = 100, 50 \times 3 = 150, 50 \times 4 = 200, 50 \times 5 = 250$
12. $4 \times 1 = 4, 4 \times 2 = 8, 4 \times 3 = 12, 4 \times 4 = 16, 4 \times 5 = 20$

2

13. no; 405 remainder 4   **14.** yes; $230 \div 5 = 46$ **15.** no, 608 remainder 2
16. yes; $1221 \div 3 = 407$ **17.** yes; $756 \div 7 = 108$ **18.** yes; 920

$\div\ 8 = 115$ **19.** yes; $621 \div 3 = 207$ **20.** yes; $426 \div 6 = 71$
21. yes; $1232 \div 2 = 616$

3

22. $1 \cdot 24,\ 2 \cdot 12,\ 3 \cdot 8,\ 4 \cdot 6$ **23.** $1 \cdot 36,\ 2 \cdot 18,\ 3 \cdot 12,\ 4 \cdot 9,\ 6 \cdot 6$
24. $1 \cdot 45,\ 3 \cdot 15,\ 5 \cdot 9$ **25.** $1 \cdot 32,\ 2 \cdot 16,\ 4 \cdot 8$
26. $1 \cdot 16,\ 2 \cdot 8,\ 4 \cdot 4$ **27.** $1 \cdot 27,\ 3 \cdot 9$ **28.** $1 \cdot 20,\ 2 \cdot 10,\ 4 \cdot 5$
29. $1 \cdot 30,\ 2 \cdot 15,\ 3 \cdot 10,\ 5 \cdot 6$ **30.** $1 \cdot 12,\ 2 \cdot 6,\ 3 \cdot 4$ **31.** $1 \cdot 8,\ 2 \cdot 4$
32. $1 \cdot 4,\ 2 \cdot 2$ **33.** $1 \cdot 15,\ 3 \cdot 5$ **34.** $1 \cdot 81,\ 3 \cdot 27,\ 9 \cdot 9$
35. $1 \cdot 64,\ 2 \cdot 32,\ 4 \cdot 16,\ 8 \cdot 8$ **36.** $1 \cdot 38,\ 2 \cdot 19$ **37.** $1 \cdot 46,\ 2 \cdot 23$
38. $1 \cdot 51,\ 3 \cdot 17$ **39.** $1 \cdot 18,\ 2 \cdot 9,\ 3 \cdot 6$
40. $1 \cdot 72,\ 2 \cdot 36,\ 3 \cdot 24,\ 4 \cdot 18,\ 6 \cdot 12,\ 8 \cdot 9$

Self-Study Exercises 3–3

1

1. $1 \cdot 14,\ 2 \cdot 7;\ 1, 2, 7, 14$ **2.** $1 \cdot 22,\ 2 \cdot 11;\ 1, 2, 11, 22$
3. $1 \cdot 11;\ 1, 11$ **4.** $1 \cdot 17;\ 1, 17$ **5.** $1 \cdot 18,\ 2 \cdot 9,\ 3 \cdot 6;\ 1, 2, 3, 6, 9, 18$
6. $1 \cdot 24,\ 2 \cdot 12,\ 3 \cdot 8,\ 4 \cdot 6;\ 1, 2, 3, 4, 6, 8, 12, 24$

2

7. $2 \cdot 5 \cdot 5$ **8.** $2 \cdot 2 \cdot 13$ **9.** $3 \cdot 3 \cdot 5 \cdot 5$ **10.** $5 \cdot 5 \cdot 5$
11. $2 \cdot 2 \cdot 5 \cdot 5$ **12.** $2 \cdot 2 \cdot 2 \cdot 5 \cdot 5$ **13.** $5 \cdot 13$ **14.** $3 \cdot 5 \cdot 5$
15. $11 \cdot 11$ **16.** $2 \cdot 2 \cdot 2 \cdot 2 \cdot 3 \cdot 3$ **17.** $2^3 \cdot 71$ **18.** $2^4 \cdot 7$
19. $2^2 \cdot 31$ **20.** $2^2 \cdot 41$ **21.** $2^3 \cdot 3^2$ **22.** $2^2 \cdot 3^2 \cdot 5^2$

Self-Study Exercises 3–4

1

1. 6   **2.** 30   **3.** 56   **4.** 12   **5.** 180   **6.** 60   **7.** 24   **8.** 18
9. 24   **10.** 700   **11.** 27   **12.** 16   **13.** 90   **14.** 120   **15.** 180
16. 60   **17.** 96   **18.** 72   **19.** 60   **20.** 300   **21.** 66   **22.** 312

2

23. 6   **24.** 5   **25.** 1   **26.** 1   **27.** 2   **28.** 2   **29.** 15   **30.** 5
31. 12   **32.** 6

Self-Study Exercises 3–5

1

1. $\dfrac{4}{5},\ \dfrac{8}{10},\ \dfrac{12}{15},\ \dfrac{16}{20},\ \dfrac{20}{25},\ \dfrac{24}{30}$ **2.** $\dfrac{7}{10},\ \dfrac{14}{20},\ \dfrac{21}{30},\ \dfrac{28}{40},\ \dfrac{35}{50},\ \dfrac{42}{60}$ **3.** $\dfrac{3}{4} = \dfrac{18}{24}$ **4.** 6
5. 12   **6.** 36   **7.** 5   **8.** 20   **9.** 21   **10.** 12

2

11. $\dfrac{1}{2}$ **12.** $\dfrac{3}{5}$ **13.** $\dfrac{3}{4}$ **14.** $\dfrac{5}{16}$ **15.** $\dfrac{1}{2}$ **16.** $\dfrac{7}{8}$ **17.** $\dfrac{5}{16}$ **18.** $\dfrac{1}{4}$
19. $\dfrac{1}{4}$ **20.** $\dfrac{6}{25}$ **21.** $\dfrac{5}{8}$ **22.** $\dfrac{1}{4}$ **23.** $\dfrac{3}{4}$ **24.** $\dfrac{3}{16}$ **25.** $\dfrac{7}{32}$

3

26. $\dfrac{1}{2}$ **27.** $\dfrac{1}{10}$ **28.** $\dfrac{1}{5}$ **29.** $\dfrac{7}{10}$ **30.** $\dfrac{1}{4}$ **31.** $\dfrac{1}{40}$ **32.** $3\dfrac{9}{10}$ **33.** $4\dfrac{4}{5}$

34. $\dfrac{189}{500}$ **35.** $\dfrac{7}{8}$ **36.** $\dfrac{3}{8}$ **37.** $\dfrac{5}{8}$ **38.** $\dfrac{3}{4}$ **39.** $\dfrac{3}{16}$ **40.** $2\dfrac{3}{8}$ **41.** $\dfrac{5}{6}$

42. $\dfrac{5}{16}$ **43.** $3\dfrac{1}{8}$

[4]

44. 0.4   **45.** 0.3   **46.** 0.875   **47.** 0.625   **48.** 0.45   **49.** 0.98
50. 0.21   **51.** 3.875   **52.** 1.4375   **53.** 4.5625   **54.** $0.\overline{6}$ or 0.66 . . .
55. $0.\overline{27}$ or 0.2727 . . .   **56.** $0.\overline{7}$ or 0.77 . . .   **57.** $0.\overline{384615}$ or
.384615. . .   **58.** $0.83\dfrac{1}{3}$ **59.** $0.58\dfrac{1}{3}$ **60.** 2.046875 in.   **61.** 4.5%
62. 0.125 in.

Self-Study Exercises 3–6

[1]

1. $2\dfrac{2}{5}$ **2.** $1\dfrac{3}{7}$ **3.** 1   **4.** $4\dfrac{4}{7}$ **5.** 4   **6.** $2\dfrac{1}{7}$ **7.** $2\dfrac{5}{9}$ **8.** $9\dfrac{2}{5}$ **9.** $9\dfrac{5}{9}$

10. $1\dfrac{17}{21}$ **11.** $3\dfrac{4}{5}$ **12.** 16   **13.** $7\dfrac{1}{5}$ **14.** $9\dfrac{1}{2}$ **15.** 9

[2]

16. $\dfrac{7}{3}$ **17.** $\dfrac{25}{8}$ **18.** $\dfrac{15}{8}$ **19.** $\dfrac{77}{12}$ **20.** $\dfrac{77}{8}$ **21.** $\dfrac{31}{8}$ **22.** $\dfrac{89}{12}$

23. $\dfrac{103}{16}$ **24.** $\dfrac{257}{32}$ **25.** $\dfrac{69}{64}$ **26.** $\dfrac{73}{10}$ **27.** $\dfrac{26}{3}$ **28.** $\dfrac{100}{3}$ **29.** $\dfrac{200}{3}$

30. $\dfrac{25}{2}$ **31.** 15   **32.** 18   **33.** 56   **34.** 32   **35.** 48

Self-Study Exercises 3–7

[1]

1. 72   **2.** 30   **3.** 50   **4.** 48   **5.** 24

[2]

6. $\dfrac{2}{3}$ **7.** $\dfrac{7}{16}$ **8.** $\dfrac{8}{9}$ **9.** $\dfrac{11}{16}$ **10.** $\dfrac{15}{32}$ **11.** $\dfrac{7}{12}$ **12.** $\dfrac{4}{5}$ **13.** $\dfrac{9}{10}$

14. $\dfrac{4}{15}$ **15.** $\dfrac{1}{2}$ **16.** No, $\dfrac{15}{64}$ is greater   **17.** Yes, $\dfrac{3}{8}$ is greater

18. Yes, $\dfrac{7}{16}$ is greater   **19.** Yes   **20.** $\dfrac{5}{8}$ **21.** No   **22.** No   **23.** Yes

24. No   **25.** too small   **26.** 0.03   **27.** 0.392   **28.** 5.38   **29.** 4.71
30. 98.6°   **31.** 0.0256 cm   **32.** 0.973 mills
33. Answers may vary. To compare fractions, find a common denominator and compare numerators. To compare decimals, compare the whole numbers. If they are equal, compare decimal digits by each place value.

Self-Study Exercises 3–8

[1]

1. $\dfrac{3}{8}$ **2.** $1\dfrac{3}{8}$ **3.** $\dfrac{5}{8}$ **4.** $\dfrac{25}{32}$ **5.** $\dfrac{9}{16}$ **6.** $1\dfrac{7}{16}$ **7.** $\dfrac{11}{64}$ **8.** $1\dfrac{19}{40}$

9. $1\frac{23}{36}$ **10.** $1\frac{1}{12}$ **11.** $\frac{15}{16}$ in. **12.** $1\frac{27}{32}$ in. **13.** $1\frac{15}{16}$ in. **14.** $1\frac{7}{8}$ in.

15. $1\frac{1}{4}$ in.

2

16. $6\frac{4}{5}$ **17.** $4\frac{1}{8}$ **18.** $16\frac{13}{16}$ **19.** $1\frac{11}{18}$ **20.** $4\frac{13}{16}$ **21.** $5\frac{17}{32}$ **22.** $4\frac{11}{16}$

23. $13\frac{1}{2}$ **24.** $4\frac{7}{15}$ **25.** $12\frac{11}{16}$ **26.** $8\frac{13}{16}$ in. **27.** $3\frac{15}{16}$ in. **28.** $11\frac{5}{8}$ gal

29. $24\frac{5}{32}$ in. **30.** $22\frac{7}{8}$ in.

Self-Study Exercises 3–9

1

1. $\frac{1}{4}$ **2.** $\frac{3}{16}$ **3.** $\frac{1}{16}$ **4.** $\frac{1}{8}$ **5.** $\frac{9}{64}$ **6.** $\frac{1}{8}$

2

7. $4\frac{11}{16}$ **8.** $17\frac{3}{4}$ **9.** $4\frac{15}{16}$ **10.** $5\frac{21}{32}$ **11.** $8\frac{13}{16}$ in. **12.** $10\frac{1}{8}$ in.

13. $3\frac{1}{10}$ lb **14.** $\frac{3}{16}$ in. **15.** $\frac{29}{32}$

Self-Study Exercises 3–10

1

1. $\frac{3}{32}$ **2.** $\frac{7}{32}$ **3.** $\frac{7}{16}$ **4.** $\frac{7}{12}$ **5.** $\frac{1}{3}$ **6.** $\frac{5}{32}$ **7.** $\frac{1}{15}$ **8.** $\frac{11}{25}$

9. $\frac{2}{25}$ **10.** $\frac{21}{32}$

2

11. $21\frac{7}{8}$ **12.** 75 **13.** $4\frac{1}{8}$ **14.** $36\frac{1}{10}$ **15.** $1\frac{21}{40}$ **16.** $2\frac{1}{6}$ **17.** $18\frac{3}{4}$ L

18. 90 in. **19.** $15\frac{5}{8}$ in. **20.** 264 kg copper, 84 kg tin, 36 kg zinc

Self-Study Exercises 3–11

1

1. $\frac{7}{3}$ **2.** $\frac{1}{8}$ **3.** $\frac{5}{11}$ **4.** $\frac{5}{1}$ **5.** $\frac{1}{7}$

2

6. $\frac{9}{10}$ **7.** $\frac{11}{12}$ **8.** 2 **9.** $1\frac{1}{20}$ **10.** $2\frac{1}{12}$ **11.** $4\frac{2}{3}$ **12.** $\frac{2}{5}$ **13.** 8

3

14. $13\frac{1}{3}$ **15.** 32 **16.** $\frac{5}{8}$ **17.** $14\frac{3}{4}$ **18.** 7 **19.** 18 2 × 4's

20. 10 lengths **21.** 12 shovels **22.** $23\frac{11}{12}$ in. **23.** 20 ft × 15 ft

24. 4 whole pieces   **25.** 7 strips   **26.** 23 straws, $3\frac{3}{4}$ in. left

27. 8 pieces

4

28. $\frac{1}{10}$ **29.** 3   **30.** $5\frac{3}{5}$ **31.** $\frac{12}{25}$ **32.** $\frac{1}{3}$ **33.** 5   **34.** 28   **35.** $\frac{2}{9}$

36. 3   **37.** $\frac{1}{3}$ **38.** $\frac{1}{6}$ **39.** $\frac{5}{6}$ **40.** $\frac{2}{3}$

Self-Study Exercises 3–12

1

1. $-\frac{-5}{8}, -\frac{5}{-8}, \frac{-5}{-8}$ **2.** $-\frac{-3}{-4}, \frac{-3}{4}, \frac{3}{-4}$ **3.** $-\frac{2}{-5}, -\frac{-2}{5}, \frac{2}{5}$

4. $-\frac{7}{8}, \frac{-7}{8}, \frac{7}{-8}$ **5.** $-\frac{-7}{8}, -\frac{7}{-8}, \frac{-7}{-8}$

2

6. $-\frac{2}{8}$ or $-\frac{1}{4}$ **7.** $-1\frac{1}{10}$ **8.** $1\frac{1}{10}$ **9.** $\frac{6}{11}$ **10.** $-\frac{25}{32}$

3

11. -26.297 **12.** -1.11 **13.** -91.44 **14.** -110.72 **15.** -59.04
16. 340.71 **17.** $-27.7\overline{3}$ **18.** 0.413 **19.** -26.6 **20.** 7.73

Self-Study Exercises 3–13

1

1. $1\frac{5}{24}$ **2.** $-2\frac{16}{35}$ **3.** $14\frac{59}{96}$ **4.** $1\frac{3}{4}$ **5.** $\frac{4}{5}$ **6.** $\frac{1}{2}$ **7.** $-\frac{1}{24}$

8. $282\frac{1}{10}$ **9.** $11\frac{7}{40}$ **10.** $-1\frac{5}{16}$

Self-Study Exercises 3–14

1

1. 40%   **2.** 70%   **3.** $62\frac{1}{2}\%$ **4.** $77\frac{7}{9}\%$ **5.** $\frac{7}{10}\%$ **6.** $\frac{2}{7}\%$ **7.** 20%

8. 14%   **9.** 0.7%   **10.** 1.25%   **11.** 500%   **12.** 800%   **13.** $133\frac{1}{3}\%$
14. 350%   **15.** 430%   **16.** 220%   **17.** 305%   **18.** 720%
19. 1510%   **20.** 3625%

2

21. $\frac{9}{25}$, 0.36   **22.** $\frac{9}{20}$, 0.45   **23.** $\frac{1}{5}$, 0.20   **24.** $\frac{3}{4}$, 0.75   **25.** $\frac{1}{16}$, 0.0625

26. $\frac{5}{8}$, 0.625   **27.** $\frac{2}{3}$, $0.66\frac{2}{3}$ **28.** $\frac{3}{500}$, 0.006   **29.** $\frac{1}{500}$, 0.002

30. $\frac{1}{2000}$, 0.0005   **31.** $\frac{1}{12}$, $0.08\frac{1}{3}$ **32.** $\frac{3}{16}$, 0.1875   **33.** 8   **34.** 4

35. $2\frac{1}{5}$, 2.5 **36.** $4\frac{1}{4}$, 4.25 **37.** $1\frac{19}{25}$, 1.76 **38.** $3\frac{4}{5}$, 3.8 **39.** $1\frac{3}{8}$, 1.375

40. $3\frac{7}{8}$, 3.875 **41.** $1\frac{2}{3}$ **42.** $3\frac{1}{6}$ **43.** 1.153 **44.** 2.125 **45.** 1.0625

46. $\frac{1}{10}$, 0.1 **47.** 25%, 0.25 **48.** 20%, $\frac{1}{5}$ **49.** $33\frac{1}{3}$%, $0.33\frac{1}{3}$

50. $\frac{1}{2}$, 0.5 **51.** 80%, 0.8 **52.** 75%, $\frac{3}{4}$ **53.** $\frac{2}{3}$, $0.66\frac{2}{3}$ **54.** 100%, $\frac{1}{1}$

55. 30%, 0.3 **56.** $\frac{2}{5}$, 0.4 **57.** 70%, $\frac{7}{10}$ **58.** 90%, 0.9 **59.** $\frac{3}{5}$, 0.6

60. 100%, 1 **61.** 25%, $\frac{1}{4}$ **62.** $66\frac{2}{3}$%, $0.66\frac{2}{3}$ **63.** $\frac{7}{10}$, 0.7 **64.** 50%, $\frac{1}{2}$

65. 60%, 0.6 **66.** 10%, $\frac{1}{10}$ **67.** $\frac{1}{5}$, 0.2 **68.** $33\frac{1}{3}$%, $\frac{1}{3}$ **69.** $66\frac{2}{3}$%, $\frac{2}{3}$

70. 75%, 0.75 **71.** $\frac{3}{10}$, 0.3 **72.** 90%, $\frac{9}{10}$ **73.** 20%, 0.2 **74.** 60%, $\frac{3}{5}$

75. $\frac{1}{1}$, 1

Assignment Exercises, Chapter 3

1. $\frac{3}{8}$ **3.** $\frac{1}{3}$ **5.** $\frac{13}{24}$

7. (a) 9 (b) 4 (c) 4 (d) 9 (e) 9, 4 (f) 4 (g) 9 (h) improper (i) >1 (j) 9, 4

9. f **11.** d **13.** e **15.** d **17.** $\frac{87}{100}$ **19.** $2\frac{3}{100}$ **21.** 4, 6, 8, 10, 12

23. 42, 63, 84, 105, 126 **25.** 14, 21, 28, 35, 42 **27.** 16, 24, 32, 40, 48
29. yes; $153 \div 3 = 51$ **31.** no, remainder of 2
33. $1 \times 48, 2 \times 24, 3 \times 16, 4 \times 12, 6 \times 8$; 1, 2, 3, 4, 6, 8, 12, 16, 24, 48
35. $1 \times 51, 3 \times 17$, 1, 3, 17, 51 **37.** $2 \cdot 2 \cdot 11$ or $2^2 \cdot 11$
39. $2 \cdot 2 \cdot 2 \cdot 3 \cdot 3 \cdot 3$ or $2^3 \cdot 3^3$ **41.** 360 **43.** 180 **45.** 2 **47.** 6

49. $\frac{15}{24}$ **51.** $\frac{25}{60}$ **53.** $\frac{10}{15}$ **55.** $\frac{24}{32}$ **57.** $\frac{11}{55}$ **59.** $\frac{1}{2}$ **61.** $\frac{1}{8}$ **63.** $\frac{1}{4}$

65. $\frac{17}{32}$ **67.** $\frac{3}{8}$ **69.** $\frac{3}{4}$ **71.** $\frac{7}{10}$ **73.** $\frac{19}{20}$ **75.** $\frac{109}{125}$ **77.** $\frac{1}{50}$

79. 0.2 **81.** 0.625 **83.** $0.\overline{81}$ or 0.8181 . . . **85.** $3\frac{3}{5}$ **87.** $4\frac{7}{8}$

89. $5\frac{3}{8}$ **91.** $87\frac{1}{2}$ **93.** $\frac{8}{1}$ **95.** $\frac{57}{8}$ **97.** $\frac{147}{16}$ **99.** $\frac{23}{5}$ **101.** $\frac{12}{1}$
103. 20 **105.** 33 **107.** 60 **109.** 16 **111.** 60 **113.** larger

115. no **117.** smaller **119.** $\frac{3}{8}$ **121.** $\frac{3}{16}$ **123.** $\frac{27}{32}$ **125.** $\frac{9}{19}$

127. $\frac{21}{64}$ **129.** $1\frac{13}{30}$ **131.** $8\frac{19}{32}$ **133.** $10\frac{9}{32}$ **135.** $18\frac{1}{16}$ in.

137. $12\frac{19}{32}$ in. **139.** $15\frac{25}{32}$ in. **141.** $4\frac{3}{4}$ in. **143.** $\frac{7}{8}$ in. **145.** $\frac{1}{3}$

147. $1\frac{5}{8}$ **149.** $5\frac{31}{32}$ **151.** $7\frac{11}{16}$ **153.** $\frac{7}{16}$ in. **155.** $1\frac{29}{64}$ in. **157.** $\frac{7}{24}$

159. $\frac{7}{24}$ **161.** $\frac{1}{2}$ **163.** $7\frac{7}{8}$ **165.** $1\frac{1}{5}$ **167.** $100\frac{1}{2}$ in. **169.** $2\frac{3}{4}$ cups

171. $1\frac{1}{6}$ **173.** $9\frac{1}{3}$ **175.** 24   **177.** 2   **179.** $\frac{3}{5}$ **181.** $2\frac{55}{64}$ in.

183. $\frac{3}{16}$ yd   **185.** $\frac{1}{18}$ **187.** $5\frac{1}{3}$ **189.** $\frac{1}{4}$ **191.** $\frac{1}{8}$ **193.** $-\frac{-3}{-8}$,

$+\frac{-3}{8}, \frac{3}{-8}$ **195.** $\frac{7}{8}, -\frac{-7}{8}, \frac{-7}{-8}$ **197.** $\frac{62}{63}$ **199.** $1\frac{5}{7}$ **201.** 0.0351

203. -62 **205.** $-1\frac{7}{24}$ **207.** $-11\frac{13}{16}$ **209.** 70%   **211.** 12,500%

213. 1730%   **215.** 0.72 or $\frac{18}{25}$ **217.** 0.125 or $\frac{1}{8}$ **219.** $0.0066\frac{2}{3}$ or $\frac{1}{150}$

221. 2.75 or $2\frac{3}{4}$ **223.** 1.125 or $1\frac{1}{8}$ **225.** 2.272   **227.** 0.09275

229. 3.4

Trial Test, Chapter 3

1. $\frac{3}{4}$ **3.** 3   **5.** $\frac{34}{7}$ **7.** $2^5 \cdot 3$ **9.** $\frac{1}{4}$ **11.** $3\frac{8}{9}$ **13.** $13\frac{1}{2}$ **15.** $1\frac{5}{12}$

17. $7\frac{13}{14}$ **19.** $\frac{1}{9}$ **21.** $5\frac{1}{10}$ **23.** $\frac{5}{16}$ **25.** 60%   **27.** $\frac{2}{7}$ **29.** $3\frac{5}{6}$ cups

31. 8 yd

Self-Study Exercises 4–1

1

1. $x = 5$ **2.** $b = 4$ **3.** $a = 9$ **4.** $c = 10$ **5.** $x = 4$ **6.** $a = 8$
7. $x = 6$ **8.** $b = 18$ **9.** $d = 1.5$ **10.** $x = 3.33$ **11.** $c = .8889$
12. $b = 0.75$

2

13. R missing; $B = 10$; $P = 2$ **14.** $P = 2$; $R = 20\%$; B missing
15. $R = 20\%$; $B = 10$; P missing   **16.** $P = 3$; R missing; $B = 4$
17. $R = 15\%$; B missing; $P = \$9$ **18.** R missing; $B = 25$; $P = 5$
19. $P = 6$; $B = 15$; R missing   **20.** P missing; $R = 20\%$; $B = 15$
21. $R = 35\%$; B missing; $P = 70$ **22.** R missing; $B = \$45$; $P = \$3.15$

3

23. 3.75   **24.** 0.625   **25.** $66\frac{2}{3}\%$ **26.** $37\frac{1}{2}\%$ **27.** 14.25   **28.** 350
29. 9.375   **30.** 220   **31.** 20%   **32.** 200

Self-Study Exercises 4–2

1

1. 75   **2.** 63   **3.** 206   **4.** 115.92   **5.** 0.675   **6.** 0.94   **7.** 155.10
8. 231   **9.** 924   **10.** 345

2

11. 25%   **12.** 30%   **13.** $33\frac{1}{3}\%$ **14.** 32.9%   **15.** 15.75%   **16.** 16%

17. 0.8%   **18.** $0.66\frac{2}{3}\%$ **19.** 500%   **20.** 111.25%

3

21. 72   **22.** 50   **23.** 344   **24.** 46   **25.** 360   **26.** 275   **27.** 75
28. 250   **29.** 18.4   **30.** 261

4

31. 1.0625 lb   **32.** 3%   **33.** 200 hp   **34.** 0.021 lb   **35.** 5%
36. $575   **37.** 1200 lb   **38.** 11%   **39.** 9625 parts   **40.** 100,880 welds

Self-Study Exercises 4–3

1

1. 108   **2.** 109.2   **3.** 10.2   **4.** 33.75

2

5. 27 in.   **6.** 1815 board feet   **7.** 2393 board feet   **8.** 25,908 bricks
9. $16,802.50   **10.** 1366.4 yd^3

3

11. 7%   **12.** 45%   **13.** 20%   **14.** 5%   **15.** 350 hp   **16.** 10.5 yd^3
17. 3019 board feet   **18.** 28.5 lb   **19.** 40 in.   **20.** 20%   **21.** 150 hp

Self-Study Exercises 4–4

1

1. $4.55, $80.38   **2.** $829.06   **3.** $23.30   **4.** 25%   **5.** 23%

2

6. $17.52   **7.** $3866.00   **8.** $944.50   **9.** $434.84   **10.** $3205.75

3

11. $9.75   **12.** $300   **13.** 1.75%   **14.** $367.50   **15.** $1752

Assignment Exercises, Chapter 4

1. 4.5   **3.** 48   **5.** 1.33   **7.** 6   **9.** 2.5   **11.** 0.67
13. R missing; $B = 25$; $P = 5$ **15.** $R = 5\%$; $B = 180$; P missing
17. $R = 45\%$; B missing; $P = \$36$ **19.** $P = 6$; R missing; $B = 25$
21. $R = 18\%$; $B = 150$; P missing   **23.** 24   **25.** 0.4375   **27.** 60%
29. 250%   **31.** 83   **33.** 152   **35.** 15.3%   **37.** 500%   **39.** 84
41. 37.5%   **43.** 1.14%   **45.** $266   **47.** 4%   **49.** 200 students
51. 10%   **53.** 7%   **55.** 14%   **57.** 142.1 kg   **59.** 62 cm, 63 cm
61. $1.69   **63.** 12.5%   **65.** 57.5 lb   **67.** $33\frac{1}{3}\%$ **69.** 4.8%
71. 289 hp   **73.** $439.84   **75.** $365.66   **77.** $1707.60   **79.** 1.75%
81. $11,019.75   **83.** $0.89   **85.** 8.5%   **87.** $369.69   **89.** $355
91. 26.6%   **93.** 28.6%

Trial Test, Chapter 4

1. 8   **3.** 12   **5.** $R = 40\%$; $B = 10$; P missing   **7.** $P = 9$; R missing;
$B = 27$ **9.** $R = 12\%$; $B = 50$; P missing   **11.** 9   **13.** 305   **15.** 115
17. 67.10   **19.** $525   **21.** $2500   **23.** 21.14%   **25.** 13.17%
27. 5%   **29.** $104.87

Self-Study Exercises 5–1

1

1. lb   **2.** oz   **3.** lb or oz   **4.** gal   **5.** qt   **6.** T   **7.** lb   **8.** qt
9. ft and in.   **10.** c   **11.** in.   **12.** yd   **13.** mi   **14.** oz   **15.** oz
16. mi   **17.** lb or oz   **18.** lb   **19.** lb and oz   **20.** oz

2

21. 48 in.   **22.** 21 ft   **23.** 4400 yd   **24.** $9\frac{1}{3}$ yd   **25.** 2 mi

26. 48 oz   **27.** $2\frac{3}{10}$ or 2.3 lb   **28.** 732.8 oz   **29.** 19.2 oz

30. 408 oz, 25.5 lb   **31.** 20 qt   **32.** 13 pt   **33.** $1\frac{1}{2}$ gal   **34.** 60 pt

35. 9 gal   **36.** 256 oz   **37.** $1\frac{1}{2}$ or 1.5 yd   **38.** 108 in.   **39.** 7040 ft

40. 24 c

3

41. 3 ft 8 in.   **42.** 2 mi 1095 ft   **43.** 3 lb $3\frac{1}{2}$ oz   **44.** 2 gal 1 qt

45. 2 gal   **46.** 2 T 500 lb   **47.** 2 yd 2 ft 11 in.   **48.** 2 qt 2 c 2 oz
49. 2 ft 10 in.   **50.** 6 lb 9 oz   **51.** 4 gal 2 qt 16 oz or 4 gal 2 qt 1pt
52. 6 qt 20 oz or 1 gal 2 qt 1 pt 4 oz   **53.** 2 mi 5 yd 2 ft 1 in.
54. 5 ft 4 in. or 1 yd 2 ft 4 in.   **55.** 3 T 600 lb 15 oz   **56.** 4 lb 5 oz
57. 2 lb 5 oz   **58.** 2 yd 2 ft 4 in.   **59.** 2 yd 2 ft 11 in.
60. 1 mi 875 ft 6 in.

Self-Study Exercises 5–2

1

1. 2 lb 5 oz   **2.** 4 ft 7 in.   **3.** 15 lb 11 oz   **4.** 18 ft 3 in.

5. 8 qt $\frac{1}{2}$ pt or 2 gal 1 c   **6.** 14 gal 1 qt   **7.** 9 yd 1 ft   **8.** 5 c 1 oz or
2 pt 1 c   **9.** 5 ft 4 in. or 1 yd 2 ft 4 in.   **10.** 7 yd 1 ft 5 in.
11. 2 ft 7 in.   **12.** 11 ft or 3 yd 2 ft   **13.** 5 lb 15 oz   **14.** 12 lb

2

15. 6 in.   **16.** 1 pt   **17.** 2 ft   **18.** 4 lb 14 oz   **19.** 1 lb 6 oz
20. 6 lb 11 oz   **21.** 11 in.   **22.** 10 lb 13 oz   **23.** 2 in.

24. 3 gal 3 qt $1\frac{1}{2}$ pt or 3 gal 3 qt 1 pt 1 c   **25.** 4 ft 2 in.   **26.** 2 ft 3 in.

27. 45 sec   **28.** 39 sec   **29.** 69 lb 7 oz   **30.** 16 lb 14 oz

Self-Study Exercises 5–3

1

1. 60 mi   **2.** 108 gal   **3.** 252 lb   **4.** 14 qt or 3 gal 2 qt
5. 57 lb 8 oz   **6.** 38 gal 2 qt   **7.** 43 gal 3 qt   **8.** 58 ft or 19 yd 1 ft
9. 36 lb   **10.** 7 qt 1 pt or 1 gal 3 qt 1 pt

2

11. 35 in.^{2} 12. 108 ft^2 13. 180 yd^2 14. 108 mi^2 15. 378 tiles
16. 47 gal 2 oz   17. 7 qt 1 pt or 1 gal 3 qt 1 pt   18. 6 lb 4 oz
19. 7 lb 8 oz

3

20. 6 gal   21. 1 day 15 hr   22. 10 yd 1 ft 3 in.   23. 1 yd 1 ft 7 in.
24. 3 qt $\frac{1}{2}$ pt or 3 qt 1 c   25. 6$\frac{2}{3}$ gal   26. 1 hr 40 min   27. 5$\frac{1}{4}$ qt or
5 qt 1 c   28. 7 ft 6 in.   29. 3 gal 2 qt   30. 9 pieces   31. 5 ft
32. 3 gal 1 qt 5 oz   33. 24 lb 3 oz

4

34. 3   35. 17   36. 8   37. 18   38. 3   39. 8 pieces   40. 9 boxes
41. 24 cans   42. 9 tickets   43. 9 pieces

5

44. 15$\frac{\text{gal}}{\text{sec}}$ 45. $\frac{3}{4}\frac{\text{lb}}{\text{min}}$ 46. 15,840$\frac{\text{ft}}{\text{hr}}$ 47. 2304$\frac{\text{oz}}{\text{min}}$ 48. 2$\frac{\text{qt}}{\text{sec}}$

49. 20$\frac{\text{oz}}{\text{hr}}$ 50. 40$\frac{\text{ft}}{\text{min}}$ 51. 440$\frac{\text{ft}}{\text{sec}}$ 52. 44$\frac{\text{ft}}{\text{sec}}$ 53. 3$\frac{\text{qt}}{\text{min}}$

54. 53$\frac{1}{3}\frac{\text{lb}}{\text{min}}$ 55. $\frac{5}{6}\frac{\text{gal}}{\text{sec}}$

Self-Study Exercises 5–4

1

1. (a) 1000 meters (b) 10 liters (c) $\frac{1}{10}$ of a gram (d) $\frac{1}{1000}$ of a meter

(e) 100 grams (f) $\frac{1}{100}$ of a liter   2. b   3. b   4. c   5. b   6. a

7. d   8. b   9. c   10. a   11. b   12. c   13. b   14. a   15. b
16. b

2

17. 40   18. 70   19. 580   20. 80   21. 2.5   22. 210   23. 85
24. 142   25. 153 mL   26. 460 m   27. 75 dkg   28. 160 mm
29. 400   30. 8000   31. 58,000   32. 800   33. 250   34. 2100
35. 102,500   36. 8330   37. 2,000,000   38. 70   39. 236 L
40. 467 cm   41. 38,000 dg   42. 13,000 cm   43. 2.8   44. 23.8
45. 10.1   46. 6   47. 2.9   48. 19.25   49. 1.7   50. 438.9 dm
51. 4.7 g   52. 0.225 dL   53. 2.743   54. 0.385   55. 0.15   56. 0.08
57. 2964.84   58. 0.2983   59. 0.0003   60. 0.004   61. 0.002857
62. 15.285   63. 0.0297 hm   64. 0.00003 L

3

65. 11 m   66. 12 hL   67. 6 cg   68. 2.4 dm or 24 cm
69. 5.9 cL or 59 mL   70. cannot add   71. 10.1 kL or 101 hL
72. 0.55 g or 55 cg   73. cannot subtract   74. 7.002 km or 7002 m
75. 1000 mL   76. 1.47 kL or 147 dkL   77. 516 m   78. 40.8 m
79. 150.96 dm   80. 969.5 m   81. 13 m   82. 9 cL   83. 163 g
84. 0.4 m or 4 dm   85. 5   86. 30   87. 16 prescriptions
88. 80 containers   89. 5.74 dL or 57.4 cL   90. 2.3 dkm or 23 m
91. 165.7 hm or 16.57 km   92. 9.1 kL or 91 hL   93. 1.25 cL

94. 70 mm **95.** 7.5 dm **96.** 50 mL **97.** 21,250 containers
98. 19 vials

Self-Study Exercises 5–5

$\boxed{1}$

1. 354.33 in. **2.** 130.8 yd **3.** 26.04 mi **4.** 6.36 liq qt **5.** 11 L
6. 59.4 lb **7.** 22.5 kg **8.** 17.78 cm **9.** 5.4 m **10.** 1.32288 oz
11. 21.85 L **12.** 30 m **13.** 27 kg **14.** 241.5 km **15.** 32.7 yd
16. 235.6 mi **17.** 7.44 mi **18.** 13.2 L **19.** 328 ft

Self-Study Exercises 5–6

$\boxed{1}$

1. $4\frac{9}{16}$ in. **2.** $4\frac{1}{16}$ in. **3.** $3\frac{13}{16}$ in. **4.** $3\frac{3}{8}$ in. **5.** $2\frac{1}{4}$ in. **6.** 2 in.

7. $1\frac{3}{4}$ in. **8.** $1\frac{3}{16}$ in. **9.** $\frac{3}{4}$ in. **10.** $\frac{3}{8}$ in.

$\boxed{2}$

11. 115 mm or 11.5 cm **12.** 102 mm or 1.02 cm **13.** 96 mm or 9.6 cm
14. 85 mm or 8.5 cm **15.** 57 mm or 5.7 cm **16.** 50 mm or 5 cm
17. 44 mm or 4.4 cm **18.** 30 mm or 3 cm **19.** 19 mm or 1.9 cm
20. 10 mm or 1 cm

Assignment Exercises, Chapter 5

1. lb or oz **3.** qt **5.** lb or oz **7.** T **9.** qt **11.** c **13.** yd

15. oz **17.** 4 yd **19.** 6336 ft **21.** 80 oz **23.** $42\frac{1}{2}$ lb

25. 304 oz or 19 lb **27.** 15 pt **29.** 24 pt **31.** 384,000 oz

33. 6600 ft **35.** 7 ft 5 in. **37.** 13 lb $1\frac{1}{2}$ oz **39.** 2 gal 1 pt

41. 4 yd 4 in. **43.** 4 ft 1 in. or 1 yd 1 ft 1 in. **45.** 2 gal 16 oz or
2 gal 1 pt **47.** 3 ft 11 in. **49.** 10 ft 11 in. or 3 yd 1 ft 11 in.
51. 8 gal 2 qt **53.** 14 oz **55.** 1 ft 9 in. **57.** 1 ft 5 in. **59.** 504 ft^2

61. 63 in.^{2} **63.** 10 yd 1 ft 3 in. **65.** $5\frac{5}{12}$ ft **67.** 4 lb 8 oz **69.** 3.5

71. $4\frac{4}{9}$ **73.** $300\frac{mi}{hr}$ **75.** $60\frac{mi}{hr}$ **77.** $1.25\frac{gal}{min}$ **79.** $1\frac{1}{2}$ qt **81.** kilo-

83. milli- **85.** centi- **87.** 10 times **89.** $\frac{1}{1000}$ of **91.** 1000 times

93. a **95.** a **97.** c **99.** b **101.** 6.71 dkm **103.** 2300 mm
105. 12,300 mm **107.** 230,000 mm **109.** 413.27 km **111.** 3.945 hg
113. 30.00974 kg **115.** Cannot add unlike measures.
117. 748 cg or 7.48 g **119.** 61.47 cg **121.** 15 **123.** 8.5 hL
125. 18.9 m **127.** 245 mL **129.** 6 m **131.** 100 servings **133.** 40
shirts **135.** 234.35 yd **137.** 15.9 liq qt **139.** 70.4 lb **141.** 22.86 cm

143. 156.88 qt **145.** 60 m **147.** 281.75 km **149.** $5\frac{1}{4}$ in.

151. $4\frac{7}{16}$ in. **153.** $3\frac{15}{16}$ in. **155.** $3\frac{9}{16}$ in. **157.** $2\frac{3}{4}$ in.

159. 117 mm or 118 mm   **161.** 99 mm   **163.** 60 mm   **165.** 45 mm
167. 20 mm   **169.** Answers will vary.

Trial Test, Chapter 5

1. 36 in.   **3.** 8 gal   **5.** $4\dfrac{\text{qt}}{\text{sec}}$ **7.** 165 yd   **9.** $80\dfrac{2}{3}\dfrac{\text{ft}}{\text{sec}}$ **11.** $1\dfrac{1}{4}$ in.
13. 2 gal 1 qt 4 oz   **15.** deci-   **17.** 0.298 km   **19.** 9.48 L or 94.8 dL
21. 120.75 km   **23.** 8.48 pt   **25.** $735\dfrac{\text{m}}{\text{sec}}$

Self-Study Exercises 6–1

1

1. $P = 12$ cm   **2.** square   **3.** 193 ft^2 **4.** 590 ft   **5.** 900 yd^2
    $A = 9$ cm^2
6. 16 mi^2 **7.** 5268 ft   **8.** 81 tiles   **9.** 36 tiles
10. perimeter $= 80$ in.
        area $= 400$ in.^{2}

2

11. $P = 10$ ft   **12.** rectangle   **13.** 42,500 ft^2 **14.** 180 ft^2 **15.** 241 ft^2
    $A = 6$ ft^2
16. 59 ft   **17.** 142.5 board feet   **18.** 156 in. or 13 ft   **19.** 17 ft
20. 130 in.

3

21. $P = 38$ in.   **22.** parallelogram   **23.** 156 in. or 13 ft   **24.** 9000 ft^2
    $A = 72$ in.^{2}
25. 124 in.   **26.** 18 tiles   **27.** 1111 parallelograms   **28.** 160 in.
29. 108 in.   **30.** 9 signs

Self-Study Exercises 6–2

1

1. 25.1 cm   **2.** 18.8 in.   **3.** 9.4 ft   **4.** 17.3 m

2

5. 0.6 m^2 **6.** 2.1 ft^2 **7.** 14.6 cm^2 **8.** 1.0 in.^{2}

3

9. 592.7 ft^2 **10.** 268.27 in.   **11.** 1178.1 mm^2 **12.** 168.5 mm^2
13. Yes, the cross-section area of the third pipe is larger than the combined area of the other two pipes.
14. Yes, the combined cross-sectional area of the two pipes is 25.1 in.^{2} which is greater than 20 in.^{2}—the area of the large pipe.
15. 2827.4 ft/min   **16.** 1570.8 ft/min   **17.** 78.5 ft^2 **18.** 0.448 in.
19. 4.4 m

Assignment Exercises, Chapter 6

1. $P = 57$ cm   **3.** $P = 210$ mm   **5.** 836 ft   **7.** 33 yd^2 **9.** $14.25
    $A = 162$ cm^2 $A = 2450$ mm^2
11. 3 rolls   **13.** 235 ft^2 **15.** $C = 25.13$ m   **17.** $C = 17.42$ cm
                    $A = 50.27$ m^2 $A = 12$ cm^2

19. 5.8 m^2 **21.** 2.4 in.   **23.** 33 ft/min   **25.** 15.0 in.   **27.** 1.4 yd^2
29. 59.1 in.   **31.** 66 in. diameter

Trial Test, Chapter 6

1. P = 91 ft   **3.** P = 12.2 in.   **5.** $52   **7.** 282 ft^2 **9.** center
    A = 514.5 ft^2 A = 6.08 in.^{2}
11. radius   **13.** 37.70 in.   **15.** 385.62 ft   **17.** 0.29 in.^{2} **19.** 0.15 in.^{2}

Self-Study Exercises 7–1

1

1. 8.6%   **2.** 35.5%   **3.** 56.3%

2

4. Debt retirement   **5.** Misc. expenses and General government
6. Social projects and Education costs

3

7. 5 Amps   **8.** 50 volts   **9.** 35 volts   **10.** 25 ohms

Self-Study Exercises 7–2

1

1. 10   **2.** 2   **3.** $\frac{1}{3}$ **4.** $\frac{1}{7}$ **5.** 12%   **6.** 28%   **7.** 35–37 and 38–40
8. 20–22 and 23–25   **9.** 7   **10.** 16

	Midpoint	Tally	Class Frequency
11.	15	‖	2
12.	12	‖‖	4
13.	9	𝍩 ‖	7
14.	6	𝍩 𝍩 𝍩 𝍩	20
	Midpoint	Tally	Class Frequency
15.	93	‖	2
16.	88	𝍩	5
17.	83	𝍩 ‖‖	8
18.	78	𝍩 ‖‖‖	9
19.	73	‖‖	3
20.	68	𝍩 ‖	6
21.	63	‖	2
22.	58	𝍩	5

2

23.

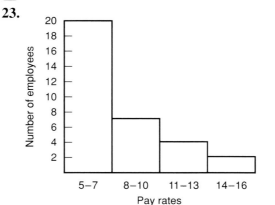

24.

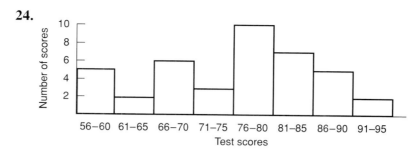

25.

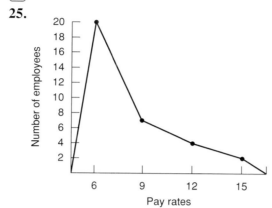

26.

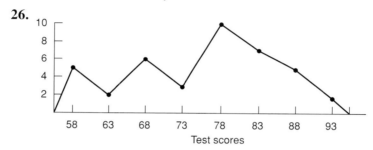

Self-Study Exercises 7–3

1

1. 16 **2.** 17 **3.** 66 **4.** 73.75 **5.** 33.7 **6.** 66.6 **7.** 42.33°F
8. 12.67°C **9.** $34.80 **10.** $39.20 **11.** 14.67 in. **12.** 8 in.
13. 20 **14.** 76 **15.** 17.3 runs **16.** 27.2 cars **17.** 24.6 mpg
18. 10.2 mpg

2

19. 44 **20.** 43 **21.** 15.5 **22.** 26.5 **23.** $30 **24.** $66
25. $5.25 **26.** $5.85 **27.** 2 **28.** 5 **29.** no mode **30.** no mode
31. $67 **32.** $32 **33.** 4 **34.** $1.97 **35.** 24 **36.** 23 **37.** 13
38. 18 **39.** $25 **40.** $33 **41.** 48°F

Self-Study Exercises 7–4

1

1. Keaton Brienne Renee
Keaton Renee Brienne
Brienne Keaton Renee
Brienne Renee Keaton
Renee Keaton Brienne

2. ABCD   BACD
ABDC   BADC
ACBD   BCAD
ADBC   BDAC
ACDB   BCDA

Renee Brienne Keaton
$3 \cdot 2 \cdot 1 = 6$ ways

ADCB	BDCA
CABD	DABC
CADB	DACB
CBAD	DBAC
CDAB	DCAB
CBDA	DBCA
CDBA	DCBA

24 ways
$4 \cdot 3 \cdot 2 \cdot 1 = 24$ ways

3. $4 \cdot 3 \cdot 2 \cdot 1 = 24$ ways   **4.** $5 \cdot 4 \cdot 3 \cdot 2 \cdot 1 = 120$ ways
5. $5 \cdot 4 \cdot 3 \cdot 2 \cdot 1 = 120$ ways

2

6. $\dfrac{1}{24}, \dfrac{1}{23}$ **7.** $\dfrac{1}{4}$ **8.** $\dfrac{11}{48}$ **9.** $\dfrac{2}{5}$ **10.** $\dfrac{1}{3}$

Assignment Exercises, Chapter 7

1. 1992, 1994, 1995   **3.** 1993, 1996, 1997   **5.** 12.9%   **7.** 17.4%

9. 7-10-97 @ 4:00 p.m.   **11.** 50   **13.** 110   **15.** $\dfrac{1}{2}$

	Midpoint	Tally	Class Frequency
17.	60.5	ⅲ ⅲ	10
19.	40.5	ⅲ ⅲ ‖	12
21.	20.5	ⅲ ‖	7

23.

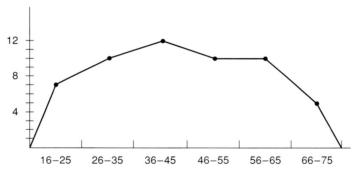

25. 36–45   **27.** 15   **29.** $\dfrac{1}{2}$ **31.** $\dfrac{10}{54} = 18.5\%$

33.

miles per gallon	Midpoint	Tally	MPG Frequency	(Answers
20–24	22	ⅲ ‖	7	may vary)
25–29	27	ⅲ ∣	6	
30–34	32	‖‖	4	

35.

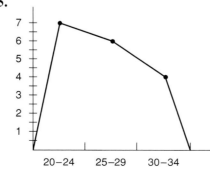

37. 12.6 cars   **39.** 13.8 mpg   **41.** $6.03   **43.** $1.85   **45.** 3.67
47. 63   **49.** 24   **51.** $\frac{10}{13}, \frac{3}{4}$ **53.** 8   **55.** $\frac{1}{5}$ **57.** $\frac{1}{4}$

Trial Test, Chapter 7

1. bar   **3.** line   **5.** 2 degrees   **7.** $65   **9.** 25%   **11.** $\frac{15}{200} = \frac{3}{40}$

13. English dept., Electronic dept.   **15.** 108.3%   **17.** $\frac{18}{20} = \frac{9}{10}$ **19.** 6

21. $\frac{1}{2}$

	Midpoint	Tally	Class Frequency
23.	5	ⅧⅢ ⅢⅢ	9

25. 14   **27.** cannot be determined   **29.** mean: 77.9   **31.** 6   **33.** $\frac{3}{5}$
median: 78
mode: 81

Self-Study Exercises 8–1

1

1. $\frac{1}{10^8}, \frac{1}{100,000,000}$ **2.** $\frac{1}{7^3}, \frac{1}{343}$ **3.** $\frac{1}{2^4}, \frac{1}{16}$ **4.** $\frac{1}{6^2}, \frac{1}{36}$ **5.** $\frac{1}{4^3}, \frac{1}{64}$
6. $\frac{1}{5^1}, 5^{-1}$ **7.** $\frac{1}{5^2}, 5^{-2}$ **8.** $\frac{1}{7^2}, 7^{-2}$ **9.** $\frac{1}{3^3}, 3^{-3}$ **10.** $\frac{1}{8^2}, 8^{-2}$
11. 8 $\boxed{x^y}$ $\boxed{-}$ 2 $\boxed{=}$ ⇒ 0.015625
    1 $\boxed{\div}$ 8 $\boxed{x^2}$ $\boxed{=}$ ⇒ 0.015625
12. 5 $\boxed{x^y}$ $\boxed{-}$ 3 $\boxed{=}$ ⇒ 8^{-03} or 0.008
    1 $\boxed{\div}$ 5 $\boxed{x^y}$ 3 $\boxed{=}$ ⇒ 8^{-03} or 0.008
13. 12 $\boxed{x^y}$ $\boxed{-}$ 7 $\boxed{=}$ ⇒ 2.7908^{-08}
    1 $\boxed{\div}$ 12 $\boxed{x^y}$ 7 $\boxed{=}$ ⇒ 2.7908^{-08}
14. 15 $\boxed{x^y}$ $\boxed{-}$ 3 $\boxed{=}$ ⇒ 2.96
    1 $\boxed{\div}$ 15 $\boxed{x^y}$ 3 $\boxed{=}$ ⇒ 2.96
15. $\boxed{(}$ 1 $\boxed{\div}$ 3 $\boxed{)}$ $\boxed{x^y}$ $\boxed{-}$ 2 $\boxed{=}$ ⇒ 9
    3 $\boxed{x^y}$ 2 $\boxed{=}$ ⇒ 9

2

16. 6   **17.** 4   **18.** −6   **19.** −35   **20.** −3   **21.** 5   **22.** −16
23. −27   **24.** −37   **25.** −60   **26.** −21°   **27.** −$61   **28.** −9
29. −10   **30.** −15   **31.** 8   **32.** $\frac{5}{16}$ **33.** 31.92   **34.** −34.84
35. −7   **36.** −6   **37.** 8   **38.** $\frac{5}{6}$ **39.** 1.26   **40.** −11.67

Self-Study Exercises 8–2

1

1. $36^{\frac{1}{2}} = 6$ **2.** $81^{\frac{1}{2}} = 9$ **3.** $64^{\frac{1}{3}} = 4$ **4.** $81^{\frac{1}{4}} = 3$ **5.** $48^{\frac{1}{7}} \approx 1.74$
6. $32^{\frac{1}{5}} = 2$ **7.** 14   **8.** 23   **9.** 14.6969   **10.** 13   **11.** 7

2

12. 2.6458   **13.** 2.8284   **14.** 1.7100   **15.** 3.4641

3

16. 2, 3   **17.** 2, 3   **18.** 6, 7   **19.** 12, 13   **20.** 1,889,568   **21.** 2401

22. 183   **23.** 1   **24.** 1   **25.** $\dfrac{1}{64}$ **26.** $\dfrac{1}{49}$ **27.** $\dfrac{1}{15}$ **28.** 3   **29.** 7

4

30. -23 **31.** 19   **32.** 50   **33.** 335   **34.** -5.944

Self-Study Exercises 8–3

1

1. $-15 = -15$ **2.** $11 = 11$ **3.** $2 = 2$ **4.** $-7 = -7$ **5.** $n = 11$
6. $m = 6$ **7.** $y = -4$ **8.** $x = 6$ **9.** $p = 9$ **10.** $b = -9$

11. $\boxed{7} + \boxed{c}$ **12.** $\boxed{4a} - \boxed{7}$ **13.** $\boxed{3x} - \boxed{2(x + 3)}$ **14.** $\boxed{\dfrac{a}{3}}$

15. $\boxed{7xy} + \boxed{3x} - \boxed{4} + \boxed{2(x + y)}$ **16.** $\boxed{14x} + \boxed{3}$ **17.** $\boxed{\dfrac{7}{(a + 5)}}$

18. $\boxed{\dfrac{4x}{7}} + \boxed{5}$ **19.** $\boxed{11x} - \boxed{5y} + \boxed{15xy}$ **20.** 5   **21.** -4 **22.** $\dfrac{1}{5}$

23. $\dfrac{2}{7}$ **24.** 6   **25.** $-\dfrac{4}{5}$ **26.** $-15c$ (Answers may vary.)

2

27. Four more than a number is seven. (Answers may vary.)
28. Five less than a number is two. (Answers may vary.)
29. Three times a number is fifteen. (Answers may vary.)
30. One more than three times a number is seven. (Answers may vary.)

3

31. $x + 5 = 12$ **32.** $\dfrac{x}{6} = 9$ **33.** $4(x - 3) = 12$ **34.** $4x - 3 = 12$

35. $12 + 7 + x = 17$ **36.** $2x + 7 = 21$ **37.** $x + 15° = 48°$

38. $x + 15 \text{ mL} = 45 \text{ mL}$ **39.** $x \cdot 5 = 45$ **40.** $\dfrac{18 - 3}{5} = x$

4

41. $-3a$ **42.** $-8x - 2y$ **43.** $7x - 2y$ **44.** $-3a + 6$ **45.** 10
46. $3a + 6b + 8c + 1$ **47.** $10x - 20$ **48.** $6x + 13$
49. $-4x + 5$ **50.** $-12a + 3$

Self-Study Exercises 8–4

1

1. $f(2) = 7, f(-3) = 2, f(0) = 5$

2. $f(-4) = -19, f(5) = 8, f\left(\dfrac{1}{6}\right) = -6\dfrac{1}{2}$

3. $f(-0.2) = -1.4$, $f\left(\dfrac{3}{4}\right) = 2.4$, $f\left(-\dfrac{3}{5}\right) = -3$

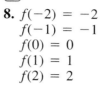

4. positive multiples of 5   **5.** integers
6. reciprocal of positive integers
7. integers less than 8

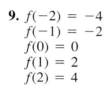

8. $f(-2) = -2$
$\quad\ f(-1) = -1$
$\quad\ f(0) = 0$
$\quad\ f(1) = 1$
$\quad\ f(2) = 2$

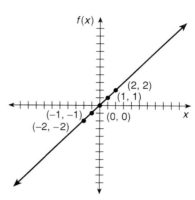

9. $f(-2) = -4$
$\quad\ f(-1) = -2$
$\quad\ f(0) = 0$
$\quad\ f(1) = 2$
$\quad\ f(2) = 4$

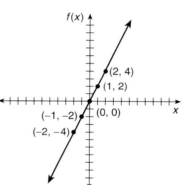

10. x is the number of days and is the independent variable. $f(x)$ is the cost of renting the wheelchair and is the dependent variable.
$f(x) = 12x + 20$

x	$f(x)$
0	20
1	32
2	44

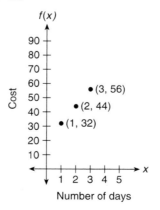

Assignment Exercises, Chapter 8

1. $\dfrac{1}{2^{-3}} = 2^3 = 8$ **3.** $\dfrac{1}{3^{-2}} = 3^2 = 9$ **5.** $\dfrac{1}{4^3}$, $\dfrac{1}{64}$ **7.** $\dfrac{1}{2^6}$, $\dfrac{1}{64}$ **9.** $\dfrac{1}{7^1}$, 7^{-1}

11. $\dfrac{1}{11^2}$, 11^{-2} **13.** 16   **15.** $-\dfrac{5}{18}$ **17.** $1\dfrac{1}{3}$ **19.** 7, 8   **21.** 11, 12

23. 12, 13   **25.** 3.317   **27.** 3.606

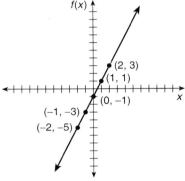

29. -38 **31.** $\boxed{15x} - \boxed{\dfrac{3x}{7}} + \boxed{\dfrac{x-7}{5}}$

33. Five more than a number is two.   **35.** A number divided by 8 is seven.

37. The product of seven more than a number times three is negative three.

39. $2x + 7 = 11$ **41.** $2(x + 8) = 40$ **43.** $f(-2)x = -5$
$$f(-1)x = -3$$
$$f(0)x = -1$$
$$f(1)x = 1$$
$$f(2)x = 3$$

45. For every increase of one unit in the independent variable, the dependent variable increases two units.   **47.**

49. $f(x) = 3x - 2$; $f(1) = 1$; $f(2) = 4$; additional data will vary

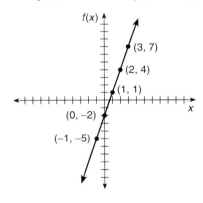

Trial Test, Chapter 8

1. $\dfrac{1}{12^2}$, $\dfrac{1}{144}$ **3.** $\dfrac{1}{10^1}$, 10^{-1}

5. To add signed numbers with unlike signs, subtract the absolute values and give the result the sign of the number having the larger absolute value.
$5 + (-7) = -2$

7. 12, 13   **9.** -287 **11.** $\boxed{5x^2}$ **13.** $x + 5 = 35$

15. Eight more than a number is twenty-one.   **17.** $f(-4) = -18$
$$f(-2) = -8$$
$$f(0) = 2$$
$$f(2) = 12$$
$$f(4) = 22$$

19.

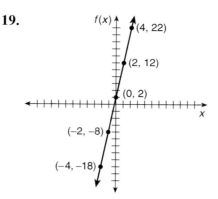

21. $f(h) = 4h - (h - 4)$ or $f(h) = 3h + 4$

Self-Study Exercises 9–1

1

1. $x = 8$ **2.** $x = 9$ **3.** $x = 5$ **4.** $n = 30$ **5.** $a = -9$

6. $b = -2$ **7.** $c = 7$ **8.** $x = -9$ **9.** $x = \dfrac{32}{5}$ **10.** $y = \dfrac{9}{2}$

11. $n = -\dfrac{5}{2}$ **12.** $b = \dfrac{28}{3}$ **13.** $x = -7$ **14.** $y = 2$ **15.** $x = 15$

16. $y = 8$ **17.** $\dfrac{10}{3}$ **18.** $-\dfrac{20}{3}$ **19.** $x = 12$ **20.** $n = 56$

2

21.–40. Each solution should check.

Self-Study Exercises 9–2

1

1. $x = 6$ **2.** $m = 7$ **3.** $a = -3$ **4.** $m = -\dfrac{1}{2}$ **5.** $y = 8$ **6.** $x = 0$

2

7. $b = -1$ **8.** $x = 8$ **9.** $t = 6$ **10.** $x = 3$ **11.** $y = 5$

12. $x = -\dfrac{7}{2}$ **13.** $a = -2$ **14.** $x = -9$ **15.** $x = \dfrac{1}{4}$ **16.** $x = -8$

17. $t = 7$ **18.** $x = -1$ **19.** $y = 11$ **20.** $y = -2$ **21.** $x = 27$

22. $y = 13$ **23.** $R = -68$ **24.** $P = \dfrac{5}{6}$ **25.** $x = 14$ **26.** $x = -\dfrac{4}{15}$

27. $m = \dfrac{49}{18}$ **28.** $s = \dfrac{8}{5}$ **29.** $m = \dfrac{8}{3}$ **30.** $T = \dfrac{68}{9}$ **31.** $x = 2$

32. $R = 0.8$ **33.** $x = 6.03$ **34.** $x = 0.08$ **35.** $R = 0.34$

Self-Study Exercises 9–3

1

1. $x = 3$ **2.** $y = -1$ **3.** $x = 1$ **4.** $t = -7$ **5.** $x = \dfrac{1}{3}$ **6.** $x = 6$

7. $a = -3$ **8.** $x = 2$ **9.** $x = \dfrac{1}{2}$ **10.** $b = \dfrac{20}{3}$

Self-Study Exercises 9–4

[1]

1. $x + 4 = 12$; $x = 8$ **2.** $2x - 4 = 6$; 5   **3.** $x + x + (x - 3) = 27$;
Two parts weigh 10 lb.
Third part weighs 7 lb.

4. $24 + x = 60$;   **5.** $4.03 - 3.97 = x$;   **6.** $x + 3x = 400$
36 gal   0.06 kg   100 ft of solid pipe
300 ft of perforated pipe

7. $x + 2x + \$70 = \$235$ **8.** $C = \$14(4)(150) = \$8400$
Spanish text: \$55
Calculator: \$110

9. $x + 2x = 325$ **10.** $A = \dfrac{1}{2}bh$; 12 ft

tank 1: $216\dfrac{2}{3}$ gal $60 = \dfrac{1}{2}(b)(10)$

tank 2: $108\dfrac{1}{3}$ gal

11. B14 = B5 + B6 + B7 + B8 + B9 + B10 + B11 + B12
D14 = D5 + D6 + D7 + D8 + D9 + D10 + D11 + D12
C5 = B5 ÷ B14 × 100
C6 = B6 ÷ B14 × 100
C7 = B7 ÷ B14 × 100
⋮
C12 = B12 ÷ B14 × 100
E5 = D5 ÷ D14 × 100
E6 = D6 ÷ D14 × 100
E7 = D7 ÷ D14 × 100
⋮
E12 = D12 ÷ D14 × 100
F5 = E5 − C5
F6 = E6 − C6
⋮
F12 = E12 − C12

12. See the figure.

	A	B	C	D	E	F
1	The 7th Inning: Budget Operating Expenses and Actual Expenses					
2						
3	Expense	Budget Amount	Percent of Total Budget	Actual Expenses	Percent of Actual Total Expense	Percent Difference from Budget
4						
5	Salaries	\$ 45,000.00	17.79%	\$ 42,000.00	16.57%	−6.7%
6	Rent	\$ 37,000.00	14.62%	\$ 36,000.00	14.20%	−2.7%
7	Depreciation	\$ 12,000.00	4.74%	\$ 14,000.00	5.52%	+16.7%
8	Utilities and phone	\$ 13,000.00	5.14%	\$ 10,862.56	4.28%	−16.4%
9	Taxes and Insurance	\$ 15,000.00	5.93%	\$ 13,583.29	5.36%	−9.4%
10	Advertising	\$ 2,000.00	0.79%	\$ 2,847.83	1.12%	+42.4%
11	Purchases	\$ 125,000.00	49.41%	\$ 132,894.64	52.41%	+6.3%
12	Other	\$ 2,000.00	0.79%	\$ 1,356.35	0.53%	−32.2%
13						
14	Total	\$ 253,000.00	99.21%	\$ 253,544.67	99.46%	

Assignment Exercises, Chapter 9

1. $x = 7$ **3.** $b = -\dfrac{15}{2}$ **5.** $y = 7$ **7.** $x = -8$ **9.** $x = 15$

11. $x = 64$ **13.** $x = -49$ **15.** $x = 84$ **17.** $x = 5$ **19.** $b = -2$
21. $x = 7$ **23.** $x = -4$ **25.** $x = 3$ **27.** $x = -1$ **29.** $a = 5$
31. $x = 5$ **33.** $x = 7$ **35.** $x = 4$ **37.** $x = 2$ **39.** $x = -3$

41. $x = \dfrac{1}{3}$ **43.** $y = -12$ **45.** $y = -3$ **47.** $x = \dfrac{9}{2}$ **49.** $y = 7$

51. $x = 9$ **53.** $x = \dfrac{20}{13}$ **55.** $x = \dfrac{7}{20}$ **57.** $c = -\dfrac{9}{32}$ **59.** $y = 1.5$

61. $R = 0.61$ **63.** $y = -1$ **65.** $x = -9$ **67.** $x = 1$ **69.** $x = 3$
71. $x = 2$ **73.** $x = -2$ **75.** $x = 2$ **77.** $x = 0$ **79.** $x = 3$

81. $x = 3$ **83.** $x = -1$ **85.** $x = -\dfrac{1}{8}$ **87.** $x - 6 = 8;\ x = 14$

89. $5(x + 6) = 42 + x;\ x = 3$ **91.** $x + (x - 3) = 51;$ 27 hr, 24 hr
93. $6 = \pi d;$ no; $D = 1.9$ in.   **95.** $A = x(2x);$ 120 ft $\times$ 60 ft

97.

1	A	B	C	D	E	F	G
2	Employee	Date	Hourly Rate	Hours Worked for Week	Regular Pay	Overtime Pay	Total Gross Pay
3	Gayden, Bertha	8/19	$ 9.25	40	$ 370.00	$ 0	$ 370.00
4	Harrover, Roy	8/19	$ 13.60	42	$ 544.00	$ 40.80	$ 584.80
5	Kearney, Claude	8/19	$ 16.50	48	$ 660.00	$ 198.00	$ 858.00
6	Oswalt, Alisa	8/19	$ 6.45	40	$ 258.00	$ 0	$ 258.00
7	Stapleton, Iven	8/19	$ 9.15	45	$ 366.00	$ 68.63	$ 434.63
8	Total Gross Payroll				$2198.00	$ 307.48	$ 2505.43

99. Answers will vary. An example would be, Three times the sum of a number and 5 is increased by 2 and the result is 80. What is the number?

Trial Test, Chapter 9

1. 6   **3.** 10   **5.** -3 **7.** -15 **9.** 3   **11.** 2   **13.** 2   **15.** 1.25
17. 2   **19.** 1500 (Henderson); 1625 (Brinks)

Self-Study Exercises 10–1

1

1. 28   **2.** -11 **3.** $\dfrac{18}{7}$ **4.** 0   **5.** $-\dfrac{27}{5}$ **6.** $-\dfrac{5}{27}$ **7.** $\dfrac{8}{5}$ **8.** $\dfrac{11}{15}$

9. -350 **10.** 48   **11.** 27   **12.** $\dfrac{1}{3}$ **13.** 13   **14.** $\dfrac{25}{8}$ **15.** 0   **16.** $\dfrac{1}{9}$

17. 576   **18.** 4   **19.** 28   **20.** -68 **21.** $\dfrac{27}{35}$ **22.** $\dfrac{108}{5}$ **23.** $\dfrac{217}{24}$

24. 8   **25.** -5 **26.** $\dfrac{3}{10}$ **27.** 70   **28.** 0   **29.** $\dfrac{32}{63}$ **30.** $\dfrac{32}{5}$

31. $-\dfrac{36}{11}$ **32.** $-\dfrac{1}{108}$ **33.** 21   **34.** $\dfrac{4}{25}$ **35.** $\dfrac{2}{21}$ **36.** $\dfrac{9}{7}$
37.–42. Answers to 12, 14, 16, 18, 25, and 32 should be checked.

43. 80 fixtures **44.** $1\frac{1}{3}$ hours **45.** $3\frac{1}{13}$ hr **46.** $3\frac{3}{7}$ days **47.** $2\frac{1}{3}$ min

Self-Study Exercises 10–2

1

1. 2 **2.** 0.8 **3.** 6.03 **4.** 16 **5.** 4.7 **6.** 0.08 **7.** 0.035
8. 0.34 **9.** -3.3 **10.** 38.8 **11.** 3.3 **12.** -0.2 **13.** 0.6
14. 1.128 **15.** 16

2

16. 87.5 lb **17.** 4.71 in. **18.** 12 hr **19.** 6.5 hr **20.** 8.3 Ω
21. $5.30 **22.** 156.25 V **23.** $264 **24.** $136 **25.** $1925

26. 12.25% **27.** 1.2% **28.** 18.5% **29.** $2777.78 **30.** $\frac{1}{2}$ yr or 6 mo

Self-Study Exercises 10–3

1

1. 3 **2.** 2 **3.** 29 **4.** $\frac{5}{8}$ **5.** $\frac{34}{5}$ **6.** $\frac{17}{14}$ **7.** 3 **8.** 21 **9.** 4

10. $-\frac{4}{5}$

2

11. $5.90 **12.** $74.13 **13.** 48 engines **14.** 15 headpieces
15. 1425 mi **16.** 3360 mi **17.** 550 lb **18.** 1.7 gal

3

19. 900 rpm **20.** 200 in.^{3} **21.** 20 in. **22.** 18 painters **23.** 10 hr
24. 50 rpm **25.** 5 in. **26.** 4 hr **27.** 5 machines **28.** 120 rpm
29. 4 helpers (5 workers total) **30.** 9 in.

4

31. $\angle A = \angle F$, $\angle B = \angle E$, $\angle C = \angle D$
32. $PQ = ST$, $\angle P = \angle S$, $\angle Q = \angle T$ **33.** $JL = MP$, $LK = NP$, $\angle L = \angle P$
34. $a = 20$, $d = 9$ **35.** $DC = 6$, $DE = 5\frac{1}{3}$ **36.** 65 ft

Assignment Exercises, Chapter 10

1. $\frac{5}{6}$ **3.** $-\frac{4}{15}$ **5.** $\frac{49}{18}$ **7.** $-\frac{60}{13}$ **9.** $\frac{8}{3}$

11.–15. Answers to 2, 6, and 10 should be checked. **17.** $-\frac{7}{4}$ **19.** $\frac{16}{7}$

21. $\frac{10}{7}$ **23.** 6 **25.** $\frac{31}{39}$ **27.** $2\frac{1}{10}$ hr **29.** 25 fixtures **31.** 1.9

33. 2 **35.** 77.2 **37.** -11.8 **39.** -0.8 **41.** 1.5 **43.** 0.44
45. 0.02 **47.** 3.4 **49.** 17 Ω **51.** 110 V **53.** 375 lb **55.** 2
57. $\frac{7}{6}$ **59.** $\frac{1}{2}$ **61.** $\frac{49}{12}$ **63.** $-\frac{21}{2}$ **65.** $\frac{3}{14}$ **67.** $-\frac{21}{23}$

69. 23 machines   **71.** $2\frac{1}{2}$ hr   **73.** 4500 women   **75.** 1500 rpm

77. 3 days   **79.** $4\frac{1}{5}$ ft   **81.** 129.7 gal   **83.** 5.0 hr   **85.** 1351 ft

87. 60 rpm   **89.** 3 machines   **91.** $\dfrac{AB}{RT} = \dfrac{BC}{TS} = \dfrac{AC}{RS}$ **93.** 10

95. 100 teeth   **97.** $2\frac{6}{7}$ gal water, $7\frac{1}{7}$ gal antifreeze

Trial Test, Chapter 10

1. 16   **3.** 21   **5.** $-\dfrac{11}{5}$ **7.** $-\dfrac{22}{7}$ **9.** $-\dfrac{56}{3}$ **11.** $\dfrac{7}{25}$ **13.** $\dfrac{10}{21}$
15. 6.17   **17.** 0.30   **19.** 254.24   **21.** 2.5   **23.** -0.15 **25.** 168.75
27. 1.33 psi   **29.** 54.7 L   **31.** 11.429 Ω **33.** 107 lb   **35.** 5

Self-Study Exercises 11–1

1
1. x^7 **2.** m^4 **3.** a^2 **4.** x^{10} **5.** y^6 **6.** a^2b

2
7. y^5 **8.** x^4 **9.** $\dfrac{1}{a}$ **10.** b **11.** 1   **12.** $\dfrac{1}{x^2}$ **13.** y^4 **14.** n^7

15. $\dfrac{1}{x^{10}}$ **16.** $\dfrac{1}{n}$ **17.** x^7 **18.** $\dfrac{1}{x^5}$ **19.** $\dfrac{1}{x^2}$

3
20. x^6 **21.** 1   **22.** $-\dfrac{1}{8}$ **23.** $\dfrac{4}{49}$ **24.** $\dfrac{a^4}{b^4}$ **25.** $8m^6n^3$ **26.** $\dfrac{x^6}{y^3}$

27. $\dfrac{9}{25}$ **28.** $4a^2$ **29.** x^6y^3

Self-Study Exercises 11–2

1
1. $7a^2$ **2.** $3x^3$ **3.** $-2a^2 + 3b^2$ **4.** $2a - 2b$ **5.** $-5x^2 - 2x$ **6.** $5a^2$
7. $-x^2 - 2y$ **8.** $2m^2 + n^2$ **9.** $9a + 2b + 10c$ **10.** $-2x + 3y + 2z$

2
11. $14x^3$ **12.** $2m^3$ **13.** $-21m^2$ **14.** $-2y^6$ **15.** $2x^2$ **16.** $-\dfrac{a}{2}$

17. $\dfrac{1}{2x^2}$ **18.** $-\dfrac{3x^3}{4}$ **19.** $3x^2 - 18x$ **20.** $12x^3 - 28x^2 + 32x$
21. $-8x^2 + 12x$ **22.** $10x^2 + 4x^3$ **23.** $3x - 2$ **24.** $4x^3 - 2x - 1$
25. $7x - \dfrac{1}{x}$ **26.** $4x^2 + 3x$

Self-Study Exercises 11–3

1
1. 37   **2.** 1820   **3.** 0.56   **4.** 1.42   **5.** 780,000   **6.** 62   **7.** 0.00046

8. 0.61   **9.** 0.72   **10.** 42   **11.** 10^{10} **12.** 10^{-7} or $\dfrac{1}{10^7}$

13. 10^{-3} or $\dfrac{1}{10^3}$ **14.** 10 **15.** 10^3 **16.** 10^2 **17.** $\dfrac{1}{10^3}$ **18.** 1

19. $\dfrac{1}{10^5}$ **20.** $\dfrac{1}{10}$

2

21. 430 **22.** 0.0065 **23.** 2.2 **24.** 73 **25.** 0.093 **26.** $83{,}000$
27. 0.0058 **28.** $80{,}000$ **29.** 6.732 **30.** 0.00589

3

31. 3.92×10^2 **32.** 2×10^{-2} **33.** 7.03×10^0 **34.** 4.2×10^4
35. 8.1×10^{-2} **36.** 2.1×10^{-3} **37.** 2.392×10^1 **38.** 1.01×10^{-1}
39. 1.002×10^0 **40.** 7.21×10^2 **41.** 4.2×10^5 **42.** 3.26×10^4
43. 2.13×10^1 **44.** 6.2×10^{-6} **45.** 5.6×10^1

4

46. 2.144×10^7 **47.** 5.6×10^3 **48.** 2.36×10^{-2} **49.** 4.73×10^{-3}
50. 7×10^2 **51.** 6.5×10^{-5} **52.** 7×10^2 **53.** 8×10^{-3}
54. 3.2285×10^{13} **55.** 4.2×10^2 or 420

Self-Study Exercises 11–4

1

1. binomial **2.** binomial **3.** monomial **4.** monomial **5.** monomial
6. monomial **7.** trinomial **8.** trinomial **9.** monomial **10.** binomial
11. binomial **12.** binomial

2

13. 1 **14.** 2 **15.** $2, 1, 0$ **16.** $3, 1, 0$ **17.** $1, 0$ **18.** $2, 0$ **19.** 0
20. 0 **21.** $1, 0$ **22.** $2, 0$ **23.** 2 **24.** 3 **25.** 6 **26.** 5 **27.** 2
28. 1

3

29. $-3x^2 + 5x$; 2; $-3x^2$; -3 **30.** $-x^3 + 7$; 3; $-x^3$; -1
31. $9x^2 + 4x - 8$; 2; $9x^2$; 9 **32.** $5x^2 - 3x + 8$; 2; $5x^2$; 5
33. $7x^3 + 8x^2 - x - 12$; 3; $7x^3$; 7 **34.** $-15x^4 + 12x + 7$; 4; $-15x^4$; -15
35. $8x^6 - 7x^3 - 7x$; 6; $8x^6$; 8 **36.** $-14x^8 + x + 15$; 8; $-14x^8$; -14

Self-Study Exercises 11–5

1

1. $\$2415.77$ **2.** $\$2050.40$ **3.** (a) 122.14 (b) 149.18 (c) 164.87
(d) 182.21 **4.** (a) $\$1197.22$ (b) $\$1576.91$ **5.** 64 **6.** 0.0041
7. $9{,}765{,}625$ **8.** 0.0020 **9.** 243 **10.** 0.0032

2

11. 2.08×10^{-87} **12.** 4.28×10^{-96} **13.** 5.58×10^{-44} **14.** 7.39
15. 0.05 **16.** 1.23 **17.** 0.03

3

18. $x = 7$ **19.** $x = -3$ **20.** $x = 2$ **21.** $x = 10$ **22.** $x = 7$

23. $x = 3$ **24.** $x = 6$ **25.** $x = -4$ **26.** $x = -6$ **27.** $x = \dfrac{7}{6}$

28. $x = \dfrac{5}{2}$ **29.** $x = 5$ **30.** $x = 8$

Self-Study Exercises 11–6

1

1. $\log_3 9 = 2$ **2.** $\log_2 32 = 5$ **3.** $\log_9 3 = \dfrac{1}{2}$ **4.** $\log_{16} 2 = \dfrac{1}{4}$

5. $\log_4 \dfrac{1}{16} = -2$ **6.** $\log_3 \dfrac{1}{81} = -4$

2

7. $3^4 = 81$ **8.** $12^2 = 144$ **9.** $2^{-3} = \dfrac{1}{8}$ **10.** $5^{-2} = \dfrac{1}{25}$

11. $25^{-0.5} = \dfrac{1}{5}$ **12.** $4^{-0.5} = \dfrac{1}{2}$ **13.** $x = 3$ **14.** $x = \dfrac{1}{81}$ **15.** $x = 2$

16. $x = -2$ **17.** $x = 625$ **18.** $x = -4$

3

19. 0.4771 **20.** 0.7782 **21.** 0.3802 **22.** 0.6232 **23.** 2.1761
24. -2.9208 **25.** 1.3863 **26.** 0.9163 **27.** -1.8971 **28.** 5.6168
29. 4.6052

4

30. 3 **31.** 6 **32.** 5 **33.** 2 **34.** 1.9358

5

35. 3 **36.** 5 **37.** 8

6

38. 2.096 **39.** 1.893 **40.** 1.631

Assignment Exercises, Chapter 11

1. x^{10} **3.** x^6 **5.** x^3 **7.** x^{12} **9.** x^{15} **11.** $-2x^4 - 4x^3 + 13x$

13. $11x^2 - 11y^2$ **15.** $21x^6$ **17.** $-\dfrac{2x^3}{3}$ **19.** $10x^3 + 15x^2 - 20x$

21. $2x^2 - 4x + 7$ **23.** 10^{12} or $1{,}000{,}000{,}000{,}000$ **25.** 10^{-3} or $\dfrac{1}{1000}$

27. 8730 **29.** 5.2×10^4 **31.** 1.7×10^{-4} **33.** 4.368×10^{126}
35. 3.38×10^{10} **37.** No. The term $-3x^{-2}$ has a negative exponent.
39. 1.25×10^{-6} **41.** 2.71×10^{-35} **43.** (a) \$2536.48 (b) \$546.84

45. $x = -2$ **47.** $x = 4$ **49.** $x = -\dfrac{5}{3}$ **51.** $x = 4$ **53.** $x = -5$

55. $x = 1$ **57.** $x = 4$ **59.** 0.01831563889 **61.** 0.00004539992976

63. $\log_3 81 = 4$ **65.** $\log_{27} 3 = \dfrac{1}{3}$ **67.** $\log_4 \dfrac{1}{64} = -3$

69. $\log_9 \dfrac{1}{3} = -\dfrac{1}{2}$ **71.** $\log_{12} \dfrac{1}{144} = -2$ **73.** $11^2 = 121$ **75.** $15^0 = 1$

77. $7^1 = 7$ **79.** $4^{-2} = \dfrac{1}{16}$ **81.** $9^{-0.5} = \dfrac{1}{3}$ **83.** 0.6990 **85.** 2.2553

87. -0.3979 **89.** 5.5984 **91.** -0.2231 **93.** 2.1133 **95.** 2
97. 343 **99.** $x = -2$ **101.** a. 2 b. 4 c. 8.1761 **103.** 3.17

105. a. $C_t = 72,000,000(0.65)^x$

    b. 30,420,000 units per cubic meter after 60 days

       8,354,092.5 units per cubic meter after 150 days

    c. approximately 500 days or 16 to 17 months

Trial Test, Chapter 11

1. x^5 **3.** x^3 **5.** $\dfrac{16}{49}$ **7.** $\dfrac{x^4}{y^2}$ **9.** $-6a^6$ **11.** $12a^3 - 8a^2 + 20a$

13. 10^6 **15.** 42,000   **17.** 4.2   **19.** 0.042   **21.** 2.4×10^2

23. 8.6×10^{-4} **25.** 7.83×10^{-3} **27.** 2.952×10^0 **29.** 3.5×10^2

31. 1.5×10^6 ohms   **33.** 58.09475019   **35.** 248,832   **37.** $x = 9$

39. $x = 2$ **41.** $\log_4 \dfrac{1}{2} = -\dfrac{1}{2}$ **43.** $3^{-3} = \dfrac{1}{27}$ **45.** 3.4657   **47.** $x = 3$

49. 2.5850   **51.** 1.4641   **53.** $S = \$224.94$ thousands   **55.** \$5596.82

Self-Study Exercises 12–1

1

1. $x^{\frac{3}{5}}; \sqrt[5]{x^3}$ **2.** $x^{\frac{3}{5}}; (\sqrt[5]{x})^3$ **3.** $x^{\frac{4}{2}}; \sqrt{x^4}$ **4.** $x^{\frac{2}{6}}; (\sqrt[6]{x})^2$ **5.** $x^{\frac{3}{4}}; \sqrt[4]{x^3}$

6. $x^{\frac{4}{2}}; (\sqrt{x})^4$ **7.** $a^{\frac{1}{4}}$ **8.** $m^{\frac{1}{3}}$ **9.** $(3x)^{\frac{1}{5}}$ **10.** $(7a)^{\frac{1}{4}}$ **11.** $(xy)^2$ **12.** $3a^2$

13. $32^{\frac{1}{3}}x^{\frac{7}{3}}$ **14.** $1296x^6y^8$ **15.** $16^{\frac{2}{3}}a^{\frac{8}{3}}b^{\frac{14}{3}}$

2

16. a^3 **17.** $b^{\frac{4}{5}}$ **18.** $64xy^{12}$ **19.** $32mn^{10}$ **20.** $a^{\frac{2}{5}}$ **21.** $\dfrac{1}{a^{\frac{1}{2}}}$ **22.** $m^{\frac{1}{8}}$

23. $x^{\frac{7}{6}}$ **24.** $a^{\frac{5}{3}}$ **25.** $2a^{\frac{17}{5}}$ **26.** $\dfrac{1}{3a^{\frac{13}{3}}}$ **27.** $\dfrac{2}{9a^{\frac{9}{8}}}$ **28.** $a^{\frac{5}{3}}b^{\frac{5}{5}} = 3.175$

29. $64ab^6$ or 2^{19} or 529,288   **30.** $c^{\frac{7}{20}} = 1.756$ **31.** $\dfrac{b^{\frac{3}{3}}}{2}$ or $2^{\frac{7}{3}}$ **32.** $a^{7.2}$

33. $a^{1.4}$ **34.** $27a^{3.6}b^6$ **35.** $4^{10}c^{13}d^3$

Self-Study Exercises 12–2

1

1. x^5 **2.** x^3 **3.** $\pm x^8$ **4.** $\dfrac{x^7}{y^{12}}$ **5.** a^3b^5 **6.** $-ab^2c^6$ **7.** $\dfrac{xy^2}{z^5}$

8. $\dfrac{a^2}{b^5c^6}$

2

9. $2\sqrt{6}$ **10.** $7\sqrt{2}$ **11.** $4\sqrt{3}$ **12.** $x^4\sqrt{x}$ **13.** $y^7\sqrt{y}$ **14.** $2x\sqrt{3x}$

15. $2a^2\sqrt{14a}$ **16.** $6ax^2\sqrt{2a}$ **17.** $2x^2yz^3\sqrt{11xz}$

18. Answers will vary.

Self-Study Exercises 12–3

1

1. $12\sqrt{3}$ **2.** $-4\sqrt{5}$ **3.** $2\sqrt{7}$ **4.** $9\sqrt{3} - 8\sqrt{5}$ **5.** $-3\sqrt{11} + \sqrt{6}$

6. $43\sqrt{2} + 6\sqrt{3}$ **7.** $12\sqrt{3}$ **8.** $13\sqrt{5}$ **9.** $11\sqrt{7}$ **10.** $-3\sqrt{6}$

11. $8\sqrt{2} - 6\sqrt{7}$ **12.** $6\sqrt{3}$ **13.** $27\sqrt{5}$ **14.** $47\sqrt{2}$ **15.** $6\sqrt{10}$

16. $-\sqrt{3}$

2

17. $15\sqrt{10}$　**18.** 240　**19.** $20x\sqrt{15x}$　**20.** $28x^5\sqrt{6x}$　**21.** $x\sqrt{3x}$

22. $\dfrac{2\sqrt{2}}{\sqrt{15y}}$ or $\dfrac{2\sqrt{30y}}{15y}$　**23.** $\dfrac{2\sqrt{2}}{x\sqrt{7}}$ or $\dfrac{2\sqrt{14}}{7x}$　**24.** $\dfrac{2}{3y\sqrt{2y}}$ or $\dfrac{\sqrt{2y}}{3y^2}$

25. $\dfrac{4x\sqrt{x}}{3}$　**26.** $x\sqrt{7}$　**27.** $4\sqrt{3}$　**28.** $\dfrac{x}{\sqrt{6}}$ or $\dfrac{x\sqrt{6}}{6}$　**29.** $\dfrac{y}{\sqrt{7}}$ or $\dfrac{y\sqrt{7}}{7}$

30. $2x\sqrt{2x}$

3

31. $2\sqrt{2}$　**32.** $\dfrac{10\sqrt{3}}{3}$　**33.** $\dfrac{8}{5\sqrt{3}}$ or $\dfrac{8\sqrt{3}}{15}$　**34.** $\dfrac{15x^3\sqrt{x}}{4}$

4

35. $\dfrac{5\sqrt{3}}{3}$　**36.** $\dfrac{6\sqrt{5}}{5}$　**37.** $\dfrac{\sqrt{2}}{4}$　**38.** $\dfrac{\sqrt{55}}{11}$　**39.** $\dfrac{\sqrt{6}}{3}$　**40.** $\dfrac{5\sqrt{3x}}{3}$

41. $\dfrac{2x\sqrt{7}}{7}$　**42.** $\dfrac{2x^3\sqrt{3x}}{3}$

Self-Study Exercises 12–4

1

1. $5i$　**2.** $6i$　**3.** $8xi$　**4.** $4y^2i\sqrt{2y}$

2

5. i　**6.** 1　**7.** 1　**8.** 1

3

9. $15 + 0i$　**10.** $0 + 33i$　**11.** $5 + 2i$　**12.** $8 + 4i\sqrt{2}$　**13.** $7 - i\sqrt{3}$
14. $0 + 7i$　**15.** $-4 + 0i$

4

16. $11 + 5i$　**17.** $(2\sqrt{3} + 2\sqrt{2}) - 8i\sqrt{3}$　**18.** $4 + i$　**19.** $12 + 20i$
20. $1 + 9i$

Self-Study Exercises 12–5

1

1. ± 5　**2.** ± 3　**3.** ± 3.464　**4.** ± 0.667　**5.** ± 3　**6.** ± 0.926
7. ± 1.155　**8.** ± 4　**9.** ± 2.449　**10.** $\pm 5i$

2

11. $x = 81$　**12.** $y = 192$　**13.** $y = 66$　**14.** $x = \pm 5.292$
15. $x = -1.979$　**16.** $x = \pm 2$　**17.** $x = 50$　**18.** $x = \pm 4.899$
19. $y = \pm 2.105$　**20.** $y = \pm 2.9$

Assignment Exercises, Chapter 12

1. $x^{\frac{7}{2}}$; $\sqrt{x^7}$　**3.** $x^{\frac{2}{3}}$; $(\sqrt[3]{x})^2$　**5.** $x^{\frac{1}{2}}$　**7.** $x^{\frac{4}{5}}$　**9.** $x^{\frac{4}{3}}y^{\frac{4}{3}}$　**11.** $\sqrt{7}$
13. $\sqrt[5]{y^3}$　**15.** $x^{\frac{1}{3}}$　**17.** $(4y)^{\frac{1}{5}}$　**19.** $2b^4$　**21.** a^2　**23.** y　**25.** $27x^{\frac{3}{4}}y^6$
27. $64a^3x^{\frac{3}{2}}$　**29.** $x^{\frac{1}{2}}$　**31.** $a^{\frac{7}{6}}$　**33.** $\dfrac{1}{x^{\frac{1}{8}}}$　**35.** $a^{\frac{8}{3}}$　**37.** $2a^{\frac{7}{2}}$; 22.627

39. $\dfrac{3}{2a^{\frac{22}{5}}}$; 0.071   **41.** $a^{6.3}$; 78.793   **43.** y^6 **45.** $-b^9$ **47.** 5   **49.** x

51. $3p\sqrt{p}$ **53.** $3a\sqrt{2b}$ **55.** $4x^2y\sqrt{2x}$ **57.** $5x^5y^4\sqrt{3y}$ **59.** $-2\sqrt{3}$

61. $-\sqrt{7}$ **63.** $11\sqrt{6}$ **65.** $-28\sqrt{3}$ **67.** $-5\sqrt{2}$ **69.** $-17\sqrt{2}$

71. $24\sqrt{3}$ **73.** $40\sqrt{21}$ **75.** $-160\sqrt{6}$ **77.** $\dfrac{3}{4}$ **79.** $\dfrac{3\sqrt{6}}{8}$

81. $\dfrac{\sqrt{15}}{10}$ **83.** $\sqrt{3}$ **85.** $\dfrac{9}{16}$ **87.** $\dfrac{5\sqrt{17}}{17}$ **89.** $\dfrac{\sqrt{21}}{6}$ **91.** $\dfrac{\sqrt{6}}{4}$

93. $\dfrac{5\sqrt{2}}{4}$ **95.** $10i$ **97.** $\pm 2y^3i\sqrt{6y}$ **99.** -1 **101.** i **103.** $0 + 15i$

105. $0 - 12i$ **107.** $7 - 4i$ **109.** $11 + i$ **111.** ± 6 **113.** ± 2

115. $\pm 2i$ **117.** ± 3 **119.** $\dfrac{3\sqrt{6}}{2}$ or 3.674   **121.** 46   **123.** ± 10.262

125. ± 8.888 **127.** ± 1 **129.** ± 7.874 **131.** $20 + 15i$

Trial Test, Chapter 12

1. $6\sqrt{14}$ **3.** $6\sqrt{3}$ **5.** $\dfrac{4\sqrt{6}}{3}$ **7.** $2\sqrt{6}$ **9.** $\dfrac{7\sqrt{15x}}{2x}$ **11.** ± 1

13. 49   **15.** 9   **17.** $\pm 7i$ or no real solutions   **19.** $3x^5$ **21.** $x^2y^3z^6$
23. $5x^{\frac{1}{6}}y^2$ **25.** $2x$ **27.** $-i$ **29.** $-3 + 5i$

Self-Study Exercises 13–1

1

1. 25   **2.** 15   **3.** 1565   **4.** 9   **5.** 4.5   **6.** $310   **7.** 14.25%
8. 3 years   **9.** $2600   **10.** 90   **11.** 8%   **12.** $3378.38   **13.** 13.6

14. 17%   **15.** 66 in.   **16.** 9.0 cm   **17.** $10.50   **18.** $\dfrac{1}{4}$ mi

19. 153.9 in.^{2} **20.** 6.25 km^2 **21.** $48.75

Self-Study Exercises 13–2

1

1. $X = E - I$ **2.** $K = M - A$ **3.** $r = \dfrac{S}{2\pi h}$ **4.** $b = \dfrac{m}{x + y}$

5. $y = \dfrac{A - mx}{2m}$ **6.** $T_2 = \dfrac{T_1 V_2}{V_1}$ **7.** $r = \sqrt{\dfrac{V}{\pi h}}$ **8.** $X = \dfrac{m^2}{Y}$

9. $R = \dfrac{100P}{B}$ **10.** $b = \dfrac{P - 2s}{2}$ **11.** $r = \dfrac{C}{2\pi}$ **12.** $l = \dfrac{A}{w}$

13. $C = \dfrac{R}{A - B}$ **14.** $s = \sqrt{A}$ **15.** $R = \dfrac{D}{T}$ **16.** $D = P - S$

17. $r = \sqrt{\dfrac{A}{\pi}}$ **18.** $C = \sqrt{a^2 + b^2}$ **19.** $I = A - P$ **20.** $T = \dfrac{I}{PR}$

Self-Study Exercises 13–3

1

1. 355   **2.** 165   **3.** 290   **4.** 0   **5.** -175 **6.** 344   **7.** 333
8. -81

2

9. 920   **10.** 558   **11.** 250   **12.** 460   **13.** 640   **14.** 672
15. -110 **16.** 140   **17.** -460 **18.** 492

3

19. 35   **20.** 0   **21.** 45   **22.** 5   **23.** 15   **24.** 10   **25.** 65   **26.** 50
27. 80   **28.** 120

4

29. 158   **30.** 59   **31.** 113   **32.** 122   **33.** 68   **34.** 419   **35.** 590
36. 770   **37.** 365   **38.** 32

Self-Study Exercises 13–4

1

1. $P = 309.3$ mm   **2.** $P = 52.9$ ft   **3.** 348 in.^{2} **4.** 260 ft^2
     $A = 5053.05$ mm^2 $A = 162$ ft^2
5. 8220 ft^2 **6.** 98 ft

2

7. $P = 28.6$ in.   **8.** $P = 48$ cm   **9.** 15.75 cm^2 **10.** 72 ft; 216 ft^2
     $A = 33$ in.^{2} $A = 96$ cm^2
11. 9.75 ft^2 **12.** 30 ft^2 **13.** 24.75 in.^{2} **14.** $175.50   **15.** 8 ft
16. 10 ft 8 in.

3

17. $AC = 24$ cm   **18.** $BC = 10$ mm   **19.** $AB = 17$ yd   **20.** 6.403 cm
21. 6.325 m   **22.** 12 ft   **23.** 30 ft   **24.** 39.783 cm   **25.** 26 ft 10 in.
26. 12.806 ft   **27.** 13 cm   **28.** 45 dm   **29.** 10.607 mm   **30.** 61.083 ft
31. $E_a = 170.294$ V

4

32. $LSA = 28$ units^2 **33.** $LSA = 848.23$ units^2 **34.** 3244 ft^2
     $TSA = 432$ units^2 $TSA = 1357.17$ units^2
35. 39.4 in.^{2} **36.** 14,700 mm^2 **37.** 37.3 in.^{2} **38.** 600 in.^{2}
39. 69.1 in.^{2}

5

40. $V = 576$ units^3 **41.** $V = 3817.04$ units^3 **42.** 23.3 in.^{3}
43. 18,850 ft^3 **44.** 320 cm^3 **45.** 55.36 in.^{3} **46.** 236 ft^3

6

47. 314.2 cm^2 **48.** 904.8 in.^{3} **49.** 6361.7 ft^2 **50.** 356,892.8 gal
51. 50.3 ft^2 **52.** 225.6 gal

7

53. $LSA = 188.5$ ft^2; $TSA = 301.6$ ft^2; $V = 301.6$ ft^3 **54.** 4712.4 ft^3
55. 589.0 cm^2 **56.** 50.3 L   **57.** 1649.3 ft^2 **58.** 10.0 ft   **59.** $347

Self-Study Exercises 13–5

1

1. 4.8 ohms   **2.** 61.6%   **3.** 35 miles per hour   **4.** 1.5 amperes
5. 6 ft^3 **6.** 3 amperes   **7.** 2 cylinders   **8.** 1600 rpm   **9.** 71.71 in.
10. 4.4 ohms

Assignment Exercises, Chapter 13

1. 12.5   **3.** 3   **5.** 1.25   **7.** 9.6%   **9.** 10.8   **11.** $193.60
13. 2 years   **15.** 50 in.   **17.** $89.50   **19.** 95.0 in.^{2}

21. $F = D + m + n$ **23.** $w = \dfrac{V}{lh}$ **25.** $a = \dfrac{c}{h + b}$ **27.** $r = \sqrt{\dfrac{B}{cx}}$

29. $B = \dfrac{A}{P}$ **31.** $r = \dfrac{I}{Pt}$ **33.** $r = s + d$ **35.** $t = \dfrac{V_0 - V}{32}$

37. $t = \dfrac{A - P}{Pr}$ **39.** $P = S + D$ **41.** 351   **43.** 472   **45.** 203

47. 104°   **49.** 185   **51.** $P = 12$ in.   **53.** $P = 339$ mm
                        $A = 6$ in.^{2} $A = 5899.5$ mm^2
55. $A = 8.125$ ft^2 **57.** 11.25 ft   **59.** 19.5 ft^2 **61.** Two sheets
63. 48 ft   **65.** 199.5 ft^2 **67.** 15 in.   **69.** 7.141 ft   **71.** 8 yd
73. 20.248 mi   **75.** 30 cm   **77.** 21.633 in.   **79.** 4.243 in.
81. 600 cm^3 **83.** 616 cm^2 **85.** 2733.186 cm^2 **87.** 102 yd^3
89. 60 ft^2 **91.** 348,962 gal   **93.** 1017.9 m^2 **95.** 7238.2 ft^3
97. 169.6 cm^2 **99.** 377.0 in.^{3} **101.** 2 gal   **103.** 116.6 ft^2
105. 243.3 in.^{2} **107.** 2094 ft^3 **109.** 2.75 amperes   **111.** 10 ft^3
113. 3.2 amperes   **115.** 1400 rpm   **117.** 2 cylinders
119. Answers will vary.

Trial Test, Chapter 13

1. $L = \dfrac{RA}{P}$ **3.** $r = \sqrt{\dfrac{d}{\pi sn}}$ **5.** 361   **7.** 383   **9.** 250 ft^2 **11.** 10 dm
13. 50 in.^{2} **15.** 6768 gal   **17.** 3848.5 ft^3 **19.** 2080.7 cm^2
21. 10.9 ft^3 **23.** 1497.05 lb   **25.** 12,000 calories

Self-Study Exercises 14–1

1

1. $7(a + b)$ **2.** $m(m + 2)$ **3.** $3x(2x + 1)$ **4.** $5(ab + 2a + 4b)$
5. $2ax(2x + 3a)$ **6.** $a(5 - 7b)$ **7.** $3(4a^2 - 5a + 2)$
8. $3x(x^2 - 3x - 2)$ **9.** $2ab(4a + 7b^2)$ **10.** $3m^2(1 - 2m)$

Self-Study Exercises 14–2

1

1. $a^2 + 11a + 24$ **2.** $x^2 + x - 20$ **3.** $y^2 - 10y + 21$
4. $2a^2 + 5a - 12$ **5.** $3a^2 - 8ab + 4b^2$ **6.** $5cx - 25xy - cy + 5y^2$
7. $6x^2 - 17x + 12$ **8.** $2a^2 - 7ab + 5b^2$ **9.** $21 - 52m + 7m^2$
10. $40 - 21x + 2x^2$ **11.** $x^2 + 11x + 28$ **12.** $y^2 - 12y + 35$
13. $m^2 - 4m - 21$ **14.** $3bx + 18b - 2x - 12$ **15.** $12r^2 - 7r - 10$
16. $35 - 22x + 3x^2$ **17.** $4 - 14m + 6m^2$ **18.** $6 + 13x + 6x^2$
19. $2x^2 + x - 15$ **20.** $20x^2 - 13xy - 21y^2$ **21.** $14a^2 + 19ab - 3b^2$
22. $30a^2 - 13ab - 10b^2$ **23.** $27x^2 + 30xy - 8y^2$
24. $20x^2 - 47xy + 24y^2$ **25.** $21m^2 + 29mn - 10n^2$

2

26. $a^2 - 9$ **27.** $4x^2 - 9$ **28.** $a^2 - y^2$ **29.** $16r^2 - 25$ **30.** $25x^2 - 4$
31. $49 - m^2$ **32.** $4 - 12x + 9x^2$ **33.** $9x^2 + 24x + 16$
34. $Q^2 + 2QL + L^2$ **35.** $a^4 + 2a^2 + 1$ **36.** $4d^2 - 20d + 25$
37. $9a^2 + 12ax + 4x^2$ **38.** $9x^2 - 49$ **39.** $36 + 12Q + Q^2$
40. $y^2 - 25x^2$ **41.** $16 - 24j + 9j^2$ **42.** $9m^2 - 4p^2$

43. $m^4 + 2m^2p^2 + p^4$ **44.** $4a^2 - 49c^2$ **45.** $81 - 234a + 169a^2$
46. $x^3 + p^3$ **47.** $Q^3 - L^3$ **48.** $27 + a^3$ **49.** $8x^3 - 64p^3$
50. $27m^3 - 8$ **51.** $216 + p^3$ **52.** $125y^3 - p^3$ **53.** $x^3 + 8y^3$
54. $Q^3 - 216$ **55.** $27T^3 + 8$

Self-Study Exercises 14–3

$\boxed{1}$

1. difference **2.** not difference **3.** difference **4.** difference
5. not difference **6.** difference **7.** $(y + 7)(y - 7)$
8. $(4x + 1)(4x - 1)$ **9.** $(3a + 10)(3a - 10)$ **10.** $(2m + 9n)(2m - 9n)$
11. $(3x + 8y)(3x - 8y)$ **12.** $(5x + 8)(5x - 8)$ **13.** $(10 + 7x)(10 - 7x)$
14. $(2x + 7y)(2x - 7y)$ **15.** $(11m + 7n)(11m - 7n)$
16. $(9x + 13)(9x - 13)$

$\boxed{2}$

17. not perfect square **18.** perfect square **19.** not perfect square
20. not perfect square **21.** perfect square **22.** perfect square
23. $(x + 3)^2$ **24.** $(x + 7)^2$ **25.** $(x - 6)^2$ **26.** $(x - 8)^2$
27. $(2a + 1)^2$ **28.** $(5x - 1)^2$ **29.** $(3m - 8)^2$ **30.** $(2x - 9)^2$
31. $(x - 6y)^2$ **32.** $(2a - 5b)^2$

$\boxed{3}$

33. difference of two cubes **34.** not a special sum or difference of cubes
35. sum of two cubes **36.** not a special sum or difference of two cubes
37. difference of two cubes **38.** sum of two cubes
39. $(2m - 7)(4m^2 + 14m + 49)$ **40.** $(y - 5)(y^2 + 5y + 25)$
41. $(Q + 3)(Q^2 - 3Q + 9)$ **42.** $(c + 1)(c^2 - c + 1)$
43. $(5d - 2p)(25d^2 + 10dp + 4p^2)$ **44.** $(a + 4)(a^2 - 4a + 16)$
45. $(6a - b)(36a^2 + 6ab + b^2)$ **46.** $(x + Q)(x^2 - xQ + Q^2)$
47. $(2p - 5)(4p^2 + 10p + 25)$ **48.** $(3 - 2y)(9 + 6y + 4y^2)$

Self-Study Exercises 14–4

$\boxed{1}$

1. $(x + 3)(x + 2)$ **2.** $(x - 7)(x - 4)$ **3.** $(x + 6)(x + 2)$
4. $(x - 3)(x - 1)$ **5.** $(x + 7)(x + 1)$ **6.** $(x + 5)(x + 2)$
7. $(x - 4)(x + 3)$ **8.** $(y - 5)(y + 2)$ **9.** $(y - 3)(y + 2)$
10. $(a + 4)(a - 3)$ **11.** $(b + 3)(b - 1)$ **12.** $(7 + b)(2 - b)$
13. $(x - 4)(x - 3)$ **14.** $(x - 6)(x + 5)$ **15.** $(x + 9)(x + 2)$
16. $(x - 6)(x - 3)$ **17.** $(x - 9)(x + 2)$ **18.** $(x + 18)(x - 1)$
19. $(x + 5)(x + 4)$ **20.** $(x - 10)(x - 2)$ **21.** $(x - 8)(x - 2)$
22. $(x - 16)(x - 1)$ **23.** $(x - 14)(x + 1)$ **24.** $(x - 7)(x + 2)$

$\boxed{2}$

25. $(x + y)(x + 4)$ **26.** $(3x + 2)(2x - y)$ **27.** $(3x + 5)(m - 2n)$
28. $(6x - 7)(5y - 6)$ **29.** $(x - 2)(x + 8)$ **30.** $(3x - 1)(2x - 7)$
31. $(x - 4)(x + 1)$ **32.** $(2x - 1)(4x + 3)$ **33.** $(x - 5)(x + 4)$
34. $(x - 2)(3x + 5)$

$\boxed{3}$

35. $(3x + 1)(x + 2)$ **36.** $(3x + 2)(x + 4)$ **37.** $(3x + 2)(2x + 3)$
38. $(x - 2)(x - 16)$ **39.** $(3x - 4)(2x - 3)$ **40.** $(2x - 5)(x - 2)$
41. $(3x - 5)(2x - 1)$ **42.** $(p + 9)(p - 4)$ **43.** $(6x - 5)(x - 1)$
44. $(4x + 3)(2x + 5)$ **45.** $(3x - 5)(5x + 1)$ **46.** $(y - 12)(y - 3)$

47. $(2x - 7)(x + 1)$ **48.** $(6x - 5)(2x + 3)$ **49.** $(5x + 3)(2x - 1)$
50. $(Q - 11)(Q + 4)$

[4]

51. $(6x - 5y)(x + 2y)$ **52.** $(3a + 2b)(2a - 7b)$ **53.** $(6x - 5)(3x + 2)$
54. $(5x - 4y)(4x + 3y)$ **55.** $(x + 8)(x - 7)$ **56.** $(a + 19)(a + 2)$

[5]

57. $4(x - 1)$ **58.** $(x + 3)(x - 2)$ **59.** $(2x + 3)(x - 1)$
60. $(x + 3)(x - 3)$ **61.** $4(x + 2)(x - 2)$ **62.** $(m + 5)(m - 3)$
63. $2(a + 2)(a + 1)$ **64.** $(b + 3)^2$ **65.** $(4m - 1)^2$
66. $(x + 7)(x + 1)$ **67.** $(2m + 1)(m + 2)$ **68.** $(2m + 1)(m - 3)$
69. $(2a - 5)(a + 1)$ **70.** $(3x - 2)(x + 4)$ **71.** $(3x + 5)(2x - 3)$
72. $(4x - 1)(2x + 3)$

Assignment Exercises, Chapter 14

1. $5(x + y)$ **3.** $4(3m^2 - 2n^2)$ **5.** $2a(a^2 - 7a - 1)$
7. $5x(3x - 4)(x + 1)$ **9.** $6a^2(3a + 2)$ **11.** $x^2 + 11x + 28$
13. $m^2 - 4m - 21$ **15.** $12r^2 - 7r - 10$ **17.** $4 - 14m + 6m^2$
19. $2x^2 + x - 15$ **21.** $14a^2 + 19ab - 3b^2$ **23.** $27x^2 + 30xy - 8y^2$
25. $21m^2 + 29mn - 10n^2$ **27.** $36x^2 - 25$ **29.** $49y^2 - 121$
31. $64a^2 - 25b^2$ **33.** $64x^2 - y^2$ **35.** $9y^2 - 16z^2$ **37.** $x^2 + 18x + 81$
39. $x^2 - 6x + 9$ **41.** $16x^2 - 120x + 225$ **43.** $64 + 112m + 49m^2$
45. $16x^2 - 88x + 121$ **47.** $g^3 - h^3$ **49.** $8H^3 - 27T^3$ **51.** $216 + i^3$
53. $z^3 + 8t^3$ **55.** $343T^3 + 8$ **57.** not difference   **59.** difference
61. difference   **63.** not perfect-square trinomial
65. perfect-square trinomial   **67.** perfect-square trinomial
69. difference of two cubes   **71.** difference of two cubes
73. sum of two cubes   **75.** $(5y - 2)(5y + 2)$
77. *NSP*, this is a sum of two squares, not a difference.   **79.** $(a + 1)^2$
81. $(4c - 3b)^2$ **83.** $(n - 13)^2$ **85.** $(6a + 7b)^2$ **87.** $(7 - x)^2$
89. *NSP*, this is a sum of two squares, not a difference.
91. *NSP*, the middle term needs a y factor.   **93.** $(7 + 9y)(7 - 9y)$
95. $(3x + 10y)(3x - 10y)$ **97.** $(3x - y)^2$ **99.** $(3xy + z)(3xy - z)$
101. $(x + 2)^2$

103. $\left(\dfrac{2}{5}x + \dfrac{1}{4}y\right)\left(\dfrac{2}{5}x - \dfrac{1}{4}y\right)$ **105.** $(T - 2)(T^2 + 2T + 4)$

107. $(d + 9)(d^2 - 9d + 81)$ **109.** $(3k + 4)(9k^2 - 12k + 16)$
111. $(x + 8)(x + 3)$ **113.** $(x + 10)(x + 3)$ **115.** $(x - 8)(x - 1)$
117. $(x - 13)(x + 2)$ **119.** $(x + 8)(x - 3)$ **121.** $(6x + 1)(x + 4)$
123. $(5x - 4)(x - 6)$ **125.** $(3x + 7)(2x - 5)$ **127.** $(7x + 8)(x - 3)$
129. $(3a + 10)(3a - 10)$ **131.** $(2x + 1)(x - 2)$ **133.** $(a + 9)(a - 9)$
135. $(y - 7)^2$ **137.** $(b + 5)(b + 3)$ **139.** $(13 + m)(13 - m)$
141. $(x - 8)(x + 4)$ **143.** $(x + 20)(x - 1)$ **145.** $2(x - 4)(x + 2)$
147. $2x(x - 6)(x + 1)$ **149.** This enables us to factor more rapidly and easily.

Trial Test, Chapter 14

1. $3x + 6y$ **3.** $14x^3 + 21x^2 - 35x$ **5.** $m^2 - 49$ **7.** $a^2 + 6a + 9$
9. $2x^2 - 11x + 15$ **11.** $x^3 - 8$ **13.** $125a^3 - 27$ **15.** $x(7x + 8)$
17. $7ab(a - 2)$ **19.** $(x + 8)(x + 1)$ **21.** $(x - 9)(x + 4)$
23. $(3x + 2)(2x - 3)$ **25.** $(3x + 5)(x + 6)$ **27.** $(a + 8b)^2$
29. $(b - 5)(b + 2)$ **31.** $(3x + 4)(x - 1)$ **33.** $(3m - 2)(m - 1)$
35. $(3a - 2)(9a^2 + 6a + 4)$

Self-Study Exercises 15–1

[1]

1. $7x^2 - 4x + 5 = 0$ **2.** $8x^2 - 6x - 3 = 0$ **3.** $7x^2 - 5 = 0$
4. $x^2 - 6x + 8 = 0$ **5.** $x^2 - 9x + 8 = 0$ **6.** $x^2 - 4x - 8 = 0$
7. $3x^2 - 6x + 5 = 0$ **8.** $x^2 - 6x - 5 = 0$ **9.** $x^2 - 16 = 0$
10. $8x^2 - 7x - 8 = 0$ **11.** $8x^2 + 8x - 10 = 0$
12. $0.3x^2 - 0.4x - 3 = 0$

[2]

13. $x^2 - 5x = 0$
$a = 1, b = -5, c = 0$
14. $3x^2 - 7x + 5 = 0$
$a = 3, b = -7, c = 5$
15. $7x^2 - 4x = 0$
$a = 7, b = -4, c = 0$
16. $3x^2 - 5x + 8 = 0$
$a = 3, b = -5, c = 8$
17. $x^2 - 5x + 6 = 0$
$a = 1, b = -5, c = 6$
18. $11x^2 - 8x = 0$
$a = 11, b = -8, c = 0$
19. $x^2 - x = 0$
$a = 1, b = -1, c = 0$
20. $9x^2 - 7x - 12 = 0$
$a = 9, b = -7, c = -12$
21. $x^2 - 5 = 0$
$a = 1, b = 0, c = -5$
22. $x^2 + 6x - 3 = 0$
$a = 1, b = 6, c = -3$
23. $5x^2 - 0.2x + 1.4 = 0$
$a = 5, b = -0.2, c = 1.4$
24. $\frac{2}{3}x^2 - \frac{5}{6}x - \frac{1}{2} = 0$
$a = \frac{2}{3}, b = -\frac{5}{6}, c = -\frac{1}{2}$
25. $1.3x^2 - 8 = 0$
$a = 1.3, b = 0, c = -8$
26. $\sqrt{3}\,x^2 + \sqrt{5}\,x - 2 = 0$
$a = \sqrt{3}, b = \sqrt{5}, c = -2$
27. $8x^2 - 2x - 3 = 0$
28. $x^2 + 3x = 0$
29. $5x^2 + 2x - 7 = 0$
30. $2.5x^2 - 0.8 = 0$

Self-Study Exercises 15–2

[1]

1. $3, -\frac{2}{3}$ **2.** $3, -4$ **3.** $\frac{11}{5}, -1$ **4.** $3, 3$ **5.** $3, -2$ **6.** $\frac{3}{4}, -\frac{1}{2}$
7. $1.84, -10.84$ **8.** $-0.18, -1.82$ **9.** $3.14, -0.64$ **10.** $2.39, 0.28$
11. $-0.75 \pm 0.97i$ **12.** $0.5 \pm 1.32i$

[2]

13. width $= 5.55$ cm, length $= 8.55$ cm
14. width $= 4$ in., length $= 10$ in.
15. 4 m **16.** width $= 11$ ft, length $= 16$ ft
17. length $= 110$ ft, width $= 70$ ft (nearest ft) **18.** 111.5 kg

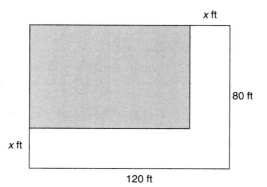

Self-Study Exercises 15–3

1

1. $x = \pm 3$ **2.** $x \pm 7$ **3.** $x = \pm\dfrac{8}{3}$ **4.** $x = \pm\dfrac{7}{4}$ **5.** $y = \pm 3$

6. $x = \pm\dfrac{9}{2}$ **7.** $x = \pm\sqrt{5}$ or ± 2.236 **8.** $x = \pm 1$ **9.** $x = \pm 2$

10. $x = \pm 1.732$ **11.** 23 ft   **12.** 82 yd

13. Isolate the squared letter, then take the square root of both sides.

14. opposites

Self-Study Exercises 15–4

1

1. $x = 3, 0$ **2.** $x = 0, \dfrac{7}{3}$ **3.** $x = 0, 2$ **4.** $x = 0, -\dfrac{1}{2}$ **5.** $x = 0, \dfrac{1}{2}$

6. $x = 0, 3$ **7.** $x = 0, 6$ **8.** $y = 0, -4$ **9.** $x = 0, -\dfrac{2}{3}$

10. $x = 0, \dfrac{4}{3}$ **11.** 8 or 0   **12.** 15 units

13. An incomplete quadratic equation is missing the constant or number term while a pure quadratic equation is missing the linear term.

14. Yes, the common factor of x will be set equal to zero.

Self-Study Exercises 15–5

1

1. $-3, -2$ **2.** $3, 3$ **3.** $7, -2$ **4.** $3, -6$ **5.** $-4, -3$ **6.** $5, 3$

7. $14, -1$ **8.** $3, 6$ **9.** $\dfrac{1}{2}, 3$ **10.** $-\dfrac{1}{3}, -4$ **11.** $\dfrac{3}{5}, -\dfrac{1}{2}$ **12.** $-\dfrac{1}{3}, -\dfrac{3}{2}$

13. $-\dfrac{3}{2}, -5$ **14.** $\dfrac{4}{3}, 2$ **15.** $\dfrac{1}{6}, -3$ **16.** $w = 5$ ft   **17.** $w = 18$ in.
$$$\phantom{-\dfrac{3}{2}, -5}$$$$\phantom{\dfrac{4}{3}, 2}$$$$\phantom{\dfrac{1}{6}, -3}$$$$x = 11$ ft $x = 21$ in.

Self-Study Exercises 15–6

1

1. $x = \dfrac{2}{3}, -1$ **2.** $x = \dfrac{3 \pm \sqrt{5}}{2}$ or $2.62, 0.38$

3. $x = \dfrac{-1 \pm \sqrt{17}}{4}$ or $0.78, -1.28$ **4.** no real solutions or $x = \dfrac{1 \pm i\sqrt{2}}{3}$

or $0.33 \pm 0.47i$ **5.** $x = \dfrac{3 \pm \sqrt{37}}{2}$ or $4.54, -1.54$

6. $x = \dfrac{-5 \pm \sqrt{97}}{6}$ or 0.81 or -2.47

7. If a is positive and c is negative, the equation will have real roots.

8. If $b^2 - 4ac$ is a perfect square and is positive, the roots will be real, rational, and unequal.

Self-Study Exercises 15–7

1

1. 2   **2.** 1   **3.** 4   **4.** 1   **5.** 4   **6.** 7   **7.** $x = 0, 2, -3$

8. $x = 0, \dfrac{1}{2}, -3$ **9.** $x = 0, \dfrac{5}{2}, \dfrac{2}{3}$ **10.** $x = 0, 2, 5$

11. $x = 0, 3, -2$ **12.** $x = 0, -\dfrac{1}{2}, -2$ **13.** $\dot{x} = 0, 3, -3$

14. $x = 0, \dfrac{1}{2}, -\dfrac{1}{2}$ **15.** $x = 0, \dfrac{3}{4}, -\dfrac{3}{4}$ **16.** $x = 0, 4, -4$ **17.** Real root: $x = 0$; Imaginary roots: $\pm\, i\, \sqrt{5}$ **18.** $w(w + 3)(w + 7) = 421$; When expanded and put in standard form, it will not factor.

Assignment Exercises, Chapter 15

1. pure   **3.** pure   **5.** incomplete   **7.** pure   **9.** complete
11. $a = 1$ **13.** $a = 1$ **15.** $a = 1$ **17.** $9, -1$ **19.** $2, -4$
  $\quad b = -2$ $\quad b = 3$ $\quad b = -3$
  $\quad c = -8$ $\quad c = -4$ $\quad c = 2$
21. $2, -\dfrac{1}{2}$ **23.** $1.78, -0.28$ **25.** $-0.23, -1.43$ **27.** $w = 11$ ft,

$l = 22$ ft   **29.** $w = 14$ in., $l = 42$ in.   **31.** $x = \pm 10$ **33.** $x = \pm\dfrac{3}{2}$

35. $y = \pm 1.740$ **37.** $x = \pm 2.828$ **39.** $x = \pm 2.236$ **41.** $x = \pm 2$
43. $x = \pm 4.123$ **45.** $y = \pm 3.055$ **47.** $x = \pm 4$ **49.** $x = \pm 8$

51. 16.4 cm   **53.** $0, 5$ **55.** $0, 2$ **57.** $0, -\dfrac{1}{2}$ **59.** $0, 5$ **61.** $0, -\dfrac{2}{3}$

63. $0, -3$ **65.** $0, 9$ **67.** $0, -8$ **69.** $0, \dfrac{5}{3}$ **71.** $0, \dfrac{1}{2}$ **73.** $x = 0$ or 4

75. $3, 1$ **77.** $-5, 2$ **79.** $-6, -1$ **81.** $4, 2$ **83.** $-\dfrac{2}{3}, \dfrac{3}{2}$ **85.** $-\dfrac{2}{5}, \dfrac{5}{2}$

87. $-\dfrac{3}{4}, -1$ **89.** $\dfrac{3}{4}, -\dfrac{1}{3}$ **91.** $-21, 2$ **93.** $\dfrac{2}{3}, -1$ **95.** $3, 2$

97. $6, -3$ **99.** $-6, -5$ **101.** width $= 12$ ft   **103.** real, rational, equal
              length $= 19$ ft
105. real, irrational, unequal   **107.** real, rational, unequal   **109.** 3
111. 1 **113.** 8 **115.** $0, 2, \dfrac{2}{3}$ **117.** $0, -2, -3$ **119.** $0, -5, \dfrac{1}{2}$

121. $0, 2$ **123.** $0, -2, -4$ **125.** $0, 5, \dfrac{-4}{?}$ **127.** $0, 3 \pm \sqrt{2}$ or
$4.414, 1.586$ **129.** $0, -1, 4$ **131.** $0, \sqrt{17}, -\sqrt{17}$ or ± 4.123

Trial Test, Chapter 15

1. pure   **3.** incomplete   **5.** ± 9 **7.** $\pm\dfrac{4}{3}$ **9.** ± 2.33 **11.** $0, 2$

13. $3, 2$ **15.** $\dfrac{3}{2}, 4$ **17.** 169.79 mils   **19.** 1.98 amps   **21.** 6

23. $0, \dfrac{3}{2}, -5$ **25.** $0, 3$

Self-Study Exercises 16–1

[1]

1. $\dfrac{4}{9}$ **2.** $\dfrac{3}{8}$ **3.** $2ab^2$ **4.** $\dfrac{x}{x + 2}$ **5.** $\dfrac{1}{2}$ **6.** $\dfrac{3(3x + 2)}{x(2x - 1)}$ or $\dfrac{9x + 6}{2x^2 - x}$

7. $\dfrac{1}{a - b}$ **8.** $\dfrac{x^2 - 3x - 10}{x^2 + 3x - 10}$ **9.** $\dfrac{x - 2}{x + 2}$ **10.** $\dfrac{2x - 4}{3}$ or $\dfrac{2(x - 2)}{3}$

11. -1 **12.** $-(x + 2)$

Self-Study Exercises 16–2

1

1. $\dfrac{10x^3}{9y^2}$ **2.** $3(a - b)$ or $3a - 3b$ **3.** 4 **4.** $\dfrac{a - 7}{b + 5}$

5. $\dfrac{3(x + 1)}{x - 1}$ or $\dfrac{3x + 3}{x - 1}$ **6.** $\dfrac{1}{12a}$ **7.** $\dfrac{1}{2}$ **8.** $\dfrac{x}{x - 2}$

9. $\dfrac{50(a - b)}{ab}$ or $\dfrac{50a - 50b}{ab}$ **10.** 1

2

11. $\dfrac{25}{27}$ **12.** $1\dfrac{1}{20}$ **13.** $\dfrac{x - 5}{12x}$ **14.** $\dfrac{2}{3x(x - 2)}$ or $\dfrac{2}{3x^2 - 6x}$ **15.** $\dfrac{x + 3}{2}$

16. $\dfrac{x - 3}{3}$ **17.** $3(x - 6)$ **18.** $\dfrac{3x^2}{4}$

Self-Study Exercises 16–3

1

1. $\dfrac{5}{7}$ **2.** $\dfrac{13}{16}$ **3.** $\dfrac{3x}{2}$ **4.** $\dfrac{23x}{24}$ **5.** $\dfrac{4 + 3x}{2x}$ **6.** $\dfrac{55}{6x}$

7. $\dfrac{9x - 1}{(x + 1)(x - 1)}$ or $\dfrac{9x - 1}{x^2 - 1}$ **8.** $\dfrac{10x + 6}{(x + 3)(x - 1)}$ or $\dfrac{10x + 6}{x^2 + 2x - 3}$

9. $\dfrac{x - 7}{(2x + 1)(x - 1)}$ or $\dfrac{x - 7}{2x^2 - x - 1}$ **10.** $\dfrac{x + 1}{(x - 6)(x - 5)}$ or $\dfrac{x + 1}{x^2 - 11x + 30}$

11. $\dfrac{5(2x + 1)}{3x}$ **12.** $\dfrac{8(3x + 2)}{5x}$ or $\dfrac{24x + 16}{5x}$ **13.** $-\dfrac{28}{9}$ or $-3\dfrac{1}{9}$ **14.** $-\dfrac{1}{7}$

Self-Study Exercises 16–4

1

1. 0 **2.** 0 **3.** $5, 0$ **4.** $0, -9$ **5.** $-8, 8$ **6.** $2, -2$ **7.** $\dfrac{3}{4}, 0$

8. $0, -3$

2

9. 35 **10.** 5 **11.** 1 **12.** $\dfrac{1}{4}$ **13.** 4 **14.** $-\dfrac{4}{5}$ **15.** $1\dfrac{5}{7}$ days

16. 8 investors **17.** 25 mph going, 40 mph returning **18.** 3 min
19. 50 lb dark-roast, 75 lb medium-roast, $2 per pound **20.** 6 ohms

Assignment Exercises, Chapter 16

1. $\dfrac{3}{4}$ **3.** $\dfrac{b^2}{2ac}$ **5.** $\dfrac{x-3}{2(x+3)}$ **7.** 1   **9.** $\dfrac{m^2-n^2}{m^2+n^2}$ **11.** $\dfrac{1}{1+y}$ **13.** 5

15. $y+1$ **17.** $\dfrac{2}{x+6}$ **19.** 3   **21.** $\dfrac{5x^3}{4y^2}$ **23.** 15   **25.** $\dfrac{4(2y-1)}{y-1}$

27. 1   **29.** $\dfrac{x+3}{x+2}$ **31.** $\dfrac{1}{4}$ **33.** $\dfrac{(y-1)^3}{y}$ **35.** $\dfrac{y+3}{y+2}$ **37.** $\dfrac{3x^2}{2}$

39. $(y+4)(y-3)$ **41.** $\dfrac{5}{4x-12}$ **43.** $\dfrac{4x^2}{3}$ **45.** $\dfrac{51}{28}$

47. $\dfrac{3x^2-30}{2x^2+12}$ **49.** $\dfrac{2}{3}$ **51.** $\dfrac{4x}{7}$ **53.** $\dfrac{19x}{12}$ **55.** $\dfrac{15-7x}{3x}$

57. $\dfrac{13}{4x}$ **59.** $\dfrac{10x+5}{(x-3)(x+2)}$ **61.** $\dfrac{6x-38}{(x+3)(x-4)}$ **63.** $\dfrac{x+3}{x-5}$

65. 0, 2   **67.** 0, $\dfrac{1}{2}$ **69.** $\dfrac{-20}{3}$ **71.** $\dfrac{1}{7}$ **73.** 3 students   **75.** $7\dfrac{1}{2}$ hr

Trial Test, Chapter 16

1. $\dfrac{1}{2}$ **3.** $\dfrac{3x-4}{x+3}$ **5.** $-\dfrac{x-2}{x+2}$ or $\dfrac{2-x}{x+2}$ **7.** $\dfrac{3a}{y}$ **9.** $\dfrac{1}{x^2(x+2y)}$

11. $\dfrac{2x+1}{x}$ **13.** $-\dfrac{5}{(x+2)(x-3)}$ **15.** $\dfrac{12+x}{4x}$ **17.** $-\dfrac{2}{3x-2}$ or

$\dfrac{2}{2-3x}$ **19.** $\dfrac{1}{x+2y}$ **21.** $\dfrac{2x^2}{x-3}$ **23.** 0, -3 **25.** $\dfrac{11}{7}$

27. $\dfrac{-5 \pm 20\sqrt{2}}{2}$ or 11.642, -16.642 **29.** 5 persons

Self-Study Exercises 17–1

1. F   **2.** F   **3.** T   **4.** T

5. [8, 12]

6. $(5, \infty)$

7. $(-\infty, -2]$

8. $(-6, \infty)$

9. (843,000, 1,000,000)

10. [4, 18]

11. $n < 12$

Self-Study Exercises 17–2

1

1. $x < 6$; $(-\infty, 6)$

$$x < 6$$
4 5 6 7 8

2. $a \geq -3$; $[-3, \infty)$

$$a \geq -3$$
−4 −3 −2 −1 0 1

3. $y \leq 8$; $(-\infty, 8]$

$$y \leq 8$$
6 7 8 9 10

4. $b > -1$; $(-1, \infty)$

$$b > -1$$
−3 −2 −1 0 1 2

5. $t < 6$; $(-\infty, 6)$

$$t < 6$$
2 3 4 5 6

6. $y < 5$; $(-\infty, 5)$

$$y < 5$$
3 4 5 6 7

7. $a \leq -2$; $(-\infty, -2]$

$$a \leq -2$$
−4 −3 −2 −1 0 1

8. $x \leq \dfrac{1}{4}$; $\left(-\infty, \dfrac{1}{4}\right]$

$$x \leq \tfrac{1}{4}$$
−3 −2 −1 0$\frac{1}{4}$ 1 2

9. $t \leq 7$; $(-\infty, 7]$

$$t \leq 7$$
5 6 7 8 9 10

10. $y > 11$; $(11, \infty)$

$$y > 11$$
9 10 11 12 13 14

11. $x \geq 3$; $[3, \infty)$

$$x \geq 3$$
0 1 2 3 4 5

12. $x < 1$; $(-\infty, 1)$

$$x < 1$$
−1 0 1 2 3 4 5

13. $x > \dfrac{1}{3}$; $\left(\dfrac{1}{3}, \infty\right)$

$$x > \tfrac{1}{3}$$
−1 0$\frac{1}{3}$ 1 2 3

14. $a \leq -3$; $(-\infty, -3]$

$$a \leq -3$$
−4 −3 −2 −1 0 1

15. $x < \dfrac{1}{2}$; $\left(-\infty, \dfrac{1}{2}\right)$

16. $2x + 196 \geq 52{,}800$
$x \geq 26{,}302$

$$x < \tfrac{1}{2}$$
−2 −1 0$\frac{1}{2}$1 2 3

Self-Study Exercises 17–3

1

1. T; all elements in U **2.** F; 2 is not in A **3.** T; $\emptyset$ is a subset of all sets.
4. F; No element of A is in B **5.** { } or $\emptyset$ **6.** $\{2, 0, -2\}$
7. $\{-4, -2, 0, 1, 2, 3, 4, 5\}$ **8.** $\{-5, -3, -1, 1, 2, 3, 4, 5\}$
9. $\{-5, -4, -3, -2, -1, 1, 2, 3, 4, 5\}$ **10.** $\{-5, -4, -3, -1, 0\}$

2

11. $2 < x < 6$; $(2, 6)$

13. No solution; ∅

12. $2 < x < 7$; $(2, 7)$

14. $x = 5$; $[5]$

15. No solution; ∅
16. $3 < x < 7$; $(3, 7)$

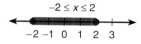

18. $-2 \leq x \leq 2$; $[-2, 2]$

20. $3 < x < 5$; $(3, 5)$

17. $0 < x < 3$; $(0, 3)$

19. $-6 < x < -3$; $(-6, -3)$

3

21. $x < 2$ or $x > 9$

$(-\infty, 2)$ or $(9, \infty)$
23. $x \leq 4$ or $x \geq 5$

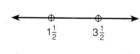

$(-\infty, 4]$ or $[5, \infty)$
25. $x \leq 1\frac{1}{2}$ or $x \geq 3\frac{1}{2}$

$\left(-\infty, 1\frac{1}{2}\right]$ or $\left[3\frac{1}{2}, \infty\right)$

22. $x \leq 3$ or $x \geq 7$

$(-\infty, 3]$ or $[7, \infty)$
24. $x \leq 1$ or $x \geq 3$

$(-\infty, 1]$ or $[3, \infty)$

26. $5.22 \leq 5.27 \leq 5.32$;
$x - 0.05 \leq x \leq x + 0.05$

Self-Study Exercises 17–4

1. $-3 < x < 2$

2. $x < -3$ or $x > \frac{1}{2}$

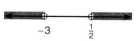

3. $\dfrac{2}{3} < x < \dfrac{5}{2}$

4. $y \leq -6$ or $y \geq 2$

5. $-4 < a < 6$

6. $2 \leq x \leq 5$

7. $x \leq -2$ or $x \geq 3$

8. $1 \leq x \leq 3$

9. $\dfrac{1}{4} \leq x \leq 5$

10. $x < -\dfrac{1}{2}$ or $x > \dfrac{1}{3}$

2

11. $-7 < x < 3$

12. $x < -8$ or $x > 3$

13. $x < \dfrac{1}{3}$ or $x > 3$

14. $-7 < x < 7$

15. $0 < x < 1$

Self-Study Exercises 17–5

1

1. ± 8 **2.** $8, -2$ **3.** $9, 5$ **4.** $6, -3$ **5.** $3, \dfrac{5}{3}$ **6.** ± 11

7. no solution   **8.** $7, 1$ **9.** $4, -3$ **10.** no solution

Self-Study Exercises 17–6

1

1. $-5 < x < 5$

2. $-1 < x < 1$

3. $-2 < x < 2$

4. $-7 \leq x \leq 7$

5. $-5 < x < -1$

6. $-7 < x < -1$

7. $-1 < x < 7$

8. $-1 < x < 5$

9. $-1 \leq x \leq 11$

10. $-6 \leq x \leq 8$

11. $-\dfrac{2}{3} < x < 4$

12. $\dfrac{1}{2} < x < \dfrac{7}{2}$

13. $-1 \leq x \leq \dfrac{1}{5}$

14. $-3 \leq x \leq -\dfrac{1}{2}$

15. $x < -2$ or $x > 2$

16. $x < -3$ or $x > 3$

17. $x \leq -5$ or $x \geq 5$

18. $x \leq -6$ or $x \geq 6$

19. $x < 3$ or $x > 7$

20. $x < -2$ or $x > 10$

21. $x \leq -8$ or $x \geq 2$

22. $x \leq -10$ or $x \geq 2$

23. $x \leq \dfrac{5}{2}$ or $x \geq \dfrac{5}{2}$ or all real numbers

24. $x \leq -\dfrac{2}{3}$ or $x \geq 2$

25. $x \leq -2$ or $x \geq -\dfrac{6}{5}$

26. $x \leq -\dfrac{3}{2}$ or $x \geq 3$

27. $\left\{ x \,\middle|\, \dfrac{1}{3} \leq x \leq 1 \right\}$

28. $\{x \mid 0 \leq x \leq 5\}$

29. $\left\{x \;\middle|\; x \leq -\dfrac{4}{5} \text{ or } x \geq 2\right\}$

30. $\left\{x \;\middle|\; x \leq 1\dfrac{1}{4} \text{ or } x \geq 1\dfrac{3}{4}\right\}$

31. no solution
{ } or Ø

32. $\{x \mid x \in R\}$ all reals

33. $x \leq 0$ or $x \geq 6$　**34.** $x < -\dfrac{1}{2}$ or $x > 0$　**35.** $1 \leq x \leq 7$

36. no solution　**37.** $-7 < x < 7$　**38.** $-11 < x < -5$
39. $x < -12$ or $x > 12$　**40.** $x < -10$ or $x > -6$

Assignment Exercises, Chapter 17

1

1. a set with no elements { } or Ø　**3.** $5 \in W$
5. $(-7, \infty)$　**7.** $[-4, 2)$

9. $(-2, \infty)$　**11.**

13.　**15.**

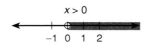

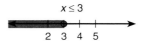

17.　**19.**

21.　**23.**

25.　**27.** { } or Ø

29. $\{-1\}$　**31.** $\{5\}$
33. $x > -2; (-2, \infty)$　**35.** $x \geq -3; [-3, \infty)$

37. $\dfrac{1}{2} < x < 4\dfrac{1}{2}; \left(\dfrac{1}{2}, \dfrac{9}{2}\right)$　**39.**

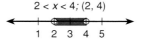

41. no solution; ∅

43.

$2 < x < 6; (2, 6)$

45.

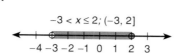

$-3 < x \le 2; (-3, 2]$

47.

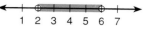

$-2 \le x \le -\frac{3}{4}; [-2, -\frac{3}{4}]$

49. $x < 2$ or $x > 8$

$(-\infty, 2)$ or $(8, \infty)$

51. $x < -9$ or $x > 8$

$(-\infty, -9)$ or $(8, \infty)$

53.

$x < 2$ or $x > 5$

55.

$-\frac{1}{3} < x < \frac{3}{2}$

57.

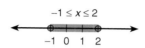

$-1 \le x \le 2$

59.

$-\frac{1}{2} \le x \le 3$

61.

$-5 < x < \frac{3}{2}$

63. $-1 < x < 7$

65. $x < -8$ or $x > 0$

67. ± 12 **69.** $4, -10$ **71.** $20, -4$ **73.** $6, -\frac{5}{2}$ **75.** $1, -\frac{23}{7}$

77. $3, -\frac{13}{7}$ **79.** $\frac{11}{3}, \frac{7}{3}$ **81.** ± 7 **83.** ± 16 **85.** $10, -4$ **87.** $2, -\frac{1}{2}$

89. $-1, -3$

91. $-1 < x < 7$

93. $-4 < x < 12$

95. no solution   **97.** $\$108,000 < I < \$250,000$

Trial Test, Chapter 17

1. $x \ge -12$

$[-12, \infty)$

3. $x > 3$

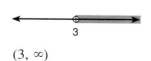

$(3, \infty)$

5. $x > 2$

$(2, \infty)$

9. $x > -2$

$(-2, \infty)$

13. no solution

7. $x \leq -6$

$(-\infty, -6]$

11. $-8 < x < 4$

$(-8, 4)$

15. $x < -\dfrac{3}{2}$ or $x > 1$

$\left(-\infty, -\dfrac{3}{2}\right)$ or $(1, \infty)$

17. $x < 2$ or $x > 6$

$(-\infty, 2)$ or $(6, \infty)$

21. $A \cap B' = \{9\}$

25. ± 15

29. $x < -18$ or $x > 2$

$(-\infty, -18)$ or $(2, \infty)$

19. $A \cup B = \{1, 2, 3, 4, 5, 6, 7, 8, 9\}$

23. $-5 < x < 2$

$(-5, 2)$

27. ± 2

Self-Study Exercises 18–1

[1] Values chosen for table of values may vary.

1.

x	$f(x)$
-2	-13
-1	-10
0	-7
1	-4
2	-1

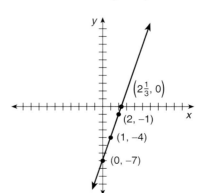

2. $f(x) = 39 + 0.1x$

3.

x	f(x)
0	39
50	44
100	49
150	54
200	59

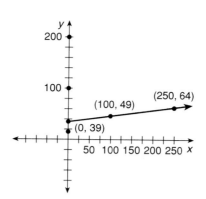

4. $79

5.

x	f(x)
−2	17
−1	10
0	5
1	2
2	1
3	2
4	5

6.

x	f(x)
−2	$\frac{1}{9}$
−1	$\frac{1}{3}$
0	1
1	3
2	9

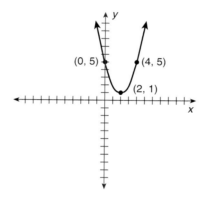

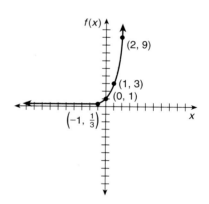

7.

x	f(x)
0.1	−1.2
0.3	0
0.5	0.4
1	1.1
2	1.8
3	2.2
4	2.5
5	2.7
8	3.2

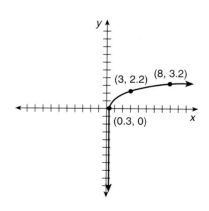

2

8. $f(x) = 3x - 5$

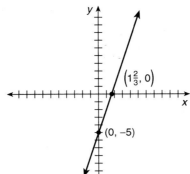

9. $f(x) = 5x - 6$

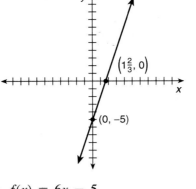

10. $f(x) = 6x - 5$

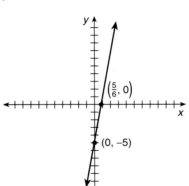

11. $f(x) = x - 1$

12. $f(x) = x + 7$

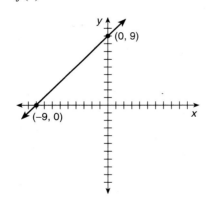

Self-Study Exercises 18–2

1

1. no   **2.** yes   **3.** yes   **4.** no   **5.** yes   **6.** yes   **7.** yes   **8.** no
9. yes   **10.** no   **11.** yes   **12.** yes   **13.** yes   **14.** no   **15.** no
16. no   **17.** yes   **18.** no

2

19. $y = 5$ **20.** $y = 5$ **21.** $x = -6$ **22.** $x = 21$ **23.** $y = 15$
24. $y = 0$ **25.** $x = 8$ **26.** $x = 3$ **27.** $y = 2$ **28.** $y = 1$
29. $x = 12$ **30.** $y = -12$ **31.** $x = 18$ **32.** $y = 12$
33. $(1, 3); (1, -2); (1, 0); (1, 1)$ **34.** $(-1, 4); (3, 4); (0, 4); (2, 4)$

35. (2, −7); (−3, −7); (0, −7); (5, −7)   **36.** (9, −2); (9, 3); (9, 0); (9, 2)
37. $12   **38.** $2   **39.** $136   **40.** $13,800

Self-Study Exercises 18–3

1

1. x-intercept, (5, 0)
y-intercept, (0, 5)

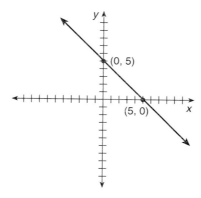

2. x-intercept, (5, 0)
y-intercept, $\left(0, \dfrac{5}{3}\right)$

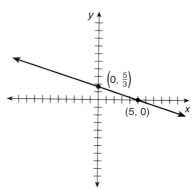

3. x-intercept, (−4, 0)
y-intercept, (0, 8)

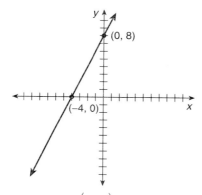

4. x-intercept, $\left(\dfrac{1}{3}, 0\right)$
y-intercept, (0, −1)

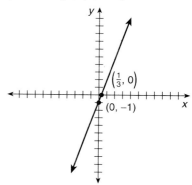

5. x-intercept, $\left(\dfrac{2}{5}, 0\right)$
y-intercept, (0, −2)

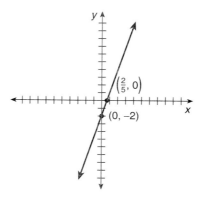

6. x-intercept, (3, 0)
y-intercept, $\left(0, \dfrac{3}{2}\right)$

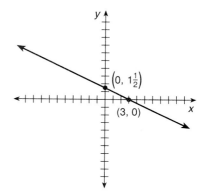

7. slope = $\frac{2}{1}$; y-intercept, (0, −3) **8.** slope = $-\frac{1}{2}$; y-intercept, (0, −2)

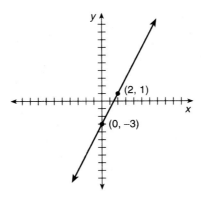

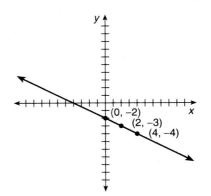

9. slope = $-\frac{3}{5}$; y-intercept, (0, 0) **10.** slope = $\frac{1}{2}$; y-intercept,

$$\left(0, \ -1\frac{1}{2}\right)$$

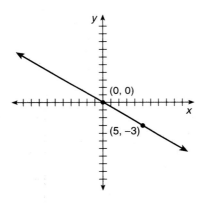

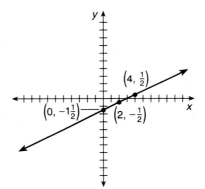

11. slope = $-\frac{2}{1}$; y-intercept, (0, 1)

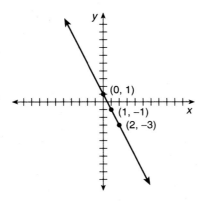

Self-Study Exercises 18–4

1.

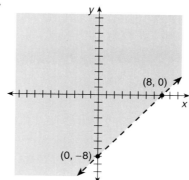

2.

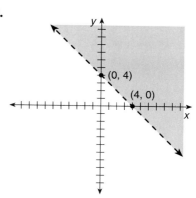

3.

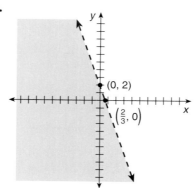

4.

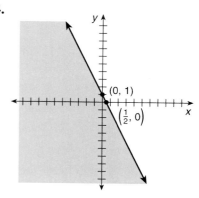

5.

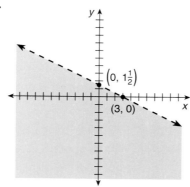

6.

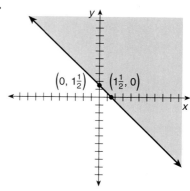

7.

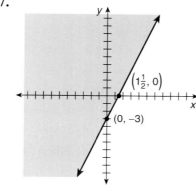

8.

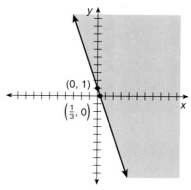

9.

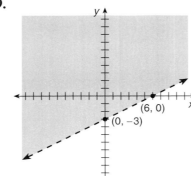

10.

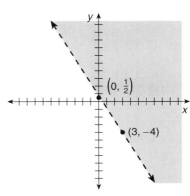

Self-Study Exercises 18–5

1. linear   **2.** quadratic   **3.** quadratic   **4.** other   **5.** linear   **6.** other

2

7. $y = x^2$

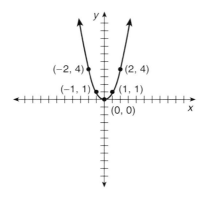

8. $y = 3x^2$

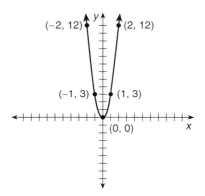

9. $y = \dfrac{1}{3}x^2$

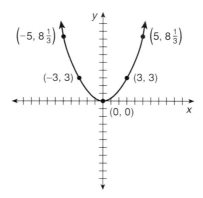

10. $y = -4x^2$

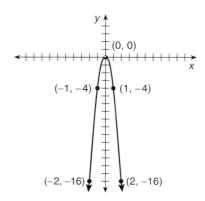

11. $y = -\dfrac{1}{4}x^2$

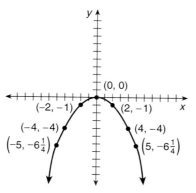

12. $y = x^2 - 4$

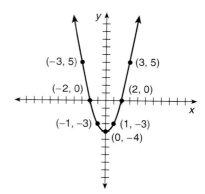

13. $y = x^2 + 4$

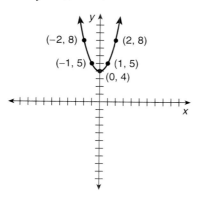

14. $y = x^2 - 6x + 9$

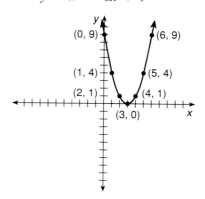

15. $y = -x^2 + 6x - 9$

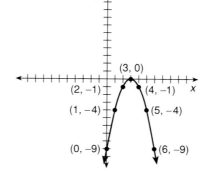

16. $y = 3x^2 + 5x - 2$

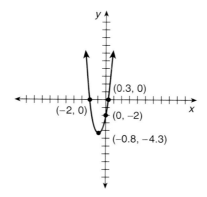

17. $y = (2x - 3)(x - 1)$

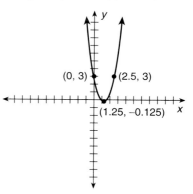

18. $y > 2x^2 - 9x - 5$

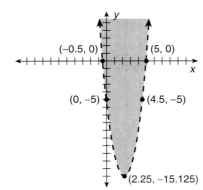

19. $y \geq -2x^2 + 9x + 5$

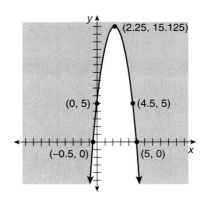

Assignment Exercises, Chapter 18

Since plotted solutions will vary, check graphs by comparing x- and y-intercepts.

1.

x	$f(x)$
-1	-5
1	-1
3	3

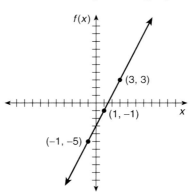

3.

x	$f(x)$
-1	-3
0	0
1	3

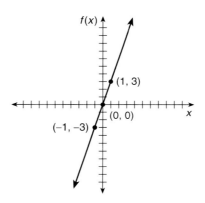

5.

x	$f(x)$
-1	3
0	0
1	-3

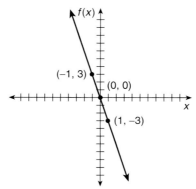

7.

x	f(x)
−1	−1
0	1
1	3

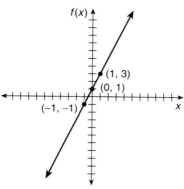

9.

x	f(x)
−5	7
−3	−5
−1	−9
1	−5
3	7

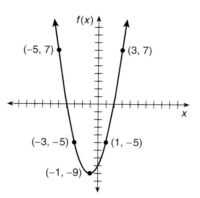

11.

x	f(x)
0.1	−2.3
1	0
3	1.1
10	2.3

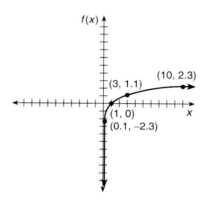

13. $f(x) = x - 6$

15. $f(x) = 3x + 6$

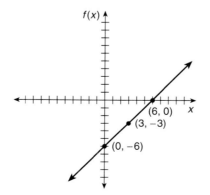

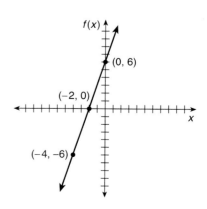

17. $f(x) = 2x - 2$

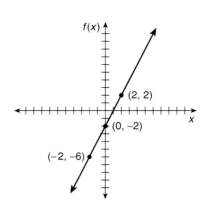

19. no **21.** yes **23.** yes **25.** $y = 1$

27. x-intercept, $(-1, 0)$; y-intercept, $\left(0, -\frac{1}{4}\right)$

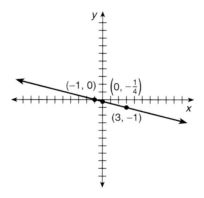

29. $3x - y = 1$; x-intercept, $\left(\frac{1}{3}, 0\right)$ **31.** $y = 5x - 2$, $m = \frac{5}{1}$, $b = -2$
y-intercept, $(0, -1)$

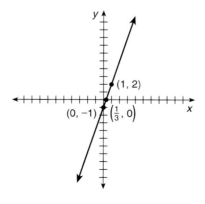

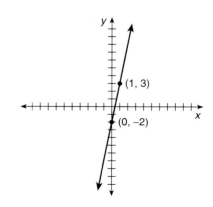

33. $y = -3x - 1$, $m = \dfrac{-3}{1}$, $b = -1$ **35.** $y = x - 4$, $m = \dfrac{1}{1}$, $b = -4$

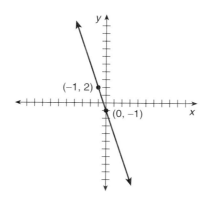 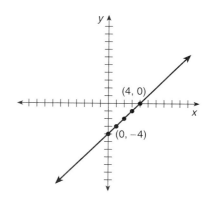

37. $y = \dfrac{1}{2}x + \dfrac{1}{2}$, $m = \dfrac{1}{2}$, $b = \dfrac{1}{2}$ **39.** $4x + y < 2$, $\left(\dfrac{1}{2}, 0\right)$, $(0, 2)$

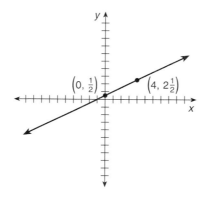 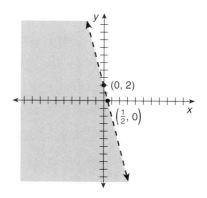

41. $3x + y \leq 2$, $\left(\dfrac{2}{3}, 0\right)$, $(0, 2)$ **43.** $x - 2y < 8$, $(8, 0)$, $(0, -4)$

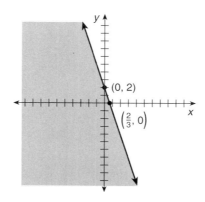 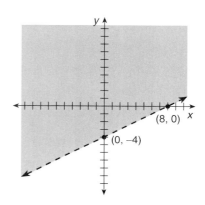

45. $y \geq 3x - 2$, $\left(\dfrac{2}{3}, 0\right)$, $(0, -2)$

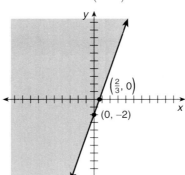

47. $y > \dfrac{2}{3}x - 2$, $(3, 0)$, $(0, -2)$

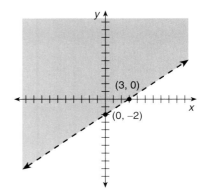

49. quadratic   **51.** other nonlinear   **53.** other nonlinear   **55.** $y = -x^2 - 1$

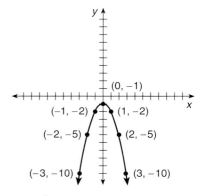

57. $y = x^2 - 6x + 8$

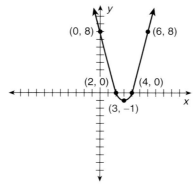

59. $y = -x^2 + 2x - 1$

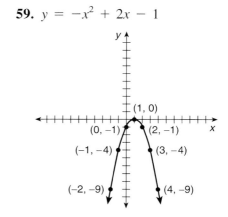

61. $y = -x^2 + 4x - 4$

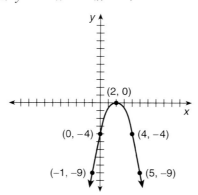

63. $y \geq -2x^2$

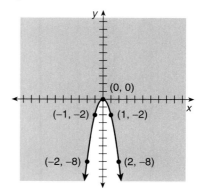

65. a)

gal	ft²
1	400
2	800
3	1200
4	1600
5	2000
6	2400
7	2800
8	3200
9	3600
10	4000

b)

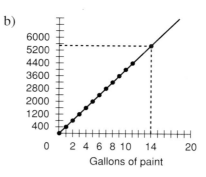

c) 14 gal
d) $279.30
e) $302.34

67.

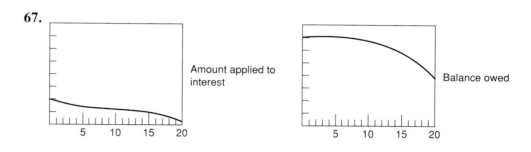

Amount applied to interest

Balance owed

Trial Test, Chapter 18

Since plotted solutions will vary, check graphs by comparing x- and y-intercepts.

1.

x	$f(x)$
-2	-1
0	0
2	1

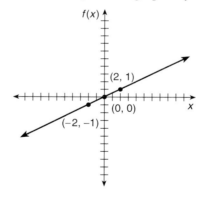

3.

x	$f(x)$
-1	-6
0	-4
1	-2
2	0

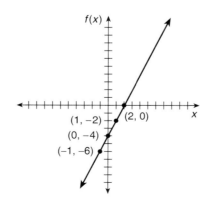

5. $f(x) = 3x - 3$; $x = 1$

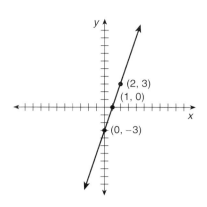

7. $(2, 5)$

9. x-intercept: $(2, 0)$

y-intercept: $\left(0, -\dfrac{1}{2}\right)$

11.

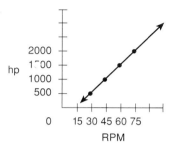

13. x-intercept $(8, 0)$

y-intercept $(0, 4)$

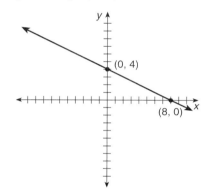

15. y-intercept $= (0, -3)$

slope $= \dfrac{-2}{1}$

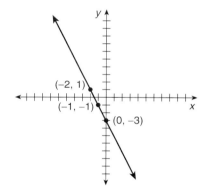

17. y-intercept $= (0, -5)$

slope $= \dfrac{1}{1}$

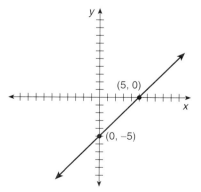

19. $2x - y \leq 2$

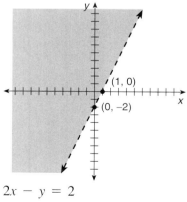

21. $x + y < 1$

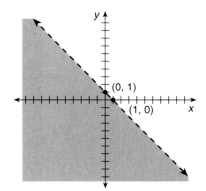

23. $y = x^2 + 2x + 1$

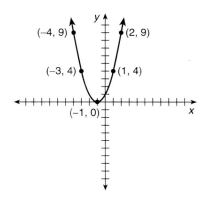

25. $y \le x^2 - 6x + 8$

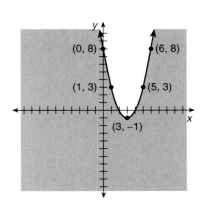

27. $y < -\frac{1}{2}x^2$

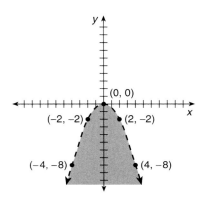

Self-Study Exercises 19–1

1

1. $\frac{1}{2}$ **2.** $-\frac{5}{3}$ **3.** 3   **4.** 1   **5.** $\frac{3}{4}$ **6.** 0   **7.** $-\frac{5}{4}$ **8.** -1
9. no slope   **10.** -3

2

11. Problem 6 is horizontal, slope is zero.
    Problem 9 is vertical, slope not defined.

Self-Study Exercises 19–2

1

1. $y = \frac{2}{3}x + \frac{25}{3}$ **2.** $y = -\frac{1}{2}x + 3$ **3.** $y = 2x + 1$ **4.** $y = x - 1$

2

5. $y = -\frac{5}{3}x + \frac{38}{3}$ **6.** $x = -1$ **7.** $y = -3$ **8.** $x = -4$ **9.** $y = 2x$
10. $y = 0$

Self-Study Exercises 19–3

1. slope $= 4$
   y-intercept $= 3$
2. slope $= -5$
   y-intercept $= 6$
3. slope $= -\dfrac{7}{8}$
   y-intercept $= -3$
4. slope $= 0$
   y-intercept $= 3$
5. $y = 2x - 5$
   slope $= 2$, y-intercept $= -5$
6. $y = \dfrac{5}{2}$

   slope $= 0$, y-intercept $= \dfrac{5}{2}$
7. $y = -2x + 6$
   slope $= -2$, y-intercept $= 6$
8. $y = 2x + 5$
   slope $= 2$, y-intercept $= 5$
9. $x = 4$
   no slope, no y-intercept
10. $x = 9$
    no slope, no y-intercept

2

11. $y = \dfrac{1}{4}x + 7$
12. $y = -8x - 4$
13. $y = -3x + 1$
14. $y = \dfrac{2}{3}x - 2$
15. $y = 4$

Self-Study Exercises 19–4

1

1. $x + y = -1$
2. $2x + y = 9$
3. $x - 3y = 4$
4. $3x - y = -7$
5. $x + 3y = 7$
6. $3x - y = -3$
7. $2x + 3y = 5$
8. $3x + 2y = 6$
9. $2x - 5y = 11$
10. $6x - 8y = 3$

Self-Study Exercises 19–5

1

1. $x - y = -1$
2. $x - 2y = -13$
3. $3x + y = 12$
4. $x + 3y = -9$
5. $3x - y = 11$
6. $2x - y = -7$
7. $3x - 2y = 11$
8. $x - 4y = 0$
9. $2x - 2y = -3$
10. $x + 5y = 4$

Self-Study Exercises 19–6

1

1. 10

2. 13

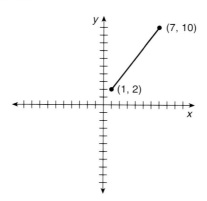

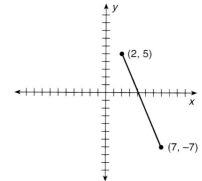

3. 7.071

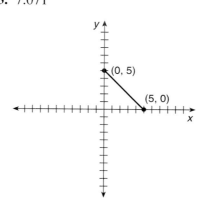

4. 5.385

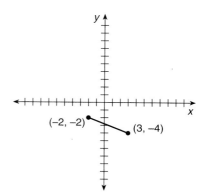

5. 11.180

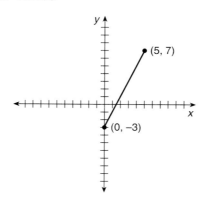

6. 13

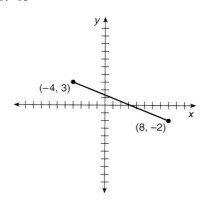

7. 10

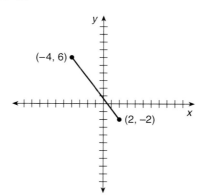

8. 5

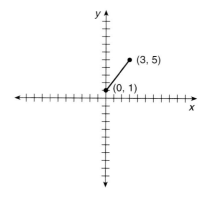

9. 15

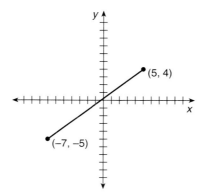

10. 10.296

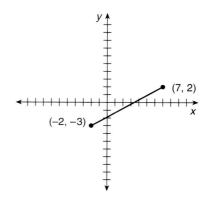

2

11. $(4, 6)$ **12.** $\left(\dfrac{9}{2}, -1\right)$ **13.** $\left(\dfrac{5}{2}, \dfrac{5}{2}\right)$ **14.** $\left(\dfrac{1}{2}, -3\right)$ **15.** $\left(\dfrac{5}{2}, 2\right)$

16. $\left(2, \dfrac{1}{2}\right)$ **17.** $(-1, 2)$ **18.** $\left(\dfrac{3}{2}, 3\right)$ **19.** $\left(-1, -\dfrac{1}{2}\right)$ **20.** $\left(\dfrac{5}{2}, -\dfrac{1}{2}\right)$

Assignment Exercises, Chapter 19

1. $\dfrac{1}{3}$ **3.** 2 **5.** $\dfrac{5}{8}$ **7.** $-\dfrac{4}{3}$ **9.** no slope **11.** $-\dfrac{4}{7}$ **13.** $\dfrac{7}{2}$ **15.** 0

17. no slope **19.** 1 **21.** $(-1, 2)(3, 2)$ Answers will vary.

23. $y = \dfrac{1}{3}x + 4$ **25.** $y = \dfrac{3}{4}x - 3$ **27.** $y = 4x - 5$

29. $y = -\dfrac{2}{3}x - \dfrac{2}{3}$ **31.** $y = -\dfrac{1}{11}x + \dfrac{17}{11}$ **33.** $y = \dfrac{7}{4}x - \dfrac{5}{4}$

35. $y = \dfrac{9}{5}x + \dfrac{3}{5}$ **37.** $y = x - 3$ **39.** $y = x - 1$ **41.** $y = -2$

43. $m = 3$ $b = \dfrac{1}{4}$ **45.** $m = -5$ $b = 4$ **47.** no slope no y-intercept **49.** $m = \dfrac{1}{8}$ $b = -5$

51. $y = -2x + 8$ $m = -2, b = 8$ **53.** $y = \dfrac{3}{2}x - 3$ $m = \dfrac{3}{2}, b = -3$ **55.** $y = \dfrac{3}{5}x - 4$ $m = \dfrac{3}{5}, b = -4$

57. $y = \dfrac{5}{3}$ $m = 0, b = \dfrac{5}{3}$ **59.** $y = 3x - 2$ **61.** $y = 2x - 2$

63. $x + y = 7$ **65.** $x - 2y = 8$ **67.** $x - 3y = 20$
69. $x + 3y = -10$ **71.** $3x - 4y = -7$ **73.** $x - y = -4$
75. $2x - y = -4$ **77.** $x - 5y = -11$ **79.** $2x + 10y = 31$
81. $x + 4y = 2$ **83.** 4.472 **85.** 10.440 **87.** 5.831 **89.** 9.434

91. 7.280 **93.** $(1, 5)$ **95.** $\left(1\dfrac{1}{2}, 3\right)$ **97.** $\left(-1\dfrac{1}{2}, 2\dfrac{1}{2}\right)$ **99.** $\left(1, -\dfrac{1}{2}\right)$

101. $\left(-1\dfrac{1}{2}, -3\right)$

103. a)

Lipsticks x	Cost y
0	5000
100	5453
1000	9530
2000	14,060

b) $(0, 5000)$; $(100, 5453)$; $(1000, 9530)$; $(2000, 14,060)$

c)

d) $f(x) = 4.53x + 5000$ e) $4.53x + 5000 = 8.99x$
$$5000 = 4.46x$$
$$x = 1121.076$$
or 1122 lipsticks

Trial Test, Chapter 19

1. $-\dfrac{2}{3}$ **3.** $\dfrac{5}{2}$ **5.** 1   **7.** $y = \dfrac{2}{3}x - 7$ **9.** $y = \dfrac{2}{3}x + \dfrac{7}{3}$

11. $y = 2$ **13.** $m = 3$
$b = -22$ **15.** $y = \dfrac{1}{4}x$

$$m = \dfrac{1}{4}, b = 0$$

17. $y = -6x + 3$ **19.** $y = \dfrac{3}{2}x + 3$ **21.** $2x + y = 5$
$m = -6, b = 3$

23. $x - 2y = 10$ **25.** $5.385, \left(\dfrac{1}{2}, 2\right)$

Self-Study Exercises 20–1

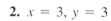

1. $x = 7, y = 5$ **2.** $x = 3, y = 3$

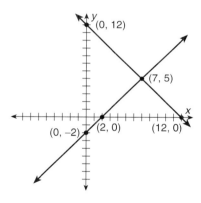   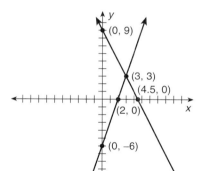

3. no solution, no intersection   **4.** many solutions, lines coincide

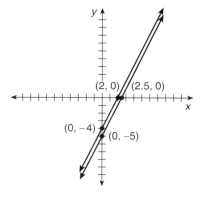   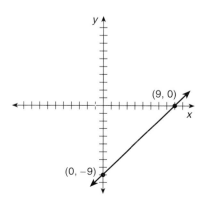

5. $x = 4$, $y = -2$

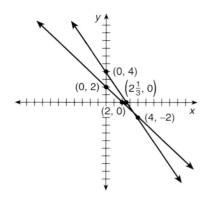

6.

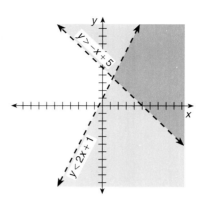

7.

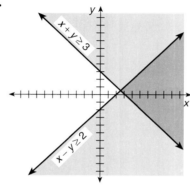

8. $x - 2y \leq -1$ $x + 2y \geq 3$

     $y > \dfrac{1}{2}x + \dfrac{1}{2}$ $y > -\dfrac{1}{2}x + \dfrac{3}{2}$

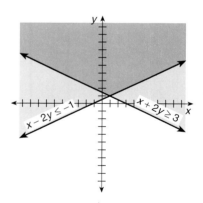

9. $x + y > 4$ $x - y > -3$

     $y > -x + 4$ $y < x + 3$

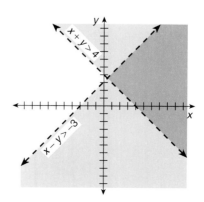

10. $3x - 2y < 8$ $2x + y \le -4$

$\qquad y > \dfrac{3}{2}x - 4$ $y \le -2x - 4$

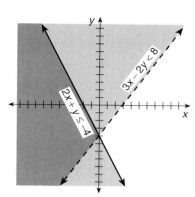

Self-Study Exercises 20–2

☐1

1. $a = 5, b = -1$ **2.** $m = 4, n = -1$ **3.** $x = 1, y = -1$
4. $a = 3, b = -3$ **5.** $x = 2, y = \dfrac{3}{2}$

☐2

6. $x = 3, y = 0$ **7.** $x = 1, y = 5$ **8.** $a = 2, b = \dfrac{8}{3}$
9. $x = -1, y = 2$ **10.** $a = 6, y = 1$

☐3

11. inconsistent; no solution   **12.** $x = -3, y = 6$
13. dependent; many solutions   **14.** dependent; many solutions
15. $a = 5, b = 1$

Self-Study Exercises 20–3

☐1

1. $a = 20, b = 10$ **2.** $r = 5, c = 7$ **3.** $x = 13, y = 11$
4. $x = 7, y = 5$ **5.** $p = \dfrac{1}{2}, k = \dfrac{1}{3}$ **6.** $x = 3, y = 1$

Self-Study Exercises 20–4

☐1

1. short 15.5 in.   **2.** $18,000 at 4%   **3.** $11 per shirt
    long 32.5 in. $17,000 at 5% $ 6 per hat
4. (7) 8-cylinder jobs   **5.** resistor $0.25   **6.** rate of plane = 130 mph
    (3) 4-cylinder jobs   capacitor $0.30   rate of wind = 10 mph
7. $3000 at 7%   **8.** rate of current = 1.67 mph   **9.** dark roast, $2.50
    $5000 at 9%   rate of motorboat = 11.67 mph   with chicory, $1.90
10. 75% mixture, 123 lb   **11.** scientific $11.00   **12.** $4000 at 10%
     10% mixture, 77 lb   graphing $30.00 $6000 at 15%
13. reserved $20.00   **14.** 4 pints at 75%
     general $15.00   4 pints at 25%

Assignment Exercises, Chapter 20

1.

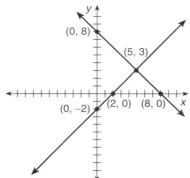

3.

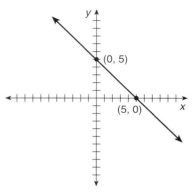

dependent; many solutions, lines coincide

5.

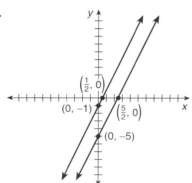

no solutions, lines parallel

7.

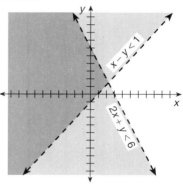

9.

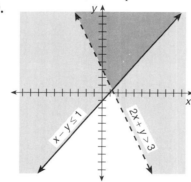

11. (3, 0) **13.** (−2, 4) **15.** (2, −2) **17.** (6, 3)

19. dependent; many solutions **21.** (5, 1) **23.** $\left(2, \dfrac{8}{3}\right)$ **25.** (2, 0)

27. (3, 2) **29.** (2, 1.5) **31.** (4, 4) **33.** (7, 5) **35.** (4, −6)

37. (3, −4) **39.** $\left(\dfrac{4}{5}, \dfrac{2}{5}\right)$ **41.** (28, 44) **43.** $\left(-3, \dfrac{5}{2}\right)$ **45.** (4, −6)

47. (12, 2) **49.** (1, −2)

51. electrician $75 **53.** shellac $2.50 **55.** 119°, 56° **57.** $2000 at 10%
apprentice $35   thinner $3.50 $3000 at 12%

59. columbian = $5 **61.** Ohio = $1.95 **63.** name brand = $320
blended = $4   Alaska = $2.10   generic = $175

65. telephone = $15,000 **67.** 4 gal. 75%
showroom = $25,000   4 gal. 25%

69. In line 4, $-6c$ should be $-9c$. **71.** Answers will vary.
$c = 3$
$d = 1$

Trial Test, Chapter 20

1. $(5, 0)$ **3.** $(14, -9)$

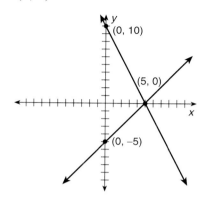
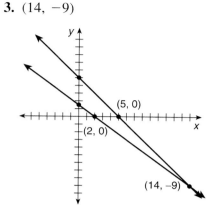

5. $(-6, -6)$ **7.** $\left(\dfrac{1}{2}, \dfrac{7}{2} \right)$

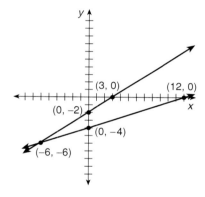

9. $(0, 0)$ **11.** $(1, 2)$ **13.** $(2, 2)$ **15.** $(-3, 15)$ **17.** $(-9, -11)$
19. $20A$, $15A$ **21.** $L = 51$ in.   **23.** \$20,000 at 12.5%
               $W = 34$ in.   \$ 5000 at 11%
25. $0.000215\ F$
     $0.000055\ F$

Self-Study Exercises 21–1

1
1. $\overleftrightarrow{PQ}$, $\overleftrightarrow{QR}$, & $\overleftrightarrow{PR}$ **2.** $\overrightarrow{PQ}$, $\overrightarrow{PR}$ **3.** $\overline{QR}$ **4.** Yes
5. No. End points are different.   **6.** No   **7.** Yes   **8.** Yes
9. Yes. $\overline{XY}$ and $\overline{YX}$ contain same points.   **10.** No

2
11. parallel   **12.** intersect   **13.** coincide   **14–15.** coincide or intersect

3
16. $\angle DEF$, $\angle FED$ **17.** $\angle E$ **18.** $\angle 1$ **19.** $\angle a$ **20.** $\angle ABC$ or $\angle CBA$
21. $\angle B$

22. 360°   23. 180°   24. 90°   25. acute   26. obtuse   27. right
28. obtuse   29. acute   30. straight   31. acute   32. obtuse   33. Yes
34. No   35. No   36. No   37. Yes   38. Yes   39. No   40. 90°
41. congruent

Self-Study Exercises 21–2

1

1. 54°   2. 46°8′18″   3. 37°52′56″   4. 74°3′12″   5. 16°59′45″
6. 54°44′12″   7. 64°24′46″

2

8. 0.7833°   9. 0.01°   10. 0.0872°   11. 0.1708°   12. 29.7°

3

13. 21′   14. 12′   15. 7′12″   16. 12′47″   17. 18′54″

4

18. 7°30′   19. 11°25′43″   20. 38.1908°   21. 12°38′50″

Self-Study Exercises 21–3

1

1. scalene   2. equilateral   3. isosceles

2

4. longest $\overline{TS}$, shortest $\overline{RS}$ 5. longest k, shortest l 6. $\angle C$, $\angle B$, $\angle A$
7. $\angle X$, $\angle Z$, $\angle Y$

3

8. $\angle A = \angle F$, $\angle C = \angle D$, $\angle B = \angle E$ 9. $\angle P = \angle S$, $\angle Q = \angle T$, $QP = TS$
10. $\angle L = \angle P$, $PM = LJ$, $LK = PN$

4

11. $BC = 12$ cm, $AB = 16.971$ cm   12. $AC = 7.071$ m, $BC = 7.071$ m
13. $AC = 9.899$ m, $AB = 14.0$ m   14. $AC = 8$ m, $BC = 8$ m
15. $AC = 14.697$ mm, $BC = 14.697$ mm   16. 25′5″   17. 37.830 ft

5

18. $AB = 12$ cm, $BC = 10.392$ cm   19. $AC = 9.0$ mm, $BC = 15.588$ mm
20. $AC = 4.619$ in., $AB = 9.238$ in.   21. $AC = 5.715$ cm, $AB = 11.431$ cm
22. $AB = 5′6″$, $BC = 4′9″$ 23. 35′   24. rafter = 18′0″, run = 15′7″
25. 4.5 cm

Self-Study Exercises 21–4

1

1. $x = 16′$, $y = 22′$ 2. $x = 6′$, $y = 25′6″$

$\boxed{2}$

3. $245'6'' = P$ **4.**   96 mm $= P$ **5.** $p = 58'8''$
     1942.5 ft$^2 = A$ 662.4 mm$^2 = A$
6. 20 yd^2 @ \$16.50 = \$330   **7.** 24 lb   **8.** 1086 ft^2, $\angle = 135°$
9. 63 in.^{2} **10.** 625 dm^2 **11.** 33 in.^{2}

$\boxed{3}$

12. equilateral triangle, $\angle 60°$ **13.** square, $\angle 90°$
14. regular pentagon, $\angle 108°$ **15.** regular hexagon, $\angle 120°$

Self-Study Exercises 21–5

$\boxed{1}$

1. 120.64 cm^2 **2.** 198.44 mm^2 **3.** 2.43 ft^2 **4.** 14.18 in.^{2} **5.** 2.75 m^2
6. 530.14 ft^2 **7.** 75.40 in.^{2} **8.** 17.0 in.^{2}

$\boxed{2}$

9. 28.27 mm   **10.** 13.09 in.   **11.** 3.20 ft   **12.** 7.06 cm   **13.** 1.64 m
14. 15.71 in.

$\boxed{3}$

15. 16.01 cm^2 **16.** 48.98 in.^{2} **17.** 54.83 in.^{2} **18.** 51.99 cm^2
19. 91.29 in.^{2} **20.** 561.81 mm^2 **21.** 257.93 ft^2

Self-Study Exercises 21–6

$\boxed{1}$

1. 60°   **2.** 30°   **3.** 90°   **4.** 11.5 in.   **5.** 11.5 in.   **6.** 13.33 in.
7. 57.5 in.^{2} **8.** 5.75 in.   **9.** 6.67 in.   **10.** 41.89 in.   **11.** 2.9 cm
12. 1 in.

$\boxed{2}$

13. 90°   **14.** 45°   **15.** 8.49 in.   **16.** 45°   **17.** 6 in.   **18.** 0.5 m
19. 0.35 m   **20.** 90°   **21.** 0.35 m   **22.** 0.29 in.   **23.** 1.06 in.
24. 63.62 dkm^2

$\boxed{3}$

25. 34.64 mm   **26.** 120°   **27.** 60°   **28.** 40 mm   **29.** 90°
30. 10 mm   **31.** 20 mm   **32.** 60°   **33.** 30°   **34.** 3 in.
35. 39.97 mm   **36.** 2.84 in.

Assignment Exercises, Chapter 21

1. $\overleftrightarrow{AB}$ **3.** $\overrightarrow{AB}$ **5.** $\overrightarrow{NO}$ **7.** No   **9.** No   **11.** $\overleftrightarrow{AB}$ & $\overleftrightarrow{CD}$
13. $\overleftrightarrow{EF}$ & $\overleftrightarrow{GH}$ **15.** Yes   **17.** $\angle EDF$ or $\angle FDE$ **19.** $\angle P$ **21.** $\angle M$
23. right   **25.** acute   **27.** straight   **29.** obtuse   **31.** neither
33. supplementary   **35.** neither   **37.** equal   **39.** vertical
41. $31°08'11''$ **43.** $34°37'56''$ **45.** $60°40'18''$ **47.** $0.4833°$
49. $0.1261°$ **51.** $45'$ **53.** $13'3''$ **55.** $23°20'$ **57.** $18°39'50.5''$
59. isosceles   **61.** longest TS, shortest RS **63.** $\angle C$, $\angle B$, $\angle A$ **65.** $\angle C =$
$\angle D$, $\angle B = \angle E$, $CB = DE$ **67.** $\angle M = \angle J$, $JL = PM$, $JK = MN$
69. $RS = 10.607$ cm, $ST = 10.607$ cm   **71.** $RS = 9.0$ hm, $ST = 9.0$ hm
73. $AB = 13.856$ dm, $BC = 6.928$ dm   **75.** $AC = 17.321$ in., $AB = 20.0$ in.

77. $AB = 46'10''$, $BC = 23'5''$ **79.** $8'4''$ **81.** 19.630 mm **83.** 115 in.,
908.5 in.^{2} **85.** regular pentagon, 108° **87.** 130 ft **89.** 83 ft^2 **91.** 72
bricks **93.** 2.62 cm^2 **95.** 1610.28 mm^2 **97.** 2.18 ft^2 **99.** 6.66 mm^2
101. 6.96 cm^2 **103.** 179.9 mm^2 **105.** 232.27 ft^2 **107.** 44.50 in.^{2}
109. 31.42 in. **111.** 36.74 mm **113.** $AE = 5$ **115.** $\angle GJO = 45°$
117. $IJ = 20$ **119.** $\angle MOP = 30°$ **121.** 5 mm **123.** 3.54 cm
125. area pentagon $<$ area circle $\pi(5)^2 = 78.54$ cm^2
area pentagon $>$ area square 50.00 cm^2

Trial Test, Chapter 21

1. $\overline{BC}$ **3.** acute **5.** parallel **7.** perpendicular **9.** $15°45'29''$
11. $18'45''$ **13.** largest $\angle A$, smallest $\angle B$ **15.** 13.657 cm **17.** $155'$
19. $135°$ **21.** 43.20 mm **23.** $25.13''$ **25.** 1.57 in.

Self-Study Exercises 22–1

[1]

1. 6.45 in. **2.** 14 cm **3.** 4.25 rad **4.** 3.49 rad **5.** 7 cm
6. 4.5 in.

[2]

7. 17.12 in.^{2} **8.** 4.32 cm^2 **9.** 2.24 cm **10.** 0.85 rad

[3]

11. 0.79 rad **12.** 0.98 rad **13.** 1.36 rad **14.** 2.44 rad **15.** 0.38 rad
16. 3.10 rad **17.** 0.78 rad **18.** 0.18 rad

[4]

19. $45°$ **20.** $30°$ **21.** $143.2394°$ **22.** $80.2141°$ **23.** $28°38'52''$
24. $22°30'$ **25.** $42°58'19''$ **26.** $63°1'31''$ **27.** 1.53 cm **28.** $61°$
29. 47.50 cm^2 **30.** $40.1070°$

Self-Study Exercises 22–2

[1]

1. $\dfrac{5}{13}$ **2.** $\dfrac{12}{13}$ **3.** $\dfrac{5}{12}$ **4.** $\dfrac{12}{13}$ **5.** $\dfrac{5}{13}$ **6.** $\dfrac{12}{5}$ **7.** $\dfrac{9}{15} = \dfrac{3}{5}$

8. $\dfrac{12}{15} = \dfrac{4}{5}$ **9.** $\dfrac{9}{12} = \dfrac{3}{4}$ **10.** $\dfrac{12}{15} = \dfrac{4}{5}$ **11.** $\dfrac{9}{15} = \dfrac{3}{5}$ **12.** $\dfrac{12}{9} = \dfrac{4}{3}$

13. $\dfrac{16}{34} = \dfrac{8}{17}$ **14.** $\dfrac{30}{34} = \dfrac{15}{17}$ **15.** $\dfrac{16}{30} = \dfrac{8}{15}$ **16.** $\dfrac{30}{34} = \dfrac{15}{17}$

17. $\dfrac{16}{34} = \dfrac{8}{17}$ **18.** $\dfrac{30}{16} = \dfrac{15}{8}$ **19.** $\dfrac{9}{16.64} = 0.5409$ **20.** $\dfrac{14}{16.64} = 0.8413$

21. $\dfrac{9}{14} = 0.6429$ **22.** $\cdot\dfrac{14}{16.64} = 0.8413$ **23.** $\dfrac{9}{16.64} = 0.5409$

24. $\dfrac{14}{9} = 1.5556$

2

25. $\sin A = \dfrac{12}{20} = \dfrac{3}{5}$ $\sin B = \dfrac{16}{20} = \dfrac{4}{5}$

   $\cos A = \dfrac{16}{20} = \dfrac{4}{5}$ $\cos B = \dfrac{12}{20} = \dfrac{3}{5}$

   $\tan A = \dfrac{12}{16} = \dfrac{3}{4}$ $\tan B = \dfrac{16}{12} = \dfrac{4}{3}$

   $\csc A = \dfrac{20}{12} = \dfrac{5}{3}$ $\csc B = \dfrac{20}{16} = \dfrac{5}{4}$

   $\sec A = \dfrac{20}{16} = \dfrac{5}{4}$ $\sec B = \dfrac{20}{12} = \dfrac{5}{3}$

   $\cot A = \dfrac{16}{12} = \dfrac{4}{3}$ $\cot B = \dfrac{12}{16} = \dfrac{3}{4}$

26. $\sin A = \dfrac{8.15}{9.73} = 0.8376$ $\sin B = \dfrac{5.32}{9.73} = 0.5468$

   $\cos A = \dfrac{5.32}{9.73} = 0.5468$ $\cos B = \dfrac{8.15}{9.73} = 0.8376$

   $\tan A = \dfrac{8.15}{5.32} = 1.5320$ $\tan B = \dfrac{5.32}{8.15} = 0.6528$

   $\csc A = \dfrac{9.73}{8.15} = 1.1939$ $\csc B = \dfrac{9.73}{5.32} = 1.8289$

   $\sec A = \dfrac{9.73}{5.32} = 1.8289$ $\sec B = \dfrac{9.73}{8.15} = 1.1939$

   $\cot A = \dfrac{5.32}{8.15} = 0.6528$ $\cot B = \dfrac{8.15}{5.32} = 1.5320$

Self-Study Exercises 22–3

1

1. 0.3584   **2.** 0.9981   **3.** 1.0724   **4.** 0.6088   **5.** 0.5225   **6.** 0.9304
7. 0.4068   **8.** 0.9336   **9.** 0.8039   **10.** 0.9163   **11.** 1.0990
12. 0.8843   **13.** 0.9757   **14.** 0.4950   **15.** 3.3191   **16.** 0.9446
17. 0.6845   **18.** 0.9377   **19.** 0.3960   **20.** 0.2199

2

21. 20.0°   **22.** 22.5°   **23.** 67°   **24.** 62.5°   **25.** 57.3°   **26.** 34.8°
27. 51.7°   **28.** 82.1°   **29.** 39.0°   **30.** 21.2°   **31.** 0.5585 rad
32. 0.6807 rad   **33.** 0.8028 rad   **34.** 1.1694 rad   **35.** 0.2537 rad
36. 1.4802 rad   **37.** 0.3061 rad   **38.** 1.1804 rad   **39.** 0.2802 rad
40. 1.1362 rad

3

41. 2.1943   **42.** 1.4945   **43.** 1.0306   **44.** 1.0187   **45.** 2.8824
46. 1.8341   **47.** 0.0524   **48.** 1.3563   **49.** 4.7174   **50.** 1.9661
51. 67.0°   **52.** 56.9°   **53.** 15.6°   **54.** 83.0°   **55.** 30.0°   **56.** 0.9903
57. 0.9659   **58.** 1.6391   **59.** 0.8001   **60.** 60°44′19.5″   **61.** 62°3′0.2″
62. 9°21′15.2″   **63.** 79°8′27.5″

Assignment Exercises, Chapter 22

1. 1.05 radians   **3.** 5.24 radians   **5.** 1.74 radians   **7.** 150°
9. 97.4028°   **11.** 67°30′   **13.** 1.61 cm   **15.** 1.71 ft   **17.** 3.38 in.

19. 2.09 in.   **21.** 15.70 cm^2 **23.** $\dfrac{15}{25} = \dfrac{3}{5}$ **25.** $\dfrac{15}{20} = \dfrac{3}{4}$ **27.** $\dfrac{25}{20} = \dfrac{5}{4}$

29. $\dfrac{20}{25} = \dfrac{4}{5}$ **31.** $\dfrac{20}{15} = \dfrac{4}{3}$ **33.** $\dfrac{25}{15} = \dfrac{5}{3}$ **35.** $\dfrac{2\text{ ft}}{2\text{ ft }2\text{ in.}} = \dfrac{24\text{ in.}}{26\text{ in.}} = \dfrac{12}{13}$

37. $\dfrac{2\text{ ft}}{10\text{ in.}} = \dfrac{24\text{ in.}}{10\text{ in.}} = \dfrac{12}{5}$ **39.** $\dfrac{10\text{ in.}}{2\text{ ft}} = \dfrac{10\text{ in.}}{24\text{ in.}} = \dfrac{5}{12}$ **41.** $\dfrac{7}{12.62} = 0.5547$

43. $\dfrac{7}{10.5} = 0.6667$ **45.** $\dfrac{12.62}{10.5} = 1.2019$ **47.** $\dfrac{10.5}{7} = 1.5000$

49. 0.4540   **51.** 0.3057   **53.** 1.0724   **55.** 0.7585   **57.** 0.8403
59. 0.1708   **61.** 52.5°   **63.** 45.3°   **65.** 16.7°   **67.** 1.2188 rad
69. 1.2101 rad   **71.** 0.3313 rad   **73.** 1.0576   **75.** 1.1326   **77.** 2.1536
79. 1.0090   **81.** 1.0592   **83.** -2.1557 **85.** $\sin\theta = \tan\theta$ for $\theta = 0°$, 180°, and 360°.

Trial Test, Chapter 22

1. 0.61 radians   **3.** 5.50 radians   **5.** $15.4167° = 0.27$ radian
7. $16.2° = 0.28$ radian   **9.** 112.5°   **11.** $68°45'18''$ **13.** 1 in.

15. 1.71 m   **17.** $\dfrac{10}{26} = \dfrac{5}{13}$ **19.** $\dfrac{26}{10} = \dfrac{13}{5}$ **21.** $\dfrac{11.5}{12.54} = 0.9171$

23. 0.7986   **25.** 0.9135   **27.** 16.0°   **29.** 19.5°   **31.** 0.8726 rad
33. 0.3150 rad   **35.** 1.3673

Self-Study Exercises 23–1

1

1. 27.8°   **2.** 6.889 in.   **3.** 28.21 m   **4.** 16.15 ft   **5.** 43.2°   **6.** 49.6°
7. $S = 51.1°$ **8.** $S = 45°$
  $U = 38.9°$ $t = 6.647$ m
  $u = 11.31$ yd $s = 4.700$ m
  $18^2 = 14^2 + (11.31)^2$ $(6.647)^2 = (4.7)^2 + (4.7)^2$
  $324 \doteq 323.9161$ $44.18 = 44.18$
9. $U = 55.5°$ **10.** $U = 74°$
  $s = 4.814$ mm $t = 50.79$ m
  $u = 7.005$ mm $u = 48.82$ m
  $(8.5)^2 \doteq (4.814)^2 + (7.005)^2$ $(50.79)^2 \doteq (48.82)^2 + (14)^2$
  $72.25 \doteq 23.17 + 49.07$ $2579.62 \doteq 2383.39 + 196$
  $72.25 \doteq 72.24$ $2579.62 \doteq 2579.39$

2

11. 35.7°   **12.** 10.05 cm   **13.** 31.50 ft   **14.** 35.49 dm   **15.** 15.88 mm
16. 14.12 yd   **17.** $R = 45.9°$ **18.** $Q = 44°$
                $Q = 44.1°$ $r = 12.23$ cm
                $r = 16.52$ ft $q = 11.81$ cm
19. $R = 16.5°$ **20.** $Q = 30.5°$
  $r = 4.147$ dkm $r = 13.58$ m
  $s = 14.60$ dkm $s = 15.76$ m

3

21. 28.6°   **22.** 4.075 m   **23.** 5.979 ft   **24.** 23.5°   **25.** 0.04663 cm
26. 0.01212 m   **27.** $D = 34.8°$ **28.** $E = 48°$
                $E = 55.2°$ $d = 6.303$ ft
                $f = 5.604$ m $f = 9.419$ ft

29. $D = 16.5°$
$d = 5.963$ in.
$f = 20.99$ in.

30. $D = 54.0°$
$E = 36.0°$
$f = 13.60$ ft

Self-Study Exercises 23–2

[1]

1. 150.1 m   **2.** 11.91 ft   **3.** 9.950 cm   **4.** 46.71 cm   **5.** 21.09 in.

[2]

6. 22.38 in.   **7.** 53.62 ft   **8.** 59.0°   **9.** 39.23 ft   **10.** 14.9°
11. 23.25 ft   **12.** 10 ohms   **13.** 17.32 ohms   **14.** 7.654 cm   **15.** 35.8°

Assignment Exercises, Chapter 23

1. $A = 38.1°$, $B = 51.9°$, $b = 13.15$ in.
3. $G = 48.5°$, $f = 1.681$ mm, $h = 2.537$ mm
5. $X = 52°$, $y = 5.469$ m, $z = 8.883$ m
7. $D = 28.3°$, $E = 61.7°$, $f = 14.76$ ft
9. $G = 53°$, $g = 25.21$ m, $k = 31.57$ m   **11.** 32.2°   **13.** 11.57 in.
15. $B = 60.5°$, $b = 3.641$ m, $c = 4.183$ m   **17.** 48.6° and 41.4°
19. $B = 56°$, $a = 88.91$ ft, $b = 131.8$ ft   **21.** 56.54 cm

Trial Test, Chapter 23

1. 48.8°   **3.** 15.41 ft   **5.** 42.9°   **7.** 6.420 cm   **9.** 62.9°
11. $B = 65.5°$, $b = 30.72$ m, $c = 33.76$ m   **13.** 38.7°   **15.** 48 in.
17. 40.06 A   **19.** 84.3°   **21.** 44.7°   **23.** $X_L = 9.64$ ohms; $R = 11.49$ ohms

Self-Study Exercises 24–1

[1]

1. 13   **2.** 15   **3.** 7.280   **4.** 8.544   **5.** 2.746

[2]

6. 67.38°   **7.** 53.13°   **8.** 20.56°   **9.** 68.20°   **10.** 75.96°

[3]

11. 19; 42°   **12.** 8; 72°   **13.** 1; 225°   **14.** 6; 75°   **15.** $12 + 5i$; 13;
22.62°   **16.** $6 + 4i$; 7.211; 33.69°

Self-Study Exercises 24–2

[1]

1. 60°   **2.** 15°   **3.** 70°   **4.** 15°   **5.** 32°   **6.** 70°   **7.** 32°   **8.** 62°
9. 0.96 rad   **10.** 0.44 rad

[2]

11. all positive   **12.** sin, csc positive; others negative
13. tan, cot positive; others negative   **14.** cos, sec positive; others negative
15. cos, sec positive; tan, sin zero; cot, csc undefined

[3]

16. -0.5000 **17.** -0.8391 **18.** -0.8011 **19.** -0.6536
20. -0.8660 **21.** -0.2910 **22.** -0.1736 **23.** 0.5000   **24.** 146.3°
25. 3.61° rad

Self-Study Exercises 24–3

[1]

1. $C = 63°$; $b = 40.00$; $c = 45.22$ **2.** $A = 50°$; $a = 6.504$; $c = 8.075$
3. $C = 33°$; $b = 56.37$; $c = 35.45$ **4.** $B = 30°$; $a = 5.543$; $b = 3.200$
5. $C = 50°$; $a = 20.84$; $c = 16.03$

2

6. $B = 21.5°$; $C = 118.5°$; $c = 57.42$ **7.** $C = 10.8°$; $A = 99.2°$; $a = 15.76$
8. $A = 43.2°$; $B = 116.8°$; $b = 23.47$ or $A = 136.8°$; $B = 23.2°$; $b = 10.35$
9. $B = 14.5°$; $C = 135.5°$; $c = 11.21$

3

10.  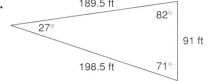   **11.** 805.9 ft

Self-Study Exercises 24–4

1

1. $A = 30.8°$
  $B = 125.1°$
  $C = 24.1°$

2

2. $D = 71.7°$
  $E = 62.9°$
  $F = 45.4°$

3. $r = 32.42$
  $S = 49.7°$
  $T = 58.3°$

4. $j = 18.68$
  $K = 37.5°$
  $L = 34°$

3

5. 49.27 ft
6. 60.8°
  44.1°
  75.1°

Self-Study Exercises 24–5

1

1. 27.93 cm^2 **2.** 90.42 m^2 **3.** 73.21 ft^2 **4.** 21.22 in.^{2} **5.** 122.6 mm^2
6. 76.31 m^2 **7.** 192.8 dm^2 **8.** 453.6 dkm^2

2

9. 14.97 m^2 **10.** 3.712 in.^{2} **11.** 24.38 cm^2 **12.** 26.72 cm^2

3

13. 3.32 cm^2 **14.** 70.4 ft^2 **15.** 5108 ft^2

Assignment Exercises, Chapter 24

1. 7.211; 56.3°   **3.** 6.083; 9.5°   **5.** 13.89; 59.7°   **7.** 0.1016 rad
9. 41°   **11.** 0.8832 rad   **13.** 32°15'10"   **15.** 0.8632   **17.** −0.3420
19. 0.3420   **21.** 1   **23.** 2.83; 2.36 rad   **25.** $A = 40°$; $b = 10.8$; $c = 4.3$
27. $A = 30.3°$; $B = 104.7°$; $b = 9.6$
29. 1st − $A = 39.4°$; $C = 112.6°$; $c = 13.4$ **31.** 42.0 ft
  2nd − $A = 140.6°$; $C = 11.4°$; $c = 2.9$
33.   **35.**   **37.**

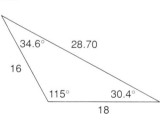      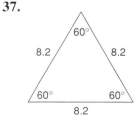

39. 97.98 ft   **41.** 343.3 ft; 665.4 ft   **43.** 99.68 m^2 **45.** 710.4 cm^2
47. 185.4 km^2 **49.** 27.71 ft^2 **51.** 708.6 in.^{2} **53.** 12,050 ft^2

Trial Test, Chapter 24

1. 0.8192   **3.** −0.9397   **5.** 135°   **7.** 101.3°   **9.** 8.991 ft   **11.** 131.8°
13. 6.830 m   **15.** 36.98 m^2 **17.** 150.9 ft   **19.** 66.88 ft^2
21. 6.4 milliamps   **23.** 30 ohms

Index